Biocontrol Agents

Entomopathogenic and Slug Parasitic Nematodes

Biocontrol Agents

Entomopathogenic and Slug Parasitic Nematodes

Edited by

Mahfouz M.M. Abd-Elgawad
Phytopathology Department, National Research Centre, El-Tahrir Street, Dokki 12622, Giza, Egypt

Tarique Hassan Askary
Division of Entomology, Sher-e-Kashmir University of Agricultural Sciences and Technology, Main Campus, Shalimar, Srinagar-190025, Jammu and Kashmir, India

James Coupland
FarmForest Research Inc, Box 910, Almonte, Ontario, Canada, K0A 1A0

CABI

CABI is a trading name of CAB International

CABI
Nosworthy Way
Wallingford
Oxfordshire OX10 8DE
UK

Tel: +44 (0)1491 832111
Fax: +44 (0)1491 833508
E-mail: info@cabi.org
Website: www.cabi.org

CABI
745 Atlantic Avenue
8th Floor
Boston, MA 02111
USA

Tel: +1 (617)682-9015
E-mail: cabi-nao@cabi.org

A catalogue record for this book is available from the British Library, London, UK.

Library of Congress Cataloging-in-Publication Data

Names: Abd-Elgawad, Mahfouz M. M., editor. | Askary, Tarique Hassan, editor.
| Coupland, James, editor.
Title: Biocontrol agents : entomopathogenic and slug parasitic nematodes/
edited by Mahfouz M.M. Abd-Elgawad, Tarique Hassan Askary, James Coupland.
Description: Boston, MA : CAB International, [2017] | Includes
bibliographical references and index.
Identifiers: LCCN 2016045645| ISBN 9781786390004 (hbk : alk. paper) | ISBN
9781786390028 (epub)
Subjects: LCSH: Nematodes as biological pest control agents.
Classification: LCC SB976.N46 B56 2017 | DDC 632/.96--dc23 LC
record available at https://lccn.loc.gov/2016045645

ISBN-13: 978 1 78639 000 4

Commissioning editor: Rachael Russell
Editorial assistant: Emma McCann
Production editor: James Bishop

Typeset by SPi, Pondicherry, India
Printed and bound by CPI Group (UK) Ltd, Croydon, CR0 4YY

Contents

Contributors

Adler R. Dillman, Department of Nematology, 900 University Avenue, University of California, Riverside, California, USA. E-mail: adler.dillman@ucr.edu

David M. Glen, Styloma Research Consulting, Cheddar, BS27 3QP, UK. E-mail: davidmglen@btopenworld.com

Ed Lewis, Department of Entomology and Nematology, University of California, Davis, California, 95616, USA. E-mail: eelewis@ucdavis.edu

Tarique Hassan Askary, Division of Entomology, Sher-e-Kashmir University of Agricultural Sciences and Technology, Main Campus, Shalimar, Srinagar-190025, Jammu and Kashmir, India. E-mail: tariq_askary@rediffmail.com

Antoinette P. Malan, Department of Conservation Ecology and Entomology, Faculty of AgriSciences, Stellenbosch University, Stellenbosch, South Africa. E-mail: apm@sun.ac.za

Claudia Dolinski, Universidade Estadual Norte Fluminense/CCTA/LEF, Campos dos Goytacazes, RJ, Brazil. E-mail: claudia.dolinski@censanet.com.br

James Coupland, FarmForest Research Inc, Box 910, Almonte, Ontario, Canada, K0A 1A0. E-mail: couplandj@hotmail.com

Mahfouz M.M. Abd-Elgawad, Phytopathology Department, National Research Center, El-Tahrir Street, Dokki 12622, Giza, Egypt. E-mail: mahfouzian2000@yahoo.com

C. Sankaranarayanan, Section of Nematology, Division of Crop Protection, ICAR-Sugarcane Breeding Institute, Coimbatore – 641 007, India. E-mail: csankar@lycos.com

Sergei E. Spiridonov, Severtsov Institute of Ecology and Evolution, Russian Academy of Sciences, Leninskii prospect, 33, Moscow, 199071, Russia. E-mail: s_e_spiridonov@rambler.ru

S.S. Hussaini, Ex Principal Scientist, National Bureau of Agriculturally Important Insects, Bangalore, India. E-mail: sshblr@yahoo.com

J. Gulsar Banu, ICAR-Central Institute for Cotton Research, Regional Station, Coimbatore – 641003, India. E-mail: gulsarsci@gmail.com

Francesca De Luca, Istituto per la Protezione Sostenibile delle Piante, Consiglio Nazionale delle Ricerche, Via Amendola 122/D, 70126 Bari, Italy. E-mail: francesca.deluca@ipsp.cnr.it

M.M.E. Saleh, Pests and Plant Protection Department, National Research Centre, El-Tahrir Street, Dokki 12622, Giza, Egypt. E-mail: mmesaleh@hotmail.com

Atef Sayed Abdel-Razek, Pests and Plant Protection Department, National Research Centre, El-Tahrir Street, Dokki 12622, Cairo, Egypt. E-mail: abdelrazek820@yahoo.com

Shahina Fayyaz, National Nematological Research Centre, University of Karachi, Karachi-75270, Pakistan. E-mail: shahinafayyaz@gmail.com

Salma Javed, National Nematological Research Centre, University of Karachi, Karachi-75270, Pakistan. E-mail: sajaved@uok.edu.pk

Christos I. Rumbos, Laboratory of Entomology and Agricultural Zoology, Department of Agriculture, Crop Production and Rural Environment, University of Thessaly, Phytokou str., 38446, N. Ionia, Magnesia, Greece. E-mail: christosrumbos@yahoo.gr

Christos G. Athanassiou, Laboratory of Entomology and Agricultural Zoology, Department of Agriculture, Crop Production and Rural Environment, University of Thessaly, Phytokou str., 38446, N. Ionia, Magnesia, Greece. E-mail: athanassiou@agr.uth.gr

Nazir Javed, Department of Plant Pathology, University of Agriculture Faisalabad, Punjab, Pakistan. E-mail: nazirpp2003@gmail.com

M. Jamal Ahmad, Division of Entomology, Sher-e-Kashmir University of Agricultural Sciences and Technology, Main Campus, Shalimar, Srinagar-190025, Jammu and Kashmir, India. E-mail: ahmad_j@rediffmail.com

Glen Stevens, School of Natural Sciences and Mathematics, Ferrum College, Ferrum, Virginia, 24088, USA. E-mail: gstevens3@ferrum.edu

Qing Yu, Canadian National Collection of Nematodes, Biodiversity Program, Ottawa Research and Development Centre, Agriculture and Agri-food Canada, Ottawa, Ontario, K1A 0C6, Canada. E-mail: Qing.Yu@agr.gc.ca

Fengcheng Sun, Canadian Food Inspection Agency, Nematology Laboratory, Ottawa Plant Laboratory, 3851 Fallowfield Road, Ottawa, Ontario, K2H 8P9, Canada. E-mail: sunfc@inspection.gc.ca

I. Cannayane, Vegetable Research Station, Palur-607102, Cuddalore Dt, India. E-mail: nematnau@gmail.com

K. Sankari Meena, Tamil Nadu Agricultural University, Coimbatore – 641003, India. E-mail: meenaa5@rediffmail.com

Achinelly María Fernanda, Centro de Estudios Parasitológicos y de Vectores, CEPAVE (CCT La Plata CONICET – UNLP), Calle 120 y 60 S/N, (1900) La Plata, Buenos Aires, Argentina. E-mail: fachinelly@cepave.edu.ar

Camino Nora, Centro de Estudios Parasitológicos y de Vectores, CEPAVE (CCT La Plata CONICET – UNLP), Calle 120 y 60 S/N, (1900) La Plata, Buenos Aires, Argentina. E-mail: nemainst@cepave.edu.ar

Muhammad Kamran, Plant Pathology Research Institute, Ayub Agricultural Research Institute, Faisalabad, Punjab, Pakistan. E-mail: mkamran.uaf.pk@gmail.com

Huma Abbas, Department of Plant Pathology, University of Agriculture Faisalabad, Punjab, Pakistan. E-mail: huma_1633@yahoo.com

Eliceche Daiana, Centro de Estudios Parasitológicos y de Vectores, CEPAVE (CCT La Plata CONICET – UNLP), Calle 120 y 60 S/N, (1900) La Plata, Buenos Aires, Argentina. E-mail: daia226@gmail.com

Luis Garrigós Leite, Istituto Biológico, Centro Experimental, Campinas, SP, Brazil. E-mail: lgleite@biologico.sp.gov.br

Vanessa Andaló, Universidade Federal de Uberlândia, Campus Monte Carmelo, MG, Brazil. E-mail: vanessaandalo@ufu.br

Alcides Moino Junior, Universidade Federal de Lavras, Lavras, MG, Brazil. E-mail: alcmoino@den.ufla.br

Elder Batista, Universidade Estadual Paulista/FCAV, Jaboticabal, SP, Brazil. E-mail: elderspb@gmail.com

Edson Tadeu Iede, Embrapa Florestas/CNPF, Estrada da Ribeira, Colombo, PR, Brazil. E-mail: edson.iede@embrapa.br

Firoza Kazi, National Nematological Research Centre, University of Karachi, Karachi-75270, Pakistan. E-mail: nnrc@uok.edu.pk

Tabassum Ara Khanum, National Nematological Research Centre, University of Karachi, Karachi-75270, Pakistan. E-mail: tabassumak@uok.edu.pk

Justin L. Hatting, South African Agricultural Research Council-Small Grain Institute (ARC-SGI), Bethlehem, South Africa. E-mail: hattingj@arc.agric.za

Eustachio Tarasco, Dipartimento del Suolo, della Pianta e degli Alimenti (DISSPA), Sez. Entomologia e Zoologia, University of Bari 'Aldo Moro', via Amendola 165/A 70126 Bari, Italy. E-mail: eustachio.tarasco@uniba.it

Adriano Ragni, Dipartimento del Suolo, della Pianta e degli Alimenti (DISSPA), Sez. Entomologia e Zoologia, University of Bari 'Aldo Moro', via Amendola 165/A 70126 Bari, Italy. E-mail: adrragni@gmail.com

Giovanna Curto, Plant Protection Service – Servizio Fitosanitario Emilia-Romagna Region, Nematology Laboratory, via di Corticella, 133, 40129 Bologna, Italy. E-mail: gcurto@regione.emilia-romagna.it

Jolanta Kowalska, Department of Ecology and Environmental Protection, Institute of Plant Protection – National Research Institute, Wladyslawa Wegorka 20 Street, 60318 Poznan, Poland. E-mail: J.Kowalska@iorpib.poznan.pl

Congli Wang, Key Laboratory of Mollisols Agroecology, Northeast Institute of Geography and Agroecology, Chinese Academy of Sciences, Harbin 150081, China. E-mail: clwangjb@yahoo.com

S. Subramanian, Department of Nematology, Tamil Nadu Agricultural University, Coimbatore, India. E-mail: subbunema@yahoo.co.in

Chunjie Li, Key Laboratory of Mollisols Agroecology, Northeast Institute of Geography and Agroecology, Chinese Academy of Sciences, Harbin 150081, China. E-mail: lichunjie@iga.ac.cn

Tiffany Baiocchi, Department of Biochemistry, 900 University Avenue, University of California, Riverside, California, USA. E-mail: tbai001@ucr.edu

M. Muthulakshmi, Department of Nematology, Tamil Nadu Agricultural University, Coimbatore, India. E-mail: lakshminem@gmail.com

Jiří Nermut', Biology Centre CAS, Institute of Entomology, Branišovská 1160/31, 370 05 České Budějovice, Czech Republic. E-mail: Jirka.Nermut@seznam.cz

Vladimír Půža, Biology Centre CAS, Institute of Entomology, Branišovská 1160/31, 370 05 České Budějovice, Czech Republic. E-mail: vpuza@seznam.cz

Timothy D. Paine, Department of Entomology, University of California, Riverside, CA, USA. E-mail: timothy.paine@ucr.edu

Irma Tandingan De Ley, Department of Nematology, University of California, Riverside, CA 92521, USA. E-mail: Irma.deley@ucr.edu

Rory McDonnell, Department of Crop and Soil Science, Oregon State University, Corvallis, OR 97331- 3006, USA. E-mail: rory.mcdonnell@oregonstate.edu

Žiga Laznik, University of Ljubljana, Biotechnical Faculty, Department of Agronomy, Jamnikarjeva 101, 1000 Ljubljana, Slovenia. E-mail: Ziga.Laznik@bf.uni-lj.si

Balachander Manohar, ICAR-National Bureau of Agriculturally Important Insects (NBAII), Hebbal, Bengaluru-560024, Karnataka, India. E-mail: bala9383@gmail.com

Shailesha Arakalagud Nanjundaiah, ICAR-National Bureau of Agriculturally Important Insects (NBAII), Hebbal, Bengaluru-560024, Karnataka, India. E-mail: anshylesha@gmail.com

Rajan, Indian Council of Agricultural Research, Krishi Bhawan, New Delhi-110001, India. E-mail: rajan.newdelhi@gmail.com

Stanislav Trdan, University of Ljubljana, Biotechnical Faculty, Department of Agronomy, Jamnikarjeva 101, 1000 Ljubljana, Slovenia. E-mail: stanislav.trdan@bf.uni-lj.si

Mushtaq Ahmad Ganai, Division of Entomology, Sher-e-Kashmir University of Agricultural Sciences and Technology, Main Campus, Shalimar, Srinagar-190025, Jammu and Kashmir, India. E-mail: Ganaimushtaq12@gmail.com

Salas Augusto, Centro de Estudios Parasitológicos y de Vectores, CEPAVE (CCT La Plata CONICET – UNLP), Calle 120 y 60 S/N, (1900) La Plata, Buenos Aires, Argentina. E-mail: augustokan@gmail.com

Rusconi Matías, Centro de Estudios Parasitológicos y de Vectores, CEPAVE (CCT La Plata CONICET – UNLP), Calle 120 y 60 S/N, (1900) La Plata, Buenos Aires, Argentina. E-mail: vlc15mat@hotmail.com

Paul De Ley, Department of Nematology, University of California, Riverside, CA 92521, USA. E-mail: paul.deley@ucr.edu

M. Nagesh, Division of Plant Protection, ICAR-CPRI, Bemloe, Shimla-171001, Himachal Pradesh, India. E-mail: nageshm.nbaii@gmail.com

Preface

Growing dissatisfaction with chemical pesticides to combat agricultural pests stems from the desperate need of humans to provide more and better food for an overpopulated world. Therefore, accelerating public concern about the overuse of synthetic chemical pesticides has urged researchers to find safe alternatives to these chemicals, with a clear aim at avoiding their health hazards and environmental pollution. Many researchers and stakeholders are addressing such a goal through the development of biocontrol agents as environmentally friendly biological alternatives. Entomopathogenic and slug parasitic nematodes rank high among these alternatives.

Insecticidal or entomopathogenic nematodes (EPNs) in the families Steinernematidae and Heterorhabditidae offer a successful example of a biological pesticide, especially against soil-dwelling and wood-boring insect pests. Actually, they have many attributes of an ideal biological control agent. They are extraordinarily lethal, with most insects being killed within 24–48 h. They a have wide host range of economically important insect pests, but non-target populations are undisturbed. This high degree of safety means that, unlike chemicals, nematode applications do not require masks or other safety equipment. Nematodes do not require specialized application equipment, as they are usually compatible with existing equipment. Application via irrigation has further improved grower acceptance. Of less recognized occurrence, but of increasing importance, the slug parasitic nematode *Phasmarhabditis hermaphrodita* is a microscopic nematode in the family Rhabditidae. It is a lethal parasite of important slug species, and therefore a commercial molluscicide containing the nematode is also used for the biological control of slug pests. In addition, other phasmarhabditid species have been discovered recently with potential as a molluscicide.

The role of different entomopathogenic and slug parasitic nematodes, along with their symbiotic bacteria, as biocontrol agents of many insect pests and molluscs, respectively, is well known, and recent years have seen an intensive worldwide search for additional species and also advances or breakthroughs in this field, such as the proper formulation of some strains and their efficacy, distribution, virulence, survival biology, safe application and commercialization. Such striking advances in the different aspects of studying these nematodes entail that there is much relevant information for interested researchers such as nematologists and insect pathologists to learn from one other and keep pace with the up-to-date knowledge in each of the relevant scopes and disciplines.

In this spirit, we targeted authors who were authorities in their discipline to compose the 32 chapters contained in this volume. The idea to compile the chapters of this book is based on our past experiences in the field of biological control. This book has a broad readership and will be a useful volume for those generally interested in the biological control of pests, but is aimed specifically at those undergraduate and postgraduate students and research practitioners having insect nematology and/or mollusc control as their disciplines or subjects of interest.

In the present volume, 63 experts from 15 countries contribute authoritative chapters that capture the full breadth of the basic and applied information on these important biocontrol agents that are used or have potential in the management of insect pests, molluscs and/or other researched targets such as plant parasitic nematodes. The information includes the remarkable developments and latest achievements in this direction. The volume is divided into seven parts. There are two chapters in Part I. They introduce comprehensive information on beneficial nematodes in general and their importance, with emphasis on crop pest management. In Part II, there are four chapters devoted to covering the different aspects of the morphology, taxonomy, biology and diversity of EPNs. Part III deals with EPNs and their symbiotic bacteria against crop insect pests and consists of seven chapters. Four chapters describe their role in the management of such insects in the orders lepidoptera, coleoptera and diptera, as well as stored grain pests. Two chapters address the toxic secretions of the EPN-mutualistic bacterial species in the two genera *Xenorhabdus* and *Photorhabdus*, and their efficacy against crop insect pests either singly or with EPNs. The last chapter of this part is devoted to the mass production, formulation and application of EPNs. Part IV addresses 11 developed and developing countries as points in case, a chapter for each country, where the role of EPNs in the integrated pest management strategies implemented in each of these countries is presented.

With the dramatic advances in the different aspects of molecular biology, Part V presents the genetics for enhancing the efficacy of EPNs. It contains two chapters dealing with nematode breeding, as well as classical and current methods to achieve such an enhancement. Part VI has three chapters organized for slug parasitic nematodes. The first chapter outlines the research on nematode biology, parasitism, production and application. The second chapter summarizes the discovery and commercialization of a slug parasitic nematode, whereas the third chapter introduces striking advances in the slug and snail parasitic nematodes in North America.

In Part VII, there are three chapters to conclude the potential commercialization and future prospects of entomopathogenic and slug parasitic nematodes. The first outlines the commercial success of the compatibility of EPNs and phytopharmaceuticals. The second analyses different strategies for making EPNs a cost-effective biocontrol agent. It presents the shortcomings and hindrances in the way of commercializing nematode biocontrol agents, whereas in the final chapter future thrusts in expanding the use of entomopathogenic and slug parasitic nematodes in agriculture are identified on the basis of the critical issues mentioned in the earlier chapters of this volume, and relevant ideas are suggested as to how to utilize these nematodes tactically to achieve maximum potential in managing their hosts or relevant pests at the field level.

We place on record our sincere thanks for the time and effort of all the contributors who helped to make this book possible. We wish to thank those who generously permitted us to use their photographs and also to those who provided the necessary information and relevant data needed for several chapters in this book; these persons are acknowledged specifically in the respective chapters. Finally, we express gratitude to our families for their support in innumerable ways during the preparation of this book.

Mahfouz M.M. Abd-Elgawad, Tarique Hassan Askary and James Coupland
April, 2017

Part I: Beneficial Nematodes and Crop Protection

1 Beneficial Nematodes in Agroecosystems: A Global Perspective

Tarique Hassan Askary[1]* and Mahfouz M.M. Abd-Elgawad[2]

[1]*Division of Entomology, Sher-e-Kashmir University of Agricultural Sciences and Technology, Srinagar, India;* [2]*Phytopathology Department, National Research Center, Giza, Egypt*

1.1 Introduction

As is generally known, there are a plenty of harmful nematode species, but many other nematodes are beneficial. Both of the opposite nematode groups are found to have natural processes to maintain and stabilize the balance of life of various organisms. Associations between nematodes and other invertebrates range from 'casual' (i.e. phoretic, commensal) to obligate parasitism and pathogenesis (Stock, 2005). There are more than 30 families in five major groups – Rhabditina, Tylenchina, Myolaimina, Spirurina and Mermithina that have associations with invertebrates. The first four belong to the class Enoplea, and the fifth to the class Chromadorea (Stock, 2005). More specifically, invertebrate parasitism pertains to four major groups of nematodes (Poinar, 1983). Accordingly, the most primitive group, the Rhabditida, gave rise to members of the Oxyurida (*c.*420 million years ago), as well as to the Steinernematidae and Heterorhabditidae (375 million years ago). A fundamental evolutionary framework of the Nematoda based on 18S ribosomal DNA sequence analysis (Blaxter *et al.*, 1998) supports Poinar's hypothesis, suggesting invertebrate parasitism arose independently at least four times (Dorris *et al.*, 1999; Stock, 2005). These data also indicate an association between invertebrate and vertebrate parasitism, with invertebrate pathogenic parasitic clades lying basal to major vertebrate parasitic ones (Dorris *et al.*, 1999; Blaxter *et al.*, 2000; Stock, 2005).

In parallel with the increasing recognition of the desirable traits of nematode groups for pest and pathogen management, there is an apparent and growing dissatisfaction with the intensive use of chemical pesticides, and therefore alternative management tactics are being sought. Researchers and growers have been improving pest management tactics in response to issues emanating from applying synthetic chemicals to suppress noxious pathogens and arthropod pests. These problems include pesticide resistance to these chemicals, unsafe application, health hazards, environmental pollution and the development of pest biotypes that break host plant resistance. Such chemical troubles made a big impact on the agricultural society and drew attention to the use of biocontrol agents as a safe and effective biopesticide alternative. Admittedly, some nematode groups, for example entomopathogenic and slug parasitic nematodes, have begun to gain common acceptance, and their commercial applications are being developed as environmentally benign alternatives to chemical pesticides. Moreover, other nematode groups, such as mermithids and fungivorous nematodes, have potential as biocontrol agents of medically important insect species and/or soilborne plant pathogens and thus should be prioritized through

*Corresponding author. E-mail: tariq_askary@rediffmail.com

improved techniques to fill pesticide/fungicide pipelines properly and safely. Hence, it is expected that the above-mentioned current pressing problems of chemical pesticides and continuous technical advances in nematode production, formulation and application will provide further opportunities to integrate the use of additional nematode groups in pest management programmes.

In addition, several groups of the beneficial nematodes have a mutualistic association with bacteria. Interestingly, *Xenorhabdus* and *Photorhabdus* are two bacterial genera mutualistically associated with infective juveniles (IJs) of entomopathogenic nematodes (EPNs) from the genera *Steinernema* and *Heterorhabditis*, respectively (Askary, 2010). Although the nematode species of the two genera are able to kill a diverse array of insect pests, their symbiotic bacteria have proved to have an effective variety of toxins and antibiotics, thus exhibiting the potential of stand-alone biocontrol agents. For example, concentrated metabolites from *Xenorhabdus* spp. and *Photorhabdus* spp. could suppress a variety of peach and pecan diseases (Shapiro-Ilan *et al.*, 2014). Another type of such beneficials are the slug parasitic nematodes, particularly *Phasmarhabditis hermaphrodita* (Rhabditidae), which have demonstrated significant impact on managing mollusc pests. The slug parasitic nematodes also associate with bacteria, *Moraxella osloensis* to kill their hosts (Askary, 2010; Askary *et al.*, 2012). With the advancement of time, entomopathogenic and slug parasitic nematodes, as primary biopesticides, have aroused interest among research workers in the exploitation of other beneficial nematode groups for effective control of different crop pests and pathogens.

This chapter outlines the historical records of insect nematology before highlighting the major groups of beneficial nematodes. These include EPNs, to use against insect pests, slug nematodes to manage mollusc pests, entomophilic nematodes such as *Thripinema* to manage thrips, predatory nematodes to manage plant parasitic nematodes (PPNs) and fungal-feeding nematodes to manage soilborne plant pathogens. Though, the first two groups are the most promising but here we shall address them briefly because other chapters of this volume deal with them in more detail. Research achievements related to these nematode groups, particularly in managing insect pests, mollusc pests and plant pathogens, are summarized. Major impediments that preclude beneficial nematodes from a significant share of the pesticide market are presented, and possible solutions for their effective utilization are discussed. To develop these nematodes as biopesticides, we will address only those nematode groups that will be financially beneficial to the biocontrol market and that have had promising results published. Future research priorities and goals are also identified.

1.2 Historical Background

Nguyen (2010) has narrated in detail the cornerstone of the history of insect parasitic nematodes. The earliest record of a nematode–insect association was reported by Aldrovandus (1602), who described in his *De Animalibus Insectis* the existence of 'worms' emerging from grasshopper cadavers. Aldrovandus was the first to use the word *Vermes* (worms) associated with insects. Reaumur (1734) described a worm that was most possibly *Sphaerularia bombi* Dufour. In 1747, Gould graphically described the worms emerging from ants. These worms were probably the mermithids. Linnaeus (1758), in his *Systema Naturae*, found eight genera in the Vermes Intestini. Out of these eight, two were truly parasitic worms, whereas the third, i.e. *Gordius*, was associated with Linnaeus. The microscopic study by Goeze (1782) helped to differentiate between the various kinds of nematodes. His work also described the emergence of mermithids from soil soon after rain. Bremser (1824) in the *New Atlas of Intestinal Worms* reported Leblond's discussion of the finding of Audouin of mermithid worms in cockchafers in France. Thereafter, several authors recorded by Nguyen (2010) have published their findings of *Mermis*.

Among beneficial nematodes, EPNs have been highly ranked as the most studied and probably most useful group worldwide. Steiner (1923) isolated and described the first EPN species, *Steinernema kraussei*, from infected sawflies in Germany. From New Jersey, and other states in the USA, outstanding scientists (e.g. Gaugler and Kaya, 1990; Gaugler, 2002; Grewal *et al.*, 2005a; Shapiro-Ilan *et al.*, 1997; Shapiro-Ilan and Gaugler, 2002) have been focusing heavily on almost all aspects of EPNs. On the other hand, Filipjev – the founder of modern nematology in Russia – was also the first Russian scientist to describe EPNs. In 1934, he published a description of the nematode *Steinernema feltiae* (Filipjev, 1934). The material was sent to him from the Udmurt Republic of the Russian Federation (northeast of the European part), where local scientists had discovered intensive infestation of armyworms

(Lepidoptera) by these nematodes. Up to now, the ubiquitous *S. feltiae* was a valid species. In Russia, it was Galina V. Veremchuk who initiated applied research in EPNs in the All-Union Plant Protection Institute, St Petersburg, Pushkin. At this stage, she was mainly active in the taxonomy of Steinernematidae, describing new forms of steinernematids, although the species she described are now considered as junior synonyms of other steinernematid species (Nguyen and Hunt, 2007). She also adapted the techniques for rearing EPNs on laboratory insects and artificial media. Her pupil – L.G. Danilov – was the first specialist in Russia who proposed commercial products based on EPNs (Abd-Elgawad and Spiridonov, 2014).

Recently, numerous scientists have filed the history of beneficial nematodes and/or their useful applications, based on the experiments conducted in different countries of the world (e.g. Gaugler, 2002; Grewal *et al.*, 2005a; Stock; 2005). This latter reviewed the historical development of insect parasitic nematodes. Poinar and Grewal (2012) checked the history of entomopathogenic nematology. The two main groups, i.e. *Steinernema* and *Heterorhabditis*, have been discussed here, but Poinar and Grewal (2012) should be consulted for more information on their early descriptions; their morphology, which originally was only used to distinguish between the species; and for descriptions of the bacteria associated with nematodes and their role in the nematode–insect complex, including antibiotic properties, phase variants and impeding host defence responses.

1.3 Nematodes and Their Beneficial Groups

Nematodes are wonderful animals. They have more than one interesting aspect for each of their seniority, diversity, usage and study. For seniority, phylogenetic relationships of their different groups are being constructed and/or refined (e.g. Nguyen and Hunt, 2007; Mráček *et al.*, 2014). The diversities are in the form of many free-living plant and animal parasites. The metabolic and behavioural activities generally determine the functional roles of nematodes. These functional roles of nematodes may be classified as ecosystem services, disservices or neutral effects (Ferris, 2010). Among the disservices attributable to nematodes is overgrazing that almost diminishes the role of prey organisms or herbivores and which reduces carbon fixation and availability to other organisms in the food web. Unfortunately, management to ameliorate the potential disservices of certain nematodes results in unintended but long-lasting diminution of the services of others. The beneficial roles of nematodes may be enhanced by environmental stewardship that fosters greater biodiversity and, consequently, complementarity and continuity of their services (Ferris, 2010). Likewise, based on their feeding habits, nematodes may have a negative or positive impact on humans. A small group of nematodes are known to have a negative effect on humans, but researchers are trying to use their (the nematodes) feeding habits in a beneficial way. It is interesting to know that some PPNs species can act as biological agents, as they control aquatic weeds (e.g. Zettler and Freeman, 1972; Smart, 1984; Gerber and Smart, 1987), although their large group is generally known to cause losses in economically important crops (Abd-Elgawad and Askary, 2015).

The main focus in the utilization of beneficial nematodes has centred on their potential as biological control agents of insects and other arthropod pests and, to a lesser degree, soil pathogens. Currently, realization of the attributes of biocontrol agents, particularly in the case of EPNs and their symbiotic bacteria, has spurred developments across a far broader scientific front (Stock, 2005). In this era of discovery, the tools of molecular biology and genetics are being used increasingly to address a range of biological questions, and this will directly serve the applied utilization of these nematodes as effective biopesticides. Also, using such tools will advance these nematodes as unique and intrinsically interesting biological model systems (see De Luca and Abd-Elgawad, Chapter 6, this volume) for basic research, as well as in applied fields such as plant health, human medicine, pharmaceutical bioprospecting and genetic engineering (Stock, 2005; Murfin *et al.*, 2012; Dillman, 2013).

There are numerous examples of association between nematodes and arthropods that range from beneficial to antagonistic (Poinar, 1983). These associations are of four different types: (i) phoretic (nematodes are transported by an insect); (ii) necromenic (nematodes obtain nutrition from insect cadavers); (iii) facultative parasitism; and (iv) obligate parasitism. Further and more detailed breakdown of these groups has been reported by Sudhaus (2008). Dillman (2013) proposed that insect parasitism evolved in this sequence, with parasites evolving from non-parasitic insect associates (Fig. 1.1). He states that nematode interaction with bacteria

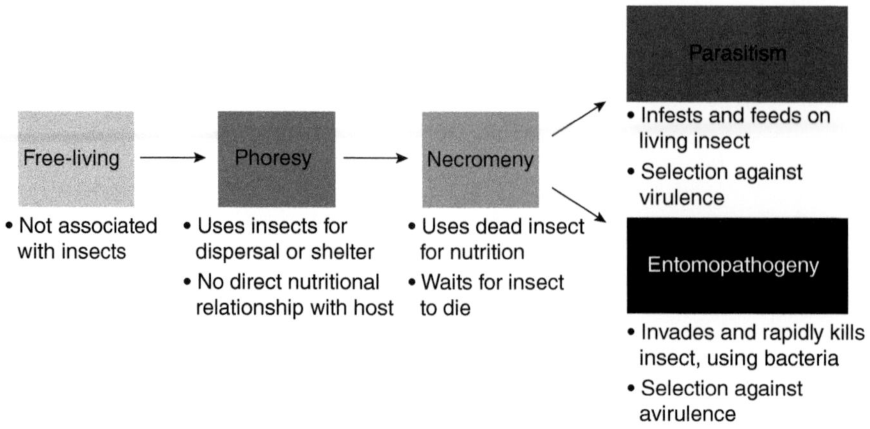

Fig. 1.1. Evolution of nematode–insect associations. Free-living: microbotrophic nematodes not known to associate with arthropods, vertebrates, plants, or fungi, only perhaps transiently associated with insects. Phoresy: nematodes adapt to use insects for dispersal or shelter but do not have a direct nutritional relationship with them. Necromeny: nematodes feed on saprophytic insect cadavers but do not participate in causing the death of the insect. Parasitism: nematodes adapt to use live insects directly for nutrition, inflict some level of harm or even cause the death of the host. Entomopathogeny: nematodes in association with insect pathogenic bacteria cause rapid disease, leading to the death of the host insect; the nematodes then feed and develop on the insect and bacterial resources. The distinction between parasitism and entomopathogeny is based on prominent features, including the use of pathogenic bacteria and the direction of selection (against virulence or avirulence), making the nematodes either more or less immediately harmful to their host (adapted from Dillman, 2013).

takes place in three ways: (i) trophism (nematodes eat bacteria); (ii) parasitism (pathogens cause nematode diseases if not resisted); and (iii) mutualism (nematodes and bacteria cooperate).

The beneficial nematodes may be grouped as entomopathogenic, slug parasitic, entomophilic, predatory and fungal-feeding forms. Herein we shall illustrate the differences among these beneficial nematode groups, besides discussing what makes these tiny creatures so interesting and useful.

1.3.1 Entomopathogenic nematodes

The two important nematode families falling in the group of EPNs are Steinernematidae Travassos, 1927, and Heterorhabditidae Poinar, 1975, which are considered as one of the most successful examples of biological pesticide used to control soil-dwelling insect pests. They possess virtually all the attributes of an ideal biological control agent. They enter the host through the natural openings such as mouth, anus or spiracles, or sometimes by abrasing the intersegmental membranes of the insect cuticle (e.g. *Heterorhabditis* spp.) (Bedding and Molyneux, 1982; Peters and Ehlers, 1994), reach the haemocoel, release there the cells of symbiotic bacteria from their intestine, which ultimately results in killing the host within 48 h (Woodring and Kaya, 1988). Because the nematode symbiotic bacterium kills insects so quickly, there is no intimate host–parasite relationship as is characteristic for other insect parasitic nematodes (Shapiro-Ilan and Gaugler, 2015). Consequently, as can be observed under laboratory tests, EPNs are very pathogenic and cause high mortality in a broad range of insect pests. However, nematodes have a limited host range in fields, due to quite narrow host specificity in some species. The efficacy of EPNs differs under field and laboratory conditions, which most possibly is due to the optimized circumstances for EPNs to function in the laboratory than in field conditions. EPNs as a group comprise of more than 90 species and are useful against numerous insect pests (Grewal *et al.*, 2005a). We list in Table 1.1 many of the insect pests (foliar, soil surface, cryptic and subterranean habitats) that are parasitized with EPNs. In future, this list is certain to expand as field

Table 1.1. Host suitability of some *Steinernema* and *Heterorhabditis* species against various insect pests of agricultural importance.

Steinernema sp./*Heterorhabditis* sp.	Host insect
Steinernema carpocapsae	Black cutworm (*Agrotis ipsilon*), diamondback moth (*Plutella xylostella*), greater wax moth (*Galleria mellonella*), tobacco caterpillar (*Spodoptera litura*), mealybug (*Centrococcus* sp.), navel orangeworm (*Amyelois transitella*), artichoke plume moth (*Platyptilia carduidactyla*), leafminers (*Liriomyza* spp.), bruchid (*Callosobruchus* sp.), sweet potato weevil (*Cylas formicarius*), grey weevil (*Myllocerus* sp.), shore flies (*Scatella* spp.), Bihar hairy caterpillar (*Diacrisia obliqua*), cat flea (*Ctenocephalides felis*), codling moth (*Cydia pomonella*), banana moth (*Opogona sachari*), blue butterfly (*Lampides boeticus*), gram pod borer (*Helicoverpa armigera*), banana root borer (*Cosmopolites sordidus*), corn earworm (*Helicoverpa zea*), rice moth (*Corcyra cephalonica*), white grub (*Holotrichia consanguinea*), cranberry girdler (*Chrysoteuchia topiaria*), mole crickets (*Scapteriscus* spp.)
S. abbasi	Tobacco caterpillar (*S. litura*)
S. feltiae	Sweet potato weevil (*C. formicarius*), tobacco caterpillar (*S. litura*), shore flies (*Scatella* spp.), leafminers (*Liriomyza* spp.), fungus gnats (*Bradysia* spp.), codling moth (*C. pomonella*), banana root borer (*Cosmopolites sordidus*), corn earworm (*H. zea*)
S. masoodi	Mealybug (*Centrococcus* sp.), bruchid (*Callosobruchus* sp.), grey weevil (*Myllocerus* sp.), Bihar hairy caterpillar (*D. obliqua*), gram pod borer (*H. armigera*), legume pod borer (*Maruca vitrata*), wheat flour beetle (*Tribolium castaneum*), greater wax moth (*G. mellonella*), blue butterfly (*L. boeticus*), rice moth (*C. cephalonica*)
S. seemae	Bruchid (*Callosobruchus* sp.), grey weevil (*Myllocerus* sp.), Bihar hairy caterpillar (*D. obliqua*), gram pod borer (*H. armigera*), legume pod borer (*M. vitrata*), wheat flour beetle (*T. castaneum*), greater wax moth (*G. mellonella*), blue butterfly (*L. boeticus*), rice moth (*C. cephalonica*)
S. thermophilum	Rice moth (*C. cephalonica*), gram pod borer (*H. armigera*), greater wax moth (*G. mellonella*)
S. glaseri	Greater wax moth (*G. mellonella*), rice moth (*C. cephalonica*), white grub (*H. consanguinea*), scarab grubs, black vine weevil (*Otiorhynchus sulcatus*), banana root borer (*C. sordidus*)
S. riobrave	Plum curculio (*Conotrachelus nenuphar*), corn earworm (*H. zea*), mole crickets (*Scapteriscus* spp.), Diaprepes root weevil (*Diaprepes abbreviatus*)
S. scapterisci	Mole crickets (*Scapteriscus* spp.)
Heterorhabditis indica	Tobacco caterpillar (*S. litura*)
H. marelata	Strawberry root weevil, (*Otiorhynchus ovatus*), black vine weevil (*O. sulcatus*)
H. bacteriophora	Scarab grubs, iris borer (*Macronoctua onusta*), fungus gnats (*Bradysia* spp.), Diaprepes root weevil (*D. abbreviatus*), corn rootworm (*Diabrotica* spp.), citrus root weevil (*Pachnaeus* spp.), black vine weevil (*O. sulcatus*), billbug (*Sphenophorus* spp.), banana moth (*O. sachari*), cabbage butterfly (*Pieris brassicae*)
H. megidis	Black vine weevil (*O. sulcatus*)
H. downesi	Large pine weevil (*Hylobius albietis*), black vine weevil (*O. sulcatus*)
H. zealandica	Scarab grubs, grape root borer (*Vitacea polistiformis*)

research progresses and correct matches of EPNs and target pests are being made.

Although dozens of important insect pests are susceptible, non-target populations are undisturbed (e.g. Gaugler and Kaya, 1990; Kaya and Gaugler, 1993; Gaugler, 2002; Dillon *et al.*, 2012). Despite their high pathogenicity to insects, their lack of threat to the environment led the US Environmental Protection Agency to exempt EPNs from registration requirements; nearly all countries provide a similar exemption. Thus, contrary to chemicals, or even the bacterial biocontrol agent *Bacillus thuringiensis*, EPN applications do not cause pollution or pollinator toxicity, and do not require masks or other safety equipment. The adverse effects on beneficial insects caused by chemicals lead to some other problems too, such as giving rise to secondary pest outbreaks, besides manifesting environmental pollution and health hazards (e.g. Luckman and Metcalf, 1982). The application of nematodes is easy, as they are compatible with existing equipment including pressurized, mist, electrostatic, fan and aerial sprayers. Application via irrigation has further improved grower acceptance (Abd-Elgawad, 1998).

EPNs have been utilized in classical, conservation and augmentative biological control programmes. Research on these beneficial nematodes is concentrated mainly on their potential as inundatively applied augmentative biocontrol agents. The importance of EPNs for the control of above and below ground insect pests was highlighted by Lacey and Georgis (2012). The development of EPNs especially in the 1980s and early 1990s was due to advancement in the mass production and formulation technology of EPNs, the discovery of numerous efficacious isolates/strains of EPNs and the desirability of reducing pesticide usage. Commercially produced EPNs are currently in use for control of scarab larvae in lawns and turf, fungus gnats in mushroom production, invasive mole crickets in lawn and turf, black vine weevil in nursery plants and Diaprepes root weevil in citrus, in addition to other insect pests (Lacey and Georgis, 2012). In the past few years, the characteristics of some commercialized species of EPN–host matching have been reported (Abd-Elgawad, 2001; Shapiro-Ilan and Gaugler, 2015). The commercially common species are *Steinernema carpocapsae*, *S. feltiae*, *S. glaseri*, *S. kushidai*, *S. riobrave*, *S. scapterisci*, *Heterorhabditis bacteriophora*, *H. indica* and *H. megidis*.

EPNs can suppress some species of PPNs (e.g. Grewal *et al.*, 1999; Somasekhar *et al.*, 2002). Jagdale *et al.* (2009) have provided biochemical and molecular evidence that EPNs induce components of systemic resistance in plants against PPNs. Nevertheless, there is numerous evidence of the success and failure of EPNs against both phytonematodes and crop insect pests of ornamental plants, trees, lawn and turf. If EPNs did not inherit the common defects of biologicals, they would surpass chemical distribution and sales in pesticide markets. The merits and demerits of EPNs are presented in Table 1.2. Lacey and Georgis (2012) are of the opinion that various desirable traits, such as recycling and the persistence ability of some EPN species in the environment, the direct and/or indirect effects on populations of PPNs and plant pathogens and compatiblity with a wide range of chemical and biological pesticides, offer good scope for the inclusion of EPNs in integrated pest management (IPM) programmes.

Moreover, mobilizing different EPN species together with naturally occurring soil microbiota has been examined. Hurley *et al.* (2014) demonstrated an increase in plant growth and health, in addition to a reduction in the use of non-specific chemical fertilizers and pesticides. They investigated the interaction between two naturally occurring taxa; i.e. EPNs (*S. feltiae*, *S. carpocapsae* and *H. bacteriophora*) and a range of bacterial endophytes. In both laboratory and

Table 1.2. Merits and demerits of entomopathogenic nematodes (Modified from Lacey and Georgis, 2012).

Merits	Demerits
Broad pest insect host range	Cost of production
Rapid speed of kill	Limited shelf life and refrigerated storage required
Can actively seek or ambush host Mass *in vivo* and *in vitro* production capability Possible to use conventional application equipment Safety: for all vertebrates, most non-target invertebrates and the food supply	Environmental limitations: requirements for adequate moisture (to enable survival and infectivity) and temperatures (above or below that required for optimal infectivity), sensitivity to UV radiation, lethal effect of several pesticides (nematicides, fumigants and others), lethal or restrictive soil chemistries (high salinity, high or low pH, etc.)
Little or no registration required	Limited spatial modelling to predict the nematodes' persistence and fates
Can be adopted and produced in relatively unsophisticated facilities	Most growers are unaware of their safe and secure usage, especially in developing countries

greenhouse trials, the bacterial endophytes had no adverse effects on the ability of the EPNs to kill *Galleria mellonella* hosts. In experiments where *G. mellonella* were exposed to increasing nematode doses, the presence of *Pseudomonas fluorescens* F113 and S118 caused *H. bacteriophora* to enter the insects in higher numbers.

1.3.2 Slug parasitic nematodes

Phasmarhabditis hermaphrodita (Family: Rhabditidae) is a lethal parasite of the slug *Deroceras reticulatum*. Moreover, the nematode is also parasitic on other slug species belonging to the families Milacidae, Limacidae and Arionidae (Wilson *et al.*, 1993; Askary *et al.*, 2012). The bacteria symbiotically associated with the IJs of this nematode is *Moraxella osloensis*. IJs move through the soil, locate the slug host and infect it by penetrating through a natural opening at the backside of the mantle. Once inside the body of the host, the IJs release bacterial cells, start feeding on multiplying bacteria and develop into self-fertilizing hermaphrodites. The death of slug takes place within 7–21 days after infection. *M. osloensis* produces a heat-stable endotoxin, which consists of a lipopolysaccharide lethal to slugs (Tan and Grewal, 2002). *M. osloensis* kill slugs only when they are carried by IJs of *P. hermaphrodita* (Tan and Grewal, 2001). Due to its commercial importance, utilization of *P. hermaphrodita* as a molluscicide is being developed, and isolating it for the first time from new regions, such as North America (see De Ley *et al.*, Chapter 29, this volume) and Egypt (Genena *et al.*, 2011) is in progress. Needless to say, indigenous *P. hermaphrodita* isolate(s) may be more adaptive and less expensive, with no risk to native fauna and flora, as can be the case with imported strains. Details on nematode biology, parasitism, production and application can be found in Chapter 27, this volume (by Nermut' and Půža), and on the discovery and commercialization of slug parasitic nematodes in Chapter 28, this volume (by Glen and Coupland). Since this chapter addresses beneficial nematodes, we would like to stress the main benefits of using *P. hermaphrodita* (http://www.slugoff.co.uk/killing-slugs/nematodes):

- they are fully organic and eco-friendly
- they do not have any adverse effects on children, pets, birds or wildlife
- they are totally safe to apply around food crops
- the slugs infected by *P. hermaphrodita* retreat underground before death: thus, there is no abundance of dead slugs on the surface
- dead slugs eaten by other creatures are non-toxic
- impossible to overapply
- the nematodes work more actively in wet weather conditions – slugs are most active and destructive in these conditions, and wet weather renders poison pellets less effective
- *P. hermaphrodita* is compatible with most chemical pesticides or fertilizers.

1.3.3 Entomophilic nematodes

Entomophilic nematodes literally means the round worms that like insects. So, they refer originally to the primary group of insect parasitic nematodes. This term may cover all nematode groups that are associated with and/or harmful to insect pests. Nevertheless, some of these groups have been demonstrating such special attributes that scientists have preferred to separate them as self-standing groups. These latter currently include EPNs, slug parasitic nematodes, predatory nematodes and fungal-feeding nematodes. The rest of the primary groups are still found in the original entomophilic nematodes. In this chapter, we adopt the classification of Grewal *et al.* (2005a), who included three subgroups as entomophilic nematodes, i.e. *Beddingia* (*Deladenus*), *Thripinema* and mermithid nematodes.

Deladenus siricidicola (= Beddingia siricidicola) (Tylenchida, Neotylenchidae)

Sirex noctilio (Hymenoptera: Siricidae), commonly known as the Sirex woodwasp, is a serious pest of pine (*Pinus* spp.) trees. It is a wood-boring wasp of pine trees found in different regions of the world (Morris, 2015). The wasp kills pine trees through the use of a symbiotic, tree-pathogenic fungus (*Amylostereum areolatum*), along with a phytotoxic mucus, which are injected into the tree along with insect eggs during oviposition.

Deladenus siricidicola has been reported as the most successful nematode being used in the control of *S. noctilio* (Stock, 2005). This was the first commercial use of nematodes to control any pest (Bedding and Iede, 2005). The nematode was discovered by Zondag (1962) and later described by Bedding (1968). *D. siricidicola* has two morphologically different orders; a fungal-feeding form that lives in the pine tree and eats the symbiotic fungus injected into the tree by an adult *S. noctilio*. By adopting this form, the nematode can reproduce in the tree for many generations in the absence of the woodwasp as it spreads throughout the tree. The nematode achieves parasitic form when it detects the larvae of *S. noctilio*. The infective juveniles (IJs) of *D. siricidicola* penetrate the larvae and reproduce. The offspring (nematode juveniles) migrate into the reproductive organs of the host, making their way into the eggs before the shells have hardened. The host, *S. noctilio*, is not killed by the nematode, but it undergoes 'parasitic castration'. The parasitized adult female of *S. noctilio* after emerging from the trees, mate and oviposit in a new pine tree, where during oviposition they inject fragments of the symbiotic fungus into the tree wood, in addition to phytotoxic mucus and eggs. These eggs are full of nematodes, instead of woodwasps. The cycle is repeated when the female host of *S. noctilio* injects into the tree wood and the nematodes are transmitted by this process. At this stage, the nematodes return to the fungal-feeding form. The nematodes cannot recycle if they penetrate *S. noctilio* males since the males cannot transmit the nematodes to females during mating.

Mass production of *D. siricidicola* for the control of *S. noctilio* can be performed under laboratory condition. The nematode stage that is fungal feeding is inoculated into flasks with established cultures of the symbiotic fungus of *S. noctilio*. The nematodes thus produced in the culture are harvested and then inoculated into trees of *S. noctilio*. Prior to the release of nematodes, trap trees are made to attract *S. noctilio*, which is achieved by injecting herbicides into the trees, which slowly kills them and makes them both attractive and highly susceptible to infestation by the woodwasps. The trees are injected 3 months before the *S. noctilio* are anticipated to emerge from the trees. Once woodwasp infestation is confirmed on the trap trees, the nematodes are injected. A proper hammer is used to punch bores into the trees, and the nematodes, suspended in a gel, are introduced into the holes in the live pine. Infected *S. noctilio* then disperse the beneficial worms to other trees.

This control programme has great success in Tasmania and Australia (Bedding, 2009; Morris, 2015), and the use of the nematode in this method saves an estimated US$16–60 million/year in Australia. Bedding (1984, 2009) provides details of the remarkable success of *D. siricidicola* and its application as a classical biocontrol agent. In Brazil, this control programme saves US $6.6 million/year. But the success is not uniform in all regions. The use of *D. siricidicola* against *S. noctilio* gave unsatisfactory results in South Africa and South America. The failures may be due to several factors, such as climate, a defective strain of the nematode or poor inoculation technique. Hurley *et al.* (2007) suggested that a single, integrated control strategy for *S. noctilio* would probably require significant local adaptation in new areas of infestation and over time. The low success in some regions may translate into poorer establishment and spread of *D. siricidicola* than was seen in areas with high initial success of *D. siricidicola* introduction. The factors responsible for this might be the low moisture content of the wood or incompatibility between nematode strain and wasp or nematode and fungus. This needs to be re-examined and further species or strains of nematode need to be evaluated to match specific conditions (Hurley *et al.*, 2007). In the past two decades, researchers have explained the cause of and response to Sirex breakdown and the strategies adopted to prevent its recurrence (Haugen, 1990; Haugen and Underdown, 1990; Bedding and Iede, 2005). Hominick and Perry (2000) suggested the biology of both the host and the parasite be studied in great detail, as well as their best matching. Moreover, it is basic to commercial success to be able to mass-culture the nematodes. Even before marketing, this is a prerequisite that will drive to large-scale experimentation.

Thripinema

The genus *Thripinema* was established by Siddiqi (1986) during a taxonomic revision of the species. Several species of nematodes belonging to the genus *Thripinema* (= *Howardula*) (Tylenchida: Allantonematidae) are known to naturally parasitize thrips (Thysanoptera). The tylenchoid ancestors of the Family Allantonematidae progressed towards mycetophagy and eventually parasitism of the insect haemocoel (Siddiqi, 2000).

We would like to cite an example of the life cycle of *Thripinema fuscum* in a female tobacco thrips, *Frankliniella fusca* host. The development of a large robust stylet and the correlated transformation of the broad corpus into a muscular organ are the keys that allow the entomoparasitic female to penetrate the exoskeleton of the insect and develop in the host haemocoel. Once inside, the spermaticized slender female goes through a radical transformation, where its development is synchronized with that of the host (Funderburk and Sims-Latsha, 2005). The parasitic female increases in size and the genital organs gradually fill most of the body cavity. The eggs are set free in the host's haemocoel. The eggs hatch and the juveniles moult to the fourth stage and exit through the anus or oviduct of the host. Supposedly, the male undergoes a final moult in the free life and mates with the female. In a single heterosexual cycle, both free-living and parasitic forms are found, the former partially free-living, as they do not feed or multiply outside the host. Sterility of the thrips host is the outcome of parasitism (Loomans *et al.*, 1997; Funderburk and Sims-Latsha, 2005). Sims *et al.* (2012) studied in detail the *in vivo* life cycle of the nematode, *T. fuscum*, and its parasitic effects on the thrips, *F. fusca*. The parasitic female of *T. fuscum* produced eggs within 4–5 days after entering into *F. fusca*, filling the host's haemocoel with eggs and developing juveniles. Mature juveniles migrated to the hindgut, while the fully developed juveniles first emerged after 9 days of parasitization from the posterior end of the male or female *F. fusca*. Parasitization induced a displacement of the host alimentary tract, an atrophy of the ovaries and fat body, and an alteration of energy and waste reserves in host tissues. Observations reveal that this parasitic nematode is well adapted to exploit the digestive, reproductive and excretory system of the host. To date, there are at least five described species of *Thripinema*: *Thripinema reniraoi*, *Thripinema aptini*, *Thripinema nicklewoodi*, *Thripinema khrustalevi* and *T. fuscum*. They obligatorily parasitize at least 13 species of thrips (Sims, 2010). Nematode infection is not lethal to thrips, but causes sterilization of adult female hosts of thrips. The biogeographical range of thrips parasitic nematodes will probably extend over time. Currently, it includes countries such as the UK, Germany, the USA, Canada, Russia and India.

Arthurs and Heinz (2002) described methods for the *in vivo* production of the nematode, *T. nicklewoodi*, for the biological control of western flower thrips, *Frankliniella occidentalis* (Pergande), an important insect pest of greenhouse flower crops. The methods were conducive for both fertilization and horizontal transmission of *T. nicklewoodi* in 1.5 ml microcentrifuge tubes (infection arenas) in the presence of 100% humidity, a temporary food source and preferably a damp substrate. *F. occidentalis* were reared after exposure to infection arenas on excised bean leaves, *Phaseolus vulgaris* (L.), in polypropylene containers for 2 weeks at 25°C. A single generation of nematodes was reproduced and developed by this method within the abdominal cavity of the infected host. It was found that the nematodes in infected thrips were first released between 12 and 14 days, with an average of 7.9 days at 25°C. On the basis of another experiment, Arthurs and Heinz (2003) suggested that *T. nicklewoodi* might help to prevent thrips outbreaks and reduce direct feeding damage and secondary virus spread during the pre-lethal period of infection. Lim (2003) observed an excretion rate of 21.4 nematodes/day by the parasitized female thrips. The sex ratio of the excreted *T. nicklewoodi* was 6:1 (female:male). Male thrips were found to be parasitized by nematodes as readily as females. Parasitism resulted in reducing the longevity of both female and male adult thrips by 26% and 61%, respectively.

Mermithid nematodes

Mermithid nematodes may be terrestrial, semi-terrestrial, or aquatic, but all are obligate endoparasites of members of the phylum Arthropoda, particularly insects. Mermithids tend to be host specific, usually to a single host species or family (Poinar, 1979). Most aquatic mermithids have seen limited study, with the exception of species parasitizing mosquitoes (Sanad *et al.*, 2013). Research focus on mosquito mermithids as biological alternatives to chemical insecticides is attributed to the nematode invasion of medically important disease vectors (Petersen, 1973; Platzer *et al.*, 2005; Sanad *et al.*, 2013).

Mermithids produce pre-parasitic juveniles that enter host species by the ingestion of eggs, as encysted juveniles within ingested paratenic hosts, or by direct penetration of the host integument and entrance into the haemocoel (Platzer *et al.*, 2005). After infection, the parasitic juvenile stage grows and moults once. After storing sufficient sustenance for its adult phase, it emerges and kills the host. Further development of the adults and reproduction take place outside the host. The role of mermithids in biological control have recorded attributes such as ease of application, environmental safety, host specificity, laboratory manipulation of life-history traits, lethality, mass rearing *in vivo* and the potential for long-term recycling in the environment (Petersen, 1973, 1985; Popiel and Hominick, 1992; Platzer *et al.*, 2005; Platzer, 2007). Other aspects of mermithids, such as the role of the surface coat of *Romanomermis culicivorax* in evasion of the host immune system (Shamseldean *et al.*, 2007), the effects of insect growth regulators on the host of the mosquito parasitic nematode *R. iyengari* (Suman *et al.*, 2013), and the penetration and emergence patterns of *R. iyengari* and *Strelkovimermis spiculatus* in laboratory exposures against the larvae of *Culex pipiens* (Sanad *et al.*, 2013), have been studied. The latter authors found that mermithid penetration and emergence behaviours and sites may be so important that they could be used as supplementary clues for identification, and that species differences could be useful in developing production and release strategies.

The main problem with mermithids is that their rearing may be costly, and the population dynamics are not compatible with the characteristics of an ideal biocontrol organism (Popiel and Hominick, 1992; Platzer *et al.*, 2005). Yet Petersen and Willis (1972) published good mass rearing procedures of *R. culicivorax* (= *Reesimermis nielseni*), which parasitize culicid mosquito larvae, and thereafter field releases were tried and monitored (Petersen, 1985). On the other hand, two mermithid species have been used to reduce noctuid and brown plant hopper populations (Chen *et al.*, 1992; Choo *et al.*, 1995). However, the same limitations that restricted the efficacy of aquatic mermithids for vector control seem also to apply for the control of agricultural pests (Hominick and Perry, 2000). Hominick and Tingle (1984) assessed the limitations of mermithids for vector control using population biology models. They found mermithids to be controlled by robust density-dependent restraints (such as environmental, sex determination, parasite-induced host mortality and reduced fecundity of crowded females), such that they could cause only moderate long-term depressions in their host populations. Also, Suman *et al.* (2013) stated that there would be minimal interest in developing mosquito mermithids as biological insecticides against mosquitoes while the microbial biopesticide, *B. thuringiensis* var. *israelensis*, remains a widely available, inexpensive and effective alternative biological agent (World Health Organization, 2006). Platzer *et al.* (2005) stressed that investigations on mermithids have been demonstrating their long-term effects on host populations (Baker and Capinera, 1997), and that their mass rearing *in vivo* in developing countries could be a viable and attractive alternative for the control of vector insects (Santamarina and Perez, 1997; Platzer *et al.*, 2005).

1.3.4 Predatory nematodes

This group of nematodes feed generally on soil microorganisms and can be found wherever their prey is. For example, *Mononchus* spp. may be found in damp soil, sand or gravel on the shore of a water source, or on the bottom of a lake, pond, stream, river, or marsh (Opperman and Salstead, 2015). Just as these nematodes prey on microorganisms, there are predators preying on this group: for example, small fish, insects and crustaceans. Our main interest in these nematodes stems from their ability to prey on PPNs. They reduce populations of PPNs in virtually all soils because of their constant association with them in the rhizosphere, and also release nutrients in plant-available forms, which may enable plants to withstand the nematode burden on their roots better (Khan and Kim, 2007).

Predatory nematodes belonging to the orders Mononchida, Diplogasterida, Dorylaimida, Aphelenchida, Enoplida and Rhabditida are categorized into three groups based on their feeding apparatus, food and feeding habits (Bilgrami, 1997; Bilgrami and Brey, 2005). Yet most of these nematodes fall under the first four orders (Khan and Kim, 2007; Kim, 2015). The nematodes belonging to the order Mononchida, Dorylaimida and Aphelenchida are good predators of PPNs, but the shortcomings with these beneficial nematodes are their longevity and stability related to shelf life (storage), which are the main attributes for the commercialization of any biocontrol agents (Askary, 2015; Askary and Martinelli, 2015). The latter authors reported that the data available in field/natural conditions were insufficient to conclude whether they were effective biocontrol agents of PPNs. However, they stressed that diplogasterids were the most suited for the biocontrol of nematodes, among the different types of predators, because of their short life cycles, easy culture, prey specificity, chemotaxis sense and resistance to adverse conditions (Bilgrami and Jairajpuri, 1990; Kim, 2015).

The mononchid predators have a strongly sclerotized buccal cavity, which is often armed with one tooth, more large puncturing teeth, numerous small grasping teeth, or both. Several of the commonly occurring mononchids feed extensively, though not exclusively, on plant parasitic and other nematodes (Bilgrami *et al.*, 1986; Bilgrami, 2008; Kim, 2015). They can be cultured on prey species in laboratory conditions. They may swallow their prey completely if it is of a smaller size, or they may feed by cutting larger prey into pieces. The duration of the life cycle varies from species to species. The life cycle of *Prionchulus punctatus* needs 45 days, but *Mononchus aquaticus* takes only 15 days at

25°C (Maertens, 1975; Grootaert and Maertens, 1976). Female *Mononchus* spp. lay eggs in the soil after mating. Some of these species are hermaphrodites, where both animals lay eggs after mating. Eggs hatch in a few hours or up to several weeks, depending on the species and the temperature. Nematode eggs are tough. They resist freezing and can survive without oxygen. Sometimes, they are transported when the mud they are in sticks to the feet of a large animal. Aquatic *Mononchus* spend most of their time crawling through the soil, or on plants, looking for food. Nematodes are important, because they help mix up the soil, which keeps a healthy environment for plants and other animals. *Mononchus* can release a sticky fluid from its tail to grab on to objects when it needs to hang on (Opperman and Salstead, 2015). Nevertheless, several researchers (e.g. Thorne, 1927; H.H. Ameen, Cairo, 2015, personal communication) found that soil populations of mononchids tended to be unstable and to change rather abruptly, sometimes disappearing or becoming drastically reduced. However, in pot trials, Cohn and Mordechai (1974) found a constant correlation between a high population level of *Mylonchulus sigmaturus* and a low population level of citrus nematode, *Tylenchulus semipenetrans*. Small (1979) also found significant reductions in the population densities of potato cyst nematode, *Globodera rostochiensis*, and root-knot nematode, *Meloidogyne incognita*, in the presence of a predatory nematode, *P. punctatus*, in pot experiments. Azmi (1983) found an increase in the abundance of *Iotonchus monhystera* and a reduction in that of spiral nematode, *Helicotylenchus dihystera*. Rama and Dasgupta (1998) reported that predator–prey relationship existed between the predator *Ironus tenuicaudatus* and two PPNs, *T. semipenetrans* and *H. dihystera*, in mandarin orange orchards. Khan and Kim (2007) gave numerous examples of the species of mononchid predators and their prey PPNs.

Dorylaimid predators have a hollow stylet, properly called odontostyle, which is used to puncture the prey organisms and to feed through (Khan and Kim, 2007). Linford and Oliviera (1937) concluded that, when feeding on other nematodes, the large odontostyle of these predators disturbs the internal organs of the prey to quickly render it immobile. The species of Nygolaimoidea and *Sectonema* have a large, slender, protrusible tooth called a mural tooth, with which they pierce or slit their prey. This tooth is not hollow and food is sucked up through the oral aperture. Dorylaimid predators are the most ubiquitous group of predatory nematodes, occurring in all types of soils, climates and habitats (Khan and Kim, 2007). The presence of two, three or more genera of dorylaimids in one place/field is quite usual, and their abundance has been estimated to be 200–500 million/acre (Thorne, 1930). In addition to their feeding on nematodes, dorylaimids are also known to feed on other soil microorganisms. Most studies on the predatory potential of dorylaimids are based on *in vitro* evaluation. For example, Bilgrami *et al.* (1985) provided a detailed account of the predation abilities of *Aquatides thornei*. Dorylaimids generally take 3–6 months to complete one life cycle, but *Labronema vulvapapillatum* completes its cycle in 36 days at 25°C (Ferris, 1968; Wood, 1973; Khan and Kim, 2007). The detection of *Eudorylaimus obtusicaudatus* feeding on eggs inside cysts of sugar-beet cyst nematode, *Heterodera schachtii* and an increase in the population of *Thornia* sp. in the presence of *T. semipenetrans* but a decrease in their absence in pot trials (Boosalis and Mankau, 1965) indicates the biocontrol potential of the nematodes. The efficient rate of predation by dorylaimid predators (Khan *et al.*, 1991; Bilgrami, 1992), the positive correlation between predation and prey trophic groups (Bilgrami, 1995), prey preference (Khan *et al.*, 1995; Khan and Jairajpuri, 1997), efficient prey searching ability, attraction and aggregation activities at the feeding site (Bilgrami and Gaugler, 2005) and the wide range of predation on PPNs (Khan and Kim, 2007) have established their credentials as efficient biocontrol agents.

The diplogasterid predators, generally found abundantly in decomposing organic manure, have a comparatively smaller buccal cavity than those of mononchids, but are armed with teeth (Khan and Kim, 2007). They complete their life cycle in 8–15 days and are the most readily cultured nematodes, being easily maintained on simple nutrient media containing bacteria (Yeates, 1969; Yeates and Bongers, 1999). Diplogasterid predators appear to be more prey selective than other groups. *Odontopharynx longicaudata* attacked and killed 6 out of 17 prey species presented in a laboratory study, and 100% of individuals of *Anguina pacificae* were killed. The ability of *O. longicaudata* to eliminate a population of *A. pacificae* suggests that a target prey–predator relationship exists in the field habitat from where the two species were collected (Chitambar and Noffsinger, 1989). A strong degree of prey preference was reported for *Mononchoides fortidens*, *M. longicaudatus* and *M. gaugleri*; they preferred juveniles of endoparasitic over ectoparasitic nematodes (Bilgrami *et al.*, 2005). Studies on the predatory behaviour of diplogasterid predators on PPNs revealed their chemotaxis behaviour towards attractants from prey nematodes (Bilgrami and Jairajpuri, 1988, 1989). Fauzia *et al.* (1998) demonstrated the ability of *M. longicaudatus* to reduce root galling by root-knot nematodes in pot tests, resulting in improved vegetative growth and increased root mass. Osman (1988) reported a decline in populations of *Meloidogyne javanica* and *T. semipenetrans* in potted soil in the presence of diplogasters. Khan and Kim (2005) reported that a pre-plant application of *M. fortidens* in potted field soil infested with *Meloidogyne arenaria*, reduced the root galling on tomato plants and suppressed the *M. arenaria* population, but on the other hand, the final population density of *M. fortidens* increased by about twofold compared to the density initially added. Thus, the diplogasterid predators may prove promising as biocontrol agents, but only further studies will confirm their utility at the commercial level. The main advantages of using diplogasterid predators lie in their chemotaxis sense, high rate of predation and easy culture, prey specificity and formation of dauer larvae under adverse conditions (Bilgrami and Jairajpuri, 1990). Diplogasterid predators and their prey nematodes have been listed by Khan and Kim (2007).

Aphelenchid predators in the genus *Seinura* have a typical aphelenchid stylet. Such a needle-like stylet penetrates the nematode cuticle and injects digestive enzymes into the prey body, which immediately paralyses the prey to ingest the contents of the prey body (Wood, 1974). Although *Seinura* spp. is generally of a small size, they can feed on larger preys (Linford and Oliviera, 1937). One generation of these species is completed in 3–6 days (Hechler and Taylor, 1966). They possess a high reproductive potential and can be cultured easily and rapidly on fungivorous nematodes under laboratory conditions (Khan and Kim, 2007).

Several issues hinder the tactful use of predatory nematodes as effective biocontrol agents of PPNs, especially concerning their mass production and application. On the other hand, Bilgrami and Brey (2005) explain factors that can contribute to the magnitude of successful control potential of PPNs by predatory nematodes. These might include resistance/susceptibility of prey nematodes to predation, prey specifity, efficacy, cannibalism, span of life cycle, ecology, culture and conservation. For instance, the numerous merits of diplogastrid over mononchid predators include their high rates of reproduction, relative ease of *in vitro* culture, predation, short life cycle and ability to detect and respond to prey attractants and rare cannibalism (Siddiqi *et al.*, 2004; Bilgrami *et al.*, 2005). Diplogastrids also possess greater tolerance to unfavourable environmental conditions (Bilgrami, 1997) but on the other hand, density-dependent predation (Bilgrami *et al.*, 2005). Bilgrami *et al.* (2008) conducted the first field release of a predatory nematode, *M. gaugleri* (Nematoda: Diplogastrida), to control PPNs. At 30 days post-application, plant parasitic and non-parasitic nematode populations decreased by up to 24.3 and 19.3%, respectively. *M. gaugleri* decreased populations of four parasitic (*Tylenchus*, *Ditylenchus*, *Tylenchorhynchus*, *Aphelenchus*) and one non-parasitic nematode genera (*Mesorhabditis*).

A general drawback in the field application of predacious nematodes for the control of PPNs is that a huge population of predators is required to execute biocontrol programmes successfully (Khan and Kim, 2007). The low usability of predatory nematodes against PPNs is also due to difficulties in the mass production and storage of the predatory nematodes (Kim, 2015). Moreover, the extent of their prey on PPNs will depend on the soil fauna, since these predators may deviate to other soil microorganisms vulnerable to predatory nematodes. Early juvenile stages cannot rely on feeding upon live prey due to their small size; they obtain nourishment from other sources such as soil particles, agar or bacteria (Grootaert and Maertens, 1976). Nelmes (1974) reared *P. punctatus* on *Aphelenchus avenae* and *Panagrellus redivivus* in soil extract agar medium and reported that *A. avenae*, a beneficial nematode species, appeared to be more suitable prey for many mononchids.Therefore, the use of predatory nematodes as biocontrol agents should be concentrated on the migratory stages of endoparasitic PPNs, as these are susceptible to predation.

1.3.5 Fungal-feeding nematodes

Nematode species of the genera *Aphelenchus*, *Filenchus*, *Tylenchus* and *Iotonchium* have been reported to feed on fungi. This number of fungivorous nematode genera is expected to increase, either through adding additional genera or splitting any of the existing genera. For instance, a remarkable abundance of GH16 (endo-1,3-beta-glucanase), which is thought to be important for the degradation of fungal cell walls, was detected in nematode *Aphelenchoides besseyi* (Wang *et al.*, 2014). This enzyme has been identified in various fungivorous nematodes and PPNs, first in the PPN, *Bursaphelenchus* species (Kikuchi *et al.*, 2005), and more recently in *A. avenae* (Karim *et al.*, 2009). The fact that *A. besseyi* harbours considerable GH16 (Wang *et al.*, 2014) confirms that it is a facultative fungal feeder. Furthermore, Wang *et al.* (2014) speculated on the possibility that these fungi-consuming nematodes acquired endo-1,3-beta-glucanase genes from bacteria to obtain their fungal feeding ability, and subsequently acquired cellulase genes from fungi, which permitted them to parasitize plants.

Fungivorous nematodes are beneficial organisms for organic matter decomposition, in the field, and as biocontrol agents of plant pathogenic fungi (Wasilewska and Bienkowski, 1985; Wang *et al.*, 2004; Ishibashi, 2005). Most of these studies have concentrated on one species, *A. avenae* Bastian 1865, which is ubiquitous in the temperate zone and has been cultured experimentally on many species of fungi (Townshend, 1964; Ishibashi, 2005). The potential of fungivorous nematodes for the control of soilborne fungal diseases of plants has been reported either explicitly or implicitly. For example, plant pathogenic fungi such as *Fusarium oxysporum* f. sp. *conglutinans*, *F. oxysporum* f. sp. *cucumerinum*, *Pythium ultimum* and *Rhizoctonia solani* have been reported as good hosts for *A. avenae* (Okada and Kadota, 2003). Having described a mass production and a long-term preservation method, Ishibashi (2005) stated a number of merits related to *A. avenae* application for the biocontrol of soilborne plant diseases; industrial wastes can be used effectively in *A. avenae* mass production, and repeated applications of the used substrates (which still contain the nematodes) can further suppress PPNs and soil fungal pathogens such as *Rhizoctonia*, *Pythium*, *Phytophthora*, *Verticillium* or *Fusarium*. Also, for nematode formulation, he described *A. avenae* condensed cakes to reduce transportation costs. Nevertheless, *A. avenae* fecundity is gradually reduced, eventually rendering the population extinct, if its stock culture is subcultured serially with the same fungus (Mankau and Mankau, 1962). Such *A. avenae* deterioration owing to subculturing can be prevented by storing dried nematodes or by subculturing on several different host fungi. When large numbers of nematodes (more than 1 × 10^4/100 ml soil) are applied to sterilized soil together with seeds, seed germination may be hindered (Kobayashi and Chikuo, 1993). Such an adverse effect may not occur in natural soils due to the existence of both other sources of food for *A. avenae* and nematode antagonists. Differences in virulence among *A. avenae* strains and application timing could result in significant variations observed in efficacy studies (Gupta, 1986; Ishibashi, 2005). An effective biocontrol of soilborne diseases by *A. avenae* and mixed application with other beneficial nematodes could materialize (Ishibashi, 1998). Yet Ishibashi stressed that *A. avenae*, could only be used at the recommended dose as a prophylactic treatment; not to treat disease-suffering plants. Also, Lagerlöf *et al.* (2011) found that *Aphelenchoides* spp. and *A. avenae* could suppress damping-off caused by *R. solani* in cauliflower seedlings and enhance the disease-suppressive effect of compost in the greenhouse. They found that *Aphelenchoides* spp. + *R. solani* had 85% healthy plants compared with 45% in *R. solani* without nematodes. *A. avenae* suppressed damping-off significantly in all substrates, from almost 100% dead plants in peat with *R. solani* to 65% healthy plants in *R. solani* + *A. avenae*. There were no additive suppressive effects (enhancement) between nematode effects and the suppressive compost (Lagerlöf *et al.*, 2011).

On the other hand, the fungi determined as good hosts for *Filenchus misellus*, based on nematode population growth rate and female body length were two Basidiomycota fungi (*Agaricus bisporus*, *Coprinus cinereus*), three Ascomycota fungi (*Chaetomium cochlioides*, *Chaetomium funicola*, *Chaetomium globosum*) and a plant pathogenic fungus (*R. solani*) (Okada and Kadota, 2003). Reciprocally, *Pleurotus ostreatus*, which is a predaceous fungus for other nematodes, was also a good host for the nematode, *F. misellus*. *A. avenae* was trapped and preyed upon by *Pleurotus hyphae.* Differences in habitat preference between *Filenchus* and *Aphelenchus* could be explained on the basis of the host status and habitat preferences of the tested fungi (Okada and Kadota, 2003). Hasna *et al.* (2007) examined the attraction of the fungivorous nematodes *A. avenae* and *Aphelenchoides* spp. to seven fungal species (*Pyrenochaeta lycopersici*, *Botrytis cinerea*, *R. solani* strains AG 3 and AG 2-1, *Verticillium dahliae*, *Pochonia bulbillosa*, *Mortierella hyalina* and *Trichoderma harzianum*) as food on agar plates 24 h after inoculation. They reported that *A. avenae* and *Aphelenchoides* spp. were attracted to all the fungi tested. *A. avenae* was preferentially attracted to *V. dahliae* ($P < 0.0001$) and *Aphelenchoides* spp. did not show any preference, except for low attraction to *R. solani*. *A. avenae* and *Aphelenchoides* spp. reproduced on all fungal species tested. After 6 weeks of incubation, the highest number of nematodes was found on *P. lycopersici* and *P. bulbillosa*, while the lowest number occurred on *R. solani* for *A. avenae* and on *T. harzianum* for *Aphelenchoides* spp. Hasna *et al.* (2007) concluded that the suitability of a fungus as a host was not clearly related to the attraction to that fungus.

1.4 Major Impediments and Possible Solutions Concerning the Expansion of Nematode Use

Although commercially formulated beneficial nematodes are used as non-chemical alternatives for pest control in a variety of crops worldwide, they have admittedly small niche markets (e.g. Gaugler and Kaya, 1990; Gaugler, 2002; Kaya *et al.*, 2006; Bilgrami *et al.*, 2008; Lacey and Georgis, 2012; Glen and Coupland, Chapter 28, and Askary and Ahmad, Chapter 13, this volume). That is because these nematodes do not properly satisfy the main criteria set up by end-users, or growers, to choose their commercialized output as a pest control product. Grewal *et al.* (2005b) nominated factors that limited the expansion of nematode use to be high product cost, limited product availability, suboptimum ease of use and suboptimum efficacy. On the contrary, they specified research needs for enhancing the use of nematodes in biocontrol programmes in terms of lowering product cost, increasing product availability, enhancing product ease of use, enhancing product efficacy and enhancing the carry-over effect. Here, keeping continuation with the above discussion it becomes essential to present crucial aspects in brief for the commercial eligibility of beneficial nematodes as biocontrol agents, though separate chapters are alloted (Askary and Ahmad, Chapter 13; Askary, Nermut, Ahmad and Ganai, Chapter 32) in this book where these issues have been discussed in detail.

1.4.1 Nematode market within the confines of product cost and availability

The product cost for any biocontrol agent plays a significant role in its successful utilization and commercialization (Askary, 2015). In the case of EPNs, the high product cost is due to the relatively expensive and lengthy processes involved in their mass production, formulation, storage and transport. In addition, nematodes as microorganisms require controlled conditions for their viability during such processes and efficacy on application. For example, even for the most sophisticated mass production technology of EPNs, known as liquid culture, Gaugler and Han (2002) reported commercial-scale (*c.*10,000-l bioreactors) production costs of US$31 for *S. carpocapsae* and US$42 for *H. bacteriophora* per hectare (2.5×10^9 nematodes/ha). Nevertheless, costs to end-users remain greater than the alternative pest management tactic in most markets. They concluded that growers seemed unlikely to pay a premium to use nematodes when there were familiar, easy-to-use, low-cost alternatives. Recently, a distinct cottage industry has emerged in some countries that utilizes the *in vivo* process for nematode mass production for sale. The larvae of the wax moth, *G. mellonella*, are most commonly used in many countries for rearing nematodes *in vivo* because of their commercial availability. Nevertheless, in other countries, like Egypt, the Egyptian cotton leafworm (*Spodoptera littoralis*) has been targeted as the prime candidate to replace *G. mellonella* for the following reasons: (i) *S. littoralis* is susceptible to most species of *Heterorhabditis* and *Steinernema*; (ii) a yield of >100,000 IJs/cadaver is produced for most nematode species; (iii) wild populations are readily available in large numbers all year; (iv) larvae are polyphagous and can be reared easily on inexpensive artificial diets or on one of many different fresh diets; (v) they have a relatively short life cycle of 3–5 weeks; and (vi) are one of the most serious economic pests in Egypt resistant to many chemical pesticides. The only defect of some *S. littoralis* populations is that the larval cadaver may disintegrate during the infection process. Using this *in vivo Galleria* process, yields between 0.5×10^5 and 4×10^5 IJs/larva, depending on the nematode species, have been obtained. The *in vivo* process, however, lacks economy of scale; the labour, equipment and material (insect) costs increase as a linear function of production capacity. Perhaps even more important is the lack of improved quality while increasing scale. Also, *in vivo* nematode production is increasingly sensitive to biological variations and catastrophes as the scale increases, thus increasing the cost of production (Grewal *et al.*, 2005b). Continued modifications of the *in vivo* process may fairly increase efficiency and profitability (Gaugler *et al.*, 2002; see Baiocchi *et al.*, Chapter 25, this volume). Efforts to date have focused on economies of scale or yield, formulation and application technology (e.g. Gaugler and Han, 2002; Grewal *et al.*, 2005b; Shapiro-Ilan *et al.*, 2012). Such approaches can provide extraordinary savings.

Moreover, another present pressing factor is the lack of nematode product availability in the retail market. This is mainly due to the relatively short shelf life of the product. Again, controlled conditions necessary for such living organisms in terms of a significant room temperature shelf life are either costly or not available. Therefore, their products are not obtainable in the typical pesticide distribution outlets such as lawn and garden stores, as well as farmer cooperatives. The only formulation, water-dispersible granules (WDGs), that possesses 5- to 6-month room temperature shelf life of EPNs was taken out of the retail market due to fungal contamination (Grewal *et al.*, 2005b). Thus, end-users have to follow special guidelines. For example, overnight, refrigerated shipping requirement is imperative for nematode transfer. Also, growers should search for nematode products in unusual places: for example, the Internet. They also have to look for and expect the arrival of the product for prompt but careful application. Although EPNs are easily mass produced, their supply depends on the production capacity and financial capability of the producers to build an inventory in anticipation of demand. Limited production capacity and poor storage stability, coupled with the seasonal nature of demand seriously limit further expansion in the use of nematodes (Grewal *et al.*, 2005b).

1.4.2 Suboptimum ease of use and efficacy

Contrary to standard chemical pesticides, applications of beneficial nematodes require a well-acquainted user for optimum outcomes. Applications should probably be made in the late evening to avoid exposure of nematodes to lethal sunlight. When applying nematodes, a post-application irrigation is always required and a pre-application irrigation is often needed (see Askary and Ahmad, Chapter 13, this volume). Humidity should be high for foliar application, and both cold and warm temperatures limit nematode efficacy. Some nematode products are quite tedious to apply and are thus not suitable for large area application. For example, EPNs held on sponges need to be hand-squeezed into water before application. Due to the labour-intensive application and constant refrigeration requirements, these formulations are only applicable in the home lawn and garden situations (Grewal *et al.*, 2005b).

Field efficacy of EPNs is limited to selected insects and conditions (see Table 1.2), though high nematode efficacy has been demonstrated in laboratory. Many reasons can contribute to low efficacy (see Table 1.2). Poor efficacy can result from an inappropriate application method, the use of a poor quality product or suboptimum application conditions. Suboptimum application conditions, such as high or low moisture and temperature, exposure to sunlight, lack of or delay in post-application irrigation, impervious soil surface and high clay content in soil, can result in low efficacy of nematodes (Grewal *et al.*, 2005b; Lacey and Georgis, 2012; also see Askary and Ahmad, Chapter 13, this volume). The quality of commercially produced nematodes is essentially self-regulated (Grewal *et al.*, 2005b). Poor quality of commercially produced nematodes may result from factors unknown to the producer, such as deterioration during transport and distribution. Experience with commercial distribution showed that in most cases in which nematodes failed to control the target pests, the reasons were either misuse or exposure to high temperature during transportation. Low efficacy may also result from the recommendation of inappropriate nematode species or strain and the lack of proper instructions on handling and application (see Askary and Ahmad, Chapter 13, this volume). Eventually, Abd-Elgawad and Spiridonov (2014) stressed that more financial, informative, regulatory and technical support was needed in order to exploit the potential for increased pest management through nematode applications.

1.4.3 Limited spatial modelling to predict nematodes' persistence and fates

Application strategies for the use of beneficial nematodes have been developed using a trial-and-error approach. These approaches have certainly seen successes in reducing pest populations. Yet, more often than not, when applications of EPNs are followed, the applied populations seem to perish within 1 or 2 weeks. Why is persistence usually so short, and how can we extend it? Further nematode research is necessary, possibly through an iterative approach to predicting and improving the efficacy and persistence of the nematodes using a series of timed releases followed by spatial modelling to describe the nematodes' fates. The models of early releases should guide subsequent application strategies. Hence, a mathematical modelling approach should be combined with field trial data to improve EPN application strategies against key pests that have shown erratic results.

1.4.4 Most growers are unaware of their use, especially in developing countries

There are many growers, especially in developing countries, who have not heard of beneficial nematodes. Other, still large, groups of farmers do not know the advantages of using biocontrol agents in general and specifically the nematodes. Very few extensionists in these countries have been assigned to transfer knowledge, technologies and methodologies in terms of nematology-related issues and their optimum application techniques. It is also possible that some growers/end-users may lack confidence in the utilization of beneficial nematodes due to lack of experience or knowledge, or they may assume that nematodes are less effective as living microorganisms.

1.4.5 Possible solutions concerning the expansion of nematode use

Our standpoint is that the use of EPNs, and/or their bacterial associates, must be a way forward in crop protection/pest management. This is especially important in developing countries in which agricultural sectors have vast tracts of arable areas and where EPNs can be adopted and produced in relatively unsophisticated facilities. Researchers in such countries should isolate indigenous EPNs and develop their locally adapted techniques that exploit the nematodes for effective pest management through integrated pest management programmes. Local EPNs may be more adaptive and less expensive, with no risk to the local fauna and flora, unlike imported strains. Therefore, local EPN production should be encouraged to eliminate or reduce transport, packaging, formulation and storage costs. Moreover, these researchers should be acquainted with the new advances in entomopathogenic nematology, possibly through joint scientific research projects with developed countries (Abd-Elgawad and Spiridonov, 2014). More importantly, proper addressing by Grewal *et al.* (2005b) of the above-mentioned four topics, which are considered major impediments to the expansion of nematode use, should be considered. Currently, a small number of end-users, especially in urban areas, use biocontrol products due to their concerns about the health and safety of their children and pets (Grewal *et al.*, 2005b). This number is lower in developing countries. Therefore, another important point is to enlighten end-users and growers about the indirect benefits gained by using such biocontrol agents, such as the safety of humans and the environment.

On the other hand, genetic improvements, using either gene manipulation or conventional methods, and enhancing the efficiency of symbiotic bacteria may be the solution to developing stable nematode formulations and improving desirable nematode traits. These latter may include searching abilities, producing more virulent nematodes and increasing their tolerance to environmental extremes (Lacey and Georgis, 2012). However, it is important to evaluate the benefits of such improvements cautiously, since novel technologies utilized for genetic manipulation could entail the withdrawal of organic approval and registration exemption status, even though they have already been set up in many countries. These possible deprivations do not negate that molecular biology and genetics will offer important services to strengthen and expand the market share of beneficial nematodes in integrated pest management programmes.

1.5 Conclusions and Future Prospects

Although numerous nematode species are notorious pests/pathogens, not all parasitic nematodes are bad, and some are currently being used as a safe and environmentally sound biopesticide alternative to synthetic chemicals for suppressing different groups of pests. Factually, these beneficial nematodes can play an important role as biocontrol agents in crop pest management against pest insects, molluscs, PPNs and soilborne fungal pathogens. Admittedly, EPNs (Rhabditida: Steinernematidae and Heterorhabditidae) rank high among beneficial nematode groups for combating noxious insects, especially soil-dwelling pests. There are many insect parasitic nematodes, but EPNs are the best studied of these and are remarkably different in their lifestyle and in their particular parasitism (Dillman,

2013). Nevertheless, demonstrated successful control of numerous insects by EPNs often has not led to the capture of a significant share of the pesticide market for these pests (Lacey and Georgis, 2012). Shapiro-Ilan *et al.* (2002) concluded that two basic elements were necessary for EPNs to be successful: a suitable nematode for the target pest and favourable economics.

Slug parasitic nematodes may be the second candidate in terms of importance among beneficial nematode groups. Realization of their practical use has been encouraging developments across a far broader scientific front. Recent years have seen an intensive worldwide search for fresh genetic material, resulting in novel isolates of slug parasitic nematodes in new regions, with potential success in commercialization against mollusc pests. Other beneficial nematodes include species of *Beddingia* (*Deladenus*) and *Thripinema*, as well as mermithid nematodes. Control of Sirex woodwasp by *B. siricidicola* was the first commercial use of nematodes to control any pest (Bedding and Iede, 2005). Yet nematode strain deterioration, inadequate distribution of nematodes and main problems, for example wind, fire and inadequate thinning, may result in the serious reproduction of sirex populations that precede control action by the nematodes. Yet, under favourable conditions for nematodes, *B. siricidicola* is expected to be used for pest control in many infested areas of the world. On the other hand, *Thripinema* spp. hold their hosts of thrips pests alive, but render the female thrips sterile. Conclusive results demonstrated that the nematode could suppress populations of thrips (Sims, 2010; Sims *et al.*, 2009, 2012). The free-living females of *Thripinema* spp. should survive to find new hosts to be effective in recycling as biocontrol agents. Arthurs and Heinz (2002) have produced *Thripinema* successfully *in vivo*, where the number of infected thrips approximately doubled in each cycle/generation. Nevertheless, the host cues necessary for parasite development need to be better understood before *in vitro* procedures for mass production are possible. Further, the host specificity of *Thripinema* sp. needs to be elucidated so that their potential against individual species of thrips pests can be evaluated accurately (Funderburk and Sims-Latsha, 2005). Mermithid nematodes have desirable attributes such as ease of application, environmental safety, host specificity, laboratory manipulation of life history traits, lethality, mass rearing *in vivo* and the potential for long-term recycling in the environment. Unfortunately, their rearing is costly. In addition, their population dynamics are not compatible with the characteristics of an ideal biocontrol agent (Platzer *et al.*, 2005). Yet Santamarina and Perez (1997) suggested that their mass rearing *in vivo* in developing countries could offer a viable and attractive alternative for the control of vector insects. Skilful application of this nematode should consider factors that can affect its efficacy adversely. For example, insect growth regulators (IGRs) have significant negative impacts on different stages of mosquito mermithid that can destabilize the balance of host–parasite population interaction. Therefore, IGRs should be used with warning in mosquito habitats in order not to disturb and affect the populations of this parasite adversely (Suman *et al.*, 2013).

Predatory nematodes are another significant group of beneficial nematodes with biological control potential against PPNs (Bilgrami, 1997). Although field experimentation of a predatory nematode, *M. gaugleri* (Nematoda: Diplogastrida), to control PPNs has demonstrated satisfactory results, more investigations are warranted to predict the field persistence as well as the biotic and abiotic factors affecting *M. gaugleri* (Bilgrami *et al.*, 2008). Generally, the edaphic factors affecting the biological control efficacy of predatory nematode species and their (predator) ability to reduce PPN populations shoud be assessed in further investigations. As for the fungivorous nematode, there is an existing possibility for their control of plant fungal pathogens. Ishibashi (2005) reported a successful mass-production method and a long-term preservation approach for the fungivorous nematode, *A. avenae*. In addition, several merits were demonstrated by this nematode as a biocontrol agent of soilborne plant diseases. Industrial wastes can be used efficiently in its mass production. The used substrates (which still contain the nematodes) can be applied again to further suppress PPNs as well as important soil fungal pathogens: for example, *Rhizoctonia*, *Pythium*, *Phytophthora*, *Verticillium* and *Fusarium*. Additionally, condensed *A. avenae* cakes are intended to reduce transportation costs. Ishibashi (1998) suggested the effective management of soilborne diseases in agriculture through the mixed application of *A. avenae* and other beneficial nematodes. Nevertheless, skilful use of fungivorous nematodes involves their application through integrated pest management programmes and at the recommended dose, as well as their single use as as a prophylactic treatment, but not to treat disease-suffering plants.

There are major barriers that prevent commercial nematode products from realizing a significant share in the pesticide market. Their products are costly, have limited availability and suboptimum ease of use and efficacy. In addition, limited mathematical modelling to predict the nematodes' persistence and fates has sometimes led to erratic results in their applications. Also, most growers are unaware of their wise utilization, especially in developing countries. Therefore, more financial, informative, regulatory and technical support is needed in order to exploit the potential for increased pest management through applications of beneficial nematodes. For example, the need for guidelines to increase the number of people interested in the activities of nematode application against pests in the hope of reaping clean and safe agroproducts for humans and the environment should be adequately addressed. More importantly, extensive research should be directed towards developing and improving the processing of beneficial nematodes, that also include their isolation, identification, mass production, formulation, packaging, transportation, application and shelf life.

Acknowledgements

This work was supported in part by US-Egypt Science and Technology Joint Fund (cycle 17; project no. 172) and Egyptian In-House, National Research Centre Project No 11030133 entitled 'Pesticide alternatives against soilborne pathogens attacking legume cultivation in Egypt'. The authors are also thankful for the support provided by Dr M. Jamal Ahmad, Incharge, Biocontrol Laboratory, Division of Entomology, SKUAST-K, Srinagar, India, during the preparation of this chapter.

References

Abd-Elgawad, M.M.M. (1998) The criteria and techniques of the entomopathogenic nematode's production and utilization. *Agriculture and Develop in the Arab Homeland* (in Arabic) 17, 39–51.

Abd-Elgawad, M.M.M. (2001) Entomopathogenic nematode–host matching. *Egyptian Journal of Agronematology* 5, 91–104.

Abd-Elgawad, M.M.M. and Askary, T.H. (2015) Impact of phytonematodes on agriculture economy. In: Askary, T.H. and Martinelli, P.R.P. (eds) *Biocontrol Agents of Phytonematodes.* CAB International, Wallingford, UK, pp. 3–49.

Abd-Elgawad, M.M.M., and Spiridonov, S.E. (2014) Entomopathogenic nematode application in Egypt and Russia: challenges and opportunities. *Egyptian Journal of Agronematology* 13, 1–22.

Aldrovandus, U. (1602) *De animalibus insectis libri septum cum singulorum iconibus ad vivum expressis.* Clementem Ferronium, Bonon, 1638.

Arthurs, S. and Heinz, K.M. (2002) *In vivo* rearing of *Thripinema nicklewoodi* (Tylenchida: Allantonematidae) and prospects as a biological control agent of *Frankliniella occidentalis* (Thysanoptera: Thripidae). *Journal of Economic Entomology* 95, 668–674.

Arthurs, S. and Heinz, K.M. (2003) Thrips parasitic nematode *Thripinema nicklewoodi* (Tylenchida: Allantonematidae) reduces feeding, reproductive fitness, and tospovirus transmission by its host, *Frankliniella occidentalis* (Thysanoptera: Thripidae). *Environmental Entomology* 32, 853–858.

Askary, T.H. (2010) Nematodes as biocontrol agents. In: Lichtfouse, E. (ed.) *Sociology, Organic Farming, Climate Change and Soil Science*. Springer, The Netherlands, pp. 347–378.

Askary, T.H. (2015) Limitations, research needs and future prospects in the biological control of phytonematodes. In: Askary, T.H. and Martinelli, P.R.P. (eds) *Biocontrol Agents of Phytonematodes.* CAB International, Wallingford, UK, pp. 446–454.

Askary, T.H. and Martinelli, P.R.P. (2015) *Biocontrol Agents of Phytonematodes.* CAB International, Wallingford, UK.

Askary, T.H., Khan, A.A., Waliullah, M.I.S., Banday, S.A., Iqbal, U. and Mir, M.M. (2012) Slug pest management through nematodes in agricultural and horticultural crops. In: Boeri, F. and Chung, J.A. (eds) *Nematodes: Morphology*, Functions and Management Strategies. Nova Publishers, New York, pp. 197–211.

Azmi, M.I. (1983) Predatory behaviour of nematodes: biological control of *Helicotylenchus dihystera* through the predaceous nematodes, *Iotonchus monhystera*. *Indian Journal of Nematology* 13, 1–8.

Baker, G.L. and Capinera, J.L. (1997) Nematodes and nematomorphs as control agents of grasshoppers and locusts. *Memoirs of the Entomological Society of Canada* 171, 157–211.

Bedding, R.A. (1968) *Delandenus wilsoni* new species and *Delandenus siricidicola* new species neotylenchidae entomophagous mycetophagous nematodes parasitic in siricid woodwasps fungi conifers. *Nematologica* 14, 515–525.

Bedding, R.A. (1984) Nematode parasites of Hymenoptera. In: Nickle, W.R. (ed.) *Manual of Agricultural Nemtaology*. Marcel-Dekker, New York, pp. 755–796.

Bedding, R.A. (2009) Controlling the pine-killing woodwasp, *Sirex noctilio*, with nematodes. In: Hajek, A.E., Glare, T.R. and O'Callaghan, M. (eds) *Use of Microbes for Control of Invasive Arthropods*. Springer, The Netherlands, pp. 213–235.

Bedding, R.A. and Iede, E.T. (2005) Application of *Beddingia siricidicola* for Sirex woodwasp control. In: Grewal, P.S., Ehlers, R.-U. and Shapiro-Ilan, D.I. (eds) *Nematodes as Biological Control Agents*. CAB International, Wallingford, UK, pp. 385–399.

Bedding, R.A. and Molyneux, A.S. (1982) Penetration of insect cuticle by infective juveniles of *Heterorhabditis* spp. (Heterorhabditidae: Nematoda). *Nematologica* 28, 354–359.

Bilgrami, A.L. (1992) Resistance and susceptibility of prey nematodes to predation and strike rate of the predators, *Mononchus aquaticus*, *Dorylaimus stagnalis*, and *Aquatides thornei. Fundamental and Applied Nematology* 15, 265–270.

Bilgrami, A.L. (1995) Numerical analysis of the relation between *Mesodorylaimus bastiani* (Nematoda: Dorylaimida) and different prey trophic categories. *Nematologia Mediterranea* 23, 81–88.

Bilgrami, A.L. (1997) *Nematode Biopesticides*. Aligarh Muslim University Press, Aligarh, India.

Bilgrami, A.L. (2008) Biological control potentials of predatory nematodes. In: Ciancio, A. and Mukerij, K.G. (eds) *Integrated Management and Biocontrol of Vegetable and Grain Crops Nematodes*. Springer, Dordrecht, The Netherlands, pp. 3–28.

Bilgrami, A.L. and Brey, C. (2005) Potential of predatory nematodes to control plant-parasitic nematodes. In: Grewal, Ehlers, R.-U. and Shapiro-Ilan, D.I. (eds) *Nematodes as Biocontrol Agents*.CAB International, Wallingford, UK, pp. 447–464.

Bilgrami, A.L. and Gaugler, R. (2005) Feeding behaviour of the predatory nematodes *Laimydorus baldus* and *Discolaimus major* (Nematoda: Dorylaimida). *Nematology* 7, 11–20.

Bilgrami, A.L. and Jairajpuri, M.S. (1988) Attraction of *Mononchoides longicaudatus* and *M. fortidens* (Nematoda: Diplogasterida) towards prey and factors influencing attraction. *Revue de Nématologie* 11, 195–202.

Bilgrami, A.L. and Jairajpuri, M.S. (1989) Aggregation of *Mononchoides longicaudatus* and *M. fortidens* (Nematoda: Diplogasterida) at feeding sites. *Nematologica* 34, 119–121.

Bilgrami, A.L. and Jairajpuri, M.S. (1990) Diplogasterid predators. In: Jairajpuri, M.S., Alam, M.M. and Ahmad, I. (eds) *Nematode Biocontrol: Aspects and Prospects*. CBS Publishers and Distributors Pvt Ltd, New Delhi, pp. 135–142.

Bilgrami, A.L., Ahmad, I. and Jairajpuri, M.S. (1985) Predatory behavior of *Aquatides thornei* (Nygolaimina: Nematoda). *Nematologica* 30, 457–462.

Bilgrami, A.L., Ahmad, I. and Jairajpuri, M.S. (1986) A study on the intestinal contents of some mononchs. *Revue de Nématologie* 17, 339–340.

Bilgrami, A.L., Gaugler, R. and Brey, C. (2005) Prey preference and feeding behaviour of diplogastrid predator *Mononchoides gaugleri* (Nematoda: Diplogasterida). *Nematology* 7, 333–342.

Bilgrami, A.L., Brey, C. and Gaugler, R. (2008) First field release of a predatory nematode, *Mononchoides gaugleri* (Nematoda: Diplogastrida), to control plant-parasitic nematodes. *Nematology* 10, 143–146.

Blaxter, M., Dorris, M. and De Ley, P. (2000) Patterns and processes in the evolution of animal parasitic nematodes. *Nematology* 2, 43–55.

Blaxter, M.L., De Ley, P., Garey, J.R., Liu, L.X., Sceldeman, P., Vierstraete, A., Vanfleteren, J.R., Mackey, L.Y., Dorris, M., Frisse, L.M., Vida, J.T. and Thomas, W.K. (1998) A molecular evolutionary framework for the phylum Nematoda. *Nature* 392, 71–75.

Boosalis, M.G. and Mankau, R. (1965) Parasitism and predation of soil microorganisms. In: Baker, K.F. and Snyder, W.C. (eds) *Ecology of Soil-borne Plant Pathogens*. University of California Press, Berkeley, California, pp. 374–391.

Bremser, J.G. (1824) Nouvel Atlas, compose de 15 planches in –4° dissinees par A. Foucaud, avec un texte explicatif renfermant des observations inedites par Charles Leblond, Paris.

Chen, G., Ren, H.F., Luo, L.W., Zhou, S.W., Zhang, Z.L., *et al.* (1992) Infection process and natural control effect of *Ovomermis sinensis* (Nematoda: Mermithidae) to *Mythimna separate* (Lep: Noctuidae). *Wuyi Science Journal* 9, 249–260.

Chitambar, J.J. and Noffsinger, E.M. (1989) Predaceous behaviour and life history of *Odontopharynx longicaudata* (Diplogasterida). *Journal of Nematology* 21, 284–291.

Choo, H.Y., Kaya, H.K. and Kim, H.H. (1995) Biological studies on *Agamermis unka* (Nematoda, Mermithidae) a parasite of the brown plant-hopper *Nilaparvata lugens*. *Biocontrol Science and Technology* 5, 209–223.

Cohn, E. and Mordechai, E. (1974) Experiments in suppressing citrus nematode populations by use of marigold and a predacious nematode. *Nematologia Mediterranea* 2, 43–53.

Dillman, A.R. (2013) Host seeking and the genomic architecture of parasitism among entomopathogenic nematodes. PhD thesis, California Institute of Technology, Pasadena, California.

Dillon, A.B., Foster, A., Williams, C.D. and Griffin, C.T. (2012) Environmental safety of entomopathogenic nematodes – effects on abundance, diversity and community structure of non-target beetles in a forest ecosystem. *Biological Control* 63, 107–114.

Dorris, M., De Ley, P. and Blaxter, M.L. (1999) Molecular analysis of nematode diversity and the evolution of parasitism. *Parasitology Today* 15, 188–193.

Fauzia, M., Jairajpuri, M.S. and Khan, Z. (1998) Biocontrol potential of *Mononchoides longicaudatus* on *Meloidogyne incognita* on tomato plants. *International Journal of Nematology* 8, 89–91.

Ferris, H. (2010) Contribution of nematodes to the structure and function of the soil food web. *Journal of Nematology* 42, 63–67.

Ferris, R.V. (1968) Biometric analysis in the genus *Labronema* (Nematoda: Dorylaimida) with a description of *L. thornei* n. sp. *Nematologica* 14, 276–284.
Filipjev, I.N. (1934) *Harmful and Useful Nematodes in Rural Economy*. Moskva, Leningrad, Russia (in Russian).
Funderburk, J.E. and Sims-Latsha, K. (2005) The entomophilic Thripinema. In: Grewal, P.S., Ehlers, R.-U. and Shapiro-Ilan, D.I. (eds) *Nematodes as Biocontrol Agents*. CAB International, Wallingford, UK, pp. 401–410.
Gaugler, R. (ed.) (2002) *Entomopathogenic Nematology*. CAB International, Wallingford, UK.
Gaugler, R. and Han, R. (2002) Production technology. In: Gaugler, R. (ed.) *Entomopathogenic Nematology*. CAB International, Wallingford, UK, pp. 289–310.
Gaugler, R. and Kaya, H.K. (1990) *Entomopathogenic Nematodes in Biological Control.* CRC Press, Boca Raton, Florida.
Gaugler, R., Brown, I., Shapiro-Ilan, D. and Atwa, A. (2002) Automated technology for *in vivo* mass production of entomopathogenic nematodes. *Biological Control* 24, 199–206.
Genena, M.A.M., Mostafa, F.A.M. and Fouly, A.H. (2011) First record for the slug parasitic nematode, *Phasmarhabditis hermaphrodita* (Schneider) in Egypt. *Archives of Phytopathology and Plant Protection* 44, 340–345.
Gerber, K. and Smart, G.C. Jr (1987) Plant-parasitic nematodes associated with aquatic vascular plants. In: Veech, J.A. and Dickson, D.W. (eds) *Vistas on Nematology*. Society of Nematologists, USA, pp. 488–501.
Goeze, J.A.E. (1782) *Versuch einer Naturgeschichte der Eingeweidewur-mer thierischer Korper, die derselben erwahnen.* Landmarks of Science, Micro-opaque, New York.
Gould, W. (1747) *An Account of English Ants*. A. Miller, London.
Grewal, P.S., Lewis, E.E. and Venkatachari, S. (1999) Allelopathy: a possible mechanism of suppression of plant-parasitic nematodes by entomopathogenic nematodes. *Nematology* 1, 735–743.
Grewal, P.S., Ehlers, R.-U. and Shapiro-Ilan, D.I. (2005a) *Nematodes as Biocontrol Agents*. CAB International, Wallingford, UK.
Grewal, P.S., Ehlers, R.-U. and Shapiro-Ilan, D.I. (2005b) Critical issues and research needs for expanding the use of nematodes in biocontrol. In: Grewal, P.S., Ehlers, R.-U. and Shapiro-Ilan, D.I. (eds) *Nematodes as Biocontrol Agents*. CAB International, Wallingford, UK, pp. 479–489.
Grootaert, P. and Maertens, D. (1976) Cultivation and life cycle of *Mononchus aquaticus*. *Nematologica* 22, 173–181.
Gupta, M.C. (1986) Biological control of *Fusarium moniliformae* Sheldon and *Pythium butleri* Subramanium by *Aphelenchus avenae* Bastian, 1865 in chitin and cellulose amended soils. *Soil Biology and Biochemistry* 18, 327–329.
Hasna, M.K., Insunza, V., Lagerlöf, J. and Rämert, B. (2007) Food attraction and population growth of fungivorous nematodes with different fungi. *Annals of Applied Biology* 151, 175–182.
Haugen, D.A. (1990) Control procedures for *Sirex noctilio* in the Green Triangle: review from detection to severe outbreak (1977–1987). *Australian Forestry* 53, 24–32.
Haugen, D.A. and Underdown, G. (1990) *Sirex noctilio* program in response to the 1987 Green Triangle outbreak. *Australian Forestry* 53, 33–40.
Hechler, H.C. and Taylor, D.P. (1966) The life histories of *Seinura celeris*, *S. olivierae*, *S. oxura* and *S. steineri* (Nematoda: Aphelenchoididae). *Proceedings of the Helminthological Society of Washington* 33, 71–83.
Hominick, W.M. and Perry, R. (2000) Biological control. In: Chen, D. (ed.) *The Biology of Nematodes*. CAB International, Wallingford, UK, pp. 115–143.
Hominick, W.M. and Tingle, G.A. (1984) Mermithid nematodes and the control of insect vectors of human disease. *Biocontrol News and Information* 5, 7–21.
Hurley B.P., Slippers, B. and Wingfield, M.J. (2007) A comparison of control results for the alien invasive woodwasp, *Sirex noctilio*, in the southern hemisphere. *Agricultural and Forest Entomology* 9, 159–171.
Hurley, M.J., Brazil, D. and Kakouli-Duarte, T. (2014) Entomopathogenic nematodes and endophytic bacteria: a novel approach with potential for plant health. In: *Proceedings, 6th International Congress of Nematology*, Cape Town, South Africa, 4–9 May, pp. 188–189.
Ishibashi, N. (1998) Possible integration of beneficial nematodes and microbial agents for biocontrol of soil pests. *Japanese Journal of Nematology* 28 (Special Issue), 22–28.
Ishibashi, N. (2005) Potential of fungal-feeding nematodes for the control of soil-borne plant pathogens. In: Grewal, P.S., Ehlers, R.-U. and Shapiro-Ilan, D.I. (eds) *Nematodes as Biocontrol Agents*. CAB International, Wallingford, UK, pp. 467–475.
Jagdale, G.B., Kamoun, S. and Grewal, P.S. (2009) Entomopathogenic nematodes induce components of systemic resistance in plants: biochemical and molecular evidence. *Biological Control* 51, 102–109.
Karim, N., Jones, J.T., Okada, H. and Kikuchi, T. (2009) Analysis of expressed sequence tags and identification of genes encoding cell-wall-degrading enzymes from the fungivorous nematode *Aphelenchus avenae*. *BMC Genomics* 10, 525.
Kaya, H.K. and Gaugler, R. (1993) Entomopathogenic nematodes. *Annual Review of Entomology* 38, 81–206.
Kaya, H.K., Aguillera, M.M., Alumai, A., Choo, H.Y., De la Torre, M., Fodor, A., Ganguly, S., Hazır, S., Lakatos, T.,

Pye, A., Wilson, M., Yamanaka, S., Yang, H. and Ehlers, R.-U. (2006) Status of entomopathogenic nematodes and their symbiotic bacteria from selected countries or regions of the world. *Biological Control* 38, 134–155.

Khan, Z. and Jairajpuri, M.S. (1997) Predatory behavior of *Paractinolaimus elongatus* on plant parasitic nematodes. *International Journal of Nematology* 8, 178–181.

Khan, Z. and Kim, Y.H. (2005) The predatory nematode, *Mononchoides fortidens* (Nematoda: Diplogasterida), suppresses the root-knot nematode, *Meloidogyne arenaria*, in potted field soil. *Biological Control* 35, 78–82.

Khan, Z. and Kim, Y.H. (2007) A review on the role of predatory soil nematodes in the biological control of plant parasitic nematodes. *Applied Soil Ecology* 35, 370–379.

Khan, Z., Bilgrami, A.L. and Jairajpuri, M.S. (1991) Some observations on the predation ability of *Aporcelaimellus nivalis* (Altherr, 1952) Heyns 1966 (Nematoda: Dorylaimida). *Nematologica* 37, 333–342.

Khan, Z., Bilgrami, A.L. and Jairajpuri, M.S. (1995) Observations on the predation abilities of *Neoactinolaimus agilis* (Dorylaimida: Actinolaimoidea). *Indian Journal of Nematology* 25, 129–135.

Kikuchi, T., Shibuya, H. and Jones, J.T. (2005) Molecular and biochemical characterization of an endo-beta-1,3-glucanase from the pinewood nematode *Bursaphelenchus xylophilus* acquired by horizontal gene transfer from bacteria. *Biochemical Journal* 389, 117–125.

Kim, Y.H. (2015) Predatory nematodes as biocontrol agents of phytonematodes. In: Askary, T.H. and Martinelli, P.R.P. (eds) *Biocontrol Agents of Phytonematodes.* CAB International, Wallingford, UK, pp. 393–420.

Kobayashi, N. and Chikuo, Y. (1993) Biological control of cucumber *Fusarium* wilt by the fungivorous nematode, *Aphelenchus avenae*. In: Ishibashi, N. (ed.) *Detection of Beneficial Nematodes in Japan and Systematized Nematode Application*. Technical Bulletin Grant-in-Aid, Ministry of Education, Japan Grant No 02506001, Saga University, Saga, Japan, pp. 116–122.

Lacey, L.A. and Georgis, R. (2012) Entomopathogenic nematodes for control of insect pests above and below ground with comments on commercial production. *Journal of Nematology* 44, 218–225.

Lagerlöf, J., Insunza, V., Lundegårdh, B. and Rämert, B. (2011) Interaction between a fungal plant disease, fungivorous nematodes and compost suppressiveness. *Acta Agriculturae Scandinavica, Section B-Soil and Plant Science* 61, 372–377.

Lim, U.T. (2003) *Thripinema nicklewoodi* (Tylenchida: Allantonematidae), a potential biological control agent of *Frankliniella occidentalis* (Thysanoptera: Thripidae). PhD thesis, University of Massachusetts, Amherst, Massachusetts.

Linford, M.B. and Oliviera, J.M. (1937) The feeding of hollow-spear nematodes on other nematodes. *Science* 85, 295–297.

Linnaeus, C. von. (1758) *Systema naturae per regna tria naturae: secundum classes, ordines, genera, species, cum characteribus differentiis, synonymis, locis* (in Latin). 10th edn. Laurentius Salvius, Stockholm.

Loomans, A.J.M., Murai, T. and Greene, I.D. (1997) Interactions with hymenopterous parasitoids and parasitic nematodes. In: Lewis, T. (ed.) *Thrips as Crop Pests*. CAB International, Wallingford, UK, pp. 355–397.

Luckman, W.H. and Metcalf, R.L. (1982) The pest-management concept. In: Metcalf, R.L. and Luckman, W.H. (eds) *Introduction to Insect Pest Management*. John Wiley and Sons, New York, pp. 1–32.

Maertens, D. (1975) Observations on life cycle of *Prionchulus punctatus* (Cobb, 1917) and culture conditions. *Biologisch Jaarboek (Dodonaea)* 43, 197–218.

Mankau, R. and Mankau, S.K. (1962) Multiplication of *Aphelenchus avenae* on phytopathogenic soil fungi. *Phytopathology* 52, 741.

Morris, E.E. (2015) *Deladenus siricidicola* (= *Beddingia siricidicola*) (Tylenchida, Neotylenchidae). In: Shelton, A. (ed.) *Biological Control: A Guide to Natural Enemies in North America*. Available at: https://biocontrol.entomology.cornell.edu/pathogens/nematodes.php (accessed 25 October 2016).

Mráček, Z., Půža, V. and Nermuť, J. (2014) Addendum to the description of *Steinernema carpocapsae* (Weiser, 1955) Wouts, Mráček, Gerdin and Bedding, 1982. *Russian Journal of Nematology* 22, 109–120.

Murfin, K.E., Dillman, A.R., Foster, J.M., Bulgheresi, S., Slatko, B.E., Sternberg, P.W. and Blair, H.G. (2012) Nematode–bacterium symbioses – cooperation and conflict revealed in the 'Omics' age. *Biological Bulletin* 223, 85–102.

Nelmes, A.J. (1974) Evaluation of feeding behaviour of *Prionchulus punctatus* (Cobb) a nematode predator. *Journal of Animal Ecology* 43, 553–565.

Nguyen, K.B. (2010) History of insect parasitic nematodes. Entomology and Nematology Department, University of Florida. Available at: http://entnemdept.ufl.edu/nguyen/FLNEM/HISTORY/entomophilic_history.htm (accessed 26 August 2016).

Nguyen, K.B. and Hunt, D.J. (2007) *Entomopathogenic Nematodes: Systematics, Phylogeny and Bacterial Symbionts. Nematology Monographs and Perspectives*, Vol 5. Brill Academic Publishers, The Netherlands.

Okada, H. and Kadota, I. (2003) Host status of 10 fungal isolates for two nematode species, *Filenchus misellus* and *Aphelenchus avenae*. *Soil Biology and Biochemistry* 35, 1601–1607.

Opperman, C. and Salstead, A. (2015) Predatory Nematode. Available at: http://www.fcps.edu/islandcreekes/ecology/predatory_nematode.htm (accessed 26 August 2016).

Osman, G.Y. (1988) Studies on the potential use of the predator *Diplogaster* sp. (Nematoda, Diplogasteridae) on certain root-parasitic nematodes. *Anzeiger für Schädlingskunde, Pflanzenschutz, Umweltschutz* 61, 70–73.

Peters, A. and Ehlers, R.-U. (1994) Susceptibility of leather jackets (*Tipula paludosa* and *Tipula oleracea*; Tipulidae: Nematocera) to the entomopathogenic nematode *Steinernema feltiae*. *Journal of Invertebrate Pathology* 63, 163–171.

Petersen, J.J. (1973) Role of mermithid nematodes in biological control of mosquitoes. *Experimental Parasitology* 33, 239–247.

Petersen, J.J. (1985) Nematodes as biological control agents: Part I. Mermithidae. *Advances in Parasitology* 24, 307–346.

Petersen, J.J. and Willis, O.R. (1972) Procedures for mass rearing of a mermithids parasite of mosquitoes. *Mosquito News* 32, 226–230.

Platzer, E.G. (2007) Mermithid nematodes. In: Floore, T.G. (ed.) *Biorational Control of Mosquitoes. Journal of the American Mosquito Control Association* (S2), Bulletin No 7, pp. 58–64.

Platzer, E.G., Mullens, B.A. and Shamseldean, M.M. (2005) Mermithid nematodes. In: Grewal, P.S., Ehlers, R.-U. and Shapiro-Ilan, D.I. (eds) *Nematodes as Biocontrol Agents*. CAB International, Wallingford, UK, pp. 411–418.

Poinar, G.O. Jr (1979) *Nematodes as Biological Control for Insects*. CRC Press, Boca Raton, Florida.

Poinar, G.O. Jr (1983) *The Natural History of Nematodes*. Prentice-Hall, Inc., Englewood Cliffs, New Jersey.

Poinar, G.O. Jr and Grewal, P.S. (2012) History of entomopathogenic nematology. *Journal of Nematology* 44, 153–161.

Popiel, I. and Hominick, W.M. (1992) Nematodes as biological control agents: Part II. *Advances in Parasitology* 31, 381–433.

Rama, K. and Dasgupta, M.K. (1998) Biocontrol of nematodes associated with mandarin orange decline by the promotion of predatory nematode *Iotonchus tenuicaudatus* (Kreis, 1924). *Indian Journal of Nematology* 28, 118–124.

Reaumur de, R.A.F. (1734) *Mémoires pour servir à l'histoire des insects*. A Paris: De l'imprimerie royale, France.

Sanad, M.M., Shamseldean, M.S.M., Elgindi, A.Y. and Gaugler, R. (2013) Host penetration and emergence patterns of the mosquito-parasitic mermithids *Romanomermis iyengari* and *Strelkovimermis spiculatus* (Nematoda: Mermithidae). *Journal of Nematology* 45, 30–38.

Santamarina, A.M. and Perez, R.P. (1997) Reduction of mosquito larval densities in natural sites after introduction of *Romanomermis culicivorax* (Nematoda: Mermithidae) in Cuba. *Journal of Medical Entomology* 34, 1–4.

Shamseldean, M.M., Platzer, E.G. and Gaugler, R. (2007) Role of the surface coat of *Romanomermis culicivorax* in immune evasion. *Nematology* 9, 17–24.

Shapiro-Ilan, D.I. and Gaugler, R. (2002) Production technology for entomopathogenic nematodes and their bacterial symbionts. *Journal of Industrial Microbiology and Biotechnology* 28, 137–146.

Shapiro-Ilan, D.I., Glazer, I. and Segal, D. (1997) Genetic improvement of heat tolerance in *Heterorhabditis bacteriophora through hybridization. Biological Control* 8, 153–159.

Shapiro-Ilan, D.I. and Gaugler, R. (2015) Nematodes (Rhabditida: Steinernematidae & Heterorhabditidae). In: Shelton, A. (ed.) *Biological Control: A Guide to Natural Enemies in North America*. Available at: https://biocontrol.entomology.cornell.edu/pathogens/nematodes.php (accessed 25 October 2016).

Shapiro-Ilan, D.I., Gouge, D.H. and Koppenhofer, A.M. (2002) Factors affecting commercial success: case studies in cotton, turf, and citrus. In: Gaugler, R. (ed.) *Entomopathogenic Nematology*. CAB International, Wallingford, UK, pp. 333–355.

Shapiro-Ilan, D.I., Han, R. and Dolinksi, C. (2012) Entomopathogenic nematode production and application technology. *Journal of Nematology* 44, 206–217.

Shapiro-Ilan, D.I., Bock, C.H. and Hotchkiss, M.W. (2014) Suppression of pecan and peach pathogens on different substrates using *Xenorhabdus bovienii* and *Photorhabdus luminescens*. *Biological Control* 77, 1–6.

Siddiqi, M.R. (1986) *Tylenchida Parasites of Plants and Insects*. CAB Commonwealth Institute of Parasitology, London.

Siddiqi, M.R. (2000) *Tylenchida Parasites of Plants and Insects*, 2nd edn. CAB International, Wallingford, UK.

Siddiqi, M.R., Bilgrami, A.L. and Tabassum, K. (2004) Description and biology of *Mononchoides gaugleri* sp. n. (Nematoda: Diplogasterida). *International Journal of Nematology* 14, 124–129.

Sims, K.R. (2010) Interactions between *Frankliniella fusca* (Thysanoptera: Thripidae), *Thripinema fuscum* (Tylenchida: Allantonematidae), and entomoparasites (tomato spotted wilt virus, *Pantoea ananatis*, and wolbachia). PhD thesis, University of Florida's North Florida Research and Education Center in Quincy, Florida.

Sims, K.R., Funderburk, J.E., Reitz, S.R. and Boucias, D.G. (2009) The impact of a parasitic nematode, *Thripinema fuscum*, on the feeding behavior and vector competence of *Frankliniella fusca*. *Entomologia Experimentalis et Applicata* 132, 200–208.

Sims, K., Becnel, J.J. and Funderburk, J. (2012) The morphology and biology of the entomophilic *Thripinema fuscum* (Tylenchida: Allantonematidae), and the histopathological effects of parasitism on the host *Frankliniella fusca* (Thysanoptera: Thripidae). *Journal of Natural History* 46, 1111–1128.

Small, R.W. (1979) The effects of predatory nematodes on populations of plant parasitic nematodes in pots. *Nematologica* 25, 94–103.

Smart, G.C. Jr (1984) Biological control of and by nematodes. *Journal of the Georgia Enomological Society* 19, 28–40.

Somasekhar, N., Grewal, P.S., de Nardo, E.A.B. and Stinner, B.R. (2002) Non-target effects of entomopathogenic nematodes on the soil nematode community. *Journal of Applied Ecology* 39, 735–744.

Steiner, G. (1923) *Aplectana kraussei* n. sp., einer in der Blattwespe Lyda sp. parasitieren de Nematodenform, nebst Bemerkungen überdas Seitenorgan der parasitischen nematoden. *Zen-tralblatt fuer Bakteriologie, Parasitenkunde, Infektionskrankheiten und Hygiene* 59, 14–18.

Stock, S.P. (2005) Insect-parasitic nematodes: from lab curiosities to model organisms. *Journal of Invertebrate Pathology* 89, 57–66.

Sudhaus, W. (2008) Evolution of insect parasitism in rhabditid and diplogastrid nematodes. In: Makarov, S.E. and Dimitrijevic, R.N. (eds) *Advances in Arachnology and Developmental Biology*. Vienna-Belgrade-Sofia, SASA, Belgrade and UNESCO MAB, Serbia, pp. 143–161.

Suman, D.S., Brey, C.W., Wang, Y., Sanad, M., Shamseldean, M.S.M. and Gaugler, R. (2013) Effects of insect growth regulators on the mosquito-parasitic nematode *Romanomermis iyengari*. *Parasitology Research* 112, 817–824.

Tan, L. and Grewal, P.S. (2001) Pathogenecity of *Moraxella osloensis,* a bacterium associated with the nematode *Phasmarhabditis hermaphrodita, to the slug Deroceras reticulatum*. *Applied Environmetal Microbiology* 67, 5010–5016.

Tan, L. and Grewal, P.S. (2002) Endotoxin activity of *Moraxella osloensis* against the grey garden slug, *Deroceras reticulatum*. *Applied Environmetal Microbiology* 68, 3943–3947.

Thorne, G. (1927) The life history, habits and economic importance of some mononchs. *Journal of Agricultural Research* 34, 265–286.

Thorne, G. (1930) Predaceous nemas of the genus *Nygolaimus* and a new genus, *Sectonema*. *Journal of Agricultural Research* 41, 445–466.

Townshend, J.L. (1964) Fungus hosts of *Aphelenchus avenae* Bastian, 1865, and *Bursaphelenchus fungivorous* Franklin and Hooper, 1962 and their attractiveness to these nematode species. *Canadian Journal of Microbiology* 10, 727–737.

Wang, F., Li, D., Wang, Z., Dong, A., Liu, L.,Wang, B., Chen, Q. and Liu, X. (2014) Transcriptomic analysis of the rice white tip nematode, *Aphelenchoides besseyi* (Nematoda: Aphelenchoididae). PLoS ONE 9(3), e91591, DOI: 10.1371/journal.pone.0091591.

Wang, K.H., McSorley, R. and Gallaher, R.N. (2004) Relationship of soil management history and nutrient status to nematode community structure. *Nematropica* 34, 83–95.

Wasilewska, L. and Bienkowski, P. (1985) Experimental study on the occurrence and activity of soil nematodes in decomposition of plant material. *Pedobiologia* 28, 41–57.

Wilson, M.J., Glen, D.M. and George, S.K. (1993) The rhabditid nematode *Phasmarhabditis hermaphrodita* as a potential biological control agent for slugs. *Biocontrol Science and Technology* 3, 503–511.

Wood, F.H. (1973) Nematode feeding relationships: relationships of soil dwelling nematodes. *Soil Biology and Biochemistry* 5, 593–601.

Wood, F.H. (1974) Biology of *Seinura demani* (Nematoda: Aphelenchoididae). *Nematologica* 20, 347–353.

Woodring, J.L. and Kaya, H.K. (1988) Steinernematid and Heterorhabditid nematodes: a handbook of biology and techniques. Southern Cooperative Series Bulletin 331, Arkansas Agricultural Experimental Station, Fayetteville, Arkansas.

World Health Organization (2006) *Pesticides and their Application for the Control of Vectors and Pests of Public Health Importance* (6th edn). WHO/CTD/NTD/WHOPES/GCDPP, Geneva, Switzerland, 113 pp.

Yeates, G.W. (1969) Predation by *Mononchoides potohikus* (Nematoda: Diplogasteridae) in laboratory culture. *Nematologica* 15, 1–9.

Yeates, G.W. and Bongers, T. (1999) Nematode diversity in agroecosystems. *Agriculture, Ecosystems and Environment* 74, 113–135.

Zettler, F.W. and Freeman, T.E. (1972) Plant pathogens as biocontrols of aquatic weeds. *Annual Review of Phytopathology* 19, 455–470.

Zondag, R. (1962) A nematode disease of *Sirex noctilio*. New Zealand Forestry Service, Interim Research Report, pp. 1–16.

2 Beneficial Nematodes and the Changing Scope of Crop Protection

JAMES COUPLAND,[1*] MAHFOUZ M.M. ABD-ELGAWAD[2] AND TARIQUE HASSAN ASKARY[3]

[1]*FarmForest Research Inc, Almonte, Canada;* [2]*Phytopathology Department, National Research Center, Giza, Egypt;* [3]*Division of Entomology, Sher-e-Kashmir University of Agricultural Sciences and Technology, Srinagar, India*

2.1 Introduction

Not since the publication of *Silent Spring* (Carson, 1962) have pesticides come under such intense public scrutiny. Over the years, there have been spikes of media attention due to the negative impacts of pesticides, such as from atrazine, a herbicide of the triazine class (Hayes *et al.*, 2010). The most recent flurry of attention has been due to the loss of pollinators and the possible implication of a new class of insecticide (namely neonicotinoids, a neuroactive insecticide) in their demise. This widespread concern and the intense public interest in how crop protection products are actually developed and used has impacted the agrochemical industry, along with the agricultural sector in general.

It is well established that the indiscriminate use of synthetic chemical pesticides has caused many issues, such as the genetic development of pest resistance and ecological damage. This has led to a gradual loss in biodiversity and the resurgence of non-target pests and secondary pest outbreaks, as well as environmental pollution, contamination of groundwater and general human and animal health hazards (e.g. Edwards, 1993; Lietti *et al.*, 2005; Peshin and Pimentel, 2014; Pimentel and Peshin, 2014). Such deleterious effects of chemical pesticides have now alarmed consumers, especially with the growing public awareness of food supplies. In fact, contributions to the current progress of using safe pesticides, such as entomopathogenic and slug parasitic nematodes, are direct reflections of the irrational applications of synthetic chemical pesticides, which are increasingly raising global, specific, social, scientific and economic issues related to their side effects on humans and their environment (e.g. Rechcigl and Rechcigl, 2000; Lacey and Shapiro-Ilan, 2008; Askary, 2010; Askary *et al.*, 2012; Lacey and Georgis, 2012; Campos-Herrera, 2015). Such issues stem not only from the dangerous chemical components of these pesticides but also from using the wrong application time, inefficient equipment, misuse in off label crops and unsafe spraying. These practices are usually capable of accelerating the rate at which the side effects of these pesticides expand.

In the agricultural and rural sector, the harmful impacts of these chemicals, along with the steady growth of the organic food movement, have changed the landscape in which farmers work to control their crop pest problems. Moreover, the intensive application of chemical pesticides is not necessarily directly proportional to enhanced yield per unit area. In the Mediterranean countries, Spanish and Italian citrus growers use fourfold and 15-fold more amounts of pesticides than do Greek growers, but both have recorded similar citrus yields per hectare (Gutiérrez *et al.*, 2005; Campos-Herrera *et al.*, 2010). On the other hand, biopesticides such as many species of beneficial nematodes, particularly entomopathogenic and slug parasitic nematodes, have demonstrated their economic importance as potential pest control products, through their wide distribution and host ranges,

*Corresponding author. E-mail: couplandj@hotmail.com

 Biocontrol Agents: Entomopathogenic and Slug Parasitic Nematodes (eds. M.M.M. Abd-Elgawad, T.H. Askary and J. Coupland)

easy formulation and successful commercial production, as well as their high efficacy against target pests of agricultural importance (Askary, 2010; Askary *et al.*, 2012). Therefore, these beneficial nematodes have shown utility in terms of insect and slug pest management; the use of which would significantly lower chemical pesticide use and be of benefit to humans and the environment. Consequently, conscious changes from the reliance on economic threshold levels and chemical pesticide-driven approaches in the past to more ecologically sustainable agroecosystem analysis-based integrated pest management (IPM) strategies are currently being developed.

Gradually, the expression 'sustainability' has become more commonly used in modern agriculture in recent years. Contrary to the over-reliance on chemical pesticides, sustainability is aimed at maintaining natural and agricultural resources and the benefits (food, commodities) they produce (Campos-Herrera, 2015). Therefore, sustainability and the increased use of IPM strategies are now the new emerging strategies in global agriculture and a direct result of the over and indiscrimate use of pesticides in the past. In this context, this chapter sheds light on beneficial nematodes as a component of environmentally friendly biopesticides that can contribute significantly in changing the scope of crop protection towards sustainable agroecosystems.

We then detail a way forward for the use of biopesticides and in particular beneficial nematodes in modern agriculture. We give an example of one of the most pressing issues of a pesticide and its impact on agriculture and society.

2.2 Neonicotinoid Pesticides: An Example

Neonicotinoids (abbreviated as neonics) are a class of neuroactive insecticides chemically similar to nicotine. In the 1980s, Shell, and in the 1990s, Bayer, began work on their development (Kollmeyer *et al.*, 1999). Nicotine is secondary plant compound that acts as an insecticide, but is also toxic to mammals. In fact, nicotine has a lower lethal dose for rats than for flies. This spurred a scientific search for compounds that retained the insecticidal properties of nicotine but were less toxic to mammals, but preliminary investigation of nicotine-related compounds (nicotinoids) as insecticides was initially unsuccessful. However, a precursor (called nithiazine) to the nicotinoid was synthesized at Purdue University. When Shell researchers were screening potential candidates, this precursor showed some insecticide potential and they refined it to develop nithiazine as an insecticide. Nithiazine was later found to be a post-synaptic acetylcholine receptor agonist, meaning it had the same mode of action as nicotine. Nithiazine does not act as an acetylcholinesterase inhibitor, in contrast to the organophosphate and carbamate insecticides. While nithiazine had the desired specificity (i.e. low mammalian toxicity), it was not photostable (it broke down in sunlight), so it was not commercially viable. Neonicotinoids were developed a decade later, with the first commercially successful neonicotinoid, imidacloprid, launched by Bayer in 1985.

Most neonicotinoids are water-soluble and break down slowly in the environment, so they can be taken up by the plant and provide protection from insects as the plant grows. During the late 1990s, this class of pesticides, primarily imidacloprid, became widely used. Beginning in the early 2000s, two other neonicotinoids, clothianidin and thiamethoxam, were in use as well. Currently, virtually all maize that is planted in the Midwestern USA is treated with one of these two insecticides and various fungicides. In addition, most soybean seeds are also treated with a neonicotinoid insecticide, usually thiamethoxam, and rapeseed with clothianidin, which is one of the neonicotinoids that is most toxic to honeybees.

This new interest in the possible negative impacts of neonicotinoids had a significant effect in the European Economic Community (EEC). In response to growing concerns about the impact of neonicotinoids on honeybees, the European Commission asked the European Food Safety Authority to study their safety. The results of the peer-reviewed study were published in January 2013. They concluded that the previous industry-sponsored studies were flawed and that the pesticide posed an unacceptably high risk to bees. In response to the study, the European Commission recommended a moratorium across the European Union (EU): the precautionary principle in accordance with the Regulation (EC) No 1107/2009 should be applied when using neonicotinoids due to their effects on honeybees.

On 29 April 2013, 15 of the 27 EU member states voted to enact a 2-year ban on the use of three neonicotinoids. Eight nations voted against the ban, while four abstained. The law prevents the use of imidacloprid, clothianidin and thiamethoxam on flowering plants for 2 years unless compelling scientific evidence comes out that says the chemicals are safe. Temporary suspensions had previously been enacted in France, Germany and Italy. In Switzerland, where neonicotinoids were never used in alpine areas, a ban was also implemented due to accidental poisonings of bee populations and the relatively low safety margin for other beneficial insects.

However, the ban remains tendentious, with environmentalists calling the move 'a significant victory for common sense and our beleaguered bee populations' and saying it is 'crystal clear that there is overwhelming scientific, political and public support for a ban'. Great Britain, which voted against the bill, disagreed: 'Having a healthy bee population is a top priority for us, but we did not support the proposal for a ban because our scientific evidence doesn't support it'. Bayer Cropscience, which makes two of the three banned products, remarked 'Bayer remains convinced neonicotinoids are safe for bees, when used responsibly and properly ... clear scientific evidence has taken a back-seat in the decision-making process'. Reaction in the scientific community overall was mixed. Biochemist Lin Field said the decision was based on 'political lobbying' and could lead to the overlooking of other factors involved in colony collapse disorder. Zoologist Lynn Dicks of Cambridge University disagreed, saying 'this is a victory for the precautionary principle, which is supposed to underlie environmental regulation'. A bee researcher, Dave Goulson, who has published recent studies (Goulson, 2015) on the deleterious effects of neonicotinoids on wild bees, has called the ban 'excellent news for pollinators', and said 'the weight of evidence from researchers clearly points to the need to have a phased ban of neonicotinoids'. It is now coming up to the 2-year mark of the ban and the EU is currently weighing evidence to see if the ban will be extended. However, recent papers have been published that seem to support the ban, due not only to the impact on honeybees but also on important wild bees (Rundlöf *et al.*, 2015).

The impact of this ban was also felt in Canada, where public concern about the impact of neonicotinoids increased significantly. Indeed, the Ontario Beekeepers Federation has launched a US$400 million class action lawsuit against Bayer Cropscience and Syngenta for bee losses for the past decade. Public anxiety has led the Ontario government's agricultural division, the Ontario Ministry of Agriculture, Food and Rural Affairs (OMAFRA), to change its policy regarding the use of neonicotinoid pesticides, especially as seed treatments. Starting on 1 July 2015, Ontario was the first jurisdiction in North America to protect bees and other pollinators through new rules to reduce the number of acres planted with neonicotinoid-treated maize and soybean seeds by 80% by 2017. To support this goal, stringent requirements were put in place for the sale and use of neonicotinoid-treated maize and soybean seed that would help ensure treated seed was only used when there was evidence of a pest problem (OMAFRA, 2015). Its aim as stated is 'reducing neonicotinoid use in these two crops presents the greatest potential to reduce pollinator exposure to the neurotoxic insecticide'.

The USA has had a divergent policy from the precautionary principle followed by the EU. Even though there has been considerable movement on this issue at the US Environmental Protection Agency (EPA), including new labelling requirements and pollinator risk assessment guidelines, as well as likely restrictions on new outdoor uses of these chemicals in the absence of additional effects data on developing honeybees, the EPA has an ongoing review to take any necessary action on current use patterns of these chemicals in general. However, a recent court ruling has revoked the EPA registration of a new neonicotinoid, Sulfoxaflor, which was given unconditional registration in 2013. The court said the US EPA gave unconditional approval of the insecticide Sulfoxaflor, even though it was clearly toxic to honeybees, and the manufacturer failed to conduct the required tests of its impact on not just adult bees, but also brood and colony strength.

In June 2014, US President Barack Obama issued a memorandum establishing a Pollinator Health Task Force, co-chaired by the Secretary of Agriculture and the Administrator of the EPA. The Pollinator Health Task Force then issued a *National Strategy to Promote the Health of Honey Bees and Other Pollinators* in May 2015 (The White House, 2015). The strategy lays out current and planned federal actions to achieve the following overarching goals:

1. Honeybees: reduce honeybee colony losses during winter (overwintering mortality) to no more than 15% within 10 years. This goal is informed by the previously released Bee Informed Partnership surveys and the newly established quarterly and annual surveys by the USDA National Agricultural Statistics Service. Based on the robust data anticipated from the national, statistically based National Agricultural Statistics Service (NASS) surveys of beekeepers, the Task Force will develop baseline data and additional goal metrics for winter, summer and total annual colony loss.

2. Monarch butterflies: increase the eastern population of the monarch butterfly to 225 million butterflies occupying an area of approximately 15 acres (6 ha) in the overwintering grounds in Mexico, through domestic/international actions and public–private partnerships, by 2020.

3. Pollinator habitat acreage: restore or enhance 7 million acres of land for pollinators over the next 5 years through federal actions and public–private partnerships.

While this strategy is to address primarily habitat loss for pollinators, a crucial component of it is to protect pollinators from exposure to pesticides. To do this, they propose to implement new harmonized guidance for assessing pesticide risks to pollinators. This is based on a guidance document produced by the EPA in June 2014, which, working in collaboration with Health Canada and the California Department of Pesticide Regulation, has released a harmonized guidance for assessing the risks posed to bees by pesticides (EPA, 2014). The guidance describes a tiered process beginning with a conservative screen (Tier 1) that uses laboratory-based acute and chronic toxicity studies of individual adult and larval honeybees. Along with this, the strategy, more importantly, is to re-evaluate the whole class neonicotinoid family of pesticides:

> Consistent with the President's requirements, EPA has further expedited its broad re-evaluation of the nitroguanidine-substituted neonicotinoid subclass (*i.e.*, imidacloprid, clothianidin, dinotefuran, thiamethoxam) under the 2015 - 2017 schedule laid out in Appendix A. As part of EPA's ongoing effort to protect pollinators, the EPA has sent letters to registrants of neonicotinoid pesticides with outdoor uses, informing them that EPA will likely not be in a position to approve most applications for new uses of these chemicals until new bee data have been submitted and pollinator risk assessments are complete.
>
> (The White House, 2015)

Along with this, the strategy outlines that seed treatments by neonicotinoids are also to be reassessed on a crop-by-crop basis. This is based on an economic analysis of the benefits of imidacloprid, clothianidin and thiamethoxam seed treatments for insect control in US soybean production.

> The assessment examines the use of neonicotinoid seed treatments in terms of the extent of use and the pests targeted. The assessment also estimates the biological and economic impacts of not allowing the use to continue on soybeans. The draft analysis was released for public comment between October 2014 and January 2015. EPA is reviewing the comments and analyzing additional information relevant to the assessment. EPA typically assesses the benefits of a chemical on a crop by crop basis. EPA's assessment of neonicotinoids' benefits on soybeans is the first completed for the neonicotinoids because some scientific publications claim that treating soybean seeds has little value. EPA will perform additional benefits assessments as part of the registration review process in which EPA will consider both risks and benefits for each of the neonicotinoids.
>
> (The White House, 2015)

More importantly is the reassessment of other registered pesticides now in use for their potential impact on pollinator health and to restrict label usage on pesticides that are known to be acutely toxic to bees. This change in direction for the EPA indicates a large change in policy in the use of synthetic pesticides in crop production.

Furthermore, recently the human health considerations of some pesticides have come under increased scrutiny. In October 2015, the EPA proposed to revoke the use of chlorpyrifos, a widely used insecticide that was sprayed on a variety of crops including oranges, apples, cherries, grapes, broccoli and asparagus (EPA, 2015). The pesticide, in use since 1965, has caused dozens of farmworkers to become ill in recent years. Traces have been found in waterways, threatening fish, and regulators say overuse could make targeted insects immune to the pesticide. If the EPA revokes all tolerances for this product, all agricultural uses will be banned (EPA, 2015).

All of this has put significant pressure on how agriculture uses synthetic pesticides, and there is a growing realization that the old system of using agrochemicals has to be revisited. There is an urgent requirement for alternative tactics to help make crop protection more sustainable. Many experts have promoted the use of IPM as the best way forward, and the EU has placed it centrally within its 2009 Sustainable Use Directive on pesticides (European Parliament, 2010).

IPM is a systems approach that combines different crop protection practices along with careful monitoring of pests and their natural enemies (Flint and van den Bosch, 1981). The idea behind IPM is that combining different practices together overcomes the shortcomings of individual practices alone. The overall aim is not to eradicate pest populations but to reduce them below levels that cause economic damage. The main IPM tactics used are:

1. Synthetic pesticides that are highly selective but which are classified by regulators as low risk, such as synthetic insect growth regulators.
2. Crop cultivars that are bred with resistance to various pests.
3. Cultivation practices such as crop rotation, intercropping and undersowing.
4. Physical methods such as mechanical weeders.
5. Natural products such as semiochemicals or biocidal plant extracts (biopesticides).
6. Biological control with natural enemies, including predatory insects and mites, parasitoids, parasites and microbial pathogens, along with microbial antagonists against plant pathogens and microbial pathogens of weeds, some of which are commercial biopesticides.
7. Finally, decision support tools for farmers to allow them to decide when it is economically beneficial to apply pesticides.

IPM can be carried out at different levels of sophistication with a basic level of targeting one pest on one crop to a whole farm level encompassing all pests and crops on the farm. The highest level encompasses all members of the agricultural community, from extension services to agrochemical industry and regulators.

Analysis of this approach over 62 research projects in 26 countries has shown that IPM leads to substantial reductions in pesticide applications (Pretty, 2008). Most farmers practice good pesticide practice (GPP), such as rotating pesticide classes, and good agronomic practice (GAP), such as crop rotation. However, the adoption of IPM principles is low in most agricultural communities, with the adoption of insect pheromones for pest monitoring quite low, except for some fruit crops, and the introduction of arthropod predators for biological control mostly used in greenhouse systems only. Indeed, in greenhouse systems, biological control now plays a central role in the production of many greenhouse crops. Pesticide resistance arising in the 1960s in some key greenhouse pests drove the development of alternative management strategies. Along with this, the adoption of bumblebees for pollination in many greenhouse vegetable crops reinforced pressure to reduce pesticide use. Indeed, there are now highly effective IPM programmes in place in most greenhouse systems which use biocontrol of insect and mite pests using predators, parasitoids, parasitic nematodes and entomopathogens. In Europe, IPM based around biological control is used in over 90% of greenhouse tomato, cucumber and sweet pepper production (Van Lenteren, 2000). In light of the aforementioned narration, there is a worldwide interest in the development of alternative strategies and the search for new types of safe insecticides.

2.3 A Sea Change in Crop Protection

The agrochemical industry has for years been relatively unchanged in how it carries out research and development of its chemical crop protection products. From the discovery of new candidate chemistries to field testing and ultimate registration, the process has been fairly straightforward though increasingly expensive. Much of the expense in registration has been, quite correctly, on the toxicological profile of the candidate compounds, both in relation to human toxicity and environment. Such studies are time-consuming and very expensive, as not only the compound itself has to be assessed but also its environmental breakdown components. Indeed, the environmental fate of the compound is now as important as mammalian toxicity and tetragenetic effects.

With this large level of scrutiny and the consequent difficulties in getting compounds to registration in a hugely onerous regulatory environment, it is unsurprising that some of the largest agrochemical producers are turning their attention to more sustainable and ecologically benign products.

This, and a new sense that when incorporated into crop production and pest management programmes, biopesticides offer the potential for more sustainable crop yields and quality than a chemical-only programme (Askary and Martinelli, 2015). Benefits in agricultural programmes include reduced chemical pesticide residue and resistance management, shorter field re-entry, biodegradability and low risk to these beneficials including predators and pollinators, including honeybees.

Along with this is a growing awareness of the human health risks and environmental issues in relation to the agricultural use of synthetic pesticides (Dunlap and Bues, 1992), and as more land is put into agricultural use, this becomes more apparent to nearby communities. The realization that pesticides could be contaminating urban areas has been one of the main components of the rural–urban divide that has begun in some communities. Also, as the organic movement has grown, even non-organic food consumers have a heightened concern about food safety as it relates to synthetic pesticides (Bruhn *et al.*, 1992). This has led to more consumer demands for pesticide-free food products and a growth in the local organic farming movement (Thakore, 2006). The public has recently shown great interest in agricultural biopesticides as the news of pollinator decline has spread. There is a realization that conservation is now paramount to our future food supply industry. These two issues may be the main movers in fostering the widespread utilization of biopesticides in the future.

Admittedly, the harmful impact of chemical pesticides consists of the environmental effects of pesticides on non-target species, while their hazardous impact on humans, birds, livestock and animals is related to their unwise usage in terms of unsafe active ingredients, dosage, timing and application methods, as well as their poor production, transport and storage practices (Table 2.1). Over 98% of applied insecticides and 95% of usable herbicides extend beyond their target species; a destination other than that intended can pollute the environment (Miller, 2004). That is usually due to their spread across entire agricultural fields and/or deep enough to cause groundwater contamination. Runoff can carry pesticides into aquatic environments, while wind can carry them to other fields, grazing areas, human settlements and undeveloped areas, potentially affecting other species. Over time, repeated application increases pest resistance, while its effects on other species can facilitate the pest's resurgence (Damalas and Eleftherohorinos, 2011).

It is well known that each pesticide or pesticide class is accompanied by specific precautions associated with environmental concerns and human safety. Nevertheless, the undesirable effects of these classes have led many pesticides to be banned, while safety-strict regulations have reduced or suspended the use of others. Consequently, the pesticides currently available have generally become less persistent and more species specific, reducing their harmful environmental impact. Moreover, the rates of pesticides applied per unit area have decreased significantly

Table 2.1. Examples of the environmental effects of pesticides. (Adapted from https://en.wikipedia.org/wiki/Environmental_impact_of_pesticides.)

Pesticide/class	Negative impact(s)	Reference
Organochlorine DDT/DDE	Eggshell thinning in raptorial birds	Rattner, 2009
DDT	Carcinogen	Turusov *et al.*, 2002
Chlordane	Interacts with vertebrate immune systems	Galloway and Depledge, 2001
Organophosphate	Acute mortality attributed to inhibition of acetylcholine esterase activity	Fleischli *et al.*, 2004
Carbamates, the phenoxy herbicide 2,4-D and atrazine	Interact with vertebrate immune systems	Galloway and Depledge, 2001
Carbamate	Thyroid disruption properties in rodents, birds, amphibians and fish	Rattner, 2009
Atrazine	Interacts with vertebrate immune systems	Galloway and Depledge, 2001
Nicotinoid	Respiratory, cardiovascular, neurological and immunological toxicity in rats and humans	Lin *et al.*, 2013
Bt maize/Cry	Reduced abundance of some insect taxa, predominantly susceptible Lepidopteran herbivores, as well as their predators and parasitoids	Kohler and Triebskorn, 2013
Spinosyns	Affect various physiological and behavioural traits of beneficial arthropods, particularly hymenopterans	Biondi *et al.*, 2012

(https://en.wikipedia.org/wiki/Environmental_impact_of_pesticides). In parallel, the importance of beneficial nematodes in crop pest management is increasing due to their advantageous attributes, and as a consequence they are contributing to the continuous decrease in the use of chemical pesticides (see Askary and Abd-Elgawad, Chapter 1, this volume).

Furthermore, continuous progress in developing formulations that enhance nematode storage stability for different types of beneficial nematodes is being achieved. For example, entomopathogenic nematodes (EPNs) can be stored for over 5 months under refrigeration in several products and for 12 months in one product. Products based on alginate gel formulations were the first to enable storage of *Steinernema carpocapsae* for 3–4 months at room temperature. With the introduction of water-dispersible granules, *S. carpocapsae* can remain efficacious for 5–6 months at 25°C. Furthermore, extending the shelf life of commercial products of many nematode species is being tried through physical trapping, metabolic inhibition, cold storage, or by the induction of anhydrobiosis to decrease nematode activity and metabolism (see Askary and Ahmad, Chapter 13, this volume). Also, polyether-polyurethane sponge as an inert carrier is very common for both a short storage period and shipping, while vermiculite and water-dispersible granular products are often used in retail markets to control insect pests of citrus, mushrooms, turfgrass, cranberries and greenhouses. In the future, more understanding of nematode behaviour and physiology will certainly lead to further development of perfect formulations. Likewise, great strides have been made in EPN mass production technology using both *in vitro* liquid and solid culture, while further innovations are still expected to overcome definite obstacles in EPN field efficacy through a sound understanding of the nematode-interacting biotic and abiotic factors (see Askary and Ahmad, Chapter 13, this volume).

Moreover, other groups of beneficial nematodes, such as species of *Beddingia* (*Deladenus*), *Thripinema* and mermithid, as well as predatory nematodes, are in developmental processes in terms of their more proper formulations, mass production and/or applications, to possibly occupy a reasonable share in the pesticide market (Grewal *et al.*, 2005a). Much more research concerning biotic and abiotic factors that favour the application of nematodes as biocontrol agents is now needed so this promising technology can be used in many diverse agricultural systems. (see Askary and Abd-Elgawad, Chapter 1, this volume).

2.4 The Difficult Rise of Biopesticides

Conventional pesticides have been under scrutiny for at least the past decade, which has more or less culminated with the recent pollinator crisis. There has also been basically a decade-long decline in the use of chemical pesticides (Bailey *et al.*, 2010a). Whether this has been due to the impact of new restrictive registration processes, especially in the EU, and a move to ban many existing pesticides, especially those that are deemed to be human endocrine disruptors (EC 91/414), is unknown. What is known is that there has been unprecedented growth in the biopesticide market, with an estimated growth rate of 3% in 2012 (Glare *et al.*, 2012), but more recently upwardly revised to 15% yearly from 2014 (Pucci, 2014). The most recent estimates are that, globally, biopesticides now account for 3.5% of all pesticides sold, or US$3 billion, up from US$1 billion 5 years ago. The upper limit on biopesticides is estimated at 10% of the crop protection market.

In recent years, there has been an almost explosive increase in the number of small companies developing sustainable or natural pest control products. The majority of these are using products such as essential oils that have been repurposed as natural pesticides for use in the control of insect pests and plant diseases. This off-the-shelf and into the crop protection business has been a successful business model and has little of the extra costs in development as a new compound. Indeed, many of the natural products are actually of food-grade level, thus rendering some toxicological tests mute.

In parallel with this, there have been some companies that have approached the biopesticide industry much like the large agrochemical corporations. They have developed methods to screen a large number of natural products and also to look at microbial, fungal and nematode potential candidates as well (Lamberth *et al.*, 2013). While this approach is much more expensive, it has the potential to find more effective commercial products that have a greater potential for large-scale agricultural use. This approach has allowed them to develop products for the agricultural industry instead of just the home and garden market (Glare *et al.*, 2012).

The interest in biopesticides can be seen in the number of mergers in the industry between the large agrochemical corporations and these smaller, biologically-based companies, the more recent ones being the investment of Monsanto in Novozymes Biologicals and the takeover of Becker Underwood by BASF. In 2015, FMC reported that it had nine biologicals in the pipeline (Agrow Reports, 2015).

It is important to realize that there are wide-ranging estimates of the actual size of the biopesticide market in the overall crop protection market. These wide-ranging estimates are due to the variation in the source of the information, with proponents often estimating a higher market share than is actually present. However, there have been many recent changes in the regulatory environment surrounding the commercialization and use of biopesticides, and this bodes well for continued growth.

Registration is often identified as the largest barrier to the commercialization of biopesticides (Sundh *et al.*, 2012; Mubyana-John and Taylor, 2015). The issues surrounding this have been discussed in recent books (Ravensburg, 2011; Sundh *et al.*, 2012; Askary and Martinelli, 2015). In general, legislatures have to understand the nature of biopesticides, and especially microbial pesticides. Most regulators have operated in the same framework that they use for chemicals, which restricts progress by basically treating them the same. While there are moves to change the regulations surrounding biopesticides, many jurisdictions have yet to implement these rules.

In Canada and the USA, there have been significant changes in the regulatory environment regarding the registration of new biopesticides. Part of this was driven by legislation banning the use of chemical pesticides in some cities and in Canada, with two provinces following suit (Bailey *et al.*, 2010b). This concurrently supported policy changes with PMRA (Pest Management Regulatory Agency, Canada), which encouraged lower-risk pesticides, of which biopesticides were a component part. This change indicated that the agency intended to facilitate the registration of biopesticides, especially ones that were already registered in the USA. This has led to a significant increase in harmonization in the registration process between the two countries, which will soon encompass all Organisation for Economic Co-operation and Development (OECD) countries (Meeussen, 2012). This has led to a reduction in the time of review for new biopesticide registrations and exemption of some fees for the scientific review of microbial pesticides. This has resulted in the registration of 22 new pesticides between 1995 and 2002, which compares favourably to the ten that were registered between 1984 and 1994 (PMRA, 2004). In 2009, the EU member states adopted a Sustainable Use of Pesticides Directive (2009/128/EC), which has further supported the adoption of biopesticides in the EU.

While there have been significant strides in the regulatory effort to get more biopesticides to market, there has not been the growth that was originally expected in the crop protection market. In North America, the current rate of adoption of biopesticides is low to moderate, with about 10% of vegetable farms and 30% of fruit farms applying biopesticides regularly (Cuddeford, 2007). In field crops, there is less than 5% use in the USA, and in Canada less than 2% (Cuddeford, 2007). While this uptake may be due to the lack of appropriate products, there are other aspects of biopesticides that make them difficult for farmers to support.

Farmers in general are used to a chemically driven concept of pest control in their operations. While advantages such as lower toxicity to the environment and worker safety are well known to many producers, they offset that with the fact that biopesticides are often seen as less effective, slow to act, have a limited pest range, a short field and shelf life and are difficult to use. This, along with the fact that they are often more expensive, makes it harder for them to determine their economic benefit. For biopesticides to be adopted in a more broad sense, there has to be some effort in developing products that have good efficacy, ease of use and are cost-effective. These advantages, along with education of the end-user through extension, will make market penetration much more effective in the future.

In recent years, there have been good commercial successes in biopesticides that have a broader pest range. Products such as Serenade target multiple fungal diseases, and Timorex Gold is now used on a wide range of crops. This, along with formulation advances and shelf-life stability, have brought some modicum of commercial success to several products in that they can be used like synthetic chemical pest control products.

While crop protection market estimates are usually based on all biopesticides and include plant-derived compounds as well as microbial agents that usually include EPNs, it is useful for the purposes of this book to mention what is the market share of microbial pesticides as opposed to all other components of biopesticides in general. Microbial pesticides only account for about 1–2% of all pesticides sold, although they have shown remarkable growth as well. It is difficult to get any real data on the amount of microbial pesticides sold as they are often lumped with plant-derived biopesticides. The most commercially available microbial pesticides are the *Bacillus thuringiensis* group of microbial pesticides, with at least 50% of the biopesticide market (CPL Business Consultants, 2010). Recent estimates from the same report indicated that in 2014, the assessed global market for microbials was US$3.3 billion, with nematodes being about US$92 million, or 3% of the global microbial biopesticide market.

2.5 Commercial Use of Entomopathogenic and Slug Parasitic Nematodes

The agricultural use of EPNs really starts in 1929 with the discovery of *Steinernema glaseri* on a golf course in New Jersey (USA). It was found to be parasitizing the grubs of the introduced pest, the Japanese beetle (*Popillae japonica* Newm.) (Glaser and Fox, 1930). While *S. glaseri* is still commercially available, it was the first widely available nematode for the suppression of an insect pest, and its characterization and original rearing techniques are still the basis of nematode rearing to this day (McCoy and Glazer, 1936). While there are numerous taxa of nematodes known to parasitize a wide variety of invertebrate pests and pathogens (see Askary and Abd-Elgawad, Chapter 1, this volume), the EPNs in the families *Steinernematidae* and *Heterorhabditidae* (Askary, 2010) followed by the slug parasitic nematodes, *Phasmarhabditis* (Coupland, 1995; Askary *et al.*, 2012) have been the most successfully commercialized nematodes as biological control agents of agricultural pests (Figs 2.1 and 2.2). They have been in use as biological pesticides against many insects and slug pests on different crops and/or home gardens, lawns and turfgrass; examples are given in (Table 2.2).

The biology of EPNs is summarized in Gaugler (2002), and their commercial use is described by Grewal *et al.* (2005b) and Lacey *et al.* (2015). Nematodes are effective biocontrol agents for many soil-dwelling pests. Indeed, many EPNs possess several of the attributes of excellent commercial biological control agents (Shapiro-Ilan and Grewal, 2008). They have a good environmental profile and are safe to non-target organisms and human health (Ehlers, 2005). This profile has allowed many regulatory agencies to reduce the requirements for the registration of these EPNs as pesticides, especially in the USA and Canada (Bailey *et al.*, 2010a,b; Lacey and Georgis, 2012)

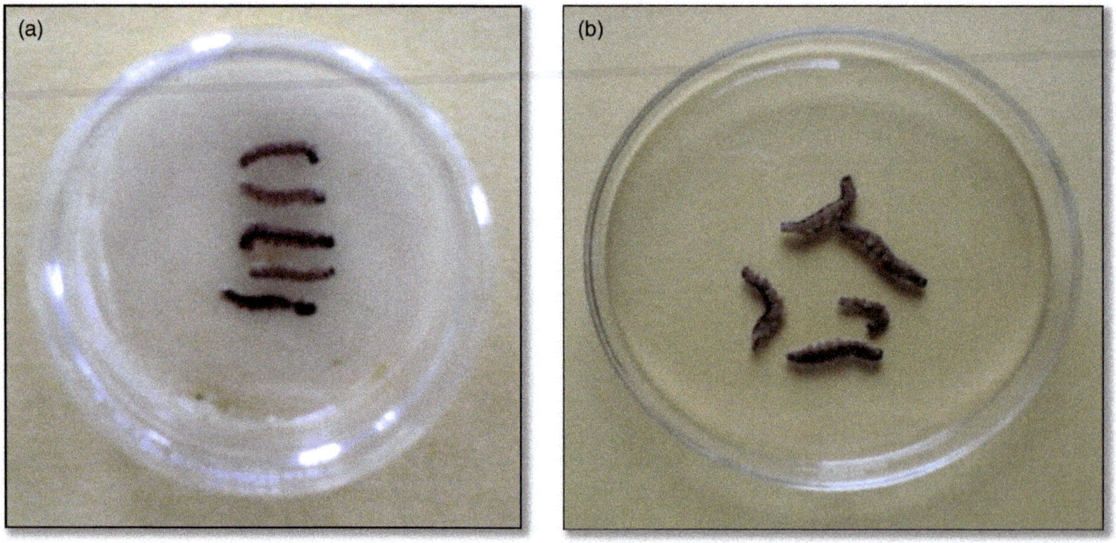

Fig. 2.1. Insect larvae killed by entomopathogenic nematode *Heterorhabditis bacteriophora*. (a) Dead larvae of armyworm, *Mythimna separata*; (b) dead larvae of cabbage butterfly, *Pieris brassicae*.

Fig. 2.2. Larva of cabbage butterfly, *Pieris brassicae*, infested with *Heterorhabditis bacteriophora*.

and somewhat less so in the EU, where there is still an onerous registration process for microbials (Grant, 2008). Furthermore, EPNs tend to have a large host range, often encompassing dozens of insect species, and some commercialized EPN species have been labelled for use against a dozen or more insect species. Along with this, they have fairly rapid speed of kill and actively seek and attack suitable hosts, and as such are dynamic control agents (Kaya and Gaugler, 1993).

EPNs have been widely commercially available for over 25 years. However, agricultural level applications have not met expectations for such good biopesticides. At the beginning of their commercialization, there were projected sales of over US$100 million, but so far the market is closer to 10% of those projections (Gaugler and Han, 2002; Shapiro-Ilan *et al.*, 2012; Campos-Herrera, 2015).

There are a number of barriers beyond acceptable efficacy that have hindered the uptake of EPNs and slug parasitic nematodes into the broader agricultural market, including the cost, shelf life, handling, mixing, coverage, competition, compatibility and profit margins for manufacturers and distributors alike. Most of the current markets are limited to specific pests of specific crops, such as citrus, turf and ornamentals.

2.5.1 Production

One of the main constraints to commercialization of EPNs was sufficient production for widespread use. For smaller amounts of EPNs, an *in vivo* culture method is sufficient to produce enough nematodes for small-scale use (Shapiro-Ilan *et al.*, 2016). This method uses a susceptible supply of inexpensive hosts on which to rear the nematodes. It has been scaled up though for mass production, recently using a LOTEK system (Gaugler *et al.*, 2002); one of the best things about *in vivo* systems is that the nematodes are run through an insect host, ensuring pathogenicity, which can be a problem in *in vitro* cultivation. While this works up to a point, there is a need for large-scale culture, which can only be done using *in vitro* methodologies.

While Glaser and Fox (1930) were able to rear *S. glaseri* successfully on a solid medium axenically, the first commercial scale monoxenic solid culture system was developed by Bedding (1984). This method entails using a culture medium (i.e. pigs' kidney) on sponge pieces and was able to produce large numbers of EPNs in relation to the amount of medium input. This system works well for small to medium-size production of EPNs in areas where labour costs are not an issue.

The main method for the commercial production of EPNs worldwide is liquid fermentation. However, the progress to this stage is often beset with difficulties, as described by Ehlers (2007). From the discovery of suitable EPNs to the assessment of the various bacterial symbionts, the commercialization process is time-consuming and difficult. The process can take many years (see Glen and Coupland, Chapter 28, this volume). Improved production, formulation and application technology will lead to improved efficacy in the field. With improved production efficiency, costs should be reduced and will lead to better economic considerations when brought to the agricultural crop protection industry.

Since information is relatively limited concerning the slug parasitic nematodes, we are satisfied with what has been reported about their production and application

Table 2.2. Use of entomopathogenic and slug parasitic nematodes as biological pesticides.[a] (Adapted from Shapiro-Ilan and Gaugler, 2015.)

Pest Common name	Pest Scientific name	Key Crop(s) targeted	Efficacious nematodes
Artichoke plume moth	*Platyptilia carduidactyla*	Artichoke	Sc
Armyworms	Lepidoptera: Noctuidae	Vegetables	Sc, Sf, Sr
Banana moth	*Opogona sachari*	Ornamentals	Hb, Sc
Banana root borer	*Cosmopolites sordidus*	Banana	Sc, Sf, Sg
Billbug	*Sphenophorus* spp. (Coleoptera: Curculionidae)	Turf	Hb, Sc
Black cutworm	*Agrotis ipsilon*	Turf, vegetables	Sc
Black vine weevil	*Otiorhynchus sulcatus*	Berries, ornamentals	Hb, Hd, Hm, Hmeg, Sc, Sg
Borers	*Synanthedon* spp. and other sesiids	Fruit trees and ornamentals	Hb, Sc, Sf
Cat flea	*Ctenocephalides felis*	Home yard, turf	Sc
Citrus root weevil	*Pachnaeus* spp. (Coleoptera: Curculionidae)	Citrus, ornamentals	Sr, Hb
Codling moth	*Cydia pomonella*	Pome fruit	Sc, Sf
Corn earworm	*Helicoverpa zea*	Vegetables	Sc, Sf, Sr
Corn rootworm	*Diabrotica* spp.	Vegetables	Hb, Sc
Cranberry girdler	*Chrysoteuchia topiaria*	Cranberries	Sc
Crane fly	Diptera: Tipulidae	Turf	Sc
Diaprepes root weevil	*Diaprepes abbreviatus*	Citrus, ornamentals	Hb, Sr
Fungus gnats	Diptera: Sciaridae	Mushrooms, greenhouse	Sf, Hb
Grape root borer	*Vitacea polistiformis*	Grapes	Hz, Hb
Iris borer	*Macronoctua onusta*	Iris	Hb, Sc
Large pine weevil	*Hylobius albietis*	Forest plantings	Hd, Sc
Leafminers	*Liriomyza* spp. (Diptera: Agromyzidae)	Vegetables, ornamentals	Sc, Sf
Mole crickets	*Scapteriscus* spp.	Turf	Sc, Sr, Scap
Navel orangeworm	*Amyelois transitella*	Nut and fruit trees	Sc
Plum curculio	*Conotrachelus nenuphar*	Fruit trees	Sr
Scarab grubs[b]	Coleoptera: Scarabaeidae	Turf, ornamentals	Hb, Sc, Sg, Ss, Hz
Shore flies	*Scatella* spp.	Ornamentals	Sc, Sf
Strawberry root weevil	*Otiorhynchus ovatus*	Berries	Hm
Small hive beetle	*Aethina tumida*	Beehives	Hi, Sr
Sweet potato weevil	*Cylas formicarius*	Sweet potato	Hb, Sc, Sf
Common grey garden slug	*Deroceras reticulatum*	Home gardens	Ph

Notes: [a]At least one scientific study reported 75% suppression of these pests using the nematodes indicated in field or greenhouse experiments. Subsequent/other studies may reveal other nematodes that are virulent to these pests. Nematode species used are abbreviated as follows: Hb = *Heterorhabditis bacteriophora*; Hd = *Heterorhabditis downesi*; Hi = *Heterorhabditis indica*; Hm = *Heterorhabditis marelata*; Hmeg = *Heterorhabditis megidis*; Hz = *Heterorhabditis zealandica*; Sc = *Steinernema carpocapsae*; Sf = *Steinernema feltiae*; Sg = *Steinernema glaseri*; Sk = *Steinernema kushidai*; Sr = *Steinernema riobrave*; Sscap = *Steinernema scapterisci*; Ss = *Steinernema scarabaei*; Ph = *Phasmarhabditis hermaphrodita*. [b]The efficacy of various pest species within this group varies according to nematode species.

(see Nermut' and Půža, Chapter 27, this volume), as well as their efficacy and commercialization (see Glen and Coupland, Chapter 28, this volume).

2.5.2 Application

One of the main issues with efficacy is the mismatch between where and how the EPNs are to be used. The biology and behaviour of the nematode and its pest target, along with the environment in which it is to be used, have to be considered carefully. This usually entails carrying out a large number of field trials to test application protocols that will achieve consistent results (Georgis and Gaugler, 1991). Factors such as soil moisture (based on rainfall and irrigation), thatch depth, soil type and seasonal temperature, along with nematode strain and application method, could be a predictor of failure or success.

Along with improved production efficiency, there are new advancements in application technologies. Some, such as the application of infected hosts, attract and kill methods, slow-release bags and plant dips, are currently being used. There is also ongoing research in the development of application equipment and the impact of sprayers on dispersal and mortality (Wright *et al.*, 2005).

2.5.3 Efficacy

Another aspect of reduced uptake of EPNs is a perceived lack of efficacy. Predictive control is necessary if EPNs are to gain any market penetration as agricultural biopesticides. However, due to the complex interplay of abiotic and biotic factors, achieving this predictability is one of the great challenges in EPN development. While many nematodes can infect many different hosts successfully, there is variation in virulence and optimal development that occurs between species and strains. It is therefore imperative that rigorous screening of different potential species and strains is carried out before the development of a commercial programme.

New EPNs are being discovered at a fairly high rate, and it is probable that more effective strains or species will be discovered (Stock and Hunt, 2005). Of the more than 100 EPN species that have been discovered to date, more than 40% have been discovered since 2001. This, along with selective breeding for pathogenicity, should improve efficacy in EPNs. This has already been the case with the discovery of *Steinernema scapterisci* for the control of mole crickets and with *Steinernema riobrave* for *Diaprepes* root weevils and other crop pests, which made a considerable impact on the EPN commercial market (Shapiro-Ilan *et al.*, 2002).

Currently, less than 20% of the more than 35 species discovered since 2001 have been tested for biocontrol efficacy in either the laboratory or the field. Admittedly, there has been an efficacy gap in the EPN performance between laboratory and field. Biotic and abiotic factors contributing to such a gap in EPN efficacy should be further investigated in the context of knowing and creating optimal conditions for maximum EPN efficacy against the insect hosts. This is especially important since it is certain that, with the continually increased number of new strains and species, there will be an opportunity for increasing virulence and more potential for successful commercialization.

2.5.4 Use of entomopathogenic and slug parasitic nematodes worldwide

Commercial use of EPNs will expand as the list of potential crop pests that are suitable for control by EPN increases worldwide. There has been increased research in expanding the use of EPNs to some new or existing targets, and this has certainly borne fruit. The subject of this book is to detail the use of EPNs and slug parasitic nematodes in different countries around the world. Alt hough commercial products of these nematodes, containing different species for use against various pests, are found on several continents, especially North America, South America, Africa, Asia and Europe, Australia (Table 2.3), pesticide markets have not been saturated yet and need a lot of these biopesticides in order to achieve the required balance. Moreover, while there is more uptake of EPNs in some regions than in others, there may be reasons for this that may be overcome in the future.

Indeed, there may have to be a whole paradigm shift in how we produce and use EPNs to make them more available to the agricultural market. This is eloquently put forward by Gaugler (1997), who mentions that the current paradigm for commercializing biopesticides is basically based on a chemical pesticide industry model. Biopesticides, with the notable exceptions of some biofungicides and bioinsecticides, fit this model poorly. If we use this model for EPNs with production costs of US$50–70/ha, these far exceed most synthetic chemicals. Add to this the low shelf-life time (3–6 months on average) and the special handling and distribution, it falls well short of the conventional farmer-friendly standard. One of the models that Gaugler proposes is that we emulate business sectors that have overcome biological limitations successfully. A model for this is the microbrewery industry, which has shown rapid growth in many areas of the world. The idea of 'small batch' custom production of a fresh product sold without preservatives has led to the commercial success of high-quality products. The concept that we can overcome the costs associated with production, storage and transport by using local production seems a reasonable alternative to the more centralized system we have now. In this way, only 'fresh' biologicals would be applied, which should ensure performance. One way to do this would be to use existing cottage industries of small biological control companies that already produce insect predators and parasitoids. There could be a dissemination

Table 2.3. Commercial entomopathogenic and slug parasitic nematode products available in various countries.

Company, country (product name)	Nematode species[a]	Key pest(s)	Reference or website
BASF, USA and Australia (Millenium®, Nemasys®, Nematac®)	Hb, Sc, Sf, Sk	Several species of Lepidoptera and Coleoptera	http://betterplants.basf.us/products/product-index.html
BioLogic, USA (Scanmask, Ecomask, Heteromask)	Hb, Sc, Sf	Several species of Lepidoptera and Coleoptera	https://biologicco.com/
Green Spot, Ltd, USA (various products)	Hb, Sc, Sf	Several species of Lepidoptera and Coleoptera	https://greenmethods.com/nematodes/
Optinem Cydia, Italy (Agrifutur)	Sf	Codling moth (*Cydia pomonella*)	http://www.agrotechnico.com/doc/optinem_CYDIA_ST_FR.pdf
Bioplanet, Italy (Nemopak HB)	Hb	Vine weevil (*Otiorhynchus sulcatus*) and other Coleoptera in the soil	http://bioplanet.it/it/heterorhabditis-bacteriophora/
CBC Biogard, Italy (Nemastar)	Sc	Codling moth (*C. pomonella*), other Lepidoptera of orchards (*Cydia* spp., *Synanthedon* spp., etc.), palm borer (*Paysandisia archon*), red palm weevil (*Rhynchophorus ferrugineus*), Noctuid and Crambid moths (*Spodoptera* spp., *Agrotis* spp., *Duponchelia fovealis*, etc.), summer chafer (*Amphimallon solstitialis*), *Rhizotrogus aestivus*, flatheaded woodborer (*Capnodis tenebrionis*), adults of mole cricket (*Gryllotalpa gryllotalpa*), crane fly (*Tipula paludosa*, *Tipula oleracea*)	http://www.biocomes.eu/partner/biogard-division-cbc-europe/
Nature Insect Control, Canada (Lawn Guardian)	Hb, *Steinernema* spp.	Grubs (e.g. *Popillia japonica*)	http://www.naturalinsectcontrol.com/product.php?id=000000416
The National Research Centre, Egypt (NOBUG)	Sc	Lepidopterous insect pests	Abd-Elgawad and Spiridonov, 2014
'Biodan' and 'Biometodika' Companies, Russia (Nemabact and EntonemF)	SC, Sf	Thrips, grubs, cabbage flies and bark beetles	Abd-Elgawad and Spiridonov, 2014
Henan Jiyuan Baiyun Industry Co Ltd, China (Naima)	Various	Chive root gnat (*Bradysia odoriphaga*) and below-ground pests	http://www.keyunnpv.cn/Products/tiandishengwunongyao/330273.html
Bio Controle, Brazil (Bio *Steinernema*)	Sb	Sugarcane weevil (*Sphenophorus levis*)	http://www.pg.fca.unesp.br/Teses/PDFs/Arq1207.pdf
BASF, UK and Australia (Nemaslug)	Ph	Common grey garden slug (*Deroceras reticulatum*)	http://www.harrodhorticultural.com/nemaslug-slug-killer-pid7623.html

Note: [a]Nematode species used are abbreviated as follows: Hb = *Heterorhabditis bacteriophora*; Sc = *Steinernema carpocapsae*; Sf = *Steinernema feltiae*; Sb = *Steinernema brazilense*; Sk = *Steinernema kushidai*; Ph = *Phasmarhabditis hermaphrodita*.

of 'do-it-yourself' technology, where local cooperative farm organizations could produce their own EPNs. This technology would also help farmers in developing countries to upgrade their own home-grown systems for improving the EPN market targeted to their specific needs. On the other hand, researchers, especially nematologists, should always be ready to cope with any new introduction(s) that entails changes in current agricultural systems. For example, Florida citrus acreage is favoured for broad EPN application to control citrus root weevils effectively (Shapiro-Ilan *et al.*, 2002; Dolinski *et al.*, 2012). Nevertheless, the recent introduction of the bacterial disease, huanglongbing (HLB or citrus greening), has been modifying the citrus production system there. Consequently, a new 'Advanced Production System' is being followed, potentially to sustain citriculture in the face of increased rates of tree loss due to HLB. The new system dramatically changed soil properties, such as pH ($P < 0.001$) compared to properties in traditional systems. Responses by soil organisms to these changes included higher infestation rates by the bacterium *Paenibacillus* sp. on the EPN *Steinernema diaprepesi* ($P = 0.05$), which may have contributed to reduced numbers of infective juveniles (IJs) of native EPNs in the new system ($P = 0.01$). Therefore, further studies of the physical causes of this effect are needed to solve this issue, which has the potential to reduce EPN populations and increase herbivory by the citrus root weevils (Campos-Herrera *et al.*, 2013). However, since the discovery and use of *S. glaseri*, there have been worldwide successes in the use of EPNs against some of the most intractable crop and tree pests. EPNs have a great future for use in the growing biopesticide market, but it will take research and some 'out-of-the-box' thinking to make this a reality.

2.6 Conclusions

The long and intensive use of chemical pesticides has caused many problems, such as developing pesticide-resistant pests, unpredictable outbreaks of non-targeted insect pests, health hazards and environmental pollution. This has pushed the crop protection industry worldwide to undertake a paradigm shift in how we use pesticides in agriculture. (Abd-Elgawad, 2008). In this context, agricultural pest control strategies are shifting from unilateral chemical methods to environmentally friendly pest management approaches such as biocontrol agents, biorational compounds, biopesticides, host plant resistance and ecological management (Rechcigl and Rechcigl, 2000; Peshin and Pimentel, 2014; Abd-Elgawad and Vagelas, 2015).

Entomopathogenic and slug parasitic nematodes are at the forefront of these environmentally friendly biocontrol agents that can be applied effectively against insect pests and slugs, respectively (Askary, 2010; Campos-Herrera, 2015). They are highly effective, safe and ecologically acceptable, as well as having a wide host range, relative ease of formulation and commercial production, and worldwide distribution, along with their host pests (e.g. Grewal *et al.*, 2005a; Askary, 2010; Askary *et al.*, 2012; Abd-Elgawad, 2014; Wilson and Rae, 2015). The use of entomopathogenic and slug parasitic nematode based pesticides has been reported for protecting agricultural crops in various agroecosystems. Their use can, on the one hand, decrease reliance on synthetic pesticides and, on the other hand, assist in maintaining a desirable agro-ecological balance. Nevertheless, their relatively low revenues and limited share in the pesticide market worldwide constitute the present pressing problems to their uptake. Therefore, current research is under way to solve critical issues and requirements for expanding their applications (Grewal *et al.*, 2005b; Abd-Elgawad and Spiridonov, 2014). This fact also applies to other numerous taxa of beneficial nematodes known to parasitize a wide range of agricultural pests and pathogens (see Askary and Abd-Elgawad, Chapter 1, this volume). There is great potential for using such newly discovered beneficial nematodes in many wide ranging agro-ecosystems.

Acknowledgements

This work was supported in part by US-Egypt Science and Technology Joint Fund (cycle 17; project no. 172) and Egyptian In-House, National Research Centre Project No 11030133 entitled 'Pesticide alternatives against soilborne pathogens attacking legume cultivation in Egypt'. The third author is also indebted to Division of Entomology, Sher-e-Kashmir University of Agricultural Sciences and Technology, Srinagar, Jammu and Kashmir, India for providing necessary facilities required in the preparation of the chapter.

References

Abd-Elgawad, M.M.M. (2008) The current status of phytonematode management in Egypt with special

reference to applicable nematicides. *Egyptian Journal of Agronematology* 6, 33–46.
Abd-Elgawad, M.M.M. (2014) Spatial patterns of *Tuta absoluta* and heterorhabditid nematodes. *Russian Journal of Nematology* 22, 89–100.
Abd-Elgawad, M.M.M. and Spiridonov, S.E. (2014) Entomopathogenic nematode application in Egypt and Russia: challenges and opportunities. *Egyptian Journal of Agronematology* 13, 1–22.
Abd-Elgawad, M.M.M. and Vagelas, I.K. (2015) Nematophagous bacteria: field application and commercialization. In: Askary, T.H. and Martinelli, P.R.P. (eds) *Biocontrol Agents of Phytonematodes*. CAB International, Wallingford, UK, pp. 276–309.
Agrow Reports (2015) Agrow Precision Farming for Crop Protection. Available at: https://store.agra-net.com/reports/agrowprecision15.html (accessed 1 September 2016).
Askary, T.H. (2010) Nematodes as biocontrol agents. In: Lichtfouse, E. (ed.) *Sociology, Organic Farming, Climate Change and Soil Science*. Springer, The Netherlands, pp. 347–378.
Askary, T.H. and Martinelli, P.R.P. (2015) *Biocontrol Agents of Phytonematodes.* CAB International, Wallingford, UK.
Askary, T.H., Khan, A.A., Waliullah, M.I.S., Banday, S.A., Iqbal, U. and Mir, M.M. (2012) Slug pest management through nematodes in agricultural and horticultural crops. In: Boeri, F. and Chung, J.A. (eds) *Nematodes: Morphology, Functions and Management Strategies*. Nova Publishers, New York, pp. 197–211.
Bailey, A., Chandler, D., Grant, W.P., Greaves, J., Prince, G. and Tatchell, M. (2010a) *Biopesticides: Pest Management and Regulation*. CAB International, Wallingford, UK.
Bailey, I.K.L., Boyetchko, S.M. and Langle T. (2010b) Social and economic drivers shaping the future of biological control: a Canadian perspective on the factors affecting the development and use of microbial pesticides. *Biological Control* 52, 221–229.
Bedding, R.A. (1984) Large scale production, storage and transport of insect-parasitic nematodes, *Neoaplectana* spp. and *Heterorhabditis* spp. *Annals of Applied Biology* 104, 117–120.
Biondi, A., Mommaerts, V., Smagghe, G., Viñuela, E., Zappalà, L. and Desneux, N. (2012) The non-target impact of spinosyns on beneficial arthropods. *Pest Management Science* 68, 1523–1536.
Bruhn, C.M., Diaz-Knauf, K., Feldman, N., Harwood, J., Ho, G., *et al.* (1992) Consumer food safety concerns and interest in pesticide-related information. *Journal of Food Safety* 12, 253–262.
Campos-Herrera, R. (2015) *Nematode Pathogenesis of Insects and Other Pests: Ecology and Applied Technologies for Sustainable Plant and Crop Protection.* Springer International Publishing, Switzerland.
Campos-Herrera, R., Stuart, R.J., EL-Borai, F.E., Gutiérrez, C. and Duncan L.W. (2010) Entomopathogenic nematode ecology and biological control in Florida citrus. In: Ciancio, A. and Mukerji, K.G. (eds) *Integrated Management of Arthropod Pests and Insect Borne Diseases*, Vol 5. Springer Science+Business Media BV, New York, pp. 101–130.
Campos-Herrera, R., Pathak, E., El-Borai, F.E., Schumann, A., Abd-Elgawad, M.M.M. and Duncan, L.W. (2013) New citriculture system suppresses native and augmented entomopathogenic nematodes. *Biological Control* 66, 183–194.
Carson, R. (1962) *Silent Spring*. Houghton Mifflin, Riverside Press, Cambridge, Boston, Massachusetts, 368 pp.
CPL Business Consultants (2010) *The 2010 Worldwide Biopesticides Market Summary*, Vol 1. CPL Business Consultants and CAB International Centre, Wallingford, UK.
Coupland, J.B. (1995) Susceptibility of helicid snails to isolates of the nematode *Phasmarhabditis hermaphrodita* from southern France. *Journal of Invertebrate Pathology* 66, 207–208.
Cudderford, V. (2007) *Biocontrol Files*. Co-published by World Wildlife Fund, the Biocontol Network and Agriculture and Agri-food Canada, pp. 4–5.
Damalas, C.A. and Eleftherohorinos, I.G. (2011) Pesticide exposure, safety issues, and risk assessment indicators. *International Journal of Environmental Research and Public Health* 8, 1402–1419.
Dolinski, C., Choo, H.Y. and Duncan, L.W. (2012) Grower acceptance of entomopathogenic nematodes: case studies on three continents. *Journal of Nematology* 44, 226–235.
Dunlap, R.E. and Bues, C.E. (1992) Understanding public concerns about pesticides: an empirical examination. *The Journal of Consumer Affairs* 26, 418–438.
Edwards, C.A. (1993) The impact of pesticides on the environment. In: Pimentel, D. and Lehman, H. (eds) *The Pesticide Question, Environment, Economics and Ethics*. Chapman & Hall, New York, pp. 13–46.
Ehlers, R.-.U. (2005) Forum on safety and regulation. In: Grewal, P.S., Ehlers, R.-U. and Shapiro-Ilan, D.I. (eds) *Nematodes as Biocontrol Agents*. CAB International, Wallingford, UK, pp. 107–114.
Ehlers, R.-U. (2007) Entomopathogenic nematodes: from science to commercial use. In: Vincent, C., Goettel, M.S. and Lazarovits, G. (eds) *Biological Control: A global perspective*. CAB International, Wallingford, UK, pp. 136–151.
EPA (Environmental Protection Agency) (2014) EPA Action to Protect Pollinators. Available at: https://www.epa.gov/pollinator-protection/epa-actions-protect-pollinators (accessed 1 September 2016).
EPA (2015) EPA Proposes to Revoke Chlorpyrifos Food Residue Tolerances. Available at: http://www2.epa.gov/pesticides/epa-proposes-revoke-chlorpyrifos-food-residue-tolerances (accessed 1 September 2016).
European Parliament (2010) Pesticides: framework for the Community action to achieve sustainable use of pesticides. Available at: http://www.europarl.europa.eu/oeil/home/home.do (accessed 25 October 2016).

Fleischli, M.A., Franson, J.C., Thomas, N.J., Finley, D.L. and Riley, W. Jr (2004) Avian mortality events in the United States caused by anticholinesterase pesticides: a retrospective summary of National Wildlife Health Center Records from 1980 to 2000. *Archives of Environmental Contamination and Toxicology* 46, 542–550.

Flint, M.L and van den Bosch, R. (1981) *Introduction to Integrated Pest Management*. Plenum Press, New York, 240 pp.

Galloway, T.S. and Depledge, M.H. (2001) Immunotoxicity in invertebrates: measurement and ecotoxicological relevance. *Ecotoxicology* 10, 5–23.

Gaugler, R. (1997) Alternative paradigms for commercializing biopesticides. *Phytoparasitica* 25, 179–181.

Gaugler, R. (2002) *Entomopathogenic Nematology*. CAB International, Wallingford, UK.

Gaugler, R. and Han, R. (2002) Production technology. In: Gaugler, R. (ed.) *Entomopathogenic Nematology*. CAB International, Wallingford, UK, pp. 289–310.

Gaugler, R., Brown, I., Shapiro-Ilan, D. and Atwa, A. (2002) Automated technology for *in vivo* mass production of entomopathogenic nematodes. *Biological Control* 24, 199–206.

Georgis, R. and Gaugler, R. (1991) Predictability in biological control using entomopathogenic nematodes. *Journal of Economic Entomology* 84, 713–720.

Glare, T.R., Caradus, J., Gelernter, W., Jackson, T., Keyhani, N., *et al*. (2012) Have biopesticides come of age? *Trends in Biotechnology* 30, 250–258.

Glaser, R.W. and Fox, H. (1930) A nematode parasite of the Japanese beetle *Popillia japonica* (Newm.). *Science* 71, 16–17.

Goulson, D. (2015) Neonicotinoids impact bumblebee colony fitness in the field: a reanalysis of the UK's Food & Environment Research Agency 2012 experiment. *PeerJ* 3:e854 https://doi.org/10.7717/peerj.854

Grant, W. (2008) More support needed for biopesticides. *Chemistry and Industry* 10, 25.

Grewal, P.S., Ehlers, R.-U. and Shapiro-Ilan, D.I. (eds) (2005a) *Nematodes as Biocontrol Agents*. CAB International, Wallingford, UK.

Grewal, P.S., Ehlers, R.-U. and Shapiro-Ilan, D.I. (2005b) Critical issues and research needs for expanding the use of nematodes in biocontrol. In: Grewal, P.S., Ehlers, R.-U. and Shapiro-Ilan, D.I. (eds) *Nematodes as Biocontrol Agents*. CAB International, Wallingford, pp. 479–489.

Gutiérrez, C., Campos-Herrera, R., Jiménez, J., Escuer, M., Labrador, S. and Rivera, Y. (2005) Control de plagas en agricultura ecológica: Eficacia de los nematodos entomopatógenos. (Pest control in ecological agriculture: efficacy of entomopathogenic nematodes.) *Alimentaría* 4, 62–67.

Hayes, T.B., Khoury, V., Narayan, A., Nazir, M., Park, A., *et al*. (2010) Atrazine induces complete femization and chemical castration of male African clawed frogs (*Xenopus laevis*). *Proceedings of the National Academy of Sciences* 107, 4612–4617.

Kaya, H. and Gaugler, R. (1993) Entomopathogenic nematodes. *Annual Review of Entomology* 38, 181–206.

Kohler, H.R. and Triebskorn, R. (2013) Wildlife ecotoxicology of pesticides: can we track effects to the population level and beyond? *Science* 341, 759, DOI: 10.1126/science.1237591.

Kollmeyer, W.D., Flattum, R.F., Foster, J.P., Powell, J.E., Schroeder, M.E. and Soloway, S.B. (1999) Discovery of the nitromethylene heterocycle insecticides. In: Yamamoto, I. and Casida, J. (eds) *Nicotinoid Insecticides and the Nicotinic Acetylcholine Receptor*. Springer-Verlag, Tokyo, pp. 71–89.

Lacey, L.A. and Georgis, R. (2012) Entomopathogenic nematodes for control of insect pests above and below ground with comments on commercial production. *Journal of Nematology* 44, 218–225.

Lacey, L.A. and Shapiro-Ilan, D.I. (2008) Microbial control of insect pests in temperate orchard systems: potential for incorporation into IPM. *Annual Review of Entomology* 53, 121–144.

Lacey, L., Grzywacz, D., Shapiro-Ilan, D.I., Frutos, R., Brownbridge, M. and Goettel, M. (2015) Insect pathogens as biological control agents: back to the future. *Journal of Invertebrate Pathology* 132, 1–41.

Lamberth, C., Jeanmart, S., Luksch, T. and Plant, A. (2013) Current challenges and trends in the discovery of agrochemicals. *Science* 341, 742–746.

Lietti, M.M., Botto, E. and Alzogaray, R.A. (2005) Insecticide resistance in Argentine populations of *Tuta absoluta* (Meyrick) (Lepidoptera: Gelechiidae). *Neotropical Entomology* 34, 113–119.

Lin, P.C., Lin, H.J., Liao, Y.Y., Guo, H.R. and Chen, K.T. (2013) Acute poisoning with neonicotinoid insecticides: a case report and literature review. *Basic and Clinical Pharmacology and Toxicology* 112, 282–286.

McCoy, E.E. and Glaser, R.W. (1936) Nematode culture for Japanese beetle control. *New Jersey Department of Agriculture Circular* 265, 3–9.

Meeussen, J. (2012) OECD Guidelines and harmonization for microbial control agents. In: Sundh, I., Wilcks, A. and Goettel, M. (eds) *Beneficial Organisms in Agriculture, Food and the Environment: Safety Assessment and Regulations*. CAB International, Wallingford, UK, pp. 308–321.

Miller, G.T. (2004) *Sustaining the Earth: An Integrated Approach*, 7th edn. Brooks Cole Publishing, California, 384 pp.

Mubyana-John, T. and Taylor, J. (2015) Nematophagous fungi: regulations and safety. In: Askary, T.H. and Martinelli, P.R.P. (eds) *Biocontrol Agents of Phytonematodes.* CAB International, Wallingford, UK, pp. 203–213.

OMAFRA (Ontario Ministry of Agriculture, Food and Rural Affairs) (2015) Ontario Introducing New Rules to

Protect Pollinators. Available at: http://news.ontario.ca/ene/en/2015/06/ontario-introducing-new-rules-to-protect-pollinators.html (accessed 1 September 2016).

Peshin, R. and Pimentel, D. (2014) *Integrated Pest Management; Experiences with Implementation, Global Overview*, Vol 4. Springer, Dordrecht, The Netherlands.

Pimentel, D. and Peshin, R. (2014) *Integrated Pest Management: Pesticide Problems*, Vol 3. Springer Science+Business Media, New York.

PMRA (Pest Management Regulatory Agency, Canada) (2004) Pest management progress report 2003. Publications, Pest Management Regulatory Agency, Catalogue Number H114-13/2003E, Ottawa Canada, 82 pp.

Pretty, J. (2008) Agricultural sustainability: concepts, principles and evidence. *Philosophical Transactions of the Royal Society B: Biological Sciences*, 363, 447–465.

Pucci, J. (2014) Biopesticides primed for growth. *Farm Chemicals International*. Available at: https://www.agribusinessglobal.com/agrichemicals/biopesticides-primed-for-growth/ (accessed 1 September 2016).

Rattner, B.A. (2009) History of wildlife toxicology. *Ecotoxicology* 18, 773–783.

Ravensburg, W.J. (2011) *A Roadmap to the Successful Development and Commercialization of Microbial Pest Control Products for Control of Arthropods*. Springer Science+Business Media BV, The Netherlands, 383 pp.

Rechcigl, J.E. and Rechcigl, N.A. (2000) *Insect Pest Management: Techniques for Environmental Protection*. Lewis Publishers, Boca Raton, Florida, 393 pp.

Rundlöf, M., Andersson, G.K.S., Bommarco, R., Fries, I., Hederström, V., *et al*. (2015) Seed coating with a neonicotinoid insecticide negatively affects wild bees. *Nature* 521, 77–80.

Shapiro-Ilan D.I. and Gaugler, R. (2015) Nematodes (Rhabditida: Steinernematidae & Heterorhabditidae). In: Shelton, A. (ed.) *Biological Control: A Guide to Natural Enemies in North America*. Available at: https://biocontrol.entomology.cornell.edu/pathogens/nematodes.php (accessed 25 October 2016).

Shapiro Ilan, D.I. and Grewal, P. (2008) Entomopathogenic nematodes and insect management. In: Capinera, J.L. (ed.) *Encyclopedia of Entomology, 2nd edition*. Springer, Dordrecht, The Netherlands, pp. 1336–1340.

Shapiro-Ilan, D.I., Gouge, D.H. and Koppenhofer, A.M. (2002) Factors affecting commercial success: case studies in cotton, turf, and citrus. In: Gaugler, R. (ed.) *Entomopathogenic Nematology*. CAB International, Wallingford, UK, pp. 333–355.

Shapiro-Ilan, D.I., Han, R. and Dolinksi, C. (2012) Entomopathogenic nematode production and application technology. *Journal of Nematology* 44, 206–217.

Shapiro-Ilan, D.I., Morales-Ramos, J.A. and Rojas, M.G. (2016) *In vivo* production of entomopathogenic nematodes. *Methods in Molecular Biology* 1477, 137–158.

Stock, S.P. and Hunt, D.J. (2005) Morphology and systematics of nematodes used in biocontrol. In: Grewal, P.S., Ehlers, R.-U. and Shapiro-Ilan, D.I. (eds) *Nematodes as Biological Control Agents*. CAB International, Wallingford, UK, pp. 3–46.

Sundh, I., Wilcks, A. and Goettel, M.S. (2012) *Beneficial Microorganisms in Agriculture, Food and the Environment: Safety Assessment and Regulation*. CAB International, Wallingford, UK.

Thakore, Y. (2006) The biopesticides market for global agricultural use. *Industrial Biotechnology* 2, 194–208.

The White House (2015) National Strategy to Promote the Health of Honey Bees and Other Pollinators: Pollinator Health Task Force. Available at: https://www.whitehouse.gov/sites/default/files/microsites/ostp/Pollinator%20Health%20Strategy%202015.pdf (accessed 1 September 2016).

Turusov, V., Rakitsky, V. and Tomatis, L. (2002) Dichlorodiphenyltrichloroethane (DDT): ubiquity, persistence, and risks. *Environmental Health Perspectives* 110, 125–128.

Van Lenteren, J.C. (2000) A greenhouse without pesticides. *Crop Protection* 19, 375–384.

Wilson, M. and Rae, R. (2015) *Phasmarhabditis hermaphrodita* as a control agent for slugs. In: Campos-Herrera, R. (ed.) *Nematodes Pathogenesis of Insects and Other Pests*. Springer International Publishing, Switzerland, pp. 509–522.

Wright, D.J., Peters, A., Schroer, S. and Fife, J.P. (2005) Application technology. In: Grewal, P.S., Ehlers, R.-U. and Shapiro-Ilan, D.J. (eds) *Nematodes as Biocontrol Agents*. CAB International, Wallingford, UK, pp. 91–106.

Part II: Entomopathogenic Nematodes – Morphology, Taxonomy, Biology and Diversity

3 Entomopathogenic Nematodes of the Families Steinernematidae and Heterorhabditidae: Morphology and Taxonomy

SERGEI E. SPIRIDONOV*

Severtsov Institute of Ecology and Evolution, Russian Academy of Sciences, Moscow, Russia

3.1 Introduction

The entomopathogenic nematodes (EPNs) are representatives of Nematoda, which are characterized by: (i) adult stages found in nature only inside the cadavers of insects killed and processed by symbiotic bacteria; (ii) infective (= dauer) juveniles (IJs) being the only free-living stage in soil. These third-stage juveniles carry in their intestines specific symbiotic bacteria. Immediately after the discovery of EPNs in the beginning of the 20th century, an attempt was made to use these pathogenic organisms in the control of insect pests (Poinar, 1979), but the working technology was developed only 20–30 years ago. It is possible now to cultivate EPNs on artificial media or on laboratory insects (Gaugler and Han, 2002). Formulations on the basis of living IJs are now widely used in Europe, the Americas, and also in some countries of Africa and Asia, to control different insect pests (Shapiro-Ilan *et al.*, 2012; Půža, 2015). However, numerous factors influence the commercial success of EPN application (Shapiro-Ilan *et al.*, 2002).

EPNs are also interesting subjects for evolutionary studies, as they illustrate the symbiosis between nematodes and bacteria. Still retaining the morphological features of their free-living ancestors, EPNs demonstrate a wide array of adaptations for such a relationship with insect hosts on the genetic/molecular level (Dillman *et al.*, 2015). A similar type of symbiosis has also been reported for *Phasmarhabditis* nematodes, pathogenic nematodes of terrestrial molluscs (Morand *et al.*, 2004), though in the latter, several associated bacterial species participate in the pathogenic process.

Presently, the representatives of two nematode families are regarded as EPNs: Steinernematidae Filipjev, 1934, and Heterorhabditidae Poinar, 1976 (Nguyen and Hunt, 2007). The family Heterorhabditidae is monotypic (i.e. contains a single genus, *Heterorhabditis*), while Steinernematidae contains two genera (*Steinernema* and exotic *Neosteinernema*), of which we will discuss only *Steinernema*. These two families belong to two different evolutionary lines of the order Rhabditida. The Heterorhabditidae family belongs to the superfamily Strongyloidea of the infraorder Rhabditomorpha, whereas Steinernematidae belongs to the superfamily Strongyloidoidea of the infraorder Panagrolaimomorpha (De Ley and Blaxter, 2002). EPNs have intimate relationships with their symbiotic bacteria (*Xenorhabdus* for *Steinernema* and *Photorhabdus* for *Heterorhabditis*). The bacteria kill the insects, provide the necessary nutrients and co-evolve with their nematode hosts, being responsible for different elements of a nematode–insect interaction (Murfin *et al.*, 2015a,b, Sicard *et al.*, 2006).

*E-mail: s_e_spiridonov@rambler.ru

 Biocontrol Agents: Entomopathogenic and Slug Parasitic Nematodes (eds. M.M.M. Abd-Elgawad, T.H. Askary and J. Coupland)

The present chapter focuses on EPN taxonomy. The chapter presents the diversity of these nematodes in condensed form and offer ways in which to navigate it.

3.2 Entomopathogenic Nematode Survey: Collection of Soil Samples and Nematode Strains

The presence of EPNs in soil can be detected through the use of baits – living insects (usually larvae of a greater wax moth or a mealworm) buried in containers filled with soil from the plot studied. Insect cadavers of evenly yellowish, brownish, violet, brick red colours and all intermediate shades should be removed from the soil (usually after 2–5 days of exposure) and transferred onto a moistened filter paper in Petri dishes. Mass migration of IJs can happen after 3–7 or even >12 days. Emerging IJs are to be collected alive as a stock for re-inoculations. These can be obtained from a decomposing insect cadaver by way of filtration through cheesecloth, cotton pads, Kleenex tissues, etc. The suspension containing IJs will be obtained after 30–60 min of filtration. This suspension can be stored in vials with a 2–3 mm layer of water in a refrigerator (3–6°C).'Falcon'-type cell culture flasks are the best for storage. Tropical EPN strains are to be stored at 12–22°C. Second-generation adults can be also collected from filtrate after prolonged filtration (>30–60 min) and fixed with hot 4–6% formalin (60–70°C) for further morphological examination. A portion can be transferred into 1.7 ml Eppendorf tubes and frozen immediately for molecular study. Freshly migrated IJs will maturate in 4–10 days at room temperature and can then be used for re-inoculation of insects. The dissection of insect cadavers with signs of successful EPN colonization (even coloration, no smell of rotting) on the 3rd to 5th day post-inoculation usually provides first-generation adults. Permanent slides can then be made (Nguyen and Hunt, 2007).

3.3 Contemporary Situation in Entomopathogenic Nematode Taxonomy

Nematode taxonomy, which was based solely on morphological data just two decades ago, is now surviving a period of transition towards a molecular epoch. EPN taxonomy is an example of new possibilities of molecular methods. In many cases, it is easier and more time-efficient to obtain DNA sequences and identify the newly found EPN isolate with a BLAST (Basic Local Alignment Search Tool) search in the National Center for Biotechnology Information GenBank (Altshul *et al.*, 1990) than to go through morphological procedures. Nowadays, the expressed nucleotide difference (e.g. on the level of 4–5% of compared sequences) is considered to be sufficient for describing new species, even when the morphological characteristics of such new taxon are not clear-cut. It is always advisable to collect material for the morphological examination. If the newly isolated EPN strain demonstrates an aberrant nucleotide sequence, properly fixed and slide-mounted material will serve for the description of a new species. An increasing amount of nucleotide data are changing the shape of nematology, but concurrently creating pitfalls like the incorrect attribution of a sequence to the taxon. It means that only congruent morphological and molecular data can provide reliable identification.

3.4 Classification

Both EPN families are classified in the order Rhabditida of the Chromadoria subclass of Nematoda (De Ley and Blaxter, 2002). In the modern understanding of the composition of the order Rhabditida, this taxon is equal to the former Secernentea subclass, within which the separate evolutionary lines are identified and considered as infraorders. According to this new approach to nematode classification, the family Steinernematidae belongs to the infraorder Panagrolaimomorpha, and the family Heterorhabditidae to the infraorder Rhabditomorpha. According to Sudhaus and Fitch (2001), heterorhabditids originated from the group of still existing saprobiotic genera.

3.5 Heterorhabditid Morphology

The morphology of the nematodes of family Heterorhabditidae retains features characteristic of free-living Rhabditidae. These features are especially obvious in males, which appear only in the second generation inside the insect cadaver. These can be found easily after the dissection of insect cadavers in physiological saline (0.9% NaCl), as these are rod-like and more transparent than the large, sausage-like females. The morphology of heterorhabditid males is very similar to that of free-living Rhabditomorpha (tubular buccal cavity, club-like pharynx with valves in the basal bulb).

A typical rhabditid *bursa copulatrix* at the male posterior end (Fig. 3.1c) consists of two sublateral alae supported by nine rays (= genital papillae). Unlike free-living nematodes, heterorhabditid females have a thick, sausage-shaped body. Under a light microscope, the darker intestine is visible, which consists of a few (about two dozen) large cells with a comparatively narrow lumen squeezed between them. Heterorhabditid females keep such ancestral morphological Rhabditomorpha features as the presence of an auxiliary excretory system (Spiridonov, 1981). Inside older females, numerous actively moving juveniles can be seen (Fig. 3.1a). The juveniles undergo up to two moults inside the female body, eventually disrupting the uterus walls. In soil, the third-stage IJ of a heterorhabditid nematode is covered with the cuticle of the second-stage juvenile (thus forming so-called 'dauer' juveniles). The third-stage cuticle surface is smooth (Fig. 3.1f). A prominent tooth is visible on the anterior end. Longitudinal ridges intersect with transversal annulation at the anterior end of the second-stage juvenile (Fig. 3.1b, d and e).

3.6 Heterorhabditid Taxonomy

The phylogenetic tree inferred from analysis of the ITS rDNA (internal transcribed spacers of ribosomal DNA) of *Heterorhabditis* species is presented in Fig. 3.3a. Heterorhabditid species are divided into clades, which are characterized with 100% bootstrap support. The names of these groupings are only conditional, proposed in an attempt to reflect a few common patterns of the geographical distribution of the nematodes of the group (Table 3.1).

3.7 Steinernematid Morphology

The most obvious difference between steinernematids and heterorhabditids lies in the morphology of the male tail, which lacks the caudal alae (Fig. 3.2c and d). Sensory papillae are arranged in the male tail in two subventral rows precloacally and in lateral, subventral and apical positions postcloacally. The characteristic feature of a steinernematid male tail is the presence of a prominent median papilla in a short distance anterior to a cloacal opening. Steinernematid females have a thick, sausage-shaped body filled with gonadal tubes and intestine. Gonadal tubes are morphologically differentiated with well-defined *receptaculum seminis* on the border between the ovary and uterus. Important morphological information can be obtained through the observation of living females (preferably first generation). Studying the junction between the ovary and uterus, it is possible to see spermatozoa inside females. Chains of amoeboid spermatozoa are characteristic for some steinernematids (e.g. *Steinernema feltiae*, *Steinernema affine*, *Steinernema intermedium*), while spermatozeugmata (Yushin *et al.*, 2007) – giant amoeboid cells (megaspermatozoon) with microspermatozoa on the surface – are characteristic for the steinernematids taxonomically related to *Steinernema glaseri*, *Steinernema carpocapsae* and *Steinernema bicornutum*. Steinernematid IJs shed their second-stage cuticular sheath soon after migration from the host cadaver. In addition to transversal annulations, the third-stage cuticle of steinernematid IJs has longitudinal ridges along both body sides – so-called lateral fields, which are quite diverse in the genus *Steinernema* (Fig. 3.2a–b and e–i). In some juveniles, the margin of annulated cuticle forms a fold on the border with the lateral field (Fig. 3.2h), which can be erroneously considered as an additional ridge.

3.8 Steinernematid Taxonomy

Phylogenetic relationships of steinernematids were also inferred from the analysis of ITS rDNA sequences (Fig. 3.3b). Nowadays, every year brings about three to five publications on the new steinernematid species, and to date, at least 91 nominal *Steinernema* species are known.

Several evolutionary lines can be detected in the genus *Steinernema*. These evolutionary lines are shown on the phylogenetic tree of the genus as clades with the bootstrap support close to 100% (which is usually lower in '*glaseri*' clade) and correspond to those revealed in the phylogenetic analysis of the smaller set of species (Spiridonov *et al.*, 2004). Recently, similar evolutionary lines in *Steinernema* genus were demonstrated after analysis of 3885 strictly homologous genes (Dillman *et al.*, 2015). This study also demonstrated genus-wide expansion in the number and diversity of proteases, protease inhibitors and fatty acid- and retinol-binding proteins in *Steinernema*.

The diversity of steinernematid species is presented in Table 3.2. Known species of steinernematids are listed according to the clades (evolutionary lines) inside the genus. Some *Steinernema* species not belonging to the main clades are presented

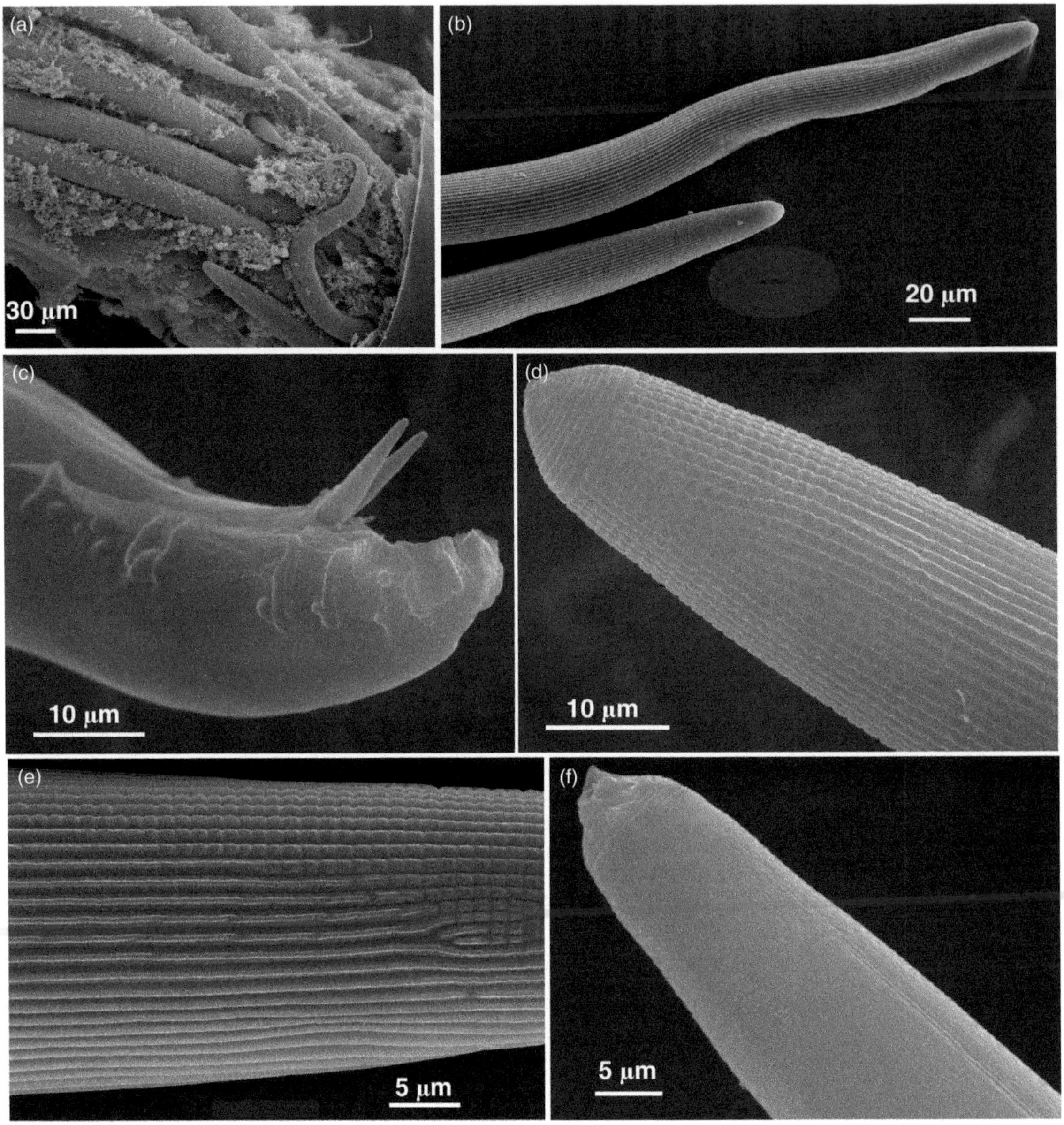

Fig. 3.1. Morphology of *Heterorhabditis* (Russian strain of *Heterorhabditis megidis*). (a) Completely formed IJs inside first-generation female; (b) longitudinal ridges as a main pattern of the IJ's cuticle surface; (c) male tail, lateral aspect; (d) surface structure on the IJ's anterior end (the cuticle of the second juvenile stage, penetration tooth is visible on the anterior end); (e) two types of cuticular ornamentation of the IJ's surface: midbody longitudinal ridges and criss-cross pattern of anterior end; (f) surface structure of the IJ's anterior end (the cuticle of the third juvenile stage). All images in lateral aspect.

below as 'independent' subclades. Some of these probably represent independent evolutionary lines with a single or a few species described. Currently, four such independent lineages have been reported from the Americas ('*rarum*' subclade, '*costaricense*' subclade, '*neocurtillae*' subclade, '*unicornum*' subclade) and one from Africa ('Cameroonian' subclade).

Table 3.1. Species diversity of *Heterorhabditis* divided into four main clades. The clades are characterized according to the structure of invasive juveniles (IJs). All the measurements are in micrometres (μm).

Species	Type locality (distribution)	Length of IJs	Male tail length	Spicule length	Gubernaculum length
'Temperate' clade unites *Heterorhabditis* species with medium-sized and long IJs (550–800 μm) described mainly from non-tropical areas.					
Heterorhabditis atacamensis Edgington, Buddie, Moore, France, Merino & Hunt, 2011	Atacama Desert, Chile	578–666	24–36	40–49	17–22
Heterorhabditis marelatus Liu & Berry, 1996	Oregon, USA	588–700	24–38	41–50	18–22
Heterorhabditis safricana Malan, Nguyen, de Waal & Tiedt, 2008	Western Cape Province, South Africa	550–676	27–49	35–54	19–27
Heterorhabditis downesi Stock, Griffin & Burnell, 2002	Wexford County, Ireland	588–692	29–34	41–47	17–20
Heterorhabditis megidis Poinar, Jackson & Klein, 1987	Ohio, USA. Reported throughout Holarctic (USA, Canada, Europe, Turkey, Israel, Japan, some parts of Russia)	736–800	35–43	46–54	17–24
Heterorhabditis zealandica Poinar, 1990	Auckland, New Zealand	570–740	30–41	48–55	19–25
***'Bacteriophora'* clade** unites *Heterorhabditis* species with medium-sized and long IJs (512–687) phylogenetically related to type species of the genus.					
Heterorhabditis bacteriophora Poinar, 1975	Brecon, Australia	512–671	22–36	36–44	18–25
Heterorhabditis beicherriana Li, Liu, Nermut, Půža & Mráček, 2012	Beijing, China	566–687	32–45	40–49	22–27
Heterorhabditis georgiana Nguyen, Shapiro-Ilan & Mbata, 2008	Georgia, USA	547–651	29–41	41–49	20–28
'Tropical' clade unites tropical species of *Heterorhabditis* with medium-sized and long IJs (332–620)					
Heterorhabditis amazonensis Andalo, Nguyen & Molino 2007	Brasilia, Amazonas state	567–612	29–41	35–45	19–23
Heterorhabditis baujardi Phan, Subbotin, Nguyen & Moens, 2003	Kon Tum Province, Vietnam	497–595	28–38	33–45	18–22
Heterorhabditis floridensis Nguyen, Gozel, Koppenhöfer & Adams, 2006	Florida, USA	554–609	29–40	36–46	17–30
Heterorhabditis mexicana Nguyen, Shapiro-Ilan, Stuart, McCoy, James & Adams, 2004	Tamaulipas State, Mexico	530–620	21–36	30–47	18–32
Heterorhabditis somsookii Maneesakorn, An, Grewal & Chandrapatya, 2015	Kanchanaburi Province, Thailand	502–565	20–30	32–45	17–23
Heterorhabditis sonorensis Stock, Rivera-Orduño & Flores-Lara, 2009	Sonora State, Mexico	495–570	25–45	31–45	20–31
Heterorhabditis taysearae Shamseldean, Abou El-Sooud, Abd-Elgawad & Saleh, 1996	Beheira Governorate, Egypt	332–499	20–29	30–42	14–21
***'Indica'* clade** unites *Heterorhabditis* species with medium-sized and long IJs (479–683) described from tropical and subtropical regions.					
Heterorhabditis indica Poinar, Karunakar & David, 1992	Tamil Nadu State, India	479–573	24–32	35–48	18–23
Heterorhabditis gerrardi Plichta, Joyce, Clarke, Waterfield & Stock, 2009	New South Wales, Australia	551–683	28–46	34–48	16–27
Heterorhabditis noenieputensis Malan, Knoetze & Tiedt, 2014	Northern Cape Province, South Africa	484–578	21–32	37–49	17–24

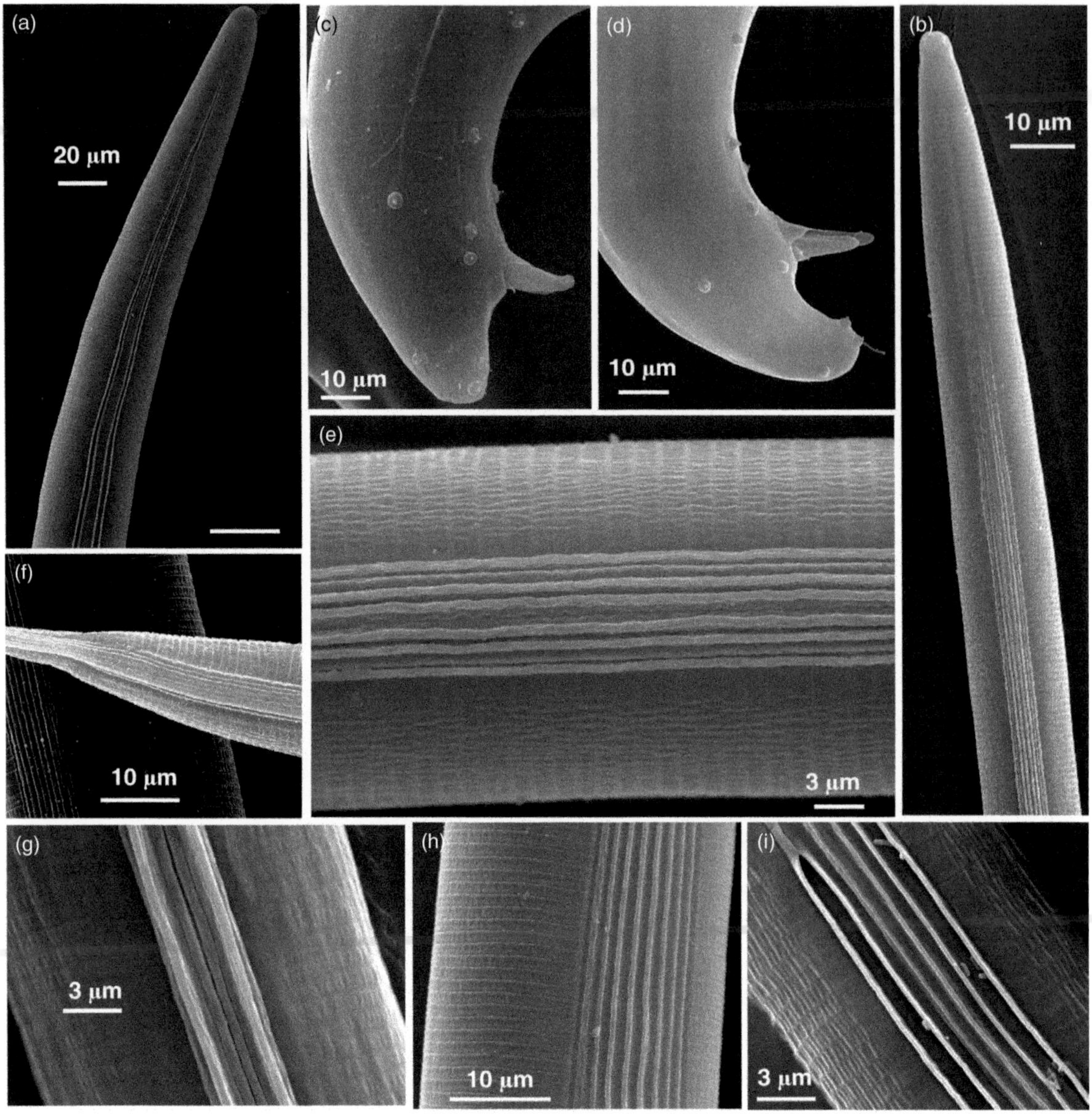

Fig. 3.2. Morphology of *Steinernema*. (a) *Steinernema cameroonense*: IJ's anterior end; (b) *Steinernema feltiae*: IJ's anterior end; (c) *Steinernema arenarium*: first-generation male, tail; (d) *Steinernema cameroonense*: first-generation male, tail; (e) *Steinernema feltiae*: IJ's lateral field; (f) *Steinernema schliemanni*: IJ's lateral field at tail region; (g) *Steinernema affine*: IJ's lateral field; (h) *Steinernema arenarium*: IJ's lateral field; (i) *Steinernema carpocapsae*: IJ's lateral field. All images are in lateral aspect.

3.9 Identification Procedure

DNA can be extracted from numerous individuals with special columns (e.g. Promega Wizard® SV Genomic DNA Purification System). If only a few individuals are available, it is reasonable to use proteinase K-based methods (e.g. Holterman *et al.*, 2006). Primers proposed by Curran *et al.* (1994) – forward 'TW81' GTTTCCGTAGGTGAACCTGC and reverse 'AB28' ATATGCTTAAGT TCAGCGGGT are good for the first attempt to amplify the ITS region of rDNA. In some cases, primers proposed

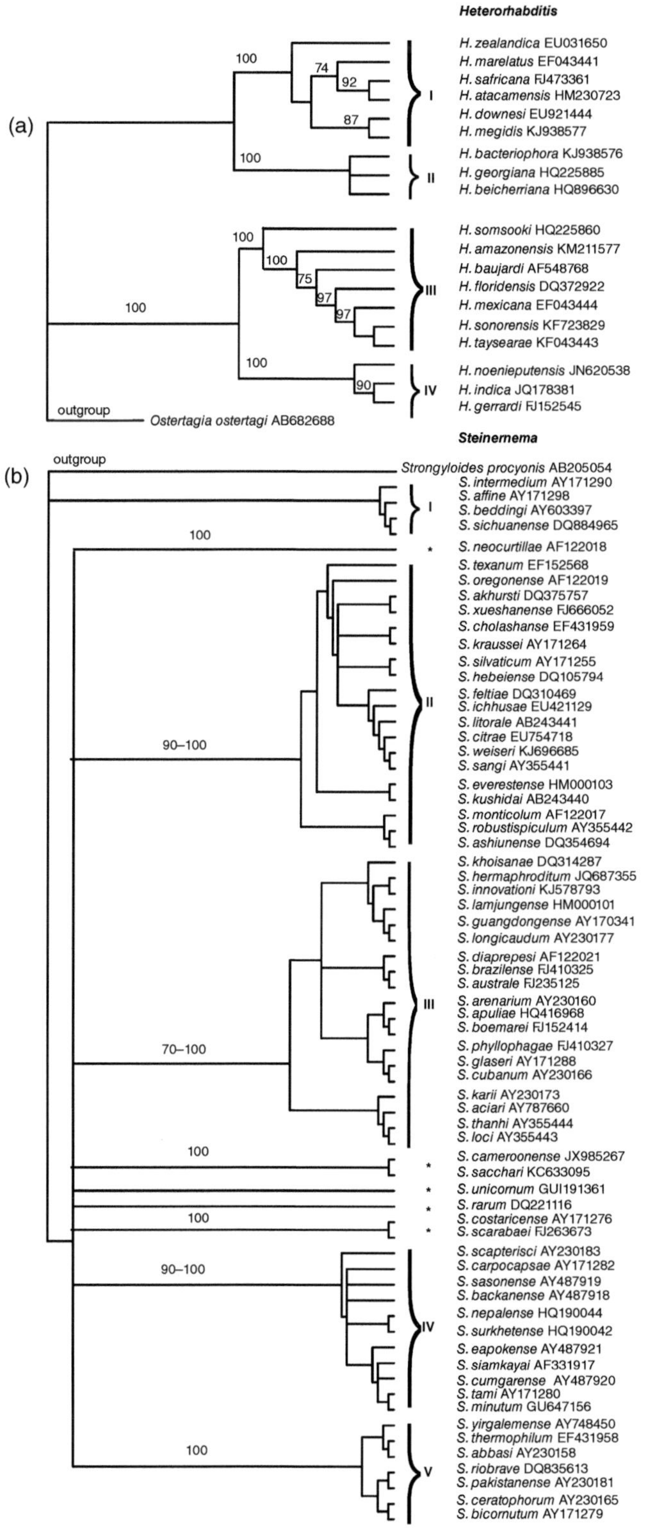

Fig. 3.3. The relationships of *Heterorhabditis* and *Steinernema* species as inferred from analysis of ITS rDNA. (a) Maximum parsimony tree of *Heterorhabditis* species; (b) maximum parsimony tree of *Steinernema* species. Not all valid species of *Steinernema* are included, but all the known clades and subclades are presented. Bootstrap (maximum parsimony, 1000 bootstrap repeats) values are indicated near the nodes. Asterisks mark subclades of uncertain position in the genus *Steinernema*.

Table 3.2. Species diversity of *Steinernema* divided into five main clades. The clades are characterized according to the structure of invasive juveniles (IJs). All the measurements are in micrometres (µm).

Species	Origin of type specimens; distribution	Length of IJs	First-generation male		
			D% (the ratio: anterior end to pharynx length)	Spicule length	Gubernaculum length
'*Affine–intermedium*' clade unites species with IJs' length 600–850. Lateral field of all species contains two central bands with two prominent submarginal ridges.					
Steinernema affine (Bovien, 1937) Wouts, Mráček, Gerdin & Bedding, 1982	Denmark, Skive. Distribution: Europe, Canada	626–788	24–49%	60–82	44–52
Steinernema arasbaranense Nikdel, Niknam & Ye, 2011	East Azerbaijan Province, Iran	693–795	43–59%	70–82	42–53
Steinernema beddingi Qiu, Hu, Zhou, Pang & Nguyen, 2005	Yunnan, China	700–790	54–63%	63–78	38–48
Steinernema intermedium (Poinar, 1986) Mamiya, 1988	South Carolina, USA	608–800	67–80%	80–106	48–96
Steinernema poinari Mráček, Puza & Nermut', 2014	Bohemia, Czech Republic. Distribution: throughout temperate Eurasia	687–848	42–64%	62–78	45–57
Steinernema sichuanense Mráček, Nguyen, Tailliez, Boemare & Chen, 2006	Sichuan Province, China	606–787	45–56%	65–72	40–51
'*Feltiae–kushidai–monticolum*' clade – IJs' length 420–1100. Lateral fields of IJs may contain eight equal ridges, or only six equal ridges, or eight unequal ridges, with submarginal ones poorly discernible. Even closely related species can have different lateral field structure (e.g. *Steinernema kraussei* with seven unequal ridges and *Steinernema silvaticum* with eight equal ridges).					
'*Feltiae*' subclade					
Steinernema feltiae (Filipjev, 1934) Wouts, Mráček, Gerdin & Bedding, 1982	Udmurt Republic, Russia. Distribution: cosmopolitan	766–928	64–72%	62–68	48–56
Steinernema akhursti Qiu, Hu, Zhou, Mei, Nguyen & Pang, 2005	Sichuan Province, China	727–909	50–85%	60–82	44–52
Steinernema citrae Stokwe, Malan, Nguyen, Knoetze & Tiedt, 2011	Western Cape Province, South Africa	623–849	49–59%	32–45	32–59
Steinernema cholashanense Nguyen, Půža & Mráček, 2008	Sichuan Province, China	727–909	50–85%	60–82	44–52
Steinernema hebeiense Chen, Li, Yan, Spiridonov & Moens, 2012	Hebei, China	610–710	48–59%	54–63	38–50

Steinernema ichnusae Tarasco, Mráček, Nguyen & Triggiani, 2008	Sardinia, Italy	767–969	59–65%	64–67	43–46
Steinernema jollietti Spiridonov, Krasomil-Osterfeld & Moens 2004	Near St Louis, USA	625–820	46–50%	55–70	45–60
Steinernema kraussei (Steiner, 1923) Travassos, 1927	Westphalia, Germany. Distribution: temperate Holarctic	797–1102	Approx. 53%	42–53	29–37
Steinernema litorale Yoshida, 2004	Aichi Prefecture, Japan	834–988	55–72%	67–89	18–25
Steinernema oregonense Liu & Berry, 1996	Oregon, USA	820–1110	64–75%	65–73	52–59
Steinernema puntauvense Uribe-Lorío, Mora & Stock, 2007	Punta-Uva Coast, Costa-Rica	631–728	45–85%	71–81	30–40
Steinernema silvaticum Sturhan, Spiridonov & Mráček, 2005	Berlin, Germany. Distribution: Western Europe	670–975	45–63%	42–64	30–43
Steinernema texanum Nguyen, Stuart, Andalo, Gozel & Rogers, 2007	Texas, USA	732–796	58–73%	55–66	39–53
Steinernema tielingense Ma, Chen, Han, Khatri-Chhetri, De Clercq & Moens, 2012	Liaoning Province, China	824–979	64–78%	79–98	49–70
Steinernema weiseri Mráček, Sturhan & Reid, 2003	Near České Budějovice, Czech Republic. Distribution: Western and Central Europe, Turkey	586–828	39–60%	62–72	46–57
Steinernema xinbinense Ma, Chen, De Clercq, Waeyenberge, Han & Moens, 2012	Liaoning Province, China	635–744	41–50%	49–62	30–41
Steinernema xueshanense Mráček, Liu & Nguyen, 2009	Yunnan, China	768–929	73–87%	66–91	41–56
'*Kushidai*' subclade					
Steinernema kushidai Mamiya, 1988	Shizuoka Prefecture, Japan	424–662	42–59%	48–72	39–60
Steinernema everestense Khatri-Chhetri, Waeyenberge, Spiridonov, Manandhar & Moens, 2011	Eastern Nepal	705–838	44–66%	65–87	39–57

Continued

Table 3.2. Continued.

Species	Origin of type specimens; distribution	Length of IJs	First-generation male: D% (the ratio: anterior end to pharynx length)	First-generation male: Spicule length	First-generation male: Gubernaculum length
Steinernema sangi Phan, Nguyen & Moens, 2001	Thanh Hoa Province, Vietnam	704–784	42–63%	58–80	34–46
'*Monticolum*' subclade					
Steinernema monticolum Stock, Choo & Kaya, 1997	Gyeongnam Province, Korea	612–821	49–61%	61–80	35–54
Steinernema ashiuense Phan, Takemoto & Futai, 2006	Kyoto Prefecture, Japan	720–800	44–56%	50–65	25–43
Steinernema changbaiense Ma, Chen, De Clercq, Han & Moens, 2012	Liaoning Province, China	730–790	52–66%	56–70	33–46
Steinernema schliemanni Spiridonov, Waeyenberge & Moens, 2010	Mecklenburg-Vorpommern, Germany	842–1008	50–58%	61–81	43–64
Steinernema robustispiculum Phan, Subbotin, Waeyenberge & Moens, 2005	Kontum Province, Vietnam	642–778	50–63%	51–65	36–44
'***Glaseri***' **clade** includes the steinernematids with the longest IJs in the genus (average length >1000 μm, but in certain species shorter IJs are present). Lateral field with eight equal ridges. Spermatozeugmata present.					
Steinernema glaseri (Steiner, 1929) Wouts, Mráček, Gerdin & Bedding, 1982	New Jersey, USA. Distribution: Western Europe, South and East Asia, Americas	864–1448	Approx. 90%	62–90	40–50
Steinernema apuliae Triggiani, Mráček & Reid, 2004	Bari area, Italy	945–1212	75–92%	64–80	46–54
Steinernema arenarium (Artyukhovsky, 1967) Wouts, Mráček, Gerdin & Bedding, 1982	Voronezh region, Russia. Distribution: South Europe, European Russia	930–1580	52–96%	63–93	45–63
Steinernema australe Edgington, Buddie, Tymo, Hunt, Nguyen, France, Merino & Moore, 2009.	Isla Magdalena, Chile	1162–1484	59–87%	55–78	36–51
Steinernema boemarei Lee, Sicard, Skeie & Stock, 2009	South France	1005–1323	68–99%	64–96	43–65

Steinernema brazilense Nguyen, Ginarte, Leite, Santos & Harakava, 2009/2010	Mato Grosso do Sul State, Brazil	1023–1284	57–80%	75–89	41–56
Steinernema cubanum Mráček, Hernández & Boemare, 1994	Pinar-del-Rio Province, Cuba	1149–1509	Approx. 70%	50–67	37–42
Steinernema diaprepesi Nguyen & Duncan, 2002	Florida, USA	880–1133	68–86%	71–90	45–61
Steinernema ethiopiense Tamiru, Waeyenberge, Hailu, Ehlers, Půža & Mráček, 2012	Western Welega Province, Ethiopia	768–1010	54–61%	69–77	46–57
Steinernema guangdongense Qiu, Fang, Zhou, Pang & Nguyen, 2004	Guangdong Province, China	987–1145	67–78%	80–94	47–73
Steinernema hermaphroditum Stock, Griffin & Chaerani, 2004	Seram Island, Indonesia	700–1000	44–50%	65–70	47–50
Steinernema innovationi Çimen, Lee, Hatting, Hazir & Stock, 2014	Free State, South Africa	1000–1103	54–63%	74–91	54–63
Steinernema jeffreyense Malan, Knoetze & Tiedt, 2015	Eastern Cape Province, South Africa	784–1043	34–68%	79–95	51–61
Steinernema khoisanae Nguyen, Malan & Gozel, 2006	Cape Province, South Africa	904–1159	71–99%	70–88	50–63
Steinernema longicaudum Shen & Wang, 1992	Shandong Province, China	929–1170	56–92%	72–108	54–65
Steinernema lamjungense Khatri-Chhetri, Waeyenberge, Spiridonov, Manandhar & Moens 2011	Western Nepal	654–786	24–49%	60–82	44–52
Steinernema puertoricense Roman & Figueroa, 1994	Puerto Rico	1057–1238	Approx. 77%	71–88	36–45
Steinernema pui Qui, Zhao, Wu, Lv & Pang, 2011	Yunnan Province, China	900–1120	70–91%	78–88	58–65
Steinernema tophus Çimen, Ming-Min Lee, Hatting, Hazir, Stock, 2014	Western Cape State, South Africa	966–1126	71–103	70–91	47–65
Steinernema vulcanicum Clausi, Longo, Rappazzo, Tarasco & Vinciguerra, 2011	Sicily, Italy	1081–1323	42–90%	46–88	32–65
'*Kari*' subclade					

Continued

Table 3.2. Continued.

Species	Origin of type specimens; distribution	Length of IJs	First-generation male: D% (the ratio: anterior end to pharynx length)	First-generation male: Spicule length	First-generation male: Gubernaculum length
Steinernema karii Waturu, Hunt & Reid, 1997	Central Province, Kenya	876–982	53–78%	73–91	42–64
Steinernema aciari Qiu, Yan, Zhou, Nguyen & Pang, 2004	Guangdong Province, China	975–1250	69–88%	75–95	48–65
Steinernema loci Phan, Nguyen & Moens, 2001	Ha Tinh Province, Vietnam	896–1072	61–80%	60–80	40–52
Steinernema leizhouense Nguyen, Qiu, Zhou & Pang, 2006	Guangdong Province, China	900–1050	67–83%	75–93	50–60
Steinernema thanhi Phan, Nguyen & Moens, 2001	Thanh Hoa Province, Vietnam	720–960	64–82%	67–78	40–56
'***Carpocapsae*' clade** includes the steinernematids with the shortest IJs in the genus (usually <600 µm). Lateral field includes six well discernible ridges – four in the central group and two marginal ones (Fig. 3.3i). Very thin submarginal ridges can be detected in some specimens. Spermatozeugmata present.					
Steinernema carpocapsae (Weiser, 1955) Wouts, Mráček, Gerdin & Bedding, 1982	Bohemia, Czech Republic	438–650	30–41%	59–72	39–56
Steinernema anatoliense Hazir, Stock & Keskin, 2003	East Anatolia, Turkey	507–580	46–55%	68–84	42–59
Steinernema asiaticum Anis, Shahina, Reid & Rowe, 2002	Karachi, Pakistan	360–450	35–57%	61–74	46–62
Steinernema backanense Phan, Spiridonov, Subbotin & Moens, 2006	Balkan Province, Vietnam	391–557	30–52%	53–62	36–51
Steinernema balochiense Fayyaz, Khanum, Ali, Solangi, Gulsher & Javed, 2015	Balochistan, Pakistan	415–528	40–51%	60–76	40–47
Steinernema colombiense López-Núñez, Plichta, Góngora-Botero & Stock, 2008	Quindío Department, Colombia	549–732	30–50%	64–77	47–63
Steinernema cumgarense Phan, Spiridonov, Subbotin & Moens, 2006	Dak Lak Province, Vietnam	384–432	35–52%	68–81	48–66

Steinernema eapokense Phan, Spiridonov, Subbotin & Moens, 2006	Dak Lak Province, Vietnam	370–434	37–48%	63–74	47–56
Steinernema huense Phan, Mráček, Půža, Nermuť & Jarošová, 2014	Bach Ma National Park, Vietnam	444–571	46–69%	60–72	46–56
Steinernema meghalayensis Ganguly, Rathour & Singh, 2011	Meghalaya State, India	430–630	29–38%	65–78	45–60
Steinernema minutum Maneesakorn, Grewal & Chandrapatya, 2010	Chumphon Province, Thailand	364–455	41–52%	71–95	52–73
Steinernema nepalense Khatri-Chhetri, Waeyenberge, Spiridonov, Manandhar & Moens, 2011	Sunsari District, Nepal	424–513	34–54%	56–78	47–60
Steinernema ritteri de Doucet & Doucet, 1990	Cordoba Province, Argentina	470–590	Approx. 50%	58–75	33–50
Steinernema sasonense Phan, Spiridonov, Subbotin & Moens, 2006	Kon Tum Province, Vietnam	437–516	34–46%	54–78	42–56
Steinernema scapterisci Nguyen & Smart, 1990	Uruguay	517–609	32–39%	72–92	59–75
Steinernema siamkayai Stock, Somsook & Reid, 1998	Phetchabun Province, Thailand	398–495	35–49%	75–80	47–65
Steinernema surkhetense Khatri-Chhetri, Waeyenberge, Spiridonov, Manandhar & Moens, 2011	Surkhet District, Nepal	393–450	37–64%	58–78	42–63
Steinernema tami Pham, Nguyen, Reid & Spiridonov, 2000	Dong Nai Province, Vietnam	400–600	30–60%	71–84	38–55
Steinernema websteri Cutler & Stock, 2003	Beijing, China	553–631	30–50%	64–72	42–56
'*Bicornutum*' clade includes the steinernematids with IJs' length 361–873. Two horn-like projections on the anterior end of IJs. Lateral field with eight ridges, but submarginal ones in some specimens are less distinct. Spermatozeugmata present.					
Steinernema bicornutum Tallosi, Peters & Ehlers, 1995	Vojvodina Province, Serbia. Distribution: Europe and Canary Island, Jamaica	648–873	50–60%	53–70	38–50

Continued

Table 3.2. Continued.

Species	Origin of type specimens; distribution	Length of IJs	First-generation male: D% (the ratio: anterior end to pharynx length)	First-generation male: Spicule length	First-generation male: Gubernaculum length
Steinernema abbasi Elawad, Ahmad & Reid, 1997	Oman	496–579	51–68%	57–74	35–50
Steinernema bifurcatum Fayyaz, Yan, Qiu, Han, Gulsher, Khanum & Javed, 2014	Punjab, Pakistan	460–590	42–58%	60–85	30–49
Steinernema ceratophorum Jian, Reid & Hunt, 1997	Liaoning Province, China	591–800	33–69%	54–90	25–45
Steinernema maqbooli Shahina, Tabassum, Mehreen & Salma, 2013	Sindh Province, Pakistan	425–603	49–66%	53–71	30–40
Steinernema pakistanense Shahina, Anis, Reid, Rowe & Maqbool, 2001	Sindh and Balochistan Provinces, Pakistan	649–716	50–60%	62–73	36–45
Steinernema papillatum San-Blas, Portillo, Nermut, Puza, Morales-Montero, 2015	Venezuela, Zulia State	572–720	43–65	42–62	23–36
Steinernema riobrave Cabanillas, Poinar & Raulston, 1994	Texas, USA	361–701	60–80%	63–75	48–56
Steinernema yirgalemense Nguyen, Tesfamariam, Gozel, Gaughler & Adams, 2005	Yirgalem, Ethiopia	548–693	50–66%	51–77	42–54
'*Costaricense*' subclade					
Steinernema costaricense Uribe-Lorío, Mora & Stock, 2007	Costa Rica. Distribution: Missouri, USA	680–1032	38–66%	55–89	40–62
Steinernema scarabaei Stock, Koppenhöfer, 2003	New Jersey, USA	890–959	53–77%	67–83	36–50
Steinernema australe Edgington, Buddie, Tymo, Hunt, Nguyen, France, Merino & Moore, 2009	Isla Magdalena, Chile	1162–1484	59–87%	55–78	36–51
'Cameroonian' subclade					
Steinernema cameroonense Kanga, Trinh, Waeyenberge, Spiridonov, Hauser & Moens, 2012	Southern Cameroon	490–694	48–76%	51–85	37–57

Steinernema nyetense Kanga, Trinh, Waeyenberge, Spiridonov, Hauser & Moens, 2012	Southern Cameroon	565–708	40–70%	67–98	40–62
Steinernema sacchari Nthenga, Knoetze, Berry, Tiedt & Malan, 2014	KwaZulu-Natal Province, South Africa	630–722	54–88%	73–89	50–68
'*Rarum*' subclade					
Steinernema rarum (de Doucet, 1986) Mamiya, 1988	Cordoba Province, Argentina. Distribution: USA	446–578	Approx. 50%	42–52	23–38
'*Neocurtillae*' subclade					
Steinernema neocurtillae Nguyen & Smart, 1992	Florida, USA	741–988	13–26%	52–64	44–59
'*Unicornum*' subclade					
Steinernema unicornum Edgington, Buddie, Tymo, France, Merino & Hunt, 2009	Tierra del Fuego, Chile	654–786	24–49%	60–82	44–52

by Vrain *et al.* (1992) – forward '18S' TGATTA CGTCCCTGCCCTTT and reverse '26S' TTTCACTCGCCGTTACTAAGG – can be tested. It is also reasonable to amplify part of the LSU (large subunit) rDNA with primer proposed by Nunn (1992) with 'D2A' ACAAGTACCGTGAGGGAA-AGTTG as forward one and 'D3B' TCGGAAGG-AACCAGCTACTA as reverse one. The 'D2A' as forward one can be replaced with primer 'LSU391' AGCGGAGGA AAAGAAACTAA (Nadler *et al.*, 2006). The amplification of ITS rDNA can provide decisive diagnostic information (more nucleotide differences with known sequences), but can also lead to the situation when direct sequencing of the polymerase chain reaction (PCR) product is impossible because of *heteroplasmy* (the presence of different copies of the ribosomal cluster in the same individual of a nematode). Being an obstacle to the identification of EPN species through the direct sequencing of PCR product, such intra-individual variability provides a deeper insight into the intraspecific structure of EPNs (Půža *et al.*, 2015).

Thus, it is always better to amplify and sequence both ITS rDNA and D2D3 LSU rDNA. The best way to analyse obtained sequences is to detect related sequences in the NCBI GenBank with BLAST (Altshul *et al.*, 1990). Phylogenetic analysis of obtained and downloaded sequences can be made with MEGA (Molecular Evolutionary Genetics Analysis) (Tamura *et al.*, 2011).

The data presented in the tables for known EPN species cannot ensure the species identification, as the morphometric parameters in related species significantly overlap. It can be expected that certain morphometric features depend on the nutritional conditions inside the host cadaver, and can be more variable than reported. In some cases, the presented morphometric data can be sufficient to exclude some of the described species from comparison with newly isolated nematodes. It is reasonable to refer to the detailed species descriptions that can be found in the monograph edited by Nguyen and Hunt (2007), at the website http://entnemdept.ufl.edu/nguyen/morph/kbnstein.htm, or to download as separate papers from resources such as ResearchGate.

3.10 Molecular Approaches and their Application in Nematode Taxonomy

3.10.1 Molecular tools

The advent of molecular methods has brought profound changes in zoological systematics. Concerning the EPNs, these methods are the only sound basis to describe existing diversity, as morphological features are obviously variable, and the extent of this variation is still understudied. The three main loci of the EPN genome being used for taxonomy are the ITS rDNA, the D2D3 expansion segment of LSU rDNA and the partial sequence of *Cox*I mitochondrial gene. ITS rDNA of steinernematids is very variable, whereas the variability of D2D3 LSU rDNA in some cases is not sufficient for the secure delimitation of species. *Cox*I mtDNA sequences are very informative, but up to now have been obtained for only a small portion of known EPN species. Modern EPN molecular taxonomy is still based on the single locus (ITS rDNA), with the only incidental addition of data for other loci. ITS rDNA data are now available in the NCBI GenBank for nearly all EPN species, and a BLAST search for the newly obtained ITS rDNA sequence will provide all the information needed. The sequencing of EPN mitochondrial genomes (Montiel *et al.*, 2006), or even complete genomes (Dillman *et al.*, 2015), is opening new possibilities for EPN taxonomy. Further comparative analysis is needed to apply such massive bioinformatics data for EPN taxonomy.

3.10.2 Target regions

The study of EPNs can be started with a very simple set of equipment and material. Indeed, only microscopes and a supply of laboratory insects (wax moth or mealworm) are necessary for any EPN survey. Though the species concept is still underdeveloped for EPNs, biodiversity studies on EPNs should be continued, and new regions ought to be explored. Thus, several subclades of *Steinernema* have been discovered in Africa and South America. These groups as yet unite few species, but we can expect that speciation in these subclades is as active as in the better-studied clades. It is an indication that these areas are promising for the search of new EPN species. The question of EPN evolution on a geological timescale is still open, and we can only wonder which region of the planet in the past played a decisive role in the establishment and evolution of this group. Surprisingly, a few, if any, endemic autochthonous forms have been described for the temperate Australasian region (New Zealand, Tasmania), which we believe deserves more intensive sampling. The EPN fauna of the vast territory of the North Palearctic looks quite uniform, with *Heterorhabditis megidis* and *S. feltiae* reported throughout Eurasia from the Atlantic up to the Pacific, but the tropical belt of the Old and New Worlds is surely hiding its undescribed richness of EPNs.

3.11 Conclusions and Future Prospects

In the past decades, the main successes in EPN application were achieved in the developed countries. However, the visible potential of these organisms for pest control permits expectations of the further spreading of these techniques and related research. The almost cosmopolitan presence of EPNs in natural habitats is a valuable prerequisite for the development of EPN-based methods of pest control in developing countries. One can find in its home area a species of EPN well suited to existence in local climatic and biotic conditions. In this respect, study of the regions for which only fragmentary EPN data are available would be fruitful in two ways: in the description of EPN diversity and the possible recruitment of new biocontrol agents. The bias toward a practical application significantly affected the content of EPN studies. Still, very little is known about the 'field' ecology of these nematodes, in pristine habitats lying outside of human intervention, as well as in agroecosystems after EPN application. It seems that, in normal conditions, some steinernematid species can exist under quite low population densities (Spiridonov *et al.*, 2007), appearing as a succession of infestations of small insect hosts like dipteran larvae. Some aspects of EPN ecology such as the survival of released IJs in soil can be studied through the use of vitally stained juveniles (Spiridonov, 1991), but the main elements of EPN life in soil are still completely enigmatic. Even progress in studies on EPN diversity is limited by an insect baiting of soil samples as the only and dominant isolation approach. It is important to remember that not all EPN species can develop successfully in *Galleria mellonella* or *Tenebrio molitor*; therefore, the insect baiting method can fail on such EPN species. It also hampers the laboratory maintenance of certain EPN species for which a suitable laboratory host has not yet been found. We expect that the new approach of metagenomic analysis will open new horizons in the study of EPN biodiversity.

Acknowledgements

The author is very grateful to Dr Tarique H. Askary, SKUAST-K, Srinagar, India, and Dr Mahfouz M.M. Abd-Elgawad, National Research Centre, Dokki, Egypt, for their invitation to contribute to this volume and for their editorial work on this chapter.

References

Altschul, S.F., Gish, W., Miller, W., Myers, E.W. and Lipman, D.J. (1990) Basic local alignment search tool. *Journal of Molecular Biology* 215, 403–410.

Curran, J., Driver, F., Ballard, J.W.O. and Milner, R.J. (1994) Phylogeny of *Metarhizium*: analysis of ribosomal DNA sequence data. *Mycological Research* 98, 547–552.

De Ley, P. and Blaxter, M.L. (2002) Systematic position and phylogeny. In: Lee, D.L. (ed.) *The Biology of Nematodes*. Taylor and Francis, London, pp. 1–30.

Dillman, A.R., Macchietto, M., Porter, C.F., Rogers, A., Williams, B., *et al.* (2015) Comparative genomics of *Steinernema* reveals deeply conserved gene regulatory networks. *Genome Biology* 16, 200–221.

Gaugler, R. and Han, R. (2002) Production technology. In: Gaugler, R. (ed.) *Entomopathogenic Nematology*. CAB International, Wallingford, UK, pp. 289–310.

Holterman, M., van der Wurff, A., van den Elsen, S., van Megen, H., Bongers, N., *et al.* (2006) Phylum-wide analysis of SSU rDNA reveals deep phylogenetic relationships among nematodes and accelerated evolution toward crown clades. *Molecular Biology and Evolution* 23, 1792–1800.

Montiel, R., Lucena, M.A., Medeiros, J. and Simoñes, N. (2006) The complete mitochondrial genome of the entomopathogenic nematode *Steinernema carpocapsae*: insights into nematode mitochondrial DNA evolution and phylogeny. *Journal of Molecular Evolution* 62, 211–225.

Morand, S., Wilson, M.J. and Glen, D.M. (2004) Nematodes (Nematoda) parasitic in terrestrial gastropods. In: Barker, G.M. (ed.) *Natural Enemies of Terrestrial Molluscs*. CAB International, Wallingford, UK, pp. 525–558.

Murfin, K.E., Whooley, A.C., Klassen, J.L. and Goodrich-Blair, H. (2015a) Comparison of *Xenorhabdus bovienii* bacterial strain genomes reveals diversity in symbiotic functions. *BMC Genomics* 16, 889.

Murfin, K.E., Lee, M.M., Klassen, J.L., McDonald, B.R., Larget, B., *et al.* (2015b) *Xenorhabdus bovienii* strain diversity impacts coevolution and symbiotic maintenance with *Steinernema* spp. nematode hosts. *MBio* 6, e00076.

Nadler, S.A., Bolotin, E. and Stock, S.P. (2006) Phylogenetic relationships of *Steinernema*. *Systematic Parasitology* 63, 161–181.

Nguyen, K.B. and Hunt, D.J. (2007) Entomopathogenic nematodes; systematics, phylogeny and bacterial symbionts. *Nematology Monographs and Perspectives*. Vol 5. Brill, Leiden-Boston, 816 pp.

Nunn, G.B. (1992) Nematode molecular evolution. PhD thesis, University of Nottingham, UK.

Poinar, G.O. Jr (1979) *Nematodes for Biological Control of Insects*. CRC Press, Boca Raton, Florida.

Půža, V. (2015) Control of insect pests by entomopathogenic nematodes. In: Lugtenberg, B. (ed.) *Principles of Plant–Microbe Interactions.* Springer International Publishing, Bern, pp. 175–183.

Půža, V., Chundelová, D., Nermuť, J., Žurovcová, M. and Mráček, Z. (2015) Intra-individual variability of ITS regions in entomopathogenic nematodes (Steinernematidae: Nematoda): implications for their taxonomy. *BioControl* 60, 547–554.

Shapiro-Ilan, D.I., Gouge, D.H. and Koppenhöfer, A.M. (2002) Factors affecting commercial success: case studies in cotton, turf and citrus. In: Gaugler, R. (ed.) *Entomopathogenic Nematology*. CAB International, Wallingford, UK, pp. 333–355.

Shapiro-Ilan, D., Han, R. and Dolinski, C. (2012) Entomopathogenic nematode production and application technology. *Journal of Nematology* 44, 206–217.

Sicard, M., Hinsinger, J., Le Brun, N., Pages, S., Boemare, N. and Moulia, C. (2006) Interspecific competition between entomopathogenic nematodes (*Steinernema*) is modified by their bacterial symbionts (*Xenorhabdus*). *BMC Evolutionary Biology* 6, 68.

Spiridonov, S.E. (1981) Auxiliary excretory system of the entomopathogenic nematode *Heterorhabditis bacteriophora* (Rhabditida, Heterorhabditidae) *Zoologicheskii Zhurnal* 60, 1887–1888 [in Russian].

Spiridonov, S.E. (1991) Vital staining of invasive larvae of Steinernematidae (Nematoda, Rhabditida). *Angewandte Parasitologie* 32, 155–156.

Spiridonov, S.E., Reid, A.P., Podrucka, K., Subbotin, S.A. and Moens, M. (2004) Phylogenetic relationships within the genus *Steinernema* (Nematoda: Rhabditida) as inferred from analyses of sequences of the ITS1-5.8S-ITS2 region of rDNA and morphological features. *Nematology* 6, 547–566.

Spiridonov, S.E., Moens, M. and Wilson, M.J. (2007) Fine scale spatial distributions of two entomopathogenic nematodes in a grassland soil. *Applied Soil Ecology* 37, 192–201.

Sudhaus, W. and Fitch, D. (2001) Comparative studies on the phylogeny and systematics of Rhabditidae (Nematoda). *Journal of Nematology* 33, 1–70.

Tamura, K., Peterson, D., Peterson, N., Stecher, G., Nei, M. and Kumar, S. (2011) MEGA: Molecular evolutionary genetics analysis using maximum likelihood, evolutionary distance, and maximum parsimony methods. *Molecular Biology and Evolution* 28, 2731–2739.

Vrain, T.C., Wakarchuk, D.A., Levesque, A.C. and Hamilton, R.I. (1992) Intraspecific rDNA restriction fragment length polymorphism in the *Xiphinema americanum* group. *Fundamental and Applied Nematology* 15, 563–573.

Yushin, V.V., Yoshida, M. and Spiridonov, S.E. (2007) Riders on the sperm: sperm dimorphism and spermatozeugmata in nematodes from the genus *Steinernema* (Rhabditida: Steinernematidae). *Nematology* 9, 61–75.

4 Entomopathogenic Nematodes: General Biology and Behaviour

J. GULSAR BANU,[1]* I. CANNAYANE[2] AND K. SANKARI MEENA[3]

[1]ICAR-Central Institute for Cotton Research, Coimbatore, India; [2]Vegetable Research Station, Palur, India; [3]Tamil Nadu Agricultural University, Coimbatore, India

4.1 Introduction

Entomopathogenic nematodes (EPNs) are round worms that are parasitic to many groups of insects. They occur naturally in the entire soil environment and infest insect hosts by means of chemical cues obtained from the insects, the concentration of carbon dioxide and vibration from the insects (Kaya and Gaugler, 1993). EPNs belong to the families Steinernematidae (*Steinernema* spp.) and Heterorhabditidae (*Heterorhabditis* spp.), possess the attributes of effective biocontrol agents and are widely used against a variety of crop insect pests (Grewal *et al.*, 2005; Koppenhöfer, 2007; Askary, 2010). A few years ago, a new genus in the EPN group Heterorhabditidoides was proposed, with associated bacteria being *Serratia* spp. (Zhang *et al.*, 2008, 2012).

Steinernema and *Heterorhabditis* have the potential to recycle and remain in the soil for a long duration. They can also cause direct or indirect effects on plant parasitic nematodes and soilborne pathogens. EPNs also improve the soil quality (Lacey and Georgis, 2012) and are compatible with many pesticides and biocontrol agents. They are specific to target pests and generally non-toxic to beneficial organisms. That is why the US Environmental Protection Agency (EPA) has exempted EPN products from registration. The application of EPNs does not need any protective equipment. They possess the optimal balance of biocontrol attributes (Cranshaw and Zimmerman, 2013). These positive attributes make them an excellent candidate to fit finely into integrated pest management (IPM) programmes.

EPNs parasitize the insect host by entering into the host body, and they lead an endoparasitic mode of life. The symbiotic bacteria associated with them play an important role in the suppression of insect pests. Approximately 30–500 symbiotic bacteria are reported to be associated with the two nematodes, *Steinernema* and *Heterorhabditis* (Cowles and Goodrich-Blair, 2004; Goetsch *et al.*, 2006). EPNs and the symbiotic bacteria control a wide range of insect pests living in the soil and in cryptic habits.

The present chapter is focused on the general biology of EPNs. The symbiotic relationships of nematodes and bacteria and the survival biology and foraging behaviour of nematodes are discussed in detail.

4.2 Entomopathogenic Nematode–Bacteria Symbiosis

Steinernema and *Heterorhabditis* have symbiotic association with the Gram-negative bacteria *Xenorhabdus* and *Photorhabdus* spp., respectively. The symbiotic bacteria are present in the epithelial surface of the anterior intestine of the infective juveniles (IJs) (Poinar, 1966; Bird and Akhurst, 1983). The anterior part of the nematode intestine is modified as a bacterial chamber to carry out the cells of the symbiotic bacterium. IJs, when entering into the insect host, release the bacteria from their intestine into the haemolymph of the host body. The bacteria produce a wide range of toxins and hydrolytic exoenzymes, which kill the host insect generally within 24–48 h. The bacteria, on finding

*Corresponding author. E-mail: gulsarsci@gmail.com

a suitable medium, multiply very rapidly. The nematodes feed on the multiplying bacteria, moult and multiply inside the host (Fig. 4.1). The nematode–bacteria symbiosis has a potential effect against a wide range of insects belonging to different orders (Kaya and Gaugler, 1993).

The symbiotic bacteria, by the production of different defensive enzymatic compounds, even control the plant parasitic nematodes. About three different types of secondary metabolites, namely ammonia, indole and stilbene derivatives, identified from these bacteria were found to be toxic to the second-stage juveniles (J_2) of the root-knot nematode, *Meloidogyne incognita*, and fourth-stage juveniles (J_4) and adult stages of the pinewood nematode, *Bursaphelanchus xylophilus* (Hu *et al.*, 1999). The nematode–bacteria complex is being used worldwide in the field of agriculture and the medicinal industries. Further research should be carried out on these symbiotic relationships at the molecular level to learn the evolution and genetics of the symbiosis.

4.2.1 Mutualism

Mutualism is a type of interaction that exists between two different organisms and which results in a beneficial effect for each individual from the activity of the other organism. A symbiotic association occurs between microorganisms and animals in every ecological niche and ranges from obligate to facultative dependence (Pontes and Dale, 2006). The benefits derived due to these interactions are distinct and include mutual influence on nutrition, defence, reproduction and development (Taylor *et al.*, 2005; Pais *et al.*, 2008). Microbial mutualism is the basis for the evolution of eukaryotic cells, and it is the ubiquitous feature of biotic systems.

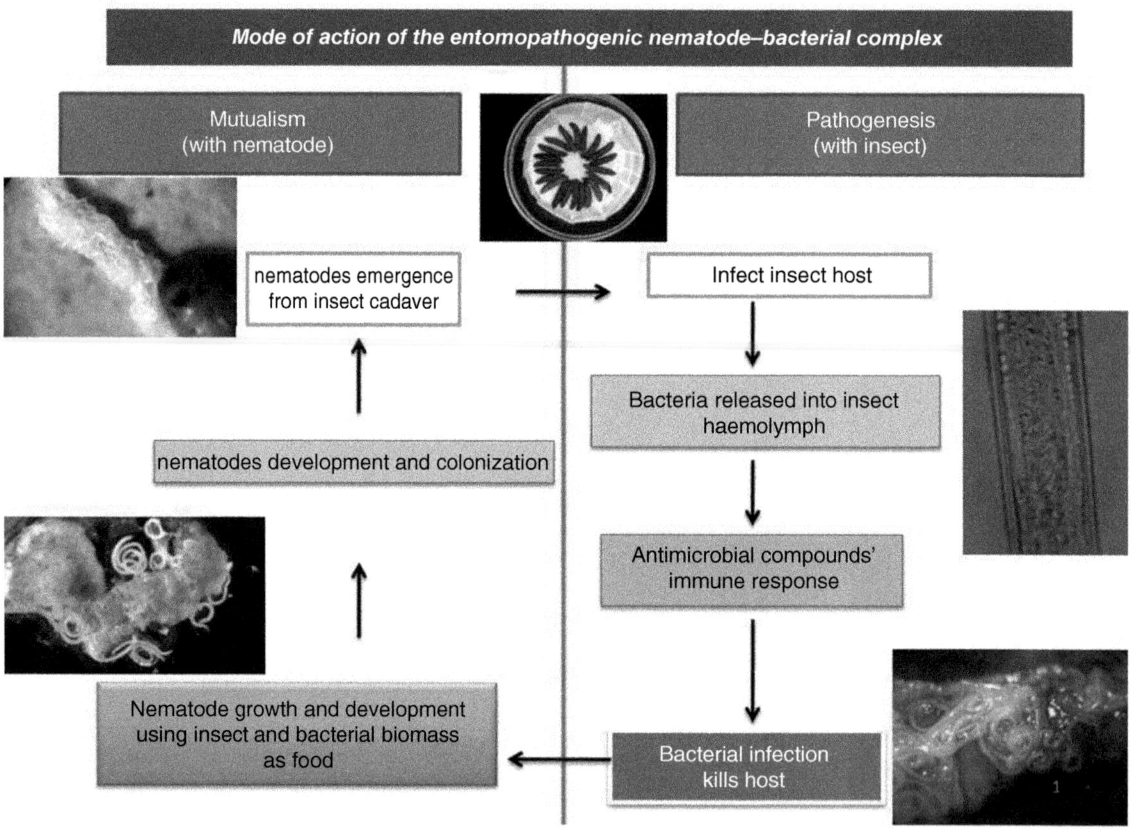

Fig. 4.1. A chart showing the EPN–bacteria relationship during insect infestation.

Insects are used as a nutritional source by a group of terrestrial nematodes that are symbiotically associated with Gamma proteobacteria. This evolution has occurred at two independent times, in the genera *Steinernema* and *Heterorhabditis* (Adams and Nguyen, 2002; Goodrich-Blair and Clarke, 2007). Though *Steinernema* spp. and *Heterorhabditis* spp. are not closely related, their symbiotic bacteria *Xenorhabdus* and *Photorhabdus*, respectively form a phylogenetic sister group (Suzuki *et al.*, 1996). The non-feeding IJ stage of EPNs present in the soil carries symbiotic bacteria into the host insect, which subsequently kill the host and use it as a nutrient source for their reproduction. The symbiotic bacteria confer to the virulence, immune modulation and reproductive fitness of the EPN (Goodrich-Blair and Clarke, 2007; Park *et al.*, 2007; Clarke, 2008).

The success of parasitism correlates with the number of bacteria carried per IJ, indicating that the trade-off is induced by symbiosis. Finally, the intensity of the trade-off depends on the parental effects and, more generally, is greater under restrictive environmental conditions (Emelianoff *et al.*, 2008). IJs of *Steinernema* carry *Xenorhabdus* bacteria in the receptacle of the nematode intestine that are essential for their survival and coexistence. IJs infect the host insects and penetrate through their natural openings and release the bacterial symbiont into the insect haemocoel (Richards and Goodrich-Blair, 2009). The toxic secretions of the bacteria kill the insect host (Ciche and Ensign, 2003; Snyder *et al.*, 2007). In *Steinernema* spp., the region known as the receptacle is an extracellular space between two epithelial cells (Bird and Akhurst, 1983; Martens *et al.*, 2003; Snyder *et al.*, 2007). Although *Photorhabdus* associated with *Heterorhabditis* spp. are clearly limited to a definite region of the intestine, the colonized intestinal region of *Heterorhabditis* spp. appears to be morphologically similar to that of the rest of the intestine (Ciche *et al.*, 2008).

IJs ambush or hunt the insect host (Campbell and Gaugler, 1992) and penetrate the insect integument. Nematodes reproduce two to three generations in the insect cadaver. When the food begins to deplete, the development of IJs takes place due to high nematode density and low nutrition cue. IJs exit the cadaver to search and infect a new insect host. Colonization of the nematode by the bacterial symbiont, followed by the development of non-feeding IJs is the critical event for maintenance of the symbiosis. During this stage, EPNs exhibit specificity for their bacterial symbiont (Ciche and Ensign, 2003; Cowles and Goodrich-Blair, 2008). Apart from mutualism, the bacterial symbiont also exhibits pathogenicity towards the insects (Clarke, 2008). These systems help us to study how pathogenicity is neutralized by the nematode host. Mutualism depends on factors such as nutritional factors (Watson *et al.*, 2005), surface structure of the EPN (Bennett and Clarke, 2005), gene regulation (Cowles and Goodrich-Blair, 2006; Cowles *et al.*, 2007; Herbert and Goodrich-Blair, 2007) and stress response (Heungens *et al.*, 2002), as well as some unknown function (Cowles and Goodrich-Blair, 2008). Due to the expression of virulence factors including the Tc proteins, the bacteria kill the insect host, which forms a source of nutrients for both the bacteria and the EPN (Forst *et al.*, 1997). The size of the host plays an important role in limiting the development of nematodes (Bastidas *et al.*, 2014). Irrespective of the interaction, bacteria have subtle effects on host development, nutrition and immunity. Molecular-level studies on these interactions will reveal the fundamental principles of pathogenicity and mutualism (Goodson *et al.*, 2005).

4.2.2 Colonization process

Colonization of the symbiotic bacteria takes place in the anterior intestinal cells of the nematodes. Initially, few bacterial cells colonize the nematode pharyngeal-intestinal valve (PIV) that is present anterior to the intestinal epithelium. During the process of colonization, due to the constriction and relaxation of the nematode intestine, the bacteria initially present in the PIV occupy the receptacle (Chaston *et al.*, 2013). The bacteria–nematode interaction is unique and was decided by events occurring at the molecular and cellular level interaction between the host and the bacteria during the colonization process. The SR1 gene of the symbiotic bacteria is species specific, which helps in the colonization of the host. For example, symbiosis region 1 (SR1) gene is involved at each stage of colonization of *Xenorhabdus nematophila*, whereas in *Xenorhabdus szentirmaii*, which lacks the SR1 gene, colonization will not occur unless SR1 is expressed ectopically (Chaston *et al.*, 2013).

IJs enter into the susceptible insect hosts either through their natural openings in the cuticle (e.g. the mouth, anus or spiracles) or by penetrating through the integument with the help of a buccal, tooth-like appendage. After entering the insect body, the IJ migrates into the insect haemolymph, where signals from the insects induce the IJ to eject the symbiotic bacteria from their gut (Wang and Bedding, 1996; Ciche and Ensign, 2003; Cagnolo *et al.*, 2004). The regurgitated bacteria show exponential growth in the insect and reach cell densities up to 10^9 cfu/insect within 48 h (Watson *et al.*, 2005). Rapid growth of the symbiotic bacteria is accompanied by the secretion of toxins and other molecules, which damage the host tissues and ultimately reduce the effectiveness of the insect innate immune system (Goodrich-Blair and Clarke, 2007).

The post-exponential growth phase of the bacteria coincides with the death of the insect. During this stage, all the internal organs and tissues of the insect will be converted into bacterial biomass. End of the pathogenic period and initiation of the mutualistic period of the nematode and bacteria is pronounced by the death of the host insect (ffrench-Constant *et al.*, 2003). For example, exponential growth of *Photorhabdus* coincides with the development of the IJ into an adult hermaphrodite. At this point, the hermaphrodite will lay about 200–300 eggs that will pass through four juvenile stages into either male or hermaphrodite adult nematodes (Ciche *et al.*, 2008). Further, no subsequent external egg-laying is observed, and intrauterine nematode reproduction continues for about 1–2 generations in response to environmental cues. The final generation of developing nematodes enters into the alternative J_3 stage, which is the infective stage of the EPN. During development, *Photorhabdus* colonize the gut region of the IJ before its emergence from the insect cadaver and enter into the soil (Ciche *et al.*, 2008). Under optimal laboratory conditions, it takes about 10–20 days for the EPN to complete its life cycle (from IJ infection to emergence). To be an effective symbiont, the bacteria should possess the following three criteria: (i) they must be a highly efficient pathogen of the insect; (ii) they must support nematode growth and development; and (iii) they must be able to colonize the IJ stage of the nematode. However, IJs can also be recovered when inoculated directly on to high densities of *Photorhabdus* grown on agar plates, implying that the bacteria also produce a signal, termed 'the food signal', which can stimulate nematode development (Strauch and Ehlers, 1998).

4.2.3 Life cycle

The EPN completes its life cycle inside the insect body, except the IJ, which is the only free-living stage found in the soil. IJs penetrate the insect host through natural openings (spiracles, mouth, anus), and also through cuticle intersegments, thereby entering into the haemecoel (insect body cavity) containing the insect blood (haemolymph). Juveniles release the symbiotic bacteria from the intestine into the haemolymph, where bacterial multiplication occurs, killing the insect within 24–48 h. Nematodes in turn feed on the dead cadaver for their growth and multiplication. Depending on the availability of insect source, the nematode has one or many generations inside the insect cadaver, multiplying several folds, and finally exits (IJs only) into the environment to continue its life cycle (Kaya and Gaugler 1993).

In *Heterorhabditis* spp., the first generation is self-fertile hermaphrodites, while the next generation produces both males and females by cross-fertilization (amphimictic). *Steinernema* spp. produces all generations as male and female, which is termed as gonochorism (Grewal *et al.*, 2005). Insects killed by *Heterorhabditis* become a red colour, while *Steinernema*-infested insects become brown or tan (Kaya and Gaugler, 1993). The life cycle of EPNs is generally completed within 12–15 days at a temperature of 25–30°C, which is the optimum temperature for nematode growth and reproduction.

Approximately 30–200 *X. nematophila* cells were found to occupy a specific region of the intestinal lumen of the infective juveniles of *Steinernema carpocapsae*, and the transportation of the bacterial symbiont into the insect's blood occurred by defecation (Herbert and Goodrich-Blair, 2007). Thus, the bacterial symbionts appear to provide specific nutritional factors to the nematodes during their development. During the stationary phase, both *Photorhabdus* and *Xenorhabdus* spp. produce intracellular protein crystals, which are considered to be a source of nutrition to the nematodes (Bowen and Ensign, 2001).

The nematode–bacteria relationship is species specific, and each species of *Heterorhabditis* or *Steinernema* isolated from nature is colonized by a specific bacterial phylotype, while nematodes grown under laboratory conditions on non-cognate bacteria produce uncolonized IJs (Ciche and Ensign, 2003; Sicard *et al.*, 2004a,b). Maternal nematodes acquire *Photorhabdus* symbionts as a persistent intestinal biofilm prior to transmission to the IJ stage offspring developing inside the maternal body (Joyce *et al.*, 2006).

4.2.4 Pathogenicity

The pathogenicity and fecundity of EPNs depend on density-dependent factors. They directly affect the number of IJs produced by infested cadavers or indirectly affect the longevity of nematodes. To establish the pathogenic effect, the EPN and its symbiotic bacteria should overwhelm the host immune response, which may affect the invasion in a density-dependent manner (Dunphy and Thurston, 1990).

Photorhabdus and *Xenorhabdus* are pathogenic to a wide range of insect larvae, and a dose of less than 5 cfu when directly injected into the blood system is sufficient to kill the insects within 48–72 h. Insect innate immunity consists of both cellular and humeral components that are triggered by surveillance proteins on recognition of invading organisms (Leulier *et al.*, 2003). Circulating haemocytes (blood cells) that recognize foreign material and recruit other haemocytes to encapsulate the invading organism were reported to be the cell-based innate immunity responses. The capsule that formed was then melanized by the action of phenoloxidase, and the melanized nodule containing the trapped organism was removed from circulation (Kanost *et al.*, 2004).

The insecticidal activities of EPNs are due mainly to the 'toxins' secreted by the bacterial symbionts. Various types of bacterial pathogens including *Photorhabdus* species produce glycosyl transferase toxins. Recent studies have discovered novel, unexpected variations in host protein targets and amino acid acceptors of toxin-catalysed glycosylation (Jank *et al.*, 2013). It is well recognized that bacteria communicate via small diffusible molecules, a process termed quorum sensing. The best understood quorum sensing systems are those that use acylated homoserine lactones (AHLs) for communication (Brameyer *et al.*, 2014).

4.2.5 Importance of bacterial symbionts

The coexistence of EPNs and their bacterial symbionts are well documented and characterized for their mutual benefits and subsequent infection/parasitism on insects. The described species and their associated bacterial symbionts are furnished in Tables 4.1–4.7. In the case of *Neosteinernema longicurvicaudatum*, only one described species is available and its bacterial symbiont has not yet been identified and characterized, and hence not mentioned in the table.

4.3 Survival Biology

EPNs adopt unique survival tactics against environmental extremities. They undergo dormant stages during unfavourable environmental conditions, and the dormant stage will prolong throughout their lifespan to withstand the fluctuating environments. Nematode dormancy is differentiated into diapause and quiescence. Diapause is the arrested developmental stage in nematodes and is controlled by hormones (Denlinger, 2002). Nematodes cannot be recovered from diapause until specific requirements are satisfied, even under suitable environmental conditions. Quiescence is a facultative metabolism that occurs under unfavourable environmental extremities and can be retrieved under favourable environmental conditions.

An unfavourable environmental condition is induced by water, oxygen, osmotic pressure and temperature. There appears to be a wide range of response to environmental stress exhibited across EPN species and strains, which includes variation in desiccation tolerance, hypoxia tolerance and freeze tolerance (Shapiro-Ilan *et al.*, 2014). Quiescence induced by these factors is termed as anhydrobiosis (lack of water), thermobiosis (high temperature), cryobiosis (low temperature), anoxybiosis (low oxygen) and osmobiosis (osmotic stress) (Barrett, 1991). These environmental stresses induce the development of unique juvenile stages in EPNs, which is termed as dauer juveniles and is the infective stage for EPNs. EPNs are able to withstand adverse environmental conditions inside their host cadavers. The IJ stage that lives in the soil is particularly vulnerable to environmental variability (Brown and Gaugler, 1997). Long-term persistence of EPNs in the field is related to those abiotic factors such as soil moisture, temperature or soil texture. Acclimation to temperature stress had both negative and positive effects on the survival of both nematode species, although the overall variation meant that many of these effects were non-significant. With improved basal thermal tolerance for either species at upper lethal temperatures, no consistent reduction in the plasticity of nematodes was recorded, while lower temperatures reduced their survival ability. The lower temperature range of 5–12.5°C lowered the survival rate of *Heterorhabditis zealandica*. Such results indicate that the thermal niche breadth of EPN species can differ significantly depending on the recent thermal conditions. Hence, thermal niche

Table 4.1. Described species of *Steinernema* and their original habitat.

Serial number	Entomopathogenic nematode species	Host/habitat and locality	Reference
1	*Steinernema abbasi* Elawad, Ahmad & Reid, 1997	Bollworm Sultanate of Oman	Elawad *et al*., 1997
2	*Steinernema aciari* Qiu, Yan, Zhou, Nguyen & Pang, 2005	Soil, China	Qui *et al*., 2005a
3	*Steinernema affine* (= *affinis*) Wouts, Mráček, Gerdin & Bedding, 1982, syn. *Neoaplectana affinis* Bovien, 1937	Bibionid fly larva, Denmark	Wouts *et al*., 1982
4	*Steinernema akhursti* Qiu, Hu, Zhou, Pang & Nguyen, 2005	Soil, China	Qiu *et al*., 2005b
5	*Steinernema anatoliense* Hazir, Stock & Keskin, 2003	Soil, Turkey	Hazir *et al*., 2003
6	*Steinernema apuliae* Triggiani, Mráček & Reid, 2004	Soil, Italy	Triggiani *et al*., 2004
7	*Steinernema arasbaranense* Nikdel, Niknam & Ye, 2011	Soil, Iran	Nikdel *et al*., 2011
8	*Steinernema arenarium* (= *anomalae* = *anomali*) (Artyukhovsky, 1967) Wouts, Mráček, Gerdin & Bedding, 1982 (Kozodoi, 1984) Curran, 1989	Soil, Central Russia	Artyukhovsky, 1967
9	*Steinernema ashiuense* Phan, Takemoto & Futai, 2006	Soil, Japan	Phan *et al*., 2006a
10	*Steinernema asiaticum* Anis, Shahina, Reid & Rowe, 2002	Soil, Pakistan	Anis *et al*., 2002
11	*Steinernema australe* Edgington, Buddie, Tymo, Hunt, Nguyen, France, Merino & Moore, 2009	Soil, Chile	Edgington *et al*., 2009a
12	*Steinernema backanense* Phan, Spiridonov, Subbotin & Moens, 2006	Soil, Vietnam	Phan *et al*., 2006b
13	*Steinernema balochiense* Fayyaz, Khanum, Ali, Solongi, Gulsher & Javed, 2015	Soil, Pakistan	Fayyaz *et al*., 2015
14	*Steinernema beddingi* Qiu, Hu, Zhou, Pang & Nguyen, 2005	Soil, China	Qui *et al*., 2005c
15	*Steinernema bicornutum* Tallosi, Peters & Ehlers, 1995	Soil, Yugoslavia	Tallosi *et al*., 1995
16	*Steinernema bifurcatum* Fayyaz, Yan, Qiu, Han, Gulsher, Khanum & Javed, 2014	Soil, Pakistan	Fayyaz *et al*., 2014
17	*Steinernema boemarei* Lee, Sicard, Skeie & Stock, 2009	Soil, France	Lee *et al*., 2009
18	*Steinernema brazilense* Nguyen, Ginarte, Leite, Santo & Harakava, 2010	Soil, Brazil	Nguyen *et al*., 2010
19	*Steinernema carpocapsae* (Weiser, 1955) Wouts, Mráček, Gerdin & Bedding, 1982	*Cydia pomonella*, Czechoslovakia	Wouts *et al*., 1982
20	*Steinernema caudatum* Xu, Wang & Li, 1991	Soil, China	Xu *et al*., 1991
21	*Steinernema ceratophorum* Jian, Reid & Hunt, 1997	Soil, China	Jian *et al*., 1997
22	*Steinernema cholashanense* Nguyen, Puza & Mráček, 2008	Soil, Eastern Tibet	Nguyen *et al*., 2008a
23	*Steinernema citrae* Stokwe, Malan, Nguyen, Knoetze & Tiedt, 2011	Soil, South Africa	Stokwe *et al*., 2011
24	*Steinernema colombiense* López-Núñez, Plichta, Góngora-Botero & Stock, 2008	Soil, Colombia	López-Núñez *et al*., 2008
25	*Steinernema costaricense* Uribe-Lorio, Mora & Stock, 2007	Soil, Costa Rica	Uribe-Lorio *et al*., 2007
26	*Steinernema cubanum* (= *cubana*) Mráček, Hernandez & Boemare, 1994	Soil, Cuba	Mráček *et al*., 1994
27	*Steinernema cumgarense* Phan, Spiridonov, Subbotin & Moens, 2006	Soil, Vietnam	Phan *et al*., 2006b

Continued

Table 4.1. Continued.

Serial number	Entomopathogenic nematode species	Host/habitat and locality	Reference
28	*Steinernema diaprepesi* Nguyen & Duncan, 2002	Citrus weevil *Diaprepes abbreviates*, USA	Nguyen and Duncan, 2002
29	*Steinernema dharanaii* Kulkarni, Rizvi, Kumar, Paunikar & Mishra, 2012	Soil, India	Kulkarni *et al.*, 2012
30	*Steinernema eapokense* Phan, Spiridonov, Subbotin & Moens, 2006	Soil, Vietnam	Phan *et al.*, 2006b
31	*Steinernema ethiopense* Tamiru, Waeyenberge, Hailu, Ehlers, Půža & Mráček, 2012	Soil, Ethiopia	Tamiru *et al.*, 2012
32	*Steinernema everestense* Chhetri, Waeyenberge, Spiridonov & Moens, 2011	Soil, Nepal	Chhetri *et al.*, 2011a
33	*Steinernema feltiae* (Filipjev, 1934) Wouts, Mráček, Gerdin & Bedding, 1982	*Bibio* sp., Russia	Wouts *et al.*, 1982
34	*Steinernema glaseri* (Steiner, 1929) Wouts, Mráček, Gerdin & Bedding, 1982	*Popillia japonica*, New Jersey, USA	Steiner, 1929
35	*Steinernema guangdongense* Qiu, Fang, Zhou, Pang & Nguyen, 2004	Soil, China	Qiu *et al.*, 2004
36	*Steinernema hebeiense* Chen, Li, Yan, Spiridonov & Moens, 2006	Soil, China	Chen *et al.*, 2006
37	*Steinernema hermaphroditum* Stock, Griffin & Chaerani, 2004	Soil, Indonesia	Stock *et al.*, 2004
38	*Steinernema huense* Phan, Mracek, Puza, Nermut and Jarosova, 2014	Soil, Vietnam	Phan *et al.*, 2014
39	*Steinernema ichnusae* Tarasco, Mracek, Nguyen & Triggiani, 2008	Soil, Italy	Tarasco *et al.*, 2008
40	*Steinernema innovationi* Çimen, Lee, Hatting, Hazir & Stock, 2015	Soil, South Africa	Çimen *et al.*, 2015
41	*Steinernema intermedium* (= *intermedia*) (Poinar, 1985) Mamiya, 1988	Soil, USA	Poinar, 1985
42	*Steinernema jollieti* Spiridonov, Osterfeld & Moens, 2004	Soil, USA	Spiridonov *et al.*, 2004
43	*Steinernema jeffreyense* Malan, Knoetza & Tiedt, 2015	Soil, South Africa	Malan *et al.*, 2016
44	*Steinernema karii* Waturu, Hunt & Reid, 1997	Soil, Kenya	Waturu *et al.*, 1997
45	*Steinernema khoisanae* Nguyen, Malan & Gozel, 2006	Soil, South Africa	Nguyen *et al.*, 2006a
46	*Steinernema kraussei* (Steiner, 1923) Travassos, 1927	Web spruce saw fly, Germany	Steiner, 1923
47	*Steinernema kushidai* Mamiya, 1988	Scarabaeid beetle larva, Japan	Mamiya, 1988
48	*Steinernema lamjungense* Chhetri, Waeyenberge, Spiridonov, Manadhar & Moens, 2011	Soil, Nepal	Chhetri *et al.*, 2011b
49	*Steinernema leizhouense* Nguyen, Qiu, Zhou & Pang, 2006	Soil, China	Nguyen *et al.*, 2006b
50	*Steinernema* litorale Yoshida, 2004	Soil, Japan	Yoshida, 2004
51	*Steinernema loci* Phan, Nguyen & Moens, 2001	Soil, Vietnam	Phan *et al.*, 2001a
52	*Steinernema longicaudum* Shen & Wang, 1991	Soil, China	Shen and Wang, 1991
53	*Steinernema meghalayensis* Ganguly, Rathour, Kumar & Singh, 2011	Soil, India	Ganguly *et al.*, 2011
54	*Steinernema monticolum* Stock, Choo & Kaya, 1996	Soil, Korea	Stock *et al.*, 1997

Continued

Table 4.1. Continued.

Serial number	Entomopathogenic nematode species	Host/habitat and locality	Reference
55	*Steinernema minutum* Maneesakorn, Grewal & Chandrapatya, 2010	Soil, Thailand	Maneesakorn *et al.*, 2010
56	*Steinernema neocurtillae* (= *neocurtillis*) Nguyen & Smart, 1992	*Neocurtilla hexadactylla*, Florida, USA	Nguyen and Smart, 1992
57	*Steinernema oregonense* (= *oregonensis*) Liu & Berry, 1996	Soil, Oregon, USA	Liu and Berry, 1996a
58	*Steinernema pakistanense* Shahina, Anis, Reid & Maqbool, 2001	Soil, Pakistan	Shahina *et al.*, 2001
59	*Steinernema papillatum* Blas, Portillo, Nermut & Puza Montero, 2015	Soil, Venezuela	Blas *et al.*, 2015
60	*Steinernema phyllophagae* Nguyen & Buss, 2011	*Phyllophaga* sp., USA	Nguyen and Buss, 2011
61	*Steinernema poinari* Mráček, Puza and Nermut, 2014	Soil, Czech Republic	Mráček *et al.*, 2014
62	*Steinernema puertoricense* (= *puertoricensis*) Roman & Figueroa, 1994	Soil, Puerto Rico	Roman and Figueroa, 1994
63	*Steinernema pui* Qui, Zhao, Wu, Lv & Pang, 2011	Soil, China	Qui *et al.*, 2011
64	*Steinernema puntauvense* Uribe-Lorio, Mora & Stock, 2007	Soil, Costa Rica	Uribe-Lorio *et al.*, 2007
65	*Steinernema rarum* (= *rara*) (de Doucet, 1986) Mamiya, 1988	*Heloithis* sp., Argentina	de Doucet, 1986
66	*Steinernema riobrave* (= *riobravis*) Cabanillas, Poinar & Raulston, 1994	*Helicoverpa zea*, USA	Cabanillas *et al.*, 1994
67	*Steinernema ritteri* de Doucet & Doucet, 1990	Soil, Argentina	de Doucet and Doucet, 1990
68	*Steinernema robustispiculum* Phan, Subbotin, Waeyenberge & Moens, 2005	Soil, Vietnam	Phan *et al.*, 2005
69	*Steinernema sacchari* Nthenga, Knoetze, Berry, Tiedt & Malan 2014	Soil, South Africa	Nthenga *et al.*, 2014
70	*Steinernema sangi* Phan, Nguyen & Moens, 2001	Soil, Vietnam	Phan *et al.*, 2001b
71	*Steinernema sasonense* Phan, Spiridonov, Subbotin & Moens, 2006	Soil, Vietnam	Phan *et al.*, 2006b
72	*Steinernema scapterisci* Nguyen & Smart, 1990	Mole cricket, America	Nguyen and Smart, 1990
73	*Steinernema scarabaei* Stock and Koppenhöfer, 2003	*Anomala* (= *Exomala*) *orientalis* and *Popillia japonica*, New Jersey, USA	Stock and Koppenhöfer, 2003
74	*Steinernema serratum* Liu, 1992	Soil, South Africa	Liu, 1992
75	*Steinernema siamkayai* Stock, Choo & Kaya, 1998	Soil, Thailand	Stock *et al.*, 1998
76	*Steinernema sichuanense* Mráček, Nguyen, Taillier, Bomoemare & Chen, 2006	Soil, China	Mráček *et al.*, 2006
77	*Steinernema silvaticum* Sturhan, Spiridonov & Mráček, 2005	Soil, Germany	Sturhan *et al.*, 2005
78	*Steinernema tami* Luc, Nguyen, Reid & Spiridonov, 2000	Soil, Vietnam	Luc *et al.*, 2000
79	*Steinernema texanum* Nguyen, Tuart, Andalo, Gozel & Michael, 2007	Soil, Texas, USA	Nguyen *et al.*, 2007
80	*Steinernema thanhi* Phan, Nguyen & Moens, 2001	Soil, Vietnam	Phan *et al.*, 2001a
81	*Steinernema thermophilum* Ganguly & Singh, 2000	Soil, India	Ganguly and Singh, 2000
82	*Steinernema tielingense* Ma, Chen, Li, Han, Chhetri, Clercq & Moens, 2012	Soil, North China	Ma *et al.*, 2012a

Continued

Table 4.1. Continued.

Serial number	Entomopathogenic nematode species	Host/habitat and locality	Reference
83	*Steinernema tophus* Çimen, Hatting, Hazir & Stock, 2014	Soil, South Africa	Çimen *et al.*, 2014
84	*Steinernema unicornum* Edgington, Buddie, Tymo, France, Merino & Hunt, 2009	Soil, Chile	Edgington *et al.*, 2009b
85	*Steinernema vulcanicum* Clausi, Longo, Rappazzo, Tarasco & Vinciguerra, 2011	Soil, Italy	Clausi *et al.*, 2011
86	*Steinernema websteri* Cutler & Stock, 2003	Soil, China	Cutler and Stock, 2003
87	*Steinernema weiseri* Mráček, Sturhan & Reid, 2003	Soil, Czech Republic	Mráček *et al.*, 2003
88	*Steinernema xinbinense* Ma, Chen, Clercq, Waeyenberge, Han & Moens, 2012	Soil, China	Ma *et al.*, 2012b
89	*Steinernema xueshanense* Mráček, Zhi and Nguyen, 2009	Soil, China	Mráček *et al.*, 2009
90	*Steinernema yirgalemense* Nguyen, Tesfamariam, Gozel, Gaugler & Adams, 2004	Soil, Ethiopia	Nguyen *et al.*, 2004a
91	*Steinernema masoodi* Ali, Shaheen, Pervez & Hussain 2005	Soil, Uttar Pradesh, India	Ali *et al.*, 2005
92	*Steinernema seemae* Ali, Shaheen, Pervez & Hussain 2005	Soil, Uttar Pradesh, India	Ali *et al.*, 2005

should be characterized across a broad range of EPN species to understand the evolution of thermal limits for the effective survival and performance of the nematodes (Hill *et al.*, 2015).

The nematode cuticle becomes a multilayered structure to withstand desiccation; nematodes close their external openings during this stage to prevent the entry of toxic chemicals and microbial antagonists. Nematodes also use their internal biochemical reserves for survival during diapause. Glycogen is the chief energy reserve for EPNs, followed by lipids in maintaining their infectivity during unfavourable conditions (Patel *et al.*, 1997). These energy reserves play a significant role in supporting the physiological and behavioural process of EPNs under environmental extremities.

4.3.1 Cold tolerance

Nematodes survive in high altitudes and temperate countries may even withstand subzero temperatures, either by the mechanism of freeze tolerance or freeze susceptible. Freeze tolerance is the mechanism by which they survive in freezing conditions, even during ice formation in their tissues. Freeze-susceptible nematodes maintain their body fluids as liquids at a temperature below their melting points by the mechanism of supercooling.

Steinernema feltiae, *Steinernema anomali* and *Heterorhabditis bacteriophora* are reported to be freeze-tolerant nematodes (Brown and Gaugler, 1996), while *H. zealandica* has been observed as a freeze-avoiding nematode (Wharton and Surrey, 1994). In this condition, sugars and polyols present in the nematode's body act as cryoprotectants. Trehalose (sugars) has great potential to enhance the supercooling ability of the nematodes. The variation contributed for cold tolerance is attributed to the different adaptive strategies of the nematodes, which include cryoprotective dehydration, anhydrobiosis and freeze avoidance or other tolerance strategies (Wharton, 2011). Much of the work related to the thermal performance of EPNs has focused on their cold tolerance, in which the overwinter survival of free-living nematodes in polar environments has been observed (Perry and Wharton, 2011). An increase in the freezing tolerance of *H. bacteriophora* and *S. feltiae* has been recorded after their acclimation to temperatures below their propagation temperature (Ali and Wharton, 2013).

4.3.2 Heat tolerance

Temperature plays a key role in the survival and activity of terrestrial invertebrates. Thermal tolerance may be characterized by traits like thermal maxima, minima and optima. The mean tolerated

Table 4.2. Described species of *Heterorhabditis* and their original habitat.

Serial number	Entomopathogenic nematode species	Host/habitat and locality	Reference
1	*Heterorhabditis amazonensis* Andalo, Nguyen & Moino, 2006	Soil, Brazil	Andalo *et al.*, 2006
2	*Heterorhabditis argentinensis* Stock, 1993 (synonym of *Heterorhabditis bacteriophora*)	*Graphognathus* sp., Argentina	Stock, 1993
3	*Heterorhabditis atacamensis* Edgington, Buddie, Moore, France, Merino & Hunt, 2011	Soil, Chile	Edgington *et al.*, 2011
4	*H. bacteriophora* Poinar, 1976	*Heliothis punctigera*, Australia	Poinar, 1976
5	*Heterorhabditis baujardi* Phan, Subbotin, Nguyen & Moens, 2003	Soil, Vietnam	Phan *et al.*, 2003
6	*Heterorhabditis beicherriana* Li, Liu, Nermut, Puza & Mracek, 2012	Soil, China	Li *et al.*, 2012
7	*Heterorhabditis brevicaudis*[a] (may be synonym of *Heterorhabditis indica*) Liu, 1994	Soil, China	Liu, 1994
8	*Heterorhabditis downesi* Stock, Griffin & Burnell, 2002	Soil, Ireland	Stock *et al.*, 2002
9	*Heterorhabditis floridensis* Nguyen, Gozel, Koppenhofer & Adams, 2006	Soil, Florida	Nguyen *et al.*, 2006c
10	*Heterorhabditis georgiana* Nguyen, Shapiro-Ilan & Mdata, 2008	Soil, Georgia	Nguyen *et al.*, 2008b
11	*Heterorhabditis gerrardi* Plichta, Joyce, Clarke, Waterfield & Stock, 2009	Soil, Australia	Plichta *et al.*, 2009
12	*Heterorhabditis hawaiiensis* Gardner, Stock & Kaya, 1994 (synonym of *H. indica*)	Soil, Hawaii, USA	Gardner *et al.*, 1994
13	*Heterorhabditis indica* (= *indicus*) Poinar, Karunaka & David, 1992	Soil, India	Poinar *et al.*, 1992
14	*Heterorhabditis marelatus* Liu & Berry, 1996	Soil, Oregon, USA	Liu & Berry, 1996b
15	*Heterorhabditis megidis* Poinar, Jackson and Klein, 1987	*Papillia japonica*, USA	Poinar *et al.*, 1987
16	*Heterorhabditis mexicana* Nguyen, Shapiro-Ilan, Stuart, James, McCoy & Adams, 2004	Soil, Mexico	Nguyen *et al.*, 2004b
17	*Heterorhabditis noenieputensis* Malan, Knoetze & Tiedt, 2014	Soil, South Africa	Malan *et al.*, 2014
18	*Heterorhabditis poinari*[a] Kakulia & Mikaia, 1997	Soil, Georgia	Kakulia and Mikaia, 1997
19	*Heterorhabditis safricana* Malan, Nguyen, De Waal & Tiedt, 2008	Soil, South Africa	Malan *et al.*, 2008
20	*Heterorhabditis sonorensis* Stock, Rivera-Orduño & Lara, 2009	*Diceroprocta ornea*, Mexico	Stock *et al.*, 2009
21	*Heterorhabditis taysearae* Shamseldean, El-Sooud, Abd-Elgawad & Saleh, 1996	Soil, Egypt	Shamseldean *et al.*, 1996
22	*Heterorhabditis zealandica* Poinar, 1990	Black beetle, New Zealand	Poinar, 1990

Note: [a]Species inquirenda.

Table 4.3. Described species of *Neosteinernema* and its original habitat.

Serial number	Entomopathogenic nematode species	Host/habitat and locality	Reference
1	*Neosteinernema longicurvicauda* Nguyen and Smart, 1994	Termite *Reticulitermes flavipes* (Koller), Florida, USA	Nguyen and Smart, 1994

Table 4.4. Described species of *Heterorhabditidoides* and its original habitat.

Serial number	Entomopathogenic nematode species	Host/habitat and locality	Reference
1	*Heterorhabditidoides chongmingensis* Zhang, Liu, Xu, Sun, Yang, An, Gao, Lin, Lai, He, Wu & Zhang, 2008	Soil, China	Zhang *et al.*, 2008
2	*Heterorhabditidoides rugaoensis* Zhang, Liu, Tan, Wang, Qiao, Yedid, Dai, Qiu, Yan, Tan, Su, Lai & Gao, 2012	Soil, China	Zhang *et al.*, 2012

temperature for non-adapted EPN species ranged from 33.3°C to 40.1°C, whereas for adapted species, 34.8°C to 39.2°C was the optimum temperature for survival and infectivity (Mukuka *et al.*, 2010). Higher temperatures normally affect the growth, infectivity and reproduction of nematodes. However, some nematode species can survive well in elevated temperature regimes. *Steinernema riobrave*, *Steinernema glaseri* and *Heterorhabditis indica* have been noted as heat-tolerant nematodes (Grewal *et al.*, 1994). Heat shock proteins (HSP) are stress-related proteins that help the nematodes to survive in high temperatures. HSP 70 has been detected in *H. bacteriophora* HP 88 (Selvan *et al.*, 1996).

4.3.3 Desiccation tolerance

Generally, nematodes need a thin film of water around them for their movement and infectivity. Desiccation or dry condition of the soil occurs by a slow rate of water loss. Nematodes can withstand desiccation by coiling their cuticle tightly and reducing the exposure of the cuticle area to the air. During dehydration, the accumulation of polyols and sugars acts as a protectant to the membranes of the cuticle. In dry conditions, the survival of EPNs generally ranges from 2 to 3 weeks (Kaya, 1990). Studies on desiccation tolerance carried out in different strains of *S. carpocapsae* revealed the maximum survival of the nematodes when the nematodes were exposed to slow drying conditions (Glazer, 1992). Also, it was found that *S. carpocapsae* strains could withstand desiccation better than *S. glaseri* (Kung *et al.*, 1991) and *S. riobravae* (Baur *et al.*, 1995). *S. carpocapsae* infective juveniles clump in large aggregates in response to desiccation (Simons and Poinar, 1973).

The sheath of the IJs helps to reduce water loss during desiccation (Womersley, 1990). However, other workers did not find any suitable evidence to support the role of the sheath in the desiccation survival of *S. carpocapsae* (Campbell and Gaugler, 1991; Patel *et al.*, 1997). Newly emerged *S. carpocapsae* survive better in desiccation, while aged IJs could not tolerate desiccation (Patel *et al.*, 1997). The survival of *S. carpocapsae* under desiccated conditions might be due to its lower rate of water loss because of the difference in cuticle permeability.

Desiccation tolerance studies were carried out in *S. feltiae* strains, namely IS-6, IS-15 and SF. Among the strains, the highest survival rate was recorded in IS-6, isolated from the desert region of Israel, followed by IS-15, isolated from northern Israel. Desiccation tolerance of IS-6 and IS-15 strains was associated with the dispersal ability of the aggregated IJs at a slow dehydration regime, followed by their entry into anhydrobiosis as individually coiled IJs. Increased trehalose levels were recorded in the IJs after 3 days of exposure to 97% RH, while an opposite trend was recorded in glycogen. On rehydration, glycogen regained its original concentration after 24 h duration, while trehalose reached only 50% from the original, which suggests that during the rehydration process, trehalose is used as an energy source or is hydrolysed for the glycogeneogenesis process (Solomon *et al.*, 1999).

4.3.4 *Endotokia matricida*

Endotokia matricida is the condition that occurs in EPNs under reduced food supply wherein intrauterine birth causes maternal death (*endotokia matricida*). It is relatively common in rhabditid nematodes and typical for EPNs. It helps in the long-term survival and subsequent infection of insects in the soil when external food supply is reduced (Bell, 1982).

In the case of *Heterorhabditis* spp., IJs consume the body cavity of the mother (Johnigk and Ehlers, 1999). Prior to this process, the mother's rectal glands are colonized by the symbiotic *Photorhabdus*, which serve as the source of inoculum for the progeny. A low number of *Photorhabdus* cells (one to two) colonize each IJ, and these bacteria reproduce within the lumen of the nematode gut to give a mature

Table 4.5. Bacterial symbionts associated with *Steinernema* species.

Serial number	Entomopathogenic nematode species	Bacterial symbiont	References
1	*Steinernema abbasi*	*Xenorhabdus indica*	a
2	*S. aciari*	*X. ishibashii*	a
3	*S. affine* (= *affinis*) syn. *Neoaplectana affinis*	*X. bovienii*	a
4	*S. akhursti*	Undescribed	a
5	*S. anatoliense*	*X. bovienii*	a
6	*S. apuliae*	*X. kozodoii*	a
7	*S. arasbaranense*	Undescribed	–
8	*S. arenarium* (= *anomalae* = *anomali*)	*X. kozodoii*	a
9	*S. ashiuense*	Undescribed	a
10	*S. asiaticum*	Undescribed	a
11	*S. australe*	*X. magdalenensis*	Tailliez *et al*., 2012
12	*S. backanense*	Undescribed	b
13	*S. balochiense*	Undescribed	–
14	*S. beddingi*	Undescribed	a
15	*S. bicornutum*	*X. budapestensis*	a
16	*S. bifurcatum*	*X. indica*	Fayyaz *et al*., 2014
17	*S. boemarei*	Undescribed	b
18	*S. brazilense*	Undescribed	b
19	*S. carpocapsae*	*X. nematophila*	a
20	*S. caudatum*	Undescribed	b
21	*S. ceratophorum*	*X. budapestensis*	a
22	*S. cholashanense*	Undescribed	b
23	*S. citrae* Stokwe	Undescribed	–
24	*S. colombiense*	Undescribed	b
25	*S. costaricense*	*X. szentirmaii*	a
26	*S. cubanum* (= *cubana*)	*X. poinarii*	a
27	*S. cumgarense*	Undescribed	–
28	*S. diaprepesi*	*X. doucetiae*	a
29	*S. dharanaii*	Undescribed	–
30	*S. eapokense*	Undescribed	b
31	*S. ethiopense*	Undescribed	–
32	*S. everestense*	Undescribed	b
33	*S. feltiae*	*X. bovienii*	a
34	*S. glaseri*	*X. poinarii*	a
35	*S. guangdongense*	Undescribed	b
36	*S. hebeiense*	Undescribed	b
37	*S. hermaphroditum*	*X. griffiniae*	a
38	*S. huense*	Undescribed	Phan *et al*., 2014
39	*S. ichnusae*	Undescribed	b
40	*S. innovationi*	Undescribed	–
41	*S. intermedium* (= *intermedia*)	*X. bovienii*	a
42	*S. jollieti*	*X. bovienii*	a
43	*S. jeffreyense*	Undescribed	–
44	*S. karii*	*X. hominickii*	a
45	*S. khoisanae*	*X. khoisanae*	Ferreira *et al*., 2013
46	*S. kraussei*	*X. bovienii*	a
47	*S. kushidai*	*X. japonica*	a
48	*S. lamjungense*	Undescribed	–

Continued

Table 4.5. Continued.

Serial number	Entomopathogenic nematode species	Bacterial symbiont	References
49	*S. leizhouense*	Undescribed	a
50	*S. litorale*	*X. bovienii*	a
51	*S. loci*	Undescribed	a
52	*S. longicaudum*	*X. ehlersii*	a
53	*S. meghalayensis*	*X. meghalayensis*	Kumar, 2011
54	*S. monticolum*	*X. hominickii*	a
55	*S. minutum*	*X. stockiae*	b
56	*S. neocurtillae* (= *neocurtillis*)	Undescribed	a
57	*S. oregonense* (= *oregonensis*)	*X. bovienii*	a
58	*S. pakistanense*	Undescribed	a
59	*S. papillatum*	Undescribed	–
60	*S. phyllophagae*	Undescribed	–
61	*S. poinari*	Undescribed	–
62	*S. puertoricense* (= *puertoricensis*)	*X. romanii*	a
63	*S. pui*	Undescribed	b
64	*S. puntauvense*	*X. bovienii*	a
65	*S. rarum* (= *rara*)	*X. szentirmaii*	a
66	*S. riobrave* (= *riobravis*)	*X. cabanillasii*	a
67	*S. ritteri*	Undescribed	a
68	*S. robustispiculum*	Undescribed	a
69	*S. sacchari*	Undescribed	–
70	*S. sangi*	Undescribed	a
71	*S. sasonense*	Undescribed	a
72	*S. scapterisci*	*X. innexi*	a
73	*S. scarabaei*	*X. koppenhoeferii*	a
74	*S. serratum*	*X. ehlersii*	a
75	*S. siamkayai*	*X.sockiae*	a
76	*S. sichuanense*	Undescribed	a
77	*S. silvaticum*	*X. bovienii*	a
78	*S. tami*	Undescribed	a
79	*S. texanum*	Undescribed	–
80	*S. thanhi*	Undescribed	a
81	*S. thermophilum*	*X. indica*	a
82	*S. tielingense*	Undescribed	–
83	*S. tophus*	Undescribed	Çimen *et al.*, 2014
84	*S. unicornum*	Undescribed	b
85	*S. vulcanicum*	Undescribed	–
86	*S. websteri*	*X. nematophila*	a
87	*S. weiseri*	*X. bovienii*	a
88	*S. xinbinense*	Undescribed	b
89	*S. xueshanense*	Undescribed	a
90	*S. yirgalemense*	*X. indica*	a

Notes: [a]Stock and Blair, 2008; [b]Lewis and Clarke, 2012.

population of 50–150 cfu/IJ (Ciche and Ensign, 2003). Hence, colonization in *Heterorhabditis* IJs consists of three stages: (i) colonization of the rectal gland of the mother; (ii) colonization of the IJ gut; and (iii) outgrowth.

4.4 Behaviour

Behaviour is the integration of sensory and motor inputs of nematodes, which are assessed in terms of locomotion, movement, feeding, mating, migration and penetration (Yeates, 2004). It involves the process

Table 4.6. Bacterial symbionts associated with *Heterorhabditis* species.

Serial number	Entomopathogenic nematode species	Bacterial symbiont	References
1	*Heterorhabditis amazonensis*	Undescribed	a
2	*H. argentinensis*	*Photorhabdus temperata*	b
3	*H. atacamensis*	Undescribed	b
4	*H. bacteriophora*	*P. luminescens* subsp. *luminescens* subspecies including *laumondii TT01*, *kayaii*, *thracensis*	a
5	*H. baujardi*	*P. luminescens*	a
6	*H. beicherriana*	Undescribed	b
7	*H. brevicaudis* (= species inquirenda, may be synonym of *Heterorhabditis indica*)	*P. luminescens* subsp. *akhurstii*	a
8	*H. downesi*	*P. temperata*	a
9	*H. floridensis*	*P. luminescens* subsp. *luminescens*	a
10	*H. georgiana*	*P. luminescens akhurstii*	b
11	*H. gerrardi*	*P. asymbiotica*	b
12	*H. hawaiiensis*	Undescribed	b
13	*H. indica* (= *indicus*)	*P. luminescens* subsp. *akhurstii*	a
14	*H. marelatus*	Undescribed	a
15	*H. megidis*	*P. temperata*	a
16	*H. mexicana*	Undescribed	a
17	*H. noenieputensis*	Undescribed	b
18	*H. poinari*	Undescribed	a
19	*H. safricana*	Undescribed	b
20	*H. sonorensis*	*P. luminenscens* subsp. *sonorensis*	b
21	*H. taysearae*	Undescribed	a
22	*H. zealandica*	*P. heterorhabditis*	a

Notes: [a]Stock and Blair, 2008; [b]Jagdale, 2013.

Table 4.7. Bacterial symbionts associated with *Heterorhabditidoides* species.

Serial number	Entomopathogenic nematode species	Bacterial symbiont	References
1	*Heterorhabditidoides chongmingensis*	*Serratia nematodiphila*	Zhang *et al.*, 2009
2	*H. rugaoensis*	*S. nematodiphila*	Zhang *et al.*, 2012

such as perception, wherein nematodes' sense variations are associated with their internal and external environment and respond to the changes they perceive. Behaviours are classified into two types: operational (voluntary) and consequential (influenced by stimuli). Operational behaviour refers to the exogenous or endogenous activities of nematodes such as movement, hatching, vulval contractions and stylet movement. It brings similar spatio-temporal organizations into a single group, which includes its body postures, wave patterns, movement in the egg, muscular movement during defecation, swarming, nictation, orientation and penetration. A behavioural process may be simple (e.g. determining the number of muscles involved in coordinating a particular activity) or complicated (e.g. the different movements involved in a single type of behaviour such as orientation, feeding or copulation). Various subcomponents of behaviour such as crawling, forward, backward, sideward or lateral movements, coiling and head and tail movements may be considered to simplify the classification. Behaviour by consequence is extremely varied in nematodes, since they are governed by various endogenous activities, physical and chemical stimuli that are sensory in nature. Changes in behaviour due to chemical (e.g. kairomones, allomones, sex pheromones) or physical (e.g. temperature, moisture, electricity, evasive actions taken to avoid predation) stimuli are examples of behaviour by consequence (Gaugler and Bilgrami, 2004; Yeates, 2004).

4.4.1 Foraging behaviour

The foraging strategy of EPNs decides which species is to be used for the management of specific target pests. Cruiser foraging species are best suited to find immobile hosts, including those present deep in soil or in other cryptic habitats, whereas ambush foragers are highly helpful in controlling mobile insects (Gaugler *et al.*, 1997).

Ambusher

Ambushers such as *S. carpocapsae* and *Steinernema scapterisci* have adapted 'sit and wait' behaviour to infect mobile insects (billbugs, webworms, cutworms, mole crickets and army worms), when they come in contact with the nematode on the surface of the soil. These nematodes do not respond to host released cues. Infective juveniles of some *Steinernema* spp. can stand on their tails (body waving (= nictate)) and infect passing insect hosts easily by jumping on them. Since highly mobile insects live in the upper soil, ambushers are generally effective in infecting insects on the surface than those present in deep soil. (Campbell and Kaya, 2000; Bal *et al.*, 2014).

Cruiser

Cruiser nematodes such as *H. bacteriophora*, *H. megidis*, *S. glaseri* and *S. kraussei* move actively in search of hosts and are distributed throughout the soil profile. They are more effective against less mobile hosts such as white grubs and black vine weevils. Cruisers never nictate, but respond to carbon dioxide released by insects as cues (Lewis *et al.*, 1993).

Intermediate

Some nematode species such as *S. feltiae* and *S. riobrave* have adapted a new strategy called the intermediate strategy, which is in between ambush and cruise strategies, to attack both the mobile and sedentary insects present at the surface or deep in the soil. *S. feltiae* is highly effective against fungus gnats and mushroom flies, whereas *S. riobrave* is effective against corn earworms, citrus root weevils and mole crickets (Lewis *et al.*, 1992; Campbell and Kaya, 2000).

The adoption of a certain foraging strategy is correlated with other aspects of parasite ecology, behaviour, physiology and anatomy, and thus influences how parasites interact with hosts (Campbell and Lewis, 2002; Ramos-Rodriguez *et al.*, 2007). Host selection by parasitic nematodes follows a hierarchial series of steps that include search for host habitat, host finding, host acceptance and host suitability (Doutt, 1964). The specific behaviours and cues used by IJs to search for the hosts vary, depending on the nematode species. Typically, ambushers that use a 'sit and wait' foraging strategy respond to volatile chemical cues from the host (when in a standing posture) by changing their behaviour to enhance contact with a passing insect (Campbell and Kaya, 2000). However, they are not attracted to host volatile cues while crawling on the substrate, unless they have already contacted a host (Lewis *et al.*, 1995). In contrast, cruise foragers that actively search for hosts respond readily to volatile chemicals cues by crawling towards the source of the cues (Lewis *et al.*, 1992, 1993; Campbell and Kaya, 2000).

4.4.2 Host recognition

Factors such as CO_2, temperature gradients and host faeces have been shown to influence the behaviour of IJs (Gaugler *et al.*, 1980; Byers and Poinar, 1982; Grewal *et al.*, 1993; Lewis *et al.*, 1993). The distribution of nematodes in soil is often clumped (Stuart and Gaugler, 1994; Glazer *et al.*, 1996; Campbell *et al.*, 1998). Therefore, IJs have the potential to encounter hosts that are already infected. The decision of whether or not to invade an already infected host has important fitness consequences for the individual IJ because survival and reproduction are determined by the host environment, and infection decisions are irreversible.

For an IJ encountering an insect, prior conspecific infection may facilitate host exploitation because the host immune system may already be compromised and potential mates may be present. Alternatively, infecting an already infected host, particularly at times late in the infection process, may lead to increased competition for diminished resources, or even the lack of sufficient nutrients in a host cadaver to complete development (Selvan *et al.*, 1993). In addition to hosts infection by conspecifics, nematodes may also encounter host infection by heterospecific individuals. These costs and benefits could lead to selection for the ability to distinguish between infected or uninfected hosts, and among hosts at different times after infection. It is hypothesized that early in an infection by conspecific nematodes, a host should be of higher quality, and therefore more attractive to an infective stage, than a host late in the infection process, which is predicted to produce cues that suppress nematode infection. There is some evidence that waste products from hosts late in an infection repel nematodes (Kunkel *et al.*, 2006).

4.4.3 Response of infective juveniles to insects

The decision to invade an insect depends not only on external factors, such as the cues from potential hosts, but also on the internal state of the IJ. Bohan and Hominick (1996, 1997) described fluctuations in infectivity of *S. feltiae* wherein a proportion of IJs switched between a non-infectious and an infectious state. An increase in the infectivity of *H. megidis* in the initial weeks after emerging from the host was attributed to a gradual change in the infection tendency of individual IJs, rather than a switch between states (Griffin, 1996; Dempsey and Griffin, 2002; Ryder and Griffin, 2003). Taking a modelling approach to the dynamics of infection, IJs were distinguished in a population at a given time based on their sensitivity in two categories: (i) risk taker and (ii) risk averse (Fushing *et al.*, 2008). Risk takers invade the host readily by overcoming the host's defence, while risk-averse individuals wait until the host has been invaded by others. In the case of *S. feltiae*, a majority of individuals were risk averse, waiting until the host was invaded. This was similar to the scenario described by Hay and Fenlon (1995), whose modified binomial model indicated that a sizeable proportion of *S. feltiae* IJs invaded the host secondarily but could not initiate infection on their own. But in the case of *S. carpocapsae*, the distinction into risk-prone and risk-averse individuals was seen only in response to certain hosts, indicating certain flexibility in infection. These models can be useful only in exploring phased infectivity. An increase in the proportion of risk takers in a population can be correlated with an increase in the overall infectivity of a population. IJs delay infectivity for some time, during which they (or their competitors) migrate away from the host from which they have emerged en masse, to avoid overcrowding (Dempsey and Griffin, 2002).

4.5 Conclusion

Entomopathogenic nematodes, a versatile group of invertebrate animals, are routinely being debated to expose the intriguing properties with respect to their target-oriented action on a range of diversified insect taxa worldwide. *Xenorhabdus* spp. and *Photorhabdus* spp. are the symbionts of the most exploited EPNs, Steinernematids and Heterorhabditids; both have mutually benefited from their existence. For the successful survival of cohabitance, nematode–bacteria require a third-party living host, insects. It is well documented that the degree of efficacy of EPNs varies with nematode species, host insect and the ecological niche where these interactions take place. EPNs adapt to a wide range of survival, reproduction, behavioural and host recognition strategies. Understanding the different aspects of the organisms is a prime criterion to be included as a component in the biomanagement programme, which requires some acceptable traits that can comply with an economically viable approach. Hence, it is suggested that the selection of an EPN for controlling a target insect is of foremost importance in the control strategy, which is based on the foraging strategies of the EPN and the mobility of the insects.

Acknowledgements

The senior author gratefully acknowledges the Director, ICAR-Central Institute for Cotton Research, Nagpur, for the facilities rendered to carry out the research on EPNs at the ICAR-Central Institute for Cotton Research, Regional Station, Coimbatore, Tamil Nadu, India, to compile this chapter. The co-authors extend their sincere gratitude to the Department of Nematology, Tamil Nadu Agricultural University, Coimbatore, and the Vegetable Research Station, Palur, in providing the facilities to compile this chapter.

References

Adams, B.J and Nguyen, K.B. (2002) Taxonomy and systematics. In: Gaugler R. (ed.) *Entomopathogenic Nematology*. CAB International, Wallingford, UK, pp. 1–34.

Ali, F. and Wharton, D.A. (2013) Cold tolerance abilities of two entomopathogenic nematodes, *Steinernema feltiae* and *Heterorhabditis bacteriophora*. *Cryobiology* 66, 24–29.

Ali, S.S., Shaheen, A., Pervez, R. and Hussain, M.A. (2005) *Steinernema masoodi* sp. n. and *Steinernema seemae* sp.n. (Nematoda: Rhabditida: Steinernematidae) from India. *International Journal of Nematology* 15, 89–99.

Andalo, V., Nguyen, K.B. and Moino, A. Jr (2006) *Heterorhabditis amazonensis* n. sp. (Rhabditida: Heterorhabditidae) from Amazonas, Brazil. *Nematology* 8, 853–867.

Anis, M., Shahina, F., Reid, A.P and Rowe, J. (2002) *Steinernema asiaticum* sp. n. (Rhabditida: Steinernematidae) from Pakistan. *International Journal of Nematology* 12, 220–231.

Artyukhovsky, A.K. (1967) *Neoaplectana arenaria* nov. sp. (Steinernematidae, Nematoda) inducing nematode disease in chafers of the Voronezh region. *Trudy Voroneszhskogo Gosudarstvennogo Zapovednika* 15, 94–100.

Askary, T.H. (2010) Nematodes as biocontrol agents. In: Lichtfouse, E. (ed.) *Sociology, Organic Farming, Climate Change and Soil Science*. Springer, The Netherlands, pp. 347–378.

Bal, H.K., Taylor, R.A.J. and Grewal, P.S (2014) Ambush foraging entomopathogenic nematodes employ 'sprinters' for long-distance dispersal in the absence of hosts. *Journal of Parasitology* 100, 422–432.

Barrett, J. (1991) Anhydrobiotic nematodes. *Agricultural Zoology Reviews* 4, 161–176.

Bastidas, B., Edgar, P. and San-Blas, E. (2014) Size does matter: the life cycle of *Steinernema* spp. in micro-insect hosts. *Journal of Invertebrate Pathology* 121, 46–55.

Baur, M.E., Kaya, H.K. and Thurston, G.S. (1995) Factors affecting entomopathogenic infection of *Plutella xylostella* on a leaf surface. *Entomologia Experimentalis et Applicata* 77, 239–250.

Bell, G. (1982) *The Masterpiece of Nature: The Evolution and Genetics of Sexuality*. University of California Press, Berkeley, California.

Bennett, H.P.J. and Clarke, D.J. (2005) The pbgPE operon in *Photorhabdus luminescens* is required for pathogenicity and symbiosis. *Journal of Bacteriology* 187, 77–84.

Bird, A.F. and Akhurst, R.J. (1983) The nature of the intestinal vesicle in nematodes of the family Steinernematidae. *International Journal of Parasitology* 13, 599–606.

Blas, S.E., Portillo, E., Nermut, J., Puza, V. and Montero, M.P (2015) *Steinernema papillatum* n. sp. (Rhabditida: Steinernematidae), a new entomopathogenic nematode from Venezuela. *Nematology* 17, 1081–1097.

Bohan, D.A. and Hominick, W.M. (1996) Investigations on the presence of an infectious proportion amongst populations of *Steinernema feltiae* (Site 76 strain) infective stages. *Parasitology* 112, 113–118.

Bohan, D.A. and Hominick, W.M. (1997) Long-term dynamics of infectiousness within the infective stage pool of the entomopathogenic nematode *Steinernema feltiae* (Site 76 strain) Filipjev. *Parasitology* 114, 301–308.

Bovien, P. (1937) Some type of association between nematodes and insects. *Videnskabelige Meddelelser Dansk Naturhistorisk Forening* 101, 1–114.

Bowen, D.J. and Ensign, J.C. (2001) Isolation and characterization of intracellular protein inclusions produced by the entomopathogenic bacterium, *Photorhabdus luminescens*. *Applied Environmental Microbiology* 67, 4834–4841.

Brameyer, S., Kresovic, D., Bode, H.B. and Heermann, R. (2014) Lux R solos in *Photorhabdus* species. *Frontiers in Cellular and Infection Microbiology* 4, 1–11.

Brown, I.M. and Gaugler, R. (1996) Cold tolerance of steinernematid and heterorhabditid nematodes. *Journal of Thermal Biology* 21, 115–121.

Brown, I.M. and Gaugler, R. (1997) Temperature and humidity influence emergence and survival of entomopathogenic nematodes. *Nematologica* 43, 363–375.

Byers, J.A. and Poinar, G.O. Jr (1982) Location of insect hosts by the nematode, *Neoaplectana*, in response to temperature. *Behaviour* 79, 1–10.

Cabanillas, H.E., Poinar, G.O. Jr and Raulston, J.R. (1994) *Steinernema riobravis* n. sp. (Rhabditida: Steinernematidae) from Texas. *Fundamental and Applied Nematology* 17, 123–131.

Cagnolo, S.R., Donari, Y.M. and Di Rienzo, J.A. (2004) Existence of infective juveniles in the offspring of first- and second-generation adults of *Steinernema rarum* (OLI strain): evaluation of their virulence. *Journal of Invertebrate Pathology* 85, 33–39.

Campbell, L.R. and Gaugler, R. (1991) Role of the sheath in desiccation tolerance of two entomopathogenic nematodes. *Nematologica* 37, 324–332.

Campbell, L.R. and Gaugler, R. (1992) Effect of exsheathment on motility and pathogenicity of two entomopathogenic nematode species. *Journal of Nematology* 24, 365–370.

Campbell, J.F. and Kaya, H.K. (2000) Influence of insect associated cues on the jumping behavior of entomopathogenic nematodes (*Steinernema* spp.). *Behaviour* 137, 591–609.

Campbell, J.F. and Lewis, E.E. (2002) Entomopathogenic nematode host-search strategies. In: Lewis, E.E., Campbell, J.F. and Sukhdeo, M.V.K. (eds) *The Behavioural Ecology of Parasites*. CAB International, Wallingford, UK, pp. 13–38.

Campbell, J.F., Orza, G., Yoder, F., Lewis, E.E. and Gaugler, R. (1998) Spatial and temporal distribution of endemic and released entomopathogenic nematode populations in turfgrass. *Entomologia Experimentalis et Applicata* 86, 1–11.

Chaston, J.M., Murfin, K.E., Heath-Heckman, E.A. and Goodrich-Blair, H. (2013) Previously unrecognized

stages of species-specific colonization in the mutualism between *Xenorhabdus* bacteria and *Steinernema* nematodes. *Cellular Microbiology* 15, 1545–1559.
Chen, S., Li, X., Yan, A., Spiridonov, S.E. and Moens, M. (2006) A new entomopathogenic nematode, *Steinernema hebeiense* sp. n. (Rhabditida: Steinernematidae), from North China. *Nematology* 8, 563–574.
Chhetri, H.B.K., Waeyenberge, L., Spiridonov, S.E., Manandhar, H.K. and Moens, M. (2011a) *Steinernema everestense* n. sp. (Rhabditida: Steinernematidae), a new species of entomopathogenic nematode from Pakhribas, Dhankuta, Nepal. *Nematology* 13, 443–462.
Chhetri, H.B.K., Waeyenberge, L., Spiridonov, S.E., Manandhar, H.K. and Moens, M. (2011b) *Steinernema lamjungense* n. sp. (Rhabditida: Steinernematidae), a new species of entomopathogenic nematode from Lamjung district, Nepal. *Nematology* 13, 589–605.
Ciche, T.A. and Ensign, J.C. (2003) For the insect pathogen *Photorhabdus luminescens*, which end of a nematode is out? *Applied Environmental Microbiology* 69, 1890–1897.
Ciche, T.A., Kim, K.S., Kaufmann-Daszczuk, B., Nguyen, K.C and Hall, D.H. (2008) Cell invasion and matricide during *Photorhabdus luminescens* transmission by *Heterorhabditis bacteriophora* nematodes. *Applied Environment Microbiology* 74, 2275–2287.
Çimen, H., Lee, M.M., Hatting, J., Hazir, S. and Stock, S.P. (2014) *Steinernema tophus* sp. n. (Nematoda: Steinernematidae), a new entomopathogenic nematode from South Africa. *Zootaxa* 3821, 337–353.
Çimen, H., Lee, M.M., Hatting, J., Hazir, S. and Stock, S.P. (2015). *Steinernema innovationi* n. sp. (*Panagrolaimomorpha*: Steinernematidae), a new entomopathogenic nematode species from South Africa. *Journal of Helminthology* 89, 415–427.
Clarke, D.J. (2008) *Photorhabdus*: a model for the analysis of pathogenicity and mutualism. *Cell Microbiology* 10, 2159–2167.
Clausi, M., Longo, A., Rappazzo, G., Tarasco, E. and Vinciguerra, M.T. (2011) *Steinernema vulcanicum* n. sp. (Rhabditida: Steinernematidae), a new entomopathogenic nematode species from Sicily (Italy). *Nematology* 13, 409–423.
Cowles, C.E. and Goodrich-Blair, H. (2004) Characterization of a lipoprotein, NilC, required by *Xenorhabdus nematophila* for mutualism with its nematode host. *Molecular Microbiology* 54, 464–477.
Cowles, C.E and Goodrich-Blair, H. (2006) H. *nilR* is necessary for co-ordinate repression of *Xenorhabdus nematophila* mutualism genes. *Molecular Microbiology* 62, 760–771.
Cowles, C.E and Goodrich-Blair, H. (2008) The *Xenorhabdus nematophila nilABC* genes confer the ability of *Xenorhabdus* spp. to colonize *Steinernema carpocapsae* nematodes. *Journal of Bacteriology* 190, 4121–4128.
Cowles, K.N., Cowles, C.E., Richards, G.R., Martens, E.C and Goodrich-Blair, H. (2007) The global regulator Lrp contributes to mutualism, pathogenesis and phenotypic variation in the bacterium *Xenorhabdus nematophila*. *Cell Microbiology* 9, 1311–1323.
Cranshaw, W.S. and Zimmerman, R. (2013) Insect Parasitic Nematodes. Colorado State University Extension, June 2013.
Cutler, G.C. and Stock, S.P. (2003) *Steinernema websteri* sp. n. (Rhabditida: Steinernematidae), a new entomopathogenic nematode from China. *Nematologia Mediterranea* 31, 215–224.
de Doucet, M.M.A. (1986) A new species of *Neoaplectana* Steiner, 1929 (Nematoda: Steinernematidae) from Cordoba, Argentina. *Revue de Nematologie* 9, 317–323.
de Doucet, M.M.A. and Doucet, M. (1990) *Steinernema ritteri* n. sp. (Nematoda: Steinernematidae) with a key to the species of the genus. *Nematologica* 36, 257–265.
Dempsey, C.M. and Griffin, C.T. (2002) Phased activity in *Heterorhabditis megidis* infective juveniles. *Parasitology* 124, 605–613.
Denlinger, D.L. (2002) Regulation of diapause. *Annual Review of Entomology* 47, 93–122.
Doutt, R.L. (1964) Biological characteristics of entomophagous adults. In: DeBach, P. (ed.) *Biological Control of Insect Pests and Weeds*. Reinhold, New York, pp. 145–167.
Dunphy, G.B. and Thurston, G.S. (1990) Insect immunity. In: Gaugler, R. and Kaya, H.K. (eds) *Entomopathogenic Nematodes in Biological Control*. CRC Press, Boca Raton, Florida, pp. 301–318.
Edgington, S., Buddie, A.G., Tymo, L. and Moore, D. (2009a) *Steinernema australe* n. sp. (Panagrolaimomorpha: Steinernematidae), a new entomopathogenic nematode from Isla Magdalena, Chile. *Nematology* 11, 699–717.
Edgington, S., Buddie, A.G., Tymo, L.M., France, A., Merino, L. and Hunt, D.J. (2009b) *Steinernema unicornum* sp. n. (Panagrolaimomorpha: Steinernematidae), a new entomopathogenic nematode from Tierra del Fuego, Chile. *Journal of Nematode Morphology and Systematics* 12, 113–131.
Edgington, S., Buddie, A.G., Moore, D., France, A., Merino, L. and Hunt, D.J. (2011) *Heterorhabditis atacamensis* n. sp. (Nematoda: Heterorhabditidae), a new entomopathogenic nematode from the Atacama Desert, Chile. *Journal of Helminthology* 85, 381–394.
Elawad, S.A., Ahmad, W. and Reid, A. (1997) *Steinernema abbasi* sp. n. (Nematoda: Steinernematidae) from the Sultanate of Oman. *Fundamental and Applied Nematology* 20, 433–442.
Emelianoff, V., Chapuis, E., Le Brun, N., Chiral, M., Moulia, C. and Ferdy, J.B. (2008) A survival-reproduction trade-off in entomopathogenic nematodes mediated by their bacterial symbionts. *Evolution* 62, 932–942.

Fayyaz, S., Yan, X., Qiu, L., Han, R., Gulsher, M., *et al.* (2014) A new entomopathogenic nematode, *Steinernema bifurcatum* n. sp. (Rhabditida: Steinernematidae) from Punjab, Pakistan. *Nematology* 16, 821–836.

Fayyaz, S., Khanum, T.A., Ali, S., Solangi, G.S., Gulsher, M. and Javed, S. (2015) *Steinernema balochiense* n. sp. (Rhabditida: Steinernematidae), a new entomopathogenic nematode from Pakistan. *Zootaxa* 3904, 387–402.

Ferreira, T., Van Reenen, C.A., Endo, A., Sprooer, C., Malan, A.P. and Dicks, L.M.T. (2013) Description of *Xenorhabdus khoisanae* sp. nov., the symbiont of the entomopathogenic nematode, *Steinernema khoisanae. International Journal of Systematic and Evolutionary Microbiology* 63, 3220–3224.

ffrench-Constant, R.,Waterfield, N., Daborn, P., Joyce, S., Bennett, H., *et al.* (2003) *Photorhabdus:* towards a functional genomic analysis of a symbiont and pathogen. *FEMS Microbiology Review* 26, 433–456.

Filipjev, J. (1934) Miscellanea Nematologica 1. Eine neue Art der Gattung *Neoaplectana* Steiner nebst Bermerkunge über die systematische Stellung der letzteren. *Parasitologicheskii Sbornik Zoologicheskogo Instituta Akademii Nauk SSSR* 4, 229–240.

Forst, S., Dowds, B., Boemare, N. and Stackebrandt, E. (1997) *Xenorhabdus* and *Photorhabdus* spp.: bugs that kill bugs. *Annual Review of Microbiology* 51, 47–72.

Fushing, H., Zhu, L., Shapiro-Ilan, D.I., Campbell, J.F. and Lewis, E.E. (2008) State-space based mass event-history model I: many decision-making agents with one target. *Annals of Applied Statistics* 2, 1503–1522.

Ganguly, S. and Singh, L.K. (2000) *Steinernema thermophilum* sp. n. (Rhabditida: Steinernematidae) from India. *International Journal of Nematology* 10, 183–191.

Ganguly, S., Rathour, K.S., Kumar, S. and Singh, M. (2011) *Steinernema meghalayensis* sp. n. (Rhabditida: Steinernematidae) from northeastern hilly region of India. *Indian Journal of Nematology* 41, 83–97.

Gardner, S.L., Stock, S.P. and Kaya, H.K. (1994) A new species of *Heterorhabditis* from the Hawaiian islands. *Journal of Parasitology* 80, 100–106.

Gaugler, R. and Bilgrami, A.L. (2004). *Nematode Behavior*. CAB International, Wallingford, UK, 439 pp.

Gaugler, R., LeBeck, L., Nakagaki, B. and Boush, G.M. (1980) Orientation of the entomogenous nematode, *Neoaplectana carpocapsae* to carbon dioxide. *Environmental Entomology* 9, 649–652.

Gaugler, R., Lewis, E. and Stuart, R.J. (1997) Ecology in the service of biological control: the case of entomopathogenic nematodes. *Oecologia* 109, 483–489.

Glazer, I. (1992) Survival and efficacy of *Steinernema carpocapsae* in an exposed environment. *Biocontrol Science and Technology* 2, 101–107.

Glazer, I., Kozodoi, E., Salame, L. and Nestel, D. (1996) Spatial and temporal occurrence of natural populations of *Heterorhabditis* spp. (Nematoda: Rhabditida) in a semiarid region. *Biological Control* 6, 130–136.

Goetsch, M., Owen, H., Goldman, B. and Forst, S. (2006) Analysis of the PixA inclusion body protein of *Xenorhabdus nematophila*. *Journal of Bacteriology* 188, 2706–2710.

Goodrich-Blair, H and Clarke, D.J. (2007) Mutualism and pathogenesis in *Xenorhabdus* and *Photorhabdus*: two roads to the same destination. *Molecular Microbiology* 64, 260–268.

Goodson, M.S., Kojadinovic, M., Troll, J.V., Scheetz, T.E., Casavant, T.L., *et al.* (2005) Identifying components of the NF-kappaB pathway in the beneficial *Euprymna scolopes–Vibrio fischeri* light organ symbiosis. *Applied Environmental Microbiology* 71, 6934–6946.

Grewal, P.S., Gaugler, R. and Selvan, S. (1993) Host recognition by entomopathogenic nematodes: behavioral response to contact with host feces. *Journal of Chemical Ecology* 19, 1219–1231.

Grewal, P.S., Selvan, S. and Gaugler, R. (1994) Thermal adaptation of entomopathogenic nematodes: niche breadth for infection, establishment, and reproduction. *Journal of Thermal Biology* 19, 245–253.

Grewal, P.S., Ehlers, R.-U. and Shapiro-Ilan, D.I. (2005) *Nematodes as Biocontrol Agents*. CAB International, Wallingford, UK.

Griffin, C.T. (1996) Effects of prior storage conditions on the infectivity of *Heterorhabditis* sp. (Nematoda: Heterorhabditidae). *Fundamental and Applied Nematology* 19, 95–102.

Hay, D.B. and Fenlon, J.S. (1995) A modified binomial model that describes the infection dynamics of the entomopathogenic nematode *Steinernema feltiae* (Steinernematidae; Nematoda). *Parasitology* 111, 627–633.

Hazir, S., Stock, S.P. and Keskin, N. (2003) A new entomopathogenic nematode, *Steinernema anatoliense* n. sp. (Rhabditida: Steinernematidae), from Turkey. *Systematic Parasitology* 55, 211–220.

Herbert, E.E. and Goodrich-Blair, H. (2007) Friend and foe: the two faces of *Xenorhabdus nematophila*. *Nature Reviews Microbiology* 5, 634–646.

Heungens, K., Cowles, C.E. and Goodrich-Blair, H. (2002) Identification of *Xenorhabdus nematophila* genes required for mutualistic colonization of *Steinernema carpocapsae* nematodes. *Molecular Microbiology* 45, 1337–1353.

Hill, M.P., Malan, A.P. and Terblanche, J.S. (2015) Divergent thermal specialisation of two South African Entomopathogenic Nematodes. *PeerJ PrePrints* https://dx.doi.org/10.7287/peerj.preprints.1152v1 (accessed 6 September 2016).

Hu, K., Li, J. and Webster, J.M. (1999) Nematicidal metabolites produced by *Photorhabdus luminescens* (Enterobacteriaceae), bacterial symbiont of entomopathogenic nematodes. *Nematology* 1, 457–469.

Jagdale, G.B. (2013) Described *Heterorhabditis* species and their symbiotic bacteria. Available at: http://blog.

bugsforgrowers.com/natural-predators/entomopathogenic-nematodes/described-heterorhabditis-species (accessed 6 September 2016).

Jank, T., Bogdanovic, X., Wirth, C., Haaf, E., Spoerner, M., *et al*. (2013) A bacterial toxin catalyzing tyrosine glycosylation of Rho and deamidation of Gq and Gi proteins. *Nature Structural and Molecular Biology* 20, 1273–1284.

Jian, B., Reid, A.P. and Hunt, D.J. (1997) *Steinernema ceratophorum* n. sp. (Nematoda: Steinernematidae), a new entomopathogenic from north east China. *Systematic Parasitology* 37, 115–125.

Johnigk, S.A. and Ehlers, R.-U. (1999) *Endotokia matricida* in hermaphrodites of *Heterorhabditis* spp. and the effect of the food supply. *Nematology* 1, 717–726.

Joyce, S.A., Watson, R.J. and Clarke, D.J. (2006) The regulation of pathogenicity and mutualism in *Photorhabdus*. *Current Opinion in Microbiology* 9, 127–132.

Kakulia, G. and Mikaia, N. (1997) New species of the nematode *Heterorhabditis poinari* sp. nov. (Rhabditida: Heterorhabditidae). *Bulletin of the Georgian Academy of Sciences* 155, 457–459.

Kanost, M.R., Jiang, H. and Yu, X.Q. (2004) Innate immune responses of a lepidopteran insect, *Manduca sexta*. *Immunological Reviews* 198, 97–105.

Kaya, H.K. (1990) Soil ecology. In: Gaugler, R. and Kaya, H.K. (eds) *Entomopathogenic Nematodes in Biological Control*. CRC Press, Boca Raton, Florida, 93–177.

Kaya, H.K. and Gaugler, R. (1993) Entomopathogenic nematodes. *Annual Review of Entomology* 38, 181–206.

Koppenhöfer, A.M. (2007) Nematodes. In: Lacey, L.A. and Kaya, H.K. (eds) *Field Manual of Techniques in Invertebrate Pathology: Application and Evaluation of Pathogens for Control of Insects and other Invertebrate Pests*. Second edn. Springer, The Netherlands, pp. 249–264.

Kozodoi, E.M. (1984) A new entomophagous nematode *Neoaplectana anomali* sp. n. (Rhabditida: Steinernematidae) and its biology. *Zoologich-esky Zhurnal* 63, 1605–1609.

Kulkarni, N., Rizvi, A.N., Kumar, V., Paunikar, S. and Mishra, V.K. (2012) Morphological and molecular characterization of *Steinernema dharanaii* sp. n.: a new entomopathogenic nematode from India. *Indian Journal of Tropical Biodiversity* 20, 107–116.

Kumar, S. (2011) Identification and molecular characterization of symbiotic bacteria (*Xenorhabdus* sp.) associated with indigenous entomopathogenic nematode (*Steinernema* sp.). PhD Thesis, Department of Biotechnology, Kurukshetra University, Haryana, India.

Kung, S.P., Gaugler, R. and Kaya, H.K. (1991) Effects of soil temperature, moisture, and relative humidity on entomopathogenic nematode persistance. *Journal of Invertebrate Pathology* 57, 242–249.

Kunkel, B.A., Shapiro-Ilan, D.I., Campbell, J.F. and Lewis, E.E. (2006) Effect of *Steinernema glaseri* infected host exudates on movement of conspecific infective juveniles. *Journal of Invertebrate Pathology* 93, 42–49.

Lacey, L.A. and Georgis, R. (2012) Entomopathogenic nematodes for control of insect pests above and below ground with comments on commercial production. *Journal of Nematology* 4, 218–225.

Lee, M.M., Sicard, M., Skeie, M. and Stock, S.P. (2009) *Steinernema boemarei* n. sp. (Nematoda: Steinernematidae), a new entomopathogenic nematode from southern France. *Systematic Parasitology* 72, 127–141.

Leulier, F., Parquet, C., Pili-Floury, S., Ryu, J.H., Caroff, M., *et al*. (2003) The *Drosophila* immune system detects bacteria through specific peptidoglycan recognition. *Nature Immunology* 4, 478–484.

Lewis, E.E. and Clarke, D.J. (2012) Nematode parasites of entomopathogens. In: Veega, F.E. and Kaya, H.K. (eds) *Insect Pathology*. Second edn. Academic Press, San Diego, California, pp. 395–443.

Lewis, E.E., Gaugler, R. and Harrison, R. (1992) Entomopathogenic nematode host finding: response to host contact cues by cruise and ambush foragers. *Parasitology* 105, 309–315.

Lewis, E.E., Gaugler, R. and Harrison, R. (1993) Response of cruiser and ambusher entomopathogenic nematodes (Steinernematidae) in host volatile cues. *Canadian Journal of Zoology* 71, 765–769.

Lewis, E.E., Grewal, P.S. and Gaugler, R. (1995) Hierarchical order of host cues in parasite foraging: a question of context. *Parasitology* 110, 207–213.

Li, X.Y., Liu, Q.Z., Nermut, J., Puza, V. and Mracek, Z. (2012) *Heterorhabditis beicherriana* n. sp. (Nematoda: Heterorhabditidae), a new entomopathogenic nematode from the Shunyi district of Beijing, China. *Zootaxa* 3569, 25–40.

Liu, J. (1992) Taxonomic study of the genus *Steinernema* Travassos and *Heterorhabditis* Poinar. *Proceedings of the XIX International Congress of Entomology*, Beijing, 318 pp.

Liu, J. (1994) A new species of the genus *Heterorhabditis* from China (Rhabditida: Heterorhabditidae). *Acta Zootaxonomica Sinica* 19, 268–272.

Liu, J. and Berry, R.E. (1996a) *Steinernema oregonensis* n. sp. (Rhabditida: Steinernematidae) from Oregon, USA. *Fundamental and Applied Nematology* 19, 375–380.

Liu, J. and Berry, R.E. (1996b) *Heterorhabditis marelatus* n. sp. (Rhabditida: Heterorhabditidae) from Oregon. *Journal of Invertebrate Pathology* 67, 48–54.

López-Núñez, J.C., Plichta, K., Góngora-Botero, C. and Stock, S.P. (2008) A new entomopathogenic nematode, *Steinernema colombiense* n. sp. (Nematoda: Steinernematidae) from Colombia. *Nematology* 10, 561–574.

Luc, P.V., Nguyen, K.B., Reid, A.P. and Spiridonov, S.E. (2000) *Steinernema tami* sp. n. (Rhabditida: Steinernematidae) from Cat Tien Forest, Vietnam. *Russian Journal of Nematology* 8, 33–43.

Ma, J., Chen, S., Li, X., Han, R., Chhetri, H.B.K., De Clercq, P. and Moens, M. (2012a) A new entomopathogenic nematode, *Steinernema tielingense* n. sp (Rhabditida: Steinernematidae), from north China. *Nematology* 14, 321–338.

Ma, J., Chen, S., De Clercq, P., Waeyenberge, L., Han, R. and Moens, M. (2012b) A new entomopathogenic nematode, *Steinernema xinbinense* n. sp. (Nematoda: Steinernematidae), from north China. *Nematology* 14, 723–739.

Malan, A.P., Nguyen, K.B., deWaal, J.Y. and Tiedt, L. (2008) *Heterorhabditis safricana* n. sp. (Rhabditida: Heterorhabditidae) an entomopathogenic nematode from South Africa. *Nematology* 10, 381–396.

Malan, A.P., Knoetze, R. and Tiedt, L. (2014) *Heterorhabditis noenieputensis* n. sp. (Rhabditida: Heterorhabditidae), a new entomopathogenic nematode from South Africa. *Journal of Helminthology* 88, 139–151.

Malan, A.P., Knoetze, R. and Tiedt, L.R. (2016) *Steinernema jeffreyense* n. sp. (Rhabditida: Steinernematidae), a new entomopathogenic nematode from South Africa. *Journal of Helminthology* 90, 262–278.

Mamiya, Y. (1988) *Steinernema kushidai* n. sp. (Nematoda: Steinernematidae) associated with scarabaeid beetle larvae from Shizuoka, Japan. *Applied Entomology and Zoology* 23, 313–320.

Maneesakorn, P., Grewal, P.S. and Chandrapatya, A. (2010) *Steinernema minutum* sp. nov. (Rhabditida: Steinernematidae): a new entomopathogenic nematode from Thailand. *International Journal of Nematology* 20, 27–42.

Martens, E.C., Heungens, K. and Goodrich-Blair, H. (2003) Early colonization events in the mutualistic association between *Steinernema carpocapsae* nematodes and *Xenorhabdus nematophila* bacteria. *Journal of Bacteriology* 185, 3147–3154.

Mráček, Z., Hernandez, E.A. and Boemare, N.E. (1994) *Steinernema cubana* sp. n. (Nematoda: Rhabditida: Steinernematidae) and the preliminary characterization of its associated bacterium. *Journal of Invertebrate Pathology* 64, 123–129.

Mráček, Z., Sturhan, D. and Reid, A. (2003) *Steinernema weiseri* sp. n. (Rhabditida: Steinernematidae), a new entomopathogenic nematode from Europe. *Systematic Parasitology* 56, 37–47.

Mráček, Z., Nguyen, K.B., Tailliez, P., Boemare, N. and Shulong, C. (2006) *Steinernema sichuanense* n. sp. (Rhabditida: Steinernematidae), a new species of entomopathogenic nematode from the province of Sichuan, east Tibetan Mts., China. *Journal of Invertebrate Pathology* 93, 157–169.

Mráček, Z., Zhi, L.Q. and Nguyen, K.B. (2009) *Steinernema xueshanense* n. sp. (Rhabditida: Steinernematidae), a new species of entomopathogenic nematode from the province of Yunnan, southeast Tibetan Mts., China. *Journal of Invertebrate Pathology* 102, 69–78.

Mráček, Z., Puza, V. and Nermut, J. (2014) *Steinernema poinari* sp n. (Nematoda: Steinernematidae) a new entomopathogenic nematode from the Czech Republic. *Zootaxa* 3760, 336–350.

Mukuka, J., Strauch, O., Waeyenberge, L., Viaene, N., Moens, M. and Ehlers, R.-U. (2010) Heat tolerance among different strains of the entomopathogenic nematode *Heterorhabditis bacteriophora*. *Biocontrol* 55, 423–434.

Nguyen, K.B. and Buss, E.A. (2011) *Steinernema phyllophagae* n. sp. (Rhabditida: Steinernematidae), a new entomopathogenic nematode from Florida, USA. *Nematology* 13, 425–442.

Nguyen, K.B. and Duncan, L.W. (2002) *Steinernema diaprepesi* n. sp. (Rhabditida: Steinernematidae), a parasite of the root weevil *Diaprepes abbreviatus* (L) (Coleoptera: Curculionidae). *Journal of Nematology* 34, 159–170.

Nguyen, K.B. and Smart, G.C. Jr (1990) *Steinernema scapterisci* n. sp. (Steinernematidae: Nematoda). *Journal of Nematology* 22, 187–199.

Nguyen, K.B. and Smart, G.C. Jr (1992) *Steinernema neocurtillis* n. sp. (Rhabditida: Steinernematidae) and a Key to Species of the Genus *Steinernema*. *Journal of Nematology* 24, 463–477.

Nguyen, K.B. and Smart, G.C. Jr (1994) *Neosteinernema longicurvicauda* n. gen. n. sp. (Rhabditida: Steinernematidae), a parasite of the termite *Reticulitermes flavipes* (Koller). *Journal of Nematology* 26, 162–174.

Nguyen, K.B., Tesfamariam, M., Gozel, U., Gaugler, R. and Adams, B.J. (2004a) *Steinernema yirgalemense* n. sp. (Rhabditida: Steinernematidae) from Ethiopia. *Nematology* 6, 839–856.

Nguyen, K.B., Shapiro-Ilan, D.I., Stuart, R.J., McCoy, C.W., James, R.J. and Adams, B.J. (2004b) *Heterorhabditis mexicana* n. sp. (Rhabditida: Heterorhabditidae) from Tamaulipas, Mexico, and morphological studies of the bursa of *Heterorhabditis* spp. *Nematology* 6, 231–244.

Nguyen, K.B., Malan, A.P. and Gozel, U. (2006a) *Steinernema khoisanae* n. sp. (Rhabditida: Steinernematidae), a new entomopathogenic nematode from South Africa. *Nematology* 8, 157–175.

Nguyen, K.B., Qiu, L., Zhou, Y. and Pang, Y. (2006b) *Steinernema leizhouense* sp. n. (Nematoda: Steinernematidae), a new entomopathogenic nematode from southern China. *Russian Journal of Nematology* 14, 101–118.

Nguyen, K.B., Gozel, U., Koppenhofer, H.S. and Adams, B.J. (2006c) *Heterorhabditis floridensis* n. sp. (Rhabditida: Heterorhabditidae) from Florida. *Zootaxa* 1177, 1–19.

Nguyen, K.B., Stuart, R.J., Andalo, V., Gozel, U. and Rogers, M.E. (2007) *Steinernema texanum* n. sp. (Rhabditida: Steinernematidae), a new entomopathogenic nematode from Texas, USA. *Nematology* 9, 379–396.

Nguyen, K.B., Puza, V. and Mracek, M. (2008a) *Steinernema cholashanense* n. sp. (Rhabditida: Steinernematidae), a new species of entomopathogenic nematodes from the province of Sichuan Cholachan mountains, China. *Journal of Invertebrate Pathology* 97, 251–264.

Nguyen, K.B., Shapiro-Ilan, D.I. and Mbata, G.N.M. (2008b) *Heterorhabditis georgiana* n. sp. (Rhabditida: Heterorhabditidae) from Georgia, USA. *Nematology* 10, 433–448.

Nguyen, K.B., Ginarte, C.M.A., Leite, L.G., dos Santo, J.M. and Harakava, R. (2010) *Steinernema brazilense* n. sp. (Rhabditida: Steinernematidae), a new entomopathogenic nematode from Mato Grosso, Brazil. *Journal of Invertebrate Pathology* 103, 8–20.

Nikdel, M., Niknam, G.R. and Ye, W. (2011) *Steinernema arasbaranense* n. sp. (Nematoda: Steinernematidae), a new entomopathogenic nematode from Arasbaran forests, Iran. *Nematologia Meditaranea* 39, 17–28.

Nthenga, I., Knoetze, R., Berry, S., Tiedt, L.R. and Malan, A.P. (2014) Steinernema sacchari n. sp. (Rhabditida: Steinernematidae), a new entomopathogenic nematode from South Africa. *Nematology* 16, 475–494.

Pais, R., Lohs, C., Wu, Y., Wang, J. and Aksoy, S. (2008) The obligate mutualist *Wigglesworthia glossinidia* influences reproduction, digestion, and immunity processes of its host, the tsetse fly. *Applied Environmental Microbiology* 74, 5965–5974.

Park, Y., Herbert, E.E., Cowles, C.E., Cowles, K.N., Menard, M.L., *et al.* (2007) Clonal variation in *Xenorhabdus nematophila* virulence and suppression of *Manduca sexta* immunity. *Cell Microbiology* 9, 645–656.

Patel, M.N., Perry, R.N. and Wright, D.J. (1997) Desiccation survival and water contents of entomopathogenic nematodes, *Steinernema* spp. (Rhabditida: Steinernematidae). *International Journal for Parasitology* 27, 61–70.

Perry, R.N. and Wharton, D.A. (2011) *Molecular and Physiological Basis of Nematode Survival.* CAB International, Wallingford, UK.

Phan, K.L., Nguyen, K.B. and Moens, M. (2001a) *Steinernema loci* sp. n. and *Steinernema thanhi* sp. n. (Rhabditida: Steinernematidae) from Vietnam. *Nematology* 3, 503–514.

Phan, K.L., Nguyen, N.C. and Moens, M. (2001b) *Steinernema sangi* sp. n. (Rhabditida: Steinernematidae) from Vietnam. *Russian Journal of Nematology* 9, 1–7.

Phan, K.L., Subbotin, S.A., Nguyen, N.C. and Moens, M. (2003) *Heterorhabditis baujardi* sp. n. (Rhabditida: Heterorhabditidae) from Vietnam and morphometric data for *H. indica* populations. *Nematology* 5, 367–382.

Phan, K.L., Subbotin, S.A., Waeyenberge, L. and Moens, M. (2005) A new entomopathogenic nematode, *Steinernema robustispiculum* n. sp. (Rhabditida: Steinernematidae), from Chumomray National Park in Vietnam. *Systematic Parasitology* 60, 23–32.

Phan, K.L., Takemoto, S. and Futai, K. (2006a) *Steinernema ashiuense* sp. n. (Nematoda: Steinernematidae), a new entomopathogenic nematode from Japan. *Nematology* 8, 681–690.

Phan, K.L., Spiridonov, S.E., Subbotin, S.A. and Moens, M. (2006b) Four new species of *Steinernema* Travassos, 1927 with short infective juvenile from Vietnam. *Russian Journal of Nematology* 14, 11–29.

Phan, K.L., Mracek, Z., Puza, V., Nermut, J. and Jarosova, A. (2014) *Steinernema huense* sp. n., a new entomopathogenic nematode (Nematoda: Steinernematidae) from Vietnam. *Nematology* 16, 761–775.

Plichta, K.L., Joyce, S.A., Clarke, D., Waterfield, N. and Stock, S.P. (2009) *Heterorhabditis* gerrardi n. sp. (Nematoda: Heterorhabditidae): the hidden host of Photorhabdus asymbiotica (Enterobacteriaceae: gamma-Proteobacteria). *Journal of Helminthology* 83, 309–320.

Poinar, G.O. Jr (1966) The presence of *Achromobacter nematophilus* in the infective stage of a *Neoplectana* sp. (Steinernematidae: Nematoda). *Nematologica* 12, 105–108.

Poinar, G.O. Jr (1976) Description and biology of a new insect parasitic rhabditoid, *Heterorhabditis bacteriophora* n. gen. n. sp. (Rhabditida; Heterorhabditidae n. fam.). *Nematologica* 21, 463–470.

Poinar, G.O. Jr (1985) *Neoaplectana intermedia* n. sp. (Steinernematidae: Nematoda) from South Carolina. *Revue de Nematology* 8, 321–327.

Poinar, G.O. Jr (1990) Taxonomy and biology of Steinernematidae and Heterorhabditidae. In: Gaugler, R. and Kaya, H.K. (eds) *Entomopathogenic Nematodes in Biological Control*. CRC Press, Boca Raton, Florida, pp. 23–61.

Poinar, G.O. Jr, Jackson, T. and Klein, M. (1987) *Heterorhabditis megidis* sp. n. (Heterorhabditidae: Rhabditida) parasitic in the Japanese beetle, *Popillia japonica* (Scarabaeidae: Coleoptera), in Ohio. *Proceedings of the Helminthological Society of Washington* 54, 53–59.

Poinar, G.O., Karunakar, G.K. and David, H. (1992) *Heterorhabditis indicus* n. sp. (Rhabditida: Nematoda) from India: separation of *Heterorhabditis* spp. by infective juveniles. *Fundamental and Applied Nematology* 15, 467–472.

Pontes, M.H. and Dale, C. (2006) Culture and manipulation of insect facultative symbionts. *Trends in Microbiology* 14, 406–412.

Qiu, L., Yan, X., Zhou, Y., Nguyen, K.B. and Pang Y. (2005a) *Steinernema aciari* sp. n. (Nematoda: Steinernematidae), a new entomopathogenic nematode from Guangdong, China. *Journal of Invertebrate Pathology* 88, 58–69.

Qiu, L., Hu, X., Zhou, Y., Mei, S., Nguyen, K.B. and Pang, Y. (2005b) *Steinernema akhursti* sp. n. (nematoda: steinernematidae) from Yunan, China. *Journal of Invertebrate Pathology* 90, 151–160.

Qiu, L., Hu, X., Zhou, Y., Pang, Y. and Nguyen, K.B. (2005c) *Steinernema beddingi* n. sp. (Nematoda: Steinernematidae), a new entomopathogenic nematode from Yunan, China. *Nematology* 7, 737–749.

Qiu, L., Fang, Y.U., Zhou, Y., Pang, Y. and Nguyen, K.B. (2004) *Steinernema guangdongense* sp. n. (Nematoda: Steinernematidae), a new entomopathogenic nematode from southern China with a note on *S. serratum* (nomen nudum). *Zootaxa* 704, 1–20.

Qiu, L., Zhao, J., Wu, Z., Lv, Z. and Pang, Y. (2011) *Steinernema pui* sp. n. (Rhabditida, Steinernematidae), a new entomopathogenic nematode from Yunnan, China. *Zootaxa* 2767, 1–13.

Ramos-Rodriguez, O., Campbell, J.F., Christen, J.M., Shapiro-Ilan, D.I., Lewis, E.E. and Ramaswamy, S.B. (2007) Attraction behaviour of three entomopathogenic nematode species towards infected and uninfected hosts. *Parasitology* 134, 729–738.

Richards, G.R. and Goodrich-Blair, H. (2009) Masters of conquest and pillage: *Xenorhabdus nematophila* global regulators control transitions from virulence to nutrient acquisition. *Cell Microbiology* 11, 1025–1033.

Roman, J. and Figueroa, W. (1994) *Steinernema puertoricensis* n. sp. (Rhabditida: Steinernematidae), a new entomopathogenic nematode from Puerto Rico. *Journal of Agriculture, University of Puerto Rico* 78, 167–175.

Ryder, J.J. and Griffin, C.T. (2003) Phased infectivity in *Heterorhabditis megidis*: the effects of infection density in the parental host and filial generation. *International Journal for Parasitology* 33, 1013–1018.

Selvan, S., Campbell, J.F. and Gaugler, R. (1993) Density-dependent effects on entomopathogenic nematodes (Heterorhabditidae and Steinernematidae) within an insect host. *Journal of Invertebrate Pathology* 62, 278–284.

Selvan, S., Grewal, P.S., Leustek, T. and Gaugler, R. (1996) Heat shock enhances thermotolerance of infective juvenile insect-parasitic nematodes, *Heterorhabditis bacteriophora* (Rhabditida: Heterorhabditidae). *Experientia* 52, 727–730.

Shahina, F., Anis, M., Reid, A.P., Rowe, J. and Maqbool, M.A. (2001) *Steinernema pakistanense* sp. n. (Rhabditida: Steinernematidae) from Pakistan. *International Journal of Nematology* 11, 124–133.

Shamseldean, M.M., Abou El-Sooud, A.B., Abd-Elgawad, M.M.M. and Saleh, M.M. (1996) Identification of a new heterorhabditid species from Egypt, *Heterorhabditis taysearae* n. sp. (Rhabditida: Heterorhabditidae). *Egyptian Journal of Biological Pest Control* 6, 129–138.

Shapiro-Ilan, D.I., Brown, I. and Lewis, E.E. (2014) Freezing and desiccation tolerance in entomopathogenic nematodes: diversity and correlation of traits. *Journal of Nematology* 46, 27–34.

Shen, C.P. and Wang, G.H. (1991) Description and study of an entomopathogenic nematode, *Steinernema longicaudum* sp. nov. *Procedings of the First National Academy Association*. Chinese Science and Technology Press, Beijing, pp. 220–231.

Sicard, M., Brugirard-Ricaud, K., Pages, S., Lanois, A., Boemare, N.E., *et al.* (2004a) Stages of infection during the tripartite interaction between *Xenorhabdus nematophila*, its nematode vector, and insect hosts. *Applied Environmental Microbiology* 70, 6473–6480.

Sicard, M., Ferdy, J.B., Pages, S., Le Brun, N., Godelle, B., *et al.* (2004b) When mutualists are pathogens: an experimental study of the symbioses between *Steinernema* (entomopathogenic nematodes) and *Xenorhabdus* (bacteria). *Journal of Evolutionary Biology* 17, 985–993.

Simons, W.R. and Poinar, G.O. (1973) The ability of *Iveoapiectana carpocapsae* (Steinemematidae: Nematodea) to survive extended periods of desiccation. *Journal of Invertebrate Pathology* 22, 228–230.

Snyder, H., Stock, S.P., Kim, S.K., Flores-Lara, Y. and Forst, S. (2007) New insights into the colonization and release processes of *Xenorhabdus nematophila* and morphology and ultrastructure of the bacterial receptacle of its nematode host, *Steinernema carpocapsae*. *Applied and Environmental Microbiology* 73, 5338–5346.

Solomon, A., Paperna, I. and Glazer, I. (1999) Desiccation survival of the entomopathogenic nematode *Steinernema feltiae*: induction of anhydrobiosis. *Nematology* 1, 61–68.

Spiridonov, S.E., Osterfeld, K.K. and Moens, M. (2004) *Steinernema jollieti* sp. n. (Rhabditida: Steinernematidae), a new entomopathogenic nematode from the American Midwest. *Russian Journal of Nematology* 12, 85–95.

Steiner, G. (1923) *Aplectana kraussei* n. sp., eine in der Blattwespe *Lyda* sp. Parasitierende Nematoden Form, nebst Bemerkungenüber das Seitenorgan der parasitischen Nematoden. *Zentralblatt fur Bakteriologie Parasitenkunde Infektionskranheiten und Hygiene Abteilung* 59, 14–18.

Steiner, G. (1929) *Neoaplectana glaseri* n. g. n. sp. (Oxyuridae), a new nemic parasite of the Japanese beetle (*Popillia japonica* Newm.). *Journal of the Washington Academy of Science* 19, 436–440.

Stock, S.P. (1993) A new species of the genus *Heterorhabditis* Poinar, 1976 (Nematoda: Heterorhabditidae) parasitizing *Graphognathus* sp. larvae (Coleoptera: Curculionidae) from Argentina. *Research and Reviews in Parasitology* 53, 103–107.

Stock, S.P. and Blair, H.D. (2008) Entomopathogenic nematodes and their bacterial symbionts: the inside out of a mutualistic association. *Symbiosis* 46, 65–75.

Stock, S.P. and Koppenhöfer, A.M. (2003) *Steinernema scarabaei* n. sp. (Rhabditida: Steinernematidae), a natural pathogen of scarab beetle larvae (Coleoptera: Scarabaeidae) from New Jersey. *Nematology* 5, 191–204.

Stock, S.P., Choo, H.Y. and Kaya, H.K. (1997) An entomopathogenic nematode, *Steinernema monticolum* sp. n. (Rhabitida: Steinernematidae) from Korea with a key to other species. *Nematologica* 43, 15–29.

Stock, S.P., Samsook, V. and Reid, A.P. (1998) A new entomopathogenic nematode *Steinernema siamkayai* sp. n. (Rhabditida: Steinernematidae) from Thailand. *Systematic Parasitology* 41, 105–113.

Stock, S.P., Griffin, C.T. and Burnell, A.M. (2002) Morphological characterisation of three isolates of *Heterorhabditis* Poinar, 1976 from the 'Irish group' (Nematoda: Rhabditida: Heterorhabditidae) and additional evidence supporting their recognition as a distinct species, *H. downesi* n. sp. *Systematic Parasitology* 51, 95–106.

Stock, S.P., Griffin, C.T. and Chaenari, R. (2004) Morphological and molecular characterization of *Steinernema hermaphroditum* n. sp. (Nematoda: Steinernematidae), an entomopathogenic nematode from Indonesia, and its phylogenetic relationship with other closely related taxa. *Nematology* 6, 401–412.

Stock, S.P., Orduño, B.R. and Lara, Y.F. (2009) *Heterorhabditis sonorensis* n. sp. (Nematoda: Heterorhabditidae), a natural pathogen of the seasonal cicada *Diceroprocta ornea* (Walker) (Homoptera: Cicadidae) in the Sonoran desert. *Journal of Invertebrate Pathology* 100, 175–184.

Stokwe, N.F., Malan A.P., Nguyen, K.B., Knoetze, R. and Tiedt, L. (2011) *Steinernema citrae* n. sp. (Rhabditida: Steinernematidae), a new entomopathogenic nematode from South Africa. *Nematology* 13, 569–587.

Strauch, O. and Ehlers, R.-U. (1998) Food signal production of *Photorhabdus luminescens* inducing the recovery of entomopathogenic nematodes *Heterorhabditis* spp. in liquid culture. *Applied Microbiology and Biotechnology* 50, 369–374.

Stuart, R. and Gaugler, R. (1994) Patchiness in populations of entomopathogenic nematodes. *Journal of Invertebrate Pathology* 64, 39–45.

Sturhan D., Spiridonov, S.E. and Mráček, Z. (2005) *Steinernema silvaticum* sp. n. (Rhabditida: Steinernematidae), a new entomopathocenic nematode from Europe. *Nematology* 7, 227–241.

Suzuki, T., Yabusaki, H. and Nishimura, Y. (1996) Phylogenetic relationships of entomopathogenic nematophilic bacteria: *Xenorhabdus* spp. and *Photorhabdus* sp. *Journal of Basic Microbiology* 36, 351–354.

Tailliez, P., Page, S., Edgington, S., Tymo, L.M. and Buddie, A.G. (2012) Description of *Xenorhabdus magdalenensis* sp. nov., the symbiotic bacterium associated with *Steinernema australe. International Journal of Systematic and Evolutionary Microbiology* 62, 1761–1765.

Tallosi, B., Peters, A. and Ehlers, R.-U. (1995) *Steinernema bicornatum* sp. n. (Rhabditida: Nematoda) from Vojvodina, Yugoslavia. *Russian Journal of Nematology* 3, 71–80.

Tamiru, T., Waeyenberge, L., Hailu, T., Ehlers, R.-U., Půža, V. and Mráček, Z. (2012) *Steinernema ethiopiense* sp. n. (Rhabditida: Steinernematidae), a new entomopathogenic nematode from Ethiopia. *Nematology* 14, 741–757.

Tarasco, E., Mracek, Z., Nguyen, K.B. and Trigiani, O. (2008) *Steinernema ichnusae* sp. n. (Nematode: Steinernematidae), a new entomopathogenic nemarode from Sardinia Island (Italy). *Journal of Invertebrate Pathology* 99, 173–185.

Taylor, M.J., Bandi, C. and Hoerauf, A. (2005) *Wolbachia* bacterial endosymbionts of filarial nematodes. *Advances in Parasitology* 60, 245–284.

Triggiani, O., Mráček Z. and Reid A. (2004) *Steinernema apuliae* sp. n. (Rhabditida: Steinernematidae): a new entomopathogenic nematode from southern Italy. *Zootaxa* 460, 1–12.

Uribe-Lorio, L., Mora, M. and Stock, S.P. (2007) *Steinernema costaricense* n. sp. and *Steinernema puntauvense* n. sp. (Rhabditida, Steinernematidae), two new entomopathogenic nematodes from Costa Rica. *Systematic Parasitology* 68, 167–172.

Wang, J.X. and Bedding, R.A. (1996) Population development of *Heterorhabditis bacteriophora* and *Steinernema carpocapsae* in the larvae of *Galleria mellonella. Fundamental and Applied Nematology* 19, 363–367.

Watson, R.J., Joyce, S.A., Spencer, G.V. and Clarke, D.J. (2005) The *exbD* gene of *Photorhabdus temperate* is required for full virulence in insects and symbiosis with the Nematode *Heterorhabditis*. *Molecular Microbiology* 56, 763–773.

Waturu, C.N., Hunt, D.J. and Reid, A.P. (1997) *Steinernema karii* sp. n. (Nematoda: Steinernematidae), a new entomopathogenic nematode from Kenya. *International Journal of Nematology* 7, 68–75.

Weiser, J. (1955) *Neoaplectana carpocapsae* n. sp. (Anguillulata, Steinernematidae) novy cizopasnic housenik obalece jablecneho, *Carpocapsa pomonella* L. *Věstník československé Zoologické Společynosti* 19, 44–52.

Wharton, D.A. (2011) Cold tolerance. In: Perry, R.N. and Wharton, D.A. (eds) *Molecular and Physiological Basis of Nematode Survival.* CAB International, Wallingford, UK, pp. 182–204.

Wharton, D.A. and Surrey, M.R. (1994) Cold tolerance mechanisms of the infective larvae of the insect parasitic nematode *Heterorhabditis zealandica* Poinar. *Cryo Letters* 15, 353–360.

Womersley, C.Z. (1990) Dehydration survival and anhydrobiotic potential. In: Gaugler, R. and Kaya. H.K. (eds) *Entomopathogenic Nematodes in Biological Control*. CRC Press, Boca Raton, Florida, pp. 117–138.

Wouts, W.M., Mracek, Z., Gerdin, S. and Bedding, R.A. (1982) *Neoaplectana* Steiner, 1929 a junior synonym of *Steinernema* Travassos, 1927 (Nematoda: Rhabditida). *Systematic Parasitology* 4, 147–154.

Xu, F.Z., Wang, G.H. and Li, X.F. (1991) A new species of the genus *Steinernema* (Rhabditida: Steinernematidae). *Zoological Research* 12, 17–20.

Yeates, G.W. (2004) Ecological and behavioural adaptations. In: Gaugler, R. and Bilgrami, A.L. (eds) *Nematode Behaviour*. CAB International, Wallingford, UK, pp. 1–24.

Yoshida, M. (2004) *Steinernema litorale* n. sp. (Rhabditida: Steinernematidae), a new entomopathogenic nematode from Japan. *Nematology* 6, 819–838.

Zhang, C., Liu, J., Xu, M., Sun, J., Yang, S., *et al.* (2008) *Heterorhabditidoides chongmingensis* gen. nov., sp. nov. (Rhabditida: Rhabditidae), a novel member of the entomopathogenic nematodes. *Journal of Invertebrate Pathology* 98, 153–168.

Zhang, C.X., Yang, S.Y., Xu, M.X., Sun, J., Liu, H., *et al.* (2009) A novel species of *Serratia*, family Enterobacteriaceae: *Serratia nematodiphila* sp. nov., symbiotically associated with entomopathogenic nematode *Heterorhabditidoides chongmingensis* (Rhabditida: Rhabditidae). *International Journal of Systematic and Evolutionary Microbiology* 59, 1603–1608.

Zhang, K.Y., Liu, X.H., Tan, J., Wang, Y., Qiao, L., *et al.* (2012) *Heterorhabditidoides rugaoensis* n. sp. (Rhabditida: Rhabditidae), a novel highly pathogenic entomopathogenic nematode member of Rhabditidae. *Journal of Nematology* 44, 348–360.

5 Entomopathogenic Nematodes: Ecology, Diversity and Geographical Distribution

S.S. Hussaini*

National Bureau of Agriculturally Important Insects, Bangalore, India

5.1 Introduction

Ecology is an essential facet for lasting management of insect pests with entomopathogenic nematodes (EPNs), besides enlisting biodiversity and geographical distribution. The EPNs (families Heterorhabditidae and Steinernematidae) have a mutualistic symbiotic association with enteric Gammaproteobacteria (*Steinernema/Xenorhabdus* and *Heterorhabditis/Photorhabdus*), which confer high virulence against insects. EPNs are a natural mortality factor for soil-dwelling arthropods. Much focus has been on aspects of taxonomy, phylogeny, biogeography, genetics, physiology, biochemistry and ecology, as well as on commercial production and application technologies. Of late, EPNs and their symbionts are viewed as model systems for advancing research in other disciplines such as soil ecology, symbiosis and evolutionary biology. Integration of existing information, particularly the accumulating information on their biology, into increasingly detailed population models is critical to improving our ability to exploit and manage EPNs as biocontrol agents. Understanding ecological processes in a changing world vis-à-vis climate change, which is a global phenomenon, is critical. Some recent advances in ecology, community ecology, interactions with various factors – edaphic and soil organisms – and their effects on the population dynamics of EPNs, the diversity of EPNs – structural and genetical – and their geographic distribution have been summarized in this chapter. The unique biological features of EPNs demand an insight into the molecular ecology to understand behaviour, ecological genetics, life-history evolution, reproduction and adaptation to extreme environmental conditions, besides an insight into evolutionary aspects. In this chapter, the focus is on some aspects of ecology that are relevant for EPN utility.

There have been reviews and focused attention on ecological considerations and model systems, heightened interest as new species are increasingly being recorded (Glazer, 2002; Hominick, 2002; Lewis, 2002; Henderson *et al.*, 2009; Hiltpold *et al.*, 2010; Kruitbos *et al.*, 2010; Qiu *et al.*, 2011; Campos-Herrera *et al.*, 2012; Ma *et al.*, 2012a,b; Tamiru *et al.*, 2012; Zhang *et al.*, 2012; Mracek and Nermut, 2014; Nthenga *et al.*, 2014; Çimen *et al.*, 2015; Malan *et al.*, 2015).

5.2 Ecology

More than a quarter of the world's population is infected with nematode parasites, and more than a hundred species of nematodes are parasites of humans. Despite extensive morbidity and mortality caused by these parasites, the biological mechanisms of host–parasite interactions are poorly understood. EPNs are soil-transmitted parasites, and their foraging strategies are believed to range from 'ambush' to 'cruise' foragers. Nematode ecology is affected by compounds produced during the decomposition of all organisms in soil. EPN infective juveniles (IJs) are likely to encounter both

*E-mail: sshblr@yahoo.com

  Biocontrol Agents: Entomopathogenic and Slug Parasitic Nematodes (eds. M.M.M. Abd-Elgawad, T.H. Askary and J. Coupland)

uninfected and infected insects in soil, and host quality depends on the stage of infection. Nematode response to infected hosts will change over the course of an infection. The host infection status on long-range attraction to host volatile cues has been documented. The attraction response of *Steinernema carpocapsae*, *Steinernema glaseri* and *Steinernema riobrave* with different foraging strategies to infected and uninfected insects (*Galleria mellonella* and *Tenebrio molitor*) has been tested at different intervals. Based on their foraging strategies, *S. carpocapsae* was not very responsive to hosts, *S. glaseri* was highly responsive and *S. riobrave* was intermediate. Generally, the level of attraction did not change with time after infection and was similar between infected and uninfected hosts. *S. glaseri* infected *T. molitor* exceptionally tended to be less attractive to *S. glaseri* than uninfected hosts. The results suggest that any influence of host infection status on infection behaviour is occurring at subsequent steps in the infection process than at host attraction, or involves non-volatile cues (Rodriguez *et al.*, 2007).

5.2.1 Host cues and attraction

Attraction to hosts is regulated largely by host cues. Parkman and Frank (1992) used song as a cue to trap nematodes in an effort to disseminate *Steinernema scapterisci* as a classical, inoculative biological control agent. *Scapteriscus* spp. was attracted to synthetic *Scapteriscus* song and was exposed to *S. scapterisci* on Florida golf courses and sod farms. The nematode was established and held in moist sand. *Steinernema* spp. are important natural regulating agents of insect populations. Nematodes respond to a variety of stimuli that aid in survival and host finding. The strategies differ along a continuum from ambush to cruiser. Directional movement too has been demonstrated in response to electrical current (Shapiro-Ilan *et al.*, 2009). *S. glaseri* moved to a higher electric potential and *S. carpocapsae* to a lower potential. *Steinernema* spp. may detect the electrical or electromagnetic fields in nature and use stimuli differentially. The host-searching behaviour of some Indian species was analysed in agar plate (petri dish scoring system) using the last instar larvae of *G. mellonella* as host. The scoring system was based on the probability of nematode distribution in the presence/absence of the insect host. The movement of nematodes in the presence of the host increased significantly as the duration increased. Among *Heterorhabditis* spp., *Heterorhabditis indica* PDBC EN 13.31 and PDBC EN 13.4 proved the best cruisers, followed by *Heterorhabditis* sp. PDBC EN 4.6 and PDBC EN 4.5 isolates. Among steinernematids, *Steinernema feltiae* was the best, followed by *Steinernema abbasi* and *S. carpocapsae* (Hussaini *et al.*, 2010). If a fixed hierarchy of cues is followed during foraging, parasites should respond most strongly to cues offered in natural order. *S. carpocapsae*, an ambush forager, aggregated at the source of volatile host cues only after attachment to the host cuticle. They parasitized hosts more efficiently after contact with the cuticle. *S. glaseri*, a cruise forager, was unaffected by exposure to combinations of host cues. *S. feltiae*, with characteristics of ambushing and cruising, was affected by cue hierarchies when either contact or volatile cues were presented first. Host-associated materials encountered out of context may not qualify as host cues for *S. carpocapsae*. The order in which cues are encountered is more predictable for ambushers than for cruisers. A more fixed contextual framework is found in an ambusher's response to host materials (Lewis *et al.*, 1995a). The competing goal of the forager often complicates foraging strategies. On feeding, IJs forage exclusively for a single host. Two questions were posed: (i) what is the relationship between metabolic rate, energy reserves and foraging strategy; and (ii) when a foraging strategy fails, will an IJ parasite switch strategies? Behaviour, metabolic rate changes, energy reserves and infectivity to *Galleria* larvae were measured for species of EPNs stored in water. *S. carpocapsae* was least active and had the lowest metabolic rate, and *Heterorhabditis bacteriophora* was more active, with a high metabolic rate. *S. glaseri* was most active and had an intermediate metabolic rate. Neither changed foraging strategy. *S. carpocapsae* decreased nictation and increased locomotory rate (Lewis *et al.*, 1995b). Host recognition is an important step in the life cycle of parasites. *H. bacteriophora* IJs are strongly attracted to 1-heptanol, 1-octanol and 1-nonanol and find the short-term alcohols to be only slightly attractive, unlike *Caenorhabditis elegans* (O'Halloran and Burnell, 2003). Ammonia present in cockroach faeces is inhibitory to nematodes. During contact with faeces of hosts, recognition by *H. bacteriophora* and *S. glaseri*, alteration in the frequency and/or duration of forward crawling, head-waving, body-waving, stopping, backward crawling, head-rubbing and head-thrusting occurred.

S. carpocapsae responded only to *Blattella germanica* faeces, and *S. scapterisci* did not respond to any of the insect species. *S. scapterisci* showed no avoidance behaviour. On contact with insect gut contents, *H. bacteriophora*, *S. carpocapsae*, *S. glaseri* and *S. scapterisci* reduced the duration of forward crawling and increased the frequency and/or duration of one or more search parameters. Infectivity was related to host recognition behaviour (Grewal *et al.*, 1993a,b). Lewis *et al.* (1993) found *S. glaseri* was attracted only to sources producing carbon dioxide, whereas *S. carpocapsae* did not, and showed little directionality to live *Galleria*. *S. glaseri* was attracted to carbon dioxide. Treatment with sodium metaperiodate or sodium hypochlorite inhibited nematode attraction towards insect plasma (Khlibsuwan *et al.*, 1992), increasing randomly wandering nematodes. Nematode migration towards the source was unaffected by temperatures below 33°C, and was impaired at 35°C and 37°C. The adverse effect of 5 mM and 10 mM NaIO4 on migratory behaviour was reversed 24 h after rinsing with buffered saline. The effect of NaOCl on behaviour was slightly reversible at 0.2 and 0.4% (v/v) concentration, but irreversible at 0.6 and 1.0%. The effect of heat treatment at 35°C and 37°C was reversible. Gaugler *et al.* (1991) evolved a strain for better host-finding abilities in *S. carpocapsae* (G13), selected after 13 rounds using wax moth and compared with two wild-type strains. The G13 strain showed a significant response to different insect hosts (*Popillia japonica* Newman, *Leptinotarsa decemlineata* (Say), *Acheta domesticus* L., *Apis mellifera* L., *B. germanica* L. and *G. mellonella*) up to 20 cm, compared with 3.5 cm in the wild types. Increasing exposure period to 8 h did not increase host finding for any strain, and was correlated positively with host carbon dioxide production. Complete inhibition of chemotaxis was observed when CO_2 was removed. The 'foraging strategy' paradigm (distinction between species with ambush and cruise strategies) as applied to EPNs is being challenged and alternative paradigms proposed (Griffin, 2012). Infection decisions are based on the condition of the potential host: already infected, long-dead hosts or healthy live hosts may be invaded. Phased increase in infectivity may occur. EPN reproductive behaviour (including sperm competition, copulation, attraction and effects of pheromones) and the role of *endotokia matricida* as parental behaviour are exploited by the symbiont for transmission. *S. glaseri* IJs orient towards the root tips of germinating tomato seedlings. This parallels that of the IJs of root-knot *Meloidogyne javanica* Chitwood, and carbon dioxide is largely responsible. *S. glaseri*, applied at high dose (5×10^6/plant) one application or at a lower dose (5×10^5) daily over 10 days, reduced the numbers and reproductive capacity of *M. javanica*. The effect was fourfold when expressed as a total egg mass count and sixfold when expressed as egg masses/g of fresh weight of root. The application of *S. glaseri* larvae also led to the production of larger plants (Bird and Bird, 1986).

Saenz-Aponte and Olivares (2008) evaluated the mobility and speed of *Steinernema* sp. SNIO198 and found the isolate to have ambushing and cruising behaviour in an oil palm plantation in Colombia. Pre-infection of one species affects subsequent infections. Pre-infection of *S. carpocapsae* and *S. glaseri* on subsequent nematode invasion were investigated at 25°C. The behaviour of *G. mellonella* larvae infected through injection with *S. carpocapsae* IJs did not change 6 h post-infection, but invasion of both nematode species decreased by 50% compared to previously uninfected hosts (Wang *et al.*, 1998). Pre-infection with *S. glaseri* suppressed subsequent invasion of both species. It was also decreased by symbiotic *Xenorhabdus* bacteria. Invasion rates were 1.8, 0 and 5.4% for *S. carpocapsae* and 1.8, 5.0 and 7.8% for *S. glaseri,* with the addition of *Xenorhabdus nematophilus* or *Xenorhabdus poinarii* and the use of hot water treatments, respectively, as against 16.5% for *S. carpocapsae* and 15.1% for *S. glaseri.* At the early stage of subsequent invasion, the sex ratios (% male) of the invading *S. carpocapsae* in the hosts with/without the pre-infections were 5.3% and 12.3%, respectively. Insects infected with a *Steinernema*/*Xenorhabdus* complex may produce suppressive substance(s) against subsequent nematode invaders. The behavioural responses to host and environmental cues and their oral or cuticle penetration ability showed that *S. glaseri* migrated 5.54 ñ 0.41 and 3.20 ñ 0.40 mm towards whole and wounded grass roots, respectively, while *H. bacteriophora* migrated 7.46 ñ 0.51 and 3.24 ñ 0.41 mm, 30 min post-inoculation; more attracted by wounded than by whole roots, both migrated 4.15 ñ 0.4 and 16.43 ñ 0.9 mm towards gut fluid of *P. japonica*. *S. glaseri* migrated towards host haemolymph at 4.76 ñ 0.33 mm, but *H. bacteriophora* was not attracted to haemolymph. Grass roots enhanced host penetration by *S. glaseri*, but penetration decreased with increase in the number of grass roots. *S. glaseri* was more capable of penetrating the host haemocoel via the gut than was *H. bacteriophora*. The penetration sites for *H. bacteriophora* on the cuticle (75 ñ 6.3%) were higher than on the gut (25.0 ñ 5.5%) (Wang and Gaugler, 1998). Sexual attraction and behaviour was differential. Virgin females of *S. carpocapsae* produced a non-strain specific sex pheromone as an attractant (Neves *et al.*, 1998). IJs respond differently to unparasitized/parasitized hosts by conspecific/heterospecific populations. Except for *S. carpocapsae*, other *Steinernema* spp. were attracted to unparasitized hosts. IJs were repelled from hosts infected for 4 h with heterospecific infections, except *Steinernema anomali*, whereas *S. glaseri* was repelled only from *S. riobrave*-infected hosts. *S. glaseri* were attracted to a few heterospecific infections and *S. anomali* and *S. riobrave* were attracted to two of five heterospecific infections. *S. anomali* and *S. glaseri* were more attracted to hosts infected with the out-group *H. bacteriophora* than those by conspecific nematodes. *S. carpocapsae*, *S. anomali* and *S. glaseri* IJs were more attracted to insects colonized by conspecific nematodes than to uninfested insects. *S. carpocapsae* IJs were repelled from the 24-h-old conspecific infections; *S. glaseri* were less attracted to 24- than to 4-h-old conspecific infections (Kaya *et al.*, 1993b; Grewal *et al.*, 1997).

Herbivore-induced volatile emissions benefit plant hosts by recruiting natural enemies of herbivorous insects. Signals were described in the subterranean/terrestrial environment. In larvae of the root weevil, *Diaprepes abbreviates* L. on citrus, the interactions between plant, insect and nematode are poorly understood. More *Steinernema diaprepesi* Nguyen and Duncan penetrated weevil-infested roots than uninfested roots. The roots damaged by weevil attracted more nematodes than mechanically damaged roots and sand. Volatiles were identified as terpene compounds. Swingle citrus roots release induced volatiles as an indirect defence in response to feeding, and some are attractive for EPNs (Ali *et al.*, 2010). *H. bacteriophora* isolates BJ and D-H-Da1 aggregated around tomato root rather than a *G. mellonella* larva when presented separately. Both isolates showed no preference for a *G. mellonella*–tomato seedling combination. Dongrae IJs showed no preference for a wax moth larva/tomato seedling presented simultaneously. In contrast, roots of bean plants (*Vicia faba* cv. Windsor) hindered the ability of BJ IJs to infect *G. mellonella* larva, but tomato roots increased host finding by D-H-Da1 juveniles (Cutler and Webster, 2003). *S. carpocapsae* individuals were attracted to germinated seeds of marigold and unsterilized root pieces of marigold and tomato. No chemotactic response was seen for germinated seeds of tomato or sterilized root pieces or root aqueous extracts of either plant. The ability of *S. glaseri* to infect wax moth through sand columns moistened with root aqueous extracts of these plants and aqueous suspension of α-terthienyl: root exudates of marigold and tomato did not affect infectivity adversely but α-terthienyl at 20–40 ppm reduced the nematodes that infected the hosts (Kanagy and Kaya, 1996).

5.2.2 Survival

A low level of persistence in many agricultural systems is a major drawback (Cheng and Hou, 1997). In a coastal prairie, *Heterorhabditis marelata* IJs has lived for a year in the absence of prey, independent of initial density, suggesting that *H. marelata* does not aggregate at high densities to reduce risk of desiccation (Preisser *et al.*, 2005). In Egypt, Shamseldean and Abd-Elgawad (1995) found survival of IJs at low levels of soil moisture (3%) and ceiling temperature (42°C), and faint electric conductivity (0.48 mmhos/cm) to be low. Increased recovery was obtained with increased time at 23°C. Recovery of *S. riobrave* was more than others. Foliar sprays of *S. carpocapsae* All strain on chrysanthemums ('Manatee Iceberg') effectively controlled second-instar larvae of *Liriomyza trifolii* Blanchard (>85% mortality) at high relative humidity (RH). Free moisture on the leaf surface and increased RH enhanced survival and activity. Glycerin was an effective adjuvant compared to Wilt Pruf, Folicote and Sunspray Ultrafine Spray oil. Enhanced efficacy (80%) was with *S. carpocapsae* as against Abamectin (99.6%) (Broadbent and Olthof, 1995). On bean (*Phaseolus vulgaris* L.), foliage persistence/activity of *S. carpocapsae* 'Mexican' showed mortality related to RH; during 6 h exposure at 60–80%, with gradual reduction and high mortality within 2 h at 45% RH. Antidessicant 'Folicote' ensured IJ survival at 60% RH, and 38–60% increase in viability during 6 h exposure. At 80% RH, 'Folicote' caused 10–20% increased viability, but not at 45% RH. On tomato and soybean leaves, survival was better than on cotton, pepper (*Capsicum frutescens*), bean leaves or filter paper.

At 60% RH, IJ movement ceased within 45–60 min of exposure, and the nematode body shrank with no change in pathogenicity for 4 h (Glazer, 1992). Poultry manure killed 70–100% *S. feltiae* All, *Steinernema bibionis* SN and *Heterorhabditis heliothidis* NC strains within 18 h. Manure slurry for 6 h killed 95% of *Musca domestica* L., and <40% larval mortality for 12 h. Most (90–97%) nematodes remained on the surface. Poor survival and limited movement of nematodes in poultry manure appears to make them unlikely candidates for biocontrol of filth flies in this habitat (Georgis *et al.*, 1987). Temperatures between 20°C and 40°C in soil had a linear effect and less significant quadratic and cubic effects on survival. Soil moisture levels of 10–30% did not affect survival. Exposure periods of 1 and 2 weeks did not affect survival (Gray and Johnson, 1983). In the laboratory, 46.6% of the populations of *Steinernema seemae*, *Steinernema masoodi* and *S. carpocapsae* survived and tolerated sublethal temperature (45°C) for 6 h, and 43.3% infectivity was observed against *Helicoverpa armigera* pre-pupae. Survivors did not infect at 25°C and 30°C (Ali *et al.*, 2007). Spray adjuvants, glycerin, Triton-X 100 and Tween 20 at 500–2000 ppm had no adverse effect on viability survival and pathogenicity of *S. glaseri* and *H. indica*. Infectivity and emergence was highest at 500 ppm of Triton-X 100, followed by Tween 20 and glycerine. For *S. glaseri*, it was highest was at 500 ppm, followed by 1000 ppm of Triton-X 100 and Tween 20 (Priya and Subramanian, 2006).

During storage, 99 and 85% survival of *S. carpocapsae* and *H. indica*, respectively, was obtained in the aqueous suspension of ascorbic acid after 60 days. Infectivity was not affected in sorbic acid and MPHB; yields were 3.1–3.3 lakh for *S. carpocapsae* and 5.1–5.4 lakh for *H. indica* (Hussaini *et al.*, 2004b). Antidesiccants glycerin, liquid paraffin, Triton-X 100, Tween 80 and castor oil tested for survival and pathogenicity of *S. carpocapsae*, *Steinernema tami*, *S. abbasi*, *H. indica* and *H. bacteriophora* against *Plutella xylostella* in leaf disc assay showed that 0.25, 0.5 and 1.0% sprayed on to cabbage leaf discs improved infectivity. Absolute mortality was obtained with 0.5% glycerin and 1% liquid paraffin for 1 and 2 h. Glycerin and liquid paraffin provided maximum IJ protection with Triton-X 100 and Tween 80 at higher concentrations. Castor oil was the least effective (Hussaini *et al.*, 2005a). IJs survive for months in soil or water with stored energy reserves. Intraspecific variation in survival exists. *Heterorhabditis megidis* isolates stored at 20°C in water up to 14 weeks with ST_{50} (50% survival time) of 5.6–12.5 weeks. Regression analysis for initial energy reserves, size (area and length), the time at which 50% of energy reserves were depleted (ET_{50}) and motility (head movements/min in week 0) was presented. Energy depletion rate, ET_{50}, is a predictor of survival, explaining the variation in ST_{50}. Depletion rate was highest in the first week. By week 5, when they begin to die, energy reserves completely depleted and death was due to starvation. Intraspecific variation is due to variation in rates of depletion of energy reserves and locomotory activity (Fitters and Griffin, 2006). It is a challenging task to store IJs in water for longer periods. Survival and infectivity of IJs of *S. tami*, *S. abbasi*, *S. carpocapsae*, *H. indica* and *H. bacteriophora* were tested on storage in distilled water. The effect of ageing with freshly hatched IJs of *Steinernema* and *Heterorhabditis* sp. monitored showed that *S. tami* and *S. carpocapsae* survived up to 40 days without mortality or loss in infectivity and *S. abbasi* with 2% mortality. *Heterorhabditis* spp. survived for 20 days and declined to 60 and 20% after 40 and 60 days storage, respectively. Infectivity to *G. mellonella* and *Spodoptera litura* was not altered in unstored and stored IJs of *Steinernema* spp., whereas it varied in *Heterorhabditis* spp. as fresh IJs were non-infective compared to stored IJs (Hussaini *et al.*, 2005b).

5.2.3 Persistence in environment

Natural and artificially infested populations of western corn rootworm, *Diabrotica virgifera virgifera* (Coleoptera: Chrysomelidae) and *Diabrotica barberi* in maize (*Zea mays*) were controlled with *S. carpocapsae* All applied after planting. Control was on a par with that of terbufos (Counter 15G @ 9.9 kg a.i./ha) and chlorpyrifos (Lorsban 4E emulsion, 1.12 kg a.i./ha). Chlorpyrifos and the nematode rates gave lower root damage ratings than terbufos. With a natural pest infestation, the high nematode rate, terbufos and chlorpyrifos provided lower root damage ratings. All treatments were less effective under higher pest pressure. Viable nematodes were found 1–3 days after application and 28 days later at low levels (Wright *et al.*, 1993). *S. carpocapsae* and *S. glaseri* persistence was greater in sandy soil than in clay. Soil pH of 4–8 had no adverse effects, but the population drastically declined in highly alkaline (pH 10) medium.

Persistence decreased as soil oxygen decreased from 20 to 1%. *S. carpocapsae* persisted better at 5–25°C temperature than at 35°C. *S. glaseri* persisted better at 15–35°C than at 5°C, attributed to the climatic origins of the species. *S. carpocapsae* and *S. glaseri* persisted better at low moistures (2 and 4%, respectively), compared with 8 and 16%. Decreased RH from 100 to 25% decreased the nematodes. *S. carpocapsae* persisted 32 days at 100% RH, 16 days at 97% RH, 8 days at 90% RH, 4 days at 75% RH and 2 days at 50 and 25% RH. *S. glaseri* persisted 32 days, 16 days, 4 days, 12 h and 4 h at 100%, 97%, 90%, 75%, 50% and 25%, respectively. *S. glaseri* appeared more susceptible to environmental conditions (Kung, 1990). Survival on foliage of bean plants indicated that *S. feltiae* 'Mexican' and 'Pye' strains withstood the conditions better than *S. feltiae* 'All'. Foliage application of the *S. feltiae* 'Pye', mixed in solutions of the antidesiccant glycerol (1% wt/wt) or Folicote (6% wt/wt), resulted in 75 and 95% control of *H. armigera* larvae, respectively. The 'All' strain was less tolerant to desiccation (Glazer and Navon, 1989). Bark compost provided the best inoculum medium for introducing nematodes into soil (Ishibashi and Kondo, 1986). Different agricultural practices – fertilization (organic and nitrogen, phosphorus and potash (NPK)), irrigation (dripping and sprinkling), plough tilling and use of the herbicide – influence nematode activity in the field. All practices together affected only *H. bacteriophora* negatively in the second year and *S. feltiae* was not affected in any year. *S. feltiae* persisted longer than *H. bacteriophora*. Shortest persistence was in the herbicide treated and tilled plots for *S. feltiae* and *H. bacteriophora*, respectively (Susurluk, 2008).

5.2.4 Non-target effects

In a classical biocontrol programme using EPNs, it is important to understand the impact of the introduced nematode on the population of target and non-target organisms. If found promising, further evaluation on the effect of augmentation of the exotic EPNs on the target pest population should be carried out to further evaluate the impact of the indigenous species on the pest population to identify the potential ones and then decide whether the indigenous species should be conserved or augmented. Under laboratory conditions, nematodes proved to be effective in controlling some foliar pests (Trdan *et al.*, 2009), but had a negative influence on beneficial organisms. Variable responses have been reported relating to the susceptibility of beneficial insects and other orthropods to EPNs. Until now, studies on the non-target effects of EPNs have been performed on species of non-target organisms (NTOs) and a large range from completely harmless to pronounced harmful effects have been established (Bathon, 1996; Farag, 2002; Powell and Webster, 2004). Field trials showed a moderate influence/no effect of EPNs on non-target arthropods (Kaya and Hotchkin, 1981; Georgis *et al.*, 1991), and Kaya (1978a,b) reported that EPNs could affect the interaction between braconid and tachinid parasitoids and their hosts. Moreover, Shannag and Capinera (2000) found *S. carpocapsae* detrimental to the development of *Cardiochiles diaphaniae* Marsh (Hymenoptera: Braconidae), a parasite of watermelon worm *Diaphania hyalinata* (L.) and pickleworm *Diaphania nitidalis* Stoll (Lepi: Pyralidae). There are no published records to show the susceptibility of mites to EPNs. Mites have been related to population reduction of EPNs. Ishibashi *et al.* (1987) observed collembolans, mites, tardigrates, mononchid and dorylaimid nematodes preyed on *S. carpocapsae* and found no deleterious effects of EPNs on them. Epsky *et al.* (1988) demonstrated that the mesostigmatid mite *Gamasellodes vermivorax* Walter reduced *S. carpocapsae* efficacy against *G. mellonella* larvae. Mesostigmatid mite *Alycus roseus* Koch (Acarina: Bimichaeliidae) completed its development from the last instar nymphs to adults and produced viable eggs while feeding on plant parasitic nematodes. *Stratiolaelaps* (*Hypoaspis*) *miles* (Berlese) predated on *S. feltiae* IJs (Brodsgaard *et al.*, 1996). Several steinernematids attach to mites and serve as phoretic hosts (Epsky *et al.*, 1988). Earthworms improve soil conditions – aeration, drainage, organic matter content – change soil structure, move large amounts of soil and affect microflora and fauna diversity (Brown, 1995; Doube and Brown, 1998). Risk identification and evaluation procedure, risk management, including mitigation or reduction, and finally risk/benefit analysis of the EPN release are important inputs. Under host range assessment, selection of appropriate non-target species is the primary and most important step. Many relations – phoretic, paratenic, intermediate or sole host – between EPN and earthworms have been reported (Poinar, 1978; Gunnardson and Rundgren, 1986; Timper and Davies, 2004); however, no adverse effects on earthworms by EPNs are on record. Non-susceptibility of earthworms to steinernematids has been recorded (Capinera *et al.*, 1982; Nguyen and Smart, 1991; Shapiro-Ilan *et al.*, 1993). Some authors feel that earthworms could be used as vectors to introduce and/or disperse beneficial organisms, including EPNs (Shapiro-Ilan *et al.*, 1993, 1995; Potter *et al.*, 1994; Eng *et al.*, 2005). Campos-Herrera *et al.* (2006) studied the interaction of *S. feltiae* with redworms, *Eisenia fetida* (Lumbricidae), and reported 20–90% infection. Ten days post-infected hosts of steinernematids contain dauers (Kaya and Stock, 1997), and were scavenged by ants (the dauers could attach to the ants). Scavenging ants impact the steinernematids, as they consume parasitic life stages of EPNs (Baur *et al.*, 1998) and cause desiccation of cadavers once one or two openings have been made through the host cuticle. In addition, the large species of ants, *Veromessor (Aphaenogaster) anderi* Forel and *Formica pacifica* Francoeur, also carry the cadavers to their nests. The ants removed four of ten cadavers containing *S. feltiae* by dragging them into the openings of the colony, as the cadavers containing *H. bacteriophora* were not disturbed. Thus, ants could play a critical role in the dispersal of steinernematids and not *Heterorhabditis* sp. Poinar (1978) reviewed the associations between earthworms and nematodes and reported over 150 nematode spp. associated with 80 earthworm spp. Cropping history not only influenced earthworm numbers but also the composition of populations. Under laboratory conditions, *H. bacteriophora* can kill spiders, woodlice and diplopods (Bathon, 1996), and a transient effect was noticed in the field. An effect on snail *Oncomelania hupensis* Gredler was observed in the laboratory, but severe impact is unlikely in the field. Hazardous effects were recorded for tadpoles, *Bufo marinus* L., Pacific treefrog, *Pseudacris* (*Hyla*) *regilla* Baird and Girard, African clawed frog, *Xenopus laevis* Wagler, and for lizard, *Anolis marmoratus* Dumeril & Bibron. For the other soil biota nematodes, EPNs can also cause non-target effects (Grewal *et al.*, 1999; Lewis *et al.*, 2001; Jagdale *et al.*, 2002, 2009).

Vertical dispersal was increased in the presence of earthworms as *S. carpocapsae* (All) was found on the exterior/interior of earthworms, *Lumbricus terrestris* L. or *Aporrectodea trapezoids* Orley (Shapiro-Ilan *et al.*, 1993). The non-target effect of *S. feltiae*, *S. carpocapsae* and *H. bacteriophora* on the larvae of the spotted lady beetle, *Adalia bipunctata* (L.), and the lacewing, *Chrysoperla carnea* (Stephens) at different concentrations and temperatures showed that larvae of *A. bipunctata* were more susceptible to nematode attack than the larvae of *C. carnea*. The lowest mortality of both was recorded at 15°C; and after 4 days, differences between 20°C and 25°C were not found (Rojht *et al.*, 2009). The application of *S. carpocapsae* to soil did not affect microbial biomass, respiration or nitrogen pools (Nardo *et al.*, 2006). Increase in ammonium was measured in the nematode plus larva treatment. EPNs can affect system-level processes in soil indirectly, and hence the importance of indirect interactions in affecting functions in soil food webs. The insecticide trichlorfon suppressed the microbial biomass and nitrification process.

5.2.5 Mulch

Mulches enhanced EPN efficacy. Compared with bare ground, mulches enhanced control by providing cocooning sites for codling moth larvae and an easy-to-treat substrate that maintained moisture and enhanced nematode activity. *S. carpocapsae* or *S. feltiae* were tested against cocooned sentinel codling moth larvae in cardboard strips, followed by 2 h of irrigation in plots that were covered with one of four mulches (clover, shredded paper, grass hay or wood chips). Mortalities of 97 and 98% were observed in paper-mulched plots, respectively, compared to 80 and 76% in bare plots. A significant portion of sentinel larvae (25, 14%) that were placed in crevices in the soil in hay- and paper-mulched plots revealed the natural presence of *Heterorhabditis* sp. (Lacey *et al.*, 2006).

5.2.6 Tritrophic interaction

EPNs infecting insect pests have evolved and function within a multitrophic context, their behaviour and physiology being influenced by elements from other trophic levels – the herbivore victim and its food plant. Effects of host plant and a secondary metabolite, cucurbitacin D, on the mortality of spotted cucumber beetle, *Diabrotica undecimpunctata howardi* (Barber), from infection and progeny production by nematodes was ascertained. In general, rootworms fed on maize suffered lower mortality than those reared on groundnuts or squash varieties. Progeny production was highest from rootworms fed on maize, lower for groundnut and lowest on squash, and those fed on bitter squash had lower production than from non-bitter squash for all strains (Barbercheck *et al.*, 1995). Similar effects on the susceptibility of pea leaf weevil, *Sitona lineatus* (L.), to *S. carpocapsae* infection was noted (Jaworska and Ropek, 1994). The mortality of *S. lineatus* was significantly greater for larvae from peas than those from faba beans. Young adults from pea-fed larvae were more susceptible to infection. Larvae of *S. lineatus* from beans appeared more favourable hosts for nematode multiplication than from peas, based on juveniles of *S. carpocapsae* that emerged from bean-fed *S. lineatus*. *Agrotis ipsilon* larvae fed on collard (kale) foliage or on artificial diet were equally infected and killed by *S. carpocapsae*. Nematode progeny production was lower in kale and artificial diet-fed larvae. The addition of commercially obtained sinigrin, a glucosinolate allelochemical found in collards, to artificial diet demonstrated the suppression of nematode progeny production (Epsky and Capinera, 1994). The mortality of *D. u. howardi* from *S. carpocapsae* and *H. bacteriophora* and progeny production varied according to the pest host plant. Mortality from *S. carpocapsae* was lower when the pest was fed on groundnut roots than those fed on squash or maize roots. *H. bacteriophora* induced mortality was lower when the pest fed on maize roots than on groundnut or squash roots. Progeny production from larvae fed on squash roots was lower than that from maize or groundnut. Host plant affects EPNs and a soil-inhabiting stage of an insect pest (Barbercheck, 1993). *D. u. howardi* susceptibility to *H. bacteriophora* (Lewiston) and *S. carpocapsae* (Mexican) has been monitored (Eben and Barbercheck, 1997). Larvae were reared on maize (Pioneer 3192), groundnuts (NC 9) and squash species, marrow (*Cucurbita pepo* L.) var. Crookneck, and a wild bitter African squash (*Citrullus colocynthis* Schrad), which produces a bitter cucurbitacine E. Larvae suffered higher mortality when reared on both squash spp. compared to those reared on maize or groundnuts, and mortality caused by *H. bacteriophora* was higher in soil than in sand. *S. carpocapsae* caused ten times greater mortality than *H. bacteriophora*. Hussaini *et al.* (2001c) found that black cutworm larvae feeding on potato, castor, pumpkin, sunflower, chickpea, cotton and artificial diet were infected differently by *S. carpocapsae* PDBCEN 6.1, *S. abbasi*, *S. tami* and *H. indica* PDBCEN 13.3. Mortality due to *S. carpocapsae* was lower in pests fed on castor and pumpkin. Progeny production was much lower in foliage-fed larvae, particularly pumpkin, compared to those fed on artificial diet. Chemical cues can also originate from the microbes associated with the host insect and the interactions between herbivore and host plant or between herbivores and associated organisms.

5.2.7 Habitat quality

Soil habitat quality influences EPN behaviour. *S. carpocapsae* remained at the point of application in sand with no taxis towards hosts, but dispersed in peat much more and showed significant taxis towards hosts. *H. megidis* dispersed well in both media, but showed taxis towards hosts in sand. In outdoor mesocosms, *S. carpocapsae* outcompeted *H. megidis* in host finding in peat; the contrary was true in sand. EPNs may be habitat specialists and highlight the difficulties of studying soil-transmitted parasites in non-soil media (Kruitbos *et al.*, 2010). Habitat heterogeneity enhances conservation of biocontrol agents (BCAs) in agricultural systems. Complex habitats may conserve beneficial organisms in the soil. The effect of maize and more complex (maize plus mixed annual plant refuge) habitats were compared for influence on the persistence and dispersal of *S. carpocapsae* in soil. Detection at source was not affected by habitat complexity but was associated with soil bulk density, plant density and soil matric potential. Maximum movement rate was 33.3 cm/day, 9 days after application, decreasing with increasing distance from the application point. Soil moisture affected dispersal as it was detected further away in complex habitats, with higher soil moisture. *S. carpocapsae* movement is not dependent on habitat complexity but responds to variation in plant density and moisture (Jabbour and Barbercheck, 2008). The effects of climate change on EPN populations in geographically dispersed databases show decrease in parasitism as climatic variability increases.

5.2.8 Soil type and texture

Soil texture affects nematode dispersal and infectivity, and dense roots affect movement and infectivity adversely. Soil type and texture are important factors that influence the efficacy and persistence of applied EPNs. Infectivity and persistence of *S. carpocapsae* and *H. indica* was affected by soil type and texture, with lighter soil types favouring infectivity and heavier soil types with higher silt fraction favouring persistence (Shakeela and Hussaini, 2009). Maximum activity occurred with 700–800 æm size particles, and not at 600 æm. Activity was limited in sand fraction (1000–1100 æm). A mixture of small particles, or large ones to fractions of the mid-range (700–1000 æm), reduced the percentage (Blackshaw and Sentamizhselvan, 1991). *Lycoriella cuprina* larvae were exposed to IJs of *Heterorhabditis* sp. D1 and *S. glaseri* KG strain in sand, loamy sand and a clay loam varying in moisture potentials. Less parasitism was observed in high clay soils and the lowest was caused by *S. glaseri*. Parasitism in loamy sand occurred at low moisture potentials below permanent wilting point of plants. In sand, both nematodes parasitized larvae at saturation (Molyneux and Bedding, 1984).

The vertical migration of *S. feltiae* IJs decreased as clay and silt increased. Nematodes placed on the soil surface remained within 2 cm of it, but some penetrated to a depth of 10 cm in silica sand and coarse sandy loam. Some pupae at the same depth were also infected in silty clay loam soil. In pure silica sand and coarse sandy loam, nematodes introduced 14 cm below the surface were able to infect pupae placed at 4–24 cm. Movement was least in clay soil and limited in silty clay loam. Nematodes showed a tendency to disperse upwards from the point of application, particularly when moth pupae were present (Georgis and Poinar, 1983). Nictating larvae did not aggregate under any of the soil moisture conditions (Kondo and Ishibashi, 1985). Diapausing cocooned larvae of codling moth, *Cydia pomonella* L., overwinter in cryptic habitats in the soil or in the bark of infested trees. Cocooned larvae were more susceptible than non-cocooned larvae (Navaneethan *et al.*, 2010). Spatial and temporal distribution of *H. bacteriophora* and *S. carpocapsae* along transects placed at different turf grass sites was investigated with *G. mellonella* as bait. *H. bacteriophora* was recovered from fewer sections in a transect than *S. carpocapsae*, and both species did not differ in patchiness of distribution. In one transect with a *H. bacteriophora* population, *S. feltiae* was also recovered, but the two populations seldom overlapped spatially. In transects with adequate scarab larval density, *H. bacteriophora* and *P. japonica* density were correlated inversely. Endemic *H. bacteriophora* populations may suppress *P. japonica* populations. In one transect, an epizootic of *H. bacteriophora* in an undetermined host may have occurred. Edaphic factors were relatively uniform along transects and were, at most, correlated weakly with nematode recovery. Uniform inoculative releases of *H. bacteriophora* tended to return to patterns of distribution typical of endemic populations (Campbell *et al.*, 1998).

The efficacy of *H. bacteriophora*, *H. megidis* and *S. feltiae* in reducing *D. v. virgifera* as a function of soil characteristics has been presented by Toepfer *et al.* (2010). Nematodes have the potential to reduce insect larvae in most soils, and efficacy is higher in heavy clay or silt clay soils than in sandy soils, which is in contrast to the common assumption that nematodes perform better in sandy soils than in heavy soils (Toepfer *et al.*, 2010). Soil texture, chemistry and moisture have a profound effect on the activity and persistence of EPNs. Nematodes' natural habitat is the soil; ticks and other arthropod pests prefer to stay on the soil surface, under stones or leaf litter, spending much of their life cycle in the humid environment. Therefore, consideration of the effect of the soil environment on nematode activity is a prerequisite for the successful use of EPNs against arthropod pests. In another study, many IJs of *S. carpocapsae* and *S. riobrave* were found in the uppermost soil layer, and the heterorhabditid strains were absent from the upper 6 cm. The IJs of *S. feltiae* and *S. carpocapsae* exhibited an intermediate behaviour. *S. carpocapsae* activity in the soil upper layer (1 cm depth) was affected strongly by soil type: the greatest number recorded from sandy loam soil, less in the lighter soils – marine sand and calcareous sandstone – and few were recovered from heavy soils. Even when the soil moisture was low and the number of nematodes found in the upper layer was low, tick mortality remained high. This demonstrated the possible use of the nematodes as an anti-tick agent and the importance of knowing the exact interaction of nematodes with the immediate environment of the pest in order to optimize pest-control activity of the nematode (Alekseev *et al.*, 2006).

The infectivity of *H. bacteriophora* to *Maladera* (*Cephaloserica*) *insanabilis* Brenske, 1894 (*Maladera matrida*, 1986) was variable in sandy, loamy sand and sandy loam. It was more in sandy soil, followed by loamy sand and sandy loam soil. Lethal infection in grubs occurred earlier in loamy sand and sandy soils than in sandy loam soil, as the lethal infection rate index (LIRI) was 103.89 and 122.52 in the latter two types, respectively, as against 100 in the former (Bhatnagar *et al.*, 2002).

5.2.9 Temperature effect

Temperature is a critical input affecting several features. A range of temperatures (5–30°C) affected the infectivity of *Heterorhabditis* sp. and *S. carpocapsae* to *Otiorhynchus sulcatus*. Increase in temperature increased the infectivity of both nematodes. Total mortality of *O. sulcatus* was obtained when *Heterorhabditis* sp. was kept at 20°C for 12 days. Maximum mortality of 65% was obtained with *S. carpocapsae* at 20°C and 25°C (Miduturi *et al.*, 1995). Thermal niche breadths for infection, establishment and reproduction are well defined. *S. riobrave* infected *G. mellonella* larvae at the widest temperature range (10–39°C), and *S. feltiae* at the narrowest (8–30°C). The thermal niche breadths for establishment within hosts were the widest for *S. glaseri* (10–37°C) and the narrowest for *S. feltiae* (8–30°C), and for reproduction it was widest for *S. glaseri* (12–32°C) and narrowest for *S. carpocapsae* (20–30°C). *S. scapterisci* (20–32°C), *S. riobrave* (20–35°C) and *Steinernema* sp. (20–32°C) were more adapted to warm temperature reproduction, and *S. feltiae* to cooler temperatures (10–25°C). Although heterorhabditids are endemic to warmer climates, the upper thermal limits and temperature optima for reproduction of *H. bacteriophora* and *H. megidis* were cooler than those of some steinernematids from South America and the Caribbean. Thermal niche breadths did not differ between conspecific populations from different localities, but varied among species from the same locality. EPN species appear to have well-defined thermal niches unaffected by their locality (Grewal *et al.*, 1994). The developmental temperatures of five geographical isolates of *S. feltiae* obtained from France, Turkey, California, Hawaii and North Carolina have been documented for infection, establishment and reproduction (Hazir *et al.*, 2001). These studies demonstrate the conserved nature of the thermal niche breadth in species of EPN. Under normal soil conditions, moisture levels are high enough to support plant growth. Similarly, temperature is an important environmental factor for the survival of nematodes and biological processes (Askary, 2010). EPNs are usually killed at temperatures above 37°C (Griffin, 1993; Grewal *et al.*, 1994; Ghally, 1995).

Hague *et al.* (1991) found the efficacy of two strains of *S. carpocapsae* to be very similar; the UK strain was more effective at lower temperatures than the All strain. *S. feltiae* strain was less infective at high temperatures (30°C), but actively penetrated the insect host at lower temperatures (10°C). The associated bacteria are sensitive to temperature fluctuations with certain optima ranges. *X. nematophilus* grows at 18–35°C (optimum 30°C); *S. feltiae* can develop to adults at 12–30°C (optimum 25°C); the infection rate rose to 81–100% at 23–30°C but lost infectivity below 12°C and over 35°C (Yang and Li, 1988). The infectivity of *Steinernema* spp. to post-feeding larvae of Australian blowfly, *Lucilia* (*Phaenicia*) *cuprina* (Wiedemann), was higher over a greater range. Parasitization range was greater than that permitting development and reproduction. Heterorhabditids took longer to complete their life cycle than did the steinernematids (Molyneux, 1986). Tolerance to high and freezing temperatures is related to propagation temperatures. *S. riobrave* had the highest LT_{50} values, *S. feltiae* the lowest, and *S. carpocapsae* All displayed an intermediate tolerance. The infectivity of *S. feltiae* freeze-surviving nematodes was 90%. The capacity to withstand freezing diminished by propagation at warmer temperatures (Jagdale and Gordon, 1997). In the host cadaver, IJs can survive adverse environmental conditions for limited periods, but low temperatures and RH prevent emergence. Wax moth larvae infected at 25°C and exposed to 75–100% RH had significantly fewer *H. bacteriophora* and *S. feltiae* emerging at low RH, whereas emergence of *S. glaseri* or *S. carpocapsae* was unaffected. There was no survival of *S. carpocapsae* within the insect cadaver at 25°C after 28 days; at 15°C, 28% survived after 40 days with 75% RH (Brown and Gaugler, 1997). IJs of *S. feltiae* (All) applied to soil in turkey houses with populations of lesser mealworm, *Alphitobius diaperinus* Panzer (Tenebrionidae), at an air temperature of 26.4°C were alive for 3 weeks but not at 5, 7 or 9 weeks. Beetle populations were unaffected. IJs did not survive in soil for more than 2 weeks at 24°C (Geden and Axtell, 1988).

Reproduction of *S. carpocapsae* (Mexican and DD136) and *H. heliothidis* usually followed parasitization by IJs aged between 1 and 32 weeks (Molyneux, 1985). *H. heliothidis* was more sensitive to temperature fluctuations, but were most virulent. *H. heliothidis* and the Mexican strain were equally virulent and more than DD136 (Dunphy and Webster, 1986). Exsheathment of IJs had no effect on survival in sand. The effect of different temperatures on the infectivity and progeny production of indigenous isolates *S. carpocapsae* PDBC EN 6.11, *S. abbasi* PDBC EN 3.1, *S. tami* PDBC EN 2.1 and *H. indica* PDBC EN 6.71, PDBC EN 13.3 were studied under laboratory conditions against *G. mellonella* and *A. ipsilon* final-instar larvae. All isolates caused absolute mortality of the test insect larvae and produced progeny at 25°C and 32°C. Low/no infection was observed at lower temperatures: 15°C and 8°C, respectively. The time required for all isolates to kill host insects varied at 15°C, 25°C and 32°C. The shortest time for mortality of *A. ipsilon* and *G. mellonella* larvae was observed at 32°C, followed by 25°C. Absolute mortality of larvae was observed with all isolates after 48 and 72 h at 25°C and 32°C, whereas the same was not achieved at 15°C, even after 96 h post-exposure. Progeny production varied with reference to temperature. A suitable temperature for infection and development of populations was 25°C (Hussaini *et al.*, 2005b). Boff *et al.* (2000) found that an increase in the storage period resulted in an increase in active nematodes; infectivity and reproduction decreased at low and high temperatures. The time to first emergence and the body length of IJs was affected by time, storage temperature and inoculum level. Infectivity and development were optimum when stored at 10°C and 15°C. Susceptibility of last-instar larvae of cabbage root fly, *Delia radicum* (L.), to *S. feltiae*, *S. carpocapsae*, *Steinernema arenarium*, *H. megidis* and *H. bacteriophora* was evaluated in the laboratory at 10°C, 15°C and 20°C. *S. feltiae* was the only species that killed the larvae at 10°C; *S. carpocapsae*, *S. arenarium* and *H. megidis* were effective at 15–20°C, whereas *H. bacteriophora* killed the maggot only at 20°C. Temperature significantly affected the host-searching ability of test species. Mobility was reduced at low temperatures. Significant effects of the host presence on nematode mobility were found for *S. feltiae*, *S. arenarium* and *H. megidis*, but not for *S. carpocapsae* and *H. bacteriophora*. The dynamics of the attachment to and penetration into the host were monitored for *S. feltiae* at 10°C, 15°C or 20°C and for *S. carpocapsae* at 20°C. In the period (6–30 h), *S. carpocapsae* attached in higher numbers at 20°C than did *S. feltiae* at all temperatures. At 20°C, *S. carpocapsae* penetrated the host only after 30 h, while *S. feltiae* penetrated after 15, 9 and 6 h at 10°C, 15°C and 20°C, respectively (Chen *et al.*, 2003).

5.2.10 Soil moisture

Different strains have varying ability to withstand low RH. Survival of *S. carpocapsae* (All) was reduced to 20% and to 0% after 4 h and 8 h, respectively, at 50–70% RH. Mexican and Pye strains survived twice as long. RH levels (30°C – 80%) showed reduced viability of *S. carpocapsae* Mexican. On rough surfaces, nematode survival was almost twofold higher than on smooth surfaces (Glazer, 1991). Survival and pathogenicity of *S. carpocapsae* were greater at 5–20°C and, conversely, for *S. glaseri* it was significantly greater at 15–35°C. *S. carpocapsae* and *S. glaseri* survived best at 2% and 4%, respectively. After rehydration, they caused lethal infection of *G. mellonella* larvae (Kung *et al.*, 1991).

Steinernema thermophilum Ganguly and Singh is well adapted to varying moisture profiles. Infectivity increased at 3–9% moisture level, and declined later. The optimum soil moisture level was 9%, and optimal range 3–16% (Ganguly and Gavas, 2004). The effects of soil moisture on the survival, nictating activity and infectivity of *S. feltiae* was that larvae nictated on soil particles at an RH of 10–50% and invaded final-instar larvae of *S. litura*. The amount of nictating activity corresponded with the infectivity of the larvae. Despite higher survival rates at RH 10, 15 and 50% than at 25–40%, nematodes did not infect cutworm larvae. Nictating larvae did not aggregate at any of the moisture levels (Kondo and Ishibashi, 1985). The virulence of *H. megidis* (UK211) and *S. carpocapsae* (All) was tested in loamy sand at gravimetric water contents of negative 0.01, 0.06, 0.3, 1.0 and 15 bars. *Curculio* larval survival decreased as moisture levels increased. The nematode effect was most pronounced at –0.06 bars. At –0.01 bars, larval survival was ≤5%, regardless of nematode presence, thus indicating that intense irrigation alone might reduce pecan weevil, *Curculio caryae* (G.H. Horn) populations with no adverse effect of surfactant (Kinetic) on *C. caryae* suppression. In the greenhouse, larval survival was lower in *S. carpocapsae* (Italian) and *S. riobrave* (7–12) than in *S. carpocapsae* (Agriotos, Mexican) and *S. riobrave* (355) treatments. It was reduced to approximately 20% in the *S. riobrave* (7–12 treatment) (Shapiro-Ilan *et al.*, 2006).

5.2.11 Osmotic stress

Most salts cause high mortality within 24 h of exposure. Potassium chloride at 0.3 M showed low mortality levels. However, 0% survival in the high temperature assay (HTA) was observed. The viability of fresh IJs of *S. carpocapsae* was reduced considerably when exposed to 45°C for 12 h in a HTA. Carbohydrates did not reduce viability, except polyethylene glycol (PEG) 300 and 600 (2.0 and 1.2, respectively). Exposure for 72 h to glycerol (2.2–3.8), PEG-300 (1.2–1.6) and PEG-600 (0.8) increased heat tolerance. The survival rate was high in nematodes stored for 72 days following an increase in glycerol concentration (0.7–3.5). Survival was lower in nematodes exposed directly to 3.5 glycerol or stored in distilled water for 36 days. After storage for 4 days, survival levels in the HTA were 87.3 and 49.2% at 4 h and 8 h exposure, respectively, which decreased to 72.7 and 26.5% after 19 days, respectively. The infectivity of *S. carpocapsae* desiccated by gradual osmotic pressure towards *T. molitor* was similar in fresh or stored IJs (Glazer and Salame, 2000).

S. carpocapsae All and *S. scapterisci* SS adapted well to the osmotic solution (a mixture of fortified artificial seawater and glycerol) at 15°C, with mortality below 2%. *S. feltiae* SN, *Steinernema longicaudum* X-7 and *H. bacteriophora* H06 were moderately tolerant (20% mortality). Osmotic treatment greatly enhanced their heat tolerance and retained their infectivity to wax moth larvae (Yan *et al.*, 2010). *Steinernema pakistanense* isolates 98 AB, 15 AM and BS 266 IJs tolerated high temperatures better in seawater salinity 38 ppt than in distilled water; 94% survived at 38°C and were infective. *S. feltiae* (AO5) did not survive at 36°C in both sea and distilled water. *H. indica* (isolate 7 ma) tolerated more salinity at 40°C in seawater, with 15% infectivity. *H. indica* (32NR) was more temperature tolerant and survived at 40°C in both media (Shahina *et al.*, 2005).

5.2.12 Pesticides, fertilizers and amendments

EPNs are often applied in conjunction with other biological or chemical pesticides, fertilizer or soil amendments and it is more economical to tank mix both for application. EPN IJs are tolerant to short exposures (2–6 h) of most agrochemicals (Hazir *et al.*, 1987; Rovesti and Deseo, 1990; Grewal *et al.*, 1992) and are tank mixed. Some pesticides can reduce nematode viability and virulence (Zimmerman and Cranshaw, 1990; Patel and Wright, 1996; Grewal *et al.*, 1998). Several studies on the effect of agrochemicals on different species of EPN include acaricides, fungicides, herbicides and insecticides. Some are synergistic in combination (Koppenhofer *et al.*, 2002). EPNs include *S. carpocapsae* (Rao *et al.*, 1975; Hara and Kaya, 1982, 1983a; Rovesti and Deseao, 1990; Zimmerman and Cranshaw, 1990), *S. feltiae* (Rovesti and Deseo, 1990; Zimmerman and Cranshaw, 1990; Vainio, 1994; Grewal *et al.*, 1998; Krishnayya and Grewal, 2002), *S. diaprepesi* (Grewal *et al.*, 2001), *Steinernema bicornutum* (Hussaini *et al.*, 2001b), *S. glaseri* (Koppenhofer and Kaya, 1997), *H. bacteriophora* (Rovesti *et al.*, 1988; Rovesti and Deseo, 1991; Koppenhofer and Kaya, 1997) and *H. indica* (Kulkarni *et al.*, 2009). Nematode species differ in their susceptibility to pesticides. Certain exposures can stimulate and interact synergistically with nematodes, but some are detrimental. The synergistic interaction of *Heterorhabditis* spp. and *S. glaseri* with neonicotinoid imidacloprid in scarab larvae is well documented (Koppenhofer and Kaya, 1998; Koppenhofer *et al.*, 2003; Anuar and Daniel, 2009). Imidachloprid reduces the grub's defensive behaviour, resulting in increased attachment and penetration (Koppenhofer *et al.*, 2000). Another one, thiomethoxam, was found compatible with *H. megidis*, *S. feltiae*, *S. glaseri* and *S. carpocapsae*, while not compatible with *H. bacteriophora* (Koppenhofer *et al.*, 2003; Alumai and Grewal, 2004). Compatibility of spinosad was also established (Alumai and Grewal, 2004) with *S. carpocapsae* and *H. bacteriophora*. Cartap, profenofos and pyraclofos were most toxic to IJs (83.4% and 57.1% mortality, respectively, in solutions of 100 æg/ml after 48 h). Diazinon, dichlorvos, fenthion, malathion, trichlorfon, propetamphos and prothiofos showed weak toxicity at 100 æg/ml. Other chemicals tested showed no toxicity at 100 æg/ml. IJs were incubated in insecticide solutions (100 æg/ml) for 24 h and then used to treat newly moulted last-instar larvae of *S. litura*. Organophosphates (OPs) except acephate, malathion and temephos, carbamate (methomyl), pyrethroids (permethrin and ethofenprox) and cartap inhibited the pathogenicity of IJs to these larvae (Zhang *et al.*, 1994). Mortality of last-instar *P. japonica* was highest with isazofos, bendiocarb and chlorpyrifos. Carbaryl and *H. bacteriophora* and *S. glaseri* were less effective treatments in these arenas. Soil characteristics affecting activity were organic matter (chlorpyrifos, bendiocarb and *S. glaseri*) and pH (carbaryl and isazofos). *S. glaseri* was the most effective treatment in arenas simulating turf grass, followed by chlorpyrifos, then bendiocarb. Irrigation improved insecticide efficacy with higher soil organic matter (Cowles and Villani, 1994).

Movement, nictation and infectivity of *S. carpocapsae* strain All were compared for ensheathed (EnJ) and desheathed (DeJ) IJs exposed to acephate, dichlorvos, methomyl, oxamyl, or permethrin. Normal sinusoidal movement, uncoordinated motion, twitching, convulsion or formation of a pretzel shape, an inactive 'S' posture with fine twitching, or a quiescent straight posture were evident. DeJs displayed these movements at lower concentrations than did EnJs. The treated EnJs caused generally lower mortality in *S. litura* than untreated, EnJs alone and greater mortality than insecticides alone. Nictation of DeJs was suppressed by the chemicals at low concentration, except for acephate and permethrin. Nictating EnJs or DeJs killed host insects faster than the non-nictating juveniles (Ishibashi and Takii, 1993).

H. bacteriophora and *S. feltiae* IJs exposed to insecticides Basudin 25 EC (diazinon), Bi 58 (dimethoate 37%), Ipofos 50 (bromfenvinfos) and Nexion EC 40 (bromophos) at recommended concentrations showed no toxicity after 24 h exposure at 20°C. Mortality for *S. feltiae* was from 12.3% (dimethoate) to 14.7% (bromophos) and for *H. bacteriophora* from 14.3% (diazinon) to 16.7% (dimethoate). Mortality rates were lower at 10°C. Surviving larvae on exposure at 20°C were infective to *Galleria* sp., but with reduced pathogenicity (Jaworska, 1990).

Butyric acid, which is produced through the fermentation of organic matter by anaerobic soil bacteria, possesses nematicidal properties. In sand amended with 0.88 mg butyric acid/g, plant parasitic and fungivorous nematodes were reduced (84–100%), whereas bacterivorous nematodes were reduced at 8.8 mg (70–98%). *Heterorhabditis* was affected adversely by 0.88 mg (59% reduction), while *Steinernema* was reduced 85% at 8.8 mg. Vapour from a 0.1 M solution reduced plant parasitic and fungivorous nematodes (89–96%), while vapour from 1 M solution reduced EPNs by 94–99%. Bacterivorous nematodes did not survive (Browning *et al.*, 2004).

The effectiveness of EPNs against the larvae of mustard beetle, *Phaedon cochleariae* (Chrysomelidae), at different temperatures showed the mortality of last-instar larvae of *P. japonica* was highest when exposed to isazofos, bendiocarb and chlorpyrifos in 30-ml soil arenas in laboratory assays (Jaworska and Gospodarek, 2009). Carbaryl and *H. bacteriophora* and *S. glaseri* were less effective treatments. Soil × treatment (insecticides and nematodes) interactions were highly significant, often changing larvicide rank order of efficacy within soils. The soil characteristics affecting insecticidal activity were organic matter (chlorpyrifos, bendiocarb and *S. glaseri*) and pH (carbaryl and isazofos). In arenas (7.6 cm × 15 cm diameter) simulating turf conditions, *S. glaseri* was most effective, followed by chlorpyrifos and bendiocarb. Soil organic matter improved insecticidal efficacy (Cowles and Villani, 1994).

Oxamyl treatment did not enhance IJ pathogenicity to *G. mellonella* larvae. Aberrant nematode movement and partial paralysis were caused at higher concentrations. IJs of *S. carpocapsae* showed a low level of locomotory activity, whereas *H. bacteriophora* IJs maintained a high level of activity. Treatment at 30 æg/ml increased sinusoidal movements and reduced host finding and pathogenicity. Higher rates induced complete paralysis. Oxamyl is incompatible with *S. carpocapsae* and *H. bacteriophora* (Gaugler and Campbell, 1991). IJs of *S. carpocapsae* and *S. feltiae* tolerate most of the 75 commercial pesticides tested. The response of the two species was very similar. Only dodine (fungicide) and alachlor and paraquat (herbicides) seriously affected the IJs. Parathion, aldicarb, methomyl, flubenzimine, metam sodium and fenamiphos were the most toxic (Rovesti and Deseo, 1990). Bentazone, ioxynil, hexaconazole, cyromazine and buprofezin did not affect *S. feltiae*, but quizalofop-ethyl, tralkoxydim, sulfur and potassium soap were toxic. Pirimicarb, cypermethrin, diazinon, simazine and metalaxyl plus mancozeb did not affect either fungus, but glyphosate, dimethoate, MCPA (2-methyl-4-chlorophenoxyacetic acid), vinclozolin, trifluralin, thiram and propiconazole inhibited at least one species (Vainio and Hokkanen, 1990). Compatibility of *H. bacteriophora*, *H. heliothidis*, *S. carpocapsae* and *S. feltiae* with commonly used pesticides was assessed. The susceptibility of the *Neoaplectana* spp. was similar, as was that of the *Heterorhabditis* spp. (Rovesti *et al.*, 1990).

The herbicide alachlor was most deleterious in assays, reducing dauer activity by 44–90% in aqueous solution. Only 19–30% of the nematodes exposed to other herbicides were inactive. Bioassays of nematodes extracted from these suspensions and tested against final-instar larvae of *G. mellonella* yielded LD_{50}s of 45.2, 1.1, 0.9 and 1.1 nematodes/larva for the alachlor, 2,4-D, glyphosate and sethoxydim treatments, respectively against an LD_{50} of 2.8 nematodes/larva in control. Infectivity in the soil was reduced by exposure to 2,4-D and alachlor. At the recommended rate, the LD_{50}s for the alachlor, 2,4-D and control treatments were 26.7,

8.8 and 2.7 nematodes/larva, respectively (Forschler *et al.*, 1990). The presence of fenamiphos in sand adversely affected the infectivity of *S. feltiae* and *S. carpocapsae* to *G. mellonella*. Fenamiphos was nematostatic in sand over a 4- to 7-day exposure period, but the nematodes were infectious (Kaya and Burlando, 1989). Insecticides (carbaryl, chlorpyrifos, DDT, dimethoate, endosulfan and malathion) and fungicides (carbendazim, mancozeb and zineb) on *S. feltiae* showed that the nematodes were active 72 h after treatment with carbaryl, dimethoate, endosulfan, malathion and zineb at field doses (Das and Diwakar, 1987). More juveniles (J_3) of *S. feltiae* were recovered from the 1,3-D-treated soil than from non-treated soil, while native nematodes in the treated soil fluctuated more with the addition of DD-136 than those in non-treated soil. The J_3 persisted longer in silt than in sandy soils. Inundative soil application of DD-136 increased native rhabditids and decreased plant parasitic nematodes. DD-136 in chemically treated soil effectively attacked the invading insect pests, and also suppressed the recovery of the plant nematodes. *S. feltiae* IJs were affected adversely by mevinphos, fenamiphos, trichlorfon, methomyl and oxamyl, causing partial paralysis with a curled or coiled posture. Affected nematodes did not infect larvae of *Spodoptera exigua* (Hb.), but when washed in distilled water they recovered and were infectious and behaviour was unaffected (Hara and Kaya, 1983a).

S. feltiae invaded and reproduced in larvae of *S. exigua* (Hb.) killed by insecticides trichlorfon, mevinphos, methomyl, fenvalerate or permethrin. When killed larvae of the noctuid were exposed to IJs 48 h after treatment, 72–100% of the dead larvae produced nematode progeny. The progeny were infectious to living *S. exigua* larvae. However, when larvae were first infected with the nematode and treated topically with fenamiphos or methomyl 48 h later, the development and reproduction of *S. feltiae* were affected adversely (Hara and Kaya, 1983b).

A high mortality (>50%) of *S. feltiae* on 5-day exposure to neem seed shell (*Azadirachta indica*) extract (5 g/100 ml water), but not after exposure to seed kernel extracts, was observed. Neemix (azadirachtin) had a slight adverse effect on the survival of IJs. Bancol 50 WP 0.75% (bensultap) caused 100% mortality in 3 days. The ability to infect *G. mellonella* was not affected, but limited their establishment. The number of established nematodes per *G. mellonella* decreased with increasing Neemix concentration (Pezowicz *et al.*, 1997). *S. glaseri* mortality with diazinon (0.040) was significant: highest (82.66%) in 0.035% concentrations, on a par with 0.04%, and highly compatible with chlorpyrifos and limnool (Prabhuraj and Viraktamath, 2004).

S. feltiae was tolerant to 8 insecticides (a.i. kinoprene, lufenuron, methomyl, metoxyfenozide, oxamyl, piperonyl-butoxide, pyriproxyfen, tebufenozide), 7 acaricides (a.i. azocyclotin, clofentezin, diafenthiuron, etoxazole, fenbutatinoxide, fenpyroximate, tebufenpyrad) and 4 fungicides (a.i. captan, fenhexamid, kresoxim-methyl, nuarimol) under laboratory conditions. Insect larval mortality caused by *S. feltiae* + insecticides and *S. feltiae* + fungicides during 72 hours varied from 2.26% to 18.68% and from 7.04% to 8.86%, respectively. Acaricides with a.i. fenpyroximate and tebufenpyrad considerably influenced the ability of *S. feltiae* to infect larvae of *T. molitor*. Tebufenpyrad caused 95% and fenpyroximate 85% reduction in *S. feltiae* virulence (Redova, 2010).

Lorsban™ (chlorpyrifos), Decis™ (deltamethrin), Match™ (lufenuron), Deltaphos™ (deltramethrin + triazophos), Dimilin™ (diflubenzuron), Stallion™ (gamacyhalothrin), Karate Zeon™ (lambdacyhalothrin), Tracer™ (spinosad), Vexter™ (chlorpyrifos), Galgotrin™ (cypermethrin), Certero™ (triflumuron) and Talcord™ (permethrin) were compatible with *H. indica*, *S. carpocapsae* and *S. glaseri* (Negrisoli *et al.*, 2010). There are a few studies on the compatibility of herbicides with EPNs (Rovesti *et al.*, 1988; Forschler *et al.*, 1990; Rovesti and Deseo, 1990, 1991). Herbicides did not have a significant effect on viability and infectivity. Glyphosate and MCPA individually had a negligible effect on *S. feltiae* and a moderate effect on *H. bacteriophora*. However, exposure to glyphosate and MCPA caused higher mortality (26–57%) than glyphosate (0.67–15%) or MCPA alone (2–19%). Krishnayya and Grewal (2002) found 100% mortality of *S. feltiae* exposed to fungicides Cinnamaldehyde and hydrogen dioxide/peroxyacetic acid mix (Zerotol) within 4 h and 120 h, respectively, while azoxystrobin (Abound) caused no mortality of nematodes. Cyromazine seed treatments, which affect insect cuticle formation, and EPNs, which must pass through openings in the insect cuticle, were applied independently and in combination to test for synergism between the two in their effects on onion maggots. In assays with *H. bacteriophora* at 10–160 nematodes/larva, significant mortality in first- and second-stadium onion maggots was caused. Cyromazine applied as a film-coated seed treatment (0.78–50 g ai/kg seed) alone or in combination with nematodes per first and second instars, respectively showed no interactions between cyromazine and nematodes. Differences among rates of cyromazine in plant and larval mortality were significant, and differed for the two larval stadia. Although cyromazine and EPNs could be used independently for onion maggot control, no synergism was found in combined use (Yildrim and Hoy, 2003). Larvae of the scarabaeid, *Cyclocephala hirta* Le Conte, are major pests of turfgrass in California. In field experiments, bendiocarb (chemical insecticide), *Bacillus popilliae* (milky disease bacterium) and two EPNs, *S. feltiae* and *H. bacteriophora* were evaluated against third instar larvae of *C. hirta*. The results showed no significant differences in population reduction among the treatments, but the larval population of *C. hirta* reduced in all plots. The population of *C. hirta* had a natural occurrence of *B. popilliae* and *Rickettsiella popilliae* (Phillip). The prevalence of *R. popilliae* during the course averaged <10%, but that of *B. popilliae* averaged 20%. Bioassays for *B. popilliae* showed that 67–90% of the larvae became infected with this bacterium and none with *R. popilliae* (Kaya *et al*., 1993).

Nitrate accumulation of lettuce varieties and effect on *S. glaseri* and cockchafer grubs (*Melolontha melolontha* L.) was investigated. An increasing amount of water supply decreased the nitrate content of lettuce in equivalent nitrogen rate treatments. Nitrogenous fertilizers had no harmful effect on cockchafer grubs without nematodes. *S. glaseri* proved to be more sensitive to N-fertilizers than cockchafer grubs (Nadasy and Nadasy, 2006).

Acephate, bifenthrin, deltamethrin, fipronil and imidacloprid in aqueous solutions were tested for compatibility with *S. scapterisci* in turf grass. The survival of the nematode was >95% in solutions of acephate, bifenthrin and imidacloprid. Infectivity of *S. scapterisci* in adult *Scapteriscus vicinus* Scudder was >60% in acephate and bifenthrin and <40% in imidacloprid. Nematodes were compatible with most insecticides without reduction in survival or infectivity (Barbara and Buss, 2005). Compatibility of *S. feltiae*, *S. glaseri*, and *H. indica* with pesticides and fertilizers was tested. All were highly compatible with urea and muriate of potash up to 400 ppm. *Steinernema* spp. were compatible with superphosphate at 100 ppm, *H. indica* up to 200 ppm, *S. glaseri* and *H. indica* with atrazine up to 10 000 ppm, whereas *S. feltiae* was highly compatible up to 5000 ppm. *S. feltiae*, *S. glaseri* and *H. indica* were compatible with 312, 625 and 1250 ppm of carbofuran; 312 ppm of phorate; 2500, 5000 and 1250 ppm of quinalphos; and 2500, <312 and <312 ppm of aldrin, respectively (Karunakar *et al*., 2002). The effects of diazinon, diflubenzuron and *S. feltiae* on a small infestation of sciarid, *Lycoriella auripila* (Winnertz), on mushrooms, *Agaricus bisporus* (J.E. Lange) Imbach, were investigated. *S. feltiae* applied at a rate of 3 × 10^6 IJs/0.56 m^2 tray caused increases of 28.5 and 19% in the mean numbers and weights of mushrooms, respectively. Diflubenzuron treatment reduced total numbers and weights by 14.6% and 6%, respectively, and a combination of diazinon and diflubenzuron reduced total numbers and weights by 18.5% and 9.4%, respectively. Nematodes delayed the onset of the first flush of mushrooms, while diflubenzuron delayed the third and fourth flushes. It is suggested that the yield benefits of the application of *S. feltiae* result mainly from effects of the nematodes on the mushrooms rather than solely from the suppression of a damaging pest population (Grewal *et al*., 1992). Sciarids are controlled successfully by *S. feltiae* in mushrooms. However, the phorids are not controlled. A combination of IJs and a low dose of insecticides caused higher mortality of *Lycoriella solani* (Diptera: Sciaridae) and *Megaselia halterata* (Diptera: Phoridae) larvae as compared to individual treatments (Sznyk-Basalyga *et al*., 2002). The efficacy of EPNs on foliage is enhanced through the addition of adjuvants and surfactants (Glazer, 1992; Baur *et al*., 1997; Mason and Wright, 1998). The addition of organic matter effectively changes the environment and can increase the diversity of organisms. Organic materials can improve bulk density, porosity and moisture-holding capacity, and can enhance plant growth and health. Bednarek and Gaugler (1997) found that the application of organic manure resulted in increased densities of native *S. feltiae*, whereas NPK fertilizers suppressed nematode densities

Metallic ions affect the activity of nematodes positively. Cadmium, copper, lead and zinc increased *S. feltiae* invasive larvae mortality, and decreased their pathogenicity and reproduction. Manganese at 400 mg Mn (II)/dm^3 concentration affected *S. feltiae* positively during storage, causing decreased mortality, improved pathogenicity and reproduction from insect bodies in comparison to distilled water (Jaworska and Gorczyca, 2002; Jaworska and Gospodarek, 2009). Stimulation of *Steinernema* sp. and *H. indica* PDBC isolates exposed to Mg and Mn ions caused increased pathogenicity and progeny production (Hussaini *et al*., 2001d). Similar results were obtained by Menti *et al*. (2000). Cadmium nitrate added to sand at concentrations of 1, 5 and

10 ppm, which refers to the first to sixth degree of soil contamination adopted by the Institute of Crops, Fertilization and Soil Science in Poland, affected the invasive capacity of IJs of *S. feltiae* immediately and after 7, 14 and 21 days (Jarmul and Kamionek, 2003). Magnesium had an effect on beneficial entomopathogenic fungi, nematodes and predatory arthropods. Magnesium fertilization of soil contaminated with heavy metals does not affect the majority of epigeal invertebrates, but may influence single species, favouring the occurrence of some (*Bembidion* sp. Latreille – Carabidae) and restricting others (*Harpalus rufipes* De Geer). Magnesium synergism increased infectiveness and pathogenicity of entomopathogenic fungi such as *Beauveria bassiana*, *Paecilomyces fumosoroseus* (Wize) A.H.S. Br. & G. Sm., *Paecilomyces farinosus* (Holmsk.) A.H.S. Br. & G.Sm. or *Metarhizium anisopliae* at 160 mg/dm^3. Increased pathogenicity of *S. carpocapsae* and *H. bacteriophora* was observed when magnesium ions were added. The effective dose differed per nematode and insect combination (450 mg/dm^3) or jointly with entomopathogenic fungi (320 mg/dm^3): a 10- to 300-fold higher pathogenicity was observed. Magnesium and Mn ions had beneficial effects on the infectivity of some Indian isolates in cutworm, *A. ipsilon* Hufnagel and *Agrotis segetum* (Hussaini *et al.*, 2001d).

5.2.13 UV protectants

Walia *et al.* (2008) tested PABA (para-aminobenzoic acid), Congo red, zinc oxide and titanium dioxide at different concentrations against *H. bacteriophora* and two isolates (Janti and RB-5) of *Steinernema* sp. nematode suspensions (1000 IJs/ml) treated with the above chemicals at different concentrations exposed to UV radiation in a laminar flow bench for 30 and 60 min. PABA and Congo red provided nearly absolute protection from UV radiation. Zinc oxide was not effective (70–90% mortality in exposed IJs). Titanium dioxide protected *H. bacteriophora* completely at 0.5% concentration only; whereas in *Steinernema* sp. Janti isolate protection was 76%, and in RB-5 isolate 40% IJs were alive even at 0.5% concentration. The IJs exposed to PABA and Congo red remained infective to *G. mellonella* larvae, and titanium dioxide hampered infectivity. PABA and Congo red at 0.25% and zinc oxide and titanium dioxide at 1% were not phytotoxic to mungbean. The effect of UV protectants was not nematode density dependent. Commercially available Blankophor fluorescent brighteners were compared with the standard Tinopal LPW as solar radiation protectants for *S. carpocapsae* (All). Blankophor BBH and Tinopal LPW were the most successful UV screens, with 95% of the original nematode infectivity to larvae of *Galleria* retained after 4 h exposure to direct sunlight. The Blankophor HRS and DML preserved 80 and 85% infectivity, and the P167 preserved 70% infectivity after sunlight exposure. Other Blankophors (RKH, LPG and BSU) were not as effective (Nickle and Shapiro, 1994).

H. bacteriophora IJs exposed to medium-wave UV radiation (302 nm) were inactivated. *S. carpocapsae* and irradiated *H. bacteriophora* showed significant delay in progeny emergence when exposed for 1.5 min, a decline in reproductive capacity at 2 min, and loss of pathogenicity at 4 min. Negative effects were not noted until 6 min for *S. carpocapsae*. Complete inactivation of both species occurred in 4 and 10 min, respectively. The IJ sheath did not provide any protection in both species. The poor UV stability of *H. bacteriophora* is consistent with the tendency of heterorhabditids to be less tolerant to environmental stress than steinernematids (Gaugler *et al.*, 1992). A stilbene fluorescent brightener, Tinopal LPW, was used as a UV protectant for *S. carpocapsae* (All). Irradiation of an aqueous suspension of nematodes produced an LC_{50} in 15.7 min under a sun lamp and in 31.7 min in direct sunlight. Irradiation by both sun lamp and sunlight did not reduce their biological activity, as measured by their ability to parasitize wax moth larvae after exposure of 8 h and 4 h, respectively. Tinopal LPW appeared promising as a radiation protectant (Nickle and Shapiro, 1992).

The effects of UV exposure on the survival of *Steinernema kushidai* (J_3) and its insecticidal activity against larvae of the cupreous chafer, *Anomala cuprea* (Hope), a humus and plant root feeder, were examined at 25°C. The UV used was from sunlight and from UV lamps. The peak wavelengths of UV A, UV B and UV C lamps were 350, 310 and 254 nm, respectively. Mortality increased after exposure to sunlight for 40 min at intensities of 4.3, 3.3 and 3.1 mW/cm^2, decreasing insecticidal activity. UV A (1.2 mW/cm^2) was not harmful to J_3 for up to 40 min exposure. Contrarily, all J_3 were killed by exposure to UV B (0.9 mW/cm^2) for 5 min and to UV C (0.9 mW/cm^2) for 30 s. The insecticidal activity of J_3 and the number of *A. cuprea* cadavers producing nematode progeny after the treatments decreased at 40 min, 5 min and 10 s exposures to sunlight, UV B and UV C, respectively, but did not decrease after exposure to UV A. Sunlight, UV B and UV C decreased the density of viable symbiotic bacterium, *Xenorhabdus japonicus*, associated with *S. kushidai* in the body of J_3 (Fujiie and Yokoyama, 1998).

The effect of optical brighteners (OBs) and PABA as UV protectants for EPNs was tested under laboratory conditions. IJs of indigenous EPN isolates of *Steinernema* sp. (SSL2), PDBC EN 13.21 and PDBC EN 14.1, and two of *H. indica*, PDBC EN 13.22 and PDBC EN 14.3, were subjected to irradiation in aqueous suspension of OBs and PABA. The addition of OBs/PABA enhanced the tolerance of IJs to UV. Concentrations of 0.05/0.1% of all protectants offered excellent protection (Hussaini *et al.*, 2003). In onion, leek and chives, a combination of *Lecanicillium muscarium* (Petch) Zare et Gams, *P. fumosoroseus* (Wize) A.H.S. Br. & G. Sm., *S. feltiae* and *H. bacteriophora* commercial formulations controlled onion thrips, *Thrips tabaci* (Linderman), successfully and were compatible (Jung and Kim, 2004).

Green manure is a standard agricultural practice. The addition of lucerne leaves (*Medicago sativa* L.) to soil increased nematode-trapping fungus, *Dactylellina candidum* (Nees) Yan Li and microbivorous nematodes. *D. candidum* did not increase when lucerne leaves were added to soil lacking nematodes, but did increase if nematodes were added to soil lacking lucerne leaves. *D. candidum* and microbivorous nematodes did not exhibit predator–prey dynamics (i.e. fungus seldom reduced numbers of resident nematodes), and after initially increasing in lucerne-amended soil, numbers of *D. candidum* then decreased, whereas numbers of nematodes continued to increase (Jaffee, 2006).

Xenorhabdus sp. and *Photorhabdus temperata* subsp. *temperata* (Ptt) are the symbiotic bacteria of *Steinernema monticolum* and *H. megidis*, respectively. When the nematodes enter target insects, the symbiotic bacteria are released into the insect haemocoel. The bacteria cause septicaemia and kill the insect host. When ingested, the bacteria cause high mortality of third-instar larvae of *S. exigua*, but show no pathogenicity to the fifth-instar larvae. The high mortalities in the third-instar larvae seem to be caused by antibiotic activity against the opportunistic human pathogen, *Bacillus cereus*, a gut symbiont required for optimal development of *S. exigua*. To increase their pathogenicity in the fifth instar, the bacteria should be delivered into the haemocoel. To this end, *Bacillus thuringiensis* as a synergist was used to facilitate entry of the bacteria from the gut lumen into the haemocoel. The bacterial mixture treatment was highly synergistic against the fifth-instar larvae. The synergistic effects were shown by the successful infection of *Xenorhabdus* sp. or Ptt in the insect haemocoel a possibility that *Xenorhabdus* sp. or *Photorhabdus* sp. can be applied to kill *S. exigua* by oral treatment in a mixture with *Bt* (Jung and Kim, 2004).

5.2.14 Other agents

In general, if the natural enemies of an insect pest live above the ground, they are not affected by nematodes. If they are living in the soil, there may be a possibility of infection, although in practice this did not seem to occur. Exposing *A. ipsilon* larvae to *S. carpocapsae* 24 h after parasitism by *Meteorus rubens* prevented the emergence of *M. rubens* from host larvae at concentrations of 15 IJs. The effect of *S. carpocapsae* decreased when exposure was delayed to 7 and 12 days after parasitism by *M. rubens*. The survival of *A. ipsilon* larvae was shorter when exposure was delayed from 24 h to 7 and 12 days (Zaki *et al.*, 1997). The combined effect of *S. abbasi*, *S. carpocapsae* and *S. feltiae* with and without arbuscular mycohrrizal fungus, *Glomus intraradices* N.C. Shenck & G.S. Sm., on the development and reproduction of root-knot nematodes, *Meloidogyne incognita*, in tomato was evaluated in microplots. Observations at harvest stage revealed that application of *S. carpocapsae* at $4 \times 10^5/m^2$ along with arbuscular mycorrhizal fungi (AMF) reduced the number of galls and egg masses per root system, number of eggs/eggmass and hatch rate of eggs. Enhancement in growth, development and yield of tomato in terms of shoot height, shoot weight, root length, root weight, number of branches/plant, number of flowers, number of fruit and fruit yield was obtained. This treatment was followed by *S. abbasi* at $2 \times 10^5/m^2$ + AMF and *H. indica* alone at $4 \times 10^5/m^2$. The persistence of *H. indica* was greater in the plot receiving a dose of $4 \times 10^5/m^2$, followed by 2×10^5 and $1 \times 10^5/m^2$; however, there was no difference in per cent mortality of *G. mellonella* larvae with and without AM fungus treatments (Hussaini *et al.*, 2009).

Soil mites are predatory on several organisms. *Sancassania* sp. Oudermans (Arachnida: Acaridae) adult female mites consumed more than 80% of

S. feltiae IJs within 24 h. The IJs showed less infectivity to *G. mellonella*. Predation was greater in sandy soil than in loamy soil; *S. feltiae* IJs were predated more than *H. bacteriophora*. No phoretic relationship was observed between mites and nematodes, and the nematodes did not infect the mites (Karagoz *et al.*, 2007). Pest control is greatly improved in a combination of bioagents with synergistic or additive effects. A synergistic effect on the virulence of *Hoplia philanthus* (Coleop: Scarabaeidae) was observed in a combination of *H. megidis* with *M. anisopliae* (Ansari *et al.*, 2004). Combining bacteria *Paenibacillus popilliae* Dutky (Eubacteriales: Bacillaceae) (= *Bacillus popilliae*) or *Bt* Buibui gave higher control of white grubs than either alone (Thurston *et al.*, 1994; Koppenhofer and Kaya, 1997). *Paenobacillus* sp. appears to be widespread in phoretic association with heterorhabditid sp. (Enright *et al.*, 2003; Enright and Griffin, 2005). Spore attachment to nematode cuticle impaired motility and reduced population growth rate of *S. diaprepesi* (El Borai *et al.*, 2005). Terrestrial isopod wood lice, *Porcellio scaber* (Latreille), and spotted snake millipede, *Blaniulus guttulatus* (Fabricius) were highly pathogenic to *S. carpocapsae* and *S. feltiae*. Pathogenicity was similar at higher doses. Nematodes multiplied in the arthropod bodies (Jaworska, 1991).

Grubs of the western masked chafer, *C. hirta*, infected with the milky disease bacterium, *B. popilliae* (*Bp*), were more susceptible to *H. bacteriophora* and *S. glaseri* than healthy grubs. Increased susceptibility was due to greater ease of nematode penetration through the midguts of *Bp*-infected insects and occurred after >10 days exposure to *B. popilliae* spores. By that time, the bacterium was proliferating rapidly in the insect haemolymph and possibly stressed the insect by physiological starvation. In the greenhouse, *S. glaseri* killed more *Bp*-infected grubs than healthy grubs, whereas no differences were noted with *H. bacteriophora*. Thus, to achieve the best grub control in the presence of *B. popilliae*, the application of *S. glaseri* is recommended (Thurston *et al.*, 1994). Mortality of the Mexican beetle, *Zygogramma bicolorata* Pallister, grubs and adults (biocontrol agent of the invasive weed *Parthenium hysterophorus* L.) due to *S. glaseri* occurred faster in the adult (67 h) than in the grub stage (72 h). Observations indicate that *Z. bicolorata* is not a promising host for EPNs (Seth *et al.*, 2007). In the field and greenhouse, the performance of nematodes was monitored in the presence of *Bt* (Koppenhofer *et al.*, 1999).

Collembolans *Folsomia candida* Willem (Isotomidae) and *Sinella caceca* (Entomobryidae) readily consumed large numbers of *S. glaseri*, *S. carpocapsae* and *S. feltiae*. However, phoresy or adverse effects of nematodes on Collembola were not noticed. *S. carpocapsae* exposed to collembolans caused less mortality of *G. mellonella*. In contrast, it did not reduce the efficacy of nematodes against *P. japonica* grubs in turfgrass plugs. It is more likely to reduce the effectiveness of nematode applications against surface-feeding insects than against subterranean root feeders (Gilmore and Potter, 1993).

Neochtina eichhorniae Warner (Curculionidae), bioagent of water hyacinth, *Eichhornia crassipes* (Mart.) Solms., was not affected adversely by *Bt*, *B. bassiana* and *M. anisopliae*. However, two strains of *S. carpocapsae* caused 60–70% mortality. Limited potential for these pathogens to control *N. eichhorniae* in resource recovery or wastewater treatment facilities is noted (Haag and Boucias, 1991). Devi (2007a) found higher larval mortality of waxmoth larvae (70%) when *H. indica* was inoculated alone than with *Arthrobotrys oligospora* combined (62%). *A. oligospora* + *S. carpocapsae* resulted in lower larval mortality (47%) than *S. carpocapsae* alone (60%). *H. indica* multiplication was greater than when combined with fungus. A similar trend was also observed with *S. carpocapsae*. Hence, *H. indica* was more effective than *S. carpocapsae* in habitats with abundant nematophagous fungi.

The addition of biocidal compound Sincocin-AG to soil reduced *S. feltiae* and *H. heliothidis* numbers when poultry manure had not been added (Ameen, 1991). In non-sterilized soil, the survival of *S. carpocapsae*, *S. glaseri* and *H. bacteriophora* was lower than in sterilized soil. *Hirsutella rhossiliensis* was the dominant antagonist in the soil. In non-sterilized soil, more *S. glaseri* than *S. carpocapsae* or *H. bacteriophora* were parasitized by *H. rhossiliensis* (Deuteromycotina: Hyphomycetes); consequently, the survival of *S. glaseri* was lower. Differential survival of nematodes in the presence of *H. rhossiliensis* may be due to differences in their retention of the J_2 cuticle, susceptibility to fungal adhesion and motility. Highly motile nematodes encountered more *H. rhossiliensis* conidia than less motile nematodes. Relative motility in soil was in the order *S. glaseri* > *H. bacteriophora* > *S. carpocapsae*. EPNs that retain their J_2 cuticle exhibit low motility and are refractory to fungal adhesion, and may be suitable for controlling insects in habitats with abundant parasitic fungi such as *H. rhossiliensis* (Timper *et al.*, 1991). This suggests that *H. bacteriophora* may be more effective than *Neoaplectana* sp. in habitats with abundant nematode parasitic fungi. *Dactylella megalospora* trapped and assimilated 18 genera of plant parasitic and nine non-plant parasitic nematodes. Details of the characterization of *D. megalospora*, the nematode/fungus interaction, trap senescence and resistance to trapping are documented. The fungus trapped and destroyed more than 99% of 300,000 *S. carpocapsae* (Esser *et al.*, 1991). Devi (2007b) found *G. mellonella* larvae previously exposed to *B. bassiana*, *Nomuraea rileyi* (Farlow) Samson and *Lecanicillium* sp. showed that combined treatment with *S. carpocapsae* and *H. indica* was efficient compared to either fungus or nematode alone. No significant difference in the infectivity of *H. indica* and *S. carpocapsae* was found, except in the combination of *B. bassiana* and *S. carpocapsae*. Combined *N. rileyi* and *H. indica* showed higher mortality, followed by combining *B. bassiana* and *H. indica* over fungus or nematode alone. For *S. carpocapsae* plus *B. bassiana*, insect mortality was not significantly different compared with *N. carpocapsae* alone. In treatments with nematode plus fungus, the percentage mortality of *S. exigua* from *S. carpocapsae* and *H. bacteriophora* was reduced relative to the treatments with nematodes alone. Inundative release of EPNs where *B. bassiana* occurs may result in greater control of soilborne insects than applying nematodes alone (Barbercheck and Kaya, 1991). Better performance has indicated the potential of commercial *Paecilomyces lilacinus* strain 251 for integration with EPNs in pest management strategies (Rumbos *et al.*, 2007).

Pea aphid, *Acyrthosiphon pisum* (Harris), and green aphid, *Megoura viciae* (Buckton) exposed to IJs of *Steinernema* sp. strain *Hylobius* on moist filter paper were parasitized, and while only a few adults of *A. pisum* were parasitized, parasitism of *M. viciae* reached 28% in 3 days. Females developing in aphids grew only to a very small size, and mated but did not lay fertile eggs. Such larvae as did hatch were later encapsulated in *Endotokia matricida*. Similarly, *Coccinella septempunctata* L. collected from pine forest debris were parasitized within 4 days (Mracek and Ruzicka, 1990). Nematodes and fungi have overlapping host ranges.

The host period of lethal infection (PLI) due to *B. bassiana* infection was impacted and was shorter than with either pathogen alone. *B. bassiana* and nematodes rarely co-produced progeny in dually infected hosts. *Xenorhabdus* was observed in host haemolymph 12 h after application of the nematodes. *Xenorhabdus* or nematodes prevented/inhibited growth of *B. bassiana* if nematodes were applied within 24 h (Barbercheck and Kaya, 1990). The braconid parasitoid, *Apanteles ultor*, did not develop when the host was infected with *S. feltiae* or *H. heliothidis*, but the parasites began to be affected within 24 h by *S. feltiae* and within 36 h by *H. heliothidis*. Death is presumed to have been caused by associated bacteria, since no braconid larvae contained any nematodes (Triggiani, 1985): the few surviving braconid larvae failed to pupate as they were killed by the nematode.

Diaprepes abbreviatus was infected with fungi, *M. anisopliae*, *B. bassiana*, *P. lilacinus* and *Aspergillus ochraceus* Wilhelm and the EPNs *S. feltiae* and *Heterorhabditis* sp. throughout the year. Fungus-infected larvae were most abundant from May to July, while nematode parasitism was predominant May to November. In simultaneous use, EPNs caused the death of parasitoids. Ectoparasitic ichneumonid species, *Mastrus ridibundus* (Gravenhorst) and *Liotryphon caudatus* (Ratzeburg), exposed to *S. carpocapsae* (10 IJs/cm^2; ~LC80–90 for codling moth larvae) within codling moth cocoons had 70.7 and 85.2% mortality, respectively. Diapausing parasitoid larvae were protected from nematode penetration within cocoons. The parasitoid females were able to detect and avoid ovipositing on nematode-infected cocooned codling moth larvae. Complementarity of EPNs and parasitoids was noticed (Lacey *et al.*, 2003). Similarly EPNs and a predatory phytoseiid mite, *Amblyseius cucumeris* (Oudemans), alone or in combination greatly reduced western flower thrips, *Frankliniella occidentalis* (Pergande), on beans. Second-instar larvae of thrips fell off the plants and pupated in soil due to the presence of mites. The introduction of 10 adult mites/plant caused 83% thrips control (Ebssa *et al.*, 2006).

Factors driving change in the population dynamics of nematodes in the soil are almost completely unknown. Top-down control by microarthropod predators may have a significant impact on nematode population dynamics. Mites and Collembola were capable of reducing nematode numbers in the laboratory and were feeding on a targeted nematode species in the field. A PCR-based approach was developed for the detection of predation on three species of slug- and insect-pathogenic nematodes:

Phasmarhabditis hermaphrodita Schneider, *H. megidis* and *S. feltiae*. The collembolan, *F. candida*, and the mesostigmatid mite, *Stratiolaelaps miles* (Berlese), were employed as model predators to calibrate post-ingestion prey DNA detection times. Detection times for nematode DNA within the guts of Collembola were longer than in mites, with half-lives of 08.75 h and 05.03 h, respectively. *F. candida* reduced numbers of *H. megidis*, with rates of predation of ~0.4 IJs per collembolan/h over 10 h (Read *et al.*, 2006). Whether the interaction between EPNs and insect pathogenic fungi is additive, antagonistic or synergistic depends on species, insect host, rate and time of application (Koppenhofer and Kaya, 1997; Ansari *et al.*, 2004, 2006). The antimicrobial nature of metabolites produced by associated bacteria is well known (Barbercheck and Kaya, 1990; Chen *et al.*, 1994; Webster *et al.*, 2002). However, diverse toxic metabolites are described in several biological control agents, including species of *Beauveria*, *Metarhizium* and *Paecilomyces* (Vey *et al.*, 2001). Ansari *et al.* (2005) investigated the mutual effects between the bacteria *Photorhabdus luminescens* and *Xenorhabdus poinarii* and different entomopathogenic fungi. *P. luminescens* was found antagonistic to *M. anisopliae*, *Beauveria bassiana*, *B. brongniartii* and *P. fumosoroseus*, inhibiting growth and conidia production, however, *X. poinarii* did not inhibited fungal growth. Henderson *et al.* (2009) stated that mustard green manures or seed meal were high in glucosinolates, and produced a natural biofumigant on incorporation into the soil. *Brassica carinata* seed meal singly and both the seed meal and *Steinernema* spp. incorporated reduced root-knot nematode damage to potato tubers and increased marketable tuber yields. Thus, mustard seed meal applications harmful to the target root-knot nematode also disrupted the ability of *Steinernema* spp. to act as biocontrol agents for the control of potato beetles following biofumigation.

Root systems of maize varieties emit (E)-beta-caryophyllene (E beta C) in response to feeding by larvae of the beetle, *D. virgifera virgifera* LeConte. The sesquiterpene was shown to attract EPNs. By comparing beetle emergence and root damage for two maize varieties, one that emits E beta C and one that does not, it was found that root protection by *H. megidis* and *S. feltiae* was higher on the emitting variety, but this was not the case for *H. bacteriophora* (Hiltpold *et al.*, 2010) The infectivity and multiplication of *S. feltiae*, *S. glaseri* and *H. indica* in granulovirus (GV)-infected *Chilo infuscatellus* and *Chilo sacchariphagus indicus* were studied. Nematodes developed and reproduced in *C. infuscatellus* and *C. s. indicus* larvae infected with GV. Multiplication of dauer juveniles (DJs) was significantly higher in healthy larvae of borers than GV-infected larvae (Karunakar *et al.*, 2003).

5.2.15 Interspecific interactions

The relative importance of competition in determining the characteristics of organisms, populations and communities has long been a major issue in ecology (Wootton, 1994; Pianka, 1999). More importantly in EPN applications are the risks of displacing endemic EPNs and other non-targets. Competition is more dramatic when it occurs between species, but also occurs within species and has important consequences. There has been little research on inter- and intraspecific competition. Numerous aspects of EPN ecology and behaviour could have evolved in this context and might enable the coexistence of certain species. Habitat heterogeneity can also facilitate coexistence and the avoidance of competition. Habitat patches exit in a matrix within which numbers, arrangement and size of patches influence the movement of organisms, food web interaction and persistence of populations (Polis *et al.*, 1997; With *et al.*, 2002). Interspecific competition can be direct or indirect. Direct is between two species, whereas indirect applies to a wide range of effects mediated by the presence of one or more species by a change in the chemical or physical environment (Pianka, 1994). Indirect effects could be positive or negative, and the latter only is considered competitive. As EPNs are associated with symbiotic bacteria in cadavers, direct competition might be rare, but indirect competition mediated by bacteria could be common. An insect is rarely infected by more than one species of EPN (Ehlers *et al.*, 1991), and heterorhabditids and steinernematids apparently cannot develop on each other's bacteria (Alatorre-Rosas and Kaya, 1990, 1991). However, *S. glaseri* and *S. carpocapsae* can co-infect and produce progeny from a single *G. mellonella* (Koppenhoffer *et al.*, 1995a). In such a case, *S. glaseri* is affected less negatively by mixed infection than is *S. carpocapsae*, perhaps by its faster development and also ability to use *S. carpocapsae* symbiont. Indirect mutualism and commensalism are positive effects involving consumer–resource interaction. Higher-order interactions are non-additive (Barbercheck *et al.*, 1995). Apparent competition occurs when two prey species share a common natural enemy. A numerical response in predatory mites and subsequent reduction in two EPN species applied as well as endemic was observed (Walter *et al.*, 1986). Free-living bacterivorous, fungivorous predatory and omnivorous nematodes are important constituents in decomposition and nutrient cycling food webs. Grewal *et al.* (1997) observed no effects of *S. carpocapsae* and *S. glaseri* on free-living nematodes. Despite a reasonable understanding of the events that lead to and culminate in infection, changes within the host affect approaching IJs, which remains unexplored. The quality of a host is dynamic, and what is critical for the parasite is the level of fecundity that is supported by the host (Grewal *et al.*, 1992; Grewal and Richardson, 1993). Infective juveniles do not find hosts via random movement; infection decisions are made based on the characteristics of the hosts. Campbell *et al.* (1999) stated that when provided enough hosts, all IJs that result from a given infection will infect a host immediately. Tillage is especially disruptive to the soil environment and can influence survival and persistence. The diversity and abundance of predators are greater under no-till, and natural control of insects in soil may be enhanced in conservation tillage systems (Stinner and House, 1990; Brust, 1991; Letorneau, 1998). A high level of crop residues in no-tillage regimes might influence the abundance of EPNs through a greater number and diversity of hosts. Under conventional tillage regimes, the soil surface tends to have a greater fluctuation in temperature/moisture than under no/reduced tillage regimes (Hummel *et al.*, 2002; Millar and Barbercheck, 2002). However, these effects vary in different species.

The occurrence of more than one species of EPN within the same location is not unusual. Several surveys have shown occurrence sympatrically (Puza and Mracek, 2005). Interactions of two sympatric EPNs, *Steinernema affine* and *S. kraussei* in single species, simultaneous and sequential infections of *G. mellonella* were performed. The invasion rate of *S. affine* was not affected by the mixed infection; the invasion rate of *S. kraussei* was strongly reduced. *S. affine* outcompeted *S. kraussei* and the progeny production of the latter species occurred sporadically. *S. affine* was able to infect and multiply in a higher number of hosts, and also in already infected hosts or killed by *S. kraussei* producing normal progeny (Chevalier and Webster, 2006; Puza and Mracek, 2009). When *S. carpocapsae* and *S. glaseri* were co-infected (Koppenhofer *et al.*, 1995b), both penetrated, established and developed into normal adults. Although *S. glaseri* and *S. carpocapsae* produced progeny from the same cadavers, *S. glaseri* was less affected by mixed infection than *S. carpocapsae*. The ratios between the two species affected progeny production. In combinations of *S. glaseri* and *S. carpocapsae*, *S. glaseri* had reductions in progeny production of 53, 9 and 30%, respectively, for ratios 8:16, 16:16 and 16:18 IJs, whereas *S. carpocapsae* had reductions of 35, 85 and 98%, respectively. The superiority of *S. glaseri* is ascribed to its faster development and also the less specific association of *S. glaseri* with its bacteria, *X. poinarii*. Vertical distribution and patchiness contribute to the coexistence (Stuart *et al.*, 2006). Kondo and Ishibashi (1989) observed *S. carpocapsae* to be a stronger competitor than *S. glaseri* in mixed infections of *S. litura*. Reduction in progeny of *S. glaseri* was 80% from co-infected *S. litura*, while *S. carpocapsae* was not affected. While studying the interactions of *S. carpocapsae*, *S. feltiae* and *H. bacteriophora* in sand, Neumann and Shields (2006) did not observe any evidence of multiple infection on *G. mellonella*, even though the nematodes infected hosts in close proximity. *Steinernema* sp. can develop together in one insect host, though one species may suffer the effect of competition more than the other (Hay and Richardson, 1995). *S. feltiae* was the most virulent species/isolates, Nemasys and Sus 94 were superior to the rest. *S. kraussei* was least infectious. Increasing the density of *S. carpocapsae* and *H. bacteriophora* within *G. mellonella* larvae declined penetration, showing the detrimental effects of overcrowding. *S. carpocapsae* and *H. bacteriophora* had optimal survival at a density of 100 nematodes/host, and *H. bacteriophora* tolerated greater densities. Fecundity was reduced at densities above and below 100 nematodes/host (Selvan *et al.*, 1993). When species with different search strategies and pest species of soil insects occurred in the same habitat (Kaya *et al.*, 1993b), soil temperature influenced the results. At 22°C, both nematode species killed their respective target insects within 1 week. At 16°C, *S. carpocapsae* killed larvae of *A. ipsilon*, but *H. bacteriophora* was not effective against larvae of *O. sulcatus*. Interspecific competition between *H. heliothidis* and *S. feltiae*, *S. bibionis* or *S. glaseri* for *G. mellonella* in sand showed that in vertical dispersal, *Neoaplectana* spp. outcompeted *H. heliothidis* for insect hosts, while *H. heliothidis* outcompeted *Neoaplectana* spp. for hosts located distally to nematode placement. In downward dispersal, *N. feltiae* infected 73% of the hosts located at 5 cm from nematode placement and *H. heliothidis* infected 20%. At 10–25 cm from placement, *H. heliothidis* was the only nematode species infecting the hosts. Similar trends were observed with *H. heliothidis* and *N. bibionis* and *H. heliothidis* and *N. glaseri* when they were in competition for insect hosts. In upward dispersal, *H. heliothidis* was dominant over *N. feltiae.* In horizontal dispersal, the steinernematids outcompeted *H. heliothidis* for hosts located distally (10–30 cm). In dual infection of *H. heliothidis* and *Neoaplectana* sp., the development of both occurred, but they eventually died (Alatorre-Rosas and Kaya, 1990, 1991).

S. abbasi and *H. indica* are found in date palm plantations throughout the whole year. *S. abbasi* caused more mortality in larvae of *G. mellonella* than those of *H. indica.* Mortality by *S. abbasi* reached 100% in March, August, October and November. The coexistence of *S. abbasi* and *H. indica* in the soil around the trunks of palm trees resulted in successful combined infection, with both species in *G. mellonella* larvae. Such infected larvae showed the symptoms of either nematode species and the combined infection was detected in the second passage, when some infected larvae showed symptoms of *S. abbasi* and others showed symptoms of *H. indica* infection (Abbas and Mousa, 2003).

Symbioses between invertebrates and prokaryotes are biological systems of particular interest in order to study the evolution of mutualism. The symbioses between *Steinernema* and *Xenorhabdus* are very tractable model systems. Previous studies have demonstrated a highly specialized relationship between each strain of nematode, and the bacteria and mutualism played a role in the important life history traits of each partner, such as access to insect host resources, dispersal and protection against biotic and abiotic factors. The impact of *Xenorhabdus* symbionts on the progression and outcome of interspecific competition between *Steinernema* spp. was addressed. This was influenced by associated bacteria: the results with aposymbiotic nematodes were totally opposite. *Xenorhabdus* provides *Steinernema* with access to food sources, and also new abilities to deal with competitors (Sicard *et al.*, 2006).

5.2.16 Vertical movement

Soil type affects the vertical movement of different isolates. *S. carpocapsae* (NY001) and *Steinernema* sp. (NY008-2E) remained primarily at depths of <15 cm, whereas heterorhabditids dispersed to 35 cm depth. Movement of *H. bacteriophora* (Oswego) was greatest in loamy sand and that of *Steinernema* sp. (BY008-2F) was greatest in sand loam. *H. bacteriophora* (Oswego) and *S. carpocapsae* (NY001) infected *G. mellonella* larvae in the field at soil temperatures of 15–18°C; *Steinernema* sp. (NY008-2E) at temperatures between 13 and 15°C (Ferguson *et al.*, 1995). The heterorhabditid isolates HUK211, K122, *H. megidis* HO1 and HL81 had superior migratory ability. Activity at low temperatures was not related to a specific taxonomic group (Westerman, 1994). Heating and cooling of soil surface creates temperature gradient fluctuations. When heat waves propagate horizontally to eliminate gravitational effects, the movement of *Ditylenchus phyllobius* (Thorne, 1934) Filipjev, 1936, *S. glaseri* and *H. bacteriophora* was rapid and random. However, the reniform nematode, *Rotylenchulus reniformis*, moved away from and the root-knot, *M. incognita* moved toward the thermal surface. Responses were the same whether heat waves propagated upward/downward; the differences in movement of the species was related to body length, with the smallest nematodes moving downward and the largest moving upward (Robinson, 1994).

Field studies were conducted to determine the vertical distribution of the larvae and pupae of curculionid, *O. sulcatus* (Fabricius), and the depth of penetration of nematode in field-grown Japanese yew, *Taxus cuspidata* Siebold & Zucc. Larvae of *O. sulcatus* were found at 30 cm but more than 90% in the top 15 cm. The distribution of pupae was similar. Soil columns may not be effective models of nematode infection in undisturbed field soils (Hanula, 1993). When *S. scapterisci* IJs were released on the soil surface in both the laboratory and field in Florida, some of them moved downwards through the soil at least 10 cm in 5 days and infected and killed mole crickets (*Scapteriscus* spp.). When released 2 cm below the soil surface, most of them moved into the upper 2 cm layer of soil, while some moved 10 cm downward. When placed at the centre of a 16-cm soil column, IJs moved in both directions, more moving downward than upward. IJs were more efficient in killing mole crickets in the field than in the laboratory (Nguyen and Smart, 1991). Ambushers employ a long-distance dispersal strategy to maximize reproductive success in the absence of hosts. Vegetation enhanced lateral dispersal, particularly of *H. bacteriophora*. The pattern of dispersal differed spatio-temporally, with an average of ~6 cm/day. *S. carpocapsae* ambushed close to the cadaver (<3.8 cm), whereas *H. bacteriophora* dispersed between 7 and 12 cm away from cadavers. *S. carpocapsae* dispersed faster than *H. bacteriophora*. This is an adaptive dispersal strategy by *S. carpocapsae* in the absence of hosts (Harit *et al.*, 2014). Phoresis is a strategy in which an organism attaches to another for dispersal. Relatively immobile larvae of the large pine weevil, *Hylobius abietis* L., are highly susceptible to EPNs and support reproduction, whereas adults do not. IJs of *H. megidis* and *S. carpocapsae* use the resistant but highly mobile adults of *H. abietis* for phoretic dispersal. The strategy depends on nematode species and habitat (sand versus peat), and no sex-biased dispersal was evident in *S. carpocapsae*. Sibling competition and inbreeding is thus avoided (Kruitbos *et al.*, 2009).

The role of the sexes in the emergence and dispersal of EPNs from hosts has rarely been addressed. *S. feltiae* (UK76 and SBI1), emerging from *G. mellonella* cadavers, were classed as non-dispersed (remaining on the cadaver for up to 9 days) and dispersed (actively moving away from the cadaver). Sex ratios within both classes were examined in infective (individuals that invaded *G. mellonella* successfully) and non-infective populations. Sex ratios differed from 1:1 in the SBI1 non-dispersed population (female bias) and SBI1 non-dispersed infective population (male bias). For each isolate, dispersed individuals were more infective than non-dispersed by 11% of SBI1 and 22% of UK76. Non-dispersed individuals were found to be mature IJs compared with 78% of SBI1 and 82% of UK76 dispersed individuals. SBI1 males survived alone in *G. mellonella* cadavers for longer periods than did females, hinting at 'male colonization' (Grewal *et al.*, 1993b; Rolston *et al.*, 2006).

5.2.17 pH

Decline of soil pH (8 to 4) decreased the survival of *S. carpocapsae* and *S. glaseri* IJs in a 16-week period, whereas it was similar at pH 4, 6 and 8 in the first 4 weeks and at pH 10 for 1 or more weeks. At oxygen:nitrogen ratios of 1:99, 5:95 and 10:90, *S. carpocapsae* survived better than *S. glaseri* in the first 2 weeks, and survival of both species declined to less than 20% after 4 weeks. Pathogenicity of *S. carpocapsae* was greater than that of *S. glaseri* in the first 2 weeks (Kung *et al.*, 1990). *H. indica* survival was maximum at pH 5 and *S. carpocapsae* at 7, and least at pH 2 and 3, respectively. *S. carpocapsae* tolerated a broader range of pH (4–9) than *H. indica*, whose survival and infectivity were not affected by varying pH levels, except at 2 and 3. Progeny production was on a par with the control (Hussaini *et al.*, 2004a).

There are still many challenges facing entomopathogenic nematologists in monitoring and predicting the impacts of climate change. Research needs to focus on pest phenotypic and genotypic flexibility, responses to global changes operating in concert, and then awareness of some patterns may become apparent in due course. We have to equip ourselves to face and tackle the consequences of climate change and its effect on potential EPNs. In this context, climatic pattern mapping of a region is important in terms of risk assessment of pests as well as for the EPN introduced. This would give an insight into species distribution, in relation to relevant climatic data, particularly the establishment of EPN species.

5.3 Biodiversity

Nematodes are one of the most highly diversified group of animals on earth. They are the most abundant metazoans in the soil, found to occupy all habitable aquatic and terrestrial habitats. Generally, they are free living in marine, freshwater or in soil. Four of every five multicellular animals on the planet are nematodes. They occupy any niche that provides an available source of organic carbon in marine, freshwater and terrestrial environments.

5.3.1 Classical studies

EPNs have been in focus of late as potential agents for biological control. Most orders of insects are associated, and natural associations involve 11 orders of nematodes and 19 of insects. Associations range from phoretic to facultative/obligate parasitism, and may lead to host death, sterility, reduced fecundity, delayed development, or aberrant behaviour. They have a unique symbiotic relationship with bacteria (*Xenorhabdus* or *Photorhabdus*) (Boemare *et al.*, 1993). Bacteria live in the intestine of the nematodes, released into the haemocoel of the host, cause septicaemia and produce conditions for nematode growth and reproduction (Askary,

2010). Specific single nematode–bacterial association is on record; however, one bacterial species may be associated with more than one nematode species. The attractive features are production of the infective stage and the symbiotic association with bacterium. Information on the population dynamics and ecology of EPNs in natural systems is critical. At spatial and temporal levels, EPNs may also regulate insect populations, impacting ecosystems and altering plant species richness, composition and productivity. Worldwide degradation of soils, particularly in desert or semi-desert ecosystems, and rational management urgently requires increased knowledge of the ecology of these soils, including knowledge of the diversity of the soil flora and fauna and their respective interactions. Nematodes and insects are among the most abundant faunal components of the soil.

Variability across seasons, habitats, geographic regions and factors that affect soil texture, moisture, temperature and the availability of susceptible hosts determine local distribution. Populations are mostly patchy. In general, populations can exhibit uniform, random, or patchy distribution, but pattern depends on the scale of measurement. The distribution of resources is also correlated with patchiness, which influences the dynamics of predation, parasitism and competition. Besides the limited dispersal capability of IJs, the variability of founding, establishment under different circumstances, passive dispersal by water or phoretic dispersal within or upon hosts or other organisms, mites or earthworms are critical inputs. Degrees of patchiness could be characteristic for a particular species or set of conditions within the soil profile relating to foraging strategies (Dutilleul and Legendre, 1993). Metapopulation structure and dynamics is an important theme in population ecology, with natural populations being viewed as a series of transient ephemeral local populations with average lifespans that are much shorter than those of the whole network. Rates of birth and death, immigration and emigration influence local populations, but the persistence of metapopulations results from a balance between recurrent colonization and extinction events, with a high turnover of local populations for some EPNs under certain conditions. The genetic diversity among patches appears to be considerable (Hanski, 2001). Patchiness has ramifications at the population and community levels by influencing gene flow and altering the dynamics of competition, predation and parasitism (McCauley, 1995; Harrison and Hastings, 1996). Longevity and tolerance to stresses like heat, UV, hypoxia and desiccation differ among isolates. Different isozyme patterns and metabolic enzymes have been recorded. With localized patches, very little genetic variation may be exhibited. Estimation of species geographic distribution is critical to biodiversity and conservation biology for all organisms, including nematodes. New tools that offer possibilities for a better understanding of biological diversity are geographic information systems (GIS) and global positioning systems (GPS), remote sensing and ecological modelling. Moreover, these tools have been applied in nematology, particularly for plant parasitic nematodes, to help make predictions for future control management strategies. However, explorations for predicting species distribution of EPNs are yet not possible as most data on EPNs are fragmented. Sampling has rarely been planned in a systematic manner, and has been restricted to countries or certain geographic regions only. Within localized patches at a given site, nematodes often exhibit very little genetic variation. In a cadaver, if the number of IJs colonizing is small, it would produce genetically related progeny, and thus effective population sizes may be relatively small, especially so for *Heterorhabditis* spp. as the first generation is hermaphroditic.

5.3.2 Molecular characterization

Molecular biology is a useful tool to elucidate the biodiversity and biogeography of EPNs. These nematodes are morphologically conservative: molecular techniques will provide for developing a robust, morphologically based taxonomy. Knowledge on biogeography and habitat specificity is necessary. Molecular approaches differentiate the species/subspecies/strains; understand their biology and evolution; and allow identification for registration, quarantine and proprietary protection (Curran, 1990), and their symbiotic bacteria (Akhurst, 1980; Boemare and Akhurst, 1988; Akhurst and Boemare, 1990).

To assign the DNA genotypes to a given taxonomic level, supporting data on cross-breeding, the use of restriction fragment length polymorphisms (RFLPs) and other DNA sequence-based procedures is required to determine in-depth inter and intraspecific variations. The ribosomal DNA repeat unit is an ideal choice for identification purposes and for differentiating species and strains of *Steinernema*/

Heterorhabditis (Reid and Hominick, 1992; Reid, 1994) and deducing phylogenetic relationships.

Further, biochemical and molecular techniques have been used in the taxonomy of *Xenorhabdus* and *Photorhabdus* (Akhurst and Boemare, 1990; Boemare and Akhurst, 1994). Commonly used techniques in bacterial taxonomy are biochemical and cultural characteristics; DNA/DNA hybridization; 16S rRNA oligonucleotide cataloguing; immunology and electrophoresis. Molecular genetics are now routinely employed to elucidate genes in the bacteria with the host–bacterial fuctions – *cip* (Bintrim and Ensign, 1998), *rpo* S (Vivas and Goodrich-Blair, 2001; He *et al.*, 2004) and in screens to identify symbiosis or nematode growth and reproduction (Ciche *et al.*, 2001), siderophore activity (Ciche *et al.*, 2003) and nematode colonization (Heungens *et al.*, 2002). The interactions between nematode and bacteria are confirmed to be intricate and complex. The formation of a colonized IJ in a multistage process in which nematode and bacteria recognize each other through a molecular dialogue is also evident (Heungens *et al.*, 2002). Molecular studies on *H. indica* isolates from India, Kenya, Indonesia and Cuba were characterized by molecular diagnostic features (Stock *et al.*, 2000). The Kenyan isolates represent a distinct subgroup of *H. indica.* These isolates lack one of the two *Hind* I restriction sites that are present in the γ DNA internal transcribed spacer (ITS) product of all the other isolates tested. They also differ from other *H. indica* isolates in γ DNA internal guide sequence (IGS) *Hae* III restriction profile. *Steinernema* and *Heterorhabditis* nematodes have similar life histories. The non-feeding IJs seek out insect hosts, especially in the soil environment. On the location of hosts, nematodes penetrate through natural body openings or thin cuticle and symbiotic bacterium is released, which multiply rapidly and cause insect death. The nematodes feed on the bacteria and liquefying insect, and mature into adults. The relationship is one of classic mutualism. Nematode growth and reproduction depend on the conditions established in the host cadaver by the bacterium. Bacterium contribute anti-immune proteins to assist the nematode in overcoming host defences and antimicrobials that suppress colonization of the cadaver by competing secondary invaders.

5.4 Geographical Distribution

There has been tremendous research and commercial interest in EPNs since the late 1970s. In addition to biological control benefits, microbiologists have found associated bacteria producing toxins and antibiotics that serve as a model system and help in understanding the symbiotic association (Boemare, 2002). The proliferation of research has been concentrated primarily in North America and Western Europe, and in Australia in the 1980s and 1990s (Commonwealth Scientific and Industrial Research Organisation, CSIRO). However, in other parts of the world, including India, it has gained impetus subsequently. Two EPN species could occupy the same continent/region, and even the same square metre, coexist as they occupy different soil depths or depend on different insects, and be widely distributed in soils throughout the world in different habitats/ecosystems, depending on their insect hosts. Some suggest steinernematids to be more common in cooler, temperate regions, while heterorhabditids are more common in warmer, tropical conditions (Hominick *et al.*, 1996). Distribution is aggregated rather than random. Prevalence and intensity can vary with time; some species are dominant while others are rare in a particular locality. In Germany, the prevalence of steinernematids was highest in woodlands (50.3%) (Sturhan, 1999), and *Steinernema affinis*, *S. feltiae*, *S. intermedium* and *Steinernema* sp. B were predominant. Similarly, heterorhabditids were equally abundant in turf and weedy habitats, but never in closed canopy forests (Stuart and Gaugler, 1994). Some steinernematids show a distinct habitat preference reflecting the distribution of suitable hosts that are adapted. Mracek and Beevao (2000) emphasized an essential impact of host aggregation on the incidence of EPNs. In Vietnam (Phan *et al.*, 2004) and Thailand, active programmes on EPNs exist, as well as in Jordan, Azerbaijan, Ukraine and Russia in varying degrees. Hungary has been an active participant of the Eurpean Commission's Cooperation in the Field of Science and Technical Research (COST), but still lags behind other European countries as the programmes are still in early phases. Hominick and Briscoe (1990) studied prevalence in Britain and found steinernematids more prevalent compared to heterorhabditids, the most abundant being *S. bibionis*. In regions of Slovakia, prevalence was highest in forest soil (58%). *S. affine*, *S. intermedium* and *S. feltiae* were most prevalent, and several species showed distinct habitat preference. In the Guadeloupe Islands, heterorhabditids compared to *Steinernema* spp. were abundant (97 and 3%, respectively). *H. indica*

(88%) occurred more frequently than *H. bacteriophora* (12%), with the highest prevalence in coastal areas (91.4%), tropical lowlands (5.7%) and tropical middle altitude areas (2.9%). Surveys in Belgium showed a prevalence of steinernematids and heterorhabditids in the region. *S. feltiae*, followed by *S. affinis*, *Steinernema* sp. B3 and *Heterorhabditis* sp. (north-west European strain) (Miduturi *et al.*, 1996), were abundant in woodlands, followed by cultivated land and grassland. Hara *et al.* (1991) reported the distribution of EPNs in the Hawaiian Islands, from six ecologically diverse habitats; 22 sites were positive for *Heterorhabditis* sp. and two for *Steinernema* sp.

EPN research in China, as of the mid-1980s and early 1990s, has been comprehensive in collaboration with CSIRO and ACIAR (Australian Centre for International Agricultural Research) (Bedding, 1990). China has been a rich source of EPNs. Species include *S. glaseri*, *S. carpocapsae*, *H. bacteriophora* (Hominick, 2002) and *H. indica* (Fang *et al.*, 2004). In addition, several species have been isolated (Xu *et al.*, 1991; Jian *et al.*, 1997; Qiu *et al.*, 2004, 2005a,b,c). In China, Zhang *et al.* (1992) found occurrence in 13.1% of the 175 samples, and incidence was affected by soil type and habitat. Per cent occurrence was 20.8, 15.1, 12.2 and 5.4% for loose sandy, sandy loam, medium loam and light loam soils, respectively. Hominick and Briscoe (1990) found steinernematids from 48.6% and only one site yielded *Heterorhabditis* sp. Most frequent was *S. bibionis*. The prevalence of steinernematids formed a continuum so that roadside verges harboured them most often, while heathland sites yielded least. Soil temperatures of 15°C at 5 cm in fields and verges, 13°C in hedgerows and heathland and 12°C in woodland were found. Prevalence in Britain was influenced by soil type, associated with calcareous soils, with a calcareous subsoil horizon. Soils periodically or seasonally waterlogged, suitably enriched by humus, and hence high in organic matter, harboured nematodes, and some had a clay content or subsoils high in clay. In different regions of Slovakia, Sturhan and Liskova (1999) found the rate of prevalence was highest in forest soil (58%). *S. affine*, *S. intermedium* and *S. feltiae* were most prevalent (74%); other species were *S. kraussei*, *S. carpocapsae* and *S. bicornutum*. Mixtures of 2–4 species were present in 30% of the population. Several species showed a distinct habitat preference: *S. affine* for arable soil and *S. intermedium* for woodland and riverbank vegetation with trees. Preference for certain soil types was less apparent. During the 1980s, Professor N. Ishibashi, Saga University, introduced a study of EPNs in Japan and organized Japanese researchers for basic and applied research (Ishibashi, 1987, 1990). *S. carpocapsae* was the first nematode registered for commercialization, and later *S. glaseri* and *S. kushidai* Mamiya were discovered (Mamiya, 1988) and commercialized. Yoshida *et al.* (1998) reported distribution in the five climatic regions, with eight *Steinernema* sp. and two *Heterorhabditis* sp. *H. indica* was found in the coastal region, from the subtropical to the warm temperature zone, and *H. megidis* in the eastern part of the warm temperature coastal region, respectively.

Surveys initiated in Korea during the 1980s found 30 isolates from forests, riparian habitats, agricultural fields, golf courses and seashores, and from insect cadavers (Choo *et al.*, 1995; Lee *et al.*, 1996a,b; Stock *et al.*, 1997a,b). Species identified include *S. carpocapsae*, *S. glaseri*, *S. longicaudum*, *H. bacteriophora* and *H. megidis* (Choo *et al.*, 1995; Stock *et al.*, 1997a,b). *S. monticolum* Stock, Choo and Kaya was described as a new species (Stock *et al.*, 1997b). Ecological characterization studies were carried out in the laboratory involving *S. carpocapsae*, *S. longicaudum*, *H. bacteriophora* and an undescribed *Heterorhabditis* sp. (Choo *et al.*, 1999, 2002; Chung, 2001). In Turkey, several surveys have been conducted, with the isolation of known species (Ozer *et al.*, 1995; Kepnekci, 2002; Hazir *et al.*, 2003b; Kepenekci and Susurluk, 2003) and Ozer *et al.* (1995) initially stated that they recovered *S. carpocapsae* from the coast of the Black Sea, which was later identified as *S. feltiae* (Hominick *et al.*, 1996). However Kepenekci (2002) isolated *S. carpocapsae* from the forest area of Anamur. From Turkey, Kepenekci and Susurluk (2003) isolated *S. feltiae* and *H. bacteriophora* (two isolates); Hazir *et al.* (2003b) isolated *H. bacteriophora*, *S. feltiae*, *S. affine* (Bovien) and Hazir *et al.* (2003a) isolated a new species, *Steinernema anatoliense*.

In Central America, the detection of indigenous isolates is promising because of the great biodiversity in these countries. *S. carpocapsae* was discovered in Chihuahua (Poinar, 1990) and *H. indica* from the subtropical region in Mexico (La Sierra, Tabasco) (Cortez-Madrigal *et al.*, 2003). *Heterorhabditis* and *Steinernema* sp. were collected from semi-desert areas at an altitude higher than 2000 m in Zacatecas (Cortez-Madrigal *et al.*, 2003). Later, 64 locations were sampled from maize, grain, forage sorghum and Sudan grass in six Mexican states (Molina-Ochoa *et al.*, 2003), and Jalisco Mexicoa *Heterorhabditis* sp. was isolated from soil (Diaz-Madiros *et al.*, 2002). In the Centro de Ciencias Medio Ambientales (El Selvador), an indigenous strain was isolated from Alto Valle. In Costa Rica, two *Steinernema* isolates and *H. indica* were recovered from soil (Uribe-Lorío *et al.*, 2005). Ruiz-Vega *et al.* (2003) recorded *S. feltiae* and a new *Heterorhabditis* sp. from Oaxaca, Mexico. Nguyen *et al.* (2004b) described *Heterorhabditis mexicana* from Tamaulipas. Most of South America remains unexplored for EPNs. According to Poinar (1990), the first report on a species of *Rhabditis* was initially described 65 years ago as *Rhabditis humbeltoni* Pereira (Pereira, 1973). Later, *S. glaseri*, initially described from North America, was recovered from the eggs of *Migdolus fryanus*, field collected from Sao Paulo (Pizano *et al.*, 1985). New species described from South America include *Steinernema rarum* (Doucet, 1986) and *Steinernema ritteri* (Doucet and Doucet, 1990), both found in Argentina, and *S. scapterisci* from mole crickets in Uruguay (Nguyen and Smart, 1990). The latest to be described is *Heterorhabditis argentinensis* from *Graphognatus* sp. (Stock, 1993). Comprehensive surveys suggest rich fauna. In Brazil, sampling from 1532 localities generated 18 isolates of *S. carpocapsae* and 13 of *Heterorhabditis* (Fowler, 1988); *S. feltiae* was also recovered (Fowler and Garcia, 1988). In Argentina, *S. feltiae*, *S. carpocapsae*, *S. scapterisci*, *H. bacteriophora* and *H. argentinensis* were found in the Pampean region (Stock, 1992, 1995). In Venezuela, surveys indicated the presence of six isolates of *Heterorhabditis* spp., including *H. indica* (Rosales and Suarez, 1998). *S. feltiae* was isolated from Colombia (Saenz, 1999) and unidentified *Heterorhabditis* from Suriname (Segeren Oever *et al.*, 1984).

The African continent represents a fertile field for EPN exploration as very few countries have been surveyed. A number of species and strains have been reported. *Steinernema karii* Waturu Hunt and Reid (Waturu *et al.*, 1997) and *Heterorhabditis taysaerae* Shamseldean, Abou El Sooud, Abd-Elgawad and Saleh (Shamseldean *et al.*, 1996) from Egypt. *H. indica* and *H. bacteriophora* were reported from Kenya and Egypt (Burnell and Stock, 2000; Stack *et al.*, 2000; Hominick, 2002). In addition, Spaull (1991) added several isolates from South Africa. These were all from a diverse range of environments. Species of *Steinernema* and *Heterorhabditis* spp. were recovered from Uganda (Spaull, 1991). India, as is the case with many other parts of the world, has a rich biodiversity resource due to varied geographic, climate and weather conditions. It is divided into 15 agroclimatic and agroecological zones, which in the most part consist of tropical and subtropical areas. Hence, the search for indigenous species/strains resulted in a number of isolates from different regions (Hussaini, 2003). Although the work on EPNs in India began in the 1960s, the natural occurrence of *H. bacteriophora* from Burliar was first reported in 1988. Later surveys throughout the country by the Project Directorate of Biological Control, Bangalore (PDBC) (Hussaini, 2003) and different workers documented the natural occurrence of EPNs. Although these surveys proved the abundance of indigenous isolates of *Steinernema* and *Heterorhabditis* in the country, the numbers of species identified are limited. Surveys for new isolates of EPNs were carried out in nine states of India, namely Andhra Pradesh, Gujarat, Haryana, Karnataka, Maharastra, Orissa, Pondichery, Tamil Nadu and Uttar Pradesh. In this survey, 92 soil samples were found to have EPNs. They were identified as species of *Heterorhabditis* and *Steinernema* (Anon., 2000). This discrepancy was not due to a lack of diversity in EPNs but was because of the difficulties faced in the taxonomic study of EPNs. Hussaini *et al.* (2001a) identified some of the native populations of *Steinernema* by RFLP analysis and analysis of the PCR amplified ITS rDNA region using 17 restriction enzymes. These results showed that *S. abbasi* Elawad Ahmed and Reid and *S. tami* Luc Nguyen Reis and Spiridonov were present in India. The genetic relatedness among the native *Steinernema* sp. was studied using 11 decamer oligonucleotide primers to generate RAPD DNA fragments. Umarao *et al.* (2002) studied the RAPD profiles and *Steinernema thermophilum* was differentiated from native *Steinernema*. So far reported are *S. carpocapsae*, from Karnataka, *S. abbasi*, *S. bicornutum* from Delhi, *S. tami* from Assam and *Steinernema* sp. SSL2 from Tamil Nadu and Lakshadweep (Hussaini *et al.*, 2001a), *S. thermophilum* from Delhi (Ganguly and Singh, 2000), *S. riobrave* from Gujarat (Ganguly *et al.*, 2002) and *S. glaseri* from Kerala (Banu, 2003). *S. seemae* n. sp., *S. masoodi* n. sp. and *Steinernema qazii* were reported from Kanpur (Ali *et al.*, 2005, 2009) and *Steinernema dharanaii* from Jabalpur, M.P. (Kulkarni *et al.*, 2012). *H. indica* was reported from many places throughout the country, namely Tamil Nadu, Karnataka and Kerala (Poinar *et al.*, 1992; Banu *et al.*, 1998; Hussaini, 2003). *H. bacteriophora* was reported from Tamil Nadu (Sivakumar *et al.*, 1988; Josephrajkumar and Sivakumar, 1997). Sreenivasan *et al.* (2012) collected 200 soil samples from cotton fields of ten different districts of Tamil Nadu and detected EPNs in arable lands from all ten cotton growing districts, with 27 positive sites (13.5%) ranging from irrigated to rain-fed cotton fields of soil types sandy to clayey, soil pH (4.9–8.4), electrical conductivity (0.11–1.25 dS/m) and organic matter content (0.21–3.8%). Based on morphological and molecular tools, from among 27 positive samples, 20 samples yielded steinernematids (74%) and seven yielded heterorhabditids (26%). The *Heterorhabditis* isolates were identified as *H. bacteriophora*. *Steinernema* sp. included *S. carpocapsae*, *Steinernema siamkayii* and *S. monticolum*. *S. carpocapsae* was predominant. The cropping patterns comprising cotton–vegetables had higher EPN prevalence (26%) than cotton–pulses (22%), cotton–banana (3.3%) and cotton–millets (2.8%). *S. carpocapsae* was recovered from cotton–pulses, cotton–millets and cotton–vegetable cropping patterns, whereas *H. bacteriophora* was isolated from cotton–pulses, cotton–vegetable and cotton–banana cropping patterns. Gokte-Narkhedkar *et al.* (2001) found 16 isolates of *Steinernema* and *Heterorhabditis* spp. from cotton growing areas of north, south and central India. Three isolates were of *H. bacteriophora*, while two were of *H. indica*. In Kerala, EPNs were found in 9.38% of the samples collected (Banu, 2003). Forty isolates of *Heterorhabditis* and 16 of *Steinernema* were documented from coastal, plain and hilly areas, distributed widely in places with annual rainfall ranging from 1500 to more than 3000 mm. Both *Heterorhabditis* and *Steinernema* spp. occurred frequently in sandy loam soils. Naturally occurring EPNs from ginger (*Zingiber officinale*) ecosystems from different agroclimatic regions in India were documented (Pervez *et al.*, 2014). Also, 202 soil samples were collected from different locations of Kozhikode, Wayanad, Kottayam, Idukki (Kerala), Kodagu (Karnataka), Coimbatore (Tamil Nadu), Kolasib (Mizoram), Faizabad (Uttar Pradesh) and Barapani (Meghalaya) districts to determine the EPN population: eight samples were positive for EPNs, three were from Kozhikode district, three from Idukki district and one each from Waynad and Faizabad districts. Out of the eight EPNs isolated, three species belonged to genus *Steinernema*, one to *Heterorhabditis* and four to *Oscheius*. These EPNs have been identified on the basis of morphometric and morphological characterization. Four hundred and twenty four pseudostems of ginger infested with shoot borer (*Conogethes punctiferalis*) larvae were collected from the different localities of Kozhikode, Wynad and Kodagu districts and 112 larvae were found dead. Among the dead larvae, only one EPN, namely *Oscheius* sp., was recorded from the shoot borer larvae. The survey reveals that *Oscheius* spp. and *Steinernema* spp. occur widely, and *Steinernema ramanai* and *Oscheius gingeri* have been reported as new species from the rhizosphere of ginger. Bamel and Waghmare (2014) recorded *S. siamkayii* from Indian Agricultural Research Institute (IARI) fields, New Delhi, and Ganguly *et al.* (2005) presented molecular characterization of this species. Raj Kumar *et al.* (2001) surveyed the Udaipur district of Rajasthan: out of 105 soil samples collected, 5 yielded nematodes. Further, Parihar *et al.* (2002) collected 477 soil samples from 13 districts of Rajasthan, and of the 176 samples collected from Udaipur, only 3 samples collected from rose and 4 from maize showed the presence of *Steinernema* sp. and *Heterorhabditis* sp., respectively. Surveys in the Ri-bhoi district of Meghalaya showed the presence of *Steinernema* spp. in samples collected randomly from dryland, wetland, forest and jhum lands. Samples from wetlands and forests, mostly from loamy and lacerite soil, showed a higher prevalence of EPNs. The countrywide surveys clearly document the prevalence of EPNs throughout the country, with a distinct distribution pattern for *Steinernema* and *Heterorhabditis*. Habitat preference of *Heterorhabditis* was apparent, as they were extracted only from coastal regions compared to *Steinernema*, which was isolated from all habitats (Prasad *et al.*, 2001; Hussaini, 2003). Earlier reports also support the specific habitat preference of *H. indica*. The association of *H. indica* with sandy coastal soils has been reported by Griffin *et al.* (2000). All organisms have specific niche requirements that will be satisfied only in particular habitats, and each species will be physiologically and behaviourally adapted for survival in particular habitats.

Details of soil temperature, moisture and RH effects on the survival and pathogenicity of EPNs are given elsewhere in the chapter. The diversity in persistence is affected by the surrounding soil flora and fauna. In addition to abiotic factors, besides spatial and temporal dynamics, it is these links between EPN populations and the surrounding soil flora and fauna that delimit the fundamental niche of EPNs in the soil community. The spatial distribution of EPNs in the soil consists of a network of local populations occupying patches (Curran, 1993; Campbell *et al.* 1995). These local populations have distinct probabilities of extinction and are connected by the dispersal of dauers (see Timper *et al.*, 1988, Parkman and Frank, 1992, and Shapiro-Ilan *et al.*, 1995, for descriptions of dauer dispersal at metapopulation scales). The local population dynamics within patches of suitable habitat are regulated by density-dependent (Selvan *et al.*, 1993; Koppenhöfer and Kaya; 1996) and stochastic density-independent processes. The boundaries of these local populations may shift in time, depending on the number and distribution of insects within the patch at any given time and the dispersal ability of dauers. These local populations expand, contract, move or perish, depending on host densities, antagonist distributions or edaphic factors. The size and shape of local populations of EPNs will delimit areas of unsuitable habitat for insect populations, the distribution of resource pools for nematophagous fungi and invertebrate predators, and competitive hindrance to the colonization of favourable microsites by other nematode species (free-living, plant parasitic and other EPNs). On all scales, distribution depends on the ability to survive unfavourable conditions. EPNs have been recovered from soils in all six of the world's biomes (Nearctic, Neotropical, Palearctic, Ethiopian, Oriental and Australian), indicating that different species and strains are adapted to persist in a wide variety of climates (Choo *et al.*, 1996). In many regions, the edaphic conditions are unlikely to be suitable at all times, and free-living, infective-stage dauers must possess physiological, morphological or behavioural adaptations to persist when conditions are unfavourable. Furthermore, throughout the geographic range of EPN species, hosts are not equally abundant and may only be available for short intervals throughout the year. The variation in persistence between species and strains of *Steinernema* and *Heterorhabditis* results from differences in physiological and behavioural adaptations. Inactivity caused by low temperature or desiccation increases the persistence of most EPNs. Therefore, inactivity may not contribute substantially to the variation in persistence among species. Soil temperature and moisture levels within the range where dauers are active, significantly affect the persistence of different species/strains. Documentation of delayed infectivity is incomplete; therefore, it is impossible to say how it contributes to variation in persistence. Differences in the soil distribution between ambusher and cruiser nematodes do explain differences in persistence; cruising dauers below the surface persist longer than ambushing dauers at the soil surface. Differences in energy usage between ambusher and cruiser species, however, do not correlate with differences in persistence. The nutritional status of dauers affects persistence significantly, and differences in nutritional status account, in part, for differences in the persistence of individual dauers. The endogenous differences between species/strains combine with exogenous factors to determine the persistence of individual dauers. Heterorhabditids were correlated with ocean beaches within 100 m of the seashore (0 m elevation). These positive sites had soils containing sand grains from coral and shells, with a moderately alkaline pH (8.0) and low organic content (12%). The steinernematid isolates came from inland areas in silty clay and silt loam soils with a higher organic content (15–35%).

EPNs are remarkably versatile in being useful against many soil insect pests in diverse cropping systems, yet they are clearly underutilized. Like other BCAs, nematodes are constrained by being living organisms that require specific conditions to be effective. Unlike pesticides, desiccation or UV light rapidly inactivates insecticidal nematodes. Similarly, nematodes are effective within a narrower temperature range than chemicals, and are more impacted by suboptimal soil type, thatch depth and irrigation frequency. Before 1995, most survey data were insufficient for correlation of habitats and populations to assess the pattern of distribution (Hominick *et al.*, 1996). However, larger sample sizes, accurate identification and habitat associations of species have become more apparent. Soil type, availability of susceptible hosts, physiological and behavioural adaptations are key factors affecting distribution. Broadly, heterorhabditids have been isolated primarily from sandy coastal soils, some taxa have greater presence in calcareous soils (*H. indica*) or more acidic soils (*H. bacteriophora* and *H. marelata*), whereas others range beyond coastal and are distributed broadly in turf and weedy habitats (*H. megidis*) and tropical forests (Phan *et al.*, 2003). The prevalence of steinernematids appears to be highest in woodlands (Hominick *et al.*, 1996). Recently in the USA and Europe, surveys have revealed habitat association for several steinernematids (Hominick *et al.*, 1995; Stock *et al.*, 1999; Sturhan, 1999; Sturhan and Liskova, 1999). *S. feltiae* are common in grassland and woodland (Hominick, 2002). *S. feltiae* and *S. affine* are the only species recorded from arable soils in Germany (Sturhan, 1999). *S. kraussei* and *S. intermedium* are mainly forest/woodland species. *S. kraussei* are found mainly in coniferous forests in the USA and Canada, and *S. dharanai* from teak forests in Jabalpur, India (Kulkarni *et al.*, 2012). Surveys for natural occurrence in all states and territories of Australia revealed four species of *Heterorhabditis* (three previously unknown), six of *Steinernema* (four previously unknown) and one of an undescribed steinernematid genus. In Tasmania, in collections made for over 10 years, nematodes were more commonly detected in winter and spring than in summer and autumn, with no difference in occurrence between forest and grassland (Akhurst and Bedding, 1986). Two distinct components of biodiversity are planned and associated (Vandemeer and Perfecto, 1995). Planned biodiversity is associated with crops or animals intentionally included by the farmer and vary as per the management system in time and space. Associated biodiversity includes the fauna and flora that colonize the ecosystem (Cook and Baker, 1983). At the global scale, the pattern of geographic distribution indicates both families of EPN are ubiquitous and found on all continents except Antarctica (Hominick *et al.*, 1996; Hominick, 2002). Table 5.1 lists the species recorded in the past 15 years (2001–2015).

Surveys on natural occurrence have been carried out throughout the world by different workers. In Thailand, Tangchitsomkid *et al.* (1998) recovered nematodes from 9 out of 42 locations (21.4%), eight isolates of *Steinernema* sp. and one of *Heterorhabditis* sp. Nematodes were collected from central plains at a temperature range of 24–27°C (10–15 cm deep from the soil surface) and 29–37°C (at the soil surface), and they were heat-tolerant isolates. Surveys in Pakistan revealed the occurrence in 12.4% of the samples (Shahina *et al.*, 1998). In the Azorean archipelago, 3.9% samples were positive, with *Heterorhabditis* spp. isolated from 30 sites on six islands and *Steinernema* spp. from 16 sites on three islands. Sao Miguel and Terceira Islands were positive for both genera, and Pico Island was positive only for *Steinernema* (Rosa *et al.*, 2000). EPNs were found from sea level up to 750 m. Seventy per cent of the samples positive for *Heterorhabditis* were collected below 150 m, whereas 62.5% of the samples positive for *Steinernema* were collected above 300 m. *Heterorhabditis* was not isolated above 450 m. *Steinernema* was collected mostly in loamy sand and sandy loam soils with a pH below 6, whereas *Heterorhabditis* was collected mostly in sandy and loamy sand soils with a pH higher than 6. Both species were found in cropland, orchards and pastures; *Heterorhabditis* was also found in woodland and native vegetation.

In Turkey, *H. bacteriophora* and *H. marelata* were found in two samples and *S feltiae* in one sample (Hazir *et al.*, 2003a,b). *H. marelata* was recorded for the first time in Turkey. Nematodes were recovered from six out of seven regions sampled, with 22 positive sites (2%) out of 1080 sites sampled. A single nematode isolate was recovered at each of the positive sites, of which 15 were steinernematids and 7 were heterorhabditids, representing a total of four species, namely *H. bacteriophora*, *S. feltiae*, *S. affine* and *Steinernema* n. sp. The most common species was *S. feltiae*, followed by *H. bacteriophora*. *S. affine* was from four sites in two regions and *Steinernema* n. sp. from one site. *H. bacteriophora* and *S. feltiae* have been found in many parts of the world, whereas *S. affine*, so far until this survey, has only been recovered in Europe. *S. affine* was isolated from the European (Marmara) as well as the Asiatic region (Middle Anatolia) of Turkey. A new undescribed *Steinernema* sp. was isolated from the eastern region (East Anatolia). Positive sites had sandy, sandy loam or loam (68.2%) and sandy clay loam or clay loam (31.8%), and the pH ranged from 5.6 to 7.9. Habitats were broadly classified as disturbed (59.1%), which included agricultural fields and poplar planted for lumber and windbreaks, and undisturbed (40.9%), which included pine forest, grassland, marsh and reed sites. *S. feltiae*, *S. affine*, and *H. bacteriophora* were recovered from both disturbed and undisturbed habitats. The new *Steinernema* sp. was recovered from grassland. Kepenekci and Susurluk (2003) studied *H. bacteriophora* and *H. marelata* from two samples and *S. feltiae* in one sample. *H. marelata* was recorded for the first time in the nematode fauna of Turkey.

In soils from various habitats in Guadeloupe (Grande Terre, Basse Terre) and neighbouring islands (Marie-Galante, La Desirade, Petite Terre, Les Saintes,

Table 5.1. Species of Steinernema and Heterorhabditis along with their authorities, type hosts and localities in the past 15 years (2001–2015).

Species	Isolated from	Locality	Author	Comments
Steinernema loci, *S. thanhi*	Soil, beach Soil, beach	Thanhoa, Vietnam Hatinh Province, Vietnam	Phan *et al*., 2001a	*S. loci* morphologically similar to *S. arenarium* and *S. glaseri* but distinguished from *S. glaseri* by a shorter IJ body L.
S. sangi	Soil	Thanhoa, Vietnam	Phan *et al*., 2001b	Similar to *S. kraussei.*
S. pakistanense	Soil	Karachi, Pakistan	Shahina *et al*., 2001	Having two prominent horn-like structures on labia. Close to *S. bicornutum* and *S. ceratophorum*.
S. diaprepesi	*Diaprepes abbreviatus*	Palk County, Florida, USA	Nguyen and Duncan, 2002	Closely related to *S. feltiae*, *S. glaseri*, and *S. oregonense*.
S. asiaticum	Soil	Karachi, Pakistan	Anis *et al*., 2002	Smallest body length of infective stage.
S. weiseri	Soil	České Budějovice, Czech Republic	Mracek *et al*., 2003	British *Steinernema* sp. D1 is considered conspecific with *S. weiseri*, most similar to *S. feltiae*, with which it did not hybridize.
S. anatoliense	Soil	Kars, Anatolia, Turkey	Hazir *et al*., 2003a	RFLP analysis of the ITS region of rDNA shows *S. anatoliense* n. sp. to be distinct from 50 other *Steinernema*.
S. websteri	Larvae of *Agrotis segetum*	China	Cutler and Stock, 2003	Resembles *S. abbasi* in the total body length of the IJs. Most similar to *S. carpocapsae*, *S. abbasi* and *S. riobrave* in general morphology of IJs.
S. scarabaei	*Anomala*, *Popillia japonica*	New Jersey, USA	Stock and Koppenhofer, 2003	Mucronated tail in both first-generation adults.
S. guangdongense	Soil	Jijia town, China	Qiu *et al*., 2004	Closest to *S. longicaudum*.
S. jollieti	Soil	Woodland, Midwest, USA	Spiridonov *et al*., 2004	Clade *S. kraussei*, *S. feltiae*, *S. weiseri*, and *S. oregonense.*
S. apuliae	Salt pan border habitat	Southern Italy	Triggiani *et al*., 2004	Belongs to long IJ group represented by *S. glaseri* and *S. arenarium*.
S. hermaphroditum	Soil	Moluccan islands, Indonesia	Stock *et al*., 2004	Presence of hermaphrodites in the first adult generation.
S. yirgalemense	Soil, area with banana and sugarcane	Yirgalem, Ethiopia	Nguyen *et al*., 2004a	Placed in the '*bicornutum* group'. It is similar to *S. riobrave*.
S. masoodi	Rhizosphere of pigeon pea	Kanpur, India	Ali *et al*., 2005	

Continued

Table 5.1. Continued.

Species	Isolated from	Locality	Author	Comments
S. seemae				
S. akhursti	Soil	Yunnan, China	Qiu *et al.*, 2005b	Belongs to *feltiae* group; closely related to *S. feltiae*, *S. oregonense*, *S. kraussei*, *S. weiseri*, *S. sangi*, *S. thanhi*, *S. monticolum*.
S. beddingi	Soil, cabbage field	Yunnan, China	Qiu *et al.*, 2005c	Close to *S. affine*, *S. intermedium*.
S. aciari	Soil	Haimen town, Guangdong, China	Qiu *et al.*, 2005a	Belongs to the *S. glaseri* group.
S. robustispiculum	Woodland	Chumamray National Park, Vietnam	Phan *et al.*, 2005	Morphometrics of IJs *S. robustispiculum* are close to those of *S. monticolum* except for position of excretory pore.
S. silvaticum	Forest soil, woodland	Berlin, Germany	Sturhan *et al.*, 2005	Belongs to '*feltiae–kraussei–oregonense*' group; easily distinguished from *S. kraussei*, *S. feltiae*, *S. oregonense*, *S. weiseri and S. jollieti* by morphological characters of the IJs and males.
S. eapokense	Soil, forest	Backan, Vietnam	Phan *et al.*, 2006b	Closely related species of the '*carpocapsae–scapterisci–tami*' group.
S. sasonense	Soils, humid, fernalie acrisols of broad-leaved forest, slopes of granite mountains	Chumomray mt. Sason, Sa Thay, Kon Tum Province, Vietnam	Phan *et al.*, 2006b	Closely related species of the '*carpocapsae–scapterisci–tami*' group.
S. cumgarense	Soil, forest	Daklak, Vietnam	Phan *et al.*, 2006b	Closely related species of the '*carpocapsae–scapterisci–tami*' group.
S. beckanense	Forest	Kon Tum, Vietnam	Phan *et al.*, 2006b	Closely related species of the '*carpocapsae–scapterisci–tami*' group.
S. leizhouense	Soil, eucalyptus plantations	Leizhou, Tangjia, Guangdong, China	Nguyen *et al.*, 2006a	Closest species *S. loci* and *S. aciari*.
S. ashiuense	Riverbank	Kyoto, Japan	Phan *et al.*, 2006a	Placed in the '*feltiae–kraussei–oregonense*' group in the clade containing *S. robustispiculum* and *S. monticolum*.
S. sichuanense	Soil	Sichuan, China	Mracek *et al.*, 2006	In *S. affine/intermedium* group, IJ of *S. sichuanense* n. sp. differs from *S. affine* by its absence of the internal tail spine.
S. hebeiense	Sandy soil	Hebei Province, China	Chen *et al.*, 2006	*Feltiae–kraussei–oregonense* group.
S. khoisanae	Soil	South Africa	Nguyen *et al.*, 2006c	*S. khoisanae* and others of *glaseri* group form a monophyletic assemblage.

Continued

Table 5.1. Continued.

Species	Isolated from	Locality	Author	Comments
S. texanum	Mixed, weedy vegetation soil	Texas, USA	Nguyen *et al.*, 2007	Related to the *feltiae*-group, including *akhursti*, *hebeiense*, *jollieti*, *kraussei*, *kushidai*, *litorale*, *monticolum*, *oregonense*, *sangi*, *silvaticum*, *weiseri*.
S. costaricense, *S. puntauvense*	Soil	Costa Rica	Uribe-Lorío *et al.*, 2007	*S. puntauvense*: the position of excretory pore in IJ (av. 25 micron).
S. ichnusae	Soil	Platamona Beach, north-west Sardinia, Italy	Tarasco *et al.*, 2008	Related species (*S. feltiae*, *S. kraussei*, *S. litorale*, *S. oregonense* and *S. cholashanense*).
S. colombiense	Soil	Colombia	Nunez *et al.*, 2008	Same clade as *S. carpocapsae*, *S. monticolum*, *S. siamkayai*, *S. scapterisci*.
S. cholashanense	Soil	Chola Mountain, Sichuan, China	Nguyen *et al.*, 2008a	Related species *S. feltiae*, *S. kraussei* and *S. oregonense*.
S. australe	Soil, beach	Isla Magdalena, Chile	Edgington *et al.*, 2009b	The *glaseri* group.
S. qazii	Soil	Hamirpur, Uttar Pradesh, India	Ali *et al.*, 2009	Resembles *S. abbasi*.
S. xueshanense	Soil, Tibet	Yunnan, China	Mracek *et al.*, 2009	Belongs to the *S. feltiae–kraussei* group.
S. boemarei	Soil	Southern France	Lee *et al.*, 2009	Belongs to the *glaseri* group.
S. unicornum	Soil	Chile	Edgington *et al.*, 2009a	
S. braziliense	Natural forest	Mato Grosso, Brazil	Nguyen *et al.*, 2010	Belongs to the *glaseri* group.
S. arasbaranense	Soil	Arasbaran forests, Kerengan village, east Azarbaijan, Iran	Nikdel *et al.*, 2011	Clustered with species in the *affine/intermedium* group but which clearly differs from these species.
S. pui	Soil	Xiaoji town, Yunnan, China	Qiu, *et al.*, 2011	Belongs to the *S. glaseri* group.
S. vulcanicum	Soil, chestnut wood	Sicily, Italy	Clausi *et al.*, 2011	Belongs to the *S. glaseri* group.
S. everestense	Soil	Pakhribas, Dhankuta District., Nepal	Khatri Chhetri *et al.*, 2011a	A post-anal swelling is developed in most first- and all second-generation females.
S. lamjungense	Soil	Tarku, Nepal	Khatri Chhetri *et al.*, 2011b	The *arenarium–glaseri–karii–longicaudum* group.
S. phyllophagae	White grub, live oak	Florida, USA	Nguyen and Buss, 2011	Closely related to *glaseri* group, includes *S. aciari*.
S. meghalayensis	Soil	Meghalaya, India	Ganguly *et al.*, 2011	
S. xinbinense	Soil	China	Ma *et al.*, 2012a	Belongs to the *feltiae* group.
S. ethiopiense	Soil	Mendi area, Western Wollega, Ethiopia	Tamiru *et al.*, 2012	The closest relative species is the afro-tropical *S. karii*.
S. changbaiense	Soil	North-eastern China	Ma *et al.*, 2012b	

Continued

Table 5.1. Continued.

Species	Isolated from	Locality	Author	Comments
S. dharanaii	Teak plantation	India	Kulkarni *et al.*, 2012	Belongs to the *glaseri*-group.
S. sacchari	Soil, sugarcane field	Kwa Zulu-Natal, South Africa	Nthenga *et al.*, 2014	Belongs to monophyletic group, 'Cameroonian' clade, of *S. cameroonense*, *S. nyetense* and *S. sacchari*. Closely related to *feltiae–kraussei–oregonense* Clade III.
S. citrae	Citrus orchard	Rietkloof farm, Piketberg Cape Province, South Africa	Stokwe *et al.*, 2011	Closely related to *feltiae* group. Body L of IJ close to *S. texanum* and *S. weiseri*, though differs in diameter.
S. poinari	Soil	South-west Bohemia, Czech Republic	Mracek and Nermut, 2014	Can be separated from *S. affine* and *S. intermedium* by cross-breeding tests.
S. tophus		South Africa	Çimen *et al.*, 2014	Within clade V, was found closely related to *S. innovationi* and *S. khoisanae*, belongs to *glaseri* group.
S. innovationi	–	South Africa	Çimen *et al.*, 2015	Belongs to *glaseri* group.
S. balochiense	*Psidium guajava* L., rhizophere soil samples	Balochistan, Pakistan	Fayyaz *et al.*, 2015	Closest species *S. nepalense*.
S. jeffreyense	*Psidium guajava* L., rhizophere soil samples	South Africa	Malan *et al.*, 2015	Belongs to the '*glaseri*' group (Clade V).
S. pwaniensis		Tanzania	Puza *et al.*, 2016	
Heterorhabditis downesi	Soil	Wexford, Ireland	Stock *et al.*, 2002	
H. boujardi	Soil	Vietnam	Phan *et al.*, 2003	Was easily distinguished from *H. downesi*, *H. marelata*, *H. megidis* and *H. bacteriophora*, was slightly separated from *H. bacteriophora* by variables of IJs, was clearly distinguished by variables of the males.
H. mexicana	Sorghum field	Tamaulipas, Mexico	Nguyen *et al.*, 2004b	A sister taxon to *H. indica. H. mexicana* n. sp. is morphologically similar to *H. bacteriophora*, *H. brevicaudis* and *H. indica*.
H. amazonensis	Soil	Brazil	Andalo *et al.*, 2006	Morphologically similar to *H. baujardi*, *H. floridensis*, *H. mexicana* and *H. indica*, and can be distinguished from these species mainly by male and female characters.
H. sanorensis		Mexico	Stock *et al.*, 2009	Sequence data placed *H. sonorensis* n. sp. as a member of the *indica* group.

Continued

Table 5.1. Continued.

Species	Isolated from	Locality	Author	Comments
H. gerrardi	Soil	Australia	Plichta *et al.*, 2009	Closely related to members in the '*indica* group'.
H. georgiana	Soil	USA	Nguyen *et al.*, 2008b	Morphologically similar to *H. bacteriophora*.
H. beicherriana	Orchards	Shunyi District Beijing, China	Li *et al.*, 2012	Belongs to the *bacteriophora* group with sister taxa *H. bacteriophora* and *H. georgiana*.
H. floridensis	Citrus groves and plants	Florida, USA	Nguyen *et al.*, 2006b	Forms a clade with *H. mexicana*, *H. baujardi* and *H. indica* and differs from these species by several nucleotide autapomorphies.
H. noenieputensis		South Africa	Malan *et al.*, 2014	Closest to *H. indica* and *H. gerrardi* in the *indica* group.
H. safricana	Peach orchards	South Africa	Malan *et al.*, 2008	Typical for species in the *megidis* group.
H. atacamensis	Atacama desert	Chile	Edgington *et al.*, 2011	The closest species, *H. safricana*.
Heterorhabditidoides rugaoensis		Rugao, China	Zhang *et al.*, 2012	Some species of *Oscheius*/Insectivora group should be reassigned to the genus *Heterorhabditidoides*. Morphologically very similar to *Heterorhabditidoides chongmingensis*.

Saint-Barthelemy, Saint-Martin), Constant *et al.* (1998), found EPNs in 35 of the 538 sites (6%): *Heterorhabditis* in 34 of the 35 sites (97%) and *Steinernema* in one site (3%). *Heterorhabditis* sp. found were *H. indica* (88%) and *H. bacteriophora* (12%); occuring in coastal (91.4%), tropical lowlands (5.7%) and tropical middle altitude (2.9%) areas, and none in mountain areas.

In five islands of Indonesia, Griffin *et al.* (2000) isolated EPNs from 79 sites; *Heterorhabditis* and *Steinernema* were equally prevalent, recovered from 11.7% of samples, (20.3%), both from coastal sites only. Nematodes were more prevalent on the Moluccan islands of Ambon and Seram than on Java or Bali, and were not detected on Sulawesi, where non-coastal sites only were sampled. RFLP analysis was used. *H. indica* and two RFLP types of *Steinernema* were identified.

Prishchepa *et al.* (2000) investigated in Belarus; a total of 15 *Steinernema* isolates (SBM1-96, SBM2-96, SBM3-96, SBS1-96, SBL-96, SBM10-96, SBMg21-96, SBMg22-96, SBI-96, SBSL-98, SBR-98, SBN-98, SBP1-98, SBP2-98 and SBG-99), 2 *Steinernema* strains (*S. feltiae* strain SBS2-96 and *S. carpocapsae* strain SBZ-97) and an isolate of *Heterorhabditis* sp. HBI-97 were identified. Distribution of the identified species in various soil communities, and the morphological characteristics of *S. feltiae* strain SBS2-96 and *S. carpocapsae* strain SBZ-97, are outlined. Miduturi *et al.* (1996) found an aggregated distribution of *S. feltiae* and *H. megidis* (NWE type) in a grassland habitat. Sand fraction and organic matter content did not always influence the distribution of the nematodes. The largest numbers of *S. feltiae* were detected in samples with a sand fraction content higher than 90%. The area in which *H. megidis* was detected was somewhat limited and had a sand fraction greater than 94%. Numbers of *S. feltiae* were influenced negatively by higher organic matter content.

EPNs were recovered from 38.1%, 28.5%, 23.8% and 9.6% of the samples taken in woodlands, roadside verges, cultivated land and grassland, respectively (Miduturi *et al.*, 1997). *S. feltiae* A1 type was isolated in all the habitats except in roadside verges, whereas *S. affinis* was not found in cultivated land. *Steinernema* sp. B3 was isolated from cultivated fields and roadside verges and *H. megidis* from grassland. *S. feltiae* was prevalent in sand to loamy sand soils with a wide range of organic matter content. *Heterorhabditis* sp. and *Steinernema* sp. B_3 were isolated in sandy loam and loam soils with an organic matter content of 5.6% and 7.5%, respectively. All the isolates were found in soils with a pH range of 3.6–7.8. However, most *S. feltiae* populations were isolated in acidic soils.

Fischer-Le-Saux *et al.* (1998) studied the genetic diversity of *Xenorhabdus* and *Photorhabdus* bacteria associated with *Steinernema* spp. and *Heterorhabditis* spp. by RFLP analysis of PCR-amplified 16S rRNA genes (rDNAs). Most of the strains were isolated from the Caribbean basin. Thirty distinctive 16S rDNA genotypes were identified, and cluster analysis was used to distinguish *Xenorhabdus* from *Photorhabdus*. The genus *Xenorhabdus* appears more diverse than the genus *Photorhabdus*, and for both genera the bacterial genotype diversity is in congruence with the host nematode taxonomy. The occurrence of symbiotic bacterial genotypes was related to the ecological distribution of host nematodes.

Surveys were carried out in five climatic regions of Japan (Yoshida *et al.*, 1998): subtropical islands, warm temperature coastal region along the Pacific, central mountainous region, temperate region and cool temperature region. Over 1400 soil samples from 266 sites were tested for the presence of nematodes; 20 steinernematids and 29 heterorhabditids were detected in 62 sites and 24 sites, respectively. Eight steinernematid species were recognized: seven undescribed species and *S. kushidai*. Each species had a characteristic distribution: three widespread common, one coastal, one subtropical and three mountainous species. In contrast, the heterorhabditids were identified as *H. indica* and *H. megidis*. The former was widely found in the coastal region from the subtropical to the warm temperature zone, while the latter was found mainly in the eastern part of the warm temperature coastal region. A natural population was studied in the Pampean region, Argentina, and in a distribution of natural habitats in California (Stock, 1995; Stock *et al.*, 1999).

Genetic diversity in seven *Heterorhabditis* species – namely *H. bacteriophora* (HP88 and E1), *H. indica* (EMS-13, Coimbatore), *Heterorhabditis hawaiiensis* (MG-13), *H. megidis* (HSH2 and HO1), *Heterorhabditis zealandica*, *H. marelata* (OH10) and *Heterorhabditis hepialius* (Bodega Bay) – has been investigated (Hashmi and Gaugler, 1998). Primer-specific bands generated with individual nematodes of each genotype were similar. Although the isolates used were from different geographical regions, they showed little variability in band patterns. On average, the similarity among isolates of *H. bacteriophora*, *H. indica* and *H. megidis* was 83.8%. Overall mean within-species similarity for these three species was 89.75%. High within-population similarity (96.2%) was observed with randomly selected individual nematodes, while interspecific similarities based on random amplified polymorphic DNA (RAPD) markers were 31.3%. The species tested were distinguished by RAPD. Similarities between species ranged from 11.1 to 95%. Four species groups identified using cluster analysis were: *H. bacteriophora* and *H. zealandica*; *H. megidis*; *H. indica* and *H. hawaiiensis*; and *H. marelata* and *H. hepialius*.

Genetic variation in laboratory-reared biocontrol agents may be reduced due to founder effect, inbreeding and selection. RAPD-PCR DNA has been used to compare genetic variation in two strains of *H. bacteriophora* (Shapiro-Ilan *et al.*, 1997). One strain (IS5) was isolated from the field and the other strain (HP88) has been reared under laboratory conditions for over 10 years. For each strain, 15 inbred lines were generated by eight cycles of selfing of a single hermaphrodite (reaching >90% homozygosity). Genomic DNA from each inbred line was screened with 14 decamer primers. Genetic variation was calculated based on average percentage similarity of DNA banding patterns and cluster analysis. The level of within-population variation detected did not differ significantly between the strains.

5.5 Genetic Diversity

Genotypic variation among IJs of *H. bacteriophora* (strain HP88) to heat, desiccation, UV tolerance and host-finding ability was assessed by comparing the performance of inbred lines in laboratory assays (Glazer *et al.*, 1991). Each line consisted of highly homozygous offspring originating from one individual obtained from a natural population. Considerable variation in all four traits was detected among the different inbred lines. The heritability values for heat or UV tolerance and for host-finding ability were high, indicating that selection should be an efficient way to improve these traits in the population. Desiccation tolerance varied considerably within each line. Heritability value was low, as this was influenced mainly by environmental variation, and selective breeding for higher desiccation tolerance would be inefficient. Improvement through the induction of mutations may be a better alternative in this population. Hashmi and Gaugler (1998) used isolates of *H. bacteriophora* (HP88 and E1), *H. indica* (EMS-13, Coimbatore), *H. hawaiiensis* (MG-13), *H. megidis* (HSH2 and HO1), *H. zealandica, H. marelata* (OH10) and *H. hepialius* (Bodega Bay) from different geographical regions and observed little variability.

In soils from 351 sites in ecologically diverse habitats from the Hawaiian Islands (MSL 4200 m), 6.8% were positive for EPNs, 6.3% for a *Heterorhabditis* sp. from Kauai (6), Oahu (5), Maui (4), Molokai (1) and Hawaii (6), and two sites had *Steinernema* sp. from Maui; none from the island of Lanai.

5.5.1 Qualitative and quantitative composition

Heterorhabditis and *Steinernema* are in different clades belonging to Rhabditidae (Clade V) and Strongylidae (Clade IV), respectively. *Photorhabdus* and *Xenorhabdus* bacteria are monophyletic clades in Enterobacteriaceae and are related to each other more closely than the non-EPN associated bacteria (Rainey *et al*., 1995). However, different species, genera and families of these parasites occur in different ecosystems/habitats, depending on their insect hosts. They occur throughout the USA in many different soil types and habitats, natural and managed. The probability of occurrence was correlated positively with below-ground insect species richness and abundance, and correlated negatively with the percentage of vegetation basal cover and soil temperature. The effect of soil temperature on survival is given elsewhere. Heterorhabditids and steinernematids were found in many areas of all continents, excluding Antarctica. Some research suggests steinernematids are more common in cooler temperate regions and heterorhabditids in warmer tropical condions (Hominick *et al*., 1996). These families inhabit most terrestrial habitats (Hominick and Briscoe, 1990). The rate of prevalence of *Steinernema* sp. was highest in woodlands in Germany (Sturhan, 1999), where *S. affinis*, *S. intermedium* and *Steinernema* sp. B were predominant. *Heterorhabditis* spp. was abundant in turf and weedy habitats, but never found in close canopy forest (Stuart and Gaugler, 1994). Hominick *et al.* (1995) demonstrated that at least some steinernematids showed a distinct preference that might reflect the distribution of suitable hosts adapted for the habitat. One of the major factors restricting the nematode host range is the foraging behaviour of IJs. The effect of cropping practices on populations of *S. carpocapsae* was examined in Taiwan. Populations in fallow and bare plots were very low in February–March, due to snow. Higher populations were found in no-tillage maize plots than in conventional-tillage plots from October to August. Populations in no-tillage and conventional tillage maize, sorghum and soybean fields decreased over time from October to April. Five months after application, no IJs were recovered from most of the plots, except for those that had received a conventional-tillage maize treatment. The numbers in soil covered with maize and sorghum debris was significantly lower after 30 days at 25°C. In the horizontal movement study, *S. carpocapsae* was capable of moving 3.5 cm/day in bare soil plots and 7.5 cm in rye mulch covered plots, indicating that the mulch enhanced movement in agriculture systems.

The seasonal population dynamics of endemic *S. carpocapsae* and *H. bacteriophora* in turfgrass and their effect on populations of Japanese beetle, *P. japonica*, larvae, and mobile arthropods associated with the soil surface and thatch layer were considered; both species were recovered from samples collected from April to December. *S. carpocapsae* tended to be more prevalent than *H. bacteriophora.* However, *H. bacteriophora* had a higher density than *S. carpocapsae.* Sections with *H. bacteriophora* had significantly lower *P. japonica* densities compared to sections without nematodes. *S. carpocapsae* did not have a measurable impact on *P. japonica* populations. Mobile surface arthropods, as measured by pitfall catches, were higher in *H. bacteriophora*-positive sections than in sections without nematodes. This difference was due to higher catches of Araneae and Lithobiomorpha. Mobile surface arthropod populations did not differ between sections with and without *S. carpocapsae.* The scope of the impact of *H. bacteriophora* on host populations is likely to be limited because of its patchy distribution. Understanding the ability of natural populations of predators, parasitoids and pathogens to suppress prey populations is important in determining their potential as biological control agents (Campbell *et al*., 1995). Many more steinernematids than heterorhabditids have been described. Genetic variation with populations of

Heterorhabditis spp. could be extremely low due to hermaphroditism (Downes and Griffin, 1996). Discoveries in unusual places modify known distribution patterns. Habitat preferences may reflect the distribution of suitable hosts, and differences in host preference can explain the coexistence of species (Sturhan, 1999).

5.6 Conclusion

The global biodiversity of EPNs and associated bacteria is tremendous, yet to date a miniscule fraction consists of described species. A stable systematic framework can be utilized by many other disciplines requiring representation of history and scientific inferences logically; for example, ecology, biogeography, behaviour, genomics of ecology, transgenic, selective breeding, host ranges, etc. More robust phylogenies and their utility as model systems for investigations in evolution, molecular genetics, biology is evident. Genome sequences to be acquired for nematode hosts that complement the symbiotic counterparts will help in better understanding and improve the performance of EPNs as biocontrol agents. Simulation studies are required to provide insight into population dynamics. EPNs are important natural enemies of insects in soil throughout the world, regulating insect populations under various circumstances. Manipulations in managed habitats through augmentation and conservation would provide a stable, benign control of soil insects. The impacts of such efforts on soil ecosystems remain largely unknown and variable. Within a broad ecological context, various interactions with biotic and abiotic environments are the varying characteristics among species, strains and habitats. The economic and social realities of modern agriculture, environmental degradation and climate change provide important backdrops for studies, so that EPNs remain prime subjects for ecological research. Organic manure/fertilizer used in the field has increased/decreased EPN establishment and recycling. Tolerance to pesticides is variable. However, they are tenable in tank mixes. Nematicides remain incompatible in general. The long-term effects on endemic nematodes need assessment. Moisture and temperature remain critical abiotic factors. Two distinct components of biodiversity are planned and associated (Vandemeer and Perfecto, 1995). Planned biodiversity is associated with crops or animals intentionally included by farmers and vary per management system in time and space. Associated biodiversity includes the fauna and flora that colonize the ecosystem (Cook and Baker, 1983). Many nematodes have physiological or behavioural adaptations, allowing the resumption of activity after prolonged inactivity. Wide ranges of pH are tolerated. Though acid deposition may be a limiting factor in some areas, no quantitative authentic records are available. Species/strains have different temperature regimes for infection, survival, activity and reproduction. Temperature tolerances are modified through natural and artificial selections. Population biology research is a challenging task of modern ecology and EPNs provide a salient case. This research is fundamental to aspects such as persistence and to develop prediction models in designing control programmes. It is expected that the value of the EPN–bacterial association as models for studying biological phenomena may enhance their value in biological control and increase our understanding manyfold.

Acknowledgements

I am thankful to my friends, colleagues and family for the help rendered in compiling this chapter.

References

Abbas, M.S.T. and Mousa, S.A. (2003) Comparative existence of *Steinernema abbasi* and *Heterorhabditis indicus* in soil of a date palm plantation. *Egyptian Journal of Agricultural Research* 81, 1073–1083.

Akhurst, R.J. (1980) Morphological and functional dimorphism in *Xenorhabdus* spp. bacteria symbiotically associated with the insect pathogenic nematodes, *Neoaplectana* and *Heterorhabditis*. *Journal of General Microbiology* 121, 303–309.

Akhurst, R.J. and Bedding, R.A. (1986) Natural occurrence of insect pathogenic nematodes (Steinernematidae and Heterorhabditidae) in soil in Australia. *Journal of the Australian Entomological Society* 25, 241–244.

Akhurst, R.J. and Boemare, B.R. (1990) Biology and taxonomy of *Xenorhabdus*. In: Gaugler, R. and Kaya H.K (eds) *Entomopathogenic Nematodes in Biological Control.* CRC Press, Boca Raton, Florida, pp. 75–90.

Alatorre-Rosas, R. and Kaya, H.K. (1990) Interspecific competition between entomopathogenic nematodes in the genera *Heterorhabditis* and *Steinernema* for an insect host in sand. *Journal of Invertebrate Pathology* 55, 179–181.

Alatorre-Rosas, R. and Kaya, H.K. (1991) Interaction between two entomopathogenic nematode species in the same host. *Journal of Invertebrate Pathology* 57, 1–6.

Alekseev, E., Glazer, I. and Samish, M. (2006) Effect of soil texture and moisture on the activity of entomopathogenic nematodes against female *Boophilus annulatus* ticks. *BioControl* 51, 507–518.

Ali, J.G., Alborn, H.T. and Stelinski, L.L. (2010) Subterranean herbivore-induced volatiles released by citrus roots upon feeding by *Diaprepes abbreviatus* recruit entomopathogenic nematodes. *Journal of Chemical Ecology* 36, 361–368.

Ali, S.S., Shaheen, A., Pervez, R. and Hussain, M.A. (2005) *S. masoodi* sp. n. and *S. seemae* sp. n. (Nematoda: Steinernematidae) from India. *International Journal of Nematology* 15, 89–99.

Ali, S.S., Pervez, R., Hussain, M.A. and Ahmad, R. (2007) Effect of temperature on survival of *Steinernema seemae*, *S. masoodi* and *S. carpocapsae* (Rhabditida: Steinernematidae) and their subsequent infectivity to prepupa of *Helicoverpa armigera* (Hubner). *Archives of Phytopathology and Plant Protection* 40, 183–187.

Ali, S.S., Shaheen, A., Asif, M. and Akhtar, M.H. (2009) *S. qazii* sp.n. (Nematoda: Steinernematidae) from Kanpur, India. *Trends in Biosciences* 2, 59–64.

Alumai, A. and Grewal, P.S. (2004) Tank-mix compatibility of the entomopathogenic nematodes, *Heterorhabditis bacteriophora* and *Steinernema carpocapsae*, with selected chemical pesticides used in turfgrass. *Biocontrol Science and Technology* 14, 725–730.

Ameen, H.H. (1991) The effects of some biocontrol agents on the predaceous nematode, *Mononchus truncatus*. *Bulletin of Faculty of Agriculture, University of Cairo* 42, 429–434.

Andalo, V., Nguyen, K.B. and Moino Junior, A. (2006) *Heterorhabditis amazonensis* n. sp. (Rhabditida: Heterorhabditidae) from Amazonas, Brazil. *Nematology* 8, 853–867.

Anis, M., Shahina, F., Reid, A.P. and Rowe, J. (2002) *Steinernema asiaticum* sp. n from Pakistan. *International Journal of Nematology* 12, 220–232.

Anon. (2000) *Annual Report*. Sugarcane Breeding Institute, Coimbatore, India.

Ansari, M.A., Tirry, L. and Moens, M. (2004) Interaction between *Metarhizium anisopliae* CLO 53 and entomopathogenic nematodes for the control of *Hoplia philanthus*. *Biological Control* 31, 172–180.

Ansari, M.A., Waeyenberge, L. and Moens, M. (2005) First record of *Steinernema glaseri* Steiner, 1929 (Rhabditida) from Belgium: a natural pathogen of *Hoplia philanthus* (Coleoptera: Scarabaeidae). *Nematology* 7, 953–956.

Ansari, M.A., Tirry, L., Shah, F.A. and Moens, M. (2006) Field trials against *Hoplia philanthus* (Coleop: Scarabaeidae) with a combination of an entomopathogenic nematode and the fungus *Metarhizium anisopliae* CLO 532006. *Biological Control* 39, 453–459.

Anuar, M.R. and Daniel, C.P. (2009) Synergies between biological and neonicotinoid insecticides for the curative control of the white grubs *Amphimallon majale* and *Popillia japonica*. *Biological Control* 51, 169–180.

Askary, T.H. (2010) Nematodes as biocontrol agents. In: Lichtfouse, E. (ed.) *Sociology, Organic Farming, Climate Change and Soil Science*. Springer, The Netherlands, pp. 347–378.

Bamel, V. and Waghmare, C. (2014) Report of Entomopathogenic Nematode, *Steinernema siamkayai* (Stock, Somsook and Reid 1998) (Steinernematidae: Nematoda) from fields of IARI, New Delhi. *Indian Journal of Nematology* 44, 255–258.

Banu, J.G. (2003) Scope of entomopathogenic nematodes for the management of pests of intercrops in coconut based cropping system. In: Hussaini, S.S., Rabindra, R.J. and Nagesh, M. (eds) *Current Status of Research on Entomopathogenic Nematodes in India*. Project Directorate of Biological Control, Bangalore, India, pp. 179–184.

Banu, J.G., Sosamma, V.K. and Koshy, P.K. (1998) Natural occurrence of an entomopathogenic nematode, *Heterorhabditis indicus* (Poinar, Karunakar and David) from Kerala, India. In: Nematology: challenges and opportunities in 21st Century. *Proceedings of IIIrd International Symposium on Afro-Asian Society of Nematologists*. Sugarcane Breeding Institute, Coimbatore, India, pp. 274–280.

Barbara, K.A. and Buss, E.A. (2005) Integration of insect parasitic nematodes (Rhabditida: Steinernematidae) with insecticides for control of pest mole crickets (Orthoptera: Gryllotalpidae: *Scapteriscus* spp.). *Journal of Economic Entomology* 98, 689–693.

Barbercheck, M.E. (1993) Tritrophic level effects on entomopathogenic nematodes. *Environmental Entomology* 22, 1166–1171.

Barbercheck, M.E. and Kaya, H.K. (1990) Interactions between *Beauveria bassiana* and the entomogenous nematodes, *Steinernema feltiae* and *Heterorhabditis heliothidis*. *Journal of Invertebrate Pathology* 55, 225–234.

Barbercheck, M.E. and Kaya, H.K. (1991) Competitive interactions between entomopathogenic nematodes and *Beauveria bassiana* (Deuteromycotina: Hyphomycetes) in soil borne larvae of *S. exigua* (Lepi: Noctuidae). *Environmental Entomology* 20, 707–712.

Barbercheck, M.E., Wang, J. and Hirsh, I.S. (1995) Host plant effects on entomopathogenic nematodes. *Journal of Invertebrate Pathology* 66, 169–177.

Bathon, H. (1996) Impact of entomopathogenic nematodes on non-target hosts. *Biocontrol Science and Technology* 6, 421–434.

Baur, M.E., Kaya, H.K., Gaugler, R. and Tabashnik, B. (1997) Effects of adjuvants on entomopathogenic nematode persistence and efficacy against *Plutella xylostella*. *Biocontrol Science and Technology* 7, 513–525.

Baur, M.E., Kaya, H.K. and Strong, D.R. (1998) Foraging ants as scavengers on entomopathogenic nematode-killed insects. *Biological Control* 12, 231–236.

Bedding, R.A. (1990) Logistics and strategies for introducing EPN technology into developing countries. In: Gaugler, R. and Kaya, H.K. (eds) *Entomopathogenic*

Nematodes in Biological Control. CRC Press, Boca Raton, Florida, pp. 223–246.

Bednarek, A. and Gaugler, R. (1997) Compatibility of soil amendments with entomopathogenic nematodes. *Journal of Nematology* 29, 220–227.

Bhatnagar, A., Bareth, S.S. and Shinde, V. (2002) Effect of soil texture on nematode infectivity to whitegrub, *Maladera insanabilis* Brenske. *Pest Management and Economic Zoology* 10, 71–75.

Bintrim, S.B and Ensign, J.C. (1998) Insertional inactivation of genes encoding the crystalline inclusion proteins of *Photorhabdus luminescens* results in mutants with pleiontropic phenotypes. *Journal of Bacteriology* 180, 1261–1269.

Bird, A.F. and Bird, J. (1986) Observations on the use of insect parasitic nematodes as a means of biological control of root-knot nematodes. *International Journal for Parasitology* 16, 511–516.

Blackshaw, R.P. and Senthamizhselvan, M. (1991) The effect of sand particle size on *Steinernema feltiae* sensu Filipjev (1934) (*N. bibionis* Bovien) activity against *Galleria mellonella* larvae. *Annals of Applied Biology* 118, 637–643.

Boemare, N. (2002) Biology, taxonomy and systematics of *Photorhabdus* and *Xenorhabdus*. In: Gaugler, R. (ed.) *Entomopathogenic Nematology*. CAB International, Wallingford, UK, pp. 35–56.

Boemare, N.E. and Akhurst, R.J. (1988) Biochemical and physiological characterization of colony form variants in *Xenorhabdus* spp. (Enterobacteriaceae). *Journal of General Microbiology* 134, 751–761.

Boemare, N.E. and Akhurst, R.J. (1994) DNA homology between *Xenorhabdus* and *Photorhabdus* spp.: convergent evolution of two different genera. In: Burnell, A.M., Ehlers, R.U. and Masson, J.P. (eds) *Genetics of Entomopathogenic Nematode-Bacterium Complexes*. Proceedings of Symposium and Workshop, St Patricks College, Maynooth, Co. Kildare, Ireland. European Commision, DG XII, Luxembourg, pp. 59–69.

Boemare, N.E., Akhurst, R.J. and Mourent, R.J. (1993) DNA relatedness between *Xenorhabdus* spp. (Enterobacteriaceae) symbiotic bacteria of entomopathogenic nematodes, and proposal to transfer *Xenorhabdus luminesense* to a new genus, *Phorhabdus* gen. nov. *International Systematic Bacteriology* 43, 249–255.

Boff, M.I.C., Wiegers, G.L. and Smits, P.H. (2000) Effect of storage time and temperature on infectivity, reproduction and development of *Heterorhabditis megidis* in *G. mellonella*. *Nematology* 2, 635–644.

Broadbent, A.B. and Olthof, T.H.A. (1995) Foliar application of *Steinernema carpocapsae* (Rhabditida: Steinernematidae) to control *Liriomyza trifolii* (Diptera: Agromyzidae) larvae in chrysanthemums. *Environmental Entomology* 24, 431–435.

Brodsgaard, H.F., Sardar, M.A. and Enkegaard, A. (1996) Prey preference of *Hypoaspis miles* (Berlese) (Acarina: Hypoaspididae) non-interference with other beneficial in glass house crops. *IOBC/WPRS Bulletin* 19, 23–26.

Brown, G.G. (1995) How do earthworms affect microfloral and faunal community diversity? *Plant Soil* 170, 209–231.

Brown, I.M. and Gaugler, R. (1997) Temperature and humidity influence emergence and survival of entomopathogenic nematodes. *Nematologica* 43, 363–375.

Browning, M., Dawson, C., Alm, S.R., Gorres, J.H., and Amador, J.A. (2004) Differential effects of butyric acid on nematodes from four trophic groups. *Applied Soil Ecology* 27, 47–54.

Brust, G.E. (1991) Augmentation of an endemic entomogenous nematode by agroecosystem manipulation for the control of a soil pest. *Agricultural Ecosystem Environment* 36, 175–184.

Burnell, A.M. and Stock, S.P. (2000) *Heterorhabditis*, *Steinernema* and their bacterial symbionts – lethal pathogen of insects. *Nematology* 2, 31–42.

Campbell, J.F., Lewis, E., Yoder, F. and Gaugler, R. (1995) Entomopathogenic nematode seasonal population dynamics and impact on insect populations in turfgrass. *Biological Control* 5, 598–606.

Campbell, J.F., Orza, G., Yoder, F., Lewis, E. and Gaugler, R. (1998) Spatial and temporal distribution of endemic and released entomopathogenic nematode populations in turfgrass. *Entomologia Experimentalis et Applicata* 86, 1–11.

Campbell, J.F., Koppenhofer, A.M., Kaya, H.K. and Chinnasri, B. (1999) Are there temporarily noninfectious daeur stages in EPN populations: a test of the phased infectivity hypothesis. *Parasitology* 118, 499–508.

Campos-Herrera, R., Trigo, D. and Gutiérrez, C. (2006) Phoresy of the entomopathogenic nematode *Steinernema feltiae* by the earthworm, *Eisenia fetida*. *Journal of Invertebrate Pathology* 92, 50–54.

Campos-Herrera, R., Barbercheck, M., Hoy, C.W. and Stock, S.P. (2012) Entomopathogenic nematodes as a model system for advancing the frontiers of ecology. *Journal of Nematology* 44, 162–176.

Capinera, J.L., Blue, S.L., and Wheeler, G.S. (1982) Survival of earthworms exposed to *Neoaplectana carpocapsae* nematodes. *Journal of Invertebrate Pathology* 39, 419–421.

Chen, G., Dunphy, G.B. and Webster, J.M. (1994) Antifungal activity of two *Xenorhabdus* species and *Photorhabdus luminescens*, bacteria associated with the nematodes *Steinernema* sp. and *Heterorhabditis megidis*. *Biological Control* 4, 157–162.

Chen, S., Li, J., Han, X. and Moens, M. (2003) Effect of temperature on the pathogenicity of entomopathogenic nematodes (*Steinernema* and *Heterorhabditis* spp.) to *Delia radicum*. *BioControl* 48, 713–724.

Chen, S., Li , X., Yan, A., Spiridonov, S.E. and Moens, M. (2006) A new entomopathogenic nematode *Steinernema hebeiense* sp. n. (Rhabditida: Steinernematide), from North China. *Nematology* 8, 563–574.

Cheng, C.C. and Hou, R.F. (1997) Effects of environmental factors on survival of the entomopathogenic nematode, *Steinernema carpocapsae*. *Chinese Journal of Entomology* 17, 120–131.

Chevalier, R. and Webster, J.M. (2006) Temporary changes in populations of soil organisms after field application of entomopathogenic nematodes. *Journal of Nematology* 38, 466–473.

Choo, H.Y., Kaya, H.K. and Stock, S.P. (1995) Isolation of entomopathogenic nematodes (Steinernematidae and Heterorhabditidae) from Korea. *Japanese Journal of Nematology* 25, 44–51.

Choo, H.Y., Kim, H.H., Lee, D.W. and Park, Y.D. (1996) Microbial control of fly maggots with entomopathogenic nematodes and fungus in outhouses of farm houses. *Korean Journal of Applied Entomology* 35, 80–84.

Choo, H.Y., Lee, D.W., Ha, P.J., Kim, H.H. and JungLee, S.M. (1999) Temperature and dose size effects on infectivity and reproduction of EPN, *S. longicaudum* Gongjun strain. *Korean Journal of Pesticide Science* 3, 60–68.

Choo, H.Y., Lee, D.W., Yoon, H.S., Lee, S.M. and Hang, D.T. (2002) Effects of temperature and nematode conc on pathogenicity and reproduction of entomopathogenic nematode, *S. carpocapsae* Poncheon. *Korean Journal of Applied Entomology* 41, 269–277.

Chung, H.J. (2001) Effects of temperature and dose on infectivity and reproduction of heterorhabditids in *Galleria mellonella* larvae. MSc thesis, Gyeogsang National University, Jinju, Korea.

Ciche, T.A., Bintrim, S.B., Horswill, A.R. and Ensign, J.C. (2001) A phosphopantetheinyl transferase homologue is essential for *P. luminescens* to support growth and reproduction of the entomopathogenic nematode *H. bacteriophora*. *Journal of Bacteriology* 183, 3117–3126.

Ciche, T.A., Blackburn, M., Carney, J.R. and Ensign, J.C. (2003) Photobactin: a catechol siderophore produced by *P. luminescens*, an entomopathogen mutually associated with *H. bacteriophora*. *Applied Environmental Microbiology* 69, 4706–4713.

Çimen, H., Lee, M.M., Hatting, J., Hazir, S. and Stock S.P. (2014) *Steinernema tophus* sp. n. (Nematoda: Steinernematidae), a new entomopathogenic nematode from South Africa. *Zootaxa* 3821, 337–353.

Çimen, H., Lee, M.M., Hatting, J., Hazir, S. and Stock, S.P. (2015) *Steinernema innovationi* n. sp. (Panagrolaimomorpha: Steinernematidae), a new entomopathogenic nematode species from South Africa. *Journal of Helminthology* 89, 415–427.

Clausi, M., Longo, A., Rappazzo, G., Tarasco, E. and Vinciguerra, M.T. (2011) *Steinernema vulcanicum* n. sp. (Rhabditida: Steinernematidae), a new entomopathogenic nematode species from Sicily (Italy). *Nematology* 13, 409–423.

Constant, P., Marchay, L., Fischer, L.S.M., Briand, P.S. and Mauleon, H. (1998) Natural occurrence of entomopathogenic nematodes (Rhabditida: Steinernematidae and Heterorhabditidae) in Guadeloupe islands. *Fundamental and Applied Nematology* 21, 667–672.

Cook, R.J. and Baker, K.F. (1983) *The Nature and Practice of Biological Control of Plant Pathogens*. American Phytopathological Society, St Paul, Michigan.

Cortez Madrigal, H., Morales-Salvador, P. and Adams, B. (2003) Primer registro en Mexico del nematode *Heterorhabditis indicus* Poinar. In: Vazquez-Garcia Peres Dominguez, J.F., Ibarra Cortes, Balpuesta Leon, C., Vazquez-Reyes, R., Cerventos Rios, J. and Ibarra Frias, N. (eds) *Memorias del XXVI Congreso Nacional de Control Biologico*. Sociedad Mexicana de Control Biologico, Texcoco, Mexico, pp. 68–70.

Cowles, R.S. and Villani, M.G. (1994) Soil interactions with chemical insecticides and nematodes used for control of Japanese beetle (Coleoptera: Scarabaeidae) larvae. *Journal of Economic Entomology* 87, 1014–1021.

Curran, J. (1990) Efficacy of entomopathogenic nematodes in field soils. In: *Proceedings and Abstracts, Vth International Colloquium on Invertebrate Pathology and Microbial Control*. Society for Invertebrate Pathology, Adelaide, Australia, pp. 224–227.

Curran, J. (1993) Post application biology of entomopathogenic nematodes in soil. In: Bedding, R., Akhurst, R. and Kaya, H.K. (eds) *Nematodes and the Biological Control of Insects*. CSIRO, East Melbourne, Australia, pp. 67–77.

Cutler, C.G. and Stock, S.P. (2003) *S. websteri* sp. n. an entomopathogenic nematode from China. *Nematologia Mediterranea* 31, 215–224.

Cutler, G.C. and Webster, J.M. (2003) Host-finding ability of three entomopathogenic nematode isolates in the presence of plant roots. *Nematology* 5, 601–608.

Das, J.N. and Diwakar, B.J. (1987) Compatibility of certain pesticides with DD-136 nematode. *Plant Protection Bulletin India* 39, 20–22.

Devi, G. (2007a) Impact of *Arthrobotrys oligospora* on infectivity of entomopathogenic nematodes (Meghalaya isolates). *Annals of Plant Protection Sciences* 15, 532–533.

Devi, G. (2007b) Compatibility of indigenous entomopathogenic nematodes with entomopathogenic fungi. *Indian Journal of Plant Protection* 35, 149–150.

Diaz Maderos, P., Najera Rincon, M.B., Lezma Gutierrez, R., Rebolledo-Domingiez, and OFlore Lopez, H.E. (2002) Hypomycetes y nematodes (Nematoda: Heterorhabditidae) entomopathogenos de gallinas ciegas (Coleop: Melolontidae), en los altos de Jalisco, Mexico. In: Baez-Sanudo, R. and Juvera-Bracemontes, J.J. (eds) *Memorias del XXV Congress Nacional de Control Biologico*. Sociedad Mexicana de Control Biologico, Texcoco, Mexico, pp. 236–238.

Doube, B.M. and Brown, G.G. (1998) Life in a complex community: functional interactions between earthworm, organic matter, microorganisms and plant growth. In: Edwards, C.A. (ed.) *Earthworm Ecology*. Lucie Press, Boca Raton, Florida, pp. 179–211.

Doucet, M.M.A. (1986) A new species of *Neoaplectana* Steiner 1929 (Nematoda: Steinernematidae) from Cardoba, Argentina. *Revue de Nematology* 9, 317–323.

Doucet, M.M.A. and Doucet, M.E. (1990) *Steinernema ritteri* n. sp. (Nematoda: Steinernematidae) with a key to the species of the genus. *Nematologica* 36, 257–265.

Downes, M.J., and Griffin, C.T. (1996) Dispersal behavior and transmission strategies of the entomopathogenic nematodes *Heterorhabditis* and *Steinernema*. *Biocontrol Science and Technology* 6, 347–356.

Dunphy, G.B. and Webster, J.M. (1986) Temperature effects on the growth and virulence of *Steinernema feltiae* strains and *Heterorhabditis heliothidis*. *Journal of Nematology*18, 270–272.

Dutilleul, P. and Legendre, P. (1993) Spatial heterogeneity against heteroscedasticity: an ecological paradigm versus a statistical concept. *Oikas* 66, 152–171.

Eben, A. and Barbercheck, M.E. (1997) Host plant and substrate effects on mortality of southern corn rootworm from entomopathogenic nematodes. *Biological Control* 8, 89–96.

Ebssa, L., Borgemeister, C. and Poehling, H.M. (2006) Simultaneous application of entomopathogenic nematodes and predatory mites to control western flower thrips *Frankliniella occidentalis*. *Biological Control* 39, 66–74.

Edgington, S., Buddie, A.G., Tymo, L.M., France, A., Merino, L. and Hunt, D.J. (2009a) *Steinernema unicornum* sp. n. (Panagrolaimomorpha: Steinernematidae), a new entomopathogenic nematode from Tierra del Fuego, Chile. *Journal of Nematode Morphology and Systematics* 12, 113–131.

Edgington, S., Buddie, A.G., Tymo, L., Hunt, D.J., Nguyen, K.B., *et al.* (2009b) *Steinernema australe* n. sp. (Panagrolaimomorpha: Steinernematidae), a new entomopathogenic nematode from Isla Magdalena, Chile. *Nematology* 11, 699–717.

Edgington, S., Buddie, A.G., Moore, D., France, A., Merino, L. and Hunt, D.J. (2011) *Heterorhabditis atacamensis* n. sp. (Nematoda), a new entomopathogenic nematode from the Atacama Desert, Chile. *Journal of Helminthology* 85, 381–394.

Ehlers, R.-U., Deseo, K.V. and Stackebrandt, E. (1991) Identification of *Steinernema* spp. (Nematoda) and their symbiotic bacteria *Xenorhabdus* spp. from Italian and German soils. *Nematologica* 37, 360–366.

El Borai, F.E., Duncan, L.W. and Preston, J.F. (2005) Bionomics of a phoretic association between *Paenibacillus* sp. and the entomopathogenic nematode, *Steinernema diaprepesi*. *Journal of Nematology* 37, 18–25.

Eng, M.S., Preisser, E.L. and Strong, D.R. (2005) Phoresy of the entomopathogenic nematode *Heterorhabditis marelatus* by a non-host organism, the isopod *Porcellio scaber*. *Journal of Invertebrate Pathology* 88, 173–176.

Enright, M.R. and Griffin, C.T. (2005) Effect of *Paenibacillus nematophilus* on the EPN *H. megidis*. *Journal of Invertebrate Pathology* 88, 40–48.

Enright, M.R., Mclnerney, J.O. and Griffin, C.T. (2003) Characterization of endospore-forming bacteria associated with entomopathogenic nematode, *Heterorhabditis* spp. and *Paenibacillus nematophilus*. *International Journal of Systematic and Evolutionary Microbiology* 53, 435–441.

Epsky, N.D. and Capinera, J.L. (1994) Influence of herbivore diet on the pathogenesis of *Steinernema carpocapsae* (Nematoda: Steinernematidae). *Environmental Entomology* 23, 487–491.

Epsky, N.D., Walter, D.E. and Capinera, J.L. (1988) Potential role of nematophagous microarthropods as biotic mortality factors of entomogenous nematodes (Rhabditida: Steinernematidae, Heterorhabditidae). *Journal of Economic Entomology* 81, 821–825.

Esser, R.P., El-Gholl, N.E. and Price, M. (1991) Biology of *Dactylella megalospora* Drechs., a nematophagous fungus. *Proceedings of Soil and Crop Science Society of Florida* 50, 173–180.

Fang, Y.Y., Zhou, Y., Yang, X., Pang, Y. and Qiu, L.H. (2004) A new record of entomopathogenic nematode from China. *Acta Zootaxonomica Sinica* 29, 692–696.

Farag, N.A. (2002) Impact of two entomopathogenic nematodes on the ladybird, *Coccinella undecimpunctata* and its pray, *Aphis fabae*. *Annals of Agricultural Science* (Cairo) 47, 431–443.

Fayyaz, S., Khanum, T.A, Ali, S., Solangi, G.S, Gulsher, M. and Javed, S. (2015) *Steinernema balochiense* n. sp. (Rhabditida: Steinernematidae) a new entomopathogenic nematode from Pakistan. *Zootaxa* 7, 387–402.

Ferguson, C.S., Schroeder, P.C. and Shields, E.J. (1995) Vertical distribution, persistence, and activity of entomopathogenic nematodes (Nematoda: Heterorhabditidae and Steinernematidae) in alfalfa snout beetle (Coleoptera: Curculionidae) infested fields. *Environmental Entomology* 24, 149–158.

Fischer-Le-Saux, M., Mauleon, H., Constant, P., Brunel, B. and Boemare, N. (1998) PCR-ribotyping of *Xenorhabdus* and *Photorhabdus* isolates from the Caribbean region in relation to the taxonomy and geographic distribution of their nematode hosts. *Applied and Environmental Microbiology* 64, 4246–4254.

Fitters, P.F.L. and Griffin, C.T. (2006) Survival, starvation, and activity in *Heterorhabditis megidis* (Nematoda: Heterorhabditidae). *Biological Control* 37, 82–88.

Forschler, B.T., All, J.N. and Gardner, W.A. (1990) *Steinernema feltiae* activity and infectivity in response to herbicide exposure in aqueous and soil environments. *Journal of Invertebrate Pathology* 55, 375–379.

Fowler, H.G. (1988) Occurrence and infectivity of entomogenous nematodes in mole crickets in Brazil. *International Rice Research Newsletter* 13, 34–35.

Fowler, H.G. and Garcia, C.R. (1988) Nematodes as natural control agents of mole crickets (Orthop: Gryllotalpidae): field and laboratory studies. *Revista Brasileira Biology* 48, 789–795.

Fujiie, A. and Yokoyama, T. (1998) Effects of ultraviolet light on the entomopathogenic nematode, *Steinernema kushidai* and its symbiotic bacterium, *Xenorhabdus japonicas*. *Applied Entomology and Zoology* 33, 263–269.

Ganguly, S. and Gavas, R. (2004) Effect of soil moisture on the infectivity of entomopathogenic nematode, *Steinernema thermophilum* Ganguly and Singh. *International Journal of Nematology* 14, 78–80.

Ganguly, S. and Singh, L.K. (2000) *Steinernema thermophilum* sp. n. (Rhabditida: Steinernematidae) from India. *International Journal of Nematology* 10, 183–191.

Ganguly, S., Singh, M., Lal, M., Singh, L.K., Vyas, R.V. and Patel, D.J. (2002) New record of an entomopathogenic nematode, *Steinernema riobrave* from Gujarat. *Indian Journal of Nematology* 32, 223.

Ganguly, S., Rathour, K.S. and Pandey, J. (2005) New records of *Steinernema siamkayai* Stock from Champawat District of Uttaranchal and its biochemical characterization. *Indian Journal of Nematology* 35, 203–204.

Ganguly, S., Rathour, S., Kumar, S. and Singh, M. (2011) *Steinernema meghalayensis* n. sp. (Rhabditida: Steinernematidae) from North Eastern Hilly region of India. *Indian Journal of Nematology* 41, 83–97.

Gaugler, R. and Campbell, J.F. (1991) Behavioural response of the entomopathogenic nematodes *Steinernema carpocapsae* and *Heterorhabditis bacteriophora* to oxamyl. *Annals of Applied Biology* 119, 131–138.

Gaugler, R., Campbell, J.F. and Gupta, P. (1991) Characterization and basis of enhanced host-finding in a genetically improved strain of *Steinernema carpocapsae*. *Journal of Invertebrate Pathology* 57, 234–241.

Gaugler, R., Bednarek, A. and Campbell, J.F. (1992) Ultraviolet inactivation of heterorhabditid and steinernematid nematodes. *Journal of Invertebrate Pathology* 59, 155–160.

Geden, C.J. and Axtell, R.C. (1988) Effect of temperature on nematode *Steinernema feltiae* (Steinernematidae) treatment of soil for control of lesser mealworm (Coleoptera: Tenebrionidae) in turkey houses. *Journal of Economic Entomology* 81, 800–803.

Georgis, R. and Poinar, G.O.Jr (1983) Effect of soil texture on the distribution and infectivity of *Neoaplectana carpocapsae* (Nematoda; Steinernematidae). *Journal of Nematology* 15, 308–311.

Georgis, R., Mullens, B.A. and Meyer, J.A. (1987) Survival and movement of insect parasitic nematodes in poultry manure and their infectivity against *Musca domestica*. *Journal of Nematology* 19, 292–295.

Georgis, R., Kaya, H. and Gaugler, R. (1991) Effect of steinernematid and heterorhabditid nematodes on non-target arthropods. *Environmental Entomology* 20, 815–822.

Ghally, S.E. (1995) Some factors affecting the activity and pathogenicity of *Heterorhabditis heliothidis* and *Steinernema carpocapsae* nematodes. *Journal of Egyptian Society of Parasitology* 25, 125–135.

Gilmore, S.K. and Potter, D.A. (1993) Potential role of Collembola as biotic mortality agents for entomopathogenic nematodes. *Pedobiologia* 37, 30–38.

Glazer, I. (1991) Ecological considerations for entomopathogenic nematodes activity under sub-optimal conditions. *Bulletin SROP* 14, 28–31.

Glazer, I. (1992) Survival and efficacy of *Steinernema carpocapsae* in an exposed environment. *Biocontrol Science and Technology* 2, 101–107.

Glazer, I. (2002) Survival biology. In: Gaugler, R. (ed.) *Entomopathogenic Nematology*. CAB International, Wallingford, UK, pp. 169–188.

Glazer, I. and Navon, A. (1989) Activity and persistence of entomoparasitic nematodes tested against *Heliothis armigera* (Lepidoptera: Noctuidae). *Journal of Economic Entomology* 83, 1795–1800.

Glazer, I. and Salame, L. (2000) Osmotic survival of the entomopathogenic nematode *Steinernema carpocapsae*. *Biological Control* 18, 251–257.

Glazer, I., Gaugler, R. and Segal, D. (1991) Genetics of the nematode *Heterorhabditis bacteriophora* strain HP88: the diversity of beneficial traits. *Journal of Nematology* 23, 324–333.

Gokte-Narkhedkar, N., Lavhe, N.V. and Mayee, C.D. (2001) Entomopathogenic nematodes – potential agent against American boll worm of cotton (*Heliothes armigera*). *SAIC Newsletter* 11, 4.

Gray, P.A. and Johnson, D.T. (1983) Survival of the nematode *Neoaplectana carpocapsae* in relation to soil temperature, moisture and time. *Journal of the Georgia Entomological Society* 18, 454–460.

Grewal, P.S. and Richardson, P.N. (1993) Effects of application rates of *Steinernema feltiae* (Nematoda: Steinernematidae) on biological control of the mushroom fly *Lycoriella auripila* (Diptera: Sciaridae). *Biocontrol Science and Technology* 3, 29–40.

Grewal, P.S., Richardson, P.N., Collins, G. and Edmondson, R.N. (1992) Comparative effects of *Steinernema feltiae* (Nematoda: Steinenematidae) and insecticides on yield and cropping of mushroom *Agaricus bisporus*. *Annals of Applied Biology* 121, 511–520.

Grewal, P.S., Tomalak, M., Keil, C.B.O. and Gaugler, R. (1993a) Evaluation of genetically selected strain of *Steinernema feltiae* against the mushroom sciarid, *Lycoriella mali*. *Annals of Applied Biology* 123, 695–702.

Grewal, P.S., Selvan, S., Lewis, E.E. and Gaugler, R. (1993b) Males as colonizing sex in insect parasitic nematodes. *Experientia* 49, 605–608.

Grewal, P.S., Selvan, S. and Gaugler, R. (1994) Thermal adaptation of entomopathogenic nematodes: niche breadth for infection, establishment, and reproduction. *Journal of Thermal Biology* 19, 245–250.

Grewal, P.S., Lewis, E.E. and Gaugler, R. (1997) Response of infective stage parasites (Nematoda: Steinernematidae) to volatile cues from infected hosts. *Journal of Chemical Ecology* 23, 503–515.

Grewal, P.S., Weber, T. and Betterley, D.A. (1998) Compatibility of the insect-parasitic nematode, *Steinernema feltiae*, with chemicals used in mushroom production. *Mushroom News* 46, 6–10.

Grewal, P.S., Converse, V. and Georgis R. (1999) Influence of production and bioassay methods on infectivity of two ambush foragers (Nematoda: Steinernematidae). *Journal of Invertebrate Pathology* 73, 40–44.

Grewal, P.S., Power, K.T. and Shetlar, D.J. (2001) Neonicotinoid insecticides alter diapause and survival of overwintering white grubs (Coleop: Scarabaeidae). *Pest Management Science* 57, 852–857.

Griffin, C.T. (1993) Temperature responses of EPN: implications for the success of biological control programmes. In: Bedding, R. Akhurst, R. and Kaya, H. (eds) *Nematodes and the Biological Control of Insects*. CSIRO, East Melbourne, Australia, pp. 15–126.

Griffin, C.T. (2012) Perspectives on the behaviour of entomopathogenic nematodes from dispersal to reproduction: traits contributing to nematode fitness and biocontrol efficacy. *Journal of Nematology* 44, 177–184.

Griffin, C.T., Chaerani, R., Fallon, D., Reid, A.P. and Downes, M.J. (2000) Occurrence and distribution of the entomopathogenic nematodes *Steinernema* spp. and *Heterorhabditis indica* in Indonesia. *Journal of Helminthology* 74, 143–150.

Gunnardson, T. and Rundgren, S. (1986) Nematode infestation and hatching failure of lumbricid cocoons in acidified and polluted soils. *Pedobiologia* 29, 165–173.

Haag, H. and Boucias, D.G. (1991) Infectivity of insect pathogens against *Neochetina eichhorniae*, a biological control agent of water hyacinth. *Florida Entomologist* 74, 128–133.

Hague, N.G.M., Haukeland, S. and Otto, AA. (1991) Effect of temperature on the efficacy of steinernematid nematodes. *Mededelingen-van-de-Faculteit-Landbouwwetenschappen, Rijksuniversiteit, Gent* 56, 1245–1251.

Hanski, I. (2001) Spatially realistic theory of metapopulation ecology *Naturissenschaften* 88, 372–381.

Hanula, J.L. (1993) Vertical distribution of black vine weevil (Coleoptera: Curculionidae) immatures and infection by entomogenous nematodes in soil columns and field soil. *Journal of Economic Entomology* 86, 340–347.

Hara, A.H. and Kaya, H.K. (1982) Effect of selected insecticides and nematicides on the *in vitro* development of entomogenous nematode *Neoaplectana carpocapsae*. *Journal of Nematology* 14, 486–491.

Hara, A.H. and Kaya, H.K. (1983a) Toxicity of selected organophosphate and carbamate pesticides to infective juveniles of the entomogenous nematode *Neoaplectana carpocapsae* (Rhabditida). *Environmental Entomology* 12, 496–501.

Hara, A.H. and Kaya, H.K. (1983b) Development of the entomogenous nematode, *Neoaplectana carpocapsae* (Rhabditida: Steinernematidae), in insecticide-killed beet armyworm (Lepidoptera: Noctuidae). *Journal of Economic Entomology* 76, 423–426.

Hara, A.H., Gaugler, R., Kaya, H.K. and Lebeck, L.M. (1991) Natural populations of entomopathogenic nematodes (Rhabditida, Heterorhabditidae, Steinernematidae) from the Hawaiian Islands. *Environmental Entomology* 20, 211–216.

Harit, K.B., Robin, A., Taylor, J. and Grewal, P.S. (2014) Ambush foraging entomopathogenic nematodes employ 'sprinters' for long-distance dispersal in the absence of hosts. *British Journal of Applied Science and Technology* 4, 16–65.

Harrison, S. and Hastings, A. (1996) Genetic and evolutionary consequences of metapopulation structure. *Trends of Ecological Evolution* 11, 180–183.

Hashmi, G. and Gaugler, R. (1998) Genetic diversity in insect-parasitic nematodes (Rhabditida: Heterorhabditidae). *Journal of Invertebrate Pathology* 72, 185–189.

Hay, D.B. and Richardson, P.N. (1995) Inter- and intraspecific variation in infectivity of *Steinernema* spp. to larvae of the mushroom fly *Lycoriella solani. Entomologia Experimentalis et Applicata* 77, 11–15.

Hazir, S., Ishibashi, N., Choi, D.R., Ishibashi, Y., Yamamoto, Y., *et al.* (1987) Integrated control of insects and nematodes by mixing application of steinernematid nematodes with chemicals. In: *Recent Advances in Biological Control of Insect Pests by Entomogenous Nematodes in Japan*. Min of Edn Grant No 59860005, Saga University, Japan, pp. 165–176.

Hazir, S., Stock, S.P., Kaya, H.K., Koppenhofer, A.M. and Keskin, N. (2001) Developmental temperature effects on five geographic isolates of the entomopathogenic nematode *Steinernema feltiae* (Nematoda: Steinernematidae). *Journal of Invertebrate Pathology* 77, 243–250.

Hazir, S., Stock, S.P. and Keskin, N. (2003a) A new entomopathogenic nematode *S. anatoliense* sp. n. from Turkey. *Systematic Parasitology* 55, 211–220.

Hazir, S., Keskin, N., Stock, S.P., Kaya, H.K. and Ozcan, S. (2003b) Diversity and distribution of entomopathogenic nematodes (Rhabditida: Steinernematidae and Heterorhabditidae) in Turkey. *Biodiversity and Conservation* 12, 375–386.

He, H., Snyder, H.A. and Forst, S. (2004) Unique organization and regulation of the mrx fimbrial operon of *X. nematophila. Microbiology* 150, 1439–1446.

Henderson, D.R., Riga, E., Ramirez, R.A., Wilson, J. and Snyder, W.E. (2009) Mustard biofumigation disrupts biological control by *Steinernema* spp. nematodes in the soil. *Biological Control* 48, 316–322.

Heungens, K., Cowles, C.E. and Goodrich-Blair, H. (2002) Identification of *X. nematophila* genes required for mutualistic colonization of *Steinernema carpocapsae* nematodes. *Molecular Microbiology* 45, 1337–1353.

Hiltpold, I., Toepfer, S., Kuhlmann, U. and Turlings, T.C.J. (2010) How maize root volatiles affect the efficacy of entomopathogenic nematodes in controlling the western corn rootworm? *Chemoecology* 20, 155–162.

Hominick, W.M. (2002) Biogeography. In: Gaugler, R. (ed.) *Entomopathogenic Nematology.* CAB International, Wallingford, UK, pp. 115–144.

Hominick, W.M. and Briscoe, B.R. (1990) Occurrence of entomopathogenic nematodes (Rhabditida: Steinernematidae and Heterorhabditidae) in British soils. *Parasitology* 100, 295–302.

Hominick, W.M., Reid, A.P. and Briscoe, B.R. (1995) Prevalence and habitat specificity of steinernematid and heterorhabditid nematodes isolated during soil surveys of the UK and The Netherlands. *Journal of Helminthology* 69, 27–32.

Hominick, W.M., Reid, A.P., Bohan, D.A. and Briscoe, B.R. (1996) Entomopathogenic nematodes: biodiversity, geographical distribution and the convention on biodiversity. *Biocontrol Science and Technology* 6, 317–331.

Hummel, R., Walgenbach, J.F., Barbercheck, M.E., Kennedy, G.G., Hoyt, G.D. and Arellano, C. (2002) Effects of production practices on soil borne entomopathogens in west North Carolina vegetable systems. *Environmental Entomology* 31, 84–91.

Hussaini, S.S. (2003) Progress of research work on entomopathogenic nematodes in India. In: Hussaini, S.S., Rabindra, R.J. and Nagesh, M. (eds) *Current Status of Research on Entomopathogenic Nematodes in India*. Project Directorate of Biological Control, Bangalore, India, pp. 27–68.

Hussaini, S.S., Ansari, M.A., Ahmad, W. and Subbotin, S.A. (2001a) Identification of some Indian populations of *Steinernema* species by RFLP analysis of the ITS region of r DNA. *International Journal of Nematology* 11, 73–76.

Hussaini, S.S., Singh, S.P. and Shakeela, V. (2001b) Compatibility of entomopathogenic nematodes with selected pesticides and their influence on some biological traits. *Entomon* 26, 37–44.

Hussaini, S.S., Singh, S.P. and Shakeela, V. (2001c) Effect of different host plants on infectivity and progeny production of entomopathogenic nematodes, (Steinernematidae, Heterorhabditidae: Rabditida) in *Agrotis ipsilon. Proceedings of the Second National Symposium on IPM in Horticultural Crops, New Molecules, Biopesticides and Environment*. Institution of Agricultural Technologists, Bangalore, India, pp. 223.

Hussaini, S.S., Satya, K.J. and Hussain, M.A. (2001d) Effect of chemically stimulated entomopathogenic nematodes for control of insect pests. *Proceedings of the Second National Symposium on IPM in Horticultural Crops: New Molecules, Biopesticides and Environment*. Institution of Agricultural Technologists, Bangalore, India, pp. 204–206.

Hussaini, S.S., Hussain, M.A. and Satya, K.J. (2003) Survival and pathogenicity of indigenous entomopathogenic nematodes in different UV protectants. *Indian Journal of Plant Protection* 31, 12–18.

Hussaini, S.S., Nagesh, M., Rajeshwari, R. and Fathima, M.S. (2004a) Effect of pH on survival, pathogenicity and progeny production of some indigenous isolates of entomopathogenic nematodes. *Indian Journal of Nematology* 34, 169–173.

Hussaini, S.S., Nagesh, M., Rajeshwari, R. and Fathima, M.S. (2004b) Effect of adjuvants on survival and pathogenicity of some indigenous isolates of entomopathogenic nematodes. *Indian Journal of Plant Protection* 32, 111–114.

Hussaini, S.S., Nagesh, M., Rajeshwari, R. and Dar, M.H. (2005a) Effect of antidesiccants on survival and pathogenicity of indigenous isolates of entomopathogenic nematodes against *Plutella xylostella* (L.). *Annals of Plant Protection Sciences* 13, 179–186.

Hussaini, S.S., Shakeela, V. and Dar, M.H. (2005b) Influence of temperature on infectivity of entomopathogenic nematodes against black cutworm, *Agrotis ipsilon* (Hufnagel) and greater wax moth, *Galleria mellonella* (Linnaeus) larvae. *Journal of Biological Control* 19, 51–57.

Hussaini, S.S., Kumar, K.C.K., Adholeya, A. and Shakeela, V. (2009) Interaction effect of *Steinernema* spp. and arbuscular mycorrhizal fungus, *Glomus intraradices* on the development and reproduction of *Meloidogyne incognita* (Kofoid and White) in tomato. *Trends in Biosciences* 2, 33–36.

Hussaini, S.S., Shakeela, V. and Kiran Kumar, K.C. (2010) Study of attraction behaviour of indigenous isolates of entomopathogenic nematodes using *Galleria mellonella. Indian Journal of Nematology* 40, 148–154.

Ishibashi, N. (1987) *Recent Advances in Biological Control of Insect Pests by Entomogenous Nematodes in Japan*. Ministry of Education, Culture, and Science, Japan. Grant No 59860005.

Ishibashi, N. (1990) *Development of Biological Integrated Control of Pests by Beneficial Nematodes*. Ministry of Education, Culture, and Science, Japan. Grant No 6286006.

Ishibashi, N. and Kondo, E. (1986) *Steinernema feltiae* (DD-136) and *S. glaseri*: persistence in soil and bark compost and their influence on native nematodes. *Journal of Nematology* 18, 310–316.

Ishibashi, N. and Takii, S. (1993) Effect of insecticides on movement, nictation, and infectivity of *Steinernema carpocapsae*. *Journal of Nematology* 25, 204–213.

Ishibashi, N., Young, F.M., Nakashima, M., Abiru, C. and Haraguchi, N. (1987) Effect of application of DD-I36 on silk worm, *Bombyx mori*, predatory insect, *Agriosphodorus dohrni*, parasitic, *Trichomal apantelocterus*, soil mites, and other non-target soil arthropods, with brief notes on feeding behaviour and predator pressure of soil mites, tardigrades, and predatory nematodes on DD-136 nematodes. In: Ishibashi, N. (ed.) *Recent Advances in Biological Control of Insect Pest*. Ministry of Education, Culture, and Science, Saga, Japan, pp. 150–164.

Jabbour, R. and Barbercheck, M.E. (2008) Soil and habitat complexity effects on movement of the entomopathogenic nematode *S. carpocapsae* in maize. *Biological Control* 47, 235–243.

Jaffee, B.A. (2006) Interactions among a soil organic amendment a nematode and nematode trapping fungus, *Dactyllellina candidum. Phytopathology* 96, 1388–1396.

Jagdale, G.B. and Gordon, R. (1997) Effect of propagation temperatures on temperature tolerances of entomopathogenic nematodes. *Fundamental and Applied Nematology* 21, 177–183.

Jagdale, G.B., Somasekhar, N., Grewal, P.S. and Klein, M.G. (2002) Supply of plant parasitic nematodes by application of live and dead IJs of EPN, *S.carpocapsae* on boxwoods, *Buxus* spp. *Biological Control* 24, 42–49.

Jagdale, G.B., Kamaoun, S. and Grewal, P.S. (2009) Entomopathogenic nematodes induce components of systemic resistence in plants: biochemical and molecular evidence. *Biological Control* 51, 102–109.

Jarmul, J. and Kamionek, M. (2003) The influence of cadmium ions on the behaviour of *Steinernema feltiae* Filipjev IJs in soil. *Papierok-B Bulletin-OILB/SROP* 26, 197–202.

Jaworska, M. (1990) Effect of some insecticides on entomophilic nematodes. *Zeszyty-Problemowe-Postepow-Nauk-Rolniczych* 91, 73–79.

Jaworska, M. (1991) Infection of the terrestrial isopod *Porcellio scaber* Latr. and millipede *Blaniulus guttulatus* Bosc. with entomopathogenic nematodes (Nematoda: Rhabditida) in laboratory conditions. *Folia Horticulturae* 3, 115–120.

Jaworska, M. and Gorczyca, A. (2002) The effect of metal ions on mortality, pathogenicity and reproduction of entomopathogenic nematodes *Steinernema feltiae* Filipjev (Rhabditida, Steinernematidae). *Polish Journal of Environmental Studies* 11, 517–519.

Jaworska, M. and Gospodarek, J. (2009) Effect of magnesium on beneficial organisms. *Journal of Elementology* 14, 257–263.

Jaworska, M. and Ropek, D. (1994) Influence of host-plant on the susceptibility of *Sitona lineatus* L. (Col., Curculionidae) to *Steinernema carpocapsae* Weiser. *Journal of Invertebrate Pathology* 64, 96–99.

Jian, B., Reid, A.P. and Hunt, D.J. (1997) *Steinernema ceratophorum* n. sp., a new EPN from North China. *Systematic Parasitology* 37, 115–125.

Josephrajkumar, A. and Sivakumar, C.V. (1997) A survey of entomopathogenic nematodes in Kanyakumari district, Tamil Nadu, India. *Indian Journal of Entomology* 59, 45–50.

Jung, S.C. and Kim, Y.G. (2004) Synergistic effect of entomopathogenic bacteria (*Xenorhabdus* sp. and *P. temperata* ssp. *temperata*) on the pathogenicity of *Bacillus thuringiensis* ssp. *aizawai* against *Spodoptera exigua* (Lepidoptera: Noctuidae). *Environmental Entomology* 35, 1584–1589.

Kanagy, K.M.N. and Kaya, H.K. (1996) The possible role of marigold roots and α-terthienyl in mediating host-finding by steinernematid nematodes. *Nematologica* 42, 220–231.

Karagoz, M., Baris, G., Ibrahim, C., Kaya, H.K. and Hazir, S. (2007) Predation of entomopathogenic nematodes by *Sancassania* sp. (Acari: Acaridae). *Experimental and Applied Acarology* 43, 85–95.

Karunakar, G., Easwaramoorthy, S. and David, H. (2002) Compatibility of *Steinernema feltiae, S. glaseri* and *Heterorhabditis indica* with certain fertilizers, herbicide and pesticides. *Sugar Technology* 4, 123–130.

Karunakar, G., Easwaramoorthy, S. and David, H. (2003) Interaction between entomopathogenic nematodes and granulosis viruses of *Chilo infuscatellus* Snellen and *Chilo sacchariphagus indicus* (Kapur). *Journal of Biological Control* 16, 153–156.

Kaya, H.K. (1978a) Infectivity of *Neoaplectana carpocapsae* and *Heterorhabditis heliothidis* to pupae of the parasite *Apanteles militaris*. *Journal of Nematology* 10, 241–244.

Kaya, H.K. (1978b) Interaction between *Neoaplectana carpocapsae* (Nematoda: Steinernematidae) and *Apanteles militaris* (Hymenoptera: Bracondiae), a parasitoid of the armyworm, *Pseudaletia unipuncta*. *Journal of Invertebrate Pathology* 31, 358–364.

Kaya, H.K. and Burlando, T.M. (1989) Infectivity of *Steinernema feltiae* in fenamiphos-treated sand. *Journal of Nematology* 21, 434–436.

Kaya, H.K. and Hotchkin, P.G. (1981) The nematode *Neoaplectana carpocapsae* Weiser and its effect on selected ichneumonid and braconid parasites. *Environmental Entomology* 10, 474–478.

Kaya, H.K. and Stock, S.P. (1997) Techniques in insect nematology. In: Lacey, L.A. (ed.) *Manual ofTechniques in Insect Pathology*. Academic Press, San Diego, California, pp. 282–324.

Kaya, H.K., Klein, M.G. and Burlando, T.M. (1993a) Impact of *Bacillus popilliae*, *Rickettsiella popilliae* and entomopathogenic nematodes on a population of the scarabaeid, *Cyclocephala hirta*. *Biocontrol Science and Technology* 3, 443–453.

Kaya, H.K., Burlando, T.M. and Thurston, G.S. (1993b) Two entomopathogenic nematode species with different search strategies for insect suppression. *Environmental Entomology* 22, 859–864.

Kepenekci, I. (2002) Entomopathogenic nematodes (Rhabditida) in the Mediterranean Region of Turkey. *Nematologia Mediterranea* 21, 13–16.

Kepenekci, I. and Susurluk, I.A. (2003) Three entomopathogenic nematodes from Turkey. *Pakistan Journal of Nematology* 21, 19–23.

Khatri-Chhetri, H.B., Waeyenberge, L., Spiridonov, S., Manandhar, H. and Moens, M. (2011a) *Steinernema everestense* n. sp. (Rhab: Steinernematidae), a new species of entomopathogenic nematode from Pakhribas, Dhankuta, Nepal. *Nematology* 13, 443–462.

Khatri-Chhetri, H.B., Waeyenberge, L., Spiridonov, S., Manandhar, H.K. and Moens, M. (2011b) *Steinernema lamjungense* n. sp. (Rhabditida: Steinernematidae), a new species of entomopathogenic nematode from Lamjung district, Nepal. *Nematology* 13, 589–605.

Khlibsuwan, W., Ishibashi, N. and Kondo, E. (1992) Effects of enzymes, chemical and temperature on *Steinernema carpocapsae* attraction to host plasma. *Journal of Nematology* 24, 482–488.

Kondo, E. and Ishibashi, N. (1985) Effects of soil moisture on the survival and infectivity of the entomogenous nematode, *Steinernema feltiae* (DD-136). *Proceedings of the Association for Plant Protection of Kyushu* 31, 186–190.

Koppenhöfer, A.M. and Kaya, H.K. (1996) Coexistence of entomopathogenic nematode species with different foraging behaviour. *Fundamental and Applied Nematology* 19, 175–183.

Koppenhöfer, A.M. and Kaya, H.K. (1997) Additive and synergistic interaction between entomopathogenic nematodes and *Bacillus thuringiensis* for scarab grub control. *Biological Control* 8, 131–137.

Koppenhöfer, A.M. and Kaya, H.K. (1998) Synergism of imidacloprid and an entomopathogenic nematode: a novel approach to white grub (Coleoptera: Scarabaeidae) control in turfgrass. *Journal of Economic Entomology* 91, 618–623.

Koppenhofer, A.M., Haya, H.K. and Taormino, S. (1995a) Infectivity of entomopathogenic nematodes infectivity of EPN (Rhabditida: Steinernematidae) at different soil depths and moisture. *Journal of Invertebrate Pathology* 6, 99–103.

Koppenhofer, A.M., Kaya, H.K., Shanmugam, S. and Wood, G.L. (1995b) Interspecific competition between steinernematid nematodes within an insect host. *Journal of Invertebrate Pathology* 66, 99–103.

Koppenhofer, A.M., Choo, H.Y., Kaya, H.K., Lee, D.W. and Gelernter, W.D. (1999) Increased field and greenhouse efficacy against scarab grubs with a combination of an entomopathogenic nematode and *Bacillus thuringiensis*. *Biological Control* 14, 37–44.

Koppenhöfer, A.M., Brown, I.M., Gaugler, R., Grewal, P.S., Kaya, H.K. and Klein, M.G. (2000) Synergism of entomopathogenic nematodes against white grubs: greenhouse and field evaluation. *Biological Control* 19, 245–252.

Koppenhofer, A.M., Cowles, R.S., Cowles, E.A., Fuzy, E.R. and Baugartner, L. (2002) Comparison of neonicotinoid insecticides as synergists for entomopathogenic nematodes. *Biological Control* 24, 90–97.

Koppenhöfer, A.M., Cowles, R.S., Cowles, E.A., Fuzy, E.M. and Kaya, H.K. (2003) Effect of neonicotinoid synergists on entomopathogenic nematode fitness. *Entomologia Experimentalis et Applicata* 106, 7–18.

Krishnayya, P.V. and Grewal, P.S. (2002) Effect of neem and selected fungicides on viability and virulence of the entomopathogenic nematode *Steinernema feltiae. Biocontrol Science and Technology* 12, 259–266.

Kruitbos, L.M., Heritage, S. and Wilson, M.J. (2009) Phoretic dispersal of entomopathogenic nematodes by *Hylobius abietis. Nematology* 11, 419–427.

Kruitbos, L.M., Heritage, S., Hapca, S. and Wilson, M.J. (2010) The influence of habitat quality on the foraging strategies of the entomopathogenic nematodes *Steinernema carpocapsae* and *Heterorhabditis megidis*. *Parasitology* 137, 303–309.

Kulkarni, N., Paunikar, S. and Hussaini, S.S. (2009) Tolerance of EPN, *Heterorhabditis indica* to some common insecticides useful for developing IPM strategy against forest insect pests. Fifth International Conference Biopesticides: Stakeholders Perspective, 26–30 April 2009, at TERI, New Delhi, India (Abstract), pp. 174.

Kulkarni, N., Rizvi, A.N., Kumar, V., Paunikar, S. and Kumar, V. (2012) Morphological and molecular characterization of *Steinernema dharanaii* sp. n: a new entomopathogenic nematode from India. *Indian Journal of Tropical Biodiversity* 20, 107–116.

Kung, S.P. (1990) Abiotic factors affecting the persistence of two entomopathogenic nematodes, *Steinernema carpocapsae*, and *Steinernema glaseri* (Nematoda: Steinernematidae) in the soil. Dissertation Abstracts International. *B. Sciences and Engineering* 51, 1620. Rutgers State University of New Jersey, New Brunswick, New Jersey, USA.

Kung, S.P., Gaugler, R. and Kaya, H.K. (1990) Influence of soil pH and oxygen on persistence of *Steinernema* spp. *Journal of Nematology* 22, 440–445.

Kung, S.P., Gaugler, R. and Kaya, H.K. (1991) Effects of soil temperature, moisture, and relative humidity on entomopathogenic nematode persistence. *Journal of Invertebrate Pathology* 57, 242–249.

Lacey, L.A., Unruh, T.R. and Headrick, H.L. (2003) Interactions of two idiobiont parasitoids (Hymenoptera: Ichneumonidae) of codling moth (Lepidoptera: Tortricidae) with the entomopathogenic nematode *Steinernema carpocapsae* (Rhabditida: Steinernematidae). *Journal of Invertebrate Pathology* 83, 230–239.

Lacey, L.A., Granatstein, D., Arthurs, S.P., Headrick, H. and Fritts, R. Jr (2006) Use of entomopathogenic nematodes (Steinernematidae) in conjunction with mulches for control of overwintering codling moth (Lepidoptera: Tortricidae). *Journal of Entomological Science* 41, 107–119.

Lee, M.M., Sicard, M., Skeie, M. and Stock, S.P. (2009) *S. boemarei* n. sp. (Nematoda: Steinernematidae), a new entomopathogenic nematode from southern France. *Systematic Parasitology* 72, 127–141.

Lee, S.M., Lee, D.W. and Choo, H.Y. (1996a) Isolation of entomopathogenic nematodes and entomopathogenic fungi in the southern parts of Korean. *Forest Research Institute Journal of Forest Science* 53, 110–116.

Lee, S.M., Lee, D.W. and Choo, H.Y. (1996b) Biology and pathogenicity of entomopathogenic nematodes, *Steinernema* spp., isolated from forest soil in southern Korea. *Forest Research Institute Journal of Forest Science* 53, 117–123.

Letourneau, D.K. (1998) Conservation biological control lessons for conserving natural enemies. In: Barbosa, P. (ed.) *Conservation Biological Control*. Academic Press, New York, pp. 9–38.

Lewis, E.E. (2002) Behavioural ecology. In: Gaugler, R. (ed.) *Entomopathogenic Nematology*. CAB International, Wallingford, UK, pp. 205–224.

Lewis, E.E., Gaugler, R. and Harrison, R. (1993) Response of cruiser and ambusher entomopathogenic nematodes (Steinernematidae) to volatile cues. *Canadian Journal of Zoology* 71, 765–769.

Lewis, E.E., Grewal, P.S. and Gaugler, R. (1995a) Hierarchical order of host cues in parasite foraging strategies. *Parasitology* 110, 207–213.

Lewis, E.E., Selvan, S., Campbell, J.F. and Gaugler, R. (1995b) Changes in foraging behaviour during the infective stage of entomopathogenic nematodes. *Parasitology* 110, 583–590.

Lewis, E.E., Grewal, P.S. and Sardanelli, S. (2001) Interactions between the *S. feltiae–X. bovienii* insect pathogen complex and root-knot nematode, *M. incognita*. *Biological Control* 21, 55–62.

Li, X.Y., Liu, Q.Z., Nermut, J., Puza, V. and Mracek, Z. (2012) *Heterorhabditis beicherriana* n. sp. (Nematoda: Heterorhabditidae), a new entomopathogenic nematode from the Shunyi district of Beijing, China. *Zootaxa* 3569, 25–40.

Ma, J., Chen, S., Clercq, P. de, Waeyenberge, L., Han, R. and Moens, M. (2012a) A new entomopathogenic nematode, *Steinernema xinbinense* n. sp. (Nematoda: Steinernematidae), from North China. *Nematology* 14, 723–739.

Ma, J., Chen, S.L., Clercq, P. de, HanR.C. and Moens, M. (2012b) *Steinernema changbalense* sp. n. (Rhabditida) a new species of entomopathogenic nematodes from northeast China. *Russian Journal of Nematology* 20, 97–112.

Malan, A.P., Nguyen, K.B., De Waal Jeanne, Y. and Tiedt, L. (2008) *Heterorhabditis safricana* n. sp. (Rhabditida: Heterorhabditidae), a new entomopathogenic nematode from South Africa. *Nematology* 10, 1163.

Malan, A.P., Knoetze, R. and Tiedt, L.R. (2014) *Heterorhabditis noenieputensis* n. sp. (Rhabditida: Heterorhabditidae), a new entomopathogenic nematode from South Africa. *Journal of Helminthology* 88, 139–151.

Malan, A.P., Knoetze, R. and Tiedt, L.R. (2015) *Steinernema jeffreyense* n. sp. (Rhabditida: Steinernematidae), a new entomopathogenic nematode from South Africa. *Journal of Helminthology* 11, 1–17.

Mamiya, Y. (1988) *Steinernema kushidai* n. sp. (Nematoda: Steinernematidae) associated with scarabaeid beetle larvae from Shizuoca, Japan. *Applied Entomology and Zoology* 23, 313–320.

Mason, J.M. and Wright, D.J. (1998) Screening and selection of adjuvants for the spray application of entomopathogenic nematodes against a foliar pest. *Crop Protection* 17, 463–470.

McCauley, D.E. (1995) Effects of population dynamics on genetics in mosaic landscapes. In: Hansson, L., Fahrig, L. and Merriam, G. (eds) *Mosaic Landscapes and Ecological Processes*. Chapman and Hall, London, pp. 178–198.

Menti, H., Wright, D.J. and Perry, R.N. (2000) Infectivity of populations of the entomopathogenic nematodes *Steinernema feltiae* and *Heterorhabditis megidis* in relation to temperature, age and lipid content. *Nematology* 2, 515–521.

Miduturi, J.S., Clercq, R.de, Casteels, H. and Grisse, A.de (1995) Effect of temperature on the infectivity of entomopathogenic nematodes against black vine weevil (*Otiorrhynchus sulcatus* F.). *Parasitica* 50, 103–108.

Miduturi, J.S., Moermans, R. and Moens, M. (1996) Distribution of entomopathogenic nematodes in a grassland. *Russian Journal of Nematology* 5, 67–70.

Miduturi, J.S., Waeyenberge, L. and Moens, M. (1997) Natural distribution of entomopathogenic nematodes (Heterorhabditidae and Steinernematidae) in Belgian soils. *Russian Journal of Nematology* 5, 55–65.

Millar, L.C. and Barbercheck, M.E. (2002) Effect of tillage practices on entomopathogenic nematodes in a corn agroecosystem. *Biological Control* 25, 1–11.

Molina-Ochoa, M., Lezama-Gutierrez, R., Gonzalez-Ramirez, M., Lopez-Edwards, M., Rodriguez-Vega, M.A., and Arceo-Palacios, F. (2003) Pathogens and parasitic nematodes associated with populations of fall armyworm (Lepidoptera: Noctuidae) larvae in Mexico. *Florida Entomologist* 86(3), 244–253.

Molyneux, A.S. (1985) Survival of infective juveniles of *Heterorhabditis* spp. and *Steinernema* spp. (Nematoda: Rhabditida) at various temperatures and their subsequent infectivity for insects. *Revue de Nematologie* 8, 165–170.

Molyneux, A.S. (1986) *Heterorhabditis* spp. and *Steinernema* (= *Neoaplectana*) spp.; temperature, and aspects of behaviour and infectivity. *Experimental Parasitology* 62, 169–180.

Molyneux, A.S. and Bedding, R.A. (1984) Influence of soil texture and moisture on the infectivity of *Heterorhabditis* sp. D1 and *Steinernema glaseri* for larvae of the sheep blowfly, *Lucilia cuprina*. *Nematologica* 30, 358–365.

Mracek, Z. and Beevao, S. (2000) Insect aggregations and entomopathogenic nematode occurrence. *Nematology* 2, 297–301.

Mracek, Z. and Nermut, J. (2014) *Steinernema poinari* sp. n. (Nematoda: Steinernematidae) a new entomopathogenic nematode from the Czech Republic. *Zootaxa* 3760, 336–350.

Mracek, Z. and Ruzicka, Z. (1990) Infectivity and development of *Steinernema* sp. strain *Hylobius* (Nematoda, Steinernematidae) in aphids and aphidophagous coccinellids. *Journal of Applied Entomology* 110, 92–95.

Mracek, Z., Sturhan, D. and Reid, A. (2003) *Steinernema weiseri* n. sp. a new entomopathogenic nematode from Europe. *Systematic Parasitology* 56, 37–47.

Mracek, Z., Nguyen, K.B., Tailliez, P., Boemare, N. and Long, C.S. (2006) *S. sichuanense* n. sp. (Rhabditida, Steinernematidae), a new species of entomopathogenic nematode from the province of Sichuan, east Tibetan Mts., China. *Journal of Invertebrate Pathology* 93, 157–169.

Mracek, Z., Liu, Q.Z. and Nguyen, K.B. (2009) *S. xueshanense* n. sp. (Rhabditida, Steinernematidae), a new species of entomopathogenic nematode from the province of Yunnan, SE Tibetan Mts., China. *Journal of Invertebrate Pathology* 102, 69–78.

Nadasy, E. and Nadasy, M. (2006) Some harmful or useful environmental effect of nitrogen fertilisers. *Cereal Research Communications* 34, 49–52.

Nardo, E.A.B.de, Grewal, P.S., McCartney, D. and Stinner, B.R. (2006) Non-target effects of entomopathogenic nematodes on soil microbial community and nutrient cycling processes: a microcosm study. *Applied Soil Ecology* 34, 250–257.

Navaneethan, T., Strauch, O., Besse, S., Bonhomme, A. and Ehlers, R.-U. (2010) Influence of humidity and a surfactant-polymer-formulation on the control potential of the entomopathogenic nematode *Steinernema feltiae* against diapausing codling moth larvae (*Cydia pomonella* L.). *BioControl* 55, 777–788.

Negrisoli Junior, A.S., Garcia, M.S. and Negrisoli, C.R.C.B. (2010) Compatibility of entomopathogenic nematodes (Nematoda: Rhabditida) with registered insecticides for *Spodoptera frugiperda* (Smith, 1797) (Lepidoptera: Noctuidae) under laboratory conditions. *Crop Protection* 29, 545–549.

Neumann, G. and Shields, E.J. (2006) Interspecific interactions among three entomopathogenic nematodes, *Steinernema carpocapsae* Weiser, *S. feltiae* Filipjev, and *Heterorhabditis bacteriophora* Poinar, with different foraging strategies for hosts in multipiece sand columns. *Environmental Entomology* 35, 1578–1583.

Neves, Simoes, N. and Mota, M. (1998) Evidence for a sex pheromone in *Steinernema carpocapsae*. *Nematologica* 44, 95–98.

Nguyen, K.B. and Buss, E.A. (2011) *Steinernema phyllophagae* n. sp. (Rhabditida: Steinernematidae), a new entomopathogenic nematode from Florida, USA. *Nematology* 13, 425–442.

Nguyen, K.B. and Duncan, L.W. (2002) *Steinernema diaprepesi* n. sp. a parasite of the citrus weevil *Diaprepes abbreviatus* (Coleoptera: Curculionidae). *Journal of Nematology* 34, 19–170.

Nguyen, K.B. and Smart, G.C.Jr (1990) Vertical dispersal of *Steinernema scapterisci. Journal of Nematology* 22, 574–578.

Nguyen, K.B. and Smart, G.C.Jr (1991) Pathogenicity of *Steinernema scapterisci* to invertebrates. *Journal of Nematology* 23, 7–11.

Nguyen, K.B., Tesfamariam, F., Gozel, U., Gaugler, R. and Adams, B.J. (2004a) *Steinernema yirgalemense* n. sp. (Rhabditida: Steinernematidae) from Ethiopia. *Nematology* 6, 839–856.

Nguyen, K.B., Shapiro-Ilan, D.I., Stuart, R.J., McCoy, C.W., James, R.R. and Adams, B.J. (2004b) *Heterorhabditis mexicana* n.sp. (Rhab: Heterorhabditidae) from Tamaulipas, Mexico, and morphological studies of the bursa of *Heterorhabditis* spp. *Nematology* 6, 231–244.

Nguyen, K.B., Qiu, L., Zhou, Y. and, Pang, Y. (2006a) *Steinernema leizhouense* sp. n. (Nematoda: Steinernematidae), a new entomopathogenic nematode from southern China. *Russian Journal of Nematology* 14, 101–118.

Nguyen, K.B., Gozel, U., Koppenhofer, H.S. and Adams, B.J. (2006b) *Heterorhabditis floridensis* n. sp. (Rhabditida: Heterorhabditidae) from Florida. *Zootaxa* 1177, 1–19.

Nguyen, K.B., Malan, A.P. and Gozel, U. (2006c) *S. khoisanae* sp. nov. (Rhabditida: Steinernematidae), a new entomopathogenic nematode from South Africa. *Nematology* 8, 157–175.

Nguyen, K.B., Stuart, R.J., Andalo, V., Gozel, U. and Rogers, M.E. (2007) *S. texanum* n. sp. (Rhabditida: Steinernematidae), a new entomopathogenic nematode from Texas, USA. *Nematology* 9, 379–396.

Nguyen, K.B., Puza, V. and Mracek, Z. (2008a) *S. cholashanense* n. sp. (Rhabditida, Steinernematidae) a new species of entomopathogenic nematode from the province of Sichuan, Chola Shan Mountains, China. *Journal of Invertebrate Pathology* 97, 251–264.

Nguyen, K.B., Shapiro-Ilan, D.I. and Mbata, G.N.M. (2008b) *Heterorhabditis georgiana* n. sp. from Georgia. *Nematology* 10, 433–448.

Nguyen, K.B., Ginarte, C.M., Leite, L.G, Santos, J.M. and Harakava, R. (2010) *Steinernema brazilense* n. sp. a new entomopathogenic nematode from Mato Grosso, Brazil. *Journal of Invertebrate Pathology* 103, 8–20.

Nickle, W.R. and Shapiro, M. (1992) Use of a stilbene brightener, Tinopal LPW, as a radiation protectant for *Steinernema carpocapsae*. *Journal of Nematology* 24, 371–373.

Nickle, W.R. and Shapiro, M. (1994) Effects of 8 brighteners as solar radiation protectants for *Steinernema carpocapsae*, All strain. *Journal of Nematology* 26, 782–784.

Nikdel, M., Niknam, G. and Ye, W. (2011) *Steinernema arasbaranense* n. sp. (Nematoda: Steinernematidae), a new entomopathogenic nematode from Arasbaran forests, Iran. *Nematologia Mediterranea* 39, 17–28.

Nthenga, I., Knoetze, R., Berry, S., Tiedt, L.R. and Malan, A.P. (2014) *Steinernema sacchari* n. sp. (Rhabditida: Steinernematidae), a new entomopathogenic nematode from South Africa. *Nematology* 16, 475–494.

Nunez, J.C.L., Plichta, K., Gongora Botero, C.E. and Stock, P.S. (2008) A new entomopathogenic nematode, *S. colombiense* n. sp. (Nematoda: Steinernematidae), from Colombia. *Nematology* 10, 561–574.

O'Halloran, D.M. and Burnell, A.M. (2003) An investigation of chemotaxix in the insect parasitic nematode *H. bacteriophora*. *Parasitology* 127, 375–385.

Özer, N., Keskin, N. and Kirbas, Z. (1995) Occurrence of entomopathogenic nematodes (Steinernematidae: Heterorhabditidae) in Turkey. *Nematologica* 41, 639–640.

Parihar, A., Siddiqui, A.U., Yadav, Y.S. and Ram, B. (2002) Occurrence of entomopathogenic nematodes in Rajasthan. *National Symposium on Biodiversity and Management of Nematodes in Cropping Systems for Sustainable Agriculture*, held at ARS, Durgapura (Jaipur), Rajasthan, India, pp. 80.

Parkman, J.P. and Frank, J.H. (1992) Use of a sound trap to inoculate *Steinernema scapterisci* (Rhabditida: Steinernematidae) into pest mole cricket populations (Orthoptera: Gryllotalpidae). *Florida Entomologist* 76, 75–82.

Patel, M.N. and Wright, D.J. (1996) The influence of neuroactive pesticides on the behaviour of entomopathogenic nematodes. *Journal of Helminthology* 70, 53–61.

Pereira, C. (1973) *Rabditis humbeltoni* n.sp. nematode apparentemente semiparasito da broco do algodoeiro (*Gasteocerdes brasiliensis*). *Archivo do Instituto Biologico* 8, 215–230.

Pervez, R., Eapen, S.J., Devasahayam, S. and Jacob, T. (2014) Natural occurrence of entomopathogenic nematodes associated with ginger (*Zingiber officinale* Rosc.) ecosystem in India. *Indian Journal of Nematology* 44, 238–246.

Pezowicz, E., Ignatowicz, S. and Kamionek, M. (1997) Comparative effects of water extracts of seeds of the Indian neem tree, *Azadirachta indica* A. Juss, a plant-derived insecticide, NeemixTM, and a synthetic insecticide, Bancol 50 WP, on the efficacy of the entomogenous nematode, *Steinernema feltiae* (Filip). *Annals of Warsaw Agricultural University SGGW, Horticulture* 18, 31–39.

Phan, K.L., Nguyen, N.C. and Moens, M. (2001a) *Steinernema loci* sp. n. and *Steinernema thanhi* sp. n. from Vietnam. *Nematology* 3, 503–514.

Phan, K.L., Nguyen, N.C. and Moens, M. (2001b) *Steinernema sangi* sp. n. from Vietnam. *Russian Journal of Nematology* 9, 1–7.

Phan, K.L., Subbotin, S.A., Nguyen, N.C. and Moens, M. (2003) *Heterorhabditis baujardi* sp. n from Vietnam and morphometric data for *H. indica* populations. *Nematology* 5, 367–382.

Phan, K.L., Tirry, L. and Moens, M. (2004) Pathogenic potential of six isolates of entomopathogenic nematodes (Rhabditida: Steinernematidae) from Vietnam. *BioControl* 50, 477–491.

Phan, K.L., Subbotin, S.A., Waeyenberg, L. and Moens, M. (2005) *Steinernema robustispiculum* from Chumomray National Park in Vietnam. *Systematic Parasitology* 60, 23–32.

Phan, K.L., Takemoto, S. and Futai, K. (2006a) *Steinernema ashiuense* sp. n. (Nematoda: Steinernematidae), a new entomopathogenic nematode from Japan. *Nematology* 8, 681–690.

Phan, K.L., Spiridonov, W.E., Subbotin, S.A. and Moens, M. (2006b) Four new species of short *Steinernema* Travasos, 1928 with infective juveniles from Vietnam. *Russian Journal of Nematology* 14, 11–29.

Pianka, E.R. (1994) Comparative ecology of *Varanus* in the Great Victoria Desert. *Australian Journal of Ecology* 19, 395–408.

Pianka, E.R. (1999) *Evolutionary Ecology*, sixth edn. Harper and Row, New York.

Pizano, M.A., Aguillera, M.M., Monteiro, A.R. and Ferraz, L.C.B. (1985) Incidence of *Neoplectana glaseri* Steiner, 1929 (Nematoda: Steinernematidae) parasitizing *Migdolus fryanus* (Westwood, 1863) (Col. Cerambycidae). *Entomological Newsletter* 17, 9–10.

Plichta, K.L., Joyce, S.A., Clarke, D., Waterfield, N. and Stock, S.P. (2009) *Heterorhabditis gerrardi* n. sp. (Nematoda: Heterorhabditidae) the hidden host of *Photorhabdus asymbiotica* (Enterobacteriaceae: gamma-Proteobacteria). *Journal of Helminthology* 83, 309–320.

Poinar, G.O. Jr (1978) Associations between nematodes and oligochaetes. *Proceedings of Helminthological Society of Washington*, pp. 268.

Poinar, G.O. Jr (1990) Taxonomy and biology of Steinernematidae and Heterorhabditidae. In: Gaugler, R. and Kaya, H.K. (eds) *Entomopathogenic Nematodes in Biological Control*. CRC Press, Boca Raton, Florida, pp. 23–61.

Poinar, G.O. Jr, Karunakar, G.K. and David, H. (1992) *Heterorhabditis indicus* n. sp. (Rhabditida, Nematoda) from India: separation of *Heterorhabditis* spp. by infective juveniles. *Fundamental and Applied Nematology* 15, 467–472.

Polis, G.A., Anderson, W.B. and Holt, R.D. (1997) Toward an integration of landscape and food web ecology: the dynamics of spatially subsidized food webs. *Annals of Ecology and Systenatics* 28, 289–316.

Potter, D.A., Spicer, P.G., Redmond, C.T. and Powell, A.J. (1994) Toxicity of pesticides to earthworms in Kentucky bluegrass turf. *Bulletin of Environmental Contamination and Toxicology* 52, 176–181.

Powell, J.R. and Webster, J.M. (2004) Interguild antagonism between biological controls: impact of entomopathogenic nematode application on aphid predator, *Aphidoletes aphidimyza* (Diptera: Cecidomyiidae). *Biological Control* 30, 110–118.

Prabhuraj, A. and Viraktamath, C.A. (2004) Compatibility of *Steinernema glaseri* with two inorganic and one neem based insecticide. *Indian Journal of Entomology* 66, 377–379.

Prasad, G.S., Ranganath, H.R. and Singh, P.K. (2001) Occurrence of EPN in parts of South Andaman. *Current Science* 80, 501–502.

Preisser, E.L., Dugaw, C.J., Denis, B. and Strong, D.R. (2005) Long-term survival of the entomopathogenic nematode, *H. marelatus. Environmental Entomology* 34, 1501–1506.

Prishchepa, L., Mikulskaya, N. and Bezruchionok, N. (2000) Study of biological diversity of entomopathogenic nematodes in Belarus. *Vestsi Akademii Agrarnykh Navuk Respubliki Belarus'* 2, 59–62.

Priya, P. and Subramanian, S. (2006) Survival, infectivity and reproduction of entomopathogenic nematodes in spray adjuvants. *Indian Journal of Nematology* 36, 129–130.

Puza, V. and Mracek, Z. (2005) Seasonal dynamics of entomopathogenic nematodes of the genera *Steinernema* and *Heterorhabditis* as a response to abiotic factors and abundance of insect host. *Journal of Invertebrate Pathology* 89, 116–122.

Puza, V. and Mracek, Z. (2009) Mixed infection of *G. mellonella* with two entomopathogenic nematode (Nematoda: Rhabditida) species: *S. affine* benefits from the presence of *S. kraussei*. *Journal of Invertebrate Pathology* 102, 40–43.

Puza, V., Nermut, J., Mracek, Z., Gengler, S. and Haukland, S. (2016) *Steinernema pwaniensis* n sp. (Nematoda: Rhabditida) entomopathogenic nematode from Tanzania. *Journal of Helminthology* 14, 1–15.

Qiu, L.H., Fang, Y.Y., Zhou, Y., Pang, Y. and Nguyen, K.B. (2004) *Steinernema guangdongemse* sp. n. – a new entomopathogenic nematode from South China with note on *S. serratum* (*nomen nudum*). *Zootaxa* 704, 1–20.

Qiu, L., Yan, X., Zhou, Y., Nguyen, K.B. and Pang, Y. (2005a) *Steinernema aciari* sp. n. (Nematoda: Steinernematidae), a new entomopathogenic nematode from Guangdong, China. *Journal of Invertebrate Pathology* 88, 58–69.

Qiu, L., Hu, X., Zhou, Y., Mei, S., Nguyen, K.B. and Pang, Y. (2005b) *Steinernema akhursti* sp. n. (Nematoda: Steinernematidae) from Yunnan, China. *Journal of Invertebrate Pathology* 90, 151–160.

Qiu, L., Hu, X., Zhou, Y., Pang, Y. and Nguyen, K.B. (2005c) *Steinernema beddingi* n. sp. (Nematoda: Steinernematidae), a new entomopathogenic nematode from Yunnan, China. *Nematology* 7, 737–749.

Qiu, L., Zhao, J., Wu, Z., Lv, Z. and Pang, Y. (2011) *Steinernema pui* sp. n. (Rhabditida, Steinernematidae), a new entomopathogenic nematode from Yunnan, China. *Zootaxa* 2767, 1–13.

Rainey, F.A., Ehlers, R.-U. and Stackebrandt, E. (1995) Inability of the polyphasic approach to sytematics to determine the relatedness of the genera *Xenorhabdus* and *Photorhabdus*. *International Journal of Systematic Bacteriology* 45, 339–381.

Raj Kumar, M., Parihar, A. and Siddiqui, A.U. (2001) Studies on entomopathogenic nematodes of Udaipur. *National Congress on Centenary of Nematology in India Appraisal and Future Plans*, held at IARI, New Delhi, India, pp. 118.

Rao, P.P.S., Das, P.K. and Padhi, G. (1975) Note on compatibility of DD-136, an insect parasitic nematode with some insecticides and fertilizers. *Indian Journal of Agricultural Sciences* 45, 275–277.

Read, D.S., Sheppard, S.K., Bruford, M.W., Glen, D.M. and Symondson, W.O. (2006) Molecular detection of predation by soil micro-arthropods on nematodes. *Molecular Ecology* 15, 1963–1972.

Redova, S. (2010) Effect of selected pesticides on the vitality and virulence of the entomopathogenic nematode *Steinernema feltiae* (Nematoda: Steinernematidae). *Plant Protection Science* 46, 83–88.

Reid, A.P. (1994) Molecular taxonomy of *Steinernema*. In: Burnell, A.M., Ehlers, R.U. and Masson, J.P. (eds) *Genetics of Entomopathogenic Nematode Bacterium Complexes*. COST 812. ECSC-EC, EACE, Brussels, pp. 49–58.

Reid, A.P. and Hominick, W.M. (1992) Restriction fragment length polymorphism within the ribosomal DNA repeats unit of British entomopathogenic nematodes (Rhabditida: Steinernematidae). *Parasitology* 105, 317–391.

Robinson, A.F. (1994) Movement of five nematode species through sand subjected to natural temperature gradient fluctuations. *Journal of Nematology* 26, 46–58.

Rodriguez, R.O., Campbell, J.F., Christen, J.M., Shapiro-Ilan, D.I., Lewis, E.E. and Ramaswamy, S.B. (2007) Attraction behaviour of three entomopathogenic nematode species towards infected and uninfected hosts. *Parasitology* 134, 729–738.

Rojht, H., Kac, M. and Trdan, S. (2009) Non-target effect of entomopathogenic nematodes on larvae of two spotted lady beetle (Coleoptera: Coccinellidae) and green lacewing (Neuroptera: Chrysopidae) under laboratory conditions. *Journal of Economic Entomology* 102, 1440–1443.

Rolston, A.N., Griffin, C.T. and Downes, M.J. (2006) Emergence and dispersal patterns of two isolates of the entomopathogenic nematode *Steinernema feltiae*. *Journal of Nematology* 38, 221–228.

Rosa, J.S., Bonifassi, E., Amaral, J., Lacey, L.A., Simoes, N. and Laumond, C. (2000) Natural occurrence of entomopathogenic nematodes (Rhabditida: *Steinernema, Heterorhabditis*) in the Azores. *Journal of Nematology* 32, 215–222.

Rosales, L.C. and Suarez, Z. (1998) Nematodos entomopatogenos como posibles agentes de control del gorgojo negro del platano *Cosmopolites sordidus* (Germar, 1824) (Coleoptera: Curculionidae). *Boletin Entomologia Venezolana* 13, 123–140.

Rovesti, L. and Deseo, K.V. (1990) Compatibility of chemical pesticides with the entomopathogenic nematodes, *Steinernema carpocapsae* Weiser and *S. feltiae* Filipjev (Nematoda: Steinernematidae). *Nematologica* 36, 237–245.

Rovesti, L. and Deseo, K.V. (1991) Compatibility of pesticides with the entomopathogenic nematode, *Heterorhabditis hetiothidis*. *Nematologica* 37, 113–116.

Rovesti, L., Heinzpeter, E.W., Tagliente, F. and Deseo, K.V. (1988) Compatibility of pesticides with entomopathogenic nematode *Heterorhabditis bacteriophora* Poinar (Nematoda: Heterorhabditidae). *Nematology* 34, 462–476.

Rovesti, L., Fiorini, T., Bettini, G., Heinzpeter, E.W. and Tagliente, F. (1990) Compatibility of *Steinernema* spp. and *Heterorhabditis* spp. with pesticides. *Informatore Fitopatologico* 40, 55–61.

Ruiz-Vega, J., Aquino-Bolaños, T., Kaya, H.K.P. and Stock, P. (2003) Colecta y evaluación de nemátodos entomopathágenos para el control de gallinas ciegas Phyllophaga vetula (Horn) en Oaxaca, México. *Folia Entomologia. Mexicana* 42, 169–175.

Rumbos, C., Mendoza, A., Kiewnick, S. and Sikora, R.A. (2007) Effect of *Paecilomyces lilacinus* strain 251 on the survival and virulence of entomopathogenic nematodes under laboratory conditions. *Nematologia Mediterranea* 35, 103–107.

Saenz, A. (1999) Evaluacion de procedimentos para et aislamentoy almacenamiento del entomonematodonativo *Steinernema feltiae* (Rhabditida: Steinernematidae). *Agronomia Colombiana* 17, 37–42.

Saenz-Aponte, A. and Olivares, W. (2008) Searching capacity of the entomopathogenic nematode *Steinernema* sp. SNIO 198 (Rhabditida: Steinernematidae). *Revista Colombiana de Entomologia* 34, 51–56.

Segeren v.d. Oever, H.A., Sanchit Bekker, M. and Sauerrs Muller, A. (1984) Biological control of *Castnia dedalus* with insect parasitic nematodes. *De Surinamese-Landbouw* 32, 45–50.

Selvan, S., Campbell, J.F. and Gaugler, R. (1993) Density-dependent effects on entomopathogenic nematodes (Heterorhabditidae and Steinernematidae) within an insect host. *Journal of Invertebrate Pathology* 62, 278–281.

Seth, R.K., Barik, T.K. and Gautam, R.D. (2007) Safety evaluation of Parthenium beetle, *Zygogramma bicolorata* Pallister (Chrysomelidae: Coleop) against entomopathogenic nematode, *S. glaseri* (Steiner). *Journal of Entomological Research* 31, 313–317.

Shahina, F., Anis, M., Zainab, S. and Maqbool, M.A. (1998) Entomopathogenic nematodes in soil samples collected from Sindh, Pakistan. *Pakistan Journal of Nematology* 16, 41–50.

Shahina, F., Anis, M., Reid, A.P., Rowe, J. and Maqbool, M.A. (2001) *Steinernema pakistanense* sp.n. from Pakistan. *International Journal of Nematology* 11, 124–133.

Shahina, F., Manzar, H. and Bhatti, M.I. (2005) Effect of salt and temperature stress on survival and infectivity of infective juveniles of entomopathogenic nematodes. *Pakistan Journal of Nematology* 23, 103–109.

Shakeela, V. and Hussaini, S.S. (2009) Influence of soil type on infectivity and persistence of indigenous isolates of entomopathogenic nematodes *H. indica* and *S. carpocapsae*. *Journal of Biological Control* 23, 63–72.

Shamseldean, M.M. and Abd-Elgawad, M.M. (1995) Survival and infectivity of entomopathogenic nematodes under environmental stress. *Anzeiger fur Schadlingskunde, Pflanzenschutz, Umweltschutz* 68, 31–33.

Shamseldean, M.M., Abou-El-Sooud, A.B., Abd-Elgawad, M.M. and Saleh, M.M. (1996) Identification of a new heterorhabditid species from Egypt, *Heterorhabditis taysaerae*. *Journal of Biological Control* 6, 129–138.

Shannag, H.K. and Capinera, J.L. (2000) Interference of *Steinernema carpocapsae* (Nematoda: Steinernematidae) with *Cardiochiles diaphaniae* (Hymenoptera: Braconidae), a parasitoid of melonworm and pickleworm (Lepidoptera: Pyralidae). *Environmental Entomology* 29, 612–616.

Shapiro-Ilan, D.I., Berry, E.C. and Lewis, L.C. (1993) Interactions between nematodes and earthworms: enhanced dispersal of *Steinernema carpocapsae*. *Journal of Nematology* 25, 189–192.

Shapiro-Ilan, D.I., Tylka, G.L., Berrye, E.C. and Lewis, L.C. (1995) Effects of earthworms on the dispersal of *Steinernema* spp. *Journal of Nematology* 27, 21–28.

Shapiro-Ilan, D.I., Glazer, I. and Segal, D. (1997) Genetic diversity in wild and laboratory populations of *Heterorhabditis bacteriophora* as determined by RAPD-PCR analysis. *Fundamental and Applied Nematology* 20, 581–585.

Shapiro-Ilan, D.I., Cottrell, T.E., Brown, I., Gardner, W.A., Hubbard, R.K. and Wood, B.W. (2006) Effect of soil moisture and a surfactant on entomopathogenic nematode suppression of the pecan weevil, *Curculio caryae*. *Journal of Nematology* 38, 474–482.

Shapiro-Ilan, D.I., Campbell, J.F., Lewis, E.E., Elkon, J.M. and Kim Shapiro, D.B. (2009) Directional movement of steinernematid nematodes in response to electrical current. *Journal of Invertebrate Pathology* 100, 134–137.

Sicard, M., Hinsinger, J., Le Brun, N., Pages, S., Boemare, N. and Moulia, C. (2006) Interspecific competition between entomopathogenic nematodes (*Steinernema*) is modified by their bacterial symbionts (*Xenorhabdus*). BMC *Evolutionary Biology* 6, 68.

Sivakumar, C.V., Jayaraj, S. and Subramanian, S. (1988) Observations on Indian population of the entomopathogenic nematode *Heterorhabditis bacteriophora*. *Journal of Biological Control* 2, 112–113.

Spaull, V.W. (1991) *Heterorhabditis* and *Steinernema* species for the control of sugarcane stalk borer in South Africa. *Phytophylactica* 23, 213–215.

Spiridonov, S.E., Krasomil-Osterfeld, K. and Moens, M. (2004) *Steinernema jollieti*, sp. n. (Rhabditida: Steinernematidae), a new entomopathogenic nematode from the American Midwest. *Russian Journal of Nematology* 12, 85–95.

Sreenivasan, N., Prabhu, S., Makesh, S. and Sivakumar, M. (2012) Natural occurrence of entomopathogenic nematode species (Rhabditida: Steinernematidae and Heterorhabditidae) in cotton fields of Tamil Nadu, India. *Journal of Natural History* 46, 45–46.

Stack, C.M., Easwaramoorthy, S.G., Mehta, U., Downes, M., Griffin, C. and Burnell, A. (2000) Molecular characterization of *Heterorhabditis indica* isolates from India, Kenya, Indonesia and Cuba. *Nematology* 2, 477–487.

Stinner, B.R. and House, G.J. (1990) Arthropods and other invertebrates in conservation tillage agriculture. *Annual Review of Entomology* 35, 299–318.

Stock, S.P. (1992) Presence of *Steinernema scapterisci* parasitizing the mole cricket *Scapteriscus borellii* in Argentina. *Nematologia Mediterranea* 20, 163–165.

Stock, S.P. (1993) A new species of the genus *Heterorhabditis* Poinar, 1975 (Nematoda: Heterorhabditidae) parasitizing *Graphognatus* sp. larvae (Coleoptera: Curculionidae) from Argentina. *Research Review Parasitology* 53, 103–107.

Stock, S.P. (1995) Natural populations of entomopathogenic nematodes in Pampean region of Argentina. *Nematropica* 25, 143–148.

Stock, S.P. and Koppenhofer, A.M. (2003) *Steinernema scarabaei* n. sp. a natural pathogen of scarab beetle larvae from New Jersey, USA. *Nematology* 5, 191–204.

Stock, S.P., Choo, H.Y. and Kaya, H.K. (1997a) First record of *Steinernema glaseri* Steiner, 1929 (Nematoda: Steinernematidae) in Asia, with notes on intraspecic variation. *Nematologica* 43, 377–381.

Stock, S.P., Choo, H.Y. and Kaya, H.K. (1997b) An entomopathogenic nematode, *Steinernema monticolum* sp. n. (Rhabditida: Steinernematidae) from Korea with a key to other species. *Nematologica* 43, 15–29.

Stock, S.P., Pryor, B.M. and Kaya, H.K. (1999) Distribution of entomopathogenic nematodes (Steinernematidae and Heterorhabditidae) in natural habitats in California, USA. *Biodiversity and Conservation* 8, 535–549.

Stock, S.P., Mracek, Z. and Webster, J.M. (2000) Morphological variation between allopatric populations of *S. kraussei* (Steiner 1923). *Nematology* 2, 143–152.

Stock, S.P., Griffin, C.T. and Burnell, A.M. (2002) Morphological characterisation of 3 isolates of *Heterorhabditis* from the Irish group (Rhabditida: Heterorhabditidae) and additional evidence supporting their recognition as a distinct sp. *H. downesi* n. sp. *Systematic Parasitology* 51, 95–106.

Stock, S.P., Griffin, C.T. and Chaerani, R. (2004) Morphological and molecular characterisation of *Steinernema hermaphroditum* n. sp. (Nematoda: Steinernematidae), an entomopathogenic nematode from Indonesia, and its phylogenetic relationships with other members of the genus. *Nematology* 6, 401–412.

Stock S.P., Rivera-Orduño, B. and Flores Lara, Y. (2009) *Heterorhabditis sonorensis* n. sp. (Nematoda: Heterorhabditidae), a natural pathogen of the seasonal cicada, *Diceroprocta ornea* (Walker) (Homoptera: Cicadidae) in the Sonoran desert. *Journal of Invertebrate Pathology* 100, 175–184.

Stokwe, N.F., Malan, A.P., Nguyen, K.B., Knoetze, R. and Tiedt, L. (2011) *Steinernema citrae* n. sp. (Rhabditida: Steinernematidae), a new entomopathogenic nematode from South Africa. *Nematology* 13, 569–587.

Stuart, R.J. and Gaugler, R. (1994) Patchiness in populations of etomopathogenic nematode. *Journal of Invertebrate Pathology* 64, 39–45.

Stuart, R.J., Barbercheck, M.E., Grewal, P.S., Taylor, R.A.J. and Hoy, C.W. (2006) Population biology of entomopathogenic nematodes: concepts, issues, and models. *Biological Control* 38, 80–102.

Sturhan, D. (1999) Prevalence and habitat specificity of entomopathogenic nematodes in Germany, In: Gwynn, R.L., Smith, P.H., Griffin, C., Ehlers, R.-U., Boemare, N.E. and Masson, J.P. (eds) COST 819, *Entomopathogenic Nematodes: Application and Persistence of Entomopathogenic Nematodes*. European Commission, Luxembourg, pp. 123–132.

Sturhan, D. and Liskova, M. (1999) Occurence and distribution of entomopathogenic nematodes in the Slovak Republic. *Nematology* 1, 273–277.

Sturhan, D., Spiridonov, S.E. and Mracek, Z. (2005) *Steinernema silvaticum* sp. n. (Rhabditida: Steinernematidae), a new entomopathogenic nematode from Europe. *Nematology* 7, 227–241.

Susurluk, I.A. (2008) Effects of various agricultural practices on persistence of the inundative applied entomopathogenic nematodes, *Heterorhabditis bacteriophora* and *S. feltiae* in the field. *Russian Journal of Nematology* 16, 23–32.

Sznyk-Basałyga, A., Bednarek, A. and Lewandowski, M. (2002) Effect of simultaneous application of entomopatogenic nematodes and insecticide Trigard on flies *Lycoriella solani* (Diptera: Sciaridae) and *Megaselia halterata* (Diptera: Phoridae). *Progress in Plant Protection* 42, 460–463.

Tamiru, T., Waeyenberge, L., Hailu, T., Ehlers, R.-U., Puza, V. and Mracek, Z. (2012) *Steinernema ethiopiense* sp. n. (Rhabditida: Steinernematidae), a new entomopathogenic nematode from Ethiopia. *Nematology* 14, 741–757.

Tangchitsomkid, N., Sontirat, S., Tangchitsomkid, N. and Sontirat, S. (1998) Occurrence of entomopathogenic nematodes in Thailand. *Kasetsart Journal Natural Sciences* 32, 347–354.

Tarasco, E., Mracek, Z., Nguyen, K.B. and Triggiani, O. (2008) *S. ichnusae* sp. n. (Nematoda: Steinernematidae), a new entomopathogenic nematode from Sardinia Island (Italy). *Journal of Invertebrate Pathology* 99, 173–185.

Thurston, G., Kaya, H.K. and Gaugler, R. (1994) Characterizing the enhanced susceptibility of milky disease-infected scarabaeid grubs to entomopathogenic nematodes. *Biological Control* 4, 67–73.

Timper, P. and Davies, K.G. (2004) Biotic interactions. In: Gaugler, R. and Bilgrami, A.L. (eds) *Nematode Behaviour*. CAB International, Wallingford, UK, pp. 277–307.

Timper, P., Kaya, H.K. and Gaugler, R. (1988) Dispersal of entomogenous nematode *Steinernema feltiae* (Rhabditida: Steinernematidae) by infected adult insects. *Environmental Entomology* 17, 546–550.

Timper, P., Kaya, H.K. and Jaffee, B.A. (1991) Survival of entomogenous nematodes in soil infested with the nematode-parasitic fungus *Hirsutella rhossiliensis* (Deuteromycotina: Hyphomycetes). *Biological Control* 1, 42–50.

Toepfer, S., Kurtz, B. and Kuhlmann, U. (2010) Influence of soil on the efficacy of entomopathogenic nematodes in reducing *Diabrotica virgifera virgifera* in maize. *Journal of Pest Science* 83, 257–264.

Trdan S., Vidrih, M., Andjus, L. and Laznik, Ž. (2009) Activity of four entomopathogenic nematode species against different developmental stages of Colorado potato beetle, *Leptinotarsa decemlineata* (Coleoptera, Chrysomelidae). *Helminthologia* 46, 14–20.

Triggiani, O. (1985) Influence of nematodes of the families Steinernematidae and Heterorhabditidae on the parasitoid *Apanteles ultor* Rhd. (Hymenoptera: Braconidae). *Difesa delle Piante* 8, 293–300.

Triggiani, O., Mracek, Z. and Reid, A. (2004) *Steinernema apuliae* sp. n., a new entomopathogenic nematode from Southern Italy. *Zootaxa* 460, 1–12.

Umarao, Ganguly, S., Vyas, R.V. and Ganguly, A.K. (2002) Molecular characterization of *Steinernema thermophilum* Ganguly and Singh (Nematoda: Steinernematidae). *International Journal of Nematology* 12, 215–219.

Uribe-Lorío, L., Mora, M. and Stock, S.P. (2005) First record of entomopathogenic nematode in Costa Rica. *Journal of Invertebrate Pathology* 88, 226–231.

Uribe-Lorío, L., Mora, M. and Stock, S.P. (2007) *Steinernema costaricense* n. sp. and *S. puntauvense* n. sp. (Rhabditida: Steinernematidae), two new entomopathogenic nematodes from Costa Rica. *Systematic Parasitology* 68, 167–182.

Vainio, A. (1994) Effect of pesticides on long-term survival of *Steinernema feltiae* in field. *Bulletin-OILB-SROP* 17, 7076.

Vainio, A. and Hokkanen, H. (1990) Side-effects of pesticides on the entomophagous nematode *Steinernema feltiae*, and the entomopathogenic fungi *Metarhizium anisopliae* and *Beauveria bassiana* in the laboratory. *Proceedings and Abstracts, Vth International Colloquium on Invertebrate Pathology and Microbial Control*, Adelaide, Australia, 20–24 August 1990, 334 pp.

Vandermeer, J. and Perfecto, I. (1995) *Breakfast of Biodiversity: The Truth About Rain Forest Destruction.* Institute for Food and Development Policy, Oakland, California.

Vey, A., Hoagland, R. and Butt, T.M. (2001) Toxic metabolites of fungal biocontrol agents. In: Butt, T.M., Jackson, C. and Morgan, N. (eds) *Fungal Biocontrol Agents, Progress, Problems and Potential*. CAB International, Wallingford, UK, pp. 311–346.

Vivas, E.I. and Goodrich-Blair, H. (2001) *X. nematophilus* as a model for host bacterium intersctions: *rpo*S is necessary for mutualism with nematodes. *Journal of Bacteriology* 183, 4687–4693.

Walia, K.K., Walia, R.K. and Nandal, S.N. (2008) Effect of ultra-violet radiation protectants on indigenous isolates of entomopathogenic nematodes. *Journal of Biological Control* 22, 417–424.

Walter, D.E., Hudgens, R.A. and Freckman, D.W. (1986) Consumption of nematodes by fungivorous mites, *Tyrophagus* spp. (Acarina: Astigmata: Acaridae). *Oecologia* 70, 357–361.

Wang, X.D., and Ishibashi, N. (1998) Effects of precedent infection of entomopathogenic nematodes (Steinernematidae) on the subsequent invasion of infective juveniles. *Japanese Journal of Nematology* 28, 8–15.

Wang, Y. and Gaugler, R. (1998) Host and penetration site location by entomopathogenic nematodes against Japanese beetle larvae. *Journal of Invertebrate Pathology* 72, 313–318.

Waturu, C.N., Hunt, D.J. and Reid, A.P. (1997) *Steinernema karii* sp.n a new entomopathogenic nematode from Kenya. *International Journal of Nematology* 7, 68–75.

Webster, J.M., Chen, G., Hu, K. and Li, J. (2002) Bacterial metabolites. In: Gaugler, R. (ed.) *Entomopathogenic Nematology*. CAB International, Wallingford, UK, pp. 99–114.

Westerman, P.R. (1994) The vertical migration of *Heterorhabditis* spp. and *Steinernema* spp. at 9°C and the relationship to efficacy against *Otiorhynchus sulcatus* at 9°C. *Bulletin OILB SROP* 17, 81–85.

With, K.A., Pavuk, D.M., Worchuk, J.L., Oates, R.K. and Fisher, J.L. (2002) Threshold effects of landscape structures on biological control in agroecosystems. *Ecological Applications* 12, 52–65.

Wootton, J.T. (1994) The nature and consequences of indirect effects of ecological communities. *Annual Review of Ecology and Systematics* 25, 443–466.

Wright, R.J., Witkowski, J.F., Echtenkamp, G. and Georgis, R. (1993) Efficacy and persistence of *Steinernema carpocapsae* (Rhabditida: Steinernematidae) applied through a center-pivot irrigation system against larval corn rootworms (Coleoptera: Chrysomelidae). *Journal of Economic Entomology* 86, 1348–1354.

Xu, F.Z., Wang, G.H. and Li, X.F. (1991) A new species of the genus *Steinernema* (Rhabditida: Steinernematidae). *Zoological Records* 12, 17–20.

Yan, X., Liu, X.J., Han, R.C., Chen, S.L., Clercq, P. de and Moens, M. (2010) Osmotic induction of anhydrobiosis in entomopathogenic nematodes of the genera *Heterorhabditis* and *Steinernema. Biological Control* 53, 325–330.

Yang, P. and Li, S.C. (1988) The effect of temperature on the development and pathogenicity of entomopathogenic nematodes. *Insect Knowledge* 25, 300–302.

Yildrim, E. and Hoy, C.W. (2003) Interaction between cyromazine and the entomopathogenic nematode *Heterorhabditis bacteriophora* Poinar 'GPS11' for control of onion maggot, *Delia antiqua* (Meigen). *Crop Protection* 22, 923–927.

Yoshida, M., Reid, A.P., Briscoe, B.R. and Hominick, W.M. (1998) National survey of entomopathogenic nematodes (Rhabditida: Steinernematidae and Heterorhabditidae) in Japan. *Fundamental and Applied Nematology* 21, 185–198.

Zaki, F.N., Awadallah, K.T. and Gesraha, M.A. (1997) Competitive interaction between the braconid parasitoid *Meteorus rubens* Nees and the entomogenous nematode, *Steinernema carpocapsae* on larvae of

Agrotis ipsilon Hufn. (Lepidoptera: Noctuidae). *Journal of Applied Entomology* 121, 151–153.

Zhang, G.Y., Yang, H.W., Zhang, S.G. and Jiang, H. (1992) Survey on the natural occurrence of entomophilic nematodes (Steinernematidae and Heterorhabditidae) in the Beijing area. *Chinese Journal of Biological Control* 8, 157–159.

Zhang, K.Y., Liu, X.H., Tan, J., Wang, Y., Qiao, L., *et al.* (2012) *Heterorhabditidoides rugaoensis* n. sp. (Rhabditida: Rhabditidae), a novel highly pathogenic entomopathogenic nematode member of Rhabditidae. *Journal of Nematology* 44, 348–360.

Zhang, L., Shono, T., Yamanaka, S. and Tanabe, H. (1994) Effects of insecticides on the entomopathogenic nematode *Steinernema carpocapsae* Weiser. *Applied Entomology and Zoology* 29, 539–547.

Zimmerman, R.J. and Cranshaw, W.S. (1990) Compatibility of three entomopathogenous nematodes (Rhabditida) in aqueous solutions of pesticides used in turf grass maintenance. *Journal of Economic Entomology* 83, 97–100.

6 Molecular Systematics and Phylogenetic Reconstruction of *Steinernema* and *Heterorhabditis*

FRANCESCA DE LUCA[1]* AND MAHFOUZ M.M. ABD-ELGAWAD[2]

[1]*Istituto per la Protezione Sostenibile delle Piante, Consiglio Nazionale delle Ricerche, Bari, Italy;* [2]*Phytopathology Department, National Research Center, Giza, Egypt*

6.1 Introduction

Entomopathogenic nematodes (EPNs) have a global spread since they form a group of obligate and lethal parasites of insects. EPNs belonging to the genera *Heterorhabditis* and *Steinernema* possess the attribute of potential biological insecticides and are therefore attractive from a commercial viewpoint. These nematodes, in general, are relatively conserved in gross morphology such as microscopic sizes or morphological similarity, thus making the task of identification difficult for taxonomists. Since the discovery of hermaphroditism in steinernematids by Griffin *et al.* (2001), diagnosis of EPNs by traditional methods has now been of limited utility in EPN taxonomy, even for trained nematologists. The diagnosis of *Heterorhabditis* needs a more sophisticated molecular analysis because an amphimictic and automictic turning may alternate in every species (Dix *et al.*, 1994). Furthermore, it is quite common that heterorhabditid isolates of the same species have different morphometrics (Abd-Elgawad and Nguyen, 2007). On one hand, morphological characters lead to an ambiguous identification, whereas on the other hand, molecular techniques offer new opportunities in solving problems related to EPN identification. In the past decade, different methods based on DNA analysis have been developed and used for the identification of species, studies on intra- and interspecific variability, population genetics, ecology and selection of the appropriate species and/or strains as biological agents (Dubey *et al.*, 2011). Moreover, DNA is not influenced by environmental conditions, or by the tissue or developmental stage of an organism.

Molecular characterization of EPNs has important implications for their proper classification, biodiversity studies, population genetics, ecology and selection of the appropriate species and/or strains in biological control programmes. One such example is *Heterorhabditis brevicaudis*, which on a morphological basis was considered as a species inquirenda, until the molecular data were produced to support its description (Hsieh *et al.*, 2009). For instance, a precise and reliable identification of EPN species/strain has important implications in controlling certain insect species and to realize optimal execution in reality (Abd-Elgawad, 2001). Specimens belonging to an EPN strain show common, but also specific, characters that may differ from other strains within the same species, e.g. virulence, pesticide resistance, persistence, heat tolerance and host ability. In conclusion, classical taxonomy and molecular diagnostics should be combined and considered, and the concept of integrative taxonomy must also be introduced for EPNs.

*Corresponding author. E-mail: francesca.deluca@ipsp.cnr.it

6.2 Molecular Techniques for Entomopathogenic Nematode Identification

6.2.1 Routine molecular methods

DNA sequence divergence between EPN species was first studied by visualization (in ethidium bromide stained agarose gels) of repetitive DNA restriction fragment length differences (RFLDs) in total genomic DNA of *Steinernema glaseri*, *Steinernema carpocapsae*, *Steinernema* spp. and *Heterorhabditis* isolates (Curran, 1990). Then, the screening of randomly selected plasmid clones of *S. carpocapsae* Eco RI genomic DNA fragments identified a species-specific DNA probe which, in Southern blots of DNA from all ten isolates, only hybridized to the genomic DNA of the three *S. carpocapsae* isolates (Curran, 1990).

The introduction of polymerase chain reaction (PCR) (Saiki *et al.*, 1985), has affected all fields of the life sciences significantly. It has brought about dramatic progress in systematics, evolutionary studies and molecular investigations (Adams, 1998). A paramount advantage of methods based on PCR is that this technique overcomes the limitations of morphological identification, as their sensitivity permits the analysis of informative DNA markers and genes from minute amounts of material. Any very small DNA piece/slice may be utilized via amplification and detection by PCR starting from individual nematodes, even single individuals within a complex mixture of nematodes from a soil community (Subbotin *et al.*, 2001; Hubschen *et al.*, 2004; Campos-Herrera *et al.*, 2014).

The general steps of DNA extraction from individual specimens (Sambrook and Russell, 2001) are as follows:

1. Extraction of gross DNA from individual nematodes.
2. PCR amplification of target regions using specific primers for identification.
3. Detection and recovery of the PCR amplicons from agarose gels.
4. Direct sequencing of purified amplicons.
5. Cloning PCR products into T vectors, conversion of qualified *Escherichia coli* cells and rapid characterization of positive clones by colony PCR.
6. Sequencing of selected clones.
7. Similarity search using the basic local alignment search tool (BLAST) to identify the EPN origins.

The PCR amplification needs a DNA template having the target region, DNA polymerase, two primers flanking the target region and four deoxynucleotide triphosphates (dATP, dGTP, dTTP and dCTP) within a buffer of magnesium ions. The primers usually used are universal primers that are complementary to definite sequences of DNA for a broad range of organisms; primers that are complementary to one species are called species specific. When such sequences, arranged in a particular order, of flanking regions of the target fragment are not conserved, PCR with degenerate primers that allow annealing and amplification of a set of related sequences can be performed.

Diagnostic restriction fragment length polymorphism models related to polymerase chain reaction (PCR-RFLP) are still useful in supporting the characterization and identification of species and subspecies of the two main EPN genera (Reid *et al.*, 1997; Stack *et al.*, 2000; Nguyen *et al.*, 2001; Phan *et al.*, 2001; Spiridonov *et al.*, 2004; Susurluk and Toprak, 2006; Dolinski *et al.*, 2008). The amplified products with variation in sequences can be differentiated by restriction enzyme digestion. Such a digestion is usually applied on the amplicons of various EPN species, subspecies or strains, and then the fragments obtained are separated by gel electrophoresis. Distinct sequences within restriction sites can clearly lead to distinct RFLP patterns via the digestion of the augmented fragments. For trustworthy diagnosis, internal transcribed spacer (ITS)-RFLP has been in use for the identification and characterization of examined species, thus revealing that they are reliable tools for the complementary morphological descriptions of uncharacterized EPNs (Griffin *et al.*, 2000; Phan *et al.*, 2001; Susurluk *et al.*, 2007; Torrini *et al.*, 2014).

The acronym RAPD refers to random amplified polymorphic DNA, because this method amplifies random DNA fragments. The RAPD technique is used to characterize/describe the intended EPN species/strains (Liu and Berry, 1995, 1996a; Stock *et al.*, 2001) to estimate variations among EPN species/strains on a genetic basis (Hashmi and Gaugler, 1998) and to assess phyletic relations among these taxa (Liu and Berry, 1996b; Stock and Hunt, 2005). In particular, RAPD was used to differentiate *Heterorhabditis marelatus* effectively from *Heterorhabditis bacteriophora* (Liu and Berry, 1996b) and to explore the divergence of various *Heterorhabditis* subspecies/populations (Shapiro-Ilan *et al.*, 1997). A single arbitrary primer, about ten nucleotides long can amplify DNA sequences under low temperature conditions of annealing to permit quick screening of a nematode genome, and consequently to detect variation via this RAPD. When annealing of the random primer with two complementary sequences found on opposite strands of the genome in the right position and close enough from each other occurs, the relevant DNA bit is subjected to amplification. When any of such circumstances are not present, for example the primer site differs by any nucleotide, then annealing does not take place. However, RAPD fingerprints have been discouraged because of inconsistent results. The quality and the concentration of DNA, and the PCR cycling conditions, can yield inconsistent output. Also, the variation due to laboratory procedures in the Material and Methods may even exceed that of the PCR cycling conditions (Adams *et al.*, 2006). Another fingerprinting technique is the amplified fragment length polymorphism (AFLP) technique used to examine and compare the genetic variability between populations/strains of EPNs (Masumu *et al.*, 2006; Dillon *et al.*, 2008). The AFLP technique is more reliable and robust than RAPD because it is not influenced by amplification conditions such as the quantity and quality of DNA template or annealing temperature. AFLP has these phases: whole DNA digestion via two restriction enzymes; ligation of double-strand adapters to the extremities of the restriction fragments; amplification using two primers integral to the adapter and one that shows one to three nucleotides neighbouring to the restriction places; finally, separation of AFLP-PCR fragments on denaturing polyacrylamide gel. Single-enzyme-AFLP, or SE-AFLP, analysis is a fingerprinting method derived from AFLP via one instead of two adapters (Wang *et al.*, 2013). The SE-AFLP method was executed to assess the diversity, on a genetic basis, among some populations of *Steinernema monticolum* from Korea, but no correlation could be detected between the genetic similarity of populations and their geographic vicinity (Dillon *et al.*, 2008; Wang *et al.*, 2013).

6.2.2 Genes committed for molecular identification of entomopathogenic nematodes

Nuclear and mitochondrial DNA are the most favourable regions in nematode phylogenesis and systematics. It is well known that genes at these regions markedly undergo the process of mutating at different rates over time. Generally, the process of mutating in non-coding regions such as introns and ITS has a higher rate than mitochondrial genes, since the latter genes evolve according to their function and structure. Ribosomal DNA contains units (cistrons) that are repeated in tandem a hundred times. The copies are located on different chromosomes and are subjected to the principle of concerted evolution that determines the homogenization of the sequences among the various cistrons. Every single cistron consists in a non-coding intergenic region, or IGS, an external transcribed spacer (ETS) and three genes, 18S, 5.8S and 28S, which are also separated from the first and second ITSs (ITS1 and ITS2). The coding regions (18S and 28S) of each cistron are conserved in a broad range of different genera. So, they can be utilized to contrast with other taxa, especially those that were separated for long time ago. But the average of evolution in external and intergenic spacers (ETS, ITS and IGS) is more than that in 18S and 28S. Therefore, ETS, ITS and IGS can be useful for resetting/rebuilding evolutionary cases, especially recent ones, and for the distinction of closely related species and subspecies.

However, even within the same gene, regions are present with different variability. In fact, the 28S gene has variable and conserved regions (D expansion). The expansion domains D are variable even between closely related species, and are helpful for recording recent evolutionary events (Stock *et al.*, 2001; Stock and Koppenhöfer, 2003; Sonnenberg *et al.*, 2007; Tarasco *et al.*, 2008). Numerous high-level phylogenetic studies have focused on the analysis of the 18S rRNA gene where results were compatible with those based on nematode morphology. Both morphological and molecular studies stressed that Steinernematidae and Heterorhabditidae families represented distinct and unrelated lineages (Blaxter *et al.*, 1998; Stock and Hunt, 2005; Sommer and Streit, 2011). For instance, at species level, the 18S marker also resulted in being too conserved in resolving phylogenetic relationships within *Heterorhabditis* or *Steinernema* (Stock *et al.*, 2001). The ITS containing region of rDNA is highly variable, is of great diagnostic value and is very beneficial in discriminating *Heterorhabditis* or *Steinernema*, at both species and strain level, is useful to infer evolutionary relationships among genera and species. Of special interest for identification of EPNs are ITS1 and ITS2 (the first and second internal transcribed spacers) because they simultaneously have both a low and a high level of intraspecific sequence variation and interspecific differences, respectively (Nguyen *et al.*, 2001; Ma *et al.*, 2012; Nthenga *et al.*, 2014). Nevertheless, the ITS1 and ITS2 regions in EPNs may exhibit significant heterogeneity, even within the same species, via different individual nematodes.

Nuclear genes encoding proteins (actin, the major sperm protein and heat shock proteins) and intron regions are often used in the diagnosis and evolutionary studies of plant parasitic nematodes (De Luca *et al.*, 2013). Targeting intron regions is a helpful approach for discriminating between closely related species that lack sufficient divergence in the ITS regions. Recently, a novel molecular marker, the heat shock protein-70 (*hsp*-70) gene was used to differentiate six *Heterorhabditis* species that were nearly morphologically identical (Saeb, 2015). This study revealed that the exon primer set of *hsp*-70 was specific of *Heterorhabditis megidis*, opening the door for many other applications including quantitative PCR analysis to facilitate the quantification of nematodes in infected insects. The intron primer set of the *hsp*-70 gene produced distinct and reproducible patterns that characterized and identified the different species of *Heterorhabditis*. The resulting intron sequences may be utilized for designing sequence characterized amplified region-polymerase chain reaction (SCAR-PCR), to reach much easier and specific identification of *Heterorhabditis* species in the future.

The extraction of mitochondrial DNA (mtDNA) can be achieved without much effort. It comprises conserved sequences that enable it to apply common primers easily and possesses a high copy number. Therefore, mtDNA broadly serves in comparing nematode populations, as well as EPN evolutionary and taxonomic status (Stock and Hunt, 2005; Montiel *et al.*, 2006). Blouin *et al.* (1999) examined *H. marelatus* isolates to locate the genetic variability via amplification of the mitochondrial ND4 gene. They found *H. marelatus* to have much lower mtDNA diversity within its populations. Additional mtDNA genes, COXII and 16S rDNA, were reported in EPNs (Szalanski *et al.*, 2000). These loci showed variability at species level and were used to differentiate various isolates of *Steinernema* spp. However, these markers were not able to reveal the intraspecific variability among populations of *Steinernema feltiae*. In addition, complete mitochondrial DNA of *S. carpocapsae* was determined (Montiel *et al.*, 2006) by using a long PCR approach. They reported that *S. carpocapsae* was more connected to *Caenorhabditis elegans* and *Ascaris suum* than to *Strongyloides stercoralis*, based on the comparison of their mitochondrial genome. Yet these relations between the four species do not match their previously investigated phylogeny, which used nuclear ribosomal DNA sequences. Montiel *et al.* (2006) assumed that this disagreement stemmed from the differential pressure of selection and/or different strategies of reproduction exerted on mitochondrial and nuclear genes.

Of recent special interest as a molecular marker for the taxonomic differentiation and evolutionary research of EPNs is the subunit I of cytochrome oxidase (COI) encoded by the mitochondrial gene. It has several attributes as a target gene. It has not only quite preserved regions but also mutable areas with a broad range of mutational rates. COI was found to be the best marker to examine the genetic diversity of *S. feltiae* isolates (Valadas *et al.*, 2013).

6.2.3 DNA barcoding

DNA barcoding is the most recent and promising approach for nematode identification in general and EPNs in particular. It is a molecular method dedicated to the identification of biological identity that depends on testing the variability of a relevant marker (Floyd *et al.*, 2002; Hebert *et al.*, 2003). Similar to barcoded merchandise, organisms are barcoded via genomic regions that contain substantial information from their ancient/recent evolutionary biology. Blaxter (2004) reported that a DNA barcode, represented by a specific sequence, could demonstrate both phylogenetic and species-specific data of its organism. The barcodes can help in locating an unknown specimen phylogenetically in its right position compared to known reference sequences, as well as in EPN identification. DNA barcoding also provides the analysis of nematodes both as individuals and as a mixture of individuals and species at different phases (bulk) that are subjected, after lysis, to the amplification of the intended region.

6.2.4 Quantitative PCR

Of recent special interest is the use of the real-time PCR technique in the detection and quantification of many EPNs, or to study cryptic species (Torr *et al.*, 2007; Campos-Herrera *et al.*, 2011, 2012, 2014, 2015). In 'end-point' PCR, the amplification product is displayed only at the end of the reaction and does not give information on the amount of template DNA in the sample. The technique of real-time PCR measures continuously the amplification of specific products during the exponential phase of PCR; that is, when the amplification efficiency is influenced minimally by the reaction variables, allowing much more accurate information to be obtained than 'end-point' PCR. Quantification of the product is achieved by adding compounds whose fluorescence emitted at each cycle is proportional to the quantity of the product amplified. Real-time PCR is achieved by using specific hybridization probes (TaqMan, Molecular Beacons and Scorpions) or intercalating dyes (Syber green) that bind to double-stranded DNA. The fluorescence changes can be determined directly in the tube where the reaction has taken place. The presence of non-specific products is highlighted at the end of the amplification when the samples are heated and changes in fluorescence generate a dissociation curve. An amplification curve is obtained for each sample, and detection of the specific product is carried out, determining the cycle at which the fluorescence intensity slightly increases compared to the background value, known as the threshold cycle (Ct). The Ct value is inversely proportional to the target DNA, and the quantity of the target DNA is calculated by comparing the Ct values of known endogenous DNA.

At the moment, more than 20 real-time qPCR probes are available to study EPN assemblages in the field (Torr *et al.*, 2007; Campos-Herrera *et al.*, 2011, 2013, 2014, 2015).

6.2.5 Sequencing technologies

Sequencing allows the identification of known species, regardless of their morphological characteristics (e.g. free-living nematodes and entomopathogenic and plant parasitic nematodes), but also new species as yet unknown. The genetic variability among different species involves differences in the genomic sequences proportionally to the evolutionary divergence among species, but still allows the correct identification of species. Analysis of the sequences obtained, by comparing with a sequence database available online, allows a rapid and reliable identification of the EPN studied.

Direct sequencing of the amplified products allows the identification of polymorphisms (one or more nucleotides in one or more of the sequence positions) and a measure of the genetic variability of the species, but does not allow sequence variants to be isolated. To overcome this problem, the PCR products are cloned into a plasmid vector and several clones sequenced to also recover sequence variants present at low frequency. The combination of PCR with the sequencing method of dideoxynucleotides, the defined method of cycle sequencing, has allowed the determination of sequences of species from small amounts of template DNA. Automation of sequencing has therefore reduced the cost and time of these procedures producing a large amount of data. Recently, the sequences of different nuclear (18S, 28S and ITS, *hsp*-70) and mitochondrial genes (ND4 and COI) have been determined, and the inter- and intraspecies variability has allowed the phylogenetic relationships of EPNS to be established at different taxonomic levels.

Sequencing technology is evolving rapidly, and over the past 5 years, the NGS approach (next-generation sequencing) has found wide use in all fields of diagnostics. The principal steps in NGS sequencing are the following: the total DNA is first fragmented into small segments that vary from 100 to 500 base pairs (bp), called library. These fragments, by the addition of adapters, are linked covalently to a solid support (flow cell) that serves as a template for enzymatic incorporation, mediated by a DNA polymerase, of nucleotides labelled with fluorophores (sequencing by synthesis) producing millions of parallel reactions. The new sequences, called reads, are then reassembled using a known reference genome.

NGS technology allows the study of genomics, transcriptomics and metagenomics, and research on EPN identification and analysis of the composition of nucleic acids from matrices of a different nature (e.g. investigation of the microbial composition of plant or soil samples, or food). Furthermore, NGS technology allows high accuracy detection of alleles or under-represented species that may be important for the protection of crops and the environment.

6.2.6 Phylogenesis

In the past decade, innovative sequencing technologies have led to an increase in database sequences that have allowed assessment of the phylogenetic relationships between different species and populations of EPNs. Several molecular markers have been explored to resolve the relationships among species of *Steinernema* and *Heterorhabditis*, including several nuclear (18S, 28S and ITS, *hsp*-70) and mitochondrial (COI, COII and ND4) genes (Peat, 2010; Saeb, 2015). The 18S rRNA gene was the first marker to be sequenced and used to infer phylogenetic relationships between *Steinernema* and *Heterorhabditis* (Liu *et al.*, 1997; Blaxter *et al.*, 1998). But Blaxter *et al.* (1998) showed in his phylogenetic tree that *H. bacteriophora* was the only EPN that grouped with the free-living nematode *C. elegans*, while another species of EPN, *S. carpocapsae* clustered with other nematodes. This result suggests that the 18S gene is highly conserved within the phylum Nematoda (Burnell, 2002) and does not contain phylogenetically informative characters (replacements) useful in determining the relationships among species of EPNs and other nematodes. Markers such as the 28S rRNA gene, the ITS region (Reid, 1994; Reid *et al.*, 1997), mitochondrial ND4 and COI are widely used to reconstruct the phylogenetic relationships between EPNs. The 28S rRNA gene is characterized by expansion regions called D domains, with different degrees of variability. The D2–D3 expansions diverge between different species, and therefore can be used to study relatively recent evolutionary events (Powers, 2004; Blaxter *et al.*, 2005; He *et al.*, 2005; Adams *et al.*, 2006; Gozel *et al.*, 2006). The mitochondrial COI and ND4 genes code for proteins, and therefore have selective constraints to maintain their function. The ribosomal genes do not encode for protein, and thus accumulate mutations (indels) more easily compared to protein-coding genes (Nguyen *et al.*, 2001; De Luca *et al.*, 2004). Recent studies have shown that the ITS, 16S, COI and ND4 are species-specific markers, because they evolve rapidly and vary between species, between populations within a species and even between individuals of the same population, and can also be used as a molecular barcode (Adams *et al.*, 1998; Spiridonov *et al.*, 2004).

The representation of phylogenetic relationships includes three steps: (i) alignment of the sequences; (ii) the construction of a phylogenetic tree; and (iii) comparison with other phylogenetic trees. The nucleotide or amino acid counterparts are aligned with each other, and three alignment modes can be distinguished: (i) matches, the same nucleotide is present in all sequences; (ii) mismatch, different nucleotides are present in the same position; and (iii) gaps, no nucleotide is present in one of the sequences in the same location. For closely related species, sequences can be aligned easily; for sequences not closely related, there are different computer programs with special algorithms that help to obtain alignments. The most used programs are Clustal, which allows a progressive sequence alignment, and BioEdit, which allows aligning and manually adjusting the alignment itself. The methods most used for the construction of phylogenetic trees are: (i) the distance matrix methods; and (ii) the discrete methods. The distance-matrix methods (UPGMA and Neighbor Joining) are based on a measure of evolutionary distance between all pairs of sequences, and the values of distance represent the distance between taxa. The distance-matrix methods can produce both rooted and unrooted trees, depending on the algorithm used. The discrete methods (maximum parsimony (MP), maximum likelihood (ML) and Bayesian inference (BI)) represent trees that minimize the number of evolutionary events accurately (Fig. 6.1). More information about the relationships between taxa can be reached by inserting a sequence (out-group) not related to all the others. When the out-group is chosen appropriately, you will have greater genetic distance, and it will be located at the basal position in the tree. The reliability of a phylogenetic analysis can be evaluated through the calculation of the bootstrap value. Bootstrapping consists of a random resampling of the multi-alignment (Phillips *et al.*, 2004). The multi-alignment is analysed many times (100–1000), and for each of them a phylogenetic tree is constructed. At the end, a number of trees equal to the number of simulated multi-alignments will be determined and tested if the same node is recovered. If the same node shows the same percentage of iterations, this means that the node is well supported. Two trees reproduce the same node when it shares the same taxonomic units.

6.3 Conclusions and Future Prospects

EPNs of the two aforementioned genera together with their symbiotic bacteria symbolize a distinct symbiotic complex for the biocontrol of insect pests (Askary, 2010). Molecular techniques are very useful for carrying out accurate identification of their species/strains, since species identification using solely morphology-based taxonomy may be a good but not sufficient criterion for separating species (Adams and Nguyen, 2002). Many cryptic species (morphologically similar but genetically different) have been reported for nematodes.

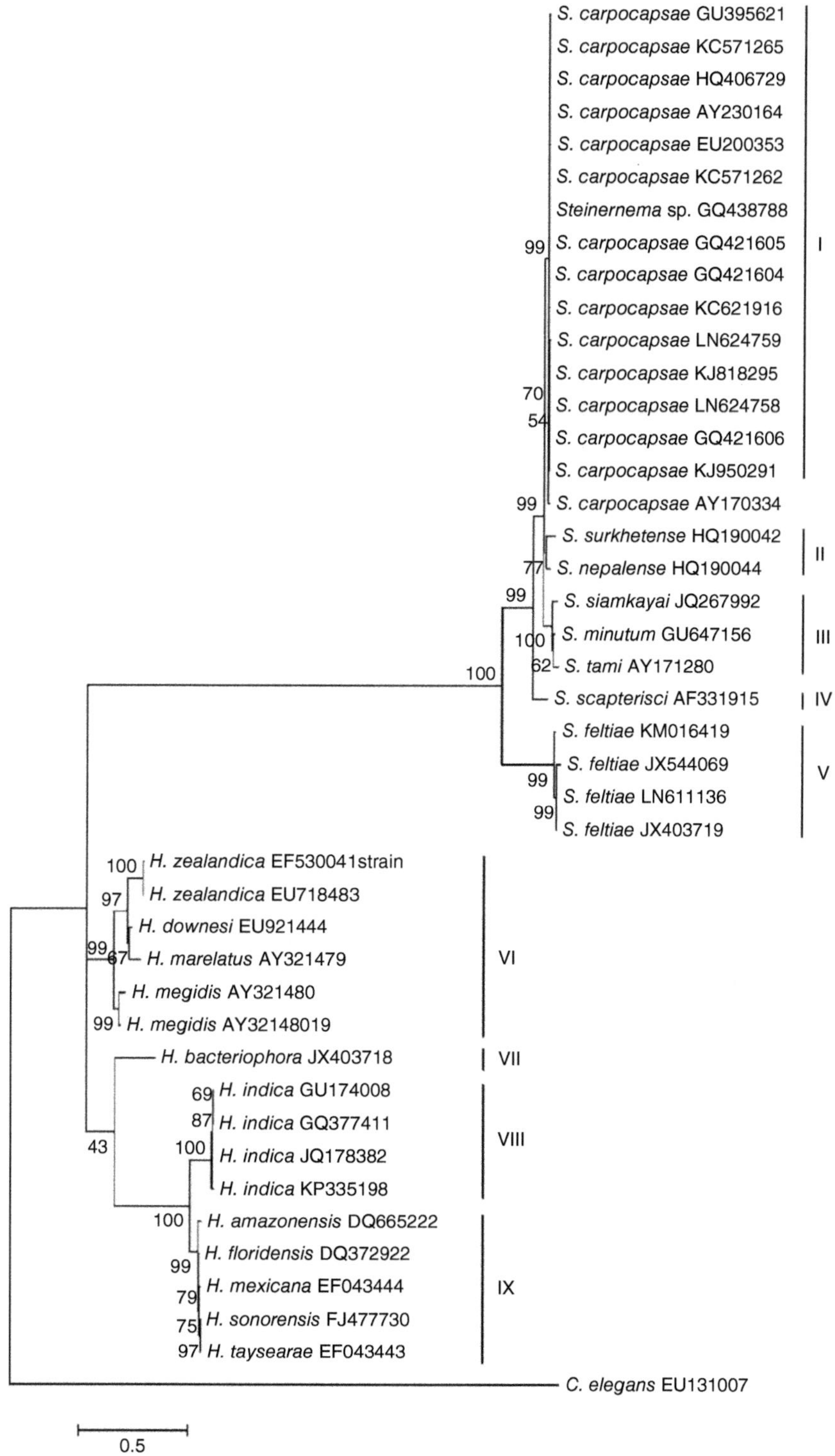

Fig. 6.1. Phylogenetic tree of ITS-containing region of *Steinernema* and *Heterorhabditis* species. Sequences were analysed using the maximum likelihood method. Numbers at nodes indicate bootstrap values.

Thus, these species may remain undetected, even if a careful morphological characterization has been carried out. Recently, for accurate and reproducible species/strain identification of EPNs, researchers have been testing a polyphasic approach: the combination of morphological, molecular and phylogenetic analyses (Dillman, 2013; Fayyaz *et al.*, 2014, 2015). In addition, the conserved flanking regions of ITSs being highly conserved have allowed the design of species-specific primers that can be used for both traditional PCR amplification and for real-time PCR for EPN mixed populations in soil samples. We have reconstructed phylogenetic tree of ITS-containing region of *Steinernema* and *Heterorhabditis* species; but it is subject to further modification especially with the discovery of new EPN species/strains and/or molecular tools.

Acknowledgements

The authors acknowledge Dr Elena Fanelli for critical reading of the manuscript before submission and the funding provided by the Italian MEF to the CNR for the project 'Innovazione e Sviluppo del Mezzogiorno-Conoscenze Integrate per Sostenibilità ed Innovazione del Made in Italy Agroalimentare' Legge n. 191/2009. This work was supported in part by US-Egypt Science and Technology Joint Fund (cycle 17; project no. 172) and Egyptian In-House, National Research Centre Project No 11030133 entitled 'Pesticide alternatives against soil-borne pathogens attacking legume cultivation in Egypt'.

References

Abd-Elgawad, M.M.M. (2001) Entomopathogenic nematode–host matching. *Egyptian Journal of Agronematology* 5, 91–104.

Abd-Elgawad, M.M.M. and Nguyen, K.B. (2007) Isolation, identification and environmental tolerance of new heterorhabditid populations from Egypt. *International Journal of Nematology* 17, 116–123.

Adams, B.J. (1998) Species concepts and the evolutionary paradigm in modern nematology. *Journal of Nematology* 30, 1–21.

Adams, B.J. and Nguyen, K.B. (2002) Taxonomy and systematics. In: Gaugler, R. (ed.) *Entomopathogenic Nematology*. CAB International, Wallingford, UK, pp. 1–33.

Adams, B.J., Burnell, A.M. and Powers, T.O. (1998) A phylogenetic analysis of *Heterorhabditis* (Nemata: Rhabditidae) based on internal transcribed spacer 1 DNA sequence data. *Journal of Nematology* 30, 22–39.

Adams, B.J., Fodor, A., Koppenhöfer, H.S., Stackebrandt, E., Stock, P.S. and Klein, M.G. (2006) Biodiversity and systematics of nematode-bacterium entomopathogens. *Biological Control* 37, 32–49.

Askary, T.H. (2010) Nematodes as biocontrol agents. In: Lichtfouse, E. (ed.) *Sociology, Organic Farming, Climate Change and Soil Science*. Springer, The Netherlands, pp. 347–378.

Blaxter, M.L. (2004) The promise of a DNA taxonomy. *Philosophical Transactions of the Royal Society B* 359, 669–679.

Blaxter, M.L., De Ley, P., Garey, J.R., Liu, L.X., Scheldeman, P., *et al.* (1998) A molecular evolutionary framework for the *phylum* Nematoda. *Nature* 392, 71–75.

Blaxter, M.L., Mann, J., Chapman, T., Thomas, F., Whitton, C., *et al.* (2005) Defining operational taxonomic units using DNA barcode data. *Philosophical Transactions of Royal Society London B Biological Sciences* 360, 1935–1943.

Blouin, M.S., Liu, J. and Berry, R.E. (1999) Life cycle variation and the genetic structure of nematode populations. *Heredity* 83, 253–259.

Burnell, A. (2002) Genetics and genetic improvement. In: Gaugler, R. (ed.) *Entomopathogenic Nematology*. CAB International, Wallingford, UK, pp. 241–264.

Campos-Herrera, R., El-Borai, F.E., Stuart, R.J., Graham, J.H. and Duncan L.W. (2011) Entomopathogenic nematodes, phoretic *Paenibacillus* spp., and the use of real time quantitative PCR to explore soil food webs in Florida citrus groves. *Journal of Invertebrate Pathology* 108, 30–39.

Campos-Herrera, R., El-Borai, F.E. and Duncan L.W. (2012) Wide interguild relationships among entomopathogenic and free-living nematodes in soil as measured by real time qPCR. *Journal of Invertebrate Pathology* 111, 126–135.

Campos-Herrera, R., Pathak, E., El-Borai, F.E., Schumann, A., Abd-Elgawad, M.M.M. and Duncan, L.W. (2013) New citriculture system suppresses native and augmented entomopathogenic nematodes. *Biological Control* 66, 183–194.

Campos-Herrera, R., El-Borai, F.E., Ebert, T.A., Schumann, A. and Duncan L.W. (2014) Management to control citrus greening alters the soil food web and severity of a pest–disease complex. *Biological Control* 76, 41–51.

Campos-Herrera, R., Jaffuel, G., Chiriboga, X., Blanco-Perez, R., Fesselet, M., *et al.* (2015) Traditional and molecular detection methods reveal intense interguild competition and other multitrophic interactions associated with native entomopathogenic nematodes in Swiss tillage soils. *Plant Soil* 389, 237–255.

Curran, J. (1990) Molecular techniques in taxonomy. In: Gaugler, R. and Kaya, H.K. (eds) *Entomopathogenic Nematodes in Biological Control*. CRC Press, Boca Raton, Florida, pp. 63–74.

De Luca, F., Fanelli, E., Di Vito, M., Reyes, A. and De Giorgi, C. (2004) Comparison of the sequences of the D3 expansion of the 26S ribosomal genes reveals different extent of heterogeneity in different populations and species of *Pratylenchus* from the Mediterranean region. *European Journal of Plant Parasitology* 110, 949–957.

De Luca, F., Vovlas, N., Lucarelli, G., Troccoli, A., Radicci, V., *et al.* (2013) *Heterodera elachista* the Japanese cyst nematode parasitizing corn in Northern Italy: integrative diagnosis and bionomics. *European Journal of Plant Pathology* 136, 857–872.

Dillman, A.R. (2013) Host seeking and the genomic architecture of parasitism among entomopathogenic nematodes. PhD thesis, California Institute of Technology, Pasadena, California.

Dillon, A.B., Rolston, A.N., Meade, C.V., Downes, M.J. and Griffin, C.T. (2008) Establishment, persistence and introgression of entomopathogenic nematodes in a forest ecosystem. *Ecological Application* 18, 735–747.

Dix, I., Koltai, H., Glazer, I. and Burnell, A.M. (1994) Sperm competition in mated 1st generation hermaphrodite females of the HP-88 strain of *Heterorhabditis* (Nematoda, Heterorhabditidae) and progeny sex-ratios in mated and unmated females. *Fundamental and Applied Nematology* 17, 17–27.

Dolinski, C., Kamitani, F.L., Machado, I.R. and Winter, C.E. (2008) Molecular and morphological characterization of heterorhabditid entomopathogenic nematodes from the tropical rainforest in Brazil. *Memorias do Instituto Oswaldo Cruz*, Rio de Janeiro, 103, 150–159.

Dubey, J., Tiwary, B.N. and Ganguly, S. (2011) Biopesticidal nematodes and molecular techniques involved for their characterization. In: Dhingra, H.K., Jha, P.N. and Bajpai, P. (eds) *Current Topics in Biotechnology and Microbiology.* Lap Lambert Academic Publishing AG & CO, KG, Dudweller Landstr, Germany, pp. 7–36.

Fayyaz, S., Yan, X., Qiu, L., Han, R., Gulsher, M., *et al.* (2014) A new entomopathogenic nematode, *Steinernema bifurcatum* n. sp. (Rhabditida: Steinernematidae) from Punjab, Pakistan. *Nematology* 16, 821–836.

Fayyaz, S., Khanum, T.A., Ali, S., Solangi, G.S., Gulsher, M. and Javed, S. (2015) *Steinernema balochiense* n. sp. (Rhabditida: Steinernematidae) a new entomopathogenic nematode from Pakistan. *Zootaxa* 3904, 387–402.

Floyd, R., Abebe, E., Papert, A. and Blaxter, M.L. (2002) Molecular barcodes for soil nematode identification. *Molecular Ecology* 11, 839–850.

Gozel, U., Lamberti, F., Duncan, L., Agostinelli, A., Rosso, L., *et al.* (2006) Molecular and morphological consilience in the characterisation and delimitation of five nematode species from Florida belonging to the *Xiphinema americanum*-group. *Nematology* 8, 521–532.

Griffin, C.T., Chaerani, R., Fallon, D., Reid, A.P. and Downes, M.J. (2000) Occurrence and distribution of the entomopathogenic nematodes *Steinernema* spp. and *Heterorhabditis indica* in Indonesia. *Journal of Helminthology* 74, 143–150.

Griffin, C.T., O'Callaghan, K.M. and Dix, I. (2001) A self-fertile species of *Steinernema* from Indonesia: further evidence of convergent evolution amongst entomopathogenic nematodes? Parasitology 122, 181–186.

Hashmi, G. and Gaugler, R. (1998) Genetic diversity in insect-parasitic nematodes (Rhabditida: Heterorhabditidae). *Journal of Invertebrate Pathology* 72, 185–189.

He, Y., Subbotin, S.A., Rubtsova, T.V., Lamberti, F., Brown, D.J.F. and Moens, M. (2005) A molecular phylogenetic approach to Longidoridae (Nematoda: Dorylaimida). *Nematology* 7, 111–124.

Hebert, P.D., Cywinska, A., Ball, S.L. and deWaard, J.R. (2003) Biological identifications through DNA barcodes. *Proceedings of Royal Society B Biological Sciences* 270, 313–321.

Hsieh, F.C., Tzeng, C.Y., Tseng, J.T., Tsai, Y.S., Meng, M.H. and Kao, S.S. (2009) Isolation and characterization of the native entomopathogenic nematode, *Heterorhabditis brevicaudis*, and its symbiotic bacteria from Taiwan. *Current Microbiology* 58, 564–570.

Hubschen, J., Kling, L., Ipach, U., Zinkernagel, V., Bosselut, N., *et al.* (2004) Validation of the specificity and sensitivity of species-specific primers that provide a reliable molecular diagnostic for *Xiphinema diversicaudatum*, *X. index* and *X. vuittenezi*. *European Journal of Plant Pathology* 110, 199–208.

Liu, J. and Berry, R.E. (1995) Determination of PCR conditions for RAPD analysis in entomopathogenic nematodes (Rhabditida: Heterorhabditidae and Steinernematidae). *Journal of Invertebrate Pathology* 65, 79–81.

Liu, J. and Berry, R.E. (1996a) Phylogenetic analysis of the genus *Steinernema* by morphological characters and randomly amplified polymorphic DNA fragments. *Fundamental and Applied Nematology* 19, 463–469.

Liu, J. and Berry, R.E. (1996b) *Heterorhabditis marelatus* n. sp. (Rhabditida: Heterorhabditidae) from Oregon. *Journal of Invertebrate Pathology* 67, 48–54.

Liu, J., Berry, R.E. and Moldenke, A.F. (1997) Phylogenetic relationships of entomopathogenic nematodes (Heterorhabditidae and Steinernematidae) inferred from partial 18S rRNA gene sequences. *Journal of Invertebrate Pathology* 69, 246–252.

Ma, J., Chen, S.L., De Clercq, P., Waeyenberge L., Han, R. and Moens, M. (2012) A new entomopathogenic nematode, *Steinernema xinbinense* n. sp. (Rhabditida: Steinernematidae), from north China. *Nematology* 14, 723–739.

Masumu, J., Geysen, D., Vansnick, E., Geerts, S. and van den Bossche, P. (2006) A modified AFLP for *Trypanosoma congolense* isolate characterization. *Journal of Biotechnology* 125, 22–26.

Montiel, R., Lucena, M.A., Medeiros, J. and Simões, N. (2006) The complete mitochondrial genome of the entomopathogenic nematode *Steinernema carpocapsae*: insights into nematode mitochondrial

DNA evolution and phylogeny. *Journal of Molecular Evolution* 62, 211–225.

Nguyen, K.B., Maruniak, J. and Adams, B.J. (2001) The diagnostic and phylogenetic utility of the rDNA internal transcribed spacer sequences of *Steinernema. Journal of Nematology* 33, 73–82.

Nthenga, I., Knoetze, R., Berry, S., Tiedt, L.R. and Malan, A.P. (2014) *Steinernema sacchari* n. sp. (Rhabditida: Steinernematidae), a new entomopathogenic nematode from South Africa. *Nematology* 16, 475–494.

Peat, S.M. (2010) Utilization of phylogenetic systematics, molecular evolution, and comparative transcriptomics to address aspects of nematode and bacterial evolution. PhD thesis. Brigham Young University, Provo, Utah.

Phan, K.L., Nguyen, N.C. and Moens, M. (2001) *Steinernema loci* sp. n. and *Steinernema thanhi* sp. n. (Rhabditida: Steinernematidae) from Vietnam. *Nematology* 3, 503–514.

Phillips, M.J., Delsuc, F. and Penny, D. (2004). Genome-scale phylogeny and the detection of systematic biases. *Molecular Biology and Evolution* 2, 1455–1458.

Powers, T.O. (2004) Nematode molecular diagnostics: from bands to barcodes. *Annual Review of Phytopathology* 42, 367–383.

Reid, A.P. (1994) Molecular taxonomy of *Steinernema*. In: Burnell, A.M., Ehlers, R.U. and Mason, J.P. (eds) *Genetics of Entomopathogenic Nematode–Bacterium Complexes*. European Commission Luxembourg City, Luxemburg, pp. 49–58.

Reid, A.P., Hominick, W.M. and Briscoe, B.R. (1997) Molecular taxonomy and phylogeny of entomopathogenic nematode species (Rhabditida: Steinernematidae) by RFLP analysis of the ITS region of the ribosomal DNA repeat unit. *Systematic Parasitology* 37, 187–193.

Saeb, A.T.M. (2015) Heat shock protein Hsp70 multigene family as a new genetic target for the differentiation and identification of entomopathogenic nematodes (Rhabditida: Heterorhabditidae). *Advances in Life Sciences and Health* 2, 16–30.

Saiki, R.K., Scharf, S.J., Faloona, F., Mullis, K.B., Erlich, H. and Arnheim, N. (1985) Enzymatic amplification of β-globin genomic sequence and restriction site analysis for diagnosis of sickle anaemia. *Science* 230, 1350–1354.

Sambrook, J. and Russell, D.W. (2001) *Molecular Cloning: A Laboratory Manual*, 3rd edn. Cold Spring Harbor Laboratory Press, New York, 2344 pp.

Shapiro-Ilan, D.I., Glazer, I. and Segal, D. (1997) Genetic diversity in wild and laboratory populations of *Heterorhabditis bacteriophora*as determined by RAPD-PCR analysis. *Fundamental and Applied Nematology* 20, 581–585.

Sommer, R.J. and Streit, A. (2011) Comparative genetics and genomics of nematodes: genome structure, development and lifestyle. *Annual Review of Genetics* 45, 1–20.

Sonnenberg, R., Nolte, A.W. and Tautz, D. (2007) An evaluation of LSU rDNA D1-D2 sequences for their use in species identification. *Frontiers in Zoology* 4, 6. DOI: 10.1186/1742-9994-4-6.

Spiridonov, S.E., Reid, A., Podrucka, K., Subbotin, S.A. and Moens, M. (2004) Phylogenetic relationships within the genus *Steinernema* (Nematoda: Rhabditida) as inferred from analyses of sequences of the ITS1–5.8S–ITS2 region of rDNA and morphological features. *Nematology* 6, 547–566.

Stack, C.M., Easwramoorthy, S.G., Metha, U.K., Downes, M.J., Griffin, C.T. and Burnell, A.M. (2000) Molecular characterization of *Heterorhabditis indica* isolates from India, Kenya, Indonesia and Cuba. *Nematology* 2, 477–487.

Stock, S.P. and Hunt, D.J. (2005) Morphology and systematics of nematodes used in biocontrol. In: Grewal, P.S., Ehlers, R.U. and Shapiro-Ilan, D.I. (eds) *Nematodes as Biocontrol Agents*. CAB International, Wallingford, UK, pp. 3–43.

Stock, S.P. and Koppenhöfer, A.M. (2003) *Steinernema scarabaei* n. sp. (Rhabditida: Steinernematidae), a natural pathogen of scarab beetle larvae (Coleoptera: Scarabaeidae) from New Jersey. *Nematology* 5, 191–204.

Stock, S.P., Campbell, J.F. and Nadler, S.A. (2001) Phylogeny of *Steinernema* Travassos, 1927 (Cephalobina: Steinernematidae) inferred from ribosomal DNA sequences and morphological characters. *Journal of Parasitology* 87, 877–889.

Subbotin, S.A., Vierstraete, A., De Ley, P., Rowe, J., Wayenberge, L., *et al.* (2001) Phylogenetic relationships within the cyst-forming nematodes (Nematoda, Heteroderidae) based on analysis of sequences from the ITS regions of ribosomal DNA. *Molecular Phylogenetics and Evolution* 21, 1–16.

Susurluk, I.A. and Toprak, U. (2006) Molecular identification of three entomopathogenic nematodes from Turkey by PCR-RFLP of the ITS regions. *Phytoparasitica* 34, 17–20.

Susurluk, I.A., Tarasco, E., Triggiani, O. and Ehlers, R.-U. (2007) Molecular characterisation of entomopathogenic nematodes isolated in Italy by PCR-RFLP analysis of the ITS region of the ribosomal DNA repeat unit. *Nematologia Mediterranea* 35, 23–28.

Szalanski, A.L., Taylor, D.B. and Mullin, P.G. (2000) Assessing nuclear and mitochondrial DNA sequence variation in *Steinernema* (Rhabditida: Steinernematidae). *Journal of Nematology*, 32, 229–233.

Tarasco, E., Mráček, Z., Nguyen, K.B. and Triggiani, O. (2008) *Steinernema ichnusae* sp. n. (Nematoda: Steinernematidae) a new entomopathogenic nematode from Sardinia Island (Italy). *Journal of Invertebrate Pathology* 99, 173–185.

Torr, P., Spiridonov, S.E., Heritage, S. and Wilson, M.J. (2007) Habitat associations of two entomopathogenic nematodes: a quantitative study using real-time quantitative polymerase chain reactions. *Journal of Animal Ecology* 76, 238–245.

Torrini, G., Landi, S., Benvenuti, C., De Luca, F., Fanelli, E., *et al.* (2014) Morphological and molecular characterization of a *Steinernema carpocapsae* (Nematoda: Steinernematidae) strain isolated in Veneto Region (Italy). *Redia* XCVII, 89–94.

Valadas, V., Laranjo, M., Mota, M. and Oliveira, S. (2013) A survey of entomopathogenic nematode species in continental Portugal. *Journal of Helminthology* 88, 327–341.

Wang, H., Jung, Y.H., Son, D. and Choo, H.Y. (2013) High level of genetic diversity among *Steinernema monticolum* in Korea revealed by single-enzyme amplified fragment length polymorphism. *Journal of Invertebrate Pathology* 113, 146–151.

Part III: Entomopathogenic Nematodes and Their Symbiotic Bacteria Against Crop Insect Pests

7 Efficacy of Entomopathogenic Nematodes Against Lepidopteran Insect Pests

M.M.E. Saleh*

Pests and Plant Protection Department, National Research Centre, Giza, Egypt

7.1 Introduction

The insects belonging to order Lepidoptera include moths and butterflies. These insects are of great economic importance as they play the role of pollinators and produce silk, but others also act as crop pests. They comprise 174,250 species that belong to 126 families (Capinera, 2008). Almost all the valuable plants essential to humans are attacked by lepidopteran pests. These include agricultural crops, horticultural crops, timber and shade trees. Larvae, after hatching from eggs, cause much damage as they often feed on live plant matter such as leaves, flowers, fruit and their seeds. However, the insects of this order do not act as the vectors of plant diseases, nor do they parasitize or cause injury to humans (Powell, 2009). Some lepidopterans also feed on the turf of grasses and sedges. Others, such as fungus moths, scavenger moths and snout moths, feed on dead and decaying plant matter, particularly mouldy debris (*Encyclopedia Britannica*, https://www.britannica.com/animal/lepidopteran).

Therefore, in view of the overall economic loss due to lepidopteran pests, they need to be managed. There are several methods to manage lepidopteran pests and one of them is the biological method wherein biocontrol agents are used. These biocontrol agents are exploited for insect pest control through inoculative, inundative and augmentative releases (Lacey *et al.*, 2001). Entomopathogenic nematodes (EPNs) belonging to the families Steinernematidae and Heterorhabditidae are the biocontrol agents that are lethal to insects in soil, cryptic foliage and exposed foliage. They are widespread in all continents except Antarctica (Hominick, 2002). Since more than 90% of insect species have life-cycle stages in soil (Klein, 1990), it is very common for them to be exposed to EPNs. EPNs have been known for decades (Poinar, 1990); however, assessment of their efficacy against target insect pests became important after the availability of commercial formulations in the 1990s (Georgis and Manweiler, 1994; Glazer and Lewis, 2000). The nematodes *Steinernema* and *Heterorhabditis* have the ability to locate, infect and kill a wide range of insect species. They have four juvenile stages, but only the third-stage juvenile (also called 'infective' or 'dauer') are free living in the soil, can survive outside the insect host and move from one insect to another. Insect mortality, due to nematode infection, is caused by a symbiotic bacterium (*Xenorhabdus* spp. for steinernematids and *Photorhabdus* spp. for heterorhabditids) that the infective juveniles (IJs) carry in their intestines and release in the insect haemolymph (Akhurst and Boemare, 1990; Askary, 2010). Invasion by the nematode occurs through natural openings (spiracles, mouth and anus) or, in some cases, directly through the cuticle of certain insects (Bedding and Molyneux, 1982; Peters and Ehlers, 1994). The bacteria cells proliferate and eventually kill the insect host, usually within 72 h (Askary, 2010). At temperatures ranging from 18–28°C, the life cycle of nematodes is completed in 6–18 days, depending on the host insect and the nematode species (Poinar, 1990). The invading IJs belonging to *Heterorhabditis* develop into hermaphrodites, and those belonging to *Steinernema* develop into females

*E-mail: mmesaleh@hotmail.com

or males. One or more generations of progeny develop within one host, and reproductions continue until host-derived nutrients are depleted. At this time, the third-stage IJs leave the insect cadaver in search of new hosts.

In the past three decades, considerable attention has been directed to *Heterorhabditis* and *Steinernema* and their respective bacterial partners, *Photorhabdus* and *Xenorhabdus* (Griffin *et al.*, 2006) in order to exploit them against crop insect pests. The development of large-scale production and ease-of-use formulations has created marketing opportunities for EPNs (Georgis, 2002). EPNs evaluated against the control of lepidopteran pests have been reported to provide encouraging results. *Steinernema carpocapsae* has been found to be effective in controlling the soil-dwelling stage of the peach fruit moth, *Carposina niponensis*. The fruit damage caused by the pest was reduced to less than 3% (Wang, 1993). EPNs applied against pests such as the carpenterworm (Cossidae), *Holcocerus insularis*, which attacks ash, Chinese scholar tree, Chinese hawthorn and willows, and *Zeuzera multistrigata*, which attacks Australian pine, *Casuarina equisetifolia*, have been found to be highly successful (Yang *et al.*, 1993). In Egypt, although there were many successful applications of EPNs against economic insect pests, namely apple borers, *Zeuzera pyrina* and/or *Synanthedon myopaeformis* (Saleh *et al.*, 1994; Abdel-Kawy *et al.*, 1996; Saleh and Abbas, 1998); however, lack of effective mass-production technology still remains the main obstacle facing the commercialization of EPNs in the country.

Thus, with the expansion of commercial interest in EPNs, the efficacy of many economically important insect pests has been evaluated in a wide range of laboratory and field assays (Glazer and Lewis, 2000). Although the host range of EPNs includes more than 200 insect species, nematodes have only been marketed successfully for a small fraction of these insects. Accordingly, in this chapter, I have selected Lepidopteran insect pests representing different agroecosystems to highlight the efficacy of EPNs against them. Here, I mean by the term 'efficacy' the qualitative as well as the quantitative measurement of the ability of EPNs to produce insect disease.

7.2 Examples of Efficacy of Entomopathogenic Nematodes against Lepidopteran Insects in Different Habitats

The efficacy of EPNs is influenced by nematode species, strain, production and storage conditions, persistence in the introduced habitat and susceptibility of target insects (Glazer, 2002; Mráček, 2002). The control efficacy of EPNs against Lepidopteran insect pests on different crops has been presented in Table 7.1.

7.2.1 Soil habitat

The soil habitat offers an excellent site for insect nematode interactions. More than 90% of insect pests spend part of their life cycle in soil, and soil is the natural reservoir of steinernematid and heterorhabditid nematodes (Gaugler, 1988; Georgis and Poinar, 1989; Begley, 1990). Shapiro-Ilan *et al.* (2005) suggested that EPNs offered a narrow window of opportunity for control of defoliating Lepidopteran insects that have soil stages. Noctuids, for example, are most vulnerable pre-pupae and pupae when they search for a soil site in which to pupate. It has been reported that pre-pupae of the beet armyworm, *Spodoptera exigua*, were highly susceptible to infection by *S. carpocapsae* (Kaya and Hara, 1980). Furthermore, *S. carpocapsae* could infect >90% of *S. exigua* pre-pupae in soil at concentrations of 5 nematodes/cm^2 of soil surface (Kaya and Griev, 1982). In field trials, *Steinernema feltiae*, applied to soil, significantly reduced the larval population of the turnip moth *Agrotis segetum*. Adult beet armyworms were also susceptible to nematode infections during emergence from the soil (Kaya and Griev, 1982). Timper *et al.* (1988) demonstrated that adults of beet armyworm infected with *S. carpocapsae* dispersed up to 11 m from the site of infection. This suggests that infected adult insects may account in part for the wide distribution of EPNs. In Spain, Garcia-del-Pino *et al.* (2013) studied the efficacy of *S. carpocapsae*, *S. feltiae* and *Heterorhabditis bacteriophora* used as soil treatments against larvae, pupae and adults of *Tuta absoluta* under laboratory conditions. The nematodes caused highest mortality to larvae that were dropped in soil and ready to pupate. Larval mortality was 100, 52.3 and 96.7% in the case of *S. carpocapsae*, *S. feltiae* and *H. bacteriophora*, respectively. There was no mortality of pupae, however, *S. carpocapsae* caused 79.1% mortality of adults emerging from soil. Also, the insecticides showed no adverse affect on the survival, infectivity and reproduction of nematodes. IJs that survived insecticide exposition caused infection on the larvae of *Galleria*, but the difference was not significant when the performance of IJs was compared with the

Table 7.1. Examples of efficacy of entomopathogenic nematodes against Lepidopteran insects.

Crop	Pest	Nematode	Method	Efficacy	Reference
Apple	*Zeuzera pyrina*	*Steinernema carpocapsae*	Spray	44%	Abdel-Kawy *et al.*, 1996
			Injection	64%	
			Nematode-wetted cotton, 800 IJs/ml	76% larval mortality	
		S. riobravis	Injection 200 IJs/ml	76% larval mortality	El-Shazly *et al.*, 1999
	Synanthedon myopaeformis	*Steinernema* sp.	Spray 3–4 × 10^4 IJs/ infested callus	32–76% mortality	Kahounova and Mrácek, 1991
		S. carpocapsae	Spray	54%	Abdel-Kawy *et al.*, 1996
			Injection	24%	
			Nematode-wetted cotton, 800 IJs/ml	63.3% mortality	
	Choristonerura rosaceana	*S. carpocapsae*	Spray 2 × 10^9 IJs/ha	13–37% control	Bélair *et al.*, 1999
	Cydia pomonella	*S. carpocapsae*	Spray on tree trunks and soil 50 IJs/cm^2	83%	Lacey and Unruh, 1998
		S. riobravis		31%	
		Heterorhabditis bacteriophora		43% reduced emergence	
Avocado	*Boarmia selenaria*	*S. carpocapsae*	Spray 250 IJs/m^2	98 ± 2% control	Glazer and Wysoki, 1990
Bentgrass	*Agrotis* spp.	*S. carpocapsae* *S. glaseri*	Spray 2.5 × 10^9 IJs/ha	Significant mortality	Buhler and Gibb, 1994
Blackcurrant	*Incurvaria capitella*	*S. carpocapsae*		58–86% larval mortality	Samersov *et al.*, 1998
Cherry	*Euzophera semifuneralis*	*H. bacteriophora*	Spray on tree bark 3.9 × 10^5 IJs/l	Insignificant control	Kain and Agnello, 1999
		S. feltiae	6.5 × 10^4/l		
Citrus	*Phyllocnistis citrella*	*S. carpocapsae*	Spray 5–30 × 10^6/l	69% mines reduction	Beattie *et al.*, 1995
Cabbage	*Plutella xylosstella*	*S. carpocapsae*	Spray 50 IJ/cm^2 soil	41% damage control	Baur *et al.*, 1998
	Pieris rapae	*S. carpocapsae*	Spray 40 IJs/cm^2 soil	35% damage control	Bélair *et al.*, 2003
Clover	*Wiseana cervinata*	*H. bacteriophora*, *S. feltiae*	Spray 4 × 10^5 IJs/m^2	25% 60% reduction	Wright and Jackson, 1992
Maize	*Ostrinia nubilalis*	*S. carpocapsae*	Spray 5 × 10^4 IJs/plant	5–20% reduced plant damage	Ben-Yakir *et al.*, 1998
	Helicoverpa zea	*S. riobravis*	Spray 2 × 10^9 IJs/ha	22–100% infection	Cabanillas and Raulston, 1995
		S. carpocapsae *S. riobravis*	Spray 2 × 10^5 IJs/m^2	82–97% pre- and pupal infection	Cabanillas and Raulston, 1995

Continued

Table 7.1. Continued.

Crop	Pest	Nematode	Method	Efficacy	Reference
		S. riobravis	Spray 3.7×10^6–1.2×10^7 IJs/m^2	79–93% mortality	Feaster and Steinkraus, 1996
Cotton	*Spodoptera littoralis*	*S. carpocapsae*	Spray 250 IJs/ml	61%	Glazer *et al.*, 1992
	Earias insulana	*S. carpocapsae*	Spray 125 IJs/ml	76% reduction	Glazer *et al.*, 1992
Peach	*Synanthedon exitiosa*	*Heterorhabditis heliothidis*	Spray	Significant reduction	Cossentine *et al.*, 1990
Squash	*Diaphania nitidatis*	*S. carpocapsae*	Spray 7.4×10^9 IJs/ha	85–97% infected larvae	Shannag *et al.*, 1994
Sugarcane	*Eoreuma loflini*	*S. riobravis*	Spray 1.24×10^9 IJs/ha 2.47×10^9 IJs/ha	Ineffective	Legaspi *et al.*, 2000
Tomato	*Tuta absoluta*	*S. carpocapsae*, *S. feltiae*, *H. bacteriophora*	Soil application	100, 52.3 and 96.7% larval mortality 79% adult mortality	Garcia-del-Pino *et al.*, 2013

control. Thus, it was concluded that larvae of *T. absoluta*, falling from leaves after insecticide application, could serve as suitable hosts for EPNs. This will also help to increase the population and persistence of nematodes in soil. Gelechid pink bollworm, *Pectinophora gossypiella*, completes five generations during the Arizona cotton growing season. During the time when the insect larvae enter the soil to pupate, IJs cause infection by entering into the larvae, thus serving as biological control agents of *P. gossypiella*. *Steinernema riobrave* have been reported to perform better than *S. carpocapsae* in controlling insect pests under field conditions (Henneberry *et al.*, 1996). Low potential of control was observed for adults of the pink bollworm, which is often considered to be an unsuitable developmental stage to control. In an experiment conducted in Texas, *S. riobrave* proved highly effective in controlling the pre-pupae and pupae of the corn earworm, *Helicoverpa zea* (Cabanillas and Raulston, 1996b). A similar result was achieved by Feaster and Steinkraus (1996) when *S. riobrave* was evaluated for inundative control of this pest in Arkansas maize fields.

7.2.2 Cryptic habitats

Cryptic habitats are considered as most favourable, enhancing the infectivity, survival and persistence of EPNs, because the cryptic environment minimizes nematode death from ultraviolet radiation and desiccation. Experiments carried out by researchers in these habitats have provided mostly consistent and efficacious results. Shapiro-Ilan *et al.* (2005) reported that many orchards have hosts available much of the time in a year, favourable soil conditions (moist, sandy) and shade, that are amenable to insect suppression by EPNs. Additionally, crops produced in orchards are often relatively high in value, which facilitates the economic feasibility of nematode applications. As a result, a number of orchard insect pests have been studied extensively for their potential to be controlled by EPNs, and some have become commercial success stories. The economic damage inflicted by Lepidopterans is significant and affects production both by direct feeding damage on the fruit and reduction in fruit quality from indirect feeding on the tree. The leopard moth borer, *Z. pyrina* (Lepidoptera: Cossidae), is a destructive pest causing partial or complete damage to apple trees. They also attack pear, pomegranate and olive trees (Ismail *et al.*, 1992). The application of different EPN species caused 20–90% mortality among larvae infesting apple and olive trees (Abdel-Kawy *et al.*, 1992, 1996). Miller and Bedding (1982) reported up to 90% mortality of the larvae of the currant borer moth, *Synathedon tipuliformis*, feeding extensively on pith in the stem of the host plants. A higher level of parasitism by *S. feltiae* was achieved in old wood, probably due to rougher bark and bigger entrance holes to tunnels, than in new wood. However, some new-generation IJs migrated from parasitized larvae to infect larvae in other tunnels. When *Steinernema* sp. were sprayed on to larvae of the apple clearwing borer moth, *Synanthedon myopaeformis* (Lepidoptera: Sessidae), the mortality of the pest varied between 32 to 76% during a 4-week period, although the weather conditions were unfavourable during the experiment (Kahounova and Mrácek, 1991). Extensive studies with the lepidopteran insects in the family Cossidae have shown that they can be controlled effectively by EPNs (Lindegren *et al.*, 1981; Forschler and Nordin 1988). In China, Wang (1993) used *S. carpocapsae* against the apple borer and the clear wing moth borer and got encouraging results in the control of these pests. Carpenterworms (*Prionoxystus robiniae* (Peck), Lepidoptera: Cossidae) damage a lot of ornamental and fruit trees in the USA and Canada. Lindegren *et al.* (1981) injected *S. carpocapsae* into the gallery openings of fig trees in California. Within 2–6 days, there was a decreased feeding activity of the carpenterworm larvae, and the final pest mortality reached 44–92%. Suppression up to 100% of the *P. robiniae* was reported by Forschler and Nordin (1988) when EPNs were injected into galleries, but spraying on bark surface provided variable results. The American plum borer, *Euzophera semifuneralis*, is one of the most important fruit tree boring pests of tart cherries that are reported to cause a significant decline in the lifespan of cherry orchards. It is associated with bark wounds that provide an entry into the cambium layer where the larvae feed. Laboratory bioassays proved high sensitivity of EPNs to larvae of *E. semifuneralis*; however, the result was not significant when *S. feltiae* and *H. bacteriophora* were tested to control the plum borer larvae, and this may be due to inadequate larval contact with the nematode spray (Kain and Agnello, 1999). Bark folds create a cryptic microhabitat for the larvae of codling moth, *Cydia pomonella*, to pupate, though EPNs can act in the space protecting them against desiccation. The application of EPNs in July resulted in a total larval mortality of 32%, whereas in October and February the mortality rate was 80 and 95%, respectively, in overwintering populations of pests settled in corrugated paper bands around tree trunks. This high efficacy of EPNs is attributed to rainfall during the autumn and winter period (Kaya *et al.*, 1984). The effective application of EPNs in the cryptic environment has been reported when *S. carpocapsae* entered the blossom of squash plants, where they persisted for days and infected pickleworm, *Diaphania nitidalis* (Crambidae), feeding on the buds, blossoms, vines and fruit of squash such as cucumbers and melons (Shannag *et al.*, 1994).

7.2.3 Foliar habitat

The larvae of Lepidoptera are the primary foliar feeders. However, the success rate of EPNs in controlling the larvae on foliage is less than in soil and cryptic habitats (Arthurs *et al.*, 2004). Field trials using nematodes as biological insecticides against foliage-feeding insects have encountered many problems. The foliar environment often exposes the nematodes to unfavourable moisture conditions that result in their rapid desiccation and death. Field applications of EPNs against foliage-feeding insect pests have resulted in varying efficacy when used in an unfavourable environment. Nematodes require high relative humidity, water drops on leaves and avoidance of direct sunshine, otherwise they desiccate rapidly. Tropical conditions and rainy periods enhance nematode survival on foliage, but in temperate and dry areas, nematode application has limited success. When weather conditions are virtually optimal, foliar spray can be effective. Moreover, high temperatures and sunlight are fatal to nematodes exposed on the leaf surface. Accordingly, nematodes applied on foliage must be protected from detrimental environmental effects. The tropic regions with heavy rainfall and glasshouses (areas with moderate temperatures and high humidity) are preferred environments for foliar applications of nematodes. Furthermore, in evening applications, the incorporation of anti-desiccants into aqueous nematode suspensions, oil formulations or anhydrobiotic nematodes offer possibilities for increasing efficacy against foliage insects. It has been noted that sprayable gel could enhance nematode efficacy in above-ground applications (Shapiro-Ilan *et al.*, 2010). Gels such as alginates or cross-linked polyacrylamides are based on moisture-holding capacities and can serve as carriers or baits in EPN formulations (Kaya and Nelsen, 1985; Georgis, 1990). When sprayable gel is used as a post-application cover, it aids in protecting nematodes. The adjuvants having anti-desiccant properties have generally been found to cause a low toxic effect on nematodes. However, in a few cases anti-desiccants increased nematode efficacy in above-ground applications (Glazer *et al.*, 1992; Head *et al.*, 2004), but other studies did not prove this (Miller and Bedding, 1982).

Nevertheless, in many cases, EPNs have reduced foliar insect populations successfully to an acceptable level. The application of *S. carpocapsae* on cherry trees effectively controlled third-instar fall webworm, *Hyphantria cunea*, in Mexico. In field trials, nematodes applied at 3000 nematodes/ml under conditions of high humidity and moderate temperature persisted on foliage for between 11 and 40 h and caused 43–100% mortality (Yamanaka *et al.*, 1987). Because the webworm larvae are gregarious and form silk webs at the feeding site, the webs may provide some protection to the nematodes by slowing the rate of nematode desiccation and reducing exposure to solar radiation (Begley, 1990). *S. carpocapsae* applied on wet foliage for 24 h in the field infected >80% of the armyworm, and high humidity for 15 h followed by evening applications and the addition of a 'viscous material' enhanced the mortality of the cabbage worm, *Pieris rapae* (Wang and Li, 1987). In Florida, *S. carpocapsae* was found to be successful against the beet armyworm *S. exigua* in a commercial nursery producing chrysanthemum stock. IJs applied at dawn controlled the pests effectively and survived by 52–84% after 3 h (Begley, 1990). Several research workers have found *S. carpocapsae* effective against diamondback moth (DBM) (Moris, 1985; Salem *et al.*, 2007). The cells of *Xenorhabdus nematophila* (the symbiotic bacteria of the nematode, *S. carpocapsae*) and its metabolites when applied against the larvae of DBM were found very promising (Mahar *et al.*, 2004). Researchers have found *S. carpocapsae* effective in controlling citrus leaf miner, *Phyllocnistis citrella*, mining immature foliage and destroying the epidermal cells of plants; however, its use is not commercially acceptable at the concentration tested (Beattie *et al.*, 1995). The oblique-banded leaf roller, *Choristoneura rosaceana*, damaging foliages of apple tree in eastern North America, was controlled by 13–37% with *S. carpocapsae*. The low larval mortality is attributed to the low survival rate of IJs, the difficulties of nematode access to larvae, because the suspension could not penetrate the tightly woven larval silk (Bélair *et al.*, 1999). In Brazil, the efficacy of 33 native and imported isolates of *S. carpocapsae*, *H. bacteriophora*, *Steinernema glaseri*, *S. riobravae*, *Steinernema rarum* and *S. feltiae* were assessed against the larvae and pupae of the apple leaf roller, *Bonagota salubricola*. Laboratory bioassays showed that all tested isolates were pathogenic to the pest larvae (Barbosa-Negrisoli *et al.*, 2010). However, the adjuvants added to the suspension significantly increased nematode efficacy in the field.

A generalized life cycle of a foliar pest is represented by *Ostrinia nubilalis*, which overwinters as a mature fifth instar in maize stalk debris. The number of generations per year ranges from one to four, depending on geographic location (Showers, 1979). *S. carpocapsae* applied to maize plants at 50,000 IJs/plant effectively reduced the number of *O. nubilalis* larvae. A similar application to ears of sweetcorn reduced the percentage of damaged ears from 20% to 5% (Ben-Yakir *et al.*, 1998). When *S. carpocapsae* was applied to maize residues at 9 × 104 IJs/m^2, few diapausing fifth-instar *O. nubilalis* were killed, even though nematodes were isolated from the soil 2 months after application (Lewis and Raun, 1978). When sprayed on vegetative maize, *S. carpocapsae* caused 33–43% infection on larvae of *Spodoptera frugiperda*; however, when sprayed on ears of sweetcorn, 71% of larvae were infected (Richter and Fuxa, 1990).

7.3 Case Studies

7.3.1 The black cutworm *Agrotis ipsilon* (Lepidoptera: Noctuidae)

The black cutworm, *Agrotis ipsilon*, is a common destructive pest of many economic vegetables and field crops, including maize (*Zea mays*). It attacks seedlings and cuts plant stems above the soil surface. The insect larvae hide in soil, from sun rays and drought or for pupation. The larvae develop through six instars, pupate in the soil and the adults emerge 7–10 days after pupation, to begin another generation, or depending on location and the time of year. The application of EPNs to control *A. ipsilon* is usually done with adjuvants in formulation, to mitigate the adverse effects of sun rays and drought (Arthurs *et al.*, 2004). Saleh *et al.* (2015) evaluated the effect of several formulation adjuvants on the performance of two Egyptian isolates of EPNs, namely *H. bacteriophora* BA1 and *S. carpocapsae* BA2, against the larvae of *A. ipsilon* and the larvae of the cotton leafworm, *Spodoptera littoralis* (Lepidoptera: Noctuidae), in semi-field and field assays. Formulation adjuvants were used individually or in combinations with both tested nematodes. The adjuvants included Tween 80, paraffin oil, glycerin, calcium alginate, carboxy methyl cellulose (CMC) and Arabic gum. They were used at a concentration of 0.1%, except glycerin (3%). The adjuvants tested had no adverse effects on nematode survival or efficacy. In semi-field assays, maize plants were cultivated in pots of approximately 300 cm^2 surface area. When the plants were 3 weeks old, they were infested artificially with the larvae of *A. ipsilon* or *S. littoralis*. The plants were then sprayed with tested nematodes with and without combinations of formulation adjuvants. The results showed that the nematode suspension alone belonging to *H. bacteriophora* BA1 and *S. carpocapsae* BA2 killed the larvae of *A. ipsilon* by 75% and 93%, respectively. The mortality of *S. littoralis* larvae was 20% and 60% when sprayed with *H. bacteriophora* BA1 and *S. carpocapsae* BA2, respectively. All infected cadavers of *A. ipsilon* were found under the soil. The added effect of formulation adjuvants for both tested nematodes was insignificant against *A. ipsilon* larvae but significant against *S. littoralis* larvae. The best effect on nematode performance resulted from the combinations of surfactants, oils and polymers. The combination CMC 0.1% + Tween80 0.1% + paraffin oil 0.1% increased the effect of *S. carpocapsae* from 60% to 85% against the larvae of *S. littoralis*. Field results revealed that formulation adjuvants were more useful against *S. littoralis* larvae infesting whorl maize leaves rather than *A. ipsilon* larvae attacking maize stems at the soil surface. The results support the findings of other researchers that adjuvants used in formulation aids in controlling foliar pests (Mason *et al.*, 1998; Schroer and Ehlers, 2005; Shapiro Ilan *et al.*, 2012). A study conducted in the USA showed that *S. carpocapsae* was highly effective (95%) against black cutworm (Georgis and Poinar, 1994). Nevertheless, EPNs are not widely applied against black cutworm due to damage thresholds on golf course trees, and particularly greens, where chemical insecticides perform even better and are more consistent than *S. carpocapsae*. The larvae were controlled significantly by *S. carpocapsae* and

S. glaseri, but after 8 days of application, the nematodes lost their persistence (Buhler and Gibb, 1994).

7.3.2 The corn earworm *Helicoverpa zea* (Boddie), (Lepidoptera: Noctuidae)

The corn earworm, *H. zea*, is a pest of many field and vegetable crops (Sparks *et al.*, 1989). It is considered the second most damaging pest in North America after the codling moth (Hardwick, 1965). Eggs are laid individually on maize (*Z. mays* L.) leaf hairs and silk. The eggs hatch in 3–4 days. Newly hatched larvae feed on silk, causing the most damage to plants. The larvae of *H. zea* damage maize by feeding in whorls, silks, tassels, or kernels. The larvae have cryptic feeding habitats (feed within maize ears), and pupation takes place in the soil. The larvae pupate 5–10 cm below the soil surface. Pupal mortality is high in wet soil where the moisture level is 18% or above. Dehydration can also lead to high death rates among pupae if soil moisture is as low as 1–2% (Ditman *et al.*, 1940). However, maize is a safe place for their development; the pests survive to adults, disperse widely and produce offspring that damage other crops such as cotton (Cabanillas and Raulston, 1995).

Nematodes sprayed with the suspension of *S. carpocapsae* on ears of sweetcorn infected 71% of larvae and caused up to a 53% reduction in larval numbers of a mixed population of *H. zea* and *S. frugiperda* (Richter and Fuxa, 1990). In experiments with soil insects, *S. carpocapsae* applied at 1.3 and 2.5 × 10^9 IJs/ha to the soil surface was effective or better than several chemical insecticides in protecting seedling maize from *A. ipsilon* (Levine and Oloumi-Sadeghi, 1992). *S. riobrave* applied to the soil surface at 200,000 IJs/m^2 infected 95% of the pre-pupae and pupae of *H. zea*. Nematodes were most effective when applied through furrow irrigation or applied to soil prior to irrigation. The application of *S. carpocapsae* was also effective against the pupae of *H. zea* (Cabanillas and Raulston, 1996a,b). In order to mange soil insects, nematodes are applied to soil with a 5-cc disposable syringe and a 21 gauge needle; 1000 nematodes suspended in 4 ml of water are injected into the soil to a depth of approximately 3 cm. The syringes are then washed with 5 ml water, which is expelled on to the soil surface. To manage corn rootworm, nematodes are applied with a Wheaton self-filling syringe and 14-gauge needle (Jackson and Hesler, 1996). In Arkansas, Feaster and Steinkraus (1996) applied *S. riobravis* to control *H. zea* in maize fields. Infection of *H. zea* was highly significant in nematode-treated plots. Mean mortalities from *S. riobravis* infections were 78.5% and 94.8% at treatment levels of 5.2 × 10^5 and 5.3 × 10^6 IJs/m^2 of soil, respectively. When irrigated and non-irrigated plots were compared, no significant differences were found in *S. riobravis* infecting *H. zea*.

7.3.3 The leopard moth borer, *Zeuzera pyrina* (Lepidoptera: Cossidae)

The leopard moth borer, *Z. pyrina*, is a destructive pest causing partial or complete damage to apple trees. They also attack pear, pomegranate and olive trees (Ismail *et al.*, 1992). The small tree branches are attacked first. Young larvae of *Z. pyrina* bore the branches and feed inside the sapwood. After the larvae become mature, they migrate into the heartwood. Due to larval infestation, branches die and entrance holes are filled with silk and frass. The process of feeding and tunnelling in older wood lasts for 2–3 years. After several migrations, larger branches and the trunk are attacked by larvae. Here, they form ascending galleries under the bark. Later, galleries are formed in the wood. The entry holes made by the larvae are marked by sap outflows, sawdust and frass (Carter, 1984).

Experiments conducted in the laboratory have shown high efficacy of *Steinernema* and *Heterorhabditis* against the larvae and pupae of *Z. pyrina* (Saleh *et al.*, 1994). Another experiment resulted in 20% larval mortality of *Z. pyrina* when *S. carpocapsae*/*H. bacteriophora* were sprayed on to apple trees (Abdel-Kawy *et al.*, 1988, 1992). Larval mortality increased to 90% when the nematode suspension was injected in insect galleries. Under laboratory and field conditions, the efficacy of five Egyptian and imported EPNs was evaluated against *Z. pyrina*. It was concluded that all the nematodes used in the study were effective in causing mortality of the larvae/pupae of the pest. However, *S. riobravae* performed the best under field conditions, causing 76% larval mortality when injected at 2000 IJs/ml into the galleries (El-Shazly *et al.*, 1999). In Egypt, Abdel-Kawy *et al.* (1996) tested four different strains/species of nematodes by using three application methods to control *Z. pyrina* larvae infesting apple orchards. Of these four nematodes, *S. carpocapsae* All and *H. bacteriophora* HP88 were exotic, whereas *S. carpocapsae* S2 and *H. bacteriophora* S1

(Shamseldean *et al.*, 1996) were native isolates. The three application techniques used were: (i) spraying water suspension of nematodes on the trunks and branches of infested trees using a portable pressure sprayer after removing its filter to ensure the easy passage of nematodes through the sprayer; (ii) injection of nematode suspension in galleries tunnelled by the pest larvae, using a 50 cc syringe ending with a plastic hose, which was inserted into the gallery opening while injecting – each gallery received 2 ml of the nematode suspension; and (iii) soaking small cotton plugs with the nematode suspension and putting them on the gallery openings. Two concentrations of nematodes were tested, i.e. 800 and 1600 IJs/ml. Twenty-four treatments were conducted with three replicates/treatment. Three infested trees containing 29–143/three trees active galleries for each treatment were assigned. The active galleries (containing live larvae) were those producing fresh frass. Control and treated trees were inspected for insect mortality after 1 week. The data were analysed and are presented in Table 7.2.

Before treatments, the infestation level of *Z. pyrina* larvae in apple trees ranged from 29 to 134 larvae/three trees. After 1 week of nematode application, mortality of larvae inside galleries ranged from 33% to 88%, depending mostly on the application technique. The cotton ball technique proved superior to the injection technique. No significant differences were observed among most of the nematodes used in the study. Also, the differences were insignificant between the concentration of 1600 and 800 IJs/ml. In order to develop an analytic pattern model for above-ground applications of *S. carpocapsae*, 136 field trials were conducted by Arthurs *et al.* (2004). The model showed differences in nematode treatment efficacy depending on the pests' target habitat (bore holes > cryptic foliage > exposed foliage) and trial location (greenhouse > field studies). The experimental results revealed that the rates of nematode infection were affected by relative humidity and temperature during and up to 8 h after the application. Contrary to this, there was no significant variation in nematode performance

Table 7.2. Larval mortality of the apple borer, *Zeuzera pyrina*, infesting apple trees after treatments with entomopathogenic nematodes.

Nematode	Method	Concentration (nematodes/ml)	Treated galleries/ three trees	Larval mortality (%)	
				Mean	SE
Steinernema carpocapsae All	Injection	1600	57	76.46	3.86
		800	31	60.95	19.54
	Cotton	1600	55	76.98	6.52
		800	46	82	9.07
	Spray	1600	98	35.59	7.40
		800	73	33.76	8.29
S. carpocapsae S2	Injection	1600	29	70	10
		800	36	71.97	4.8
	Cotton	1600	31	88.89	6.41
		800	36	87.12	7.23
	Spray	800	45	45.01	24.09
Heterorhabditis bacteriophora HP88	Injection	1600	65	55.31	9.88
		800	97	50.35	7.44
	Cotton	1600	74	75.22	7.62
		800	37	58.22	5.41
	Spray	1600	74	34.8	3.72
		800	88	50.11	11.03
H. bacteriophora S1	Injection	1600	69	57.99	7.95
		800	52	56.72	7.34
	Cotton	1600	65	72.14	7.01
		800	143	64.81	17.57
	Spray	1600	37	65.28	1.39
		800	74	53.77	4.6
Control			29	0	

due to spray adjuvants (wetting agents and anti-desiccants) and nematode dosage. The mortality rate was highest in those insect larvae that bored into layers of bark and cambium. The tree holes created as a result of larval boring provide environments for nematodes to locate and infect host larvae while protected from hostile abiotic conditions. Direct gallery injections apply suspension to gallery openings with a steam nozzle or squirt bottle and were effective (≥80% control) but laborious. Saleh (2014) investigated the effects of Egyptian isolates of *S. carpocapsae* and *H. bacteriophora* injected in pomegranate branches infested with *Z. pyrina*. The nematodes were injected at a concentration of 500 IJs/ml. The nematodes were found to reach the target site and caused infection to the larvae of 66.67% and 83.33% in the case of *S. carpocapsae* and *H. bacteriophora*, respectively. When infected larvae were placed in White traps, all infected larvae supported nematode development and produced new offspring (Figs 7.1 and 7.2).

7.3.4 The codling moth, *Cydia pomonella* L. (Lepidoptera: Tortricidae)

The codling moth, *C. pomonella*, is a serious pest in almost all the countries where apple and pear are grown. The pest overwinters in cryptic habitats as cocooned diapausing larvae (Lacey *et al.*, 2006a), which, if controlled, would reduce or eliminate damage to fruit, early in the following growing season. The nematode *S. carpocapsae* has been used to control cocooned codling moth larvae in Europe and North America where the larvae are located near the base of trees, close to the soil (Lacey *et al.*, 2000). Evaluation of EPNs has been carried out to a limited extent against cocooned codling moth larvae (Dutky, 1959; Kaya *et al.*, 1984; Nachtigall and Dickler, 1992; Lacey and Unruh, 1998; Lacey and Chauvin, 1999). *S. carpocapsae* and *S. feltiae* when tested against codling moth proved very much effective. Under optimal conditions, the application of nematodes at 1 × 10^6 IJs/tree resulted in controlling over 95% of codling moth larvae (Kaya *et al.*, 1984). Low moisture and extreme temperature are considered as the major limiting factors for the activity of nematodes in the orchard agroecosystem. Habitat modification, particularly the use of irrigation before and after treatment, application timing and formulation with anti-desiccants are some strategies that warrant attention to enhance or extend the activity of IJs in orchards and other cryptic habitats. Selecting and evaluating nematodes that are cold active would also prove beneficial in controlling overwintering larvae in the late autumn and early spring. The correct choice of nematode species and other operational factors warrant investigation to provide growers with practical control options. Lacey *et al.* (2006b) conducted trials in apple and pear orchards to control cocooned codling moth larvae with two EPNs, namely *S. carpocapsae* and *S. feltiae*. The study was conducted in late summer, autmn and early spring. Nematodes were applied to apple trees at 10^6 IJs/tree, plus supplemental wetting to aid the survival of the IJs. In September, the larval mortality caused by both species of nematode varied between 94% and 95% (mortality in sentinel codling moth larvae). In the following month (i.e. in October), *S. feltiae* was much effective, causing 90% larval mortality, whereas *S. carpocapsae* proved much less effective than *S. feltiae*, with 58% larval mortality, ostensibly due to the cooler conditions. In the same location in mid-October 2001, *S. feltiae* (10^6 IJs/tree), when applied in combination with a wetting agent (Silwet L77) or a humectant (Stockosorb), were most effective in controlling cocooned sentinel codling moth larvae in cardboard strips (80% mortality) and logs (34–47%). The trees were misted for 4 h after treatment. Larval mortality also increased in strips without post-application wetting when either of the above adjuvants was used with IJs. However, the efficacy of the nematodes did not improve significantly on

Fig. 7.1. Larvae of *Zeuzera pyrina* after application of entomopathogenic nematodes: upper – infected with *Steinernema carpocapsae*; lower – infected with *Heterorhabditis bacteriophora.*

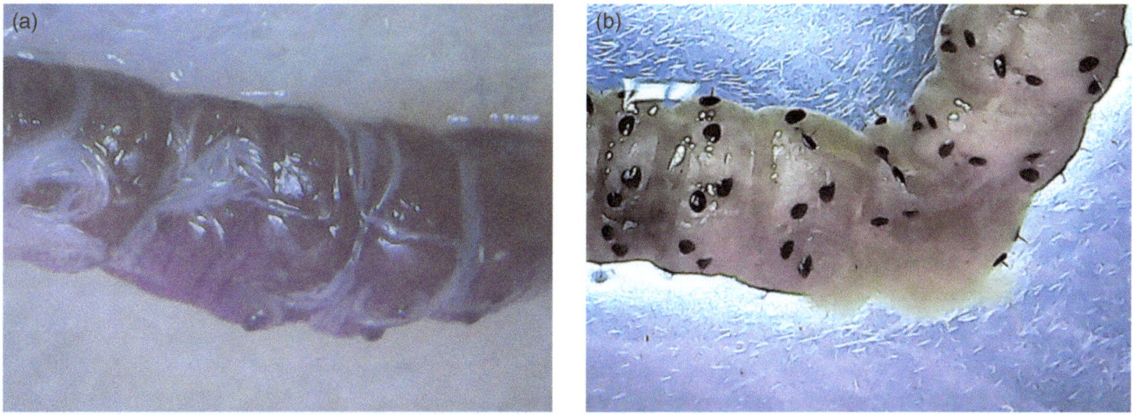

Fig. 7.2. Different developmental stages and recycling of entomopathogenic nematodes in larvae of *Zeuzera pyrina*. (a) Different developmental stages of *Heterorhabditis bacteriophora*. (b) Recycling of *Steinernema carpocapsae*.

logs. Several workers have conducted such experiments in the USA and Europe to evaluate the efficacy of EPNs against *C. pomonella* (Lacey *et al.*, 2005, 2006a,b; Unruh and Lacey, 2001). Lacey and Unruh (1998) tested three EPN species, namely *S.carpocapsae*, *S. riobrave* and *H. bacteriophora*, against cocooned larvae of codling moth, *C. pomonella*, in leaf litter. The best performance was shown by *S. carpocapsae*, causing 99% larval mortality, followed by *H. bacteriophora* (83%) and *S. riobrave* (80%). The efficacy of the nematodes decreased when applied on tree logs; it was 83% for *S. carpocapsae*, 43% for *S. riobrave* and 31% for *H. bacteriophora*.

7.3.5 Other fruit tree borers

Steinernema and *Heterorhabditis* have been reported to control a variety of sesiid borers (Miller and Bedding, 1982; Deseo and Miller, 1985; Kaya and Brown, 1986; Cossentine *et al.*, 1990; Nachtigall and Dickler, 1992; Williams *et al.*, 2002; Shapiro-Ilan *et al.*, 2009). Among these, the clear wing borer, *S. myopaeformis*, is considered an important pest in Europe, the USA and some parts of Africa. The larvae of *S. myopaeformis* make galleries under the bark of fruit trees, especially old ones with damaged trunks. In Egypt, three application methods (spray, injection or cotton balls wetted with nematode suspension) using a concentration of 800 IJs (of either *S. carpocapsae* or *H. bacteriophora*)/ml of water were applied against larvae of *S. mayopaeformis* confined under the bark of apple trees. The nematodes applied by using cotton ball method were very effective, causing 66.7% larval mortality, followed by the injection and spray methods (Abdel-Kawy *et al.*, 1996).

The lesser peach tree borer, *Synanthedon pictipes* (Grote & Robinson) (Lepidoptera: Sesiidae), is an important pest of peach. Eggs are laid by adult moths on the trunk and scaffold limbs, in cracks of the tree's bark, in the crotch or near injured areas (Bobb, 1959; Johnson *et al.*, 2005; Cottrell *et al.*, 2008). Larvae bore into the inner bark and cambium. They feed and develop there and the overwintering of the second-generation larvae takes place in the tunnels. The damage caused due to larval feeding reduces tree vigour. In the case of severe infestation, there is loss of scaffold limbs and/or premature loss of trees and orchard productivity (Johnson *et al.*, 2005). Shapiro-Ilan and Cottrell (2006) reported high mortality in *S. pictipes* caused by steinernematids under laboratory tests; however, heterorhabditids were less virulent. EPNs have been found less effective when applied on foliage to control insect pests, and the reason behind this is the harmful effects of ultraviolet (UV) radiation and desiccation of nematodes (Gaugler and Boush, 1978; Begley, 1990; Shapiro-Ilan *et al.*, 2006). Despite this, several studies confirm high levels of pest mortality in above-ground applications of EPNs (Arthurs *et al.*, 2004), and that too includes against several species of *Synanthedon* (Miller and Bedding, 1982; Deseo and Miller, 1985; Kaya and Brown, 1986; Nachtigall and Dickler, 1992). An improved formulation or application technique may be helpful in increasing the efficacy of EPNs during above-ground applications

(Shapiro-Ilan *et al.*, 2006; Lacey and Shapiro-Ilan, 2008). There are many reports confirming that addition of anti-desiccants or other adjuvants were useful in providing improved above-ground control of different foliar pests such as *Plutella xylostella* and *S. littoralis* (Glazer *et al.*, 1992; Baur *et al.*, 1997; Head *et al.*, 2004; Schroer and Ehlers, 2005). As compared to foliar applications, relatively little attention has been paid by scientists to improve EPN formulations that are applied against borer pests. Miller and Bedding (1982) tested the efficacy of a wetting agent, Arlatone, in combination with *S. feltiae* in order to control currant borer, *S. tipuliformis*. The adjuvant was less effective when compared with the application of nematodes alone. Shapiro-Ilan *et al.* (2010) tested different formulations of the nematode *S. carpocapsae* with an aim to improve its efficacy in controlling *S. pictipes*. It was concluded that nematode treatments followed by the application of a sprayable gel could enhance the control of *S. pictipes*, and also other above-ground pests. *S. carpocapsae*, when applied in combination with sprayable gel, reduced the population of *S. pictipes* in fields significantly, but application of nematodes alone did not affect the survival of *S. pictipes*.

7.4 Conclusion

In the above-mentioned sections, examples of the field efficacy of EPNs against lepidopterans have indicated that most of them are at least susceptible to nematodes in soil, cryptic and foliar habitats. The high susceptibility of the greater wax moth, *Galleria mellonella*, renders it as the reference laboratory insect for measuring quality traits of EPNs, as well as one of the best *in vivo* hosts for nematode production. Additionally, *G. mellonella* is often used as a bait insect for nematode extraction from the soil and in experiments (Woodring and Kaya, 1988; Gaugler, 2002; Grewal *et al.*, 2005). Because of the generally high susceptibility of many lepidopterans to nematodes, the efficacy of these bioinsecticides is usually determined by insect habitat and environment rather than by nematode species and dose of application. Therefore, further research should be carried out on targeted EPN traits, such as including UV-radiation tolerance and resistance to environmental extremes such as high/low temperature stress and desiccation, as well as IJ longevity to enhance EPN efficacy (Glazer, 2014; see Baiocchi *et al.*, Chapter 25, this volume). Indeed, the soil habitat, especially sandy soil, offers an excellent site for insect–nematode interactions. More than 90% of these insect species have life-cycle stages in soil, which is the natural EPN habitat. Thus, exposure to EPNs should be common where lepidopteran soil stages including larvae, pre-pupae, pupae and even emerging adults are easy targets. There are many reports in the literature of successful attempts to control the soil stages of lepidopteran insects with both steinernematids and heterorhabditids, including the black cutworm, *A. ipsilon*, the corn earworm, *H. zea*, and the beet armyworm, *S. exigua* (e.g. Kaya and Griev, 1982; Feaster and Steinkraus, 1996; Grewal *et al.*, 2005; Saleh *et al.*, 2015).

Lepidoptera in cryptic habitats, like stem and root borers, are also considered favourable for infectivity, survival and persistence of EPNs, because the cryptic environment minimizes nematode death from ultraviolet radiation and desiccation. Currently, field experiments in these habitats have provided consistent and efficacious results. In frequent cases, over 70% control of the borers *Z. pyrina* or *S. myopaeformis* was attained by injecting EPNs in their galleries. Spraying the trunks of apple trees with these nematodes during winter resulted in over 80% mortality in overwintering larvae of the apple moth, *C. pomonella*. Field applications of nematodes as biological insecticides against foliage-feeding Lepidoptera are problematic. The foliar environment often exposes the nematodes to unfavourable conditions that result in their rapid desiccation and death. This subject needs further studies, especially because adjuvants with anti-desiccant properties usually enhance nematode control in such above-ground applications. As mentioned above, some successful cases of controlling lepidopteran insects with EPNs on foliage have been reported, including the diamondback moth, *P. xylostella*, the cabbage worm, *P. rapae*, and the beet armyworm, *S. exigua*, infesting the foliage of chrysanthemum plants.

The future of expanding the use of EPNs against lepidopteran pests will depend on the strategies that need to be adopted, such as lowering the cost of production, proper formulation, stable shelf life (at least 1 year) of the product, efficient application and moderate to high nematode efficacy under field conditions (see Askary *et al.*, Chapter 32, this volume). The production of EPNs in developing countries on a small scale, such as in cottage industries, and their use against lepidopteran pests infesting vegetable and field crops will boost the expanding use of nematodes.

Acknowledgement

This work was achieved through the framework of the 10th Research Plan of the National Research Centre in Egypt, Internal Project No. 10120610.

References

Abdel-Kawy, A.G.M., El-Kifl, T.A.H. and Abdel-Bary, N.A. (1988) Controlling the leopard moth *Zeuzera pyrina* by the nematode *Neoaplectana carpocapsae* under field conditions. *Bulletin of Zoological Society of Egypt* 36, 108–120.

Abdel-Kawy, A.G.M., El-Bishry, M.H. and El-Kifl, T.A.H. (1992) Controlling the leopard moth *Zeuzera pyrina* by three entomopathogenic nematode species in the field. *Bulletin of the Faculty of Agriculture, University of Cairo* 43, 769–780.

Abdel-Kawy, A.M., Saleh, M.M.E. and El-Azzazi, A.M. (1996) Using entomopathogenic nematodes in the control of two apple borers. *Egyptian Journal of Biological Pest Control* 6, 147–151.

Akhurst, R.J. and Boemare, N.E. (1990) Biology and taxonomy of *Xenorhabdus*. In: Gaugler, R. and Kaya, H.K. (eds) *Entomopathogenic Nematodes in Biological Control*. CRC Press, Boca Raton, Florida, pp. 75–90.

Arthurs, S., Heinz, K.M. and Prasifca, J.R. (2004) An analysis of using entomopathogenic nematodes against above-ground pests. *Bulletin of Entomological Research* 94, 297–306.

Askary, T.H. (2010) Nematodes as biocontrol agents. In: Lichtfouse, E. (ed.) *Sociology, Organic Farming, Climate Change and Soil Science*. Springer, The Netherlands, pp. 347–378.

Barbosa-Negrisoli, C.R.C., Negrisoli, A.S, Dolinski, C. and Bernardi, D. (2010) Efficacy of entomopathogenic nematodes (Nemata: Rhabditida) to control Brazilian apple leafroller *Bonagota salubricola* (Meyrick, 1937) (Lepidoptera: Tortricidae). *Crop Protection* 29, 1274–1279.

Baur, M.E., Kaya, H.K., Gaugler, R. and Tabashnik, B.E. (1997) Effects of adjuvants on entomopathogenic nematode persistence and efficacy against *Plutella xylostella*. *Biocontrol Science and Technology* 7, 513–525.

Baur, M.E., Kaya, H.K., Tabashnik, B.E. and Chilcutt, C.F. (1998) Suppression of the diamond-back moth (Lepidotera: Plutellidae) with an entomopathogenic nematode (Rhabditidae: Steinernematidae) and *Bacillus thuringiensis* Berliner. *Journal of Economic Entomology* 91, 1089–1095.

Beattie, G.A.C., Somsook, V., Watson, D.M., Clift, A.D. and Jiang, L. (1995) Field evaluation of *Steinernema carpocapsae* (Weiser) (Rhabditida: Steinernematidae) and selected pesticides and enhancers for control of *Phyllocnistis citrella* Stainton (Lepidoptera: *Gracillariidae). Journal of Australian Entomological Society* 34, 335–342.

Bedding, R.A. and Molyneux, A.S. (1982) Penetration of insect cuticle by infective juveniles of *Heterorhabditis* spp. *Nematologica* 28, 354–359.

Begley, J.W. (1990) Efficacy against insects in habitats other than soil. In: Gaugler, R. and Kaya, H.K. (eds) *Entomopathogenic Nematodes in Biological Control*. CRC Press, Boca Raton, Forida, pp. 215–232.

Bélair, G., Vincent, C., Lemire, S. and Coderre, D. (1999) Laboratory and field assays with entomopathogenic nematodes for the management of oblique banded leafroller *Choristoneura rosaceana* (Harris) (Tortricidae). *Journal of Nematology* 31(4S), 684–689.

Bélair, G., Fournier, Y. and Dauphinais, N. (2003) Efficacy of steinernematid nematodes against three pests of crucifers in Quebec. *Journal of Nematology* 35, 259–265.

Ben-Yakir, D., Efron, D., Chen, M. and Glazer, I. (1998) Evaluation of entomopathogenic nematodes for biocontrol of the European corn borer, *Ostrinia nubilalis* on sweet corn in Israel. *Phytoparasitica* 26, 101–108.

Bobb, M.L. (1959) The biology of the lesser peach tree borer in Virginia. *Journal of Economic Entomology* 52, 634–636.

Buhler, W.G. and Gibb, T.J. (1994) Persistence of *Steinernema carpocapsae* and *S. glaseri* (Rhabditida: Steinernematidae) as measured by their control of black cutworm (Lepidoptera: Noctuidae) larvae in bent grass. *Journal of Economic Entomology* 87, 638–642.

Cabanillas, H.E. and Raulston, J.R. (1995) Impact of *Steinernema riobravis* (Rhabditida: Steinernematidae) on the control of *Helicoverpa zea* (Lepidoptera: Noctuidae) in corn. *Journal of Economic Entomology* 88, 58–64.

Cabanillas, H.E. and Raulston, J.R. (1996a) Effects of furrow irrigation on the distribution and infectivity of *Steinernema riobravis* against corn earworm in corn. *Fundamental and Applied Nematology* 19, 273–281.

Cabanillas, H.E. and Raulston, J.R. (1996b) Evaluation of *Steinernema riobravis*, *S. carpocapsae*, and irrigation timing for the control of corn earworm, *Helicoverpa zea*. *Journal of Nematology* 28, 75–82.

Capinera, J.L. (2008) Butterflies and moths. *Encyclopedia of Entomology*, 2nd edn. Springer, Dordrecht, The Netherlands, pp. 626–672.

Carter, D.J. (1984) Pest Lepidoptera of Europe with special reference to the British Islands. *Series Entomologica 31*. Dr W. Junk Publishers, Dordrecht, The Netherlands, 431 pp.

Cossentine, J.E., Banham, F.L. and Jensen, L.B. (1990) Efficacy of the nematode, *Heterorhabditis heliothidis* (Rhabditida: Heterorhabditidae) against the peach tree borer, *Synanthedon exitiosa* (Lepidoptera: Sesiidae) in peach trees. *Journal of the Entomological Society of British Columbia* 87, 82–84.

Cottrell, T.E., Fuest, J. and Horton, D.L. (2008) Influence of *Prunus* spp., peach cultivars, and bark damage on oviposition choices by the lesser peachtree borer (Lepidoptera: Sesiidae). *Environmental Entomology* 37, 1508–1513.

Deseo, K.V. and Miller, L.A. (1985) Efficacy of entomogenous nematodes, *Steinernema* spp., against clearwing moths, *Synanthedon* spp. in north Italian apple orchards. *Nematologica* 31, 100–108.

Ditman, L.P., Weiland, G.S. and Guil, J.H. Jr (1940) The metabolism in the corn earworm III: weight, water, and diapause. *Journal of Economic Entomology* 23, 282–295.

Dutky, S.R. (1959) Insect microbiology. *Advances in Applied Microbiology* 1, 175–200.

El-Shazly, E.A., Saleh, M.M.E. and Semeada, A.M. (1999) Laboratory and field evaluation of some entomopathogenic nematodes against *Zeuzera pyrina* L. (Lepidoptera: Cossidae) on apple trees. *Journal of Agricultural Society of Mansoura University* 24, 3673–3679.

Encyclopaedia Britannica (2014) Available at: http://www.britannica.com/Bchecked/topic/336811/lepidopteran (accesssed 10 September 2014).

Feaster, M.A. and Steinkraus, D.C. (1996) Inundative biological control of *Helicoverpa zea* (Lepidoptera: Noctuidae) with the entomopathogenic nematode *Steinernema riobravis* (Rhabditida: Steinernematidae). *Biological Control* 7, 38–43.

Forschler, B.T. and Nordin, G.L. (1988) Comparative pathogenicity of selected entomopathogenic nematodes to hardwood borers, *Prionoxystus iobiniae* (Lepidoptera: Cossidae) and *Megacyllene robinae* (Coleoptera: Cyrambycidae). *Journal of Invertebrate Pathology* 52, 343–349.

Garcia-del-Pino, F., Alabern, X. and Morton, A. (2013) Efficacy of soil treatments of entomopathogenic nematodes against the larvae, pupae and adults of *Tuta absoluta* and their interaction with the insecticides used against this insect. *BioControl* 58, 723–731.

Gaugler, R. (1988) Ecological considerations in the biological control of soil inhabiting insects with entomopathogenic nematodes. *Agriculture, Ecosystems and Environment* 24, 351–360.

Gaugler, R. (ed.) (2002) *Entomopathogenic Nematology*. CAB International, Wallingford, UK, 388 pp.

Gaugler, R. and Boush, G.M. (1978) Effects of ultraviolet radiation and sunlight on the entomogenous nematode *Neoaplectana carpocapsae*. *Journal of Invertebrate Pathology* 32, 291–296.

Georgis, R. (1990) Formuulation and application technology. In Gaugler, R. and Kaya, H.K. (eds) *Entomopathogenic Nematodes in Biological Control*. CRC Press, Boca Raton, Florida, pp. 173–191.

Georgis, R. (2002) The Biosys experiment: an insiders perspective. In: Gaugler, R. (ed.) *Entomopathogenic Nematodes*. CAB International, Wallingford, UK, pp. 357–372.

Georgis, R. and Manweiler, S.A. (1994) Entomopathogenic nematodes: a developing biological control technology. *Agricultural Zoology Reviews* 6, 63–94.

Georgis, R. and Poinar, G.O. Jr (1989) Field effectiveness of entomophilic nematodes *Neoaplectana* and *Heterorhabditis*. In: Leslie, A.R. and Metcalf, R.L. (eds) *Integrated Pest Management for Turf Grass and Ornamentals*. Environmental Protection Agency, Washington, DC, pp. 213–224.

Georgis, R. and Poinar, G.O. Jr (1994) Nematodes as bioinsecticides in turf and ornamentals. In: Leslie, A. (ed.) *Handbook of Integrated Pest Management for Turf and Ornamentals*. CRC Press, Boca Raton, Florida, pp. 477–489.

Glazer, I. (2002) Survival biology. In: Gaugler, R. (ed.) *Entomopathogenic Nematology*. CAB International, Wallingford, UK, pp. 169–187.

Glazer, I. (2014) Genetic improvement and breeding of EPN: the race for the 'super nematode'. In: *Proceedings, 6th International Congress of Nematology*, Cape Town, South Africa, 4–9 May 2014, 30 pp.

Glazer, I. and Lewis, E.E. (2000) Bioassays for entomopathogenic nematodes. In: Navon, A. and Ascher, K.R.S. (eds) *Bioassays of Entomopathogenic Microbes and Nematodes*. CAB International, Wallingford, UK, pp. 229–247.

Glazer, I. and Wysoki, M. (1990) Steinernematid and heterorhabditid nematodes for biological control of the giant looper, *Boarmia selenaria*. *Phytoparasitica* 18, 9–16.

Glazer, I., Klein, M., Navon, A. and Nakache, Y. (1992) Comparison of efficacy of entomopathogenic nematodes combined with anti-desiccants applied by canopy sprays against three cotton pests (Lepidoptera: Noctuidae). *Journal of Economic Entomology* 85, 1636–1641.

Grewal, P.S., Ehlers, R.-U. and Shapiro-Ilan, D.I. (eds) (2005) *Nematodes as Biocontrol Agents*. CAB International, Wallingford, UK, 505 pp.

Griffin, C.T., Boemare, N.E. and Lewis, E.E. (2006) Biology and behavior. In: Grewal, P.S., Ehlers, R.-U. and Shapiro-Ilan, D.I. (eds) *Nematodes as Biocontrol Agents*. CAB International, Wallingford, UK, pp. 47–64.

Hardwick, D.F. (1965) The corn earworm complex. *Memoirs of the Entomological Society of Canada* 40, 1–247.

Head, J., Lawrence, A.J. and Walters, K.F.A. (2004) Efficacy of the entomopathogenic nematode, *Steinernema feltiae*, against *Bemisia tabaci* in relation to plant species. *Journal of Applied Entomology* 128, 543–547.

Henneberry, T.J., Jech, L.F., Burke, R.A. and Lindegren, J.E. (1996) Temperature effects on infection and mortality of *Pectinophora gossypiella* (Lepidoptera: Gelechiidae) larvae by two entomopathogenic nematode species. *Environmental Entomology* 25, 179–183.

Hominick, W.M. (2002) Biogeography. In Gaugler, R. (ed.) *Entomopathogenic Nematology*. CAB International, Wallingford, UK, pp. 115–143.

Ismail, I.I., Abu-Zied, N.A. and Abdallah, F.F. (1992) Population dynamics of the leopard moth, *Zeuzera pyrina* L., and its control on olive trees in Egypt. *Journal of Plant Diseases and Protection*, 99, 519–524.

Jackson, J.J. and Hesler, S.L. (1996) Placement and application rate of the nematode *Steinernema carpocapsae* (Rhabditida: Steinernematidae) for suppression of the western corn rootworm (Coleoptera: Chrysomelidae). *Journal of the Kansas Entomological Society* 68, 461–467.

Johnson, D., Cottrell, T. and Horton, D. (2005) Peach tree borer. In: Horton, D. and Johnson, D. (eds) *Southeastern Peach Grower's Handbook*. University of Georgia, Georgia Cooperative Extension Services, Handbook No 1, pp. 266–269.

Kahounova, L. and Mrácek, Z. (1991) Larval mortality in *Synanthedon myopaeformis* (Lepidoptera) in apple trees sprayed with *Steinernema* sp. (Nematoda: Steinernematidae) strain *Hylobius*. *Acta Entomologica Bohemoslovaca* 88, 205–210.

Kain, D.P. and Agnello, A.M. (1999) Pest status of American plum borer (Lepidoptera: Pyralidae) and fruit tree borer control with synthetic insecticides and entomopathogenic nematodes in New York state. *Journal of Economical Entomology* 92, 193–200.

Kaya, H.K. and Brown, L.R. (1986) Field application of entomogenous nematodes for biological control of clear-wing moth borers in alder and sycamore trees. *Journal of Arboriculture* 12, 150–154.

Kaya, H.K. and Griev, B.J. (1982) The nematode *Neoaplectan carpocapsae* and the beet armyworm *Spodoptera exigua:* infectivity of prepupae and pupae in soil and of adults during emergence from soil. *Journal of Invertebrate Pathology* 39, 192–197.

Kaya, H.K. and Hara, A.H. (1980) Differential susceptibility of lepidopterous pupae to infection by the nematode *Neoaplectana carpocapsae. Journal of Invertebrate Pathology* 36, 389–393.

Kaya, H.K. and Nelsen, C.E. (1985) Encapsulation of steinernematid and heterorhabditid nematodes with calcium alginate: a new approach for insect control and other applications. *Environmental Entomology* 14, 572–574.

Kaya, H.K, Joos, J.L., Falcon, L.A. and Berlowitz, A. (1984) Suppression of the codling moth (Lepidoptera: Olethreutidae) with the entomogenous nematode, *Steinernema feltiae* (Rhabditida: Steinernematidae). *Journal of Economic Entomology* 77, 1240–1244.

Klein, M.G. (1990) Efficacy against soil-inhibiting insect pests. In: Gaugler, R. and Kaya, H.K. (eds) *Entomopathogenic Nematodes in Biological Control*. CRC Press, Boca Raton, Florida, pp. 195–214.

Lacey, L.A. and Chauvin, R.L. (1999) Entomopathogenic nematodes for control of codling moth in fruit bins. *Journal of Economic Entomology* 92, 104–109.

Lacey, L.A. and Shapiro-Ilan, D.I. (2008) Microbial control of insect pests in temperate orchard systems: potential for incorporation into IPM. *Annual Review of Entomology* 53, 121–144.

Lacey, L.A. and Unruh, T.R. (1998) Entomopathogenic nematodes for control of codling moth: effect of nematode species, dosage, temperature and humidity under laboratory and simulated field conditions. *Biological Control* 13, 190–197.

Lacey, L.A., Knight, A. and Huber, J. (2000) Microbial control of lepidopteran pests of apple orchards. In: Lacey, L.A. and Kaya, H.K. (eds) *Field Manual of Techniques in Invertebrate Pathology: Application and evaluation of pathogens for control of insects and other invertebrate pests.* Kluwer Academic Publishers, Dordrecht, The Netherlands, pp. 557–576.

Lacey, L.A., Frutos, R., Kaya, H.K. and Vail, P. (2001) Insect pathogens as biological control agents: do they have a future? *Biological Control* 21, 230–248.

Lacey, L.A., Neven, L.G., Headrick, H.L. and Fritts, J.R., (2005) Factors affecting entomopathogenic nematodes (Steinernematidae) for control of overwintering codling moth (Lepidoptera: Tortricidae) in fruit bins. *Journal of Economic Entomology* 98, 1863–1869.

Lacey, L.A., Granatstein, D., Arthurs, S.P., Headrick, H. and Fritts, J.R. (2006a) Use of entompathogenic nematodes (Steinernematidae) in conjuction with mulches for control of overwintering codling moth (Lepidoptera: Tortricidae). *Journal of Entomological Science* 41, 107–119.

Lacey, L.A., Arthurs, S.P., Unruth, T.R., Headrick, H. and Fritts, J.R. (2006b) Entomopathogenic nematodes for control of codling moth (Lepidoptera: Tortricidae) in apple and orchards: effect of nematode species and seasonal temperatures, adjuvants, application equipment, and post-application irrigation. *Biological Control* 37, 214–223.

Legaspi, J.C., Legaspi, B.C. and Saldana, R.R. (2000) Evaluation of *Steinernema riobravis* (Nematoda: Steinernematidae) against the Mexican rice borer (Lepidoptera: Pyralidae). *Journal of Entomological Science* 35, 141–149.

Levine, E. and Oloumi-Sadeghi, H. (1992) Management of diabroticite rootworms in corn. *Annual Review of Entomology* 36, 229–255.

Lewis, L.C. and Raun, E.S. (1978) Laboratory and field evaluation of the DD-136 strain of *Neoaplectana carpocapsae* (nematodes) for control of the European corn borer, *Ostrinia nubilalis. Iowa State Journal of Research* 52, 391–396.

Lindegren, J.E., Yamashita, T.T. and Barnett, W.W. (1981) Parasitic nematodes may control carpenter worm in fig trees. *California Agriculture* 35, 25–30.

Mahar, A.N., Munir, M., El-Awad, S., Gowen, S.R. and Hauge, N.G.M. (2004) Microbial control of the diamondback moth *Plutella xylostella* L. (Lepidoptera: Yponmeutidae) using bacteria from the entomopathogenic nematode *Steinernema carpocapsae. Journal of Zhejiang University Science* 5, 1183–1190.

Mason, J.M., Matthews, G.A. and Wright, D.J. (1998) Screening and selection of adjuvants for the spray application on entomopathogenic nematodes against a foliar pest. *Crop Protection* 17, 463–470.

Miller, L.A. and Bedding, R.A. (1982) Field testing of the insect parasitic nematode, *Neoaplectana bibionis* (Nematoda: Steinernematidae) against currant borer moth, *Synanthedon tipuliformis* (Lep.: Sesiidae) in blackcurrants. *Entomophaga* 27, 109–114.

Moris, O.N. (1985) Susceptibility of 31 species of agricultural insect pests to entomopathogenic nematodes *Steinernema carpocapsae* and *Heterorhabditis bacteriophora. Canadian Entomologist* 117, 401–407.

Mráček, Z. (2002) Use of entomoparasitic nematodes (EPANs) in biological control. In: Upadhyay, R.K. (ed.) *Advances in Microbial Control of Insect Pests*. Kluwer Academic Publishers, New York, pp. 235–264.

Nachtigall, G. and Dickler, E. (1992) Experiences with field applications of entomoparasitic nematodes for biological control of cryptic living insects in orchards. *Acta Phytopathologica et Entomologica Hungarica* 27, 485–490.

Peters, A. and Ehlers, R.-U. (1994) Susceptibility of leatherjackets (*Tipula paludosa* and *T. oleracae*, Tipulidae: Nematocera) to the entomopathogenic nematode *Steinernema feltiae*. *Journal of Invertebrate Pathology* 63, 163–171.

Poinar, G.O. Jr (1990) Taxonomy and biology of steinernematidae and heterorhabditidae. In: Gaugler, R. and Kaya, H.K. (eds) *Entomopathogenic Nematodes in Biological Control*. CRC Press, Boca Raton, Florida, pp. 23–61.

Powell, J.A. (2009) Lepidoptera (moths and butterflies). In: Resh, V.H. and Cardé, R.T. (eds) *Encyclopedia of Insects.* Academic Press, San Diego, pp. 631–664.

Richter, A.R. and Fuxa, J.R. (1990) Effect of *Steinernema feltiae* on *Spodoptera frugiperda* and *Heliothis zea* (Lepidoptera: Noctuidae) in corn. *Journal of Economic Entomology* 83, 1286–1291.

Saleh, M.M.E. (2014) Utilization of entomopathogenic nematodes in the biological control of some fruit tree borers. First Annual Report 10th Research Plan of the National Research Centre, In-House Project No 10120610, Egypt.

Saleh, M.M.E. and Abbas, M.S.T. (1998) Suitability of certain entomopathogenic nematodes for controlling *Zeuzera pyrina* L. (Lepidoptera: Cossidae) in Egypt. *International Journal of Nematology* 8, 126–130.

Saleh, M.M.E., Semeada, A.M. and El-Shazly, E.A. (1994) Virulence of native and foreign entomopathogenic nematodes to the apple borer, *Z. pyrina*. *Egyptian Journal of Biological Pest Control* 4, 107–112.

Saleh, M.M.E., Mona, A.H., Gehan, A.H., Hussein, M.A., Salem, H.A. and Hala, M.S.M. (2015) Foliar application of entomopathogenic nematodes for controlling *Spodoptera littoralis* and *Agrotis ipsilon* (Lepidoptera: Noctuidae) on corn (*Zea mays*) plants. *Advances in Applied Agricultural Science* 3, 51–61.

Salem, S.A., Abdel-Rahman, H.A., Zebitz, C.P.W., Saleh, M.M.E., Ali, F.I. and El-kholy, M.Y. (2007) Evaluation of entomopathogenic nematodes in controlling some cabbage pests. *Journal of Applied Science Research* 3, 323–328.

Samersov, V.F., Yarchakovskaya, S.I. and Bezruchenok, N.N. (1998) Ecologically safe protection of black currant. *Zashchita i Karantin Rastenii* 9, 11–12.

Schroer, S. and Ehlers, R.-U. (2005) Foliar application of the entomopathogenic nematode *Steinernema carpocapsae* for biological control of diamondback moth larvae (*Plutella xylostella*). *Biological Control* 33, 81–86.

Shamseldean, M.M., Abo E1-Sooud, A.B. and Saleh, M.M.E. (1996) Isolation of *Steinernema carpocapsae* (Weiser, 1955) Poinar, 1990 (Steinemematidae), *Heterorhabditis bacteriophora* Poinar, 1976 and *Heterorhabditis indicus* Poinar *et al.*, 1992 (Heterorhabditidae) as a first record from Egypt. *Egyptian Journal of Biological Pest Control* 6, 185–199.

Shannag, H.K., Webb, S.E. and Capinera, J.L. (1994) Entomopathogenic nematode effect on pickleworm (Lepidoptera: Pyralidae) under laboratory and field conditions. *Journal of Economic Entomology* 87, 1205–1212.

Shapiro-Ilan, D.I. and Cottrell, T.E. (2006) Susceptibility of the lesser peach tree borer (Lepidoptera: Sesiidae) to entomopathogenic nematodes under laboratory conditions. *Environmental Entomology* 35, 358–365.

Shapiro-Ilan, D.I., Duncan, L.W., Lacey, L.A. and Han, R. (2005) Orchard applications. In: Grewal, P.S., Ehlers, R.-U. and Shapiro-Ilan D.I. (eds) *Nematodes as Biocontrol Agents*. CAB International, Wallingford, UK, pp. 215–229.

Shapiro-Ilan, D.I., Gouge, G.H., Piggott, S.J. and Fife, J. (2006) Application technology and environmental considerations for use of entomopathogenic nematodes in biological control. *Biological Control* 38, 124–133.

Shapiro-Ilan, D.I., Cottrell, T.E., Mizell III, R.F., Horton, D.L. and Davis, J. (2009) A novel approach to biological control with entomopathogenic nematodes: prophylactic control of the peach tree borer, *Synanthedon exitiosa*. *Biological Control* 48, 259–263.

Shapiro-Ilan, D.I., Cottrell, T.E., Mizell III, R.F., Horton, D.L., Behle, R.W. and Dunlap, C.A. (2010) Efficacy of *Steinernema carpocapsae* for control of the lesser peach tree borer, *Synanthedon pictipes*: improved above ground suppression with a novel gel application. *Biological Control* 54, 23–28.

Shapiro-Ilan, D.I., Han, R. and Dolinksi, C. (2012) Entomopathogenic nematode production and application technology. *Journal of Nematology* 44, 206–217.

Showers, W.B. (1979) Effect of diapause on the migration of the European corn borer into the southeastern United States. In: Rabb, R.L. and Kennedy, G.G. (eds) *Movement of Highly Mobile Insects: Concepts and Methodology in Research.* North Carolina

State University Graphics, Raleigh, North Carolina, pp. 420–430.

Sparks, A.N., Westbrook, J.K., Wolf, W.W., Raulston, J.R. and Pair, S.D. (1989) The potential importance of insect migration to management strategies of maize arthropods. *Acta Phytopathologica et Entomologica Hungarica* 24, 195–206.

Timper, P., Kaya, H.K. and Gaugler, R. (1988) Dispersal of the entomogenous nematode *Steinernema feltiae* (Rhabditida; Steinernematidae) by infected adult insects. *Environmental Entomology* 17, 546–550.

Unruh, T.R. and Lacey, L.A. (2001) Control of codling moth, *Cydia pomonella* (Lepidoptera: Tortricidae) with *Steinernema carpocapsae*: effects of supplemental wetting and pupation site on infection rate. *Biological Control* 20, 48–56.

Wang, J. (1993) Control of the peach fruit moth, *Carposina niponensis*, using entomopathogenic nematodes. In: Bedding, R., Akhurst, R. and Kaya, H. (eds) *Nematodes and the Biological Control of Insect Pests*. CSIRO Publications, East Melbourne, Australia, pp. 59–65.

Wang, J.X. and Li, Y.L. (1987) Entomogenous nematode research in China. *Revue de Nématologie* 10, 483–489.

Williams, R.N., Fickle, D.S., Grewal, P.S. and Meyer, J.R. (2002) Assessing the potential of entomopathogenic nematodes to control the grape root borer *Vitacea polistiformis* (Lepidoptera: Sesiidae) through laboratory and greenhouse bioassays. *Biocontrol Science and Technology* 12, 35–42.

Woodring, J.L. and Kaya, H.K. (1988) Steinernematid and heterorhabditid nematodes: a handbook of techniques. Southern Cooperative Series Bulletin No 331, Arkansas Agricultural Experiment Station, Fayetteville, Arkansas, 30 pp.

Wright, P.J. and Jackson, T.A. (1992) Efficacy of entomogenous nematodes for control of porina (*Wiseana cervinata*) in Canterbury pastures during winter. *New Zealand Journal of Agricultural Research* 35, 435–439.

Yamanaka, S., Seta, K. and Yasuda, M. (1987) Evaluation of the use of entomogenous nematode, *Steinernema feltiae* (str. Mexican) for the biological control of the fall webworm *Hyphantria cunea* (Lepodoptera: Arctidae). *Japanese Journal of Nematology* 16, 26–31.

Yang, H.W., Jian, H. and Zhang, S.G. (1993) Discussion of a simple method for measurement of infectivity of Steinernema carpocapsae. In: *Proceedings of International Symposium on the Use of Biological Control Agent Under Integrated Pest Management*. Fukuoka, Japan, pp. 107–111.

8 Efficacy of Entomopathogenic Nematodes Against Coleopteran Pests

J. Gulsar Banu*

Central Institute for Cotton Research, Regional Station, Coimbatore, India

8.1 Introduction

Among the different insect orders prevailing in the world, Coleoptera consist of the largest insect order, with close to 390,000 described species in almost 30,000 genera of 176 families worldwide (Slipinski *et al.*, 2011), representing about 40% of known insect species. This order consists of the biggest and most conspicuous insects, namely beetles and weevils, which occur in almost all climatic conditions and cause huge economic loss to agricultural crops (Gressitt, 2015).

Conventionally used organophosphate and carbamate insecticides have increased crop yields and brought about a 70% reduction in coleopteran grubs. But the subsequent application of chemical insecticides has resulted in lowering the efficacy of the chemicals used in the control of white grubs (Georgis *et al.*, 2006). Other drawbacks associated with conventional insectides are soil toxicity, its lethal effect on non-target organisms and on beneficial soil microorganisms, pest resurgence and environmental pollution. In addition to the above problems, poor storage facilities associated with conventional insecticides in developing countries make them unsuitable for insect control. Hence, emphasis should be given to an alternative and ecofriendly management strategy for coleopteran insect control.

Entomopathogenic nematodes (EPNs) are among one of the safer biocontrol agents used for the control of coleopteran insects. When applied under favourable environmental conditions, EPNs have been found to be as effective as chemical insecticides in the management of coleopteran insects. Moreover, increased demand for organic fruit and vegetables in recent years has provided opportunities for nematode-based commercial products. Many EPN-based commercial products are currently available worldwide for the control of these insect pests. EPNs are widely used as biocontrol agents against various coleopteran insects (Lacey and Georgis, 2012) (Table 8.1). The efficacy of EPNs is influenced by several factors like host-finding mechanism, adaptability to environment and the presence of host range, etc. The application of proper species of EPNs in their proper stage will result in the effective control of coleopteran insect pests dwelling in the subterranean and epigeal (surface) habitat.

The present chapter is focused on the success of EPNs against various coleopteran insect pests, the various biotic and abiotic factors that reduce the efficacy of EPNs, the application of EPNs in integrated management programmes and the genetic improvement methods to increase the efficacy of EPNs against various insect hosts.

8.2 Entomopathogenic Nematodes for the Control of White Grubs

White grubs (Coleoptera: Scarabaeidae) are the root-feeding larva of the scarab beetles. They are serious soilborne pests of many agricultural and horticultural plants worldwide (Guo *et al.*, 2015). *Steinernema glaseri* was the first EPN used for the control of Japanese beetle, *Popillia japonica*. However, a large-scale colonization programme of

*E-mail: gulsarsci@gmail.com

 Biocontrol Agents: Entomopathogenic and Slug Parasitic Nematodes (eds. M.M.M. Abd-Elgawad, T.H. Askary and J. Coupland)

Table 8.1. *Steinernema* and *Heterorhabditis* species commonly used in the biocontrol of coleopteran insects.

Crop	Target insects	Insect stages controlled by entomopathogenic nematodes	Nematodes	Reference
Turfgrass	Japanese beetle (*Popillia japonica*)	Larva	*Sg*	Gaugler *et al.*, 1992
Turfgrass	Billbug (*Sphenophorus* spp.)	Larva/adult	*Hb*, *Sc*[a]	Shapiro-Ilan and Gaugler, 2010
Berries and ornamentals	Blackvine weevil (*Otiorhynchus sulcatus*)	Larva/adult	*Hb*, *Hd*, *Hm*, *Hmeg*, *Sc*, *Sg*[a]	
Citrus and ornamentals	Citrus root weevil (*Pachnaeus* spp.)	Larva/adult	*Sr*, *Hb*[a]	
Citrus and ornamentals	*Diaprepes* root weevil (*Diaprepes abbreviates*)	Larva/adult	*Hb*, *Sr*[a]	
Forest planting	Large pine weevil (*Hylobius albietis*)	Grub	*Hd*, *Sc*[a]	
Turf and ornamentals	Scarab beetle	Larva	*Hb*, *Sc*, *Sg*, *Hz*[a]	
Berries	Strawberry root weevil (*Otiorhynchus ovatus*)	Larva/pupae	*Hm*[a]	
Beehives	Small hive beetle (*Aethina tumida*)	Larva/pupae	*Hi*, *Sr*[a]	
Sweet potato	Sweet potato weevil (*Cylas formicarius*)	Larva	*Hb*, *Sc*, *Sf*[a]	
Laboratory assay	May beetle chafer (*Meliolontha hippocastani*)	Larva	*Sc*	Lortkipanidze *et al.*, 2011
Laboratory assay	Banana weevil (*Cosmopolites sordidus*)	Larva	*Steinernema* and *Heterorhabditis*	Mwaitulo *et al.*, 2011
Laboratory assay	Mustard beetle (*Phaedon cochleariae*)	Pupae	*Hb*, *Sf*, *Sc*	Mahar *et al.*, 2012
Maize	Western corn rootworm (*Diabrotica virgifera*)	Larva	*Hb*, *Hm*, *Sf*, *Sk*	Toepfer *et al.*, 2012
Laboratory assay	*Stethobaris nemesis*	Larva and weevil	*Sc*, *Sf*	Shapiro-Ilan *et al.*, 2012
Cabbage	Striped flea beetle (*Phyllotreta striolata*)	Larva	*Sc*, *Hi*	Yan *et al.*, 2013
Laboratory assay	Red palm weevil (*Rhynchophorus ferrugineus*)	Larva	*Sc*	Manachini *et al.*, 2013
Groundnut	White grub (*Holotrichia paralella*)	Grub	*Sl*, *Hb*	Guo *et al.*, 2013
Laboratory assay	Spotted Oriental cucumber beetle (*Epilachna chrysomelina*)	Larva and adult	*Hb*, *Sc*	Ebeid *et al.*, 2014
Laboratory assay	*Temnorhynchus baal*	Larva	*Sg*	Atwa and Hassan, 2014
Laboratory assay	European cock chafer (*Melolontha melolontha*)	Larva	*Sf*, *Hb*	Erbas *et al.*, 2014
Groundnut	White grub (*Holotrichia olita*)	Grub	*Sl*, *Hb*	Guo *et al.*, 2015
Conifer trees	Pine weevil (*Hylobius abietis*)	Larva and pupae	*Hd*, *Sg*	Williams *et al.*, 2013

Notes: *Sc* = *Steinernema carpocapsae*; *Sf* = *Steinernema feltiae*; *Sg* = *Steinernema glaseri*; *Sl* = *Steinernema longicaudatum*; *Sk* = *Steinernema kraussei*; *Hb* = *Heterorhabditis bacteriophora*; *Hd* = *Heterorhabditis downesi*; *Hm* = *Heterorhabditis mareletus*; *Hmeg* = *Heterorhabditis megidis*; *Sr* = *Steinernema riobrave*; *Hz* = *Heterorhabditis zealandica*; *Hi* = *Heterorhabditis indica*.

[a]Application of the nematodes mentioned in the table provided about 75% insect control under greenhouse and field conditions.

this EPN against *P. japonica* failed, mainly because of a lack of awareness of the symbiotic bacteria associated with the nematodes. This initial effort built a base for the use of EPNs in biological control programmes (Gaugler *et al.*, 1992). Later, other EPNs, namely *Steinernema scarabaei*, *Heterorhabditis bacteriophora* and *Heterorhabditis zealandica* (Koppenhofer *et al.*, 2006), were used against this beetle and were found to be effective in its management. Among them, *H. bacteriophora* and *H. zealandica* are commercially available products for the control of grubs of the beetle (Grewal *et al.*, 2005) (Table 8.2).

Based on cost–benefit analysis, two EPN species, namely *Steinernema longicaudum* X-7 and *H. bacteriophora* H06 were effective biocontrol agents for the management of *Holotrichia oblita* larvae in groundnut fields (Guo *et al.*, 2015).

Generally, white grub species differ in their susceptibility to different nematode species (Fig. 8.1). Similarly, the virulence of the EPN is also affected by the different larval stages of the host. The inability of a nematode to persist/survive in the particular larval stage of the insect makes it an unsuccessful candidate in insect control. For example, young instars of white grub Japanese beetle are more susceptible to EPNs (Klein *et al.*, 2007) than the fully grown grub stages. Other factors such as size, immune response of the host and its behaviour contribute to the differences in the susceptibility of nematodes to the various host developmental stages. The portal of entry of nematodes was found to be too small in younger stages of the grubs, which hindered the entry of EPNs, and the immune system developed by the later stages of the grubs

Table 8.2. Efficacy of entomopathogenic nematodes against some white grub species.

White grub species	Entomopathogenic nematode strains[a]	Control under field conditions (%)	References
Popillia japonica	*Steinernema scarabaei*	100	Koppenhofer and Fuzy, 2003a
	Heterorhabditis bacteriophora (GPS 11)	34–97	Grewal *et al.*, 2004
	H. bacteriophora (TF)	65–92	Koppenhofer and Fuzy, 2003a,b
	H. zealandica (X1)	73–98	Grewal *et al.*, 2004
Anomala orientalis	*S. scarabaei*	100	Koppenhofer and Fuzy, 2003a,b
Mordellistena castanea	*S. scarabaei*	71–86	Koppenhofer and Fuzy, 2003b
Cyclocephala borealis	*H. zealandica* (X1)	72–96	Grewal *et al.*, 2004;
	S. scarabaei	84	Koppenhofer and Fuzy,
	H. bacteriophora (GPS 11)	47–83	2003a

Note: [a]EPN strains were applied at the rate of $\leq 2.5 \times 10^9$ IJ/ha.

Fig. 8.1. Entomopathogenic nematode infection in sugarcane white grub.

becomes stronger in resisting the EPNs. These differences in the virulence of EPNs and the larval stages of the grubs should be taken into account while developing the biocontrol strategies of EPNs against white grub management.

In some cases, EPNs were found to be more effective than insecticides. It was found that the application of EPNs increased groundnut yield by 70% over the control, which was greater than treatment with the insecticide Chlorpyrifos, in which only a 40% increase in yield was obtained. With the optimum dose of 7.5×10^8 infective juveniles (IJs)/ha, the net profit obtained in the control of white grubs in groundnut fields using *S. longicaudatum* (X-7) and *H. bacteriophora* (H06) was, respectively, about US$1017.49 and US$1432.21 more than the application of Chlorpyrifos (Guo *et al.*, 2013).

Favourable environmental conditions such as soil moisture, texture, temperature and the prevailing white grub species associated with the groundnut field made the nematodes perform better, with good persistence in the field. The sandy soil of the groundnut field, with a soil moisture content of 17% and a temperature of 17–27°C, favoured the easy dispersion and infection of the IJs of these nematodes (Guo *et al.*, 2015).

Regarding production cost and stability, it was found that less cost was involved in the production of *H. bacteriophora* (H06) as solid culture than *S. longicaudatum* (X-7). Under unfavourable conditions, *H. bacteriophora* (H06) was found to be more stable than *S. longicaudatum* (X-7) (Yan *et al.*, 2010). The development of *Heterorhabditis* as liquid culture increased the production efficacy of the culture, making them effective against white grub control. Thus, the profit factor and efficacy of EPNs in the management of white grubs in groundnut fields make it superior to the standard insecticide in the management of the insect pest.

Although the white grubs of June beetle, margined vine chafer and garden chafer are controlled effectively with *H. bacteriophora* in integrated pest management (IPM) strategies, their application needs a greater quantity of water, which is not economically viable for small farmers. Moreover, the activity of the product is affected by inadequate weather conditions prevailing during the application of the EPNs, and hence repeated EPN application is needed for effective control of the grubs, which thus incurs a high cost for the grub management process.

White grubs have developed various defensive mechanisms such as morphological, ecological and behavioural barriers to prevent or delay the entry of EPN species into the host. These factors have conferred on them varying degrees of resistance against EPNs. Ecological and behavioural barriers, such as the infrequent release of CO_2, rather than continuous release, and brushing of the grub's body with its legs or mouthparts (Gaugler *et al.*, 1994), prevent the infection by EPNs. Morphological barriers, including covering the spiracles with sieve plates (Grewal *et al.*, 2005), dense peritrophic membrane of the grubs and frequent defecation, delay the penetration of EPNs into the grub's body. The immune response created by the melanotic encapsulation of the grubs prevents EPN infection into the haemocoel of the host. *H. bacteriophora* was found to be killed by the melanotic encapsulation of the grub, while *S. glaseri* escaped from the encapsulation (Wang *et al.*, 1995).

Generally, LC_{50} values for white grubs are found to be higher than for other insect species. The defensive and evasive mechanisms of the white grubs against EPNs are the reason for these increased LC_{50} values. This is one of the major obstacles involved in the inundative control of white grubs with EPN species (Wang *et al.*, 1994). To overcome this, more virulent nematode strains should be selected through genetic engineering and/or a selective breeding approach. The collection of indigenous strains may provide isolates more suitable for inundative release against local insect pests because of adaptation to the local climate (Banu *et al.*, 2004).

Another major factor that hinders the use of nematodes against white grubs is the competition from chemical insecticides, as insecticides are cheaper compared to the cost of EPNs and, moreover, are easy to apply in the field. The cost of *H. bacteriophora* was four times higher than the similarly effective organophosphate and carbamate compounds, and twice as much as the insecticides imidacloprid, clothianidin and halofenazide. This limits the commercial use of EPNs against white grubs (Georgis *et al.*, 2006).

EPNs can be applied successfully in areas with restricted use of chemicals. For example, in Japan, *S. glaseri* was used successfully to control white grubs in golf courses, as there was restricted use of chemicals in those areas. Another case where EPNs replaced chemicals was in Germany, where no insecticides were available for the control of white grubs in golf courses. *H. bacteriophora* was used successfully over a commercial formulation to control the insect pest (Shapiro-Ilan *et al.*, 2002).

S. glaseri was superior to *Heterorhabditis indica* and *H. bacteriophora* (1000 IJs/pupae) in the control of pupae of the white grub, *Holotrichia serrata*. The nematodes *S. glaseri* and *H. indica* were also found to be pathogenic to the adults of *H. serrata*. The injection of *S. glaseri* (1000 IJs/beetle) into the beetle caused 100% mortality of the beetle in 3 days, while the same nematode when applied through soil caused mortality of the beetle in 6 days. Similarly, the application of *H. indica* through injection and soil application caused mortality of the beetle in 4 and 5 days, respectively (Sankaranarayanan *et al.*, 2006).

The soil-dwelling stages of the striped flea beetle, *Phyllotreta striolata* (Coleoptera: Chrysonelidae), infesting cabbage fields was controlled effectively with *S. carpocapsae* and *H. indica* (0.75×10^9 IJs/ha), which ultimately reduced the adult population in the field. Increased yield of the crops over the control were observed with the application of these EPNs (Yan *et al.*, 2013).

Steinernema spp. and *Heterorhabditis* sp. also exhibit efficient management of coleopteran pupae. The larvae of the plum curculio, *Conotrachelus nenuphar* (Coleoptera: Curculionidae), exit the fruit and enter into the soil for pupation. This is controlled effectively by the application of *Steinernema riobrave* and *H. indica* (Kim and Alston, 2008; Shapiro-Ilan *et al.*, 2008; Pereault *et al.*, 2009).

8.3 Entomopathogenic Nematodes for the Control of Weevils

Weevils are a type of beetle that belongs to the superfamily, Curculionoidea. Weevils are destructive insect pests of both field and horticultural crops. Both larval and adult stages are the infective stages in the weevils. Successful control of weevils like the black vine weevil, carrot weevil, sweet potato weevil, citrus root weevil, banana root weevil and large pine weevil by EPNs has been reported by several workers.

8.3.1 Black vine weevil (*Otiorhynchus sulcatus* Fabricius)

One of the most important weevil pest species is the black vine weevil, *Otiorhynchus sulcatus*, infesting annual and perennial ornamentals and small soft fruit crops in many parts of the world. It is a major pest on strawberry. Infestation of single larvae of this insect at the base of the stem can cause more damage to the plant than the presence of several larvae at the peripheral region of the plant. In cranberry, the application of *Steinernema carpocapsae* at 30 IJs/cm^2 reduced the infestation of black vine weevil by 96 or 100% (Booth *et al.*, 2002). Similarly, in strawberry, the application of *Heterorhabditis megidis* provided effective control of this larvae at temperatures above 10°C (Haukeland and Lola-Luz, 2010). *Heterorhabditis* species were found to be more effective than *Steinernema* species in the management of this weevil (Van Tol and Raupp, 2006). *Heterorhabditis* species have an active foraging strategy and move deeper into the soil, which makes them an ideal nematode for the management of black vine weevil.

Temperature plays a key role in the larval development of the insect, as larvae younger than its third-instar stages do not survive in low temperatures. The insect has adapted to this by the production of vigorous eggs early in the season, and the larvae enter into the overwintering stages. Control of the larvae can be achieved by reducing the larval population during autumn before they enter into the overwintering stage, or by targeting the larvae during the spring season before they resume feeding (Haukeland and Lola-Luz, 2010).

As temperature and soil type are the major critical factors for EPNs to control this larvae (Van Tol *et al.*, 2004), multiple applications of *H. megidis* in outdoor crops has resulted in the effective control of the larvae (Lola-Luz *et al.*, 2005; Lola-Luz and Downes, 2007). But this is not economically viable for all growers. A cold autumn and spring season are critical for nematodes, because the nematodes' efficacy is reduced during these seasons. The threshold temperature for *O. sulcatus* larvae activity ranges from 2 to 6°C (Stenseth, 1979), which reduces nematode infectivity in the larvae and hence takes longer to control the insect larvae (Long *et al.*, 2000). In the summer season, these nematodes exhibit poor virulence because of lack of available hosts and poor persistence in the soil (Van Tol and Raupp, 2006). Even when they did persist in the soil, after 3 months there appeared to be no significant reduction in the developing larvae of *O. sulcatus*.

8.3.2 Citrus root weevil (*Diaprepes abbreviates* L.)

Citrus root weevil, *Diaprepes abbreviates*, is the major pest of citrus trees in Florida. Grubs of this weevil are subterranean in nature and infest the roots of the trees severely, which results in their death. They lay their eggs on the leaves of the trees and, after hatching, enter into the soil. The application of infective stage EPNs during the entry of the larvae into the soil provides effective control of this insect. *S. carpocapsae* was the first nematode shown to be pathogenic to *D. abbreviates* and developed commercially for citrus root weevil control. Later, *S. carpocapsae* was replaced by *S. riobrave*, which caused greater mortality of *D. abbreviatus* under field conditions. The application of *S. riobrave* at 200 million nematodes per acre resulted in the effective control (85–95%) of this insect (Duncan and McCoy, 1996). Similarly, a single annual application of 2 million *S. riobrave*/tree reduced *D. abbreviatus* population by 85% over a period of 2 years (Bullock *et al.*, 1999).

S. riobrave and *H. indica* are commercially marketed in USA against this insect. These nematodes are highly effective at the temperature range of 27 ± 2°C under coarse sandy soil (McCoy *et al.*, 2007; Stuart *et al.*, 2008). Efficient management of pests can be achieved only when the EPNs are applied at the soil surface while the insects are transiting over it. The irrigation system also plays a major role in the delivery of EPN IJs into the zone where the larvae enter into the soil (Raja *et al.*, 2015).

In most studies, field application rates of EPNs have been relatively high compared to the label rate (Shapiro- Ilan *et al.*, 2002), but Stuart *et al.* (2008) reported that the application of *S. riobrave* twice a year at the label rate of 20,000 IJs/ft^2 (= 22 IJs/cm^2) significantly reduced root weevil emergence and enhanced the yield. But the result varied with the product quality, its application rates, suitability of edaphic condition for EPNs and the experimental methods used.

8.3.3 Red palm weevil (*Rhynchophorus ferrugineus* Olivier)

The red palm weevil, *Rhynchophorus ferrugineus*, is a most destructive pest affecting palm trees, including the coconut palm, date palm and oil palm. The weevil larvae make holes in the trunks of the trees and weaken the trees, which eventually kills the host trees. Even though the adults of the weevil cause damage to the trees, mortality of the trees occurs due to the burrowing of the larvae into the tree trunk, which ultimately leads to the death of the trees. The cryptic feeding habit exhibited by the larvae makes them difficult to control.

EPNs are suitable candidates for the management of *R. ferrugineus*. Natural occurrence of three nematodes, namely *Rhabditis* sp., *Teratorhabditis* sp. and *Mononchoides* sp., has been reported from the adult of *R. ferrugineus* (Banu and Rajendran, 2003a). Two native EPNs (*H. indica* and *Steinernema* sp.) and an exotic isolate, *S. glaseri*, were reported pathogenic to both grubs and adults of *R. ferrugineus* (Banu *et al.*, 2003). *H. indica* SA, isolated from Arabic Gulf countries, reduced the weevil population significantly, under both laboratory and field conditions. The two indigenous EPN isolates, namely *Steinernema* sp. SA (Saudi isolate), extracted from naturally infested red palm weevil adults, and *H. indica* HSA (Saudi isolate), extracted from soil samples, were tested for their virulence against the adults of red palm weevil in date palm trees in Qatar, Saudi Arabia. The nematode suspensions were sprayed at 2 million nematodes/5 l/tree. The spray was applied to the base and soil region around the tree. The weevils caught in the trap after the spray were counted. A reduction in emergence of about 32.87 and 35.5% was recorded with the application of *Steinernema* sp. SA and *H. indica* HSA, respectively, with the single spray. Among the two isolates, *Heterorhabditis* recorded better field performance over *Steinernema* species in the control of the weevils. The temperature tolerance nature of *Heterorhabditis* species, with its deeper reduction of weevil population, made this a suitable strain for the management of red palm weevils in date palm trees of Arabic countries (Saleh *et al.*, 2001).

Though various degrees of control have been achieved with EPNs against red palm weevils, it is still difficult to control this weevil with biocontrol agents. There are various reasons for this. One of the main reasons is the use of humoral and cellular responses of the insect against the biocontrol agents. Parasitic bacteria and fungi are killed by humoral defence responses, while EPNs are encapsulated by haemocytes (Schmidt *et al.*, 2001). A haemocyte-mediated immune response is part of the innate defence response that eliminates foreign materials through various mechanisms such as endocytosis, phagocytosis or encapsulation. Phagocytosis is the process of engulfing particles of the external environment into the cytoplasmic vesicles of the host insect, but this process is influenced by several factors like the type of foreign particles, biological inhibitors and the quantity of haemocytes. Changes in the quantity of haemocytes can inhibit the insect cellular immune defence system.

EPNs have evolved active and passive mechanisms to avoid the host immune response that prevents the host from recognizing the nematode

(Brivio *et al.*, 2002). Infectivity of *S. carpocapsae* to the late-instar larvae of *R. ferrugineus* recorded only a small number of live nematodes (4/20 ul) in the haemolymph of the insect larvae, The nematodes did not produce any offspring inside the host body (Triggiani and Tarasco, 2011) and were found not to be encapsulated by the haemocytes of the host (Manachini *et al.*, 2013). Parasitization of the nematode resulted in a decreased number of haemocytes, plasmatocytes and granulocytes of the host larva by which the EPNs evade the host immune response. Although nematodes suppress the host immune response to a certain extent, the inability of the EPN to replicate inside the host larvae suggests the poor potential of *S. carpocapsae* against the host.

8.3.4 Annual bluegrass weevil (*Listronotus maculicollis* Dietz)

EPNs are an important biocontrol agents of the blue grass weevil, *Listronotus maculicolis*, on golf courses in north-eastern USA. An acceptable level of control of the larvae of annual blue grass weevils in golf course turf has been achieved with the application of *S. carpocapsae*, *H. bacteriophora* and *H. megidis* (Vittum, 2005). Combined application of EPNs with the insecticide imidacloprid improved the efficacy of the EPNs against this pest. The application of *S. carpocapsae* (2.5 billion IJs/ha) and imidacloprid (336 g ai/ha) combinations resulted in a 77–78% reduction of blue grass weevil, whereas a nematode application alone recorded only a 50–58% reduction of the pest under field conditions in USA (Kostromytska *et al.*, 2015).

8.3.5 Bluegrass billbug (*Sphenophorus parvulus* Gyllenhal)

The bluegrass billbug, *Sphenophorus parvulus*, infests Kentucky grasses and perennial rye grasses in cool seasons and Bermuda grass and zoysia grass in warm seasons. *S. carpocapsae* and *H. bacteriophora* are equally effective in the management of larvae of this insect. Soil temperature is the critical factor in controlling the efficacy of the EPN against this pest. The application of EPNs in hot summer was found to be ineffective against newly hatched larvae of the host. Similarly, EPN application during the autumn season limits its efficacy because of the low prevailing temperature. Late spring was found to be the optimum period for EPNs to control the larvae of the weevil. These deficiencies have been overcome by the development of new EPN formulations that were found to be effective even at temperatures as low as 12°C (Georgis *et al.*, 2006).

8.4 Induced Defence Mechanism

In the subterranean environment, interactions between plants and insects induce signals in the plants that, in turn, attract EPNs. Thus, tritrophic interaction between plants, insects and nematodes in certain plants induce indirect defence mechanisms against the insects infesting the plants.

In citrus, infestation of root weevil larvae causes serious damage to the roots. The bioassay carried out in the root weevil infested citrus rootstock Swingle citrumelo revealed the emission of volatile compounds from its roots, which were identified as terpene compounds. These volatiles play a key role in the induction of indirect defence mechanisms by acting as anti-herbivore substances. In turn, these volatiles attract the EPN *Steinernema diaprepesi* towards them, which can control the *D. abbreviatus* larvae effectively (Ali *et al.*, 2010). Similarly, Thuja plants (*Thuja occidentalis*) infested with vine weevil larvae produced exudates that were found to attract *H. megidis* (Van Tol and Van der Sommen, 2001). Maize roots infested with Western corn rootworm (*Diabrotica virgifera*) were found to release the terpeniod compound, (E)-ß-caryophyllene, which attracts *H. megidis* towards them (Rasmann *et al.*, 2005).

Release of volatiles from the above plants was observed only during the infestation of the insects and not observed during the mechanical damage of the plants. These volatiles act as a signal clues for the attraction of EPN towards the plants. Thus, they play an indirect defence mechanism in insect management. By the exploitation of the natural compounds produced by the plants during the insect damage, the efficacy of EPN can be enhanced in order to get better control. Application of (E)-ß-caryophyllene (extracted from the Western corn rootworm damaged maize roots) directly into the soil was found to reduce the rootworm damage in the maize plants through enhanced activity of the EPN (Rasmann *et al.*, 2005). Further, advanced genetic researches were made to engineer the cultivable maize to release the compound (E)-ß-caryophyllene to protect them from insect damage by the recruitment of EPNs (Hiltpold *et al.*, 2010).

Green pea (*Pisum sativum*) secretes root exudates from the root cap cells to defend themselves against the invasion of plant pathogens and nematodes.

These root exudates in turn induce reversible quiescence in the EPN species. On storage of the root exudates of the plant at a lower concentration, significant increase in the activity and infectivity of the EPN was observed. Formulation of these root exudates at the appropriate concentration will favour the long-term storage of EPNs with increased effectiveness (Hiltpold *et al.*, 2015).

8.5 Factors Decreasing the Efficacy of Entomopathogenic Nematodes

The survival of nematodes in the host and in the soil (in the absence of an insect host) depends on several biotic and abiotic factors. These factors play a major role in interfering with the activity of EPNs against insects. Generally, the defence response of an EPN–bacterial complex against a wide range of insect orders varies significantly based on their symbiotic relationship, and the target pest should be fixed based on this complex reaction. Developmental stages of the host and recycling potential of the nematodes inside the host should be given importance to improve the effectiveness of the nematodes. Optimal environmental conditions are essential to get the maximum effect of the nematodes.

The following are the factors responsible for mitigating the efficacy of nematodes. These factors should be given priority to minimize their deleterious effects against the nematodes and to enhance the nematodes' defence reaction against the insects.

8.5.1 Biotic factors

The survival of EPNs is affected greatly by several biotic factors. The foraging behaviour of the nematodes is an important factor in determining their specificity against insects. Host suitability of the nematodes and its recycling potential in the host are the major biotic factors decreasing the nematodes' efficacy. The survival of nematodes in the soil is affected greatly by the presence of natural enemies in the soil. They reduce the EPN population in the soil drastically. The different biotic factors mentioned below are responsible for decreasing nematode efficacy against insects.

Host suitability

This is the predominant factor affecting the pathogenicity of EPNs. Although EPNs possess a wide spectrum of insecticidal activity under laboratory conditions, their virulence to particular insect hosts under field conditions is narrowed because of the prevailing ecological barriers.

Behaviour and physiological adaptation

For the effective management of coleopteran insects, the important criterion to be considered is the appropriate selection of nematode species to match with the target insects (Shapiro-Ilan *et al.*, 2012). Ambushing nematodes (stationary nematodes) could not be used to control sedentary subterranean insects like white grubs, as both nematodes and the host are relatively sedentary and inhabit different parts of the soil profile. In such cases, cruiser nematodes are better adapted to parasitize the white grubs. This was confirmed by the study conducted by Georgis and Gaugler (1991) with 82 EPNs against the larvae of the Japanese beetle. The nematode species used for the study did not reduce the larvae of the beetle because they were unsuitable. Thus, understanding nematode behaviour is very much essential to optimize host–parasite matches.

Physiological adaptation is one of the major factors affecting the efficacy of EPNs. For example, *S. carpocapsae* enters the insect body through the spiracles. But in the case of white grubs, their spiracles are covered with sieve plates (Forschler and Gardner, 1991) and it is not possible for the nematodes to enter into the insect host.

Host developmental stages

Nematode efficacy varies with the developmental stages of the host insect. In the case of white grubs, early instars are more susceptible to nematodes, and this susceptibility decreases when the insect grub grows larger, from the first instar to the successive stages. For example, decreased efficacy of *H. bacteriophora* was observed as *Anomala orientalis* grew from the first instar to further developmental stages. But *S. glaseri* and *S. scarabaei* were equally lethal to all the instars of white grubs (Koppenhofer and Fuzy, 2004). Thus, to get the maximum efficacy of the EPN, the selection of nematode species and the host's developmental stages are the factors that need to be given consideration.

Intra- and interspecific competition

Intraspecific competition occurs if too many IJs of the same species of EPN are present in the single host.

Regarding interspecific competition, Steinernematid and Heterorhabditid species cannot coexist in the same host because of the association of the symbiotic bacteria, so there is no advantage in using the Steinernematid and Heterorhabditid nematode for suppression of the single insect host (Wu, 2013).

Recycling potential

Generally, the effects of nematode treatment last in the soil for 2–6 weeks, and no long-term control of insects has been reported. The recycling potential of EPNs under field conditions is hampered by several biotic and abiotic factors, such as antagonism, predation, depletion of energy, soil temperature, moisture and humidity. This problem can be overcome by recycling the nematodes in host insects, which enables them to persist for longer.

High reproduction capacity of *S. carpocapsae* was obtained when they were reared on *Galleria mellonella* than on *Curculio caryae*. Poor nutritional factors of *C. caryae* may have contributed to the reduced reproduction rate of the nematodes in the host. The lipid content of the nematode also reduces their reproductive potential, as *S. carpocapsae* cultured in *C. caryae* were found to contain more lipid content in their body than *S. carpocapsae* cultured in *G. mellonella*. Nematodes obtained from one round of recycling in *C. caryae* larvae can be capable of killing more weevils, but their potential decrease on continuous recycling in the *C. caryae* larvae due to the reduced reproduction rate of the nematodes in the host (Shapiro-Ilan *et al.*, 2005).

The application of *H. bacteriophora* (HC1) to coffee soil managed the coffee berry borer (*Hypothenemus hampei*) population effectively. The nematode applied to the coffee soil was found to persist and was recycled in the soil for about 3 months after its application. The nematode was extracted from soil samples by the Baermann funnel method and inoculated in the last-instar larvae of *G. mellonella* to observe the virulence of the nematode in its post-application persistence. About 14–24% death of the host larvae was observed with the application of the nematode (Rodriguez *et al.*, 2011).

Foraging strategies

It is very important to understand the foraging strategies of EPNs in order to include them in biocontrol programmes. Nematodes exhibiting cruise foraging strategies can be used to control immobile and cryptic habit insects, while nematodes with ambush foraging strategies cannot be used to control the above-mentioned insects. A few cases may be an exception to this. For example, *S. carpocapsae* (ambushing nematode) can parasitize the larvae and pupae of the large pine weevil, *Hylobius abietis*, which lives under the bark of the tree (Dillon *et al.*, 2006).

Interaction with other soil biota

Other biotic factors that reduce the survival and infectivity of EPNs include the presence of nematophagous fungi and other predators, namely predatory nematodes, mites, protozoans, turbellarians, tardigrades, insects, etc. (Kaya and Koppenhofer, 1996). Depending on the interaction with other soil biota, it results in synergistic, antagonistic or additive interactions. EPN interaction with *Metarhizium anisopliaea* and *Paenibacillus papillae* resulted in a synergistic interaction, while its interaction with *Beauveria bassiana* resulted in an antagonistic interaction.

The presence of the saprobic fungus, *Fusarium oxysporum*, had a significant impact on the virulence and reproduction of *Heterorhabditis sonorensis*. Simultaneous application of the fungus and nematode resulted in an antagonistic interaction between them, which reduced the progeny of the EPN to half. The presence of fungal mycelia in the soil affects the ability of the nematode juvenile to locate the host insect, which results in reduced insect mortality. Simultaneous inoculation of the fungus and nematode gives the fungal spores a better chance to attach to the nematode cuticle, which results in the death of the nematode (Navarro *et al.*, 2014).

Interaction between the EPN, *S. feltiae*, and the earthworm, *Eisenia fetida*, resulted in reduced vigour index of the nematodes. Earthworms act as a vector in the dispersal of nematodes. Although nematodes had no deleterious effect on earthworms, the virulence of the nematodes was affected when they passed through the gut of the earthworm. The mortality of *Spodoptera littoralis* with *S. feltiae* was observed to be 0% after the passage of the nematode through the gut of the earthworm (Herrera *et al.*, 2006).

8.5.2 Abiotic factors

Generally, the survival of nematodes ranges from weeks to months. Abiotic factors play a significant role in affecting the survival and infectivity of nematodes against the insect host. Environmental factors such as ultraviolet (UV) radiation, soil moisture, soil temperature, desiccation, soil texture, soil pH and the agrochemicals used in the field have a greater impact in affecting the field efficacy of EPNs. The presence of organic matter interferes with the activity of the nematode and reduces its efficacy. Under favourable environmental conditions, EPNs produce a long-term effect in the management of insects (Susurluk, 2011). It has been observed that sandy soil/sandy loam soil having the temperature range of 15–25°C with low moisture content supports the maximum survival of EPNs (Smart, 1995). The following abiotic factors frequently interfere with the activity of nematodes.

Ultraviolet radiation

EPNs and their symbiotic bacteria are very sensitive to sunlight, especially their UV radiation (Banu and Rajendran, 2003b). A minimum of 4 min exposure of *H. bacteriophora* to medium-wave UV light caused significant loss in the pathogenicity of the nematode, whereas *S. carpocapsae* lost their efficacy after 6 min exposure to UV light. Moreover, application of the nematode in direct sunlight reduced their virulence by about 6.6 and 94.9% on an exposure period of 30 and 60 min, respectively. Hence, it is advisable to apply the EPN formulation for the control of soilborne pests in the field either in the morning or in the evening, to avoid exposure to UV radiation (Kaya and Gaugler, 1993; Fujiie and Yokoyama, 1998). The inability of EPNs to withstand UV radiation is the main drawback associated with their use against foliar insect pests. To expand the use of nematodes against a wide range of pests, the problem of their sensitivity to UV radiation needs to be solved.

Tinopal LPW, a stilbene fluorescent brightener, was proved to be an effective UV protectant for the EPN, *S. carpocapsae*. Tinopal LPW at 1% concentration was found to be non-toxic to nematodes. Nematodes added in 1% Tinopal LPW maintained their original activity remaining (OAR) over a period of 8 h under a sun lamp in laboratory conditions and for 4 h in direct sunlight under field conditions. Tinopal LPW is commonly used in soaps and detergents. It absorbs UV light and converts it into visible light. On application, this product was found to be harmless to nematodes and their symbiotic bacteria (Nickle and Shapiro, 1992).

Soil moisture

Soil moisture is an important critical factor for nematode mobility and infection (Koppenhofer and Fuzy, 2007). Infectivity of the nematode starts to decline in low soil moisture, which prevents the nematode from movement and host infection. Insecticidal activity of *Steinernema kushidai* against the white grub *Anomala caprea* infesting turfgrass was increased with the increase in soil moisture from 10 to 40% (Fujiie *et al.*, 1996). Saturated soil creates anaerobic conditions and prevents nematode movement. In field conditions, soil should be kept moist after the application of EPNs for the effective functioning of the nematode against insects.

Soil temperature

Both high and low temperatures are found detrimental to EPNs. Most *S. feltiae* are found to be effective at the temperature range of 2–30°C, while *Heterorhabditis* can infect the host at a temperature range of 7–35°C. At a temperature of 10°C, *S. carpocapsae* will become inactive (Georgis *et al.*, 2006; Lacey *et al.*, 2006). Similarly, at a soil temperature below 20°C, *H. bacteriophora* become ineffective in white grub control (Georgis and Gaugler, 1991). But a few exceptional nematodes are more cold or heat tolerant than other nematodes. For example, *S. feltiae* is more freeze tolerant and can infect insects even at a soil temperature below 10°C (Chen *et al.*, 2003). A temperature above 30°C reduces the efficacy of most EPNs, while some species like *S. riobrave* are effective even at temperatures above 35°C (Lacey and Unruh, 1998).

Under field conditions, nematode efficacy declines below 16°C. *S. carpocapsae* and *H. bacteriophora* can cause a more significant reduction of June beetle populations at a temperature of 25°C than at 12°C (Townsend *et al.*, 1998). The storage of EPNs at a very low temperature prior to their application reduces their infectivity (Kaya and Gaugler, 1993).

Soil texture and pH

Clay soils limit nematode movement, since they have a small pore size. EPNs occur more frequently in sandy loam soil than in sandy soil (Banu *et al.*, 2002). Nematode efficacy decreases with an increase in the clay content in soil, whereas in fine-textured soil, nematode efficacy increases. It has been observed that *H. bacteriophora* is more effective against *P. japonica* in fine-textured soil, as fine-textured soil retains moisture better than coarse-textured soil (Georgis and Gaugler, 1991).

A pH value between 4 and 8 had little or no impact on nematode survival (Kaya, 1990). Soil salinity and acidity beyond this range will have a negative impact on the survival and infectivity of nematodes. Infection of *H. bacteriophora*, *H. zealandica*, *S. glaseri* and *S. scarabaei* was found to be lowest in acidic sandy soil with a pH of 3.9 under laboratory conditions (Koppenhofer and Fuzy, 2006).

Soil organic matter

The accumulation of organic matter between the soil and turfgrass foliage restricts the downward movement of nematodes (Georgis and Gaugler, 1991). The thickness of the organic matter is correlated negatively with nematode efficacy. Organic matter increases the populations of predators and pathogens of EPNs in the soil, which in turn decreases the nematode population in the soil. On the other hand, a report of Kaya (1990) suggested that increased abundance of arthropod hosts in organic soil sustained or enhanced the EPN population in the soil. Regarding movement, *H. bacteriophora* was found to be more motile in organic soil than *S. carpocapsae* (Barbercheck and Kaya, 1991).

Volume and frequency of irrigation

Soil moisture is one of the important critical factors for the movement of nematodes, which can be retained by the supply of an appropriate quantity of water through irrigation at regular intervals. Generally, nematodes need a thin film of water in the interstitial spaces of the soil for their effective propulsion. Irrigation frequency is correlated positively with the efficacy of EPNs. Post-application irrigation of a minimum 0.74 cm is required for the establishment of nematodes in turfgrass (Shetlar *et al.*, 1988). Pre-EPN application irrigation is generally required for the effective infectivity of EPNs. The amount of post-EPN release irrigation water is a crucial factor affecting EPN activity. The application of too much water makes the nematodes drain away, and too little water allows them to remain on the soil surface. Hence, the appropriate quantity of water is a must for the effective functioning of EPNs.

Improving the efficacy of entomopathogenic nematodes against coleopteran pests

The selection of the proper species of EPN with a higher virulence and persistence will result in the effective control of insects. The efficacy of EPN strains can also be improvised by genetic engineering methods, alteration of the soil environment and the addition of adjuvants to the EPN formulation.

Persistence of entomopathogenic nematodes in coleopteran hosts

EPNs have the potential to persist in the environment when used as biocontrol agents against insect pests. But their long-term persistence is limited. A laboratory experiment conducted on the longevity of 29 strains of 11 species of EPN revealed the maximum survival capability of *S. carpocapsae* (Sal) and *S. diaprepesi*, followed by *H. bacteriophora* (Lewiston), *H. megidis*, *S. feltiae* and *S. riobrave* (3–3 and 355). Lower levels of survivability were observed in other *H. bacteriophora* strains (Hb, HP88 and Oswego), as well as in *S. glaseri* and *S. rarum*. Relative to *S. glaseri* and *S. rarum*, a lower range of longevity was observed in *H. indica*, *H. marelatus* and *H. mexicana*, respectively. *S. glaseri* was re-isolated from the field after 5 years from the rhizosphere of cardamom, where it was introduced for the management of *Basilepta felvicorne* (Banu *et al.*, 2002). Although nematode persistence can vary in laboratory and field conditions, basic data on their longevity will be more helpful in choosing the best match of EPN for a particular target pest (Shapiro-Ilan *et al.*, 2006). A single application of EPN can impact more than one generation of white grubs (Grewal, 2001), as *S. glaseri* has been reported to maintain in the field in New Jersey, USA, for about 14 years, provided the density of Japanese beetle larva in the field is less than five per square foot.

8.6 Integrated Management of Coleopteran Pests

To obtain better control and management of crop pests, we need to utilize EPNs in combination with insecticides. A deep knowledge on the compatibility of EPNs with insecticides is indeed needed for the effective utilization of the combined products. Compatibility is not only species specific but also strain specific. Although EPNs are compatible with most pesticides, fungicides, herbicides, fertilizers, adjuvants, surfactants, plant extracts and biocontrol agents (Nishimatsu and Jackson, 1998; Scheepmaker *et al.*, 1998), they are sensitive to certain insecticides, wetting agents and surfactants.

Increased efficacy in the control of white grubs was observed in the combined application of EPNs (*S. glaseri* and *H. bacteriophora*) with *Bacillus thuringiensis* (Koppenhofer *et al.*, 1997) and the insecticide imidacloprid (Koppenhofer *et al.*, 2000). *S. feltiae* showed a level of direct compatibility with the insecticide pirimicarb, whereas *S. carpocapsae* and *S. kraussei* are sensitive to pirimicarb and Azadirachtin (Laznik and Trdan, 2014). Dimethoate and imidacloprid were compatible with *S. feltiae*, while cypermethrin was proven to be non-compatible for direct mixing with *S. feltiae* (Cuthbertson *et al.*, 2007).

The nature of the interactions between EPNs and other entomopathogens varies depending on the nematode species and the relative timing or rate of application (Koppenhofer and Kaya, 1997). A synergistic interaction was observed between EPNs and other entomopathogens such as *Paenibacillus popilliae*, *B. thuringiensis* and *Metarhizium anisopliae* (Ansari *et al.*, 2004), while antagonistic interaction was observed between EPNs and *B. bassiana* or *Paecilomyces fumosoroseus* (Shapiro-Ilan *et al.*, 2004). Application of the fungus, *M. anisopliae*, 1 or 2 weeks prior to the application of an EPN (*H. bacteriophora*, *S. feltiae* or *S. kraussei*) provided 100% mortality of the third-instar larvae of *O. sulcatus* (Ansari *et al.*, 2008). A good combination of EPNs, insecticides and other biocontrol agents will surely provide a better control of insects in IPM packages.

8.7 Selective Breeding of Entomopathogenic Nematodes for Insect Pest Control

Breeding of EPNs selectively to increase their responsiveness to foraging cues will surely enhance the efficacy of EPNs in insect management. In maize, damage of the roots by the Western corn rootworm (Coleoptera: Chrysomelidae) emits a compound, (E)-ß-caryophyllene, which attracts *H. bacteriophora* towards the roots. It is the most infectious nematode against the worm, but its attraction towards the roots was found to be very weak. To overcome this drawback, artificial selection of *H. bacteriophora* strain was made. After six generations of selection, the resultant strain was found to be twice as effective as its original form, which revealed that the manipulation of EPNs would give a greater potential in the control of insects (Hiltpold *et al.*, 2010).

Through artificial selection, many beneficial traits of EPNs, such as host finding, enhanced virulence and tolerance to environmental extremities (Grewal *et al.*, 1996; Ehlers *et al.*, 2005), have been improved. The enhancement of the beneficial traits in EPNs under laboratory conditions may alter some other traits negatively, such as storage stability under field conditions. Hence, while breeding to increase the virulence of a strain, other important traits should also be considered.

The virulence of EPNs and their symbiotic bacteria can be enhanced through genetic engineering methods. Physiological and behavioural traits of EPNs can be modified/improved by these methods. The main target in the genetic improvement of EPNs is to increase their efficacy to compete with insecticides, resistance to environmental extremes and to increase their suitability to culture in liquid formulations. Symbiotic bacteria associated with EPNs also need to be improved for the above traits.

S. glaseri can infest Japanese beetle larvae easily, but it is difficult to mass produce them, while *H. bacteriophora* has poor efficacy against the beetle but can be cultured easily. For successful infection by EPN, it needs to disable or overwhelm the insect's immune response towards it. The incorporation of anti-immune, surface-coat protein gene from *S. glaseri* into immune-susceptible *H. bacteriophora* increases its pathogenicity towards the Japanese beetle, which is possible by genetically engineered *H. bacteriophora* in laboratory experiments. This functional genomics approach will enhance the basic understanding of the specificity of the nematode–host interaction that is mediated by the insect's response.

By combining the genetic engineering and artificial selection strategies, improvement in the infectivity of the nematode and its persistence in the soil can be achieved, which results in effective, economical and ecologically safer methods of insect management.

8.8 Conclusion and Future Prospects

The application of EPNs are highly lethal to coleopteran insect pests as they kill the insect hosts within a short period of 24–48 h of infection when compared to other biocontrol agents, which take about days to a week to kill the insect hosts. Moreover, EPNs were found to be compatible with other biocontrol agents, fungicides, insecticides and fertilizers. The application of EPNs is an environmentally friendly and cost-effective method in insect control. They can be mass produced on a large scale through *in vitro* and *in vivo* methods (Askary, 2010).

The differences in virulence of EPNs with the larval stages of grubs should be considered when developing the biocontrol formulation of EPNs against white grubs. Advances in mass production technologies of EPNs may create an interest in the commercialization of the products. Many EPN-based commercial products are available in countries like the USA, Australia and Europe against various insects like fruit weevil, mealybug, codling moth, etc. (Van Zyl and Malan, 2014). The application of EPNs has potential benefit over their chemical counterparts by reducing non-target impacts, increasing resistance management and safety applications. Considering the above, EPNs should be added as one of the important components in sustainable IPM programmes.

The potential use of EPNs with reduced production costs and more pathogenicity is very bright for the near future. Advanced research is needed in general on nematode physiology and biochemistry to develop novel formulations of EPNs with enhanced virulence under field conditions by exploiting the various strains of EPNs with good potential in insect control. Also, attention should be given on the compatibility of EPNs with other chemical insecticides. Genetic engineering studies on nematodes will have to be improved to combat the environmental extremities and to obtain more virulent strains in insect control. Studies on the tritrophic interaction between plants, insects and nematodes at the molecular level will aid in the development of effective formulations with efficient EPN strains to manage/control insects. Thus, focus should be give in near future for the development of insect specific nematode strains. The impact of EPNs through defence volatile production in plants should also be improved for the management of coleopteran insects in perennial crops, which will be rather more advantageous than in annual crops.

Regarding commercial products in EPNs, efforts should be taken to optimize formulations for longer shelf life. Emphasis should be placed on low-cost application techniques for the release of nematodes into the soil and to protect them during storage. Despite the regular formulations, concern should be given to the development of new, attractive and cost-effective formulations like encapsulation and nano products of EPNs, which might be an effective alternative to conventional pest control methods and will increase the efficacy of EPNs in the long run. The development of EPNs in the control of insects under complex environmental situations with low-cost application methods, along with long-term storage and persistence, is the need of the hour and should be fulfilled in the near future through genetic engineering methods.

Acknowledgements

I am grateful to the Director, ICAR-Central Institute for Cotton Research, Nagpur, for permitting me to compile this chapter. Thanks are also due to Dr K. Sankari Meena, Senior Research Fellow, ICAR-Central Institute for Cotton Research, Regional Station, Coimbatore, Tamil Nadu, India, for the timely help rendered in the preparation of this chapter.

References

Ali, J.G., Alborn, H.T. and Stelinski, L.L. (2010) Subterranean herbivore-induced volatiles released by citrus roots upon feeding by *Diaprepes abbreviatus* recruit entomopathogenic nematodes. *Journal of Chemical Ecology* 36, 361–368.

Ansari, M.A., Tirry, L. and Moens, M. (2004) Interaction between *Metarhizium anisopliae* CLO53 and entomopathogenic nematodes for the control of *Hoplia philanthus*. *Biological Control* 31, 172–180.

Ansari, M.A., Shah, F.A. and Butt, T.M. (2008) Combined use of entomopathogenic nematodes and *Metarhizium anisopliae* as a new approach for black vine weevil, *Otiorhynchus sulcatus*, control. *Entomologia Experimentalis et Applicata* 129, 340–347.

Askary, T.H. (2010) Nematodes as biocontrol agents. In: Lichtfouse, E. (ed.) *Sociology, Organic Farming, Climate Change and Soil Science*. Springer, The Netherlands, pp. 347–378.

Atwa, A.A. and Hassan, S.H. (2014) Bioefficacy of two entomopathgenic nematodes against *Spodoptera littoralis* Boisduval (Lepidoptera) and *Temnorhynchus baal* Reiche (Coleoptera) larvae. *Journal of Biopesticides* 7, 104–109.

Banu, J.G. and Rajendran, G. (2003a) Nematodes associated with *Rhynchophorus ferrugineus* in Kerala. *Journal of Experimental Zoology (India)* 6, 99–100.

Banu, J. and Rajendran, G. (2003b) Effect of UV radiation on the survival and infectivity of three entomopathogenic nematodes. *Indian Journal of Nematology* 33, 86–87.

Banu, J.G., Khuong, B., Nguyen, K.B. and Rajendran, G. (2002) Occurrence and distribution of entomopathogenic nematodes in Kerala. *International Journal of Nematology* 15, 9–16.

Banu, J.G., Rajendran, G. and Subramanian, S. (2003) Susceptibility of red weevil *Rhynchophorus ferrugineus* (Oliv.) to entomopathogenic nematodes. *Annals of Plant Protection Sciences* 11, 104–106.

Banu, J.G., Subaharan, K. and Iyer, R. (2004) Occurrence and distribution of entomopathogenic nematodes in white grub endemic areas of Kerala. *Journal of Plantation Crops* 32(S), 333–334.

Barbercheck, M.E. and Kaya, H.K. (1991) Effect of host condition and soil texture on host finding by the entomogenous nematodes *Heterorhabditis bacteriophora* (Rhabditida: Heterorhabditidae) and *Steinernema carpocapsae* (Rhabditida: Steinernematidae). *Environmental Entomology* 20, 582–589.

Booth, S.R., Tanigoshi, L.K. and Shanks, C.H. (2002) Evaluation of entomopathogenic nematodes to manage root weevil larvae in Washington state cranberry, strawberry, and red raspberry. *Environmental Entomology* 31, 895–902.

Brivio, M.F., Pagani, M. and Restelli, S. (2002) Immune suppression of *Galleria mellonella* (Insecta, Lepidoptera) humoral defenses induced by *Steinernema feltiae* (Nematoda Rhabditida): involvement of the parasite cuticle. *Experimental Parasitology* 101, 149–156.

Bullock, R.C., Pelosi, R.R. and Killer, E.E. (1999) Management of citrus root weevils (Coleoptera: Curculionidae) on Florida citrus with soil-applied entomopathogenic nematodes (Nematoda: Rhabditida). *Florida Entomologist* 82, 1–7.

Chen, S., Li, J., Han, X. and Moens, M. (2003) Effect of temperature on the pathogenicity of entomopathogenic nematodes (*Steinernema* and *Heterorhabditis* spp.) to *Delia radicum*. *Biological Control* 48, 713–724.

Cuthbertson, A.G.S., Mathers, J.J., Northing, P., Luo, W. and Walters, K.F.A. (2007) Establishing effect of commonly used insecticides for aphid control on the infectivity of the entomopathogenic nematode, *Steinernema feltiae* using a streamlined screening method. *Estudos de Biologia* 29, 17–21.

Dillon, A.B., Ward, D., Downes, M.J. and Gri, C.T. (2006) Suppression of the large pine weevil, *Hylobius abietis* (L.) (Coleoptera: Curculionidae) in pine stumps by entomopathogenic nematodes with different foraging strategies. *Biological Control* 38, 217–226.

Duncan, L.W. and McCoy, C.W. (1996) Vertical distribution in soil, persistence, and efficacy against citrus root weevil (Coleoptera: Curculionidae) of two species of entomogenous nematodes (Rhabditida: Steinernematidae: Heterorhabditidae). *Environmental Entomology* 25, 174–178.

Ebeid, A.R., Abdel-Rahman, A.A. and Gesrah, M.A. (2014) Virulence of native entomopathogenic nematodes against the spotted oriental cucumber beetle, *Epilachna chrysomelina* (F.) (Coleoptera: Coccinellidae) in laboratory and semi-field experiments. *Egyptian Journal of Biological Pest Control* 24, 53–56.

Ehlers, R.U., Oestergaard, J., Hollmer, S., Wingen, M. and Strauch, O. (2005) Genetic selection for heat tolerance and low temperature activity of the entomopathogenic nematode –bacterium complex *Heterorhabditis bacteriophora–Photorhabdus luminescens*. *Biocontrol* 50, 699–716.

Erbas, Z., Gokce, C., Hazir, S., Demirbag, Z. and Demir, I. (2014) Isolation and identification of entomopathogenic nematodes (Nematoda: Rhabditida) from the Eastern Black Sea region and their biocontrol potential against *Melolontha melolontha* (Coleoptera: Scarabaeidae) larvae. *Turkish Journal of Agriculture and Forestry* 38, 187–197.

Forschler, B. and Gardner, W. (1991) Parasitism of *Phyllophaga hirticula* by *Heterorhabditis heliothidis* and *Steinernema carpocapsae*. *Journal of Invertebrate Patholo*gy 58, 396–407.

Fujiie, A. and Yokoyama, T. (1998) Effects of ultraviolet light on the entomopathogenic nematode, *Steinernema kushidai* and its symbiotic bacterium, *Xenorhabdus japonicus*. *Applied Entomology Zoology* 33, 263–269.

Fujiie, A., Takata, Y., Tachibana, M. and Yokoyama, T. (1996) Insecticidal activity of an entomopathogenic nematode, *Steinernema kushidai* (Nematoda: Steinernematidae) against *Anomala cuprea* (Coleoptera: Scarabaeidae) larvae under different soil moisture conditions. *Applied Entomology Zoology* 31, 453–455.

Gaugler, R., Campbell, J.F., Selvan, S. and Lewis, E.E. (1992) Large-scale inoculative releases of the entomopathogenic nematode *Steinernema glaseri*: assessment 50 years later. *Biological Control* 2, 181–187.

Gaugler, R., Wang, Y. and Campbell, J. (1994) Aggressive and evasive behaviors in *Popillia japonica* (Coleoptera: Scarabaeidae) larvae: defense against entomopathogenic nematode attack. *Journal of Invertebrate Pathology* 64, 193–199.

Georgis, R. and Gaugler, R. (1991) Predictability in biological control using entomopathogenic nematodes. *Journal of Economic Entomology* 84, 713–720.

Georgis, R., Koppenhofer, A.M., Lacey, L.A., Belair, G., Duncan, L.W., *et al.* (2006) Successes and failures in the use of parasitic nematodes for pest control. *Biological Control* 38, 103–123.

Gressitt, J.L. (2015) Coleopteran insects. Available at: www.britannica.com/animal/beetle (accessed 14 September 2016).

Grewal, P.S. (2001) Using entomopathogenic nematodes for the turf grass pest management. *Turfgrass Trends* 10, 1–7.

Grewal, P.S., Gaugler, R. and Wang, Y. (1996) Enhanced cold tolerance of the entomopathogenic nematode, *Steinernema feltiae* through genetic selection. *Annals of Applied Biology* 129, 335–341.

Grewal, P.S., Power, K.T., Grewal, S.K., Suggars, A. and Haupricht, S. (2004) Enhanced consistency in biological control of white grubs (Coleoptera: Scarabaeidae) with new strains of entomopathogenic nematodes. *Biological Control* 30, 73–82.

Grewal, P.S., Koppenhofer, A.M. and Choo, H.Y. (2005) Lawn, turfgrass, and pasture applications. In: Grewal, P.S., Ehlers, R.-U. and Shapiro-Ilan, D.I. (eds) *Nematodes as Biocontrol Agents.* CAB International, Wallingford, UK, pp. 115–146.

Guo, W., Yan, X., Zhao, G. and Han, R. (2013) Efficacy of entomopathogenic *Steinernema* and *Heterorhabditis* nematodes against white grubs (Coleoptera: Scarabaeidae) in peanut fields. *Journal of Economic Entomology* 106, 1112–1117.

Guo, W., Yan, X., Zhao, G., Chen, J. and Han, R. (2015) Efficacy of entomopathogenic *Steinernema* and *Heterorhabditis* nematodes against *Holotrichia oblita*. *Journal of Pest Science* 88, 359–368.

Haukeland, S. and Lola-Luz, T. (2010) Efficacy of the entomopathogenic nematodes *Steinernema kraussei* and *Heterorhabditis megidis* against the black vine weevil *Otiorhynchus sulcatus* in open field-grown strawberry plants. *Agricultural and Forest Entomology* 12, 363–369.

Herrera, R.C., Trigo, D. and Gutierrez, C. (2006) Phoresy of the entomopathogenic nematode, *Steinernema feltiae* by the earthworm, *Eisenia fetida. Journal of Invertebrate Pathology* 92, 50–54.

Hiltpold, I., Baroni, M., Toepfer, S., Kuhlmann, U. and Turlings, T.C.J. (2010) Selection of entomopathogenic nematodes for enhanced responsiveness to a volatile root signal helps to control a major root pest. *Journal of Experimental Biology* 213, 2417–2423.

Hiltpold, I., Jaffuel, G. and Turlings, T.C.J. (2015) The dual effects of root-cap exudates on nematodes: from quiescence in plant-parasitic nematodes to frenzy in entomopathogenic nematodes. *Journal of Experimental Botany* 66, 603–611.

Kaya, H.K. (1990) Soil ecology. In: Gaugler, R. and Kaya, H.K. (eds) *Entomopathogenic Nematodes in Biological Control*. CRC Press, Boca Raton, Florida, pp. 93–115.

Kaya, H.K. and Gaugler, R. (1993) Entomopathogenic nematodes. *Annual Review of Entomology* 38, 181–206.

Kaya, H.K. and Koppenhofer, A.M. (1996) Effects of microbial and other antagonistic organism and competition on entomopathogenic nematodes. *Biocontrol Science and Technology* 6, 357–371.

Kim, H.G. and Alston, D.G. (2008) Potential of two entomopathogenic nematodes for suppression of *Plum curculio* (*Conotrachelus nenuphar*, Coleoptera: Curculionidae) life stages in northern climates. *Environmental Entomology* 37, 1272–1279.

Klein, M.G., Grewal, P.S., Jackson, T.A. and Koppenhofer, A.M. (2007) Lawn, turf and grassland pests. In: Lacey, L.A. and Kaya, H.K. (eds) *Field Manual of Techniques in Invertebrate Pathology: Application and Evaluation of Pathogens for Control of Insects and other Invertebrate Pests*. Springer, Dordrecht, The Netherlands, pp. 655–675.

Koppenhofer, A.M. and Fuzy, E.M. (2003a) *Steinernema scarabaei* for the control of white grubs. *Biological Control* 28, 47–59.

Koppenhofer, A.M. and Fuzy, E.M. (2003b) Biological and chemical control of the Asiatic garden beetle, *Maladera castanea* (Coleoptera: Scarabaeidae). *Journal of Economic Entomology* 96, 1076–1082.

Koppenhofer, A.M. and Fuzy, E.M. (2004) Effect of white grub developmental stage on susceptibility to entomopathogenic nematodes. *Journal of Economic Entomology* 97, 1842–1849.

Koppenhofer, A.M. and Fuzy, E.M. (2006) Effect of soil type on infectivity and persistence of the entomopathogenic nematodes *Steinernema scarabaei*, *Steinernema glaseri*, *Heterorhabditis zealandica* and *Heterorhabditis bacteriophora*. *Journal of Invertebrate Pathology* 92, 11–22.

Koppenhofer, A.M. and Fuzy, E.M. (2007) Soil moisture effects on infectivity and persistence of the entomopathogenic nematodes *Steinernema scarabaei*, *Steinernema glaseri*, *Heterorhabditis zealandica*, and *Heterorhabditis bacteriophora*. *Applied Soil Ecology* 35, 128–139.

Koppenhofer, A.M. and Kaya, H.K. (1997) Additive and synergistic interaction between entomopathogenic nematodes and *Bacillus thuringiensis* for scarab grub control. *Biological Control* 8, 131–137.

Koppenhofer, A.M., Baur, M.E., Stock, S.P., Choo, H.Y., Chinnasri, B. and Kaya, H.K. (1997) Survival of entomopathogenic nematodes within host cadavers in dry soil. *Applied Soil Ecology* 6, 231–240.

Koppenhofer, A.M., Grewal, P.S. and Kaya, H.K. (2000) Synergism of imidacloprid and entomopathogenic nematodes against white grubs: the mechanism. *Entomologia Experimentalis et Applicata* 94, 283–293.

Koppenhofer, A.M., Grewal, P.S. and Fuzy, E.M. (2006) Virulence of the entomopathogenic nematodes *Heterorhabditis bacteriophora*, *Heterorhabditis zealandica*, and *Steinernema scarabaei* against five white grub species (Coleoptera: Scarabaeidae) of economic importance in turfgrass in North America. *Biological Control* 38, 397–404.

Kostromytska, O., Albrecht, M., Koppenhofer, A.M. and Saona, C.R. (2015) Advancing integrated management of annual bluegrass weevil. *Turfgrass and Environmental Research Online* 14, 19–21.

Lacey, L.A. and Georgis, R. (2012) Entomopathogenic nematodes for control of insect pests of above and below ground with comments on commercial production. *Journal of Nematology* 44, 218–225.

Lacey, L.A. and Unruh, T.R. (1998) Entomopathogenic nematodes for control of coding moth: effect of nematode species, dosage, temperature and humidity under laboratory conditions. *Biological Control* 13, 190–197.

Lacey, L.A., Arthurs, S.P., Unruh, T.R., Headrick, H. and Fritts, R. (2006) Entomopathogenic nematodes for control of codling moth (Lepidoptera: Tortricidae) in apple and pear orchards: effect of nematode species and seasonal temperatures, adjuvants, application equipment and post-application irrigation. *Biological Control* 37, 214–223.

Laznik, Z. and Trdan, S. (2014) The influence of insecticides on the viability of entomopathogenic nematodes (Rhabditida: Steinernematidae and Heterorhabditidae) under laboratory conditions. *Pest Management Science* 70, 784–789.

Lola-Luz, T. and Downes, M. (2007) Possibilities for biological control of black vine weevil, *Otiorhynchus sulcatus* in Ireland using entomopathogenic nematodes. *Biological Control* 40, 314–319.

Lola-Luz, T., Downes, M. and Dunne, R. (2005) Control of black vine weevil larvae *Otiorhyncus sulcatus* (Fabricius) (Coleoptera:Curculionidae) in grow bags outdoors with nematodes. *Agricultural and Forest Entomology* 7, 121–126.

Long, S.J., Richardson, P.N. and Fenlon, J.S. (2000) Influence of temperature on the infectivity of entomopathogenic nematodes (*Steinernema* and *Heterorhabditis* spp.) to larvae and pupae of the vine weevil, *Otiorhynchus sulcatus* (Coleoptera: urculionidae). *Nematology* 2, 309–317.

Lortkipanidze, M., Gorgadze, O., Kokhia, M., Melashvili, N. and Kuchava, M. (2011) Effectiveness of entomopathogenic nematodes (*Steinernemacarpocapsae*) against the *Melolontha hippocastani* (Coleoptera: Scarabaeidae). *Bulletin of the Georgian National Academy of Sciences* 5, 155–157.

McCoy, C.W., Stuart, R.J., Duncan, L.W. and Shapiro-Ilan, D.I. (2007) Application and evaluation of entomopathogens for citrus pest control. In: Lacey, L.A. and Kaya, H.K. (eds) *Field Manual of Techniques in Invertebrate Pathology: Application and Evaluation of Pathogens for Control of Insects and other Invertebrate Pests*, 2nd edn. Springer, Dordrecht, The Netherlands, pp. 567–581.

Mahar, A.N., Jan, N.D. and Mahar, A.Q. (2012) Comparative effectiveness of entomopathogenic nematodes against the pupae of mustard beetle, *Phaedon cochleariae* F. (Chrysomelidae: Coleoptera). *Pakistan Journal of Zoology* 44, 517–523.

Manachini, B., Schillaci, D. and Arizza, V. (2013) Biological responses of *Rhynchophorus ferrugineus* (Coleoptera: Curculionidae) to *Steinernema carpocapsae* (Nematoda: Steinernematidae). *Journal of Economic Entomology* 106, 1582–1589.

Mwaitulo, S., Haukeland, S., Saehre, G., Laudisoit, A. and Maerere, A.P. (2011) First report of entomopathogenic nematodes from Tanzania and their virulence against larvae and adults of the banana weevil, *Cosmopolites sordidus* (Coleoptera: Curculionidae). *International Journal of Tropical Insect Science* 31, 154–161.

Navarro, P.D., McMullen, J.G. and Stock, S.P. (2014) Interactions between the entomopathogenic nematode, *Heterorhabditis sonorensis* (Nematoda, Heterorhabditidae) and the saprobic fungus, *Fusarium oxysporum* (Ascomycota: Hypocreales). *Journal of Invertebrate Pathology* 115, 41–47.

Nickle, W.R. and Shapiro, M. (1992) Use of a stilbene brightener, tinopal LPW, as a radiation protectant for *Steinernema carpocapsae. Journal of Nematology* 24, 371–373.

Nishimatsu, T. and Jackson, J.J. (1998) Interaction of insecticides, entomopathogenic nematodes and larvae of the Western corn rootworm (Coleoptera: Chrysomellidae). *Journal of Economic Entomology* 91, 410–418.

Pereault, R.J., Whalon, M.E. and Alston, D.G. (2009) Field efficacy of entomopathogenic fungi and nematodes targeting caged last-instar plum curculio (Coleoptera: Curculionidae) in Michigan cherry and apple orchards. *Environmental Entomology* 38, 1126–1134.

Raja, R.K., Hazir, C., Gumus, A., Asan, C., Karagoz, M. and Hazir, S. (2015) Efficacy of the entomopathogenic nematode, *Heterorhabditis bacteriophora* using different application methods in the presence or absence of a natural enemy. *Turkish Journal of Agriculture and Forestry* 39, 277–285.

Rasmann, S., Kollner, T.G., Degenhardt, J., Hiltpold, I., Toepfer, S., *et al*. (2005) Recruitment of entomopathogenic nematodes by insect-damaged maize roots. *Nature* 434, 732–737.

Rodríguez, M.G., Hernández, M., Borrero, Y., Gómez, L. and Enrique, R. (2011) Recycle of entomopathogenic nematodes in soils grower coffee crop (*Coffea* spp.) in Buey Arriba. *Revista de Proteccion Vegetal*. Available at: https://www.scienceopen.com/document?vid=2b29bde1-e776-4975-a555-1d1e5d6e4e40 (accessed 22 September 2016).

Saleh, M.M.E., Hanounik, S.B., Al-Muhanna, U.E., Al-Dhaher, H. and Al-Garrash, Z.H. (2001) Distribution of *Heterorhabditis indica* (Nematoda: Heterorhabditidae) in Eastern Saudi Arabia. *International Journal of Nematology* 11, 215–218.

Sankaranarayanan, C., Somasekhar, N. and Singaravelu, B. (2006) Biocontrol potential of entomopathogenic nematodes, *Heterorhabditis* and *Steinernema* against pupae and adults of white grub *Holotrichia serrata* F. *Sugar Tech* 8, 268–271.

Scheepmaker, J.W.A., Geels, F.P., Van Griensven, L.J.L.D. and Smits, P.H. (1998) Susceptibility of larvae of the mushroom fly, *Megaselia halterata* to the entomopathogenic nematode *Steinernema feltiae* in bioassays. *Biocontrol* 43, 201–214.

Schmidt, O., Theopold, U. and Strand, M. (2001) Innate immunity and its evasion and suppression by hymenopteran endoparasitoids. *Bioessays* 23, 344–351.

Shapiro-Ilan, D.I. and Gaugler, R. (2010) Nematodes: Rhabditida: Steinernematidae and Heterorhabditidae. In: Shelton, A. (ed.) *Biological Control: A Guide to Natural Enemies in North America*. Available at: https://www.biocontrol.entomology.cornell.edu/pathogens/nematodes.html (accessed 28 January 2015).

Shapiro-Ilan, D.I., Gouge, D.H. and Koppenhofer, A.M. (2002) Factors affecting commercial success: case studies in cotton, turf and citrus. In: Gaugler, R. (ed.) *Entomopathogenic Nematology.* CAB International, Wallingford, UK, pp. 333–356.

Shapiro-Ilan, D.I., Jackson, M., Reilly, C.C. and Hotchkiss, M.W. (2004) Effects of combining an entomopathogenic fungi or bacterium with entomopathogenic nematodes on mortality of *Curculio caryae* (Coleoptera: Curculionidae). *Biological Control* 30, 119–126.

Shapiro-Ilan, D.I., Dutcher, J.D. and Hatab, M. (2005) Recycling potential and fitness of Steinernematid nematodes cultured in *Curculio caryae* and *Galleria mellonella. Journal of Nematology* 37, 12–17.

Shapiro-Ilan, D.I., Stuart, R.J. and McCoy, C.W. (2006) A comparison of entomopathogenic nematode longevity in soil under laboratory conditions. *Journal of Nematology* 38, 119–129.

Shapiro-Ilan, D.I., Mizell, R.F., Cottrell, T.E. and Horton, D.H. (2008) Control of plum curculio, *Conotrachelus nenuphar*, with entomopathogenic nematodes: effects of application timing, alternate host plant, and nematode strain. *Biological Control* 44, 207–215.

Shapiro-Ilan, D.I., Han, R. and Dolinksi, C. (2012) Entomopathogenic nematode production and application technology. *Journal of Nematology* 44, 206–217.

Shetlar, D.J., Suleman, P.E. and Georgis, R. (1988) Irrigation and use of entomogenous nematodes, *Neoaplectana* spp. and *Heterorhabditis heliothidis* (Rhabditida: Steinernematidae and Heterorhabditidae), for control of Japanese beetle (Coleoptera: Scarabaeidae) grubs in turfgrass. *Journal of Economic Entomology* 81, 1318–1322.

Slipinski, S.A., Leschen, R.A.B. and Lawrence, J.F. (2011) Order Coleoptera Linnaeus, 1758. In: Zhang, Z.Q. (ed.) *Animal Biodiversity: An Outline of Higher-level Classification and Survey of Taxonomical Richness. Zootaxa* 3148, 203–208.

Smart, G.C. Jr (1995) Entomopathogenic nematodes for the biological control of insects. *Journal of Nematology* 27(4S), 529–534.

Stenseth, C. (1979) Effect of temperature on development of *Otiorhynchus sulcatus* (Coleoptera: Curculionidae). *Annals of Applied Biology* 91, 179–185.

Stuart, R.J., Borai, E.F.E. and Duncan, L.W. (2008) From augmentation to conservation of entomopathogenic nematodes: trophic cascades, habitat manipulation and enhanced biological control of *Diaprepes abbreviatus* root weevils in Florida citrus groves. *Journal of Nematology* 40, 73–84.

Susurluk, I.A. (2011) Potential of *S. feltiae* (Rhabditida: Steinernematidae) as a biological control agent against cabbage maggot, *Delia radicum* (Diptera: Anthomyiidae) in oil seed rape. *Turkish Journal of Agriculture and Forestry* 35, 413–419.

Toepfer, S., Knuth, P., Glas, M. and Kuhlmann, U. (2012) Successful application of entomopathogenic nematodes for the biological control of Western corn rootworm larvae in Europe – a mini review. *Proceedings of International Conference on the German Diabrotica Research Program*, Berlin, pp. 59–66.

Townsend, M.L., Johnson, D.T. and Steinkraus, D.C. (1998) Laboratory studies of the interactions of environmental conditions on the susceptibility of green June beetle (Coleoptera: Scarabaeidae) grubs to entomopathogenic nematodes. *Journal of Entomology Science* 33, 40–48.

Triggiani, O. and Tarasco, E. (2011) Evaluation of the effects of autochthonous and commercial isolates of *Steinernematidae* and *Heterorhabditidae* on *Rhynchophorus ferrugineus. Bulletin of Insectology* 64, 175–180.

Van Tol, R.W.H.M. and Raupp, M.J. (2006) Nursery and tree applications. In: Grewal, P.S., Ehlers, R.-U. and Shapiro-Ilan, D.I. (eds) *Nematodes as Biological Control Agents*. CAB International, Wallingford, UK, pp. 167–190.

Van Tol, R.W.H.M. and Van der Sommen, A.T.C. (2001) Plants protect their roots by alerting the enemies of grubs. *Ecology Letters* 4, 292–294.

Van Tol, R.W.H.M., Van Dijk, N. and Sabelis, M.W. (2004) Host plant preference and performance of the vine weevil, *Otiorhynchus sulcatus*. *Agriculture and Forest Entomology* 6, 267–278.

Van Zyl, C. and Malan, A.P. (2014) The role of entomopathogenic nematodes as biological control agents of insect pests, with emphasis on the history of their mass culturing and *in vivo* production. *African Entomology* 22, 235–249.

Vittum, P. (2005) Annual bluegrass weevil: a metropolitan pest on the move. *Golf Course Management* (May, 2005) pp. 105–108.

Wang, Y., Gaugler, R. and Cui, L. (1994) Variations in immune response of *Popillia japonica* and *Acheta domesticus* to *Heterorhabditis bacteriophora* and *Steinernema* species. *Journal of Nematology* 26, 11–18.

Wang, Y., Campbell, J.F. and Gaugler, R. (1995) Infection of entomopathogenic nematodes *Steinernema glaseri* and *Heterorhabditis bacteriophora* against *Popillia japonica* (Coleoptera: Scarabaeidae) larvae. *Journal of Invertebrate Pathology* 66, 178–184.

Williams, C.D., Dillon, A.B., Harvey, C.D., Hennessy, R., Namara, L. and Griffin, C.T. (2013) Control of a major pest of forestry, *Hylobius abietis* with entomopathogenic nematodes and fungi using eradicant and prophylactic strategies. *Forest Ecology and Management* 305, 212–222.

Wu, S. (2013) Efficacy of entomopathogenic nematodes and entomopathogenic fungi against masked chafer white grubs, *Cyclocephala* spp. (Coleoptera: Scarabaeidae). PhD thesis. Virginia Polytechnic Institute and State University, Blacksburg, Virginia.

Yan, X., Liu, X., Han, R., Chen, S., De Clercq, P. and Moens, M. (2010) Osmotic induction of anhydrobiosis in entomopathogenic nematodes of the genera, *Heterorhabditis* and *Steinernema*. *Biological Control* 53, 325–330.

Yan, X., Han, R., Moens, M., Chen, S. and Clercq, P.D. (2013) Field evaluation of entomopathogenic nematodes for biological control of striped flea beetle, *Phyllotreta striolata* (Coleoptera: Chrysomelidae). *Biological Control* 58, 247–256.

9 Efficacy of Entomopathogenic Nematodes against Dipteran Pests

Atef Sayed Abdel-Razek*

Pests and Plant Protection Department, National Research Centre, Cairo, Egypt

9.1 Introduction

Entomopathogenic nematodes (EPNs) such as *Steinernema* and *Heterorhabditis* are considered important biological control agents that are being exploited against a variety of crop insect pests of economic importance. More than 70 species of *Steinernema* and 20 of *Heterorhabditis* have been reported so far (Campos-Herrera, 2015), and this number is increasing continuously with the passage of time. This shows the beneficial effect of these nematodes in agriculture, which is drawing the attention of researchers working in different laboratories of the world. As indicated in other relevant chapters of this volume, Steinernematidae and Heterorhabditidae, the two families of EPNs, have been receiving extensive efforts, including fundamental and applied research, workshops and meetings, writing and issuing books and periodicals and application methodologies to exploit their superior attributes, especially as effective biological control agents (e.g. Gaugler and Kaya, 1990; Grewal *et al.*, 2005; Campos-Herrera, 2015). These nematodes occur naturally in soils of all the continents except Antarctica (Hominick, 2002; Adams *et al.*, 2006) and are being exploited successfully by many growers and stakeholders against soil-dwelling insects in numerous crops around the world (Campos-Herrera, 2015).

Dipteran insect pests have a wide range of habitats. On the other hand, EPNs also have variable host-searching behaviour, different styles of survivability and overall wonderful nematode–host matching for insects, in the order Diptera as well (Abd-Elgawad, 2001; Lewis *et al.*, 2006; Campos-Herrera, 2015). For example, Kepenekci *et al.* (2015) have recently documented that the nematode species *Steinernema feltiae* is adaptable to dipterous larvae. Generally speaking, once the nematode infective juveniles (IJs) are in the haemocoel, the nematodes release the mutualistic bacterium to overcome the host's immune response (Simões and Rosa, 1996; Forst *et al.*, 1997; Abdel-Razek, 2006; Askary, 2010) and kill the insect quickly in 48–72 h (Abdel-Razek, 2002; Boemare, 2002). Inside the cadaver, the nematodes cycle through several generations until the cadaver is fully exploited as a food resource for the nematodes and their symbiotic bacteria.

Like nematode hosts of other insect orders, many dipterous species can show different degrees of susceptibility to EPNs. Strictly speaking, many biotic and abiotic factors can also modify the efficacy of dipterous insects. These factors include EPN species and strains, insect species and developmental stages, and environmental stresses (Eidt and Thurston, 1995; Simões and Rosa, 1996; Campos-Herrera, 2015). On the other hand, the nematode's ability can overcome the barriers of its susceptible insect hosts through factors such as its physical force (e.g. rupture thin tracheae by body thrusting or, as in *Heterorhabditis*, use a tooth situated terminally on their mouth). The IJs may also pierce holes in the tissue with proteolytic secretions (Abu Hatab *et al.*, 1993; Peters and Ehlers, 1994).

EPNs have been reported to control a variety of dipteran insect pests in different habitats (Langford *et al.*, 2014; Campos-Herrera, 2015; Kepenekci *et al.*, 2015). The present chapter is an attempt in this direction to focus on such examples in a more explanatory manner. Factors affecting the success and failures of EPNs in the control of dipteran pests are also summarized.

*E-mail: abdelrazek820@yahoo.com

9.2 General Biology and Characteristics of Dipterans

The order Diptera is one of the four largest groups of living organisms. The common members of the order Diptera are flies, gnats, maggots, midges and mosquitoes. All members of diptera possess a pair of wings and go through complete metamorphosis (egg, larva, pupa and adult). The larva of diptera differs from other insects in having no legs and a head with chewing mouthparts (culiciform), or no head with only rudimentary mouth hooks for feeding (veriform). Culiciform-type larvae are typical in mosquitoes, while veriform-type larvae are typical in true flies. All diptera larvae live in aquatic or moist environments, where they feed on plant and animal material (Chapman, 1998; Daly *et al.*, 1998). Some destroy our food, especially grains, vegetables and fruit, others act as biological control agents of weeds and insects, whereas a few are helpful in solving crimes or pollinating plants (Chapman, 1998; Blagoderov *et al.*, 2002).

9.3 General Control Methods of Dipterans

Dipteran insects are usually controlled by the following methods.

9.3.1 Cultural control

This includes the use of various farm practices that discourage or prevent the development of fly pest problems and which directly or indirectly reduce pest populations, such as crop rotation, ploughing, planting time, pruning, thinning, etc.

9.3.2 Mechanical/physical control

In this method, the simple fly swat, fly screen, mosquito nets or light traps are used. Mechanical control, such as the use of barriers or hand picking insects, is more practical on a small scale than on a large scale.

9.3.3 Genetic control

Male sterilization, selective breeding and genetic modification are the techniques used in genetic control. Male sterilization techniques involve the mass rearing of a pest, laboratory sterilization of males and their release into the field with the purpose of increasing the wild male population, leading to infertile egg laying.

9.3.4 Biological control

This can be performed by direct or indirect use of parasites, predators or pathogens (bacteria, viruses, fungi and protozoa) that are natural enemies of a particular insect pest. The objective is to bring down the insect pest population and avoid economic losses. The biocontrol agents used to control dipteran pests are *Bacillus thuringiensis*, *Beauveria bassiana*, EPNs, *Trichogramma* wasp, green lacewings, praying mantis and lady beetles.

9.3.5 Chemical control

Chemical control includes both behaviour modifiers and insecticides.

9.4 Dipteran Habitats Targeted by Entomopathogenic Nematodes

Generally, dipterans are controlled by chemical insecticides, bait sprays, or ground cover sprays of potent organophosphorus pesticides. These methods are hazardous for both the environment and beneficial organisms. Therefore, an eco-friendly method of pest control is needed by growers (Roessler, 1989). In the past two decades, this target has been fulfilled to some extent by EPNs belonging to the families Steinernematidae and Heterorhabditidae. These nematodes are pathogenic to a wide range of agriculturally important dipteran pests and therefore are considered an alternative to chemical insecticides (Gaugler and Kaya, 1990). Infective juveniles (IJs) enter their host via the mouth, anus and spiracles (Steinernematidae and Heterorhabditidae), and sometimes also by penetrating the host cuticle (Heterorhabditidae) (Shapiro-Ilan and Lewis, 1999). They kill their hosts with the help of their symbiotic bacteria, i.e. *Xenorhabdus* spp. (in the case of *Steinernema*) or *Photorhabdus* spp. (in the case of *Heterorhabditis*); the bacteria release toxic metabolites in the haemocoel of the insect host, killing them within 48 h (Woodring and Kaya, 1988; Askary, 2010).

Steinernematids and heterorhabditids have been found pathogenic against a broad range of insect pests in various habitats (Kaya and Gaugler, 1993), and are therefore used inundatively in numerous high-value cropping systems (Koppenhöfer *et al.*, 2000). These nematodes have been found successful against soil-inhabiting insects (as soil application), as well as above-ground insects (foliar spray) and in cryptic habitats (Shapiro-Ilan *et al.*, 2006). The pathogenicity, host-searching behaviour and survivability of different nematode species varies, which makes them a suitable agent in biological control programmes (Kaya, 1990). Nematodes such as *Steinernema carpocapsae* and *Steinernema scapterisci* may search for their hosts at or near the soil profile, whereas *Heterorhabditis bacteriophora* and *Steinernema glaseri* search deeper in the soil for their hosts. The former group is referred to as 'ambushers', as they remain at a place and wait for a host that is moving at or in the soil surface. The latter group is referred to as 'cruisers', which are highly mobile and best adapted to find sedentary hosts. This group responds strongly to long-range host chemical cues (Lewis *et al.*, 1993). Eventually, dipteran habitats targeted by EPNs could be classified as follows.

9.4.1 Subterranean habitats

EPNs are soil dwelling and therefore isolated from soil habitats. These beneficial nematodes can control effectively fungus gnats, *Lycoriella* sp., subterranean pests in mushroom production and *Bradysia* spp., found in the cultivation of greenhouse plants. *S. feltiae* and *Heterorhabditis* spp. have been found effective in the control of *Lycoriella auripilla*, *Lycoriella mali*, *Lycoriella solani*, *Bradysia coprophila* and *Bradysia difformis* (Scheepmaker *et al.*, 1998a,b; Jagdale *et al.*, 2004, 2007; Jess *et al.*, 2005; Tomalak *et al.*, 2005; Grewal, 2007). *S. feltiae*, when applied at 1.0–1.5 × 10^6 IJs/m^2, resulted in significant control of *Lycoriella* spp. that was comparable to or better than the insecticides commonly used in mushroom production (Jagdale *et al.*, 2004; Grewal, 2007). The effectiveness of *S. feltiae* is longer than chemical insecticides, namely diflubenzuron and methoprene. This is due to the fact that the nematode recycles in gnat larvae (Grewal and Richardson, 1993).

9.4.2 Soil surface habitats

EPNs are adapted to a wide range of climatic conditions, i.e. hot, temperate and cold zones, and even from lowlands to high alpine altitudes (Steiner, 1996). They occur naturally in soils of all the continents except Antarctica (Griffin *et al.*, 1990; Hominick, 2002; Adams *et al.*, 2006). The larvae and pupae of a large number of dipteran families such as Tipulidae, Tabanidae, Leptidae, Asilidae, Bombyliidae, Therevidae, Dolichopodidae, Mycetophilidae and Bibionidae are found in soil, especially soil rich in organic matter. Researchers from different parts of the world have reported the successful application of nematodes against soil-dwelling insects in a variety of crops (Georgis *et al.*, 2006; Kaya *et al.*, 2006; Campos-Herrera, 2015).

9.4.3 Foliar application

S. carpocapsae are the most common nematode species that are widely used for the control of foliar and other above-ground dipteran pests. They are characterized with an ambushing, host-finding strategy, which makes them ideal candidates for insect pests encountered on the soil surface. There are several reports confirming *S. carpocapsae* and *S. feltiae* as potential candidates for the control of *Liriomyza trifolii* (Hara *et al.*, 1993; LeBeck *et al.*, 1993; Sher *et al.*, 2000; Tomalak *et al.*, 2005), *Liriomyza huidobrensis* (Williams and Walters, 2000) and *Tuta absoluta* (Batalla-Carrera *et al.*, 2010). However, the efficacy of nematodes was affected when applied on foliage insect pests. Field trials were conducted to evaluate the efficacy of EPNs against insect pests in above-ground habitats. The results revealed the lowest nematode efficacy against the insect pests of foliar habitats (Arthurs *et al.*, 2004). This was supposed to be due to rapid desiccation of the IJs on the leaf surface. However, use of anti-desiccants has been found to increase the effectiveness of IJs during foliar application.

9.4.4 Cryptic habitats

EPNs are reported to control dipteran pests of orchards found in cryptic habitats (Lacey and Chauvin, 1999; Cossentine *et al.*, 2002; Lacey *et al.*, 2005; de Waal *et al.*, 2010). A good example in this context is Mediterranean fruit flies that can be controlled successfully by nematodes. The insect larval population thus controlled results in reduced emergence of adult flies in the following spring. The most evaluated nematode species for such activity are *S. carpocapsae*, *S. feltiae*, *H. bacteriophora* and

Heterorhabditis zealandica. However, larvicidal activity of nematodes applied against the insects of cryptic habitat faces the greatest influence of abiotic factors such as temperature, moisture and type of habitat (Kaya *et al.*, 1984; Lacey and Unruh, 1998; Unruh and Lacey, 2001; Lacey *et al.*, 2006; Navaneethan *et al.*, 2010; de Waal *et al.*, 2011).

9.5 Factors Affecting the Success/ Failure of Entomopathogenic Nematodes

I list below the factors affecting the successes and failures of EPNs in the control of dipteran pests. These are categorized under two subheadings.

9.5.1 Success factors of entomopathogenic nematodes

1. EPNs are safe for humans, plants and animals, and are relatively safe for the environment, especially when compared to traditional pesticides.
2. The application of EPNs does not require personal protective equipment, such as safety masks and re-entry intervals. Their applications do not cause residues or groundwater contamination.
3. EPNs harbour bacteria that cannot survive independently, and are thus unable to infect humans (Crow, 2009; Barbercheck, 2011).
4. EPNs do not cause infection on non-target insects that are found above the soil, such as foliar-inhabiting insects.
5. Mortality caused by EPNs in soil-inhabiting non-target arthropods is temporarily spatially restricted, and affects only a part of the population (Ehlers, 2005).
6. The dispersion of EPNs does not take place beyond the treated plot (Booth *et al.*, 2002).
7. The introduction of exotic EPNs can result in reducing the native nematode species (biological pollution), but generally, due to the low mobility of nematodes, large parts of the native nematode populations will survive (Ehlers, 2005).
8. EPNs when in the soil around plant roots are protected from the harmful effect of UV light and high temperatures as soil in these locations is generally moist and humid (Polavarapu, 1999; Cowles *et al.*, 2005).
9. EPNs have a broad insect host range and a rapid speed of kill through actively seek or ambush foraging behaviour.
10. EPNs are easy to mass produce. They can either be mass produced by using *in vivo* and/ or *in vitro* techniques.
11. There is the possibility to use them with conventional application equipment.
12. They are free from registration in many countries.

9.5.2 Failure factors of entomopathogenic nematodes

1. EPNs are sensitive to environmental conditions, such as UV rays, high temperatures and rainfall.
2. EPNs need special environmental limitations such as adequate moisture for their survival and pathogenicity.
3. EPNs may not be compatible with some chemical pesticides used to control other insect pests.
4. EPNs are generally more expensive as compared to traditional chemical insecticides (Wright *et al.*, 2005).
5. EPNs need reapplication where the natural balance of the insect hosts is overloaded with nematodes. Under such cases, nematodes, after killing insects in a population, die off themselves. The small number of insects that survive build up again, and therefore to kill this growing insect population, the next application of EPNs is required (Crow, 2009).
6. EPN formulations have a short shelf life and they need refrigeration for storage.
7. EPNs have an expensive production cost.

Fertilizers and chemical pesticides can also have an impact on the efficacy of EPNs, which can be positive, neutral, or negative (Shapiro-Ilan *et al.*, 1996). The interaction of EPNs with other microorganisms, chemical pesticides and other EPNs varies, based on the specific chemical and nematode species or strain, dosages and timing of application (Benz, 1971; Koppenhöfer and Grewal, 2005). Hence, it is suggested that combinations should be tested on a case-by-case basis.

9.6 Entomopathogenic Nematodes Against Dipteran Insect Pests

Several reports confirm that EPNs are effective against a large number of insect pests (Begley, 1990; Klein, 1990; Bedding *et al.*, 1993). Nematode efficacy is governed by factors such as foraging behaviour, the temperature requirements of a particular nematode species and the pest's accessibility in its habitat. Successful application of EPNs against pests has been reported in habitats that provide protection from environmental extremes (especially in soil), their natural habitat and cryptic habitats. The efficacy in manure varied from low to high. IJs are exposed to detrimental conditions on foliage or on similar habitats, which can only be remedied marginally by adjuvants. Some dipteran pests, along with the crops in which they have been controlled successfully with EPNs, are mentioned in Table 9.1.

9.6.1 Entomopathogenic nematodes against leaf miners, *Liriomyza* spp. (Diptera: Agromyzidae)

There are more than 330 described species of *Liriomyza*, and many are economically important pests of field crops, ornamentals and vegetables (Spencer, 1973; Kang, 1996). More than 20 species of *Liriomyza* are of economic importance; this includes six that are polyphagous. These are *Liriomyza sativae* (Branchard), *L. trifolii* (Burgess), *L. huidobrensis* (Branchard), *Liriomyza bryoniae* (Kaltenbach), *Liriomyza strigata* (Meigen) and *Liriomyza longei* Frick (Morgan *et al.*, 2000; van der Linden, 2004). As a result of anthropogenic activities, these *Liriomyza* species now occur in most of the temperate and tropical regions of the world (Minkenberg and van Lenteren, 1986). Farm growers generally use synthetic and natural insecticides for their management (Parrella *et al.*, 1984; Ferguson, 2004), which has resulted adversely in the effectiveness of these insecticides on *Liriomyza* and its natural enemies (Lange *et al.*, 1980; Parrella *et al.*, 1984).

Nematode species such as *Heterorhabditis heliothidis*, *Heterorhabditis megidis*, *Heterorhabditis* sp. (strain UK 211), *S. carpocapsae* and *S. feltiae* have been found to cause a pathogenic effect on *Liriomyza* spp. (Harris *et al.*, 1990; Williams and Walters, 2000). IJs enter the leaf mine via the punctures created by female flies during egg laying or host feeding. On contact, nematodes are more likely to infect the host larva via the anus rather than the mouth, but not through spiracles, due to their small size and morphology (LeBeck *et al.*, 1993). The nematodes kill the larvae, the pre-puparia and the early puparia. On the basis of an experiment, it was reported that first- and second-instar larvae of

Table 9.1. Entomopathogenic nematodes in the control of some dipteran pests.

Nematode	Pest common name	Pest scientific name	Family	Targeted crops
Steinernema spp., *Heterorhabditis* spp.	Cabbage maggot	*Delia radicum*	Diptera: Anthomyiidae	Cabbage
Steinernema feltiae, *Heterorhabditis bacteriophora*	Fungus gnats	*Bradysia* spp.	Diptera: Sciaridae	Mushrooms
S. carpocapsae, *S. feltiae*	Leaf miners	*Liriomyza* spp.	Diptera: Agromyzidae	Vegetables, ornamentals
S. carpocapsae, *S. feltiae*	Shore flies	*Scatella* spp.	Diptera: Ephydridae	Ornamentals
S. carpocapsae	Crane fly	*Tipula* sp.	Diptera: Tipulidae	Turf
S. weiseri, *S. carpocapsae*, *S. riobrave*, *S. feltiae*, *Heterorhabditis* spp.	Mediterranean fruit fly (Medfly)	*Ceratitis capitata*	Diptera: Tephritidae	Vegetables, stone fruit, pome fruit, citrus, loquats and guavas
S. feltiae, *S. carpocapsae*, *S. riobrave*, *S. glaseri*, *H. bacteriophora*, *H. marelatus*	Olive fruit fly	*Bactrocera oleae*	Diptera: Tephritidae	Olive
S. feltiae, *H. bacteriophora*	Melon fruit fly	*Dacus ciliatus*	Diptera: Trypaneidae	Cucurbits
S. feltiae, *H. marelatus*, *H. bacteriophora*, *S. carpocapsae*	Cherry fruit fly	*Rhagoletis cerasi*	Diptera: Tephritidae	Cherry

Liriomyza soon die, i.e. 0.25–0.66 h after penetration by one nematode, whereas pre-puparia die after 15 h (LeBeck *et al.*, 1993). A low larval mortality of 4% (Hara *et al.*, 1993) and a high mortality of 85–97% was observed under laboratory or greenhouse conditions at high humidity (Broadbent and Olthof, 1995; Head *et al.*, 2002). The ambient relative humidity affects the survival and virulence of nematodes. Hara *et al.* (1993) reported that at least 92% relative humidity was needed to achieve even moderately high (>65%) levels of parasitism. Interestingly, temperature is not as critical as humidity. Williams and MacDonald (1995) inoculated *S. feltiae* on second-instar larvae. The nematode was found equally effective throughout the temperature range of 10–30°C. However, in a few cases, a special adjuvant such as glycerin may be required to improve the efficacy of the nematodes.

The control of *Liriomyza* by EPNs has proven impractical because the mortality caused by the nematodes has been found to be <80%, and also the nematodes are sensitive to humidity (Colombo and Locatelli, 1985). The high costs and variable effectiveness of EPNs (as compared to other control agents) on *Liriomyza* have restricted their use to high-value crops in small niche markets. However, there is scope for future research as a few nematodes cause relatively rapid death of *Liriomyza* instars, having a wide range of susceptibility. But it is suggested that EPNs be integrated with other practices to achieve an effective control (Sher *et al.*, 2000; Williams and Walters, 2000).

9.6.2 Entomopathogenic nematodes against cherry fruit fly, *Rhagoletis cerasi* L. (Diptera: Tephritidae)

Rhagoletis cerasi is considered a serious pest of cherry in Europe, the Americas and other cherry growing areas of the world. Eggs are laid individually by *R. cerasi*, each inserted singly under the skin of a ripening fruit. The pitted area around the oviposition punctures becomes discoloured. Maggots feed on the flesh of developing fruit (Alford, 2007). Fruits fed on by maggots often rot, and heavy infestations result in reduced market value. In recent years, the production of sweet cherries has faced severe yield loss due to high infestation levels of cherry fruit fly and the consequent refusal of the market. Generally, traps and crop netting are used to control this key pest of sweet cherries in Europe. But in the future, the problem may become worse as the chemical pesticide dimethoate has been removed from registration in conventional cherry growing areas, which will probably result in increasing the pest population in a given area when not controlled adequately during the coming years. Under this situation, EPNs are the best alternative strategy as they do not require registration and can be applied in organic growing. They are commercially available and are offered in formulations that are easy to use and compatible with the orchard environment. Thus, if their control potential is proved practically against *R. cerasi*, they can be implemented directly in organic farming and conventional cherry growing areas.

Temperature and nematode concentration have a great impact on the efficacy of native EPN species when applied against the larvae of *R. cerasi*. *S. feltiae* is one of the most efficacious nematode species tested at all temperatures and concentrations (Kepenekci *et al.*, 2015). *S. feltiae* is adaptable to lower temperatures (Grewal *et al.*, 1994; Hazir *et al.*, 2001). Even in the tropics, isolate of *S. feltiae* (MG-14) does well at low temperatures (Hazir *et al.*, 2001). *S. feltiae* caused more than 40% mortality at a temperature of 10 and 15°C when 1000 IJs/larva were applied. On the other hand, *Heterorhabditis* spp. are adaptable to warm temperatures (Grewal *et al.*, 1994; Karagoz *et al.*, 2009). In a study conducted by Kepenekci *et al.* (2015), *S. carpocapsae* was found to be the least virulent species against the larvae of *R. cerasi* evaluated at different temperatures. Wouts *et al.* (1982) tested the efficacy of *Steinernema bicornutum* and *Heterorhabditis* sp. against the larvae and pupae of the European cherry fruit fly and found that *S. carpocapsae* and *S. feltiae* were the most successful species in laboratory, semi-field and field conditions, with 88%, 78% and 88% larval mortality, respectively. Stark and Lacey (1999), on the basis of an experiment, reported that *S. carpocapsae* caused more infectivity (65%) than *S. feltiae* (35%), *H. bacteriophora* (50%) and *Heterorhabditis marelatus* (15%) against the larvae of the western cherry fruit fly, *Rhagoletis indifferens* Curran. They conducted the experiments in 100 × 15 mm petri dishes at 25°C with a nematode concentration of 6360 IJs/20 larvae. When the results were compared to intermediate concentration (500 IJs) at 25°C, *S. feltiae* showed the highest larval mortality (76%), followed by *H. marelatus* (72%), *H. bacteriophora* (66%) and *S. carpocapsae* (37%).

Nematode efficacy to host cues is influenced by the foraging behaviour and responses to the host

(Lewis, 2002; Griffin *et al.*, 2005). The larvae of *R. cerasi* emerge from the fruit to pupate in the top few centimetres of the soil surface (Özdem and Kılınçer, 2008). *S. feltiae* has an intermediate foraging strategy, seeking its hosts on or just beneath the soil surface (Lewis *et al.*, 2006), and can infect a number of Tephritidae larval species (Stark and Lacey, 1999; Sirjani *et al.*, 2009). A study conducted in the field showed that *S. feltiae* caused 79% mortality against larvae of *R. indifferens* (Yee and Lacey, 2003). In another study, Herz *et al.* (2006) applied *S. feltiae* by tractor-mounted spray boom under the canopy area of cherry trees and assessed the persistence of nematodes in the field after application. After application, *S. feltiae* was found highly active in soil samples, but dropped to 60% of the initial activity within 1 week.

9.6.3 Entomopathogenic nematodes against Mediterranean fruit fly, *Ceratitis capitata* (Wiedemann) (Diptera: Tephritidae)

The Mediterranean fruit fly (Medfly), *Ceratitis capitata*, is a native to Africa but has established in many parts of the world (Fimani, 1989; Liquido *et al.*, 1991). It is considered a major pest of a wide range of fruit, vegetables and nuts. The female Medfly lays eggs beneath the skin of fruit at the time when it starts ripening. The eggs hatch within 2–4 days and the maggots develop inside the fruit. They feed for 1–2 weeks on the fleshy part of the host fruit, and sometimes make a tunnel through it, thus reducing the infected portion to a juicy, inedible mass. When fully grown, maggots stop feeding and exit the fruit to pupate in the soil. The emergence of adults from the puparia takes place after 1–2 weeks, and thus the life cycle is completed (Fimani, 1989).

During the period when the maggots leave the fruit and enter the soil, they are more likely at this stage to get infected by EPNs (Lindegren and Vail, 1986; Lindegren, *et al.*, 1990; Gazit *et al.*, 2000; Bazman *et al.*, 2008). Laboratory studies revealed that the Turkish isolate of *Steinernema weiseri* was found effective against Medfly maggots, causing 100% mortality when the maggots were held individually in wells with 50 IJs (Bazman *et al.*, 2008). Gazit *et al.* (2000) tested 12 EPN isolates against the last instar of Medfly. The highest mortality was caused by *Steinernema riobrave* (82.5%), followed by *Heterorhabditis* sp. (81.5%) and *S. feltiae* (33.9%). Karagoz *et al.* (2009) tested the efficacy of 5 EPN species (Turkish isolates) against the last instar of Medfly under laboratory conditions. The isolate *S. feltiae* 09-31 Aydin was highly virulent and caused 78% mortality, whereas the mortality caused by *S. weiseri* and *S. carpocapsae* was 50 and 56%, respectively.

9.6.4 Entomopathogenic nematodes against cabbage maggot, *Delia radicum* Linnaeus (Diptera: Anthomyiidae)

The cabbage maggot, *Delia radicum*, damages cabbage and other brassica crops grown in the temperate regions of the world. Another species, *Delia platura*, is cosmopolitan and attacks more than 40 species of plants such as spinach (*Spinacia oleracea* L.), artichoke (*Cynara scolymus* L.), beet (*Beta vulgaris* L.) and wild cabbage (*Brassica oleracea* L.) (Darvas and Szappanos, 2003; Valenciano *et al.*, 2004). *D. platura* is widespread and considered a major pest in Europe, North and South America, and is very common in northern Africa, Japan, India, Australia and New Zealand (Valenciano *et al.*, 2004). The pest attacks at the time of seedling emergence. The life cycle of the fly is synchronized with the sowing and harvesting stages. During sowing, the females are attracted to the organic matter, where they lay approximately 100 eggs in the soil near the stems (Capinera, 2001; Gouinguené and Stadler, 2006). During the time of seedling germination, the eggs hatch and the first-instar larvae feed on seedlings. Feeding by the larvae may result in yield losses of up to 30% (Valenciano *et al.*, 2004).

EPNs are considered a potential and safe alternative measure to control *Delia* species (Chen and Moens, 2003; Chen *et al.*, 2003; Nielsen, 2003). Due to their wide host range and high reproduction capacity, they can be exploited against *Delia* spp. in the field (Valenciano *et al.*, 2004; Gil *et al.*, 2007; Delgado and Sáenz, 2012). Laboratory tests demonstrated that different doses of *Steinernema* spp. and *Heterorhabditis* spp. were efficient in managing *D. radicum* up to 40–60%. However, the results are inconsistent in terms of the maggot stage, nematode species, strain, dose and experimental conditions (Bracken, 1990; Royer *et al.*, 1996; Willmott *et al.*, 2002). Bracken (1990) evaluated seven strains/species of *Steinernema* and *Heterorhabditis* against first-instar *D. radicum*. A dose of 98–341 IJs/host was found effective in causing >50% mortality of the pest. At the higher dosages, surface application of nematodes was more effective than subsurface application in preventing injury caused by *D. radicum*. When the eggs of *D. radicum* were placed on the soil surface near the root (where eggs of *D. radicum* are laid naturally), nematode deposition on the surface led to more nematodes closer to *D. radicum* maggots as compared to subsurface application. The ability of the nematodes to remain active or persist after application influences the effectiveness of the nematode treatments (Fuxa, 1987). However, surface application may leave a portion of IJs on or near the surface, thus exposing them to desiccation or the harmful effects of ultraviolet light (Gaugler and Boush, 1978). Although nematode applications did not eradicate *D. radicum* populations in cabbage, treatments reduced significantly the amount of deep feeding scars caused by *D. radicum* that may lead to the loss of plants or reduced yield. Therefore, it is suggested that the application of EPNs should be done at the time of transplanting by equipping mechanical transplanters with automatic and metered devices for injecting nematodes on to or into the soil around the base or root of each plant (Schroeder *et al.*, 1996).

9.6.5 Entomopathogenic nematodes against the melon fruit fly, *Dacus ciliatus* (Loew) (Diptera: Trypaneidae)

The melon fruit fly, *Dacus ciliatus*, is considered one of the serious pests of most cucurbitaceous crops in Africa and Asia (Maher, 1957; Azab *et al.*, 1970). Heavy infestations of cucurbits caused by *D. ciliatus* have been reported in parts of Egypt and in South Africa (White and Elson-Harris, 1994; Fetoh, 2003). Adult flies lay the eggs into the cucurbit fruit, where the larvae develop. As a result of oviposition, larval activity and probably infection by fungi, the injured fruit deforms and rots (Fletcher, 1987). Several research workers have reported that fruit fly larvae are highly susceptible to EPNs; however, pupae are generally less susceptible (Stark and Lacey, 1999; Gazit *et al.*, 2000; Yee and Lacey, 2003).

Five species of EPNs comprising the genus *Heterorhabditis* and *Steinernema* were screened against *D. ciliatus*. It was noted that *S. feltiae* and *H. bacteriophora* caused a significant level of mortalities on the larvae and pupae of melon fruit fly. The most promising was *S. feltiae*, followed by *H. bacteriophora* (BA1 and HB), *Steinernema abbasi* (ABB) and *Heterorhabditis indica* (SAA2) (Hussein *et al.*, 2006). In Iran, Kamali *et al.* (2013) isolated native EPNs and evaluated them against *D. ciliatus* under laboratory and greenhouse conditions. The native strain *H. bacteriophora* was compared with the commercial strain of *S. carpocapsae*. Under laboratory conditions, the larvae and adult of *D. ciliatus* were susceptible to nematode infection; however, mortality was low on the pupae. Both nematode strains reproduced successfully and emerged from the larvae of *D. ciliatus*. Under greenhouse conditions, larval mortality due to nematodes was higher in sandy loam and sand soils than in clay loam. Nematode efficacy was higher at 25 and 30°C. Fetoh *et al.* (2011) evaluated *H. bacteriophora* and *S. carpocapsae* All strains on the different stages of *D. ciliatus*. The results showed *H. bacteriophora* was more virulent than *S. carpocapsae* in causing mortality in larvae and adults of *D. ciliatus*. However, the nematodes were less effective on pupae.

9.6.6 Entomopathogenic nematodes against the olive fruit fly, *Bactrocera oleae* (Rossi) (Diptera: Tephritidae)

Olive fruit fly, *Bactrocera oleae*, is a serious pest of olive that poses a virtual threat to all commercial and fruit-bearing ornamental olive plantings (Daane and Johnson, 2010). The adult female uses its ovipositor to cut the fruit skin and then lays eggs beneath it. The maggots feed inside the fruit and destroy the pulp, resulting in degraded fruit. They pupate within the olive fruit or in the soil. The infested fruits dry out and fall from the trees (Nardi *et al.*, 2003). The commercial value of such infested fruits that are meant for table consumption and oil production is reduced (Rice, 2000).

The susceptibility of olive fruit flies to various species/strains of *Steinernema* and *Heterorhabditis* have been noticed, but not many reports are available on this aspect. Sirjani *et al.* (2009) evaluated six species of EPN, namely *S. carpocapsae*, *S. feltiae*, *S. riobrave*, *S. glaseri*, *H. bacteriophora* and *H. marelatus* against third-instar larvae of olive fruit fly. When IJs were sprayed over naturally infested fallen olives, it caused larval mortality within the treated olives as well as in the soil. Larval mortality was highest (67.9%) when treated with IJs of *S. feltiae*, which indicates that this nematode may have the potential to suppress overwintering populations of *B. oleae*.

9.6.7 Entomopathogenic nematodes against some other dipteran pests

Heterorhabditis and *Steinernema* have been used very successfully against the peach fruit fly, *Bactrocera zonata* (Attala *et al.*, 2002; Soliman, 2007a,b), the cucurbit fly, *D. ciliatus* (Fetoh and El-Gendi, 2006), and the Queensland fruit fly, *Bactrocera tryoni* (Langford *et al.*, 2014). Langford *et al.* (2014) demonstrated the efficacy of three nematode species, i.e. *H. bacteriophora*, *S. carpocapsae* and *S. feltiae*, against *B. tryoni* under laboratory conditions. All three nematode species effectively killed maggots of *B. tryoni*; however, pupae were unaffected. The efficacy of cold-adapted *S. feltiae* against *B. tryoni* was comparatively more than *H. bacteriophora*, *S. carpocapsae* at all concentrations (50–1000 IJs/cm^2) and at different moisture levels and temperatures. However, the increased mortality of maggots caused by the nematodes was concentration dependent.

Several studies have addressed the efficacy of EPNs against the house fly, *Musca domestica* L. (Diptera: Muscidae) in the laboratory and in confined poultry environments (Renn *et al.*, 1985; Geden *et al.*, 1986; Belton *et al.*, 1987; Mullens *et al.*, 1987a,b). Georgis *et al.* (1987) reported that poultry manure was toxic to IJs. Manure in feedlots is frequently mixed with soil and grain; these differences may make the filth fly substrate in the cattle feedlot environment more hospitable to EPNs than that in confined poultry systems. Despite these differences, the efficacy of EPNs for the control of flies in the confined cattle environment has not been explored (Petersen, 1993), and only a few species and strains of nematodes have been screened against filth flies, which gave conflicting results for some nematode species, possibly because of differences in strain pathogenicity (Laumond *et al.*, 1979; Gaugler and Kaya, 1990).

In a laboratory experiment, third-instar larvae of *M. domestica* were exposed to four different strains of *S. feltiae*, two of *Steinernema bibionis*, and one of *H. heliothidis* on moist filter paper in petri dishes. All the strains caused infectivity on flies; however, strains of *S. feltiae* were more promising than *S. bibionis* or *H. heliothidis* (Mullens *et al.*, 1987a). Geden *et al.* (1986) tested the efficacy of three EPN species, i.e. *S. feltiae*, *S. glaseri* and *H. heliothidis*, against the larval, pupal and adult stages of the house fly under different habitat conditions. A high susceptibility was noted when second- and third-instar larvae and adult flies were inoculated with *S. feltiae* and *H. heliothidis* in petri dishes containing nematode-treated filter paper. In rearing medium, the mortality of second- and third-instar larvae when exposed to nematodes *S. feltiae* and *H. heliothidis* was 55–61 and 11–26%, respectively, at the highest dosage of 5000 IJs/host. When a suspension of *S. feltiae* was prepared in a 5% sucrose bait on cotton balls, the mortality of adult flies ranged from 53 to 67% at a dose ranging from 1000 to 100,000 IJs/ml of bait. Renn *et al.* (1985) evaluated *H. heliothidis* (NZ strain) and *S. feltiae (Agriotos* strain) in the laboratory against immature and adult stages of a strain (G) of house fly, *M. domestica*. Both species of nematodes killed larval stages of *M. domestica* on inoculated filter papers. The most pathogenic was *S. feltiae*, killing >90% of all larval stages at the two highest doses of 25,000 and 50, 000 IJs/0.5 ml tap water.

9.7 Entomopathogenic Nematodes in Integrated Pest Management Strategies Against Dipterans

EPNs, when in combination with other nematode species, fungi and viruses, often lead to high pest mortality, whereas nematode–bacteria interactions range from antagonistic to synergistic, depending on the agents, application timing or physico-chemical characters of the system (Koppenhöfer and Grewal, 2005). Generally, EPNs are applied to ecosystems that usually receive other inputs such as chemical pesticides, surfactants (e.g. wetting agents), fertilizers and soil amendments and it is therefore desirable to tank mix one or more inputs to save time and money. IJs are tolerant to short exposures (2–6 h) to most agrochemicals, including herbicides, fungicides, acaricides and insecticides (Rovesti and Deseo, 1990; Ishibashi, 1993). It has been reported that azadirachtin (Stark, 1996), *B. thuringiensis* products, pesticidal soap (Kaya *et al.*, 1995), herbicides, fungicides, acaricides, insecticides, nematicides and many other pesticides have limited to strong toxic effects on IJs (Rovesti and Deseo, 1990). Synergistic interactions between EPNs and other control agents have been observed for various insecticides (Koppenhöfer and Kaya, 1998; Nishimatsu and Jackson, 1998) and pathogens (Thurston *et al.*, 1994; Koppenhöfer *et al.*, 1999). Inorganic fertilizer, composted manure or urea do not have negative effects on nematode virulence, but fresh manure does (Bednarek and Gaugler, 1997; Shapiro-Ilan *et al.*, 1996).

Abd-Elgawad *et al.* (2012) sprayed orange trees with kaolin (alone), *S. riobrave* (alone) and kaolin + *S. riobrave* in an attempt to check the fruit infestation by Medfly, *C. capitata*. The results showed a reduction in larval infestation by 67.0, 78.0 and 99.8% in treatments of kaolin, *S. riobrave* and kaolin + *S. riobrave*, respectively. Mahmoud (2007) evaluated botanical insecticides based on azadirachtin and the nematode *S. feltiae* for the control of peach fruit fly, *B. zonata*. Of the 25 treatment combinations between azadirachtin from Neem Seed Kernel (NSK) extract and *S. feltiae*, 18 gave synergistic responses, 4 were additive, none antagonistic and 3 without any response. The same number of combinations with NeemAzal T 5% showed 19 synergistic responses, 1 additive, none antagonistic and 5 without any response. Combinations of Neemix 4.5% and *S. feltiae* showed 11 synergistic responses, 5 additive, 3 antagonistic and 6 without any response. Botanical insecticides based on azadirachtin, especially NSK extract and NeemAzal T 5%, in combination with *S. feltiae* may result in increasing the efficacy of control of the peach fruit fly, *B. zonata*.

9.8 Conclusion and Future Prospects

The health and environmental problems associated with chemical pesticides have resulted in a resurgence of biological control alternatives for pest management. EPNs rank high among such biocontrol agents, being second only to *B. thuringiensis* (Bt). They can control a wide variety of insect pests of different orders effectively. This applies to many dipteran insect pests in different habitats (e.g. Lewis *et al.*, 2006; Lacey and Georgis, 2012; Langford *et al.*, 2014; Kepenekci *et al.*, 2015). Nevertheless, in some experimental hosts such as mosquitoes, a rapid melanization reaction will kill the nematode IJs before they can release their symbiotic bacteria (Bronskill, 1962). Also, the survival of IJs can generally be affected by abiotic (UV light, temperature extremes and desiccation) and biotic (predators and pathogens) factors.

Valid as they are against numerous dipterous pests, these nematodes may be more effective as inundative agents against a number of dipteran insect pests and as inoculative agents for classical biological control if significant advances are attained, such as that made with the incorporation of bacterial insecticidal genes of Bt into plants for insect control. The main target traits for future improvement in EPNs (see Baiocchi *et al.*, Chapter 25, this volume) are quite applicable against pest members of the order diptera.

It is important, however, to assess carefully the return of investment when new technologies are used for genetic modifications that could result in the loss of organic approval and registration exemption status as a prerequisite for application in many countries. Genetic improvements will play a major role in strengthening and expanding the market share of EPNs in the current crop–dipterous pest complex. Moreover, other advancements such as the mass production, formulation and application of nematodes needs to be strengthened in order to increase the efficacy and field host range beyond the current dipterous insects, crops and environment. If attained, the use of EPNs in biological pest control would abandon its marginal position in the pesticide market.

Acknowledgements

This work is supported by In-House Project No 10120609 entitled 'Insect pests control for some of exporting vegetable and fruit crops produced organically'. The author thanks NRC administrative departments for financial support. Many thanks also to colleagues at the Department of Pests and Plant Protection, NRC, Egypt.

References

Abd-Elgawad, M.M.M. (2001) Entomopathogenic nematode–host matching. *Egyptian Journal of Agronematology* 5, 91–104.

Abd-Elgawad, M.M.M., Abdel-Razek, A.S and Abd El-Wahab, A.E. (2012) Efficacy of kaolin, entomopathogenic nematode and fungi on the Mediterranean fruit fly infesting citrus. *Egyptian Journal of Agronematology* 11, 178–192.

Abdel-Razek, A.S. (2002) Pathogenicity of bacteria symbiotically associated with insect pathogenic nematodes against the greater wax moth, *Galleria mellonella* (L.). *Archives of Phytopathology and Plant Protection* 35, 53–60.

Abdel-Razek, A.S. (2006) The role of the different fermentation media and infection processes on the infectivity of the bacterial symbionts to *Spodoptera littoralis* (Biosduval) (Lepidoptera: Noctuidae). *Archives of Phytopathology and Plant Protection* 39, 353–363.

Abu Hatab, M.A., Selvan, S. and Gaugler, R. (1993) Role of proteases in penetration of insect gut by the entomopathogenic nematode *Steinernema glaseri* (Nematoda: Steinernematidae). *Journal of Invertebrate Pathology* 66, 125–130.

Adams, B.J., Fodor, A., Koppenhofer, H.S., Stackebrandt, E., Stock, S.P. and Klein, M.G. (2006) Biodiversity and systematics of nematode–bacterium entomopathogens. *Biological Control* 37, 32–49.

Alford, D.V. (2007) *Pests of Fruit Crops*. Academic Press, Boston, Massachusetts.

Arthurs, S., Heinz, K.M. and Prasifka, J.R. (2004) An analysis of using entomopathogenic nematodes against above-ground pests. *Bulletin of Entomological Research* 94, 297–306.

Askary, T.H. (2010) Nematodes as biocontrol agents. In: Lichtfouse, E. (ed.) *Sociology, Organic Farming, Climate Change and Soil Science*. Springer, The Netherlands, pp. 347–378.

Attala, A., Fatima, A. and Eweis, M.A. (2002) Preliminary investigation on the utilization of entomopathogenic nematodes as biological control agents against the peach fruit fly, *Bactrocera zonata* (Saunders) (Diptera: Tephritidae). *Egyptian Journal of Agricultural Research* 80, 1045–1053.

Azab, A.K., El-Nahal, A.K.M. and Swailem, S.M. (1970) The immature stages of the melon fruit fly, *Dacus ciliatus. Bulletin de la Société Entomologique d'Egypte* 54, 243–247.

Barbercheck, M. (2011) Insect-parasitic nematodes for the management of soil-dwelling insects. Penn State College of Agricultural Sciences, University Park, Pennsylvania. Available at: http://ento.psu.edu/extension/factsheets/parasitic-nematodes (accessed 15 September 2016).

Batalla-Carrera, L., Morton, A. and García-del-Pino, F. (2010) Efficacy of nematodes against the tomato leafminer *Tuta absoluta* in laboratory and greenhouse conditions. *Biological Control* 55, 523–530.

Bazman, I., Ozer, N. and Hazir, S. (2008) Bionomics of the entomopathogenic nematode, *Steinernema weiseri* (Rhabditida: Steinernematidae). *Nematology* 10, 735–742.

Bedding, R.A., Akhurst, R.J. and Kaya, H.K. (1993) *Nematodes and the Biological Control of Insect Pests.* CSIRO Publishing, Australia.

Bednarek, A.J. and Gaugler, R. (1997) Compatibility of soil amendments with entomopathogenic nematodes. *Journal of Nematology* 29, 220–227.

Begley, J.W. (1990) Efficacy against insects in habitats others than soil. In: Gaugler, R. and Kaya, H.K. (eds) *Entomopathogenic Nematodes in Biological Control*. CRC Press, Boca Raton, Florida, pp. 215–231.

Belton, P., Rutherford, T.A., Trotter, D.B. and Webster, J.M. (1987) *Heterorhabditis heliothidis*: potential biological control agent of house flies in caged-layer poultry barns. *Journal of Nematology* 19, 263–266.

Benz, G. (1971) Synergism of micro-organisms and chemical insecticides. In: Burgess, H.D. and Hussey, N.W. (eds) *Microbial Control of Insects and Mites*. Academic Press, London, pp. 327–355.

Blagoderov, V.A., Lukashevich, E.D. and Mostovski, M.B. (2002) 'Order Diptera Linné, 1758. The true flies' (= Muscida Laicharting, 1781). In: Rasnitsyn, A.P. and Quick, D. (eds) *History of Insects*. Kluwer Academic Publishers, Dordrecht, The Netherlands, pp. 227–240.

Boemare N.E. (2002) Biology, taxonomy and systematics of *Photorhabdus and Xenorhabdus*. In: Gaugler, R. (ed.) *Etomopathogenic Nematology*. CAB International, Wallingford, UK, pp. 35–56.

Booth, S.R., Tanigoshi, L.K. and Shanks, C.H. (2002) Evaluation of entomopathogenic nematodes to manage root weevil larvae in Washington State cranberry, strawberry and red raspberry. *Environmental Entomology* 31, 895–902.

Bracken, G.K. (1990) Susceptibility of first instar cabbage maggot, *Delia radicum* (Anthomyiidae: Diptera), to strains of the entomogenous nematodes *Steinernema feltiae, S. bibionis* (Bovien), *Heterorhabditis bacteriophora Poinar* and *H. heliothidis. Canadian Entomologist* 122, 633–639.

Broadbent, A.B. and Olthof, T.H.A. (1995) Foliar application of *Steinernema carpocapsae* (Rhabditida: Steinernematidae) to control *Liriomyza trifolii* (Diptera: Agromyzidae) larvae in chrysanthemums. *Environmental Entomology* 24, 431–435.

Bronskill, J.F. (1962) Encapsulation of Rhabditoid nematodes in mosquitoes. *Candian Journal of Zoology* 40, 1269–1275.

Campos-Herrera, R. (2015) *Nematode Pathogenesis of Insects and Other Pests*. Series: ecology and applied technologies for sustainable plant and crop protection. Springer International Publishing, Switzerland.

Capinera, J.L. (2001) Order Diptera – flies and maggots. In: Capinera, J.L. (ed.) *Handbook of Vegetable Pests*. Academic Press, San Diego, California, pp. 197–241.

Chapman, R.F. (1998) *The Insects: Structure and Function*, 4th edn. Cambridge University Press, New York.

Chen, S. and Moens, M. (2003) Susceptibility of cabbage root maggot, *Delia radicum,* to entomopathogenic nematodes (Steinernematidae and Heterorhabditidae). *Nematology Mediterranea* 31, 157–162.

Chen, S., Han, X. and Moens, M. (2003) Biological control of *Delia radicum* (Diptera: Anthomyiidae) with entomopathogenic nematodes. *Applied Entomology Zoology* 38, 441– 448.

Colombo, M. and Locatelli, D.P. (1985) Laboratory evaluation of the activity of *Steinernema feltiae* (Filip.) and *Heterorhabditis* spp. on *Liriomyza trifolii* (Burgess) and *Opogona sacchari* (Bojer) infesting cultivated flowering plants. *La Difesa delle Piant* 8, 263–269.

Cossentine, J.E., Jensen, L.B. and Moyls, L. (2002) Fruit bins washed with *Steinernema carpocapsae* (Rhabditida: Steinernematidae) to control *Cydia pomonella* (Lepidoptera: Tortricidae). *Biocontrol Science and Technology* 12, 251–258.

Cowles, R.S., Polavarapu, S., Williams, R.M., Thies, A. and Ehlers, R.-U. (2005) Soft fruit applications. In:

Grewal, P.S., Ehlers, R.-U. and Shapiro-Ilan, D.I. (eds) *Nematodes as Biocontrol Agents.* CAB International, Wallingford, UK, pp. 231–254.

Crow, W.T. (2009) Using nematodes to control insects: overview and frequently asked questions. University of Florida, Institute of Food and Agricultural Sciences, Gainesville, Florida. Available at: http://www.vammibeneficials.com/Univ_of_Florida_Nematode_FAQ.pdf (accessed 20 September 2016).

Daane, K.M. and Johnson, M.W. (2010) Olive fruit fly: managing an ancient pest in modern times. *Annual Review of Entomology* 55, 151–169.

Daly, H.V., Doyen, J.T. and Purcell, A.H. (1998) *Introduction to Insect Biology and Diversity*, 2nd edn. Oxford University Press, New York.

Darvas, B. and Szappanos, A. (2003) Male and female morphology of some central European *Delia* (Anthomyiidae) pests. *Acta Zoologica Academiae Scientiarum Hungaricae* 49, 87–101.

Delgado, O.Y. and Sáenz, A. (2012) Virulencia, producción y desplazamiento de nematodos entomopatógenos sobre larvas del picudo de la guayaba *Conotrachelus psidii* Marshall (Coleoptera: Curculionidae) en laboratorio. *Universitas Scientiarum* 17, 283–290.

de Waal, J.Y., Malan, A.P., Levings, J. and Addison, M.F. (2010) Key elements in the successful control of diapausing codling moth, *Cydia pomonella* (lepidoptera: tortricidae) in wooden fruit bins with a South African isolate of *Heterorhabditis zealandica* (Rhabditida: heterorhabditidae). *Biocontrol Science and Technology* 20, 489–502.

de Waal, J.Y., Malan, A.P. and Addison, M.F. (2011) Evaluating mulches together with *Heterorhabditis zealandica* (Rhabditida: Heterorhabditidae) for the control of diapausing codling moth larvae, *Cydia pomonella* (L.) (Lepidoptera: Tortricidae). *Biocontrol Science and Technology* 21, 255–270.

Ehlers, R.-U. (2005) Forum on safety regulations. In: Grewal, P.S., Ehlers, R.-U. and Shapiro-Ilan, D.I. (eds) *Nematodes as Biocontrol Agents.* CAB International, Wallingford, UK, pp. 107–114.

Eidt, D.C., and Thurston, G.S. (1995) Physical deterrents to infection by entomopathogenic nematodes in wire worms (Coleoptera: Elateridae) and other soil insects. *Candian Entomologist* 127, 423–429.

Ferguson, J.S. (2004) Development and stability of insecticide resistance in the leafminer *Liriomyza trifolii* (Diptera: Agromyzidae) to cyromazine, abamectin, and spinosad. *Journal of Economic Entomology* 97, 12–119.

Fetoh, B. (2003) The pumpkin fly, *Dacus ciliatus* (Loew) as new and ancient pest in Egypt. *The Agriculture Magazine* 46, 54–55.

Fetoh, B.E.A. and El-Gendi, S.S. (2006) Impact of entomopathogenic nematodes on different stages of the pumpkin fly, *Dacu ciliatus* (loew) as a new approach in its biological control. *Fayoum Journal of Agricultural Research and Development* 20, 19–25.

Fetoh, B.E.A., Abd-Elgawad, M.M.M., Shalaby, F.F. and Elyme, M.F. (2011) Pathogenic and lethal effects of the entomopathogenic nematodes on the peach fruit fly, *Bactrocera zonata* (saunders) and the cucurbit fruit fly, *Dacus ciliatus* (loew) (Diptera: Tephritidae). *Egyptian Journal of Agricultural Research* 89, 463–476.

Fimani, P. (1989) Mediterranean region. In: Robinson, A.S. and Hooper, G. (eds) *Fruit Flies: Their Biology, Natural Enemies and Control*. Elsevier, Amsterdam, pp. 39–50.

Fletcher, B.S. (1987) The biology of dacine fruit flies. *Annual Review of Entomology* 32, 115–144.

Forst, S., Dowds, B., Boemare, N. and Stackebrandt, E. (1997) *Xenorhabdus* and *Photorhabdus* spp.: bugs that kill bugs. *Annual Reviews in Microbiology* 51, 47–72.

Fuxa, J.R. (1987) Ecological considerations for the use of entomopathogens in IPM. *Annual Review of Entomology* 32, 225–251.

Gaugler, R. and Boush, G.M. (1978) Effects of ultraviolet radiation and sunlight on the entomogenous nematode, *Neoaplectana carpocapsae. Journal of Invertebrate Pathology* 32, 291–296.

Gaugler, R. and Kaya, H.K. (1990) *Entomopathogenic Nematodes in Biological Control.* CRC Press, Boca Raton, Florida, 365 pp.

Gazit, Y., Rossler, Y. and Glazer, I. (2000) Evaluation of entomopathogenic nematodes for the control of Mediterranean fruit fly (Diptera: Tephritidae). *Biocontrol Science and Technology* 10, 157–164.

Geden, C.J., Axtell, R.C. and Brooks, W.M. (1986) Susceptibility of the house fly, *Musca domestica* (Diptera: Muscidae), to the entomogenous nematodes *Steinernema feltiae*, *S. glaseri* (Steinernematidae), and *Heterorhabditis heliothidis* (Heterorhabditidae). *Journal of Medical Entomology* 23, 326–332.

Georgis, R., Mullens, B.A. and Meyer, J.A. (1987) Survival and movement of insect parasitic nematodes in poultry manure and their infectivity against *Musca domestica*. *Journal of Nematology* 19, 292–295.

Georgis, R., Koppenhöfer, A.M., Lacey, L.A., Bélair, G., Duncan, L.W., *et al.* (2006) Successes and failures in the use of parasitic nematodes for pest control. *Biological Control* 38, 103–123.

Gil, R.C., Carillo, D.Q. and Jimenez, J.G. (2007) Determinación de las principales plagas de la espinaca *(Spinacia oleracea)* en Cota, Colombia. *Revista Colombiana de Entomología* 33, 124–128.

Gouinguené, S.P. and Stadler, E. (2006) Oviposition in *Delia platura* (Diptera, anthomyiidae): the role of volatile and contact cues of bean. *Journal of Chemical Ecology* 32, 1399–1413.

Grewal, P.S. (2007) Mushroom pests. In: Lacey, L.A. and Kaya, H.K. (eds) *Field Manual of Techniques in Invertebrate Pathology: Application and Evaluation of Pathogens for Control of Insects and Other Invertebrate Pests*, 2nd edn. Springer, Dordrecht, The Netherlands, pp. 457–461.

Grewal, P.S. and Richardson, P.N. (1993) Effects of application rates of *Steinernema feltiae* (Nematoda: Steinernematidae) on control of the mushroom sciarid fly *Lycoriella auripilla*. *Biocontrol Science and Technology* 3, 29–40.

Grewal, P.S., Selvan, S. and Gaugler, R. (1994) Thermal adaptation of entomopathogenic nematodes: niche breadth for infection, establishment, and reproduction. *Journal of Thermal Biology* 19, 245–253.

Grewal, P.S., Ehlers, R.-U. and Shapiro-Ilan, D.I. (eds) (2005) *Nematodes as Biocontrol Agents*. CAB International, Wallingford, UK, 505 pp.

Griffin, C.T., Downes, M.J. and Block, W. (1990) Tests of Antarctic soils for insect parasitic nematodes. *Antarctic Science* 2, 221–222.

Griffin, C.T., Boemare, N.E. and Lewis, E.E. (2005) Biology and behaviour. In: Grewal, P.S., Ehlers, R.-U. and Shapiro-Ilan, D.I. (eds) *Nematodes as Biocontrol Agents*. CAB International, Wallingford, UK, pp. 47–59.

Hara, A.H., Kaya, H.K., Gaugler, R., Lebeck, L.M. and Mello, C.L. (1993) Entomopathogenic nematodes for biological control of the leafminer, *Liriomyza trifolii* (Dipt., Agromyzidae). *Entomophaga* 38, 359–369.

Harris, M.A., Begley, J.W. and Warkentin, D.L. (1990) *Liriomyza trifolii* (Diptera: Agromyzidae) suppression with foliar applications of *Steinernema carpocapsae* (Rhabditida: Steinernematidae) and abamectin. *Journal of Economic Entomology* 83, 2380–2384.

Hazir, S., Stock, S.P., Kaya, H.K., Koppenhofer, A.M. and Keskin, N. (2001) Developmental temperature effects on five geographic isolates of the entomopathogenic nematode *Steinernema feltiae* (Nematoda: Steinernematidae). *Journal of Invertebrate Pathology* 77, 243–250.

Head, J., Palmer, L.F. and Walters, K.F.A. (2002) Development of an integrated control strategy for leafminers in leafy salads with potential for extrapolation to other cropping systems. *Bulletin of International Organization for Biological Control of Noxious Animals and Plants* 25, 97–100.

Herz, A., Koppler, K., Vogt, H., Elias, E., Katz, P. and Peters, A. (2006) Biological control of the cherry fruit fly, *Rhagoletis cerasi* L. (Diptera, Tephritidae) by use of entomopathogenic nematodes: first experiences towards practical implementation. In: Boos, M. (ed.) *Proceedings of the 12th International Conference on Cultivation Technique and Phytopathological Problems in Organic Fruit Growing* held in Weinsberg, Germany from 31 January to 2 February 2006, pp. 67–72.

Hominick, W.M. (2002) Biogeography. In: Gaugler, R. (ed.) *Entomopathogenic Nematology*. CAB International, Wallingford, UK, pp. 115–143.

Hussein, M.A., El-Wakeil, N. and El-Sebai, T. (2006) Susceptibility of melon fruit fly, *Dacus ciliatus*, to entomopathogenic nematodes (Rhabditida) and to insecticides. *International Journal of Nematology* 16, 13–18.

Ishibashi, N. (1993) Integrated control of insect pests by *Steinernema carpocapsae*. In: Bedding, R., Akhurst, R. and Kaya, H. (eds) *Nematodes and the Biological Control of Insects*. CSIRO, East Melbourne, Australia, pp. 105–113.

Jagdale, G.B., Casey, M.L., Grewal, P.S. and Lindquist, R.K. (2004) Application rate and timing, potting medium, and host plant effects on the efficacy of *Steinernema feltiae* against the fungus gnat, *Bradysia coprophila*, in floriculture. *Biological Control* 29, 296–305.

Jagdale, G.B., Casey, M.L., Canas, L. and Grewal, P.S. (2007) Effect of entomopathogenic nematode species, split application and potting medium on the control of the fungus gnat, *Bradysia difformis* (Diptera: Sciaridae), in the greenhouse at alternating cold and warm temperatures. *Biological Control* 43, 23–30.

Jess, S., Schweizer, H. and Kilpatrick, M. (2005) Mushroom applications. In: Grewal, P.S., Ehlers, R.-U. and Shapiro-Ilan, D.I. (eds) *Nematodes as Biocontrol Agents.* CAB International, Wallingford, UK, pp. 191–213.

Kamali, S., Karimi, J., Hosseini, M., Campos-Herrera, R. and Duncan, L.W. (2013) Biocontrol potential of the entomopathogenic nematodes *Heterorhabditis bacteriophora* and *Steinernema carpocapsae* on cucurbit fly, *Dacus Ciliatus* (Diptera: Tephritidae). *Biocontrol Science and Technology* 23, 1307–1323.

Kang, L. (1996) *Ecology and Sustainable Control of Serpentine Leafminers*. Science Press, Beijing.

Karagoz, M., Gulcu, B., Hazir, C., Kaya, H.K. and Hazir, S. (2009) Biological control potential of Turkish entomopathogenic nematodes against the Mediterranean fruit fly *Ceratitis capitata*. *Phytoparasitica* 37, 153–159.

Kaya, H.K. (1990) Soil ecology. In: Gaugler, R. and Kaya, H.K. (eds) *Entomopathogenic Nematodes in Biological Control*. CRC Press, Boca Raton, Florida, pp. 93–115.

Kaya, H.K. and Gaugler, R. (1993) Entomopathogenic nematodes. *Annual Review Entomology* 38, 181–206.

Kaya, H.K., Joos, J.L., Falcon, L.A. and Berlowitz, A. (1984) Suppression of the codling moth (Lepidoptera: Olethreutidae) with the entomogenous nematode, *Steinernema feltiae* (Rhabditida: Steinernematidae). *Journal of Economic Entomology* 77, 1240–1244.

Kaya, H.K., Burlando, T.M., Choo, H.Y., and Thurston, G.S. (1995) Integration of entomopathogenic nematodes with *Bacillus thuringiensis* or pesticidal soap for control of insect pests. *Biological Control* 5, 432–441.

Kaya, H.K., Aguillera, M.M., Alumai, A., Choo, H.Y., De la Torre, M., Fodor, A., Ganguly, S., Hazır, S., Lakatos, T., Pye, A., Wilson, M., Yamanaka, S., Yang, H. and Ehlers, R.-U. (2006) Status of entomopathogenic nematodes and their symbiotic bacteria from selected countries or regions of the world. *Biological Control* 38, 134–155.

Kepenekci, I., Hazir, S. and Ozdem, A. (2015) Evaluation of native entomopathogenic nematodes for the control of the European cherry fruit fly *Rhagoletis cerasi* L. (Diptera: Tephritidae) larvae in soil. *Turkish Journal of Agriculture and Forestry* 39, 74–79.

Klein, M.G. (1990) Efficacy against soil-inhabiting insect pests. In: Gaugler, R. and Kaya, H.K. (eds) *Entomopathogenic Nematodes in Biological Control*. CRC Press, Boca Raton, Florida, pp. 195–214.

Koppenhöfer, A.M. and Grewal, P.S. (2005) Compatibility and interactions with agrochemicals and other biocontrol agents. In: Grewal, P.S., Ehlers, R.-U. and Shapiro-Ilan, D.I. (eds) *Nematodes as Biological Control Agents*. CAB International, Wallingford, UK, pp. 363–381.

Koppenhöfer, A.M., and Kaya, H.K. (1998) Synergism of imidacloprid and entomophathogenic nematodes: a novel approach to white grub control in turfgrass. *Journal* of *Economic Entomology* 91, 618–623.

Koppenhöfer, A.M., Choo, H.Y., Kaya, H.K., Lee, D.W. and Gelernter, W. (1999) Improved field and greenhouse efficacy with a combination of entomopathogenic nematode and *Bacillus thuringiensis* against scarab grubs. *Biological Control* 14, 37–44.

Koppenhöfer, A.M., Brown, I.M., Gaugler, R., Grewal, P.S., Kaya, H.K. and Klein, M.G. (2000) Synergism of entomopathogenic nematodes and imidacloprid against white grubs: greenhouse and field evaluation. *Biological Control* 19, 245–251.

Lacey, L.A. and Chauvin, R.L. (1999) Entomopathogenic nematodes for control of codling moth in fruit bins. *Journal of Economic Entomology* 92, 104–109.

Lacey, L.A. and Georgis, R. (2012) Entomopathogenic nematodes for control of insect pests above and below. *Journal of Nematology* 44, 218–225.

Lacey, L.A. and Unruh, T.R. (1998) Entomopathogenic nematodes for control of codling moth: effect of nematode species, dosage, temperature and humidity under laboratory and simulated field conditions. *Biological Control* 13, 190–197.

Lacey, L.A., Neven, L.G., Headrick, H.L. and Fritts, R. Jr (2005) Factors affecting entomopathogenic nematodes (Steinernematidae) for the control of overwintering codling moth (Lepidoptera: Tortricidae) in fruit bins. *Journal of Economic Entomology* 98, 1863–1869.

Lacey, L.A., Arthurs, S.P., Unruh, T.R., Headrick, H. and Fritts R. Jr (2006) Entomopathogenic nematodes for control of codling moth (Lepidoptera: Tortricidae) in apple and pear orchards: effect of nematode species and seasonal temperatures, adjuvants, application equipment and post-application irrigation. *Biological Control* 37, 214–223.

Lange, W.H., Agosta, G.G., Goh, K.S. and Kishiyama, J.S. (1980) Field effect of insecticides on chrysanthemum leafminer and a primary parasitoid, *Chrysocharis ainsliei* (Crawford), on artichokes in California. *Environmental Entomology* 9, 561–562.

Langford, E.A., Nielsen, U.N., Johnson, S.N. and Riegler, M. (2014) Susceptibility of Queensland fruit fly, *Bactrocera tryoni* (Froggatt) (Diptera: Tephritidae), to entomopathogenic nematodes. Biological Control 69, 34–39.

Laumond, D., Mauleon, H. and Kermarrec, A. (1979) Donnes nouvelles sur le spectre d'hotes et le parasitisme du nematode entomophage *Neoplectana carpocapsae*. *Entomophaga* 24, 113–127.

LeBeck, L.M., Gaugler, R., Kaya, H.K., Hara, A.H. and Johnson, M.W. (1993) Host stage suitability of the leafminer, *Liriomyza trifolii* (Diptera: Agromyzidae) to the entomopathogenic nematode *Steinernema carpocapsae* (Rhabditida: Steinernematidae). *Journal of Invertebrate Pathology* 62, 58–63.

Lewis, E.E. (2002) Behavioural ecology. In: Gaugler, R. (ed.) *Entomopathogenic Nematology*. CAB International, Wallingford, UK, pp. 205–224.

Lewis, E.E., Gaugler, R. and Harrison, R. (1993) Response of cruiser and ambusher entomopathogenic nematodes (Steinernematidae) in host volatile cues. *Canadian Journal of Zoology* 71, 765–769.

Lewis, E.E., Campbell, J., Griffin, C., Kaya, H.K. and Peters, A. (2006) Behavioral ecology of entomopathogenic nematodes. *Biological Control* 38, 66–79.

Lindegren, J.E. and Vail, P.V. (1986) Susceptibility of Mediterranean fruit fly, melon fly, and oriental fruit fly (Diptera: Tephritidae) to the entomogenous nematode *Steinernema feltiae* in laboratory tests. *Environmental Entomology* 15, 465–468.

Lindegren, J.E., Wong, T.T. and McInnis, D.O. (1990) Response of Mediterranean fruit fly (Diptera: Tephritidae) to the entomogenous nematode *Steinernema feltiae* in field tests in Hawaii. *Environmental Entomology* 19, 383–386.

Liquido, N.J., Shinoda, L.A. and Cunningham, R.T. (1991) Host plants of the Mediterranean fruit fly (Diptera: Tephritidae): an annotated world review. Miscellaneous publications of the *Entomological Society of America* 77, pp. 52.

Maher, A. (1957) On the bionomics of *Dacus ciliatus* Loew (Diptera, Tryphetidae). *Bulletin of Society of Entomology Egypt* 41, 527–533.

Mahmoud, M.F. (2007) Combining the botanical insecticides NSK extract, NeemAzal T 5%, Neemix 4.5% and the entomopathogenic nematode *Steinernema feltiae* Cross N 33 to control the peach fruit fly, *Bactrocera zonata* (Saunders). *Plant Protection Science* 43, 19–25.

Minkenberg, O.P.J.M. and van Lenteren, J.C. (1986) The leafminers *Liriomyza bryoniae* and *L. trifolii* (Diptera: Agromyzidae), their parasites and host plants: a review. *Agriculture University of Wageningen Papers* 86-2, 1–50.

Morgan, D.J.W., Reitz, S.R., Atkinson, P.W. and Trumble, J.T. (2000) The resolution of California populations of *Liriomyza huidobrensis* and *Liriomyza trifolii* (Diptera: Agromyzidae) using PCR. *Heredity* 35, 53–61.

Mullens, B.A., Meyer, J.A. and Cyr, T.L. (1987a) Infectivity of insect-parasitic nematodes (Rhabditida: Steinernematidae, Heterorhabditidae) for larvae of some manure breeding flies (Diptera: Muscidae). *Environmental Entomology* 16, 769–773.

Mullens, B.A., Meyer, J.A. and Georgis, R. (1987b) Field tests of insect-parasitic nematodes (Rhabditida: Steinernematidae, Heterorhabditidae) against larvae of manure breeding flies (Diptera: Muscidae) on caged-layer poultry facilities. *Journal of Economic Entomology* 80, 438–442.

Nardi, F., Carapelli, A., Dallai, R. and Frati, F. (2003) The mitochondrial genome of the olive fly *Bactrocera oleae*: two haplotypes from distant geographical locations. *Insect Molecular Biology* 12, 605–611.

Navaneethan, T., Strauch, O., Besse, S., Bonhomme, A. and Ehlers, R.-U. (2010) Influence of humidity and a surfactant-polymer formulation on the control potential of the entomopathogenic nematode *Steinernema feltiae* against diapausing codling moth larvae (*Cydia pomonella* L.) (Lepidoptera: Tortricidae) *BioControl* 55, 777–788.

Nielsen, O. (2003) Susceptibility of *Delia radicum* to steinernematid nematodes. *BioControl* 48, 431–446.

Nishimatsu, T. and Jackson, J.J. (1998) Interaction of insecticides, entomopathogenic nematodes, and larvae of the Western corn rootworm (Coleoptera: Chrysomelidae). *Journal of Economic Entomology* 91, 410–418.

Özdem, A. and Kılınçer, N. (2008) The biology of the European cherry fruit fly [*Rhagoletis cerasi* L. (Diptera: Tephritidae)]. *Acta Horticulturae* 795, 897–904.

Parrella, M.P., Keil, C.B. and Morse, J.G. (1984) Insecticide resistance in *Liriomyza trifolii*. *California Agriculture* 38, 22–23.

Peters, A. and Ehlers, R.-U. (1994) Susceptibility of leatherjackets (*Tipula paludosa* and *T. oleracea*; Tipulidae: Nematocera) to the entomopathogenic nematode *Steinernema feltiae*. *Journal of Invertebrate Pathology* 63, 163–171.

Petersen, J.J. (1993) Biotic agents to control house flies and stable flies. In: Thomas, G.D. and Skoda, S.R. (eds) *Rural Flies in the Urban Environment*. University of Nebraska-Lincoln, Research Bulletin No 317, pp. 70–82.

Polavarapu, S. (1999) Insecticidal nematodes for cranberry pest management. In: Polavarapu, S. (ed.) *Proceedings at the National Workshop on Optimal Use of Insecticidal Nematodes in Pest Management held at Rutgers University*, Chatsworth, New Jersey, 28–30 August 1999, pp. 79–90.

Renn, N., Barson, G. and Richardson, P.N. (1985) Preliminary laboratory tests with two species of entomophilic nematodes for control of *Musca domestica* in intensive animal units. *Annals of Applied Biology* 106, 229–233.

Rice, R. (2000) Bionomics of the olive fruit fly, *Bactrocera* (*Dacus*) *oleae*. *University of California Plant Protection Quarterly* 10, 1–5 (14 December 2015).

Roessler, Y. (1989) Control; insecticides; insecticidal bait and cover sprays. In: Robinson, A.S. and Hooper, G. (eds) *Fruit Flies, Their Biology, Natural Enemies and Control. World Crop Pests*, 3 (B). Elsevier, Amsterdam, pp. 329–336.

Rovesti, L. and Deseö, K.V. (1990) Compatibility of chemical pesticides with the entomophathogenic nematodes, *Steinernema carpocapsae* Weiser and *S. feltiae* Filipjev (Nematoda: Steinernematidae). *Nematologica* 36, 237–245.

Royer, L., Belair, G. and Fournier, Y. (1996) Attractiveness of cabbage maggot (Diptera:Anthomyiidae) to entomopathogenic Steinernematid nematodes *Journal of Economic Entomology* 89, 614–620.

Scheepmaker, J.W.A., Geels, F.P., Rutjens, A.J., Smits, P.H. and Van Griensven, L.J.L.D. (1998a) Comparison of the efficacy of entomopathogenic nematodes for the biological control of the mushroom pests *Lycoriella auripila* (Sciaridae) and *Megaselia halterata* (Phoridae). *Biocontrol Science and Technology* 8, 277–287.

Scheepmaker, J.W.A., Geels, F.P., Smits, P.H. and van Griensven, L.J.L.D. (1998b) Influence of *Steinernema feltiae* and diflubenzuron on yield and economics of the cultivated mushroom *Agaricus bisporus* in Dutch mushroom culture. *Biocontrol Science and Technology* 8, 269–275.

Schroeder, P.C., Ferguson, C.S., Shelton, A.M., Wilsey, W.T., Hoffman, M.P. and Petzoldt, C. (1996) Greenhouse and field evaluations of entomopathogenic nematodes (Nematoda: Heterorhabditidae and Steinernematidae) for control of cabbage maggot (Diptera: Anthomyiidae) on cabbage. *Journal of Economic Entomology* 89, 1109–1115.

Shapiro-Ilan, D.I. and Lewis, E.E. (1999) Comparison of entomopathogenic nematode infectivity from infected hosts versus aqueous suspension. *Environmental Entomology* 28, 907–911.

Shapiro-Ilan, D.I., Tylka, G.L. and Lewis, L.C. (1996) Effects of fertilizers on virulence of *Steinernema carpocapsae*. *Applied Soil Ecology* 3, 27–34.

Shapiro-Ilan, D.I., Gouge, D.H., Piggot, S.J. and Fife, J. (2006) Application technology and environmental considerations for use of entomopathogenic nematodes in biological control. *Biological Control* 38, 124–133.

Sher, R.B., Parrella, M.P. and Kaya, H.K. (2000) Biological control of the leafminer *Liriomyza trifolii* (Burgess): implications for intraguild predation between *Diglyphyus begini* Ashmead and *Steinernema carpocapsae* (Weiser). *Biological Control* 17, 155–163.

Simões, N. and Rosa, J.S. (1996) Pathogenicity and host specificity of entomophathogenic nematodes. *Biocontrol Science Technology* 6, 403–411.

Sirjani, F.O., Lewis, E.E. and Kaya, H.K. (2009) Evaluation of entomopathogenic nematodes against the olive fruit fly, *Bacterocera oleae* (Diptera: Tephritidae). *Biological Control* 48, 274–280.

Soliman, N.A. (2007a) Efficacy of the entomopathogenic nematodes *Steinernema riobravis* Cabanillas and *Heterorhabditis bacterophora* (Native strain) against the peach fruit fly, *B. zonata* and the Mediterranean fruit fly, *C. capitata*. *Egyptian Journal of Biological Pest Control* 17, 77–82.

Soliman, N.A. (2007b) Pathogenicity of the entomopathogenic nematodes to the peach fruit fly, *B. zonata* and the Mediterranean fruit fly, *C. capitata*. *Egyptian Journal Biological Pest Control* 17, 121–124.

Spencer, K.A. (1973) *Agromyzidae (Diptera) of Economic Importance*. Springer, The Netherlands.

Stark, J.D. (1996) Entomopathogenic nematodes (Rhabditida: Steinernematidae): toxicity of neem. *Journal of Economic Entomology* 89, 68–73.

Stark, J.E.P. and Lacey, L.A. (1999) Susceptibility of Western fruit fly (Diptera: Tephritidae) to five species of entomopathogenic nematodes in laboratory studies. *Journal of Invertebrate Pathology* 74, 206–208.

Steiner, W.A. (1996) Distribution of entomopathogenic nematodes in the Swiss Alps. *Revue Suisse de Zoologie* 103, 439–452.

Stock, S.P. (2015) Diversity, biology and evolutionary relationships. In: Campos-Herrera, R. (ed.) *Nematode Pathogenesis of Insects and Other Pests. Series: Sustainability in Plant and Crop Protection*. Springer International Publishing, Switzerland, pp. 3–27.

Thurston, G.S., Ni, Y. and Kaya, H.K. (1994) Influence of salinity on survival and infectivity of entomopathogenic nematodes. *Journal of Nematology* 26, 345–351.

Tomalak, M., Piggott, S. and Jagdale, G.B. (2005) Glasshouse applications. In: Grewal, P.S., Ehlers, R.-U. and Shapiro-Ilan, D.I. (eds) *Nematodes as Biological Control Agents.* CAB International, Wallingford, UK, pp. 147–166.

Unruh, T.R. and Lacey, L.A. (2001) Control of codling moth, *Cydia pomonella* (Lepidoptera: Tortricidae) with *Steinernema carpocapsae*: effects of supplemental wetting and pupation site on infection rate. *Biological Control* 20, 48–56.

Valenciano, J.B., Casquero, P.A. and Boto, J.A. (2004) Evaluation of the occurrence of bean plants *(Phaseolus vulgaris* L.) affected by bean seed fly, *Delia platura* (Meigen), grown under different sowing techniques and with different forms of pesticide application. *Field Crop Research* 85, 103–109.

van der Linden, A. (2004) Biological control of leafminers on vegetable crops. In: Heinz, K.M., van Driesche, R.G. and Parrella, M.P. (eds) *Biocontrol in Protected Culture.* Ball Publishing, USA, pp. 239–252.

White, I.M. and Elson-Harris, M.M. (1994) *Fruit Flies of Economic Significance: Their Identification and Bionomics.* CAB International, Wallingford, UK.

Williams, E.C. and MacDonald, O.C. (1995) Critical factors required by the nematode *Steinernema feltiae* for the control of the leafminers, *Liriomyza huidobrensis*, *Liriomyza bryoniae* and *Chromatomyia syngenesiae*. *Annals of Applied Biology* 127, 329–341.

Williams, E.C. and Walters, K.F.A. (2000) Foliar application of the entomopathogenic nematode *Steinernema feltiae* against leafminers on vegetables. *Biocontrol Science and Technology* 10, 61–70.

Willmott, D.M., Hart, A.J., Long, S.J., Richardson, P.N. and Chandler, D. (2002) Susceptibility of cabbage root fly *Delia radicum,* in potted cauliflower *(Brassica oleracea* var. botrytis) to isolates of entomopathogenic nematodes (*Steinernema* and *Heterorhabditis* spp.) indigenous to the UK. *Nematology* 4, 965–970.

Woodring, J.L. and Kaya, H.K. (1988) Steinernematid and Heterorhabditid nematodes: a handbook of biology and techniques. Southern Cooperative Series Bulletin 331, Arkansas Agricultural Experimental Station, Fayetteville, Arkansas, 30 pp.

Wouts, W.M., Mracek, Z., Gerdin, S. and Bedding, R.A. (1982) *Neoplectana Steiner*, 1929 a junior synonym of *Steinernema Travassos*, 1927 (Nematoda: Rhabditida). *Systematic Parasitology* 4, 147–154.

Wright, D.J., Peters, A., Schroer, S. and Fife, J.P. (2005) Application technology. In: Grewal, P.S., Ehlers, R.-U. and Shapiro-Ilan, D.I. (eds) *Nematodes as Biocontrol Agents.* CAB International, Wallingford, UK, pp. 91–106.

Yee, W. and Lacey, L.A. (2003) Stage-specific mortality of *Rhagoletis indifferens* (Diptera: Tephritidae) exposed to three species of *Steinernema* nematodes. *Biological Control* 27, 349–356.

10 Control of Stored Grain Pests by Entomopathogenic Nematodes

Shahina Fayyaz,[1]* Salma Javed,[1] Christos I. Rumbos[2] and Christos G. Athanassiou[2]

[1]*National Nematological Research Centre, University of Karachi, Pakistan;*
[2]*Laboratory of Entomology and Agricultural Zoology, Department of Agriculture, Crop Production and Rural Environment, University of Thessaly, Magnesia, Greece*

Background Issues and Aim

Grain production suffers considerable quantitative and qualitative losses during storage due to insect infestations. Current control strategies are based mainly on chemical control, i.e. fumigants and residual grain protectants. However, the development of insect resistance to commonly used insecticides, as well as the public demand for residue-free products and environmentally friendly approaches in food processing and storage urge for reliable, low-risk alternatives for the management of insects in a storage environment. The present chapter reviews the progress achieved so far on the exploitation of entomopathogenic nematodes (EPNs) for the control of stored-product insects. Recent findings on the control of stored grain pests by EPNs are summarized and the factors affecting the virulence of EPNs against storage insects are elucidated. Finally, the limitations of EPNs, as well as their future prospects and possible role in control strategies against stored-product insects, are discussed.

10.1 Introduction

Grain commodities, for example cereals and legumes, are subject to insect infestation during post-harvest processing and storage. Approximately 10–20% of the total production of durable dry commodities is lost or degraded due to stored-product insects, i.e. the direct loss due to consumption and the indirect impact due to contamination with insect fragments, faeces and webbing (Neethirajan *et al.*, 2007; Phillips and Throne, 2010). Grain protectants and fumigants are widely used in warehouses and storage facilities for stored grain protection, especially in the case of insects and mites (Arthur, 1996; Zettler and Arthur, 2000). However, several stored-product insect species have developed a significant level of resistance to chemical residual insecticides and fumigants (Collins *et al.*, 1993; Chaudhry, 2000; Guedes and Pereira, 2008; Opit *et al.*, 2012; Nguyen *et al.*, 2015). For instance, Opit *et al.* (2012) reported a considerable increase in phosphine resistance in American populations of the red flour beetle, *Tribolium castaneum* (Herbst) (Coleoptera: Tenebrionidae), and the lesser grain borer, *Rhyzopertha dominica* (F.) (Coleoptera: Bostrychidae), in the past two decades. Similarly, high levels of pyrethroid resistance were observed in Brazilian populations of the maize weevil, *Sitophilus zeamais* Motschulsky (Coleoptera: Curcolionidae) (Guedes and Pereira, 2008). Moreover, public concerns about the potential risks that derive from the consumption of pesticide-treated commodities have created a strong demand for environmentally friendly approaches in food processing and products that are free of pesticide residues. As a result, there has been a switch of scientific interest for the development and integration in stored-product protection of alternative, ecologically safe methods for the disinfestation of stored commodities (Donahaye, 2000).

*Corresponding author. E-mail: shahinafayyaz@gmail.com

Biological control of postharvest infestations has been studied extensively in the last decades and is based on the application of macro- (e.g. predators, parasitoids) or microbiological control agents (e.g. bacteria, fungi, nematodes) (Brower *et al.*, 1996; Schöller *et al.*, 1997; Moore *et al.*, 2000). Currently, several species of antagonists are on the market for application on stored grains against stored-product insects in Europe (Niedermayer and Steidle, 2013) and in the USA (Flinn and Schöller, 2012). The bacterial pathogen *Bacillus thuringiensis* Berliner (Bacillales: Bacillaceae) has been registered in the USA as grain protectant for the control of the Indian meal moth, *Plodia interpunctella* (Hübner) (Lepidoptera: Pyralidae) larvae (Lord *et al.*, 2007), while entomopathogenic fungi have been evaluated successfully against most of the major postharvest insect pests (Kavallieratos *et al.*, 2014). Among the biological control agents that have been investigated for the control of postharvest moths and beetles are EPNs. In total, 16 species of EPNs have been evaluated against a wide range of stored-product insect species (Table 10.1).

10.2 Entomopathogenic Nematodes as Biological Control Agents

EPNs are lethal obligate pathogens of insects that have a mutualistic association with bacteria of the genera *Xenorhabdus* Thomas and Poinar and *Photorhabdus* Boemare, Akhurst and Mourant (Poinar, 1990; Griffin *et al.*, 2005). The first EPN, *Steinernema kraussei* (*Aplectana kraussei*) Steiner (Rhabditida: Steinernematidae), was described by Steiner in 1923; however, their potential for insect pest management was discovered later by systematic studies in the 1970s and 1980s (Gaugler, 2002). Since then, considerable attention has been directed to the families Steinernematidae (Rhabditida: Nematoda) and Heterorhabditidae (Rhabditida: Nematoda), and particularly the genera *Steinernema* Travassos and *Heterorhabditis* Poinar (Shapiro-Ilan *et al.*, 2012). Nowadays, EPNs are used for biological control of a wide range of insect pests; i.e. soil-inhabiting insects (Kaya and Gaugler, 1993; Smart, 1995), as well as foliar (Arthurs *et al.*, 2004) and urban insect pests (Morton and García-del-Pino, 2013).

Their virulence mechanism and the infection process are based on the free-living infective juveniles (IJs), which are highly mobile and responsible for the dispersal of the nematodes, as well as the identification and orientation to the host (Ishibashi and Kondo, 1990). Infective juveniles enter into the insect cuticle mainly through natural body openings, e.g. mouth, anus, spiracles or wounds (Smart, 1995). After penetration, the nematodes release their symbiotic mutualistic bacteria (*Xenorhabdus* in *Steinernema* and *Photorabdus* in *Heterorhabditis*) into the insect's haemocoel (Poinar, 1990; Ciche *et al.*, 2006), which multiply quickly in the haemolymph, causing the death of the insect, usually within 24–48 h after application (Kaya and Gaugler, 1993; Stock and Blair, 2008).

10.3 Control of Stored Grain Pests by Entomopathogenic Nematodes

Due to their requirements for moist conditions (Kung *et al.*, 1991), EPNs appear to be a poor fit for the dry stored-product environment and therefore application of nematodes in stored commodities has received, until lately, very limited attention. Indeed, applications of EPNs to bulk grain bins or storage areas are expected to be highly problematic due to environmental limitations. Recently, improvement of nematode formulations, application equipment or approaches, as well as the discovery of highly virulent strains, has enhanced the effectiveness of EPNs (Shapiro-Ilan *et al.*, 2012) and renewed interest in using them in storage environments. As a result, several researchers have investigated the potential of EPNs to control stored-product insects. The nematode species that have been evaluated against different postharvest insect species belong exclusively to the genera *Steinernema* and *Heterorhabditis* and are presented in Table 10.1.

10.3.1 Coleoptera

Sitophilus spp.

The granary weevil, *Sitophilus granarius* (L.) (Coleoptera: Curculionidae), and the rice weevil, *Sitophilus oryzae* (L.) (Coleoptera: Curculionidae), are primary pests of stored wheat, causing severe losses to cereal grains throughout the world through direct feeding on grain kernels (Daglish *et al.*, 1996). Their immature stages develop inside the wheat kernels, rendering their control difficult (Flinn and Schöller, 2012). Early studies evaluated the virulence of *Steinernema feltiae* Filipjev (formerly *Neoplectana carpocapsae*) (Alikhan *et al.*, 1985) and *Steinernema carpocapsae* Weiser (Wójcik, 1986a,b) against adults of *S. granarius*, whereas,

Table 10.1. Entomopathogenic nematode species that have been evaluated for the control of stored-product insect pests.

Entomopathogenic nematode family	Entomopathogenic nematode species	Target insect pest	Insect stage	References
		Alphitobius diaperinus	Larva, adult	Geden *et al.*, 1985
		Callosobruchus chinensis	Larva, adult	Shahina and Salma, 2009
		Ephestia kuehniella	Larva	Morris, 1985
			Larva, pupa, adult	Ramos-Rodríguez *et al.*, 2006
			Larva	Athanassiou *et al.*, 2008
			Larva	Athanassiou *et al.*, 2010
		Lasioderma serricorne	Larva, adult	Rumbos and Athanassiou, 2012
			Larva, pupa, adult	Shahina and Salma, 2011b
		Oryzaephilus surinamensis	Larva, pupa, adult	Ramos-Rodríguez *et al.*, 2006
			Adult	Trdan *et al.*, 2006
		Plodia interpunctella	Larva, pupa, adult	Ramos-Rodríguez *et al.*, 2006
		Rhyzopertha dominica	Adult	Ramos-Rodríguez *et al.*, 2006
	Steinernema feltiae		Adult	Athanassiou *et al.*, 2010
Steinernematidae		*Sitophilus oryzae*	Adult	Ramos-Rodríguez *et al.*, 2006
			Adult	Roodaki *et al.*, 2012
			Adult	Laznik *et al.*, 2010
			Pupa, adult	Shahina and Salma, 2010
		Sitophilus granarius	Adult	Athanassiou *et al.*, 2010
			Adult	Trdan *et al.*, 2005
			Adult	Alikhan *et al.*, 1985
		Tenebrio molitor	Adult	Trdan *et al.*, 2006
			Larva	Susurluk, 2006
		Tribolium castaneum	Larva	Morris, 1985
			Larva, pupa, adult	Ramos-Rodríguez *et al.*, 2006
		T. confusum	Larva, pupa, adult	Ramos-Rodríguez *et al.*, 2006
			Larva, adult	Shahina and Salma, 2011a
			Adult	Alikhan *et al.*, 1985
			Larva, adult	Athanassiou *et al.*, 2008
		Trogoderma granarium	Larva, adult	Athanassiou *et al.*, 2010
		T. variabile	Larva, adult	Rumbos and Athanassiou, 2012
		Typhaea stercorea	Larva	Alikhan *et al.*, 1985
			Larva, pupa, adult	Ramos-Rodríguez *et al.*, 2006
			Larva, adult	Svendsen and Steenberg, 2000
		Acanthoscelides obtectus	Adult	Barbosa-Negrisoli *et al.*, 2013
		E. kuehniella	Larva, pupa, adult	Ramos-Rodríguez *et al.*, 2006
			Larva	Athanassiou *et al.*, 2010
			Larva	Barbosa-Negrisoli *et al.*, 2013
		L. serricorne	Larva, adult	Rumbos and Athanassiou, 2012

	O. surinamensis	Larva, pupa, adult	Ramos-Rodríguez *et al.*, 2006
		Adult	Trdan *et al.*, 2006
	P. interpunctella	Larva, pupa, adult	Ramos-Rodríguez *et al.*, 2006
	R. dominica	Adult	Ramos-Rodríguez *et al.*, 2006
		Adult	Athanassiou *et al.*, 2010
	S. oryzae	Adult	Ramos-Rodríguez *et al.*, 2006
S. carpocapsae		Adult	Athanassiou *et al.*, 2010
		Adult	Barbosa-Negrisoli *et al.*, 2013
	S. granarius	Adult	Wójcik, 1986a,b
		Adult	Trdan *et al.*, 2005, 2006
	Sitophilus zeamais	Adult	Barbosa-Negrisoli *et al.*, 2013
	T. molitor	Larva, pupa, adult	Ramos-Rodríguez *et al.*, 2006
		Larva	Ramos-Rodríguez *et al.*, 2007b
		Larva	Barbosa-Negrisoli *et al.*, 2013
	T. castaneum	Larva, pupa, adult	Ramos-Rodríguez *et al.*, 2006
	T. confusum	Adult	Wójcik, 1986a,b
		Larva, adult	Athanassiou *et al.*, 2010
		Larva, adult	Rumbos and Athanassiou, 2012
	T. variabile	Larva, pupa, adult	Ramos-Rodríguez *et al.*, 2006
	T. stercorea	Larva, adult	Svendsen and Steenberg, 2000
	A. obtectus	Adult	Barbosa-Negrisoli *et al.*, 2013
	O. surinamensis	Larva, pupa, adult	Ramos-Rodríguez *et al.*, 2006
	R. dominica	Adult	Ramos-Rodríguez *et al.*, 2006
	S. oryzae	Adult	Ramos-Rodríguez *et al.*, 2006
		Adult	Barbosa-Negrisoli *et al.*, 2013
	S. zeamais	Adult	Barbosa-Negrisoli *et al.*, 2013
	T. molitor	Larva, pupa, adult	Ramos-Rodríguez *et al.*, 2006
S. riobrave		Larva	Ramos-Rodríguez *et al.*, 2007b
		Larva	Barbosa-Negrisoli *et al.*, 2013
	T. castaneum	Larva, pupa, adult	Ramos-Rodríguez *et al.*, 2006
		Larva, pupa, adult	Ramos-Rodríguez *et al.*, 2007a
	T. variabile	Larva, pupa, adult	Ramos-Rodríguez *et al.*, 2006
	P. interpunctella	Larva, pupa, adult	Ramos-Rodríguez *et al.*, 2006
		Larva, pupa, adult	Ramos-Rodríguez *et al.*, 2007a
	E. kuehniella	Larva, pupa, adult	Ramos-Rodríguez *et al.*, 2006
		Larva, pupa	Barbosa-Negrisoli *et al.*, 2013
S. glaseri	*A. diaperinus*	Larva, adult	Geden *et al.*, 1985
	T. molitor	Larva	Ramos-Rodríguez *et al.*, 2007b

Continued

Table 10.1. Continued.

Entomopathogenic nematode family	Entomopathogenic nematode species	Target insect pest	Insect stage	References
		A. obtectus	Adult	Barbosa-Negrisoli *et al.*, 2013
		A. diaperinus	Larva, adult	Geden *et al.*, 1985
	S. rarum	*S. oryzae*	Adult	Barbosa-Negrisoli *et al.*, 2013
		S. zeamais	Adult	Barbosa-Negrisoli *et al.*, 2013
		T. molitor	Larva	Barbosa-Negrisoli *et al.*, 2013
		E. kuehniella	Larva, pupa	Barbosa-Negrisoli *et al.*, 2013
		C. chinensis	Larva, adult	Shahina and Salma, 2009
	S. asiaticum	*L. serricorne*	Larva, pupa, adult	Shahina and Salma, 2011b
		S. oryzae	Pupa, adult	Shahina and Salma, 2010
		T. castaneum	Larva, adult	Shahina and Salma, 2011a
		C. chinensis	Larva, adult	Shahina and Salma, 2009
	S. abbasi	*L. serricorne*	Larva, pupa, adult	Shahina and Salma, 2011b
		S. oryzae	Pupa, adult	Shahina and Salma, 2010
		T. castaneum	Larva, adult	Shahina and Salma, 2011a
		C. chinensis	Larva, adult	Shahina and Salma, 2009
	S. siamkayai	*L. serricorne*	Larva, pupa, adult	Shahina and Salma, 2011b
		S. oryzae	Pupa, adult	Shahina and Salma, 2010
		T. castaneum	Larva, adult	Shahina and Salma, 2011a
		C. chinensis	Larva, adult	Shahina and Salma, 2009
	S. pakistanense	*L. serricorne*	Larva, pupa, adult	Shahina and Salma, 2011b
		S. oryzae	Pupa, adult	Shahina and Salma, 2010
		T. castaneum	Larva, adult	Shahina and Salma, 2011a
		A. obtectus	Adult	Barbosa-Negrisoli *et al.*, 2013
		C. chinensis	Larva, adult	Shahina and Salma, 2009
		E. kuehniella	Larva	Morris, 1985
			Larva	Athanassiou *et al.*, 2010
			Larva, pupa	Barbosa-Negrisoli *et al.*, 2013
		L. serricorne	Larva, adult	Rumbos and Athanassiou, 2012
			Larva, pupa, adult	Shahina and Salma, 2011b
Heterorhabditidae	*Heterorhabditis bacteriophora*	*O. surinamensis*	Adult	Trdan *et al.*, 2006
		P. interpunctella	Larva, adult	Mbata and Shapiro-Ilan, 2005
		R. dominica	Adult	Athanassiou *et al.*, 2010
		S. oryzae	Pupa, adult	Shahina and Salma, 2010
			Adult	Athanassiou *et al.*, 2010
			Adult	Barbosa-Negrisoli *et al.*, 2013
		S. granarius	Adult	Trdan *et al.*, 2005, 2006
		S. zeamais	Adult	Barbosa-Negrisoli *et al.*, 2013
		T. molitor	Larva	Morris, 1985

			Larva	Susurluk, 2006
		T. castaneum	Larva, pupa	Ngoma *et al.*, 2012
		T. confusum	Larva	Barbosa-Negrisoli *et al.*, 2013
			Larva, adult	Shahina and Salma, 2011a
		T. stercorea	Larva, adult	Athanassiou *et al.*, 2010
			Larva, adult	Rumbos and Athanassiou, 2012
			Larva, adult	Svendsen and Steenberg, 2000
		L. serricorne	Larva, adult	Rumbos and Athanassiou, 2012
		O. surinamensis	Adult	Trdan *et al.*, 2006
	H. megidis	*P. interpunctella*	Larva, adult	Mbata and Shapiro-Ilan, 2005
		S. granarius	Adult	Trdan *et al.*, 2005, 2006
		T. confusum	Larva, adult	Rumbos and Athanassiou, 2012
		T. stercorea	Larva, adult	Svendsen and Steenberg, 2000
		C. chinensis	Larva, adult	Shahina and Salma, 2009
		L. serricorne	Larva, pupa, adult	Shahina and Salma, 2011b
		P. interpunctella	Larva, adult	Mbata and Shapiro-Ilan, 2005
	H. indica		Larva	Mbata and Shapiro-Ilan, 2010
		S. oryzae	Pupa, adult	Shahina and Salma, 2010
		T. castaneum	Larva, adult	Shahina and Salma, 2011a
	H. heliothidis	*A. diaperinus*	Larva, adult	Geden *et al.*, 1985
	H. georgiana	*P. interpunctella*	Larva	Shapiro-Ilan *et al.*, 2009
		T. molitor	Larva	Shapiro-Ilan *et al.*, 2009
	H. marelatus	*P. interpunctella*	Larva, adult	Mbata and Shapiro-Ilan, 2005
	H. zealandica	*P. interpunctella*	Larva, adult	Mbata and Shapiro-Ilan, 2005
		T. molitor	Larva, pupa	Ngoma *et al.*, 2012

more recently, Trdan *et al.* (2006) also studied the effect of two heterorhabditid species, *Heterorhabditis bacteriophora* Poinar and *Heterorhabditis megidis* Poinar, Jackson and Klein, with *H. megidis* being the least efficient. Different EPN species have also been evaluated against *S. oryzae*, with variable success (Ramos-Rodríguez *et al.*, 2006; Athanassiou *et al.*, 2010; Laznik *et al.*, 2010; Shahina and Salma, 2010; Barbosa-Negrisoli *et al.*, 2013). Most studies suggest that *S. oryzae* adults are not highly susceptible to EPNs, with mortality levels of 9% (Athanassiou *et al.*, 2010), 26% (Barbosa-Negrisoli *et al.*, 2013) or 40% (Ramos-Rodríguez *et al.*, 2006). In contrast, Laznik *et al.* (2010) noted higher mortalities (40–72%) at the highest nematode suspension concentration (2000 IJs/insect) of three *S. feltiae* strains, whereas Shahina and Salma (2010) reported increased susceptibility of *S. oryzae* adults (65–72%) to six Pakistani EPN strains (Fig. 10.1). These contradicting results indicate that the various nematode and host strains used, as well as the different experimental protocols followed, may have an impact on the results. Low susceptibility to EPNs was shown also for *S. zeamais* (Barbosa-Negrisoli *et al.*, 2013).

Tribolium spp.

Tribolium species are cosmopolitan insect pests, common in flour mills and wherever flour and related amylaceous products, such as semolina, bran, etc., are processed or stored (Aitken, 1975).

Various studies have also been conducted for the evaluation of the virulence of EPNs against *T. castaneum* (Ramos-Rodríguez *et al.*, 2006; Shahina and Salma, 2011a) and the confused flour beetle, *Tribolium confusum* (Coleoptera: Tenebrionidae) (Alikhan *et al.*, 1985; Wójcik, 1986a,b; Athanassiou *et al.*, 2008, 2010; Rumbos and Athanassiou, 2012). Larvae of *T. castaneum* were highly susceptible to three steinernematid species, i.e. *S. feltiae*, *S. carpocapsae* and *S. riobrave*, whereas pupae and adult susceptibility was species dependent (Ramos-Rodríguez *et al.*, 2006). Six heat-tolerant Pakistani nematode strains also showed successful results against *T. castaneum* larvae and adults (Shahina and Salma, 2011a), as shown in Fig. 10.2. For *T. confusum* larvae, different responses to EPNs have been demonstrated, varying from low (Alikhan *et al.*, 1985; Rumbos and Athanassiou, 2012) to high susceptibility (Athanassiou *et al.*, 2008), depending mainly on the nematode species and strain. However, most studies agree on the poor infectivity of EPNs against *T. confusum* adults (Athanassiou *et al.*, 2008, 2010; Rumbos and Athanassiou, 2012).

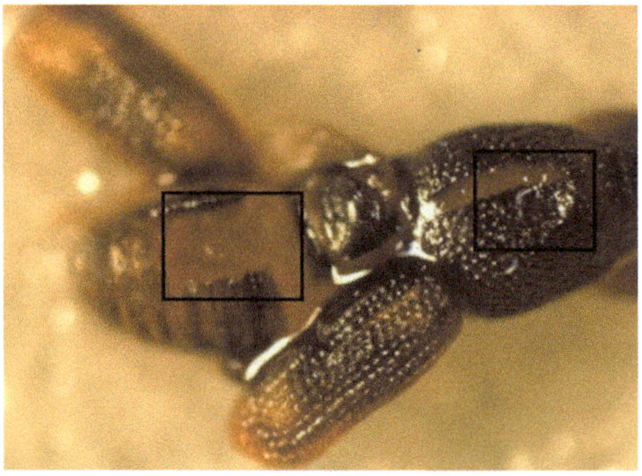

Fig. 10.1. Emergence of progenies of *Steinernema pakistanense* from adult of *Sitophilus oryzae*.

Rhyzopertha dominica (F.)

R. dominica is a primary pest of stored grain in many regions of the world and is injurious to cereals, mainly wheat, maize and rice, and in other substrates containing starch (Edde, 2012). As an internal feeder, *R. dominica* larvae develop inside the kernel; hence, studies with EPNs have focused on the susceptibility of the free-living adult stage, which has exhibited low to moderate susceptibility (Ramos-Rodríguez *et al.*, 2006; Athanassiou *et al.*, 2010).

Other stored-product beetles

A number of studies tested the infectivity of EPNs against mainly the larvae of the mealworm beetle, *Tenebrio molitor* (L.) (Coleoptera: Tenebrionidae), and high mortality levels were achieved in most cases (Morris, 1985; Ramos-Rodríguez *et al.*, 2006, 2007a; Shapiro-Ilan *et al.*, 2009; Barbosa-Negrisoli *et al.*, 2013). For instance, only 2 days after treatment, Shapiro-Ilan *et al.* (2009) reported complete control of *T. molitor* larvae for most of the steinernematid and heterorhabditid species tested. For the sawtoothed grain beetle, *Oryzaephilus surinamensis* (L.) (Coleoptera, Silvanidae), another typical secondary pest of stored grains (Trematerra *et al.*, 2000), sensitivity to EPNs varied considerably depending on the life

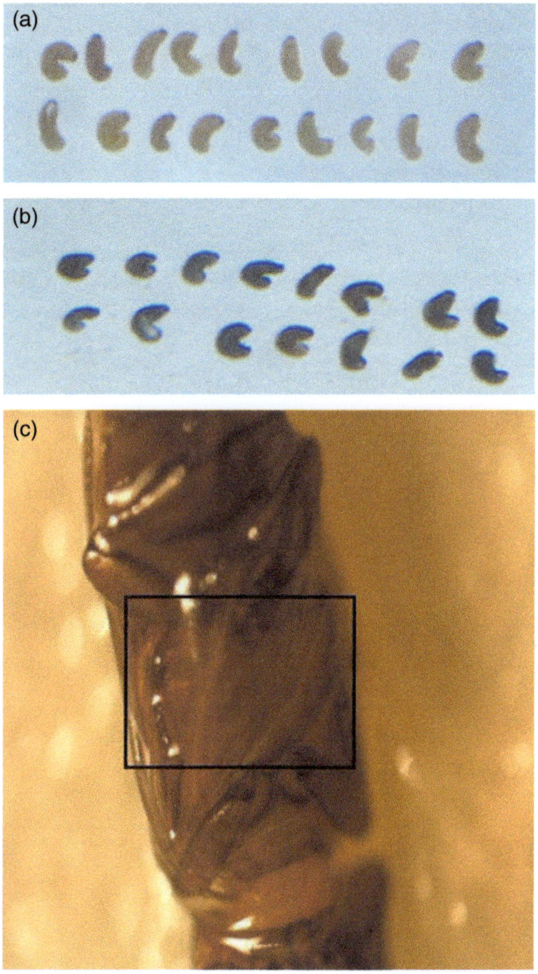

Fig. 10.2. Cadavers of *Tribolium castaneum*: (A) Larvae infested by *Heterorhabditis bacteriophora* (B) Larvae infested by *Steinernema pakistanense* (C) Emergence of new progeny of entomopathogenic nematodes from the adult beetle.

stage and the nematode species (Ramos-Rodríguez *et al.*, 2006; Trdan *et al.*, 2006). Contradicting results have been reported for the susceptibility of the cigarette beetle, *Lasioderma serricorne* (F.) (Coleoptera, Anobiidae), to EPNs. Shahina and Salma (2011b) reported for the first time that the larvae, pupae and adults of *L. serricorne* were highly susceptible to six Pakistani EPN strains (Fig. 10.3). In contrast, Rumbos and Athanassiou (2012) reported poor susceptibility of larval and adult stages of *L. serricorne* to four EPN species. Finally, for the pulse beetle, *Callosobruchus chinensis* (L.) (Coleoptera, Bruchidae), the most serious pest of pulses in tropical and subtropical countries, Shahina and Salma (2009) investigated the effect of seven Pakistani nematode strains, namely *Steinernema pakistanense* Shahina, Anis, Reid, Rowe and Maqbool, *S. asiaticum* Anis, Shahina, Reid and Rowe, *S. abbasi* Elawad, Ahmad and Reid, *S. siamkayai* Stock, Somsook and Reid, *S. feltiae* Filipjeve, 1934, *Heterorhabditis bacteriophora* and *H. indica* Poinar, Karunakar and David against last instar and adult stages of the beetle. In this study, mortality was influenced significantly by nematode density and species, whereas for *H. bacteriophora*, *S. siamkayai*, *S. pakistanense* and *H. indica* complete control (100%) was achieved at the highest density (Fig. 10.4) (Shahina and Salma, 2009).

10.3.2 Lepidoptera

Plodia interpunctella Hübner

A considerable number of reports are available on the virulence of EPNs against *P. interpunctella* (Mbata and Shapiro-Ilan, 2005, 2010; Ramos-Rodríguez *et al.*, 2006; Shapiro-Ilan *et al.*, 2009). *P. interpunctella* is a worldwide insect pest that can infect a variety of stored products and food commodities and is considered among the most economically important insect pest of processed food (Mohandass *et al.*, 2007). Mbata and Shapiro-Ilan (2005) investigated the susceptibility of *P. interpunctella* adults and larvae against seven heterorhabditid species and strains and considered *H. indica*, *H. megidis* and *Heterorhabditis marelatus* Liu and Berry as the most virulent. Moreover, they noted that egg laying was reduced significantly for adults that survived exposure to the nematodes (Mbata and Shapiro-Ilan, 2005). Likewise, steinernematid species, i.e. *S. carpocapsae*, *S. feltiae* and *S. riobrave*, were highly pathogenic to *P. interpunctella* larvae, with over 80% host mortality at doses higher than 10 IJs/individual (Ramos-Rodríguez *et al.*, 2006).

Ephestia kuehniella Zeller

Extensive work has also been conducted with another pyralid, the Mediterranean flour moth, *Ephestia kuehniella* (Zeller) (Lepidoptera: Pyralidae) (Morris, 1985; Ramos-Rodríguez *et al.*,

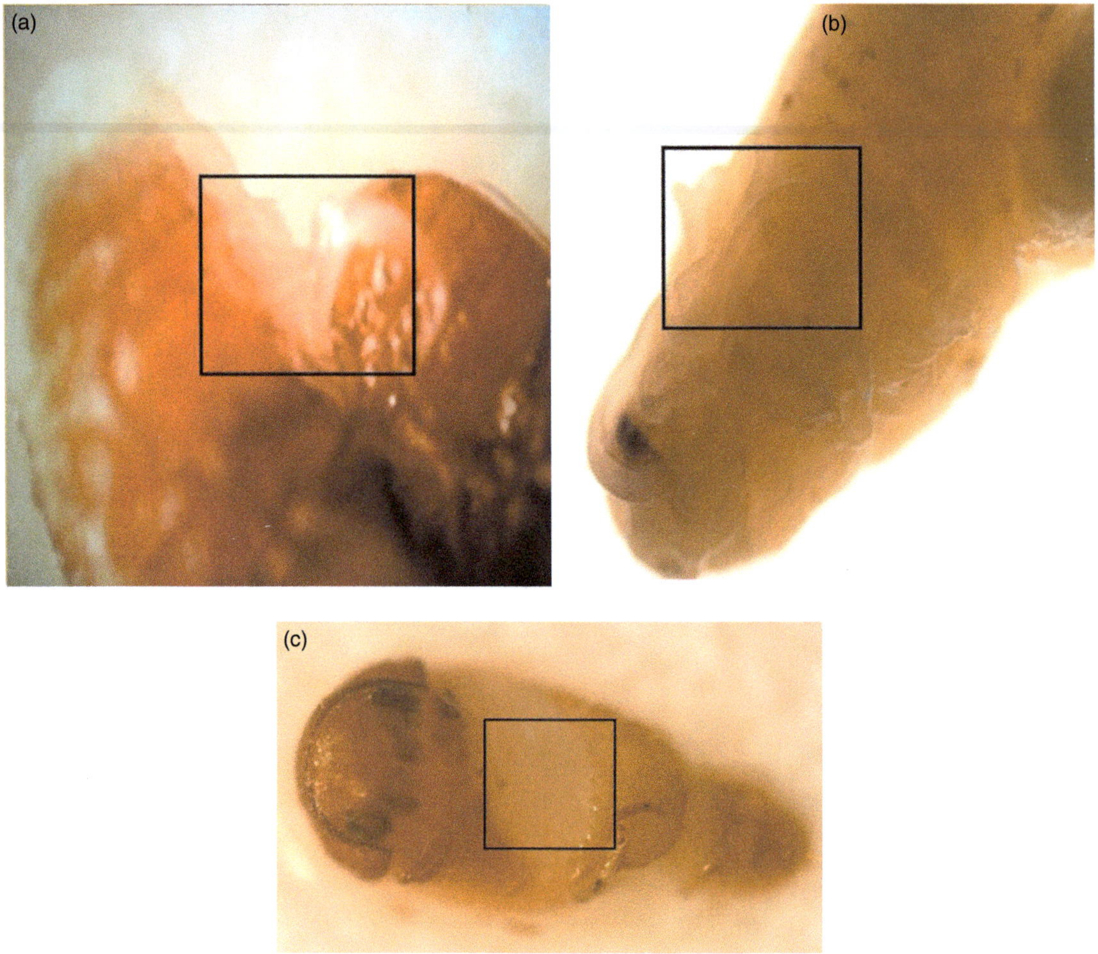

Fig. 10.3. Emergence of progenies of *Steinernema pakistanense* from adult (a), larval (b) and pupal (c) stage of *Lasioderma serricorne*.

2006; Athanassiou *et al.*, 2008, 2010; Barbosa-Negrisoli *et al.*, 2013), a serious pest of amylaceous products and dried fruits, which, apart from direct infestation, degrades the infested products with faeces and webbings (Sedlacek *et al.*, 1995). Larvae of *E. kuehniella* were highly susceptible to *H. bacteriophora*, *S. carpocapsae*, *S. riobrave* and *Steinernema rarum* Doucet, with survival rates lower than 30% at doses higher than 20 IJs/insect (Barbosa-Negrisoli *et al.*, 2013). Using similar dose rates, Ramos-Rodríguez *et al.* (2006) also reported high susceptibility of *E. kuehniella* larvae and adults to *S. carpocapsae*, *S. feltiae* and *S. riobrave*. In contrast, mortality of *E. kuehniella* larvae exposed to wheat treated with three *S. feltiae* strains varied in relation to the nematode strain and the dose, and never exceeded 70%, even at the high dose of 900 IJs/insect (Athanassiou *et al.*, 2008).

10.4 Factors Affecting the Virulence of Entomopathogenic Nematodes against Stored-product Insects

The virulence of EPNs is strongly affected by a number of biotic and abiotic factors. For example, the target insect host has a significant impact on their effectiveness. *S. granarius* adults were more susceptible to *S. feltiae* and *S. carpocapsae* than *O. surinamensis* adults (Trdan *et al.*, 2006).

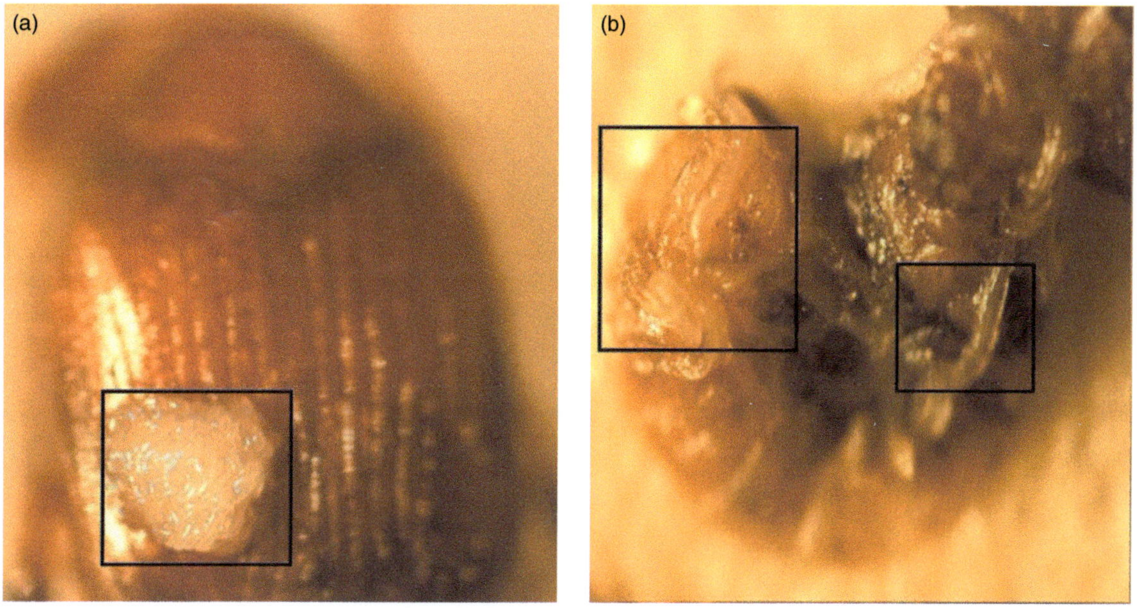

Fig. 10.4. Emergence of progenies of *Steinernema pakistanense* from adult (a) and larval stage of *Callosobruchus chinensis* (b).

Similarly, *H. bacteriophora* controlled to a certain extent *R. dominica* adults, whereas *T. confusum* and *S. oryzae* adults remained unaffected by the EPNs (Athanassiou *et al.*, 2010). The variable sensitivity of postharvest insects to EPNs has been demonstrated by a number of studies (Ramos-Rodríguez *et al.*, 2006; Trdan *et al.*, 2006; Athanassiou *et al.*, 2008, 2010; Shahina and Salma, 2009, 2011b; Barbosa-Negrisoli *et al.*, 2013) and one should always take this into account when designing management strategies. In warehouses and storage facilities in particular, where the coexistence of several postharvest insect species is the rule, whether the application of EPNs will be equally effective against all the insect species present should always be taken into consideration.

The results of various studies have shown that the developmental stage of the target insect host exerts a significant impact on the efficacy of EPNs. For instance, the application of Pakistani EPN strains (Shahina and Salma, 2011a) and *S. carpocapsae* (Ramos-Rodríguez *et al.*, 2006) was highly effective against larvae of *T. castaneum*, but not against their adults and pupae. Furthermore, *T. confusum* adults were less susceptible to the tested nematodes than their larvae (Athanassiou *et al.*, 2008, 2010). In general, the reduced susceptibility of adults to EPNs compared to larvae counts for several postharvest insect species (Ramos-Rodríguez *et al.*, 2007a,b; Shahina and Salma, 2009, 2011a,b; Athanassiou *et al.*, 2010; Rumbos and Athanassiou, 2012) and has been attributed to the structural and behavioral characteristics of the different life stages. For instance, adult beetles are protected by the hard, chitinous exoskeleton, which acts as a mechanical hurdle to the invasion of EPNs, in contrast to the soft body of the larvae, which is more easily penetrable (Gaugler, 1988). Additionally, beetle adults are generally fast moving, making their approach by EPNs more difficult (Athanassiou *et al.*, 2010). Larval stages have been found more susceptible than adults for quite a number of insect pests (Fuxa *et al.*, 1988; Shapiro-Ilan *et al.*, 2002; Trdan *et al.*, 2009). However, there are exceptions to the rule of increased larval susceptibility. Adults of *P. interpunctella* and *E. kuehniella* were more or equally susceptible as larvae to the tested nematodes (Mbata and Shapiro-Ilan, 2005; Ramos-Rodríguez *et al.*, 2006).

Temperature can also be a limiting factor for the infectivity and efficacy of EPNs. *S. feltiae*, *S. carpocapsae*, *H. bacteriophora* and *H. megidis* were more effective in controlling *S. granarius* at 20 and 25°C than at 15°C (Trdan *et al.*, 2006). Likewise, high mortality rates of *S. oryzae* adults after application of three strains of *S. feltiae* were recorded at 25°C, while at 30°C the same strains fell short of providing sufficient control (Laznik *et al.*, 2010). Generally, high temperatures affect the survival and virulence of EPNs negatively, due to the faster exhaustion of the insect's energy reservoirs (Kung *et al.*, 1991; Hsiao and All, 1996; Menti *et al.*, 2000). Therefore, it is essential for the applicator to be aware of the specific thermal requirements and the optimum temperature for infection of each nematode species. For example, *H. indica*, *Steinernema glaseri* Steiner, *S. pakistanense*, *S. asiaticum* and *S. riobrave* are relatively heat tolerant, whereas *H. megidis* and *S. feltiae* are more tolerant to cooler temperatures (Shahina and Salma, 2009, 2010, 2011a,b; Shapiro-Ilan *et al.*, 2012).

The virulence of EPNs is also affected by relative humidity (Kung *et al.*, 1991; Navaneethan *et al.*, 2010). In the only study available on the effect of humidity on the pathogenicity of EPNs against stored-product insects, the efficacy of *S. riobrave* against larvae, pupae and adults of *T. castaneum* was not affected significantly by the four relative humidity levels tested (Ramos-Rodríquez *et al.*, 2007a). However, since humidity requirements vary between nematode species and strains (Koppenhöfer *et al.*, 1995), further research should determine the efficacy of various nematode species and strains against postharvest insects under different moisture regimes. Generally, moisture is necessary for the survival and movement of EPNs towards the host, whereas their activity is affected negatively by low humidity levels, which leads to their desiccation and subsequent death (Kung *et al.*, 1991; Navaneethan *et al.*, 2010; Shapiro-Ilan *et al.*, 2012). This is particularly important for storage environments, where dry conditions usually prevail. Since the control of relative humidity in storage facilities is not always feasible, improvements in the application and formulation of EPNs should, therefore, take into account the limitation of the low humidity levels in the storage environment.

10.5 Limitations and Future Prospects

Even though information on the efficacy of EPNs against stored-product insects is restricted, the research conducted so far provides strong evidence that EPNs are a promising tool for the control of postharvest insects. However, until now, all studies carried out describe laboratory bioassays. Therefore, there is a substantial gap in field studies, where the efficacy of the large-scale application of EPNs will be evaluated under 'real world' conditions. One more weak point of the integration of EPNs in stored-product insect management programmes is their application strategies. Inundative applications of EPNs in bulked grains stored in large storage facilities are vulnerable to much criticism. The low humidity conditions that prevail in these facilities, especially in warm regions, could affect the survival and virulence of EPNs negatively, and subsequently reduce their efficacy. Moreover, the direct spraying of EPNs on edible stored commodities may raise concerns for public health issues and not meet current consumer demands.

To overcome these limitations, the development and evaluation of new, alternative methods of EPN formulations and application approaches becomes crucial. In the past decade, there has been a considerable enhancement of formulations of EPNs that improves their post-application survival and efficacy (Grewal, 2002). For instance, the encapsulation of EPNs has enhanced their efficacy against insect pests of soil (Kaya and Nelsen, 1985; Hiltpold *et al.*, 2012; Vemmer and Patel, 2013; Kim *et al.*, 2015) and could potentially improve their performance under the hostile conditions of storage environments. Improved application methods can also enhance the effectiveness of EPNs. In the lure and kill technique, an attractant (pheromone or kairomone) draws the target insect into a bait station, where it is exposed to the insecticidal agent. This technique has been evaluated recently for applications of entomopathogenic fungi against stored-product insects in storage environments (Baxter, 2008; Baxter *et al.*, 2008) and could be also investigated for the application of EPNs.

10.6 Conclusions

So far, a considerable amount of research has been conducted on the evaluation of EPNs against stored-product insects, providing evidence that EPNs can, under conditions, provide adequate control of postharvest insects, and could be included in future integrated insect management strategies in storage facilities. However, all of the more than 20 published scientific papers on the topic reviewed in this chapter present results from laboratory bioassays. Field studies on the investigation of the efficacy of EPNs against storage insects under 'real world' conditions are missing. Large-scale applications of EPNs in soils, warehouses and related storage facilities will definitely boost research on the topic and should be included in future studies. Finally, new formulations and application methods of EPNs that have already been shown to be effective against other pests merit further evaluation, as they could increase their efficacy against insects in storage environments.

Acknowledgements

Research on Pakistani strains is part of a dissertation submitted by Salma Javed in partial fulfilment of the requirement for a PhD degree in Nematology at the National Nematological Research Centre (NNRC), University of Karachi. Thanks to the Pakistan Science Foundation, the Government of Pakistan (PSF) and the Benaki Phytopathological Institute, Greece, for financial support for the research projects.

References

Aitken, A.D. (1975) Insect travelers I: Coleoptera. *Technical Bulletin* 31. HMSO London.

Alikhan, M.A., Bednarek, A. and Grabiec, S. (1985) The physiological and morphological characteristics of *Neoaplectana carpocapsae* (Nematoda: Steinernematidae) in two insect hosts. *Journal of Invertebrate Pathology* 45, 168–173.

Athanassiou, C.G., Palyvos, N.E. and Kakouli-Duarte, T. (2008) Insecticidal effect of *Steinernema feltiae* (Filipjev) (Nematoda: Steinernematidae) against *Tribolium confusum* du Val (Coleoptera: Tenebrionidae) and *Ephestia kuehniella* (Zeller) (Lepidoptera: Pyralidae) in stored wheat. *Journal of Stored Product Research* 44, 52–57.

Arthur, F.H. (1996) Grain protectants: current status and prospects for the future. *Journal of Stored Product Research* 32, 293–302.

Arthurs, S., Heinz, K.M. and Prasifka, J.R. (2004) An analysis of using entomopathogenic nematode against above ground pests. *Bulletin of Entomological Research* 94, 297–306.

Athanassiou, C.G., Kavallieratos, N.C., Menti, H. and Karanastasi, E. (2010) Mortality of four stored product pests in stored wheat when exposed to doses of three entomopathogenic nematodes. *Journal of Economic Entomology* 103, 977–984.

Barbosa-Negrisoli, C.R.D.C., Negrisoli, A.S.J., Bernardi, D. and Garcia, M.S. (2013) Activity of eight strains of entomopathogenic nematodes (Rhabditida: Steinernematidae, Heterorhabditidae) against five stored product pests. *Experimental Parasitology* 134, 384–388.

Baxter, I.H. (2008) *Entomopathogen based autodissemination for the control of Plodia interpunctella (Hübner) – an examination of the critical components*. PhD thesis, University of Southampton, UK.

Baxter, I.H., Howard, N., Armsworth, C.G., Barton, L.E.E. and Jackson, C. (2008) The potential of two electrostatic powders as the basis for an auto dissemination control method of *Plodia interpunctella* (Hübner). *Journal of Stored Product Research* 44, 152–161.

Brower, J.H., Smith, L., Vail, P.V. and Flinn, P.W. (1996) Biological control. In: Subramanyam, B. and Hagstrum, D.W. (eds) *Integrated Management of Insects in Stored Products*. Marcel Dekker, New York, pp. 223–286.

Chaudhry, M.Q. (2000) Phosphine resistance: a growing threat to an ideal fumigant. *Pesticide Outlook* 11, 88–91.

Ciche, T.A., Darby, C., Ehlers, R.-U., Forst, S. and Goodrich-Blair, H. (2006) Dangerous liaisons: the symbiosis of entomopathogenic nematodes and bacteria. *Biological Control* 38, 22–46.

Collins, P.J., Lambkin, T.M., Bridgeman, B.W. and Pulvirenti, C. (1993) Resistance to grain-protectant insecticides in coleopterous pests of stored cereals in Queensland, Australia. *Journal of Economic Entomology* 86, 239–245.

Daglish, G.J., Eelkema, M. and Harrison, L.M. (1996) Control of *Sitophilus oryzae* (L.) (Coleoptera: Curculionidae) in paddy rice using chlorpyrifos-methyl or fenitrothion in combination with several other protectants. *Journal of Stored Product Research* 32, 247–253.

Donahaye, E.J. (2000) Current status of non-residual control methods against stored product pests. *Crop Protection* 19, 571–576.

Edde, P.A. (2012) A review of the biology and control of *Rhyzopertha dominica* (F.) the lesser grain borer. *Journal of Stored Products Research* 48, 1–18.

Flinn, P.W. and Schöller, M. (2012) Biological control: insect pathogens, parasitoids, and predators. In: Hagstrum, D.W., Phillips, T.W. and Cuperus, G. (eds) *Stored Product Protection. S156*. Kansas State University, Manhattan, New York, pp. 203–212.

Fuxa, J.F., Agudelo-Silva, F. and Richter, A.R. (1988) Effect of host age and nematode strain on susceptibility of *Spodoptera frugiperda* to *Steinernema feltiae. Journal of Nematology* 20, 91–95.

Gaugler, R. (1988) Ecological considerations in the biological control of soil-inhabiting insects with entomopathogenic nematodes. *Agriculture, Ecosystems and Environment* 24, 351–360.

Gaugler, R. (2002) *Entomopathogenic Nematology.* CAB International, Wallingford, UK.

Geden, C.J., Arends, J.J. and Brooks, W.M. (1985) Susceptibility of the lesser mealworm, *Alphitobius diaperinus* (Coleptera: Tenebrionidae) to the entomogenous nematodes *Steinernema feltiae*, *S. glaseri* (Steinernematidae) and *Heterorhabditis heliothidis*

(Heterorhabditidae). *Journal of Entomology Science* 20, 331–339.

Grewal, P.S. (2002) Formulation and application technology. In: Gaugler, R. (ed.) *Entomopathogenic Nematology*. CAB International, Wallingford, UK, pp. 265–287.

Griffin, C.T., Boemare, N.E. and Lewis, E.E. (2005) Biology and behavior. In: Grewal, P.S., Ehlers, R.-U. and Shapiro-Ilan, D.I. (eds) *Nematodes as Biocontrol Agents*. CAB International, Wallingford, UK, pp. 47–64.

Guedes, R.N.C. and Pereira, E.J.G. (2008) Maize weevil × insecticides: pyrethroid resistance, associated fitness costs and mitigation, and management implications. In: Burton, E.N. and Williams, P.V. (eds) *Crop Protection Research Advances*. Nova Science Publishers, New York, pp. 125–150.

Hiltpold, I., Hibbard, B.E., Wade, F.B. and Turlings, T.C.J. (2012) Capsules containing entomopathogenic nematodes as a Trojan horse approach to control the Western corn rootworm. *Plant and Soil* 358, 11–25.

Hsiao, W.F. and All, J.N. (1996) Effects of temperature and placement site on the dispersal of the entomopathogenic nematode, *Steinernema carpocapsae* in four soils. *Chinese Journal of Entomology* 16, 95–106.

Ishibashi, N. and Kondo, E. (1990) Behavior of infective juveniles. In: Gaugler, R. and Kaya, H.K. (eds) *Entomopathogenic Nematodes in Biological Control*. CRC Press, Boca Raton, Florida, pp. 139–152.

Kavallieratos, N.G., Athanassiou, C.G., Aountala, M. and Kontodimas, D.C. (2014) Evaluation of the entomopathogenic fungi *Beauveria bassiana*, *Metarhizium anisopliae*, and *Isaria fumosorosea* for control of *Sitophilus oryzae*. *Journal of Food Protection* 77, 87–93.

Kaya, H.K. and Gaugler, R. (1993) Entomopathogenic nematodes. *Annual Review of Entomology* 38, 181–206.

Kaya, H.K. and Nelsen, C.E. (1985) Encapsulation of steinernematid and heterorhabditid nematodes with calcium alginate: a new approach for insect control and other applications. *Environmental Entomology* 14, 572–574.

Kim, J., Jaffuel, G. and Turlings, T.C.J. (2015) Enhanced alginate capsule properties as a formulation of entomopathogenic nematodes. *Biocontrol* 60, 527–535.

Koppenhöfer, A.M., Kaya, H.K. and Taormino, S.P. (1995) Infectivity of entomopathogenic nematodes (Rhabditida: Steinernematidae) at different soil depths and moistures. *Journal of Invertebrate Pathology* 65, 193–199.

Kung, S.P., Gaugler, R. and Kaya, H.K. (1991) Effects of soil temperature, moisture, and relative humidity on entomopathogenic nematode persistence. *Journal of Invertebrate Pathology* 57, 242–249.

Laznik, Z., Tóth, T., Lakatos, T., Vidrih, M. and Trdan, S. (2010) The activity of three new strains of *Steinernema feltiae* against adults of *Sitophilus oryzae* under laboratory conditions. *Journal of Food Agriculture and Environment* 8, 150–154.

Lord, J.C., Campbell, J.F., Sedlacek, J.D. and Vail, P.V. (2007) Application and evaluation of entomopathogens for managing insects in stored products. In: Lacey, L.A. and Kaya, H.K. (eds) *Field Manual of Techniques in Invertebrate Pathology*, Second edn. Springer, Dordrecht, The Netherlands, pp. 677–693.

Mbata, G.N. and Shapiro-Ilan, D.I. (2005) Laboratory evaluation of virulence of Heterorhabditid nematodes to *Plodia interpunctella* Hübner (Lepidoptera: Pyralidae). *Environmental Entomology* 34, 676–682.

Mbata, G.N. and Shapiro-Ilan, D.I. (2010) Compatibility of *Heterorhabditis indica* (Rhabditida: Heterorhabditidae) and *Habrobracon hebetor* (Hymenoptera: Braconidae) for biological control of *Plodia interpunctella* (Lepidoptera: Pyralidae). *Biological Control* 54, 75–82.

Menti, H., Wright, D.J. and Perry, R.N. (2000) Infectivity of populations of the entomopathogenic nematodes *Steinernema feltiae* and *Heterorhabditis megidis* in relation to temperature, age and lipid content. *Nematology* 2, 515–521.

Mohandass, S., Arthur, F.H., Zhu, K.Y. and Throne, J.E. (2007) Biology and management of *Plodia interpunctella* (Lepidoptera: Pyralidae) in stored products. *Journal of Stored Products Research* 43, 302–311.

Moore, D., Lord, J.C. and Smith, S.M. (2000) Pathogens. In: Subramanyam, B. and Hagstrum, D.W. (eds) *Alternatives to Pesticides in Stored Product IPM*. Kluwer Academic Publishers, Boston, Massachusetts, pp. 193–227.

Morris, O.N. (1985) Susceptibility of 31 species of agricultural pests to entomogenous nematodes *Steinernema feltiae* and *Heterorhabditis bacteriophora*. *Canadian Entomologist* 122, 309–320.

Morton, A. and García-del-Pino, F. (2013) Sex-related differences in the susceptibility of *Periplaneta americana* and *Capnodis tenebrionis* to the entomopathogenic nematode *Steinernema carpocapsae*. *Journal of Invertebrate Pathology* 112, 203–207.

Navaneethan, T., Strauch, O., Besse, S., Bonhomme, A. and Ehlers, R.U. (2010) Influence of humidity and a surfactant-polymer-formulation on the control potential of the entomopathogenic nematode *Steinernema feltiae* against diapausing codling moth larvae (*Cydia pomonella* L.) (Lepidoptera: Tortricidae). *Biocontrol* 55, 777–788.

Neethirajan, S., Karunakaran, C., Jayas, D.S. and White, N.D.G. (2007) Detection techniques for stored-product insects in grain. *Food Control* 18, 157–162.

Ngoma, L., Nyamboli, M.A., Gray, V.M. and Babalola, O.O. (2012) Virulence of two entomopathogenic nematodes (*Heterorhabditis bacteriophora*, *Heterorhabditis zealandica*) to *Galleria mellonella* (Lepidoptera: Pyralidae), *Tenebrio molitor* (Coleoptera: Enebrionidae) and pupae in the laboratory. *Life Science Journal* 9, 2572–2579.

Nguyen, T.T., Collins, P.J. and Ebert, P.R. (2015) Inheritance and characterization of strong resistance to phosphine in *Sitophilus oryzae* (L.). *Plos One* 10(4), e0124335.

Niedermayer, S. and Steidle, J.L.M. (2013) The Hohenheimer Box – a new way to rear and release *Lariophagus distinguendus* to control stored product pest insects. *Biological Control* 64, 263–269.

Opit, G.P., Phillips, T.W., Aikins, M.J. and Hasan, M.M. (2012) Phosphine resistance in *Tribolium castaneum* and *Rhyzopertha dominica* from stored wheat in Oklahoma. *Journal of Economic Entomology* 105, 1107–1114.

Phillips, T.W. and Throne, J.E. (2010) Biorational approaches to managing stored-product insects. *Annual Review of Entomology* 55, 375–397.

Poinar, G.O. (1990) Biology and taxonomy of Steinernematidae and Heterorhabditidae. In: Gaugler, R. and Kaya, H.K. (eds) *Entomopathogenic Nematodes in Biological Control*. CRC Press, Boca Raton, Florida, pp. 23–62.

Ramos-Rodríguez, O., Campbell, J.F. and Ramaswamy, S.B. (2006) Pathogenicity of three species of entomopathogenic nematodes to some major stored-product insect pests. *Journal of Stored Product Research* 42, 241–252.

Ramos-Rodríguez, O., Campbell, J.F. and Ramaswamy, S.B. (2007a) Efficacy of the entomopathogenic nematode *Steinernema riobrave* against the stored-product insect pests *Tribolium castaneum* and *Plodia interpunctella*. *Biological Control* 40, 15–21.

Ramos-Rodríguez, O., Campbell, J.F., Christen, J.M., Shapiro-Ilan, D.I., Lewis, E.E. and Ramaswamy, S.B. (2007b) Attraction behaviour of three entomopathogenic nematode species towards infected and uninfected hosts. *Parasitology* 134, 729–738.

Roodaki, M., Haghani, M. and Abdollahi, M. (2012) Isolation of *Steinernema feltiae* and its pathogenicity on rice weevil, *Sitophilus oryzae* (Coleoptera: Curculionidae). *Iranian Journal of Entomological Research* 5, 83–93.

Rumbos, C.I. and Athanassiou, C.G. (2012) Insecticidal effect of six entomopathogenic nematode strains against *Lasioderma serricorne* (F.) (Coleoptera: Anobiidae) and *Tribolium confusum* Jacquelin du Val (Coleoptera: Tenebrionidae). *Journal of Stored Product Research* 50, 21–26.

Schöller, M., Prozell, S., Al-Kirshi, A.G. and Reichmuth, C. (1997) Towards biological control as a major component of integrated pest management in stored product protection. *Journal of Stored Products Research* 33, 81–97.

Sedlacek, J.D., Weston, P.A. and Barney, R.J. (1995) Lepidoptera and Psocoptera. In: Subramanyam, B. and Hagstrum, D.W. (eds) *Integrated Management of Insects in Stored Products*. Marcel Decker, New York, pp. 41–70.

Shahina, F. and Salma, J. (2009) Laboratory evaluation of seven Pakistani strains of entomopathogenic nematode against a stored product insect pest, pulse beetle (*Callosobruchus chinensis* L.). *Journal of Nematology* 41, 255–260.

Shahina, F. and Salma, J. (2010) Laboratory evaluation of seven Pakistani strains of entomopathogenic nematode against a stored product insect pest, rice weevil (*Sitophilus oryzae* L.). *Pakistan Journal of Nematology* 28, 295–305.

Shahina, F. and Salma, J. (2011a) Pakistani strains of entomopathogenic nematode as a biological control agent against stored grain pest, *Tribolium castaneum. Pakistan Journal of Nematology* 29, 25–34.

Shahina, F. and Salma, J. (2011b) Virulence of Pakistani isolates of entomopathogenic nematode against a stored insect species, *Lasioderma serricorne. International Journal of Nematology* 21, 79–85.

Shapiro-Ilan, D.I., Mizell, R.F. and Campbell, J.F. (2002) Susceptibility of the plum curculio, *Conotrachelus nenuphar*, to entomopathogenic nematodes. *Journal of Nematology* 34, 246–249.

Shapiro-Ilan, D.I., Mbata, G.N., Nguyen, K.B., Peat, S.M., Blackburn, D. and Adams, B.J. (2009) Characterization of biocontrol traits in the entomopathogenic nematode *Heterorhabditis georgiana* (Kesha strain), and phylogenetic analysis of the nematode's symbiotic bacteria. *Biological Control* 51, 377–387.

Shapiro-Ilan, D.I., Han, R. and Dolinksi, C. (2012) Entomopathogenic nematode production and application technology. *Journal of Nematology* 44, 206–217.

Smart, G.C. (1995) Entomopathogenic nematodes for the biological control of insects. *Journal of Nematology* 27, 529–534.

Steiner, G. (1923) *Aplectana kraussei* n. sp., eine in der Blattwespe *Lyda* sp. parasitierende Nematodenform, nebst Bemerkungen über das Seitenorgan der parasitischen Nematoden. Zbl. Bakt. Parasitenk. Infektionskrank. Hyg. Abt. 2, 59: 14–18.

Stock, S.P. and Blair, H.G. (2008) Entomopathogenic nematodes and their bacterial symbionts: the inside out of a mutualistic association. *Symbiosis* 46, 65–75.

Susurluk, A. (2006) Effectiveness of the entomopathogenic nematodes *Heterorhabditis bacteriophora* and *Steinernema feltiae* against *Tenebrio molitor* (Yellow mealworm) larvae in different soil types at different temperatures. *Turkish Journal of Biology* 30, 199–205.

Svendsen, T.S. and Steenberg, T. (2000) The potential use of entomopathogenic nematodes against *Typhaea stercorea*. *BioControl* 45, 97–111.

Trdan, S., Vidric, M., Urek, G. and Milevoj, L. (2005) Concentration of suspension and temperature as factors of pathogenicity of entomopathogenic nematodes for the control of granary weevil, *Sitophilus granarius* (L.) (Coleoptera: Curculionidae). *Acta Agriculturae Slovenica* 85, 117–124.

Trdan, S., Vidric, M. and Valic, N. (2006) Activity of four entomopathogenic nematodes against young adult of *Sitophilus granarious* (Coleoptera: Curculionidae) and *Oryzaephilus surinomensis* (Coleoptera: Silvanidae) under laboratory condition. *Journal of Plant Diseases and Protection* 113, 168–173.

Trdan, S., Vidric, M., Andjus, L. and Laznik, Z. (2009) Activity of four entomopathogenic nematode species against different developmental stages of Colorado potato beetle, *Leptinotarsa decemlineata* (Coleoptera: Chrysomelidae). *Helminthology* 46, 1–14.

Trematerra, P., Sciarreta, A. and Tamasi, E. (2000) Behavioural responses of *Oryzaephilus surinamensis*, *Tribolium castaneum* and *Tribolium confusum* to naturally and artificially damaged durum wheat kernels. *Entomologia Experimentalis et Applicata* 94, 195–200.

Vemmer, M. and Patel, A.V. (2013) Review of encapsulation methods suitable for microbial biological control agents. *Biological Control* 67, 380–389.

Wójcik, W.F. (1986a) Influence of the size of host on the growth of the *Neoplectana carpocapsae* Weiser, 1955 nematodes. *Annals of Warsaw Agricultural University SGGW-AR Animal Science* 20, 75–85.

Wójcik, W.F. (1986b) Effect of worsening of physiological state of insects on the growth of the nematodes, *Neoplectana carpocapsae* Weiser, 1955. *Annals of Warsaw Agricultural University SGGW-AR Animal Science* 20, 109–111.

Zettler, J.L. and Arthur, F.H. (2000) Chemical control of stored product insects with fumigants and residual treatments. *Crop Protection* 19, 577–582.

11 Toxic Secretions of *Xenorhabdus* and Their Efficacy Against Crop Insect Pests

Nazir Javed,[1]* Muhammad Kamran[2] and Huma Abbas[1]

[1]*Department of Plant Pathology, University of Agriculture Faisalabad, Pakistan;* [2]*Plant Pathology Research Institute, Ayub Agricultural Research Institute, Faisalabad, Pakistan*

11.1 Introduction

Entomopathogenic nematodes (EPNs) such as *Steinernema* and *Heterorhabditis* are multicellular organisms that harbour bacteria in their intestine. They have the potential to provide effective control of some commercially important insect pests belonging to orders lepidoptera, coleoptera and diptera (Burnell and Stock, 2000). They are established all over the world in diverse ecological habitats and are being developed as biological control agents of insect pests (Cranshaw and Zimmerman, 2013).

EPNs carry symbiotic bacteria (*Steinernema* carry *Xenorhabdus*, *Heterorhabditis* carry *Photorhabdus*) in their intestine (Thomas and Poinar, 1979; Boemare *et al.*, 1993; Forst *et al.*, 1997; Nagesh *et al.*, 2002; Askary, 2010); the third-stage juveniles harbour the bacteria monoxenically within a bi-lobed vesicle in the anterior portion of the intestine (Bird and Akhurst, 1983; Martens *et al.*, 2003). These are motile, Gram-negative bacteria that belong to the family Enterobacteriaceae. The nematode gains entry into the insect via natural openings. On reaching the haemocoel, the nematode releases bacteria from its intestine into the insect haemolymph. The bacteria multiply rapidly there and release toxic secretions that cause septicaemia (Forst and Nealson, 1996; Ali *et al.*, 2005), killing the host insect within 48 h (Woodring and Kaya, 1988). The nematode feeds on the multiplying bacteria, moults and becomes adult. This represents a symbiotic association between nematode and bacteria.

Some strains of *Xenorhabdus* have been found highly virulent against insect pests. Even <10 cells of bacterium are sufficient to kill susceptible insects such as *Manduca sexta* or *Galleria mellonella* when injected into their haemocoel (Poinar and Thomas, 1967; Forst *et al.*, 1997). *Xenorhabdus*, when cultured in liquid medium, have been found to secrete insecticidal toxins of a highly virulent nature (Jarrett *et al.*, 1997). Enzymes such as lipase, protease and several broad-spectrum antibacterial and antifungal antibiotics are released by bacteria as they enter the stationary phase of their life cycle (Akhurst and Boemare, 1990; Forst and Nealson, 1996). The enzymes break down the insect tissues, and as a result there is a bioconversion of insect larvae into a nutrient soup, ideal for the growth and reproduction of nematodes (Ali *et al.*, 2005).

In this chapter, we discuss in detail the toxic secretions of *Xenorhabdus* that, besides possessing antimicrobial properties, also have the potential to kill insect pests of agricultural importance.

11.2 Dependency of *Steinernema* on *Xenorhabdus*

Steinernema spp. require living insect hosts to complete their life cycle. The infective stage of *Steinernema*, i.e. the third-stage juveniles, also called infective juveniles (IJs), finds a suitable host insect by following various entry points such as

*Corresponding author. E-mail: nazirpp2003@gmail.com

the mouth, anus and spiracles. Shapiro-Ilan *et al.* (2012) observed that complexities in finding the host were due to various environmental factors playing a crucial role in the movement of nematodes. These may be low temperature as well as a low oxygen level. The reproduction of *Steinernema* depends on its symbiotic relationship with the bacteria, *Xenorhabdus*; the bacteria secrete toxic compounds and hydrolytic exoenzymes, which kill the host insect. As a result, there is a bioconversion of insect larva into a nutrient soup that is ideal for the growth and reproduction of nematodes. The bacteria also secrete some substances that prevent the other pathogens or secondary microorganisms from attacking the insect. Ansari *et al.* (2005) observed the growth of fungal and other bacterial entrants inhibited by *Xenorhabdus*. Thus, the toxic secretions are meant to kill not only the host insect but also other competitors. The toxins may be antibacterial, antifungal or insecticidal, which make the prey specific to the nematodes (Khandelwal and Banerjee-Bhatnagar, 2003).

11.3 Dependency of *Xenorhabdus* on *Steinernema*

Xenorhabdus depends on *Steinernema* because it cannot survive in a natural habitat. There is no evidence of *Xenorhabdus* isolated from soil, and it is assumed, therefore, that they cannot exist in the soil environment without their nematode associates. *Xenorhabdus* establishes a symbiotic relationship with the nematode for survival. The nematode acts as a vector and carries the bacteria into the insect haemocoel. Thus, the bacteria are protected by the nematode from the external environment. The bacteria, on reaching the insect haemolymph, multiply very rapidly as they find a suitable medium to proliferate. Different species of *Xenorhabdus* associated with *Steinernema* are presented in Table 11.1.

In the following sections, we have discussed three species of *Xenorhabdus*, i.e. *Xenorhabdus nematophila*, *Xenorhabdus indica* and *Xenorhabdus bovienii*, and their toxic secretions that are lethal to insect pests. However, the main emphasis has been laid on *X. nematophila*, due to their association with the nematode *Steinernema carpocapsae*, one of the most successful EPNs exploited against crop insect pests and which are commercialized in the biopesticide market.

Table 11.1. Different species of *Xenorhabdus* symbiotically associated with *Steinernema* species. (From Ganguly, 2006; Askary, 2010).

Bacterium	Entomopathogenic nematode
Xenorhabdus nematophila	*Steinernema carpocapsae*
X. poinarii	*S. cubanum*
X. poinarii	*S. glaseri*
X. bovienii	*S. kraussei*
X. bovienii	*S. affine*
X. bovienii	*S. feltiae*
X. bovienii	*S. intermedium*
X. beddingii	*S. longicaudatum*
X. indica	*S. thermophilum*
X. innexi	*S. scapterisci*
X. japonica	*S. kushidai*
X. szentirmaii	*S. rarum*
X. ehlersii	*S. serratum*
X. budapestensis	*S. bicornutum*

11.4 *Xenorhabdus nematophila* corrig. (Poinar and Thomas 1965) Thomas and Poinar 1979

11.4.1 Significance of *Xenorhabdus nematophila* as a biopesticide

The most appropriate bacterium that could be used as a biopesticide is *X. nematophila*. *X. nematophila* is a Gram-negative, facultative anaerobic, non-spore-forming bacterium with numerous peritrichous flagella mutualistically associated with the nematode (Akhurst and Boemare, 1990). Approximately 30–200 cells of *X. nematophila* occupy a specific region of the intestinal lumen of IJs of *S. carpocapsae*.

X. nematophila multiply by attacking the tissues and immune system of the insect, which leads to insect death. *X. nematophila* has been found virulent even without a nematode when injected alone into insects (Herbert et al., 2007). The life cycle of *X. nematophila* consists of three stages: (i) insect infection; (ii) nematode and bacterial reproduction; and (iii) transmission, i.e. reassociation. The bacteria have potential virulence features, including an assortment of toxins, degradative enzymes and regulators of these compounds. Both the organisms, i.e. the nematode and the bacteria, produce a variety of metabolites throughout their life cycle, helping them to inhabit and reproduce within the insect's body. The production of metabolites by both the nematode and the bacteria contributes to the success of their symbiotic association against different insect hosts (Collmer *et al.*, 2002); they help evade the insect immune system, enzymes such as proteases, lipases and phospholipases to preserve the food supply required during reproduction, and to prevent the colonization of the insect carcass from other fungal and bacterial agents during reproduction (Akhurst and Dunphy, 1993).

There are two phases of bacteria, i.e. phase I and phase II. Phase I cell-type of *X. nematophilus* is normally associated with the nematode, whereas phase II can form spontaneously under stationary phase conditions. Phase II cells do not elaborate products normally associated with the phase I cell type. In order to achieve new phenotypic traits, several strains (19061, AN6, F1 and N2-4) of *X. nematophilus* bacterium were analysed. The pathogenicity in the larvae of *M. sexta* revealed that the phase II form of AN6 (AN6/II) was significantly less virulent in comparison to the phase I form (AN6/I); however, cells of both phase I and phase II supported the growth and development of the nematode. Although certain phenotypic traits are common to all phase II cells, virulence and motility are variable and are governed by environmental conditions (Volgyi *et al.*, 1998). In the secretions of *X. nematophila*, the particulate fraction of the insecticidal proteins is more active than the soluble fraction. The secretions are sensitive to heat. Khandelwal and Banerjee-Bhatnagar (2003) reported that the outer membrane vesicles when heated to 80°C for 15 min resulted in the deactivation of insecticidal activity. Global regulators that play a role in numerous phases of the life cycle of *X. nematophila* are the leucine-responsive regulatory protein Lrp, the two-component system CpxRA and the LysR-type regulator LrhA. CpxR is the response regulator of a two-component system (Herbert *et al.*, 2007) like LrhA and Lrp that affects motility positively. *X. nematophila* creates a plethora of putative virulence factors that include toxins, lipases, haemolysins and proteases, which may attribute to disease and intonation of host immunity (Forst *et al.*, 1997). Unlike LrhA and Lrp, CpxR has a negative effect on haemolysin activity, protease and antibiotic activities. Park *et al.* (2004, 2007) observed that *X. nematophila* restrained transcription of antimicrobial peptides (AMPs) and could check the formation of nodules around bacterial cells. The formation of nodules could be mediated by an inhibitor of insect phospholipase A (Kim *et al.*, 2005). But the mechanism through which AMP gene expression is suppressed is still unclear. *cpxR* and *lrp* mutants have flaws in restraining AMP induction (Cowles *et al.*, 2007), while *lrp* mutants are also imperfect in suppressing nodule formation. But this inability still has not been measured for *cpxR*.

The symbiotic bacteria may work effectively without their vector. *X. nematophila* has been found to inhibit the antibacterial humoral immune reaction, as well as the cellular immune reaction in beet armyworm, *Spodoptera exigua* (Ji and Kim, 2004). The inhibition of *X. nematophila* on the expression of the antibacterial peptide is not associated with inhibition of the eicosanoid pathway. There is a significant effect of media on the toxicity of bacteria. The secretions of *X. nematophila* are more lethal to *G. mellonella* larvae when applied in sand media than in filter substrate. Under laboratory conditions, a bacterial concentration of 4×10^6 cells/ml in broth caused mortality of all the larvae of *Galleria*. Thus, incursion of *Xenorhabdus* and *S. carpocapsae* is lethal when into the haemocoel of insects (Samaliev *et al.*, 2000). The bacterial cells of *X. nematophila* penetrated into the body of *Galleria* larva within 4 h in a broth culture, while in water it took 16 h for penetration. The bacterial cells possibly play the role of a parasite, or produce toxins to break the immunity of the invading insect. *X. nematophila* is supposed to enter the insect body via body openings such as the mouth, anus or spiracles, but it is still not clear how it gains entry into the haemocoel (Boemare *et al.*, 1997).

Under moist conditions, the motile bacterial cells probably gain entry into the host insect via the spiracle, the only organ besides the mouth or anus having direct entry to the external environment. In comparison to stored secretions, the application of fresh bacterial secretions could be more lethal to insect pests. Likewise, the efficacy of the dried form of secretion is more than that of the wet. Cell suspension of *X. nematophila* has been found to be toxic to *Solenopsis invicta* (fire ant) (Dudney, 1997) and *S. exigua* (larvae and pupae) (Elawad *et al.*, 1999).

Thus, *X. nematophila* has great potential as a biocontrol agent for harmful insects in obscure environments. Toxins of *X. nematophila* could feasibly be used to protect plants by the transformation of the toxin gene or genes (Bowen and Ensign, 1998) or by application of the toxin directly, an approach for the control of pests.

11.4.2 Toxic secretions of *Xenorhabdus nematophila*

X. nematophila produce toxins that are linked to pathogenicity and virulence. Toxins may cause direct toxicity to the host (aiding directly to cause disease and death), immunosuppression (aiding in overcoming host defence; antibiotic activity), or refrain competition from other pathogens or saprobes.

Xenocin

Under environmental stress conditions such as high population density and nutrient depletion, bacteriocins (antimicrobial peptides) are produced by bacteria (Cascales *et al.*, 2007). Xenocin operon of *X. nematophila* consists of *xciA* and *ximB* genes encoding a 64-kDa xenocin and a 42-kDa immunity protein that are able to kill the microbes in the insect larva. Catalytic domain of xenocin has RNase activity that causes cytotoxicity. The immunity protein is produced by a second constitutive promoter and targeted to the periplasm in a flagellum-independent manner (Singh *et al.*, 2013). The stable expression of xenocin depends on co-expression of the immunity protein.

The physiological importance of bacteriocin export through the flagellar route rather than cell lysis is reflected in the way that *X. nematophila* maintain a high population density and are adventitious over competing species. Besides, dissemination of secretion through flagella makes clear that it is effective in the environment. However, regarding *X. nematophila*, further investigation is needed as to whether the flagella can transport xenocin directly to the target host (Roy *et al.*, 2009).

42-kDa protein

A variety of protein that causes insect mortality is generated by *X. nematophila*. Screening of a cosmid library constructed from *X. nematophila* strain A24 has identified a gene that encodes a novel protein toxic to insects. 42-kDa protein encoded by the toxin gene was expressed and purified from a recombinant system. The protein killed the larvae of *G. mellonella* and *Helicoverpa armigera* when inserted at doses of around 30–40 ng/g larva. Preliminary histological analysis revealed that toxin primarily proceeded on the midgut of the insect larva (Brown *et al.*, 2004).

Txp40

Toxin gene *txp40* was identified in 59 strains of *Xenorhabdus* and *Photorhabdus* bacteria, illustrating that it is equally highly conserved and widespread. Recombinant toxin protein was found active against a large number of insect species when injected in larvae of the greater wax moth, *G. mellonella*, the Indian meal moth, *Plodia interpunctella*, the cotton bollworm, *H. armigera*, and the Australian sheep blowfly, *Lucilia cuprina*. Protein exhibited significant cytotoxicity against two dipteran cell lines and two lepidopteran cell lines, but not against a mammalian cell line. Histological data from the larvae of *H. armigera* into which the toxin was injected indicated that the primary site of action of the toxin is the midgut, though some harm to the fat body was also observed (Brown *et al.*, 2006). A total of 47 *Xenorhabdus* strains and 12 *Photorhabdus* strains were proven to contain a gene closely related to *txp40* A24. Recombinant Txp40 toxin was found to be active against a wide range of lepidopteran pests such as *G. mellonella*, *P. interpunctella* and *H. armigera*.

This report suggests that the bacteria occupy a niche in the extracellular matrix surrounding the midgut. Thus, the bacteria are associated intimately with the midgut epithelial cells. Bacterial colonization begins in the anterior midgut and haemolymph, develops to the posterior midgut and finally spreads to other tissues. Despite the speculation about the toxin's mode of action, it is obvious that Txp40 is highly conserved in *Xenorhabdus*.

17-kDa pilin

Khandelwal *et al.* (2004) reported oral toxicity on insect larvae caused by 17-kDa pilin subunit. The pervasive damage was found on the midgut epithelial membrane of fourth- and fifth-stage larvae of *H. armigera*. Phylogenetic comparison of the deduced amino acid sequence of the 17-kDa protein with other fimbrial sequences identified by position-specific iterated basic local alignment search tool (PSI-BLAST) exhibited that *Xenorhabdus* protein formed a distinct fixed cluster with the structural subunits of mannose-resistant fimbrial subunits (including MrpA of *Proteus mirabilis* and SfmA of *Serratia marcescens*). The sugar derivatives inhibit the cytotoxicity of cultured haemocytes. This indicates a specific recognition event between the pilin protein and insect cells. The association of oral larvicidal activity and concomitant damage of the gut membrane with the structural subunit of the pilin protein is considered an important finding.

A number of endosymbiotic bacterial genomes when sequenced showed that obligate host-associated bacteria have some of the negligible genomes known in nature (Moran, 2002). The production of proteins with double functions may be to minimize the genome size in *X. nematophila* so as to gain a competitive edge in the natural habitat. The abundance of pilin protein on the cell surface, which is primarily used for recognition of the host, also functions to intoxicate the insect larva rapidly. Thus, with the indication that protein can presume a structure capable for biological activity without the support of any other factor may be encouraging for the discovery of new biopesticides in future programmes.

Fimbrial shaft protein

Banerjee *et al.* (2006) purified a fimbrial shaft protein (MrxA) of *X. nematophila*. The soluble monomeric protein lysed larval haemocytes of *H. armigera*. Osmotic protection of the cells with polyethylene glycol indicated that the 17-kDa MrxA subunit made pores (inner diameter >2.9 nm) in the target cell membrane. The large, voltage-gated passive-diffusion channels in lipid bilayers were formed by MrxA protein. The sanitized/purified MrxA protein caused inflammation of larval haemocytes, trailed by cytolysis in a dose-dependent manner.

Outer membrane vesicles (OMV)

The outer membrane of Gram-negative bacteria such as *X. nematophilus* performs many specialized functions under adverse environmental conditions. They are known to secrete protein when the bacterial pathogens are inside their hosts (Pugsley, 1993). The first report of blebbing of the outer membrane from the surface of *X. nematophilus* was given by Khandelwal and Banerjee-Bhatnagar (2003). They purified OMVs obtained from the culture supernatant of *X. nematophilus* and analysed it. It was found that live cells of *X. nematophilus* and the OMV proteins derived from them had oral insecticidal activity against neonatal larvae of *H. armigera*. The OMV proteins showed chitinase activity.

GroEL protein

X. nematophila secretes an insecticidal GroEL protein toxic to insect larvae, as this protein has chitin binding activity (Joshi *et al.*, 2008). GroEL protein belongs to the relatives of molecular chaperones and is obligatory for proper folding of cellular proteins. Oral intake of XnGroEL protein has been found to be poisonous to *H. armigera* as it interacts with the larval peritrophic lining, which leads to cessation of growth and larval development. Kumari *et al.* (2013) tested the insecticidal potential of XnGroEL against *H. armigera* in transgenic tobacco plants expressing the protein. A 1.7-kb gene encoding the 58-kDa XnGroEL protein was integrated into the tobacco genome through *Agrobacterium*-mediated transformation. It was observed that the transgenic lines were having healthy growth and were normal in their phenotypes. None of the insect larvae survived, and plant damage was reduced to 55–77% in comparison to the untransformed and vector control plants. It was concluded that XnGroEL was a potential candidate that could impart insect resistance against *H. armigera*.

11.5 *Xenorhabdus indica* Somvanshi *et al.* 2009, sp. nov.

Somvanshi *et al.* (2006) isolated three strains of *X. indica* (Validation List No 130; 2009 www.bacterio.net/albidiferax.html) from three independent isolation approaches by crushing mixture of IJs of the nematode *Steinernema thermophilum*. The strains were 28(T), DSM 17382(T) and CIP 108830(T). The association of *X. indica* with the nematode *Steinernema yirgalemense* was reported for the first time by Ferreira *et al.* (2016). They identified bacterial strain 157-C isolated from *S. yirgalemense*. The strain identification was based on 16S rDNA, *recA*, *dnaN*, *gltX*, *gyrB* and *infB* gene sequence analysis, and its relation to the strain type DSM 17 382 (of *X. indica*) was confirmed by DNA–DNA hybridization.

11.5.1 Toxic secretions of *Xenorhabdus indica*

Abou El-Hag and El-Sadawy (2008) reported Xenoprotec, a protein of high molecular weight 220 kDa, derived from *X. indica*. Xenoprotec are antimicrobial proteins belonging to either bacteriocins or antimicrobial peptides. Bacteriocins are proteinaceous compounds lethal to bacteria other than the producing strain. Antimicrobial peptides are amphiphilic. Xenoprotec produced from *X. indica* was found to be highly effective against the bacteria *Staphylococcus* and *Enterococcus faecalis*. However, the mode of action of Xenoprotec still needs investigation.

Kalia *et al.* (2014) evaluated phase I and phase II of *X. indica* against neonates of *H. armigera* and *Spodoptera litura*. Both the phases were found pathogenic to the target pests. The level of virulence was unaffected by phase variation. It was concluded that the application of *X. indica* had the potential to control key lepidopteran pests. Pranaw et al. (2013) reported the effectiveness of proteases produced by *X. indica* to control the cotton bollworm, *H. armigera*. The purified protease resulted in 67.9 ± 0.64% mortality of *H. armigera* within one week.

11.6 *Xenorhabdus bovienii* (Akhurst 1983) Akhurst and Boemare 1993, comb. nov.

11.6.1 Toxic secretions of *Xenorhabdus bovienii*

X. bovienii CS03 (*Xb* CS03) strain was found to be non-virulent when injected directly into lepidopteran insects (Bisch *et al.*, 2016). The genome of *Xb* CS03 was sequenced and compared with the genome of *X. bovienii* SS-2004 (virulent strain). The two strains differed in genome size and content. *Xb* CS03 had a greater antimicrobial activity, as well as more pseudogenes than *Xb* SS-2004. Sugar *et al.* (2012) characterized *X. bovienii* (SS-2004), the bacterial symbiont of *Steinernema jollieti*. IJs harboured *X. bovienii* that was present in a membrane-bound envelope, which also enclosed the nematode-derived intravesicular structure.

The primary and secondary isolates of *X. bovienii* were pathogenic to *M. sexta* and several other insects. However, primary form stocks showed attenuated virulence, which indicated that *X. bovienii* might undergo virulence modulation. Rahoo *et al.* (2011) tested the toxicity of two bacteria, i.e. *X. bovienii* and *Photorhabdus luminescens*, at different concentrations against *G. mellonella*. Efficacy was assessed on different abiotic factors such as substrates, temperatures and moisture levels. The bacteria were found most effective on sand, causing maximum larval mortality as compared to filter paper used as substrate. The mortality of the larva was maximum when the temperature and moisture was 30°C and 20%, respectively.

11.7 Conclusions and Future Prospects

On the basis of the foregone review, it can be concluded that secretions from bacterial symbiont *Xenorhabdus* sp. may prove helpful in defending plants against insect pests. *Xenorhabdus* sp. represents a potential source for new genes encoding potent insecticidal toxins that can be transformed into plants as a defence against insect pests. The tripartite interaction, i.e. *Xenorhabdus*–nematode–insect, represents a model system wherein mutualistic and pathogenic processes can be studied in a single bacterial species. *X. nematophila* is among the few most suitable bacteria, the secretion of which can be used as a biopesticide. Furthermore, in future, the bacterial symbionts of EPNs can be exploited successfully with other management practices to control the advancing population of insect pests attacking different crops and posing economical barriers in crop production.

Acknowledgements

We, the authors, would like to express gratitude to all the nematologists of the laboratory, Department of Plant Pathology, University of Agriculture Faisalabad, Pakistan, and especially to Dr Sajid Aleem Khan for reviewing the manuscript.

References

Abou El-Hag, H.A. and El-Sadawy, H.A. (2008) Xenoprotec: antimicrobial agent derived from *Xenorhabdus indica*. *American-Eurasian Journal of Agricultural and Environmental Sciences* 3, 568–576.

Akhurst, R.J. and Boemare, N.E. (1990) Biology and taxonomy of *Xenorhabdus*. In: Gaugler, R. and Kaya, H.K. (eds) *Entomopathogenic Nematodes in Biological Control*. CRC Press, Boca Raton, Florida, pp. 75–90.

Akhurst, R.J. and Dunphy, G.B. (1993) Tripartite interactions between symbiotically associated entomopathogenic bacteria, nematodes and their insect hosts. In: Beckage, N., Thompson, S. and Federici, B. (eds) *Parasites and Pathogens of Insects*, Vol 2. Academic Press, New York, pp. 1–23.

Ali, S.S., Ahmad, R., Hussain, M.A. and Parvez, R. (2005) *Pest Management in Pulses through Entomopathogenic Nematodes*. Indian Institute of Pulses Research, Kanpur, India.

Ansari, M.A., Tirry, L. and Moens, M. (2005) Antagonism between entomopathogenic fungi and bacterial symbionts of entomopathogenic nematodes. *Biological Control* 50, 465–475.

Askary, T.H. (2010) Nematodes as biocontrol agents. In: Lichtfouse, E. (ed.) *Sociology, Organic Farming, Climate Change and Soil Science*. Springer, The Netherlands, pp. 347–378.

Banerjee, J., Singh, J., Joshi, M.C., Ghosh, S. and Banerjee, N. (2006) The cytotoxic fimbrial structural subunit of *Xenorhabdus nematophila* is a pore-forming toxin. *Journal of Bacteriology* 188, 7957–7962.

Bird, A.F. and Akhurst, R.J. (1983) The nature of the intestinal vesicle in nematodes of the family Steinernematidae. *International Journal of Parasitology* 13, 599–606.

Bisch, G., Ogier, J.C., Médigue, C., Rouy, Z., Vincent, S., *et al.* (2016) Comparative genomics between two *Xenorhabdus bovienii* strains highlights differential evolutionary scenarios within an entomopathogenic bacterial species. *Genome Biology and Evolution* 8, 148–160.

Boemare, N.E., Akhurst, R.J. and Mourant, R.G. (1993) DNA relatedness between *Xenorhabdus* spp. (Enterobacteriaceae), symbiotic bacteria of entomopathogenic nematodes, and a proposal to transfer *Xenorhabdus luminescens* to a new genus, *Photorhabdus* gen. nov. *International Journal of Systematic Bacteriology* 43, 249–255.

Boemare, N.E., Givaudan, A., Brehelin, M. and Laumond, C. (1997) Symbiosis and pathogencity of nematode bacterium complexes. *Symbiosis* 22, 21–45.

Bowen, D.J. and Ensign, J.C. (1998) Purification and characterization of the high-molecular-weight insecticidal protein complex produced by the entomopathogenic bacterium, *Photorhabdus luminescens*. *Applied and Environmental Microbiology* 64, 3029–3035.

Brown, S.E., Cao, A.T., Hines, E.R., Akhurst, R.J. and East, P.D. (2004) A novel secreted protein toxin from the insect pathogenic bacterium *Xenorhabdus nematophila*. *The Journal of Biological Chemistry* 279, 14595–14601.

Brown, S.E., Cao, A.T., Dobson, P., Hines, E.R., Akhurst, R.J. and East, P.D. (2006) Txp40, a ubiquitous insecticidal toxin protein from *Xenorhabdus* and *Photorhabdus* bacteria. *Applied and Environmental Microbiology* 72, 1653–1662.

Burnell, A.M. and Stock, P. (2000) *Heterorhabditis*, *Steinernema* and their bacterial symbionts – lethal pathogens of insects. *Nematology* 2, 31–42.

Cascales, E., Buchanan, S.K., Duché, D., Kleanthous, C., Lloubès, R., *et al.* (2007) Colicin biology. *Microbiology and Molecular Biology Review* 71, 158–229.

Collmer, A., Lindeberg, M., Petnicki-Ocwieja, T., Schneider, D.J. and Alfano, J.R. (2002) Genomic mining type III secretion system effectors in Pseudomonas syringae yields new picks for all TTSS prospectors. *Trends in Microbiology* 10, 462–469.

Cowles, K.N., Cowles, C.E., Richards, G.R., Martens, E.C. and Goodrich-Blair, H. (2007) The global regulator Lrp contributes to mutualism, pathogenesis and phenotypic variation in the bacterium *Xenorhabdus nematophila*. *Cell Microbiology* 9, 1311–1323.

Cranshaw, W.S. and Zimmerman, R. (2013) *Insect Parasitic Nematodes*. Fact Sheet No 5.573. Colorado State University Extension. Available at: http://extension.colostate.edu/docs/pubs/insect/05573.pdf (accessed 19 September 2016).

Dudney, R.A. (1997) Use of *Xenorhabdus nematophilus* Im/I and 1906/1 for fire ant control. US Patent No 5616318.

Elawad, S.A., Gowen, S.R. and Hague, N.G.M. (1999) Efficacy of bacterial symbionts from entomopathogenic nematodes against the beet army worm (*Spodoptera exigua*). Test of Agrochemicals and Cultivars No. 20. *Annals of Applied Biology (Supplement)* 134, 66–67.

Ferreira, T., van Reenen, C.A., Tailliez, P., Pagès, S., Malan, A.P. and Dicks, L.M.T. (2016) First report of the symbiotic bacterium *Xenorhabdus indica* associated with the entomopathogenic nematode *Steinernema yirgalemense*. *Journal of Helminthology* 90, 108–112. Available at: http://dx.doi.org/10.1017/S0022149X14000583 (accessed 19 September 2016).

Forst, S. and Nealson, K. (1996) Molecular biology of the symbiotic-pathogenic bacteria *Xenorhabdus* spp. and *Photorhabdus* spp. *Microbiological Reviews* 60, 21–43.

Forst, S., Dowds, B., Boemare, N. and Stackbrandt, E. (1997) *Xenorhabdus* and *Photorhabdus* spp.: bugs that kill bugs. *Annual Review of Microbiology* 51, 47–72.

Ganguly, S. (2006) Recent taxonomic status of entomopathogenic nematodes: a review. *International Journal of Nematology* 36, 158–176.

Herbert, E.E., Cowles, K.N. and Goodrich-Blair, H. (2007) CpxRA regulates mutualism and pathogenesis in *Xenorhabdus nematophila*. *Applied and Environmental Microbiology* 73, 7826–7836.

Jarrett, P.D., Ellis, D. and Morgan, J.A.W. (1997) World Intellectual Property Patent GB 97/02284.

Ji, D. and Kim, Y. (2004) An entomopathogenic bacterium, *Xenorhabdus nematophila*, inhibits the expression of an antimicrobial peptide, cecropin, of the beet armyworm, *Spodoptera exigua*. *Journal of Insect Physiology* 50, 489–496.

Joshi, M.C., Sharma, A., Kant, S., Birah, A., Gupta, G.P., *et al.* (2008) An insecticidal GroELProtein with chitin binding activity from *Xenorhabdus nematophila*. *Journal of Biological Chemistry* 283, 28287–28296.

Kalia, V., Sharma, G., Shapiro-Ilan, D.I. and Ganguly, S. (2014) Biocontrol potential of *Steinernema thermophilum* and its symbiont *Xenorhabdus indica* against lepidopteran pests: virulence to egg and larval stages. *Journal of Nematology* 46, 18–26.

Khandelwal, P. and Banerjee-Bhatnagar, N. (2003) Insecticidal activity associated with the outer membrane vesicles of *Xenorhabdus nematophilus*. *Applied and Environmental Microbiology* 69, 2032–2037.

Khandelwal, P., Choudhury, D., Birah, A., Reddy, M.K., Gupta, G.P. and Banerjee, N. (2004) Insecticidal pilin subunit from the insect pathogen *Xenorhabdus nematophila*. *Journal of Bacteriology* 186, 6465–6476.

Kim, Y., Ji, D., Cho, S. and Park, Y. (2005) Two groups of entomopathogenic bacteria, *Photorhabdus* and *Xenorhabdus*, share an inhibitory action against phospholipase A_2 to induce host immune depression. *Journal of Invertebrate Pathology* 89, 258–264.

Kumari, P., Kant, S., Zaman, S., Mahapatro, G.K., Banerjee, N. and Sarin, N.B. (2013) A novel insecticidal GroEL protein from *Xenorhabdus nematophila* confers insect resistance in tobacco. *Transgenic Resistance* 23, 99–107.

Martens, E.C., Heungens, K. and Goodrich-Blair, H. (2003) Early colonization events in the mutualistic association between *Steinernema carpocapsae* nematodes and *Xenorhabdus nematophila* bacteria. *Journal of Bacteriology* 185, 3147–3154.

Moran, N.A. (2002) Microbial minimalism: genome reduction in bacterial pathogens. *Cell* 108, 583–586.

Nagesh, M., Hussaini, S.S. and Singh, S.P. (2002) Isolation and characterization of symbiotic bacteria from *Heterorhabditis* spp. and *S. carpocapsae* Weiser. *Pest Management in Horticultural Ecosystems* 8, 38–42.

Park, Y., Kim, Y., Tunaz, H. and Stanley, D.W. (2004) An entomopathogenic bacterium, *Xenorhabdus nematophila*, inhibits hemocytic phospholipase A2 (PLA2) in tobacco hornworms *Manduca sexta*. *Journal of Invertebrate Pathology* 86, 65–71.

Park, Y., Herbert, E.E., Cowles, C.E., Cowles, K.N., Menard, M.L., *et al*. (2007) Clonal variation in *Xenorhabdus nematophila* virulence and suppression of *Manduca sexta* immunity. *Cell Microbiology* 9, 645–656.

Poinar, G.O. Jr and Thomas, G.M. (1967) The nature of *Achromobacter nematophilus* as an insect pathogen. *Journal of Invertebrate Pathology* 9, 510–514.

Pranaw, K., Singh, S., Dutta, D., Singh, N., Sharma, G., *et al*. (2013) Extracellular novel metalloprotease from *Xenorhabdus indica* and its potential as an insecticidal agent. *Journal of Microbiology and Biotechnology* 23, 1536–1543.

Pugsley, A.P. (1993) The complete general secretary pathway in gram-negative bacteria. *Microbiological Reviews* 57, 50–108.

Rahoo, A.M., Mukhtar, T., Gowen, S.R. and Pembroke, B. (2011) Virulence of entomopathogenic bacteria *Xenorhabdus bovienii* and *Photorhabdus luminescens* against *Galleria mellonella* larvae. *Pakistan Journal of Zoology* 43, 543–548.

Roy, K., Hilliard, G.M., Hamilton, D.J., Luo, J., Ostmann, M.M. and Fleckenstein, J.M. (2009) Enterotoxigenic *Escherichia coli* EtpA mediates adhesion between flagella and host cells. *Nature* 457, 594–598.

Samaliev, H.Y., Andreoglou, F.I., Elawad, S.A., Hague, N.G.M. and Gowen, S.R. (2000) The nematicidal effects of the bacteria *Pseudomonas oryzihabitans* and *Xenorhabdus nematophilus* on the root-knot nematode *Meloidogyne javanica*. *Nematology* 2, 507–514.

Shapiro-Ilan, D.I., Han, R. and Dolinksi, C. (2012) Entomopathogenic nematode production and application technology. *Journal of Nematology* 44, 206–217.

Singh, P., Park, D., Forst, S. and Banerjee, N. (2013) Xenocin export by the flagellar type III pathway in *Xenorhabdus nematophila*. *Journal of Bacteriology* 195, 1400–1410.

Somvanshi, V.S., Lang, E., Ganguly, S., Swiderski, J., Saxena, A.K. and Stackebrandt, E. (2006) A novel species of *Xenorhabdus*, family Enterobacteriaceae: *Xenorhabdus indica* sp. nov., symbiotically associated with entomopathogenic nematode *Steinernema thermophilum* Ganguly and Singh, 2000. *Systematic and Applied Microbiology* 29, 519–525.

Sugar, D.R., Murfin, K.E., Chaston, J.M., Andersen, A.W., Richards, G.R., *et al*. (2012) Phenotypic variation and host interactions of *Xenorhabdus bovienii* SS-2004, the entomopathogenic symbiont of *Steinernema jollieti* nematodes. *Environmental Microbiology* 14, 924–939.

Thomas, G.M. and Poinar, G.O. Jr (1979) *Xenorhabdus* gen. nov., a genus of entomopathogenic nematophilic bacteria of the family *Enterobacteriaceae*. *International Journal of Systematic Bacteriology* 29, 352–360.

Volgyi, A., Fodor, A., Szentirmai, A. and Forst, S. (1998) Phase variation in *Xenorhabdus nematophilus*. *Applied Environmental Microbiology* 64, 1188–1193.

Waterfield, N., Hares, M., Yang, G., Dowling, A. and ffrench-Constant, R. (2005) Potentiation and cellular phenotypes of the insecticidal toxin complexes of *Photorhabdus* bacteria. *Cell Microbiology* 7, 373–382.

Woodring, J.L. and Kaya, H.K. (1988) Steinernematid and heterorhabditid nematodes: a handbook of techniques. Southern Cooperative Series Bulletin No 331, Arkansas Agricultural Experiment Station, Fayetteville, Arkansas.

12 Toxic Secretions of *Photorhabdus* and Their Efficacy Against Crop Insect Pests

Mahfouz M. M. Abd-Elgawad*

Phytopathology Department, National Research Center, Giza, Egypt

12.1 Introduction

Photorhabdus and *Xenorhabdus* are two genera of bacteria that mutualistically associate with entomopathogenic nematodes (EPNs) in the genera *Heterorhabditis* and *Steinernema*, respectively. Fundamental and applied research on the nematode–bacterium pairs, which are capable of invading and killing many agricultural pests, has been in progress in order to explore further their biology, attributes and genetic constitution, as well as to improve their efficacy as biocontrol agents (e.g. Gaugler and Kaya, 1990; Gaugler, 2002; Grewal *et al.*, 2005; Abd-Elgawad and Vagelas, 2015; Campos-Herrera, 2015). Both bacterial genera are motile Gram-negative bacteria of the family Enterobacteriaceae (Boemare and Akhurst, 1988; Boemare *et al.*, 1993). Their life cycles are basically similar (Fig. 12.1) in terms of inhabiting the gut of their symbiotic nematode infective juveniles (IJs), and demonstrating pathogenic potential for growing in the haemolymph of insect hosts invaded by the IJs. Thus, the matchless life cycle of the two bacterial genera involves the formation of a mutualistic symbiosis with one host, i.e. the respective nematode, and a strong pathogenic attack against a distinct host, the target insect (Forst and Clarke, 2002). The bacteria profit from this relationship by being protected within the nematode body, from the competitive/unfavourable environment, and transported by the nematodes, which, after attacking their hosts, release the bacteria to the nutrient-rich haemolymph of the invaded insects. The released bacteria proliferate and generate a wide range of toxins and hydrolytic exoenzymes that bioconverse the insect host into a nutrient soup; thus offering an ideal diet for benefitting nematode development and reproduction. Eventually, together, the nematodes and the symbionts kill the arthropod host, although in most interactions the bacteria alone are highly virulent (e.g. Akhurst and Dunphy, 1993; Bowen and Ensign, 1998; Abdel-Razek, 2003; Uma *et al.*, 2010; Jallouli *et al.*, 2013; Kumar *et al.*, 2014). The nematodes multiply within insect cadavers until the nutrient supply becomes almost unavailable, at which time they develop into IJs that are recolonized by the symbiotic bacteria. This marked interdependent reproductive cycle is the output of a highly evolved interaction between the bacterium and the nematode. Boemare (2002) stated that the close relatedness of the two taxonomic structures for the bacterial genera revealed the concurrent formation of distinct species in the course of evolution, i.e. cospeciation, between the bacterial symbionts and the nematode hosts. In fact, taxonomic uncertainties do not invalidate this concept of cospeciation. He also reported that the correlation found between symbiont and host heat tolerances might support cospeciation. In this respect, the mutualism between *Heterorhabditis* and *Photorhabdus* was basically thought to be strictly one-to-one in terms of cospeciation. More recently, with the discoveries of more new EPNs and their symbionts, it has been found that some *Photorhabdus* species may have different *Heterorhabditis* hosts; for example, *Photorhabdus temperata* has the nematode hosts *Heterorhabditis*

*E-mail: mahfouzian2000@yahoo.com

bacteriophora, *Heterorhabditis zealandica* and *Heterorhabditis downesii* (Boemare, 2002; Adams *et al.*, 2006; An and Grewal, 2010; Stock, 2015). Maneesakorn *et al.* (2011) examined coevolutionary links between *Heterorhabditis* and *Photorhabdus* using a single gene approach. The ribosomal internal transcribed spacer (ITS) region was considered for *Heterorhabditis* but the housekeeping gyrase B (*gyrB*) gene was utilized for *Photorhabdus* (Stock, 2015).

Recent studies in *Xenorhabdus* and *Photorhabdus* have allowed us to learn more about their mechanisms of persistence, pathogenicity and toxins, as well as to identify genetic systems that play important roles in their mutualism and pathogenicity (e.g. Forst and Clarke, 2002; Clarke, 2008; An and Grewal, 2010; Murfin *et al.*, 2012; Gatsogiannis *et al.*, 2013; ffrench-Constant and Dowling, 2014; Hurst *et al.*, 2015). For example, to maintain sound associations between *Photorhabdus* and their host nematodes, An and Grewal (2010) found that the bacteria endured substantial transcriptional reshaping in the nematode gut. In addition to general mechanisms for starvation, the bacteria stimulate cellular acidification to deliberate growth, turn to the pentose phosphate pathway to cope with oxidative stress and insufficient nutrition, and abandon movement but promote biofilm to continue in the nematode gut until being liberated into the insect haemolymph. Such bacterial tactics indicate how the mutualistic bacteria, *Photorhabdus* spp., decrease their nutritional reliance on their nematode partner to secure efficient transmission of the pair to the insect host (An and Grewal, 2010). It is also interesting that the EPNs and their symbionts can be cultivated independently or together, and molecular techniques are being developed to examine the bacteria and/or EPNs accurately (e.g. Chaston *et al.*, 2011; Ogier *et al.*, 2014; Fayyaz *et al.*, 2015). This technical tractability has enabled the use of EPNs and bacteria as models of mutualism, pathogenesis, evolution, behaviour, ecology and drug discovery (Murfin *et al.*, 2012; Gatsogiannis *et al.*, 2013; ffrench-Constant and Dowling, 2014; Campos-Herrera, 2015; Hurst *et al.*, 2015).

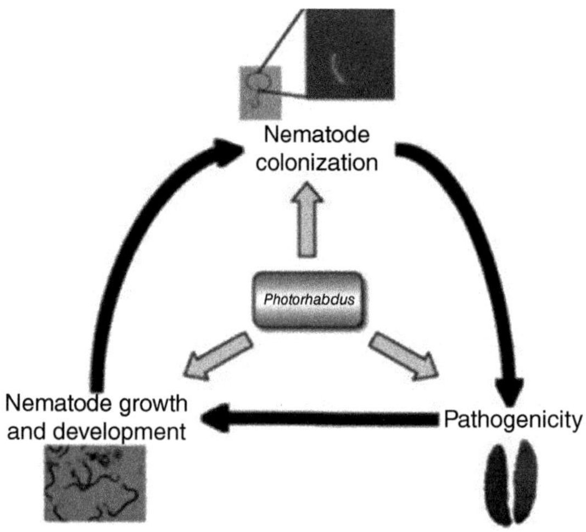

Fig. 12.1. The multiple personalities of *Photorhabdus*. The *Photorhabdus* life cycle can be defined in terms of the three distinct roles that the bacteria has during its interaction with the insect and the nematode. First, *Photorhabdus* must be able to colonize the gut of the IJ in a highly specific process. A population of approximately 100 cfu is maintained in the IJ (for periods of several months to 1 year) and the top micrograph shows green fluorescent protein (GFP)-labelled *Photorhabdus* in the foregut of the IJ. Second, after regurgitation into the insect haemolymph, *Photorhabdus* must be able to kill the insect larvae. The cuticle of the insect cadaver remains intact after death and *Photorhabdus*-infected insects often change colour due to the production of an anthraquinone pigment by the bacteria. Finally, *Photorhabdus* must provide conditions within the cadaver that will support nematode growth and development. This includes the provision of nutrients to the increasing nematode population and protecting the cadavar from predation by saprophytic organisms. (Adapted from Clarke, 2008).

Despite the above-mentioned comparable lifestyles of *Xenorhabdus* and *Photorhabdus*, Forst and Nealson (1996) reported several distinctive traits of the life cycle of each bacterial genus that are clearly different from the other (Fig. 12.1). *Xenorhabdus* spp. are carried in specialized vesicles created by an outpocketing in the *Steinernema* spp. IJ gut, but *Photorhabdus* spp. are found throughout the gut of *Heterorhabditis* spp. IJs (Bird and Akhurst, 1983; Goodrich-Blair and Clarke, 2007). The reproductive cycle of the respective nematodes is also quite different. The IJs of *Steinernema* exist as amphimictic males and females, so that both a male and a female must infect the host haemocoel to mate and reproduce inside the insect. On the contrary, the IJs of *Heterorhabditis* develop into hermaphroditic females and subsequently into amphimictic males and females. Therefore, infection of the host by a single IJ of hermaphroditic *Heterorhabditis* can reproduce inside the insect haemocoel. Moreover, bacterial colonization of maternal rectal glands and hatching of the offspring within the mother (*endotokia matricida*) is needed for the conveyance of *Photorhabdus luminescens* to IJ progeny of *H. bacteriophora* (Ciche *et al.*, 2008). This style of colonization has not been found in *Steinernema carpocapsae* nematodes, and *endotokia matricida* is not necessary for IJ colonization (Chaston *et al.*, 2011). An attractive and visible trait that differentiates *Photorhabdus* from *Xenorhabdus* is its ability to fluoresce, so much so that the whole infected insect cadavers usually glow in the dark. Therefore, its genus name started with the prefix '*Photo*'. The two bacterial genera show variations also in their molecular interactions to multiply on their host insects (Goodrich-Blair and Clarke, 2007; Chaston *et al.*, 2011). For example, mechanisms in *Photorhabdus* differ from *Xenorhabdus* in order to cope with the immune system, such as antimicrobial peptide production, of their invaded insect hosts. While *Photorhabdus* employs modification in lipopolysaccharide to tolerate the effect of the host-derived antimicrobial peptides, *Xenorhabdus nematophila* blocks the induction of insect antimicrobial peptide expression. Also, in characterizing mutants defective in colonizing their IJs, no interference in genetic determinants sought after for colonization was observed between the two genera (Goodrich-Blair and Clarke, 2007; Easom *et al.*, 2010; Somvanshi *et al.*, 2010; Chaston *et al.*, 2011). The above-mentioned basic variations in lifestyle and microscopic characters of the two bacterial genera affirm their molecular and genetic distinctions. A genomic study revealed that 16S rRNA genes of *Xenorhabdus* and *Photorhabdus* were over 94% identical, but each genus had genomic disruption by many insertions, inversions, deletions and translocations (Chaston *et al.*, 2011). They found that the two *Photorhabdus* and *Xenorhabdus* genomes shared more genes exclusive to each other (893 and 409, respectively) than between *Xenorhabdus–Photorhabdus* pairs (62 genes for *X. nematophila* and *P. luminescens*; 76 for *X. nematophila* and *Photorhabdus asymbiotica*; 155 for *Xenorhabdus bovienii* and *P. luminescens*; 170 for *X. bovienii* and *P. asymbiotica*). On the other hand, species of both genera can be grown as free-living organisms on artificial media; the relationship is not obligate as both *Photorhabdus* and *Xenorhabdus* can multiply under controlled conditions without their host nematodes. Rich nutrient media and the absence of competition are certainly responsible for this *in vitro* growth (Forst and Nealson, 1996). Interestingly, on entering the stationary phase, the bacteria secrete numerous extracellular products, including a lipase(s), phospholipase(s), protease(s) and various broad-spectrum antibiotics. Such outputs are normally transferred to the insect haemolymph when the bacteria come in the stationary phase. While these enzymes degrade the macromolecules of the insect tissues to supply the nematodes with nutrients necessary for their development, the antibiotics prevent contamination of the insect cadaver by other invading species.

This chapter is devoted to reporting the toxic secretions of *Photorhabdus* bacteria and their efficacy against crop insect pests. Yet, because these bacteria are fairly similar to members of the bacterial genus *Xenorhabdus*, the two genera have been introduced together to show aspects of their similarities and differences. It is now well established that both genera of bacteria supply similar services to their different nematode hosts using matchless physiological and metabolic mechanisms (Chaston *et al.*, 2011). Distinctive phenotypic attributes for both *Xenorhabdus* and *Photorhabdus* are presented in Table 12.1.

Identification of these bacteria requires highly trained personnel in various fields of morphology, physiology, biochemistry, serology and molecular biology. Initially, phenotypic screening is conducted for differentiation of the bacterial strains, since this

Table 12.1. Distinguishing traits of *Photorhabdus* and *Xenorhabdus*. (Adapted from Forst and Clarke, 2002).

Phenotypic trait	*Photorhabdus luminescens*	*Xenorhabdus nematophila*
Catalase	+	–
Bioluminescence	+	–
Pigments	Anthraquinones	–
Antibiotics	Hydroxystilbenes	Xenocoumacins Xenorhabdins Indole derivatives
Crystal proteins	11.6 kDa 11.3 kDa	26 kDa 22 kDa
Urease activity	d	–
Indole production	d	–
Aesculin hydrolysis	d	–

Note: d = depending on strain.

gives a preliminary comparison and designation into similar or different groups. Nevertheless, similar to their nematode partners, the necessity for prompt and accurate detection of bacterial species and/or isolates has been introducing and developing molecular approaches. Therefore, it is believed that molecular biology has revolutionized the systematics and classification of all bacteria (Stock, 2015). Approaches such as DNA reassociation and riboprinting are used for the characterization and identification of *Xenorhabdus* and *Photorhabdus*, replacing boring phenotypic-related techniques. Also, sequence data of single and multigene data sets are being adopted for evolutionary and identification studies of species/strains (e.g. Tailliez *et al.*, 2010; Kuwata *et al.*, 2013; Ferreira *et al.*, 2013, 2014; Stock, 2015). Researchers have been using comparative sequence analyses of the 16S rRNA gene and several housekeeping genes for characterizing *Photorhabdus* and *Xenorhabdus* isolates/species and investigating their phylogenetic relationships (e.g. Chaston *et al.*, 2011; Ferreira, *et al.*, 2013, 2014).

It is well established that the symbionts, *Xenorhabdus* spp. and *Photorhabdus* spp., replicate and produce a variety of primary and secondary metabolites and, in conjunction with the nematode, they induce host mortality of many insect pests in a short time, which appraises them as valuable biocontrol agents. Described species of *Xenorhabdus* are more than 20 species, but they correspond to only 4 species of *Photorhabdus* (Ferreira *et al.*, 2014; Stock, 2015). These bacteria can produce antimicrobial compounds (Akhurst, 1982), hydrolytic enzymes (Boemare and Akhurst, 1988) and insecticidal toxins (Bowen and Ensign, 1998; ffrench-Constant and Bowen, 2000; ffrench-Constant *et al.*, 2007; ffrench-Constant and Dowling, 2014). Dozens of bioactive secondary metabolites belonging to these diverse chemical classes, such as puromycin and madumycin II (Webster *et al.*, 2002), have been reported from cultures of *Xenorhabdus* spp. and *Photorhabdus* spp. Several species of *Photorhabdus* and *Xenorhabdus* yield more than one array of bioactive secondary metabolites. Webster *et al.* (2002) reported that the metabolites from *Xenorhabdus* species were more diverse than those from *Photorhabdus*. Xenorhabdins are commonly produced by *X. bovienii*, xenocoumacins by *X. nematophila* and hydroxystilbenes and anthraquinones by *Photorhabdus* spp. (Webster *et al.*, 2002). Such metabolites have both different chemical structures and a broad range of activities of agricultural and medicinal advantage, such as antibiotic, insecticidal, nematicidal, antimycotic, antiulcer, antineoplastic and antiviral. The structure and bioactivity of these metabolites have been summarized by Webster *et al.* (2002), and recent revision of such metabolites has been reviewed recently (e.g. Fuchs *et al.*, 2011; Easom and Clarke, 2012; Eom *et al.*, 2014). For example, out of five *P. temperata* strains, only two strains, J4 and J5, produced heat-stable toxins. The supernatants extracted from the culture of these strains preserved up to 95% of insecticidal activity after heat treatment for 30 min at 80°C (Jang *et al.*, 2011). Different aspects of these metabolites and their potential for controlling crop insect pests have been reported in detail for *Xenorhabdus* (see Javed *et al.*, Chapter 11, this volume). Generally, further biochemical studies are needed that also address molecular events relevant to understanding and describing the processes involved in such attributes of these symbionts, as well as the efficacy and mode of action of their secreted compounds. In this chapter, more important toxins, compared to those of other bacteria, encoded by *Photorhabdus* genomes will be presented to prove their qualification as bioinsecticides (Rodou *et al.*, 2010; ffrench-Constant and Dowling, 2014).

Some references suppose that steinernematids are generally temperate or cold-adapted species, but heterorhabditids tend to favour a tropical climate. Nevertheless, Hominick (2002) argued that the data do not support such a generalization. He speculated that accommodation was probably more a species than a generic attribute. If so, the distribution of the symbiotic bacteria *Xenorhabdus* and *Photorhabdus* is not influenced by climate. Interestingly, numerous surveys in Egypt, for example, have shown that heterorhabditids occur more frequently than steinernematids (e.g. Shamseldean and Abd-Elgawad, 1994; Salama and Abd-Elgawad, 2001; Abd-Elgawad and Nguyen, 2007; Abd-Elgawad, 2008, 2014; Shamseldean, 2010; Abu-Shady *et al.*, 2011), whereas in temperate climate more steinernematids than heterorhabditids have been reported (Hominick, 2002; Thanwisai *et al.*, 2012). On the other hand, the occurrence of some heterorhabditid nematode species, such as *Heterorhabditis indica* and *H. bacteriophora*, has been documented on all continents, with the exception of Antarctica (Hominick, 2002; Stock, 2015). Eventually, many biotic and abiotic factors such as

EPN dispersal mechanisms, soil and vegetation type, as well as the distribution of suitable hosts and natural enemies, are key factors in affecting the distribution of EPN species, and conseqently their mutualistic bacteria (Hominick, 2002; Stock, 2015; Stuart *et al.*, 2015).

Geographically, the most widespread heterorhabditid nematode is *H. bacteriophora*, which has been recorded in zones having continental and Mediterranean climates. *H. indica* is very common in the subtropics and tropics (Burnell and Stock, 2000; Stack *et al.*, 2000; Hominick, 2002). Also, *Heterorhabditis megidis* has been isolated from the northern hemisphere. It usually has a more northerly and more restricted distribution than *H. bacteriophora* (Stack *et al.*, 2000). *H. megidis* has also been isolated in Greece (Menti *et al.*, 1997), Israel (Glazer *et al.*, 1993), Japan (Yoshida *et al.*, 1998), Russia (Fischer-Le Saux *et al.*, 1998), Canada (Mracek and Webster, 1993) and Turkey (Yilmaz *et al.*, 2009). Eventually, researchers have been surveying EPNs to discover new species and/or populations adapted to local conditions and insect pests. They have found many EPN isolates with aggregated distribution on a local scale (e.g. Abd-Elgawad, 2014), although these nematodes are omnipresent on a global scale. Stock (2015) reported that EPNs have been found in 49 countries (19 from Europe, 9 from Asia, 3 from North America, 7 from Central America and the Caribbean, 8 from South America and 3 from Africa). Nevertheless, the available data of their geographic records are thought to be an artefact of sampling endeavours by nematologists interested in EPNs. Therefore, voluminous and impressive references in the field of their applications as a nematode–bacteriun complex against insect pests are available (e.g. Gaugler and Kaya, 1990; Gaugler, 2002; Grewal *et al.*, 2005; Vega and Kaya, 2012; Campos-Herrera, 2015), but this chapter reports mainly on *Photorhabdus* bacteria when applied to kill insects pests independently of their nematode hosts.

12.2 Entomopathogenic Nematodes and Their Mutualistic Bacteria-related Terminology

Confusion afflicted the mutualistic framework of EPNs as a nematode–bacterial complex when Glaser and Fox (1930) recognized only the nematodes, but not their mutualistic bacteria, infecting larvae of *Popillia japonica* (the Japanese beetle) on a New Jersey golf course in the USA. Their nematodes were identified as *Neoaplectana* (now *Steinernema* based on the first EPN species erected by Steiner (1923) as *Aplectana kraussei* (now *Steinernema kraussei*)) *glaseri* (Glaser and Fox, 1930).

Dutky (1937) was the first to identify the bacteria as symbiotic partners of the nematodes, but his research was conducted with *S. carpocapsae*. In this respect, the term 'symbiosis' has generally been somewhat debatable; whether it refers to persistent mutualisms only or it may indicate any type of persistent biological interaction, i.e. commensalistic, mutualistic, or parasitic. Martin and Schwab (2013) found that 100% of general biology textbooks surveyed used symbiosis to refer to mutualism, commensalism and parasitism, whereas the restrictive definition (i.e. symbiosis = mutualism) has ultimately withered. In order to comply with such survey results, the term 'mutualistic', not 'symbiotic', *Xenorhabdus* and *Photorhabdus* bacteria should be used. Therefore, when this term 'symbiosis' and/or its derivatives, i.e. symbiotic and symbiont, are used for the association of *Photorhabdus* and/or *Xenorhabdus* species with EPNs in textbooks (e.g. Gaugler, 2002; Grewal *et al.*, 2005; Campos-Herrera, 2015; Eissa and Abd-Elgawad, 2015) and published papers (e.g. Chaston *et al.*, 2011; Poinar and Grewal, 2012; Abd-Elgawad and Spiridonov, 2014; Nollmann *et al.*, 2015), they definitely mean mutualistic bacteria.

Had Glaser *et al.* (1942) developed an axenic (i.e. without any other organism) culture of *Steinernema glaseri* for biological control, they would have found a bacterial contaminant in the culture that could be tolerated by the nematodes; it was probably *Xenorhabdus poinari*, the symbiont of *S. glaseri*. Although Glaser's team could culture the nematodes axenically, biocontrol of *P. japonica* by this nematode met limited success (Gaugler *et al.*, 1992). The reason behind such a limited success is that the functional pathogen must consist of both symbiotic partners, nematode–bacterial complex, to manage the targeted insect pests. This is especially important in assigning the efficacy of a particular EPN species against a given pest (Lewis and Clarke, 2012). Eventually, the efficacy of EPNs relies both on the nematodes' capability to locate, recognize and attack a host, and on the virulence of the mutualistic bacteria to this host (Griffin *et al.*, 2005). Therefore, it is critical for commercially growing EPNs to retain bacteria as nematode infective juveniles, and in turn for the bacteria to possess their virulence to insects. The symbiosis is so important that when the term 'entomopathogenic nematode' is used in the scientific literature, it is often referring to the nematode–bacterium complex (or pair) responsible for the functional pathogen. So, unless indicated otherwise, the term 'entomopathogenic nematode' herein and elsewhere actually refers to the nematode–bacterium complex (e.g. Lewis and Clarke, 2012).

Caution should also be exercised in using some terms especially to report commercial production techniques and/or laboratory bioassays of EPNs and their symbionts. The term 'axenic' describes the state of a culture in which only a single species, variety, or strain of organism is present and entirely free of all other contaminating organisms (https://en.wikipedia.org/wiki/Axenic). The term 'xenic' culture refers to a culture of parasites grown in association with an unknown microbiota (http://www.medilexicon.com/medicaldictionary.php?t=21860). Also, 'monoxenic culture' is a culture of parasites grown in association with a single known bacterium (http://www.medilexicon.com/medicaldictionary.php?t=21841); e.g. an EPN with the symbiotic bacterium.

Factually, the more research we do on these bacteria, the better we will learn useful terms for the identity of their species/strains, biology and genetics, modelling, symbiosis and pathogenicity. For example, definite terms may be useful to characterize special event(s) in the life cycle of EPNs, which passes through eggs, four juvenile stages and the adults. The non-feeding, third-stage juveniles are the only free stage that can usually be found in soil; outside the insect host. When these non-feeding IJs invade their insect hosts, 'nematode recovery' occurs. This term means that the nematodes change to feeding third-stage juveniles, feed on the bacteria and moult in succession to the fourth stage, where they become hermaphrodite adults of the first generation (Adams and Nguyen, 2002). The process of exiting from the non-feeding IJ stage and resuming development (inside the insect haemolymph) is known as, or termed, 'recovery' (Golden and Riddle, 1982). *Heterorhabditis* IJs recover and develop into self-fertilizing hermaphrodites with a female phenotype. In this respect, recovery is also very important from a commercial point of view. Therefore, Moshayov *et al.* (2013) tried to characterize the recovery in *H. bacteriophora* by identifying genes involved in this process and found that most of the genes known to differentially express between recovering and developmentally arrested IJs belonged to metabolic pathways. The heterorhabditid nematode second generation consists of amphimictic females and males, which mate to provide either the third generation or IJs. In other words, nematode development and reproduction within the insect host continues until nutrition material in the cadaver is consumed (when bacteria stop multiplying), usually allowing for two to three generations (Adams and Nguyen, 2002). The *Heterorhabditis* hermaphrodite deposits some eggs externally. However, second-generation *Heterorhabditis* frequently hatch, feed and develop inside the hermaphrodite. Thus, the adults die and the IJs exit from the dead insect host, a phenomenon known as *endotokia matricida* (Johnigk and Ehlers, 1999). This also occurs in second-generation amphimictic females of *Heterorhabditis* (Wright and Perry, 2002). These new IJs colonized with the *Photorhabdus* bacterium grow, before emerging from the carcass into the soil. Also, some bacteriologists suggest that the terminology of *Photorhabdus* 'phenotypic conversion' should be more appropriate than *Photorhabdus* 'phase variation'. That is because phase variation is always known as a genetic event that is able to be reversed; usually arranged by lack of DNA stability (Boemare, 2002). In the future, additional terms may further serve in commercial aspects, such as biopesticidal development and application, and pharmaceutical industry.

A specific system of nematode food signals may dictate different stages of worm growth. *In vivo* tests have demonstrated that the IJs of *Heterorhabditis* can grow to adults even in the non-attendance of *Photorhabdus* bacteria (Han and Ehlers, 2000). Therefore, it is likely that the dominant recovery signal stems from the haemolymph of the arthropod host. The bacteria can make signals that control nematode development. Strauch and Ehlers (1998) state that *Photorhabdus* produces a food signal that is transmitted to the medium and urges IJ recovery. The signal reaches its maximum output during the late exponential phase of the *Photorhabdus* bacteria. Recovery of *H. bacteriophora* IJs is noted in the existence of cells or crystalline inclusion proteins (Bintrim and Ensign, 1998). Signal of *Xenorhabdus* bacteria does not cause *Heterorhabditis* recovery (Han and Ehlers, 1998). On the other hand, *S. carpocapsae*

IJs react to signals released by *P. luminescens*. Therefore, Forst and Clarke (2002) conclude that it is improbable that the bacterial food signal is strictly engaged in IJ recovery, but it may be implicated in commanding some later stage in EPN growth (Forst and Clarke, 2002). Moshayov *et al.* (2013) used RNAi to silence genes that were presumably related to the nematode recovery process from the infective to parasitic stage in order to know its genetic basis, which may guide to improvement in nematode recovery. Glazer (2015) speculated that such an improvement might serve in enhancing commercial EPN production and could increase nematode efficiency.

12.3 Species of *Photorhabdus*

So far, all species of the genus *Photorhabdus* are motile Gram-negative, nematode mutualistic, bioluminescent bacteria (Woodring and Kaya, 1988; Boemare, 2002; Forst and Clarke, 2002). Even *P. asymbiotica*, which was first isolated from human wounds, was later recognized as a symbiont of *Heterorhabditis gerrardi* (Plichta *et al.*, 2009). Four bacterial species have been recognized hitherto: *P. luminescens*, *P. temperata*, *P. asymbiotica* and *Photorhabdus heterorhabditis*. Other new *Photorhabdus* species, from previously described EPN isolates are still in the pipeline since they require characterization. A recent phylogenetic scheme is that of Ferreira *et al.* (2014), which recorded 17 subspecies (9 of *P. luminescens* + 6 of *P. temperata* + 2 of *P. asymbiotica*) and many more strains. For example, *P. heterorhabditis* had three strains; i.e. SF41, SF783 and Q614 (Ferreira *et al.*, 2014).

Like many other biologicals, *Photorhabdus* identification is passing through a state of flux as techniques of molecular biology are upgraded and accompanied phylogenies turn progressively into relevant tools. Therefore, the taxonomy and biological systematics of *Photorhabdus* are fairly fluid, with proposals for new species/subspecies/strains and the joint linking of categories that had previously been regarded as distantly related (Boemare, 2002; Waterfield *et al.*, 2009; Tailliez *et al.*, 2010; Ferreira *et al.*, 2013, 2014). One current direction is utilizing a multigene tactic to assort novel *Photorhabdus* isolates to avoid miscategorizations related to genes that have undergone various evolutional events (Ferreira *et al.*, 2013, 2014; Stock, 2015).

Admittedly, the more the research on these bacteria and their host nematodes, the more is the possibility of finding new *Photorhabdus* species. There are numerous described and undescribed species and strains of the bacteria-host nematodes *Heterorhabditis* spp. worldwide (Poinar and Grewal, 2012) that support this assumption. Many undescribed species of EPNs have been extracted during surveys, but species without formal descriptions are usually not referred to in texts (Hominick, 2002; Abd-Elgawad and Spiridonov, 2014), because accurate identification is substantial to their sound utilization and understanding their geographical distribution, habitat specificity and designation of their role in the biological control of arthropod pests. Furthermore, while *Xenorhabdus ishibashii* was erected as a new species since its protein sequence had less than 97% similarity with the protein sequences of its closely related bacterial species (Kuwata *et al.*, 2013), *Photorhabdus* strains or isolates that shared less than 97% nucleotide identity were separated to different subspecies (Tailliez *et al.*, 2010). Therefore, such splitting of a species to a subspecies, as well as the identification of novel *Photorhabdus* bacteria, may work towards future upgrading of some subspecies to the species level, as their characteristics, transmissible to their progeny, can designate them above subspecific rank.

The life cycle of *Photorhabdus* species and their mutualistic heterorhabditid nematodes is divided into four (I–IV) stages (Table 12.2). Initially, in stage I, the bacteria are found in the midgut of the nematode IJs. These latter search for and enter their insect host. In the following stage (II), the bacteria are liberated from the nematode into the insect haemolymph, where bacterial toxins, conducive to the insect's death, are secreted. Then, nematode development and reproduction occur, in stage III, using the decomposed insect corpse. So, the bacterial secretions of exoenzymes, antibiotics and crystal proteins support nematode reproduction via decomposing the corpse and repelling any other competing bacterial species. In the final stage (IV), new nematode generations are created where maternal nematodes acquire *Photorhabdus* cells as a constant intestinal biofilm prior to transmission to IJs' offspring developing inside the maternal body (Somvanshi *et al.*, 2010). Eventually, the bacteria colonize intestines of the IJs before exiting the insect cadaver or remains and moving on to the next insect host (Forst and Clarke, 2002; Gerdes *et al.*, 2015).

Table 12.2. Different stages of the mutualistic relationship between *Photorhabdus* species and their *Heterorhabditis* species (adapted from Forst and Clarke, 2002; Gerdes *et al.*, 2015).

Stage	Bacteria life cycle	Nematode life cycle
I	Bacteria retained in nematode gut	Infective juvenile in the soil Search for insect host Infective juvenile enters into insect haemocoel
II	Bacteria released into haemolymph Production of virulence factors Death of insect	Recovery in the haemocoel
III	Bacteria in stationary phase Production of antibiotics, exoenzymes, crystal protein Bioconversion of insect	Nematode reproduction
IV	Colonization of the intestine of infective juveniles	Development of new infective juveniles

A prominent attribute of *Photorhabdus* species is that the bacteria undergo phase variation (Forst *et al.*, 1997; Joyce *et al.*, 2006), known as phase I and phase II; two phenotypic variants of the bacterial species. From a bacteriological standpoint, phase variation is generally known as a reversible genetic event usually arranged by lack of DNA stability; for example DNA inversion taking place with considerable frequency (Boemare, 2002). For *Xenorhabdus* and *Photorhabdus*, 'phase variation' is usually utilized to indicate an alternative equilibrium in the expression/non-expression of physiological characters from the same genome. For every trait that can be quantified (e.g. antibiotic production, luminescence), the difference between phases is quantitative, and it is highly probable that this is also true for other phase-related characters emphasizing the occurrence of a genetic regulation (Boemare, 2002). Phase variation of *Photorhabdus* species can be readily examined morphologically and functionally via morphological characters, dye adsorption and antibiotic realization (Akhurst, 1980; Woodring and Kaya, 1988) for any mutualistic strain. Under natural conditions, the primary cells of the symbionts have been exclusively isolated from wild nematodes (Boemare, 2002). Yet it should be noted that *in vitro* subculturing of different EPN-mutualistic bacterial species usually influences an inconstant portion of the bacterial colony in terms of cell morphologies (Boemare and Akhurst, 1988), manufacture of different types of enzymes and respiratory enzymatic activities (Smigielski *et al.*, 1994), secondary metabolites (Akhurst, 1980, 1982) and motility (Givaudan *et al.*, 1995). Interestingly, *in vitro* subculturing may also turn a bacterial portion of phase I to phase II. In *Photorhabdus* species, phase I, also called the primary variant or phase, accompanies the heterorhabditid nematode IJs and possesses numerous traits that distinguish it from phase II (e.g. Woodring and Kaya, 1988; Forst and Clarke, 2002; Gerdes *et al.*, 2015). The second phase, also known as the secondary variant or secondary phase, is usually the output of *in vitro* cultures. Generally, phase II cells are the variant cells in which the traits of the primary phase cells are present at faint grades, or even absent. Consequently, phase II has scarce or reduced levels of the properties of phase I variants: its occurrence is probably a response to stressed conditions such as extended incubation. Nevertheless, both of the variants demonstrate similarities to their pathogenicity. On the contrary, the two variants have dissimilarities in their capability to support their host (the nematode) reproduction, which can play a central role for the *in vitro* and commercial production of EPNs. The characteristics faded or lost in the phase II cells comprise the lack of dye-binding ability, slighter pigmentation, scarcer antibiotic output and very faint bioluminescence. Despite such a loss in traits, the secondary cells are virulent; but not as virulent as primary cells (Dowds and Peters, 2002).

Boemare (2002) explained the reflection of intraspecific biodiversity of *Photorhabdus* bacteria on both their physiological significance and consequences for taxonomic studies. He reported that *Photorhabdus* cells of the primary phase were responsible for supplying EPNs with essential nutrients via killing and bioconversing the insect host and furnishing a range of antimicrobial agents. Although *Photorhabdus* cells of the secondary phase may also put the insect host to death, they are less efficient in furnishing optimum conditions for nematode growth. Moreover, some

Photorhabdus cells of the secondary phase may be harmful to their EPN partner. On the other hand, such phase variants have been the origin of conflicts between data published by various laboratories. It is found at an intraspecific level for all *Photorhabdus* but does not affect the phenotypic characters that distinguish the species level. For example, on testing a bacteriological character, even if only one variant is positive among others of a definite strain that are negative, the result should be reported as positive for this strain (Boemare, 2002). Factually, consistency should be apparent in all the potential expression of the bacterial strain for taxonomic studies.

12.3.1 *Photorhabdus luminescens* (synonym: *Xenorhabdus luminescens*)

P. luminescens is the type species of the genus *Photorhabdus*. It was first described by Khan and Brooks (1977) and deposited with the American Type Culture Collection as an unidentified bacterium with accession No 29304. Then, Thomas and Poinar (1979) described this symbiotic bacterium and assigned it as a new genus *Xenorhabdus* (now *Photorhabdus*), where the new species *Xenorhabdus luminescens* (now *P. luminescens*) associated with *H. bacteriophora* was erected. They demonstrated the importance of the bacteria in the development of the nematodes. Then, based on their DNA relatedness, *Xenorhabdus* spp. were divided into two genera, where *X. luminescens* was transferred to a new genus, *Photorhabdus* gen. nov. on the proposal of Boemare *et al.* (1993).

As the other three *Photorhabdus* spp., *P. luminescens* originally inhabits the IJ nematode gut of the family Heterorhabditidae. Species such as *H. bacteriophora* and *H. indica* are the nematode hosts of *P. luminescens* (Table 12.3). It apparently has a worldwide distribution. The bacterial length is 5 microns and width is 1 micron (Gerdes *et al.*, 2015). This species secretes many virulence factors, such as toxin complexes (TCs) and antimicrobial compounds, which can put their insect hosts to death, while at the same time saving itself and its *H. bacteriophora* partner from other competitor bacteria. Due to such bacterial properties and the nematode ability to infect a wide range of insect hosts, the *P. luminescens–H. bacteriophora* pair has been formulated in a commercial product to control many different species of crop insect pests effectively via a single natural product rather than multiple chemical products. The cadaver of the insect host turns bioluminescent as *P. luminescens* infection advances. The bacterial application has been deemed safe towards humans and the environment (Gerdes *et al.*, 2015). *P. luminescens* genome could be sequenced (Duchaud *et al.*, 2003) and compared to that of *P. asymbiotica* (Wilkinson *et al.*, 2009). It encodes a great number of toxins, adhesins, haemolysins, proteases and lipases. It also comprises a broad range of antibiotic synthesizing genes. Such proteins are important in the protection of the attacked insects from competitors, host colonization, invasion and bioconversion of the insect cadaver. These attributes offer *P. luminescens* as a favourable model for investigating different aspects of symbiosis and host–pathogen relationships. Interestingly, Duchaud *et al.* (2003) stressed that its newly characterized insecticidal proteins could be promising alternatives for suppressing insect pests. It has a membrane attack complex/perforin family (MACPF) protein but this molecule is non-lytic (Rosado *et al.*, 2007). Its genome also has the *gcvB* RNA gene that encodes a small non-coding RNA included in the organization of a quantum of amino acid transport systems and biosynthetic genes. Eventually, its genome is responsible for potency to: (i) synthesize high molecular weight TCs that knock down the host quickly (within 12 h to 48 h); (ii) produce enzymes to decompose the insect tissues into nutrients; (iii) secrete antibiotics to suppress competing bacteria; and (iv) form pigmentation and bioluminescence (Gerdes *et al.*, 2015). Research is in progress to search for more attributes of *P. luminescens* subspecies/strains (e.g. Strain TTO1) (http://biocyc.org/PLUM243265/organism-summary?object=PLUM243265).

Wee *et al.* (2000) showed that *P. luminescens* secondary cells secreted a protease inhibitor that helped to justify why protease activity was found in primary but not secondary cells. The bacterial production of protease is important because it is widely believed that protease is actively involved in the breakdown of proteins in the attacked host insects, which would permit the bacteria and its symbiotic nematode to grow and reproduce inside the insect before moving on to the following host (Forst and Clarke, 2002). Furthermore, Zhang *et al.* (2012b) found that expression of a nematode symbiotic bacterium-derived protease inhibitor protein in tobacco could enhance tolerance against *Myzus persicae*.

Table 12.3. Original[a] species associations of *Heterorhabditis* and *Photorhabdus*. (Adapted from Boemare, 2002; Plichta *et al.*, 2009; Ferreira *et al.*, 2014).[b]

Nematode species/subspecies	Bacterial species/subspecies
Heterorhabditis bacteriophora subgroup Brecon	*Photorhabdus luminescens luminescens*
H. bacteriophora subgroup HP88	*P. luminescens laumondii*
H. indica	*P. luminescens akhurstii*
H. zealandica	*P. temperata*
H. zealandica South Africa strains	*P. heterorhabditis* strains SF41 and SF783
H. bacteriophora subgroup NC	*P. temperata*
H. megidis Nearctic group (Ohio, Wisconsin)	*P. temperata*
H. megidis Palaearctic group	*P. temperata temperata*
H. gerrardi	*P. asymbiotica*

Notes: [a]Recent records (e.g. Ferreira *et al.*, 2013, 2014) displayed nine subspecies of *P. luminescens*, six subspecies of *P. temperata* and two subspecies of *P. asymbiotica*. The nine subspecies of *P. luminescens* are: *P. luminescens* subsp. *Akhurstii* and *P. luminescens* subsp. *laumondii* (Fischer-Le Saux *et al.*, 1999), *P. luminescens* subsp. *Caribbeanensis* and *P. luminescens* subsp. *Hainanensis* (Tailliez *et al.*, 2010), *P. luminescens* subsp. *kayaii* (Hazir *et al.*, 2004), *P. luminescens* subsp. *kleinii* (An and Grewal, 2010), *P. luminescens* subsp. *luminescens* (Thomas and Poinar, 1979; Boemare *et al.*, 1993), *P. luminescens* subsp. *noenieputensis* (Ferreira *et al.*, 2013) and *P. luminescens* subsp. *sonorensis* (Orozco *et al.*, 2013). The six known subspecies of *P. temperata* are: *P. temperata* subsp. *cinerea* (Tóth and Lakatos, 2008), *P. temperata* subsp. *khanii* and *P. temperata* subsp. *tasmaniensis* (Tailliez *et al.*, 2010), *P. temperata* subsp. *stackebrandtii* (An and Grewal, 2010), *P. temperata* subsp. *temperata* (Fischer-Le Saux *et al.*, 1999) and *P. temperata* subsp. *thracensis* (Hazir *et al.*, 2004, Tailliez *et al.*, 2010). The two recognized subspecies of *P. asymbiotica* are *P. asymbiotica* subsp. *australis* (Akhurst *et al.*, 2004) and *P. asymbiotica* subsp. *asymbiotica* (Fischer-Le Saux *et al.*, 1999; Akhurst *et al.*, 2004). [b]Classification and identification of *Photorhabdus* species/subspecies is currently in a state of flux as molecular characterization and phylogenies become increasingly relevant (e.g. Tailliez *et al.*, 2010; Chaston *et al.*, 2011; Ferreira *et al.*, 2013, 2014).

Being the type species of the genus *Photorhabdus*, with a global distribution and supreme insecticidal activities, *P. luminescens* received much research concerning its effects against insect pests (Table 12.4). For example, its pathogenic capability against diamondback moth (*Plutella xylostella*) pupae was better than that of *X. nematophila*. Both bacterial species killed 60% and 40% of *P. xylostella* pupae. Moreover, their LC_{50} values, 5×10^4 and 5.5×10^5 cells/ml respectively, favoured the use of *P. luminescens* (Abdel-Razek, 2003). The rate of the pupal mortality was impacted by the number of *P. luminescens* cells. A significant correlation was found between mortality of the pupae, percentage of deformed adults and the bacterial dose. A negative correlation existed between the no. of eggs laid/female, percentage of hatching, adult survival and the *P. luminescens* dose.

12.3.2 *Photorhabdus temperata*

Due to high divergence, low genomic relatedness, deep phylogenetic relationship, and differential phenotypic characters in *Photorhabdus* isolates examined by Fischer-Le Saux *et al.* (1999), two new species were erected; *P. temperata* and *P. asymbiotica*. *P. temperata* is native to temperate regions (maximum growth temperature 33-35°C). *P. luminescens* and *P. temperata* are heterogeneous genomic groups, but because their DNA-DNA hybridization values were relatively high and because each of them constituted a coherent phenotypic group based on its maximum growth temperature, Fischer-Le Saux *et al.* (1999) suggested delineating only two species. They found that sub-groups of strains within these two species shared very high DNA-DNA hybridization values and could be distinguished by stable 16s rDNA branching. Generally, a close correspondence is found between the taxonomic grouping of the symbiotic bacteria and that of their nematode hosts (Fischer-Le Saux *et al.*, 1999; Boemare, 2002).

A few strains of *P. temperata* exhibited toxicity independently from their host nematode larvae against important insect pests (Table 12.4). For example, the strain K122 of *P. temperata*, symbiotic of *Heterorhabditis downesii* K122 (Stock *et al.*, 2002), demonstrated an appropriate level of pest control when ingested by *Ephestia kuehniella* larvae (Jallouli *et al.*, 2008, 2013). Reaching medium optimization for its biomass production (Jallouli *et al.*, 2008) and overcoming the production limitations of this strain as bioinsecticide in low-cost medium (Jallouli *et al.*, 2011) were successful through adequate fermentation technology (Jallouli *et al.*, 2012).

Kwak and Shin (2015) presented the complete genome sequence of *P. temperata* subsp. thracensis 39-8, a symbiotic bacterium from *Heterorhabditis bacteriophora*, as a novel bacterium accommodating insect pathogenicity. Its genome contains one circular chromosome of 5,147,098 bp with a guanine-cytosine content of 44.10%. Kwak and Shin (2015) could foresee different virulence factors associated with insect pathogenicity. That is because the genome of *Photorhabdus temperata subsp. thracensis* 39-8 (abbreviated as *P. t. thracensis* 39-8) had various genes encoding proteins for virulence factors of insect pathogenicity; e.g. luciferases, siderophores and bacteriocin production, extracellular polysaccharides production, and type III secretion system. So, these genes in *P. temperata* strain 39-8 genome (Kwak and Shin, 2015) may open new avenues for advanced technological implementations of *Photorhabdus* insecticidal toxins which promote commercial bioinsecticide in crop pest control. Eventually, *P. temperata* genome sequence as well as the ongoing research for *P. temperata* applications as bioinsecticides should significantly support and advance these examples listed in Table 12.4.

12.3.3 *Photorhabdus heterorhabditis*

Out of four recognized species, *P. heterorhabditis* is the one most recently described (Ferreira *et al.*, 2014). Its isolates SF41T and SF783 were extracted from *H. zealandica* populations found in South Africa. Both isolates are closely related to strain Q614 isolated from a population of *Heterorhabditis* sp. from soil in Australia in the 1980s. Sequence analysis based on DNA–DNA hybridization data, a multigene approach and phenotypic traits demonstrated that strains Q614, SF41T and SF783 belonged to the same species; i.e. *P. heterorhabditis* strains SF41, SF783 and Q614. Molecular analyses revealed that *P. temperata* subsp. *cinerea* was the taxon most closely related with a DNA–DNA hybridization value of 68% (Ferreira *et al.*, 2014). Due to its recent recognition, very little work has explored its potential against agricultural insect pests.

12.3.4 *Photorhabdus asymbiotica*

This species was first recognized from clinical samples. Therefore, I would like to warn from the outset that *P. asymbiotica* must be dealt with cautiously. Since it was isolated originally from human wounds, there is probably a role for *P. asymbiotica* as an emerging human pathogen. However, *P. asymbiotica* was found later in a symbiotic association with, like the other species of the genus, a heterorhabditid nematode host species; i.e. definitely *H. gerrardi* for *P. asymbiotica* (Plichta *et al.*, 2009). Yet an uncommon enzyme, glycosyltransferase, was found in *P. asymbiotica* toxin (PaTox). As will be discussed later, among *Photorhabdus* spp., the gene encoding PaTox is only found in *P. asymbiotica* (Gerrard *et al.*, 2004). Although recent reports (e.g. Jank *et al.*, 2015a; Alamalakala *et al.*, 2015) have not ascertained whether this toxin is in charge of the wider host spectrum that threatens human health, future research is expected to clarify its pathogenic range. Jank *et al.* (2013) report that such a toxin shelters two enzymatic domains: a glycosyltransferase (PaToxG) and a deamidase domain (PaToxD). They explain how unique adjustment takes place at protected tyrosine remains (Tyr32/34) found in the switch I region of Rho proteins, so that GlcNAc of Rho proteins prevents signalling of the downstream effector. Subsequently, glycosylation of Rho by PaTox directs the dissociation of the actin cytoskeleton, sealing off phagocytosis, cell rounding and eventually leading to the death of cell and arthropod host. The PaTox seems to impact the activity of various types of cells, especially insect cells (e.g. haemocytes of *Galleria mellonella* and SF9 cells) and mammalian cells such as murine (J774) macrophages and human HeLa cells (Jank *et al.*, 2013).

The complete genome sequences of *P. asymbiotica* ATCC43949 have been reported (Wilkinson *et al.*, 2009). *P. asymbiotica* genome was reported to have a few gene clusters. They are envisaged to be engaged in the biosynthesis of unparalleled secondary metabolites that are presumed to increase the pathogenic capacities of *P. asymbiotica*. However, recent reports on *Photorhabdus* secondary metabolite production have indicated that many of its genes are silent under laboratory culture conditions (Theodore *et al.*, 2012). Using a special recipe for *P. asymbiotica* growth under different fermentation circumstances, Theodore *et al.* (2012) could attain a series of novel glidobactin/luminmycin derivatives from *P. asymbiotica*. These comprised glidobactin A (1), luminmycin A (2) and luminmycin D (3). The same chemicals were also gained when live crickets were infected with *P. asymbiotica*. Luminmycin D showed cytotoxicity against human pancreatic cells with the half maximal inhibitory concentration

Table 12.4. Examples of bioinsecticidal activity where *Photorhabdus* species provided acceptable insect pest control.

Photorhabdus species	Insect species (common name)	Comments	Reference
Photorhabdus luminescens	*Manduca sexta* (tobacco or tomato hornworm)	Bacterial toxin causes disruption of the midgut epithelium of the insect larvae. Toxin was introduced orally or via injection.	Blackburn *et al.*, 1998; Bowen and Ensign, 1998
P. luminescens	*Galleria mellonella* (greater wax moth larvae)	The 50% lethal dose of bacterium injected into insect larvae was less than five cells.	Milstead, 1979; Bowen and Ensign, 1998; Mahar *et al.*, 2005
P. luminescens	*Tenebrio molitor* (mealworm)	Bacterial HPLC-purified toxin attained >30 and 100% mortality via oral and injection introduction, respectively.	Bowen and Ensign, 1998.
P. luminescens	*Monomorium pharaonis* (Pharaoh ant)	Bacterial HPLC-purified toxin attained >30 mortality via oral introduction.	Bowen and Ensign, 1998
P. luminescens	*Blattella germanica* (German cockroach)	Bacterial HPLC-purified toxin attained >30 and 100% mortality via oral and injection introduction, respectively.	Bowen and Ensign, 1998
P. luminescens	*Periplaneta americana* (American cockroach)	Bacterial HPLC-purified toxin attained 100% mortality via injection introduction.	Bowen and Ensign, 1998
P. luminescens	*Thrips palmi* (melon thrips)	Cell-free supernatant of the bacterial forms was more virulent than cells causing high mortality within 24–48 h.	Uma *et al.*, 2010
P. luminescens	*Pieris brassicae* (cabbage white butterfly)	100% mortality of the larvae was recorded within 24 h compared to no mortality in the control plot.	Mohan *et al.*, 2003
P. luminescens	*Drosicha mangiferae* (mango mealy bug)	Spray over bug-infested mango twigs resulted in 92.5% mortality after 48 h, which was significantly higher than that made 1 h prior to the release of bugs (82.3%).	Mohan *et al.*, 2004
P. luminescens	*Plutella xylostella* (diamondback moth)	Insect mortality was attributed to different toxic bacterial enzymes that damage the haemocytes and inhibit activation of the humoral immune system of the insect.	Abdel-Razek, 2003
P. luminescens	*Aphis gossypii* (melon and cotton aphid)	Two bacterial isolates, Z-8-1 and Z-3-1, caused 100% mortality of *A. gossypii* nymphs and *Tetranychus macferlanei* after 24 and 36 h of exposure, respectively.	Kumar *et al.*, 2014
P. luminescens	*Leptinotarsa decemlineata* (Colorado potato beetle)	The 48-h LC_{50} for toxin complex a (Tca) against neonate insects was 2.7 ppm. Growth of second instar exposed to Tca for 72 h was almost entirely inhibited at >0.5 ppm.	Blackburn *et al.*, 2005
P. luminescens	*Bemisia tabaci* (sweet potato whitefly)	Tca at 0.1 and 0.2 ppm reduced nymph numbers reaching second instar by 50%.	Blackburn *et al.*, 2005
P. luminescens	*Phenacoccus solenopsis* (Solenopsis mealybug)	Caused 58% nymph mortality after 48 h and complete oviposition inhibition. A constraint is that *P. luminescens* kill coccinellids that are key mealybug predators. So, integration of biopesticides that may induce bioecological hazards needs to be used cautiously with other biocontrol agents.	Fand *et al.*, 2012

Continued

Table 12.4. Continued.

Photorhabdus species	Insect species (common name)	Comments	Reference
P. temperata	*Ephestia kuehniella* (Mediterranean flour moth)	At 12 × 10^8 cells/ml, 100% mortality of insect larvae could be reached. Insect mortality is due to toxaemia.	Jallouli *et al.*, 2013
P. temperata	*Prays oleae* (olive moth)	The bacteria exhibited oral toxicity with an LC_{50} of 58.1 × 10^6 cells/ml.	Tounsi *et al.*, 2006
P. temperata	*Frankliniella occidentalis* (western flower thrips) and *Thrips tabaci* (onion thrips)	After 7 days of drinking from bacterial supernatant, a mortality of 90% could be reached. Thrips were also killed after sucking from leaves covered with the bacterial toxins. Toxins affected thrip fecundity.	Gerritsen *et al.*, 2005
P. temperata	*Diatraea saccharalis* (sugarcane stalk borer)	LD_{50} was 16.2 bacterial cells at 48 h post-infection. Infected larvae started dying as soon as 30 h post-infection with a LT_{50} value of 33.8 h and an LT_{90} value of 44.8 h.	Carneiro *et al.*, 2008
P. luminescens	*Schistocerca gregaria* (desert locust)	Nymphs were killed when the bacteria were injected into the abdomen, applied orally, mixed with bran or applied to the foliage of food plants as compared to broth alone (control).	Mahar *et al.*, 2004
P. temperata	*Tribolium castaneum* (red flour beetle)	*P. temperata* exhibited more than six times higher pathogenicity than Xenorhabdus nematophila. The aqueous extracts of the two bacteria contained proteins that showed different profiles.	Shrestha and Kim, 2010
P. luminescens	*Luciaphorus perniciosus* (mushroom mite)	Bacterial cultures of 1 × 10^8 cells/ml resulted in 83% mortality of the female mite within 3 days after application.	Bussaman *et al.*, 2009
P. luminescens	*Choristoneura fumiferana* (the eastern spruce budworm)	By the binary PaxA and PaxB toxin on a midgut-derived cell line FPMI-CF-203/2.5 or CF-203.	Zhang *et al.*, 2014
P. temperata	*Culex pipiens pallens*	Two anthraquinones isolated from the bacteria were highly lethal to the insect larvae.	Ahn *et al.*, 2013
P. temperata	*Spodoptera exigua* (beet armyworm)	When ingested, the bacteria caused high mortality of third; but not fifth-instar larvae via antibiotic activity against *Bacillus cereus* (a gut symbiont for optimal insect growth). *Bacillus thuringiensis* as a synergist was used to enforce mortality of fifth-instar larvae.	Jung and Kim, 2006
P. luminescens	*Spodoptera litura* (Oriental leafworm moth)	The bacteria encapsulated in sodium alginate beads (2.5 × 10^7 cells/bead) and mixed with sterilized soil resulted in 100% larval mortality in 48 h.	Rajagopal *et al.*, 2006

(IC_{50}) as 0.11 μM. Also, its proteasome inhibition was recorded (IC_{50} of 0.38 μM) (Theodore *et al.*, 2012). On the other hand, Wilkinson *et al.* (2009) found *P. asymbiotica* strain ATCC43949 genome (5,064,808 bp) to be smaller than that of *P. luminescens* strain TT01 (5,688,987 bp). Yet each of the two strains has about one megabase of DNA that is distinctive for each of them. The decreased genome size of the strain ATCC43949 is linked to a smaller variation in insecticidal genes; for example, those responsible for the toxin complexes (Tcs), makes caterpillars floppy (Mcf) toxins and the *Photorhabdus* virulence cassettes (PVCs). Moreover, the *P. asymbiotica* genome demonstrated the addition of a plasmid related to pMT1 from *Yersinia pestis* and several novel pathogenicity islands, including a novel Type Three Secretion System (TTSS) encoding island. Thus, Wilkinson *et al.* (2009) suggested that strain ATCC43949 might be virulent against humans due to acquisition of the pMT1-like plasmid and specific effectors, e.g. SopB, which promoted its stability within human macrophages. The loss of insecticidal genes in the *P. asymbiotica* strain was so intriguing that it did not reflect a loss in insect pathogenicity. They concluded that *P. asymbiotica* strain ATCC43949 had gained virulence against humans by obtaining a plasmid and specific virulence traits like those in other human pathogenic bacteria. Therefore, caution should also be exercised with any new species/strain of *Photorhabdus* whereby human hazard could be introduced with such new biologicals (Abouelhag and El-Sadawy, 2012). All bacteria linked to clinical infection seem to have one or more plasmids. On the contrary, the bacteria used as biocontrol agents for crop protection do not. Therefore, it is wise to keep away from using plasmid-carrying bacteria against crop insect pests (ffrench-Constant and Dowling, 2014). If any of the toxins associated with *P. asymbiotica* are found to own potential utility in crop pest management, it is useful to investigate their role in virulence in order to develop wise statement about their mode of action and safety as well (ffrench-Constant and Dowling, 2014).

12.4 Toxins of *Photorhabdus* species

It is now well established that the pathogenicity of *Photorhabdus* bacteria is the result of the presence of pathogenicity islands on its chromosome that carry a large number of genes. They encode for different important compounds such as toxins, enzymes, bacteriocins and antibiotics. Because of their plentiful and renewable discoveries, diversified toxins from these bacteria have been identified, characterized and categorized into different groups. For example, these toxins were formerly classified into four major groups: the toxin complexes (Tcs), the *Photorhabdus* insect-related (Pir) proteins, the makes caterpillars floppy (Mcf) toxins and the *Photorhabdus* virulence cassettes (PVC) (Yang *et al.*, 2006; ffrench-Constant *et al.*, 2007; Rodou *et al.*, 2010). More recently, ffrench-Constant and Dowling (2014) grouped these toxins as Tcs, PVC, Mcf, Patox and Photox, and Binary Toxins.

On the other hand, different approaches to isolate toxin genes from *Photorhabdus* bacteria were followed (ffrench-Constant and Dowling, 2014). First, to isolate Tc genes, Tc proteins were purified, anti-Tc antibodies raised and then the Tc-encoding genes were extracted from expression libraries. Second, for PVCs, phage-like loci were expressed in recombinant *Escherichia coli* and found to produce visible phage tail-like structures. Third, concerning Mcf, gain-of-function screens could be utilized to characterize novel toxins as phenotypically expressed by *Photorhabdus*-infected caterpillars. Fourth, bioinformatic analyses may be used simply to search for motifs found in other toxins, given the large amount of genomic sequence now available for different strains and species of *Photorhabdus* (Rodou *et al.*, 2010; ffrench-Constant and Dowling, 2014).

Moreover, scientists have been trying to elucidate fully the secretion mechanisms of these toxins that may occur through different secretion systems. Rodou *et al.* (2010) utilized bioinformatics analysis and comparison against the elements of recognized secretion systems throughout the entire *P. luminescens* genome. Consequently, they could identify various copies of components of all familiar systems for secretion to be found in *P. luminescens*, except those assembling a type IV secretion system. Gatsogiannis *et al.* (2013) proposed a syringe-like injection mechanism in *P. luminescens* toxins that could serve as a robust base for the development of biopesticides.

12.4.1 The toxin complexes (Tcs)

There are four toxin complexes, namely TcA, TcB, TcC and TcD, existing in various loci. Two of these loci, TcA and TcC loci, encode for several open reading frames (ORFs), thus manufacturing multiple products per locus (Bowen *et al.*, 1998; ffrench-Constant and Bowen, 1999, 2000). The other two loci, TcB and TcD, consist of a single long ORF (Rodou *et al.*, 2010). Such protein complexes with large molecular weight could be resolved into a range of different polypeptides (30–200 kDa) on a native gel under denaturing conditions (Bowen and Ensign, 1998). Four such million molecular weight complexes were extracted from *P. luminescens* strain W14 and named toxin complex (Tc) A, B, C and D, or TcA, TcB, TcC and TcD (Bowen *et al.*,

1998). Nevertheless, further findings revealed that this nomenclature was not practical because each Tc locus might be further divided into more separate ORFs, and eventually there was a countless number of potential Tc loci in any one *P. luminescens* W14 genome (Bowen *et al.*, 1998; ffrench-Constant and Dowling, 2014). Certainly, expression of A, B and C subunit encoding genes (e.g. TcaA, TcaB and TcaC) together was substantial enough to reconstruct recombinant oral toxicity (Yang and Waterfield, 2013). Moreover, expression of TcaB and TcaC (encoding B and C) in the same bacterial cytoplasm was also needed (Waterfield *et al.*, 2001a,b), but the TcaA ORF (encoding A) could be expressed alone and then mixed with the outputs of TcaB/TcaC (B and C) to reconstruct full oral toxicity (Waterfield *et al.*, 2001a). Different aspects concerning the structure, mechanism of action and potential role of the Tcs as 'polymorphic' toxins in the biology of infection have been discussed elsewhere (Zhang *et al.*, 2012a; ffrench-Constant and Dowling, 2014).

Apparently, insecticidal activity of these toxins is of a proteinaceous nature, and their oral activity to insect pests may be obtained via the cell-free supernatant of certain *Photorhabdus* strains (e.g. Bowen and Ensign, 1998; Bowen *et al.*, 1998; Abd El-Zaher *et al.*, 2008, 2012). On the other hand, the direct delivery of Tcs into the insect's haemocoel, through the infection of the symbiotic nematodes and releasing the bacteria into the insect haemolymph, does not clarify why any *Photorhabdus* toxins are definitely toxic in the insect midgut, if they are not somehow able to extend their activities from either side of the gut (insect haemocoel or gut lumen). As a compromise, the available data propose that various entomopathogens do indeed utilize the Tcs to disturb the insect midgut, presumably either from the gut lumen itself (*Yersinia entomophaga* or *Bacillus thuringiensis*) or from the outer haemocoel (*Photorhabdus* and *Xenorhabdus*). In this respect, ffrench-Constant and Dowling (2014) confirmed that some insect pathogens that carry Tc-like genes (e.g. *Y. entomophaga*) have demonstrated its ability to pass from the gut lumen into the insect haemocoel, suggesting the Tc-like genes may somehow ease this transition across the gut.

Gatsogiannis *et al.* (2013) have demonstrated that TcA in *P. luminescens* (TcdA1) forms a minute opening in the transmembrane. They reported its construction in the pre-pore and pore constitution as viewed by cryo-electron microscopy (cryo-EM), where the TcdA1 pre-pore assembled as a pentamer forming an α-helical, vuvuzela-shaped channel less than 1.5 nanometres in diameter, encompassed by a major outer shell. Membrane insertion is triggered not only at low pH, as expected, but also at high pH; justifying Tc action directly through the midgut of insects. Meusch *et al.* (2014) interpreted Tc toxin alterations, from its pre-pore to pore state, through a molecular model where the crystal structure of the A pre-pore was fitted into the cryo-EM structure. They demonstrated in detail the exact association of the pore formation, with spectacular conformational alterations that unlocked the shell and transferred the main channel by 12 nm. Such a striking alteration in the structure of the A pore complex is expedited by three defined hinge zones and leads to a wide reorder of the four receptor domains (termed 'receptor-binding domain A, B, C and D'), offering a potential change in their binding geometry and/or affinity. Also, comparative constructions of the TcdA1 pore inserted into a membrane and in combination with TcdB2 and TccC3 display large conformational changes during membrane insertion, indicating a novel, syringe-like mechanism of protein translocation. So, Gatsogiannis *et al.* (2013) proved how ABC-type toxin complexes bridged a membrane to inject their lethal components into the cell cytoplasm of the insect host. They concluded that this mechanism was distinctive to the whole ABC-type toxin family. Their interpretation of toxin translocation can help to unravel the complexity involved in the host–pathogen interaction and the complicated life cycle of many pathogens, including *Photorhabdus* and human pathogenic bacteria (Gatsogiannis *et al.*, 2013). Moreover, the above-mentioned explanation of the syringe-like injection mechanism in *P. luminescens* toxins may pave the way to develop biopesticides in terms of adopting adequate *Photorhabdus* formulation and application approaches.

12.4.2 *Photorhabdus* virulence cassettes (PVCs)

This set of assumed toxins was specified via homology to antifeeding genes in *Serratia entomophila* (Yang *et al.*, 2006; ffrench-Constant *et al.*, 2007). Many pro-phage-like loci, known as the PVCs, were discovered in the sequenced genome of each *P. luminescens* TT01 and *P. asymbiotica* ATCC43949. The location and number of these modified phage tail loci varied between the sequenced genomes of these two *Photorhabdus* strains (Wilkinson *et al.*,

2009). For instance, four PVC-encoding loci were recognized in *P. luminescens* in a tandem array between a type IV DNA conjugation pilus operon and a *mukB* replicon partitioning gene. However, at the corresponding location in *P. asymbiotica* genome, there is just one PVC locus, and in *P. temperata* K122, the PVC loci are substituted by a Mu-like phage (Yang *et al.*, 2006). This supports the assumption that the PVC loci are not only phage-like in sequence but also they can move easily around and between bacterial genomes. The PVC loci have a specific general structure with 12–15 ORFs encoding a conserved 'PVC unit', followed by 2–4 loci encoding putative effector proteins. This inclusive form proposes that the PVC-secretion system (PVCSS) is similar to that of type VI in being generated from the tails of pro-phage, but differs in its related AAA+ATPase, which in the type VI system is a member of the ClpB clade (ClpV). Therefore, it seems that both the type VI-SS and PVC-SS have evolved independently from pro-phage tails using different AAA+ATPases to recycle the insertion system after deployment (Zhang *et al.*, 2012a). ffrench-Constant and Dowling (2014) reported that the assumed PVC encoded effectors manifested correspondence to different proteins. So, these effectors showed similarity to a range of various toxins, including TccC, from the toxin complexes; part of makes caterpillars floppy 1 (Mcf1) toxin; LopT (a homologue of the *Yersinia* cysteine protease YopT), an effector delivered by the secretion order of type III in other bacteria; and the *Photorhabdus* necrosis factor (Pnf; a homologue of Cnf). PVCs can be expressed readily in recombinant *E. coli*. The PVCs are 30 nm wide but show a huge variety in length (up to 800 nm long), depending on whether they are found in a contracted (needle within the sheath) or extended state (needle protruding from the sheath). Polyclonal antibodies elevated against the putative effector proteins do appear to label proteins brought out from the PVC when anti-effector antibody-conjugated gold particles are examined by transmission electron microscopy (ffrench-Constant and Dowling, 2014). Injection of *G. mellonella* larvae with recombinant PaPVCpnf killed the larvae and caused dramatic rearrangement of the actin cytoskeleton of haemocytes recovered from injected larvae (Yang *et al.*, 2006). Such insecticidal activities vanished on inserting transposons into the ORFs encoding the effector proteins (Yang *et al.*, 2006). Based on such data, some researchers (ffrench-Constant and Dowling, 2014) postulated that the PVCs were phage-tail-like transmission tools that foreworded nominated effectors (encoded at the end of each of the individual PVC-encoding loci) into their intended eukaryotic cells.

The nature of such transmitted effectors and precisely how they join the intended cell is still inconspicuous. Nevertheless, PVC-like particles display distinct characteristics such as cessation of feeding, death, or the triggering of metamorphosis. So, it might be that the PVCs are adhering to specific receptors in their host tissues, where they are able to 'recognize' particular structures. This recognition will enable them to display their toxicity; for example, through breaking down midgut cells or via specific species such as the settling *Hydroides* tubeworm larva (ffrench-Constant and Dowling, 2014). The latter authors supposed that this group of toxins possessed specific receptors eligible to recognize displayed sugars or proteins and that their transmitted effectors were potentially to intervene with sound physiological processes that changed the developmental course of the intended tissue. Since the bacterial genomes of *Photorhabdus* have numerous PVC loci, it is possible that the delivered effectors may be engaged both in destroying the insect midgut (like the Afp particle from *Serratia*) or in inducing developmental alternations in their nematode partners; by analogy with the function of metamorphosis-associated contractile structures in targeting tubeworm metamorphosis (ffrench-Constant and Dowling, 2014). Rodou *et al.* (2010) conclude that the protein products of the PVCs do destruct insect haemocytes without direct antibacterial activity.

12.4.3 Makes caterpillars floppy (Mcf) toxins

Daborn *et al.* (2002) found a single, high molecular weight *Photorhabdus* gene, makes caterpillars floppy (*mcf*), which was adequate to allow *E. coli* to persist within and to kill an insect. Mcf toxin is mostly dissimilar to other recognized protein sequences but holds a BH3 domain and triggers apoptosis in the haemocytes and the midgut epithelium of the invaded insect host. In this group, toxins 1 (Mcf1) and 2 (Mcf2) operate via injection (Daborn *et al.*, 2001, 2002; Yang *et al.*, 2006; ffrench-Constant *et al.*, 2007) and they are encoded by PAI II (Waterfield *et al.*, 2002, 2004), along with other haemagglutinin-like proteins (Rodou *et al.*,

2010). Mcf1 enhances apoptosis in the midgut, producing a distinct 'floppy' characteristic in the attacked insect host and in mammalian cells (Dowling *et al.*, 2004). It simulates BH3 domain proteins formed in mitochondria and possesses proapototic functions (Yang *et al.*, 2006). Mcf2 N-terminus shows a similarity to a type III secreted protein of a plant pathogen (*Pseudomonas syringae*), but it lacks the BH3-like domain (Waterfield *et al.*, 2003). Mcf1 and Mcf2 possess C-terminal domains that back their transmission via a type I secretion system (Wilkinson *et al.*, 2009). Recombinant Mcf1 and Mcf2 toxins in *E. coli* enable the bacterium to live within the insect and enhance its killing. Eventually, these toxins are possibly developed to furnish *Photorhabdus* with toxicity in order to kill different insect species quickly (ffrench-Constant and Dowling, 2014).

12.4.4 Patox and photox

Jank *et al.* (2013) utilized the basic local alignment search tool (BLAST) and bioinformatic approaches to find a gene in *P. asymbiotica* genome encoding a putative glycosylating toxin. They named it *P. asymbiotica* toxin, or 'PaTox'. The gene was cloned and the PaTox protein was purified. When injected into *G. mellonella* larvae, 100% mortality of the injected insects occurred by the full-length toxin after 88 h. Jank *et al.* (2013) also reported that on changing the DxD motif in PaTox to asparagine-X-asparagine (NxN), 61% of the larvae survived. They found PaTox to be responsible for two tasks. It targets Rho proteins (like TccC5, which is delivered by the toxin complex secretion system) by glycosylation of tyrosine (GlcNAcylates) at Y32. It also charges heterotrimeric G proteins by deamidation (Jank *et al.*, 2013, 2015a,b). Jank *et al.* (2013) found tyrosine GlcNAcylation only during *P. asymbiotica* infection of insect cells but not following infection with *P. luminescens*, probably because the former bacterium species did not have a copy of TccC5. This might imply that PaTox is somehow distinctive of, or demanded by, *Photorhabdus* strains capable of infecting humans. Nevertheless, such an assumption needs further examinations. Much like Mcf, its exact mechanism of accessing the cell of the host insect is as unclear as Mcf, though it possibly comes into the cell from early endosomes (Jank *et al.*, 2013). The perfect structure–function analysis of PaTox reveals that this toxin brings about both activation (via PaToxD) and inactivation (via PaToxG) of the same Rho protein target; it is prepared to modulate exactly the balance of Rho activity. In parallel with such a balance task, other large toxins of *Photorhabdus*, such as Mcf, may also hold more than one enzymatic activity. ffrench-Constant and Dowling (2014) speculated that this assumption might thus begin to interpret the complex phenotypes noted following the application of Mcf to different types of cells.

Another bioinformatic analysis resulted in identifying a different toxin, termed '*Photorhabdus* toxin' or 'Photox'. The *plu0822* gene was identified as the putative gene encoding mART toxin protein (Visschedyk *et al.*, 2010). The specific site of Photox modification of actin was analysed *in vivo* using a yeast-based assay system and showed that Photox specifically ADP-ribosylates actin at Arg177. Photox has an unusual domain structure with the N-terminal domain (residues 1–185) dissimilar to other toxin domains and being foreseen to be disturbed in structure. Visschedyk *et al.* (2010) presumed that this N-terminal domain might be included in accessing Photox into the intended cell, potentially utilizing VgrG as a translocator during type VI secretion. In conclusion, Photox is clearly a novel mART that ADP-ribosylates actin, but the definite point at which it is expressed during *Photorhabdus* infection, and therefore its potential interaction with other toxins that change actin assembly, is still inconspicuous (Visschedyk *et al.*, 2010; ffrench-Constant and Dowling, 2014). On the other hand, the intracellular deamidase activity of PaTox relies on the existance of the membrane localization domain (Jank *et al.*, 2015a).

12.4.5 Binary toxins

These toxins, discovered in either *Photorhabdus* or *Xenorhabdus*, are obviously smaller and more convenient to work with than the above-mentioned large *Photorhabdus* toxins and toxin delivery systems. Thus, they may also be simpler to clone and apply in the control of insect pests (ffrench-Constant and Dowling, 2014). The outputs of two loci, plu 4093-plu 4092 and plu 4437-plu 4436, in *P. luminescens* TT01 genome proved oral insecticidal activity to control wax moth, *G. mellonella*, and mosquito larvae (Waterfield *et al.*, 2005). The proteins encoded by these loci were termed

'*Photorhabdus* insect-related' (Pir) proteins A and B (or PirAB). Since their bioinsecticidal activity was demonstrated only when injected together and not when injected individually, both PirA and PirB were necessary as a binary toxin (ffrench-Constant and Dowling, 2014) to cause insecticidal activity. In addition, PirA and PirB had no oral activity against another moth, *Manduca sexta* (Waterfield *et al.*, 2005). These proteins demonstrated resemblance both to δ-endotoxins from *B. thuringiensis* and a developmentally regulated protein from a beetle, *Leptinotarsa decemlineata.* Duchaud *et al.* (2003) suggest PirAB to be mimics of insect juvenile hormone (JH) esterases that disrupt insect metamorphosis by metabolizing the insect growth regulator juvenile hormone. Yet Waterfield *et al.* (2005) found that Pir proteins were not able to metabolise JH.

The binary toxin is beneficial as mosquito larvicides, though its mode of action is still inconspicuous. PirAB demonstrated bioinsecticidal activity against the larvae of *Aedes aegypti* (the vector of Dengue fever) and *Aedes albopictus*. The toxin proved to be safe against a beneficial insect species; the *Mesocyclops thermocyclopoides* predator (Ahantarig *et al.*, 2009). Therefore, PirAB may represent a good alternative to toxins from *B. thuringiensis* and *Bacillus sphaericus* to manage mosquito larvae.

Pairs of genes encoding proteins, similar to the XaxAB cytotoxin from *X. nematophila*, are found in *P. luminescens*, where presumably they would be termed PaxAB. Vigneux *et al.* (2007) found these genes not only in various entomopathogenic bacteria of the genera *Xenorhabdus* and *Photorhabdus* but also in other pathogenic bacteria. Based on amino acid sequences, locus organization and activity features, this toxin did not fit in any family of cytotoxins (Vigneux *et al.*, 2007). It is, therefore, probably the prototype of a new family of binary toxins. Both PaxA and PaxB are required for insecticidal activity, and in combination showed effects on a Spruce budworm, *Choristoneura fumiferana*. On the other hand, maximal haemolytic activity could be recorded with precisely equal concentrations of both XaxA and XaxB. ffrench-Constant and Dowling (2014) annotated also that it was hard to understand why adding XaxA before XaxB was required for toxicity. The activity of the XaxAB-like toxin in *P. luminescens* TT01 (known here as PaxAB) has also been reported (Zhang *et al.*, 2014). Yet the XaxAB-like binary toxins need more research work to explore fully their specific mode of action (ffrench-Constant and Dowling, 2014).

12.5 Secretion systems of *Photorhabdus* bacteria

Almost all known secretion systems are represented in *P. luminescens* except for those of the type IV secretion system (Rodou *et al.*, 2010). The wide diversity of secretion machinery found in the genomes of *Photorhabdus* is essentially due to the fact that proteins that are formed in the cytoplasm must go out to the extracellular space (protein secretion). Yet *Photorhabdus* cells are encompassed by a cell envelope that involves the cell (cytoplasmic) membrane and a cell wall. In Gram-negative bacteria, such as *Photorhabdus*, protein secretion involves the crossing of the newly formed protein through both the inner membrane (IM) and the outer membrane (OM), as well as the periplasmic space spanning between them (Rodou *et al.*, 2010).

Briefly, secretion of the above-mentioned toxins can be done in one or two steps, thus splitting the secretion pathways into two groups: (i) one-step or Sec-independent pathways, where the protein moves via a path across both membranes and the periplasm; and (ii) two-step or Sec-dependent pathways, where the protein first moves into the periplasm across the IM with the help of a protein complex embedded in it, then bridges the periplasm, where it is transmitted through the OM to the extracellular environment. The one-step secretion systems of *Photorhabdus* bacteria involve type I secretion system, type III secretion system and type VI secretion system. The two-step secretion systems of *Photorhabdus* bacteria include IM translocation and OM translocation. The outer membrane translocation system is further divided into type II secretion system, type V secretion system, two-partner secretion and chaperone-usher secretion. Bioinformatics analysis using elements of the recognized secretion systems indicated the presence of multiple copies of such secretion system components throughout the *P. luminescens* genome (Rodou *et al.*, 2010). The *P. luminescens* type III is utilized for the direct delivery of effectors in the host cell. Interestingly, this type of secretion systems offers a valuable help in processes such as immune system evasion during the establishment of infection (Wilkinson *et al.*, 2009). This is important during the life cycle of the bacterium for its survival and infestation. On the other hand, type VI secretion systems were identified in *Photorhabdus* bacteria; type VI secretion system also has similar adventageous secretion apparatus. Multiple copies of other secretion systems could be related to the array of toxins, virulence factors and other proteins, substantial for survival and infestation, secreted from *Photorhabdus* bacteria (Rodou *et al.*, 2010; ffrench-Constant and Dowling, 2014).

12.6 Use and Development of *Photorhabdus* as Biocontrol Agents

It is well established that a *Photorhabdus* bacteria–heterorhabditid nematode pair has been an attractive biocontrol approach to many insect pest species, especially against soil-dwelling and wood-boring insect pests. The *Photorhabdus–Heterorhabditis* complex possesses virtually many attributes of an ideal biological control agent (Gaugler, 2002; Grewal *et al.*, 2005; Campos-Herrera, 2015). They are extraordinarily lethal to most insects, causing mortality within 36 h. Dozens of important insect pests are susceptible, yet non-target populations are undisturbed. Despite their high pathogenicity to insects, their lack of threat to the environment led the US Environmental Protection Agency to exempt EPNs from registration requirements. Simultaneous applications of this pair, in terms of EPNs and their associated bacteria, do not require masks or other safety equipment and re-entry time; residues, pollution and pollinator toxicity are not issues. Also, this complex does not require specialized application equipment, as it is compatible with existing equipment, including pressurized, mist, electrostatic, fan and aerial sprayers. Application via irrigation has further improved grower acceptance. Recently, Gerdes *et al.* (2015) stressed that the *H. bacteriophora–P. luminescens* complex could serve as a highly probable candidate to solve the issue of insect resistance. They also advocated that applying the bacteria–nematode partner as an insecticide has demonstrated no adverse effects to humans, plants, wildlife, or the environment, and consequently is a highly favourable choice.

Additionally, from the above-mentioned attributes of *Photorhabdus* bacteria, it has been realized that these bacteria have all the substantial means for the secretion of virulence factors. Therefore, *Photorhabdus* bacteria are able to establish a microbial infection when applied independently from their host nematode larvae against important insect pests (e.g. Blackburn *et al.*, 1998; Mohan *et al.*, 2003; Brown *et al.*, 2006; Rodou *et al.*, 2010; Jang *et al.*, 2012; Kumar *et al.*, 2014; Gerdes *et al.*, 2015). In the wake of many promising results for the use of these bacteria as bioinsecticides (Table 12.4), *Photorhabdus* bacteria, or its formulations, are being taken into account as potential biocontrol agents in pest management strategies. On the other hand, numerous biotic and abiotic factors give rise to doubts regarding their practical consideration and ecological safety against beneficial/non-target fauna in general, and insects in particular (Mohan and Sabir, 2005). It is expected, however, that renewed discoveries and advances in biotechnology will consider these bacteria as bioinsecticides in future agricultural practices.

12.6.1 Potential for using *Photorhabdus* as biocontrol agents

Morgan *et al.* (1997) detected the survival of *Photorhabdus* bacteria in the environment during 1 week, both in soil and in fresh water. This has probably set a time frame for applying the bacteria, independent of their mutualistic nematodes, against a target pest. As a result, attempts to utilize different formulations based on just the bacteria and/or bacterial metabolites have been reported (Table 12.4). Initially, the pathogenicity of *Photorhabdus* alone has been demonstrated by injecting them directly into the insect haemocoel by artificial means (Dunphy and Webster, 1988), although the bacteria have not been detected in free-living form in nature, raising doubts of their competence to survive and infect an arthropod host in the absence of the mutualistic EPN. On the contrary, *P. asymbiotica* is known to be a human pathogen, although is also infectious to insects (Plichta *et al.*, 2009). A distinguished difference with the other *Photorhabdus* spp. is the typical coloration acquired by the cadaver. In *P. asymbiotica*, the cadavers turn grey with pink spots, whereas the colour for other species of the same genus is, in most cases, reddish (Plichta *et al.*, 2009; Piedra-Buena *et al.*, 2015).

In fact, numerous important issues have urged scientists to explore the availability of novel pesticides and insecticides as alternatives to meet the growing dissatisfaction with chemicals and their related problems; their exploration has brought them to the *Photorhabdus* bacteria. As is widely known, the inferior performance of biopesticides compared to chemical insecticides in many situations constitutes an apparent issue. Also, the maladjusted efficacy of the *Heterorhabditis–Photorhabdus* complex, due to product quality control or environmental disagreement, is the greatest hindrance to the adoption of this complex in various management programmes against arthropod pests (e.g. Georgis, 2002; Grewal *et al.*, 2005; Abd-Elgawad and Spiridonov, 2014). On the other hand, issues of human safety and health and environmental pollution leading to the banning of some effective fumigant and granular pesticides, and outgrowth of pest biotypes that break host plant resistance, have set up real scope to integrate the utilization of the bacteria–nematode partner in management programmes of crop insect pests. For example, the growing dissatisfaction with *B. thuringiensis* (Bt) as a pesticide, due to developing pest resistance, has spurred researchers to examine more pesticides and insecticides as efficient alternatives. Following this line of thinking, a few reports (e.g. Brown *et al.*, 2006; Jang *et al.*, 2012; Jallouli *et al.*, 2013; Gerdes *et al.*, 2015) have found such alternative insecticides as *P. luminescens* or *P. temperata* bacteria, either alone or with their mutualistic heterorhabditid nematodes listed in Table 12.4.

There are thus far limited reports in the literature for the exclusive use of these symbiotic bacteria or their metabolites as a tool to control crop insect pests compared to the application of nematode–bacterium pairs, and due to this relative lack of data, a substantial amount of investigation remains to be achieved before the efficacy or commercial application of this approach, that is, the rising generation of these mutualistic bacteria as biopesticides, can be established. In this respect, technical improvements in the production and application of EPNs can supply addional opportunities to integrate the use of these symbionts, within commercial EPN products, in insect control tactics. Likewise, technical issues in the formulation of these bacteria alone (i.e. without their nematode hosts), their shelf life, production and application should be considered, explored and solved, especially to provide further scope to integrate their utilization in integrated pest management (IPM) programmes, if applicable. For example, Mohan *et al.* (2003) tested the insecticidal capability of *P. luminescens*, isolated form *H. indica*-IARI strain, against cabbage butterfly (*Pieris brassicae*), which is a polyphagous pest of crops belonging to the Brassicae family. The bacterial cultures were grown overnight on nutrient broth at 28°C up to a concentration of 10^8 colony-forming units (CFU)/ml. Before spraying, the bacterial culture was mixed with paraffin oil at 10μl/ml, Tween-20 at 0.5 μl/ml and sucrose at 0.5% as a phagostimulant adjuvant using sterile water as base. Initially, Mohan *et al.* (2003) tested the survival and retention of bacterial pathogenicity in combination with various components used in the formulation. When the formulated *P. luminescens* stored at 28°C was sprayed uniformly on the foliage of ornamental nasturtium, *Tropaeolum majus*, heavily infested with the 3–4 instar of *P. brassicae*, 100% mortality of the larvae was recorded within 24 h, compared to no mortality in the control plot. Re-isolation of the bacteria from the dead insects and comparison with the original culture proved Koch's postulates. These results proved the direct toxicity of *P. luminescens* to the insects under natural conditions, when used as a foliar spray (Mohan *et al.*, 2003). Also, Jallouli *et al.* (2013) recorded promising insecticidal activity of *P. temperata* K122 against a stored grain pest, the Mediterranean flour moth, *E. kuehniella*. At a high concentration of 12×10^8 cells/ml, 100% mortality of *E. kuehniella* larvae could be reached. Jallouli *et al.* (2013) speculated that insect mortality was due to toxaemia, as confirmed by the absence of variant small colonies or *P. temperata* colonies in *E. kuehniella* tissue. The histopathological effect of *P. temperata* toxins on the gut of infected *E. kuehniella* larvae showed destruction of the gut epithelium, the appearance of large cavities and cellular disintegration.

Moreover, such bacterial application may also target a twofold goal; i.e. controlling plant pathogens as well as insect pests simultaneously. These symbiotic bacteria have already demonstrated suppressive effects on a variety of important plant pathogens such as fungi (e.g. Akhurst, 1982; Li *et al.*, 1995; Webster, *et al.*, 2002; Abd El-Zaher *et al.*, 2012; Bock *et al.*, 2014). Also, six out of ten *P. luminescens* isolates tested showed antibacterial activities; each against one, two, or three of the tested bacterial species (Abd El-Zaher and Abd-Elgawad, 2012). Moreover, metabolites or cell filtrates of some *Photorhabdus* isolates may have superior toxicity relative to others tested (Abd El-Zaher *et al.*, 2008, 2012; Shapiro-Ilan *et al.*,

2009). Abd El-Zaher *et al.* (2012) reported that physical parameters (i.e. temperature, pH and sodium chloride) variably affected the metabolite-induced mortality percentage for the greater wax moth (*G. mellonella*). Yet achieving such a twofold goal will necessitate optimal application tactics to maximize field effectiveness of the *Photorhabdus* bacteria, e.g. a delivery system that is most conducive for favourable and effective control of crop pests and pathogens via the biocontrol agents. For example, various media/additions were found to potentiate *Photorhabdus* cell suspensions and cell-free filtrates against insect pests (e.g. Abd El-Zaher *et al.*, 2008, 2012). Shapiro-Ilan *et al.* (2014) reported that implementations of the bacteria and fermentation broth, to suppress pecan and peach pathogens, could be useful in reducing costs and/or avoiding regulatory issues compared to applying concentrated metabolites of these bacteria. However, they reported that applications of bacterial broth could only inhibit the lesion growth induced by *Phytophthora cactorum* of pecan leaves, but its efficacy to the two other pathogens examined (*Fusicladium effusum* and *Armillaria tabescens*) was not apparent. Therefore, they perceived that more broth rates would stimulate a response. Alternatively, toxicity of active compounds within the bacterial cell suspension, broth or cell-free filtrates could be enhanced through medium optimization (e.g. Abd El-Zaher *et al.*, 2008, 2012; Wang *et al.*, 2011).

Therefore, it is likely that the sophisticated management of crop pests and pathogens via these symbionts should not only consider all the above-mentioned factors but also decide management programmes on a case-by-case basis for attaining the best *Photorhabdus* species or strain–pest (or pathogen) matching. Factually, information is still scanty concerning the biotic and abiotic factors influencing the efficacy of these bacteria. Consequently, some efficacy results may appear unclear, or even conflicting. For example, Kumar *et al.* (2014) found that two *P. luminescens* isolates failed to infect both the diamondback moth and the Oriental leafworm moth (*Spodoptera litura*). On the contrary, Abdel-Razek (2003) and Rajagopal *et al.* (2006) found other isolates of the same bacterial species to be effective against the diamondback moth and the Oriental leafworm moth, respectively. Interestingly, *P. luminescens akhurstii* strain K-1 encapsulated in sodium alginate beads (2.5×10^7 cells/bead) and mixed with sterilized soil resulted in 100% *S. litura* larval mortality in 48 h, whereas *H. indica* IJs (having the bacteria in their gut) resulted in 40% mortality after 72 h. Koch's postulates, for the bacteria independent of the symbiont nematode, were proved since the bacteria could be re-isolated from the dead insect. The LC_{50} dose was recorded by Rajagopal *et al.* (2006) as 1010 cells/larva for killing *S. litura* sixth-instar larvae in 48 h.

In addition to using *Photorhabdus* concentrated metabolites or bacterial broth treatments, another avenue for pest or disease suppression is the development of bioactive compounds that are responsible for bacterial toxicity. Various active compounds in *Xenorhabdus* spp. and *Photorhabdus* spp. have been identified (Li *et al.*, 1995; Webster *et al.*, 2002; Böszörményi *et al.*, 2009; Xiao *et al.*, 2012), yet thus far none have been commercialized. For example, transcinnamic acid (TCA) was recently reported to be a major active compound in *P. luminescens*' suppressive activity against *F. effusum*, and thus further research to develop TCA as a potential control agent was suggested (Bock *et al.*, 2014). More research on the identification and activity of such bioactive compounds is warranted. Furthermore, regardless of the type of treatment (bioactive chemicals, metabolites and/or bacterial treatments), field testing and economic feasibility analysis will be needed, as various biotic and abiotic factors outside the laboratory may reduce the potency and longevity of the materials.

Apart from the above-mentioned concerns related to *P. asymbiotica*, as well as occasional episodes of allergy processes experienced by people who have had very close contact with *Photorhabdus* species and their symbiotic nematodes, these bacteria, like nematodes, have a very low risk to human health (Piedra-Buena *et al.*, 2015). Mohan and Sabir (2005) and Fand *et al.* (2012) reported other constraints where *P. luminescens* harm *Trichogramma* parasitized insect eggs and do kill coccinellids (key mealybug predators), respectively. These results suggested that the integration of *Photorhabdus* with other biocontrol agents should be utilized cautiously. Advances in biotechnological approaches and consolidation of the *Photorhabdus* data with their regional and temporal efficacy and dynamics might offer the comprehensive and sound knowledge necessary to establish and evaluate the real benefits and risks of EPNs and/or their mutualistic bacteria in nature in the long term.

12.6.2 Application development and genetic engineering of *Photorhabdus*

Recent developments in *Photorhabdus* experimentations and applications make these bacteria attractive tactics, as bioinsecticides, against many insect pests (e.g. Milstead, 1979; Akhurst and Dunphy, 1993; Bowen and Ensign, 1998; Abdel-Razek, 2003; Mahar *et al.*, 2004; Gerritsen *et al.*, 2005; Rajagopal *et al.*, 2006; Tounsi *et al.*, 2006; Abd El-Zaher *et al.*, 2008, 2012; Carneiro *et al.*, 2008; Jallouli *et al.*, 2008, 2011, 2012, 2013; Shrestha and Kim, 2010; Uma *et al.*, 2010; Kumar *et al.*, 2014; Sicard *et al.*, 2014; Mohan, 2015; Ruiu, 2015). Nevertheless, further reseach is required to make their application less expensive while more adequate and effective. For example, the development of new bacterial formulations is needed to serve for their better efficacy, ease of handling and application procedures, safe storage and transport, and reduction in overall costs of usage.

In fact, different approaches for the applications of *Photorhabdus* bacteria through genetic engineering are in progress to enhance their efficacy as bioinsecticides. For example, when *P. temperata* strains expressed the cry1Aa and/or cry1Ia genes of *B. thuringiensis*, via recombinant gene construction, an apparent enhancement of their oral toxicity against the lepidopteran olive tree pest, *Prays oleae*, was achieved (Tounsi *et al.*, 2006). This refinement was of 6.2-, 6.6- and 14.6-fold for the strains K122(pBCcry1Aa), K122(pBScry1Ia) and K122(pBCcry1Aa + pBScry1Ia), respectively. Moreover, Tounsi *et al.* (2006) proved that cry1Aa and cry1Ia interacted synergistically for better control of *P. oleae*. The authors concluded that such heterologous manner of *B. thuringiensis* cry genes in *P. temperata* should be exploited to promote and develop its host range for the control of crop insect pests.

It is also possible to engineer either the nematode or the bacteria to produce a resistant line for certain purposes (Glazer, 2015), or to combine certain genes of these bacteria for building transgenic plants (Piedra-Buena *et al.*, 2015). Some genes from *Photorhabdus* have been recombined into plants to confer them resistant to insects (Liu *et al.*, 2003). Thus, insecticidal toxins of *P. luminescens* can be considered as a potential substitute to the delta endotoxins (BT toxin) produced by *B. thuringiensis*. BT toxin accounts for 90% of the bioinsecticide market; however, many crop pests have been developing resistance towards BT toxins. This includes economically important insect pests: *Ostrinia nubilalis* (European corn borer), *Heliothis virescens* (tobacco budworm), *Pectonophora gossypiella* (pink bollworm moth), *Culex quinquefasciatus* (mosquito), *A. aegypti* (yellow fever mosquito), *Trichloro plusiani* (tiger moth), *L. decemlineata* (Colorado potato beetle), *Spodoptera exigua* (beet armyworm), *Spodoptera littoralis* (Egyptian cotton leaf worm) and *Chryosomela scripta* (cottonwood leaf beetle) (Liu *et al.*, 1999; Maithri *et al.*, 2013). Moreover, the insecticidal toxin complex protein, TcaA toxin, showed toxicity against a wide range of agricultural pests, and the phylogenetic tree constructed for TcaA revealed that this toxin did not have any ancestral relationship with BT toxins (Maithri *et al.*, 2013). If so, the expected transgenic plants will be regulated by legislation of transgenic plants to be certain that such plants with genes of *Photorhabdus* bacteria can be produced without health or environmental risks (Piedra-Buena *et al.*, 2015).

12.7 Conclusions and Future Prospects

Photorhabdus spp. are entomopathogenic bacteria that mutualistically associate with heterorhabditid nematodes. These bacteria, members of the γ-subclass of Proteobacteria, are well prepared to cause insect pathogenicity. The growing interest and intensive experimentation on *Photorhabdus* bacteria in order to use them, at least partially, as bioinsecticidal alternatives to hazardous chemicals are in progress. Most advances in this direction to improve *Photorhabdus* spp. application against insect pests will rely on our sound understanding of the bacterial capacity and interactions with the surrounding environment. Therefore, a variety of approaches should be followed concerning their proper application, improved shelf life, inexpensive mass culture, optimized interaction with other biotic and abiotic factors and wise integration with other agricultural management techniques. Researchers have also been analysing the molecular structures and events of *Photorhabdus* bacteria, which will upgrade our knowledge of genetic bases for a wide array of toxins and secondary metabolites that are responsible for insecticidal capacity. Such research efforts should be combined with academic–industry partnerships for developing more effective *Photorhabdus* bioinsecticides with promising molecular and biotechnological engineering prospects in the management of crop insect pests.

Acknowledgements

This work was supported in part by US-Egypt Science and Technology Joint Fund (cycle 17; project no. 172) and Egyptian In-House, National Research Centre Project No 11030133 entitled 'Pesticide alternatives against soilborne pathogens attacking legume cultivation in Egypt'. I would like to thank Drs Mohamed F.M. Eissa, Tarique H. Askary, Mostafa M.A. Hammam, Larry W. Duncan, Fahiem E. El-Borai, Raquel Campos-Herrera, David I. Shapiro-Ilan, Edwin E. Lewis and Moawad M.M. Mohamed for the information provided.

References

Abd-Elgawad, M.M.M. (2008) Economic production of biopesticides using techniques adequate for local conditions. Final report (Febraury), Basic Scientific Research Council, Academy of Scientific Research and Technology, Egypt, 88 pp.

Abd-Elgawad, M.M.M. (2014) Spatial patterns of *Tuta absoluta* and heterorhabditid nematodes. *Russian Journal of Nematology* 22, 89–100.

Abd-Elgawad, M.M.M. and Nguyen, K.B. (2007) Isolation, identification and environmental tolerance of new heterorhabditid populations from Egypt. *International Journal of Nematology* 17, 116–123.

Abd-Elgawad, M.M.M. and Spiridonov, S.E. (2014) Entomopathogenic nematode application in Egypt and Russia: challenges and opportunities. *Egyptian Journal of Agronematology* 13, 1–22.

Abd-Elgawad, M.M.M. and Vagelas, I.K. (2015) Nematophagous bacteria: field application and commercialization. In: Askary, T.H. and Martinelli, P.R.P. (eds) *Biocontrol Agents of Phytonematodes*. CAB International, Wallingford, UK, pp. 276–309.

Abd El-Zaher, F.H. and Abd-Elgawad, M.M.M. (2012) Assessment of the antagonistic potential for the bacterial symbionts of entomopathogenic nematodes. *Journal of Applied Sciences Research* 8, 4590–4599.

Abd El-Zaher, F.H., Abd El-Maksoud, H.K. and Abd-Elgawad, M.M.M. (2008) Use of *Photorhabdus* as a biopesticides. A- Cell suspension from *Photorhabdus* sp. against *Galleria mellonella* insect. *Catrina* 3, 125–129.

Abd El-Zaher, F.H., Abd-Elgawad, M.M.M. and Abd El-Maksoud, H.K. (2012) Use of the entomopathogenic nematode symbiont *Photorhabdus luminescens* as a biocontrol agent. B- Factors affecting the cell-free filtrates from the bacterium. *Journal of Applied Sciences Research* 8, 4600–4614.

Abdel-Razek, A.S. (2003) Pathogenic effects of *Xenorhabdus nematophilus* and *Photorhabdus luminescens* (Enterobacteriaceae) against pupae of the Diamondback Moth, *Plutella xylostella* (L.). *Anzeiger für Schädlingskunde* 76, 108–111.

Abouelhag, H.A. and El-Sadawy, H.A. (2012) Natural associations between symbionts *Photorhabdus* spp. and *Xenorhabdus* spp. and bacteria related to *Ochrobactrum anthropi*, *Bacillus pumilus* and *Enterobacter cloacae*. *Life Science Journal* 9, 4783–4790.

Abu-Shady, N.M., Shamseldean, M.M., Abd-Elbary, N.A. and Stock, S.P. (2011) Diversity and distribution of entomopathogenic nematodes (Heterorhabditidae and Steinernematidae) in Egypt. *Journal of Nematology* 43, 224.

Adams, B.J. and Nguyen, K.B. (2002) Taxonomy and systematics. In: Gaugler, R. (ed.) *Entomopathogenic Nematology*. CAB International, Wallingford, UK, pp. 1–34.

Adams, B.J., Fodor, A., Koppenhöfer, H.S., Stackenbrandt, E., Stock, S.P. and Klein, M.G. (2006) Biodiversity and systematics of nematode-bacterium entomopathogens. *Biological Control* 38, 4–21.

Ahantarig, A., Chantawat, N., Waterfield, N.R., ffrench-Constant, R. and Kittayapong, P. (2009) PirAB toxin from *Photorhabdus asymbiotica* as a larvicide against dengue vectors. *Applied and Environmental Microbiology* 75, 4627–4629.

Ahn, J., Lee, J., Yang, E., Lee, Y., Koo, K., *et al.* (2013) Mosquitocidal activity of anthraquinones isolated from symbiotic bacteria *Photorhabdus* of entomopathogenic nematode. *Journal of Asia-Pacific Entomology* 16, 317–320.

Akhurst, R.J. (1980) Morphological and functional dimorphism in *Xenorhabdus* spp., bacteria symbiotically associated with the insect pathogenic nematodes *Neoaplectana* and *Heterorhabditis*. *Journal of General Microbiology* 121, 303–309.

Akhurst, R.J. (1982) Antibiotic activity of *Xenorhabdus* spp., bacteria symbiotically associated with insect pathogenic nematodes of the families *Heterorhabditidae* and *Steinernematidae*. *Journal of General Microbiology* 128, 3061–3065.

Akhurst, R.J. and Dunphy, G. (1993) Tripartite interactions between symbiotically associated entomopathogenic bacteria, nematodes, and their insect hosts. In: Beckage, N., Thompson, S. and Federici, B. (eds) *Parasites and Pathogens of Insects*, Vol 2. Academic Press, New York, pp. 1–23.

Akhurst, R.J., Boemare, N.E., Janssen, P.H., Peel, M.M., Alfredson, D.A. and Beard, C.E. (2004) Taxonomy of Australian clinical isolates of the genus *Photorhabdus* and proposal of *Photorhabdus asymbiotica* subsp. *asymbiotica* subsp. nov. and *P. asymbiotica* subsp. *australis* subsp. nov. *International Journal of Systematic and Evolutionary Microbiology* 54, 1301–1310.

Alamalakala, L., Parimi, S., Dangat, S. and Char, B.R. (2015) Non-Bt soil microbe derived insecticidal proteins. In: Sree, K.S. and Varma, A. (eds) *Biocontrol of Lepidopteran Pests: Use of Soil Microbes and their Metaboloites*. Springer International Publishing, Switzerland, pp. 89–122.

An, R. and Grewal, P.S. (2010) Molecular mechanisms of persistence of mutualistic bacteria *Photorhabdus* in the entomopathogenic nematode host. *PLoS One* 5(10), e13154 (PubMed: 20957199).

Bintrim, S.B. and Ensign, J.C. (1998) Insertional inactivation of genes encoding the crystalline inclusion proteins of *Photorhabdus luminescens* results in mutants with pleiotropic phenotypes. *Journal of Bacteriology* 180, 1261–1269.

Bird, A.F. and Akhurst, R.J. (1983) The nature of the intestinal vesicle in nematodes of the family Steinernematidae. *International Journal of Parasitology* 16, 511–518.

Blackburn, M., Golubeva, E., Bowen, D. and ffrench-Constant, R.H. (1998) A novel insecticidal toxin from *Photorhabdus luminescens*: histopathological effects of toxin complex A (Tca) on the midgut of *Manduca sexta*. *Applied and Environmental Microbiology* 64, 3036–3041.

Blackburn, M.B., Domek, J.M., Gelman, D.B. and Hu, J.S. (2005) The broadly insecticidal *Photorhabdus luminescens* toxin complex a (Tca): activity against the Colorado potato beetle, *Leptinotarsa decemlineata*, and sweet potato whitefly, *Bemisia tabaci*. *Journal of Insect Science* 5, 11 pp. DOI: 10.1093/jis/5.1.32.

Bock, C.H., Shapiro-Ilan, D.I., Wedge, D.E. and Cantrell, C.L. (2014) Identification of the antifungal compound, trans-cinnamic acid, produced by *Photorhabdus luminescens*, a potential biopesticide against pecan scab. *Journal of Pest Science* 87, 155–162.

Boemare, N. (2002) Biology, taxonomy and systematics of *Photorhabdus* and *Xenorhabdus*. In: Gaugler, R. (ed.) *Entomopathogenic Nematology*. CAB International, Wallingford, UK, pp. 35–56.

Boemare, N.E. and Akhurst, R.J. (1988) Biochemical and physiological characterization of colony form variants in *Xenorhabdus* spp. (*Enterobacteriaceae*). *Journal of General Microbiology* 134, 751–761.

Boemare, N.E., Akhurst, R.J. and Mourant, R.G. (1993) DNA relatedness between *Xenorhabdus* spp. (Enterobacteriaceae), symbiotic bacteria of entomopathogenic nematodes, a proposal to transfer *Xenorhabdus luminescens* to a new genus, *Photorhabdus* gen. nov. *International Journal of Systematic Bacteriology* 43, 249–255.

Böszörményi, E., Ersek, T., Fodor, A., Fodor, A.M., Földes, L.S., *et al.* (2009) Isolation and activity of *Xenorhabdus* antimicrobial compounds against the plant pathogens *Erwinia amylovora* and *Phytophthora nicotianae*. *Journal of Applied Microbiology* 107, 746–759.

Bowen, D.J. and Ensign, J.C. (1998) Purification and characterization of a high-molecular weight insecticidal protein complex produced by the entomopathogenic bacterium *Photorhabdus luminescens. Applied and Environmental Microbiology* 64, 3029–3035.

Bowen, D., Rocheleau, T.A., Blackburn, M., Andreev, O., Golubeva, E., *et al.* (1998) Insecticidal toxins from the bacterium *Photorhabdus luminescens*. *Science* 280, 2129–2132.

Brown, S.E., Cao, A.T., Dobson, P., Hines, E.R., Akhurst, R.J. and East, P.D. (2006) Txp40, a ubiquitous insecticidal toxin protein from *Xenorhabdus* and *Photorhabdus* bacteria. *Applied and Environmental Microbiology* 72, 1653–1662.

Burnell, A.M. and Stock, S.P. (2000) *Heterorhabditis*, *Steinernema* and their bacterial symbionts – lethal pathogens of insects. *Nematology* 2, 31–42.

Bussaman, P., Sobanboa, S., Grewal, P.S. and Chandrapatya, A. (2009) Pathogenicity of additional strains of *Photorhabdus* and *Xenorhabdus* (Enterobacteriaceae) to the mushroom mite *Luciaphorus perniciosus* (Acari: Pygmephoridae). *Applied Entomology and Zoology* 44, 293–299.

Campos-Herrera, R. (2015) *Nematodes Pathogenesis of Insects and Other Pests*. Springer International Publishing, Switzerland.

Carneiro, C.N.B., DaMatta, R.A., Samuels, R.I. and Silva, C.P. (2008) Effects of entomopathogenic bacterium *Photorhabdus temperata* infection on the intestinal microbiota of the sugarcane stalk borer *Diatraea saccharalis* (Lepidoptera: Crambidae). *Journal of Invertebrate Pathology* 99, 87–91.

Chaston, J.M., Suen, G., Tucker, S.L., Andersen, A.W., Bhasin, A., *et al.* (2011) The entomopathogenic bacterial endosymbionts *Xenorhabdus* and *Photorhabdus*: convergent lifestyles from divergent genomes. *PLoS ONE* 6 (11), e27909. DOI: 10.1371/journal.pone.0027909.

Ciche, T.A., Kim, K.S., Kaufmann-Daszczuk, B., Nguyen, K.C. and Hall, D.H. (2008) Cell invasion and matricide during *Photorhabdus luminescens* transmission by *Heterorhabditis bacteriophora* nematodes. *Applied Environmental Microbiology* 74, 2275–2287.

Clarke, D.J. (2008) *Photorhabdus*: a model for the analysis of pathogenicity and mutualism. *Cellular Microbiology* 10, 2159–2167.

Daborn, P., Waterfield, N., Blight, M.A. and Ffrench-Constant, R.H. (2001) Measuring virulence factor expression by the pathogenic bacterium *Photorhabdus luminescens* in culture and during insect infection. *Journal of Bacteriology* 183, 5834–5839.

Daborn, P.J., Waterfield, N., Silva, C.P., Au, C.P., Sharma, S. and ffrench-Constant, R.H. (2002) A single *Photorhabdus* gene, makes caterpillars floppy (mcf), allows *Escherichia coli* to persist within and kill insects. *Proceedings of the National Academy of Sciences of the United States of America* 99, 10742–10747.

Dowds, B.C.A. and Peters, A. (2002) Virulence mechanisms. In: Gaugler, R. (ed.) *Entomopathogenic Nematology*. CAB International, Wallingford, UK, pp. 79–98.

Dowling, A.J., Daborn, P.J., Waterfield, N.R., Wang, P., Streuli, C.H. and ffrench-Constant, R.H. (2004) The insecticidal toxin Makes caterpillars floppy (Mcf) promotes apoptosis in mammalian cells. *Cell Microbiology* 6, 345–353.

Duchaud, E., Rusniok, C., Frangeul, L., Buchrieser, C., Givaudan, A., *et al.* (2003) The genome sequence of the entomopathogenic bacterium *Photorhabdus luminescens*. *Nature Biotechnology* 21, 1307–1313.

Dunphy, G.B. and Webster, J.M. (1988) Virulence mechanisms of *Heterorhabditis heliothidis* and its bacterial associate, *Xenorhabdus luminescens*, in non-immune larvae of the greater wax moth, *Galleria mellonella*. *International Journal of Parasitology* 18, 729–737.

Dutky, S.R. (1937) Investigation of disease of the immature stages of the Japanese beetle. PhD thesis. Rutgers University, New Brunswick, New Jersey.

Easom, C.A. and Clarke, D.J. (2012) HdfR is a regulator in *Photorhabdus luminescens* that modulates metabolism and symbiosis with the nematode *Heterorhabditis*. *Environmental Microbiology* 14, 953–966.

Easom, C.A., Joyce, S.A. and Clarke, D.J. (2010) Identification of genes involved in the mutualistic colonization of the nematode *Heterorhabditis bacteriophora* by the bacterium *Photorhabdus luminescens*. *BMC Microbiology* 10, 45.

Eissa, M.F.M. and Abd-Elgawad, M.M.M. (2015) Nematophagous bacteria as biocontrol agents of phytonematodes. In: Askary, T.H. and Martinelli, P.R.P. (eds) *Biocontrol Agents of Phytonematodes*. CAB International, Wallingford, UK, pp. 217–243.

Eom, S., Park, Y. and Kim, Y. (2014) Sequential immunosuppressive activities of bacterial secondary metabolites from the entomopahogenic bacterium *Xenorhabdus nematophila*. *Journal of Microbiology* 52, 161–168.

Fand, B.B., Gautam, R.D., Kamra, A., Suroshe, S.S. and Mohan, S. (2012) Bioefficacy of aqueous garlic extract and a symbiotic bacterium, *Photorhabdus luminescens* against *Phenacoccus solenopsis* Tinsley (Homoptera: Pseudococcidae). *Biopesticides International* 8, 38–48.

Fayyaz, S., Khanum, T.A., Ali, S., Solangi, G.S., Gulsher, M. and Javed, S. (2015) *Steinernema balochiense* n. sp. (Rhabditida: Steinernematidae) a new entomopathogenic nematode from Pakistan. *Zootaxa* 3904, 387–402.

Ferreira, T., van Reenen, C., Pagès, S., Tailliez, P., Malan, A.P. and Dicks, L.M.T. (2013) *Photorhabdus luminescens* subsp. *Noenieputensis* subsp. nov., a symbiotic bacterium associated with a novel *Heterorhabditis* species related to *Heterorhabditis indica*. *International Journal of Systematic and Evolutionary Microbiology* 63, 1853–1858.

Ferreira, T., van Reenen, C.A., Endo, A., Tailliez, P., Pagès, S., *et al.* (2014) *Photorhabdus heterorhabditis* sp. nov., a symbiont of the entomopathogenic nematode *Heterorhabditis zealandica*. *International Journal of Systematic and Evolutionary Microbiology* 64, 1540–1545.

ffrench-Constant, R. and Bowen, D. (1999) *Photorhabdus* toxins: novel biological insecticides. *Current Opinion in Microbiology* 2, 284–288.

ffrench-Constant, R.H. and Bowen, D.J. (2000) Novel insecticidal toxins from nematode-symbiotic bacteria. *Cellular and Molecular Life Sciences* 57, 828–833.

ffrench-Constant, R.H. and Dowling, A.J. (2014) *Photorhabdus* toxins. In: Dhadialla, T.S. and Gill, S.S. (eds) *Advances in Insect Physiology, Vol 47: Insect Midgut and Insecticidal Proteins*. Elsevier Academic Press, London, pp. 343–388.

ffrench-Constant, R.H., Dowling, A. and Waterfield, N.R. (2007) Insecticidal toxins from *Photorhabdus* bacteria and their potential use in agriculture. *Toxicon* 49, 436–451.

Fischer-Le Saux, M., Mauleon, H., Constant, P., Brunel, B. and Boemare, N. (1998) PCR-ribotyping of *Xenorhabdus* and *Photorhabdus* isolates from the Caribbean region in relation to the taxonomy and geographic distribution of their nematode hosts. *Applied and Environmental Microbiology* 64, 4246–4254.

Fischer-Le Saux, M., Viallard, V., Brunel, B., Normand, P. and Boemare, N.E. (1999) Polyphasic classification of the genus *Photorhabdus* and proposal of new taxa: *P. luminescens* subsp. *luminescens* subsp. nov., *P. luminescens* subsp. *akhurstii* subsp. nov., *P. luminescens* subsp. *laumondii* subsp. nov., *P. temperata* sp. nov., *P. temperata* subsp. *temperata* subsp. nov. and *P. asymbiotica* sp. nov. *International Journal of Systematic Bacteriology* 49, 1645–1656.

Forst, S. and Clarke, D. (2002) Bacteria-nematode symbiosis. In: Gaugler, R. (ed.) *Entomopathogenic Nematology*. CAB International, Wallingford, UK, pp. 57–78.

Forst, S. and Nealson, K.H. (1996) Molecular biology of the symbiotic-pathogenic bacteria *Xenorhabdus* spp. and *Photorhabdus* spp. *Microbiological Reviews* 60, 21–43.

Forst, S., Dowds, B., Boemare, N. and Stackebrandt, E. (1997) *Xenorhabdus* and *Photorhabdus* spp.: bugs that kill bugs. *Annual Review Microbiology* 51, 47–72.

Fuchs, S.W., Proschak, A., Jaskolla, T.W., Karas, M. and Bode, H.B. (2011) Structure elucidation and biosynthesis of lysine-rich cyclic peptides in *Xenorhabdus nematophila*. *Organic & Biomolecular Chemistry* 9, 3130–3132.

Gatsogiannis, C., Lang, A.E., Meusch, D., Pfaumann, V., Hofnagel, O., *et al.* (2013) A syringe-like injection mechanism in *Photorhabdus luminescens* toxins. *Nature* 495, 520–523.

Gaugler, R. (ed.) (2002) *Entomopathogenic Nematology*. CAB International, Wallingford, UK, 388 pp.

Gaugler, R. and Kaya, H.K. (1990) *Entomopathogenic Nematodes in Biological Control*. CRC Press, Boca Raton, Florida, 365 pp.

Gaugler, R., Campbell, J.F., Selvan, S. and Lewis, E.E. (1992) Largescale inoculative release of the entomopathogenic nematode *Steinernema glaseri*: assessment 50 years later. *Biological Control* 2, 181–187.

Georgis, R. (2002) The Biosys Experiment: an insider's perspective. In: Gaugler, R. (ed.) *Entomopathogenic Nematology*. CAB International, Wallingford, UK, pp. 357–372.

Gerdes, E., Upadhyay, D., Mandjiny, S., Bullard-Dillard, R., Storms, M., *et al.* (2015) *Photorhabdus luminescens*: virulent properties and agricultural applications. *American Journal of Agriculture and Forestry* 3, 171–177.

Gerrard, J., Waterfield, N., Vohra, R. and ffrench-Constant, R. (2004) Human infection with *Photorhabdus asymbiotica*: an emerging bacterial pathogen. *Microbes and Infection* 6, 229–237.

Gerritsen, L.J.M., Georgieva, J. and Wiegers, G.L. (2005) Oral toxicity of *Photorhabdus* toxins against thrips species. *Journal of Invertebrate Pathology* 88, 207–211.

Givaudan, A., Baghdiguian, S., Lanois, A. and Boemare, N.E. (1995) Swarming and swimming changes concomitant with phase variation in *Xenorhabdus nematophilus*. *Applied and Environmental Microbiology* 61, 1408–1413.

Glaser, R.W. and Fox, H. (1930) A nematode parasite of the Japanese beetle (*Popillia japonica* Newm.). *Science* 71, 16–17.

Glaser, R.W., McCoy, E.E. and Girth, H.B. (1942) The biology and culture of *Neoaplectana chresima*, a new nematode parasitic in insects. *Journal of Parasitology* 28, 123–126.

Glazer, I. (2015) Improvement of entomopathogenic nematodes: a genetic approach. In: Campos-Herrera, R. (ed.) *Nematodes Pathogenesis of Insects and Other Pests*. Springer International Publishing, Switzerland, pp. 29–56.

Glazer, I., Liran, N., Poinar, G.O. and Smits, P.H. (1993) Identification and biological activity of newly isolated heterorhabditid populations from Israel. *Fundamental and Applied Nematology* 16, 467–472.

Golden, J.W. and Riddle, D.L. (1982) A pheromone influences larval development in the nematode *Caenorhabditis elegans*. *Science* 218, 578–580.

Goodrich-Blair, H. and Clarke, D.J. (2007) Mutualism and pathogenesis in *Xenorhabdus* and *Photorhabdus*: two roads to the same destination. *Molecular Microbiology* 64, 260–268.

Grewal, P.S., Ehlers, R.-U. and Shapiro-Ilan, D.I. (2005) *Nematodes as Biocontrol Agents*. CAB International, Wallingford, UK.

Griffin, C.T., Boemare, N.E. and Lewis, E.E. (2005) Biology and behaviour. In: Grewal, P.S., Ehlers, R.-U. and Shapiro-Ilan, D.I. (eds) *Nematodes as Biocontrol Agents*. CAB International, Wallingford, UK, pp. 47–64.

Han, R.C. and Ehlers, R.-U. (1998) Cultivation of axenic *Heterorhabditis* spp. dauer juveniles and their response to non-specific *Photorhabdus luminescens* food signals. *Nematologica* 69, 1–12.

Han, R. and Ehlers, R.-U. (2000) Pathogenicity, development, and reproduction of *Heterorhabditis bacteriophora* and *Steinernema carpocapsae* under axenic *in vivo* conditions. *Journal of Invertebrate Pathology* 75, 55–58.

Hazir, S., Stackebrandt, E., Lang, E., Schumann, P., Ehlers, R.U. and Keskin, N. (2004) Two new subspecies of *Photorhabdus luminescens*, isolated from *Heterorhabditis bacteriophora* (Nematoda: Heterorhabditidae): *Photorhabdus luminescens* subsp. *Kayaii* subsp. nov. and *Photorhabdus luminescens* subsp. *Thracensis* subsp. nov. *Systematic and Applied Microbiology* 27, 36–42.

Hominick, W.M. (2002) Biogeography. In: Gaugler, R. (ed.) *Entomopathogenic Nematology*. CAB International, Wallingford, UK, pp. 115–144.

Hurst IV, S., Rowedder, H., Michaels, B., Bullock, H., Jackobeck, R., *et al.* (2015) Elucidation of the *Photorhabdus temperata* genome and generation of a transposon mutant library to identify motility mutants altered in pathogenesis. *Journal of Bacteriology* 197, 2201–2216.

Jallouli, W., Hammami, W., Zouari, N. and Jaoua, S. (2008) Medium optimization for biomass production and morphology variance overcome of *Photorhabdus temperate* ssp. *temperata* strain K122. *Process Biochemistry* 43, 1338–1344.

Jallouli, W., Jaoua, S. and Zouari, N. (2011) Overcoming the production limitations of *Photorhabdus temperata* ssp. *temperata* strain K122 bioinsecticides in low-cost medium. *Bioprocess and Biosystems Engineering* 34, 1039–1047.

Jallouli, W., Jaoua, S. and Zouari, N. (2012) Improvement of *Photorhabdus temperate* bioinsecticides production in low-cost media through adequate fermentation technology. *Biotechnology Progress* 28, 1278–1284.

Jallouli, W., Abdelkefi-Mesrati, L., Tounsi, S., Jaoua, S. and Zouari, N. (2013) Potential of *Photorhabdus temperata* K122 bioinsecticide in protecting wheat flour against *Ephestia kuehniella*. *Journal of Stored Products Research* 53, 61–66.

Jang, E., Ullah, I., Kim, M., Lee, K. and Shin, J. (2011) Isolation and characterization of the entomopathogenic bacterium, *Photorhabdus temperate* producing a heat stable insecticidal toxin. *Journal of Plant Diseases and Protection* 118, 178–184.

Jang, E.K., Ullah, I., Lim, J.H., Lee, I.J., Kim, J.G. and Shin, J.H. (2012) Physiological and molecular characterization of a newly identified entomopathogenic bacteria, *Photorhabdus temperata* M1021. *Journal of Microbiology and Biotechnology* 22, 1605–1612.

Jank, T., Bogdanovic, X., Wirth, C., Haaf, E., Spoerner, M., *et al.* (2013) A bacterial toxin catalyzing tyrosine glycosylation of Rho and deamidation of Gq and Gi proteins. *Nature Structural and Molecular Biology* 20, 1273–1280.

Jank, T., Trillhaase, C., Brozda, N., Steinemann, M., Schwan, C., *et al*. (2015a) Intracellular plasma membrane guidance of *Photorhabdus asymbiotica* toxin is crucial for cell toxicity. *The Journal of the Federation of American Societies for Experimental Biology* 29, 2789–2802.

Jank, T., Lang, A.E. and Aktories, K. (2015b) Rho-modifying bacterial protein toxins from *Photorhabdus* species. *Toxicon* DOI: 10.1016/j.toxicon.2015.05.017.

Johnigk, S.-A. and Ehlers, R.-U. (1999) *Endotokia matricida* in hermaphrodites of *Heterorhabditis* spp. and the effect of food supply. *Nematology* 1, 717–726.

Joyce, S.A., Watson, R.J. and Clarke, D.J. (2006) The regulation of pathogenicity and mutualism in *Photorhabdus*. *Current Opinion in Microbiology* 9, 127–132.

Jung, S. and Kim, Y. (2006) Synergistic effect of entomopathogenic bacteria (*Xenorhabdus* sp. and *Photorhabdus temperata* ssp. *temperata*) on the pathogenicity of *Bacillus thuringiensis* ssp. *aizawai* against *Spodoptera exigua* (Lepidoptera: Noctuidae). *Environmental Entomology* 35, 1584–1589.

Khan, A. and Brooks, W.M. (1977) A chromogenic bioluminescent bacterium associated with the entomophilic nematode *Chromonema heliothidis*. *Journal of Invertebrate Pathology* 29, 253–261.

Kumar, K.S.V., Vendan, K.T. and Nagaraj, S.B. (2014) Isolation and characterization of entomopathogenic symbiotic bacterium, *Photorhabdus luminescens* of *Heterorhabditis indica* from soils of five agro climatic zones of Karnataka. *Biosciences Biotechnology Research Asia* 11, 129–139.

Kuwata, R., Qiu, L.H., Wang, W., Harada, Y., Yoshida, M., *et al*. (2013) *Xenorhabdus ishibashii* sp nov., isolated from the entomopathogenic nematode *Steinemema aciari*. *International Journal of Systematic and Evolutionary Microbiology* 63, 1690–1695.

Kwak, Y. and Shin, J.-H. (2015) Complete genome sequence of *Photorhabdus temperata* subsp. thracensis 39-8T, an entomopathogenic bacterium for the improved commercial bioinsecticide. *Journal of Biotechnology* 214, 115–116.

Lewis, E.E. and Clarke, D.J. (2012) Nematode parasites and entomopathogens. In: Vega, F.E. and Kaya, H.K. (eds) *Insect Pathology*, 2nd edn. Academic Press, San Diego, California, pp. 395–424.

Li, J., Chen, G. and Webster, J.M. (1995) Antimicrobial metabolites from a bacterial symbiont. *Journal of Natural Products* 58, 1081–1086.

Liu, D., Burton, S., Glancy, T., Li, Z.-S., Hampton, R., *et al*. (2003) Insect resistance conferred by 283-kDa *Photorhabdus luminescens* protein TcdA in *Arabidopsis thaliana*. *Nature Biotechnology* 21, 1222–1228.

Liu, Y.B., Tabashnik, B., Dennehy, T., Patin, A. and Bartlet, A. (1999) Development time and resistance to Bt crops. *Nature* 400, 519–. CrossRef, PubMed.

Mahar, A.N., Munir, M. and Mahar, A.Q. (2004) Studies of different application methods of *Xenorhabdus* and *Photorhabdus* cells and their toxin in broth solution to control locust (*Schistocerca gregaria*). *Asian Journal of Plant Sciences* 3, 690–695.

Mahar, A.N., Munir, M., Gowen, S.R. and Hague, N.G.M. (2005) Role of entomopathogenic bacteria, *Photorhabdus luminescens* and its toxic secretions against *Galleria mellonella* larvae. *Journal of Entomology* 2, 69–76.

Maithri, S.K., Ramesh, K.V. and Mutangana, D. (2013) Theoretical structure prediction of TcaA from *Photorhabdus luminescens* and aminopeptidase N receptor from *Helicoverpa armigera*. *Research Journal of Recent Sciences* 2, 40–49.

Maneesakorn, P., An, R., Daneshvar, H., Taylor, K., Bai, X., *et al*. (2011) Phylogenetic and cophylogenetic relationships of entomopathogenic nematodes *Heterorhabditis* (Rhabditida) and their symbiotic bacteria *Photorhabdus* (Enterobacteriaceae). *Molecular Phylogenetics and Evolution* 59, 271–280.

Martin, B.D. and Schwab, E. (2013) Current usage of symbiosis and associated terminology. *International Journal of Biology* 5, 32–45.

Menti, H., Wright, D.J. and Perry, R.N. (1997) Desiccation survival of populations of the entomopathogenic nematodes *Steinernema feltiae* and *Heterorhabditis megidis* from Greece and the UK. *Journal of Helminthology* 71, 41–46.

Meusch, D., Gatsogiannis, C., Efremov, R.G., Lang, A.E., Hofnagel, O., *et al*. (2014) Mechanism of Tc toxin action revealed in molecular detail. *Nature* 508, 61–65.

Milstead, J.E. (1979) *Heterorhabditis bacteriophora* as a vector for introducing its associated bacterium into the hemocoel of *Galleria mellonella* larvae. *Journal of Invertebrate Pathology* 33, 324–327.

Mohan, S. (2015) Entomopathogenic nematodes and their bacterial symbionts as lethal bioagents of lepidopteran pests. In: Sree, K.S. and Varma, A. (eds) *Biocontrol of Lepidopteran Pests: Use of Soil Microbes and their Metaboloites*. Springer International Publishing, Switzerland, pp. 273–288.

Mohan, S. and Sabir, N. (2005) Biosafety concerns on the use of *Photorhabdus luminescens* as biopesticide: experimental evidence of mortality in egg parasitoid *Trichogramma* spp. *Current Science* 89, 1268–1272.

Mohan, S., Raman, R. and Gaur, H.S. (2003) Foliar application of *Photorhabdus luminescens*, symbiotic bacteria from entomopathogenic nematode *Heterorghabditis indica*, to kill cabbage butter fly *Pieris brassicae*. *Current Science* 84, 1397.

Mohan, S., Sirohi, A. and Gaur, H.S. (2004) Successful management of mango mealy bug, *Drosicha mangiferae*

by *Photorhabdus luminescens*, a symbiotic bacteria from entomopathogenic nematode, *Heterorhabditis indica*. *International Journal of Nematology* 14, 195–198.
Morgan, J.A.W., Kuntzelmann, V., Tavernor, S., Ousley, M.A. and Winstanley, C. (1997) Survival of *Xenorhabdus nematophilus* and *Photorhabdus luminescens* in water and soil. *Journal of Applied Microbiology* 83, 665–670.
Moshayov, A., Koltai, H. and Glazer, I. (2013) Molecular characterisation of the recovery process in the entomopathogenic nematode *Heterorhabditis bacteriophora*. *International Journal for Parasitology* 43, 843–852.
Mracek, Z. and Webster, J.M. (1993) Survey of Heterorhabditidae and Steinernematidae (Rhabditida: Nematoda) in Western Canada. *Journal of Nematology* 25, 710–717.
Murfin, K.E., Dillman, A.R., Foster J.M., Bulgheresi, S., Slatko, B.E., *et al.* (2012) Nematode-bacterium symbioses – cooperation and conflict revealed in the 'Omics' age. *Biological Bulletin* 223, 85–102.
Nollmann, F.I., Dauth, C., Mulley, G., Kegler, C., Kaiser, M., *et al.* (2015) Insect-specific production of new gameXpeptides in *Photorhabdus luminescens*TTO1, widespread natural products in entomopathogenic bacteria. *ChemBioChem* 16, 205–208.
Ogier, J.C., Pagès, S., Bisch, G., Chiapello, H., Médigue, C., *et al.* (2014) Attenuated virulence and genomic reductive evolution in the entomopathogenic bacterial symbiont species, *Xenorhabdus poinarii*. *Genome Biology and Evolution* 6, 1495–1513.
Orozco, R.A., Hill, T. and Stock, S.P. (2013) Characterization and phylogenetic relationships of *Photorhabdus luminescens* subsp. *Sonorensis* (c-Proteobacteria: Enterobacteriaceae), the bacterial symbiont of the entomopathogenic nematode *Heterorhabditis sonorensis* (Nematoda: Heterorhabditidae). *Current Microbiology* 66, 30–39.
Piedra-Buena, A., López-Cepero, J. and Campos-Herrera, R. (2015) Entomopathogenic nematode production and application: regulation, ecological impact and non–target effects. In: Campos-Herrera, R. (ed.) *Nematodes Pathogenesis of Insects and Other Pests*. Springer International Publishing, Switzerland, pp. 255–282.
Plichta, K.L., Joyce, S.A., Clarke, D., Waterfield, N. and Stock, S.P. (2009) *Heterorhabditis gerrardi* n. sp. (Nematoda: Heterorhabditidae): the hidden host of *Photorhabdus asymbiotica* (Enterobacteriaceae:g-Proteobacteria). *Journal of Helminthology* 83, 309–320.
Poinar, G.O. and Grewal, P.S. (2012) History of entomopathogenic nematology. *Journal of Nematology* 44, 153–161.
Rajagopal, R., Mohan, S. and Bhatnagar, R.K. (2006) Direct infection of *Spodoptera litura* by *Photorhabdus luminescens* encapsulated in alginate beads. *Journal of Invertebrate Pathology* 93, 50–53.
Rodou, A., Ankrah, D.O. and Stathopoulos, C. (2010) Review: Toxins and secretion systems of *Photorhabdus luminescens*. *Toxins* 2, 1250–1264.
Rosado, C.J., Buckle, A.M., Law, R.H., Butcher, R.E., Kan, W.T., *et al.* (2007) A common fold mediates vertebrate defense and bacterial attack. *Science* 317, 1548–1551.
Ruiu, L. (2015) Insect pathogenic bacteria in integrated pest management. *Insects* 6, 352–367.
Salama, H.S. and Abd-Elgawad, M.M.M. (2001) Isolation of heterorhabditid nematodes from palm tree planted areas and their implication in the red palm weevil control. *Anzeiger für Schädlingskunde Pflanzenschutz Umweltschutz* 74, 43–45.
Shamseldean, M.M. (2010) Nematodes and their use in biological control of insect pests in the Arab countries. (In Arabic) In: Abu-Gharbieh, W.A., Al-Hazmi, A.S., Stephan, Z.A. and Dawabah, A.A. (eds) *Plant Nematodes in Arab Countries*. Darwael for Publishing, Arab Society of Plant Protection, Amman, pp. 927–969.
Shamseldean, M.M. and Abd-Elgawad, M.M. (1994) Natural occurrence of insect pathogenic nematodes (Rhabditida: Heterorhabditidae) in Egyptian soils. *Afro-Asian Journal of Nematology* 4, 151–154.
Shapiro-Ilan, D.I., Reilly, C.C. and Hotchkiss, M.W. (2009) Suppressive effects of metabolites from *Photorhabdus* and *Xenorhabdus* spp. on phytopathogens of peach and pecan. *Archives of Phytopathology and Plant Protection* 42, 715–728.
Shapiro-Ilan, D.I., Bock, C.H. and Hotchkiss, M.W. (2014) Suppression of pecan and peach pathogens on different substrates using *Xenorhabdus bovienii* and *Photorhabdus luminescens*. *Biological Control* 77, 1–6.
Shrestha, S. and Kim, Y. (2010) Differential pathogenicity of two entomopathogenic bacteria, *Photorhabdus temperate* subsp. *temperata* and *Xenorhabdus nematophila* against the red flour beetle, *Tribolium castaneum*. *Journal of Asia-Pacific Entomology* 13, 209–213.
Sicard, M., Raimond, M., Gaudriault, S., Lanois, A., Pages, S., *et al.* (2014) Putative toxins from the entomopathogenic bacterium *Photorhabdus luminescens* kill *Armadillidium vulgare* (Terrestrial isopod). *Biological Control* 69, 40–44.
Smigielski, A.J., Akhurst, R.J. and Boemare, N.E. (1994) Phase variation in *Xenorhabdus nematophilus* and *Photorhabdus luminescens*: differences in respiratory activity and membrane energization. *Applied and Environmental Microbiology* 60, 120–125.
Somvanshi, V.S., Kaufmann-Daszczuk, B., Kim, K.S., Mallon, S. and Ciche, T.A. (2010) *Photorhabdus* phase variants express a novel fimbrial locus, made essential for symbiosis. *Molecular Microbiology* 77, 1021–1038.
Stack, C.M., Easwaramoorthy, S.G., Metha, U.K., Downes, M.J., Griffin, C.T. and Burnell, A.M. (2000)

Molecular characterisation of *Heterorhabditis indica* isolates from India, Kenya, Indonesia and Cuba. *Nematology* 2, 477–487.

Steiner, G. (1923) *Aplectana kraussei* n. sp., einer in der Blattwespe Lyda sp. parasitierende Nematodenform, nebst Bemerkungen über das Seitenorgan der parasitischen nematoden. *Zentralblatt fuer Bakteriologie, Parasitenkunde, Infektionskrankheiten und Hygiene* 59, 14–18.

Stock, S.P. (2015) Diversity, biology and evolutionary relationships. In: Campos-Herrera, R. (ed.) *Nematodes Pathogenesis of Insects and Other Pests*. Springer International Publishing, Switzerland, pp. 3–28.

Stock, S.P., Griffin, C.T. and Burnell, A.M. (2002) Morphological characterization of three isolates of *Heterorhabditis poinar*, from the 'Irish group' (*Nematoda*; *Rhabditida*; *Heterorhabditidae*) and additional evidence supporting their recognition as a distinct species *H. downesi* n.sp. *Systematic Parasitology* 51, 95–106.

Strauch, O. and Ehlers, R.-U. (1998) Food signal production of *Photorhabdus luminescens* inducing the recovery of entomopathogenic nematodes *Heterorhabditis* spp. in liquid culture. *Applied and Microbiological Biotechnology* 50, 369–374.

Stuart, R.J., Barbercheck, M.E. and Grewal, P.S. (2015) Entomopathogenic nematodes in the soil environment: distributions, interactions and the influence of biotic and abiotic factors. In: Campos-Herrera, R. (ed.) *Nematodes Pathogenesis of Insects and Other Pests*. Springer International Publishing, Switzerland, pp. 97–138.

Tailliez, P., Laroui, C., Ginibre, N., Paule, A., Pagès, S. and Boemare, N. (2010) Phylogeny of *Photorhabdus* and *Xenorhabdus* based on universally conserved protein-coding sequences and implications for the taxonomy of these two genera. Proposal of new taxa: *X. vietnamensis* sp. nov., *P. luminescens* subsp. caribbeanensis subsp. nov., *P. luminescens* subsp. hainanensis subsp. nov., *P. temperate* subsp. khanii subsp. nov., *P. temperata* subsp. tasmaniensis subsp. nov., and the reclassification of *P. luminescens* subsp. thracensis as *P. temperata* subsp. thracensis. *International Journal of Systematic and Evolutionary Microbiology* 60, 1921–1937.

Thanwisai, A., Tandhavanant, S., Saiprom, N., Waterfield, N.R., Ke Long, P., *et al.* (2012) Diversity of *Xenorhabdus* and *Photorhabdus* spp. and their symbiotic entomopathogenic nematodes from Thailand. *PLoS ONE* 7(9), e43835. DOI: 10.1371/journal.pone.0043835.

Theodore, C.M., King, J.B. and You, J. and Cichewicz, R.H. (2012) Production of cytotoxic glidobactins/luminmycins by *Photorhabdus asymbiotica* in liquid media and live crickets. *Journal of Natural Products* 75, 2007–2011.

Thomas, G.M. and Poinar, G.O. Jr (1979) *Xenorhabdus* gen. nov., a genus of entomopathogenic, nematophilic bacteria of the family Enterobacteriaceae. *International Journal of Systematic Bacteriology* 29, 352–360.

Tóth, T. and Lakatos, T. (2008) Photorhabdus temperate subsp. Cinerea subsp. nov., isolated for Heterorhabditis nematodes. *International Journal of Systematic and Evolutionary Microbiology* 58, 2579–2581.

Tounsi, S., Aoun, A.E., Blight, M., Rebai, A. and Jaoua, S. (2006) Evidence of oral toxicity of *Photorhabdus temperata* strain K122 against *Prays oleae* and its improvement by heterologous expression of *Bacillus thuringiensis* cry 1Aa and cry 1Ia genes. *Journal of Invertebrate Pathology* 91, 131–135.

Uma, G., Prabhuraj, A. and Vimala, S. (2010) Bioefficacy of *Photorhabdus luminescens*, a symbiotic bacterium against *Thrips palmi* Karny (Thripidae: Thysanoptera). *Journal of Biopesticides* 3, 458–462.

Vega, F.E. and Kaya, H.K. (2012) *Insect Pathology*, 2nd edn. Academic Press, San Diego, California.

Vigneux, F., Zumbihl, R., Jubelin, G., Ribeiro, C., Poncet, J., *et al.* (2007) The xaxAB genes encoding a new apoptotic toxin from the insect pathogen *Xenorhabdus nematophila* are present in plant and human pathogens. *Journal of Biological Chemistry* 282, 9571–9580.

Visschedyk, D.D., Perieteanu, A.A., Turgeon, Z.J., Fieldhouse, R.J., Dawson, J.F. and Merrill, A.R. (2010) Photox, a novel actin-targeting mono-ADP-ribosyltransferase from *Photorhabdus luminescens*. *Journal of Biological Chemistry* 285, 13525–13534.

Wang, Y., Fang, X., An, F., Wang, G. and Zhang, X. (2011) Improvement of antibiotic activity of *Xenorhabdus bovienii* by medium optimization using response surface methodology. *Microbial Cell Factories* 10, 98–112.

Waterfield, N., Dowling, A., Sharma, S., Daborn, P.J., Potter, U. and ffrench-Constant, R.H. (2001a) Oral toxicity of *Photorhabdus luminescens* W14 toxin complexes in *Escherichia coli*. *Applied and Environmental Microbiology* 67, 5017–5024.

Waterfield, N.R., Bowen, D.J., Fetherston, J.D., Perry, R.D. and ffrench-Constant, R.H. (2001b) The *tc* genes of *Photorhabdus*: a growing family. *Trends in Microbiology* 9, 185–191.

Waterfield, N.R., Daborn, P.J. and ffrench-Constant, R.H. (2002) Genomic islands in *Photorhabdus*. *Trends in Microbiology* 10, 541–545.

Waterfield, N.R., Daborn, P.J., Dowling, A.J., Yang, G., Hares, M. and ffrench-Constant, R.H. (2003) The insecticidal toxin Makes caterpillars floppy 2 (Mcf2) shows similarity to HrmA, an avirulence protein from a plant pathogen. *Federation of European Microbiological Societies Microbiology Letters* 229, 265–270.

Waterfield, N.R., Daborn, P.J. and ffrench-Constant, R.H. (2004) Insect pathogenicity islands in the insect pathogenic bacterium *Photorhabdus*. *Physiological Entomology* 29, 240–250.

Waterfield, N., Kamita, S.G., Hammock, B.D. and ffrench-Constant, R. (2005) The *Photorhabdus* Pir toxins are similar to a developmentally regulated insect protein but show no juvenile hormone esterase activity. *Federation of European Microbiological Societies Microbiology Letters* 245, 47–52.

Waterfield, N.R., Ciche, T. and Clarke, D. (2009) *Photorhabdus* and a host of hosts. *Annual Review of Microbiology* 63, 557–574.

Webster, J.M., Chen, G., Hu, K. and Li, J. (2002) Bacterial metabolites. In: Gaugler, R. (ed.) *Entomopathogenic Nematology*. CAB International, Wallingford, UK, pp. 99–114.

Wee, K.E., Yonan, C.R. and Chang, F.N. (2000) A new broad-spectrum protease inhibitor from the entomopathogenic bacterium *Photorhabdus luminescens*. *Microbiology* 146, 3141–3147.

Wilkinson, P., Waterfield, N.R., Crossman, L., Corton, C., Sanchez-Contreras, M., *et al.* (2009) Comparative genomics of the emerging human pathogen *Photorhabdus asymbiotica* with the insect pathogen *Photorhabdus luminescens*. *BMC Genomics* 10, 302. DOI: 10.1186/1471-2164-10-302.

Woodring, J.L. and Kaya, H.K. (1988) Steinernematid and heterorhabditid nematodes: a handbook of biology and techniques. Southern Cooperative Series Bulletin No 331, Arkansas Agricultural Experiment Station, Fayetteville, Arkansas, 30 pp.

Wright, D.J. and Perry, R.N. (2002) Physiology and biochemistry. In: Gaugler, R. (ed.) *Entomopathogenic Nematology*. CAB International, Wallingford, UK, pp. 145–168.

Xiao, Y., Meng, F., Qiu, D. and Yang, X. (2012) Two novel antimicrobial peptides purified from the symbiotic bacteria *Xenorhabdus budapestensis* NMC-10. *Peptides* 35, 253–260.

Yang, G. and Waterfield, N.R. (2013) The role of TcdB and TccC subunits in secretion of the *Photorhabdus* Tcd toxin complex. *PLoS Pathogens* 9, e1003644.

Yang, G., Dowling, J., Gerike, U., ffrench-Contant, R.H. and Waterfield, N.R. (2006) *Photorhabdus* virulence cassettes confer injectable insecticidal activity against the wax moth. *Journal of Bacteriology* 188, 2254–2261.

Yilmaz, H., Waeyenberge, L., Demir, I., Moens, M. and Demirbag, Z. (2009) A new entomopathogenic nematode species for Turkey, *Heterorhabditis megidis* Poinar, Jackson & Klein 1987 (Rhabditida: Heterorhabditidae). *Turkish Journal of Agriculture and Forestry* 33, 385–391.

Yoshida, M., Reid, A.P., Briscoe, B.R. and Hominick, W.M. (1998) Survey of entomopathogenic nematodes (Rhabditida: Steinernematidae and Heterorhabditidae) in Japan. *Fundamental and Applied Nematology* 21, 185–198.

Zhang, D., De Souza, R.F., Anantharaman, V., Iyer, L.M. and Aravind, L. (2012a) Polymorphic toxin systems: comprehensive characterization of trafficking modes, processing, mechanisms of action, immunity and ecology using comparative genomics. *Biology Direct* 7, 18. DOI: 10.1186/1745-6150-7-18.

Zhang, H., Mao, J., Liu, F. and Zeng, F. (2012b) Expression of a nematode symbiotic bacterium-derived protease inhibitor protein in tobacco enhanced tolerance against *Myzus persicae*. *Plant Cell Reports* 31, 1981–1989.

Zhang, X., Hu, X., Li, Y., Ding, X., Yang, Q., *et al.* (2014) XaxAB-like binary toxin from *Photorhabdus luminescens* exhibits both insecticidal activity and cytotoxicity. *Federation of European Microbiological Societies Microbiology Letters* 350, 48–56.

13 Entomopathogenic Nematodes: Mass Production, Formulation and Application

TARIQUE HASSAN ASKARY* AND M. JAMAL AHMAD

Division of Entomology, Sher-e-Kashmir University of Agricultural Sciences and Technology, Srinagar, India

13.1 Introduction

This is an era of biological control and agricultural scientists are working with an aim to check crop losses in order to obtain maximum profit without or least disturbing the ecological balance (Askary and Martinelli, 2015). Crop insect pests are one of the major constraints in agricultural production throughout the world. Efforts made by scientists in developing some resistant varieties against particular insect pests has provided some insulation and stability in crop production, however, management of these pests efficiently and economically by least disturbing the ecological balance is still a challenge for research practitioners. Entomopathogenic nematodes (EPNs) belonging to the family Steinernematidae and Heterorhabditidae are obligate parasites of insects and have many attributes that make them a good and promising biocontrol agent. The parasitism due to these beneficial nematodes results in suppressing the insect host immune system (Dowds and Peters, 2002; Lewis and Clarke, 2012). The nematode gains entry into the insect host through natural openings such as the mouth, anus, or spiracles, or by penetrating through the intersegmental membranes of the insect cuticle (Bedding and Molyneux, 1982; Peters and Ehlers, 1994), and reaches the haemocoel, where the third-stage juveniles, also called infective juveniles (IJs), release cells of bacterial symbionts from their intestine (Askary, 2010). The nutrient-rich haemolymph of the insect host helps in the rapid multiplication of these bacteria. The bacteria release a wide range of toxins and hydrolytic exoenzymes that cause septicaemia or blood poisoning, ultimately resulting in the death of the host within 48–72 h. The bioconversion of insect larvae into a nutrient soup takes place with the advancement of time, which is ideal for nematode growth and reproduction (Ali *et al.*, 2005). All life stages of nematode, except DJs, are found exclusively inside the insect host (Grewal *et al.*, 2005; Onstad *et al.*, 2006).

EPNs are safe to plants and vertebrates, and despite worldwide application in the field, garden and pasture, no significant acute or chronic toxicity to human or other vertebrate populations has been established. That is why the Environmental Protection Agency (EPA) in the USA, India, Australia and many European countries has exempted them from all registration requirements and related regulations (Ehlers and Hokkenen, 1996; Ehlers, 2005; Dreves and Lee, 2015).

The critical factors in making EPNs successful biocontrol agents of crop insect pests are their production and application technology (Shapiro-Ilan *et al.*, 2012). EPNs can be mass produced *in vivo* (baiting technique on insects) and commercially *in vitro* (solid/liquid media) and utilized against a variety of crop insect pests of a soil-dwelling and above-ground nature, i.e. foliar, epigeal or cryptic habitats. While planning for the successful utilization of EPNs against target insect pests, some points need to be taken into consideration, which include the mass production of EPNs, their proper

*Corresponding author. E-mail: tariq_askary@rediffmail.com

formulation with an aim to achieve a longer shelf life and the appropriate application technology, whether as soil application or injection/spraying. Besides, quality tests of nematodes under laboratory and field conditions and post-field transportation need to be assessed. This also includes the nematode's viability and virulence capability. Above all, economy is the prime consideration, and the production of EPNs at low cost is a prerequisite for research workers (Yadav *et al.*, 2015).

The present chapter focuses on two EPNs, namely *Steinernema* and *Heterorhabditis*, and on recent techniques in their mass production and formulation, as well as application, both in soil and on the foliage surface or cryptic habitats for the management of crop insect pests.

13.2 Mass Production

Steinernema and *Heterorhabditis* are symbiotically associated with bacteria belonging to the genera *Xenorhabdus* and *Photorhabdus*, respectively (Poinar, 1990; Griffin *et al.*, 2005; Askary, 2010; Lewis and Clarke, 2012). The first attempt in the exploitation of EPNs was taken against Japanese beetle (Glaser, 1932). Later, the programme failed due to lack of knowledge about EPN–bacteria symbiosis (Gaugler *et al.*, 1992). The mystery of the symbiotic relationship between the nematode *Steinernema*, and the bacterium *Achromobacter nematophilus* (now *Xenorhabdus nematophilus*) was revealed in 1966 by Poinar and Thomas and again one decade later, i.e. in 1976, by Poinar for the nematode *Heterorhabditis* and the bacterium *Xenorhabdus luminescens* (later on transferred to the genus *Photorhabdus*). These discoveries proved to be a major breakthrough in insect nematology that was later exploited by other research workers and which ultimately led to the mass production and commercialization of EPNs.

EPNs are being mass produced in several countries of North America, Europe and Asia, on both a small scale and a large scale, using bioreactors (Shapiro-Ilan and Gaugler, 2002). They can be mass-produced in two ways: (i) *in vivo* (Dutky *et al.*, 1964); and (ii) *in vitro* (Glaser, 1940). In the case of *in vivo*, insects serve as the bioreactor, whereas the *in vitro* process is carried out in artificial media (Sharma *et al.*, 2011). Each approach has certain advantages and disadvantages that may be reflected in the form of technical expertise, production costs, economy of scale and quality of end product.

13.2.1 *In vivo*

This is a low technology method with low start-up costs (such as a cottage industry) that involves the production of EPNs by using live insects, which are highly susceptible and easily available at a lower cost. The insects used under this method are the larvae of the greater wax moth, *Galleria mellonella*, the rice moth, *Corcyra cephalonica*, or the mealworm, *Tenebrio molitor*, which are reared in the laboratory (Fig. 13.1). Generally, the last instar of *G. mellonella* is preferred, due to its high susceptibility, easy availability and high yield of IJs (Woodring and Kaya, 1988). However, we conducted our experiment using *C. cephalonica* and therefore presenting figures of this insect in the chapter (Fig. 13.2a and b). The approach is based on two-dimensional systems that rely on nematode production in trays and shelves (Friedman, 1990; Shapiro-Ilan and Gaugler, 2002; Ehlers and Shapiro-Ilan, 2005). The method involves four steps: inoculation, harvest, concentration and decontamination.

Inoculation

Insects are inoculated with IJs on a tray or dish lined with filter paper or another substrate conducive to nematode infection, such as soil or plaster of Paris. The nematode dosage and host density should be optimized for maximum yield. Too low a dosage of IJs may result in low host mortality, whereas too high a dosage may result in failed infections due to competition with secondary invaders. Approximately 25–200 IJs are sufficient to cause infection on one insect larva of *G. mellonella*.

Harvest

This step is performed by using a technique based on the White trap (White, 1927), wherein after 2–5 days, the host insects killed by nematodes are placed above a water reservoir (Fig. 13.3a and b). The nematode produced by this method is harvested by placing moist filter paper on a concave side up watch glass surrounded with water in a large Petri dish. The progeny IJs migrate from the depleted host cadaver into the water reservoir, where they are trapped and subsequently harvested.

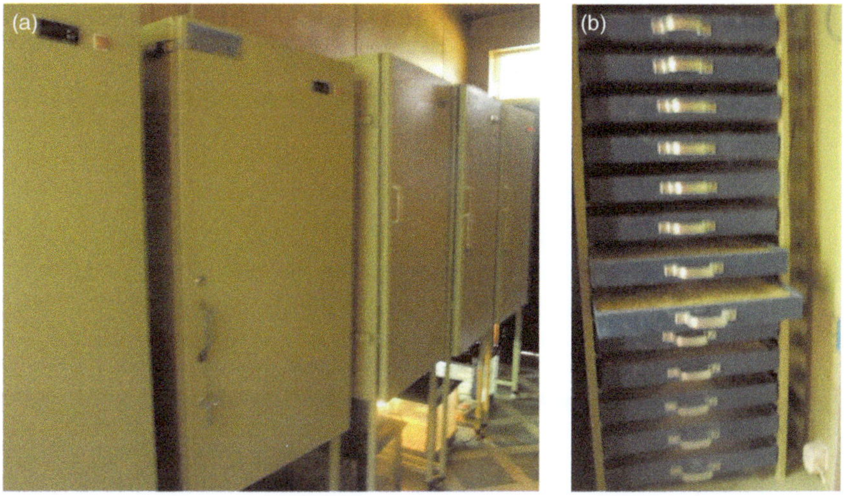

Fig. 13.1. Rearing of rice moth, *Corcyra cephalonica*, in the laboratory: (a) *Corcyra* rearing apparatus (closed door); (b) *Corcyra* rearing apparatus (door open showing movable racks).

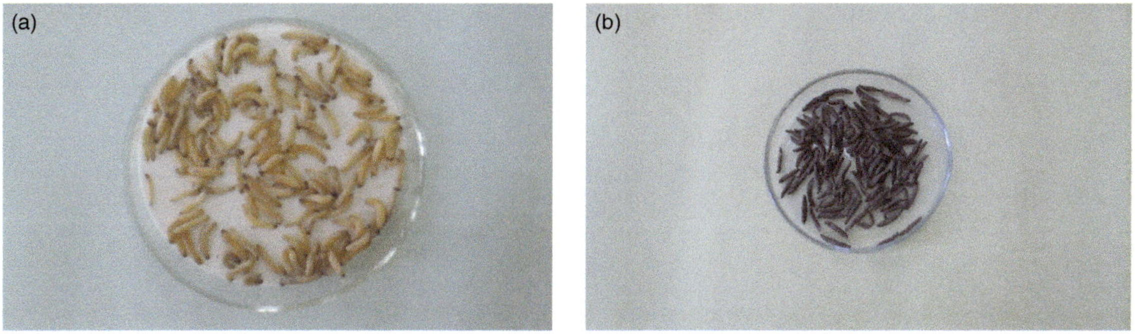

Fig. 13.2. (a) Live *Corcyra cephalonica* larvae ready for infection with EPN; (b) EPN-infected *C. cephalonica* kept for incubation.

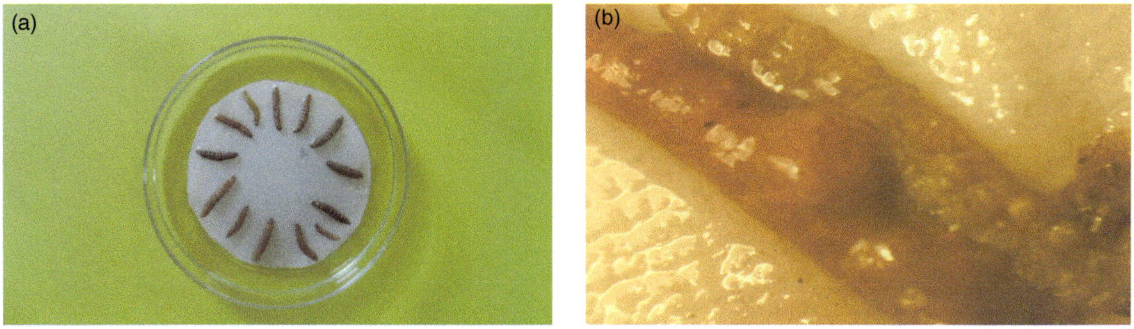

Fig. 13.3. (a) White trap method for harvesting EPNs emerged from insect cadavers; (b) infected juveniles (IJs) inside the bodies of EPN-infected insect larvae.

Precaution should be taken that there should not be too long a delay in the transfer of the infected insect cadaver to the White trap, as it may cause a negative effect on the reproductive stages of the nematodes, as well as a chance for the insect cadaver to rupture.

Concentration

IJs are decanted, transferred to a beaker and then kept in biological oxygen demand (BOD) incubator at 10–15°C. During the process, care should be taken that settling for a prolonged period may prove detrimental to the nematodes, as this often causes a lack in oxygen content (Burman and Pye, 1980). Although this may be accomplished by vacuum filtration (Lindegren *et al.*, 1993) or centrifugation (Kaya and Stock, 1997), for commercial *in vivo* operations, the total cost will be much higher for a centrifuge of sufficient capacity (Shapiro-Ilan *et al.*, 2004).

Decontamination

There is a chance of host material or microbial contamination on nematodes while migrating away from the cadaver. Therefore, the nematodes harvested by this method are washed repeatedly. This can be accomplished by gravity settling (Dutky *et al.*, 1964), wherein antimicrobial compounds such as streptomycin-sulfate, Hyamine® (methylbenzethonium chloride), merthiolate, NaOCl and $HgCl_2$ are used (Lunau *et al.*, 1993). These compounds have not been found to have any detrimental effect on nematodes during commercial application (Shapiro-Ilan *et al.*, 2004).

The yield of nematode is proportional to the size of the host cadaver (Blinova and Ivanova, 1987; Flanders *et al.*, 1996); however, it is inversely proportional to the size of a particular nematode species (Hominick *et al.*, 1997; Grewal *et al.*, 1999). Nematode yield/mg insect within host species and host susceptibility to nematode infection are inversely proportional to the size and age of the host (Dutky *et al.*, 1964; Blinova and Ivanova, 1987; Shapiro-Ilan *et al.*, 1999). However, nematode species such as *Heterorhabditis indica* and *Steinernema riobrave* have innately high reproduction capacities (Shapiro-Ilan *et al.*, 2014a). In an experiment, the yield of *S. riobrave* was recorded at more than 300,000 IJs/insect (*G. mellonella*), whereas in the case of *S. glaseri*, the yield was up to 50,000 IJs only (Grewal *et al.*, 1994; Stock and Hunt, 2005).

The nematodes produced by *in vivo* methods are of high quality, particularly when the nematodes cultured in hosts fall within the natural host range of that particular nematode species (Abu Hatab *et al.*, 1998; Abu Hatab and Gaugler, 2001; Shapiro-Ilan *et al.*, 2004). It has been noted that nematodes adapt the host on which they are reared (Stuart and Gaugler, 1996) which may result in reducing the field efficacy of the nematodes in case the target pest is unmatched (Shapiro-Ilan *et al.*, 2004). Therefore, prior to *in vivo* production, the correct selection of nematode species, as well as the host insect, must be taken into consideration for higher yields and efficacy of nematodes. Besides, host diet also plays a significant role in the quality and production of nematodes (Shapiro-Ilan *et al.*, 2014a).

The method described above is simple and cheap. It is appropriate for laboratory use and small-scale applications, such as for those growers who cannot afford large investments in *in vitro* production. However, for large-scale nematode production, this method is not suitable as it demands a large number of labourers, because the two different organisms, i.e. host insect and EPN, are cultured simultaneously, which makes the entire process costly. In spite of this, the above method of nematode production is sustainable as a cottage industry (Gaugler *et al.*, 2000; Gaugler and Han, 2002).

LOTEK technique

An improved technique for *in vivo* mass production of EPNs is LOTEK (Gaugler and Brown, 2001). This method requires less labourers and the nematode is collected through rinsing with a mist of water rather than the migration of active nematodes to a water reservoir (Brown *et al.*, 2004). Nematode production by using this system involves five steps wherein a reusable holding tray is used (except harvesting/separation), which makes the task of handling the insects easy.

INOCULATION. The prime objective of inoculation is to infect the maximum number of insects (>95%) so that the labour costs may be minimized (Brown *et al.*, 2004). This can be achieved by the immersion method used for nematode–insect combinations. Inoculum delivery on insects is performed by two ways:

1. *Immersion.* The insect larvae are placed in a tray and the tray is dipped in a concentrated suspension of IJs. Five hundred insect larvae can be placed in a tray measuring 30 × 26 × 4 cm. The concentration of IJs varies according to the nematode species and insect larva. The concentration of IJs of *S. carpocapsae* in a suspension has been optimized 2.1×10^4/ml in the case of *Tenebrio* and 8×10^3/ml in the case of *Galleria* (Managanvi *et al.*, 2012). When the nematode suspension is *H. bacteriophora*, the concentration is optimized 1.2×10^4/ml for inoculating *Galleria*.

2. *Pipette transfer*. In this step, IJs are pipetted on to an inoculation dish to holding trays, generally 3–5 days after the infection is complete. The major drawbacks of the need for a substrate are: (i) substantially increased handling time; and (ii) damage to the cadavers (Brown *et al.*, 2004).

CONDITIONING. Conditioning is performed in a humidity controlled chamber that consists of an air pump, a humidifier, a water trap and a vented chamber. The main objective of conditioning is to incubate the insect larvae killed by IJs in a controlled environment and to synchronize the emergence of IJs from the larvae. In this step, the trays containing insect larvae inoculated with IJs are transferred immediately to a humidity controlled chamber for conditioning. The trays are stacked tightly in the chamber so that the IJ-inoculated insect larvae may not escape from the tray. This method is advantageous in the sense that it reduces the amount of temperature-controlled rearing space. It requires minimal space as trays are stockpiled and stacked tightly. In case, if the procedure is done without conditioning, it becomes too much expensive due to the requirement of seven times more space to be allocated for the harvest units (due to the time in the harvester being seven times longer) and the cost of constructing additional harvesters (Brown *et al.*, 2004). Therefore, conditioning is a major cost-saving method.

HARVESTING. When the development of nematodes is complete in the incubator, the trays are shifted to a harvester. Two harvesters have been designed to fulfil the objectives:

1. *Misting chamber*. In this chamber, each tray is suspended beneath two rigid plastic pipes, each equipped with a mist nozzle delivering water at 37 ml/min and 1.78 l of water/tray. The plastic pipes have a dual function, i.e. supporting the trays and delivering water via misting nozzles. The nematodes thus emerge due to the mist, get rinsed through the holding tray perforations into the collection pan and ultimately collect into the central storage tank. Thus, the flow of water in this system aids in removing the nematodes. This is contrary to the White trap method where the nematodes themselves migrate to the water reservoir.

2. *Drip irrigation*. In the drip irrigation system, the harvester is suspended above a sloping tray that is constructed from a single piece of plastic-coated wire mesh and lined with a cotton cloth substrate. The trays are placed in between the two supports. Each support bears a row of 2.5 cm long pins and the trays are hung in between the rows of pins at an angle of 20 degrees. Water is delivered at 1 l/day on to the upper edge of each tray via 30 cm lengths of soaker hose connected to the water source. The dripping of water takes place on the trays with insect cadavers. The water seeps through the cotton cloth, gently washing the emerged nematodes off the lower edge of the tray and into a storage tank. This system has resulted in the harvesting of *H. bacteriophora* by 84.2% and 96.3% after 24 and 48 h, respectively (Managanvi *et al.*, 2012).

SEPARATION. The harvested IJs are washed and concentrated. IJs suspended from the holding tanks are pumped into a conical reservoir tank and gravity fed to a distribution manifold that divides the flow into several small streams. The reservoir tank acts as a buffer, ensuring a constant uninterrupted flow through the manifold. Streams are directed to a stainless steel screen oriented at a 38-degree angle. The water as a waste passes through the screen and is discarded. In the process, the majority of nematodes are deflected and collected on the screen surface and thereafter nematode slurry is generated, flowing slowly by gravitation through the screen to a collection gutter and finally into a temporary storage tank (Brown *et al.*, 2004). The nematode slurry is further washed and concentrated by multiple passages through the separator. The bacteria, cadaver residues, pigments, as well as other wastes, are measured by the optical density of the harvested nematode suspension before and after one pass through the separator. This method resulted in increasing nematode concentration by eight times, removing water and associated wastes by 87%, with only 4.5% loss of IJs after a single pass through the separator. Three such passes removed up to 97% of water and concentrated the IJs heavily, thus making the harvested nematodes ready for formulation or aqueous storage (Managanvi *et al.*, 2012).

CLEAN UP. The final step, i.e. the process of cleaning the harvester, is the most tedious and laborious step. All the trays of spent cadavers are emptied, placed back into the harvester and cleaned by introducing detergents and disinfectants into the water line. The process of washing and disinfection is carried out by the harvester misting nozzle.

QUALITY. The quality of nematodes produced by this method is equal to or even greater than the nematodes produced by other approaches. Since the LOTEK system is automated, the production of nematodes is achieved with reduced labour costs. Scalability is also increased, as there is no need for nematodes to migrate to the reservoir.

The mass production of EPNs by the LOTEK system is simple and incurs low construction costs, and therefore is best suited for those areas where capital and expertise are limiting factors. Another advantage of this system is the production of a large number of nematodes of one strain, sometimes required by laboratories for field use, which is not available commercially. Using this method, grower cooperatives can gain a lead by producing and applying their nematodes immediately (Gaugler, 1997), thereby evading formulation, storage, transport and distribution costs.

Cadaver application method

This method involves the culture and delivery of IJs in their infected host cadavers (Jansson *et al.*, 1993; Shapiro-Ilan and Glazer, 1996; Del Valle *et al.*, 2008). EPN-infected host insect cadavers are placed on the target site. The progeny IJs emerge from the cadaver, parasitize the target insect pests and thus cause their suppression. Under laboratory conditions, the cadaver application method proved better than the aqueous suspension method in terms of nematode dispersal (Shapiro-Ilan and Glazer, 1996), infectivity (Shapiro-Ilan and Lewis, 1999) and survival (Perez *et al.*, 2003). An additional advantage of this method is the minimum utilization of labour, and ultimately labour costs, but on the other hand, the disadvantages are the fragile condition of the nematode-infected insect cadavers, particularly soft-bodied insects, or the cadavers sticking together. However, these problems may be overcome by clay coating or encapsulating the cadavers. This will protect the cadaver from rupture and also facilitate handling (Shapiro-Ilan *et al.*, 2001, 2010; Del Valle *et al.*, 2009). Also, such problems can be avoided if hard-bodied insect hosts like *T. molitor* are used (Shapiro-Ilan *et al.*, 2008), and further by wrapping the cadaver of *T. molitor* between two pieces of masking tape. In this case, the emergence of IJs takes place between two rolls of tape, with no yield loss (Morales-Ramos *et al.*, 2010; Shapiro-Ilan *et al.*, 2010). The distribution of host cadavers on a large scale in the field can be accomplished by mechanized equipment (Zhu *et al.*, 2011). In another approach, the efficacy and persistence of IJs were found to be high when nematode-infected insect cadavers were added to bags of potting media and subsequently distributed to target pest sites (Deol *et al.*, 2011).

13.2.2 *In vitro*

EPNs when needed on a large scale but with reasonable quality and cost are produced by solid or liquid fermentation. Both these techniques are used to culture nematodes monoxenically so as to maintain consistency in quality and achieve predictability (Lunau *et al.*, 1993; Fig. 13.4).

In vitro solid culture method

The prerequisite in this method is the establishment of monoxenic culture, i.e. bacteria-free nematodes, which is accomplished with the isolation of the nematode's symbiotic bacteria from nematode-infected

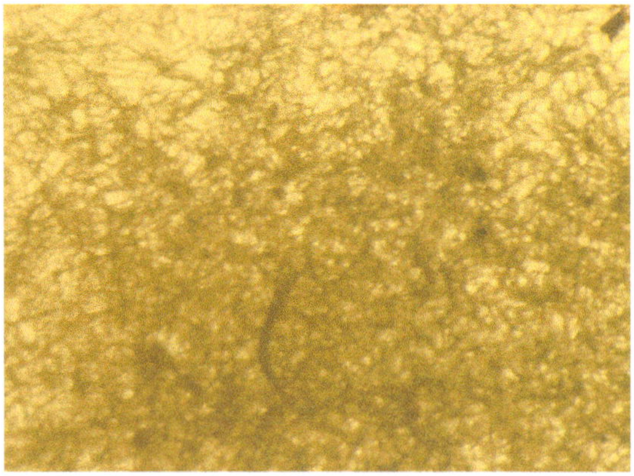

Fig. 13.4. Mass production of EPNs. A giant female appearing in the middle of IJ masses.

insect larvae or surface-sterilized IJs. After the sterile nematode eggs are prepared, they are applied to the medium pre-inoculated with bacterial symbiont. There are two phase variants of bacteria – primary and secondary. These two variants differ in characteristics such as dye absorption, response to biochemical tests and antibiotic production (Shapiro-Ilan *et al.*, 2014a). EPNs are reported to grow on secondary variant symbionts, although primary variants are more conducive to growth and are the only variant retained by IJs (Han and Ehlers, 2001). However, monoxenicity is somewhat difficult to establish by mere surface sterilization of IJs because contaminating bacteria survive beneath the nematode's cuticle (Lunau *et al.*, 1993). Therefore, in order to obtain axenic nematode eggs, gravid nematode females (Figs 13.5 and 13.6) are ruptured in an

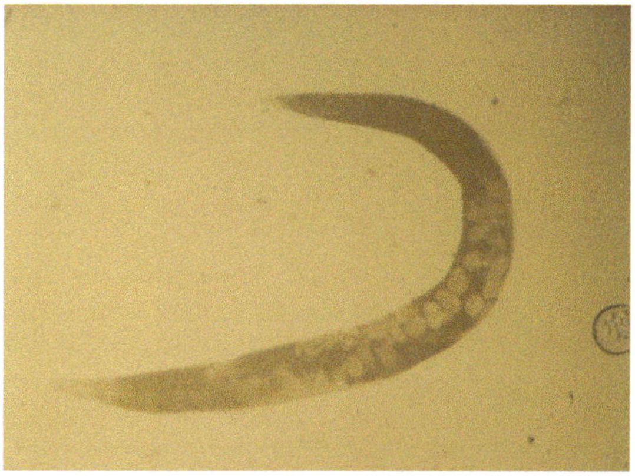

Fig. 13.5. An adult of *Heterorhabditis bacteriophora* showing its body full of eggs.

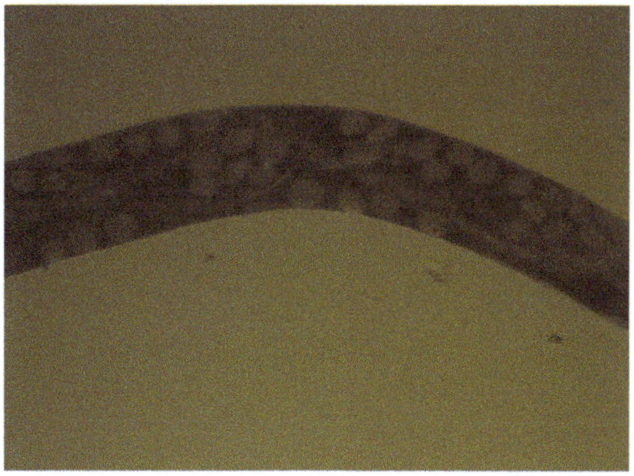

Fig. 13.6. A body portion of adult *Heterorhabditis bacteriophora* showing eggs and newly hatched juveniles.

alkaline solution (Lunau *et al.*, 1993; Han and Ehlers, 1998). The eggs obtained from the gravid female of *Steinernema* spp. or hermaphrodite *Heterorhabditis* spp. are collected in Ringer's solution (per 1 l distilled water: NaCl 9.0 g, KCl 0.42 g, $CaCl_2$ 0.48 g, $NaHCO_3$ 0.2 g) and surface sterilized in a solution of 2.5 ml 4M NaOH, 0.5 ml 12% NaOCl and 21.5 ml distilled water. The eggs are rinsed twice in sterile Ringer's solution and then transferred to sterile media, wherein the hatching of J_1 takes place. After a couple of days, J_1s are combined with their symbiotic bacteria grown on the fortified lipid agar (1.6% nutrient broth, 1.2% agar and 1% maize oil; Wouts 1981). Thus, after the formation of monoxenic cultures, the nematode culture can be scaled up in solid phase or liquid phase (Shapiro-Ilan *et al.*, 2014a).

Solid culture is performed by using various media in two-dimensional arenas such as Petri dishes. The method involves four steps: (i) preparation of solid medium; (ii) inoculation with bacteria; (iii) inoculation with IJs; and (iv) harvesting of IJs. Glaser (1932) was the first to grow EPNs *in vitro* on a solid medium axenically; however, a major breakthrough in this direction was achieved when it was prepared commercially by using chicken offal medium on a porous polyether polyurethane foam substrate (Bedding, 1981, 1984). In this three-dimensional rearing system, the liquid medium was mixed with an inert carrier, i.e. pieces of polyurethane foam. The nematodes were cultured on a crumbed polyether polyurethane sponge impregnated with emulsified beef fat and pig's kidney, along with symbiotic bacteria. However, the production of EPNs by solid media requires a high labour cost, and therefore this technology is suitable for those countries where labour costs are minimal (Bedding, 1990). Initially, the media was based on animal products, such as kidney products, but it was later improved by including various other ingredients such as yeast extract, peptone, eggs, soya flour. The media mixed with foam pieces are first autoclaved for sterilization. The sterilized medium is inoculated with bacteria, followed by nematodes 3 days later. Harvesting of nematodes is done within 15–35 days by placing the foam on to sieves immersed in water. IJs migrate out of the foam, settle downward and are pumped to a collection tank. The cleaning of IJs is performed by washing them repeatedly with water, i.e. sedimentation and decanting (Bedding, 1981, 1984; Sharma *et al.*, 2011). In order to scale up production, the technology was further refined wherein bags with a gas permeable Tyvac strip were included for ventilation automated mixing and autoclaving, and harvested through centrifugal sifters. These developments to scale up the production of nematodes reportedly have been taking place at Biotech Australia and Ecogen-Sylvan, USA (Gaugler and Han, 2002).

Nematode production by the solid culture method depends on several factors, such as media composition (Dunphy and Webster, 1989; Han *et al.*, 1992, 1995), inoculum size, culture time, temperature (Han *et al.*, 1992, 1993) and the innate reproductive capacities of different nematode species and strains (Gaugler and Han, 2002). Han *et al.* (1995) observed a significant affect on the yield of *S. carpocapsae* by the addition of water, yeast extract and egg yolk, and *H. bacteriophora* by the addition of water and lard. Inoculum size also has an impact on nematode growth. Han *et al.* (1992) reported an increase in nematode growth and a decrease in culture time when the inoculum size was increased. Salma and Shahina (2012) cultured some nematode species, namely *S. pakistanense*, *S. carpocapsae*, *S. abbasi*, *S. asiaticum*, *S. siamkayai*, *H. bacteriophora* and *H. indica* on soy flour medium and got maximum yield of these nematodes at a temperature of 32 ± 2°C; however, yield of *S. feltiae* was maximum at a lower temperature, i.e. 20 ± 2°C. The average yield of *H. bacteriophora* strain HbNJ was reported to be 1.4 × 10^9/bag (Gaugler and Han, 2002), whereas for *S. scapterisci* the average yield was 4.8 × 10^8/bag (Bonifassi *et al.*, 1999), which indicates the innate reproductive capacities of different nematode species and strains.

In vitro liquid culture method

The first liquid medium based on kidney extract for the axenic culture of *S. glaseri* was developed by Glaser (1940). Stoll (1952) was the first to develop flask-scale liquid culture of EPNs. EPNs were grown axenically in liquid culture that contained raw liver extract, and the culture was incubated on a shaker. This method resulted in the production of nematodes of approximately 400 IJs/ml. Buecher and Hansen (1971) advanced this technique by supplying sterile air through liquid media bottles. The bubbling thus formed supplied aeration to the nematodes without affecting the shear effect due to forced aeration. However, this method of producing nematodes axenically was costly and with low yield. Another drawback with this method was the absence of symbiotic bacteria in the culture, thus making it inappropriate for biocontrol purposes (Ehlers *et al.*, 1997). Now, however, the technique has been modified and mass production of nematodes is done at less cost. Yadav *et al.* (2015) mass-produced *S. carpocapsae* in different artificial media, i.e. nutrient broth, wheat flour, maize flour, lipid and modified Wout's media. The maximum production of IJs was obtained from nutrient broth media, followed by lipid media, maize flour media, modified Wout's media and wheat flour media. Although nutrient broth media produced the highest population of IJs, it was not as economical as compared to Wout's and wheat flour media.

The *in vitro* liquid method is a complex method that requires medium development, study of the biology of the nematode-bacteria complex, the development of bioreactors and understanding and control of the process parameters. Above all, *in vitro* technology needs substantial capital investment and technical expertise. Experiments have been carried out in different laboratories of the world and researchers have tested various synthetic media to mass-culture EPNs (Table 13.1). The ingredients used for liquid culture include soy flour, maize flour, yeast extract, milk powder, casein peptone, cholesterol, liver oil, thistle oil and egg yolk (Buecher and Popiel, 1989; Friedman *et al.*, 1989). The use of bioreactors in the mass production of EPNs was described for the first time by Pace *et al.* (1986). The commercial production of EPNs in liquid culture was established by a team of researchers led by Friedman in 1990 at Biosys, Inc, in Palo Alto, California, USA (Grewal, 2012). Afterwards, a UK-based company, MicroBio, established liquid production of *S. feltiae*. Liquid culture of *Heterorhabditis* was reported for the first time by Surrey and Davies (1996); however, the first commercial-scale liquid production of this nematode was by Ehlers *et al.* (1998). Liquid culture is accomplished in bioreactors having the capacity of up to 80,000 l (Georgis *et al.*, 1995). The process is carried out in large fermentors wherein enough oxygen is supplied and excessive shearing of nematodes is prevented. The time of culturing depends on the media and species, and it may vary between 2 and 3 weeks to reach maximum production. When the preparation of culture is complete, the nematodes are removed from the medium through centrifugation. One major advantage of liquid culture is that it is the most cost-effective method, and currently most

Table 13.1. Different synthetic media recommended for *in vitro* mass production of entomopathogenic nematodes. (Adapted from Hussaini, 2002.)

Synthetic medium	Nematode species	Incubation period	Temperature	Nematode harvested	Reference
Wheat bran	*Steinernema feltiae*	3 weeks	25°C	10^4	Abe, 1987
Wheat bran + salad oil	*S. feltiae*	3 weeks	25°C	10^7	Abe, 1987
Dog food agar medium	*S. feltiae*	–	–	10^5/g of medium	Hara *et al.*, 1981
Beef extract, peptone, cornmeal, water on sponge	*S. feltiae*	–	–	–	Li, 1984
Nutrient broth, yeast extract, vegetable oil, flour coated on sponge	*Heterorhabditis heliothids*	4 weeks	25°C	10×10^6/250 ml flask	Wouts, 1981
Soyapeptone 3% + yeast extract 3% + chick embryo extract 10% medium	*S. glaseri*	–	–	10,051/week for 93 days	Tarakanov, 1980
Kidney/fat homogenate	*S. feltiae* *S. bibionis* *S. glaseri* *H. bacteriophora* *H. heliothids*	2–3 weeks	25°C	38×10^6/30 flasks 29×10^6/73 flasks 8×10^6/11 flasks 36×10^6/10 flasks 32×10^6/15 flasks	Bedding, 1981

EPN products are prepared by this method. Among the various nutritional components, 60% of the total energy is derived from metabolizing lipids for the non-feeding IJs (Abu Hatab and Gaugler, 1997). It has also been reported that the suboptimal yield of nematodes is the result of lipid sources with high proportions of saturated fatty acids (Abu Hatab and Gaugler, 2001). The world's major producers who are adopting this technique for the mass production of EPNs are Andermatt Biocontrol (Switzerland), Koppert BV (The Netherlands), E-Nema GmbH (Germany), BioLogic (USA) and Becker Underwood (USA).

Steinernema and *Heterorhabditis* when produced by the *in vitro* liquid method differ in yield, due to different life cycles and reproductive biology (Shapiro-Ilan *et al.*, 2014a). It has been noted that the design of bioreactor and regulation of aeration play a significant role in maximizing mating (Neves *et al.*, 2001). All the other species of *Steinernema* except one occur as male and female and are capable of mating in maximization in liquid culture (Strauch *et al.*, 1994). Contrary to this, *Heterorhabditis* sp. does not undergo maximization of mating due to their exclusive hermaphrodite first generation. Although subsequent generations are amphimictic but cannot mate in liquid culture (Strauch *et al.*, 1994), yields of *Heterorhabditis* sp. therefore depend on the degree of recovery, i.e. when IJs moult to initiate completion of their life cycle. The recovery of *Steinernema* or *Heterorhabditis* is governed by several factors, which include nutrition (Jeffke *et al.*, 2000; Chavarria-Hernandez and de la Torre, 2001; Cho *et al.*, 2011), aeration, CO_2 (Jessen *et al.*, 2000), lipid content (Yoo *et al.*, 2000) and temperature (Ehlers *et al.*, 2000).

Undoubtedly, the *in vitro* liquid culture method is helpful in producing high-quality EPNs; however, factors like media composition and environmental conditions in the bioreactor or downstream processing can have negative affects on quality. Beneficial traits such as infectivity, environmental tolerance and fecundity can reduce by repeated culturing of nematodes (Shapiro-Ilan *et al.*, 2012). Therefore, precautionary measures against trait (mainly quality and production efficacy) deterioration should be adopted, such as improvements in inoculums and bacterial cell density (Hirao and Ehlers, 2010), timing of inoculation (Johnigk *et al.*, 2004), improvement in heat and desiccation tolerant character in the case of *Heterorhabditis* (Mukuka *et al.*, 2010; Anbesse *et al.*, 2013), optimizing media and bioprocess kinetics through modelling (Chavarria-Hernandez *et al.*, 2006, 2010), minimization of serial passages, the introduction of fresh genetic material, the cryopreservation of stock cultures (Bai *et al.*, 2004) and the creation of homozygous inbred lines that are more resistant to trait deterioration (Bai *et al.*, 2005; Chaston *et al.*, 2011).

13.3 Formulation and storage

An important consideration for the success of EPNs against target insect pests is their formulation into a stable product, which in turn is a prerequisite for successful commercialization, and this can be achieved by maximum survival of IJs in a formulation for a longer period before their utilization in the field. Nematode mortality may vary from 70 to 100% if storage life of the formulated product has expired (Grewal, 2000a). The first report in the encapsulation of EPNs with calcium alginate based on immobilizing the nematode came in 1985 by Kaya and Nelson. This success ultimately led to the development of a commercial nematode product, wherein thin sheets of calcium alginate were spread over a plastic screen in order to trap nematodes (Georgis, 1990a). Another approach in this direction was alginate and flowable gel formulations to trap nematodes physically in order to reduce their movement. These were a few major breakthroughs that later resulted in many agrochemical companies introducing the formulations into different market segments, such as for turfgrass, greenhouse, berries, homes and gardens (Goud *et al.*, 2010). In the past three decades, significant progress has been made in the formulation technology of EPNs. Besides classical aqueous suspension, EPNs can be formulated in water-dispersable granules, nematode wool, gels, vermiculite, clay peat and sponge. The nematodes formulated on to inert solid carriers like clays or vermiculite are partially dehydrated (Grewal, 2000b). The induction of partial anhydrobiosis reduces nematode activity and metabolism. Bedding (1988) described 'sandwich formulation', which consisted of a layer of nematode between two layers of clay. In this method, nematodes were mixed in clay to remove excess surface moisture and induce partial anhydrobiosis. In fact, Bedding's 'sandwich formulation' seems to be based on the findings of Simons and Poinar (1973), who reported that slow desiccation of the IJs of *S. carpocapsae* resulted in the entry of nematodes into partial anhydrobiosis. Bedding and Butler (1994) prepared a formulation by mixing nematode slurry (concentrated nematodes) with the powder of anhydrous polyacrylamide. The nematodes formulated by this method were partially desiccated but with low survival capability at room temperature. Strauch *et al.* (2000) prepared a formulation by mixing the nematode slurry with attapulgite or bentonite clay. The nematodes used in the formulation were *H. bacteriophora* (hybrid strain) and *H. indica* (LN2 strain), which survived for 2 weeks and 1 week, respectively, at 25°C. The survival of both nematodes also varied in sponge and clay at different temperatures. Granular formulation of nematodes was first developed by Capinera and Hibbard (1987), wherein the nematodes were partially encapsulated in lucerne meal and wheat flour. Six years later, Connick *et al.* (1993) described an extruded or formed granule. In this method, the nematodes were distributed throughout a wheat gluten matrix. The granules were dried to lower the moisture in order to check nematode migration and reduce the risk of contamination. But the method resulted in poor nematode survival as the granules rapidly dried out during storage. A significant progress in this direction was achieved when water-dispersible granules (WDGs) were used to encase the IJs of EPNs. The granules used in this method were of a diameter ranging between 10 and 20 mm, and consisted of a mixture of various types of silica, clays, cellulose, lignin and starches (Silver *et al.*, 1995). The water activity of the formulations was controlled in order to induce partial anhydrobiosis (Silver *et al.*, 1995; Grewal, 2000b). Water activity is a measure of how tightly the water is bound, structurally or chemically, to the nematodes (Sharma *et al.*, 2011). The use of WDGs increases the storage ability of IJs to several months at 15–25°C, besides enhancing the tolerance capacity of the nematodes to extreme temperature, enabling easier and less expensive transport and labour-intensive preparation steps. Above all, this method is cost-effective (Grewal, 2000b, 2002).

The shelf life of EPN-based products depends on the formulation and nematode species. Grewal (2002) reported the storage of *S. carpocapsae* for 3–4 months at 25°C and *S. feltiae* for 2–4 weeks in alginate gel formulations. In gel formulations, the mobility of EPNs becomes reduced, but under such conditions they can remain infective for up to 9 months (Ganguly *et al.*, 2008). However, no EPN formulation has been reported to meet the 2-year shelf-life requirement of a standard chemical pesticide (Table 13.2). Under various refrigeration ranges, actively moving nematodes can remain alive and infective for 1–6 months (Fig. 13.7). Dreves and Lee (2015) reported the storage temperature of formulated nematode varying between 39–46°F for Steinernematids and 50–60°F for Heterorhabditids. Umamaheswari *et al.* (2006) formulated three EPNs, namely *S. siamkayai*, *S. glaseri* and *H. indica*, in alginate gel capsules in an attempt to test their storage and infectivity under *in vitro* conditions at two different temperatures, i.e. 5°C and 25°C, against the larva of rice moth, *C. cephalonica*. The survival of *S. glaseri* was observed up to 24 weeks, but there was 100% survival and infectivity up to 4 weeks at 5°C. The survival of *S. siamkayai* was up to 22 and 14 weeks at 5°C and 25°C, respectively. The infectivity of *H. indica* was 100% up to 2 weeks at 5°C and 25°C. Accumulation of trehalose at low temperature is very common among EPNs (Qiu and Bedding, 1999) and may be a component of survival strategy during environmental stress (Gaugler, 2002). Chen and Glazer (2005), in an attempt to enhance the storage ability of nematodes at 23°C in distilled water containing an osmotic solution of 18% glycerol and 2% sodium alginate was dropped in 0.5% $CaCO_3.2H_2O$, which led to the formation of calcium alginate granules. The granules thus formed were placed in 5 cm Petri dishes and transformed into sealed 2.5 l desiccator under 85% relative humidity, or sealed into plastic boxes having a high relative humidity due to the presence of a piece of sponge soaked in distilled water. This resulted in 99% survival of nematodes in granules containing osmotic solution in desiccators at 100% relative humidity for up to 6 months. Recently, Kim *et al.* (2015), in an attempt to improve EPN retainment, adjusted capsule properties by changing the reaction temperature for capsule formation and by adopting post-treatment of alginate capsules with excessive Ca^{2+}. Such manufactured capsules were evaluated for EPN escape under different conditions. The hardest capsule shells were produced at 4°C with the Ca^{2+} post-treatment. The retainment of EPNs in the capsule was enhanced by 40%, only lowering the reaction temperature. Since nematode metabolism is governed by temperature, they can be immobilized by maintaining them in aqueous suspension at low temperature ranging between 5 and 15°C (Hussaini, 2002). It has been reported that the storage temperature range of steinernematids is 5–10°C, whereas for heterorhabditids it is 10–15°C (Georgis 1990a). The metabolic activities of nematode increases at 20–30°C, but on the other hand viability of nematode decreases (Georgis, 1990b). The viability of IJs depends on retaining enough water inside and around their bodies (Chen and Glazer, 2005). Therefore, continuous refrigeration is required to maintain the viability of IJs when carried on moist substrates such as sponge, vermiculite and peat (Georgis, 1990a). Fan and Hominick (1991) recorded a positive influence of cold storage on the survival of IJs. However, it has also been observed that exhaustion of energy reserves takes place much faster as storage time increases (Jung, 1996; Georgis and Kaya, 1998). Wilson and Ivanova (2004) described a non-viscous, non-adhesive and non-toxic liquid formulation for EPN storage and transport based on neutral density colloidal silica suspensions. The pathogenicity and survival of stored nematodes in this formulation without aeration was better than those stored in aerated quarter-strength Ringer's solution. Shapiro-Ilan *et al.* (2002) formulated the nematode infected insect cadavers in starches and clays, which resulted in more resistance to rupturing and sticking during agitation in comparison to non-formulated cadavers. Nematode-infected insect cadavers have also been found showing resistance to freezing (Lewis and Shapiro-Ilan, 2002). A novel method for the effective use of EPNs is their formulation in tape-enclosed cadavers (Shapiro-Ilan *et al.*, 2010). However, in all the formulations, an important consideration is the guaranteed survival of EPNs up to the period required to market the product. Such stable formulations are generally achieved by immobilizing and/or partially desiccating IJs (Hussein and Abdel-Atey, 2012), which allows the EPN products to reach the market segments with an acceptable shelf life (Grewal, 2000a).

Thus, the shelf life of nematodes in a formulated product can be enhanced by reducing nematode activity and metabolism through physical trapping, metabolic inhibition, cold storage, or by the induction of anhydrobiosis (Grewal, 2002). The shelf life can also be improved by genetic selection approaches, which may also lead to an improved persistence of these beneficial organisms in the soil.

13.4 Quality

The main hindrance in successful utilization of as biocontrol agents is the variable quantity and quality of the nematodes in the commercial products (Gaugler *et al.*, 2000). It is a well-known fact that

Table 13.2. Formulation of entomopathogenic nematodes with expected shelf life. Source: (Grewal and Peters, 2005).

Formulation	Nematode species	Strain	Shelf life (months)	
			22–25°C	2–10°C
Wettable powder	*Steinernema carpocapsae*	All	2.0–3.5	6.0–8.0
	S. feltiae	UK	2.5–3.0	5.0–6.0
		ENO2	0.5–1.0	3.0–4.0
	S. glaseri	NJ43	0.03–0.06	1.0–1.5
	Heterorhabditis indica	LN2	0.25–0.50	0
	H. bacteriophora	ENO1	0.5–1.0	2.0–3.0
Vermiculite	*S. carpocapsae*	All	0.1–0.2	5.0–6.0
	S. feltiae	SN	0.03–1.0	4.0–5.0
Water-dispersible granules	*S. carpocapsae*	All	4.0–5.0	9.0–12.0
	S. feltiae	SN	1.5–2.0	5.0–7.0
Alginate gel	*S. carpocapsae*	All	3.0–4.0	6.0–9.0
	S. feltiae	SN	0.5	4.0–5.0
Flowable gel	*S. carpocapsae*	All	1.0–1.5	3.0–4.0
	S. glaseri	NJ43	0.16–0.2	0.4–0.5
Liquid concentrate	*S. carpocapsae*	All	0.16–0.2	0.4–0.5
	S. riobrave	RGV	0.1–0.13	0.23–0.03

Fig. 13.7. Petri dishes with a suspension of infective juveniles in distilled water, placed on different shelves of a refrigerator for storage.

regardless of the type of procedure adopted in formulations, the quality of IJs declines with time (Askary, 2010). The quality of EPNs during their storage in water is influenced by high oxygen demand, sensitivity to low temperature, susceptibility to microbial contamination and the toxicity of antimicrobial agents (Grewal, 2002). The virulence capability of nematodes is assessed by conducting insect bioassays, whereas shelf life is predicted from the storage energy lipid reserve of the IJs. In extreme environmental conditions, the quality and shelf life, as well as the field performance of an EPN product are very likely to be affected (Ehlers *et al.*, 2005). The first quality control method was developed by Miller in 1989. This was an attempt to assess the pathogenic capability of commercially produced *S. carpocapsae.* The method is described as a one-on-one *G. mellonella* bioassay (Converse and Miller, 1999). However, the quality of commercially produced EPNs was reported for the first time by Gaugler *et al.* (2000). This was a significant contribution that brought awareness of the effective quality control of EPNs during commercialization. Researchers from different laboratories of the world have reported that nematodes produced by *in vivo* techniques are superior to those produced by *in vitro* techniques in terms of quality and field efficacy (Gaugler and Georgis, 1991; Yang *et al.*, 1997; Abu Hatab *et al.*, 1998; Abu Hatab and Gaugler, 1999; Shapiro-Ilan and McCoy, 2000). It has sometimes been observed that nematodes escape from the inert carriers and dry out. To prevent this, such formulations have been developed wherein the mobility and metabolism of nematodes is minimized by physical trapping; by the inclusion of metabolic inhibitors or by the induction of physical anhyrobiosis (Grewal, 2012). For a routine quality assessment of EPN species at low concentrations, the sand-well method was developed by Grewal *et al.* (1999).

It has generally been found that IJs in any formulation are susceptible to temperature extremes, ultraviolet light, anoxic conditions and contamination (Lewis and Perez, 2004). The parameters involved in manufacturing a quality EPN product are correct identity of species, total number of live nematodes, ratio of live and dead nematodes, matching of host-finding behaviour to the target pest, age of nematodes used in the formulation, virulence and reproduction ability of nematodes in the target pest, storability, tolerance to temperature extremes and cold or warm temperature activity. Other requirements are size and packaging, reliable instructions for the consumers, ease of transportation, free from contaminants, price, availability and efficacy under field conditions (Grewal and Peters, 2005).

Nematologists working in different laboratories are now paying attention to molecular approaches to improve the quality of EPNs by genetic modifications. Gaugler *et al.* (1997) demonstrated the approach of improvement in heat tolerance in populations of *H. bacteriophora*. The trait was transformed to *H. bacteriophora* by using a heat shock protein that originated from *Caenorhabditis elegans*. The deterioration trait in EPNs can also be checked by creating homozygous inbred lines (Bai *et al.*, 2005; Shapiro-Ilan *et al.*, 2014b). The frequent subculturing of nematodes in the laboratory results in a deterioration in beneficial traits. These traits may be in the form of the virulence capacity, environmental tolerance, reproductive capacity and host-finding ability of a particular species (Wang and Grewal, 2002; Bai *et al.*, 2005; Bilgrami *et al.*, 2006). The two factors responsible for trait deterioration are: (i) genetic, such as inbreeding drift, inadvertent selection; or (ii) non-genetic, such as disease or nutrition (Tanada and Kaya, 1993; Hopper *et al.*, 1993; Chaston *et al.*, 2011). The 'traits lost' occur both in nematodes and their associated/symbiont bacteria (Bilgrami *et al.*, 2006; Wang *et al.*, 2007). The causes of loss in traits are genetically based (in part), with inbreeding depression being more prominent an issue than inadvertent selection (Bai *et al.*, 2005; Adhikari *et al.*, 2009; Chaston *et al.*, 2011). Shapiro-Ilan *et al.* (1997) were the first to be successful in transferring beneficial traits, i.e. heat tolerance, from one *H. bacteriophora* population to another using a non-molecular approach, i.e. hybridization. Mukuka *et al.* (2010) combined two molecular approaches, selection and hybridization, to develop EPN populations with superior environmental tolerance.

13.5 Application Technology

The application of EPNs (IJs) is done mainly for the control of larval or pupal stages of insect pests in the soil, plant surface or cryptic habitats. The prime care during application should be minimum loss of IJs from sprayer tank to target insect (Askary, 2010). EPNs can be applied by using the equipment of other control agents; however, in order to achieve a cost-effective control, techniques required for their application should be optimized (Wright *et al.*, 2005). Nearly all agronomic and horticultural ground equipment, such as mist blowers, pressurized sprayers, electrostatic sprayers or aerial sprays, can be used in the application of EPNs (Wright *et al.*, 2005; Shapiro-Ilan *et al.*, 2006). While handling application equipment, the points to be taken into consideration are volume, nozzle type, agitation, pressure, recycling time, system environmental conditions and spray distribution pattern (Lara *et al.*, 2008). In most cases, spray volume ranging between 750 and 1900 l/ha has been found sufficient to reach the target insect pest in the soil (Grewal, 2002). During the pulverization process, the tank mixture should be stirred in order to make it homogenized. Properly homogenized tank mixture results in the better application of EPNs, due to proper distribution of the active ingredient. But on the other hand, friction with the inner parts of the hydraulic circuit may cause physical damage to the nematodes (Garcia *et al.*, 2005). Fife *et al.* (2004) are of the opinion that the cuticle structure of IJs can vary among species, and therefore, IJs can respond differently to hydrodynamic stress. Also, due to stirring, the temperature of the tank mixture increases, which ultimately results in reducing the viability of IJs. An increase in temperature during pump recirculation may prove more detrimental to EPNs than mechanical stress (Fife *et al.*, 2007). Therefore, pumps with lower capacity, such as diaphragm or roller pumps, are considered better for application. Lara *et al.* (2008) reported that the temperature of the spray tank mixture in a tank can increase from 22 to 43°C in 1 hour agitation with a centrifugal pump, and from 22 to 27°C in a tank with a diaphragm or peristaltic pumps. Grewal (2002) is of the opinion that if the temperature rises above 30°C in the application equipment for nematodes, it should be avoided. Nilsson and Gripwall (1999) reported a 10% reduction in the viability of *S. feltiae* when the suspension was sprayed with a high-pressure spray pump piston after 20 min of stirring. They concluded that the reduction in viability was due to mechanical stress, the mesh used (50 mesh) or by the increase in temperature from 19 to 23°C, although there was no reduction in nematode viability when a manual backpack sprayer was used and the suspension was not stirred. Hence, temperature and stirring of the tank mixture are also important factors that need to be considered during nematode application.

The droplet size of the spraying nematode suspension is optimized on the basis of the characteristics of the target pest. Droplet size depends on multiple factors, which include the model of the spray nozzle, the spray pressure used, the opening angle and the filter mesh (Camara *et al.*, 2008). It has been reported that the size of the droplet should be >200 μm to control the target pest present in soil (Cunha *et al.*, 2003). Small droplets lower the final velocity (Miller *et al.*, 2008), which increases the deposition time on the target pest, thus making it more susceptible to evaporation and change of trajectory. In a small plot area, water cans or backpack sprayers may be appropriate, whereas for larger plots boom sprayers should be considered for application. Air induction nozzles produce large droplets (Matthews, 2000) and are recommended for horizontal targets such as soil pests. Settling or sedimentation of IJs in the spray tank should also be taken into account during application. If the density of IJs in the spray tank is 1.05 g/cm^2, it will become heavier than the water and settle down (Wright *et al.*, 2005). IJs bigger in size settle faster than smaller ones. The distribution of nematodes becomes unequal due to sedimentation, particularly when applied under an irrigation system. However, the sedimentation speed of IJs can be made slower by increasing the viscosity of the water by adding carboxymethyl cellulose (Peters and Backes, 2003). Furthermore, the viability of nematodes is size dependent (Fife *et al.*, 2005). IJs belonging to the genus *Steinernema* are generally larger in size than *Heterorhabditis*. The medium size of *S. feltiae* has been reported as 849 μm, whereas for *H. megidis* it is 768 μm (Adams and Nguyen, 2002). This is the reason why cone nozzles are recommended over fan nozzles, because the latter can cause damage to EPNs. Moreira *et al.* (2013) evaluated different models of spray nozzles with air induction; three spray pressures (207, 413 and 720 kPa), four different additives for tank mixtures (cane molasses, mineral oil, vegetable oil and glycerin) and the influence of tank mixture stirring time to observe their effect on the viability and pathogenicity of *S. feltiae*. It was concluded that vegetable oil, mineral oil and molasses were compatible adjuvants for *S. feltiae*. The different nozzles, at pressures of up to 620 kPa, were compatible with *S. feltiae*, and the stirring of nematodes in a motorized backpack sprayer for 30 min did not cause any negative impact on the viability or pathogenicity of the nematodes. Fife *et al.* (2003) evaluated the survival of three different aged nematode species, i.e. *S. carpocapsae*, *H. bacteriophora* and *H. megidis*, in suspensions by using different pressures (1283–10,690 kPa). It was observed that nematode species differed in tolerance to high pressure: the maximum operate pressure recommended was 1380 kPa. However, the age of the nematodes did not show any impact of pressure tolerance for up to 3 weeks. Garcia *et al.* (2005) found more than 80% viability in *S. glaseri*, using pressures up to 1379 kPa during spraying. In an experiment, *S. feltiae* was sprayed at different pressures, but no significant difference in the viability of nematodes was observed at pressures up to 2000 kPa (Nilsson and Gripwall, 1999). Pressure ranging between 200 and 600 kPa is ideal for spraying. If spraying is performed at higher pressure, it may result in a finer and more airborne mist, increase wear on the nozzle material and increase energy consumption during the spraying operation. Therefore, spraying at lower pressure is suggested as it is environmentally safe and aids in conserving both resources and equipment.

It has been observed that the inherent limitations with biocontrol agents are their inconsistency and slow performance in comparison to chemicals (Askary, 2015). The success of EPNs depends on their virulence capability, host finding/foraging ability, environment tolerance, as well as persistence. The virulence capability of a single nematode strain has also been reported to vary with respect to different insect host and inoculum levels used (Khan *et al.*, 2007a,b). The post-application survival of an EPN is a major obstacle in expanding its use as a substitute to chemicals. It has also been noted that the post-application efficacy (of a higher level) of an EPN generally persists for only 2–6 weeks, and therefore the EPN is applied in an inundative or pesticidal manner, where almost no recycling is expected and reapplication is required annually or on seasonal basis (Shapiro-Ilan *et al.*, 2014a).

Several environmental factors are also responsible to limit the survival of IJs in soil (Glazer, 1996). The abiotic factors on which the successful application of EPNs depend include temperature, ultraviolet radiation, soil moisture, soil texture, optimum soil pH range (4–8) and relative humidity. Since nematodes are sensitive to desiccation, adjuvants are therefore used in the suspension to be applied for spraying. Adjuvants improve the characteristics of the spray and change the surface tension of the sprayed liquid, which consequently slows down the evaporation process (Cunha *et al.*, 2010). It is better to prefer adjuvants of natural origin, such as sugarcane molasses or vegetable oil, because petroleum-based adjuvants, like glycerin and mineral oil, are liable to cause groundwater and environmental contamination. Temperature is another factor that influences the motility and survival of IJs, and consequently the rate at which food reserves are utilized (Molyneux, 1984). However, the foremost consideration should be that appropriate nematode species must match the target pest species. Different species and strains of EPNs differ in having an optimum temperature for infection and reproduction (Grewal *et al.*, 1994). The relatively heat-tolerant nematodes are *S. glaseri*, *S. riobrave* and *H. indica*, whereas cold tolerant are *S. feltiae* and *H. megidis* (Kung *et al.*, 1991; Grewal *et al.*, 1994; Berry *et al.*, 1997; Shapiro-Ilan and McCoy, 2000). The half-time of an EPN is between a few days and 1 month (Strong, 2002). Smits (1996) reported that after the inundative release of nematodes at 0.5 million/m^2, the population density of nematodes declined rapidly, followed by a period of about 2 weeks with lower rates of decline. After this period, the nematode population reached background levels of about 10,000/m^2. Therefore, EPNs are needed to reproduce in a suitable host in order to establish and cause a long-term effect in insect pest control, although it is still debatable as to how long the nematodes can survive in soil without a host. The population density of EPNs is often correlated with insect population density, which, on the other hand, is often dependent on the food resources available (Strong, 2002). Thus, recycling of EPNs in the host insect pest determines their density and distribution in the field.

IJs mixed with water are applied with the help of commercially available ground or aerial spray equipment like mist blowers, sprayers, drip or sprinkler irrigation systems. Drip irrigation is more effective against soil-dwelling pests. Light soil irrigation prior to the application of IJs yields the best results, as this may help the nematodes to remain active and enable them to move easily through the soil pores. There should be adequate soil moisture, even after several days of IJ application. The persistence of IJs is enhanced when infected host cadavers are applied (Shapiro-Ilan *et al.*, 2001). The application of nematodes in capsules with insect attractant (also called 'lure and kill approach') may result in reducing the number of nematodes necessary for insect pest control (Hiltpold *et al.*, 2012).

The following are recommendations for applying a commercial product of EPNs in the field:

1. The EPN supplier/distributor should be given the order only a few days prior to the application of EPNs. Shipping of EPN products by overnight delivery and their application within 1–2 days after arrival is expected to provide a satisfactory result.
2. The expiry date of the commercial product of EPNs needs to be checked before application.
3. EPNs should be kept in a cooler temperature until ready for use. The seller should store the product in a refrigerator and the buyer/user should also do the same until ready for application. However, storage for a longer period will increase the shelf life but will, on the other hand, decrease the infectivity of the juveniles.
4. The nematode species to be applied should be selected precisely and matched with the most susceptible pest stage.
5. Before application, the product should be unpacked to check the colour and odour of the nematodes. The nematodes on a sponge formulation will appear as a light tan or grey paste, whereas those on vermiculite or liquid suspension will not be discernible from the carrier material (Miles *et al.*, 2012). In the case of live nematodes, the odour of the product in the container will be mild, while a strong smell like ammonia indicates dead nematodes.
6. To check the viability of the nematodes, a little amount of sponge- or vermiculite-based product should be first soaked in water and a droplet of this sample should be placed on a slide or small, clear glass bowl. The sample is viewed with 15X hand lens or microscope. The motile nematodes will appear as an 'S' shape, live but not motile as a 'J'

shape (*S. carpocapsae*). The dead nematode will appear as straight and unmoving (Miles *et al.*, 2012).

7. Soil temperature should be checked before application of EPNs. It should be >55°F but <90°F (Dreves and Lee, 2015).

8. EPNs require a thin film of water around soil particles to move through the soil particle in search of a host. Therefore, the soil should be kept moist by a light irrigation before and after the application of EPNs in the soil. Light irrigation just after the application of EPNs helps in washing the nematodes into the soil and decreasing the surface temperature of the soil. Also, the soil should not be over-irrigated, as nematode activity is inhibited in saturated soil due to lack of oxygen.

9. An appropriate method to use EPNs is to apply them in the morning or evening hours so that the nematode may be protected from desiccation and ultraviolet rays.

10. A recent report by Shapiro-Ilan *et al.* (2015) in the context of enhancing the efficacy of nematodes is the use of a gel that can be tank-mixed directly with the nematodes at a lower concentration (<2% prevents nozzle clogging). This helps to protect the EPNs from desiccation (in lieu of irrigation) and ultraviolet rays, when applied to the soil surface.

11. When the application of EPNs is to be done by sprayer, fine-meshed filters (<50 mesh) should be removed and the nozzle of the sprayer should be checked for any clogging during application.

12. The spray equipment should be clean and free from pesticide residues.

13. In the case of the combined application of nematodes with insecticides/chemical fertilizers, the compatibility of the nematode should be tested with the particular insecticide/chemical fertilizer that is to be used. It is better to follow the application instructions given on the EPN product label regarding the specific listing of chemicals that are toxic to nematodes.

14. Nematodes in a spray tank without agitation tend to sediment fast. Therefore, EPN suspension in a sprayer or watering can should be agitated from time to time in order to keep the nematodes in suspension. However, the sprayer tank should not be given excessive agitation as this may cause damage to the nematodes and decrease their viability. Since EPNs are sensitive to temperature, mechanical stress due to repeated recirculation may also lead to a rise in the temperature of the pump and nozzle, which if above 86°F proves lethal to the nematodes.

15. The field wherein EPNs have been applied needs to be kept moist for at least 7 days to obtain a satisfactory result.

Foliar application of IJs requires some careful consideration, as both solar ultraviolet radiation and wind are detrimental due to which the nematodes may fail to remain active on site, and also it may result in rapid desiccation of the nematodes. To combat this situation, anti-desiccants are added to the suspension used for foliar sprays. This method has proven helpful in retarding evaporation in the suspension, and ultimately in preventing the desiccation of nematodes (Glazer and Navon, 1990). Schroer and Ehlers (2005) reported that the addition of surfactant and polymer to the nematode suspension greatly increased the efficacy of nematodes in controlling the population of diamondback moth. Recently, Rezaei *et al.* (2015) found a promising result when evaluating the two EPNs, *S. feltiae* and *H. bacteriophora*, in combination with an adjuvant Triton X-100 (0.1% v/v), against greenhouse whitefly, *Trialeurodes vaporariorum*, infesting cucumber and pepper. Using leaf flooding and the addition of surfactants to increase leaf coverage resulted in increasing the efficacy of the EPNs (Williams and Walters, 2000; Head *et al.*, 2004). The addition of sprayable fire gel (the gel is commonly used for protecting structures from fire) and wood flour foam enhanced the efficacy of EPNs when applied to apple tree trunks for controlling codling moth, *Cydia pomonella* (Lacey *et al.*, 2010). Zeba, a formulation of super absorbent polymer, when combined with *H. zealandica* enhanced the survival and infection ability of the nematodes in the field when applied against insect pests of cryptic habitats (de Waal *et al.*, 2013).

High humidity and calm atmosphere supports nematode activity when applied as a foliar spray against insect pests. The two important considerations prior to foliage application of IJs are: (i) size of the droplet; and (ii) spray distribution system (Grewal, 2002). Depending on the EPN species, the nozzle openings of the sprayer for the application of IJs should not be <50 microns, and screens in the system should have an opening of at least 50 mesh or larger. This will make an easy and free passage of nematodes through the system. Then nematodes can withstand application pressures of approximately 300 psi, although operating pressures ranging between 20 and 60 psi are considered sufficient (Dreves and Lee, 2015). Solid cone nozzles and flat fan nozzles lead to a greater number of EPNs being deposited on leaves, which in turn results in higher mortality of the target insect pests (Lello *et al.*, 1996). Brusselman (2011) reported that spray volume is an important parameter that has an effect on EPN infectivity. The effects can vary from nematode death, loss of nematodes, difficulties in keeping the nematodes in suspension while in the spray tank to reach the target pest and covering the target site evenly. The persistence of EPNs is very low on exposed surfaces and to apply them in an excessive amount is also expensive. On the basis of these shortcomings, and other related experimental findings, it has been observed that foliar applications are less successful as compared to soil applications, due to the susceptibility of IJs to desiccation and ultraviolet rays. However, frequent low rate nematode application on foliage may provide a good result in suppressing greenhouse pests such as thrips (Shapiro-Ilan *et al.*, 2006). Hence, further research is required in this direction to find the most appropriate technique of nematode application so that the maximum number of nematodes is delivered efficiently on the target site with minimum loss in the process.

13.6 Conclusion and Future Prospects

EPNs are excellent biocontrol agents that can play an important role in sustainable farming; however, their exploitation is still limited to home gardening or small-scale farming. The hindrances to utilizing EPNs in large-scale farming are the lack of cost-effectiveness and durable formulations, as well as appropriate application methods (Kim *et al.*, 2015). Mass multiplication of EPNs to meet market demand requires both *in vivo* and *in vitro* technologies (*in vitro* solid culture method and *in vitro* liquid culture method). The critical issues encountered during the mass production of EPNs are capital outlay, technical expertise, nematode quality, and labour and cost-efficiency (Shapiro-Ilan *et al.*, 2014a). As discussed in the above review, the *in vivo* technique requires lower capital investment and expertise in comparison to the *in vitro* technique (whether solid or liquid), and is therefore most suitable for small start-up companies, but particularly in those countries or regions where labourers are cheap and easily available. Contrary to this, the *in vitro* liquid technique requires high start-up costs, or a high capital investment in the beginning, good expertise and the requirement for labour is low in comparison to the *in vivo* technique of EPN production. However, the *in vitro* solid technique stands midway between the *in vivo* and *in vitro* liquid techniques in terms of capital investment, technical expertise, labour requirement and economy of scale (cost-efficiency). One major drawback in this production system is the contamination of nematodes during upstream and downstream processing, which is difficult to monitor online (Shapiro-Ilan *et al.*, 2014a). The cost-efficiency of nematode production is highest in the *in vitro* liquid method, followed by the *in vitro* solid method. It is lowest in the *in vivo* method. Unlike the *in vitro* technique, where the nematodes are produced cheaply and commercially on a large scale and can be applied in large field areas to suppress crop insect pests, the production of nematodes by the *in vivo* technique is comparatively less cheaper and on a low scale, and therefore is suitable for nursery soil treatment or application in a small plot. In spite of this, the *in vivo* technique will likely continue, as it is appropriate for certain niche markets and can be sustained as a cottage industry (Shapiro-Ilan *et al.*, 2014a). Recently, the *in vitro* liquid method has globally been the most popular method of nematode production, as it is the most economical approach and is capable of accomplishing the task of crop pest management commercially on a large scale, but simultaneously the shortcomings of this method are its success, which is more inclined to *Heterorhabditis* because inbred lines for *Steinernema* need to be outside of liquid culture to avoid free mating. Besides, quality is always in question for the nematodes produced by this method. However, developing countries still adopt the *in vitro* solid method (Ehlers *et al.*, 2000).

Undoubtedly, it is a challenging task to use EPNs in the biological control of crop insect pests, because the potentiality of IJs depends on a series of factors, the most important being temperature, formulation, quality (viability, i.e. live and active IJs), anhydrobiosis, shelf life and finally the application technology, which is still under development (Piggot and Wardlow, 2002). The technologies used by researchers for mass production, formulation, storage and transportation have been developed all over the world with the passage of time, but a considerable level of information on the biology and potential efficacy of indigenous strains is still needed. Indigenous strains have the capability to adapt to local environmental conditions. Besides, virulence and recycling potential within the host may also be higher (Istkhar *et al.*, 2016). The identification of virulent strains by conducting bioassays under laboratory conditions and evaluating them for their efficacy against crop insect pests under simulated field conditions to learn of the difference between EPN species and isolates would help to discover the best one to be used in crop insect pest management programmes. Manufacturers must find an approach that can reduce the deterioration of beneficial traits and aid the production of high-quality nematodes. The production of EPNs in liquid culture bioreactors and their application with conventional sprayers makes them a good candidate for large-scale manufacturing at reasonable cost. However, anhydrobiosis and the poor survival of EPNs when stored at room temperature are hindrances to the realization of this beneficial microorganism as a potential biopesticide. Since IJs depend on the reserve food present in their body, which therefore prolongs their shelf life in commercial EPN-based products, the role of energy conservation in IJs becomes vital (Patel and Wright, 1997a,b; Patel *et al.*, 1997; Qiu *et al.*, 2000). Finally, it is also the responsibility of distributors/suppliers to ensure that the EPN products that are being supplied and received by end-users are in good quality condition so that performance failure of the products may be avoided. This can be solved by the development of a rapid evaluation technique so that the quality of the EPN can be judged quickly in advance by the end-user, and thus as to whether or not the product should be applied. An explicit instruction to end-users with contact information (if delivered by EPN product distributors) for returning damaged products would prove helpful in contacting the companies regarding problems in shipments and to expedite receipt of a new product (Caamano *et al.*, 2008). Since EPN quality can only be judged by their performance under field conditions prior to laboratory evaluation, it therefore becomes an uphill task for commercial producers to assure EPN users that the product they are supplying will surely control the targeted pests effectively. In the case of the EPN product failing to perform under field conditions as per the expectation of the end-users, they may go for another alternative, and that may more likely be chemical pesticides. Under such circumstances, it will also be very difficult to convince them to use EPN products again to combat insect pest problem in their fields. Hence, further research should emphasize enhanced shelf life, resistance of IJs to environmental extremes, the environmental conditions in different formulation materials, as well as proper packaging technology, so as to maintain the viability of IJs during transportation. In addition, the economics of the systems described above should always be on the priority list of manufacturers. Keeping these considerations in mind, a high-quality EPN product can be prepared and commercialized.

Acknowledgements

The work was supported in part by the All India Coordinated Research Project (AICRP), ICAR, on Biological Control of Crop Pests and Weeds running in the Division of Entomology, SKUAST-K, Srinagar, India. We are also thankful to Dr Mahaveer P. Sharma, Directorate of Soybean Research, Indore, India, for providing some important information related to the topic.

References

Abe, Y. (1987) Culture of *Steinernema feltiae* on bran media. *Japanese Journal of Nematology* 17, 31–34.

Abu Hatab, M. and Gaugler, R. (1997) Growth-mediated variations in fatty acids of *Xenorhabdus* sp. *Journal of Applied Microbiology* 82, 351–358.

Abu Hatab, M. and Gaugler, R. (1999) Lipids of *in vivo* and *in vitro* cultured *Heterorhabditis bacteriophora. Biological Control* 15, 113–118.

Abu Hatab, M. and Gaugler, R. (2001) Diet composition and lipids of *in vitro*-produced *Heterorhabditis bacteriophora. Biological Control* 20, 1–7.

Abu Hatab, M., Gaugler, R. and Ehlers, R.-U. (1998) Influence of culture method on *Steinernema glaseri* lipids. *Journal of Parasitology* 84, 215–221.

Adams, B.J. and Nguyen, K.B. (2002) Taxonomy and systematics. In: Gaugler, R. (ed.) *Entomopathogenic Nematology*. CAB International, Wallingford, UK, pp. 1–34.

Adhikari, B.N., Chin-Yo, L., Xiaodong, B., Ciche, T.A., Grewal, P.S., *et al.* (2009) Transcriptional profiling of trait deterioration in the insect pathogenic nematode *Heterorhabditis bacteriophora. BMC Genomics* 10, 609.

Ali, S.S., Ahmad, R., Hussain, M.A. and Parvez, R. (2005) *Pest Management in Pulses through Entomopathogenic Nematodes*. Indian Institute of Pulse Research, Kanpur, India.

Anbesse, S., Sumaya, N.H., Dorfler, A.V., Strauch, S. and Ehlers, R.-U. (2013) Selective breeding for desiccation tolerance in liquid culture provides genetically stable inbred lines of the entomopathogenic nematode *Heterorhabditis bacteriophora. Applied Microbiology and Biotechnology* 97, 731–739.

Askary, T.H. (2010) Nematodes as biocontrol agents. In: Lichtfouse, E. (ed.) *Sociology, Organic Farming,*

Climate Change and Soil Science. Springer, The Netherlands, pp. 347–378.

Askary, T.H. (2015) Limitations, research needs and future prospects in the biological control of phytonematodes. In: Askary, T.H. and Martinelli, P.R.P. (eds) *Biocontrol Agents of Phytonematodes.* CAB International, Wallingford, UK, pp. 446–454.

Askary, T.H. and Martinelli, P.R.P. (2015) *Biocontrol Agents of Phytonematodes.* CAB International, Wallingford, UK.

Bai, C., Shapiro-Ilan, D.I., Gaugler, R. and Yi, S. (2004) Effect of entomopathogenic nematode concentration on survival during cryopreservation in liquid nitrogen. *Journal of Nematology* 36, 281–284.

Bai, C., Shapiro-Ilan, D.I., Gaugler, R. and Hopper, K.R. (2005) Stabilization of beneficial traits in *Heterorhabditis bacteriophora* through creation of inbred lines. *Biological Control* 32, 220–227.

Bedding, R.A. (1981) Low cost *in vitro* mass production of *Neopectana* and *Heterorhabditis* sp. for field control of insect pests. *Nematologica* 27, 109–114.

Bedding, R.A. (1984) Large scale production, storage and transport of insect parasitic nematodes (*Neopectana* sp. and *Heterorhabditis* sp.). *Annals of Applied Biology* 104, 117–120.

Bedding, R.A. (1988) Storage of entomopathogenic nematodes. Patent No WO88/08668.

Bedding, R.A. (1990) Logistics and strategies for introducing entomopathogenic nematode technology into developing countries. In: Gaugler, R. and Kaya, H.K. (eds) *Entomopathogenic Nematodes in Biological Control*. CRC Press, Boca Raton, Florida, pp. 233–248.

Bedding, R.A., and Butler, K.L. (1994) Storage and transport of entomopathogenic nematodes. Australian Patent No 608852.

Bedding, R.A. and Molyneux, A.S. (1982) Penetration of insect cuticle by infective juveniles of *Heterorhabditis* spp. (Heterorhabditidae: Nematoda). *Nematologica* 28, 354–359.

Berry, R.E., Liu, J. and Groth, E. (1997) Efficacy and persistence of *Heterorhabditis marelatus* (Rhabditida: Heterorhabditidae) against root weevils (Coleoptera: Curculionidae) in strawberry. *Environmental Entomology* 26, 465–470.

Bilgrami, A.L., Gaugler, R., Shapiro-Ilan, D.I. and Adams, B.J. (2006) Source of trait deterioration in entomopathogenic nematodes *Heterorhabditis bacteriophora* and *Steinernema carpocapsae* during *in vivo* culture. *Nematology* 8, 397–409.

Blinova, S.L. and Ivanova, E.S. (1987) Culturing the nematode–bacterial complex of *Neoaplectana carpocapsae* in insects. In: Sonin, M.D. (ed.) *Helminths of Insects.* American Publishing Company, New Delhi, pp. 22–26.

Bonifassi, E., Fischer-Le Saux, M., Boemare, N., Lanois, A., Laumond, C. and Smart, G. (1999) Gnotobiological study of infective juveniles and symbionts of *Steinernema scapterisci*: a model to clarify the concept of the natural occurrence of monoxenic associations in entomopathogenic nematodes. *Journal of Invertebrate Pathology* 74, 164–172.

Brown, I., Gaugler, R. and Shapiro-Ilan, D.I. (2004) LOTEK – An improved method for *in vivo* production of entomopathogenic nematodes. *International Journal of Nematology* 14, 9–12.

Brusselman, E. (2011) Optimizing the spray application of entomopathogenic nematodes in vegetables: from spray tank to nematode deposition. Available at: https://biblio.ugent.be/publication/1320275 (accessed on 23 September 2016).

Buecher, E.J. and Hansen, E.L. (1971) Mass culture of axenic nematodes using continuous aeration. *Journal of Nematology* 3, 199–200.

Buecher, E.J. and Popiel, I. (1989) Liquid cultures of the entomogenous nematode, *Steinernema feltiae* with its bacterium symbiont. *Journal of Nematology* 21, 500–504.

Burman, M. and Pye, A.E. (1980) *Neoaplectana carpocapsae*: respiration of infective juveniles. *Nematologica* 26, 214–219.

Caamano, E.X., Cloyd, R.A., Solter, L.F. and Fallon, D.J. (2008) Quality assessment of two commercially available species of entomopathogenic nematodes: *Steinernema feltiae* and *Heterorhabditis indica*. *HortTechnology* 18, 84–89.

Camara, F.T., Santos, J.L., Silva, E.A. and Ferreira, M.C. (2008) Distribuição volumétrica e espectro de gotas de bicos hidráulicos de jato plano de faixa expandida XR11003. *Engenharia Agrícola* 28, 740–749. DOI: 10.1590/s0100-69162008000400014.

Capinera, J.L. and Hibbard, B.E. (1987) Bait formulations of chemical and microbial insecticides for suppression of crop-feeding grasshoppers. *Journal of Agricultural Entomology* 4, 337–344.

Chaston, J.M., Dillman, A.R., Shapiro-Ilan, D.I., Bilgrami, A.L., Gaugler, R., Hopper, K.R. and Adams, B.J. (2011) Outcrossing and crossbreeding recovers deteriorated traits in laboratory cultured *Steinernema carpocapsae* nematodes. *International Journal of Parasitology* 41, 801–809.

Chavarria-Hernandez, N. and de la Torre, M. (2001) Population growth kinetics of the nematode, *Steinernema feltiae*, in submerged monoxenic culture. *Biotechnology Letters* 23, 311–315.

Chavarria-Hernandez, N., Espino-Garcia, J.J., Sanjuan-Galindo, R. and Rodriguez-Hernandez, A.I. (2006) Monoxenic liquid culture of the entomopathogenic nematode *Steinernema carpocapsae* using a culture medium containing whey kinetics and modelling. *Journal of Biotechnology* 25, 75–84.

Chavarria-Hernandez, N., Ortega-Morales, E., Vargas-Torres, A., Chavarria-Hernandez, J. and Rodriguez-Hernandez, A.I. (2010) Submerged monoxenic culture of the entomopathogenic nematode, *Steinernema carpocapsae* CABA 01, in a mechanically agitated bioreactor: evolution of the hydrodynamic and mass transfer conditions. *Biotechnology and Bioprocess Engineering* 15, 580–589.

Chen, S. and Glazer, I. (2005) A novel method for long term storage of the entomopathogenic nematode *Steinernema feltiae* at room temperature. *Biological Control* 32, 104–110.

Cho, C.H., Whang, K.S., Gaugler, R. and Yoo, S.K. (2011) Submerged monoxenic culture medium development for *Heterorhabditis bacteriophora* and its symbiotic bacterium *Photorhabdus luminescens*: protein sources. *Journal of Microbiology and Biotechnology* 21, 869–873.

Connick, W.J., Nickle, W.R. and Vinyard, B.T. (1993) Pesta: new granular formulations for *Steinernema carpocapsae*. *Journal of Nematology* 25, 198–203.

Converse, V. and Miller, R.W. (1999) Development of the one-on-one quality assessment assay for entomopathogenic nematodes. *Journal of Invertebrate Pathology* 74, 143–148.

Cunha, J.P.A.R., Teixeira, M.M., Coury, J.R. and Ferreira, L.R. (2003) Avaliação de estratégias para redução da deriva de agrotóxicos em pulverizações hidráulicas. *Planta Daninha* 21, 325–332.

Cunha, J.P.A.R., Bueno, M.R. and Ferreira, M.C. (2010) Espectro de gotas de pontas de pulverização com adjuvantes de uso agrícola. *Planta Daninha* 28, 1153–1158.

Del Valle, E.E., Dolinski, C., Barreto, E.L.S., Souza, R.M. and Samuels, R.I. (2008) Efficacy of *Heterorhabditis baujardi* LPP7 (Nematoda: Rhabditida) applied in *Galleria mellonella* (Lepidoptera: Pyralidae) insect cadavers to *Conotrachelus psidii*, (Coleoptera: Curculionidae) larvae. *Biocontrol Science and Technology* 18, 33–41.

Del Valle, E.E., Dolinski, C., Barreto, E.L.S. and Souza, R.M. (2009) Effect of cadaver coatings on emergence and infectivity of the entomopathogenic nematode *Heterorhabditis baujardi* LPP7 (Rhabditida: Heterorhabditidae) and the removal of cadavers by ant. *Biological Control* 50, 21–24.

De Waal, J.Y., Malan, A.P. and Addison, M.F. (2013) Effect of humidity and a superabsorbent polymer formulation on the efficacy of *Heterorhabditis zealandica* (Rhabditida: Heterorhabditidae) to control codling moth, *Cydia pomonella* (L.) (Lepidoptera: Tortricidae) *Biocontrol Science and Technology* 23, 62–78.

Deol, Y.S., Jagdale, G.B., Canas, L. and Grewal, P.S. (2011) Delivery of entomopathogenic nematodes directly through commercial growing media via the inclusion of infected host cadavers: a novel approach. *Biological Control* 58, 60–67.

Dowds, B.C.A. and Peters, A. (2002) Virulence mechanisms. In: Gaugler, R. (ed.) *Entomopathogenic Nematology*. CAB International, Wallingford, UK, pp. 79–98.

Dreves, A.J. and Lee, J. (2015) Entomopathogenic nematodes. Available at: http://insect.pnwhandbooks.org/ipm/entomopathogenic-nematodes (accessed 21 September 2016).

Dunphy, G.B. and Webster, J.M. (1989) The monoxenic culture of *Neoaplectana carpocapsae* DD136 and *Heterorhabditis heliothidis*. *Revue de Nématologie* 12, 113–123.

Dutky, S.R., Thompson, J.V. and Cantwell, G.E. (1964) A technique for the mass production of the DD-136 nematode. *Journal of Insect Pathology* 6, 417–422.

Ehlers, R.-U. (2005) Forum on safety and regulation. In: Grewal, P.S., Ehlers, R.-U. and Shapiro-Ilan, D.I. (eds) *Nematodes as Biocontrol Agents*. CAB International, Wallingford, UK, pp. 107–114.

Ehlers, R.-U. and Hokkanen, H.M.T. (1996) Insect biocontrol with non-endemic entomopathogenic nematodes (*Steinernema* and *Heterorhabditis* spp.): conclusions and recommendations of a combined OECD and COST workshop on scientific and regulatory policy issues. *Biocontrol Science and Technology* 6, 295–302.

Ehlers, R.-U. and Shapiro-Ilan, D.I. (2005) Mass production. In: Grewal, P.S., Ehlers, R-U. and Shapiro-Ilan, D.I, (eds) *Nematodes as Biocontrol Agents*. CAB International, Wallingford, UK, pp. 65–78.

Ehlers, R.-U., Wulff, A. and Peters, A. (1997) Pathogenicity of axenic *Steinernema feltiae*, *Xenorhabdus bovienii* and bactohelminthic complex to larvae of *Tipula oleracea* and *Galleria mellonella*. *Journal of Invertebrate Pathology* 69, 212–217.

Ehlers, R.-U., Lunau, S., Krasomil-Osterfeld, K. and Osterfeld, K.H. (1998) Liquid culture of the entomopathogenic nematode–bacterium complex *Heterorhabditis megidis/Photorhabdus luminescens*. *Biocontrol* 43, 77–86.

Ehlers, R.-U., Niemann, I., Hollmer, S., Strauch, O., Jende, D., *et al*. (2000) Mass production potential of the bacto–helminthic biocontrol complex *Heterorhabditis indica–Photorhabdus luminescens*. *Biocontrol Science and Technology* 10, 607–616.

Ehlers, R.-U., Oestergaard, J., Hollmer, S., Wingen, M. and Strauch, O. (2005) Genetic selection for heat tolerance and low temperature activity of the entomopathogenic nematode–bacterium complex *Heterorhabditis bacteriophora–Photorhabdus luminescens*. *Biological Control* (Dordrecht) 50, 699–716.

Fan, X. and Hominick, W.M. (1991) Effects of low storage temperature on survival and infectivity of two *Steinernema* species (Nematoda: Steinernematidae). *Revue de Nématologie* 14, 407–412.

Fife, J.P., Derksen, R.C., Ozkan, H.E. and Grewal, P.S. (2003) Effects of pressure differentials on the viability and infectivity of entomopathogenic nematodes. *Biological Control* 27, 65–72.

Fife, J.P., Derksen, R.C., Ozkan, H.E., Grewal, P.S., Chalmers, J.J. and Krause, C.R. (2004) Evaluation of a contraction flow field on hydrodynamic damage to entomopathogenic nematodes – a biological pest control agent. *Biotechnology and Bioengineering* 86, 96–107.

Fife, J.P., Ozkan, H.E., Derksen, R.C., Grewal, P.S. and Krause, C.R. (2005) Viability of a biological pest control agent through hydraulic nozzles. *Transactions of the American Society of Agricultural Engineers* 48, 45–54.

Fife, J.P., Ozkan, H.E., Derksen, R.C. and Grewal, P.S. (2007) Effects of pumping on entomopathogenic nematodes and temperature increase within a spray system. *Applied Engineering in Agriculture* 23, 405–412.

Flanders, K.L., Miller, J.M. and Shields, E.J. (1996) *In vivo* production of *Heterorhabditis bacteriophora* 'Oswego' (Rhabditida: Heterorhabditidae), a potential biocontrol agent for soil inhabiting insects in temperate regions. *Journal of Economic Entomology* 89, 373–380.

Friedman, M.J. (1990) Commercial production and development. In: Gaugler, R. and Kaya, H.K. (eds) *Entomopathogenic Nematodes in Biological Control*. CRC Press, Boca Raton, Florida, pp. 153–172.

Friedman, M.J., Langston, S. and Pollitt, S. (1989) Mass production in liquid culture of insect killing nematodes. International Patent No W0 89/04602, US Patent 5,023,183.

Ganguly, S., Anupama Kumar, A. and Parmar, B.S. (2008) Nemagel – a formulation of the entomopathogenic nematode *Steinernema thermophilum* mitigating the shelf-life constraint of the tropics. *Nematologia Mediterranea* 36, 125–130.

Garcia, L.C., Raetano, C.G., Wilcken, S.R.S., Ramos, H.H., Leite, L.G., *et al.* (2005) Pressurização da calda de pulverização na viabilidade de microrganismos entomopatogênicos. *Engenharia Agrícola* 25, 783–790.

Gaugler, R. (1997) Alternative paradigms for biopesticides. *Phytoparasitica* 25, 179–182.

Gaugler, R. (2002) *Entomopathogenic Nematology*. CAB International, Wallingford, UK, 388 pp.

Gaugler, R. and Brown, I. (2001) LOTEK – An automated apparatus for production of insecticidal nematodes. USSN 09/845.816.

Gaugler, R. and Georgis, R. (1991) Culture method and efficacy of entomopathogenic nematodes (Rhabditida: Steinernematidae and Heterorhabditidae). *Biological Control* 1, 269–274.

Gaugler, R. and Han, R. (2002) Production technology. In: Gaugler, R. (ed.) *Entomopathogenic Nematology*. CAB International, Wallingford, UK, pp. 289–310.

Gaugler, R., Campbell, J.F., Selvan, S. and Lewis, E.E. (1992) Large-scale inoculative releases of the entomopathogenic nematode *Steinernema glaseri*: assessment 50 years later. *Biological Control* 2, 181–187.

Gaugler, R., Wilson, M. and Shearer, P. (1997) Field release and environmental fate of a transgenic entomopathogenic nematode. *Biological Control* 9, 75–80.

Gaugler, R., Grewal, P.S., Kaya, H.K. and Smith-Fiola, D. (2000) Quality assessment of commercially produced entomopathogenic nematodes. *Biological Control* 17, 100–109.

Georgis, R. (1990a) Formulation and application technology. In: Gaugler, R. and Kaya, H.K. (eds) *Entomopathogenic Nematodes in Biological Control*. CRC Press Boca Raton, Florida, pp. 173–191.

Georgis, R. (1990b) Commercialization of Steinernematid and Heterorhabditid entomopathogenic nematodes. *Proceedings of Brighton Crop Protection Conference – Pests and Diseases* 1, 275–280.

Georgis, R. and Kaya, H.K. (1998) Formulation of entomopathogenic nematodes. In: Burges, H.D. (ed.) *Formulation of Microbial Biopesticides: Beneficial Microorganisms, Nematodes and Seed Treatments.* Kluwer Academic Publishers, Dordrecht, The Netherlands, pp. 289–308.

Georgis, R., Dunlop, D.B. and Grewal, P.S. (1995) Formulation of entomopathogenic nematodes. In: Hall, F.R. and Barry, J.W. (eds) *Biorational Pest Control Agents: Formulation and Delivery.* American Chemical Society, Washington, DC, pp. 197–205.

Glaser, R.W. (1932) Apathogenic nematode of the Japanese beetle (*Popillia japonica*). *Journal of Parasitology* 18, 119.

Glaser, R.W. (1940) Continued culture of a nematode parasitic in the Japanese beetle. *Journal of Experimental Zoology* 84, 1–12.

Glazer, I. (1996) Survival mechanisms of entomopathogenic nematodes. *Biocontrol Science and Technology* 6, 373–378.

Glazer, I. and Navon, A. (1990) Activity and persistence of entomopathogenic nematodes tested against *Heliothis armigera. Journal of Economic Entomology* 83, 1795–1800.

Goud, S., Hugar, P.S. and Prabhuraj, A. (2010) Effect of temperature, population density and shelf life of EPN *Heterorhabditis indica* (RCR) in sodium alginate gel formulation. *Journal of Biopesticides* 3, 627–632.

Grewal, P.S. (2000a) Anhydrobiotic potential and long term storage of entomopathogenic nematodes (Rhabditida: Steinernematidae). *International Journal for Parasitology* 30, 995–1000.

Grewal, P.S. (2000b) Ehanced ambient storage stability of an entomopathogenic nematode through anhydrobiosis. *Pest Management Science* 56, 401–406.

Grewal, P.S. (2002) Formulation and application technology. In: Gaugler, R. (ed.) *Entomopathogenic Nematology.* CAB International, Wallingford, UK, pp. 265–287.

Grewal, P.S. (2012) Entomopathogenic nematodes as tools in integrated pest management. In: Abrol, D.P. and Shankar, U. (eds) *Integrated Pest Management: Principles and Practice.* CAB International, Wallingford, UK, pp. 162–236.

Grewal, P.S. and Peters, A. (2005) Formulation and quality. In: Grewal, P.S., Ehlers, R.U. and Shapiro-Ilan, D.I.

(eds) *Nematodes as Biocontrol Agents*. CAB International, Wallingford, UK, pp. 79–90.

Grewal, P.S., Selvan, S. and Gaugler, R. (1994) Thermal adaptation of entomopathogenic nematodes: niche breadth for infection, establishment, and reproduction. *Journal of Thermal Biology* 19, 245–253.

Grewal, P.S., Converse, V. and Georgis, R. (1999) Influence of production and bioassay methods on infectivity of two ambush foragers (Nematoda: Steinernematidae). *Journal of Invertebrate Pathology* 73, 40–44.

Grewal, P.S., Ehlers, R.-U. and Shapiro-Ilan, D.I. (2005) *Nematodes as Biological Control Agents.* CAB International, Wallingford, UK.

Griffin, C.T., Boemare, N.E. and Lewis, E.E. (2005) Biology and behaviour. In: Grewal, P.S., Ehlers, R.-U. and Shapiro-Ilan, D.I. (eds) *Nematodes as Biocontrol Agents*. CAB International, Wallingford, UK, pp. 47–64.

Han, R.C. and Ehlers, R.-U. (1998) Cultivation of axenic *Heterorhabditis* spp. dauer juveniles and their response to non-specific *Photorhabdus luminescens* food signals. *Nematologica* 44, 425–435.

Han, R.C. and Ehlers, R.-U. (2001) Effect of *Photorhabdus luminescens* phase variants on the *in vivo* and *in vitro* development and reproduction of the entomopathogenic nematodes *Heterorhabditis bacteriophora* and *Steinernema carpocapsae*. *FEMS Microbiology Ecology* 35, 239–247.

Han, R., Cao, L. and Liu, X. (1992) Relationship between medium composition, inoculums size, temperature and culture time in the yields of *Steinernema* and *Heterorhabditis* nematodes. *Fundamental and Applied Nematology* 15, 223–229.

Han, R., Cao, L. and Liu, X. (1993) Effects of inoculums size, temperature and time on *in vitro* production of *Steinernema carpocapsae* Agriotos. *Nematologica* 39, 366–375.

Han, R., Pang, X. and Li, L. (1995) Optimization of the medium components for the solid culture of entomopathogenic *Steinernema* and *Heterorhabditis* nematodes. *Natural Enemies of Insects* 17, 153–164.

Hara, A.H., Linderen, J.E. and Kaya, H.K. (1981) Monoxenic mass production of the entomogenous nematode, *Neoaplectana carpocapsae* (Wieser), on dog food/agar medium. United States Department of Agriculture, Science and Education Administration, Advances in Agriculture Technology-W-16, Oakland, California, 8 pp.

Head, J., Lawrence, A.J. and Walters, K.F.A. (2004) Efficacy of the entomopathogenic nematode, *Steinernema feltiae* against *Bemisia tabaci* in relation to plant species. *Journal of Applied Entomology* 128, 543–547.

Hiltpold, I., Hibbard, B.E., French, B.W. and Turlings, T.C.J. (2012) Capsules containing entomopathogenic nematodes as a Trojan horse approach to control the western corn rootworm. *Plant and Soil* 358, 11–25.

Hirao, A. and Ehlers, R.-U. (2010) Influence of inoculums density on population dynamics and dauer juvenile yields in liquid culture of biocontrol nematodes *Steinernema carpocapsae* and *S. feltiae* (Nematoda: Rhabditidae). *Applied Microbiology and Biotechnology* 85, 507–575.

Hominick, W.M., Briscoe, B.R., del Pino, F.G., Heng, J., Hunt, D.J., *et al.* (1997) Biosystematics of entomopathogenic nematodes: current status, protocols and definitions. *Journal of Helminthology* 71, 271–298.

Hopper, K.R., Roush, R.T. and Powell, W. (1993) Management of genetics of biological control introductions. *Annual Review of Entomology* 38, 27–51.

Hussaini, S.S. (2002) Entomopathogenic nematodes for the control of crop pests. In: Upadhyay, R.K. (ed.) *Advances in Microbial Control of Insect Pests*. Kluwer Academic/Plenum Publishers, New York, pp. 265–296.

Hussein, M.A. and Abdel-Atey, M.A. (2012) Formulation of two native entomopathogenic nematodes at room temperature. *Journal of Biopesticides* 5 (supplementary), 23–27.

Istkhar Chaubey, A.K., Aasha and Bhat, A.H. (2016) Virulence and recycling potential of entomopathogenic nematodes (Nematode: Steinernematidae, Heterorhabditidae) from Saharanpur district, western Uttar Pradesh, India. *Journal of Entomology and Zoology Studies* 4, 27–32.

Jansson, R.K., Lecrone, S.H. and Gaugler, R. (1993) Field efficacy and persistence of entomopathogenic nematodes (Rhabditida: Steinernematidae, Heterorhabditidae) for control of sweet potato weevil (Coleoptera: Apionidae) in southern Florida. *Journal of Economic Entomology* 86, 1055–1063.

Jeffke, T., Jende, D., Matje, C., Ehlers, R.-U. and Berthe-Corti, L. (2000) Growth of *Photorhabdus luminescens* in batch and glucose fed-batch culture. *Journal of Applied Microbiology and Biotechnology* 54, 326–330.

Jessen, P., Strauch, O., Wyss, U., Luttmann, R. and Ehlers, R.-U. (2000) Carbon dioxide triggers dauer juvenile recovery of entomopathogenic nematodes (*Heterorhabditis* spp.). *Nematology* 2, 310–324.

Johnigk, S.-A., Ecke, F., Poehling, M. and Ehlers, R.-U. (2004) Liquid culture mass production of biocontrol nematodes, *Heterorhabditis bacteriophora* (Nematoda: Rhabditida): improved timing of dauer juvenile inoculation. *Applied Microbiology and Biotechnology* 64, 651–658.

Jung, K. (1996) Storage of entomopathogenic nematodes of the genus *Heterorhabditis* at two temperatures. Effect on infectivity, energy reserve and number of bacteria. *IOBC/WPRS Bulletin* 19, 103–106.

Kaya, H.K. and Nelson, C.E. (1985) Encapsulation of steinernematid and heterorhabditid nematodes with calcium alginate; a new approach for insect control and other applications. *Environmental Entomology* 14, 572–574.

Kaya, H.K. and Stock, S.P. (1997) Techniques in insect nematology. In: Lacey, L.A. (ed.) *Manual of Techniques in Insect Pathology*. Academic Press, San Diego, California, pp. 281–324.

Khan, M.R., Khan, U., Askary, T.H., Mohiddin, F.A. and Khan, M.M.(2007a) Pathogenicity and host suitability for *in vivo* mass production of *Steinernema masoodi* AMU EPN-1. *International Journal of Nematology*, 17, 151–157.

Khan, M.R., Khan, U. and Askary, T.H. (2007b) Occurrence of *Steinernema masoodi* in Aligarh and its pathogenicity against six economically important insect pests. *Indian Journal of Nematology* 37, 215–216.

Kim, J., Jaffuel, G. and Turlings, T.C.J. (2015) Enhanced alginate capsule properties as a formulation of entomopathogenic nematodes. *BioControl* 60, 527–535.

Kung, S., Gaugler, R. and Kaya, H.K. (1991) Effects of soil temperature, moisture, and relative humidity on entomopathogenic nematode persistence. *Journal of Invertebrate Pathology* 57, 242–249.

Lacey, L.A., Shapiro-Ilan, D.I. and Glenn, G.M. (2010) Post application of anti-desiccant agents improves efficacy of entomopathogenic nematodes in formulated host cadavers or aqueous suspension against diapausing codling moth larvae (Lepidoptera: Tortricidae). *Biocontrol Science and Technology* 20, 909–921.

Lara, J.C., Dolinski, C., Fernandes De Sousa, E. and Figueiredo Daher, E. (2008) Effect of mini-sprinkler irrigation system on *Heterorhabditis baujardi* LPP7 (Nematoda: Heterorhabditidae) infective juvenile. *Scientia Agricola* 65, 433–437.

Lello, E.R., Patel, M.N., Matthews, G.A. and Wright, D.J. (1996) Application technology for entomopathogenic nematodes against foliar pests. *Crop Protection* 15, 567–574.

Lewis, E.E. and Clarke, D.J. (2012) Nematode parasites and entomopathogens. In: Vega, F. and Kaya, H.K. (eds) *Insect Pathology*, 2nd edn. Elsevier, Amsterdam, pp. 395–424.

Lewis, E.E. and Perez, E.E. (2004) Formulation and storage of entomopathogenic nematodes. *International Journal of Nematology* 14, 30–34.

Lewis, E.E. and Shapiro-Ilan, D.I. (2002) Host cadaver protect entomopathogenic nematodes during freezing. *Journal of Invertebrate Pathology* 81, 25–32.

Li, S.C. (1984) A new method for mass rearing of parasitic nematode, *Neoaplectana* spp. *Plant Protection Bejing, China* 10, 36–37.

Lindegren, J.E., Valero, K.A. and Mackey, B.E. (1993) Simple *in vivo* production and storage methods for *Steinernema carpocapsae* infective juveniles. *Journal of Nematology* 25, 193–197.

Lunau, S., Stoessel, S., Schmidt-Peisker, A.J. and Ehlers, R.-U. (1993) Establishment of monoxenic inocula for scaling up *in vitro* cultures of the entomopathogenic nematodes *Steinernema* spp. and *Heterorhabditis* spp. *Nematologica* 39, 385–399.

Managanvi, K., Srinivasaraghavan, E.A. and Jagdish, J. (2012) Mass production of entomopathogenic nematodes for plant protection – a review. *Environment and Ecology* 30, 1416–1421.

Matthews, G.A. (2000) *Pesticide Application Methods*, third edn. Blackwell Science, London, 432 pp.

Miles, C., Blethen, C., Gaugler, R., Shapiro-Ilan, D.I. and Murray, T. (2012) Using entomopathogenic nematodes for crop insect pest control. Pacific North West Extension Publications, USA, pp. 1–9.

Miller, P.C.H., Tuck, C.R., Murphy, S. and Ferreira, M. (2008) Measurements of the droplet velocities in sprays produced by different designs of agricultural spray nozzle. In: *Proceedings of 22nd European Conference on Liquid Atomization and Spray Systems* held at Como Lake, Italy, 8–10 September 2008. Available at: www.ilasseurope.org/ICLASS/ILASS2008_COMO/file/papers/8-5.pdf (accessed 23 September 2016).

Miller, R.W. (1989) Novel pathogenicity assessment technique for *Steinernema* and *Heterorhabditis* entomopathogenic nematodes. *Journal of Nematology* 21, 574.

Molyneux, A.S. (1984) The influence of temperature on the infectivity of heterorhabditid and steinernematid nematodes for larvae of the sheep blowfly, *Lucilia cuprina.* In: Bailey, P.T. and Swincer, D.E. (eds) *Proceedings of the 4th Australian Applied Entomological Research Conference.* H. Smith, Australian Department of Agriculture, Adelaide, Australia, pp. 344–351.

Morales-Ramos, J.A., Rojas, M.G., Shapiro-Ilan, D.I. and Tedders, W.I. (2010) Developmental plasticity in *Tenebrio molitar* (Coleoptera: Tenebrionidae): analysis of instar variation in number and development time under different diets. *Journal of Entomological Science* 45, 75–90.

Moreira, G.F., Batista, E.S.P., Campos, H.B.N., Lemos, R.E. and Ferreira, M.C. (2013) Spray nozzles, pressures, additives and stirring time on viability and pathogenicity of entomopathogenic nematodes (Nematoda: Rhabditida) for greenhouses. *PLoS ONE* 8, e65759. Available at: http://dx.doi.org/10.1371/journal.pone.0065759 (accessed 23 September 2016).

Mukaka, J., Strauch, O., Hoppe, C. and Ehlers, R.-U. (2010) Improvement of heat and desiccation tolerance in *Heterorhabditis bacteriophora* through cross-breeding of tolerant strains and successive genetic selection. *BioControl* 55, 511–521.

Neves, J.M., Teixeira, J.A., Simoes, N. and Mota, M. (2001) Effect of airflow rate on yield of *Steinernema carpocapsae* Az 20 in liquid culture in an external-loop airlift bioreactor. *Biotechnology and Bioengineering* 72, 369–373.

Nilsson, U. and Gripwall, E. (1999) Influence of application technique on the viability of the biological control agents *Verticillium lecanii* and *Steinernema feltiae*. *Crop Protection* 18, 53–59.

Onstad, D.W., Fuxa, J.R., Humber, R.A., Oestergaard, J., Shapiro-Ilan, D.I., Gouli, V.V., Anderson, R.S., Andreadis, T.G. and Lacey, L.A. (2006) *An Abridged*

Glossary of Terms Used in Invertebrate Pathology, 3rd edn. *Society for Invertebrate Pathology.* Available at: http://www.sipweb.org/resources/glossary.html (accessed 23 September 2016).

Pace, W.G., Grote, W., Pitt, D.E. and Pitt, J.M. (1986) Liquid culture of nematodes. International Patent WO 86/01074.

Patel, M.N. and Wright, D.J. (1997a) Glycogen: its importance in the infectivity of aged *Steinernema carpocapsae. Parasitology* 114, 591–596.

Patel, M.N. and Wright, D.J. (1997b) Fatty acid composition of neutral lipid energy reserves from infective juveniles of entomopathogenic nematodes. *Composition of Biochemical Physiology* 118, 341–348.

Patel, M.N., Stolinski, M. and Wright, D.J. (1997) Neutral lipids and the assessment of infectivity in entomopathogenic nematodes: observations on four *Steinernema* species. *Parasitology* 114, 489–496.

Perez, E.E., Lewis, E.E. and Shapiro-Ilan, D.I. (2003) Impact of host cadaver on survival and infectivity of entomopathogenic nematodes (Rhabditida: Steinernematidae and Heterorhabditidae) under desiccating conditions. *Journal of Invertebrate Pathology* 82, 111–118.

Peters, A. and Backes, J. (2003) Impact of substrate conditions and application method on the efficacy of *Steinernema feltiae. IOBC/ WPRS Bulletin* 26, 151–158.

Peters, A. and Ehlers, R.-U. (1994) Susceptibility of leather jackets (*Tipula paludosa* and *Tipula oleracea*; Tipulidae: Nematocera) to the entomopathogenic nematode *Steinernema feltiae*. *Journal of Invertebrate Pathology* 63, 163–171.

Piggot, S. and Wardlow, L. (2002) A fresh solution for the control of western flower thrips: dramatic results in trials. *Commercial Greenhouse Grower* February, 20–23.

Poinar, G.O. Jr (1976) Description and biology of a new insect parasitic rhabditoid. *Heterorhabditis bacteriophora* n. gen., n. sp. (Heterorhabditidae, n. fam.) (Rhabditida [Oerley]). *Nematologica* 21, 463–470.

Poinar, G.O. (1990) Biology and taxonomy of Steinernematidae and Heterorhabditidae. In: Gaugler, R. and Kaya, H.K. (eds) *Entomopathogenic Nematodes in Biological Control*. CRC Press, Boca Raton, Florida, pp. 23–62.

Poinar, G.O. Jr and Thomas, G.M. (1966) Significance of *Achromobacter nematophilus* Poinar and Thomas (Achromobacteraceae: Eubacteriales) in the development of the nematode, DD-136 (*Neoaplectana* sp. Steinernematidae). *Parasitology* 56, 385–390.

Qiu, L.H. and Bedding, R.A. (1999) Low temperature induced cryoprotectant synthesis by the infective juveniles of *Steinernema carpocapsae*: its biological significance and the mechanisms involved. *Cryo Letters* 20, 393–404.

Qiu, L.H., Lacey, M.J. and Bedding, R.A. (2000) Energy metabolism and its relation to survival and infectivity of infective juveniles of *Steinernema carpocapsae* under aerobic conditions. *Nematology* 2, 551–559.

Rezaei, N., Karimi, J., Hosseini, M., Goldani, M. and Campos-Herrera, R. (2015) Pathogenicity of two species of entomopathogenic nematodes against the greenhouse whitefly, *Trialeurodes vaporariorum* (Hemiptera: Aleyrodidae), in laboratory and greenhouse experiments. *Journal of Nematology* 47, 60–66.

Salma, J. and Shahina, F. (2012) Mass production of eight Pakistani strains of entomopathogenic nematodes (Steinernematidae and Heterorhabditidae). *Pakistan Journal of Nematology* 30, 1–20.

Schroer, S. and Ehlers, R.-U. (2005) Foliar application of the entomopathogenic nematode *Steinernema carpocapsae* for biological control of diamondback moth larvae (*Plutella xylostella*). *Biological Control* 6, 238–244.

Shapiro-Ilan, D.I. and Gaugler, R. (2002) Production technology for entomopathogenic nematodes and their bacterial symbionts. *Journal of Industrial Microbiology and Biotechnology* 28, 137–146.

Shapiro-Ilan, D.I. and Glazer, I. (1996) Comparison of entomopathogenic nematode dispersal from infected host versus aqueous suspension. *Environment and Ecology* 25, 1455–1461.

Shapiro-Ilan, D.I. and Lewis, E.E. (1999) Comparison of entomopathogenic nematode infectivity from infected hosts versus aqueous suspension. *Environment and Ecology* 28, 907–911.

Shapiro-Ilan, D.I. and McCoy, C.W. (2000) Effect of culture method and formulation on the virulence of *Steinernema riobrave* (Rhabditida: Steinernematidae) to *Diaprepes abbreviatus* (Curculionidae). *Journal of Nematology* 32, 281–288.

Shapiro-Ilan, D.I., Glazer, I. and Segal, D. (1997) Genetic improvement of heat tolerance in *Heterorhabditis bacteriophora* through hybridization. *Biological Control* 8, 153–159.

Shapiro-Ilan, D.I., Cate, J.R., Pena, J., Hunsberger, A. and McCoy, C.W. (1999) Effect of temperature and host range on suppression of *Diaprepes abbreviates* (Coleoptera: Curculionidae) by entomopathogenic nematodes. *Journal of Economic Entomology* 92, 1086–1092.

Shapiro-Ilan, D.I., Lewis, E.E., Behle, R.W. and McGuire, M.R. (2001) Formulation of entomopathogenic nematode infected cadavers. *Journal of Invertebrate Pathology* 78, 17–23.

Shapiro-Ilan, D.I., Gouge, D. and Koppenhofer, A. (2002) Factors affecting commercial success: case studies in cotton, turf and citrus. In: Gaugler, R. (ed.) *Entomopathogenic Nematology*. CAB International, Wallingford, UK, pp. 333–355.

Shapiro-Ilan, D.I., Gaugler, R. and Lewis, E.E. (2004) *In vivo* production of entomopathogenic nematodes. *International Journal of Nematology* 14, 13–18.

Shapiro-Ilan, D.I., Gouge, D.H., Piggot, S.J. and Fife, J. (2006) Application technology and environmental considerations for use of entomopathogenic nematodes in biological control. *Biological Control* 38, 124–133.

Shapiro-Ilan, D.I., Tedders, W.L. and Lewis, E.E. (2008) Application of entomopathogenic nematode infected cadavers from hard-boiled arthropods for insect suppression. US Patent No 7, 374,773.

Shapiro-Ilan, D.I., Morales-Ramos, J.A., Rojas, M.G. and Tedders, W.L. (2010) Effects of a novel entomopathogenic nematode-infected host formulation on cadaver integrity, nematode yield and suppression of *Diaprepes abbreviates* and *Aethina tumida*. *Journal of Invertebrate Pathology* 103, 103–108.

Shapiro-Ilan, D.I., Han, R. and Dolinksi, C. (2012) Entomopathogenic nematode production and application technology. *Journal of Nematology* 44, 206–217.

Shapiro-Ilan, D.I., Han, R. and Qiu, X. (2014a) Production of entomopathogenic nematodes. In: Morales-Ramos, J.A., Rojas, M.G. and Shapiro-Ilan, D.I. (eds) *Mass Production of Beneficial Organisms: Invertebrates and Entomopathogens.* Academic Press, San Diego, California, pp. 321–355.

Shapiro-Ilan, D.I., Brown, I. and Lewis, E.E. (2014b) Freezing and desiccation tolerance in entomopathogenic nematodes. Diversity and correlation of traits. *Journal of Nematology* 46, 27–34.

Shapiro-Ilan, D.I., Cottrell, T.E., Mizell, R.F. III, Horton, D.L. and Abdo, Z. (2015) Field suppression of the peach tree borer, *Synanthedon exitiosa*, using *Steinernema carpocapsae*: effects of irrigation, a sprayable gel, and application method. *Biological Control* 82, 7–12.

Sharma, M.P., Sharma, A.N. and Hussaini, S.S. (2011) Entomopathogenic nematodes, a potential microbial biopesticide: mass production and commercialisation status – a mini review. *Archives of Phytopathology and Plant Protection* 44, 855–870.

Silver, S.C., Dunlop, D.B. and Grove, D.I. (1995) Granular formulations of biological entities with improved storage stability. World Patent No WO 95/0577.

Simons, W.R. and Poinar, G.O. Jr (1973) The ability of *Neoaplectana carpocapsae* (Steinernematidae: Nematodea) to survive extended periods of desiccation. *Journal of Insect Pathology* 22, 228–230.

Smits, P.H. (1996) Post application persistence of entomopathogenic nematodes. *Biocontrol Science and Technology* 6, 379–387.

Stock, S.P. and Hunt, D.J. (2005) Morphology and systematics of nematodes used in biocontrol. In: Grewal, P.S., Ehlers, R.-U. and Shapiro-Ilan, D.I. (eds) *Nematodes as Biocontrol Agents.* CAB International, Wallingford, UK, pp. 3–43.

Stoll, N.R. (1952) Axenic culture of the parasitic nematode, *Neoaplectana glaseri*, in a fluid medium containing raw liver extract. *Journal of Parasitology* 39, 422–444.

Strauch, O., Stoessel, S. and Ehlers, R.-U. (1994) Culture conditions define automictic or amphimictic reproduction in entomopathogenic rhabditid nematodes of the genus *Heterorhabditis*. *Fundamental and Applied Nematology* 17, 575–582.

Strauch, O., Niemann, I., Neumann, A., Schmidt, A.J., Peters, A. and Ehlers, R.-U. (2000) Storage and formulation of the entomopathogenic nematodes *Heterorhabditis indica* and *Heterorhabditis bacteriophora*. *Biocontrol* 54, 483–500.

Strong, D.R. (2002) Population of entomopathogenic nematodes in food webs. In: Gaugler, R. (ed.) *Entomopathogenic Nematology*. CAB International, Wallingford, UK, pp. 225–240.

Stuart, R.J. and Gaugler, R. (1996) Genetic adaptation and founder effect in laboratory populations of the entomopathogenic nematode *Steinernema glaseri*. *Canadian Journal of Zoology* 74, 164–170.

Surrey, M.R. and Davies, R.J. (1996) Pilot scale liquid culture and harvesting of an entomopathogenic nematode, *Heterorhabditis bacteriophora*. *Journal of Invertebrate Pathology* 67, 92–99.

Tanada, Y. and Kaya, H.K. (1993) *Insect Pathology*. Academic Press, San Diego, California.

Tarakanov, V.I. (1980) Methods of continuous axenic cultivation of the insect nematode *Neoaplectana glaseri*. *Turdy Vsesoyuznogo Instituta Gel'mintologii* 25, 106–110.

Umamaheswari, R., Sivakumar, M. and Subramanian, S. (2006) Survival and infectivity of entomopathogenic nematodes in alginate gel formulations against rice meal moth larva, *Corcyra cephalonica* Stainton. *Natural Product Radiance* 5, 95–98.

Wang, X. and Grewal, P.S. (2002) Rapid genetic deterioration of environmental tolerance and reproductive potential of an entomopathogenic nematode during laboratory maintenance. *Biological Control* 23, 71–78.

Wang, Y., Bilgrami, A.L., Shapiro-Ilan, D.I. and Gaugler, R. (2007) Stability of entomopathogenic bacteria *Xenorhabdus nematophila* and *Photorhabdus luminescens* during *in vitro* culture. *Journal of Industrial Microbiology and Biotechnology* 34, 73–81.

White, G.F. (1927) A method for obtaining infective nematode larvae from cultures. *Science* 66, 302–303.

Williams, E.C. and Walters, K.F.A. (2000) Foliar application of the entomopathogenic nematode *Steinernema feltiae* against leaf miners on vegetables. *Biocontrol Science and Technology* 10, 61–70.

Wilson, M. and Ivanova, E. (2004) Neutral density liquid formulations for nematode based pesticides. *Biotechnology Letters* 26, 1167–1171.

Woodring, J.L. and Kaya, H.K. (1988) Steinernematid and Heterorhabditid nematodes: a handbook of biology and techniques. Southern Cooperative Series Bulletin 331, Arkansas Agricultural Experimental Station, Fayetteville, Arkansas, 30 pp.

Wouts, W.M. (1981) Mass production of the entomogenous nematodes *Heterorhabditis heliothidis* (Nematoda: Heterorhabditidae) on artificial media. *Journal of Nematology* 13, 467–469.

Wright, D.J., Peters, A., Schroer, S. and Fife, J.P. (2005) Application technology. In: Grewal, P.S., Ehlers, R.-U. and Shapiro-Ilan, D.I. (eds) *Nematodes as Biocontrol Agents*. CAB International, Wallingford, UK, pp. 91–106.

Yadav, S., Sharma, H.K., Siddiqui, A.U. and Sharma, S.K. (2015) *In vitro* mass production of *Steinernema carpocapsae* on different artificial media. *Indian Journal of Nematology* 45, 123–124.

Yang, H., Jian, H., Zhang, S. and Zhang, G. (1997) Quality of the entomopathogenic nematode *Steinernema carpocapsae* produced on different media. *Biological Control* 10, 193–198.

Yoo, S.K., Brown, I. and Gaugler, R. (2000) Liquid media development for *Heterorhabditis bacteriophora*: lipid source and concentration. *Applied Microbiology and Biotechnology* 54, 759–763.

Zhu, H., Grewal, P.S. and Reding, M.E. (2011) Development of a desiccated cadaver delivery system to apply entomopathogenic nematodes for control of soil pests. *Applied Engineering in Agriculture* 27, 317–324.

Part IV: Role of Entomopathogenic Nematodes in Integrated Pest Management

14 Status of Entomopathogenic Nematodes in Integrated Pest Management Strategies in the USA

Glen Stevens[1]* and Ed Lewis[2]

[1]School of Natural Sciences and Mathematics, Ferrum College, Ferrum, Virginia, USA; [2]Department of Entomology and Nematology, University of California, Davis, California, USA

14.1 Introduction

One of the most significant challenges to the field of entomopathogenic nematology is to identify and broadly disseminate conditions under which the use of insect parasitic nematodes (EPNs) represents a cost-effective, value-added approach to integrated pest management (IPM). For example, when are EPNs the optimal tool in the IPM arsenal, and when can they accomplish things that other approaches cannot? The recent work we discuss suggests answers to this very question. By virtue of their mobility, EPNs may be able to delve deep into soils where most application methods for chemical pesticides are not feasible or where chemicals are environmentally unacceptable (e.g. against root-feeding weevils in citrus, or against flies in mushroom culture), or where they can penetrate into cryptic habitats such as wounds made by stem- or root-boring insects (e.g. against peach tree borers or grape root borers). Their biology may allow them to be applied under conditions (environmental or pest-life-stage) when chemical approaches are not useful for targeting particular pest life stages (navel orangeworm in pistachio orchards, codling moth on fruit bins, etc.). In situations where the potential exposure of workers or crops to pesticides is of particular concern, for example in glasshouses and nurseries, EPNs have safety advantages in addition to excellent efficacy (e.g. control of black vine weevil, fungus gnats, or western flower thrips). Some formulations, though not all, may meet certification requirements for organic agriculture. They may even be able to contribute to the maintenance of existing IPM approaches (e.g. reducing the development of *Bacillus thuringiensis* (Bt) resistance in crops modified to produce the Bt protein (Gassmann *et al.*, 2008)). As no snail and slug parasitic nematodes are presently licensed for use in the USA, this chapter will focus solely on the use of nematodes that parasitize insects (EPN); however, the recent discovery of *Phasmarhabditis hermaphrodita* in California (De Ley *et al.*, 2014) may allow for the expansion of existing products (e.g. Nemaslug®) in the USA in the near future.

EPNs have been studied for nearly a century in the USA; while a century seems a long time, both a thorough understanding of EPN ecology, as well as the development of a substantial economic market for EPNs, remains elusive. As obligate parasites of insects, EPNs can be key top-down forces in both agricultural (Duncan *et al.*, 2003, 2007; Denno *et al.*, 2008) and unmanaged ecosystems (Ram *et al.*, 2008). This strong top-down force, the ability to greatly reduce pest populations, is the trait we attempt to exploit in IPM. Laboratory, greenhouse and field studies are all used to assess the potential benefit of EPNs in IPM, but laboratory and greenhouse studies may not translate to successful pest management in the field. Indeed, EPNs may not control many insect species in the field that show susceptibility to EPNs in the laboratory (Spence *et al.*, 2008; Karban

*Corresponding author. E-mail: gstevens3@ferrum.edu

et al., 2011). Despite this still developing understanding of EPNs, they are utilized successfully for biological control in many US crops.

Techniques for culturing them for application as inundative biological control agents are well developed (as reviewed in Shapiro-Ilan *et al.*, 2012), but EPNs occupy only a small niche in the pest solutions market, and have failed to reach the market potential expected (Lacey *et al.*, 2015). Indeed, over the first decade of the 21st century, the global market for EPNs remained flat, even as the broader biopesticide market and the chemical pesticide market expanded (Fig. 14.1).

The successes and failures of EPNs have been a subject for many reviews and case studies, including Georgis *et al.* (2006) and Dolinski *et al.* (2012). There are multiple barriers to the broader implementation of EPNs in integrated pest management, including cost relative to chemical pesticides, shelf life, environmental limitations and the poor knowledge of potential users regarding EPNs as a pest control option (Georgis *et al.*, 2006; Askary, 2010). 'Poor knowledge' here means two things: (i) how hard it is for growers to find out EPNs are an IPM option; and (ii) they offer multiple opportunities for end-user error.

EPNs are a different sort of pest control option for growers, and for scientists, from the chemicals they may replace. They are animals, which dictates particular abilities and requirements, and also implies that they may survive and establish for some time within a field. It also means that EPNs make decisions: understanding these decisions can improve our chances of effective pest management using EPNs (Lewis *et al.*, 2006). Early failures in effective biological control due to poor host–EPN matches (e.g. efforts to use *Steinernema carpocapsae*, a sit-and-wait-style ambush predator, against the fairly sedentary white grub complex) led to an appreciation of the role of behaviour in the effective utilization of EPNs, an area of research that continues to expand. Recent work has examined the role of EPNs in a complex community: EPNs are key insect pathogens, but they also compete within the insect host with each other (Zenner *et al.*, 2014) and with free-living bacterivorous nematodes (Duncan *et al.*, 2003). EPNs are generally protected within the host cadaver, though the thousands of infective juveniles (IJs) that emerge from the host are prey for a range of nematode-feeding fauna (fungi, nematodes, mites) and respond to a host of cues to determine how far and in what direction they disperse from the host (Lewis *et al.*, 2015). Understanding EPN behavioural responses and EPN effects on the soil community promises to promote our understanding of how EPNs influence (Duncan *et al.*, 2007; El-Borai *et al.*, 2007; Hodson *et al.*, 2012) and are influenced by (Stuart *et al.*, 2008; Campos-Herrera *et al.*, 2013) the soil ecosystem. This knowledge informs both inundative releases of EPNs and conservation biological control.

Certainly, the study of EPNs is expanding rapidly (San-Blas, 2013). In the USA, laboratories and

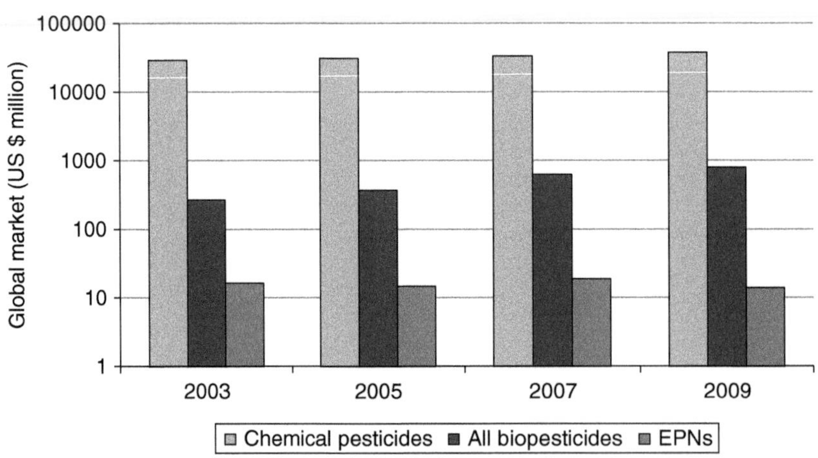

Fig. 14.1. Entomopathogenic nematode, biopesticide and chemical pesticide markets from 2003 to 2009. (Adapted from Lacey and Georgis (2012), based on data reported from the CPL Business Consultant report (2008) and the BCC Research report (2009).)

research centres around the country are working to answer the questions we laid out above, in increasingly close collaboration with scientists from around the globe (San-Blas, 2013). They are also attempting to place EPNs in an appropriate ecological context, as both predators and prey (Lewis *et al.*, 2015), acting within a physical soil framework that profoundly affects their efficacy (El-Borai *et al.*, 2007; Kaspi *et al.*, 2010; Nielsen and Lewis 2012). One of the more intriguing (and elusive) aspects of EPN ecology is the role that manipulations of the growing environment may have on the likelihood of effective 'top-down' control by EPNs of agricultural pests. The notion of conservation biological control of EPNs has received relatively scarce attention in the past decade or so (Shields 2015, but see Duncan *et al.*, 2013, for an example in citrus). This may represent an important opportunity for these ubiquitous soil organisms.

Our focus in this chapter is to document the uses for EPNs in the USA, and to pay special attention to examples where EPNs accomplish something that other pest management strategies do not.

14.2 Integrated Pest Management (IPM) in the USA

Broadly, integrated pest management is a way to make economically based decisions about reducing pest populations. Many specific definitions exist, including this one from the University of California Statewide IPM Program (found on the Web at www.ipm.ucdavis.edu/GENERAL/ipmdefinition.html):

> *Integrated pest management (IPM) is an ecosystem-based strategy that focuses on long-term prevention of pests or their damage through a combination of techniques such as biological control, habitat manipulation, modification of cultural practices, and use of resistant varieties.*
>
> *Pesticides are used only after monitoring indicates they are needed according to established guidelines, and treatments are made with the goal of removing only the target organism.*
>
> *Pest control materials are selected and applied in a manner that minimizes risks to human health, beneficial and nontarget organisms, and the environment.*

Working from this definition, we can move line by line to ask our guiding question, 'when are EPNs the optimal tool in the IPM arsenal?'

14.3 Application of Biological Control Agents (Entomopathogenic Nematodes) in Insect Pest Management in the USA

14.3.1 What can biological methods (entomopathogenic nematodes) offer to insect pest management?

In a framework based on the first sentence in the IPM definition above, EPNs have potential as biological control agents in IPM. Their effectiveness as a biological control option may be influenced by both habitat manipulations and cultural practices (Duncan *et al.*, 2007; Stuart *et al.*, 2008; Nielsen and Lewis 2012); indeed, the use of conservation biological control to maintain and even promote EPN populations is an appreciated but often understudied aspect of the field (Lewis *et al.*, 1998; Shields, 2015). If conservation biocontrol can be done effectively, however, it may be possible to maintain levels of pest injury below economic thresholds without regular inundative releases. This has been acknowledged as one of the long-term goals of the field (Lewis *et al.*, 1998) – despite this, however, successful conservation control remains elusive, and EPNs are typically applied in an inundative fashion, as a 'biological insecticide'.

The 'resistant varieties' aspect of the definition may be feasible, if the advances in chemical ecology can translate to the selection of host plant varieties that are more effective at recruiting EPNs to areas of damage within the root system. There have been some tantalizing suggestions, including recent work (Rasmann *et al.*, 2005; Ali *et al.*, 2013) that attaches chemical specificity (and thus plant breeding targets) to the longer-term knowledge that EPNs are attracted to the sites of pest damage (Choo *et al.*, 1989; Choo and Kaya, 1991; van Tol *et al.*, 2001; Ennis *et al.*, 2010).

Incorporating EPNs into the pesticide use criteria of the IPM definition (the second line in the quoted definition above) is more complex; it amounts, essentially, to asking when EPNs (or any microbial pest control formulation) are the most economically and biologically sensible option. EPNs are, in general, more expensive than chemical control options, so unless chemical use is for some reason restricted, they are often not attractive. Under some conditions, though, EPNs (and other microbial control methods) can 'outperform' chemical methods. Key settings where these conditions are met appear to be: (i) when chemical pesticide use is restricted economically or legally; (ii) when worker safety is a major concern; or (iii) when the habitat of the pest makes chemical intervention ineffective or less desirable. Effective intervention under these conditions is what biological methods offer to IPM.

When the appropriate chemical is either too expensive or when increasing regulation removes chemicals from the IPM toolkit, the relative attractiveness of EPNs increases (see Bélair *et al.*, 2010, for an example in turfgrass). Another example is the interest in increasing regulatory restrictions on chlorpyrifos in the USA due to the risks it poses to human health and environmental contamination; increasing regulation or increasing cost may increase the market attractiveness of EPN products that can also control soil pests. Similarly, pesticides that have environmental and human health concerns (e.g. water source contamination, application on school grounds and/or chemical pesticide residue on food) face an increasingly restrictive regulatory future in the USA.

The use of EPNs in an IPM framework is attractive from the perspective of worker and workplace safety. EPNs may present an ideal control product for use in a glasshouse setting, due to their ability to target key greenhouse pests such as black vine weevil (*Otiorhynchus sulcatus*), fungus gnats (*Orfelia* and *Bradysia* spp.) and western flower thrips (*Frankliniella occidentalis*), key glasshouse pests in the USA. They can be used without restrictions up to the date of harvest, have few if any worker exposure issues and no re-entry restrictions, and are compatible with other biologicals that may be in use. These benefits also apply to their use in the field, but the confined nature of glasshouse work (and the often high-value crops grown there) boosts their relative value in that setting.

EPNs currently reach their greatest use when the pest habitat makes intervention with chemical pesticides ineffective. There are many examples of this in the recent literature from the USA. These can be grouped into soil pests, wood-boring insects (both stem and root borers) and a broader category of fruit and nut pests. In general, though, where there is an aspect of pest phenology that cannot be targeted effectively with chemical insecticides, EPNs are a valuable aspect of IPM.

In the soil, EPNs are particularly valuable against root-feeding weevils (e.g. the citrus root weevil, *Diaprepes abbreviatus*, the pecan weevil, *Curculio caryae*). In the case of *D. abbreviatus*, which typically has multiple life stages present in the soil at a given time, the development of chemical resistance is a key pest life history variable that favours the use of EPNs; in addition, the relatively shallow water table and sandy soils in Florida, where EPNs are most commonly used, make groundwater contamination with chemical insecticides a significant concern. Biologicals (primarily EPNs) have replaced the use of persistent chlorinated hydrocarbons (organochlorines) in Florida citrus (Campos-Herrera *et al.*, 2015), and continue to form the foundation of the programme to control major root-feeding insect pests (*D. abbreviatus* and other root weevils) and their associated disease complexes (e.g. *Phytophthora* spp.).

EPNs are also proving to be effective options for targeting insects that bore into the roots or stems of crop plants. Well-developed examples can be found in peaches in the USA, where EPNs have proven effective at controlling both the root-feeding peach tree borer, *Synanthedon exitiosa* (Fig. 14.2), as well as the branch- and stem-feeding lesser peach tree borer, *Synanthedon pictipes* (Shapiro-Ilan *et al.*, 2010a, 2015). These species are both key pests in peaches and other *Prunus* spp., and have been controlled effectively using EPNs. The species and/or the techniques used may be applied to other Sesiid moths such as grape root borer (*Vitacea polistiformis*, Williams *et al.*, 2010), clearwing moth borers (family Sesiidae, Kaya and Brown, 1986) and the raspberry crown borer, *Pennisetia marginata* (Cowles *et al.*, 2005). The economics of using EPNs may be improved in an orchard system, particularly if wounds are obvious and/or if effective control can be obtained by applying EPNs solely to the base of orchard trees or vines (Shapiro-Ilan *et al.*, 2015).

Fig. 14.2. Application of *Steinernema carpocapsae* with a hand sprayer to an obvious wound on a peach tree caused by *Synanthedon exitosa*. Photo credit: United States Department of Agriculture (USDA).

Using EPNs in IPM programmes targeting other fruit- and nut-feeding pests has shown economic viability. The key factors in their success appear to be the ability to target a cryptic life stage of the pest during a part of the year compatible with EPN viability. In pistachio, for example, *S. carpocapsae* can be applied through irrigation sprinklers targeting overwintering larvae in February and March to provide up to 75% control of navel orangeworm (*Amyelois transitella*) in fallen pistachio (*Pistacia vera*) nuts (Siegel *et al.*, 2004, 2006); in olive (*Olea europaea*), *Steinernema feltiae* were able to kill olive fruit fly (*Bactrocera oleae*) both in treated olives and in the soil around them, achieving a mortality rate of 67.9% (Sirjani *et al.*, 2009). In apple (*Malus* sp.) the incorporation of EPNs into an attract-and-kill strategy to control plum curculio (*Conotrachelus nenuphar*) in north-eastern USA seems promising, as *Steinernema riobrave* reduced field populations of ground-dwelling *C. nenuphar* by more than 95% (Shapiro-Ilan *et al.*, 2013).

Many of these issues combine in the management of novel pests using IPM. When a new pest is found, determining whether EPNs represent a valid and valuable tool requires understanding whether there is a particular life stage in the pest that would: (i) be difficult to control with chemicals; and (ii) exist during a part of the year when EPNs can survive outside. If there is an aspect of the pest's phenology that is particularly compatible with EPN biology, and particularly if that life stage is incompatible with chemical pesticide use, EPNs may be a valuable element in the IPM toolkit for that pest species.

14.3.2 A list of available products containing entomopathogenic nematodes in the USA

EPNs are marketed in ways that target commercial growers, smaller-scale producers (market gardeners) and home gardeners. In situations where biological control options result in a value added to the crop being grown, they make sense; this valuation can be (and often will be) strictly economical, but biological options often offer more nuanced values such as perceptions of safety or ecological harmony. Companies selling EPNs in the USA (major manufacturers as well as resellers) and the products they advertise (at the time of publication) are listed in Table 14.1, and were drawn from the web page maintained at https://biocontrol.entomology.cornell.edu/pathogens/nematodes.php.

In the USA, different institutions are carrying out research on EPNs. We have listed some of them in Table 14.2.

14.4 Applications of Entomopathogenic Nematodes in Vegetable Crop Production

14.4.1 Field vegetable crop production

While a great deal of research has been focused on the control of vegetable crop pests in the field using EPNs, and both laboratory and field results show significant promise (Georgis *et al.*, 2006), thc commercial market is generally flat (Lacey and Georgis, 2012), particularly in the USA. EPNs do not compare favourably to chemical pesticides; regardless, there are results from field and laboratory assessments that suggest potential markets for EPNs should chemical insecticides become less attractive due to increased cost or unavailability due to regulatory restrictions.

Control of diamondback moth larvae using EPNs is possible, though the level of control is lower than would be needed to promote their adoption even in European markets. Similar results are seen for cutworms; although good control has been seen (e.g. Shapiro-Ilan *et al.*, 2002; Ebssa and Koppenhöfer, 2011; Bélair *et al.*, 2013), they are not utilized broadly (Lacey and Georgis, 2012).

Control of cabbage root maggot (*Delia radicum*) is still not up to commercial standards, though treating plant trays with *S. feltiae* reduces maggot numbers (Beck *et al.*, 2014) and EPNs do parasitize larvae successfully in their brief soil stage (Georgis *et al.*, 2006). Control of the carrot weevil (*Listronotus oregonensis*) is possible, though they are still not economically competitive against chemical insecticides and thus see little use (Georgis *et al.*, 2006).

In the USA, similar passages could be written about the Colorado potato beetle (*Leptinotarsa decemlineata*) and sweet potato weevil (*Cylas formicarius*): EPNs are potentially highly effective biological control agents, but their use is cost prohibitive (Georgis *et al.*, 2006). Increasing acreages of organic crops may result in increases in EPN utilization, but EPNs can be more expensive than other comparably effective microbial control formulations. If strategies for synergistically incorporating them with chemicals or other non-chemical control arise and are broadly disseminated, this may open up markets that currently remain underdeveloped.

Table 14.1. The current status of commercial formulations of entomopathogenic nematodes available in the USA.

Company name	Product name	Entomopathogenic nematode species available	Website
ARBICO, Inc	NemAttack™ NemaSeek™	Hb, Sc, Sf and blends	http://www.arbico-organics.com/category/beneficial-nematodes
BASF (formerly Becker Underwood)	Millenium®, Nemasys®, Nematac®	Hb, Sc, Sf, Sk	http://betterplants.basf.us/products/product-index.html
BioLogic Company	Scanmask, Ecomask, Heteromask	Hb, Sc, Sf	https://biologicco.com/
Crop King Inc	Larvanem, Entonem	Hm, others	https://www.cropking.com/catalog/beneficial-insects-pest-and-fungus-management/larvanem50-million
Gardens Alive, Inc	Beneficial Sf Nematodes, Grub-Away® Nematodes	Sf, Hb	http://www.gardensalive.com/category/insect-control
Gardener's Supply Company	Gnat Guard™, Grub Guard™	Steinernema sp., Sc and Hb blend	http://www.gardeners.com/search?q=nematode
Green Spot, Ltd	Various	Hb, Sc, Sf	https://greenmethods.com/nematodes/
Harmony Farm Supply & Nursery	Scanmask™	Sf	http://www.harmonyfarm.com/steinernema-feltiae-10-milllion-ct/
Hydro-Gardens, Inc	Field Guardian™, Guardian™, Gnat Patrol™, Lawn Patrol™	Steinernema (Sc, Sf) and Heterorhabditis blends	https://hydro-gardens.com/product-category/beneficials/nematodes/
IPM Laboratories, Inc	ScanMask™	Sf	https://biocontrol.entomology.cornell.edu/pathogens/nematodes.php
Koppert Biological Systems	Capsanem, Entonem, Larvanem, Terranem, Terranem-NAm	Hb, Sc, Sf blends	http://www.koppert.com/products/products-pests-diseases/
Nature's Control	Predator Nematode 'Double-Death' Mix	Sc, Hb blends	http://www.naturescontrol.com/thrip.html#pn
Peaceful Valley Farm Supply	Various, including Nemaglobe™ grub buster	Sc, Sf, Sg, Hb	https://www.groworganic.com/weed-and-pest-control/beneficial-insects/beneficial-nematodes.html
Do My Own Pest Control – Nematodes	Various products (Koppert, Orcon)	Various species	http://www.domyownpestcontrol.com/beneficial-nematodes-c-31_519.html
Rincon-Vitova Insectaries Inc	Millenium, Nemasys, Nemasys G, Nemasys L, Capsanem	Hb, Hi, Hm, Sc, Sk	http://www.rinconvitova.com/nematode.htm
Southeastern Insectaries, Inc	In-house gel formulations	Hb, Hi, Sc	http://www.southeasterninsectaries.com/nematodes.php
Territorial Seed Company	Predatory nematodes	Steinernema/Heterorhabditis mix	http://www.territorialseed.com/category/pest_control_insects
Worm's Way Inc	Orcon Beneficial Nematodes	Sf	http://www.wormsway.com/en/bnt700

Notes: Hb = *Heterorhabditis bacteriophora*; Hi = *Heterorhabditis indica*; Hm = *Heterorhabditis marelatus*; Sc = *Steinernema carpocapsae*; Sf = *Steinernema feltiae*; Sg = *Steinernema glaseri*; Sk = *Steinernema kraussei*.

Meloidogyne spp., including *Meloidogyne incognita* and *Meloidogyne arenaria*, are important and significant root-knot nematodes (RKN) in the USA; although there is no direct linkage between EPNs and RKNs, many researchers have reported reductions in RKNs when either EPNs or their bacterial symbionts are applied (Lewis and Grewal, 2006). In general, the degree of control has not been sufficient for them to be relied on in a commercial setting. Recently, though, researchers observed that

Table 14.2. Research centres in the USA involved in entomopathogenic nematode research.

Institution	Location
Cornell University	Ithaca, New York
Iowa State University	Ames, Iowa
Michigan State University	East Lansing, Michigan
Ohio Agricultural Research and Development Center	Wooster, Ohio
Rutgers, the State University of New Jersey	New Brunswick, New Jersey
University of California, Davis	Davis, California
University of California, Riverside	Riverside, California
University of Florida	Gainesville, Florida
USDA Agricultural Research Service	Various places including Byron, Georgia and Yakima, Washington

treatments of tomato seedlings with *Xenorhabdus bovienii* supernatant just prior to transplanting into infested soil provided an effective and economically feasible means for reducing pressure from RKNs in field-grown tomato seedlings (Kepenekci *et al.*, 2016).

Work on the use of EPN-infected insect cadavers may translate to increased utilization of EPNs in the USA, particularly on the scale of market gardens or other smaller venues (Dolinski *et al.*, 2015). One of the concerns of these infected cadaver formulations is the ease of application (Dolinski *et al.*, 2015) – application by hand may be feasible, but not preferable; desiccating and/or coating the cadavers may ease utilization without greatly reducing efficacy (Shapiro-Ilan *et al.*, 2001; Spence *et al.*, 2011). Also, selecting a host that is hard-bodied, such as *Tenebrio molitor*, has seen some research and success (e.g. Shapiro-Ilan *et al.*, 2010b). The use of a modified crop seed planter to deliver desiccated and/or coated cadavers into furrows in the soil (e.g. Zhu *et al.*, 2011) may improve the effectiveness of application.

14.4.2 Nursery and glasshouse production of vegetable and ornamental crops

EPNs are useful in both the nursery and glasshouse, having demonstrated effectiveness against major pests such as black vine weevil (*O. sulcatus*), strawberry root weevil (*Otiorhynchus ovatus*), white grubs (family Scarabaeidae), fungus gnats (family Sciaridae, *Orfelia* and *Bradysia* spp.), leaf miners (*Liriomyza* spp.) and western flower thrips (*F. occidentalis*) (Georgis *et al.*, 2006; Lacey *et al.*, 2015).

Nursery and greenhouse production of ornamentals is a multi-billion dollar industry in the USA, estimated by van Tol and Raupp (2006) at more than US$6 billion in annual crop sales. Given the benefits of EPNs in this context (limited worker exposure issues, no re-entry interval, etc.), this is one area in which significant expansion of EPN utilization seems promising. The incorporation of EPNs into potted nursery plants may prevent or cure pest infestations; in potted citrus seedlings, El-Borai *et al.* (2007) showed that adding the EPN *Steinernema diaprepesi* to pots could control a root-feeding weevil (in this case, the citrus root weevil, *D. abbreviatus*), leading to increased plant weight. They also showed that EPNs applied in this case could sometimes persist in pots for more than a year.

Black vine weevil control has been investigated in some detail, with the general conclusion that effective control is heavily influenced by the temperature limitations of the EPN species utilized (van Tol *et al.*, 2004; van Tol and Raupp, 2006). Recent work on the interactions between EPNs and entomopathogenic fungi suggests significant synergism, with up to 100% control of *O. sulcatus* when an application of *Metarhizium anisopliae* is followed by EPNs within 1–2 weeks (Ansari *et al.*, 2009).

The most recent work on white grubs, primarily work led by Koppenhöfer's research group at Rutgers University in New Jersey, USA, suggests that synergistic combinations of EPNs with chemical insecticides may be able to control these destructive pests while both (i) reducing the cost of EPN application, and (ii) reducing chemical pesticide dosage and frequency (reviewed in detail in Section 14.7 below, and in Koppenhöfer *et al.*, 2015). The identification of novel EPN species that are particularly effective at scarab control (e.g. *S. scarabaei*) shows the benefit of continuing to identify new EPN species and strains.

Fungus gnats (*Bradysia* spp.) can be controlled successfully using EPNs; *S. feltiae* in particular is used in many operations in Europe (Jagdale *et al.*, 2004; Georgis *et al.*, 2006). There has been little adoption of EPNs for the control of fungus gnats in the USA, however, where there are numerous cheaper chemical insecticides that can be utilized (Georgis *et al.*, 2006).

If potting media can be modified in a way that improves EPN persistence, the economics of using them in US glasshouses may improve (Dolinski *et al.*, 2012); this persistence is one of the factors that underlies EPN adoption in mushroom culture (Grewal and Richardson, 1993) and successes in the control of an invasive fungus gnat (*Bradysia agrestis*) in Korea (Kim *et al.*, 2004; Kaya *et al.*, 2006). The use of bark-based media increased the control of *Bradysia* sp. in potting soil (Nielsen *et al.*, 2008; Nielsen and Lewis, 2012). Some media (e.g. redwood sawdust) had strong negative effects on EPNs when incorporated into soilless media; peat moss, plant compost and hardwood bark improved EPN performance (Nielsen and Lewis, 2012). Regardless, the high temperatures observed in many US glasshouses suggest an area for future research to improve EPN efficacy here.

Western flower thrips (*F. occidentalis*) are very important pests, but have been difficult to control using EPNs (Ebssa *et al.*, 2001; Georgis *et al.*, 2006). High doses and multiple applications assist in effective control (Wardlow *et al.*, 2000); unfortunately, this added expense reduces their attractiveness in US applications. New work suggests that the use of EPNs in combination with other control techniques may reduce the EPN dose required for effective control. Combining applications of *Heterorhabditis bacteriophora* or *Heterorhabditis indica* with a predatory mite (*Amblyseius cucumeris*) led to up to 83% control of *F. occidentalis* (Ebssa *et al.*, 2006), and combining *S. carpocapsae* with either a Neem extract formulation (at a low EPN dose) or the entomopathogenic fungus *M. anisopliae* produced up to 74% control of *F. occidentalis* in the glasshouse (Otieno *et al.*, 2015).

Many different species of leaf miners (family Agronyzidae) can be controlled effectively by either *S. carpocapsae* or *S. feltiae* (Hara *et al.*, 1993; LeBeck *et al.*, 1993; Sher *et al.*, 2000; Williams and Walters, 2000; Tomalak *et al.*, 2005; Georgis *et al.*, 2006; Batalla-Carrera *et al.*, 2010). Despite these successes, an investigation of EPNs for use in the biocontrol of an invasive leaf miner in Alaska (the amber-marked birch leaf miner, *Profenusa thomsoni*) showed no evidence of control by *S. carpocapsae* (Progar *et al.*, 2015), so the effects of EPNs on leaf miners are certainly not universal.

14.5 Arable Crop Production

Although EPNs are being used successfully against major acreage arable crops elsewhere around the world (for example, against *Diabrotica* sp. in *Zea mays* in Europe, Toepfer *et al.*, 2014), their relative expense compared to chemical insecticides or other biological formulations (e.g. *B. thuringiensis*) means that EPN formulations are not presently used on major acreage row crops in the USA.

Despite their current failure to penetrate the US market, recent results (some from the USA, but most from elsewhere around the world) suggest conditions under which EPNs may be utilized effectively in the USA if economic conditions become more favourable for such approaches. Commercial formulations are well developed and effective for control of *Diabrotica virgifera virgifera* in Europe, although they are cost prohibitive in the USA. Interestingly, recent results suggest that they may be a part of an IPM approach to slow the development of Bt resistance in both pink bollworm (a major pest in cotton) and *D. virgifera virgifera* (Gassmann *et al.*, 2008; Petzold-Maxwell *et al.*, 2012).

The only arable crop where a significant adoption of EPNs has taken place as a biological control approach is the control of the naturalized pest alfalfa snout weevil (*Otiorhynchus ligustici*) in lucerne in New York. Researchers from Cornell University have developed strategies whereby growers can isolate, raise and apply their own EPN; these EPNs are able to control *O. ligustici* and persist within a field for multiple years across crop changes (Shields *et al.*, 2009; Shields, 2015).

14.6 Other Applications of Entomopathogenic Nematodes in the USA

14.6.1 Parks, turfgrass and urban and municipal areas

The management of parks, urban and municipal areas in the USA principally means managing turfgrass. This is true in golf courses, parks and home and office complexes; the major difference in the management of turf in the various environments is tolerance for pest damage. Turfgrass covers approximately 20,000,000 ha in the USA, and management of turf is an approximately US$40 billion/year industry (National TurfGrass Federation, http://turfresearch.org/pdf/turfinitiative.pdf)

The control of turfgrass pests using EPNs is one of the more well-studied research systems in the USA, and has provided a great deal of insight in a number of factors related to the use of EPNs in broader IPM. Namely, the benefits of specific pairings between EPNs and the pest species; EPN control of 'white grubs' in the broad sense is considered more accurately at the fine scale of specific pairings between EPN and white grub species. An EPN that controls a particular white grub well may have far less effect on another species in the same habitat, or even on different instars of the same species (Koppenhöfer *et al.*, 2015).

White grubs are major turfgrass pests; both the damage that the grubs do and the secondary damage by grub-feeding vertebrates (e.g. raccoons and wild pigs) are of major concern, particularly on golf courses and other sites with low damage thresholds. Many species of EPN have demonstrated good control of scarabs in the field (Georgis *et al.*, 2006). The availability of relatively inexpensive chemicals in the USA for use on turf means that EPNs have made limited inroads into turfgrass; however, research demonstrating significant synergism between chemicals such as imidacloprid and EPNs (Koppenhöfer and Fuzy 2002), or between EPNs and microbial agents such as Bt (Kaya and Koppenhöfer, 1996; Koppenhöfer and Kaya, 1997) may make the economics of EPN application far more attractive (Koppenhöfer *et al.*, 2015). In addition, the identification of EPNs that are highly virulent to white grubs (e.g. *Steinernema scarabaei*) promises to improve the potential benefits of EPNs in managed turf (Koppenhöfer and Fuzy, 2003, 2007, 2009).

Mole crickets (*Scapteriscus* spp.) were introduced into the USA accidentally, and were very important turfgrass pests in the south-eastern region of the USA. *Steinernema scapterisci* was a very effective classical biological control agent for mole crickets; it was originally identified from the native range of mole crickets, and successfully established after a combination of inundative applications and attract-and-infect approaches (Parkman *et al.*, 1996). Commercial formulations were used in both pastures (at low application rates) and some turfgrass sites (Georgis *et al.*, 2006) for inundative release with some success; however, *S. scapterisci* availability has come and gone from the market as products have periodically been discontinued (Koppenhöfer *et al.*, 2015).

Turfgrass weevils, including *Sphenophorus* spp. (billbugs) and the annual bluegrass weevil (*Listronotus maculicollis*), are important root-feeding pests in turfgrass. Billbugs appear vulnerable to both steinernematid and heterorhabditid EPNs (e.g. Georgis and Poinar, 1994; Smith, 1994; Kinoshita and Yamanaka, 1998), as are annual bluegrass weevils (McGraw and Koppenhöfer, 2008; McGraw *et al.*, 2010), but the results of field trials have been variable, and adoption in the USA has been limited (Koppenhöfer *et al.*, 2015).

Black cutworm (*Agrotis ipsilon*) is a significant pest in turf that can be targeted effectively with EPNs. Ebssa and Koppenhöfer (2011) examined four different EPN species (*H. bacteriophora*, *S. carpocapsae*, *S. feltiae* and *S. riobrave*) in turf managed like a golf course fairway. Of the four species, *S. carpocapsae* provided the best and most consistent control (83% reduction in pest populations). When the labelled EPN application rate was divided into separate applications, control was improved (unpublished results described in Koppenhöfer *et al.*, 2015). *A. ipsilon* is still not highly utilized in managed turf in the USA, particularly on golf courses, due to the availability of lower-cost insecticides that provide better and more consistent control.

Aside from turfgrass settings, the other places within urban and municipal landscapes where EPNs may play a role are in landscape plantings (e.g. ornamental pots and/or public gardens). Many of the same polyphagous pests found in the turf or vegetable setting are also problems in landscape plantings (e.g. black vine weevil, citrus root weevil, etc.).

Limited research has investigated the potential for using EPNs to control pests in these managed plantings. Black vine weevil can be controlled successfully in landscape pots; Burlando *et al.* (1993) demonstrated an approximately 85% reduction in *O. sulcatus* populations 1 year after adding *S. feltiae* or *H. bacteriophora* IJs to commercial planters in California. The ability to control invasive pests in ornamental plantings in Florida might also be successful using EPNs; laboratory studies of several EPN species demonstrate good control of both *Corythucha ciliata*, a native lace bug that is a pest in sycamore, and *Stethobaris nemesis*, an invasive weevil that is a pest of amaryllis (Shapiro-Ilan and Mizell, 2012). In this latter study, there were several promising species of EPNs identified, including both steinernematids and heterorhabditids.

14.6.2 Small fruit (berries, vineyards)

A number of small fruit crops in the families Rosaceae, Ericaceae, Vitaceae and Grossulariaceae (formerly Saxifragaceae) have been shown to potentially benefit from applications of EPNs to control pests (Cowles *et al.*, 2005). Grossulariaceae (*Ribes* sp., gooseberries and currants) are a very minor crop in the USA, likely owing to the long-term effects of the federal and state bans on their cultivation. The others represent significant small fruit crops.

Strawberries (*Fragaria* sp.) are a major crop in the USA; in fact, the USA leads the world in strawberry production, producing 36 billion pounds in 2012, 29% of global strawberry production (Ag Marketing Research Center, http://www.agmrc.org/commodities-products/fruits/strawberries/). EPNs are not widely used in strawberry production; however, they are used often in greenhouse and nursery production against many of the insects that can reduce production. EPNs have shown effectiveness in strawberries against black vine weevil, strawberry root weevil and strawberry crown moth; however, effective control appears to vary significantly with soil temperature and soil and crop characteristics (Booth *et al.*, 2002). Both steinernematid and heterorhabditid EPNs have shown efficacy against strawberry pests; most researchers have examined *S. carpocapsae*, *Steinernema kraussei*, *H. bacteriophora*, *Heterorhabditis marelatus* or *Heterorhabditis megidis*.

Control has generally been good against both black vine weevil and strawberry root weevil, with 75% or higher reductions in pest populations on treated plots (Berry *et al.*, 1997; Booth *et al.*, 2002; Lola-Luz and Downes, 2007). Some species appear more effective than others (Haukeland and Lola-Luz, 2010), particularly at low temperatures (below 10°C). Early spring application of *H. megidis* in Ireland reduced black vine weevil populations by nearly 90%, however, so effective control is possible even at low soil temperatures (Lola-Luz and Downes, 2007).

Some of the most exciting work targeting root weevils in strawberry is recent work showing synergism between EPNs and between EPNs and entomopathogenic fungi. Ansari *et al.* (2009) found that combining *S. kraussei* and *M. anisopliae* resulted in 100% control of overwintering larvae in strawberry bags grown in the glasshouse. The use of a combined *S. carpocapsae*/*S. feltiae* formulation applied to field plots reduced *O. sulcatus* populations in strawberry fields (and in cranberry fields) from outbreak conditions to subeconomic threshold levels (Shields, 2015).

The strawberry crown moth (*Synanthedon bibionipennis*) is a major root-boring Sesiid pest of strawberry. Laboratory assays against *S. bibionipennis* using *S. carpocapsae* and *H. bacteriophora* suggest that these pests are extremely susceptible in the laboratory; control in the field is much lower, with only 51% and 33% infection rates in the field after late autumn application (Bruck *et al.*, 2008). This points to the significance of abiotic factors, particularly temperature, in effective control using EPNs.

Cane fruit

The raspberry crown borer (*Pennisettia marginata*, Sesiidae) is a widespread pest of brambles in the USA, owing in part to the abundance of wild brambles that establish on unmanaged land across the country. As *P. marginata* is a root-boring pest, it can be difficult to detect (Cowles *et al.*, 2005), despite the potentially catastrophic effect on individual plants and plantings. Field applications of *S. feltiae* to control *P. marginata* reduced pest populations by 50% or more (Capinera *et al.*, 1986). More recent work comparing EPNs to chemical insecticides observed 46%, 53% and 33% reductions in larvae by *S. feltiae*, *S. carpocapsae* and *H. bacteriophora*, respectively; these ratings were much lower than the more effective insecticides, which averaged nearly 90% reduction in larvae (McKern *et al.*, 2007).

Cranberry

Several pests of cranberry production are susceptible to EPN infection, including beetle larvae (Koppenhöfer and Fuzy, 2008), several root weevils (Simser and Roberts, 1994) and the cranberry girdler (Simard *et al.*, 2002). These insects are all highly susceptible to infection by EPNs, but it is the growing conditions of cranberry that should particularly encourage their use. Cranberries grow in highly specialized soils. Conditions must be moist but well drained and have low pH, and they do best in soils that have high levels of peat. EPNs are not widely used in cranberry production currently, but research shows the promise of effective control if EPNs are adapted as a control option.

In the case of white grubs, the loss of organochlorine insecticides led to a major re-emergence of formerly well-controlled scarab pests such as the cranberry white grub (*Phyllophaga georgiana*). Koppenhöfer *et al.* (2008) showed that a steinernematid EPN that is well adapted to white grubs, *S. scarabaei*, provided 76–100% control of *P. georgiana*; other species were less effective at standard application rates (2.5×10^9 IJ/ha). Applications seem to be most effective when timed to target early-instar larvae.

Cranberries also face pressure from both black vine weevil and strawberry root weevil (Cowles *et al.*, 2005). Given the optimal texture (moist, well-drained soils), much of the work examining the potential benefits of EPNs against these pests should translate well to cranberry. While the low pH of the soils may interfere with long-term establishment from inundative release (Stuart *et al.*, 1997), locally adapted EPNs found in cranberry habitats may be able to be managed in a conservation biocontrol framework (Shields, 2015).

EPNs are also effective against another beetle pest in cranberry, the cranberry rootworm (*Rhabdopterus picipes*, Chrysomelidae). An assessment of the effectiveness of *H. bacteriophora* showed a reduction in its field populations of 54–79% (Polavarapu *et al.*, 2000); however, *H. bacteriophora* may be even more effective in some cases: unpublished data cited in Cowles *et al.* (2005) from Polavarapu and Koppenhöfer reported 100% mortality of cranberry rootworm 14 days after EPN application.

Cranberry girdler (*Chrysoteuchia topiaria*, Pyralidae) is a major lepidopteran pest in cranberries, and has been the focus of several investigations using EPNs. Field investigations in the Pacific Northwest using *S. carpocapsae* have shown significant reductions in *C. topiaria* populations (up to 92%, Smith *et al.*, 1993). Many different EPN species may be effective at *C. topiaria* control, including *H. marelatus*, *H. bacteriophora*, *H. megidis* and *Steinernema glaseri* (Berry and Liu, 1998; Henderson and Singhai, 1999; Simard *et al.*, 2002); these different species may vary significantly in their ability to establish long-term populations.

Blueberry

The culture of blueberries (*Vaccinium* spp.) is similar in many ways to that of cranberries. Low pH (4.5–4.8) and well-drained soils are required for effective commercial production, and significant additions of organic matter are a typical approach for achieving these characteristics. White grubs and root mealybug (*Dysmicoccus vaccinii*) are major pest concerns in blueberry culture. Results from white grub EPN assessments in turf and other small fruits translate well to blueberries (e.g. the use of *S. scarabaei*, which can reduce grub densities by 80% or more, Koppenhöfer and Fuzy, 2004) due to the well-drained texture of blueberry soils, although inundative releases may need to be repeated due to the low pH (Fischer and Fuhrer, 1990).

Laboratory examinations using excised roots were conducted to examine the efficacy of multiple different EPN species against root mealybug. In general, while Steinernematids failed to kill mealybugs, Heterorhabditids both killed and reproduced within root mealybugs. The effectiveness of different *Heterorhabditis* species and strains varied from ineffective to highly effective in sand column assays, with *H. indica* and *H. bacteriophora* showing >90% control at high doses (Stuart *et al.*, 1997).

Grapes

Grapes are a high-value crop grown across the USA. They are grown increasingly for commercial wine production in the eastern states, well within the range of a native pest of *Vitis* spp., the grape root borer (*V. polistiformis*). This Sesiid moth excavates tunnels in the roots of vines, and can lead to large-scale vine decline and mortality; much like the raspberry crown borer described above, infestations are particularly difficult to observe and diagnose before extensive damage is already present (Bergh, 2012). Given the ability many EPNs have demonstrated to penetrate the galleries produced by sesiids, it is no surprise that EPNs have shown potential use against *V. polistiformis*. To date, field trial results for two EPN species have been published. Applications of *H. bacteriophora* and *Heterorhabditis zealandica* resulted in variable control of grape root borer, ranging between 55% to as high as 92% reduction in *V. polistiformis* abundance, and good persistence of *H. bacteriophora* post-application (Williams *et al.*, 2010). Recent research characterizing the abiotic and biotic factors that are associated with grape root borer incidence in Virginia suggests that EPNs may have value as either an inundative or classical biological control option (Rijal *et al.*, 2014).

Honeybees

Finally, because of the need for effective pollination in many small fruit and orchard crops, the use of EPNs to control the small hive beetle (*Aethina tumida*) seems appropriate to include here. *A. tumida* appears to be very susceptible to EPNs in the laboratory and in the field (Ellis *et al.*, 2010; Shapiro-Ilan *et al.*, 2010b). The application of EPN-infested insect cadavers (*T. molitor*, infected with *H. indica*) achieved up to 78% control of *A. tumida* in the laboratory; it may be a straightforward control technique to apply a small number of infected *T. molitor* cadavers (Dolinski *et al.*, 2015) beneath hives to control mature larvae moving from the hive into the soil.

14.6.3 Orchard (fruit and nut tree pests)

The use of EPNs in orchard systems is one of the most common in the USA (Lacey and Shapiro-Ilan, 2008). Several aspects of orchard pest management make EPNs an ideal addition to an IPM system. First, orchards are not subject to tillage annually, so the level of physical disruption is much reduced compared with row crops, or even berry crops. Often, orchards are in place for more than a decade, so soil conditions are more amenable to EPN survival, with some exceptions. Another advantage of orchards for EPN use is that many of the pests have soil stages that are not readily accessible for more traditional pest management materials (for example, overwintering larvae within fallen fruit, or larval stages that reside deep in the soil profile). This is especially important in that there are limited materials in competition with EPN use against such pests.

The EPN *S. riobrave* has been a mainstay for IPM in citrus in Florida for decades. The target is the diaprepes root weevil in citrus and several other fruit and ornamental crops, since this insect is highly polyphagous. Other commercially available EPN species have been tested, but *S. riobrave* has proven the most effective in many field trials (Duncan and McCoy, 1996; Bullock *et al.*, 1999). *D. abbreviatus* larvae can be found very deep in the soil horizon, where chemical insecticides are unlikely to reach, and the larvae are present for much of the year in Florida, making them ideal targets for EPNs. A recent review of the use of EPNs against weevil pests in Florida (Campos-Herrera *et al.*, 2015) summarizes this work effectively.

The citrus–*D. abbreviatus* system is also one of the most studied for EPN use in the field. Thus, much has been learned about the limiting factors of EPN success as well. The impact of soil type on EPN efficacy was obviated in a number of careful studies begun by McCoy, Duncan and colleagues. EPNs provide the best levels of control in sandy, well-drained soils (McCoy, 1999). This finding has been exploited in two ways. First, EPNs are recommended for use in soil where they are likely to be successful. Analysing soil types has been suggested as a way to avoid applying EPNs to areas where they have no chance of working (Kaspi *et al.*, 2010). The second research area dealing with soil type suggests that orchard planting sites may be modified to make them more appropriate for EPN application (Duncan *et al.*, 2013; Campos-Herrera *et al.*, 2015). Another area of study that was spurred by work in citrus was the impact of natural enemies and possible EPN competitors on EPN applications (Campos-Herrera *et al.*, 2012).

Pistachio production in California is plagued with infestation of the navel orangeworm, *A. transitella*). The moths oviposit during the summer months and the larvae develop within the nuts where they overwinter inside. Most control is achieved via foliar applications of organophosphates, insect growth regulators, or pyrethroid insecticides. These materials are not effective against the overwintering larvae, which are inside nuts at the base of the trees. This situation again is one where there is no product to use against the overwintering stage of the pest to offer EPN competition. *S. carpocapsae* has been shown to kill more than 70% of larvae in fallen nuts (Siegel *et al.*, 2004). As pistachio production increases and the chemicals available for controlling this pest decrease, EPN applications have become more widely used by growers.

Root weevils comprise a group of thelytokous weevils that are highly polyphagous, and their larvae and pupae reside in the soil for extended periods. Most are univoltine, with their adults eclosing in late spring and eggs being deposited throughout most of their adult lives. These are pests in many fruit crops where EPNs can be used, including not only citrus (Fuller rose weevil) (Campos-Herrera *et al.*, 2015) but also small fruits such as strawberry (black vine weevil and strawberry root weevil, see small fruit section above) (Booth *et al.*, 2002). Again, the weevils are all susceptible to EPN infection, and their use depends on the conditions of the crop.

The plum curculio, *Conotrachelus nenuphar*, is a serious pest of many stone and pome fruits. Shapiro-Ilan *et al.* (2008) showed that *S. riobrave* reduced populations of this insect at levels between 78 and 100%, and they could be used to reduce both larval and adult populations. EPNs are not in wide use in the eastern USA, but research continues. Most studies now target how to optimize conditions in orchards for EPNs, such as irrigation practices (Shapiro-Ilan *et al.*, 2013). Studies on the pecan weevil, *C. caryae*, have also shown that EPNs have great potential as part of an IPM strategy. Shapiro-Ilan and Gardner (2012) show that a series of three autumn applications of *S. carpocapsae* can provide up to 81% control of this insect. Again, as with the work with plum curculio, research is now focusing on the impact of habitat and cultural practices more than on the susceptibility of the hosts (e.g. Shapiro-Ilan *et al.*, 2006).

Boring insects, namely the peach tree borer and the lesser peach tree borer, are receiving attention as targets for EPNs, especially with the recent work on materials to extend IJ lifespan on surfaces. One of the main shortcomings of EPNs against boring pests is that they need to be able to reach the entry to the gallery made by the target pest. Previously, they could be injected directly into the hole, but this is not practical for most situations. Shapiro-Ilan *et al.* (2010a) were the first to use a fire-retardant gel called Barricade to extend the viability of EPN IJs on bark surfaces (Fig. 14.3). With this new method for application, perhaps EPN will enjoy more use against these and other boring pests.

Fig. 14.3. Application of *Steinernema carpocapsae* to the soil near the base of a peach tree using the fire-retardant gel, Barricade, to protect and extend the viability of the nematodes, and to reduce the need for post-application irrigation. Photo credit: United States Department of Agriculture (USDA).

14.6.4 Mushrooms

Mushroom cultivation is an area where EPN species are currently utilized for effective biological control of sciarid flies (*Lycoriella* spp.) in the USA. Applications of EPNs (typically *S. feltiae* is used) are comparably effective to chemical insecticides (Jagdale *et al.*, 2004; Grewal, 2007; Lacey and Georgis, 2012), with a longer residual effectiveness in compost beds (Grewal and Richardson, 1993). EPNs have proven less effective against phorid flies (e.g. *Megaselia halterata*). While some steinernematid isolates have shown good control in the laboratory (Long *et al.*, 2000), others have failed to control the pest (Scheepmaker *et al.*, 1998; Navarro and Gea, 2014).

14.7 Interactions of Entomopathogenic Nematodes with Pesticides

Data on interactions between EPNs and pesticides can be classified into three different categories, all based on the effect of the combination on pest control effectiveness. These categories are negative (pest control and/or EPNs are less effective), neutral (no significant/consistent change in control or efficacy – EPNs are compatible but do not add value to chemical use alone) and positive/synergistic, which is when control is greater than would be expected from the simple addition of the two compounds.

There are many examples of either negative or neutral interactions between EPNs and chemical pesticides. Pesticides such as abamectin, acephate, aldicarb, dodinc, fenamiphos, methomyl, parathion and teflubenzuron are generally toxic to EPNs (Shapiro-Ilan *et al.*, 2012). Other chemicals are generally compatible, including chlorpyrifos, diazinon, endosulfan and halofenozide among others (Shapiro-Ilan *et al.*, 2006; Koppenhöfer *et al.*, 2015).

The area of synergism is of increasing interest in the USA and elsewhere, as a synergistic interaction has the benefit of reducing the necessary dose of both EPNs and the chemical pesticide. This can reduce the relative cost of using EPNs, which opens up new markets and new possibilities. As a result, studies of the possible effects of combining EPNs with either chemical pesticides or microbial control approaches continue to increase. The list of potentially synergistic chemicals includes carbaryl, chlorantraniliprole, chlorpyrifos, dimethoate, endosulfan, fonofos, tefluthrin and imidacloprid (Shapiro-Ilan *et al.*, 2012; Koppenhöfer *et al.*, 2015). The extent (or even existence) of the synergism can be very dependent on the specific combination of pest, EPN species and application (Koppenhöfer and Fuzy, 2002; Shapiro-Ilan *et al.*, 2004), and further research is necessary to develop system-specific synergisms that are broad based. Some of the better-described synergisms were identified in the turfgrass–white grub system, and were reviewed in detail most recently by Koppenhöfer *et al.* (2015). To summarize, adding multiple species of EPNs simultaneously to field plots generally results in additive mortality effects on white grubs; some synergistic interactions are observed with EPNs and *B. thuringiensis* ssp. *japonensis* and with EPNs and the fungus *M. anisopliae*; finally, several EPN species demonstrate synergistic interactions with neonicotinoid insecticides such as imidacloprid, resulting from decreased defensive behaviours on the part of white grubs exposed to the neonicotinoids, and thus increased host attachment and penetration (Koppenhöfer *et al.*, 2000, 2015). Despite these successes, the combined use of chemical pesticides and EPNs is not practised on a commercial basis (Lacey and Georgis, 2012).

14.8 Technical Restrictions for Nematode Applications in the USA

14.8.1 Regulatory restrictions

EPNs are generally considered to be safe to human end-users, non-target organisms and the environment. This has freed them from many technical regulatory requirements placed on chemical insecticides; indeed, EPNs are not subject to pesticide registration requirements in the USA (Ehlers, 2005). While in the past the importation of exotic EPN species to the USA for commercial use was straightforward, importation or introduction of new EPN species into the USA is now regulated by the Animal and Plant Health Inspection Service (APHIS) and other agencies (Rizvi *et al.*, 1996; Akhurst and Smith, 2002). This limitation largely explains the lack of slug parasitic nematode products in the USA: despite a potential market for product(s) containing the slug-pathogenic nematode *P. hermaphrodita* (which was recently isolated from slugs in California, De Ley *et al.*, 2014) and the presence of other slug-pathogenic nematodes (e.g. *Phasmarhabditis californica* n. sp. and *Phasmarhabditis papillosa*, reviewed elsewhere in this book) in the USA, no nematode products for the control of slugs are marketed at present in the USA.

14.8.2 Organic farming restrictions

Use of EPNs in certified organic farming faces significant technical limitations, despite the potential value of EPNs in an organic setting. Many commercial formulations are stabilized with fungicides to improve shelf life; these fungicides are typically not certified by the Organic Materials Review Institute (OMRI), and thus many commercial formulations of EPNs do not meet the technical standard for organic agriculture. Small producer cooperatives may be able to isolate and propagate local EPN species and apply infected cadavers (Dolinski *et al.*, 2015) or develop management practices that promote native EPNs (Shields and Testa, 2015). At present, though, there is limited outreach in the USA regarding this sort of approach (Shields, 2015).

14.8.3 Technological restrictions for successful application

Technological restrictions for successful applications vary greatly, based on the formulation, species and application environment (reviewed most recently in Shapiro-Ilan *et al.*, 2006, 2012); as a result, anyone planning to use EPNs in an IPM context should consult the range of sources and review papers that discuss approaches for a specific combination of environment, target pest and EPN species. Some generalities are possible, however, in terms of application rate, abiotic effects and protective agents.

An application rate of 25 IJs/cm^2 (2.5×10^9 IJs/ha) is considered a minimum for effective control (Shapiro-Ilan *et al.*, 2012). The cost of application at this level can be significant, and is the major factor underlying the failure to adopt EPNs for the control of otherwise appropriate pests in arable crops in the USA (e.g. *D. virgifera virgifera* in maize). In many settings, however, growers can treat smaller areas within the overall site, reducing the overall cost of application – for example, the use of an attract-and-kill approach in apple orchards in north-eastern USA (Shapiro-Ilan *et al.*, 2010a) or the application to the base of infested trees in peach orchards in south-eastern USA (Shapiro-Ilan *et al.*, 2015), or prophylactic application (Shapiro-Ilan *et al.*, 2009).

Abiotic factors that can influence the effectiveness of EPN application include UV radiation, desiccation risk and temperature; in general, increased UV, increased desiccation and extremes in temperature reduce EPN efficacy. This generally limits the value of EPN applications to above-ground plant parts (leaves or stems), with some notable exceptions. Arthurs *et al.* (2004) conducted a meta-analysis of 136 field and greenhouse trials of *S. carpocapsae* against above-ground plant pests. They found that control declined along a continuum from bore holes to cryptic foliage (e.g. curled leaves, flowers, flower buds) to exposed foliage (i.e. the leaf surface), and that control in the greenhouse was greater than that observed in the field. This observation of increased *S. carpocapsae* efficacy in bore holes reflects some of the recent successes observed in treating boring moth larvae in peach orchards (Cottrell and Shapiro-Ilan, 2006; Shapiro-Ilan *et al.*, 2009, 2010a, 2015).

Desiccation risk is significant for EPNs post-application; this may underlie many of the failures in the above-ground application of EPNs, but is also a challenge for soil application. In general, irrigation is typically recommended post-application. Irrigation is common in experimental applications as well, perhaps assuming that most growers will irrigate if possible after EPN application. Recent work by Shapiro-Ilan *et al.* (2013) in apple orchards in north-eastern USA, however, showed that in at least some cases irrigation was not necessary to achieve efficacy. Thus, the need for irrigation may be influenced by other factors (soil texture, temperature at application, etc.); this is a good thing, as many end-users may not have the ability to irrigate post-application. Further research into the specific conditions under which irrigation provides a value-added service to EPN application may be very useful as water availability and cost increase across much of the USA.

Temperature effects on EPN efficacy are very much dependent on the species of EPN utilized and its native range (Kaya, 1990; Lacey and Georgis, 2012). In general, temperatures cooler than 10°C at the time of application can reduce EPN survival and infectivity. The upper temperature range is likely influenced by the probability of desiccation or UV exposure, both of which can be expected to increase with increasing temperature. Assuming high humidity and low UV exposure, EPN efficacy up to at least 30°C is a reasonable expectation, with some species exhibiting a broader range of tolerance.

It is these major limitations (desiccation and temperature) that can interfere with effective utilization by the smaller-scale user – for example, someone who purchases EPNs from one of the websites that will ship the nematodes to the mailbox of the purchaser. If those EPNs are then moved to the dashboard of a car, or even if they are stored too long before application, the end-user's experience will be impacted negatively. These restrictions are typically included in the instructions, but may not be appreciated fully or even read carefully.

14.9 Entomopathogenic Nematodes – Non-target Effects and Effects on Threatened and Endangered Insects

At present, EPNs are considered to be of little or no threat to non-target organisms (including humans) or the environment in general (Akhurst and Smith, 2002), and therefore are not subject to pesticide registration requirements in the USA and elsewhere (Ehlers, 2005). Despite this, it is sensible to consider the possible impacts of EPNs on non-target organisms, and specifically the potential effects on threatened and endangered insect species.

Non-target studies of the effect of EPN applications have focused on the effects on microarthropods, insects (including beneficials) and spiders. In general, studies have not found consistent effects on these species groups, although some changes have been documented. Hodson *et al.* (2012) found that the application of *S. carpocapsae* to control navel orangeworm increased populations of some microarthropod and spider species, while reducing populations of earwigs and some tenebrionid beetles. Greenwood *et al.* (2011) found similar results in their examination of the non-target effects on microinvertebrates: the application of Steinernematid EPNs caused some mite groups to become more common, while the application of Heterorhabditids caused some mites and collembolans to become less common. Other studies of the effect of EPN applications have found no effect on groups such as ground-dwelling or ground-emerging arthropods (e.g. Jabbour and Barbercheck, 2011; Dillon *et al.*, 2012; Babendreier *et al.*, 2015). Perhaps importantly, though, even where studies observed small changes in the arthropod community, it is the proportional make-up of the non-target community that changes; species have not been shown to be 'lost' from even the local community, even if they appear to respond negatively to EPN applications.

Some recent work has indicated the potential effects on non-targets, including beneficial species. Harvey *et al.* (2012) indicated that EPNs might affect saproxylic species (i.e. insects that bore into downed woody debris), thus slowing down rates of woody debris breakdown, an important ecosystem service. The authors suggest that careful applications of EPNs to stumps may reduce non-target effects (in this case, EPNs are applied to control an invasive forest pest not found in the USA, limiting the relevance to US silviculture). Perhaps more concerning were laboratory assays indicating that EPNs reduced populations of both two-spotted lady beetle and green lacewings (40– 90% reduction, Rojht *et al.*, 2009). These may be of limited relevance to EPN applications, as these beneficials do not have a ground-dwelling stage and are most often found in the canopy; however, in situations where EPNs are applied to canopies (e.g. using an airblast sprayer in the control of codling moth, Lacey *et al.*, 2006), such concerns may be valid, although laboratory results are difficult to translate to field applications (Spence *et al.*, 2008). It may be valuable to investigate the non-target effects in field conditions where exposure is likely and of a particular concern.

Additional non-target issues may be related to federally designated species of concern. In 2015, the US Fish and Wildlife Service (USFWS) listed 78 insect species as threatened or endangered within the USA or its territories (http://www.fws.gov/endangered/species/us-species.html). This list comprises 18 species of Coleoptera (including 4 *Cicindela* spp.), 36 species of Lepidoptera, 7 species of Odonata, 15 of Diptera (including 12 *Drosophila* spp. from Hawaii), one Orthopteran (the Zayante band-winged grasshopper, *Trimerotropis infantilis*), and one Hemipteran (the Ash Meadows naucorid, *Ambrysus amargosus*). Many of these are not continental; that is, they are Hawaiian or Pacific Island species. Most, if not all, are found primarily in native habitat, not agriculturally developed areas; this greatly reduces potential threats from the use of EPNs. Further, the USFWS lists particular counties within the USA where each species has been identified, greatly reducing the area where the issues may arise. Despite the low risk, as many of the listed species could be vulnerable to EPNs, any growers wishing to use EPNs in counties containing these threatened or endangered species may wish to check with their local Extension agents or USFWS authorities to reduce the already small chance that EPNs may influence species of concern in the areas around their fields.

14.10 Conclusion

EPNs are used on a limited basis in the USA. They are used on crops of relatively high value, such as fruit and nuts, in production facilities where chemicals may be problematic, such as greenhouses, or around areas where concern of public exposure is significant, such as school grounds. We propose that they should not be seen as direct competitors with chemical insecticides, for several reasons that include cost, ease of use, ease of storage and speed of kill. Throughout this chapter, we have highlighted ways that EPNs can be included in IPM programmes in ways that make them complementary or superior to chemical pest management methods.

Organic production schemes are limited in their pest control options, thereby making EPNs a primary management tool, since traditional chemicals are not competitors. Other situations where chemical pest management is restricted, either due to safety concerns or where there is a paucity of registered chemicals for use, present further opportunities to increase EPN use. For example, insecticides used against soil insect pests are under scrutiny for a number of reasons that include water source contamination or risk to people using the treated resources (e.g. golf courses and school grounds). Indeed, these venues represent areas where EPNs are increasingly relied on as pest management options.

Worker safety is an area of increasing concern and regulation. Greenhouse production is ideal for EPN use for the reason that there is no post-treatment entry restriction. Workers can use EPNs to treat for root weevils, fungus gnats, etc., and essentially continue uninterrupted after treatment. As EPN applications have led to reduced populations of pests such as black vine weevil by 75% to as much as 100%, depending on conditions, they are attractive options for greenhouse operators. Boosting the utilization of EPNs in these conditions may simply require increasing the general awareness of them as an effective pest control option.

Finally, there are situations where EPNs may be effective against a life stage of a pest that would not be amenable to control by chemicals. This would include insects that spend part of their life cycle in a cryptic habitat, such as inside fallen fruit or nuts, or boring into trunks or stems. The mobility of EPN IJs makes them able to reach locations that sprayed non-mobile materials cannot. This leads to results such as a 70% reduction in navel orangeworm larvae within fallen pistachio nuts.

Even better results have been reported when EPNs are used to control the moths (e.g. Sesiidae) and weevils (e.g. *O. sulcatus*) that are pests of stone fruit, pome fruit, grapes and berries; in these settings, reductions in pest populations by 75% or more are typical. IPM programmes seek to challenge as many life stages of pests as possible, and EPNs can be used in stages where chemical insecticides would be ineffective.

Overall, EPNs are an effective pest control option in many situations in the USA; the limit to the broader utilization of EPNs is driven primarily by their increased cost relative to other chemical or biological options. In that vein, a compelling case for future research is the ability of EPNs to contribute synergistically to other chemical or biological control approaches. Effective examples of this synergism have been reported in both turfgrass (for the control of white grubs) and in greenhouses, nurseries or field settings where the black vine weevil is a threat. Identifying situations where EPNs can contribute to IPM either by providing better control, by providing it at a reduced cost, or by contributing synergistically to existing control options should continue to be a major focus moving forward.

Acknowledgements

The authors would like to acknowledge the many scientists who have worked on or continue to research EPNs, both in the USA and around the world. These individuals and their laboratories continue to advance the basic and applied knowledge of EPNs. The Society of Nematologists, the Society for Invertebrate Pathology and the Entomological Society of America, among others, continue to host high-quality meetings for discussions relating to EPNs. For that we extend our profound appreciation.

References

Akhurst, R. and Smith, K. (2002) Regulation and safety. In: Gaugler, R. (ed.) *Entomopathogenic Nematology*. CAB International, Wallingford, UK, pp. 311–332.

Ali, J.G., Campos-Herrera, R., Alborn, H.T., Duncan, L.W. and Stelinski, L.L. (2013) Sending mixed messages: a trophic cascade produced by a belowground herbivore-induced cue. *Journal of Chemical Ecology* 39, 1140–1147.

Ansari, M.A., Hussain, M.A. and Moens, M. (2009) Formulation and application of entomopathogenic nematode-infected cadavers for control of *Hoplia philanthus* in turfgrass. *Pest Management Science* 65, 367–374.

Arthurs, S., Heinz, K.M. and Prasifka, J.R. (2004) An analysis of using entomopathogenic nematodes against above-ground pests. *Bulletin of Entomological Research* 94, 297–306.

Askary, T.H. (2010) Nematodes as biocontrol agents. In: Lichtfouse, E. (ed.) *Sociology, Organic Farming, Climate Change and Soil Science*. Springer, The Netherlands, pp. 347–378.

Babendreier, D., Jeanneret, P., Pilz, C. and Toepfer, S. (2015) Non-target effects of insecticides, entomopathogenic fungi and nematodes applied against western corn rootworm larvae in maize. *Journal of Applied Entomology* 139, 457–467.

Batalla-Carrera, L., Morton, A. and Garcıa-del-Pino, F. (2010) Efficacy of nematodes against the tomato leafminer *Tuta absoluta* in laboratory and greenhouse conditions. *BioControl* 55, 523–530.

Beck, B., Spanoghe, P., Moens, M., Pollet, S., Temmerman, F. and Nuyttens, D. (2014) Improving control of the cabbage root fly with entomopathogenic nematodes. *Aspects of Applied Biology* 122, 125–134.

Bélair, G., Koppenhöfer, A.M., Dionne, J. and Simard, L. (2010) Current and potential use of pathogens in the management of turfgrass insects as affected by new pesticide regulations in North America. *International Journal of Pest Management* 56, 51–60.

Bélair, G., Simard, L. and Dionne, J. (2013) Canadian entomopathogenic nematode isolates: virulence against black cutworm (Lepidoptera: Noctuidae). *Phytoprotection* 93, 43–46.

Bergh, J.C. (2012) Grape root borer. In: Bostanian, N.J., Vincent, C. and Isaacs, R. (eds) *Arthropod Management in Vineyards: Pests, Approaches, and Future Directions.* Springer, New York, pp. 383–402.

Berry, R.E. and Liu, J. (1998) *Insect Control in Cranberry with Indigenous Entomopathogenic Nematodes*. The Cranberry Institute, Wareham, Massachusetts.

Berry, R.E., Liu, J. and Groth, E. (1997) Efficacy and persistence of *Heterorhabditis marelatus* (Rhabditida: Heterorhabditidae) against root weevils (Coleoptera: Curculionidae) in strawberry. *Environmental Entomology* 26, 465–470.

Booth, S.R., Tanigoshi, L.K. and Shanks, C.H. Jr (2002) Evaluation of entomopathogenic nematodes to manage root weevil larvae in Washington State cranberry, strawberry, and red raspberry. *Environmental Entomology* 31, 895–902.

Bruck, D.J., Edwards, D.L. and Donahue, K.M. (2008) Susceptibility of the strawberry crown moth (Lepidoptera: Sesiidae) to entomopathogenic nematodes. *Journal of Economic Entomology* 101, 251–255.

Bullock, R.C., Pelosi, R.R. and Killer, E.E. (1999) Management of citrus root weevils (Coleoptera: Curculionidae) on Florida citrus with soil-applied entomopathogenic nematodes (Nematoda: Rhabditida). *Florida Entomologist* 82, 1–7.

Burlando, T.M., Kaya, H.K. and Timper, P. (1993) In outdoor planters in commercial centers, insect-parasitic nematodes are effective against black vine weevil. *California Agriculture* 47, 16–18.

Campos-Herrera, R., El-Borai, F.E. and Duncan, L.W. (2012) Wide inter-guild relationships among entomopathogenic and free-living nematodes in soil as measured by real time qPCR. *Journal of Invertebrate Pathology* 111, 126–135.

Campos-Herrera, R., Pathak, E., El-Borai, F.E., Stuart, R.J., Gutierrez, C., *et al.* (2013) Geospatial patterns of soil properties and the biological control potential of entomopathogenic nematodes in Florida citrus groves. *Soil Biology and Biochemistry* 66, 163–174.

Campos-Herrera, R., El-Borai, F.E. and Duncan, L.W. (2015) It takes a village: entomopathogenic nematode community structure and conservation biological control in Florida (US) orchards. In: Campos-Herrera, R. (ed.) *Nematode Pathogenesis of Insects and Other Pests*. Springer International Publishing, Switzerland, pp. 329–351.

Capinera, J.L., Cranshaw, W.S. and Hughes, H.G. (1986) Suppression of raspberry crown borer, *Pennisetia marginata* (Harris) (Lepidoptera: Sesiidae) with soil applications of *Steinernema feltiae* (Rhabditida: Steinernematidae). *Journal of Invertebrate Pathology* 48, 257–258.

Choo, H.Y. and Kaya, H.K. (1991) Influence of soil texture and presence of roots on host finding by *Heterorhabditis bacteriophora. Journal of Invertebrate Pathology* 58, 279–280.

Choo, H.Y., Kaya, H.K., Burlando, T.M. and Gaugler, R. (1989). Entomopathogenic nematodes: host-finding ability in the presence of plant roots. *Environmental Entomology* 18, 1136–1140.

Cottrell, T.E. and Shapiro-Ilan, D.I. (2006) Susceptibility of the peachtree borer, *Synanthedon exitiosa*, to *Steinernema carpocapsae* and *Steinernema riobrave* in laboratory and field trials. *Journal of Invertebrate Pathology* 92, 85–88.

Cowles, R.S., Polavarapu, S., Williams, R.N., Thies, A. and Ehlers, R.-U. (2005) Soft fruit applications. In: Grewal, P.S., Ehlers, R.U. and Shapiro-Ilan, D.I. (eds) *Nematodes as Biocontrol Agents*. CAB International, Wallingford, UK, pp. 231–254.

De Ley, T.I., McDonnell, R., Lopez, S., Paine, T.D. and De Ley, P. (2014) *Phasmarhabditis hermaphrodita* (Nematoda: Rhabditidae), a potential biocontrol agent isolated for the first time from invasive slugs in North America. *Nematology* 16, 1129–1138.

Denno, R.F., Gruner, D.S. and Kaplan, I. (2008) Potential for entomopathogenic nematodes in biological control: a meta-analytical synthesis and insights from trophic cascade theory. *Journal of Nematology* 40, 61–72.

Dillon, A.B., Foster, A., Williams, C.D. and Griffin, C.T. (2012) Environmental safety of entomopathogenic nematodes – effects on abundance, diversity and community structure of non-target beetles in a forest ecosystem. *Biological Control* 63, 107–114.

Dolinski, C., Choo, H.Y. and Duncan, L.W. (2012) Grower acceptance of entomopathogenic nematodes: case studies on three continents. *Journal of Nematology* 44, 226–235.

Dolinski, C., Shapiro-Ilan, D.I. and Lewis, E.E. (2015) Insect cadaver applications: pros and cons. In: Campos-Herrera, R. (ed.) *Nematode Pathogenesis of Insects and Other Pests*. Springer International Publishing, Switzerland, pp. 207–229.

Duncan, L.W. and McCoy, C.W. (1996) Vertical distribution in soil, persistence, and efficacy against citrus root weevil (Coleoptera: Curculionidae) of two species of entomogenous nematodes (Rhabditida, Steinernematidae, Heterorhabditidae). *Environmental Entomology* 25, 174–178.

Duncan, L.W., Graham, J.H., Dunn, D.C., Zellers, J., McCoy, C.W. and Nguyen, K. (2003) Incidence of endemic entomopathogenic nematodes following application of *Steinernema riobrave* for control of *Diaprepes abbreviatus*. *Journal of Nematology* 35, 178–186.

Duncan, L.W., Graham, J.H., Zellers, J., Bright, D., Dunn, D.C., *et al.* (2007) Food web responses to augmenting the entomopathogenic nematodes in bare and animal manure-mulched soil. *Journal of Nematology* 39, 176–189.

Duncan, L.W., Stuart, R.J., El-Borai, F.E., Campos-Herrera, R., Pathak, E. and Graham, J.H. (2013) Modifying orchard planting sites conserves entomopathogenic nematodes, reduces weevil herbivory and increases citrus tree growth, survival and fruit yield. *Biological Control* 64, 26–36.

Ebssa, L. and Koppenhöfer, A.M. (2011) Efficacy and persistence of entomopathogenic nematodes for black cutworm control in turfgrass. *Biocontrol Science and Technology* 21, 779–796.

Ebssa, L., Borgemeister, C., Berndt, O. and Poehling, H.M. (2001) Efficacy of entomopathogenic nematodes against soil-dwelling life stages of western flower thrips, *Frankliniella occidentalis* (Thysanoptera: Thripidae). *Journal of Invertebrate Pathology* 78, 119–127.

Ebssa, L., Borgemeister, C. and Poehling, H.M. (2006) Simultaneous application of entomopathogenic nematodes and predatory mites to control western flower thrips *Frankliniella occidentalis. Biological Control* 39, 66–74.

Ehlers, R.-U. (2005) Forum on safety and regulation. In: Grewal, P.S., Ehlers, R.-U. and Shapiro-Ilan, D.I. (eds) *Nematodes as Biocontrol Agents*. CAB International, Wallingford, UK, pp. 107–114.

El-Borai, F.E., Brentu, C.F. and Duncan, L.W. (2007) Augmenting entomopathogenic nematodes in soil from a Florida citrus orchard: non-target effects of a trophic cascade. *Journal of Nematology* 39, 203–210.

Ellis, J.D., Spiewok, S., Delaplane, K.S., Buchholz, S., Neumann, P. and Tedders, W.L. (2010) Susceptibility of *Aethina tumida* (Coleoptera: Nitidulidae) larvae and pupae to entomopathogenic nematodes. *Journal of Economic Entomology* 103, 1126–1134.

Ennis, D.E., Dillon, A.B. and Griffin, C.T. (2010) Simulated roots and host feeding enhance infection of subterranean insects by the entomopathogenic nematode *Steinernema carpocapsae*. *Journal of Invertebrate Pathology* 103, 140–143.

Fischer, P. and Führer, E. (1990) Effect of soil acidity on the entomophilic nematode *Steinernema kraussei* Steiner. *Biology and Fertility of Soils* 9, 174–177.

Gassmann, A.L., Stock, S.P., Sisterson, M.S., Carrière, Y. and Tabashnik, B.E. (2008) Synergism between entomopathogenic nematodes and *Bacillus thuringiensis* crops: integrating biological control and resistance management. *Journal of Applied Ecology* 45, 957–966.

Georgis, R. and Poinar, G.O. Jr (1994) Nematodes as bioinsecticides in turf and ornamentals. In: Leslie, A.R. (ed.) *Integrated Pest Management for Turf and Ornamentals*. CRC Press, Boca Raton, Florida, pp. 477–489.

Georgis, R., Koppenhöfer, A.M., Lacey, L.A., Bélair, G., Duncan, L.W., *et al.* (2006) Successes and failures in the use of parasitic nematodes for pest control. *Biological Control* 38, 103–123.

Greenwood, C.M., Barbercheck, M.E. and Brownie, C. (2011) Short term response of soil microinvertebrates to application of entomopathogenic nematode-infected insects in two tillage systems. *Pedobiologia* 54, 177–186.

Grewal, P.S. (2007) Mushroom pests. In: Lacey, L.A. and Kaya, H.K. (eds) *Field Manual of Techniques in Invertebrate Pathology: Application and Evaluation of Pathogens for Control of Insects and Other Invertebrate Pests*, 2nd edn. Springer, The Netherlands, pp. 457–461.

Grewal, P.S. and Richardson, P.N. (1993) Effects of application rates of *Steinernema feltiae* (Nematoda: Steinernematidae) on control of the mushroom sciarid fly *Lycoriella auripilla*. *Biocontrol Science and Technology* 3, 29–40.

Hara, A.H., Kaya, H.K., Gaugler, R., LeBeck, L.M. and Mello, C.L. (1993) Entomopathogenic nematodes for biological control of the leafminer, *Liriomyza trifolii* (Dipt., Agromyzidae) (Rhabditida: Steinernematidae). *Entomaphaga* 38, 359–369.

Harvey, C.D., Alameen, K.M. and Griffin, C.T. (2012) The impact of entomopathogenic nematodes on a non-target, service-providing longhorn beetle is limited by targeted application when controlling forestry pest *Hylobius abietis*. *Biological Control* 62, 173–182.

Haukeland, S. and Lola-Luz, T. (2010) Efficacy of the entomopathogenic nematodes *Steinernema kraussei* and *Heterorhabditis megidis* against the black vine weevil *Otiorhynchus sulcatus* in open field-grown strawberry plants. *Agricultural and Forest Entomology* 12, 363–369.

Henderson, D. and Singhai, R. (1999) Efficacy study of three species of parasitic nematodes for control of cranberry girdler, *Chrysoteuchia topiaria*, in British Columbia, Canada. In: Deziel, G. and Hogan, M. (eds) *Cranberry Research Compilation. Progress and Final Reports on Cranberry Research Conducted in* 1998. Cranberry Institute, Wareham, Massachusetts, pp. 167–176.

Hodson, A.K., Siegel, J.P. and Lewis, E.E. (2012) Ecological influence of the entomopathogenic nematode, *Steinernema carpocapsae*, on pistachio orchard soil arthropods. *Pedobiologia* 55, 51–58.

Jabbour, R. and Barbercheck, M.E. (2011) Soil microarthropod response to the application of entomopathogenic nematode-killed insects in maize and flower strip habitats. *Pedobiologia* 54, 243–251.

Jagdale, G.B., Casey, M.L., Grewal, P.S. and Lindquist, R.K. (2004) Application rate and timing, potting medium, and host plant effects on efficacy of *Steinernema feltiae* against the fungus gnat, *Bradysia coprophila*, in floriculture. *Biological Control* 29, 296–305.

Karban, R., Holdson, A.K., Gruner, D.S., Lewis, E.E., Karban, J., *et al.* (2011) Lack of susceptibility of soil-inhabiting *Platyprepia virginalis* caterpillars, a native arctiid, to entomopathogenic nematodes in nature. *Entomologia Experimentalis et Applicata* 140, 28–34.

Kaspi, R., Ross, A., Hodson, A., Stevens, G.N., Kaya, H.K. and Lewis, E.E. (2010) Foraging efficacy of the entomopathogenic nematode *Steinernema riobrave* in different soil types from California citrus groves. *Applied Soil Ecology* 45, 243–253.

Kaya, H.K. (1990) Soil ecology. In: Gaugler, R. and Kaya, H.K. (eds) *Entomopathogenic Nematodes in Biological Control*. CRC Press, Boca Raton, Florida, pp. 93–115.

Kaya, H.K. and Brown, L.R. (1986) Field application of entomogenous nematodes for biological control of clear-wing moth borers in alder and sycamore trees. *Journal of Arboriculture* 12, 150–154.

Kaya, H.K. and Koppenhöfer, A.M. (1996) Effects of microbial and other antagonistic organism and competition on entomopathogenic nematodes. *Biocontrol Science and Technology* 6, 357–372.

Kaya, H.K., Aguillera, M.M., Alumai, A., Choo, H.Y., de la Torre, M., *et al.* (2006) Status of entomopathogenic nematodes and their symbiotic bacteria from selected countries or regions of the world. *Biological Control* 38, 134–155.

Kepenekci, I., Hazir, S. and Lewis, E.E. (2016) Evaluation of entomopathogenic nematodes and the supernatants of the *in vitro* culture medium of their mutualistic bacteria for the control of the root-knot nematodes *Meloidogyne incognita* and *M. arenaria*. *Pest Management Science* 72, 327–334.

Kim, H.H., Choo, H.Y., Kaya, H.K., Lee, D.W., Lee, S.M. and Jeon, H.Y. (2004) *Steinernema carpocapsae* (Rhabditida: Steinernematidae) as a biological control agent against the fungus gnat *Bradysia agrestis* (Diptera: Sciaridae) in propagation houses. *Biocontrol Science and Technology* 14, 171–183.

Kinoshita, M. and Yamanaka, S. (1998) Development and prevalence of entomopathogenic nematodes in Japan. *Japanese Journal of Nematology* 28, 42–45.

Koppenhöfer, A.M. and Fuzy, E.M. (2002) Comparison of neonicotinoid insecticides as synergists for entomopathogenic nematodes. *Biological Control* 24, 90–97.

Koppenhöfer, A.M. and Fuzy, E.M. (2003) Ecological characterization of *Steinernema scarabaei*, a scarab-adapted entomopathogenic nematode from New Jersey. *Journal of Invertebrate Pathology* 83, 139–148.

Koppenhöfer, A.M. and Fuzy, E.M. (2004) Effect of white grub developmental stage on susceptibility to entomopathogenic nematodes. *Journal of Economic Entomology* 97, 1842–1849.

Koppenhöfer, A.M. and Fuzy, E.M. (2007) Soil moisture effects and persistence of the entomopathogenic nematodes *Steinernema scarabaei*, *S. glaseri*, *Heterorhabditis zealandica*, and *H. bacteriophora*. *Applied Soil Ecology* 35, 128–139.

Koppenhöfer, A.M. and Fuzy, E.M. (2008) Attraction of four entomopathogenic nematode species to four white grub species. *Journal of Invertebrate Pathology* 99, 227–234.

Koppenhöfer, A.M. and Fuzy, E.M. (2009) Long-term effects and persistence of *Steinernema scarabaei* applied for suppression of *Anomala orientalis* (Coleoptera: Scarabaeidae). *Biological Control* 48, 63–72.

Koppenhöfer, A.M. and Kaya, H.K. (1997) Additive and synergistic interaction between entomopathogenic nematodes and *Bacillus thuringiensis* for scarab grub control. *Biological Control* 8, 131–137.

Koppenhöfer, A.M., Grewal, P.S. and Kaya, H.K. (2000) Synergism of imidacloprid and entomopathogenic nematodes against white grubs: the mechanism. *Entomologia Experimentalis et Applicata* 94, 283–293.

Koppenhöfer, A.M., Rodriguez-Saona, C.R., Polavarapu, S. and Holdcraft, R.J. (2008) Entomopathogenic nematodes for control of *Phyllophaga georgiana* (Coleoptera: Scarabaeidae) in cranberries. *Biocontrol Science and Technology* 18, 21–31.

Koppenhöfer, A.M., Kostromytska, O.S., McGraw, B.A. and Ebssa, L. (2015) Entomopathogenic nematodes in turfgrass: ecology and management of important insect pests in North America. In: Campos-Herrera, R. (ed.) *Nematode Pathogenesis of Insects and Other Pests.* Springer International Publishing, Switzerland, pp. 309–327.

Lacey, L.A. and Georgis, R. (2012) Entomopathogenic nematodes for control of insect pests above and below ground with comments on commercial production. *Journal of Nematology*, 44, 218–225.

Lacey, L.A. and Shapiro-Ilan, D.I. (2008) Microbial control of insect pests in temperate orchard systems: potential for incorporation into IPM. *Annual Review of Entomology* 53, 121–144.

Lacey, L.A., Arthurs, S.P., Unruh, T.R., Headrick, H. and Fritts, R. Jr (2006) Entomopathogenic nematodes for control of codling moth (Lepidoptera: Tortricidae) in apple and pear orchards: effect of nematode species and seasonal temperatures, adjuvants, application equipment and post-application irrigation. *Biological Control* 37, 214–223.

Lacey, L.A., Grzywacz, D., Shapiro-Ilan, D.I., Frutos, R., Brownbridge, M. and Goettel, M.S. (2015) Insect pathogens as biological control agents: back to the future. *Journal of Invertebrate Pathology* 132, 1–41.

LeBeck, L.M., Gaugler, R., Kaya, H.K., Hara, A.H. and Johnson, M.W. (1993) Host stage suitability of the leafminer *Liriomyza trifolii* (Diptera: Agromyzidae) to the entomopathogenic nematode *Steinernema carpocapsae* (Rhabditida: Steinernematidae). *Journal of Invertebrate Pathology* 62, 58–63.

Lewis, E.E. and Grewal, P.S. (2006) Interactions between plant-parasitic nematodes and entomopathogenic nematodes. In: Grewal, P.S., Ehlers, R.-U. and Shapiro-Ilan, D.I. (eds) *Nematodes as Biocontrol Agents*. CAB International, Wallingford, UK, pp. 349–362.

Lewis, E.E., Campbell, J.F. and Gaugler, R. (1998) A conservation approach to using entomopathogenic nematodes in turf and landscapes. In: Barbosa, P. (ed.) *Conservation Biological Control.* Academic Press, San Diego, California, pp. 235–254.

Lewis, E.E., Campbell, J., Griffin, C., Kaya, H.K. and Peters, A. (2006) Behavioral ecology of entomopathogenic nematodes. *Biological Control* 38, 66–79.

Lewis, E.E., Hazir, S., Hodson, A. and Gulcu, B. (2015) Trophic relationships of entomopathogenic nematodes in agricultural habitats. In: Campos-Herrera, R. (ed.) *Nematode Pathogenesis of Insects and Other Pests*. Springer International Publishing, Switzerland, pp. 139–163.

Lola-Luz, T. and Downes, M. (2007) Biological control of black vine weevil *Otiorhynchus sulcatus* in Ireland using *Heterorhabditis megidis*. *Biological Control* 40, 314–319.

Long, S.J., Richardson, P.N., Wilmott D.M. and Edmondson, R.N. (2000) Infectivity of entomopathogenic nematodes (Steinernematidae, Heterorhabditidae) to mushroom phorid fly (*Megaselia halterata*) larvae. *Nematology* 2, 451–459.

McCoy, C.W. (1999) Arthropod pests of citrus roots. In: Timmer, L.W. and Duncan, L.W. (eds) *Citrus Health Management.* APS Press, St Paul, Minnesota, pp. 149–156.

McGraw, B.A. and Koppenhöfer, A.M. (2008) Evaluation of two endemic and five commercial entomopathogenic nematode species (Rhabditida: Heterorhabditidae and Steinernematidae) against annual bluegrass weevil (Coleoptera: Curculionidae) larvae and adults. *Biological Control* 46, 467–475.

McGraw, B.A., Vittum, P.J., Cowles, R.S. and Koppenhöfer, A.M. (2010) Field evaluation of entomopathogenic nematodes for the biological control of the annual bluegrass weevil, *Listronotus maculicollis* (Coleoptera: Curculionidae), in golf course turfgrass. *Biocontrol Science and Technology* 20, 149–163.

McKern, J.A., Johnson, D.T. and Lewis, B.A. (2007) Biology and control of the raspberry crown borer (Lepidoptera: Sesiidae). *Journal of Economic Entomology* 100, 398–404.

Navarro, M.J. and Gea, F.J. (2014) Entomopathogenic nematodes for the control of phorid and sciarid flies in mushroom crops. *Pesquisa Agropecuária Brasileira* 49, 11–17.

Nielsen, A.L. and Lewis, E.E. (2012) Designing the ideal habitat for entomopathogen use in nursery production. *Pest Management Science* 68, 1053–1061.

Nielsen, A.L., Spence, K.O. and Lewis, E.E. (2008) Efficacy patterns of biopesticides used in potting media. *Biopesticides International* 4, 87–101.

Otieno, J.A., Pallmann, P. and Poehling, H.M. (2015) The combined effect of soil-applied azadirachtin with entomopathogens for integrated management of western flower thrips. *Journal of Applied Entomology* 140, 174–186.

Parkman, J.P., Frank, J.H., Walker, T.J. and Schuster, D.J. (1996) Classical biological control of *Scapteriscus* spp. (Orthoptera: Gryllotalpidae) in Florida. *Environmental Entomology* 25, 1415–1420.

Petzold-Maxwell, J.L., Cibils-Stewart, X., French, B.W. and Gassmann, A.J. (2012) Adaptation by western corn rootworm (Coleoptera: Chrysomelidae) to Bt maize: inheritance, fitness costs, and feeding preference. *Journal of Economic Entomology* 105, 1407–1418.

Polavarapu, S., Stuart, R.J. and Schiffhauer, D.E. (2000) Evaluation of insecticides against cranberry rootworm, 1998. *Arthropod Management Tests* 25, 66–67.

Progar, R.A., Kruse, J.J., Lundquist, J.E., Zogas, K.P. and Rinella, M.J. (2015) An entomopathogenic fungus and nematode prove ineffective for biocontrol of an invasive leaf miner *Profenusa thomsoni* in Alaska. *Biocontrol Science and Technology* 25, 373–382.

Ram, K., Gruner, D.S., McLaughlin, J.P., Preisser, E. and Strong, D.R. (2008) Dynamics of a subterranean cascade in space and time. *Journal of Nematology* 40, 85–92.

Rasmann, S., Köllner, T.G., Degenhardt, J., Hiltpold, I., Toepfer, S., *et al.* (2005) Recruitment of entomopathogenic nematodes by insect-damaged maize roots. *Nature* 434, 732–737.

Rijal, J.P., Brewster, C.C. and Bergh, J.C. (2014) Effects of biotic and abiotic factors on grape root borer (Lepidoptera: Sesiidae) infestations in commercial vineyards in Virginia. *Environmental Entomology* 43, 1198–1208.

Rizvi, S.A., Hennessey, R. and Knott, D. (1996) Legislation on the introduction of exotic nematodes. *Biocontrol Science and Technology* 6, 477–480.

Rojht, H., Kač, M. and Trdan, S. (2009) Nontarget effect of entomopathogenic nematodes on larvae of twospotted lady beetle (Coleoptera: Coccinellidae) and green lacewing (Neuroptera: Chrysopidae) under laboratory conditions. *Journal of Economic Entomology* 102, 1440–1443.

San-Blas, E. (2013) Progress on entomopathogenic nematology research: a bibliometric study of the last three decades: 1980–2010. *Biological Control* 66, 102–124.

Scheepmaker, J.W.A., Geels, F.P., Griensven, L.J.L., Van, D. and Smits, P.H. (1998) Susceptibility of larvae of the mushroom fly *Megaselia halterata* to the entomopathogenic nematode *Steinernema feltiae* in bioassays. *BioControl* 43, 201–214.

Shapiro-Ilan, D.I. and Gardner, W. (2012) Improved control of *Curculio caryae* (Coleoptera: Curculionidae) through multi-stage pre-emergence applications of *Steinername carpocapsae*. *Journal of Entomological Science* 47, 27–34.

Shapiro-Ilan, D.I. and Mizell, R.F. III (2012) Laboratory virulence of entomopathogenic nematodes to two ornamental plant pests, *Corythucha ciliata* (Hemiptera: Tingidae) and *Stethobaris nemesis* (Coleoptera: Curculionidae). *Florida Entomologist* 95, 922–927.

Shapiro-Ilan, D.I., Lewis, E.E., Behle, R.W. and McGuire, M.R. (2001) Formulation of entomopathogenic nematode-infected cadavers. *Journal of Invertebrate Pathology* 78, 17–23.

Shapiro-Ilan, D.I., Gouge, D.H. and Koppenhöfer, A.M. (2002) Factors affecting commercial success: case study in cotton, turf and citrus. In: Gaugler, R. (ed.) *Entomopathogenic Nematology*. CAB International, Wallingford, UK, pp. 333–353.

Shapiro-Ilan, D.I., Jackson, M., Reilly, C.C. and Hotchkiss, M.W. (2004) Effects of combining an entomopathogenic fungi or bacterium with entomopathogenic nematodes on mortality of *Curculio caryae* (Coleoptera: Curculionidae). *Biological Control* 30, 119–126.

Shapiro-Ilan, D.I., Gouge, D.H., Piggott, S.J. and Fife, J.P. (2006) Application technology and environmental considerations for use of entomopathogenic nematodes in biological control. *Biological Control* 38, 124–133.

Shapiro-Ilan, D.I., Mizell, R.F., Cottrell, T.E. and Horton, D.L. (2008) Control of plum curculio, *Conotrachelus nenuphar*, with entomopathogenic nematodes: effects of application timing, alternate host plant, and nematode strain. *Biological Control* 44, 207–215.

Shapiro-Ilan, D.I., Cottrell, T.E., Mizell, R.F., Horton, D.L. and Davis, J. (2009) A novel approach to biological control with entomopathogenic nematodes: prophylactic control of the peach tree borer, *Synanthedon exitiosa*. *Biological Control* 48, 259–263.

Shapiro-Ilan, D.I., Cottrell, T.E., Mizell, R.F., Horton, D.L., Behle, R.W. and Dunlap, C.A. (2010a) Efficacy of *Steinernema carpocapsae* for control of the lesser peach tree borer, *Synanthedon pictipes*: improved aboveground suppression with a novel gel application. *Biological Control* 54, 23–28.

Shapiro-Ilan, D.I., Morales-Ramos, J.A., Rojas, M.G. and Tedders, W.L. (2010b) Effects of a novel entomopathogenic nematode-infected host formulation on cadaver integrity, nematode yield, and suppression of *Diaprepes abbreviatus* and *Aethina tumida*. *Journal of Invertebrate Pathology* 103, 103–108.

Shapiro-Ilan, D.I., Han, R. and Dolinksi, C. (2012) Entomopathogenic nematode production and application technology. *Journal of Nematology* 44, 206–217.

Shapiro-Ilan, D.I., Wright, S.E., Tuttle, A.F., Cooley, D.R. and Leskey, T.C. (2013) Using entomopathogenic nematodes for biological control of plum curculio, *Conotrachelus nenuphar*: effects of irrigation and species in apple orchards. *Biological Control* 67, 123–129.

Shapiro-Ilan, D.I., Cottrell, T.E., Mizell, R.F., Horton, D.L. and Zaid, A. (2015) Field suppression of the peachtree borer, *Synanthedon exitiosa*, using *Steinernema carpocapsae*: effects of irrigation, a sprayable gel and application method. *Biological Control* 82, 7–12.

Sher, R.B., Parrella, M.P. and Kaya, H.K. (2000) Biological control of the leafminer *Liriomyza trifolii* (Burgess): implications for intra-guild predation between *Diglyphus begini* Ashmead and *Steinernema carpocapsae* (Weiser). *Biological Control* 17, 155–163.

Shields, E.J. (2015) Utilizing persistent entomopathogenic nematodes in a conservation or a more classical biological control approach. In: Campos-Herrera, R. (ed.) *Nematode Pathogenesis of Insects and Other Pests*. Springer International Publishing, Switzerland, pp. 165–184.

Shields, E.J. and Testa, A.M. (2015) New York Case Study: Biological control of *Otiorhynchus ligustici* with native persistent entomopathogenic nematodes using a more classical approach. In: Campos-Herrera, R. (ed.) *Nematode Pathogenesis of Insects and Other Pests*. Springer International Publishing, Switzerland, pp. 285–307.

Shields, E.J., Testa, A., Neumann, G., Flanders, K.L. and Schroeder, P.C. (2009) Biological control of alfalfa snout beetle with a multi-species application of locally-adapted persistent entomopathogenic nematodes: the first success. *American Entomologist* 55, 250–257.

Siegel, J., Lacey, L.A., Fritts, R. Jr, Higbee, B.S. and Noble, P. (2004) Use of steinernematid nematodes for post harvest control of navel orangeworm (Lepidoptera: Pyralidae, *Amyelois transitella*) in fallen pistachios. *Biological Control* 30, 410–417.

Siegel, J.P., Lacey, L.A., Higbee, B.S., Noble, P. and Fritts, R. Jr (2006) Effect of application rates and abiotic factors on *Steinernema carpocapsae* for control of overwintering navel orangeworm (Lepidoptera: Pyralidae, *Amyelois transitella*) in pistachios. *Biological Control* 36, 324–330.

Simard, L., Bélair, G. and Brodeur, J. (2002) Susceptibility of cranberry girdler to entomopathogenic nematodes. *The Canadian Entomologist* 134, 329–330.

Simser, D. and Roberts, S. (1994) Suppression of strawberry root weevil, *Otiorhynchus ovatus*, in cranberries by entomopathogenic nematodes (Nematoda, Steinernematidae and Heterorhabditidae). *Nematologica* 40, 456–462.

Sirjani, F.O., Lewis, E.E. and Kaya, H.K. (2009) Evaluation of entomopathogenic nematodes against the olive fruit fly, *Bactrocera oleae* (Diptera: Tephritidae). *Biological Control* 48, 274–280.

Smith, K., Fisher, G.C., Poole, A. and Jackson, D. (1993) Effectiveness of Biosafe-N, *Steinernema carpocapsae* for control of cranberry girdler, 1992. Cranberry Agricultural Research, Ocean Spray Cranberries, Lakeville-Middleboro, Massachusetts.

Smith, K.A. (1994) Control of weevils with entomopathogenic nematodes. In: Smith, K.A. and Hatsukade, M. (eds) *Control of Insect Pests with Entomopathogenic Nematodes*. Food and Fertilizer Technology Center, Republic of China in Taiwan, pp. 1–13.

Spence, K.O., Lewis, E.E. and Perry, R.N. (2008) Host-finding and invasion by entomopathogenic and plant-parasitic nematodes: evaluating the ability of laboratory bioassays to predict field results. *Journal of Nematology* 40, 93–98.

Spence, K.O., Stevens, G.N., Arimoto, H., Ruiz-Vega, J., Kaya, H.K. and Lewis, E.E. (2011) Effect of insect cadaver desiccation and soil water potential during rehydration on entomopathogenic nematode (Rhabditida: Steinernematidae and Heterorhabditidae) production and virulence. *Journal of Invertebrate Pathology* 106, 268–273.

Stuart, R.J., Polavarapu, S., Lewis, E.E. and Gaugler, R. (1997) Differential susceptibility of *Dysmicoccus vaccinii* (Homoptera: Pseudococcidae) to entomopathogenic nematodes (Rhabditida: Heterorhabditidae and Steinernematidae). *Journal of Economic Entomology* 90, 925–932.

Stuart, R.J., El-Borai, F.E. and Duncan, L.W. (2008) From augmentation to conservation of entomopathogenic nematodes: trophic cascades, habitat manipulation and enhanced biological control of *Diaprepes abbreviatus* root weevils in Florida citrus groves. *Journal of Nematology* 40, 73–84.

Toepfer, S., Knuth, P., Glas, M. and Kuhlmann, U. (2014) Successful application of entomopathogenic nematodes for the biological control of western corn rootworm larvae in Europe – a mini review. *Julius-Kühn-Archiv* 444, 59–66.

Tomalak, M., Piggott, S. and Jagdale, G.B. (2005) Glasshouse applications. In: Grewal, P.S., Ehlers, R.-U. and Shapiro-Ilan, D.I. (eds) *Nematodes as Biocontrol Agents*. CAB International, Wallingford, UK, pp. 147–166.

van Tol, R.W.H.M. and Raupp, M.J. (2006) Nursery and tree applications. In: Grewal, P.S., Ehlers, R.-U., and Shapiro-Ilan, D.I. (eds) *Nematodes as Biocontrol Agents*. CAB International, Wallingford, UK, pp. 167–190.

van Tol, R.W.H.M., van der Sommen, A.T.C., Boff, M.I.C., van Bezooijen, J., Sabelis, M.W. and Smits, P.H. (2001) Plants protect their roots by alerting the enemies of grubs. *Ecology Letters* 4, 292–294.

van Tol, R.W.H.M., van Dijk, N. and Sabelis, M.W. (2004) Host plant preference and performance of the vine weevil *Otiorhynchus sulcatus*. *Agricultural and Forest Entomology* 6, 267–278.

Wardlow, L.R., Piggott, S. and Goldsworthy, R. (2000) Foliar application of *Steinernema feltiae* for the control of flower thrips. *Mededelingen (Rijksuniversiteit te Gent. Fakulteit van de Landbouwkundige en Toegepaste Biologische Wetenschappen)* 66, 285–291.

Williams, E.C. and Walters, K.F.A. (2000) Foliar application of the entomopathogenic nematode *Steinernema feltiae* against leaf miners on vegetables. *Biocontrol Science and Technology* 10, 61–70.

Williams, R.N., Fickle, D.S., Grewal, P.S. and Dutcher, J. (2010) Field efficacy against the grape root borer *Vitacea polistiformis* (Lepidoptera: Sesiidae) and persistence of *Heterorhabditis zealandica* and *H. bacteriophora* (Nematoda: Heterorhabditidae) in vineyards. *Biological Control* 53, 86–91.

Zenner, A.N.R.L., O'Callaghan, K.M. and Griffin, C.T. (2014) Lethal fighting in nematodes is dependent on developmental pathway: male–male fighting in the entomopathogenic nematode *Steinernema longicaudum*. *PLoS ONE* 9(2), e89385. DOI: 10.1371/journal.pone.0089385.

Zhu, H., Grewal, P.S. and Reding, M.E. (2011) Development of a desiccated cadaver delivery system to apply entomopathogenic nematodes for control of soil pests. *Applied Engineering in Agriculture* 27, 317–324.

15 Status of Entomopathogenic Nematodes in Integrated Pest Management Strategies in Canada

Qing Yu[1]* Fengcheng Sun[2] and James Coupland[3]

[1]*Canadian National Collection of Nematodes, Ottawa Research and Development Centre, Ottawa, Canada;* [2]*Canadian Food Inspection Agency, Nematology Laboratory, Ottawa, Canada;* [3]*FarmForest Research Inc, Almonte, Canada*

15.1 Introduction

Entomopathogenic nematodes (EPNs) commonly refer to species of *Steinernema* and *Heterorhabditis*. In recent years, several species have been developed commercially for controlling insect pests in agriculture and in turf. In Canada, research on EPNs began in the early 60s at the Belleville Research Institute, and in the 90s, the mode of action was the focus. The EPN species on the market were mostly of foreign origin. Only recently has more research shifted towards discovering and understanding the indigenous species, and evaluating the efficacies of those indigenous species (Bélair *et al.*, 2013).

Although Canada is a large country, its agriculture is largely concentrated in the region not exceeding 100 km from the US northern border. Canada is one of the largest agricultural producers and exporters in the world. Agriculture has contributed over CAN$100 billion annually to Canada's GDP since 2000, and its importance is increasing (AAFC, 2016). Canada has more than 50.5 million hectares of land classified as dependable agricultural land (Statistics Canada, 2014). A large greenhouse industry is also concentrated in southern Ontario.

No crop escapes insect pests, which cause enormous loss to the farmers. Pest control has been one of the top priorities for research since the very beginning of the federation. Canadian agricultural, forest and natural ecosystems are threatened also by invasive alien species (IAS) of plant, animal and microbes. In 2004, the government of Canada developed a national strategy to deal with the IAS, and an action plan was implemented in the following year. Canada is a member of the North American Plant Protection Organization (NAPPO), which is a regional organization under the auspices of the International Plant Protection Convention (IPPC). Canada is a member of the Convention on Biological Diversity (CBD), which has been signed by 150 countries.

Agricultural research, other than that conducted at universities, is pursued primarily by the federal department of agriculture, which has had several different names; currently, Agriculture and Agri-Food Canada (AAFC). The Canadian Food Inspection Agency (or CFIA), the national regulatory agency, was created from the related inspection services of three separate federal government departments: Agriculture and Agri-Food Canada (AAFC), Fisheries and Oceans Canada, and Health Canada. It had been under the minister of AAFC since its creation and now is under the minister of Health Canada. The Health Canada Pest Management Regulatory Agency (PMRA) is responsible for pesticide regulation in Canada.

Nematology research in Canada has a long history (Yu and Potter, 2006). The main focus has been

*Corresponding author. E-mail: Qing.Yu@agr.gc.ca

 Biocontrol Agents: Entomopathogenic and Slug Parasitic Nematodes (eds. M.M.M. Abd-Elgawad, T.H. Askary and J. Coupland)

on plant parasitic nematodes that cause serious diseases in crops. It is worth noting the significant contributions to entomopathogenic nematology made by Dr John Webster at the Simon Fraser University, who led studies on secondary metabolites produced by symbiotic bacteria. Currently, nematology in Canada is represented by: the senior author, the only nematode taxonomist with AAFC in Ottawa, Ontario; Dr Tom Forge, focusing on the role of nematodes in nutrient cycles in soil, at AAFC in Agassi, British Columbia; Mr Guy Bélair and Dr Benjamin Mimee on nematode controls at AAFC in St Jean-sur-Richelieu, Quebec; Mr Tom Welacky on soybean cyst nematodes at AAFC in Harrow, Ontario; Dr Fengcheng Sun is the head of the nematode diagnostic laboratory with CFIA in Ottawa, Ontario; and Dr Mario Tenuta, the only active university nematologist, is located at the University of Manitoba, working on nematode ecology. The Canadian National Collection of Nematodes at AAFC in Ottawa is the major repository of its kind in the world, and is curated by Dr Yu.

The present chapter provides a general overview of EPN research and development in Canada, focusing on the specific knowledge gap in EPN biodiversity.

15.2 Integrated Pest Management in Canada

Pesticides are applied to agricultural crops to prevent losses from weeds, insects, fungi and nematodes. In 2011, 69% of Canadian farms growing crops reported applying herbicides, 15% employed insecticides and 23% used fungicides. Saskatchewan and Manitoba crop farms applied the highest share of herbicide application in 2011, while insecticide application was more common in the Atlantic Provinces and British Columbia. Fungicide application was most frequent in Manitoba and least common in Quebec. Pesticides often cause environmental damage (Statistics Canada, 2014).

To reduce the use of pesticides, farmers also use a number of alternative pest control methods. The management of pests, weeds and diseases using biological agents such as insects, mites, fungi and nematodes is an important part of the integrated pest management strategy. The biological control of insects, weeds and plant diseases has a long and successful record, not only in research but also in application (Mason and Gillespie, 2013). The biological control of insects in greenhouses in Canada has been particularly successful. Predatory beetles, parasitic wasps, mites and nematodes have all been widely used in greenhouse vegetables and flowers.

Integrated pest management (IPM) is a process that uses all necessary techniques to suppress pests effectively, economically and in an environmentally sound manner. Pest monitoring is a key component of IPM. Canada, like other industrialized nations, has gone through a painful process from relying solely on pesticides to control pests to IPM. Canada has a long history of IPM research and applications. Now Canadians have been at the forefront in trying to reduce pesticide use since the 1940s. The pesticide-only era is over and IPM is the must-have in pest management. Three types of IPM activities are carried out in Canada. AAFC has a mandate for national programmes that focus on basic research. The provincial departments of agriculture have the mandate for extension activities to disseminate the research from the federal department. Grower groups and crop associations provide some funds for federal and provincial scientists to carry out some of the research, and they also hire consultants that monitor pest development.

15.3 Application of Biological Control Agents in IPM in Canada

Biological control is the use of parasites/parasitoids, predators, pathogens, antagonists or competitors (agents) to suppress a targeted pest population. Using biological control agents, resistant varieties and culture practices such as crop rotations are the three major techniques for pest control in Canada. Researches and applications of biological control agents cover a complete spectrum of all kinds of agents in controlling insects, diseases and weeds; for example, as on 31 January 2011, over 15 species of predatory mites and 50 species of insect parasites and predators had been approved by CFIA for commercial release (AAFC, 2006). EPN commercial products currently available in Canada are presented in Table 15.1. These products are readily available at major retail stores, gardening shops and hardware stores across the country.

15.4 Native Entomopathogenic Nematode Species in Canada

No national survey has taken place for species of EPNs. A few regional surveys, three in western Canada (Mrácek and Webster, 1993; Stock *et al.*,

Table 15.1. A list of commercially available products containing entomopathogenic nematodes in Canada.

Brand	Producer	Nematode species	Pests	Crop
Lawn Guardian	Nature Insect Control (NIC)	*Heterorhabditis bacteriophora*, *Steinernema* spp.	Grubs (*Rhizotrogus majalis* and *Popillia japonica*)	Turf
Birch Leafminer Guard	NIC	*Steinernema feltiae*, *S. carpocapsae* and *H. bacteriophora*	Leaf miners (*Fenusa pusilla*)	Greenhouse crops
Grub Busters	GreenEarth	Not known	Grubs (*R. majalis* and *P. japonica*)	Turf
Nemasys L.	Becker Underwood acquired by BASF in 2012	*S. kraussei*	Weevils (*Otiorhynchus sulcatus*, *O. ovatus*, *O. singularis* and *Sciopithes obscurus*	Blueberries

2000; Mrácek *et al.*, 2005) and one each in Ontario and Quebec (Simard *et al.*, 2007), reported a total of seven species of EPN, five of which were *Steinernema* and the remaining two were *Heterorhabditis*. Their distributions are presented in Fig. 15.1. The species reported include: *Steinernema feltiae* Filipjev, 1934, in British Columbia (BC), Ontario and Quebec; *Steinernema kraussei* Steiner, 1923, BC and Quebec; *Steinernema affine* Bovien, 1937, BC; *Steinernema carpocapsae* Quebec; an unidentified *Steinernema* sp. in Alberta and Newfoundland; *Heterorhabditis megidis* Poinar, Jackson and Klein, 1987, Ontario; and an unidentified *Heterorhabditis* sp. in Quebec (Mrácek and Webster, 1993; Stock *et al.*, 2000; Mrácek *et al.*, 2005; Simard *et al.*, 2007; Bélair *et al.*, 1984, 2001; Finney-Crawley, 1985). Naturally, infestation rates varied depending on species and location. For example, in western Canada, the infestation rate by EPNs was 22%; in BC, 17 sites out of 107 site samples were infested with steinernematids, 7 out of 107 were heterorhabditids, representing 22.4% and 6.1%, respectively (Mrácek and Webster, 1993); and in Ontario, 27% of the golf courses studied were infested with steinernematids, and 21% in Quebec (Simard *et al.*, 2007). Interestingly, EPN species were only found in habitats where human influence was substantial, including agricultural, golf courses, forested and bush hedgerows: none were found in the natural grasslands or virgin forests, and where insect infestations were visible.

Steinernema species were found to be more diverse and more common than *Heterorhabditis* species. The species ratio between the two genera and the infestation rates were correlated positively. The species ratio of *Steinernema*/*Heterorhabditis* (3.5) was similar to the ratio of the numbers of the infestation sites by *Steinernema* and by *Heterorhabditis* (3.6), which was similar to the species worldwide ratio. There are about 54 nominal species of *Steinernema*, and 15 of *Heterorhabditis* (Nguyen and Hunt, 2007). The ratio is three. This phenomenon can also be found elsewhere in the northern hemisphere. For example, in Slovakia, the species ratio is three (Sturhan and Liskova, 1999). In Spain, five species of *Steinernema* were recovered against one species of *Heterorhabditis* (Garcia and Palomo, 1996). In Jordan, three species of *Steinernema* and one species of *Heterorhabditis* were found; a ratio of three (Stock *et al.*, 2008).

Surveys of EPNs are generally hampered by the fact that only the infective juvenile stage can be found in soil, all other stages taking place inside the insect hosts. Proper identification has to involve molecular methods and baiting for nematodes with insects (Bedding and Akhurst, 1975). Both are time-consuming. Obvious gaps are the diversity of EPNs in the Prairie Provinces (Alberta, Saskatchewan and Manitoba), the Maritime Provinces (New Brunswick, Nova Scotia and Prince Edward Island) and the three territories (Northwest Territories, Nunavut and Yukon).

15.5 An Outline of Studies on Entomopathogenic Nematodes in Canadian Research Centres

15.5.1 Department of Biological Sciences, Simon Fraser University, Burnaby, British Columbia

The most significant Canadian contributions to the science of entomopathogenic nematology were made by the team led by Professor Dr John M. Webster

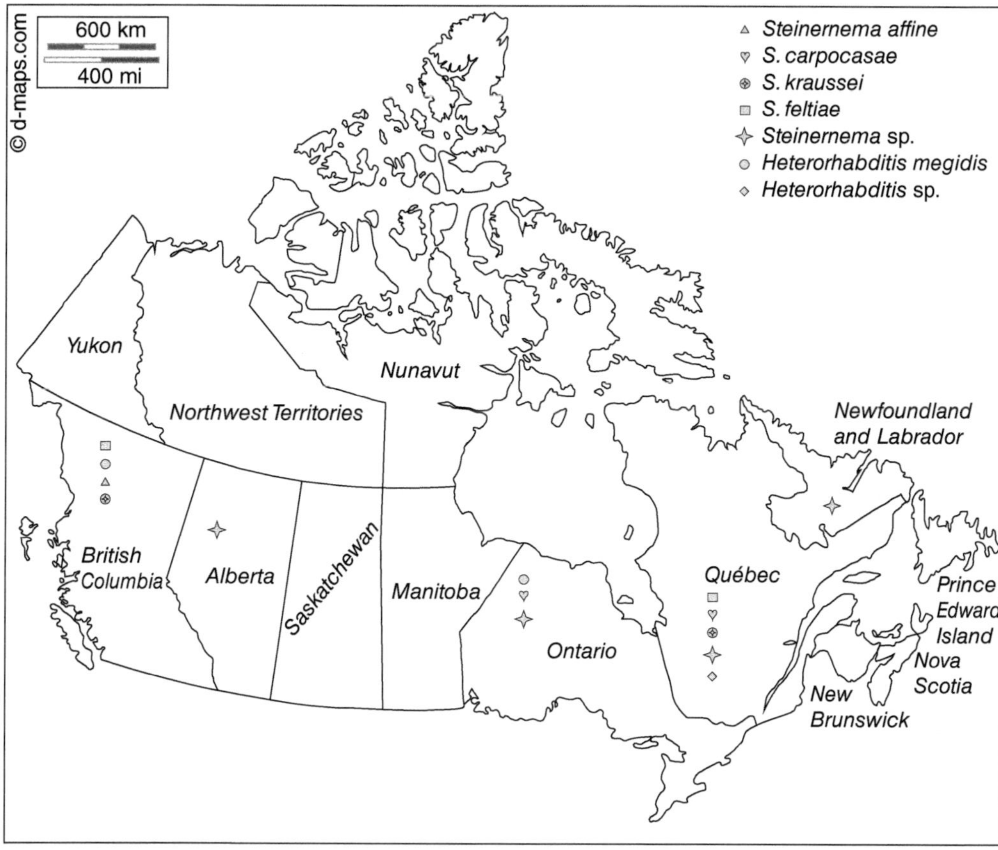

Fig. 15.1. Distribution of the native species of EPNs in Canada.

(now retired), Department of Biological Sciences, Simon Fraser University, Burnaby, British Columbia. The research explored the potential and enhancement of *Steinernema* spp. and *Heterorhabditis* spp. and their associated bacterial symbionts as biological control agents of pest insects (Saunders, 1984; Hayes, 1990; Webster *et al.*, 1998; Hayes *et al.*, 1999), and the factors in the soil that influence the behaviour of these nematodes in their search for potential insect hosts (Hui and Webster, 2000). Also, the bioactivity of the bacterial metabolites has been explored and their role in insect pathology identified. The studies on secondary metabolites of the symbiotic bacteria of EPNs have been the main focus of this group. The EPNs *Heterorhabditis* and *Steinernema*, together with their symbiont bacteria *Photorhabdus* and *Xenorhabdus*, respectively, are obligate and lethal parasites of insects (Clarke and Eberl, 2006). The bacteria are Gram-negative, belonging to the family of Enterobacteriaceae, yet they are unique both phenotypically and genotypically (Murfin *et al.*, 2012). The bacteria are pathogenic to insects. The nematodes serve as vectors: once inside the haemocoel of a host, the bacteria proliferate, release endo- and exotoxins, septicaemia ensues and then insect death (Poinar, 1990). Several species of bacteria symbionts have been studied for their insecticidal property, and their virulence factors have been determined. Twenty species of *Xenorhabdus* and three species of *Photorhabdus* have been described from different species of *Steinernema* and *Heterorhabditis*, respectively.

More than 30 bioactive secondary metabolites have been discovered (Webster *et al.*, 2002), and they have a wide range of bioactivities, such as: antibiotic (Li *et al.*, 1995a,b, 1997; Hu *et al.*, 1997, 1998, 1999; Hu and Webster, 2000); antimycotic (Chen *et al.*, 1994, 1996; Isaacson, 1994; Chen, 1996; Li *et al.*, 1997, 1998; Ng and Webster, 1997;

Isaacson and Webster, 2002); insecticidal, nematicidal (Hu *et al.*, 1995, 1996, 1997, 1998; Hu and Webster, 2000; Hu, 1999); anti-ulcer (McInerney *et al.*, 1991a,b); anticancer (Webster *et al.*, 2000); and antiviral. Among the six compounds with insecticidal toxicity, xenorhabdins were the most toxic on *Heliothis punctigera* (McInerney *et al.*, 1991a). The nematicidal properties of metabolites produced by *Xenorhabdus* and *Photorhabdus* are somewhat surprising in view of the close symbiotic relationship between the bacteria and EPNs, and no biological role has been identified. Most of the metabolites are for competing with other bacteria, and most of the compounds are known to be produced by other soil bacteria and fungi as well.

15.5.2 Department of Biology, Memorial University, St John's Newfoundland

The research team led by Dr R. Gordon at the Department of Biology, Memorial University, St John's, Newfoundland focused on finding cold-tolerant strains of EPN (Finney and Bennett, 1984; Finney-Crawley, 1985), understanding the mechanism of cold tolerance (Gordon *et al.*, 1979; Gordon and Burford, 1984; Jagdale and Gordon, 1994a,b,c,d, 1997a,b,c, 1998; Gordon *et al.*, 1996; Jagdale *et al.*, 1996; Jagdale, 1997) and acclimation in EPNs. With the retirement of Dr R. Gordon, the research activities at the department have shifted to marine nematodes.

15.5.3 Department of Natural Resource Sciences, Macdonald Campus of McGill University, Ste Anne de Bellevue, Quebec

The research team is led by Dr G.B. Dunphy and the group has made significant contribution to the EPN knowledge bank (Dunphy and Webster, 1984, 1991; Dunphy *et al.*, 1997) and has published extensively (not cited due to limitation of space) on the nematode–insect relationship, i.e. resistance mechanisms of the insect and how the nematodes and bacteria overcome the insect's defence systems in general.

15.5.4 Research laboratories in the Federal Department, Agriculture and Agri-Food Canada, Ottawa Research and Development Centre, Ottawa, Ontario

This group, led by Dr Q. Yu of the invertebrate biodiversity team of the national environmental health programme for the department, conducts taxonomic research and provides the resulting expert information on the identity of the nematodes that benefit or pose a threat to Canadian agriculture and its biological resources, including EPNs. Together with the botany and mycology taxonomy groups, one of the largest clusters of taxonomists in a world where taxonomy is a declining discipline, the biodiversity team used to be called the Biosystematics Institute, which during the 1980s was led by Dr Bill Mountain, who was transferred from the Vineland Station, Ontario. Nematode taxonomists who have worked there for a significant period include Dr A. Baker, Mr R. Mulvey, Mr B. Hopper, Dr L-Y Wu, Dr R. Anderson and Dr B. Ebsery.

A novel insecticidal metalloprotease gene was discovered in the symbiotic bacteria from an unidentified EPN species isolated in Ontario using the baiting method (Tambong, 2013; Tambong *et al.*, 2014). This research also addresses regulatory obstacles and promotes regulatory cooperation and modernization, both domestically and with key trading partners, fulfilling national obligations and those related to international treaties (IPPC, CBD). The research and technology transfer activities in nematode biosystematics improves the national capacity to identify, classify and utilize nematodes of importance to Canadian agriculture. The group provides authoritative nematode identifications and advice as part of the National Identification Service (NIS) to federal, provincial and foreign agencies and scientists for research and regulatory compliance purposes. The senior author also curates and expands the Canadian National Collection of Nematodes (CNC), and develops data on nematode species, their hosts, occurrence and distribution.

15.5.5 Vineland Research Station, Vineland, Ontario

The Vineland Research Station was considered during the 1970s–1980s to have the biggest cluster of field nematologists in north-east North America. At the time when Dr Bill Mountain, himself a nematologist, was the director, there was a cluster of five nematologists, including Mr J. Townshend, Dr F. Marks, Dr T. Olthof, Dr J. Potter and Dr W. Johnson. By the late 1990s, when only Dr John Potter remained, Dr Q. Yu and Dr K. Riga joined the team for a short period, until the nematology programme was completely terminated in 2001. Dr Theo H.A. Olthof, and his team at the station, made significant contributions in evaluating the efficacy of EPNs for controlling Sciaridae (Diptera) flies in mushroom production in Ontario (Olthof *et al.*, 1991; Rinker *et al.*, 1995), for controlling leaf miner, *Chromatomyia syngenesiae* Hardy, 1849 (Diptera: Liriomyzidae), on chrysanthemum (Olthof and Broadbent, 1991, 1992; Broadbent and Olthof, 1995), and the plum curculio, *Conotrachelus nenuphar* Herbst, 1797 (Coleoptera: Curculionidae) (Olthof and Hagley, 1993). Dr K. Riga also evaluated EPN efficacies on controlling the European corn borer, *Ostrinia nubilalis* Hübner, 1796 (Lepidoptera: Crambidae) (Riga *et al.*, 2001). The good news is that one nematologist position will be staffed.

15.5.6 Horticulture Research and Development Center, Saint-Jean, Quebec

Significant contributions have been made by G. Bélair and his team towards understanding the biodiversity of EPNs in central Canada (Bélair *et al.*, 1984, 2000; Simard, 2006; Simard *et al.*, 2007). This group has also made significant contributions in evaluating EPN efficacies on carrot weevil, *Listronotus oregonensis* LeConte, 1857 (Coleoptera: Curculionidae) (Bélair and Boivin, 1985; Boivin and Bélair, 1989), and against insect pests in orchards (Bélair *et al.*, 1984, 1998, 1999, 2000; Vincent and Bélair, 1992; Royer *et al.*, 1996; Georgis *et al.*, 2006) and on other pests (Simard *et al.*, 2001, 2002, 2006; Bélair *et al.*, 2002, 2005). This group is still active in EPN research.

15.5.7 Research Institute, Belleville, Ontario

A group led by Dr H.E. Welch, who later became professor at the University of Manitoba, was the first research group focused on EPNs in Canada. A series of papers were published on the effectiveness of *S. feltiae* (DD-136) on a few insect pests (Welch and Briand 1961a,b; Welch and Bronskill, 1962; Welch, 1965). The centre was closed in 1971. At the University of Manitoba, Dr Welch shifted his research on Mermethida.

15.5.8 Natural Resource Canada (NRC), Forestry Canada

At Natural Resource Canada (NRC), Canada Forestry Service – Maritime Regions, Fredericton, New Brunswick, a research group led by Dr Douglas C. Eidt focused on applications of EPNs in managing forest insect pests such as the seedling debarking weevil of conifer, *Hylobius congener* Dalla Torre *et al.*, 1943 (Coleoptera: Curculionidae), and the spruce bud moth, *Zeiraphera canadensis* Muturra and Freeman, 1967 (Lepidoptera: Tortricidae) (Eidt and Dunphy, 1991; Zervos *et al.*, 1991, 1994; Eidt and Thurston, 1994, 1995; Sweeney and Gesner, 1994; Eidt *et al.*, 1995a,b; Sweeney *et al.*, 1998). This research activity had been suspended with the passing of Dr Eidt in 2011.

15.5.9 Vineland Research and Innovation Centre, Vineland, Ontario (private)

The Vineland Research and Innovation Centre, a private or semi-private entity, was established in 2005 with the aim to serve the horticulture industry in Canada's Niagara region. Its researchers in Applied Genomics, Consumer Insights and Horticultural Production Systems work together to deliver innovative products and production solutions that address the needs of the horticulture industry and advance Canada's research and commercialization agenda. It has an active research and development (R&D) programme on EPNs.

15.6 Application of Entomopathogenic Nematodes in Crop Protection in Canada

15.6.1 Control of white grub on turf

White grubs are common pests of lawns and turfgrasses in Canada, and cause more problems in central, eastern and the west coast of Canada than in the Prairie Provinces. White grubs are the larval stage of scarab beetles (Coleoptera, Scarabaeidae), which feed on the roots of grasses and other plants. This feeding results in the wilting and death of the grass in irregular-shaped patches on the lawn. Complete losses of grasses are possible if measures are not applied. The three most common species are the June beetles, *Phyllophaga* spp., the European chafer, *Rhizotrogus majalis* Razoumowsky, 1789 (Coleoptera: Melolonthidae) and the Japanese beetle, *Popillia japonica* Newman, 1841 (Coleoptera: Scarabaeidae). The latter two are invasive species, introduced into Canada accidently, and are found mainly in southern Ontario and part of Quebec.

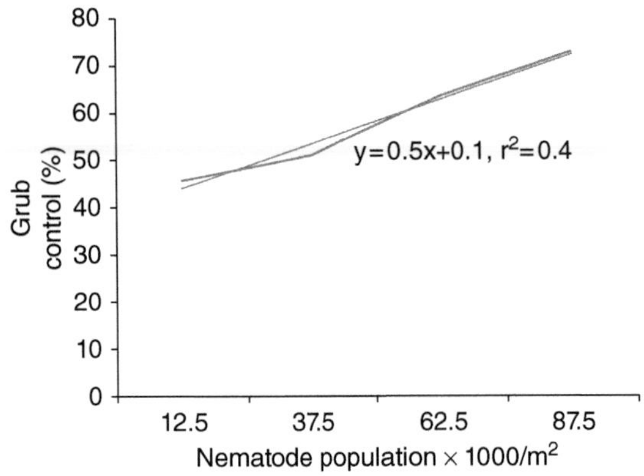

Fig. 15.2. Relationship of EPN population applied to control the level of grubs on turf.

Currently, there are few insecticides approved for use on commercial turfgrass, and virtually none are allowed for home lawns. To make management more complicated and difficult, each species causes greatest damage to grasses at different times of the year, and from different generations within species. Two to three EPN products are on the market. Lawn Guardian™, consisting of *Heterorhabditis bacteriophora* and *Steinernema* spp., was developed by Nature Insect Control (NIC), and Grub Buster was developed by Green Earth, although the identity of the nematode has not been revealed. Major retail stores, hardware stores and local garden centres carry them. The EPN products listed in Table 15.1 do not appear to differentiate on targeting different species, which may have been a factor in their failure to control the pests. Not much has been done to evaluate the efficacy of the various species of EPN on different species of white grubs in Canada. One study compared the efficacy of three species of EPN on Japanese beetles and European chafers (Barton *et al.*, 1996). Overall, the average control levels for European chafers and Japanese beetles ranged from 46% to 73%, respectively, lower than the level of control from the chemical pesticide imidacloprid. *S. feltia* at 200,000/m^2 was more effective than *H. bacteriophora* (Table 15.2). There was a positive correlation between the numbers of EPNs applied and their efficacies (Fig. 15.2). No differences in the efficacy between EPN species, nor in the mortality of the different species of grubs, were found in the study. Many questions remain to be answered. For example, most of the EPNs in these products are from foreign countries and have been in laboratory mass-production conditions for many generations, which is known to reduce the virulence of EPNs (Shapiro-Ilan *et al.*, 2006, 2014). The origin of the EPN also matters, as different species/strains vary in virulence (Jagdale, 1997). For Canadian situations, it is foreseeable that the native species would be more effective. More studies are necessary on how to reinvigorate EPNs after generations of recycling conditions in the laboratory.

As the grubs continue their expansion in invading and colonizing new regions, the need for EPNs to control the pests will increase. This market will remain the largest for EPNs in Canada.

15.6.2 Control of leaf miners in greenhouse crops

Currently, there is only one EPN product (Birch Leafminer Guard, developed by NIC) targeting leaf miners on greenhouse crops. Several species of leaf miner attack a number of greenhouse ornamental and vegetable crops. The most commonly affected ornamental crops are chrysanthemum and gerbera; however, occasionally there is damage to other crops such as gypsophila, marigolds and asters. In Ontario, vegetable crops are less commonly attacked, but certain leaf-miner species can damage tomato and cucumber crops.

The most common species is the American serpentine leaf miner, *Liriomyza trifolii* Burgess, 1880 (Diptera: Agromyzidae), almost always the pest species found in ornamental crops. The vegetable leaf miner, *Liriomyza sativae* Blanchard, 1938; Spencer, 1990 (Diptera: Agromyzidae), is most often found in vegetable crops but is an uncommon pest in general for the greenhouse vegetable industry. The pea leaf miner, *Liriomyza huidobrensis* Blanchard, 1926 (Diptera: Agromyzidae), was found for the first time in Ontario in the mid-1990s in greenhouse bedding plant crops and subsequently became a pest of field vegetables. However, their pest status has declined since the mid-2000s, with few reports since then of presence or damage.

Nematodes (*S. feltiae* and *S. carpocapsae*) have been shown to be effective in controlling leaf-miner larvae (Olthof and Broadbent, 1991; Broadbent and Olthof, 1995). When applied as a foliar spray, the nematodes enter the mines (presumably through egg-laying wound sites) and attack the larvae in the mines. To optimize the efficacy of this approach, nematodes are applied in the evening or on cloudy,

Table 15.2. Control of European chafer and Japanese beetle on turf in Ontario, Canada. (Adapted from Barton *et al.*, 1996.)

		Site 1	Site 2	Site 3	Site 4	
Treatment	Rate	European chafer	Japanese beetle	European chafer	Japanese beetle	Average control (%)
Untreated		21	12.5 a	11.5 a	8.8 a	
NTN 33893 (imidacloprid)	335 g ai/ha	0 (100%)	0.3 (100%) c	0.5 (92%) c	3.3 (67%) b	86
Scan Mask (*Steinernema feltiae*)	200,000/m^2	–	1.5 (85%) c	3.8 (67%) b	3.3 (67%) b	73
Microkil (*S. feltiae*)	50,000/m^2	–	4.5 (62%) b	10.8 (8%) a	3.3 (67%) b	46
Lawn Guardian (*S. carpocapsae* and *Heterorhabditis bacteriophora*)	5400/m^2	–	3.5 (69%) b	5.8 (50%) b	5.8 (34%) b	51
Lawn Guardian (*S. carpocapsae* and *H. bacteriophora*)	3636/m^2	–	2.8 (77%) b	5 (58%) b	3.8 (56%) b	64

Note: to indicate the statistical difference, the same letter (a, b or c) suggests no significant difference.

cool days. This has two benefits: it limits the harmful effects of UV radiation on the nematodes and it allows the leaves to stay moist for several hours, prolonging the life of the nematodes and giving them more time to enter the mines. Ideally, nematodes should be applied twice weekly, especially for the control of heavy populations. Because leaf-miner larvae can develop from egg to pupa in less than a week, a single application per week would leave a gap in control, whereby some eggs could hatch following an application of nematodes and mature through to pupation before the next application. Nematodes can play an important role in biocontrol programmes, especially in transitioning from a pesticide to a biocontrol programme. For example, in resistant leaf-miner populations against which persistent pesticides may have been applied previously, residues can compromise biocontrol programmes that use parasitic wasps. In such cases, nematodes can be used to help control leaf miners until such time as other biocontrols can be introduced safely. They can continue to be used with parasitic wasps. Nematodes on their own will not likely reduce the pest population to non-damaging levels, but can help buy some time until other biocontrols can be used.

15.6.3 Control of weevils on blueberries

Nemasys L. (consists of *S. kraussei*), developed by Becker Underwood, has been used as the only EPN product for controlling weevil on blueberries. A complex of weevil species, including the black vine weevil, *Otiorhynchus sulcatus* Fabricius, 1775 (Coleoptera: Curculionidae), strawberry root weevil, clay coloured weevil, *Otiorhynchus singularis* Linnaeus, 1767 (Coleoptera: Curculionidae), and obscure root weevil, *Sciopithes obscurus* Horn 1876 (Coleoptera: Curculionidae), are a threat to the long-term vigour of blueberry plantings, although black vine weevils are the most common weevil in blueberries in BC. Weevil larvae feed on roots, rootlets and the basal crown area, and are harmful to young plants. Adults feed at night on the leaves and cause notching on the leaf edges. Affected bushes may become stunted, yield poorly and may die. The presence of weevils is increasing in highbush blueberry and there are no effective controls. Nemasys L. has shown some consistent levels of control of the weevils in BC. There are no data to show how widespread is use of the product in the province. One study (Hayes, 1990) on dry-pick cranberries in BC revealed that the population of the black vine weevil was reduced significantly after the application of *S. carpocapsae*.

15.7 Other Areas with Advanced Stages of Research and Development of Entomopathogenic Nematodes

15.7.1 Controlling insect pests of maize

The European corn borer, *O. nubilalis* Hübner, 1796 (Lepidoptera: Crambidae), the fall armyworm, *Spodoptera frugiperda* J.E. Smith, 1797 (Lepidoptera: Noctuidae), the western corn rootworm, *Diabrotica virgifera* LeConté, 1868 (Coleoptera: Chrysomelidae), and the seed corn maggot, *Delia platura* Meigen, 1826 (Diptera: Anthomyiidae), are among the major insect pests of field maize, sweetcorn and seed maize in the province of Ontario, Canada. Not many studies have been done to evaluate EPN efficacies against insect pests of the field crop under field conditions. A series of experiments was performed in the laboratory and in the greenhouse to assess the efficacy of the EPNs *Steinernema glaseri* and *S. feltiae* in decreasing the numbers of European corn borer, fall armyworm, western corn rootworm and the seed corn maggot in maize (Riga *et al.*, 2001). Both nematode species effectively controlled the four insect pest species (Table 15.3). During greenhouse experiments, the number of plants protected with EPNs against these insect pests was significantly higher than in the untreated controls. Similar results were obtained in a microplot study of maize infected with the European corn borer, the western corn rootworm and the seed corn maggot. The application of a single nematode species against several insect pests has economic advantages. In addition, both *S. glaseri* and *S. feltiae* overwintered and survived in the field until the next growing season.

15.8 Safety and Regulations

EPNs as biological control agents if not formulated are regulated under the Plant Protection Act (PPA) by CFIA, and if formulated EPN products, which are rare, are regulated under the Plant Protection Products Act (PPPA) by the Pest Management Regulatory Agency (PMRA).

15.8.1 Importation

A researcher or other person wishing to import nematodes into Canada sends an application for an import permit to the Plant Health and Protection Permit Office of the CFIA. After initial assessment, the application is forwarded to the Plant Risk Assessment Unit. The nematologists at CFIA and AAFC are consulted to provide advice on the impact of the proposed nematode species on plant health and the environment. The permit form requires the applicant to provide information on the nematode species and origin, the condition of the organisms intended for import and the destination and known distribution within Canada.

If the nematodes to be imported consist of species known to occur in North America, as listed in Table 15.4 (CFIA internal, not published), and are produced by suppliers of established reputation (so far, only European), then importers receive a permit without condition, essentially a courtesy permit. Also, US suppliers of nematodes and nematode products do not require a permit.

A good discussion of the detailed procedure for importing biocontrol agents into Canada has been published (AAFC, 2005; De Clerck-Floate *et al.*, 2006; Hunt *et al.*, 2008). If the nematodes to be imported do not meet these criteria, the application is evaluated, like other applications for the importation of arthropod biocontrol agents, as shown in Fig. 15.3, that is, on the basis of the potential for both direct and indirect impacts on plant health, and includes a consideration of the hazards associated with plant pests, parasites and pathogens, host materials or other means of transport, and indirect impacts such as nematode agents that influence pollinator populations negatively. Once the permit application is received, it is evaluated by the Plant Risk Assessment Unit; recommendations are then forwarded from the

Table 15.3. Control of insect pests of maize using entomopathogenic nematodes. (Adapted from Riga *et al.*, 2001).

		Nematode species	
		Steinernema glaseri	*Steinernema feltiae*
Laboratory study	Insects	Control (%)	
	Seed corn maggot	72	80
	Corn rootworm	88	92
	European corn borer	68	72
	Fall armyworm	88	96
Greenhouse study			
	Seed corn maggot	60	70
	Corn rootworm	80	85
	European corn borer	72	62
	Fall armyworm	84	88

Plant Risk Assessment Unit to the Plant Health and Protection Permit Office, who are responsible for issuing permits with conditions of importation indicated, maintaining records of permits issued, providing access to permits issued to ports of entry and Import Service Centres, and accepting fees. Permits are available at http://www.inspection.gc.ca/DAM/DAM-plants-vegetaux/STAGING/text-texte/c5256_1331652913719_eng.pdf.

There are three possible outcomes of the permitting process: permits may be granted, they may be granted with conditions of entry intended to mitigate the risks identified during assessment, or they may be denied. Importers may also receive permits for importation of other nematode species under conditions of containment and control. The process for importation into containment is the same as for other import permits. However, CFIA's policy in this case is to err on the side of caution; that is, if the presence of a species in Canada is unconfirmed, introduction will be discouraged by requiring containment conditions that would prevent environmental release, and commercial intentions will not be issued a permit. It is also recommended that the applicant contact the PMRA for information on current policy with regard to the commercial use and registration of nematode-based products. Invertebrates and microorganisms (including nematodes) that have been determined through pest risk analysis not to pose a quarantine risk, listed in Appendix 1 of Policy Directive 96-1452, are exempted from the requirement for an import permit, as are organisms registered as fertilizers and biological control agents. There are no defined reviews or processing timelines for import permits. Rather, assessors try to provide recommendations to the permit office as quickly as possible. An application for imports destined for scientific research

Table 15.4. List of entomopathogenic nematodes that could be imported into Canada without a permit in 2014.

1	*Heterorhabditis bacteriophora* (Poinar)
2	*Heterorhabditis indica* (Poinar, Karunakar and David)
3	*Heterorhabditis maurelatus* (Liu and Berry)
4	*Heterorhabditis megidis* (Poinar, Jackson and Klein)
5	*Steinernema carpocapsae* (Weiser)
6	*Steinernema feltiae* (Filipjev)
7	*Steinernema glaseri* (Steiner)
8	*Steinernema kraussei* (Steiner)

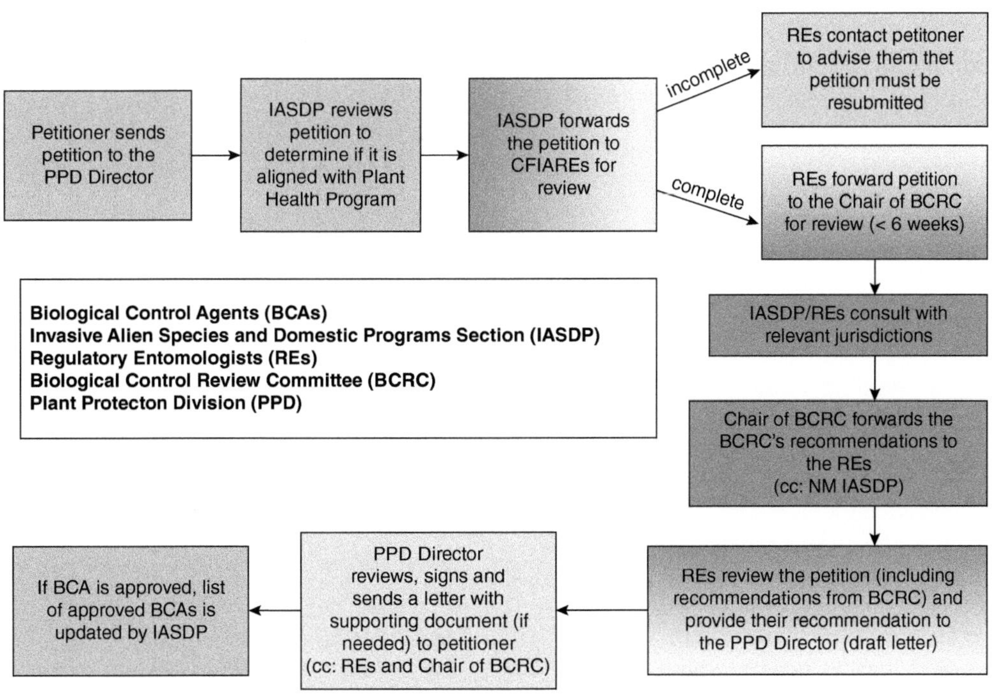

Fig. 15.3. Petition process of invertebrate biological control agents including EPNs in Canada.

costs CAN$15, while applications for other purposes cost CAN$35 (AAFC, 2006).

15.8.2 Registration

The EPNs are not required to be registered if they are alive and not formulated. Registration of formulated products is required. The impact of the imported EPN on the biodiversity of the native species needs to be considered. But discussions are ongoing to have EPNs fully federally registered, citing a precedent set with another bioinsecticide, *Bacillus thuringiensis*, which was once embraced by the environmental movement but has since been deemed unacceptable from a safety standpoint. Another important factor to be considered is the fact that the symbiotic bacteria inside EPNs are known to have a wide spectrum of other biological functions, including toxicity towards humans.

15.9 Conclusions

The status of EPNs in integrated pest management strategies in Canada is undergoing changes; on the one hand, demands for EPNs for pest control are on the rise as people are becoming increasingly environmentally conscious, but on the other hand, researches are stagnating as resources for nematode research decline. Historically, Canadian nematologists have made significant contributions to entomological nematology. Researches on bacteria metabolites of the symbiotic bacteria have been outstanding. More attention needs to be focused on the biodiversity of the native species, and their symbiotic bacteria. Commercial use of EPNs to control insect pests in Canada has been on the rise, especially in greenhouses and on turf. Most of those EPNs are imported from other countries. In some cases, efficacies have not been evaluated properly. Canada has a cold climate, and in most cases those imported EPNs are not known to be cold tolerant. Research is needed to select and evaluate local populations that are naturally more adapted to the local environmental conditions as more effective biological control agents. It is strongly suggested that the situation concerning regulation is reassessed.

Acknowledgements

Sincere thanks to Drs Peter Mason and John Webster for their critical review of this manuscript.

References

AAFC (Agriculture and Agri-Food Canada) (2005) Critical regulatory path for biocontrol agents. Available at: http://publications.gc.ca/collections/collection_2009/agr/A52-166-2005E.pdf (accessed 26 September 2016).

AAFC (2006) Guide for the Importation and Release of Arthropod Biological Control Agents. Available at: http://publications.gc.ca/collections/collection_2007/agr/A42-105-2006E.pdf (accessed 28 September 2016).

AAFC (2016) An Overview of the Canadian Agriculture and Agri-Food System 2016. Available at: https://caes.usask.ca/members/_pdf/Overview%202016-Final_eng.pdf (accessed 28 October 2016).

Barton, W., Goudy, H. and Van Eerd, L. (1996) Control of European chafer *Rhizotrogus majalis*, and Japanese beetle *Popillia japonica*. *Guelph Turfgrass Institute Annual Research Report* 1996, 33–38.

Bedding, R.A. and Akhurst, R.J. (1975) A simple technique for the detection of insect parasitic rhabditid nematodes in soil. *Nematologica* 21, 109–110.

Bélair, G. and Boivin, B. (1985) Susceptibility of the carrot weevil to *Steinernema feltiae, S. bibionis* and *Heterorhabditis heliothidis*. *Journal of Nematology* 17, 363–366.

Bélair, G. and Boivin, G. (1995) Evaluation of *Steinernema carpocapsae* Weiser for control of carrot weevil adults, *Listronotus oregonensis* (LeConte) (Coleoptera: Curculionidae), in organically grown carrots. *Biocontrol Science and Technology* 5, 225–231.

Bélair, G., Anderson, R.V. and Boivin, G. (1984) Characteristics of Listronotus strain of *Steinernema carpocapsae* from a carrot weevil host. *Proceedings of the First International Congress of Nematology,* Guelph, Canada, pp. 9.

Bélair, G., Vincent, C. and Chouinard, G. (1998) Foliar sprays with *Steinernema carpocapsae* against early-season apple pests. *Journal of Nematology* 30, 599–606.

Bélair, G., Vincent, C., Lemire, S. and Coderre, D. (1999) Laboratory and field assays with entomopathogenic nematodes for the management of oblique banded leafroller *Choristoneura rosaceana* (Harris) (Tortricidae). *Journal of Nematology* 31, 684–689.

Bélair, G., Fournier, Y. and Dauphinais, N. (2000) Survey on plant-parasitic and entomopathogenic nematodes in vineyards of Quebec. *Phytoprotection* 82, 49–55.

Bélair, G., Dauphinais, N., Fournier, Y. and Mauleon, H. (2001) Survey of plant-parasitic and entomopathogenic nematodes in vineyards of Québec. *Phytoprotection* 82, 49–55.

Bélair, G., Fournier, Y. and Dauphinais, N. (2002) Efficacy of Steinernematid nematodes against three insect pests of Crucifers in Quebec. *Journal of Nematology* 35, 259–265.

Bélair, G., Wright, D.J. and Curto, G. (2005) Vegetable and tuber crop applications. In: Grewal, P.S., Ehlers, R.-U. and Shapiro-Ilan, D.I. (eds) *Nematodes as Biocontrol Agents*. CAB International, Wallingford, UK, pp. 255–264.

Bélair, G., Simard, L. and Dionne, J. (2013) Canadian entomopathogenic nematode isolates: virulence against black cutworm (Lepidoptera: Noctuidae). *Phytoprotection* 93, 43–46.

Boivin, G. and Bélair, G. (1989) Infectivity of two strains of *Steinernema feltiae* (Rhaditida: Steinernematidae) in relation to temperature, age and sex of carrot weevil (Coleoptera: Curculionidae). *Journal of Economic Entomology* 82, 762–765.

Broadbent, A.B. and Olthof, T.H.A. (1995) Foliar application of *Steinernema carpocapsae* (Rhabditida: Steinernematidae) to control *Liriomyza trifoli* (Diptera: Agromyzidae) larvae in chrysanthemums. *Biological Control* 24, 431–435.

Chen, G. (1996) Antimicrobial activity of the nematode symbionts, *Xenorhabdus* and *Photorhabdus* (Enterobacteriaceae) and the discovery of two novel groups of antimicrobial substances, nematophin and xenorxides. PhD thesis, Simon Fraser University, British Columbia, Canada.

Chen, G., Dunphy, G.B. and Webster, J.M. (1994) Antifungal activity of two *Xenorhabdus* species and *Photorhabdus luminescens*, bacteria associated with the nematodes *Steinernema* species and *Heterorhabditis megidis*. *Biological Control* 4, 157–162.

Chen, G., Maxwell, P., Dunphy, G.B. and Webster, J.M. (1996) Culture conditions for *Xenorhabdus* and *Photorhabdus* symbionts of entomopathogenic nematodes. *Nematologica* 42, 124–127.

Clarke, D.J. and Eberl, L. (2006) Interactions between bacteria and nematodes. In: Konig, H. and Varma, A. (eds) *Intestinal Microorganisms of Termites and Other Invertebrates. Vol 6, Soil Biology*. Springer Verlag, Berlin, Heidelberg, Germany, pp. 55–64.

De Clerck-Floate, R.A., Mason, P.G., Parker, D.J., Gillespie, D.R., Broadbent, A.B. and Boivin, G. (2006) Guide for the importation and release of arthropod biological control agents in Canada. *Agriculture and Agri-Food Canada* A42-105/2006E, 55 pp.

Dunphy, G.B. and Webster, J.W. (1984) Interaction of *Xenorhabdus nematophilus* sub sp. nematophilus with the haemolymph of *Galleria mellonell*. *Journal of Insect Physiology* 30, 883–889.

Dunphy, G.B. and Webster, J.M. (1991) Antihemocytic surface components of *Xenorhabdus nematophilus* var. dutki and their modification by serum of nonimmune larvae of *Galleria mellonella*. *Journal of Invertebrate Pathology* 58, 40–51.

Dunphy, G.B., Miyamoto, C. and Meighen, E. (1997) A homoserine lactone autoinducer regulates virulence of an insect-pathogenic bacterium, *Xenorhabdus nematophilus* (Enterobacteriaceae). *Journal of Bacteriology* 179, 5288–5291.

Eidt, D.C. and Dunphy, G.B. (1991) Control of spruce bud moth, *Zeiraphera canadensis* Muturra and Freeman, in white spruce plantations with entomopathogenic nematodes, *Sternernema* spp. *The Canadian Entomologist* 123, 379–385.

Eidt, D.C. and Thurston, G.S. (1994) *Entomopathogenic Nematodes for Insect Pest Management in a Cold Climate.* Canadian Forest Service, Fredericton, New Brunswick, Canada, 66 pp.

Eidt, D.C. and Thurston, G.S. (1995) Physical deterrent to infection by entomopathogenic nematodes in wireworms (Coleoptera: Elateridae) and other soil insects. *The Canadian Entomologist* 127, 423–429.

Eidt, D.C., Zervos, S., Pyea, A.E. and Finney-Crawley, J. (1995a) Control of *Hylobius congener* Dalle Torre, Shenkling, and Marshall (Coleoptera: Curculionidae) using entomopathogenic nematodes. *The Canadian Entomologist* 127, 431–438.

Eidt, D.C., Zervos, S. and Finney-Crawley, J. (1995b) Susceptibility of *Hylobius congener* Dalle Torre, Shenkling, and Marshall to entomopathogenic nematodes. *The Canadian Entomologist* 127, 439–441.

Finney, J.R. and Bennett, G.F. (1984) *Heterorhabditis*: a potential biocontrol agent of agricultural and forest pests in Newfoundland. *Journal of Agricultural Entomology* 1, 287–295.

Finney-Crawley, J.R. (1985) Isolation of cold tolerant steinemematid nematodes in Canada. *Journal of Nematology* 17, 496.

Garcia, D.P.F. and Palomo, A. (1996) Natural occurrence of entomopathogenic nematodes (Rhabditida: Steinernematidae and Heterorhabditidae) in Spanish soils. *Journal of Invertebrate Pathology* 68, 84–90.

Georgis, R., Koppenhöfer, A.M., Lacey, L.A. Bélair, G., Duncan, L.W., *et al*. (2006) Successes and failures in the use of parasitic nematodes for pest control. *Biological Control* 38, 103–123.

Gordon, R. and Burford, I.R. (1984) Transport of palmitic acid across the tegument of the entomophilic nematode *Romanomermis culicivorm. Journal of Nematology* 16, 14–21.

Gordon, R., Finney J.R., Condon, W.J. and Rusted, T.N. (1979) Lipids in the storage organs of three mennithid nematodes and in the hemolymph of their hosts. *Comparative Biochemistry and Physiology* 64, 369–374.

Gordon, R., Chippett, J. and Tilley, J. (1996) Effects of two carbarmates on infective juveniles of *Steinernema carpocapsae* A11 strain and *Steinernema feltiae* Umea strain. *Journal of Nematology* 28, 310–317.

Hayes, A.E. (1990) The entomopathogenic nematode *Steinernema carpocapsae* (All strain), for control of black vine weevils in dry-pick cranberries in British Columbia. Master degree thesis, Simon Fraser University, British Columbia, Canada.

Hayes, A.E., Fitzpatrick, S.M. and Webster, J.M. (1999) Infectivity, distribution and persistence of the entomopathogenic nematode *Steinernema carpocapsae* All strain (Rhabditida: Steinernematidae) applied by sprinklers or boom sprayers to dry-pick cranberries. *Journal of Economic Entomology* 92, 539–546.

Hu, K. (1999) Nematicidal properties of *Xenorhabdus* spp. and *Photorhabdus* spp., bacterial symbionts of entomopathogenic nematodes. PhD thesis, Simon Fraser University, British Columbia, Canada.

Hu, K. and Webster, J.M. (2000) Antibiotic production in relation to bacterial growth and nematode development in *Photorhabdus-Heterorhabditis* infected *Galleria mellonella* larvae. *FEMS Microbiology Letters* 189, 219–223.

Hu, K., Li, J. and Webster, J.M. (1995) Mortality of plant-parasitic nematodes caused by bacterial (*Xenorhabdus* spp. and *Photorhabdus luminescens*) culture media. *Journal of Nematology* 27, 502–503.

Hu, K., Li, J. and Webster, J.M. (1996) 3,5-Dihydroxy-4-isopropylstilbene: a selective nematicidal compound from the culture filtrate of *Photorhabdus luminescens*. *Canadian Journal of Plant Pathology* 18, 104.

Hu, K., Li, J. and Webster, J.M. (1997) Quantitative analysis of a bacteria-derived antibiotic in nematode-infected insects using HPLC-UV and TLC-UV methods. *Journal of Chromatography B Biomedical Applications* 703, 177–183.

Hu, K., Li, J., Wang, W., Wu, H., Lin, H. and Webster, J.M. (1998) Comparison of metabolites produced in vitro and in vivo by *Photorhabdus luminescens*, a bacterial symbiont of the entomopathogenic nematode *Heterorhabditis megidis*. *Canadian Journal of Microbiology* 44, 1072–1077.

Hu, K., Li, J. and Webster, J.M. (1999) Nematicidal metabolites produced by *Photorhabdus luminescens* (Enterobacteriaceae), bacterial symbionts of entomopathogenic nematodes. *Nematology* 1, 457–469.

Hui, E. and Webster, J.M. (2000) Influence of insect larvae and seedling roots on the host-finding ability of *Steinernema feltiae* (Nematoda: Steinernematidae). *Journal of Invertebrate Pathology* 75, 152–162.

Hunt, E.J., Kuhlmann, U., Sheppard, A., Qin, T.-K., Barratt, B.I.P., *et al.* (2008) Review of invertebrate biological control agent regulation in Australia, New Zealand, Canada and the USA: recommendations for a harmonized European system. *Journal of Applied Entomology* 132, 89–123.

Isaacson, P.J. (1994) Antimicrobial activity of *Xenorhadbus* sp. (Enterobacteriaceae) symbiont of the entomopathogenic nematode, *Steinernema riobrave* (Rhabditida: Steinernematidae). Master degree thesis, Simon Fraser University, British Columbia, Canada.

Isaacson, P.J. and Webster J.M. (2002) Antimicrobial activity of *Xenorhadbus* sp. R10 (Enterobacteriaceae) symbiont of the entomopathogenic nematode, *Steinernema riobrave* (Rhabditida: Steinernematidae). *Journey of Invertebrate Pathology* 79, 146–153.

Jagdale, G.B. (1997) Adaption to temperature in entomopathogenic nematodes. PhD thesis, Memorial University of Newfoundland, Canada, 175 pp.

Jagdale, G.B. and Gordon, R. (1994a) Distribution of catecholamines in the nervous system of a mermithid nematode, *Romanomermis culicivorax*. *Parasitology Research* 80, 459–466.

Jagdale, G.B. and Gordon, R. (1994b) Distribution of FMRF-amide-like peptide in the nervous system of a mermithid nematode, *Romanomermis culicivorax*. *Parasitology Research* 80, 467–473.

Jagdale, G.B. and Gordon, R. (1994c) Role of catecholamines in the reproduction of *Romanomermis culicivorax*. *Journal of Nematology* 26, 40–45.

Jagdale, G.B. and Gordon, R. (1994d) Caudal papillae in *Romanomermis culicivorax*. *Journal of Nematology* 26, 235–237.

Jagdale, G.B. and Gordon, R. (1997a) Effect of temperature on the activities of glucose-6-phosphate dehydrogenase and hexokinase in entomopathogenic nematodes (Nematoda: Steinernematidae). *Comparative Biochemistry and* Physiology 118A, 1151–1156.

Jagdale, G.B. and Gordon, R. (1997b) Effect of temperature on the composition of fatty acids in total lipids and phospholipids of entomopathogenic nematodes. *Journal of Thermal Biology* 22, 245–251.

Jagdale, G.B. and Gordon, R. (1998) Variable expression of isozymes in entomopathogenic nematodes follows laboratory recycling. *Fundamental and Applied Nematology* 21, 147–155.

Jagdale G.B., Gordon, R. and Vrain, T.C. (1996) Use of cellulose acetate electrophoresis in the taxonomy of steinernematids (Rhabditida, Nematoda). *Journal of Nematology* 28, 301–309.

Li, J., Chen, G. and Webster, J.M. (1995a) Antimicrobial metabolites from a bacteria symbiont. *Journal of Natural Products* 58, 1081–1086.

Li, J., Chen, G., Wu, H. and Webster, J.M. (1995b) Identification of two pigments and a hydroxylstilbene antibiotic from *Photorhabdus luminescens*. *Applied and Environment Microbiology* 61, 4329–4333.

Li, J., Chen, G. and Webster, J.M. (1997) Nematophin, a novel antimicrobial substance produced by *Xenorhabdus nematophilus* (Enterobacteraceae). *Canadian Journal of Microbiology* 43, 770–773.

Li, J., Hu, K. and Webster, J.M. (1998) Antibiotics from *Xenorhabdus* spp. and *Photorhabdus* spp. (Enterobacteriaceae). *Chemistry of Heterocyclic Compounds* 34, 1561–1570.

Mason, P.G. and Gillespie, D.R. (2013) *Biological Control Programmes in Canada 2001–2012*. CAB International, Wallingford, UK.

McInerney, B.V., Gregson, R.P., Lacey, M.J., Akhurst, R.J., Lyons, G.R., *et al.* (1991a) Biologically active metabolites from *Xenorhabdus* spp. Part 1. Dithiolopyrrolone derivatives with antibiotic activity. *Journal of Natural Products* 54, 774–784.

McInerney, B.V., Taylor, W.C., Lacey, M.J., Akhurst, R.J. and Gregson, R.P. (1991b) Biologically active metabolites from *Xenorhabdus*spp. Part 2. Benzopyran-1-one derivatives with gastroprotective activity. *Journal of Natural Products* 54, 785–795.

Mrácek, Z. and Webster, J.M. (1993) Survey of Heterorhabditidae and Steinernematidae (Rhabditida, Nematoda) in Western Canada. *Journal of Nematology* 25, 710–717.

Mrácek, Z., Kindlmann, P. and Webster, J.M. (2005) *Steinernema affine* (Nematoda: Steinernematidae), a new record for North America and its distribution relative to other entomopathogenic nematodes in British Columbia. *Nematology* 7, 495–501.

Murfin, K.E., Dillman, A.R., Foster, J.M., Bulgheresi, S., Slatko, B.E., *et al.* (2012) Nematode–bacterium symbioses – cooperation and conflict revealed in the 'omics' age. *The Biological Bulletin* 223, 85–102.

Ng, K.K. and Webster, J.M. (1997) Antimycotic activity of *Xenorhabdus bovienii* (Enterobacteriaceae) metabolites against *Phytophthora infestans* on potato. *Canadian Journal of Plant Pathology* 19, 125–132.

Nguyen, K.B. and Hunt, D.D. (2007) *Entomopathogenic Nematodes: Systematics, Phylogeny and Bacterial Symbionts*. Brill, Koninklijke Brill NV, The Netherlands.

Olthof, T.H.A. and Broadbent, A.B. (1991) Control of a chrysanthemum leafminer, *Liriomyza trifolii*, with the entomophilic nematode, *Heterorhabditis heliothidis*. *Nematologica* 36, 379.

Olthof, T.H.A. and Broadbent, A.B. (1992) Evaluation of steinernematid nematodes for control of a leafminer, *Liriomyza trifolii*, in greenhouse chrysanthemums. *Journal of Nematology* 24, 612.

Olthof, T.H.A. and Hagley, E.A.C. (1993) Laboratory studies of the efficacy of Steinememattid nematodes against the Plum Curculio (Coleoptera: Curculionidae). *Journal of Economic Entomology* 86, 1078–1082.

Olthof, T.H.A., Rinker, D.L. and Dano, J. (1991) The effect of entomophilic nematodes on a sciarid fly, *Lycoriella mali*, and on growth and yield of mushroom, *Agaricus bisporus*. *Journal of Nematology* 23, 543.

Poinar, G.O. Jr (1990) Biology and taxonomy of Steinernematidae and Heterorhabditidae. In: Gaugler, R. and Kaya, H.K. (eds) *Entomopathogenic Nematodes in Biological Control*. CRC Press, Boca Raton, Florida, pp. 23–61.

Riga, E., Whistlecraft, J. and Potter, J. (2001) Potential of controlling insect pests of corn using entomopathogenic nematodes. *Canadian Journal of Plant Science* 81, 783–787.

Rinker, D.L., Olthof, T.H.A., Dano, J. and Alm, G. (1995) Effects of entomopathogenic nematodes on control of a mushroom-infesting Sciarid Fly and on mushroom production. *Biocontrol Science and Technology* 5, 109–120.

Royer, L., Bélair, G., Boivin, G. and Fournier, Y. (1996) Attractiveness of *Delia radicum* (Diptera: Anthomyiidae) to entomopathogenic steinernematid nematodes. *Journal of Economic Entomolology* 89, 614–620.

Saunders, J.M. (1984) Evaluation of the potential of *Heterorhabditis megidis* H90 (Nematoda) and its associated bacteria *Photorhabdus luminescens* in the control of some forest insect pests. PhD thesis, Waterloo University, Ontario, Canada.

Shapiro-Ilan, D.I, Gouge, D.H, Piggott, S.J. and Patterson, F.J. (2006) Application technology and environmental considerations for use of entomopathogenic nematodes in biological control. *Biological Control* 38, 124–133.

Shapiro-Ilan, D.I, Brown, I., and Lewis, E.E. (2014) Freezing and desiccation tolerance in entomopathogenic nematodes: diversity and correlation of traits. *Journal of Nematology* 46, 27–34.

Simard, L. (2006) Distribution, abondance et écologie saisonnièredes principaux insectes ravageurs du gazon sur les terrainsde golf du Québec et évaluation du potentiel de contrôlede nématodes entomopathogènes indigènes. PhD thesis, Université Laval Québec, Canada.

Simard, L., Bélair, G. and Brodeur, J. (2001) Susceptibility of the European chafer (Coleoptera: Scarabaeidae) to entomopathogenic nematodes (Rhabditida: Steinernematidae, Heterorhabditidae). *Journal of Nematology* 33, 297–301.

Simard, L., Bélair, G. and Brodeur, J. (2002) Susceptibility of the cranberry girdler (Lepidoptera: Pyralidae) to entomophagenic nematodes. *Canadian Entomologist* 134, 329–330.

Simard, L., Bélair, G. and Dionne, J. (2006) Virulence of entomopathogenic nematodes (Rhabditida: Steinernematidae, Heterorhabditidae) against *Tipula paludosa* (Diptera: Tipulidae), a new turfgrass insect pest on golf courses in Quebec. *Biocontrol Science and Technology* 16, 789–801.

Simard, L., Bélair, G., Stock, S., Mauléon, H. and Dionne, J. (2007) Natural occurrence of entomopathogenic nematodes (Rhabditida: Steinernematidae) on golf courses in eastern Canada. *Nematology* 9, 325–332.

Statistics Canada (2014) Study: Agriculture in Canada. Available at: http://www.statcan.gc.ca/daily-quotidien/141113/dq141113a-eng.pdf (accessed 26 September 2016).

Stock, S.P., Mracek, Z. and Webster, J.M. (2000) Morphological variation between allopatric populations of *Steinernema kraussei* (Steiner, 1923) (Rhabditida: Steinernematidae). *Nematology* 2, 143–152.

Stock, S.P., Banna, A., Darwish, L. and Katbeh, R.A. (2008) Diversity and distribution of entomopathogenic nematodes (Nematoda: Steinernematidae, Heterorhabditidae) and their bacterial symbionts (c-Proteobacteria: Enterobacteriaceae) in Jordan. *Journal of Invertebrate Pathology* 98, 228–234.

Sturhan D. and Liskova, M. (1999) Occurrence and distribution of entomopathogenic nematodes in the Slovak Republic. *Nematology* 1, 273–277.

Sweeney, J.D. and Gesner, G. (1994) Field trials for suppression of spruce cone maggots, *Strobilomyia appalachensis* Michelsen, using entomopathogenic nematodes. Natural Resources Canada, Canadian Forest Service, Maritimes Region, Fredericton, New Brunswick. R and D Technical Note No 35, 2 pp.

Sweeney, J.D., Gesner, G.N., Bennett, R. and Vrain, T. (1998) Effect of mulches on persistence of entomopathogenic nematodes (*Steinernema* spp.) and infection of *Strobilomhia neanthracina* (Diptera: Anthomyiidae) in field trials. *Journal of Economic Entomology* 91, 1320–1330.

Tambong, J.T. (2013) Phylogeny of bacteria isolated from Rhabditis sp. (Nematoda) and identification of novel entomopathogenic *Serratia marcescens* strains. *Current Microbiology* 66, 138–144.

Tambong, J.T., Xu, R., Sadiku, A., Chen, Q. and Yu, Q. (2014) Molecular detection and analysis of a novel metalloprotease gene of entomopathogenic *Serratia marcescens* strains in infected *Galleria mellonella*. *Canadian Journal of Microbiology* 60, 203–209.

Vincent, C. and Bélair, G. (1992) Biocontrol of the apple sawfly, *Hoplocampa testudinea*, with entomogenous nematodes. *Entomophaga* 37, 575–582.

Webster, J.M., Chen, G. and Li, J. (1998) Parasitic worms: an ally in the war against the superbugs. *Parasitology Today* 14, 161–163.

Webster, J.M., Li, J. and Chen, G. (2000) Anticancer property of dithiolopyrrolones. US Patent No 6,020,360.

Webster, J.M., Chen, G., Hu, K. and Li, J. (2002) Bacterial metabolites. In: Gaugler, R. (ed.) *Entomopathogenic Nematology*. CAB International, Wallingford, UK, pp. 75–92.

Welch, H.E. (1965) Entomophilic nematodes. *Annual Review of Entomology* 10, 275–302.

Welch, H.E. and Briand, L.J. (1961a) Tests of the nematode DD 136 and an associated bacterium for control of the Colorado potato beetle, *Leptinotorso decemlineata* (Say). *Canadian Entomologist* 93, 759–763.

Welch, H.E. and Briand, L.J. (1961b) Field experiment on the use of a nematode for the control of vegetable crop insects. *Proceedings of the Entomological Society of Ontario* 91, 197–202.

Welch, H.E. and Bronskill, J.F. (1962) Parasitism of mosquito larvae by the nematode, DD 136. *Canadian Journal of Zoology* 40, 1263–1268.

Yu, Q. and Potter, J. (2006) A selective history of nematology in Canada. Available at: http://nematologia.com.br/wp-content/uploads/2013/09/nemincana.pdf (accessed 26 September 2016).

Zervos, S., Johnson, S.C. and Webster, J.M. (1991) Effect of temperature and inoculum size on reproduction and development of *Heterorhabditis heliothidis- and Steinernema glaseri* (Nematoda: Rhabditoidea) in *Galleria melionella*. *Canadian Journal of Zoology* 69, 1261–1264.

Zervos, S., Eidt, D.C., Weaver, C.A.A. and Tosh, K.J. (1994) Control of black vine weevil in forest nurseries with entomopathogenic nematodes. Natural Resources Canada, Canadian Forest Service, Maritimes Region, Fredericton, New Brunswick. Technical Note No 289, 5 pp.

16 Status of Entomopathogenic Nematodes in Integrated Pest Management Strategies in Argentina

María Fernanda Achinelly,* Nora Camino, Daiana Eliceche, Augusto Salas and Matías Rusconi

Centro de Estudios Parasitológicos y de Vectores, CEPAVE (CCT La Plata CONICET - UNLP), Buenos Aires, Argentina

16.1 Introduction

Entomopathogenic nematodes (EPNs) are one of the highly influential biocontrol agents regulating the population dynamics of insect pests through association with their hosts in relationships ranging from fortuitous to parasitic. Knowledge of nematode community is useful to interpret the host–parasite relationships, and in many cases to illustrate co-evolution phenomena and to determine those potential biocontrol agents against major pests of health and agricultural interest.

The Republic of Argentina is located in the Southern Cone of South America. Due to its large area, the latitude is characterized by distinct forms of relief: the mountains to the west and the plateaus, plains and depressions eastwards. This country presents a great diversity of climates and soils where the livestock and agriculture constitute the main economic activities. The quantity and quality of agricultural performance place Argentina as an important food producer that corresponds mostly with the sown area of cereals. The most widespread are wheat and maize, besides oats, barley, rice and sorghum. Also, oil and fibre plants, sugarcane, grapes, mate, tea, snuff, fruit and vegetables are grown.

Within this context, several studies have been performed in Argentina with the aim of isolating entomonematodes as biocontrol agents of pests of public health and agricultural importance. A great diversity of species has been reported as the result of research carried out during the past 35 years. The studies focused mainly on the description of numerous species, determination of biological cycles, ecology of pathogenic species and dynamics in natural conditions, mainly for Mermithidae, Thelastomatidae, Rhabditidae and EPNs Heterorhabditidae and Steinernematidae (Doucet, 1986; Doucet and Bertolotti, 1996; Lax *et al.*, 2011; Achinelly and Micieli, 2013; Del Valle *et al.*, 2013; Camino *et al.*, 2014; Belaich *et al.*, 2015; Di Battista *et al.*, 2015). However, the research of this group of parasites was not commensurate to its potential. The lack of legislation regarding the regulation and registration requirements of different groups of bioproducts, the development of criteria or policy recommendations for studies to field or releases of new bioproducts, along with the weakening of the economy, have restricted the growth of this topic in Argentina. For these reasons, the studies have been limited mainly to laboratory stage or small field scale, leaving the development of entomonematodes as biocontrol agents in this country still incipient.

This chapter presents the results of research undertaken in Argentina and the current status of the potential use of entomonematodes as biologic control agents. Our main emphasis is inclined towards the two EPNs, *Steinernema* and *Heterorhabditis*, and their status in integrated pest management (IPM) in Argentina. The information presented joins the most relevant bibliographic information from the

*Corresponding author. E-mail: fachinelly@cepave.edu.ar

personal experience of the authors and other specialists in the country.

16.2 Integrated Pest Management in Argentina

IPM began to take hold in the 1990s as a decision system based on the selection and use of tactical control combined singly or harmoniously into a strategy related to the analysis of the cost/benefit ratio. The interests of farmers and the impacts on human health, society and the environment have been taken into account. Now, IPM is more related to the combination of technical expertise and economic analysis.

Currently, various government agencies such as INTA (National Agricultural Technology Institute) encourage the creation of consortia of integrated pest management (ConMIP) through projects in extensive crops.

16.3 Application of Biological Control Agents in Argentina

The growing interest in environmental quality has enhanced in recent years the search for new tools to control insects of agricultural importance that are safe, healthy and compatible with the environment. In Argentina, the study of parasites and pathogens began in the early 1980s, currently with a growing demand for bioinsecticides based on entomopathogens to control agricultural pests and disease vectors. The application of these agents was not commensurate with their potential, with successful results using the bacterium *Bacillus thuringiensis* to control agricultural and forestry pests, the microsporidium *Paranosema locustae* against locusts and the granulosis virus (*Betabaculovirus*; Baculoviridae) to control lepidopteran pest of fruit and nut trees (Sciocco-Cap, 2013). However, there are no registered products based on EPNs, and only a few nematodes have been evaluated at the small field scale.

16.4 An Outline of Studies on Entomonematodes (Parasites, Parasitoids and Pathogenic) in Argentinian Research Centres

A great diversity of nematode species with different associations with insects (parasites, parasitoids and pathogens) have been described in Argentina through research carried out during the past 40 years. A few groups can be mentioned: the first initiated by Dr Magdalena Agüera, at Centro de Zoología Aplicada, National University of Córdoba, continuing with the research of Drs Marcelo Doucet and Paola Lax and their working group in the province of Cordoba; a second group in Buenos Aires province, which began with Drs Stock and Camino following this subject uninterruptedly since 1980 until the present, at CEPAVE, CONICET, National University of La Plata; and a third in Santa Fe province with Dr Eleodoro Del Valle, who trained with researchers such as Dr Claudia Dolinski from Brazil, a specialist in EPNs.

Among the prominent contributions are the description of numerous species, the determination of biological cycles, the ecology of the pathogenic species and the dynamics under natural conditions – mainly for the Mermithidae, Thelastomatidae and Rhabditidae, as well as the EPNs Heterorhabditidae and Steinernematidae.

16.4.1 Entomopathogenic nematodes in insects impacting public health

Although there are no reports of EPNs isolated from insects impacting public health from Argentina, several studies have tested their efficacy. Infectivity was evaluated against mosquitoes, flies and lice. Two isolates of EPNs from Cordoba province were tested under laboratory conditions. A mortality of 84% was produced by *Heterorhabditis bacteriophora* in *Aedes aegypti* mosquito larvae at a dose of 750:1 infective juveniles (IJs)/larvae (Peschiutta *et al.*, 2014). In the same way, *Culex apicinus* mosquito larvae were susceptible to *Steinernema rarum*, with higher mortality (75%) at a dose of 400:1 IJs/larvae (Cagnolo and Almirón, 2010). Nematodes of the two entomopathogenic families, Steinernematidae and Heterorhabditidae, produced mortality in lice. Strains of *H. bacteriophora* and *S. rarum* were able to infect specimens of *Pediculus humanus capitis*. Other mosquito species were susceptible to EPNs, such as *Culex quinquefasciatus* and *Ae. aegypti* larvae (*S. rarum*), and *Culex pipiens* (*Steinernema feltiae*). Pathogenicity by *S. feltiae* and *S. rarum* to *Musca domestica* larvae has also been reported (Doucet *et al.*, 2008).

Parasitoids

Parasitoids are organisms that spend a significant portion of their life history in a single host in a relationship that is in essence parasitic; but they can also sterilize, kill and sometimes consume the host.

Mermithidae, a group of nematode parasites, principally of insects, and almost always lethal to the host, are considered parasitoids, where the host undergoes behavioural changes and is usually killed on mermithid emergence (Wise de Valdez, 2007). Several characteristics of these mermithid nematodes, such as environmental safety, host specificity, mass rearing *in vivo*, lethality, facile establishment and ready recycling, make them attractive biological control agents for mosquitoes (Platzer, 2007).

One of the first records of entomonematodes in Argentina was made by Berg (1898), who reported the presence of nematodes in locusts from Buenos Aires and Córdoba, the descriptions of which may well correspond to the family Mermithidae.

MERMITHID NEMATODES ISOLATED FROM MOSQUITOES (DIPTERA: CULICIDAE). Mermithid parasites of mosquitoes in Argentina were isolated from two species, constituting the only reports for the Neotropical Region: Poinar and Camino (1986) for *Strelkovimermis spiculatus*, with the floodwater mosquito, *Aedes albifasciatus* (Macquart), and Camino (1989) for *Hydromermis* sp. with *Psorophora ferox* larvae.

Several studies tending to use *S. spiculatus* as bioregulator of mosquito populations have been carried out in Argentina, and these are summarized below.

Strelkovimermis spiculatus *Poinar and Camino, 1986.* *S. spiculatus* was found to infect the host mosquito *Ae. albifasciatus* in rain-flooded ponds (Micieli *et al.*, 2012b) and *Cx. pipiens* in house-drainage ditches (Garcia and Camino, 1990; Muttis *et al.*, 2013) in the Buenos Aires province, Argentina. These insects are implicated in the transmission of arboviral and parasitic diseases worldwide, including extremely serious illnesses in humans, a reality that stimulates the study and development of strategies oriented around the control of these two mosquitoes. Because of the insect lethality of this nematode and the capacity to tolerate environments of elevated salinity and organic pollution, *S. spiculatus* has been identified as a possible biological control agent for culicine mosquitoes.

Studies have been performed on the life cycle and production of *S. spiculatus* (Camino and Reboredo, 1994, 1996), on the effect of certain biotic and abiotic conditions on the parasitism of *S. spiculatus* (Camino and García, 1991; Achinelly and García, 2003; Achinelly *et al.*, 2003; Micieli *et al.*, 2012a) and on the host–parasite interaction (Micieli and García, 1999; Campos and Sy, 2003; Micieli *et al.*, 2012b).

The life cycle of *S. spiculatus*, at an approximate duration of 35 days, is similar to that of other aquatic mermithids (Camino and Reboredo, 1994). Pre-parasitic but infective second-stage juveniles (J_2s) hatch from eggs and actively seek to penetrate the mosquito larvae. The third-stage juvenile (J_3) then develops in the mosquito larvae for 6–8 days, at which time the post-parasitic fourth-stage juvenile (J_4) emerges, killing the host. Finally, the females that become adults lay eggs in the aquatic substrate to complete the reproductive cycle (Fig. 16.1).

S. spiculatus was found to be parasitic on 16 mosquito species in Argentina under natural and laboratory conditions (Table 16.1). Epizootic levels were registered on only natural populations of mosquito, *Ae. albifasciatus*, with levels of infection reaching between 80 and 100% (Micieli and García, 1999; Micieli *et al.*, 2012b). Studies on this mermithid under laboratory conditions have indicated a wide range of susceptible mosquito species. This mermithid was observed to have completed development inside mosquito larvae and then emerge, resulting in lethality to all the mosquito larvae thus infected. The main difference in the host susceptibility was the degree of infection with respect to the instar phase, being notably more intense in the younger larvae (i.e. the first and second instars) at an overall mean infectivity of 80%, and ranging from 13 to 100%; much lower in the older larvae (i.e. the third and fourth instars) at a mean value of 38%, and ranging from 0 to 100% (Achinelly *et al.*, 2004).

S. spiculatus infections were not recorded with the majority of the non-target aquatic organisms exposed to the pre-parasites, with the sole exception of dipterans of the family Chironomidae: in which species, however, nematode penetration was detected in only 7% of the larvae exposed to the pre-parasites (Achinelly *et al.*, 2004). All nematodes that penetrated Chironomidae larvae died at 24–48 h post-infection through host resistance in the form of melanization. This nematode was also introduced in natural populations of culicids: *Ae. aegypti*, *Cx. apicinus* and *Cx. pipiens* larvae (Achinelly and Micieli, 2009). The levels of parasitism fluctuated greatly between instar stages, sites and parasite doses. The overall parasitism observed for *Ae.*

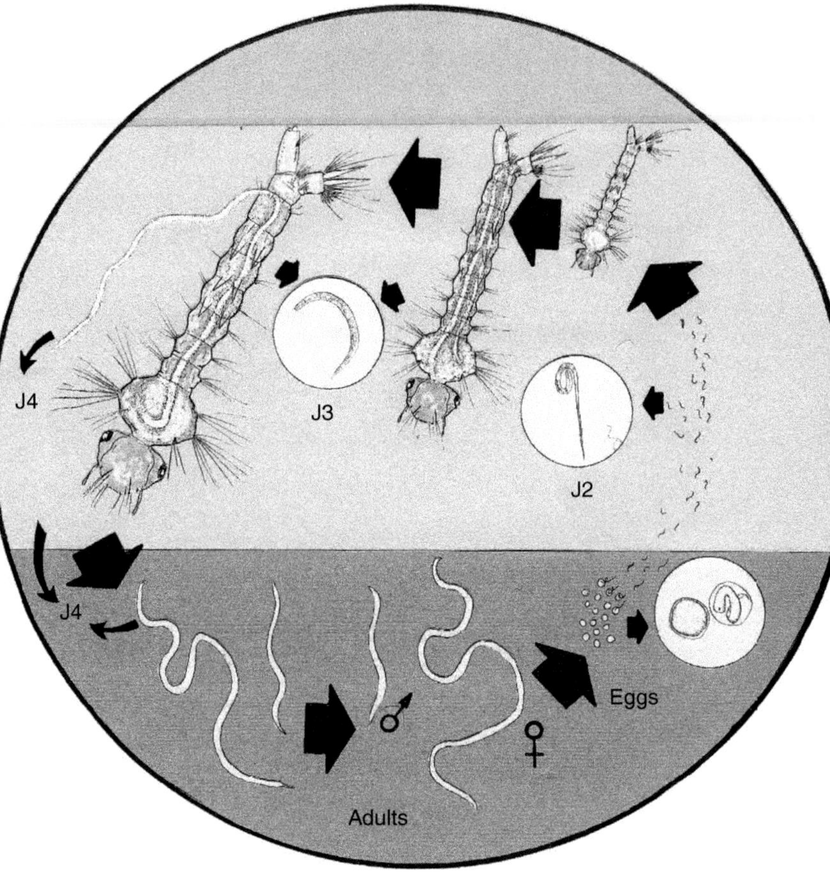

Fig. 16.1. Biological cycle of the EPN *Strelkovimermis spiculatus*: J_1, pre-parasitic juvenile inside an egg; J_2, pre-parasitic but infectious stage; J_3, parasitic stage; J4: free-living post-parasitic state.

aegypti, *Cx. pipiens* and *Cx. apicinus* was between 11–100%, 14–70% and 20–80% at dose ratios of 10:1, 50:1 and 100:1 J_2s/larvae, respectively; but the per cent infectivity among the three mosquito species was not significantly different at any given dose. The J_2s of *S. spiculatus* were observed in the field for up to approximately 10 days; which survival period agrees with the data of Micieli and García (1999), who had observed *S. spiculatus* J_2s alive for more than 7 days. Thus, mosquitoes with overlapping generations could be infected at different times with just one application of the nematodes in the breeding sites. This study constituted the first field evaluation of *S. spiculatus* with natural mosquito populations (Achinelly and Micieli, 2009).

From the above, *S. spiculatus* appears to be a promising potential biological control agent because of the mermithid's epizootic capabilities, its wide range of mosquito hosts and its innocuousness to non-target aquatic organisms. Nevertheless, certain parameters, such as the effect of this nematode's presence on water quality, should be considered before inoculating this entomoparasite in the field.

MERMITHID NEMATODES ISOLATED FROM SIMULIDS AND CHIRONOMIDS. Research in Argentina has included the characterization and description of new genera and species of mermithids that parasitize black flies in different wetlands in the following nine provinces of the country: Buenos Aires, Córdoba, Entre Ríos, Misiones, Tucumán, Neuquén, Río Negro, Jujuy and Mendoza. The resulting records from different locations in these provinces comprised two species of *Ditremamermis* Camino and Poinar, 1988; eight of *Mesomermis* Daday, 1911; four of *Gastromermis* Micoletzky, 1923; two of

Table 16.1. Spectrum of culicid hosts for the nematode *Strelkovimermis spiculatus* from Argentina. (Adapted from Achinelly and Micieli, 2013).

Mosquito larvae	Natural infections	Experimental infections	Laboratory infections
Aedes aegypti		x	x
A. albifasciatus	x		x
Anopheles albitarsis			x
Culex apicinus		x	x
C. castroi			x
C. chidesteri	x		x
C. dolosus	x		x
C. maxi	x		x
C. mollis	x		
C. pipiens	x	x	x
C. renatoi			x
Isostomya paranensis			x
Ochlerotatus crinifer	x		x
Psorophora ferox			x
P. ciliata	x		
P. cyanescens	x		

Octomyomermis Johnson, 1963; two of *Isomermis* Coman, 1953; two of *Limnomermis* Daday, 1911; one of *Hydromermis* Corti, 1902; and one of *Bathymermis* Daday, 1911.

Mermithids were also the only nematode family isolated from chironomid larval haemocoels. Three species of *Octomyomermis* genus and two of *Isomermis* sp. were recorded in the Buenos Aires province (Camino *et al.*, 2013b). They were always lethal for their hosts.

Parasites

Nematode parasites would not be as efficient as other families mentioned for use as agents of biological control of insects. However, it is worth mentioning an extensive study that was carried out in Argentina as a result of surveying and prospecting for entomonematodes.

NEMATODE PARASITES OF COCKROACHES. A survey of entomonematodes in cockroaches was conducted in order to find a nematode able to control this pest in Argentina (Table 16.2). Nematodes were isolated from *Periplaneta americana*, *Periplaneta brunnea*, *Periplaneta fuliginosa*, *Blatella vaga* and *Blatella germanica* (Lax *et al.*, 2008; Blanco *et al.*, 2012; Camino and González, 2012; Camino and de Villalobos, 2013; Camino *et al.*, 2013a; Gutierrez, 2015). All parasites isolated belong to the Oxyurida order, presenting in a location in the digestive tract without causing major changes in their hosts. *P. americana* presented the greatest diversity of parasite species. Parasitism in this host reached 70% in some cases.

16.4.2 Entomonematodes in insects of agricultural impact

Parasites and parasitoids

NEMATODES ISOLATED FROM GRASSHOPPERS, CRICKETS AND MOLE CRICKETS (ORTHOPTERA). The pampean region in Argentina is a major farming area where agricultural products are obtained for both domestic consumption and export, and where different species of grasshoppers exhibit outbreaks of significant magnitude, generating substantial economic losses. This environment was found to present a wide diversity of mermithid nematodes that might exert a form of regulation on the insect populations (Table 16.3). Entomonematode parasites of crickets and mole crickets were also registered in the Córdoba, Entre Ríos and Chaco provinces of Argentina (Camino and Stock, 1994; Camino *et al.*, 2013a), but only the Mermithidae family was lethal.

NEMATODES ISOLATED FROM SOIL PEST WHITE GRUBS (SCARABAEIDAE). A survey of nematode parasites and pathogens of white grubs in wheat fields of the pampas, Argentina, identified 14 species of nematode belonging to 5 families (Mermithidae,

Thelastomatidae, Travassonematidae, Diplogasteridae and Rhabditidae (Table 16.4). The information on the nematode community was useful for interpreting host–parasite relationships. This study provides for the first time a list of parasitic nematodes of white worms of the pampas region of the country (Camino *et al.*, 2014).

Table 16.2. Nematodes (Oxyurida) parasitizing cockroaches in Argentina.

Host	Parasite
Periplaneta americana	*Thelastoma domesticus*
	Thelastoma sp.
	Hammerschmidtiella laplatae
	Hammerschmidtiella sp.
	Blattophila sp.
	Leidynema sp.
P. brunnea	*H. ettalaensis*
P. fuliginosa	*Hammerschmidtiella* sp.
	Leidynema appendiculata
Blatella vaga	*Protellus blatta*
B. germanica	*Blatticola* sp.

Entomopathogenic nematodes

EPNs (Rhabditida: Steinernematidae, Heterorhabditidae) are obligate parasites of a wide range of soil insects. Specimens of both families naturally regulate populations of various insects, behave as obligate parasites and kill their hosts.

They are considered pathogens because of a symbiotic relationship with bacteria causing death to insects. EPNs have been found all over the world in a range of diverse habitats. The cycle of life is characterized by a free-stage infective juvenile (IJ), which locates and penetrates the host body in a passive way, *Steinernema* spp., or in an active way, *Heterorhabditis* spp., through a tooth. When reaching the haemocoel, IJs release their symbiotic bacteria,

Table 16.3. Biodiversity of insect parasitic nematodes in soil pest (Orthoptera) in Buenos Aires, Argentina.

Host		Parasite	
Acrididae	*Laplatacris dispar*	Mermithidae	*Agamermis decaudata*
	L. dispar	Mermithidae	*Amphimermis bonaerensis*
	Dichroplus elongatus	Mermithidae	*A. dichroplusi*
	Metaleptera brevicornis	Mermithidae	*A. ronderosi*
	D. elongatus	Mermithidae	*Hexamermis cochlearius*
	Staurorhectus longicornis	Mermithidae	*H. ovistriata*
	L. dispar	Mermithidae	*Longimermis acridophila*
Gryllidae	*Gryllodes laplatae*	Mermithidae	*Amphimermis* sp.
		Mermithidae	*H. macrostomata*
		Thelastomatidae	*Blatticola cristovata*
		Thelastomatidae	*Cameronia laplatae*
		Cephalobiidae	*Cephalobium bidentata*
		Cephalobiidae	*C. dispar*
		Cephalobiidae	*C. laplata*
		Cephalobiidae	*C. magdalensis*
		Cephalobiidae	*C. tridentata*
		Cephalobiidae	*C. odontolateralis*
		Thelastomatidae	*Neyraiella distinctus*
		Diplogasteridae	*Mikoletzkya* sp.
		Rhabditidae	*Alloionema* sp.
		Rhabditidae	*Pelodera* sp.
		Rhabditidae	*Cruznema lincolnensis*
Gryllotalpidae	*Neocurtilla claraziana*	Mermithidae	*Agamermis decaudata*
		Thelastomatidae	*Cephalobellus cyclocephalae*
		Thelastomatidae	*Euryconema brevicauda*
		Thelastomatidae	*Fontanema gracilis*
		Thelastomatidae	*Gryllophila cephalobulata*
		Thelastomatidae	*Schwenkiella tetradentatum*
		Travassosinematidae	*Binema bonaerensis*

Table 16.4. Entomonematode parasitises of white grubs (Scarabaeidae), an important pest in crops of Buenos Aires, Argentina.

	Scarabaeidae			
	Cyclocephala signaticolis		*Diloboderus abderus*	
Parasites	Prevalence %	Intensity (number of parasites per infected insect)	Prevalence (%)	Intensity (number of parasites per infected insect)
Mermithidae				
Amphidomermis sp.	–	–	12	5.6
Hexamemermis gracilis	–	–	10	5.5
Thelastomatidae				
Cephalobellus cyclocephalae	12	1.2	–	–
Cranifera robustum	10			
Severianoia brevicauda	–	–	60	1
Thelastoma modestus	–	–	65	30
T. rara	–	–	80	2500
Travassosinematidae				
Mirzaiella americana	–	–	32	1
Diplogasteridae				
Diplogaster octodontus	–	–	26	2.46
Noteodiplogaster papillosa	–	–	36	2.7
Patanodontus acaudatum	–	–	26	2.5
Rhabditidae				
Cruznema campestris	–	–	26	30
Parasitorhabditis platidontus	–	–	30	22
Rhabditis bonaerensis	–	–	26	17

which multiply rapidly, leading to septicaemia that kills the host within 24–48 h post-infection. The IJs are associated with symbiotic bacteria, *Xenorhabdus* spp. in Steinernematidae and *Photorhabdus* spp. in Heterorhabditidae, which multiply and generate metabolites that kill the insect and serve as a source of food for nematodes. Inside the host, IJs feed on bacteria, until reaching the adult stage. In case of Heterorhabditids, first generation is hermaphroditic but the following generations are amphimictics. But in Steinernematids all generations are always amphimictic, and they do not have hermaphrodictic generations.

In every generation, new JIs are produced, and the reproduction of the nematodes continues as long as the body of the insect continues to provide the necessary nutrients to the parasite. After leaving the body of the insect, they will move on to the ground to contact a new host. For both nematode genres, the bodies undergo a particular process, preventing invasion by other microorganisms and the consequent putrefaction by the antibiotics production for symbiotic bacteria. These nematodes provide an environmentally safe and economically reasonable alternative against a variety of important insect pests, and some of them are used commercially for biocontrol.

In Argentina seven species of the genus *Steinernema* and two of *Heterorhabditis* have been cited and studied, with more than 30 different isolations (Fig. 16.2). EPNs are widely distributed in this country, with records for the provinces of Córdoba, Neuquén, Río Negro, Santa Fe, La Pampa and Buenos Aires (Doucet, 1986; Doucet and Bertolotti, 1996; Doucet and Laumond, 1996; Doucet *et al.*, 1992, 2008; Stock, 1993a,b; Giayetto and Cichón, 2006). Researches have been oriented mainly on the biology and ecology (Doucet and Bertolotti, 1996; Lax *et al.*, 2011; Eliceche *et al.*, 2012; Achinelly *et al.*, 2013), the intraspecific variability (Doucet *et al.*, 1992, 1996) and the description of new species (Doucet, 1986; de Doucet and Doucet, 1990).

Field and laboratory assays demonstrated that insects from 11 orders and 42 families were susceptible to these nematodes. Studies on the parameters involved in EPN parasitism determined that the efficiency depended on the search strategies of the

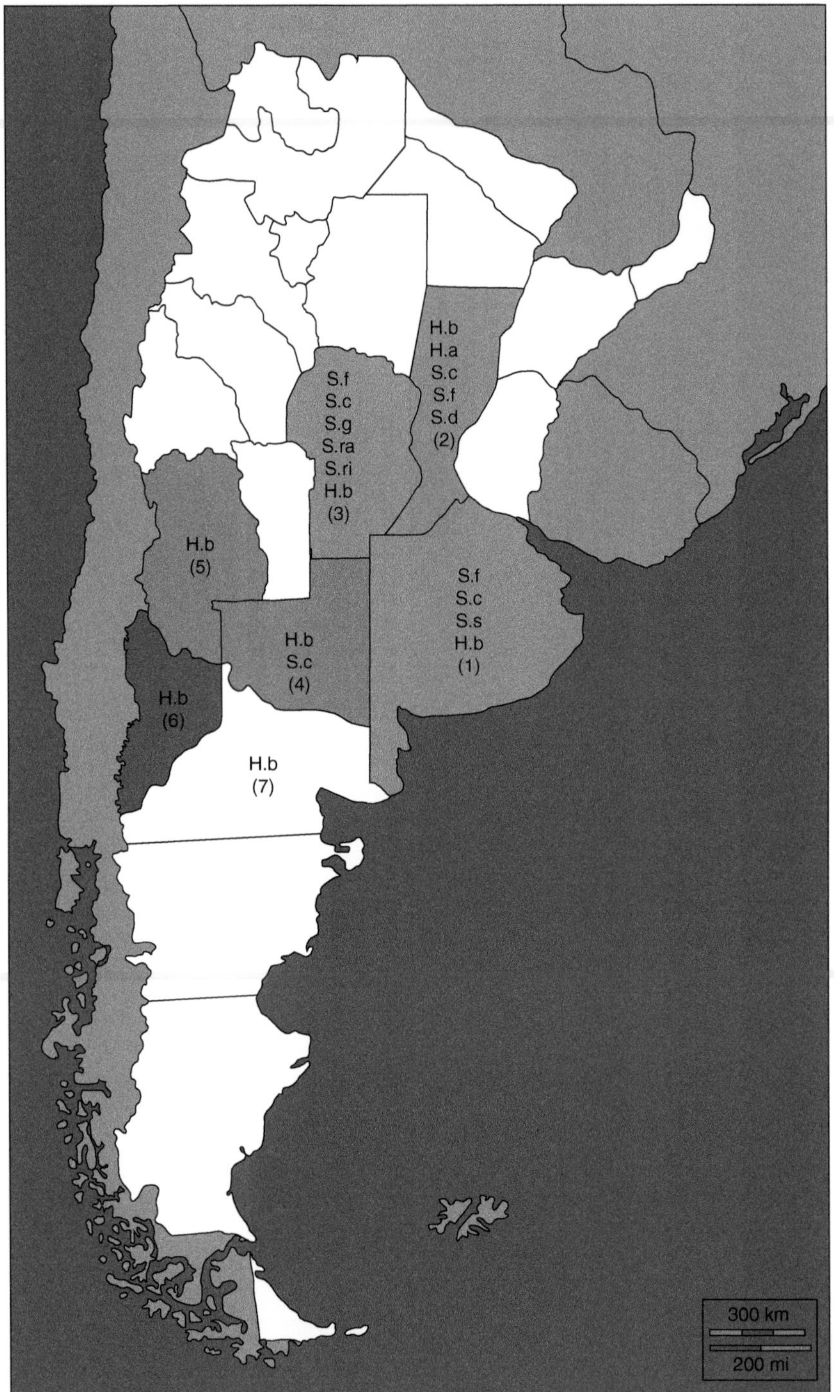

Fig. 16.2. Entomopathogenic nematodes from provinces of Argentina. 1 = Buenos Aires; 2 = Santa Fe; 3 = Córdoba; 4 = La Pampa; 5 = Mendoza; 6 = Neuquen; 7 = Rio Negro. H.a = *Heterorhabditis argentinensis*; H.b = *Heterorhabditis bacteriophora*; S.c = *Steinernema carpocapsae*; S.d = *Steinernema diaprepesi*; S.f = *Steinernema feltiae*; S.g = *Steinernema glaseri*; S.ra = *Steinernema rarum*; S.ri = *Steinernema ritteri*; S.s = *Steinernema scapterisci*.

IJs, along with their ability to remain in the environment, withstand stress conditions and become established in specific ambiences by adapting to the life cycle of their hosts (Doucet *et al.*, 1992; Doucet and Bertolloti, 1996).

STEINERNEMATIDAE. The *Steinernema* genus includes adults with an excretory pore usually opening posterior to the basal bulb, males with copulatory papillae typically comprising a single midventral, precloacal papilla plus 10–14 paired papillae, bursa absent, females can present epitygmata and IJs with symbiotic bacteria usually visible in the intestinal pouch just posterior to the basal bulb. Both males and females are necessary for reproduction (amphimictic generations). Insects parasitized by *Steinernema* spp. usually show brown and eventually darker black colours, although they have also been observed with pink, yellow and green grey colours.

Seven species of the genus *Steinernema* have been isolated from Argentina: *Steinernema carpocapsae* (Weiser, 1955) with nine isolates; *S. feltiae* (Filipjev, 1934) Wouts, (seven isolates); *Steinernema diaprepesi* Nguyen and Duncan (one isolate); *Steinernema glaseri* (Steiner, 1929) (one isolate); *Steinernema ritteri* (de Doucet and Doucet, 1990) (two isolates); *S. rarum* (Doucet, 1986) Mamiya, 1988 (nine isolates) and *Steinernema scapterisci* (Nguyen and Smart, 1990) (two isolates); with records from Córdoba, Santa Fe and Buenos Aires provinces (Doucet, 1986; de Doucet and Doucet, 1990; Stock, 1992, 1993a, 1995; Lax *et al.*, 2011). Among these citations, *S. rarum* and *S. ritteri* were described for the first time in Argentina (Doucet, 1986; de Doucet and Doucet, 1990). These *Steinernerma* species were all found in cultivated soils, with the sole exception of *S. glaseri*, with that specimen having been isolated from mountain forest sod (Doucet *et al.*, 2008).

Pathogenicity and host range. Several research workers have evaluated steinernematid nematodes against a wide range of insect pests, presented below (Table 16.5).

Recent studies have demonstrated the efficacy against lepidopterans of economic significance. An *S. rarum* isolate (OLI strain) produced mortalities of 96% and 98% in *Anticarsia gemmatalis* and *Crocidosema aporema* larvae, respectively, at a dose of 500 IJs/insect. The life cycle was shorter in *C. aporema*, whereas production was longer (4–7 days) in *A. gemmatalis* (Gianfelici *et al.*, 2014). Adults of *A. gemmatalis* were also susceptible to this nematode species with a high mortality of 80% at a dose of 500 IJs/adult (Cagnolo *et al.*, 2010).

S. diaprepesi was also evaluated against the lepidopterans *Spodoptera frugiperda* and *Helicoverpa gelotopoeon*. Infections were performed individually in Petri dishes (35 mm diameter) by applying IJs immersed in 0.5 ml water at 50 and 100 IJs/dish, each treatment contained 15 larvae. Mortality of 100% and 93% was obtained with the highest dose in *S. frugiperda* and *H. gelotopoeon*, respectively (Caccia *et al.*, 2014). In addition, studies on the pathogenicity of *S. carpocapsae* and *S. rarum* indicated the potential of these species as biocontrol agents for insects belonging to the orders Hemiptera, Diptera and Coleoptera (Bertolotti *et al.*, 2007; Doucet *et al.*, 2008; Lax *et al.*, 2011).

HETERORHABDITIDAE. The genus *Heterorhabditis* includes males usually with nine pairs of genital papillae, bursa present, IJs with dorsal tooth in the cephalic region, symbiotic bacteria retained in intestine, first-generation adults hermaphroditic, second-generation adults amphimictic.

Insects killed by *Heterorhabditis* spp. usually have colorations ranging from reddish-brown, orange and purple, and may have luminescence in the dark, depending on the bacteria of the genus *Photorhabdus*.

Two nematode species belonging to the genus *Heterorhabditis* have been registered in Argentina (see Fig. 16.2): one population of *Heterorhabditis argentinensis* Stock, 1993, in Rafaela, Santa Fe province, plus several populations of *H. bacteriophora* Poinar, 1976 – along with additional records of that species for the provinces of Córdoba (OLI, RIV strains), valleys of the Patagonian region, Neuquén and Río Negro provinces (RN, Nq, Biodiv., INTA A-V; INTA 11-12; ROCA strains), Mendoza (Rama Caída), Santa Fe (Juan and Mo strains), La Pampa and Buenos Aires (VELI) (Stock, 1993b; Doucet and Bertollotti, 1996; Giayetto *et al.*, 1998; Doucet *et al.*, 2008; Del Valle *et al.*, 2013).

The genetic proximity of *H. bacteriophora* to *H. argentinensis* has been discussed by several authors. A careful analysis of the rDNA internal transcribed spacer ITS-1 characters revealed that these closely related sister taxa would be conspecific, differing by a single transition and deletion (Adams *et al.*, 1998).

Pathogenicity and host range. A wide range of susceptible insects to heterorhabditid nematodes have been registered under laboratory and field conditions, including eight orders and 27 families (Table 16.6). Mortality was achieved in all organisms tested, completing the life cycle with rare exceptions (Thysanoptera)

Table 16.5. Spectrum of experimental and naturally infected susceptible hosts to entomopathogenic nematodes (Steinernematidae). (Adapted from Doucet *et al.*, 2008).[a]

Host range			Nematode	Isolate	References
Anoplura	Pediculidae	*Pediculus humanus capitis*	*Steinernema rarum*	NOE	Doucet *et al.*, 2008
Coleoptera	Coccinelidae	*Eriopis connexa*	*S. rarum*	NOE	Doucet *et al.*, 2008
			S. feltiae	LCHOR	Doucet *et al.*, 2008
		Hippodamia convergens	*S. rarum*	NOE	Doucet *et al.*, 2008
			S. feltiae	LCHOR	Doucet *et al.*, 2008
	Chrysomelidae	*Chrysodina* sp.	*S. rarum*	NOE	Doucet *et al.*, 2008
			S. feltiae	LCHOR	Doucet *et al.*, 2008
	Curculionidae	*Rhigopsidius piercei*	*S. rarum*	ACAB, NOE	Doucet *et al.*, 2008
		Naupactus cinereidorsum	*S. rarum*	NOE	Doucet *et al.*, 2008
			S. feltiae	LCHOR	Doucet *et al.*, 2008
	Meloidae	*Epicauta adspersa*	*S. rarum*	NOE	Doucet *et al.*, 2008
			S. feltiae	LCHOR	Doucet *et al.*, 2008
	Melyridae	*Atylus astromaculatus*	*S. feltiae*	LCHOR	Doucet *et al.*, 2008
	Scarabaeidae	*Cyclocephalla signaticolis*	*S. carpocapsae*[a]		Stock, 1995
			S. feltiae[a]		Stock, 1995
		Diloboderus abderus	*S. carpocapsae*[a]		Stock, 1995
			S. feltiae[a]		Stock, 1995
	Tenebrionidae	*Tenebrio molitor*	*S. rarum*	ACAB, NOE, OLI	Doucet *et al.*, 2008
			S. feltiae	LCHOR	Doucet *et al.*, 2008
Diptera	Culicidae	*Aedes aegypti*	*S. rarum*	OLI	Doucet *et al.*, 2008
		Culex apicinus	*S. rarum*	OLI	Cagnolo and Almirón, 2010
		C. quinquefasciatus	*S. rarum*	OLI	Doucet *et al.*, 2008
		C. pipiens	*S. feltiae*	LCHOR	Doucet *et al.*, 2008
	Muscidae	*Musca domestica*	*S. rarum*	NOE	Doucet *et al.*, 2008
			S. feltiae	LCHOR	Doucet *et al.*, 2008
	Trypetidae	*Ceratitis capitata*	*S. rarum*	ACAB, NOE	Doucet *et al.*, 2008
			S. feltiae	LCHOR	Doucet *et al.*, 2008
Hemiptera	Aphidae	*Acyrthosiphon kondoi*	*S. rarum*	NOE	Doucet *et al.*, 2008
			S. feltiae	LCHOR	Doucet *et al.*, 2008
	Coreidae	*Pachylis argentinus*	*S. feltiae*	LCHOR	Doucet *et al.*, 2008
	Nabidae	*Nabis* sp.	*S. feltiae*	LCHOR	Doucet *et al.*, 2008
	Lecanidae	*Ceroplastes grandis*	*S. rarum*	NOE	Doucet *et al.*, 2008
			S. feltiae	LCHOR	Doucet *et al.*, 2008
	Pentatomidae	*Dichelops furcatus*	*S. rarum*	NOE	Doucet *et al.*, 2008
			S. feltiae	LCHOR	Doucet *et al.*, 2008
		Nezara viridula	*S. rarum*	OLI	Doucet *et al.*, 2008
	Reduviidae	*Dipetalogaster maximus*	*S. rarum*	NOE	Doucet *et al.*, 2008
Hymenoptera	Apidae	*Apis mellifera*	*S. rarum*	NOE	Doucet *et al.*, 2008

Table 16.5. Continued.

	Formicidae	*Acromyrmex lundi*	*S. rarum*	NOE	Doucet *et al*., 2008
Isopoda	Armadillidiidae	*Armadillidium vulgare*	*S. rarum*	OLI	Doucet *et al*., 2008
Lepidoptera	Noctuidae	*Anticarsia gemmatalis*	*S. feltiae*	LCHOR	Doucet *et al*., 2008
			S. rarum	NOE, OLI	Doucet *et al*., 2008; Gianfelici *et al*., 2014
		Spodoptera frugiperda	*S. feltiae*	LCHOR	Doucet *et al*., 2008
			S. rarum	NOE	Doucet *et al*., 2008
			S. diaprepesi	SRC	Caccia *et al*., 2014; Del Valle *et al*., 2014
		Rachiplusia nu	*S. rarum*	NOE	Doucet *et al*., 2008
			S. feltiae	LCHOR	Doucet *et al*., 2008
		Heliothis sp.	*S. feltiae*	LCHOR	Doucet *et al*., 2008
			S. rarum[a]	NOE	Doucet, 1986; Doucet *et al*., 2008
		Helicoverpa gelotopoeon	*S. diaprepesi*	SRC	Caccia *et al*., 2014
	Pieraustidae	*Laxostege bifidalis*	*S. rarum*	NOE	Doucet *et al*., 2008
	Pieridae	*Colias lesbia*	*S. rarum*	NOE	Doucet *et al*., 2008
			S. feltiae	LCHOR	Doucet *et al*., 2008
	Pyralidae	*Galleria mellonella*	*S. rarum*	NOE, OLI	Doucet *et al*., 2008
			S. carpocapsae	PAMPA	Stock, 1995
			S. scapterisci	COLON	Stock, 1995
			S. diaprepesi	SRC	Lax *et al*., 2011; Del Valle *et al*., 2014
			S. ritteri		de Doucet and Doucet, 1990
			S. feltiae	LCHOR, ISABEL	Doucet *et al*., 2008; Stock, 1995
		Diatraea saccharalis	*S. rarum*	NOE	Doucet *et al*., 2008
			S. feltiae	LCHOR	Doucet *et al*., 2008
	Tortricidae	*Crocidosema aporema*	*S. rarum*	OLI	Gianfelici *et al*., 2014
Orthoptera	Gryllotalpidae	*Scapteriscus borellii*	*S. scapterisci*[a]		Stock, 1995

Table 16.6. Spectrum of experimental and naturally infected susceptible hosts to entomopathogenic nematodes (Heterorhabditidae). (Adapted from Doucet *et al.*, 2008).[a]

Host range			Nematode	Isolate	References
Anoplura	Pediculidae	*Pediculus humanus capitis*	*Heterorhabditis bacteriophora*	OLI	Doucet *et al.*, 2008
Arachnida	Tetranychidae	*Panonichus ulmi*	*H. bacteriophora*	Nq, INTA A-V, INTA 11-12, ROCA	Giayetto and Cichón, 2006
Blattodea	Blattidae	*Blatta orientalis*	*H. bacteriophora*	RIV	Doucet *et al.*, 2008
		Periplaneta americana	*H. bacteriophora*	RIV	Doucet *et al.*, 2008
Coleoptera	Coccinelidae	*Hippodamia convergens*	*H. bacteriophora*	Nq	Giayetto and Cichón, 2006
	Chrysomelidae	*Diabrotica speciosa*	*H. bacteriophora*	RIV	Doucet *et al.*, 2008
		Plagiodera erythroptera	*H. bacteriophora*	RIV	Doucet *et al.*, 2008
		Xanthogaleruca luteola	*H. bacteriophora*	BIODIV., Nq, INTA A-V, INTA 11-12, ROCA	Giayetto and Cichón, 2006
	Curculionidae	*Pantomorus leucoloma*	*H. bacteriophora*	RIV	Doucet *et al.*, 2008
		Rhigopsidius piercei	*H. bacteriophora*	ACAB	Doucet *et al.*, 2008
		Naupactus xanthographus	*H. bacteriophora*	RIV	Doucet *et al.*, 2008
		Graphognathus leucoloma	*Heterorhabditis argentinenins*[a]		Stock, 1993b
	Melyridae	*Atylus astromaculatus*	*H. bacteriophora*	RIV	Doucet *et al.*, 2008
	Nitidulidae	*Lobiopa insularis*	*H. bacteriophora*	VELI	Achinelly *et al.*, 2013
	Platypodidae	*Platypus sulcatus*	*H. bacteriophora*	BIODIV., Nq, INTA A-V, INTA 11-12, ROCA	Giayetto and Cichón, 2006
	Scarabaeidae	*Diloboderus abderus*	*H. bacteriophora*	VELI	Camino *et al.*, 2013b
	Tenebrionidae	*Tenebrio molitor*	*H. bacteriophora*	VELI	Salas *et al.*, 2013
Diptera	Culicidae	*Aedes aegypti*	*H. bacteriophora*		Peschiutta *et al.*, 2014
	Trypetidae	*Ceratitis capitata*	*H. bacteriophora*	ACAB	Doucet *et al.*, 2008
Hemiptera	Cercopidae	*Zulia entreriana*	*H. bacteriophora*	RIV	Doucet *et al.*, 2008
	Cicadellidae	*Empoasca* sp.	*H. bacteriophora*	RIV	Doucet *et al.*, 2008
		Edwarsiana crataegi	*H. bacteriophora*	BIODIV., INTA A-V, INTA 11-12	Giayetto and Cichón, 2006
	Margarodidae	*Icerya* sp.	*H. bacteriophora*	RIV	Doucet *et al.*, 2008
	Membracidae	*Ceresa* sp.	*H. bacteriophora*	RIV	Doucet *et al.*, 2008
	Pentatomidae	*Piezodorus guildinii*	*H. bacteriophora*	RIV	Doucet *et al.*, 2008
	Reduviidae	*Triatoma infestans*	*H. bacteriophora*	RIV	Doucet *et al.*, 2008
Hymenoptera	Apidae	*Apis mellifera*	*H. bacteriophora*	RIV	Doucet *et al.*, 2008
	Formicidae	*Acromyrmex lundi*	*H. bacteriophora*	Biodiv., Nq, INTA A-V, INTA 11-12, ROCA	Giayetto and Cichón, 2006
Lepidoptera	Danaidae	*Diogas erippus*	*H. bacteriophora*	RIV	Doucet *et al.*, 2008

Table 16.6. Continued.

	Nymphalidae	*Agraulis vanillae*	*H. bacteriophora*	RIV	Doucet *et al.*, 2008
	Noctuidae	*Anticarsia gemmatalis*	*H. bacteriophora*	RIV, N4, N82	Doucet *et al.*, 2008; Gianfelici *et al.*, 2014
		Heliotis sp.	*H. bacteriophora*[a]	RIV	Agüera de Doucet and Doucet, 1986
		Spodoptera frugiperda	*H. bacteriophora*	RIV, VELI	Doucet *et al.*, 2008; Camino *et al.*, 2013b
		S. praefica	*H. bacteriophora*	RIV	Doucet *et al.*, 2008
	Pieridae	*Colias lesbia*	*H. bacteriophora*	RIV	Doucet *et al.*, 2008
	Pyralidae	*Galleria mellonella*	*H. bacteriophora*	BIODIV., VELI, RIV, OLI, RN, Nq, INTA A-V, INTA 11-12, ROCA	Achinelly *et al.*, 2011; Giayetto and Cichón, 2006
	Psychidae	*Oiketicus kirbyi*	*H. bacteriophora*	RIV	Doucet *et al.*, 2008
		O. platensis	*H. bacteriophora*	BIODIV., Nq	Giayetto and Cichón, 2006
	Saturniidae	*Automeris coresus*	*H. bacteriophora*	RIV	Doucet *et al.*, 2008
		Rothschildia jacobaeae	*H. bacteriophora*	RIV	Doucet *et al.*, 2008
		Hylesia nigricans	*H. bacteriophora*	VELI	Camino *et al.*, 2013b
	Sphingidae	*Phlegetonthius sextapaphus*	*H. bacteriophora*	RIV	Doucet *et al.*, 2008
	Tortricidae	*Crocidosema aporema*	*H. bacteriophora*	N4, N82	Gianfelici *et al.*, 2014
		Cydia pomonella	*H. bacteriophora*	BIODIV., RIV, Nq, INTA A-V, INTA 11-12, ROCA	Giayetto and Cichón, 2006; Doucet *et al.*, 2008
		Grapholita molesta	*H. bacteriophora*	BIODIV., Nq, INTA A-V, INTA 11-12, ROCA	Giayetto and Cichón, 2006
Orthoptera	Acridiidae	*Xyleus modestus*	*H. bacteriophora*	RIV	Doucet *et al.*, 2008
Thysanoptera	Thripidae	*Frankliniella occidentalis*	*H. bacteriophora*	BIODIV., Nq, INTA 11-12, ROCA	Giayetto and Cichón, 2006

(Giayetto and Cichón, 2006). The susceptibility of the lepidopterans *C. aporema* and *A. gemmatalis* larvae to *H. bacteriophora* (N4 and N82 strains) was tested, indicating levels of parasitism higher than 75% at a dose of 50 IJs/insect (Gianfelici *et al.*, 2014).

H. bacteriophora (strain VELI) isolated from organic gardens constituted the second record of this species in the Buenos Aires province (Achinelly *et al.*, 2011). The Buenos Aires province is a principal area where farming is developed, and products are obtained for both domestic consumption and export. For this reason, investigations have been undertaken on the potential of this isolate as a biocontrol agent. The susceptibility to infection against lepidopterans produced over 80% parasitism in *Galleria mellonella*, *S. frugiperda* and *Hylesia nigricans*. This strain was characterized by a delay in mortality, unlike other *H. bacteriophora* strains from Argentina where nematodes killed the insects between 48 h (57.14%) and 72 h (42.85%) after contact (Camino *et al.*, 2013b; Salas *et al.*, 2013).

Efficacy was also determined against the generalist herbivore sap beetle, *Lobiopa insularis* (Coleoptera: Nitidulidae), which has recently become registered as an emerging pest in the cultivation of strawberries. Results demonstrated the susceptibility of that species to the native strain, thus pointing to a valuable and environmentally friendly alternative for the control of this beetle (Fig. 16.3). Mortality was higher in the larvae and pupae at 79 and 63%, respectively, than in adults at 10% (Achinelly *et al.*, 2013; Camino *et al.*, 2013b). The combination of cultural practices and the use of juveniles of the EPN *H. bacteriophora* in baited pitfall traps is being evaluated to control this strawberry sap beetle by the Centro de Estudios Parasitológicos y de Vectores (CEPAVE) in crops of the city of La Plata, Buenos Aires province.

16.5 Application of Entomopathogenic Nematodes in Vegetable Crop Protection

16.5.1 Field conditions

Although studies under laboratory conditions have been made in Argentina, the evaluation of EPNs under field conditions is only in the initial stages. *Steinernema riobravis* Cabanillas, Poinar and Raulston, 1994, Thermo Trilogy strain (Nematoda, Steinernematidae) was used for application to a small-scale field containing *Neocurtilla claraziana* (Saussure, 1874; Orthoptera, Gryllotalpidae), a pest of the soil in three gardens of Greater La Plata, Buenos Aires, Argentina (Camino and Reboredo, 2014). The highest dose (80,000 nematodes/3 m^2) produced a parasitism of 83%. Although high levels of infection were observed at all doses for the first 3 days post-treatment, effectiveness became significantly lower on day 7 and thereafter decreased progressively down to 0% on day 31.

16.5.2 Greenhouse environment

The effects of insect cadavers infected with three isolates of *H. bacteriophora* and one isolate of

Fig. 16.3. Adult of the strawberry sap beetle, *Lobiopa insularis*, parasitized by the entomopathogenic nematode, *Heterorhabditis bacteriophora* strain VELI.

S. diaprepesi, were evaluated on a population of *Meloidogyne incognita* in pepper (*Capsicum annuum*) and summer squash (*Cucurbita maxima*) in greenhouse experiments carried out in Santa Fe (Argentina). Two 6-day-old insect cadavers per pot were placed below the soil surface, and the soil was inoculated with 100 second-stage juveniles of *M. incognita*.

Sixty days after inoculation, the following parameters were recorded for each plant: number of leaves; dry weight of aerial parts; and numbers of galls, egg masses and eggs.

In pepper, the only variable affected was the number of eggs, in treatments with the application of *Tenebrio molitor* cadavers infected with *H. bacteriophora* Rama Caída strain.

In summer squash, several treatments using infected cadavers resulted in a decrease in the numbers of galls and egg masses. Only the treatment involving *G. mellonella* cadavers infected with the *H. bacteriophora* Rama Caída strain proved to be efficient in reducing the number of *M. incognita* eggs (Del Valle *et al.*, 2013).

The effect of the application of two such native isolates – *S. rarum* from the province of Córdoba and *H. bacteriophora* from the province of Mendoza – to a *Nacobbus aberrans* population on tomato plants under greenhouse conditions was evaluated by Caccia *et al.* (2012). The inoculation of IJs (25 IJ/cm^2) on to the soil surface compromised the reproductive ability of the target nematode, causing a decrease in the subsequent reproduction of *N. aberrans* on the plants by 57 and 53% by the two biocontrol nematodes, respectively.

Even the use of *H. bacteriophora*-infected cadavers of *G. mellonella* caused a significant decrease in the number of galls and egg masses on pepper plants parasitized by *Meloidogne javanica*, in a growth chamber (Del Valle and Doucet, 2014). However, the use of EPNs does not always reduce plant parasitic nematode (PPN) populations, so the outcomes of their interactions vary with the EPN and PPN species, the host crop and the method used to evaluate the impact on PPNs.

16.6 Other Applications of Entomopathogenic Nematodes in Argentina

16.6.1 Entomopathogenic nematodes against Diptera (Culicidae, Muscidae)

Although Steinernematidae and Heterorhabditidae are terrestrial organisms that move in the interstitial water between soil particles, several authors have evaluated the aquatic habitat as an excellent environment for the survival of these nematodes. As was mentioned under Section 16.4.1, EPNs were tested against some insects of health importance. The infectivity and life cycle of an Argentine isolate of the nematode *H. bacteriophora* was registered on *Ae. aegypti* larvae under laboratory conditions (Peschiutta *et al.*, 2014). The higher percentages of parasitism, 77.5%, 78.3% and 84.2%, corresponded to the dose 1500:1, 500:1 and 750:1 IJs/larva, respectively. The mortality of mosquito larvae began to be registered 48 h after exposition to the nematodes. *H. bacteriophora* completed its life cycle within this host successfully up to the adult stage and the emergence of IJs in few days (3–8 days). The largest number of individuals was obtained with doses 100:1 (168.2 IJs/larva).

One generation of nematodes was observed in 70% of the parasitized larvae, while a second generation was observed in a variable percentage of larvae according to the dose.

Melanization was registered from a dose of 15:1 IJs/larva, in the abdomen (50.3%), thorax (39.0%), head (6.4%) and neck (4.3%) of the mosquito larvae. The number of melanized nematodes increased with higher doses.

Cagnolo and Almirón (2010) reported for the first time parasitism by *S. rarum* in larvae of *Cx. apicinus*. They exposed mosquito larvae at 1:1, 5:1, 10:1, 15:1, 100:1 and 400:1 IJs/larva, obtaining higher mortality at the major doses (75%). Although auspicious results were obtained, efficacy has not been tested on a large scale. The results demonstrated the presence of IJs in the offspring of the first- and second-generation adults, although with a greater virulence in the first-generation IJs. The authors considered that these generational differences might be exploited for biocontrol purposes through the use of a specific generation of IJs for inoculative releases.

16.7 Role of Entomopathogenic Nematodes in Integrated Pest Management in Argentina

IPM includes preventive tasks that should be applied in a continuous and organized way to minimize the hazards caused by the presence of economic pests. It is integrated by practices, tools and the control of materials, such as natural enemies of pests, crop rotation and the use of resistant crop varieties, not excluding the use of pesticides. Unlike traditional pest control (reagent system), IPM is a proactive system that anticipates the impact of the incidence of pests in production processes.

In Argentina, IPM began to take hold in the 1990s as a decision system based on the selection and use of tactical control, singly or combined harmoniously into a strategy related to analysis of the cost/benefit ratio. It is estimated that these practices have grown considerably in recent years, supported by increasingly important professional advice, which is directly related to the health of crops and the high yields achieved (SAGPyA, 2002).

16.7.1 Tasks include

Preventive methods

In order to reduce the chance of an attack: identify risk areas; choose a suitable site for planting; prepare or correct the soil; use pest-free plants; use good sanitary quality seeds and seedlings; choose cultivars well adapted to local conditions; select resistant or tolerant varieties.

Monitoring

Monitoring is a fundamental tool in deciding which strategy to follow.

For effective management, observation of the crop throughout the year should be made frequently, determining which insects are present, their abundance, stages and economic injury level.

These tasks also include maintenance and hygiene, since insect pests need environments that provide them with air, moisture, food or shelter. Some of these are removal from the production area of plants or severely infested plant organs and weeds that are pest hosts, which could affect the crops. The use of mechanical methods to remove insects or mites, remove caterpillars manually, use of high-pressure water to remove aphids or thrips, border crops, trap crops, pruning, irrigation, fertilization, harvesting time and frequency. For crops under cover, exclusion nets or other physical barriers, mainly for aphids and thrips, are used. While many herbs or their extracts are used to control insects, mites and other harmful organisms, these crops also have their own pests, so it should be important to control them.

16.7.2 Application of products

Chemicals or curative

Low toxicity when the pest exceeds the threshold of economic damage.

Biological: natural control, with conservation of natural enemies

Once you know the type of pest to be controlled, it is necessary to plan implementation of the products. The decision on the form of application and the equipment used is crucial to obtaining good results. All insecticide used must be registered with the National Health Service, SENASA.

Verification (management control)

This task is very important and works directly on an analysis of the evolution of IPM, and helps significantly in detecting the origin of the presence of pests. In Argentinean agriculture, it is common to find family subsistence models, many of which apply pesticides at random, depending not only on the presence of pests but also on the availability of resources and the cost of pesticides. This often generates an application process that is not coordinated with neighbours, is often done in adverse weather conditions and there is widespread ignorance of the phenology of crops and pest status, as well as the toxic effects of pesticides.

The application of bioinsecticides based on entomopathogenic control in Argentina includes mainly the use of bacteria (*B. thuringiensis*; *Bacillus subtilis*) and viruses (Betabaculovirus; Baculoviridae) to control lepidopteran pests of fruit and nut trees (Sciocco-Cap, 2013). The generation, production and introduction of biopesticides is considered a cost-effective alternative compared to the application of chemicals, especially for small organic farmers who often develop management systems with low budgets. Currently, there are no records of EPNs in the National Register of Biopesticides of the *Servicio Nacional de Sanidad* (SENASA), so the applications and checks carried out have been done with native agents on a small scale, mainly for research and development (Huerga and San Juan, 2005).

In this context, the introduction of a native isolate, the nematode *H. bacteriophora* VELI strain, to control the sap beetle, *L. insularis*, pest of strawberry crops in large areas of the provinces of Buenos Aires, Santa Fe and Corrientes, achieved good results by inoculation of IJs as the strategy. The cultivated area consisted of seven ridges divided into 15–16 plots each, each parcel with a total of 12 strawberry plants. The application was performed in six plots established at random, whose dose was 10,000 IJs/plant. The results suggested that

7 months post-introduction in the environment, IJs were spread almost evenly over the cultivated space, where nematodes were isolated from both plots (applied as well as non-applied). We considered that such distribution was influenced strongly by the irrigation system supplied to plants (D. Eliceche, La Plata, 2015, personal communication). Sap beetles are strongly attracted to certain volatile plant compounds in ripening or decaying fruit, and themselves produce pheromones/kairomones that elicit an aggregating behaviour. Baits using such material can be effective in trapping and monitoring sap beetle populations, and hence determining when treatment is necessary.

The application of IJs of the nematode *H. bacteriophora* along with fermented fruit in baited pitfall traps to control *L. insularis* could be an appropriate strategy, which is currently being tested. The introduction of *H. bacteriophora* VELI strain was performed in aubergine crops in organic farming in the province of Buenos Aires, to control the attack of the weevil, *Phyrdenus muriceus*. Nematodes introduced in an area of 1000 m^2 produced a reduction of the damage caused by the weevil at a rate of 51% when IJs were inoculated (2000 IJs/lineal metre) and 69% with inoculated cadavers (1 each 10 lineal metre) compared to the control (D. Eliceche, La Plata, 2015, personal communication). The knowledge and awareness of the producers of the benefits of the optimum time for application of a biopesticide, as is the early introduction of the nematode for pest control, is relevant. Early introduction of *H. bacteriophora* to control this pest prior to planting and/or postharvest would improve the performance and effectiveness of this biological agent, avoiding the attack of crops and the establishment of the insects.

The control of plant parasitic nematodes through the use of EPNs has recently been initiated in Argentina. As mentioned above, the strategy of applying insect cadavers infected with EPNs was tested by Del Valle *et al.* (2013) on the plant parasitic nematode *M. incognita* in pepper and summer squash, where treatments involving *T. molitor* and *G. mellonella* cadavers infected with the *H. bacteriophora* Rama Caída strain affected the number of *M. incognita* eggs produced on the pepper and summer squash, respectively. In the same way, Caccia *et al.* (2012) observed a reduction of the reproduction factor when evaluating the effects of isolates of *S. rarum* and *H. bacteriophora* against the phytophagus nematode *N. aberrans* in tomato plants.

IPM in Argentina has currently become not only a philosophy of control but also a reality that must be demonstrated, in an effort to ensure that the control of pests in a sustainable manner is achieved, obtaining quality products that are safe and environmentally friendly. The latter concept includes, as the main aim, the health of the farmer and his family, and the protection of natural resources.

16.8 Technical Restrictions

Bioproducts available in the market are scarce in Argentina and null for nematodes. This is due to the high production costs compared to chemical insecticides, lack of legislation regarding the regulation and registration requirements of different groups of bioproducts, the development of criteria or policy recommendations for field studies or releases of new bioproducts and the identification of mechanisms to promote the development, marketing and use of bioproducts by the producers, as well as the control of the collection, introduction, use and/or transit of such products.

However, recently the Advisory Committee on Bio-inputs for Agricultural Use (CABUA) has been created, in the area of the National Biotechnology Advisory Committee (NBAC), to provide technical advice on the quality, efficacy and biosafety requirements necessary in the use of agricultural bioproducts, as well as to establish an appropriate regulatory framework for their use, handling and disposal in the ecosystem. This constitutes an important stimulus for the growth of this line of research in Argentina, and increases awareness of the importance of the search, use and production of bioproducts, as well as the preparation of control plans, IPM and, likewise, the need for production at the national level.

16.9 Entomopathogenic Nematodes – Potential Threat to the Ecosystem?

Studies on the impact of EPNs on beneficial soil fauna are limited to the introduction in the field of a native strain of nematode *H. bacteriophora* (VELI) on beetles in strawberry crops. The insect survey pre- and post-inoculation showed no significant differences in the number and type of registered species. Parasitism and mortality by this nematode on beneficial insects were not observed in the field. In order of importance, individuals of Staphylinidae, Carabidae and Lathrididae families were collected (Camino *et al.*, 2013b).

16.10 Conclusion

A comprehensive study of entomonematodes has been carried out in Argentina, from the 1980s to the present. Researches were directed to surveys, diversity, identification, pathogenicity and life cycles, tending mostly to find new biocontrol tools.

Eight families and 43 genera of nematodes associated to insects were recorded, with different kinds of association (parasites, pathogens and parasitoids), three of which with a potential as biocontrol agents of insects (Mermithidae, Steinernematidae and Heterorhabditidae). A wide range of insects were recorded as hosts covering eleven orders and 44 families, most of agricultural and public health importance. Entomonematodes presented a wide distribution in a variety of environments: aquatic, terrestrial, arid and humid climates, cultivated and non-cultivated, plains, mountains, valleys, steppes and forests. Some highlights of the research included the study of the nematode *S. spiculatus* in the biocontrol of mosquitoes of public health importance, with research on optimizing production, the effect of abiotic and biotic factors on infectivity, and parasitic activity and population dynamics in the field. About 30 isolates of heterorhabditid and steinernematid nematodes have been cited by researchers from Argentina over more than 35 years of study. *S. rarum* and *S. ritteri* were described as new species for first time in Argentina. The pathogenicity of EPNs included laboratory and field studies on the susceptibility of pests from horticultural crops such as tomatoes, peppers, squash, aubergine and strawberries. The isolation and study of nine species of EPNs, namely *S. carpocapsae*, *S. feltiae*, *S. scapterisci*, *S. rarum*, *S. ritteri*, *S. glaseri*, *S. diaprepesi,* and *H. bacteriophora* and *H. argentinensis* (both synonymized by Adams *et al.*, 1998), with different isolates, and records for the provinces of Córdoba, Neuquén, Río Negro, Santa Fe, La Pampa and Buenos Aires has been reported so far.

In Argentina, IPM has grown considerably in recent years, supported by increasingly important professional advice and directly related to the health of crops and the high yields achieved. The application of bioinsecticides based on entomopathogens for control currently includes the use of bacteria and viruses against lepidopteran pests of fruit and nut trees. The generation, production and introduction of biopesticides is considered a cost-effective alternative compared to the application of chemicals, especially for small organic farmers who frequently develop management systems on low budgets. So far, no company has formalized the registration of products based on EPNs in Argentina, with application and checks having only been carried out on a small scale with native agents, mainly for research and development. However, the use of nematodes in IPM, although incipient, has shown encouraging results, reducing the damage caused by pests of horticultural crops such as insects of the Coleoptera and Lepidoptera order and the root-knot nematodes, *N. aberrans*, *M. incognita* and *M. javanica*.

Although an extensive study has been conducted on entomonematodes in Argentina, the future of these investigations must involve further field trials and refinements in the techniques of mass production, along with long-term studies to ascertain the dynamics and development of these entomoparasites within natural populations of insects that are relevant to both agriculture and human health.

Acknowledgements

This work was supported by research funds from ANPCyT (Agencia Nacional de Promoción Científica y Tecnológica, Argentina), CONICET (Consejo Nacional de Investigaciones Científicas y Técnicas, Argentina), CIC (Comisión de Investigaciones Científicas, Buenos Aires, Argentina) and UNLP (Universidad Nacional de La Plata, Argentina).

References

Achinelly, M.F. and García, J.J. (2003) Efecto de la temperatura sobre la longevidad e infectividad de los juveniles preparásitos de *Strelkovimermis spiculatus* (Nematoda. Mermithidae) parásito de culícidos. *Revista de Biología Tropical* 51, 753–758.

Achinelly, M.F. and Micieli, M.V. (2009) Experimental releases of *Strelkovimermis spiculatus* (Nematoda: Mermithidae) against three mosquito species in Argentina. *Nematology* 11, 151–154.

Achinelly, M.F. and Micieli, M.V. (2013) Host range of the parasite *Strelkovimermis spiculatus* (Nematoda: Mermithidae) in Argentina mosquitoes. *Journal of Vector Ecology* 1, 1–5.

Achinelly, M.F., Micieli, M.V and Garcia, J.J. (2003) Preparasitic juveniles of *Strelkovimermis spiculatus* Poinar & Camino, 1986 (Nemata: Mermithidae) predated upon by freshwater copepods. *Nematology* 5, 885–888.

Achinelly, M.F., Micieli, M.V., Marti, G. and García, J.J. (2004) Susceptibility of neotropical mosquito larvae

(Diptera: Culicidae) and non-target aquatic organisms to the entomopathogenic nematode *Strelkovimermis spiculatus* Poinar and Camino 1986 (Nematoda: Mermithidae). *Nematology* 6, 299–302.

Achinelly, M.F., Eliceche, D. and Camino, N. (2011) Pathogenicity, biology and production of a new isolate of *Heterorhabditis bacteriophora* (Poinar, 1976) (Nematoda: Heterorhabditidae) from Argentina. *XLIV Annual Meeting Society Invertebrate Pathology* Halifax, Canada, pp. 61 (Abstract).

Achinelly, M.F., Cluigt, N. and Greco, N. (2013) Primer registro de un potencial agente de biocontrol sobre el coleóptero *Lobiopa insularis* (Nitidulidae), plaga del cultivo de frutilla en argentina. [First record on the potential biocontrol agent on the beetle *Lobiopa insularis* (Nitidulidae), plague of strawberry crop in Argentina.] *Nematrópica* 43, 277.

Adams, B.J., Burnell, A.M. and Powers, T.O. (1998) A phylogenetic analysis of *Heterorhabditis* (Nemata: Rhabditidae) based on internal transcribed spacer 1 DNA sequence data. *Journal of Nematology* 30, 22–39.

Agüera de Doucet, M.M. and Doucet, M.E. (1986) Nuevos datos para el conocimiento de *Heterorhabditis bacteriophora* Poinar, 1975. *Revista de Investigaciones, Agropecuarias INTA* XXI, 1–10.

Belaich, M.N., Buldain, D., Ghiringhelli, P.D., Hyman, B., Micieli, M.V. and Achinelly M.F. (2015) Nucleotide sequence differentiation of argentine isolates of the mosquito parasitic nematode *Strelkovimermis spiculatus* (Nematode: Mermithidae). *Journal of Vector Ecology* 40, 415–418.

Berg, C. (1898) Sobre los enemigos pequenos de la langosta peregrina *Schislocerca paranensis* (Burm.). *Comunicación Museo Nacional de Buenos Aires* 1, 25–30.

Bertolotti, M.A., Doucet, M.E. and Kambic, A. (2007) Susceptibility of *Ceratitis capitata* (Diptera: Tephritidae) to two isolates of *Steinernema rarum* (Doucet, 1986) Mamiya, 1988 (Nematoda: Steinernematidae) from Córdoba province, Argentina. *Nematropica* 37, 140.

Blanco, M.V., Lax P., Rondan Dueñas, J.C., Gardenal, C. and Doucet, M. (2012) Morphological and molecular characterisation of the entomoparasitic nematode *Hammerschmidtiella diesingi* (Nematoda, Oxyurida, Thelastomatidae). *Acta Parasitologica* 57, 302–310.

Caccia, M., Lax, P. and Doucet, M.E. (2012) Effect of entomopathogenic nematodes on the plant-parasitic nematode *Nacobbus aberrans*. *Biology and Fertility of Soils*, DOI: 10.1007/s00374-012-0724-z.

Caccia, M.G., Del Valle, E., Doucet, M.E. and Lax, P. (2014) Susceptibility of *Spodoptera frugiperda* and *Helicoverpa gelotopoeon* (Lepidoptera: Noctuidae) to the entomopathogenic nematode *Steinernema diaprepesi* (Rhabditida: Steinernematidae) under laboratory conditions. *Chilean Journal of Agricultural Research* 74, 123–126.

Cagnolo, S.R. and Almirón, W.R. (2010) Capacity of the terrestrial entomopathogenic nematode *Steinernema rarum* (Rhabditida: Steinernematidae) to parasite *Culex apicinus* larvae (Diptera: Culicidae). *Revista de la Sociedad Entomológica Argentina* 69, 141–145.

Cagnolo, S.R., Peschiutta, M.L. and Bertolotti, M.A. (2010) Susceptibility of adults of *Anticarsia gemmatalis* Hübner, 1818 (Lepidoptera: Noctuidae) to the entomopathogenic nematode *Steinernema rarum* (Doucet, 1986) Mamiya, 1988 (Rhabditida: Steinernematidae) under laboratory conditions. *Nematology* 13, 373–376.

Camino, N.B. (1989) Primer registro de culícidos (Diptera:Culicidae) parasitados por *Hydromermis* sp. (Nematoda:Mermithidae). *Neotrópica* 35, 67–70.

Camino, N.B. and de Villalobos, C. (2013) A new species of *Protellus* Cobb, 1920 (Nematoda, Thelastomatidae) parasite of the field cockroach *Blatella vaga* Hebard, 1919 (Blattodea, Blattidae) from Catamarca, Argentina. *Acta Parasitologica* 58, 98–104.

Camino, N.B. and García, J.J. (1991) Influencia de la salinidad y el pH en el parasitismo de *Strelkovimermis spiculatus* Poinar y Camino, 1986 (Nematoda: Mermithidae) en larvas de *Culex pipiens* Wied. (Diptera: Culicidae). *Neotropica* 37, 107–112.

Camino, N.B. and González, S.E. (2012) Four sympatric species of Thelastomatidae (Nematoda) parasites of the american cockroach *Periplaneta americana* (L.) (Blattodea, Blattidae) in La Plata, Argentina. *Neotropical Helminthology* 6, 239–246.

Camino, N.B. and Reboredo, G.R. (1994) Biology of *Strelkovimermis spiculatus* Poinar y Camino, 1986 (Nematoda: Mermithidae) parasite of mosquitoes (Diptera: Culicidae) under laboratory conditions. *Geotrópica* 40, 45–48.

Camino, N.B. and Reboredo, G.R. (1996) Producción de *Strelkovimermis spiculatus* Poinar y Camino, 1986 (Nematoda: Mermithidae). *Neotropica* 42, 47–50.

Camino, N.B. and Reboredo, G.R. (2014) Application to small-scale field of *Steinernema riobravis* Cabanillas, Poinar & Raulston, 1994 (Nematoda, Steinernematidae), Thermo Trilogy strain against *Neocurtilla claraziana* (Saussure, 1874) (Orthoptera, Gryllotalpidae) in gardens of La Plata, Argentina. *Neotropical Helminthology* 8, 77–83.

Camino, N.B. and Stock, S.P. (1994) *Hexamermis macrostoma* n.sp. (Nemata: Mermithidae) parasitizing the cricket *Gryllodes laplatae* (Orthoptera: Gryllidae) in Argentina. *Fundamental and Applied Nematology* 17, 397–399.

Camino, N.B., Ricuzzi, M. and López Costa, L. (2013a) A new Thelastomatidae, *Hammerschmidtiella laplatae* n. sp. (Nematoda) parasite of the cockroach *Periplaneta Americana* (L.) (Blattodea, Blattidae) from Argentina. *Neotropical Helminthology* 7, 95–103.

Camino, N.B., de Villalobos, C., Zanca, F., Achinelly, M.F., Micieli, M.V., *et al.* (2013b) Contribución al Estudio de Nemátodos y Nematomorfos Parásitos de Insectos.

Situación Actual del Conocimiento. Laboratorio de Entomonematodos. In: Berón, C. and Consolo, F. (eds) *I Taller Argentino de Patología de Insectos y sus Aplicaciones en Biocontrol*. Universidad Nacional de Mar del Plata, Mar del Plata, Argentina, pp. 21–29.

Camino, N.B., Achinelly, M.F., Rusconi, J.M., Eliceche, D. and Salas, A. (2014) Biodiversity of nematodes parasitizing soil pest white grubs (Scarabaeidae) of Buenos Aires, Argentina. *Neotropical Helminthology* 8, 227–233.

Campos, R.E. and Sy, V.E. (2003) Mortality in immatures of the floodwater mosquito *Ochlerotatus albifasciatus* (Diptera: Culicidae) and effects of parasitism by *Strelkovimermis spiculatus* (Nematoda: Mermithidae) in Buenos Aires Province, Argentina. *Memórias do Instituto Oswaldo Cruz* 98, 199–208.

Del Valle, E.E. and Doucet, M.E. (2014) Effects of *Galleria mellonella* cadavers infected with *Heterorhabditis bacteriophora* and *Steinernema rarum*, their macerates and dead infective juveniles on *Meloidogyne javanica* suppression. *Revista de la Facultad de Ciencias Agrarias, Universidad Nacional de Cuyo,* 46, 205–211.

Del Valle, E., Lax, P., Rondán Dueñas, J. and Doucet, M. (2013) Effects of insect cadavers infected by *Heterorhabditis bacteriophora* and *Steinernema diaprepesi* on *Meloidogyne incognita* parasitism in pepper and summer squash plants. *Ciencia e Investigación Agraria* 40, 109–118.

Del Valle, E., Balbi, E., Lax, P., Rondán Dueñas, J. and Doucet , M. (2014) Ecological aspects of an isolate of *Steinernema diaprepesi* (Rhabditida: Steinernematidae) from Argentina. *Biocontrol Science and Technology* 24, 690–704.

Di Battista, C.M., Fischer, S. and Campos, R.E. (2015) Parasitism prevalence and survival time of adult *Ochlerotatus albifasciatus* (Diptera: Culicidae) parasitized by *Strelkovimermis spiculatus* (Nematoda: Mermithidae. *Journal of Vector Ecology* 40, 393–397.

Doucet, M.M.A. de (1986) A new species of *Neoapleclana* Steiner, 1929 (Nematoda: Steinernematidae) from Cordoba, Argentina. *Revue de Nematologie* 9, 317–323.

Doucet, M.M.A. de and Bertolotti, M.A. (1996) Una nueva población de *Heterorhabditis bacteriophora* Poinar, 1975 para Argentina. Caracterización y acción sobre el huésped. *Nematologia mediterranea* 24,169–174.

Doucet, M.M.A. de and Doucet, M.E. (1990) *Steinernema ritteri* n. sp. (Rhabditida: Steinernematidae) with a key to the species of the genus. *Nematologica* 5, 257–265.

Doucet, M.M.A. de and Laumond, C. (1996) Uso de nematodos entomopatógenos a campo. In: Lecuona, R. (ed.) *Microorganismos patógenos empleados en el control microbiano de plagas*. Editorial Sudamericana, Buenos Aires.

Doucet, M.M.A. de, Doucet, M.E. and Nienstedt, K. (1992) Diferencias inter e intraespecíficas en la capacidad infectiva de poblaciones de *Heterorhabditis* y *Steinernema* aislados en Argentina. *Nematropica* 22, 237–242.

Doucet, M.M.A. de, Cagnolo, S.R. and Bertolotti, M.A. (1996) New *Heterorhabditis bacleriophora* Poinar, 1975 from Argentina. Variability and description of hermaphrodites from second and third generation. *Fundamental and Applied Nematology* 19, 415–420.

Doucet, M.E., Bertolotti, M.A., Cagnolo, S. and Lax, P. (2008) Nematodos entomofílicos de la provincia de Córdoba, Argentina. *Anales de la Academia Nacional de Agronomía y Veterinaria* LXII, 263–267.

Eliceche, D., Salas, A., Achinelly, M.F. and Camino, N.B. (2012) Evaluación de la temperatura de almacenamiento de la forma infectiva del nematodo *Heterorhabditis bacteriophora* sobre la supervivencia y parasitismo. *Revista Argentina de Parasitología* 1, 249.

Filipjev, J. (1934) Miscellanea Nematologica 1. Eine neue Art der Gattung *Neoaplectana* Steiner nebst Bermerkunge über die systematische Stellung der letzteren. *Parasitologicheskii Sbornik Zoologicheskogo Instituta Akademii Nauk SSSR* 4, 229–240.

García, J.J. and Camino, N.B. (1990) Primera cita para la Argentina de infecciones naturales en larvas de *Culex pipiens* (L.) (Diptera: Culicidae). *Neotropica* 36, 83–86.

Gianfelici, M., Bertolotti, M.A. and Cagnolo, S.R. (2014) Susceptibilidad de larvas de *Crocidosema aporema* (Walsingham, 1914) y *Anticarsia gemmatalis* Hübner, 1818, a tres aislados de nematodos entomopatógenos. *Revista Facultad de Ciencias Exactas, Físicas y Naturales* 1, 71–76.

Giayetto, A.L. and Cichón, L.I. (2006) Distribución, gama de huéspedes y especificidad de cinco poblaciones de *Heterorhabditis bacteriophora* (Nematoda: Heterorhabditidae) del Alto Valle de Río Negro y Neuquén, Argentina. *RIA* 35, 163–183.

Giayetto, A.L., Doucet, M.M.A. de and Bertolotti, M.A. (1998) Entomopathogenic nematodes (Rhabditidae: Heterorhabditidae and Steinernematidae) in North Patagonia, Argentina. *Nematropica* 28, 132–133.

Gutierrez, A.C. (2015) Búsqueda de patógenos de cucarachas (Blattodea) urbanas en la Provincia de Buenos Aires. PhD thesis, University of La Plata, Argentina.

Huerga, M. and San Juan, S. (2005) *El control de las plagas en la agricultura Argentina. Estudio Sectorial Agrícola Rural,* FAO, Buenos Aires.

Lax, P., Blanco M.V. and Doucet, M.E. (2008) Estudio del intestino de tres especies de cucarachas en la ciudad de Córdoba y la detección de nematodos. *VII Congreso Argentino de Entomología* Huerta Grande, pp. 208–281 (Abstract).

Lax, P., Del Valle, E., Rondan Dueñas, J., Caccia, M., Gardenal, C. and Doucet, M.E. (2011) First record of *Steinernema diaprepesi* (Steinernematidae) in Argentina. *Nematropica* 41, 377.

Mamiya, Y. (1988) *Steinernema kushidai* n. sp. (Nematoda: Steinernematidae) associated with scarabaeid beetle larvae from Shizuoka, Japan. *Applied Entomology and Zoology* 23, 313–320.

Micieli, M.V. and García, J.J. (1999) Estudios epizootiologicos de *Strelkovimermis spiculatus* Poinar and Camino, 1986 (Nematoda, Mermithidae) en una población natural de *Aedes albifasciatus* (Diptera, Culicidae) en la Argentina. *Miscelánea Zoologica* 22, 31–37.

Micieli, M.V., Risso, P., Achinelly, M.F. and Tarquini, J. (2012a) Effect of temperature, photoperiod and flooding–desiccation cycles on *Strelkovimermis spiculatus* (Nematoda: Mermithidae) preparasites under culture conditions. *Journal of Invertebrate Pathology* 110, 114–117.

Micieli, M.V., Risso, P., Achinelly, M.F., Villar, M. and Muttis, E. (2012b) Population dynamics between the mermithid *Strelkovimermis spiculatus* (Nematoda: Mermithidae) and the floodwater mosquito *Ochlerotatus albifasciatus* (Diptera: Culicidae) over time. *Biological Control* 61, 55–63.

Muttis, E., Micieli, M.V. and Garcia, J.J. (2013) *Culex pipiens* affected by joint infection of a Mosquito Iridescent Virus and *Strelkovimermis spiculatus. Journal of Invertebrate Pathology* 114, 295–297.

Nguyen, K.B. and Smart, G.C. Jr (1990) *Steinernema scapterisci* n. sp. (Steinernematidae: Nematoda). *Journal of Nematology* 22, 187–199.

Peschiutta, M., Cagnolo, S.R. and Almirón, W.R. (2014) Susceptibilidad de larvas de *Aedes aegypti* (Linnaeus) (Diptera: Culicidae) al nematodo entomopatógeno *Heterorhabditis bacteriophora* (Poinar) (Rhabditida: Heterorhabditidae). *Revista de la Sociedad Entomológica Argentina* 73, 99–108.

Platzer, E.G. (2007) Mermithid nematodes. In: Floore, T.G. (ed.) Biorational Control of Mosquitoes. *Journal of American Mosquito Control Association. Bulletin* 7, 58–64.

Poinar, G.O. Jr and Camino, N.B. (1986) *Strelkovimermis spiculatus n. sp. (*Mermithidae:Nematoda) parasitizing *Aedes albifasciatus* Mac. (Culicidae: Diptera) in Argentina. *Journal Nematologia* 18, 317–319.

SAGPyA, (2002) *Manejo integrado de plagas en el sector agroalimentario – Boletin de difusión*. Secretaría de Agricultura, Ganadería, Pesca y Alimentos, Buenos Aires.

Salas, A., Eliceche, D., Belaich, M. and Achinelly, M.F. (2013) Biología, patogenicidad y multiplicación de un aislamiento nativo del nematodo entomopatógeno *Heterorhabditis bacteriophora* proveniente de huertas de cultivo orgánico en Argentina. *Nematropica* 43, 323.

Sciocco-Cap, A. (2013) Patología de Invertebrados y Control Microbiano de Plagas en Argentina. In: Berón, C. and Consolo, F. (eds) *I Taller Argentino de Patología de Insectos y sus Aplicaciones en Biocontrol.* Universidad Nacional de Mar del Plata, Mar del Plata, Argentina, pp. 3–11.

Steiner, G. (1929) *Neoaplectana glaseri* n. g. n. sp. (Oxyuridae), a new nemic parasite of the Japanese beetle (*Popillia japonica* Newm.). *Journal of the Washington Academy of Science* 19, 436–440.

Stock, S.P. (1992) *Steinernema scapterisci* Nguyen et Smart, 1990 (Nematoda, Steinernematidae) parasitizing the mole cricket *Scapteriscus borelli* in Argentina. *Nematologia Mediterranea* 20, 7–9.

Stock, S.P. (1993a) Description of an Argentinean population of *Steinernema feltiae* (Filipjev, 1934) (Nematoda, Steinernematidae). *Nematologia Mediterranea* 21, 279–283.

Stock, S.P. (1993b) A new species of the genus Heterorhabditis Poinar, 1976 (Nematoda: Heterorhabditidae) parasitizing Graphognathus sp. Larva (Coleoptera: Curculionidae) from Argentina. *Research and Reviews in Parasitology* 53, 103–107.

Stock, S.P. (1995) Natural populations of entomopathogenic nematodes in the Pampean Region of Argentina. *Nematropica* 25, 143–148.

Weiser, J. (1955) *Neoaplectana carpocapsae* n. sp. (Anguillulata, Steinernematidae) novy cizopasnic housenek obalece jablecneho, *Carpocapsa pomonella* L. *Věstník Československé Zoologické Společynosti* 19, 44–52.

Wise de Valdez, M.R. (2007) Predator avoidance behavior of *Aedes aegypti* mosquito larvae infected with mermithid nematodes (Nematoda: Mermithidae). *Journal of Vector Ecology* 32, 150–153.

17 Status of Entomopathogenic Nematodes in Integrated Pest Management Strategies in Brazil

LUIS GARRIGÓS LEITE,[1*] VANESSA ANDALÓ,[2] CLAUDIA DOLINSKI,[3] ALCIDES MOINO JUNIOR,[4] ELDER BATISTA[5] AND EDSON TADEU IEDE[6]

[1]*Istituto Biológico, Centro Experimental, Campinas, SP, Brazil;* [2]*Universidade Federal de Uberlândia, Campus Monte Carmelo, MG, Brazil;* [3]*Universidade Estadual Norte Fluminense/CCTA/LEF, Campos dos Goytacazes, RJ, Brazil;* [4]*Universidade Federal de Lavras, Lavras, MG, Brazil;* [5]*Universidade Estadual Paulista/FCAV, Jaboticabal, SP, Brazil;* [6]*Embrapa Florestas/CNPF, Estrada da Ribeira, Colombo, PR, Brazil*

17.1 Introduction

Agriculture in Brazil is subject to many challenges, including the need to reduce the use of chemical pesticides. Four thousand agrochemicals are produced worldwide in approximately 15,000 different formulations, and 8000 of these are licensed for use in Brazil (Uso de Agrotóxicos na Agricultura, 2016). Mitigation strategies include farmer education and the development of resistant varieties and better farming techniques, such as the use of organic crops and biological pest control. In this context, integrated pest management (IPM) has been gaining prominence in the search for sustainable agriculture with lower environmental impact, greater cost reduction and added value.

Microbial control agents offer alternatives to chemical pesticides that are safe and provide selective options for IPM and sustainable agriculture. Unfortunately, microbial control agents are often compared with broad-spectrum pesticides only in terms of cost and spectra of target insects. However, their benefits are far more numerous and include efficacy, safety to non-target organisms (including natural enemies of pests), farm workers, the food supply and the environment (Lacey, 2008). Currently, more than 3 million ha in Brazil are treated with microbial agents, which have mainly targeted insect pests found above ground or on the surface of the soil. Insect pests that live below ground or in cryptic habitats pose specific problems in pest management (Jackson, 1999). They often lie undetected until the sudden appearance of damage creates the demand for rapid, curative treatments to allow the plants to recover from attack. The soil, being rich in microbes, is an ideal environment to practise microbial control of insect pests, but relatively few organisms have been developed into successful microbial control agents for soil pests. The difficulty arises in the implementation of microbial control, as the soil is a highly complex, competitive environment where the survival of applied microbes is by no means assured. In addition, soil-dwelling pests appear to show a high level of resistance to broad-spectrum pathogens and microbial toxins. However, some significant successes have been achieved with bacteria, fungi and nematodes for the control of a range of soil-inhabiting pest species.

A group of entomopathogens that has been recognized as potential agents to target insects in the soil or in cryptic environments are entomopathogenic

*Corresponding author. E-mail: lgleite@biologico.sp.gov.br

 Biocontrol Agents: Entomopathogenic and Slug Parasitic Nematodes (eds. M.M.M. Abd-Elgawad, T.H. Askary and J. Coupland)

nematodes (EPNs) from the genera *Steinernema* and *Heterorhabditis* (Rhabditida: Steinernematidae and Heterorhabditidae) (Askary, 2010). In Brazil, new strains have been selected with the potential to control several insect pests. However, the production process still requires improvement to meet growing demand, to provide products with good qualities and to suit use in various cropping systems. Another nematode highlighted in Brazil is the facultative parasite *Deladenus siricidicola* Bedding (Tylenchida: Neotylenchidae), which is used for the management of the European wood wasp, *Sirex noctilio* Fabricius (Hymenoptera: Siricidae), an important pest of *Pinus* spp. (Ferraz, 1998).

This chapter presents the current research status on EPNs in Brazil, emphasizing their use in IPM programmes.

17.2 Integrated Pest Management in Brazil

In the 1960s, there was a growing concern among the global population about the hazardous effect of chemical pesticides. In Brazil, the government reduced the use of chemical inputs through incentives that resulted in the adoption of IPM programmes by growers of different crops (Carson, 1962; Van den Bosch, 1978). This was the beginning of IPM in Brazil, which has now been popularized and is considered a major technological breakthrough in Brazilian agriculture. This management practice is now very common in the country for the control of crop insect pests, particularly those infesting soybean crops (Bueno *et al.*, 2013).

17.3 Application of Biological Control Agents in Integrated Pest Management in Brazil

IPM in Brazil was implemented for the first time for the soybean crop, which proved to be one of the most successful programmes of pest management in the world. It was established during the 1970s, and was highly adopted by farmers, causing a decrease in insecticide use of more than 50% (Panizzi, 2013). The programme was supported by two biocontrol agents: (i) NPVAg, for the control of the velvet bean caterpillar, *Anticarsia gemmatalis* Hübner (Lepidoptera: Noctuidae), which is considered a major defoliator of plants; and (ii) egg parasitoids for the control of seed-sucking stink bugs, particularly the southern green stink bug, *Nezara viridula* (L.) (Hemiptera: Pentatomidae). NPVAg proved too popular, and its use had reached 2 million ha by 2003–2004 (Moscardi *et al.*, 2002; Moscardi, 2007).

Until the 2000s, IPM in soybean was a durable programme; however, after that it started declining, due to various reasons such as a change in the landscape, the adoption of the no-tillage cultivation system and the introduction of more intense multiple cropping, and a lower input to divulge and adapt the IPM programme. Nowadays, IPM in soybean is almost a forgotten technology (Panizzi, 2013). Throughout this period, several other biological control programmes had also been implemented (Table 17.1), especially in sugarcane, which has been the biggest market for use against insect pathogens.

17.3.1 What can biological methods offer to integrated pest management?

Biopesticides are considered good components of IPM as they possess a variety of attributes. They are mostly selective, produce little or no toxic residue and their production costs are significantly lower as compared to conventional synthetic chemical pesticides (Hajek, 2004). They are self-perpetuating in nature, and have the ability to reproduce on or in close vicinity to the target pest. Also, they are easy to apply and in most cases can be applied with the farmer's existing spray equipment. However, contrary to this, they have a slower effect on targeted pests in comparison to chemical pesticides, low persistence and varied efficacy under different environmental conditions (Askary, 2015a). Therefore, biopesticides are not advised for use as a single treatment but in integration with other control measures, and their selectivity and safety can contribute meaningfully to incremental improvements in pest control (Lacey and Shapiro-Ilan, 2008; Askary, 2015a). A list of available products containing EPNs in Brazil is presented in Table 17.2.

17.3.2 An outline of studies on entomopathogenic nematodes in Brazilian research centres

In Brazil, the first study with EPNs was conducted by Travassos in 1927, wherein he transferred the nematode species *Aplectana kraussei* to a new genus called *Steinernema*. However, this issue did not receive attention for more than 50 years, i.e. in

Table 17.1. Some biological control agents used in Brazil. (Adapted from Vasconcelos, 2012.)

Agent	Pest	Crop	Treated area (ha)	Cost (US$/ha)[a]
Cotesia flavipes (wasp)	Sugarcane borer	Sugarcane	3,000,000	8.00
Neoseiulus barkeri (mite)	Broad mite and thrips	Vegetables and fruit trees	500	66.00–133.00
N. californicus (mite)	Two-spotted spider mite	Vegetables and fruit trees	500	100.00–133.00
Orius insidiosus (beetle)	Thrips	Vegetables and fruit trees	500	133.00–266.00
Phytoseiulus macropilis (mite)	Two-spotted spider mite	Vegetables and fruit trees	500	100.00–133.00
Deladenus siricidicola (nematode)	Wood wasp	Pinus forest	1,000,000	2.00[b]
Stratiolaelaps scimitus (mite)	Fungus gnats and thrips	Vegetables and fruit trees	500	166.00–266.00
Telenomus podisi (wasp)	Bug	Soybean	8000	15.00
Bacillus thuringiensis (bacteria)	Caterpillars and mosquitoes	Several	800,000	23.00–33.00
Metarhizium anisopliae (fungus)	Sugarcane spittlebug	Sugarcane	2,000,000	8.00–33.00
Trichogramma galloi (wasp)	Sugarcane borer	Sugarcane	500,000	16.00
	Fall armyworm	Maize and sorghum	5000	
T. pretiosum (wasp)	Caterpillars and borers	Tomato	3000	10.00–20.00
	Sugarcane borer	Maize and sorghum	15,000	
	Caterpillars	Soybean	10,000	

Notes: [a]Product + freight + application; [b]product cost only.

Table 17.2. Nematode mass production in Brazil for commercial or experimental use.

Entomopathogenic nematode species	Target crop	Target pest	Formulation	Dosage/acre	Name of product	Nematodes/dose	Name of company	Cost of product/ha
Deladenus siricidicola	Pine	*Sirex noctilio*	Water	Inoculative application	*Deladenus siricidicola*	Dose of 20 ml – with 1 million nematodes[a]	Embrapa Floresta	~US$2
Heterorhabditis indica LPP30 or *Heterorhabditis baujardi* LPP7	Guava	*Conotrachelus psidii*	Infected larva	2.5 × 10^8 IJs/ha	Infected larvae of *Galleria mellonella*	2.5 × 10^8 IJs	Universidade Estadual Norte Fluminense	~US$30
Steinernema brazilense	Sugarcane	*Sphenophorus levis*	Sponge	1–2 × 10^8 IJs/ha	Bio Steinernema	1–2 × 10^8 IJs	Bio Controle	~US$30

Note: [a]Each dose that is applied with gelatin is enough to treat ten infected trees.

1985, when the nematode species *Steinernema glaseri* (Steiner) was found parasitizing the sugarcane pest, *Migdolus fryanus* (Westwood) (Coleoptera: Vesperidae) (Pizano *et al*., 1985). The first field test with a nematode in Brazil was performed by Arrigoni *et al*. (1986), who evaluated an imported formulation of *Steinernema carpocapsae* (Weiser) to control *M. fryanus* in sugarcane. Six years later,

Schmitt *et al.* (1992) evaluated another imported formulation of this nematode against the banana root borer, *Cosmopolites sordidus* (Germar) (Coleoptera: Curculionidae). These tests were supported by Biosys, California, USA, carrying an intense campaign for the dissemination and evaluation of their products. Several other studies have been carried out in Brazil; however, EPNs only began to gain prominence in biological control programmes from 2000 onward, with the formation of several research groups working on various issues, including biodiversity, taxonomy, *in vivo* production, *in vitro* production, stable formulation, selection of isolates, compatibility tests with pesticides and field tests of different isolates. Now, in recent years, bacterial symbionts of EPNs have become an area of research, particularly for virulence tests and molecular studies.

Diversity of entomopathogenic nematodes

The advancement of research with EPNs in Brazil is coupled with the discovery of new species with potential use; thus, a broad field is available for investigation, given the vast diversity of species in Brazil. Pizano *et al.* (1985) recorded the occurrence of *S. glaseri* in *M. fryanus* found in sugarcane plantations in the Amalia Industrial Complex, São Paulo State. Fowler (1988) conducted surveys and covered 1532 collection points, where he got mole crickets from the genus *Scapteriscus* and *Neocurtilla* and obtained 18 isolates of *S. carpocapsae* and 13 isolates of *Heterorhabditis* spp. The nematode *Steinernema feltiae* (Filipjev) was isolated from cadavers of mole crickets (Orthoptera: Gryllotalpidae) found in São Paulo State (Fowler and Garcia, 1988). More recently, Dolinski *et al.* (2008b) isolated *Heterorhabditis baujardi* (Phan, Subbotin, Nguyen and Moens) from soil collected in Rondônia State, while Machado *et al.* (2005) isolated *Heterorhabditis indica* (Poinar, Karunakar and David) from soil collected in Itapetininga Municipality, São Paulo State. In the Benjamin Constant region, Amazonas State, a new species of *Heterorhabditis* was isolated from a soil sample using the larvae of *Galleria mellonella* L. (Lepidoptera: Pyralidae). Andaló *et al.* (2006a) identified it as *Heterorhabditis amazonensis*. Later, Andaló *et al.* (2009a) conducted a survey in the Lavras region, Minas Gerais State, and obtained other populations of the same species from fields growing sorghum (*Sorghum* sp.) and garlic (*Allium sativum*). *H. indica* LPP30 was isolated from *G. mellonella* larvae in a guava orchard in Campos dos Goytacazes, RJ (Minas, 2012). In Rio Grande do Sul, Barbosa-Negrisoli *et al.* (2011) isolated *S. glaseri*, seven strains of *Steinernema rarum* and two other species of *Steinernema* (unidentified, possibly new). In this study, four strains of *Heterorhabditis bacteriophora* and one of *H. amazonensis* were also isolated. A new species of the genus *Steinernema* was isolated from native vegetation soil in Mato Grosso do Sul State by using *G. mellonella* larvae. Nguyen *et al.* (2010) identified it as *Steinernema brazilense*.

Biology and ecology

The efficiency of EPNs in the field depends on several parameters, including those related to the behaviour, biology and ecology of the agent. Variations in these characteristics may occur across genera and species, and even among populations of the same species (Molina *et al.*, 2005; Andaló *et al.*, 2009a). The low mobility of EPNs in the soil exposes them to unfavourable biotic and abiotic factors, and may reduce their chances to find target insects for reproduction (Portillo-Aguilar *et al.*, 1999). Unfavourable factors include their natural enemies, such as fungal predators, which produce mycelial structures intended to capture nematodes (Askary, 2015b; Askary and Martinelli, 2015). Andaló *et al.* (2008a) assessed the susceptibility of *H. amazonensis* to the predatory fungus *Arthrobotrys oligospora* Fresenius, *Arthrobotrys conoides* Drechsler and *Duddingtonia flagrans* Cooke. They observed the predation of infective juveniles (IJs) by the fungus after just 8 days of exposure, which is the time needed for the fungus to stimulate vegetative growth and form traps. Dolinski *et al.* (2010) evaluated the horizontal mobility of *H. baujardi* LPP7 in soil with different texture classes, and found higher rates of mortality and infectivity at a distance of 30 cm for sand and sandy clay loam classes, with dead insects being located at distances up to 60 and 90 cm, respectively. There was no mortality in clay loam at 60 and 90 cm. Infection was not detected in any of the distances evaluated in clay soils. In another study, the dispersion of *H. baujardi* LPP7 in a guava orchard was assessed. It was observed that nematodes were present up to a depth of 10–90 cm in the soil from the point of release of the cadavers containing EPNs (Del Valle *et al.*, 2008). Andaló *et al.* (2012) evaluated the horizontal and vertical mobility of *H. amazonensis* RSC2 while searching for the fall armyworm, *Spodoptera frugiperda* (Smith) (Lepidoptera: Noctuidae). The results revealed insect mortality up to 60 cm horizontally and 20 cm vertically in the soil. Since the ability of EPNs to travel long distances is limited, they use certain strategies to reach hosts in different localities, such as attaching to a body that will act as a dispersal agent from one location to another. Mertz *et al.* (2014) studied the phoretic dispersion of *H. amazonensis* (IJs) and *Calosoma granulatum* (Coleoptera: Carabidae) (adults), and observed the horizontal transport of IJs to distances greater than 40 cm.

17.3.3 Production

The first work on the production of EPNs in Brazil focused on *in vivo* multiplication, using mainly *G. mellonella* larvae as the host insect. Other hosts were also used, such as the larvae of *Diatraea saccharalis* (F.) (Lepidoptera: Crambidae) and *Tenebrio molitor* L. (Coleoptera: Curculionidae) (Folegatti *et al.*, 1988; Leite *et al.*, 1990). *In vivo* production of EPNs was also performed using *G. mellonella* larvae as the main host, because this insect is suitable for the multiplication of the majority of species from the genera *Steinernema* and *Heterorhabditis*, providing higher yields (100,000–350,000 IJs/larva) as compared to other hosts (Dutky *et al.*, 1964; Milstead and Poinar, 1978). However, rearing *G. mellonella* is considered costly, raising costs for the production and marketing of EPNs. In an attempt to search for alternative hosts for the *in vivo* production of EPNs, Molina *et al.* (2004) evaluated the multiplication of *S. carpocapsae*, *S. glaseri* and *Steinernema arenarium* (Artyukhovsky) in the last larval instar of *G. mellonella*, *T. molitor*, *S. frugiperda* and *Bombyx mori* (L.) (Lepidoptera: Bombycidae). The highest yield of IJs was recorded in the first 3 days. The yield of IJs obtained from larvae of *G. mellonella* was highest, with 302,124 and 149,213 IJs/larva for the nematodes *S. carpocapsae* and *S. arenarium*, respectively. *B. mori* and *S. frugiperda* produced comparatively lower yields of the tested nematodes. To optimize the production of *H. baujardi* LPP7 in *G. mellonella* larvae, different concentrations of nematode inoculum with different sizes of inoculated larvae were tested. The highest yields were obtained at concentrations of 100 and 300 IJs/larva (Dias *et al.*, 2008).

In the past few years in Brazil, initiatives toward developing an EPN-based commercial product have gained attention. Leite (2006) highlighted the development of media for the *in vitro* mass production of nematodes. A medium based on beef liver soaked in sponge provided high production rates of the nematode *Heterorhabditis* sp., with a yield of up to 180,000 juveniles/ml of medium. These and other initiatives are being developed with funding from the Brazilian government in partnership with private companies. Some companies have started the small-scale production of EPNs to supply products for field tests.

17.3.4 Formulation

Various studies have been conducted to elucidate the most favourable conditions and formulations for the storage and application of EPNs. Factors such as temperature and oxygen have a direct impact on the survival and infectivity of IJs stored in aqueous suspension. Isolates of *H. amazonensis* remained viable for 6 months at a temperature of 16°C with aeration, but did not survive when stored at 8°C and 28°C (Andaló *et al.*, 2006b; Molina *et al.*, 2006). The survival of IJs may be extended by the addition of glycerin. IJs of *S. carpocapsae* were found alive and infective for longer periods in glycerin as compared to water (Andaló *et al.*, 2008b). In addition, some substrates may assist and favour the storage of nematodes, such as the sponge that preserves IJs of *H. amazonensis* JPM4 at 16°C with 50% viability after 6 months of storage (Andaló *et al.*, 2010). An agar-based medium was also tested for the transport and storage of IJs. Laboratory and field work evaluating the ability of nematodes to infect and kill *G. mellonella* larvae indicated that agar concentrations at 2 and 4% remained effective for 7 and 45 days, respectively, for the storage, transport and formulation of *H. baujardi* LPP7 IJs, with no difference to water and with the advantage of using a lower volume (Dolinski *et al.*, 2008a).

Covering is recommended for the application of EPNs with the help of EPN-infected host insect cadavers, because the insect cadaver is often subject to physical damage during manipulation and application. For instance, on the ground, insect cadavers are subject to the action of organisms, such as ants, that may alter the effectiveness of the application. Del Valle *et al.* (2009) determined the effects of different coverage of insect cadavers of *G. mellonella* infected by *H. baujardi* LPP7, and evaluated the behaviour of ants, *Ectatomma* spp., (Hymenoptera:

Formicidae) towards the cadavers. The best coverage treatments were talc and gelatine capsules, as they did not affect the emergence of IJs. However, when the formulated cadavers were placed close to the openings of ant nests, ants carried the insect cadavers with different coatings up to 20 cm away from the nest openings, with the exception of those formulated in gelatine capsules, which were not removed.

17.3.5 Application

Nematodes are applied in the field by conventional sprayers that are also used for the application of other fungicides or insecticides. However, other methods such as irrigation water may also be used for nematode application. Most of the equipment used for spraying does not generate enough pressure to damage the IJs. Lara *et al.* (2008) evaluated a microspray system for the application of IJs in the field, and suggested that this system did not affect the viability, infectivity and search capability of *H. baujardi* LPP7 IJs for the targeted pest. Garcia *et al.* (2005, 2008) studied the effect of various technologies on the application of *H. indica* (IBCB-n5) and *Steinernema* sp. IBCB-n6. The results revealed no adverse effects on the concentration, viability and effectiveness of nematodes in the control of *S. frugiperda*. The concentration of IJs decreased only when a 100 mesh screen was used, with a reduction of 28% for *H. indica* and 53% for *S. brazilense*. Moreira *et al.* (2013) reported that IJs of *S. feltiae* might be applied in mixtures with cane molasses, mineral oil, vegetable oil and glycerin without any loss of viability and pathogenicity.

Vegetable crops

GREENHOUSE. Fungus gnats (Diptera: Sciaridae: *Bradysia* spp.), western flower thrips, *Frankliniella occidentalis* (Pergande) (Thysanoptera: Thripidae), and shore flies, *Scatella stagnalis* Fallen (Diptera: Ephydridae), are major pests of greenhouse flowers in Brazil. Fungus gnats, *Bradysia* spp., are considered an important pest that occurs in greenhouse crops in Brazil. Larvae feed on the roots, causing damage to the plants, especially in ornamental, citrus, tobacco and strawberry seedlings, and also mushrooms (Gallo *et al.*, 2002). Larvae cause damage by feeding on the roots of young seedlings (Hungerford, 1916; Hamlen and Mead, 1979). Larval feeding on roots opens the avenue for other pathogens to attack, thus causing indirect damage to the plant. Adults of *Bradysia* spp. have also been reported to disseminate pathogens among plants (Kalb and Millar, 1986; Jarvis *et al.*, 1993; Harris *et al.*, 1994).

EPNs have been found to control greenhouse pests, especially fungus gnats because they spend most of their life in soil in the pot medium (Leite *et al.*, 2007; Tavares *et al.*, 2012). In Brazil, the use of native and exotic nematodes has been evaluated on *Bradysia mabiusi* (Lane) larvae in organic substrates. The nematode *H. indica* IBCB-n5 proved most virulent, providing insect mortality ranged from 67–100% when applied at 7.9–31.5 IJs/cm^2. In the case of commercial crops, the reduction in the insect population was 50–70% when nematodes were applied at 10–50 IJs/cm^2 (Leite *et al.*, 2007; Tavares *et al.*, 2012).

Arable crops

In Brazil, soil-dwelling insect pests are found in many arable crops, especially in sugarcane. Pests such as spittlebugs, *Mahanarva fimbriolata* (Stål) (Hemiptera: Cercopidae), and sugarcane weevils, *Sphenophorus levis*, cause significant losses in sugarcane production. EPNs have been tested against these pests, which led to promising results. Under field conditions, the nematode *H. indica* was applied on straw and on the ground, which caused more than 50% reduction in the population of *M. fimbriolata* (Leite *et al.*, 2005). However, the mortality of the pest was 100% when evaluated under pot conditions (Leite *et al.*, 2003). Laboratory and field tests confirmed the pathogenic action of EPNs on the larvae and adults of the sugarcane weevil, *S. levis*, causing high levels of larval mortality and reducing crop losses in the field (Tavares *et al.*, 2007, 2009; Leite *et al.*, 2012). The use of the *S. brazilense* nematode to control the larva of *S. levis* facilitated gains of up to 17 tons in sugarcane production. This value could be further increased if EPNs were used in combination with synthetic insecticides registered for the control of this pest (Gionetti *et al.*, 2011; Leite *et al.*, 2012). Since the results were encouraging, a commercial product based on EPNs was registered in Brazil for the control of *S. levis* (Agrofit, 2016) and has already been used for some time to treat the pests in approximately 500 ha of land growing sugarcane (Fig. 17.1).

Fig. 17.1. Application of *Steinernema brazilense* in a sugarcane field. The close-up image shows damage to the rhizome caused by a *Sphenophorus levis* larva.

17.4 Other Applications of Entomopathogenic Nematodes in Brazil

17.4.1 Parks, urban tree plantings and municipal forests

At the end of the 1980s, the pine plantations in Brazil began to be threatened by the presence of the European wood wasp, *S. noctilio* Fabricius (Hymenoptera: Siricidae). After the wood wasp was first detected in 1988, efforts were focused on establishing an IPM programme with an emphasis on biological control, which was based mainly on the Australian experience, using the nematode *D. siricidicola* Bedding (Nematoda: Neotylenchidae) as the main agent (Iede *et al.*, 2012).

The first nematode used against the wood wasp was introduced in Brazil in 1989–1990. It originated from CSIRO, Australia, and developed in Paraná, at Embrapa Florestas. However, the strain that was introduced and released in 1990 became defective, resulting in low rates of parasitism (2–6%). Two strategies were used to solve the problem: (i) the importation of the Kamona lineage in 1994, which was reisolated by CSIRO on the island of Tasmania, Australia, and released in the field in 1995; and (ii) new, reisolated strains from field collections in Brazil, where the nematode displayed good efficiency (with emphasis on the strain isolated in the Encruzilhada do Sul Municipality, RS), probably due to its ability to adapt to local conditions. New strains were recovered and released annually from 1995 onward, with assessments indicating parasitism levels of up to 91% in 1999–2000. During this period, 68 farms from one company were monitored, obtaining an average level of 77% parasitism for 3036 evaluated insects. Isolation of the fungus and nematode from wasps collected in the field is carried out each year to confirm the continued ability of the nematode to parasitize wood wasps.

Because several female wasps lay their eggs on the same tree, the nematodes find several live wasp larvae and also feed on the fungus during their dispersal through the stem. On finding the larvae, the nematodes receive signals to become parasitic, and actively penetrate the tegument of the host larvae in the form of parasitic females that have already been fertilized by males. Once inside the host, the female nematode waits for the larvae to metamorphose into pupae and adult wasps. At this time, each female nematode that invaded the insect may release up to 10,000 eggs, which hatch and the juveniles migrate to the reproductive system of the wasp and parasitize its developing eggs, making them infertile; however, the eggs are laid normally by the wasp. Following oviposition by the wasp, the nematodes colonize another tree and seek new hosts (Bedding and Iede, 2005). *D. siricidicola* is the only nematode capable of infecting the wood wasp, with *S. noctilio* being the only species capable of serving as a host for this nematode in Brazil. Hence, the insect is responsible for the dispersal of its parasite in the environment.

The existence of two distinct cycles facilitates the generation of *D. siricidicola* under laboratory conditions without the need of the host insect. However, the search for and isolation of field populations that have passed through the parasitic cycle are needed, because the nematode may lose its ability to change cycles and parasitize wood wasp larvae when the inoculum is maintained under laboratory conditions alone (Bedding and Iede, 2005).

17.4.2 Orchards of fruit trees

Brazil is among the world's largest producers of fruit, but only exports fruit in small quantities because of the high incidence of insect pests. *Anastrepha* Schiner, *Ceratitis capitata* (Wiedemann), *Grapholita molesta* (Busck) and *Bonagota salubricola* (Meyrick) are the main insect pests that cause significant losses to fruit crops in Brazil.

Various laboratory and field tests confirm the pathogenicity of nematodes against *C. capitata* (Silva *et al.*, 2010; Minas *et al.*, 2011; Rohde *et al.*, 2012a,b). Barbosa-Negrisoli *et al.* (2009)

tested the efficacy of *H. bacteriophora* RS88 against *Anastrepha fraterculus* under field conditions and found 51.3% larval mortality. Barbosa-Negrisoli *et al.* (2010) evaluated the effectiveness of *H. bacteriophora* RS107 against the larvae of *B. salubricola* under field conditions. The nematodes caused 70.2% larval mortality. However, the highest mortality (97%) was recorded with the application of *H. bacteriophora* RS33 against *G. molesta* in peach orchards (Barbosa-Negrisoli *et al.*, 2013).

The guava weevil, *Conotrachelus psidii* Marshall (Coleoptera: Curculionidae) is considered an important pest of guava in Brazil. Both larva and adult cause damage to fruits (Orlando *et al.*, 1974; Bailez *et al.*, 2003). In 2008, researchers from the Universidade Estadual do Norte Fluminense Darcy Ribeiro in Rio de Janeiro state, Brazil, started working with a small group of farmers with the objective to establish an IPM programme in their guava orchards. Approximately 20 farmers organized in an association named GOIACAM (Associação do Produtores de Goiabas de Cachoeiras de Macacu) paid for the nematode registration and were trained to rear *G. mellonella*. *G. mellonella* larvae were brought to the university and infected there with *H. baujardi* LPP7 and then returned to the farms as infected cadavers for application. When 20 cadavers were used in a plot of 9 m^2, the emergence of adults decreased by 40–70% as compared to trees without nematode application (Fig. 17.2). Thus, the collection of IJs is

Fig. 17.2. Field application of *Heterorhabditis baujardi* LPP7 formulated in the larvae of *Galleria mellonella* (cadaver covered by 10% talc) in a guava orchard. The close-up image shows fourth-instar larva of guava weevil, *Conotrachelus psidii*.

not required during *in vivo* production, because the IJs leave the insect cadaver directly and move into the soil in search of hosts. This technique has great potential, especially in small areas and around the canopy of fruit trees (Dolinski *et al.*, 2012).

Lara *et al.* (2008) studied the effect of irrigation application on the IJs of *H. baujardi* LPP7. The results showed that the three parameters, i.e. viability, infectivity and host-searching ability, of nematodes assessed in the experiment were not affected adversely by irrigation. In another experiment, the population fluctuation of the nematode and the occurrence of the pest on guava were studied for 1 year. The results showed a reduction in the weevil population, with lower rates of infestation and crop losses (Dolinski *et al.*, 2012; Minas, 2012). Under field conditions, *C. psidii* was found susceptible to EPNs, as the larval population reduced by up to 50% when treated with *H. indica* IBCB05 (Silva *et al.*, 2010).

The banana borer, *C. sordidus*, is considered to be the main insect pest of banana in Brazil. Direct damage is caused by the larvae, which penetrate and drill into the rhizome, building galleries in all directions. Galleries facilitate the entry of pathogenic microorganisms, including *Fusarium oxysporum* f. sp. *cubense*, which is responsible for the disease known as Panama Disease. An average reduction of 30% production has been estimated in Brazil due to insect attack (Batista Filho *et al.*, 2013). In southern São Paulo State, the nematode *S. carpocapsae* was tested against adults of *C. sordidus* through direct soil application and in two types of bait that were attractive to the insect. The baits were prepared with the pseudostem of the banana plant, at a dose of 5 × 10^6 IJs/m^2, using a volume of 0.4 l/bait. After 7 days of application, insect mortality by the nematode ranged from 50 to 70%. After 21 days of application, the mortality rate fell, ranging from 22% (applied to the soil) to 40% (application in bait) (Schmitt *et al.*, 1992).

17.4.3 Poultry production

The mealworm (*Alphitobius diaperinus*), an important pest of poultry, reproduces mainly under the feeders of the chicken shed in the sawdust covering the floor. They are benefitted by sawdust and the food falling from the feeders (Alves *et al.*, 2005). Rodrigueiro *et al.* (2008) applied *H. indica* IBCBn5 on sawdust and on the floor (previously moist) at three doses, i.e. 0.4, 4.0 and 40 IJs/cm^2, against larvae and adults of *A. diaperinus*. After 12 days of application, the nematode reduced the population of larvae on sawdust by 94, 97 and 100%, respectively, and adults by 70, 77 and 89%, respectively. However, on floor, the nematodes caused a reduction in insect population by 15, 31 and 35%, respectively.

17.4.4 Coffee pests

In Brazil, the coffee root mealybug, *Dysmicoccus cryptus* (Tinsley) (Hemiptera: Pseudococcidae), also known as cochineal floury, is frequently found in coffee plantations. The insect sucks the sap from the roots and causes severe damage and even death to plants (Santa-Cecília *et al.*, 2000). *D. cryptus* is a subsoil insect and is therefore vulnerable to the action of EPNs, as observed by Alves *et al.* (2009a) against *Dysmicoccus texensis* in laboratory tests. In an experiment, Alves *et al.* (2009b) tested the efficacy of *H. amazonensis* JPM3 against *D. cryptus* by two different methods, i.e. nematode-infected *G. mellonella* cadaver and nematode in aqueous suspension. The best results were obtained when the EPN was applied in aqueous suspension, as it caused insect mortality by 70 and 65% in the greenhouse and field, respectively.

17.5 Interactions of Entomopathogenic Nematodes with Pesticides and Biological Products

17.5.1 Compatibility with pesticides

IJs are compatible with most pesticides and inorganic fertilizers when exposed for short periods (2–24 h), which facilitates their use with these products in tank mixes (Koppenhöfer and Grewal, 2005; see Laznik and Trdan, Chapter 30, this volume). However, antagonistic, additive or synergistic effects may occur mutually. Some pesticides reduce the infectivity and survival of nematodes (Grewal *et al.*, 1998), and generally heterorhabdits are more affected than steinernematids (Grewal, 2002).

Several studies have been conducted in Brazil to evaluate the selectivity of pesticides against EPNs. Negrisoli Jr *et al.* (2008) presented a protocol (based on the International Organisation for Biological and Integrated Control West Palaearctic Regional Section (IOBC-WPRS) protocol; Vainio, 1992), which involved shortening the evaluation period for viability and infectivity, i.e. 48 h of exposure to the product using *G. mellonella* larvae as the target species in standard infectivity assessments. Chemicals may also interfere with the lipid reserves, which are the main sources of energy for IJs (Barrett and Wright, 1998). Some chemicals facilitate the survival of *H. amazonensis* RSC5, *H. amazonensis* JPM4 and *S. carpocapsae* All, but at the same time reduce the amount of lipids and infectivity in them, as was observed when evaluated against *G. mellonella* larvae (Andaló *et al.*, 2009b; Sabino *et al.*, 2014). It has also been reported that some insecticides used in low doses act synergistically with EPNs and increase the effectiveness of these organisms in inundative applications (Nishimatsu and Jackson, 1998). The nematodes *H. indica* IBCBn5 and *S. brazilense* IBCBn6 were evaluated, in combination with low doses of chemical insecticides, on the mortality of larvae and adults of the sugarcane weevil, *S. levis* Vaurie (Coleoptera: Curculionidae) under laboratory conditions. The results showed synergistic effects under various combinations, with the most effective mixture being *S. brazilense* + thiamethoxam, causing adult mortality levels above 70% (Tavares *et al.*, 2009).

17.5.2 Association with plant products

Rohde *et al.* (2013) tested the combination of EPNs with plant extracts obtained from the leaf, branch and fruit of chinaberry, *Melia azedarach* L. (Meliaceae); the leaf of rue, *Ruta graveolens* L. (Rutaceae); the root of ginger, *Zingiber officinale* Roscoe (Zingiberaceae); and the clove of garlic, *A. sativum* L. (Alliaceae), on the different developmental stages of fruit fly, *C. capitata* (Wiedemann) (Diptera: Tephritidae). It was noted that the extracts were incompatible with *S. carpocapsae* ALL and *H. amazonensis* JPM4 when the concentration was 40% w/v.

17.5.3 Association with other biological control agents

The combination of nematodes with fungi, viruses and other species of nematodes generally results in additive effects on insect mortality, whereas the combination of nematodes with bacteria produces results that range from antagonism to synergism (Koppenhöfer and Grewal, 2005). Molina *et al.* (2007) evaluated the association of the nematode *H. bacteriophora* with the fungus *Metarhizium anisopliae* against *D. saccharalis* larvae. This combination accelerated insect mortality, but lowered the production of IJs/cadaver.

17.6 Conclusions

In Brazil, EPNs are still considered a recent field of research, but one that is rapidly progressing in various aspects, such as the exploitation of biodiversity and commercial production, and the formulation of these agents. Some field experiments have been conducted, with very promising results. Other factors expected to be critical in the pursuit of commercial success include the final cost of the product, the ease in obtaining registration with governmental technical bodies and the value and shares of the agricultural market occupied by the target crop. Brazilian legislation for the regulation of biological products includes nematodes in the group of natural enemies, which makes it easy to obtain registration. Registration requires the presentation of data biologically characterizing the organisms, in addition to information about the possible effects on human and animal health, data on the identification of potential risks to the environment, and information on the quality control of organisms produced in the laboratory.

The bioinsecticide market in Brazil is directed mainly at major crops, such as soybean, maize, cotton, sugarcane and coffee. According to ABCBIO (Brazilian Association of Biological Control Companies), there are currently 107 biopesticides registered by 50 companies in Brazil. Among the products registered, 78 are microbial pesticides (fungi, bacteria and viruses), 7 are pheromones and 22 are macroorganism-based pesticides (parasites, predators and parasitoids), wherein one EPN is included. The organization also points out that 'large-scale production of biocontrol agents developed in Brazil needs to evolve the growing demand and to establish specific standards for production, quality control and inspection as well as invest in studies to suit the use in various systems'.

All the major crops have been explored extensively, with large areas infested by pests, generating great demand for nematodes. Consequently, many companies are needed to produce nematodes on a large scale. Alternatively, greenhouse crops represent a promising area for the use of EPNs in Brazil because the cultivated areas are smaller and the economic value of the products is considerably higher, providing more opportunities for companies for market supply and more opportunities for producers to invest in the product. More recently, information generated by Brazilian scientists, especially related to nematode selection and mass production, may help to leverage the establishment of commercial mass production of EPNs by companies.

Acknowledgements

The authors would like to thank the editors of this book, Drs. Tarique Hassan Askary and Mahfouz M.M. Abd-Elgawad, for reviewing the chapter.

References

Agrofit (2016) *Sistema de agrotóxicos fitossanitários*. Available at: http://www.agricultura.gov.br/servicos-e-sistemas/sistemas/agrofit (accessed 3 October 2016).

Alves, L.F.A., Rohde, C. and Alves, V.S. (2005) Patogenicidade de *Steinernema glaseri* e *S. carpocapsae* (Nematoda: Rhabdita) contra o cascudinho, *Alphitobius diaperinus* (Panzer) (Coleoptera: Tenebrionidae). *Neotropical Entomology* 34, 139–141.

Alves, V.S., Moino, A. Jr, Santa-Cecília, L.V.C., Andaló, V. and Souza, G.C. (2009a) Patogenicidade de nematoides entomopatogênicos a cochonilha-da-raiz-do-cafeeiro *Dysmicoccus texensis* (Tinsley) (Hemiptera: Pseudococcidae) em laboratório. *Arquivos do Instituto Biológico* 76, 67–73.

Alves, V.S., Moino, A. Jr, Santa-Cecilia, L.V.C., Rohde, C. and Silva, M.A.T. (2009b) Testes em condições para o controle de *Dysmicoccus texensis* (Tinsley) (Hemiptera, Pseudococcidae) em cafeeiro com nematoides entomopatogênicos do gênero *Heterorhabditis* (Rhabditida, Heterorhabditidae). *Revista Brasileira de Entomologia* 53, 139–143.

Andaló, V., Nguyen, K.B. and Moino, A. Jr (2006a) *Heterorhabditis amazonensis* n. sp. (Rhabditida: Heterorhabditidae) from Amazonas, Brazil. *Nematology* 8, 853–867.

Andaló, V., Cavalcanti, R.S., Molina, J.P., Moino, A. Jr and Magalhães, F.H.L. (2006b) Influência da aeração no armazenamento de nematoides entomopatogênicos (Rhabditida) em condições de laboratório. *Nematologia Brasileira* 30, 45–50.

Andaló, V., Moreira, G.F., Maximiniano, C., Moino, A. Jr and Campos, V.P. (2008a) Suscetibilidade de *Heterorhabditis amazonensis* (Rhabditida: Heterorhabditidae) a fungos predadores de nematodes. *Nematologia Brasileira* 32, 177–184.

Andaló, V., Cavalcanti, R.S., Molina, A.J.P. and Moino, A. Jr (2008b) Avaliação de substâncias na preservação de nematoides entomopatogênicos (Rhabditida: Heterorhabditidae, Steinernematidae) em diferentes temperaturas. *Arquivos do Instituto Biológico* 75, 301–312.

Andaló, V., Moreira, G.F., Cavalcanti, R.S. and Moino, A. Jr (2009a) Observations on the life cycle and pathogenicity of *Heterorhabditis amazonensis* (Rhabditida:

Heterorhabditidae). *Nematologia Brasileira* 33, 194–197.

Andaló, V., Moreira, G.F., Maximiniano, C., Moino, A. Jr and Campos, V.P. (2009b) Influence of herbicides on lipid reserves, mortality and infectivity of *Heterorhabditis amazonensis* (Rhabditida: Heterorhabditidae). *Nematologia Mediterranea* 37, 11–15.

Andaló, V., Cavalcanti, R.S., Molina, J.P. and Moino, A. Jr (2010) Substrates for storing entomopathogenic nematodes (Rhabditida: Steinernematidae, Heterorhabditidae). *Scientia Agricola* 67, 342–347.

Andaló, V., Santos, V., Moreira, G.F., Moreira, C.C., Freire, M. and Moino, A. Jr (2012) Movement of *Heterorhabditis amazonensis* and *Steinernema arenarium* in search of corn fall armyworm larvae in artificial conditions. *Scientia Agricola* 69, 226–230.

Arrigoni, E.B., Dinardo, L.L., Conde, A.J. and Terán, F.O. (1986) Aplicação de *Neoplectana carpocapsae* Weiser, 1955 em condições de campo para o controle de *Migdolus* spp. (Coleoptera: Cerambycidae). *Nematologia Brasileira* 8, 181–189.

Askary, T.H. (2010) Nematodes as biocontrol agents. In: Lichtfouse, E. (ed.) *Sociology, Organic Farming, Climate Change and Soil Science*. Springer, the Netherlands, pp. 347–378.

Askary, T.H. (2015a) Limitations, research needs and future prospects in the biological control of phytonematodes. In: Askary, T.H. and Martinelli, P.R.P. (eds) *Biocontrol Agents of Phytonematodes.* CAB International, Wallingford, UK, pp. 446–454.

Askary, T.H. (2015b) Nematophagous fungi as biocontrol agents of phytonematodes. In: Askary, T.H. and Martinelli, P.R.P. (eds) *Biocontrol Agents of Phytonematodes.* CAB International, Wallingford, UK, pp. 81–125.

Askary, T.H. and Martinelli, P.R.P. (2015) *Biocontrol Agents of Phytonematodes.* CAB International, Wallingford, UK.

Bailez, O.E., Viana-Bailez, A.M.M., Lima, J.O.G. de and Moreira, D.D.O. (2003) Life history of the guava weevil, *Conotrachelus psidii* Marshall (Coleoptera: Curculionidae), under laboratory conditions. *Neotropical Entomology* 32, 203–207.

Barbosa-Negrisoli, C.R.C., Garcia, M.S., Dolinski, C.M., Negrisoli, A.S. Jr, Bernardi, D. and Nava, D.E. (2009) Efficacy of indigenous entomopathogenic nematodes (Rhabditida: Heterorhabditidae, Steinernematidae), from Rio Grande do Sul Brazil, against *Anastrepha fraterculus* (Wied.) (Diptera: Tephritidae) in peach orchards. *Journal of Invertebrate Pathology* 102, 6–13.

Barbosa-Negrisoli, C.R.C., Negrisoli, A.S. Jr, Dolinski, C.M. and Bernardi, D. (2010) Efficacy of entomopathogenic nematodes (Nemata: Rhabditida) to control Brazilian apple leafroller *Bonagota salubricola* (Meyrick, 1937) (Lepidoptera: Tortricidae). *Crop Protection* 29, 1274–1279.

Barbosa-Negrisoli, C.R.C., Garcia, M.S., Dolinski, C.M., Negrisoli, A.S. Jr, Bernardi, D. and Santos, F.J. (2011) Survey for entomopathogenic nematodes (Rhabditida: Heterorhabditidae, Steinernematidae) in Rio Grande do Sul, Brazil. *Nematologia Brasileira* 34, 189–197.

Barbosa-Negrisoli, C.R.C., Negrisoli, A.S. Jr, Garcia, M.S., Dolinski, C.M. and Bernardi, D. (2013) Control of *Grapholita molesta* (Busck, 1916) (Lepidoptera: Tortricidae) with entomopathogenic nematodes (Rhabditida: Heterorhabditidae, Steinernematidae) in peach orchards. *Experimental Parasitology* 135, 466–470.

Barrett, J. and Wright, D.J. (1998) Intermediary metabolism. In: Perry, R.N. and Wright, D.J. (eds) *The Physiology and Biochemistry of Free-living and Plant-parasitic Nematodes*. CAB International, Wallingford, UK, pp. 331–353.

Batista Filho, A., Takada, H.M., Leite, L.G., Raga, A., Sato, M.E., *et al.* (2013) Brocas da bananeira (Coleoptera: Curculionidae). In: Nogueira, E.M.C., Almeida, I.M.G., Ferrari, J.T. and Beriam, L.O.S. (eds) *Bananicultura: manejo fitossanitário e aspectos econômicos e sociais da cultura*. Instituto Biológico, São Paulo, Brazil, pp. 179–193.

Bedding, R.A. and Iede, E.T. (2005) Application of *Beddingia siricidicola* for sirex woodwasp control. In: Grewal, P.S., Ehlers, R.U. and Shapiro-Ilan, D.I. (eds) *Nematodes as Biocontrol Agents*. CAB International, Wallingford, UK, pp. 385–400.

Bueno, A.F., Panizzi, A.R., Corrêa-Ferreira, B.S., Hoffmann-Campo, C.B., Sosa-Gómez, D.R., *et al.* (2013) Histórico e evolução do manejo integrado de pragas da soja no Brasil. In: Hoffmann-Campos, C.B., Correa-Ferreira, B.S. and Moscardi, F. (eds) *Soja: manejo integrado de insetos e outros artrópodes-praga*. Embrapa Soja, Londrina, Brazil, pp. 37–74.

Carson, R. (1962) *Silent Spring.* Fawcett Crest, New York.

Del Valle, E.E., Dolinski, C. and Souza, R.M. (2008) Dispersal of *Heterorhabditis baujardi* LPP7 (Nematoda: Rhabditida) applied to the soil as infected host cadavers. *International Journal of Pest Management* 54, 115–122.

Del Valle, E.E., Dolinski, C., Barreto, E.L.S. and Souza, R.M. (2009) Effect of cadaver coatings on emergence and infectivity of the entomopathogenic nematode *Heterorhabditis baujardi* LPP7 (Rhabditida: Heterorhabditidae) and the removal of cadavers by ants. *Biological Control* 50, 21–24.

Dias, P.V., Dolinski, C. and Molina, A.J.P. (2008) Influência da dose de juvenis infectantes e da massa de larvas de *Galleria mellonella* (Lepidoptera: Pyralidae) na produção *in vivo* de *Heterorhabditis baujardi* LPP7 (Rhabditida: Heterorhabditidae). *Nematologia Brasileira* 4, 317–321.

Dolinski, C., Barreto, E.L.S., Del Valle, E.E. and Stuart, R.J. (2008a) Agar media: an alternative formulation for storage, transport and application of the entomopathogenic nematode *Heterorhabditis baujardi* LPP7 Phan *et al.* (Nematoda: Heterorhabditidae). *BioAssay* 3, 1–5.

Dolinski, C., Kamitani, F.L., Machado, I.R. and Winter, C.E. (2008b) Molecular and morphological characterization of heterorhabditid entomopathogenic nematodes from the tropical rainforest in Brazil. *Memórias do Instituto Oswaldo Cruz* 103, 150–159.

Dolinski, C., Pinto, C.C.S., Robaina, R.R. and Bellini, L.L. (2010) Efeito de substratos com diferentes classes texturais na mobilidade do nematoide entomopatogenicos *Heterorhabditis baujardi* LPP7 (Rhabditida: Heterorhabditidae). *Nematologia Brasileira* 34, 123–128.

Dolinski C., Choo, H.Y. and Duncan, L.W. (2012) Grower acceptance of entomopathogenic nematodes: case studies on three continents. *Journal of Nematology* 44, 226–235.

Dutky, S.R., Thompson, J.V. and Cantwell, G.E. (1964) A technique for the mass propagation of the DD-136 nematode. *Journal of Insect Pathology* 6, 417–422.

Ferraz, L.C.C.B. (1998) Nematodes entomopatogênicos. In: Alves, S.B. (ed.) *Controle microbiano de insetos*. FEALQ, Piracicaba, Brazil, pp. 541–569.

Folegatti, M.E.G., Alves, S.B., Hawai, P.C. and Botelho, P.C.M. (1988) Nova metodologia para produção in vivo de *Neoaplectana carpocapsae*. *Nematologia Brasileira* 12, 76–83.

Fowler, H.G. (1988) Occurrence and infectivity of entomogenous nematodes in mole crickets in Brazil. *International Rice Research Newsletter* 13, 34–35.

Fowler, H.G. and Garcia, C.R. (1988) Nematodes (Rhabditida: *Steinernema feltiae*) as natural control agents of mole crickets (Orthoptera: Gryllotalpidae): field and laboratory studies. *Revista Brasileira de Biologia* 48, 789–795.

Gallo, D., Nakano, O., Silveira Neto, S., Carvalho, R.P.L., Batista, G.C. de, *et al.* (2002) *Entomologia Agrícola*, 2nd edn. FEALQ, São Paulo, Brazil.

Garcia, L.C., Raetano, C.G., Wilcken, S.R., Ramos, H.H., Leite, L.G., Batista Filho, A. and Moscardi, F. (2005) Pressurização da calda de pulverização na viabilidade de microrganismos entomopatogênicos. *Engenharia Agrícola* 25, 783–790.

Garcia, L.C., Raetano, C.G. and Leite, L.G. (2008) Application technology for the entomopathogenic nematodes *Heterorhabditis indica* and *Steinernema* sp. (Rhabditida: Heterorhabditidae and Steinernematidae) to control *Spodoptera frugiperda* (Smith) (Lepidoptera: Noctuidae) in corn. *Neotropical Entomology* 37, 305–311.

Gionetti, F.H.C., Leite, L.G., Tavares, F.M., Batista Filho, A. and Dell´Acqua, R. (2011) Virulência de nematoides entomopatogênicos (Nematoda: Rhabditida) à *Sphenophorus levis* (Coleoptera: Curculionidae). *Bragantia* 70, 81–86.

Grewal, P.S. (2002) Formulation and application technology. In: Gaugler, R. (ed.) *Entomopathogenic Nematology*. CAB International, Wallingford, UK, pp. 265–288.

Grewal, P.S., Webber, T. and Batterley, D.A. (1998) Compatibility of *Steinernema feltiae* with chemicals used in mushroom production. *Mushroom News* 46, 6–10.

Hajek, A. (2004) *Natural Enemies: An Introduction to Biological Control*. Cambridge University Press, Cambridge, UK.

Hamlen, R.A. and Mead, R.W. (1979) Fungus gnat larval control in greenhouse plant production. *Journal of Economic Entomology* 72, 269–271.

Harris, M.A., Oetting, R.D. and Moody, E. (1994) Transmission of *Thielaviopsis basicola* and *Fusarium prodiferatum* by fungus gnats. *Phytopathology* 84, 2380–2384.

Hungerford, H.B. (1916) Sciara maggots injurious to potted plants. *Journal of Economic Entomology* 9, 538–549.

Iede, E.T., Penteado, R.C.S. and Filho, W.R. (2012) The woodwasp *Sirex noctilio* in Brazil: monitoring and control. In: Slippers, B., Groot, P. and Wingfield, M.J. (eds) *The Sirex Woodwasp and its Fungal Symbiont: Research and Management of a Worldwide Invasive Pest*. Springer, New York, pp. 217–228.

Jackson, T.A. (1999) Factors in the success and failure of microbial control agents for soil dwelling pests. *Integrated Pest Management Reviews* 4, 281–285.

Jarvis, W.R. Shipp, J.L. and Gardiner, R.B. (1993) Transmission of *Pythium aphanidermatum* to greenhouse cucumber by the fungus gnat *Bradysia impatiens* (Diptera: Sciaridae). *Annals of Applied Biology* 122, 23–29.

Kalb, D.W. and Millar, R.L. (1986) Dispersal of *Verticillium alboatrum* by the fungus gnat (*Bradysia impatiens*). *Plant Disease* 70, 752–753.

Koppenhöfer, A.M. and Grewal, P.S. (2005) Compatibility and interactions with agrochemicals and other biocontrol agents. In: Grewal, P.S., Ehlers, R.-U. and Shapiro-Ilan, D.I. (eds) *Nematodes as Biological Control Agents*. CAB International, Wallingford, UK, pp. 363–381.

Lacey, L.A. (2008) Foreword. In: Alves, S.B. and Lopes, R.B. (eds) *Controle microbiano de pragas na América Latina*. FEALQ, Piracicaba, Brazil, pp. 19–20.

Lacey, L.A. and Shapiro-Ilan, D.I. (2008) Microbial control of insect pests in temperate orchard systems: potential for incorporation into IPM. *Annual Review of Entomology* 53, 121–144.

Lara, J.C., Dolinski, C., Fernandes, E. and Daher, R.F. (2008) Effects of a mini-sprinkler irrigation system on the viability, infectivity and search capability of *Heterorhabditis baujardi* LPP7 (Nematoda: Heterorhabditidae) infective juveniles. *Scientia Agricola* 65, 433–437.

Leite, L.G. (2006) Nematodes entomopatogênicos. In: Batista Filho, A. (ed.) *Controle biológico de insetos e ácaros*. Boletim Técnico, Instituto Biológico, São Paulo, Brazil, pp. 42–51.

Leite, L.G., Batista Filho, A. and Prada, W.L.A. (1990) Produção de *Neoplectana glaseri* em lagartas vivas e mortas de *Galleria mellonella*. *Revista de Agricultura* 65, 225–232.

Leite, L.G., Machado, L.A., Aguillera, M.M., Rodrigues, R.C.D. and Negrisoli, A.S. Jr (2003) Patogenicidade de *Steinernema* e *Heterorhabditis* (Nematoda: Rhabditida) contra ninfas da cigarrinha das raízes da cana-de-açúcar *Mahanarva fimbriolata* (Hemiptera: Cercopidae). *Revista de Agricultura* 78, 139–148.

Leite, L.G., Machado, L.A., Goulart, R.M., Tavares, F.M. and Batista Filho, A. (2005) Screening of entomopathogenic nematodes (Nematoda: Rhabditida) and the efficiency of *Heterorhabditis* sp. against the sugarcane root spittlebug *Mahanarva fimbriolata* (Fabr.) (Hemiptera: Cercopidae). *Neotropical Entomology* 34, 785–790.

Leite, L.G., Tavares, F.M., Bussola, R.A., Amorim, D.S., Ambrós, C.M. and Harakava, R. (2007) Virulência de nematoides entomopatogênicos (Nemata: Rhabditida) contra larvas da mosca dos fungos, *Bradysia mabiusi* (Lane, 1959) e persistência de *Heterorhabditis indica* Poinar *et al.* 1992, em substratos orgânicos. *Arquivos do Instituto Biológico* 74, 337–342.

Leite, L.G., Tavares, F.M., Botelho, P.S.M., Batista Filho, A., Polanczyk, R.A. and Schmidt, F.S. (2012) Eficiência de nematodes entomopatogênicos e inseticidas químicos contra *Sphenophorus levis* e *Leucothyreus* sp. em cana-de-açúcar. *Pesquisa Agropecuária Tropical* 42, 40–48.

Machado, L.A., Habib, M., Leite, L.G., Calegari, L.C., Goulart, R.M. and Tavares, F.M. (2005) Patogenicidade de nematoides entomopatogênicos a ovos e larvas de *Migdolus fryanus* (Westwood, 1863) (Coleoptera: Vesperidae). *Arquivos do Instituto Biológico* 72, 221–226.

Mertz, N.R., Agudelo, E.J.G., Sales, F.S., Rohde, C. and Moino, A. Jr (2014) Phoretic dispersion of the entomopathogenic nematode *Heterorhabditis amazonensis* (Rhabditida: Heterorhabditidae) by the beetle *Calosoma granulatum* (Coleoptera: Carabidae). *Phytoparasitica* 42, 179–187.

Milstead, J.E. and Poinar, G.O. (1978) A new entomogenous nematode for pest management systems. *California Agriculture* 32, 324–327.

Minas, R.S. (2012) Caracterização biológica de uma linhagem de nematoide entomopatogênico visando o controle do gorgulho da goiaba (*Conotrachelus psidii*) em dois sistemas de cultivo. Tese (Doutorado), Universidade Estadual do Norte Fluminense, Campos dos Goytacazes, RJ, Brazil.

Minas, R.S., Dolinski, C., Carvalho, R.S. and Souza, R.M. (2011) Controle biológico da mosca-do-Mediterrâneo *Ceratitis capitata* utillizando nematoides entomopatogênicos em laboratório. *Scientia Agraria* 12, 115–119.

Molina, A.J.P., Moino, A. Jr and Cavalcanti, R.S. (2004) Produção *in vivo* de nematoides entomopatogênicos em diferentes insetos hospedeiros. *Arquivos do Instituto Biológico* 71, 347–354.

Molina, A.J.P., Moino, A. Jr, Cavalcanti, R.S., Dolinski, C. and Carvalho, F.A. (2005) Patogenicidade, multiplicação e biologia de isolados nativos de nematoides entomopatogênicos (Rhabditida: Heterorhabditidae) provenientes de Lavras, MG. *Nematologia Brasileira* 29, 25–30.

Molina, A.J.P., Moino, A. Jr, Cavalcanti, R.S., Andaló, V. and Mendonça, L.A. (2006) Efecto da temperatura, concentración y tiempo de almacenamiento en la supervivencia de nematodos entomopatógenos. *Revista Colombiana de Entomología* 32, 24–30.

Molina, A.J.P., Samuels, I.R., Machado, I.R. and Dolinski, C. (2007) Interaction between isolates of *Metarhizium anisopliae* and the entomopathogenic nematode *Heterorhabditis* sp. JPM4 during infection of sugar cane borer *Diatraea saccharalis* (Lepidoptera: Pyralidae). *Journal of Invertebrate Pathology* 96, 187–192.

Moreira, G.F., Batista, E.S.P., Campos, H.B.N., Lemos, R.E. and Ferreira, M.C. (2013) Spray nozzles, pressures, additives and stirring time on viability and pathogenicity of entomopathogenic nematodes (Nematoda: Rhabditida) for greenhouses. *Plos One* 8, p. e65759.

Moscardi, F. (2007) A nuclear polyhedrosis for control of the velvetbean caterpillar in Brazilian soybeans. In: Vincent, C., Goettel, M.S. and Lazarovits, G. (eds) *Biological Control – A Global Perspectives*. CAB International, Wallingford, UK, pp. 344–352.

Moscardi, F., Morales, L. and Santos, B. (2002) The successful use of AgMNPV for the control of the velvetbean caterpillar *Anticarsia gemmatalis*, in soybean in Brazil. In: *Proceedings of the VIII International Colloquium on Invertebrate Pathology and Microbial Control.* Embrapa Soja, Londrina, Brazil, pp. 86–91.

Negrisoli, A.S. Jr, Barbosa-Negrisoli, C.R.C. and Moino, A. Jr (2008) Comparação entre metodologias de avaliação da compatibilidade de produtos fitossanitários com nematoides entomopatogênicos (Nematoda: Rhabditida). *Nematologia Brasileira* 32, 65–75.

Nguyen, K.B., Ginarte, C.M.A., Leite, L.G., Santos, J.M. and Harakava, R. (2010) *Steinernema brazilense* n. sp. (Rhabditida: Steinernematidae), a new entomopathogenic nematode from Mato Grosso, Brazil. *Journal of Invertebrate Pathology* 103, 8–20.

Nishimatsu, T. and Jackson, J.J. (1998) Interaction of insecticides, entomopathogenic nematodes, and larvae of the western corn rootworm (Coleoptera: Chrysomelidae). *Journal of Economic Entomology* 91, 410–418.

Orlando, A., Sampaio, A.S., Carvalho, A., Sacaranari, H.J. and Arruda, H.V. (1974) Notas sobre o gorgulho das goiabas *Conotrachelus psidii*, Marshall, 1922 (Coleoptera: Curculionidae), e experimentos de combate. *Biológico* 40, 251–269.

Panizzi, A.R. (2013) History and contemporary perspectives of the integrated pest management of soybean in Brazil. *Neotropical Entomology* 42, 119–127.

Pizano, M.A., Aguillera, M.M., Monteiro, A.R. and Ferroy, L.C.C.B. (1985) Incidência de *Neoaplectana glaseri* Steiner, 1929 (Nematoda: Steinernematidae) parasitando ovo de *Migdolus fryanus* (Westwood, 1863) (Col: Cerambycidae). Sociedade Brasileira de Nematologia, 9a Reunião, Piracicaba, 1.

Portillo-Aguilar, C., Villani, M.G., Tauber, M.J., Tauber, C.A. and Nyrop, J.P. (1999) Entomopathogenic nematode (Rhabditida: Heterorhabditidae and Steinernematidae) response to soil texture and bulk density. *Environmental Entomology* 28, 1021–1035.

Rodrigueiro, T.S.C., Ginarete, C.M.A., Leite, L.G., Tavares, F.M., Goulart, R.M. and Giacometti, F.H.C.

(2008) Eficiencia de *Heterorhabditis indica* IBCB-N05 (Rhabditida: Heterorhabditidae) no controle de *Alphitobius diaperinus* (Coleoptera: Tenebrionidae) sob comedouros de granja avícola. *Arquivos do Instituto Biológico* 75, 279–284.

Rohde, C., Moino, A. Jr, Carvalho, F.D. and Silva, M.A.T. (2012a) Selection of entomopathogenic nematodes for the control of the fruit fly *Ceratitis capitata* (Diptera: Tephritidae). *Agrária* 7, 797–802.

Rohde, C., Moino, A. Jr, Silva, M.A.T. and Carvalho, F.D. (2012b) Effect of *Heterorhabditis* sp. and *Steinernema carpocapsae* applied in different periods of soil infestation with larvae of *Ceratitis capitata* (Wiedemann) (Diptera: Tephritidae). *Pesquisa Aplicada e Agrotecnologia* 5, 79–84.

Rohde, C., Moino, A. Jr, Mertz, N.R., Krupa, P. and Ramalho, K.R.O. (2013) Compatibilidade de nematoides entomopatogênicos e extratos vegetais aquosos visando o controle da mosca-das-frutas *Ceratitis capitata* (Wiedemann) (Diptera: Tephritidae). *Semina. CiênciasAgrárias* 34, 1033–1042.

Sabino, P.H.S., Moino, A. Jr and Andaló, V. (2014) Effects of some insecticides on the neutral lipid percentage, survival and infectivity of *Steinernema carpocapsae* ALL and *Heterorhabditis amazonensis* JPM 4. *Nematoda* 1, e02014.

Santa-Cecília, L.V.C., Souza, J.C. de and Reis, P.R. (2000) Novas constatações da cochonilha-da-raiz *Dysmicoccus cryptus* em lavouras de café no Sul de Minas Gerais. EPAMIG, Lavras, Brazil, Circular Técnica, n. 130.

Schmitt, A.T., Gowen, S.R. and Hague, N.G.M. (1992) Baiting techniques for the control of *Cosmopolites sordidus* Germar (Coleoptera: Curculionidae) by *Steinernema carpocapsae* (Nematoda: Steinernematidae). *Nematropica* 22, 159–163.

Silva, A.C., Batista Filho, A., Leite, L.G., Tavares, F.M., Raga, A. and Schmidt, F.S. (2010) Efeito de nematoides entomopatogênicos na mortalidade da mosca-do-mediterrâneo, *Ceratitis capltata*, e do gorgulho-da-goiaba, *Conotrachelus psidii*. *Nematologia Brasileira* 34, 31–40.

Tavares, F.M., Batista Filho, A., Leite, L.G., Almeida, L.C., Silva, A.C. and Ambrós, C.M. (2007) Efeito de *Heterorhabditis indica* e *Steinernema* sp. (Nemata: Rhabditida) sobre larvas do bicudo da cana de açúcar, *Sphenophorus levis* (Coleoptera: Curculionidae), em laboratório e casa de vegetação. *Nematologia Brasileira* 31, 12–19.

Tavares, F.M., Batista Filho, A., Leite, L.G., Almeida, L.C. and Goulart, T.M. (2009) Efeito sinérgico de combinações entre nematoides entomopatogênicos (Nemata: Rhabditida) e inseticidas químicos na mortalidade de *Sphenophorus levis* (Vaurie) (Coleoptera: Curculinidae). *BioAssay* 4, 1–10.

Tavares, F.M., Batista Filho, A., Leite, L.G. and Tavares, G.M. (2012) Avaliação de isolados de nematoides entomopatogênicos sobre a mosca-dos-fungos, *Bradysia mabiusi* (Diptera: Sciaridae), praga em estufas. *BioAssay* 7, 1–6.

Travassos, L. (1927) Sobre o genera. *Oxysomatium Boletim Biologico* 5, 20–21.

Uso de Agrotóxicos na Agricultura (2016) Available at: http://www.gestaonocampo.com.br/biblioteca/uso-de-agrotoxicos-na-agricultura (accessed 4 October 2016).

Vainio, A. (1992) Guideline for laboratory testing of the side-effects of pesticides on entomophagous nematodes *Steinernema* spp. *IOBC/WPRS Bulletin* 15, 145–147.

Van den Bosch, R. (1978) *The Pesticide Conspiracy*. Doubleday & Co. Inc, New York.

Vasconcelos, Y. (2012) Inseto contra inseto. *Pesquisa Fapesp* 194, 68–73.

18 Status of Entomopathogenic Nematodes in Integrated Pest Management Strategies in India

C. Sankaranarayanan[1*] and Tarique Hassan Askary[2]

[1]*Section of Nematology, Division of Crop Protection, ICAR-Sugarcane Breeding Institute, Coimbatore, India;* [2]*Division of Entomology, Sher-e-Kashmir University of Agricultural Sciences and Technology, Srinagar, India*

18.1 Introduction

Agricultural production in India is constrained by a number of biotic and abiotic factors. India is a tropical country and has a climate that is conducive to insect pests, which is one of the major constraints in agricultural production, processing and storage. Annual crop losses due to insect pests and diseases in India are estimated to be 18% of the total agricultural output. India has very diverse agroclimatic conditions, ranging from the humid, high rainfall north-eastern zone to the north-western semi-arid and arid zones. Based on rainfall, geomorphological features, vegetation, cropping systems and land use, 20 agroclimatic zones are recognized within the country (Sehgal *et al.*, 1992). Insect pests are widespread in all the agroclimatic zones, and large quantities of insecticides are used to control these pests. Globally, India ranks 12th as a manufacturer of basic pesticides; however, in Asia, it stands second in this context. There has been a steady growth in the production of technical-grade pesticides in India, which has increased from 5000 t in 1958 to 102,240 t in 1998 (Mathur, 1999). Currently, the consumption of pesticides is showing a slight declining trend, probably due to the shift of farmers towards biopesticides, natural plant sources and other alternative methods (Das *et al.*, 2002). Indiscriminate use of insecticides, lack of adequate knowledge on proper application techniques and dosages, and frequent use of broad-spectrum insecticides are common in India. However, dependence on pesticides is not sustainable because of the associated environmental problems, development of resistance in insects against insecticides, pest resurgence and outbreak of secondary pests. Thus, there is a need to identify suitable alternative methods for the management of insect pests that would be non-hazardous, eco-friendly and cost-effective.

In the past few years, there has been an intensive worldwide search for novel biocontrol agents, and researchers have gained advances or breakthroughs in this field by the proper formulation and commercialization of some strains of biocontrol agents (Askary and Martinelli, 2015). Among these, nematodes belonging to the families Steinernematidae and Heterorhabditidae drew the attention of research practitioners and growers, as these nematodes have a short life cycle and a wide insect host range, as well as the ability to persist under unfavourable conditions and environmental extremes (Askary, 2010). These beneficial nematodes are called entomopathogenic nematodes (EPNs), which are obligate parasites of insects and kill their hosts with the aid of symbiotic bacteria (*Xenorhabdus* or *Photorhabdus*) carried in their alimentary canal. These lethal obligate insect parasites are distributed ubiquitously without pathogenicity to mammals and beneficial non-target insects, and exempted from registration requirements.

*Corresponding author. E-mail: csankar@lycos.com

Non-feeding infective juveniles (IJs) of EPNs (Fig. 18.1 & 18.2) are motile, virulent, possessing chemoreceptors, have a high reproductive potential and the ability to seek out hosts, besides having the potential for long-term establishment through recycling and compatibility with other management strategies, such as chemical pesticides. They are found to be associated with most insect orders. Association ranges from phoretic to obligate, and includes host death, sterility, reduced fecundity, delayed development or aberrant behaviour. Currently, the total reported species of *Steinernema* is 92, while that of *Heterorhabditis* is 22 (see Banu, Cannayane and Meena, chapter 4, this volume). Soil surveys conducted in divergent areas of the Indian sub continent have demonstrated a high diversity of EPNs across seasons, habitats and geographic regions (see Table 18.1). There are various target crop insect pests in India that can be managed with EPNs. Several workers have used EPNs against cutworms, stem borers, white grubs and other insect pests in laboratory and field conditions, and have found EPNs compatible with existing integrated pest management (IPM) programmes.

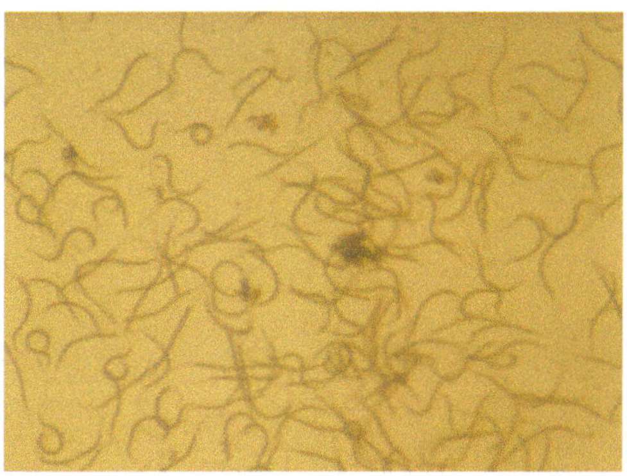

Fig. 18.1. Infective juveniles (IJs) of *Heterorhabditis bacteriophora*.

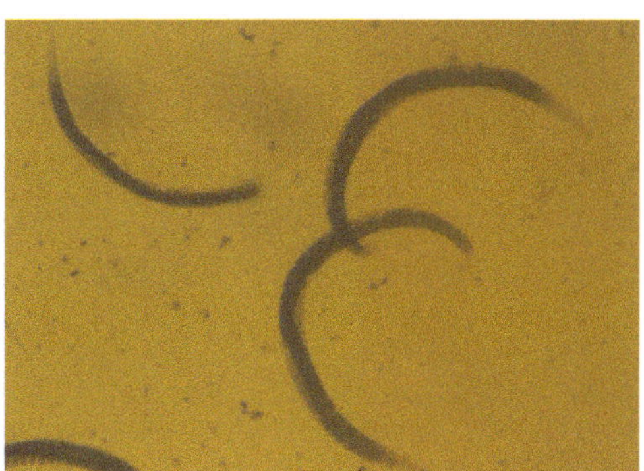

Fig. 18.2. First generation adults of *Heterorhabditis bacteriophora*.

This chapter provides an overview on the status of EPN research in India, which includes their biodiversity, efficacy against different crop insect pests, compatibility with existing IPM practices, recent development in mass production, formulation and application, and future directions for their effective utilization in the Indian farming system.

18.2 Diversity and Distribution of Entomopathogenic Nematodes in India

EPNs are used successfully to control a variety of agricultural and horticultural insect pests in different countries. Accordingly, there is an intense interest to isolate these nematodes from different parts of the world where they are adapted climatically and have the potential to control the insect pests in that particular area. Hence, surveys have been conducted in many parts of the world to isolate these beneficial nematodes, with a future aim to exploit them in the management of crop insect pests. Hominick *et al.* (1996) summarized the results of these surveys conducted prior to 1996. Since then, a number of other surveys have been conducted and new species and strains of EPNs have been isolated. In India, the initial research with EPNs was conducted primarily with exotic species/strains of *Steinernema carpocapsae*, *S. glaseri*, *S. feltiae* and *Heterorhabditis bacteriophora*, imported by researchers (Kaya *et al.*, 2006). In many cases, these exotic EPNs yielded inconsistent results in field trials, probably because of their poor adaptability to the local agroclimatic conditions. There was concern that the introduction of exotic EPNs might also have a negative impact on non-target organisms (Kaya *et al.*, 2006). Hence, the isolation of the native species and populations of EPNs was emphasized and considered a valuable resource, not only from a biodiversity perspective but also from an environmental perspective because of their adaptability and utility in biological control. Since 1990, Indian soils have been explored for the presence of EPNs, with Poinar *et al.* (1992) identifying *H. indica* from Coimbatore; Ganguly and Singh (2000) isolating *S. thermophilum* from New Delhi and Ali *et al.* (2005) isolating *S. masoodi* and *S. seemae* from Kanpur. Other species identified as indigenous isolates include *S. carpocapsae*, *S. bicornutum*, *S. glaseri*, *S. riobrave*, *S. feltiae*, *S. abbasi*, *S. tami*, *S. siamkayai* and *H. bacteriophora* (Ganguly, 2003, 2006). In India, researches on EPNs are being carried out at different research laboratories on various aspects such as survey, bioefficacy testing, mass production and formulation. Many surveys have been undertaken (Josephrajkumar and Sivakumar, 1997; Ganguly and Singh, 2000; Gokte Narkhedkar *et al.*, 2001; Hussaini *et al.*, 2001; Ganguly *et al.*, 2002, 2005, 2010; Parihar *et al.*, 2003; Umamaheswari *et al.*, 2004; Ali *et al.*, 2005, 2009; Sosamma, 2006; Varadarasan *et al.*, 2006; Ali and Shaheen, 2011; Seenivasan *et al.*, 2012) for the isolation of EPNs from various cropping systems (Fig. 18.3).

Fig. 18.3. Collection of soil sample for isolation of EPNs during a survey.

We present the current status of EPN diversity from different parts of India in Table 18.1.

18.3 Utilization of Entomopathogenic Nematodes in India against Various Insect Pests

In India, utilization of EPNs began during 1966. Rao and Manjunath (1966) made the first attempt to evaluate DD-136 against insect pests of rice, sugarcane and apple in field conditions. In the 1970s, Singh and Bardhan (1974) and Singh (1977) worked on the mortality of cutworms, their life cycle, the field efficacy of DD-136 and its compatibility with insecticides and fertilizers. High mortality of insects *viz.*, *Pseudolatia separata* and *Cirphis compacta* was obtained with *S. carpocapsae* (Israel *et al.*, 1969). Narayanan and Gopalakrishnan (1987) reported the effect of *S. feltiae* to pre-pupae, pupae and adults of *Spodoptera litura*. Work on heterorhabditids was started during the 1990s. After the mid-1980s, on an average, one research article on EPNs was published from India each year. Since 1990s, there has been great progress in the research work on EPNs, initiated from different parts of the country, and different laboratories have been actively involved in various aspects of EPN research (Table 18.2).

18.3.1 White grubs

Among the various species of white grubs infesting sugarcane in India, *Holotrichia serrata* and *Leucopholis lepidophora* are widely distributed. Much work has been done at the ICAR-Sugarcane Breeding Institute, Coimbatore, to control white grubs with EPNs. The new species of EPN (*H. indica* sp. nov.) isolated from the Coimbatore region of Tamil Nadu, India, was tested for its pathogenicity to the scarabaeids, *H. serrata* and *L. lepidophora* and was compared with that of *S. feltiae* and *S. glaseri* in the laboratory. *H. indica* and *S. glaseri* were equally efficient in producing high mortality of all first, second and third instars of *H. serrata* and *L. lepidophora*. *S. feltiae* was less effective (Karunakar, 1990). Karunakar *et al.* (2000a) observed that *S. feltiae*, *S. glaseri* and *H. indica* were not infective to the eggs of the white grub, *H. serrata*. *S. glaseri* caused significantly higher mean mortality of larvae of both *H. serrata* and *L. lepidophora*, followed by *H. indica*. *S. feltiae* caused significantly less mortality of first-instar grubs, while no mortality was observed in second and third instars. Positive correlation was observed between the dosage levels of *S. glaseri* and *H. indica* and the mortality of *H. serrata* and *L. lepidophora* in both laboratory and pot culture experiments. The mortality caused by both the species of nematodes decreased with an increase in age of the white grubs. In another experiment, Karunakar *et al.* (2000b) tested two EPNs, *S. glaseri* and *H. indica*, against different instars of white grubs, *H. serrata* and *L. lepidophora*. In laboratory conditions, *H. indica* took less time (3.80 days) than *S. glaseri* (4.37 days) to cause mortality of *H. serrata*; however, the trend was reversed in the case of *L. lepidophora*. The

Table 18.1. Diversity and distribution of entomopathogenic nematodes in India.

Entomopathogenic nematode	Place	Reference
Steinernema		
Neoaplectana sp.	Himachal Pradesh, Punjab	Singh, 1977
Steinernema feltiae	ICRISAT, Hyderabad	Singh *et al.*, 1992
Steinernema sp.	Tamil Nadu	Josephrajkumar and Sivakumar, 1997
Steinernema sp.	Gujarat	Vyas *et al.*, 1999a
S. thermophilum	ICAR-IARI, New Delhi	Ganguly and Singh, 2000
Steinernema sp.	Assam, Delhi, Gujarat, Himachal Pradesh, Uttar Pradesh	Kaushal *et al.*, 2000
S. carpocapsae, *Steinernema* sp. SSL2	Bangalore, Minicoy, Aligarh	Hussaini *et al.*, 2001
S. abbasi	IARI, New Delhi	Hussaini *et al.*, 2001
S. bicornutum	IARI, New Delhi	Hussaini *et al.*, 2001
S. tami	Jorhat, Assam	Hussaini *et al.*, 2001
Steinernema spp.	Nagpur	Gokte-Narkhedkar *et al.*, 2001
Steinernema sp.	Kerala	Sosamma and Rasmi, 2002
S. riobrave	ICAR-IARI, New Delhi	Ganguly *et al.*, 2002
Steinernema sp.	Rajasthan	Parihar *et al.*, 2003
S. riobrave	Gujarat	Ganguly *et al.*, 2002
Steinernema sp.	Rajasthan	Rajkumar *et al.*, 2003
S. carpocapsae	Uttaranchal	Misra and Khan, 2003
Steinernema sp.	Rajasthan	Parihar *et al.*, 2003
Steinernema sp.	Meghalaya	Sylesha, 2003
S. glaseri	Tamil Nadu	Banu, 2003
Steinernema spp.	Tamil Nadu	Umamaheswari *et al.*, 2004
S. masoodi	Kanpur, Uttar Pradesh	Ali *et al.*, 2005
S. seemae	Kanpur, Uttar Pradesh	Ali *et al.*, 2005
S. siamkayai	New Delhi	Ganguly *et al.*, 2005
S. siamkayi, *S. glaseri*	Kerala	Banu *et al.*, 2005
S. qazii	Kanpur, Uttar Pradesh	Ali *et al.*, 2009
Steinernema sp.	Uttar Pradesh	Shah and Azmi, 2006
Steinernema spp.	Kerala	Varadarasan *et al.*, 2006
S. pakistanensis	Haryana	Walia *et al.*, 2006
S. feltiae	Himachal Pradesh	Singh and Gupta, 2006
S. sayeedae	Koshambhi, Uttar Pradesh	Ali and Shaheen, 2011
Steinernema spp.	Andaman and Nicobar Islands	Prasad *et al.*, 2001a
S. carpocapsae	North Gujarat	Shinde *et al.*, 2010
Steinernema spp.	Karnataka	Vanitha *et al.*, 2010
S. meghalayensis	Meghalaya	Ganguly *et al.*, 2011
S. carpocapsae, *S. monticolum*, *S. siamkayai*	Tamil Nadu	Seenivasan *et al.*, 2012
S. dharanaii	Madhya Pradesh	Kulkarni *et al.*, 2012
S. carpocapsae, *S. riobrave*, *S. siamkayai*, *S. asiaticum*	Tamil Nadu	Chandrasekar *et al.*, 2013
Heterorhabditis		
Heterorhabditis indica	Tamil Nadu	Poinar *et al.*, 1992
Heterorhabditis sp.	Tamil Nadu	Josephrajkumar and Sivakumar, 1997
H. bacteriophora	Tamil Nadu	Sivakumar *et al.*, 1989
Heterorhabditis sp.	Assam, Delhi, Gujarat, Himachal Pradesh, Uttar Pradesh	Kaushal *et al.*, 2000
Heterorhabditis spp.	Andaman and Nicobar Islands	Prasad *et al.*, 2001a

Continued

Table 18.1. Continued.

Entomopathogenic nematode	Place	Reference
H. indica	Minicoy, Tamil Nadu, Uttar Pradesh, New Delhi	Hussaini *et al.*, 2001
Heterorhabditis sp.		
H. bacteriophora, *H. indica*	Nagpur	Gokte-Narkhedkar *et al.*, 2001
H. indica	Gujarat	Ganguly *et al.*, 2002
Heterorhabditis spp.	Rajasthan	Parihar *et al.*, 2003
H. indica	Kerala	Sosamma and Rasmi, 2002
H. indica	Tamil Nadu	Ambika and Sivakumar, 2002
Heterorhabditis sp.	Rajasthan	Rajkumar *et al.*, 2003
H. indica	Meghalaya	Sylesha, 2003
Heterorhabditis sp.	Rajasthan	Parihar *et al.*, 2003
H. indica	Karnataka	Prabhuraj and Patil, 2004
H. indica	Kerala	Banu *et al.*, 2005
H. bacteriophora	Himachal Pradesh	Singh and Gupta, 2006
H. bacteriophora	Tamil Nadu	Seenivasan *et al.*, 2012
H. indica	Tamil Nadu	Chandrasekar *et al.*, 2013

Table 18.2. List of different institutes involved in entomopathogenic nematode research in India.

Name of institute	Location
Indian Agricultural Research Institute (IARI)	Pusa, New Delhi
Central Plantation Crops Research Institute (CPCRI)	Kayamkulam, Kerala
Sugar Breeding Institute (SBI)	Coimbatore, Tamil Nadu
Tamil Nadu Agricultural University (TNAU)	Coimbatore, Tamil Nadu
Rajasthan Agricultural University (RAU)	Udaipur, Rajasthan
Gujarat Agricultural University (GAU)	Anand, Gujarat
National Bureau of Agricultural Insect Resources (NBAIR)	Bengaluru, Karnataka
Indian Institute of Pulses Research (IIPR)	Kanpur, Uttar Pradesh
Directorate of Rice Research (DRR)	Hyderabad, Andhra Pradesh
Central Institute for Cotton Research (CICR)	Nagpur, Maharashtra

mean time taken to cause host mortality of white grubs increased as the instar stages increased. Sankaranarayanan *et al.* (2006) reported that the pupae and adult beetles of *H. serrata* were susceptible to EPN infection.

18.3.2 Sugarcane shoot and internode borers (*Chilo* spp.)

Karunakar (1990) observed the mortality of sugarcane shoot borer, internode borer and white grubs caused due to *S. feltiae*, *S. glaseri* and *H. indica*. Karunakar *et al.* (1992) conducted a study to observe the influence of inoculum and temperature on the infectivity and multiplication of *S. carpocapsae*, *S. glaseri* and *H. indica* on sugarcane internode borer, *Chilo sacchariphagus indicus*. Inoculation of 40 IJs/larva resulted in highest yield of nematodes. The multiplication of *S. glaseri* and *H. indica* was higher at this inoculum level, while *S. carpocapsae* multiplied to the maximum at an average dosage of 10 IJs/larva. Karunakar *et al.* (1999a) screened nine lepidopteran hosts under laboratory conditions and found *Galleria mellonella*, *Corcyra cephalonica*, *Helicoverpa armigera*, *S. litura* and *Spodoptera excerptalis* highly susceptible to all EPNs used in the study. Sankaranarayanan and Easwaramoorthy (2003) evaluated 29 EPN isolates against the larva of *C. sacchariphagus indicus* and found all the EPNs used in the study caused mortality to the insect. In another experiment, Sankaranarayanan *et al.* (2003) observed that *H. indica* (LN2), *H. bacteriophora* (German isolate) and *S. glaseri* (Australian isolate) caused 100% mortality to the larvae of shoot borer *Chilo infuscatellus* in 22.8, 28.4 and 26.5 h, respectively.

18.3.3 Rice insect pests

Israel *et al.* (1969) observed high mortality of paddy cutworm (*P. separata*) and leaf folder caterpillar (*C. compacta*) when DD 136 strain (*S. carpocapsae*) was applied against them in laboratory, glasshouse and field conditions. Other important rice pests such as the paddy stem borer (*Tryporyza incertulas*), Asian rice borer (*Chilo suppressalis*), ragi pink borer (*Sesamia inferens*), paddy butterfly (*Melanitis ismene*), rice skipper (*Paranara mathias*), rice leaf folder (*Cnaphalocrocis medinalis*) and gall midge (*Pachydiplosis oryzae*) were also highly susceptible to the nematodes. The DD 136 strain completed its life cycle in 5 ± 7 days on *T. incertulas*. All the stages of *T. incertulas* except the egg were susceptible to nematodes, although at higher temperatures, the nematodes were ineffective in controlling *T. incertulas* (Rao *et al.*, 1971). Srinivas and Prasad (1991) reported 98% mortality of the fifth-instar larva of the rice leaf roller by *S. carpocapsae*. The addition of 2% glycerine and agar solutions to the nematode spray suspension has been found effective in enhancing the efficacy of nematodes (Yadava and Rao, 1970). The DD 136 strain was tolerant to concentrations of fertilizers and some common insecticides that prevailed in rice fields (Rao *et al.*, 1975).

18.3.4 Other crop insect pests

Mathur *et al.* (1966) reported that caterpillars of the maize stem borer (*Chilo partellus* Swinhoe) parasitized by *Steinernema* sp. were distinctly sluggish and inactive, and stopped feeding in the later stages of infection. Singh (1977) reported 100% mortality of *Agrotis ipsilon* and *A. segetum*, and 75% mortality of white grubs due to *Steinernema* sp. The DD 136 strain of *S. carpocapsae* was reported to cause 100% mortality of potato chafer grub, *Anomala* sp. (Rajeswari Sunderababu *et al.*, 1984). Three strains of *H. bacteriophora* and DD 136 were tested by Bhaskaran *et al.* (1994) on red hairy caterpillar, *Amsacta albistriga*, in the field. The results showed DD 136 to be more effective in comparison to other nematode strains. The LC_{50} values ranged between 6.5 and 19.4 IJs/larva. Sezhian *et al.* (1996) enhanced the effectiveness of *S. carpocapsae* by adding glycerine 0.1% and Triton X-100 against fourth instar of *S. litura* invading sunflower heads. Patel and Vyas (1995) observed 24.5% mortality of *H. armigera* on chickpea caused by *S. glaseri*. Razia and Sivaramakrishnan (2016) evaluated three species of EPNs, *S. siamkayai*, *S. pakistanense* and *H. indica* against subterranean termites, *Reticulitermes flavipes* and *Odontotermis hornei* under laboratory conditions. In sand assay method, *S. pakistanense* showed significant results, causing 100% mortality of both the pests within 24 hrs followed by *S. siamkayai* and *H. indica* which caused cent percent mortality at 48hrs when applied @ 250 IJs/ml. The bioefficacy of EPNs utilized in India against various insect pests in different crops are presented in Table 18.3.

18.4 Environmental Considerations in Using Entomopathogenic Nematodes in Indian Agriculture

Entomopathogenic nematodes possess an impressive list of positive attributes for the biological control of insect pests, including high virulence, ease of mass production, broad host range and safety (exemption from registration by the US Environmental Protection Agency). However, despite tremendous advances in research and development, the implementation of EPNs under field conditions still remains hampered by the lack of predictability in the efficacy of control. EPNs are highly sensitive to environmental stresses like heat, UV, hypoxia and desiccation, and their successful use under field conditions is highly influenced by these factors (Kaya, 1990). Normally, they should be applied in the late evening to avoid exposure to intense sunlight and UV radiation, with suitable adjuvants like glycerol or liquid paraffin.

18.4.1 Soil moisture

Soil moisture has a significant impact on the prevalence of EPNs. The prevalence of *Heterorhabditis* sp. was found to be correlated significantly and positively with rainfall, but correlated negatively with relative humidity and temperature. Prevalence increased from May, reached its peak in July and dropped gradually thereafter until January, indicating the significant influence of soil moisture on the prevalence of *Heterorhabditis* sp. (Prasad *et al.*, 2001b). *S. thermophilum* is well adapted to varying moisture profiles experienced throughout the cropping season in India. It can access the host at all the soil moisture levels tested, varying from 1 to 19%. Infectivity increased significantly at 3% and then increased gradually up to 9% moisture level, after which it started declining. The optimum soil moisture level for *S. thermophilum* was found 9%, with

Table 18.3. Bioefficiency of entomopathogenic nematodes (EPNs) utilized in India against various crop insect pests.

Pest	Entomopathogenic nematode	Application rate	Laboratory/field	Effect	References
Paddy					
Tryporyza incertulas	*Steinernema carpocapsae*		Field	EPN was effective	Rao and Manjunath, 1966
T. incertulas, *Chilo supressalis*, *Cirphis compacta*, *Pseudolatia separata*	*S. carpocapsae*		Laboratory and field	EPN was effective	Israel *et al.*, 1969
Cnaphalocrosis medinalis	*Heterorhabditis indica*, *S. asiaticum*	75 IJs/larva	Laboratory	60% mortality with *H. indica* and 40% mortality with *S. asiaticum*	Sankar *et al.*, 2009a
Scirpophaga incertulas	*H. indica*	25 IJs/egg mass; 60 IJs/larva or pupa	Laboratory	Egg hatchability affected and mortality of early-stage (20-day-old larvae) to final-instar larvae and pupae was recorded after 18 h	Sankar *et al.*, 2009b
Leptispa pygmaea	*H. indica*	5IJs to 9IJs	Laboratory	66.67–91.67% mortality of grubs	Karthikeyan and Sosamma, 2009
Orseolia oryzae	*H. indica*	2500 IJs/ml	Field	Reduction of gall midge incidence at 25 days after treatment	Karthikeyan and Sosamma, 2010
Cnaphalocrocis medinalis	*H. indica*	3000 IJs/ml	Field	Significant control of leaf folder incidence	Karthikeyan and Sosamma, 2010
Maize					
Agrotis ipsilon	*H. indica* + *Metarhizium anisopliae*		Field	Higher yield observed	Bhagat *et al.*, 2008
Cabbage and cauliflower					
Plutella xylostella	*H. indica*, *S. glaseri*	10,000 IJs/pot	Pot culture and microplot	50–58.35% larval mortality at potted and 36.6% mortality at microplot conditions	Saravanapriya and Subramanian, 2006
P. xylostella	*S. carpocapsae* (PDBC strain), *H. indica* (PDBC strain), *S. carpocapsae* (JMU strain)	2 billion/ha	Field	52.3% control of larvae	Gupta *et al.*, 2011
P. xylostella	*S. carpocapsae*	1000 IJs/pot	Pot culture	41.25% mortality of insect larvae	Sunanda *et al.*, 2013
Cardamom, tea, areca nut and ginger					
Basilepta fulvicorne	*H. indica*, *S. glaseri*		Laboratory	78.6–95.3% mortality of grub	Cannayane *et al.*, 2007
Leucopholis lepidophora	*H. indica*	20 lakhs juveniles/palm	Field	44–55% mortality of grubs in arecanut	Murthy *et al.*, 2010

Laspeyresia bipunctata	*H. indica*	2×10^9 IJs/ha	Field	Tea yield increased and effective control of insect	Devrajan *et al.*, 2010
Aetherastis circulata	*Heterorhabditis* sp.	320 IJs/5 larvae	Laboratory and field	84–92% mortality in laboratory and no significant effect in field	Thankamony and Jose, 2013
Conogethes punctiferalis	*Heterorhabditis* sp., *Steinernema* sp.		Laboratory	Shoot borer larva was susceptible to *Steinernema*	Pervez *et al.*, 2014
Chickpea					
Helicoverpa armigera	*Steinernema riobrave*, *Heterorhabditis* sp.	50,000 IJs/m^2	Field	37–49% reduction in pod damage	Vyas *et al.*, 2003
H. armigera	*Steinernema* sp.	100 million IJs/ha	Field	Increased yield	Yadav *et al.*, 2004
H. armigera	*S. carpocapsae*	600 IJs/plant	Field	41.67% mortality of larvae	Yadav *et al.*, 2008a
H. armigera	*S. masoodi*	3×10^9 IJs/ha	Field	Lowest pod damage recorded and a 68.6% increase in yield	Ahmad *et al.*, 2009
H. armigera	*S. seemae*	1×10^9 IJs/ha	Field	Lowest pod damage recorded and a 63.05% increase in yield	Ali and Asif, 2011
H. armigera	*S. masoodi*, *S. carpocapsae*, *H. indica*	$3–5 \times 10^9$ IJs/ha	Laboratory and field	Reduced pod damage up to 11–12 %	Abid *et al.*, 2014
Blackgram					
Spodoptera litura	*H. indica*, *S. glaseri*	5 × 109 IJs/ha	Pot culture and microplot	50.6–75.6% mortality	Umamaheswari *et al.*, 2006
Groundnut					
Amsacta albistriga	*S. carpocapsae*, *H. bacteriophora*		Field	*S. carpocapsae* was the most effective	Bhaskaran *et al.*, 1994
A. albistriga	*H. indica*	100,000 IJs/ml	Microplot	Effective control	Prabhu and Sudheer, 2008
Potato					
Anamola sp.	*Neoaplectana carpocapsae*		Laboratory and field	Infection and EPN development observed	Rajeswari Sunderababu *et al.*, 1984
Agrotis ipsilon	*S. riobrave*	100,000 IJs/m^2	Field	Suppression of larval population	Mathasoliya *et al.*, 2004
Brahmina coriacaea	*H. indica*		Laboratory and field	Reduction in grub population observed	Chandel *et al.*, 2005
B. coriacea	*H. indica*, *S. carpocapsae*	1, 3 and 6 × 105 IJs/m^2	Laboratory and field	Up to 80% reduction of grub population	Sharma *et al.*, 2009
Brinjal					
Myllocerus discolor	*H. indica*, *S. glaseri*	10,000 IJs/m^2	Field	92–93% reduction of grub population	Prabhuraj *et al.*, 2000
Leucinodes orbonalis	*S. carpocapsae*		Field	Reduction in borer incidence	Hussaini *et al.*, 2002a

continued

Table 18.3. Continued.

Pest	Entomopathogenic nematode	Laboratory/field	Effect	References
Sugarcane				
Chila sacchariphagus indicus	*S. carpocapsae*, *S. glaseri*, *H. indica*	Laboratory	Susceptibility of sugarcane internode borers observed	Karunakar *et al.*, 1992
Holotrichia serrata	*S. feltiae*, *S. glaseri*, *H. indica*	Laboratory	All the instars were susceptible to EPNs	Karunakar *et al.*, 2000a,b
H. serrata	*H. indica*, *S. glaseri*	Laboratory	Pupae and adults were susceptible to EPNs	Sankaranarayanan *et al.*, 2006
Chila sacchariphagus indicus	*H. indica*, *H. bacteriophora*, *S. glaseri*, *S. riobrave*, *S. feltiae*	Laboratory	*H. indica* was found superior among all EPNs, with an LD_{50} of 14.7 IJs/larva	Sankaranarayanan *et al.*, 2008
Chilo infuscatellus	*H. indica*, *H. bacteriophora*, *S. carpocapsae*, *S. glaseri*, *S. riobrave*, *S. feltiae*	Laboratory	Mortality of the early shoot borer larvae increased with increase in EPN dosage level	Sankaranarayanan *et al.*, 2011

the optimal range of 3–16% (Ganguly and Gavas, 2004). Raveendranath *et al.* (2007) reported on the basis of an experiment that the median lethal concentrations (LC_{50}) of *H. indica* were lower than those of *S. carpocapsae* in sand, red and black soils at 10% and 15% moisture after 48 h post-infection in third-instar larvae of *S. litura*. In another experiment on sand, red soil, black soil with 10% moisture content and black soil with 15% moisture content, *H. indica* was more virulent to the pupae of *S. litura* than *S. carpocapsae* (Raveendranath *et al.*, 2008).

Yadav and Lalramliana (2012) investigated the effect of soil moisture on the activity (as measured by number of nematodes established in hosts) of three EPN species (*H. indica*, *S. thermophilum* and *S. glaseri*) isolated from forest soils in Meghalaya, India. The results revealed that soil moisture had a marked influence on the establishment of IJs of different nematode species in the insect host. *S. thermophilum* showed establishment at 4% and above soil moisture level, whereas *H. indica* and *S. glaseri* showed establishment at 5% and above soil moisture level. The optimum soil moisture for different nematode species were noted as: *H. indica*, 8–18%; *S. thermophilum*, 6–20%; and *S. glaseri*, 8–25%. Further, a minimum of 6% soil moisture level was noted to be essential for achieving 100% host mortality for all the three nematode species.

18.4.2 Temperature

Karunakar *et al.* (1999b) studied the effect of temperature at eight different levels on infection, penetration and multiplication of *S. feltiae*, *S. glaseri* and *H. indica* on *G. mellonella*. No host mortality was observed at 10 and 35°C. Steinernematids showed significantly higher penetration into the host at 25°C, whereas more *H. indica* penetrated into the host at 27.5°C. The multiplication of IJs was significantly higher at 27.5 and 25°C, followed by 30°C. Hussaini *et al.* (2000) investigated the viability of *S. bicornutum* and *H. indica* suspensions in distilled water or 0.1% formalin stored at 8 and 30°C for 6 weeks. At both temperatures, suspensions of the three isolates were more viable in distilled water than in 0.1% formalin. The percentage reduction in viability under storage at 30°C was 40–50% in *S. bicornutum* isolates and 20–30% in *H. indica*. Prasad *et al.* (2001b) studied the population dynamics of *Heterorhabditis* sp. in South Andaman, India. The prevalence of *Heterorhabditis* sp. was significantly and positively with rainfall, but negatively with relative humidity and temperature. Ganguly and Singh (2001) demonstrated that *S. thermophilum* could infect and develop at a wide range of temperature (20–35°C), the optimum being 25–35°C, with better performance at 30–35°C. In another study, the life cycle of *S. thermophilum* under various temperatures was studied under laboratory conditions. The juveniles infected the host through natural openings and induced 100% mortality within 35 h at 35°C and 48 h at 25 and 30°C. The results indicated that *S. thermophilum* emerged earlier and developed faster at higher temperature (Ganguly and Singh, 2002). Moisture levels above or below 8% and temperatures above or below 24°C accompanied with 100% relative humidity decreased the survival and virulence of *Heterorhabditis* sp. (Prasad and Singh, 2003; Prasad and Ranganath, 2006). Sankaranarayanan *et al.* (2003) observed infectivity of 29 EPN isolates against the sugarcane shoot borer, *C. infuscatellus*. All the EPN isolates took less time to cause mortality of larvae at 27°C as compared to 18°C. The multiplication of EPNs in the dead cadaver was comparatively more at 27°C than at 18°C. At a relative humidity of 57–69% and room temperature of 27.0–30.1°C over a continuous period of 12 h, the IJs of *H. indica* and *S. glaseri* survived only 2 and 6%, respectively (Subramanaian and Rajeswari, 2003). Hussaini *et al.* (2004) studied the effect of heat-shock treatment on the survival and viability of some isolates of *S. abbasi*, *S. tami*, *S. carpocapsae*, *H. indica* and *H. bacteriophora* at 35, 37 and 40°C. Survival, infectivity and multiplication were unaffected at 35 and 37°C. Exposure to 40°C affected the survival and infectivity of both *Steinernema* spp. and *Heterorhabditis* spp. Gokte-Narkhdkar *et al.* (2005) studied the effects of humidity and temperature on the expression of symptoms in the larvae of rice moth, *C. cephalonica* infected with *H. indica* for 3 years. Symptom expression was pronounced in all years when the maximum humidity was more than 82%, minimum humidity was more than 60%, maximum temperature was less than 39.7°C and minimum temperature was 23.2–24.4°C. At 25°C temperature and 100% relative humidity, the production of *H. indica* was found highest (Lalramliana *et al.*, 2005). Jothi and Mehta (2007) observed that *H. indica*, *H. bacteriophora*, *H. zealandica* and *S. glaseri* caused 100% mortality of *G. mellonella* at temperatures ranging between 30 and 40°C at 24 h after inoculation. At 48 h after inoculation, *H. indica* and *H. bacteriophora* caused 100% mortality between 20 and 27.5°C, whereas *H. zealandica* was effective between 22.5 to 27.5°C. *S. glaseri* was found to be virulent even at 15°C and continued up to 27.5°C at 48 h after inoculation causing 100% mortality.

Walia *et al.* (2008) studied the effect of temperature on the mortality caused by different EPN species on *G. mellonella*. The mortality caused by *H. indica* on *G. mellonella* was 100% at 30°C, and 33.3% at 20°C after 48 h of exposure. *S. asiaticum* resulted in 93.3% mortality at 25 and 30°C, and 86.6% mortality at 20°C after 120 h. *S. siamkayai* caused 100% mortality after 72 h at all temperatures. Gokte-Narkhedkar *et al.* (2009) also observed that temperature and humidity influenced nematode multiplication. A humidity level of 60% was a critical level, and there was no nematode reproduction and multiplication below this level. A temperature of 28°C and humidity >60% were found favourable for nematode multiplication. There was little difference in the infectivity of the nematode produced at different temperature and moisture regimes to the larvae of *H. armigera*.

18.5 Integration of Entomopathogenic Nematodes in Integrated Pest Management Programmes

The use of EPNs in IPM programmes was suggested by Dutkey (1974). Ideally in an IPM approach, chemical insecticides and other biocontrol agents should not be toxic to EPNs. Compatibility must be determined before the implementation of an IPM strategy. However, compatibility per se does not necessarily ensure effective insect control. The combination should result in additive or supplemental mortality of the insect pest, reduce the amount of chemicals needed for insect control and allow the nematode to recycle. Certainly, an important consideration is the cost–benefit ratio derived from such a strategy (Kaya, 1985). In India, several compatibility studies were conducted with different EPN species. Trials were conducted on the compatibility of EPNs with insecticides such as the compatibility of DD 136 with dimethoate, malathion, mancozeb and zineb at the field recommended dosage have been reported (Das and Divakar, 1987). Studies have also been conducted to discover the compatibility of EPNs with fertilizers and pesticides. *S. glaseri*, *S. feltiae* and *H. indica* were highly compatible with urea and muriate of potash (MoP) up to a concentration of 400 ppm. Steinernematids were more compatible with 100 ppm of super phosphate, while the heterorhabditid was highly compatible up to 200 ppm. *S. glaseri* and *H. indica* were highly compatible with the herbicide, atrazine up to 10000 ppm, while *S. glaseri* was compatible with carbofuran, phorate, quinalphos and aldrin (Karunakar, 1990). Gupta and Siddiqui (1999) observed that *S. carpocapsae* was compatible with endosulfan, phosphamidon, cypermethrin, malathion, monocrotophos, dithane M-45, copper oxychloride, agallol and 2,4-D sodium salt, but incompatible with phorate and bavistin. Priya and Subramanian (2008) reported *S. glaseri* compatible with carbofuran, carbosulfan, imidacloprid, phorate and dimethoate, but incompatible with endosulfan at higher concentration (2000 ppm). Similarly, *H. indica* was compatible with carbofuran, carbosulfan and imidacloprid, but incompatible with phorate, dimethoate and endosulfan at higher concentration (2000 ppm). Maru *et al.* (2007) found *Steinernema* sp. (isolate STSIRO) and *Heterorhabditis* sp. (HMABU) compatible with endosulfan, malathion, fenvalerate and dimethoate. In an experiment, *S. mushtaqi* was found more compatible with insecticide (endosulfan and monocrotophos), fungicide (Mancozeb) and weedicide (Pendimenthilene) as compared to *S. masoodi* and *S. seemae*. S. *carpocapsae* was found least compatible with the pesticides tested (Pervez and Ali, 2011). Cypermethrin-treated *S. litura* larvae could act as an acceptable and suitable host that permitted invasion and proliferation of *S. glaseri*. Further, the efficacy of F1 EPNs indicated the potential of inoculative releases of EPNs in conjunction with cypermethrin treatment in an integrated strategy (Dhiman and Seth, 2012).

18.5.1 Compatibility of entomopathogenic nematodes with other biocontrol agents in integrated pest management programmes

EPNs can be combined with existing biological control agents to achieve better control of the target pests. Prabhuraj *et al.* (2004) evaluated the combined effect of *H. indica* and *Bacillus thuringiensis* sub sp. *kurstaki* against *H. armigera*. *H. indica* at 150 IJs + *B. thuringiensis* at 0.264 IU/mg caused 100% mortality of third-instar larva at 36 h after inoculation, which was significantly higher than the other treatments. The combined effect of

H. indica and *B. thuringiensis* at lethal doses resulted in quicker mortality of the pest as compared to individual treatments, indicating the presence of synergism. In another study, Prabhuraj *et al.* (2005) evaluated the effect of *H. indica* in combination with other entomopathogens and botanicals against *H. armigera* in the chickpea ecosystem. Pooled data on per cent larval reduction after two sprays revealed that the highest reduction (47.63%) of *H. armigera* was achieved in chlorpyrifos/quinalphos (0.04/0.05%) treatment at 7 days after spraying. However, sequential application of *H. indica* + *Pongamia pinnata* (1.0 lakh IJs + 2.5%) and *H. indica* + *Prosopis juliflora* (1.0 lakh IJs + 10%) resulted in maximum yield (1.96 and 1.83 kg/plot, respectively) with minimum pod damage (10.9 and 11.5%, respectively). Thus, there is scope for the integration of *H. indica* with *P. pinnata* and *P. juliflora* for the effective management of the chickpea pod borer. The first-instar white grub, *H. consanguinea*, was found susceptible to the nematode, *H. bacteriophora*, the fungus, *Metarhizium anisopliae*, and a combination of both. The nematode and fungus combinations were found more effective than either nematode or fungus alone. The combination of nematode with fungus required a shorter time to cause lethal infection than either of the biocontrol agents alone. Grubs exposed to combined application of nematode and fungus showed no apparent visible external symptoms of fungal infection, but nematode multiplication was observed in them (Jat and Choudhary, 2006). Devi (2007) assessed the combined efficacy of local isolates of *S. carpocapsae* and *H. indica* against *G. mellonella* larvae previously exposed to local isolates of *Beauveria bassiana*, *Nomuraea rileyi* and *Lecanicillium* sp. The larval mortality of *G. melonella* reached up to 80% when the treatments were applied in combination, but the results were less effective when the treatments were applied singly.

Bhagat *et al.* (2008) reported that different combinations of biopesticides i.e., *S. carpocapsae* + *B. bassiana*, *S. carpocapsae* + *M. anisopliae* and *H. indica* + *M. anisopliae*, provided less protection against *A. ipsilon* at earlier stages of seedling growth of maize, whereas at later stages, they were effective and registered higher yield as compared to the control. *H. indica* + *M. anisopliae* treated plots recorded higher yield as compared to other biopesticides. A successful management strategy was developed against the chickpea pod borer, *H. armigera*, by integrating locally isolated EPN *H. indica* (RCR) with other entomopathogens like *H. armigera*, nuclear polyhedrosis virus (HaNPV), *B. thuringiensis* and botanicals. Prabhuraj *et al.* (2008) reported that sequential application of *H. indica* (1 lakh IJs/l) + *P. juliflora* (10%) at 50 and 75 days after sowing resulted in highest larval reduction (23.47%), minimum pod damage (11.27%) and maximum seed yield.

18.6 Mass Production of Entomopathogenic Nematodes

For a biocontrol agent to be successful, it should be amenable for mass production on a large scale. Nematodes of both the genera i.e. *Steinernema* and *Heterorhabditis* can be mass produced in both solid and liquid cultures. Currently, commercial nematodes are produced by the solid media process or the liquid fermentation methods. In India, where labour costs are minimal, the solid media technology is adaptable and can be developed into a cottage industry utilizing the locally available nematode population and materials. The liquid fermentation process is highly efficient and economically suitable for industrialized countries. In India, efforts are being made on the mass production of EPNs by *in vivo*, using *G. mellonella* as the host, or by *in vitro* methods using the Bedding's or modified Bedding's method.

18.6.1 *In vivo*

EPNs can be mass produced by the *in vivo* technique in which the insects serve as small biological reactors. Karunakar *et al.* (1992) reared *S. feltiae*, *S. glaseri* and *H. bacteriophora* on *C. sacchariphagus indicus* and *S. excerptalis*. Prabhuraj *et al.* (2003) observed the mulberry silkworm *Bombyx mori* as a most appropriate host for the production of *S. glaseri* and *Heterorhabditis* sp. *G. mellonella* was the next best alternative for the production of *S. glaseri* and *Heterorhabditis* sp. It is suggested that mulberry silkworm, which is available throughout the year in Karnataka, India, is more suitable for *in vivo* production of EPNs as it can be obtained directly from the rearing room at low cost; in other states, *G. mellonella* may be a better effective host. The *in vivo* production of *Heterorhabditis* sp. per unit body weight was always greater than that of *Steinernema* sp. in

G. mellonella larvae (Kumar *et al.*, 2003). Singh and Gupta (2006) compared the mass production of *H. bacteriophora*, *S. feltiae* and an unidentified *Steinernema* on larvae of the rice moth, *C. cephalonica*, *G. mellonella* and on artificial medium (chicken offal homogenate). It was concluded that the IJs production of *H. bacteriophora* on the rice moth was 3.2 and 9.5 times higher than that of *S. feltiae* and *Steinernema* sp., respectively. The wax moth larvae supported 4.5, 3.0 and 2.5 times more production of *H. bacteriophora*, *S. feltiae* and *Steinernema* sp., respectively, than rice moth larvae. On the artificial medium, *H. bacteriophora* produced 54.3% more IJs than *S. feltiae*. Gokte-Narkhedkar *et al.* (2008) explored the feasibility of rearing *H. indica* on various cotton insect pests on a cottage industry scale with *C. cephalonica*, *G. mellonella*, cotton bollworms (*H. armigera*, *Earias vitella* and *Pectinophora gossypiella*), *S. litura*, *Sylepta derogata* and *Anomis* sp.

18.6.2 *In vitro*

Mass production is a key issue in EPN commercialization. Nematodes production by *in vivo* technique, where insect serve as media for nematode-bacterium growth and multiplication requires a constant source of host insects and high production costs. Alternatively, efficient *in vitro* methods have been developed by solid and liquid culturing of EPNs. EPNs are being mass produced on various substances, such as homogenized animal tissue, peptone-glucose agar and pork kidney, potato mash, dog food, chicken offal homogenate, modified dog biscuit, egg yolk, soy flour and cholesterol medium. In USA and some European countries, *Heterorhabditis* and *Steinernema* spp. are now being produced commercially in monoxenic liquid culture systems by using fermenters. In India, the mass production of EPNs is done on solid substrates on a small scale for laboratory and field experiments.

Indigenous mass production of EPNs has been demonstrated by different workers. Thinakaran Ponniah (1997) mass produced *S. glaseri* on polyurethane foam coated with yeast extract, beef extract, chickpea flour and coconut oil. Vyas *et al.* (1999b) produced *S. glaseri* with modified Bedding's media on polyurethane sponge, and obtained a yield of 50,000 IJs/sponge. Attempts were made to mass produce the EPNs *H. indica*, *H. bacteriophora* and *S. glaseri* on different artificial media at ICAR-SBI, Coimbatore, as per the method described by Wouts (1981). Hussaini *et al.* (2002b) evaluated different plant and animal protein media for the *in vitro* production of indigenous EPN isolates. The maximum yield of IJs of *S. carpocapsae* was observed on modified dog biscuit medium, *S. bicornutum* on modified egg yolk medium and *H. indica* on Wout's medium. Dog biscuit medium favoured the *in vitro* mass production of *H. indica* (Shabana and Prabhuraj, 2007). Prasad and Jayakumar (2007) mass produced *H. indica* by *in vitro* culture and obtained a maximum of 22×10^6 IJs/flask after 5 weeks. Wout's medium containing soy flour and maize oil recorded maximum populations of *H. indica* and *S. siamkayai* (Umamaheswari *et al.*, 2008). Yadav *et al.* (2008b) harvested the highest number of IJs of *H. bacteriophora* (11.65×10^5) with Wout's medium coated on polyether polyurethane chips incubated at 25 ± 1°C for 30 days. Among different substrates, namely sorghum flour, chickpea flour, groundnut flour, Parle-G biscuits, boiled potato, used tea powder, banana pseudo stem and spider lily flower stalk used for the mass production of *S. carpocapsae*, chickpea was found to be superior to others, producing 10.4×10^5 IJs/10 g media, followed by sorghum (Shinde *et al.*, 2011).

18.7 Formulation of Entomopathogenic Nematodes

Commercial formulations are developed to maintain product stability during storage and transportation. The successful market acceptance of nematode-based products depends heavily on their stability during shipping and storage, as well as their ease of use and consistent performance under field conditions. Significant progress has been made in the nematode formulation technology in developed countries but in India, EPN formulations are limited to 'talc' based for field release.

In India, the latest formulation developed by Multiplex Biotech Pvt Ltd is comprised of hydrogel-based semi-solid cream, with IJs suspension containing 1 million nematodes in a polythene pouch that can be readily mixed in a water spray tank without blocking the nozzles. A nematode shelf life of 2–3 months was achieved at room temperature (27–28°C), with a maximum of 80% survival by increasing the nematodes' metabolic rate mixing silica powder (0.1%). Currently, the company is marketing EPN products under the name of 'Soldier' (contains *Heterorhabditis indica*), recommended as soil application for soil-dwelling insect pests and 'Bouncer' (contains *Steinernema carpocapsae*) for foliar application and borer insect pests of fruit, pod and stem (Divya and Sankar, 2009). Gokte-Narkhedkar *et al.* (2003) standardized three formulations i.e. slurry, granules and suspension. Viability was found to be best with suspension, while for infectivity, both suspension and slurry were at par. Hussaini *et al.* (2003a) studied the efficacy of different formulations (talc based, alginate capsules and water) of *S. carpocapsae*, *S. abbasi* and *H. indica* against *A. ipsilon*. The results showed that alginate formulations caused the maximum mortality (33–80%). The survival of IJs was highest in water, followed by talc and alginate for *Steinernema* spp., and talc, followed by water for *H. indica*. Sitaramaiah *et al.* (2003) applied talc-based formulation of *S. carpocapsae* for the control of *S. litura* on tobacco plants in a nursery. Hussaini *et al.* (2005) used talc-based formulations of *H. indica*, *H. bacteriophora*, *S. carpocapsae* and *S. abbasi* in turfgrass for the control of white grubs, *Holotrichia longipennis*. *Steinernema* sp. and *H. indica* were prepared in five different formulations: (i) 0.1% formaldehyde solution; (ii) charcoal; (iii) 0.1% formaldehyde solution + charcoal; (iv) 2% alginate capsule; and (v) 0.1% formaldehyde + 2% alginate capsule. The survival of IJs was highest in the formulation containing 0.1% formaldehyde and alginate capsule, followed by 0.1% formaldehyde and charcoal, for *H. indica* and *Steinernema* sp. (Devi, 2007). NemaGel, a novel formulation for EPNs, based on a newly developed hydrogel, has been found to enhance the shelf life of the indigenous nematode *S. thermophilum* (Ganguly *et al.*, 2008).

18.8 Application Technology of Entomopathogenic Nematodes

EPNs can be applied with nearly all agronomic equipment, including pressurized sprayers, mist blowers and electrostatic sprayers (Shapiro-Ilan *et al.*, 2006). EPNs should be applied at the first sign a pest population is initializing to cause damage. The reapplication of nematodes depends on the success of the first nematodes released. Their survivorship and success are based on environmental conditions, moisture and soil type and percentage of live nematodes actually released during the first application. Nematodes should be reapplied at 7-day intervals if damage continues. In order to ensure maximum effectiveness, it is recommended to apply them at the optimum environmental conditions needed for their better survival. Therefore, it is best to irrigate the target site, both before and after application, because they need moist conditions to prevent desiccation and to aid with movement to find hosts. Also, the best results are obtained when the relative humidity is high, ambient temperature is neither extremely hot nor cold, soil temperature is between 10 and 35°C, soil is moist and direct sunlight is minimal. All these factors help to prevent the nematodes from drying out and increase their survival and virulence. Of paramount importance, for EPNs to be effective, is that they must be applied to soil at minimum rates of 2.5 × 109 IJs/ha (= 25/cm^2) or higher (Shapiro-Ilan *et al.*, 2002). Besides, the recycling potential of EPNs should also be considered. Generally, as long as environmental conditions are conducive, nematode populations will remain high enough to provide effective pest control for 2–8 weeks after application (Kaya, 1990).

18.8.1 Enhancement of efficacy

The most important factor that limits the use of EPNs is their poor persistence on the above-ground plant parts and consequent efficacy against foliar insects. The vulnerability of IJs to desiccation, solar radiation and higher temperature is considered as the major impediment. This has led workers to try anti-desiccants and surfactants in foliar sprays. Efficacy of *S. carpocapsae* was enhanced against *S. litura* in both laboratory and field condition when phagostimulant (0.032%), glycerine (0.1%) and a surfactant, Triton (0.05%) was added to the nematode suspension (Sezhian *et al.*, 1996). Hussaini *et al.* (2003b) studied the effect of some optical brighteners (OBs) and para-aminobenzoic acid (PABA) as UV protectants, and the results indicated the protective effect of optical brighteners and PABA in reverting the deleterious effect of UV radiation on EPNs; there was no adverse effect of the optical brighteners on the nematodes. The effects of different anti-desiccants, i.e. glycerine, liquid paraffin, Triton X-100, Tween 80 and castor oil, were tested on the survival and pathogenicity of different EPNs (*S. carpocapsae*, *S. tami*, *S. abbasi*, *H. indica* and *H. bacteriophora*) against *Plutella xylostella*. All the treatments used in the study caused no adverse effect on the survival of IJs. The infectivity of IJs improved significantly against *P. xylostella*. Absolute mortality was obtained with 0.5% glycerine and 1% liquid paraffin for 1 and 2 h. Both glycerine and liquid paraffin provided maximum protection to IJs against desiccation. Gokte-Narkhedkar *et al.* (2003) standardized eight materials (animal fat, guar gum, glycerol, aloe gel, sunflower oil, carboxy methyl cellulose (CMC), acacia gum and agar-agar) possessing anti-evaporant properties, and tested against five isolates of *H. indica* under laboratory conditions. Guar gum, glycerol, aloe gel, sunflower oil and CMC at 10% were effective. Foliar application of *S. carpocapsae* at 3×10^9 IJs/ha + glycerine + ujala (a brightener) and *H. indica* at 5×10^9 IJs/ha + glycerine + teepol (a brightener) reduced pod damage by up to 12 and 11%, respectively. Maximum grain yields of 26.85 and 21.50 q/ha were obtained by applying the above dosages, i.e. 42 and 47% increase in yield over the untreated control (Abid *et al.*, 2014). Dust and liquid formulation containing *S. seemae* was effective against *H. armigera* in pigeopea (Ali *et al.*, 2010) and chickpea (Ali and Asif, 2011). Anes and Ganguly (2012) exposed IJs of *S. thermophilum* to eight surfactants/adjuvants, i.e. Apsa 80, CMC, glycerol, liquid paraffin, starch, Triton X-100, Tween 20 and Tween 80 for a period of 1 month. All these surfactants/adjuvants were found to be compatible with *S. thermophilum* as they did not affect either the viability or virulence of the nematode, even after their exposure to the same nematode for a period of 1 month. Devi (2011) observed that Triton X-100 and Tween 20 were a little toxic, but liquid paraffin and sunflower oil were highly toxic to IJs of *S. carpocapsae* and *H. indica*.

18.9 Conclusion and Future Prospects

On the basis of foregone review, it can be concluded that the future of EPNs in the control of crop insect pests seems to be promising in India. Intensive surveys were undertaken from different parts and agroclimatic regions of the country, but there is still need for a considerable level of information on the biology and potential efficacy of indigenous strains. Proper application technology has to be developed against foliar pests using anti-desiccants and UV protectants. Above all, these beneficial nematodes always require a low temperature for their survival and application. This requires an additional expense, and ultimately makes the final product costly (Askary *et al.*, 2012). Hence, research in this quarter should be emphasized. Intensive research is needed in solid and liquid fermentation technology for the mass production and commercialization of EPNs in the Indian market. Much effort needs to be directed towards further optimizing more stable formulations that are simpler to use and have longer than a 1-year shelf life. Improvements in production technology, distribution and application will be key factors in reducing nematode costs and ensuring product quality. Initially, a biocontrol agent may be expensive, as it requires capital in setting up a biological control system; however, it may be cost-effective in the long run, due to their self-perpetuating nature (Askary, 2015). There are several companies producing EPNs commercially in foreign countries, but in India only a few companies are producing EPNs. There is a need to set up local-level cooperatives, cottage industries and farmers' entrepreneurship programmes to produce nematodes cheaply and effectively for their inclusion in integrated pest management and then for on-site use.

Acknowledgements

The authors are grateful to Dr Bakshi Ram, Director, ICAR-Sugarcane Breeding Institute, Coimbatore, for the encouragement and facilities provided. The second author is also indebted to Division of Entomology, Sher-e-Kashmir University of Agricultural Sciences and Technology (SKUAST), Srinagar, India for providing necessary facilities required to prepare the chapter.

References

Abid, M.A., Prasad, C.S., Ahmad, R., Ahmad, W. and Singh, R. (2014) Foliar application of entomopathogenic nematodes, *Steinernema masoodi*, *S. carpocapsae* and *Heterorhabditis indica* for management of legume pod borer, *Helicoverpa armigera* infesting chickpea. *Vegetos* 27, 195–199.

Ahmad, R., Ali, S.S. and Pervez, R. (2009) Field efficacy of *Steinernema masoodi* based biopesticide against *Helicoverpa armigera* (Hübner) infesting chickpea. *Trends in Biosciences* 1, 23–24.

Ali, S.S. and Asif, M. (2011) Comparative field efficacy of dust formulation and liquid formulation of *Steinernema seemae* based biopesticide and other IPM options against *Helicoverpa armigera* (Hübner) infesting chickpea. *Trends in Biosciences* 4, 35–37.

Ali, S.S. and Shaheen, A. (2011) *Steinernema sayeedae* sp. n. a heat tolerant EPN from banana rhizosphere of Koshambhi district, U.P. India. *Trends in Biosciences* 4, 123–125.

Ali, S.S, Shaheen A., Pervez, R. and Hussain, M.A. (2005) *Steinernema masoodi* sp. n. and *Steinernema seemae* sp. n. (Nematoda: Rhabditida: Steinernematidae) from India. *International Journal of Nematology* 15, 89–99.

Ali, S.S., Shaheen, A., Asif, M. and Akhtar, M.H. (2009) *Steinernema qazii* sp.n. (Nematoda: Rhabditidae: Steinernematidae) from Kanpur, India. *Trends in Biosciences* 2, 59–64.

Ali, S.S., Tickle, A.N., Nema, K.K., Pervez, R., Shaheen, A. and Ahmad, R. (2010) Field efficacy of *Steinernema masoodi* based biopesticide against *Helicoverpa armigera* (Hubner) infesting early pigeon pea. *Trends in Biosciences* 1, 63–65.

Ambika, S. and Sivakumar, C.V. (2002) Natural occurrence of entomopathogenic nematodes in three northern districts of Tamil Nadu. *Indian Journal of Entomology* 64, 288–291.

Anes, K.M. and Ganguly, S. (2012) Compatibility of entomopathogenic nematode, *Steinernema thermophilum* (Nematoda: Rhabditida) with different surfactants/adjuvants commonly used in spray mixtures. *Indian Journal of Nematology* 42, 20–23.

Askary, T.H. (2010) Nematodes as biocontrol agents. In: Lichtfous, E. (ed.) *Sociology, Organic Farming, Climate Change and Soil Science*. Springer, The Netherlands, pp. 347–378.

Askary, T.H. (2015) Limitations, research needs and future prospects in the biological control of phytonematodes. In: Askary, T.H. and Martinelli, P.R.P. (eds) *Biocontrol Agents of Phytonematodes.* CAB International, Wallingford, UK, pp. 446–454.

Askary, T.H. and Martinelli, P.R.P. (2015) *Biocontrol Agents of Phytonematodes.* CAB International, Wallingford, UK.

Askary, T.H., Khan, A.A., Waliullah, M.I.S., Banday, S.A., Iqbal, U. and Mir, M.M. (2012) Slug pest management through nematodes in agricultural and horticultural crops. In: Boeri, F. and Chung, J.A. (eds) *Nematodes: Morphology, Functions and Management Strategies.* Nova Publishers, New York, pp. 197–211.

Banu, G. (2003) Scope of entomopathogenic nematodes for the management of pest of intercrops in coconut based cropping ecosystem. In: Hussaini, S.S., Rabindra, R.J. and Nagesh, M. (eds) *Current Status of Research on Entomopathogenic Nematodes in India.* PDBC, ICAR, Bangalore, India, pp. 179–184.

Banu, J.G. Nguyen, K.B. and Rajendran, G. (2005) Occurrence and distribution of entomopathogenic nematodes in Kerala, India. *International Journal of Nematology* 15, 9–16.

Bhagat, R.M., Khajuria, M.K., Uma Shankar, Monobrullah, M. and Kaul, V. (2008) Efficacy of biopesticides and insecticides in controlling maize cutworm in Jammu. *Journal of Biological Control* 22, 99–106.

Bhaskaran, R.K.M., Sivakumar, C.V. and Venugopal, M.S. (1994) Biocontrol potential of entomopathogenic nematode in control of red hairy caterpillar, *Amsacta albistriga* of groundnut. *Indian Journal Agriculture Sciences* 64, 655–657.

Cannayane, I., Banu, J.G., Subramanian, S. and Rajavel, D.S. (2007) Preliminary evaluation of the entomopathogenic nematodes on the root grub, *Basilepta fulvicorne* in cardamom. *Indian Journal of Nematology* 37, 213–214.

Chandel, R.S., Chandla, V.K. and Dhiman, K.R. (2005) Vulnerability of potato white grub to entomogenous fungi and nematodes. *Potato Journal* 32, 193–194.

Chandrasekar, S.D., Sundararaj, P., Balu, A., Krishnak, N. (2013) Morphometric and molecular characterisation of entomopathogenic nematodes in the Western Ghats of South India. *International Journal of Nematology* 23, 175–181.

Das, J.N. and Divakar, B.J. (1987) Compatibility of certain pesticides with DD-136 nematode. *Plant Protection Bulletin* 39, 20–22.

Das, P., Das, S.K., Arya, H.P.S., Reddy, S.G. and Mishra, A. (2002) Pest and disease management. In: Das, P., Das, S.K., Arya, H.P.S., Reddy, S.G. and Mishra, A. (eds) *Inventory of Indigenous Technical Knowledge in Agriculture, Document I.* Mission Unit, Division of Agricultural Extension, Indian Council of Agricultural Research, New Delhi, pp. 48–94.

Devi, G. (2007) Compatibility of indigenous entomopathogenic nematodes with entomopathogenic fungi. *Indian Journal of Plant Protection* 35, 149–150.

Devi, G. (2011) Compatibility of entomopathogenic nematodes (Meghalaya isolates) with spray adjuvants. *Indian Journal of Nematology* 41, 52–55.

Devrajan, K., Subramanian, S. and Prabhu, S. (2010) Biological control of tea flush worm, *Laspeyresia bipunctata* using an entomopathogenic nematode, *Heterorhabditis indica*. *Indian Journal of Nematology* 40, 145–147.

Dhiman, A. and Seth, R.K. (2012) Compatibility of entomopathogenic nematode, *Steinernema glaseri* with cypermethrin. *Indian Journal of Entomology* 74, 16–23.

Divya, K. and Sankar, M. (2009) Entomopathogenic nematodes in pest management. *Indian Journal of Science and Technology* 2, 53–59.

Dutkey, S.R. (1974) Nematode parasites. In: Maxwell, F.G. and Harris, F.A. (eds) *Proceedings of Summer Institute on Biological Control of Plant Insects and Diseases*. University Press, Jackson, Mississipi, pp. 579–590.

Ganguly, S. (2003) Taxonomy of entomopathogenic nematodes and work done in India. In: Hussaini, S.S., Rabindra, R.J. and Nagesh, M. (eds) *Current Status of Research on Entomopathogenic Nematodes in India*. PDBC, ICAR, Bangalore, India, pp. 69–108.

Ganguly, S. (2006) Recent taxonomic status of entomopathogenic nematodes: a review. *Indian Journal of Nematology* 36, 158–176.

Ganguly, S. and Gavas, R. (2004) Effect of soil moisture on the infectivity of entomopathogenic nematode, *Steinernema thermophilum* Ganguly & Singh. *International Journal of Nematology* 14, 78–80.

Ganguly, S. and Singh, L.K. (2000) *Steinernema thermophilum* sp. n. (Rhabditida: Steinernematidae) from India. *International Journal of Nematology* 10, 183–191.

Ganguly, S. and Singh, L.K. (2001) Optimum thermal requirements for infectivity and development of an indigenous entomopathogenic nematode, *Steinernema thermophilum*. *Indian Journal of Nematology* 31, 148–152.

Ganguly, S. and Singh, L.K. (2002) A note on the life cycle of entomopathogenic nematode, *Steinernema thermophilum*. *Indian Journal of Nematology* 32, 98–100.

Ganguly, S., Singh, M., Lal, M., Singh, L.K., Vyas, R.V. and Patel, D.J. (2002) New record of an entomopathogenic nematode, *Steinernema riobrave* Cabanillas, Poinar & Raulston, 1994 from Gujarat. *Indian Journal of Nematology* 32, 223.

Ganguly, S., Rathour, K.S. and Pandey, J. (2005) New record of *Steinernema siamkayai* from Champawat District of Uttaranchal, and its biochemical characterisation. *Indian Journal of Nematology* 35, 203–204.

Ganguly, S., Anupama, Kumar, A. and Parmar, B.S. (2008) Nemagel – a formulation of the entompathogenic nematode *Steinernema thermophilum* mitigating the shelf-life constraint of the tropics. *Nematologia Mediterranea* 36, 125–130.

Ganguly, S., Kumar, S. and Rathour, K.S. (2010) Availability of biopesticidal nematodes to suit different agro-climatic region of India for commercialisation. *Indian Journal of Nematology* 40, 261–264.

Ganguly, S., Rathour, K.S., Kumar, S. and Singh, M. (2011) *Steinernema meghalayensis* sp. n. (Rhabditida: Steinernematidae) from northeastern hilly region of India. *Indian Journal of Nematology* 41, 83–97.

Gokte-Narkhedkar, N., Lavhe, N.V. and Mayee, C.D. (2001) Entomopathogenic nematodes – potential agent against American bollworm (*Helicoverpa armigera*) of cotton. *SAIC Newsletter* 11, 4.

Gokte-Narkhedkar, N., Banerjee, S.K., Lavhe, N.V. and Mayee, C.D. (2003) Utilization of entomopathogenic nematodes for control of cotton pests and progress of work at Central Institute for Cotton Research, Nagpur. In: Hussaini, S.S., Rabindra, R.J. and Nagesh, M. (eds) *Current Status of Research on Entomopathogenic Nematodes in India*. PDBC, ICAR, Bangalore, India, pp. 175–178.

Gokte-Narkhedkar, N., Lavhe, N.V. and Banerjee, S.K. (2005) Factors influencing expression of symptoms in *Corcyra cephalonica* by entomopathogenic nematode, *Heterorhabditis indica*. *Indian Journal of Plant Protection* 33, 273–276.

Gokte-Narkhedkar, N., Lavhe, N.V., Panchbhai, P.R. and Khadi, B.M. (2008) Cottage industry scale *in vivo* production of *Heterorhabditis indica* for the control of *Helicoverpa armigera* on cotton in India. *International Journal of Nematology* 18, 79–82.

Gokte-Narkhedkar, N., Lavhe, N.V., Mukewar, P.M. and Khadi, B.M. (2009) Optimization of conditions for maximum population index of *Heterorhabditis indica* in modified Wout's substrate for the control of *Helicoverpa armigera*. *Indian Journal of Plant Protection* 37, 213–215.

Gupta, P. and Siddiqui, M.R. (1999) Compatibility studies on *Steinernema carpocapsae* with some pesticidal chemicals. *Indian Journal of Entomology* 61, 220–225.

Gupta, S., Kaul, V., Monobrullah, M. and Kumar, S. (2011) Field evaluation of Steinernematid and Heterorhabditid nematodes against *Plutella xylostella* (L.) on cauliflower. *Annals of Plant Protection Sciences* 19, 418–422.

Hominick, W.M., Reid, A.P., Bohan, D.A. and Briscoe, B.R. (1996) Entomopathogenic nematodes: biodiversity, geographical distribution and the convention on biological diversity. *Biocontrol Science and Technology* 6, 317–331.

Hussaini, S.S., Satya, K.J. and Hussain, M.A. (2000) Mass production of a native *Steinernema* sp. (SSL2) PDBC EN 13.21 (Nematoda : Steinernematidae) on different artificial media. *Indian Journal of Plant Protection* 28, 94–96.

Hussaini, S.S., Ansari, M.A., Ahmad, W. and Subbotin, S.A. (2001) Identification of some Indian populations of *Steinernema* species (Nematoda) by RFLP analysis of the ITS region of rDNA. *International Journal of Nematology* 11, 73–76.

Hussaini, S.S., Singh, S.P. and Nagesh, M. (2002a) *In vitro* and field evaluation of some indigenous isolates of *Steinernema* and *Heterorhabditis indica* against shoot and fruit borer, *Leucinodes orbonalis*. *Indian Journal of Nematology* 32, 63–65.

Hussaini, S.S., Singh, S.P., Parthasarathy, R. and Shakeela, V. (2002b) *In vitro* production of entomopathogenic nematodes in different artificial media. *Indian Journal of Nematology* 32, 44–46.

Hussaini, S.S., Singh, S.P. and Shakeela, V. (2003a) *Variable Efficacy of Different Formulations of Entomopathogenic Nematodes against Black Cutworm* Agrotis ipsilon *(Hufnagel) Larvae*. Project Directorate of Biological Control (ICAR), Bangalore, India, pp. 193–197.

Hussaini, S.S., Hussain, M.A. and Satya, K.J. (2003b) Survival and pathogenicity of indigenous entomopathogenic nematodes in different UV protectants. *Indian Journal of Plant Protection* 31, 12–18.

Hussaini, S.S., Nagesh, M. and Shakeela, V (2004) Selection for high temperature tolerance in some native *Steinernema* and *Heterorhabditis* species. *Indian Journal of Nematology* 34, 185–192.

Hussaini, S.S., Nagesh, M., Dar, M.H. and Rajeshwari, R. (2005) Field evaluation of entomopathogenic nematodes against white grubs (Coleoptera: Scarabaeidae) on turf grass in Srinagar. *Annals of Plant Protection Sciences* 13, 190–193.

Israel, P., Rao, Y.R., Rao, V.J., Rao, P.S.P. and Verma, A. (1969) Control of paddy cutworms by DD-136, a parasitic nematode. *Current Science* 16, 390–391.

Jat, B.L. and Choudhary, R.K. (2006) Bioefficacy of entomopathogenic nematode, *Heterorhabditis bacteriophora* and fungus, *Metarhizium anisopliae* and their combinations against white grub, *Holotrichia consanguinea* Blanch. *Indian Journal of Entomology* 68, 221–224.

Josephrajkumar, A. and Sivakumar, C.V. (1997) A survey for entomopathogenic nematodes in Kanyakumari district, Tamil Nadu, India. *Indian Journal of Entomology* 59, 45–50.

Jothi, B.D. and Mehta, U.K. (2007) Impact of different temperatures on the infectivity and productivity of entomopathogenic nematodes on *Galleria mellonella*. *International Journal of Nematology* 17, 158–162.

Karthikeyan, K. and Sosamma, J. (2009) Bioefficacy of white muscardine fungus, *Beauveria bassiana* (Balsamo) Vuillemin and entomopathogenic nematode, *Heterorhabditis indica* (Poinar) against rice blue beetle, *Leptispa pygmaea* Baly. *Journal of Biological Control* 23, 79–81.

Karthikeyan, K. and Sosamma, J. (2010) *Pseudomonas flourescens* and *Heterorhabditis indica* Poinar for the management of major insect pests of rice. *Journal of Biopesticides* 3, 96–99.

Karunakar, G. (1990) Studies on entomophilic nematodes on sugarcane. MSc thesis submitted to Tamilnadu Agricultural University, Coimbatore, India, p. 180.

Karunakar, G., David, H. and Easwaramoorthy, S. (1992) Influence of dosage of *Steinernema carpocapsae*, *S. glaseri* and *Heterorhabditis indicus* on mortality of the host and multiplication of infective juveniles in sugarcane internode borer, *Chilo sacchariphagus indicus*. *Journal of Biological Control* 6, 26–28.

Karunakar, G., Easwaramoorthy, S. and David H. (1999a) Influence of temperature on infectivity, penetration and multiplication of *Steinernema feltiae, S. glaseri* and *Heterorhabditis indicus*. *International Journal of Nematology* 9, 126–129.

Karunakar, G., Easwaramoorthy, S. and David, H. (1999b) Susceptibility of nine lepidopteran insects to *Steinernema glaseri*, *S. feltiae* and *Heterorhabditis indicus* infection. *International Journal of Nematology* 9, 68–71.

Karunakar, G., Easwaramoorthy, S. and David, H. (2000a) Pathogenicity of steinernematid and heterorhabditid nematodes to whitegrubs infesting sugarcane in India. *International Journal of Nematology* 10, 19–26.

Karunakar, G., Easwaramoorthy, S. and David, H. (2000b) Host parasite interaction between two species of white grubs infesting sugarcane and two species of entompathogenic nematodes. *Sugar Tech* 2, 12–16.

Kaushal, K.K., Rawat, R., Naved, S.H. and Nand, S. (2000) Survey of entomopathogenic nematodes in India. *Annals of Plant Protection Sciences* 8, 119–121.

Kaya, H.K. (1985) Entomogenus nematodes for insect control in IPM systems. In: Hoy, M.A. and Herzog, D.C. (eds) *Biological Control in Agricultural IPM Systems*. Academic Press Inc, New York, pp. 283–302.

Kaya, H.K. (1990) Soil ecology. In: Entomopathogenic nematodes. *Annual Review of Entomology* 38, 181–206.

Kaya, H.K., Aguillera, M.M., Alumai, A., Choo, H.Y., de la Torre, M., *et al.* (2006) Status of entomopathogenic nematodes and their symbiotic bacteria from selected countries or regions of the world. *Biological Control* 38, 134–155.

Kulkarni, N., Rizvi, A.N., Kumar, V., Paunikar, S. and Mishra, V.K. (2012) Morphological and molecular characterization of *Steinernema dharanaii* sp. n.: a new entomopathogenic nematode from India. *Indian Journal of Tropical Biodiversity* 20, 107–116.

Kumar, M.R., Parihar, A. and Siddiqui, A.U. (2003) *In-vivo* culturing of indigenous entomopathogenic nematodes from Udaipur. *Indian Journal of Nematology* 33, 179.

Lalramliana, Yadav, A.K. and Shylesha, A.N. (2005) Effects of temperature and relative humidity on the emergence of infective juveniles of *Heterorhabditis indica* from Meghalaya, India. *Indian Journal of Nematology* 35, 157–159.

Maru, A.K., Siddiqui, A.U., Parihar, A. and Sharma, S.K (2007) Occurrence of entomopathogenic nematodes in varied agro-climatic zones of Rajasthan. *Indian Journal of Nematology* 37, 123–126.

Mathasoliya, J.B., Maghodia, A.B. and Vyas, R.V. (2004) Efficacy of *Steinernema riobrave* against *Agrotis ipsilon* Hufnagel (Lepidoptera: Noctuidae) on potato. *Indian Journal of Nematology* 34, 177–179.

Mathur, S.B., Srivastava, R.P. and Subba Rao, B.R.S. (1966) Records of nematodes parasitising *Chilo zonellus* Swinhoe stem borer of maize. *Indian Journal of Entomology* 28, 414–415.

Mathur, S.C. (1999) Future of Indian pesticides industry in next millennium. *Pesticide Information* 24, 9–23.

Misra, R.M. and Khan, H.R. (2003) A new nematode parasite *Steinernema carpocapsae* L. of *Hypsipyla robusta* moore (Pyralidae: Lepidoptera) from Doon Valley (Uttaranchal). *Indian Journal of Forestry* 26, 177–179.

Murthy, H.C., Naik, M.I. and Manjunath, M. (2010) Evaluation of certain new chemicals, bio-agents and plant product for the management of arecanut root grub, (*Leucopholis lepidophora*) Blanch. *Mysore Journal of Agricultural Sciences* 44, 653–655.

Narayanan, K. and Gopalakrishnan, C. (1987) Effect of entomogenous nematode, *Steinernema feltiae* to the pre-pupae, pupa and adult of *Spodopter litura*. *Indian Journal of Nematology* 17, 273–276.

Parihar, A., Siddiqui, A.U., Yadav, Y.S. and Budh Ram (2003) Occurrence of entomopathogenic nematodes in Rajasthan. *Indian Journal of Nematology* 65, 190–193.

Patel, M.C. and Vyas, R.V. (1995) Efficacy of *Stienernema glaseri* against *Helicoverpa armigera* on chickpea in pots. *International Chickpea and Pigeonpea Newsletter* 2, 39–40.

Pervez, R. and Ali, S.S. (2011) Efficacy, penetration and *in vivo* production of entomopathogenic nematodes against legume pod borer, *Maruca vitrata* Fabricius (Lepidoptera: Pyralidae). *Trends in Biosciences* 4, 103–105.

Pervez, R., Devasahayam, S. and Eapen, S.J. (2014) Determination of LD50 and LT50 of entomopathogenic nematodes against shoot borer (*Conogethes punctiferalis* Guen.) infesting ginger (*Zingiber officinale* Rosc.). *Annals of Plant Protection Sciences* 22, 169–173.

Poinar, G.O. Jr, Karunakar, G.K and David, H. (1992) *Heterorhabditis indicus* n.sp. (Rhabditida : nematoda) from India: separation of *Heterorhabditis* spp. by infective juveniles. *Fundamental and Applied Nematology* 15, 467–472.

Prabhu, S. and Sudheer, M.J. (2008) Evaluation of two native isolates of entomopathogenic nematodes *Steinernema* sp. and *Heterorhabditis indica* from Andhra Pradesh against *Amsacta albistriga* walk in groundnut. *Journal of Biopesticides* 1, 140–142.

Prabhuraj, A. and Patil, B.V. (2004) Pathogenicity of a virulent strain of *Heterorhabditis indica* against some crop pests. *Insect Environment* 10, 28–29.

Prabhuraj, A., Viraktamath, C.A. and Kumar, A.R.V. (2000) Field evaluation of entomopathogenic nematodes against brinjal root weevil, *Myllocerus discolor* Boh. *Pest Management in Horticultural Ecosystems* 2, 149–151.

Prabhuraj, A., Viraktamath, A.C.A. and Kamar, A.R.V. (2003) Evaluation of silkworm larva and root grub for the *in vivo* mass production of entomopathogenic nematodes. *Indian Journal of Entomology* 65, 34–37.

Prabhuraj, A., Shivaleela, Girish, K.S. and Patil, B.V. (2004) Effect of combination of *Heterorhabditis indica* (Rhabditida: Heterorhabditidae) and Bt against *Helicoverpa armigera* (Hübner). *Indian Journal of Entomology* 66, 369–372.

Prabhuraj, A., Patil, B.V., Girish, K.S. and Shivaleela, (2005) Field evaluation of an insect parasitic nematode, *Heterorhabditis indica* (RCR) in combination with other, entomopathogens and botanicals against chickpea podborer, *Helicoverpa armigera* (Hübner). *Journal of Biological Control* 19, 59–64.

Prabhuraj, A., Girish, K.S., Shivaleela and Patil, B.V. (2008) Integration of *Heterorhabditis indica* with other biorationals for managing chickpea pod borer, *Helicoverpa armigera* (Hüb.). *Journal of Biological Control* 22, 433–448.

Prasad, G.S. and Jayakumar, V. (2007) Population development of *Heterorhabditis indica* under *in vitro* culture. *Indian Journal of Plant Protection* 35, 155–156.

Prasad, G.S. and Ranganath, H.R. (2006) Effect of temperature, moisture and relative humidity on survival and infectivity of *Heterorhabditis indica* (Rhabditida: Heterorhabditidae) native to South Andaman's, India. *Pakistan Journal of Nematology* 24, 105–116.

Prasad, G.S. and Singh, P.K. (2003) Survival and pathogenicity of native *Heterorhabditis* sp. (Rhabditida: Heterorhabditidae) as influenced by soil temperature, moisture and relative humidity. *Indian Journal of Nematology* 33, 16–20.

Prasad, G.S., Ranganath, H.R. and Singh, P.K. (2001a) Occurrence of the entomopathogenic nematode in parts of south Andamans. *Current Science* 80, 501–502.

Prasad, G.S., Singh, P.K. and Ranganath, H.R. (2001b) Population fluctuation of entomopathogenic nematode, *Heterorhabditis* sp. in South Andaman as influenced by weather parameters. *Current Science* 80, 923–924.

Priya, P. and Subramanian, S. (2008) Compatibility of entomopathogenic nematodes *Heterorhabditis indica* (Poinar, Karunakar and David) and *Steinernema glaseri* (Stainer) with insecticides. *Journal of Biological Control* 22, 225–230.

Rajeswari Sunderababu, R., Menon, P.P.V. and Vadivelu, S. (1984) Preliminary tests with DD-136 for the control of potato chafer grub, *Anomola* sp. *Indian Journal of Nematology* 14, 187–189.

Rajkumar, Parihar, A. and Siddiqui, A.U. (2003) Isolation of indigenous entomopathogenic nematodes from Udaipur. *Indian Journal of Nematology* 33, 176–177.

Rao, P.S.P., Das, P.K. and Padhi, G. (1975) Note of compatibility of DD-136 (*Neoaplectana dutkyi*), an insect parasitic nematode with some insecticides and fertilisers. *Indian Journal of Agricultural Sciences* 45, 275–277.

Rao, V.P. and Manjunath, T.M. (1966) DD-136 nematodes that can kill many insect pests. *Indian Farming* 16, 143–144.

Rao, Y.R., Prakash, V.J., Rao, P.S., Verma, A. and Israel, P. (1971) Tests with an insect parasitic nematode DD-136 against the rice stem borer, *Tryporyza incertulas*. *Indian Journal of Entomology* 33, 215–217.

Razia, M. and Sivaramakrishnan, S. (2016) Evaluation of entomopathogenic nematodes against termites. *Journal of Entomology and Zoology Studies* 4, 324–327.

Raveendranath, S., Krishnayya, P.V., Rao, P.A., Murthy, K.V.M.K. and Hussaini, S.S. (2007) Bioefficacy of entomopathogenic nematodes, *Steinernema carpocapsae* and *Heterorhabditis indica* against third instar larvae of *Spodoptera litura*. *Indian Journal of Nematology* 37, 145–148.

Raveendranath, S., Krishnayya, P.V., Rao, P.A., Murthy, K.V.M.K. and Hussaini, S.S. (2008) Bioefficacy of entomopathogenic nematodes, *Steinernema carpocapsae* and *Heterorhabditis indica* against pupae of *Spodoptera litura*. *Indian Journal of Plant Protection* 36, 288–291.

Sankar, M., Prasad, J.S., Padmakumari, A.P., Katti, G. and Divya, K. (2009a) Combined application of two entomopathogenic nematodes, *Heterorhabditis indica* and *Steinernema asiaticum* to control the rice leaf folder, *Cnaphalocrosis medinalis* (Goen.). *Journal of Biopesticides* 2, 135–140.

Sankar, M., Prasad, J.S., Padmakumari, A.P. and Katti, G. (2009b) Efficacy of entomopathogenic nematode, *Heterorhabditis indica* Poinar (Rhabditida: Heterorhabditidae) against the rice yellow stem borer, *Scirpophaga incertulas* (Walker) (Lepidoptera: Pyralidae). *Biopesticides International* 5, 68–74.

Sankaranarayanan, C. and Easwaramoorthy, S. (2003) Bioefficacy of entomopathogenic nematode isolates against sugarcane internode borer, *Chilo sacchariphagus indicus* (Kapur) at two constant temperature. In: Ignacimuthu, S. and Jayaraj, S. (eds) *Biological Control of Insect Pests*. Phoenix Publishing House Pvt Limited, New Delhi, pp. 73–78.

Sankaranarayanan, C., Easwaramoorthy, S. and Somasekhar, N. (2003) Infectivity of entomopathogenic nematodes *Heterorhabditis* and *Steinernema* spp. to sugarcane shoot borer (*Chilo Infuscatellus* Snellen) at different temperatures. *Sugar Tech* 5, 167–171.

Sankaranarayanan, C., Somasekhar, N. and Singaravelu, B. (2006) Biocontrol potential of entomopathogenic nematodes *Heterorhabditis* and *Steinernema* against pupae and adults of white *grub Holotrichia serrata* F. *Sugar Tech* 8, 268–271.

Sankaranarayanan, C., Somasekhar, N., Singaravelu, B. and Shunmugasundaram, M. (2008) Pathogenicity of *Heterorhabditis* and *Steinernema* to sugarcane internode borer *Chilo sacchariphagus* indicus Kapur (Lepidoptera: Crambidae). *Journal of Biological Control* 22, 1–5.

Sankaranarayanan, C., Singaravelu, B., Somasekhar, N. and Santhalakshmi, G. (2011) Penetration and pathogenicity of entomopathogenic nematodes to sugarcane early shoot borer *Chilo infuscatellus* Snellen (Lepidoptera: Crambidae). *Journal of Biological Control* 25, 1–4.

Saravanapriya, B. and Subramanian, S. (2006) Biological control efficacy of entomopathogenic nematodes against *Plutella xylostella* (L.) on cabbage. *Annals of Plant Protection Sciences* 14, 516–517.

Seenivasan, N., Prabhu, S., Makesh, S. and Sivakumar, M. (2012) Natural occurrence of entomopathogenic nematode species (Rhabditida: Steinernematidae and Heterorhabditidae) in cotton fields of Tamil Nadu, India. *Journal of Natural History* 46, 2829–2843.

Sehgal, J., Mandal, D.K., Mandal, C. and Vadivelu, S. (1992) *Agro-Ecological Regions of India*. 2nd edn. Technical Bulletin No 24, National Bureau of Soil Survey and Land use Planning, Nagpur, India, pp. 130.

Sezhian, N., Sivakumar, C.V. and Venugopal, M.S. (1996) Alteration of effectiveness of *Steinernema carpocapsae* against *Spodoptera litura* larvae on sunflower by addition of an insect phagostimulant. *Indian Journal of Nematology* 26, 77–81.

Shabana, A. and Prabhuraj, A. (2007) Mass production of *Heterorhabditis indicus* in different artificial media. *Karnataka Journal of Agricultural Sciences* 20, 859–860.

Shah, N.K. and Azmi, M.I. (2006) Virulence of *Heterorhabditis indica* to the grubs of Lucerne weevil, *Hypera postica* (Coleoptera: Curculionidae). *Indian Journal of Nematology* 36, 285–286.

Shapiro-Ilan, D.I., Gouge, D.H. and Koppenhöfer, A.M. (2002) Factors affecting commercial success: case studies in cotton, turf and citrus. In: Gaugler, R. (ed.) *Entomopathogenic Nematology*. CAB International, Wallingford, UK, pp. 333–356.

Shapiro-Ilan, D.I., Gouge, D.H., Piggott, S.J. and Fife, J. (2006) Application technology and environmental considerations for use of entomopathogenic nematodes in biological control. *Biological Control* 38, 124–133.

Sharma, A., Thakur, D.R. and Chandla, V.K. (2009) Use of *Steinernema* and *Heterorhabditis* nematodes for control of white grubs, *Brahmina coriacea* Hope (Coleoptera: Scarabaeidae) in potato crop. *Potato Journal* 36, 160–165.

Shinde, S.V., Patel, H.V., Chavan, S.M., Patel, K.G., Purohit, M.S., *et al.* (2011) *In vitro* mass multiplication media and its impact assessment on entomopathogenic nematode *Steinernema carpocapsae* (Weiser) (Nematoda: Steinernematidae). *Biopesticides International* 7, 64–67.

Shinde, V., Patel, J.I. and Purohit, M.S. (2010) Report of entomopathogenic nematode, *Steinernema carpocapsae* (Weiser) (Nematoda: Steinernematidae) from North Gujarat. *Journal of Biological Control* 24, 75–76.

Singh, J. (1977) Isolation of entomophilic nematode from potato cutworms. *Current Science* 46, 454–455.

Singh, J. and Bardhan, V. (1974) Effectiveness of DD-136, an entomophilic nematode against insect pests of agricultural importance. *Current Science* 43, 662.

Singh, M. and Gupta, P.R. (2006) Occurrence of entomopathogenic nematodes in Himachal Pradesh, India and their pathogenicity against various insect species. *Pest Management and Economic Zoology* 14, 179–189.

Singh, M., Sharma, S.B. and Ranga Rao, G.V. (1992) Occurrence of entomopathogenic nematodes at ICRISAT center. *International Arachis Newsletter* 12, 15–16.

Sitaramaiah, S., Gunneswara Rao, S., Hussaini, S.S., Venkateswarulu, P. and Nageswara Rao, S. (2003) Use of entomopathogenic nematode, *S. carpocapsae* against *S. litura* in tobacco nursery. In: Tandon, P.L., Ballal, C.R., Jalali, S.K. and Rabindra, R.J. (eds) *Biological Control of Lepidopteran Pests*. PDBC, Bangalore, India, pp. 211–214.

Sivakumar, C.V., Jayaraj, S. and Subramanian, S. (1989) Observations on Indian population of the entomopathogenic nematodes *Heterorhabditis bacteriophora*. *Journal of Biological Control* 2, 112–113.

Sosamma, V.K. (2006) Distribution of entomopathogenic nematodes in coconut gardens of Kerala state and pathogenicity of *Steinernema feltiae* to red weevil, *Rhynchophorus* ferrugineus. *Journal of Plantation Crops* 34, 429–431.

Sosamma, V.K. and Rasmi, B. (2002) Survey of entomophilic nematodes in Kerala. *Indian Journal of Nematology* 32, 184–185.

Srinivas, P.R. and Prasad, J.S. (1991) Record of DD-136 nematode infection on rice leaf folder, *Cnephalocrosis medinalis*. *Indian Journal of Agricultural Sciences* 61, 348–349.

Subramanian, S. and Rajeswari, S. (2003) Foliar persistence of entomopathogenic nematodes. In: Singh, R.V., Pankaj, Dhawan, S.C. and Gaur, H.S. (eds) *Proceedings of National Symposium on Biodiversity and Management of Nematodes in Cropping Systems for Sustainable Agriculture held in Jaipur*, India, pp. 193–195.

Sunanda, B.S., Siddiqui, A.U. and Sharma, S.K. (2013) Bio-efficacy of entomopathogenic nematode *Steinernema carpocapsae* against diamondback moth (*Plutella xylostella*) infesting cabbage in India. *International Journal of Nematology* 23, 106–110.

Sylesha, (2003) Potential of entomopathogenic nematodes for insect control in north eastern hill region of India. In: Hussaini, S.S., Rabindra, R.J. and Nagesh, M. (eds) *Current Status of Research on Entomopathogenic Nematodes in India*. PDBC, ICAR, Bangalore, India, pp. 127–134.

Thankamony, S. and Jose, V.T. (2013) Evaluation of entamopathogenic nematodes against *Aetherastis circulata*. *Rubber Science* 26, 123–126.

Thinakaran Ponniah, T. (1997) Mass multiplication of entomopathogenic nematodes *Steinernema glaseri* and *Heterorhabditis indica*. MSc (Ag) dissertation, Department of Nematology, Tamil Nadu Agricultural University, Coimbatore, India.

Umamaheswari, R., Sivakumar, M. and Subramanian, S. (2004) Host range of native entomopathogenic nematodes from Tamil Nadu. *Insect Environment* 10, 151–152.

Umamaheswari, R., Sivakumar, M. and Subramanian, S. (2006) Biocontrol efficacy of entomopathogenic nematodes on *Spodoptera litura* (Lepidoptera: Noctuidae) in blackgram. *Indian Journal of Nematology* 36, 19–22.

Umamaheswari, R., Sivakumar, M. and Subramanian, S. (2008) *In vitro* production of native isolates of *Heterorhabditis indica* and *Steinernema siamkayai*. *Indian Journal of Nematology* 38, 134–137.

Vanitha, L.S., Kumari, M., Jayappa, J. and Chandrashekar, S.C. (2010) Distribution of entomopathogenic nematodes in natural habitats. *International Journal of Plant Protection* 3, 179–181.

Varadarasan, S., Sooravan, T., Chandrasekar, S.S., Ali, M.A. and Thomas, J. (2006) A survey for entomopathogenic nematodes in cardamom growing areas of Kerala and Tamil Nadu. *Journal of Plantation Crops* 34, 398–400.

Vyas, R.V., Yadav, P., Choudhary, R. and Patel, N.S. (1999a) Natural incidence of entomopathogenic nematodes in central north Gujarat. *Indian Journal of Nematology* 30, 86–88.

Vyas, R.V., Patel, N.S. and Patel, D.J. (1999b) Mass production technology for entomopathogenic nematodes, *Steinernema* spp. *Indian Journal of Nematology* 29, 178–181.

Vyas, R.V., Patel, N.B., Yadav, P., Ghelani, Y.H. and Patel, D.J. (2003) Performance of entomopathogenic nematodes for management of gram pod borer, *Helicoverpa armigera*. *Annals of Plant Protection Sciences* 11, 107–109.

Walia, K.K., Walia, R.K. and Bajaj, H.K. (2006) Occurrence, host range, pathogenicity and population build-up of *Steinernema pakistanense* Shahina *et al*., 2001 (Nematoda: Rhabditida). *International Journal of Nematology* 16, 164–168.

Walia, K.K., Walia, R.K. and Nandal, S.N. (2008) Effect of ultra-violet radiation protectants on indigenous isolates of entomopathogenic nematodes. *Journal of Biological Control* 22, 417–424.

Wouts, W.M. (1981) Mass production of the entomogenous nematode *Heterorhabditis heliothids* on artificial media. *Journal of Nematology* 13, 467–469.

Yadav, A.K. and Lalramliana (2012) Soil moisture effects on the activity of three entomopathogenic nematodes (Steinernematidae and Heterorhabditidae) isolated from Meghalaya, India. *Journal of Parasitic Diseases* 36, 94–98.

Yadav, P., Maghodia, A.B. and Vyas, R.V. (2004) Efficacy of microbial bioagents against *Helicoverpa armigera* on chickpea. *International Chickpea and Pigeonpea Newsletter* 11, 41–43.

Yadav, Y.S., Parihar, A., and Siddiqui, A.D. (2008a) Studies on bioefficacy of entomopathogenic nematode against *Helicoverpa armigera*. *Journal of Phytological Research* 21, 131–134.

Yadav, Y.S., Parihar, A. and Siddiqui, A.U. (2008b) Mass multiplication of entomopathogenic nematodes in artificial medium. *Journal of Phytological Research* 21, 157–158.

Yadava, C.P. and Rao, Y.S. (1970) On the effectiveness of entomophilic nematode DD-136 in the biological control of insect pests of rice. *Oryza* 7, 131–136.

19 Status of Entomopathogenic Nematodes in Integrated Pest Management Strategies in Pakistan

Shahina Fayyaz,* Firoza Kazi and Tabassum Ara Khanum

National Nematological Research Centre, University of Karachi, Pakistan

19.1 Introduction

The agricultural sector is the keystone of Pakistan's economy that contributes about 25% to the national gross domestic product (GDP) and which is facing a problem with the low production of many crops. Low productivity is due mainly to the insect pests that cause an annual loss of 15–20 billion rupees in agricultural crops (Manzoor, 1991). Globally, the losses of the total agricultural production due to pests have been estimated at about 14% (Oerke *et al.*, 1994). An enormous increase in the use of pesticides has strengthened crop protection (Oerke, 2006). However, about US$31 billion are spent annually on crop safety (James, 2007). Among the several pest control techniques, chemical pesticides remain the most common method to control crop pests (Khan *et al.*, 2002). In 2006, Pakistan consumed 90,676 t of chemical pesticides at a total cost of US$183 million, as per the exchange rate (Tariq *et al.*, 2007). About 80% of the insecticides used in the country were for the cotton crop only, and their indiscriminate use has disturbed the agroecosystem at a cost of US$195 million yearly (FAO, 2001). The massive application of pesticides has created numerous problems, including adverse effects on beneficial organisms, pesticide residues in the food chain, a hazard to the applicators and causing environmental pollution and groundwater contamination. Moreover, a number of insects have developed resistance to insecticides (Mason and Wright, 1997). In developing countries, pesticide poisoning has been reported in nearly 25 million agricultural workers, including women. Moreover, due to exposure to chemical pesticides, about 37,000 cases of cancer occur annually, besides creating other ecological problems (WHO, 1990).

To minimize the negative effects of synthetic chemical pesticides, plant protection institutions of the country have been undertaking intensive research and have introduced other alternative pest control options such as utilizing biopesticides and setting up biocontrol laboratories. Globally, biopesticides are becoming the mainstream as chemicals and biocontrol agents gain more of the market share every year, because some of these biocontrol measures prevent the economic disruption of food crops. Moreover, biocontrol agents are usually specific to a particular pest. They are an eco-friendly alternative to synthetic chemical insecticides and can also be used when pests have shown resistance to common pesticides.

For sustainable agriculture, biocontrol agents are considered an important tool in pest management strategies. Biological control can play an expected, even more substantial role (Ehlers, 1996). Developments in biotechnology have generated interest and the means necessary to displace the synthetic chemical insecticides that currently dominate the marketplace. Globally, entomopathogenic nematodes (EPNs) are proven to be exceptionally safe, effective and non-polluting biocontrol agents against pests of major crops like eggs, larvae, caterpillars, borers, moths, grubs, beetles, weevils, thrips, maggots, flies, crickets and grasshoppers (Denno *et al.*, 2008). EPNs have acquired significant fame

*Corresponding author. E-mail: shahinafayyaz@gmail.com

due to their non-hazardous impact on the environment, as well as on humans.

Much significance has been placed on EPNs for insect pest control, largely due to their virulence and stimulating characteristics: their broad range of insecticidal activity, their ability to kill hosts within 24–48 h, ease of culture on a large scale, ease of application in the field, and also they are not liable for registration (Kaya *et al.*, 1999; Askary, 2010). In integrated pest management (IPM) programmes, EPNs play a significant role, being non-toxic to non-target animals, specific to their target pest(s) and can be applied with ordinary tools (Shapiro-Ilan *et al.*, 2006). To regulate insect pest populations, EPNs have also been used in diverse agricultural activities (Mráček, 2002; Shapiro-Ilan *et al.*, 2002; Georgis *et al.*, 2006). However, the correct nematode strain for the particular target pest is the basis of successful biological control, and also the application mode of EPNs (Shapiro-Ilan *et al.*, 2002). Their successful use and commercialization makes them effective biological control agents against pests in North America, Australia, Europe and Asia (Koppenhöfer, 2000). Among the 80 EPN species, only 12 species have so far been commercialized globally (Kaya and Koppenhöfer, 1999; Vashisth *et al.*, 2013). Shahina and Maqbool (1996) started initial EPN work at the National Nematological Research Centre (NNRC), University of Karachi. A total of 13 species (6 new species and 7 new records) have so far been described.

The present chapter provides detailed information on indigenous EPNs regarding their distribution, identification, mass production, biology, virulence, formulation, application and commercialization. Their efficacy as a biocontrol agent or biopesticide against several insect pests is also highlighted, emphasizing the need for their inclusion in IPM strategies in Pakistan.

19.2 Integrated Pest Management in Pakistan

In 1971, the Pakistan Agricultural Research Council (PARC) initiated IPM-based research in Pakistan at the Centre for Agriculture and Biosciences International (CABI), Rawalpindi (formerly known as IIBC-PARC research station), on cotton bollworm and cotton whitefly research projects, fruit and vegetable projects in Balochistan and Khyber Pakhtunkhwa and a sugarcane project in Sindh. A ‘Cotton IPM Implementation’ project, based on the training of trainers (TOT) and farmers’ field schools (FFS), was carried out in Punjab by CABI (Poswal and Williamson, 1998).

The Government of Pakistan (2000) developed a National Integrated Pest Management Programme (Nat-IPM) that was operated and guided by the National Agricultural Research Centre (NARC) and Pakistan Agricultural Research Council (PARC), Islamabad, respectively. IPM emphasizes natural pest control strategies for healthy crop production with minimum disturbance of the agroecosystem. IPM is disseminated through the training of facilitators (ToF) and FFS activities that are based on field observations and participation. An FFS-led IPM model has been implemented in Pakistan (Khan *et al.*, 2007). Several national and international organizations related to research and development programmes have participated in the Nat-IPM activities (Ahmed, 2009).

19.3 Biological Control through Integrated Pest Management in Pakistan

Biological control is an element of the IPM strategy and can be defined as a decline in the population of pests, namely insects, mites, weeds and plant diseases, using living organisms or natural enemies. The major objective is to promote and enhance the helpful or beneficial organisms that suppress the target pest’s population by parasitizing them.

IPM technologies for cotton, sugarcane, rice, sunflower and chickpea crops have been developed in Pakistan by the National Agricultural Research System (NARS) and related institutes. Cotton and sugarcane pests have been controlled successfully by rearing and augmenting the biocontrol agents *Trichogramma* sp. and *Chrysoperla* sp. Rice pests have been managed by IPM technology developed and dispersed on farmers’ fields that results in a marked reduction of pesticide application in IPM fields. *Helicoverpa* sp. was managed on a small scale with *Trichogramma* on chickpea, sunflower and cotton. Botanical pesticides have also been developed and used to control important pests (http://parc.gov.pk/).

In the IPM system, insect parasitic nematodes, also called entomopathogenic nematodes, are regarded as promising biocontrol agents against insect pests of agriculturally important crops. These biological agents have many desirable properties; namely, they are highly toxic to the target organisms; can be mass produced easily on a large scale; have a long shelf life; and can be transported safely.

19.3.1 What can biological methods offer to integrated pest management?

The use of biological methods in an integrated management has a number of benefits and has proved their efficacy in IPM in some agricultural crops. These benefits are: reduction in chemical usages; lower production costs; prevention of excessive residue levels in products; and less risk of pests developing resistance to chemicals, minimizing the possibility of pest reappearance.

IPM has become the most widely used tactic for pest suppression among farmers in the developed countries (Norton and Mullen, 1994), and is now also implemented in Pakistan The public sector in Pakistan has concentrated its efforts on promoting IPM strategies for pest control, and has created awareness among farmers by field demonstrations. Farmers are now trained well enough to make the right decision for pest control (Ahmad *et al.*, 2005; Khaliq, 2005).

19.3.2 A list of products available in Pakistan containing entomopathogenic nematodes

A milestone towards biological control activities by EPNs has been established by the National Nematological Research Centre (NNRC) of the University of Karachi. This is the only research centre devoted to biological control through EPNs in Pakistan. At NNRC, the search for new and known species and strains of EPNs has resulted in the identification of 13 species (6 new species and 7 new records) of EPNs on the morphological and molecular level. Accessions numbers (123) were obtained from Genbank of the National Centre for Biotechnology Information (NCBI) in the USA. Mass production, formulation and storage technology has been developed. Moreover, one international patent and eight national patents have been obtained regarding EPN species. The marketing and commercialization of NNRC-EPN products as biopesticides in the field under IPM is in process. A list of NNRC-EPN products is given in Table 19.1 (see also Fig. 19.1).

19.4 An Outline of Studies on Entomopathogenic Nematodes in Pakistan Research Centres

In Pakistan, Shahina and Maqbool (1996) initiated the research for native isolates of insect parasitic nematodes or EPNs. The first evidence of the presence of EPN genera, namely *Steinernema* and *Heterorhabditis*, from soil samples was reported by the University of Karachi. Thereafter, intensive research on every aspect of EPNs, including their extraction, occurrence, geographical distribution, identification through morphological and molecular characterization, biology, virulence, selection of improved strains, mass scale production and formulations, has been carried out at the National Nematological Research Centre, University of Karachi. Besides the NNRC, some basic research on EPNs has also been conducted at the University of Agriculture, Faisalabad.

Table 19.1. Some available products of indigenous Pakistani strains of entomopathogenic nematodes.

Formulation	Nematode species	Product	Company/availability
Sponge	*Steinernema pakistanense*	PAK NEMA-1	NNRC[a]
	S. maqbooli	PAK NEMA-2	NNRC
	S. bifurcatum	PAK NEMA-3	NNRC
	S. balochiense	PAK NEMA-4	NNRC
	Heterorhabditis pakistanense	PAK NEMA-5	NNRC
	S. abbasi	PAK NEMA-7	NNRC
	S. siamkayai	PAK NEMA-20	NNRC
Aliginate gel	*S. siamkayai*	PAK NEMA Gel-22	NNRC
	S. pakistanense	PAK NEMA Gel-25	NNRC

Note: [a]NNRC = National Nematological Research Centre.

Fig. 19.1. Products of indigenous Pakistani strains of entomopathogenic nematodes.

19.4.1 Current developmental status of entomopathogenic nematodes

Of the 13 EPN species detected so far, 6 new species have been described: *Steinernema pakistanense* Shahina *et al*., 2001, *Steinernema asiaticum* Anis *et al*., 2002a, *Steinernema maqbooli* Shahina *et al*., 2013a, *Steinernema bifurcatum* Shahina *et al*., 2014a, *Steinernema balochiense* Shahina *et al*., 2015, and *Heterorhabditis pakistanense* Shahina *et al*., 2016; whereas seven species have been reported as new records: *Steinernema siamkayai* Stock *et al*., 1998, *Steinernema abbasi* Elawad *et al*., 1997, *Steinernema carpocapsae* (Weiser, 1955) Wouts *et al*., 1982, *Steinernema feltiae* (Filipjev, 1934) Wouts *et al*.,

1982, *Steinernema litorale* Yoshida, 2004, *Heterorhabditis bacteriophora* Poinar, 1976, and *Heterorhabditis indica* Poinar *et al.*, 1992, from Pakistan. Moreover, three species of symbiotic bacteria of these EPNs have also been reported, namely *Photorhabdus luminescens* (Boemare *et al.*, 1993), *Xenorhabdus nematophila* and *Xenorhabdus bovenii* (Akhurst, 1980).

Genetic diversity and phylogenetic analysis of all indigenous isolates of *Steinernema* and *Heterorhabditis* on the basis of ITS-1-5.8 S-ITS-2, 12 S rDNA mitochondrial genes, D2-D3 and 28 S was also undertaken at the molecular level. Molecular research of EPNs began with the aim of identifying and exploiting the significant potential as biological agents for the control of insect pests of agriculturally important crops. These endeavours are a step towards the development of biopesticides, which may be produced on a commercial scale for the use of farmers, after successful results.

The current developmental status of EPNs in Pakistan involves extensive research being carried out by the NNRC, University of Karachi, on a morphological and molecular basis to recognize EPNs as significant biocontrol agents against insect pests. The bioassay analysis verified the potential of native EPN strains to manage vital insect pests. Different indigenous isolates of *Steinernema* and *Heterorhabditis* have been evaluated as biological agents against insect pests of important crops, stored grain pests, as well as plant parasitic nematodes in *in vitro* and *in vivo* conditions.

Due to overlapping and variations in the morphological features of EPN species, molecular tools provide considerable help in proper identification, and the assimilation of molecular and morphological characterization offers a more reliable means of assessing species validity. Molecular markers, namely ITS1.5.8S.ITS2, D2-D3, 28S (LSU) rDNA region and 12S rDNA mitochondrial gene, were assessed for their potential in identifying EPNs, which resulted in 123 *Steinernema* and *Heterorhabditis* isolates being amplified successfully by polymerase chain reaction (PCR), purified and sequenced in the molecular laboratory at the NNRC. These sequences were used to construct a phylogenetic tree, based on maximum evolution (ME), neighbour joining (NJ), maximum likelihood (ML), maximum parsimony (MP) and Bayesian inference (BI), to resolve their genetic relationship (Shahina *et al.*, 2010a; Shahina and Mehreen, 2014b, 2015). Shahina *et al.* (2010a) began the molecular characterization of EPNs in Pakistan. As a result, 123 accessions numbers were obtained from Genbank of NCBI in the USA (Table 19.2).

Moreover, one US international patent (US 2010/0197289) and eight national patents (Pak Patent No 140579, 140580, 140622, 141095, 141447, 141641, 141515 and 141837) through the Intellectual Property Organization (IPO) have been achieved so far, and also developed in the genomic library at NNRC. The technology for the mass production, formulation and storage of effective native bioagents (EPNs) as biopesticides has been developed. The commercialization, marketing and utilization of EPNs in the field as a part of IPM are in progress. During the EPN research period, seven scholars completed their dissertations on various aspects of EPNs (Khan, 2009; Tabassum, 2010; Bhatti, 2011; Kazmi, 2012; Aatif, 2013; Mehreen, 2013; Salma, 2013).

19.4.2 *Galleria mellonella* L. cultures

Progeny of nematode isolates were produced on the greater wax moth larvae, *Galleria mellonella*, which were obtained from beehives and reared in the insect rearing laboratory at the NNRC. The insects were reared on modified artificial diet (Bhatti, 2011). A national patent on artificial diet (IPO Patent-141035) was obtained for this research in 2011.

19.4.3 Isolation of entomopathogenic nematodes

EPNs from soil samples were extracted through a *G. mellonella* baiting method (Bedding and Akhurst, 1975). Last-instar larvae of the greater wax moth were placed in 250-ml plastic boxes or containers with moistened soil and kept at room temperature (25 ± 2°C). After 7 days, dead larvae were placed on conventional White traps for culture (White, 1927). The infective juveniles (IJs) emerged from the cadaver. On emergence IJs were harvested, rinsed to remove insect debris and stored in a storage chamber at 17–20°C in a 100-ml flask containing distilled water with a drop of Triton X-100.

19.4.4 Occurrence and distribution of entomopathogenic nematodes

EPNs have been found to occur in all four provinces of Pakistan. Many researchers have undertaken surveillance and prevalence studies of EPNs

Table 19.2. Accession numbers of entomopathogenic nematode strains from Pakistan.

Genus/species	Vegetation	ITS	D2-D3	12S
Steinernema bifurcatum	*Rosa indica*	JX144741	JX068825	JX068826
S. bifurcatum	*R. indica*	JN571088	JQ838179	–
Steinernema maqbooli	*Codiaeum variegatum*	JN157769	JQ838176	JQ838183
S. maqbooli	*C. variegatum*	JQ894824	–	–
S. maqbooli	*C. variegatum*	JX232285	–	–
S. maqbooli	*C. variegatum*	JN571097	JQ838177	–
Steinernema balochiense	*Lycopersicon esculentum*	JX232284		
S. balochiense	*L. esculentum*	JX135547	JX068821	JN571100
Steinernema siamkayai	*Cynodon dactylon*	GU130183	–	–
S. siamkayai	Ornamental	GU130184	–	–
S. siamkayai	*L. esculentum*	GU130185	–	–
S. siamkayai	*Capsicum annuum*	GU130186	–	–
S. siamkayai	*C. annuum*	GU130187	–	–
S. siamkayai	*C. annuum*	JQ894821	–	JN571104
S. siamkayai	*C. annuum*	JX144736	–	–
S. siamkayai	*Psidium guajava*	–	–	JN511707
S. siamkayai	*P. guajava*	JN599145	–	JN599139
S. siamkayai	*P. guajava*	–	–	JN599145
S. siamkayai	*Abelmoschus esculentus*	JQ894822	–	JN599141
S. siamkayai	*R. indica*	JQ894823	–	JN599142
S. siamkayai	*C. dactylon*	JX135554	–	JN599143
S. siamkayai	*Triticum aestivum*	KC625525	–	–
S. siamkayai	*Mangifera indica*	KC625529	–	–
S. siamkayai	*Brassica rapa*	JF892544	–	–
S. siamkayai	*B. rapa*	JF892545	–	JN599138
S. siamkayai	*Momordica charantia*	JN157772		–
S. siamkayai	*P. guajava*	JN571085		JN571102
S. siamkayai	*M. charantia*	JQ894828	–	–
S. siamkayai	*C. dactylon*	JX068815	–	–
S. siamkayai	*C. dactylon*	JX135557	–	–
S. siamkayai	*Achrus zapota*	JX068816	–	–
S. siamkayai	*A. zapota*	JX135556	–	–
S. siamkayai	*A. zapota*	JX135555	–	–
S. siamkayai	Ornamental	KC625529	–	–
S. siamkayai	*L. esculentum*	JX110444	–	–
S. siamkayai	*C. dactylon*	JX068812	–	–
S. siamkayai	*R. indica*	JX110445	–	–
S. siamkayai	*R. indica*	JX110446	–	–
Steinernema abbasi	*Malus domestica*	EF469773	–	–
S. abbasi	*P. guajava*	JN571086	–	JN599140
S. abbasi	*Ficus carica*	JN571096	–	–
S. abbasi	*C. dactylon*	JX068813	–	–
S. abbasi	*C. dactylon*	JX144735	–	–
S. abbasi	*C. dactylon*	JX232283	–	–
S. abbasi	*C. dactylon*	JX135553	–	–
Steinernema pakistanense	*C. annuum*	KC625524	–	–
S. pakistanense	*Gossypium hirsutum*	GU130180	–	JX144739
S. pakistanense	*G. hirsutum*	GU130181	–	–
S. pakistanense	*Saccharum officinarum*	JF775374	–	–
S. pakistanense	*S. officinarum*	JF775375	–	–
S. pakistanense	*C. dactylon*	JN571091	JX068822	JX144738
S. pakistanense	*C. variegatum*	JN157768	–	–
S. pakistanense	*C. dactylon*	JQ894827	–	–

Continued

Table 19.2. Continued.

Genus/species	Vegetation	ITS	D2-D3	12S
S. pakistanense	*C. dactylon*	JX135558	–	–
S. pakistanense	*C. dactylon*	JX232282	–	–
S. pakistanense	Cucurbit	KC625526	–	–
S. pakistanense	*C. dactylon*	KC625527	–	–
S. pakistanense	*C. dactylon*	JX135550	–	–
S. pakistanense	*C. dactylon*	JX068824	–	–
S. pakistanense	*S. officinarum*	JX135545	–	–
S. pakistanense	*C. variegatum*	JX135552	–	–
Steinernema carpocapsae	*Musa acuminata*	GU130182	–	–
Steinernema litorale	*B. rapa*	JF892546	JQ795723	JN599147
Steinernema spp.	*R. indica*	JN157770	–	JQ838181
Steinernema spp.	*C. variegatum*	JN157771	JQ838178	JQ838182
Steinernema spp.	*R. indica*	JN571089	JX068817	JX068819
Steinernema spp.	*Sesamum indicum*	JN571092	–	–
Steinernema spp.	*S. indicum*	–	JX068820	JN599146
Steinernema spp.	*R. indica*	JN571095	–	–
Steinernema spp.	*C. dactylon*	JX135546	–	–
Steinernema spp.	*C. dactylon*	JX144744	–	–
Steinernema spp.	*C. variegatum*	JX135549	–	–
Steinernema spp.	*C. dactylon*	JX144734	–	–
Heterorhabditis bacteriophora	*R. indica*	JX144742	–	–
H. bacteriophora	*T. aestivum*	EF469774	–	JN571106
Heterorhabditis indica	*A. esculentus*	GU130179	–	–
H. indica	*Mangifera indica*	JQ894826	–	–
H. indica	*C. dactylon*	JX135551	–	–
H. indica	*R. indica*	JN157774	–	–
H. indica	*R. indica*	JN571090	–	–
H. indica	*C. dactylon*	JX068814	–	–
H. indica	*C. dactylon*	JX144733	–	–
H. indica	*R. indica*	JN157773	–	–
Heterorhabditis spp.	*R. indica*	JN571087	–	–
Heterorhabditis spp.	*Brassica* spp.	JN571093	–	JN571103
Heterorhabditis spp.	*G. hirsutum*	JN571094	–	JN571099
Heterorhabditis pakistanense	*M. indica*	JX144743	–	–
H. pakistanense	*C. dactylon*	JX144740	KP096496	KM216180

over time (Shahina and Maqbool, 1996; Shahina *et al.*, 1998; Shahina, 2002; Soomro *et al.*, 2002; Tabassum *et al.*, 2005; Tabassum, 2010; Bhatti, 2011; Kazmi, 2012; Mehreen, 2013; Salma, 2013; Shahina *et al.*, 2014b; Shahina and Mehreen, 2014a).

19.4.5 Biodiversity and biogeography of entomopathogenic nematodes

Pakistan has been a rich source of EPNs. In this regard, over 9500 soil samples (from 1996 to 2013) have been collected from varied climatic regions and habitats of Pakistan. Soil analysis showed 17.2% positive soil samples with EPNs. The prevalence and species diversity of nematodes varied in different regions and was influenced by soil types and habitats. EPNs were isolated from coastal areas from Sindh and Balochistan, to the high altitude areas of Khyber Pakhtunkhwa. Initial surveys of the Sindh region showed that *Heterorhabditis* was more prevalent than *Steinernema*, which preferred sandy soils near coastal areas. However, from further extensive and systematic surveys of diverse agroclimatic regions of Pakistan, it has been observed that steinernematids are more prevalent (82.4%) than heterorhabditids (17.6%). Climate, habitat, geographical region and different edaphic factors play an important role in the distribution and abundance of EPNs (Hominick *et al.*, 1996; Stock *et al.*, 1999).

Detailed studies of EPNs have revealed that *Steinernema* is the dominant and diverse genus, represented by ten species (five new and five known species), with *S. pakistanense* the most common and dominant species, while in the genus *Heterorhabditis*, *H. pakistanense*, *H. bacteriophora* and *H. indica* are the only EPN species recorded in Pakistan. These new species and strains are temperature-/heat-tolerant species, except *S. feltiae*, and are suitable to be used in pest management programmes (Shahina and Mehreen, 2014b).

19.4.6 Identified fauna of entomopathogenic nematodes

During the comprehensive taxonomic studies, 13 species of EPNs have been identified on the morphological and molecular level. The identified fauna of EPNs comprised of six species that are new to the scientific world, while seven species are new records in Pakistan.

19.4.7 New species of entomopathogenic nematodes

Indigenous isolates of EPNs have been described from Pakistan: two new species, namely *S. pakistanense* Shahina *et al*., 2001, and *S. asiaticum* Anis *et al*., 2002a, respectively, and later, three new species of *Steinernema*, namely *S. maqbooli* Shahina *et al*., 2013a, *S. bifurcatum* Shahina *et al*., 2014a, and *S. balochiense* Shahina *et al*., 2015, and a new species of *Heterorhabditis*, *H. pakistanense* Shahina *et al*., 2016, were described on the basis of morphological features and molecular characteristics.

19.4.8 *Steinernema asiaticum* Anis, Shahina, Reid and Rowe, 2002a

S. asiaticum was distinguished from all other species in the genus by having the smallest IJ body length (425 (360–450) µm). This places it in the *carpocapsae* group and close to *S. siamkayai* (L= 446 (398–495) µm), *Steinernema tami* (L= 530 (400–600) µm) and *S. carpocapsae* (L = 558 (438–650) µm). *S. asiaticum* is characterized by: very small IJ length (425 (360–450) µm); EP (32 (28–34) µm); tail length (40 (28–46) µm); and E% (78 (60–90)). There are six ridges (i.e. seven incisures) in the lateral field, and horn-like structures on the cephalic region are absent. The first-generation male is characterized by spicule length (68 (61–74) µm), D% (44 (35–57)) and tail mucron, very fine and usually only visible with a scanning electron microscope (SEM) (Fig. 19.2). The female has a vulva with moderately developed double-flapped epiptygma, and the bluntly conical to dome-shaped tail bears a short terminal spine.

Molecular characterization

S. asiaticum was characterized by the restriction fragment long polymorphism (RFLP) profile of the internal transcribed spacer (ITS) region. Isolate 211 from Pakistan had an identical RFLP profile to isolate SSL2 from Sri Lanka (Anis *et al*., 2002a).

Distribution

S. asiaticum was isolated from sandy soil samples collected from soil of grass and ornamental plants in Clifton, Malir and Gulshan-e-Maymar, Karachi, Sindh, Pakistan (Anis *et al*., 2002a) and in sandy soils from the South-west coastal areas of Sri Lanka (Amarasinghe *et al*., 1994) by baiting with *G. mellonella* larvae. No natural hosts were detected in both countries.

19.4.9 *Steinernema balochiense* Shahina, Tabassum, Ali, Solangi, Mehreen and Salma, 2015

A funnel-shaped gubernaculum is the diagnostic character of *S. balochiense*. It belongs to the *carpocapsae* group by having small juveniles. It is also diagnosed by the morphological characters of the IJs and males. The IJ is characterized by a combination of characters: small body length (455 (415–528) µm); distance from anterior end to excretory pore (35 (32–38) µm); pharynx length (90 (85–98) µm); tail length (44.3 (40–51) µm); D% and E% values (39 (36–44)) and (80 (70–92)), respectively. The male is characterized by total body length (1330 (1135–1632) µm), gubernaculum length (44.4 (40–47) µm) (Fig. 19.3), D% (44 (40–51)) and ratio of GS (63.8 (53–75)). *S. balochiense* comes close to *Steinernema nepalense*, but differs from it by having seven lines versus eight lines in the lateral field of infective stage juveniles.

Molecular characterization

S. balochiense is characterized genetically by ITS sequences (Accession No JX135547), D2-D3

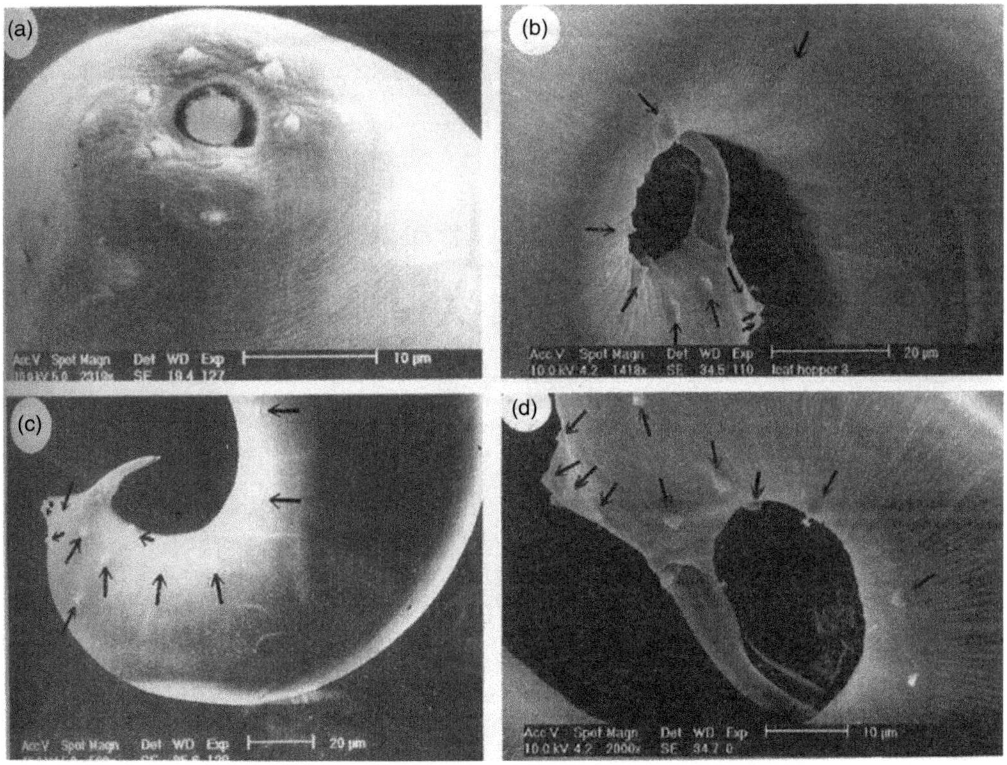

Fig. 19.2. *Steinernema asiaticum* sp. n. (a) Female face view showing six labial and four cephalic papillae; (b, c and d) tail of first-generation male, in lateral view, showing six of the paired precloacal papillae, single midventral papilla and one pair of sublateral papillae; four paired post-cloacal papillae: one pair subventral, one pair subterminal in subventral position, one pair subdorsal, one pair subterminal in subdorsal position and the extended spicules.

region (Accession No JX068821) of rDNA and 12S (Accession No. JN571100) mtRNA gene, which are deposited in the NCBI GenBank (Shahina *et al.*, 2015).

Distribution

S. balochiense was isolated from soil of guava *Psidium guajava* L. in Balochistan, Pakistan (Shahina *et al.*, 2015).

19.4.10 *Steinernema bifurcatum* Shahina, Xun, Qiu, Han, Mehreen, Tabassum and Salma, 2014a

The diagnostic character of *S. bifurcatum* is a bifurcated gubernaculum at both ends in males of the first generation. *S. bifurcatum* belongs to the *bicornutum* group as the IJ has two horn-like structures on the labial region. It can be recognized by the IJ body diameter (22 (20–24) µm), pharynx length (114 (102–130) µm), ratio a (24 (22–25)) and D% (39.7 (33–47)). *S. bifurcatum* is further recognized by the male characters of D% (48 (42–58)) and genital papillae (23 (22 + 1)) (Fig. 19.4d–f).

Molecular characterization

S. bifurcatum is characterized genetically by sequence lengths of ITS (Accession No JX989267), D2-D3 (Accession No JQ838179) of rDNA and 12S (Accession No JX068826) mtDNA. The total sequence length of the ITS region is the smallest among the six closely related species, namely *S. abbasi*, *S. bicornutum*, *S. ceratophorum*, *S. pakistanense*, *S. riobrarave* and *S. yirgalemense* The sequences of ITS, D2-D3 regions of rDNA and 12S

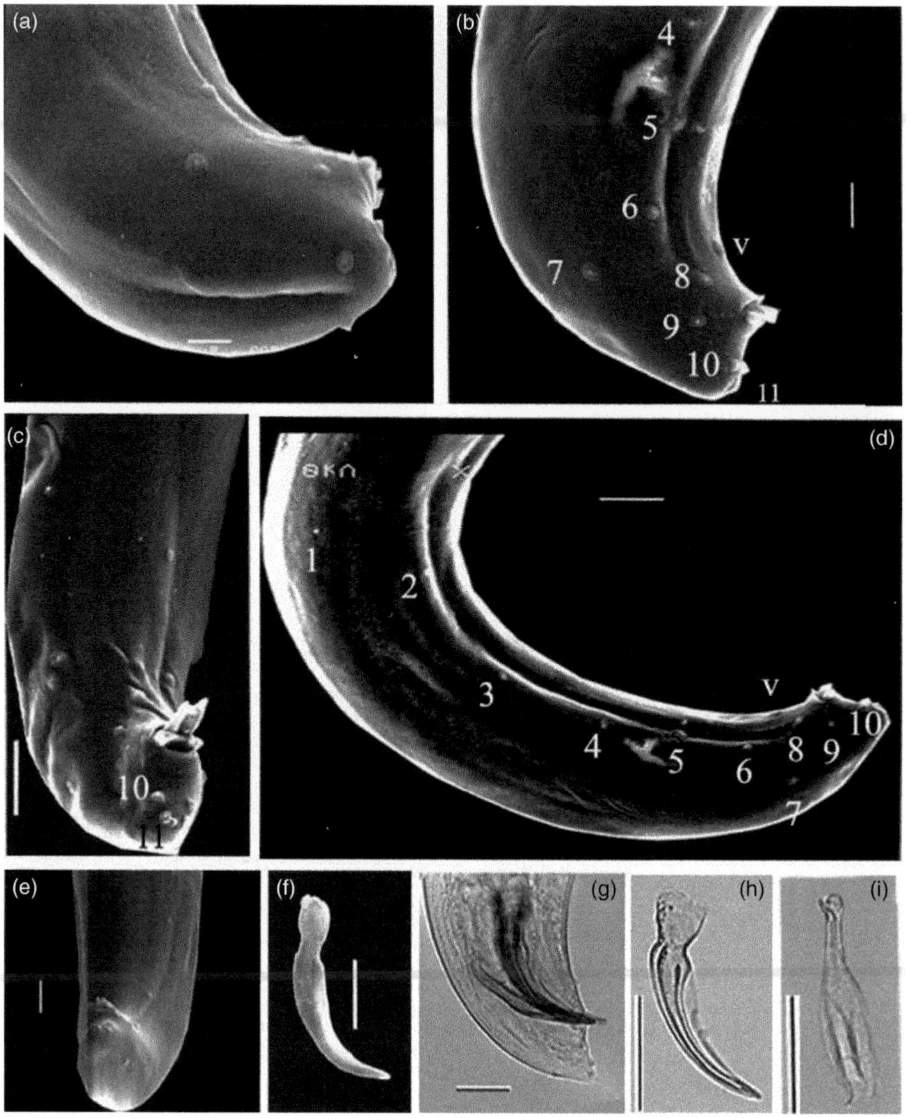

Fig. 19.3. Light and scanning electron microscope photographs of first-generation male specimens of *Steinernema balochiense* n. sp. (a–e) Tail region showing papillae; (f, h) spicules; (g) tail showing mucro; (i) gubernaculum proximal end funnel shaped. (Scale: (a) = 5 µm; (b, c, d, f) = 10 µm; (e) = 20 µm; (g) = 30 µm; (h, i) = 50 µm).

mtDNA are deposited in the NCBI GenBank (Shahina *et al.*, 2014a).

Distribution

S. bifurcatum was collected from soil around the roots of a rose plant (*Rosa indica* L.) at Sheikhupura and Lahore, Punjab, Pakistan (Shahina *et al.*, 2014a).

19.4.11 *Steinernema maqbooli* Shahina, Tabassum, Mehreen and Salma, 2013a

S. maqbooli belongs to the *bicornutum-ceratophorum-riobrave* group on the basis of a horn-like structure on the cephalic region of the third-stage IJ. *S. maqbooli* is characterized morphologically by the distinct features of the IJs and males. The IJs are

(a)

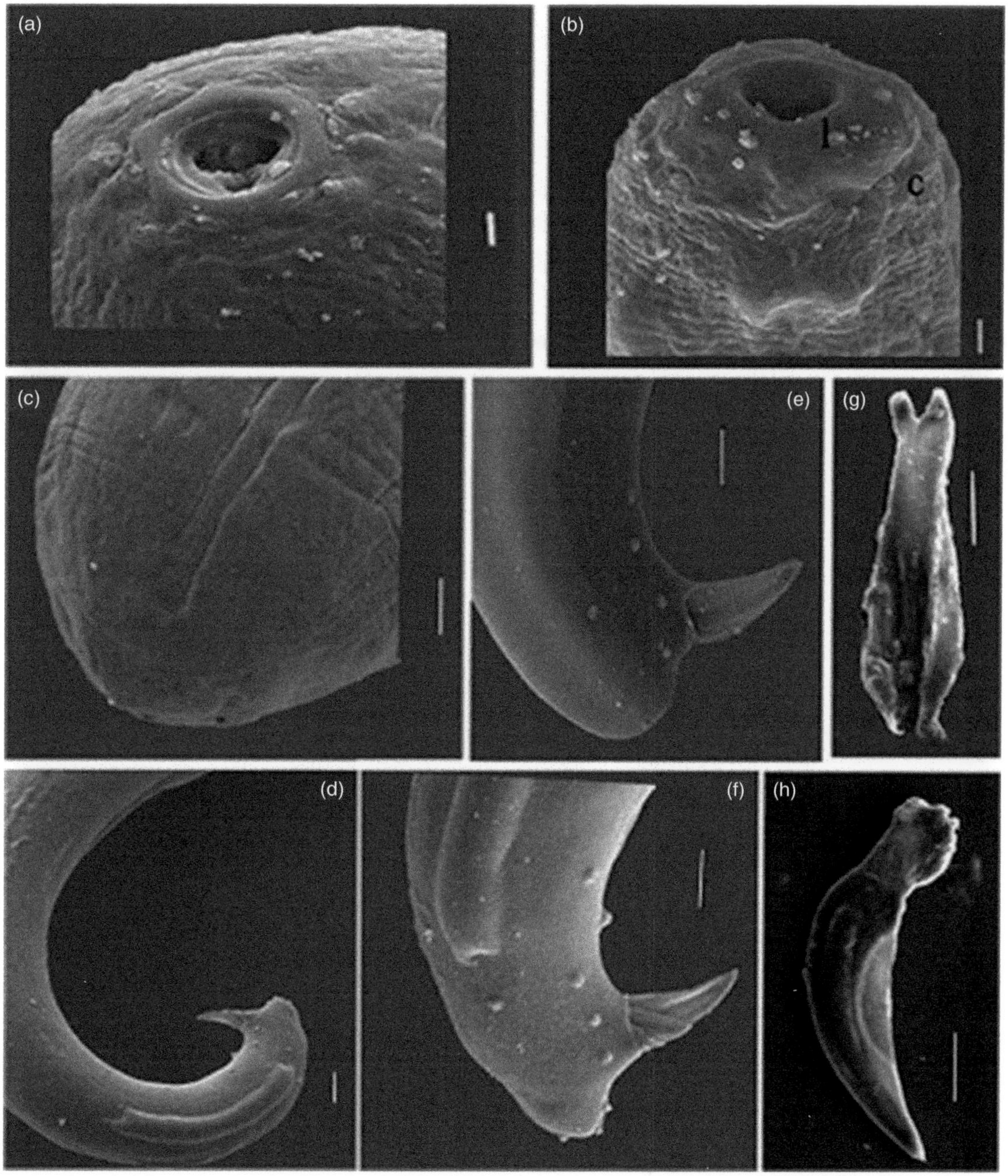

Fig. 19.4. (a) *Steinernema bifurcatum* n. sp. SEM of female and male of first generation. Female. (a, b) Anterior region showing labial (l) and cephalic papillae (c); (c) tail region. Male. (d–f) tail region showing papillae; (g) bifurcated gubernaculum; (h) spicule. (Scale bar: (a, b) = 1 μm; (c–h) = 10 μm.) (b) SEM of *Steinernema bifurcatum* n. sp. Infective juvenile. (a) Entire body; (b, c) anterior region showing labial (l), cephalic (c) papillae and horn-like structures; (d) origin of lateral field in anterior portion; (e, f) lateral field with eight ridges; (g, h) tail region showing gradual decrease in number of lateral ridges. (Scale: (a) = 100 μm; (b–d), (g) = 2 μm; (e, f) = 5 μm; (h) = 10 μm).

(b)

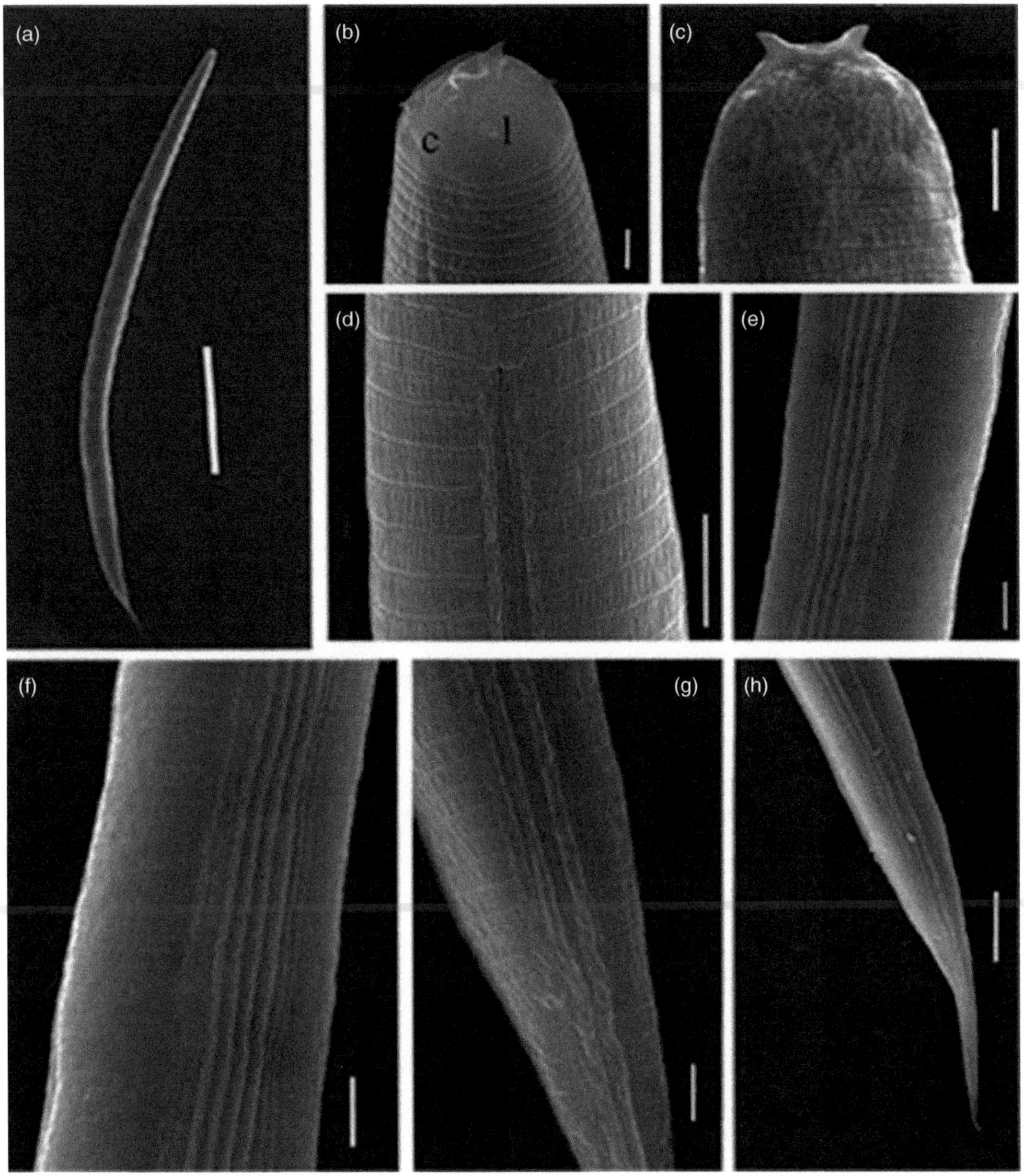

Fig. 19.4. Continued.

characterized by body length (491 (425–603) µm), distance from anterior end to excretory pore (46.6 (40–53) µm), pharynx length (99.3 (90–115) µm), tail length (49.4 (48–53) µm) and D% and E% values (46.8 (45–52)) and (94.3 (88–100)), respectively. The male is recognized by spicule length (61 (53–71) µm); D% (58.4 (49–66)) and SW% (171 (120–210)) (Fig. 19.5).

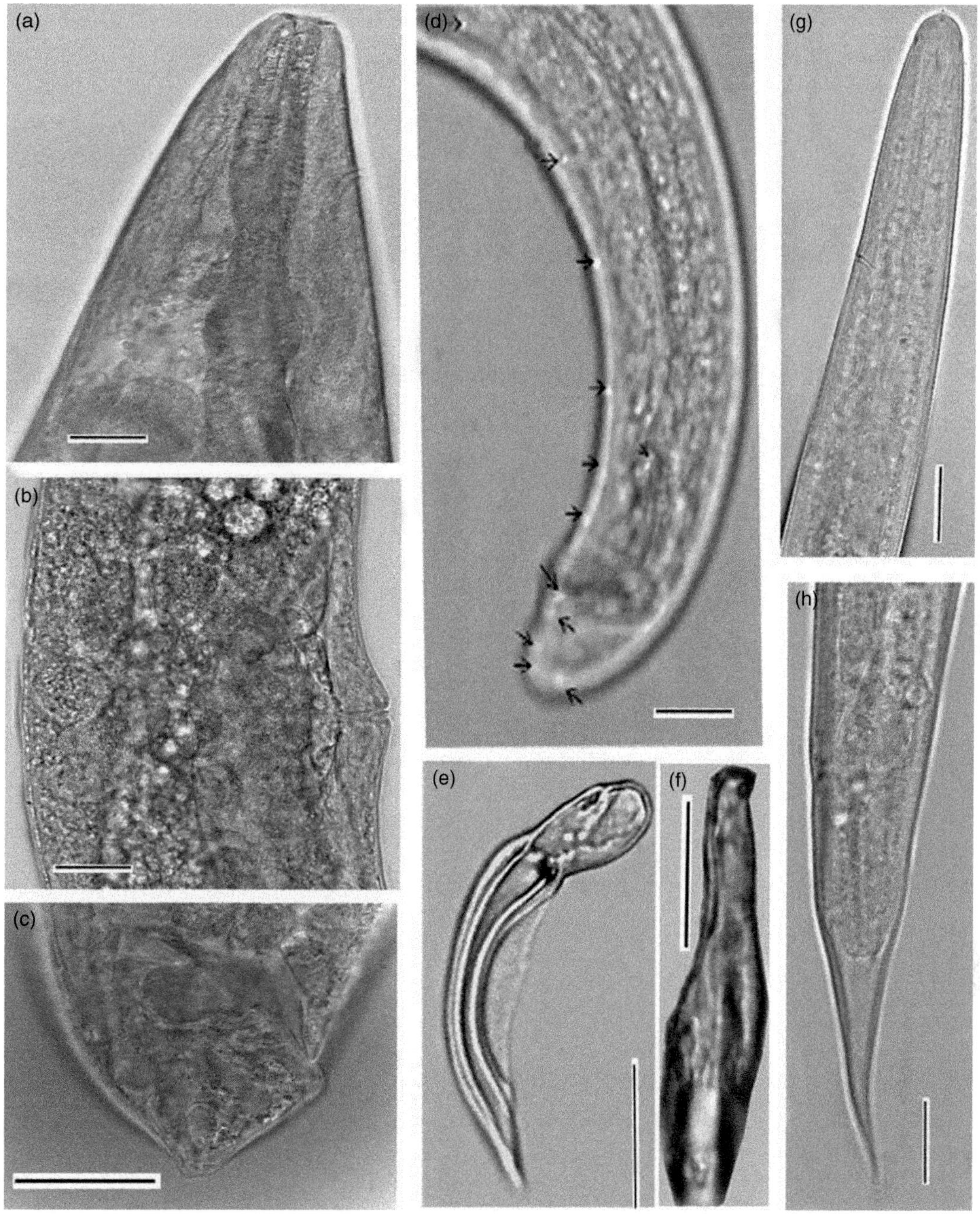

Fig. 19.5. Light micrograph of *Steinernema maqbooli* n. sp. (a–c) Female first generation: (a) pharyngeal region; (b) vulva; (c) tail. (d–f) Male first generation: (d) tail showing papillae; (e) spicules; (f) gubernaculum; (g, h) infective stage juveniles; (g) pharyngeal region; (h) tail region. (Scale (a, b) = 30 µm; (c–e) = 20 µm; (f) = 10 µm; (g, h) = 10 µm).

Molecular characterization

S. maqbooli is characterized genetically by the sequences of ITS (Accession No JN157769), D2-D3 rDNA (Accession No JQ838176) regions and 12S rDNA gene (Accession No JQ838183) sequences, which depict this *Steinernema* species as a distinct and unique entity (Shahina *et al.*, 2013a).

Distribution

S. maqbooli was encountered from *Codiaeum variegatum* L. rhizospheric soil from Quaid-e-Azam Park along the Clifton beach at Karachi, Sindh, Pakistan (Shahina *et al.*, 2013a).

19.4.12 *Steinernema pakistanense* Shahina, Anis, Reid, Rowe and Maqbool, 2001

S. pakistanense is one of a small nominal species characterized by two prominent horn-like structures on the cephalic region on the IJs, the others being *S. abbasi*, *S. bicornutum*, *S. ceratophorum*, *S. riobrave* and *S. yirgalemense*. It can be separated by the body length of the IJ, 683 (649–716) μm as compared to 635 μm or less for *S. yirgalemense*, *S. riobrave* and *S. abbasi* and 760 μm or more for *S. ceratophorum* and *S. bicornutum*. *S. pakistanense* is also characterized by the spicule shape, which differs from that of other species. Other diagnostic characters include: lateral field with eight lines, oesophageal length, excretory pore distance from anterior end in first and second male generations and spicules of a light golden yellow colour (Fig. 19.6).

Molecular characterization

The unique RFLP profile of the ITS region of rDNA confirmed that the species was distinct when compared to a database of 40 other steinernematid species. The sequence of ITS region is deposited at GenBank under Accession No AY48449. This sequence is characterized by its length (flanked by the primers 18S and 26S), 998 bp, ITS1 = 291, ITS2 = 300 bp and its composition: A = 0.22545; C = 0.16734; G = 0.22545; T = 0.31663.

Distribution

Specimens of *S. pakistanense* were encountered from soil samples of Karachi, Sindh, Pakistan. In addition to the type locality, the species has also been isolated from grass and ornamental plants collected at Sonmiani beach and Mubarak village, Balochistan, Pakistan (Shahina *et al.*, 2001). Later, it was also collected from chilli (*Capsicum annuum* L.) at Hamdard University, Sindh, Pakistan. Further, it was collected from all over Pakistan (Shahina and Mehreen, 2014a).

19.4.13 *Heterorhabditis pakistanense* Shahina, Tabassum, Salma and Mehreen, 2016

H. pakistanense belongs to species in the *indica* group on a morphological (average length of IJs) and molecular basis. *H. pakistanense* is morphologically distinct by the male characters: male body length = 819 (720–1013) μm, GS% = 58 (48–65) and the arrangement of genital papillae (on the right side the arrangement of papillae = 1 + 2 + 3 + 3, whereas on the left side the 7th, 8th and 9th papillae are absent); as well as by the IJ characters: length of oesophagus = 117 (113–125) μm, ensheathed tail length = 99 (95–110) μm, E% = 99 (95–107) μm. Hermaphrodites of *H. pakistanense* can be separated from all other species of *Heterorhabditis* except *Heterorhabditis downesi* Stock, Griffin and Burnell, 2002, by having a mucronate tail (Fig. 19.7).

Molecular characterization

H. pakistanense is characterized genetically by the sequences of ITS (Accession No JX144740), D2-D3 rDNA (Accession No KP096496) regions and 12S rDNA gene (Accession No KM216180) sequences. The molecular data of *H. pakistanense* are represented as a new species from Pakistan (Shahina *et al.*, 2016).

Distribution

H. pakistanense was collected from soil around the roots of bermuda grass (*Cynodon dactylon* L.) at Malir, Karachi, Sindh, Pakistan.

19.4.14 New records of entomopathogenic nematodes

Besides new species of EPNs, seven species of EPNs (*Steinernema* and *Heterorhabditis*) were also encountered and these were the first records of EPNs from Pakistan.

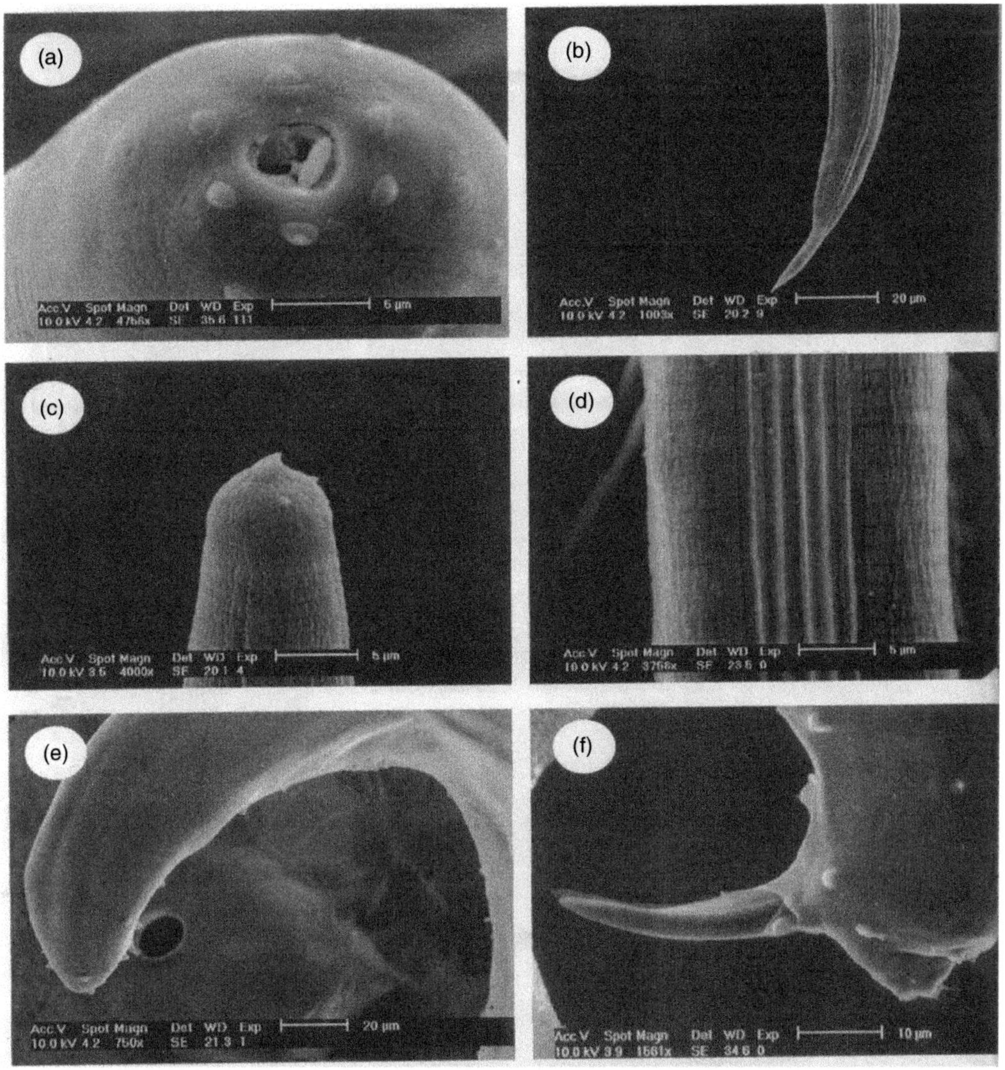

Fig. 19.6. *Steinernema pakistanense* sp. n. Scanning electron micrographs. (a) Female face view; (b) dauer juvenile tail; (c) dauer juvenile anterior end; (d) lateral field at mid-body of dauer juvenile; (e) male posterior end; (f) male tail showing protruded spicules, and genital papillae.

19.4.15 *Steinernema abbasi* Elawad, Ahmad and Reid, 1997

Morphological observations, morphometric data and DNA sequencing of the indigenous isolates come close to the description of *S. abbasi* Elawad *et al.* (1997). It has been re-described from Pakistan by Shahina *et al.* (2007).

Molecular characterization

On a molecular basis (ITS region of the rDNA), it is confirmed that the species was similar to *S. abbasi* by comparing it with the database of other steinernematid species. The sequence of the ITS region is deposited at GenBank under Accession No EF469773 (Shahina *et al.*, 2007).

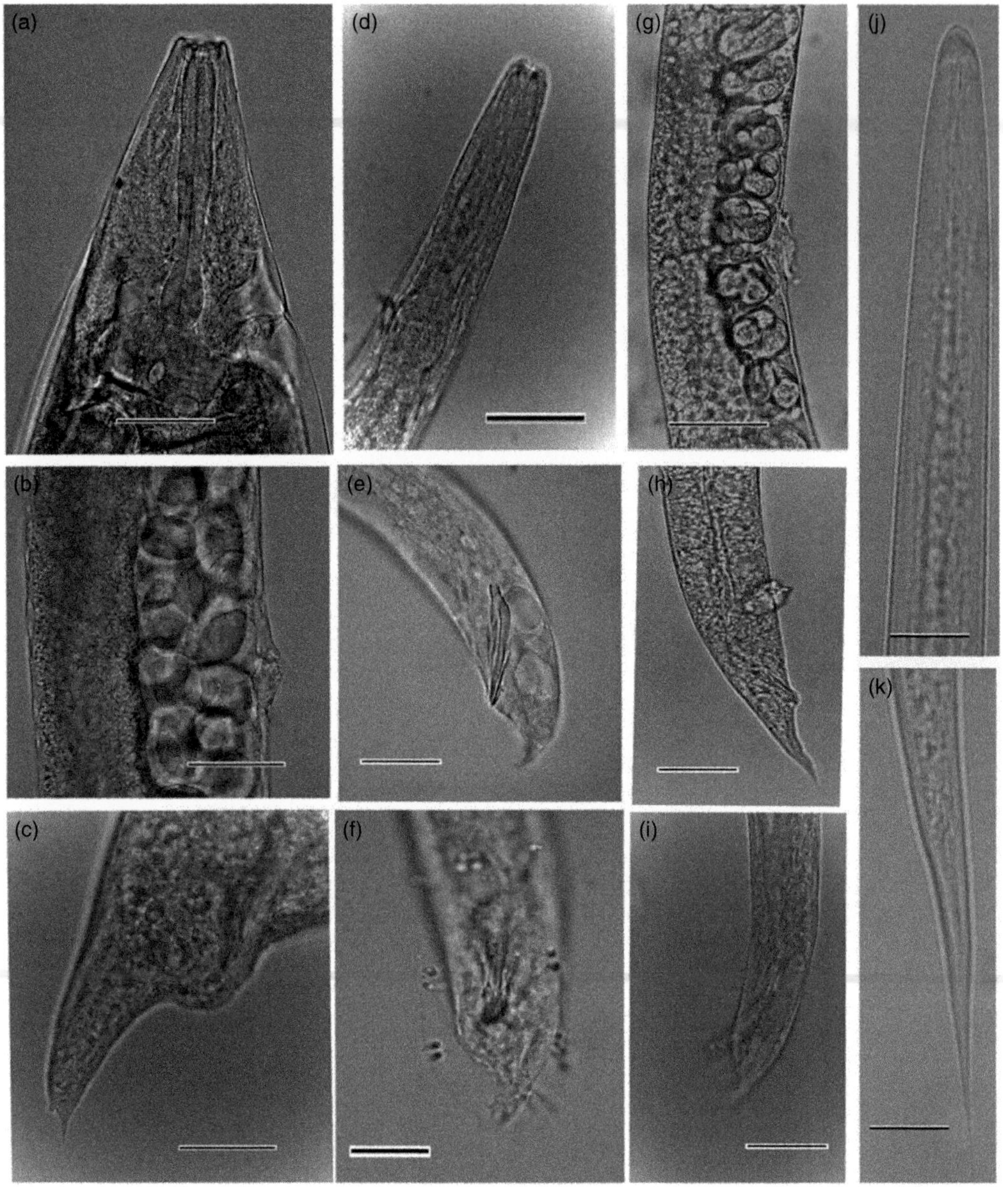

Fig. 19.7. Light photomicrographs of *Heterorhabditis pakistanense* n. sp. (a–c) Hermaphrodite: (a) pharyngeal region; (b) vulval region; (c) tail region showing mucro. (d–f, i) Male: (d) pharyngeal region; (e) tail; (f, i) tail showing papillae. (g, h) Amphimictic female: (g) vulval region; (h) tail. (j, k) Infective stage juvenile: (j) pharyngeal region; (k) tail region. (Scale: (a, d, g) = 20 μm; (e, f, h, i) = 10 μm; (j, k) = 20 μm.)

Distribution

S. abbasi was found in soil around apple trees (*Pyrus malus* L.) collected from Balakot, Khyber Pakhtunkhwa, Pakistan (Shahina *et al.*, 2007). It was also collected from ornamental plants at Hamdard University, Sindh, Pakistan (Bhatti, 2011).

19.4.16 *Steinernema carpocapsae* (Weiser, 1955) Wouts, Mracek, Gerdin and Bedding, 1982

Morphological observations, morphometric data and DNA sequencing of the indigenous isolates were similar to the original description of *S. carpocapsae*

(Weiser, 1955) Wouts *et al.*, 1982. Shahina and Tabassum (2010a) re-described *S. carpocapsae* representing the first report from Pakistan.

Molecular characterization

The sequence of the ITS region of the rDNA of indigenous isolates showed that the species was similar to *S. carpocapsae* when compared to the database of other steinernematid species. The sequence of the ITS region is deposited at GenBank under Accession No GU130182 (Shahina and Tabassum, 2010a).

Distribution

S. carpocapsae was collected from different habitats of Thatta, Sindh, Pakistan (Shahina and Tabassum, 2010a). Later, it was also collected from agricultural fields of sugarcane (*Saccharum officinarum* L.) Thatta, Sindh, Pakistan (Bhatti, 2011).

19.4.17 *Steinernema feltiae* (Filipjev, 1934) Wouts, Mracek, Gerdin and Bedding, 1982

Morphological observations and morphometric data of the specimens were similar to the original description of *S. feltiae*. It was re-described from Pakistan by Tabassum and Shahina (2004a).

Distribution

Anis *et al.* (2002b) encountered the population of *S. feltiae* from *Brassica rapa* from Zarobi, Khyber Pakhtunkhwa, and ornamental fields at Tarbella Dam, Punjab, Pakistan. Bhatti (2011) collected this species from soil around the citrus plantations of Swat, Khyber Pakhtunkhwa, Pakistan.

19.4.18 *Steinernema siamkayai* Stock, Somsook and Reid, 1998

Morphological observations, morphometric data and DNA sequencing of the indigenous isolates were close to the description of *S. siamkayai* given by Stock *et al.* (1998). It was re-described from Pakistan by Tabassum *et al.* (2007).

Molecular characterization

The sequence of the ITS region of the rDNA was compared to all the identified sequence of the *carpocapsae* group deposited in the Genbank. The species was confirmed through DNA sequencing by Dr K.B. Nguyen, University of Florida, USA.

Distribution

S. siamkayai was found from wheat (*Triticum aestivum* L.) fields of Malakand, Khyber Pakhtunkhwa, Pakistan (Tabassum *et al.*, 2007). It was also collected from agricultural fields of cabbage (*Brassica oleracea* L.) of Malir, Karachi, Sindh, Pakistan (Bhatti, 2011).

19.4.19 *Steinernema litorale* Yoshida, 2004

Morphological, molecular and cross-breeding observations indicated that indigenous isolate was similar in the characteristics of S. *litorale* described by Yoshida (2004) and belonged to the *feltiae* group of the *Steinernema* species. It has been re-described by Shahina *et al.* (2012) from Pakistan.

Molecular characterization

S. litorale has the same length of ITS and partial 26S rDNA of *S. litorale*. This is the first sequence of the mitochondrial region of *S. litorale*. The sequence was deposited in Genbank under Accession No JN599147 (Shahina *et al.*, 2012).

Distribution

S. litorale Yoshida, 2004, isolate was collected from vegetable fields of Punjab (Shahina *et al.*, 2012).

19.4.20 *Heterorhabditis bacteriophora* Poinar, 1976

H. bacteriophora is re-described from Pakistan. Morphological characters along with the molecular data are quite close to the description of Poinar, 1976.

Molecular characterization

The sequence of the ITS region of the rDNA of the indigenous isolates represented that the species was similar to *H. bacteriophora* after comparison of the database of all *Heterorhabditis* species. The sequence of the ITS region is deposited at GenBank under Accession No EF469774 (Shahina *et al.*, 2007).

Distribution

H. bacteriophora was collected from wheat (*T. aestivum* L.) plantations at Tando Allahyar, Sindh, Pakistan (Shahina *et al.*, 2007). It was also collected from agricultural fields of guava (*P. guajava* L.) at Winder, Balochistan, Pakistan (Bhatti, 2011). *H. bacteriophora* is a worldwide species. It is also prevalent in Pakistan and is found distributed in different agroclimatic regions of the country (Shahina and Mehreen, 2014a).

19.4.21 *Heterorhabditis indica* Poinar, Karunakar and David, 1992

H. indica has been re-described from indigenous isolates of Pakistan recovered by the soil-baiting technique (Anis *et al.*, 2000). Laboratory studies revealed the high virulence of *H. indica* to larvae of the greater wax moth, *G. mellonella*. The morphological and molecular (RFLP) data showed that this isolate was quite similar to the original description given by Pionar *et al.* (1992).

Molecular characterization

Heterorhabditis isolates from Pakistan were identified as *H. indica* on the basis of RFLPs in the ITS region after comparison to the database of other *Heterorhabditis* species (Anis *et al.*, 2000).

Distribution

In Pakistan, *H. indica* was found widely distributed in different habitats of Karachi, Sindh, Pakistan (Shahina *et al.*, 1998; Anis *et al.*, 2000). Later, it was recovered from mango (*Mangifera indica* L.) at Hamdard University, Sindh (Shahina *et al.*, 2007), and guava (*P. guajava* L.) at Winder, Balochistan, Pakistan (Bhatti, 2011). Further, it has been collected from many localities of Pakistan (Shahina and Mehreen, 2014a).

19.4.22 Isolation, identification and characterization of symbiotic bacteria of entomopathogenic nematodes

The indigenous isolates of EPNs were associated with symbiotic bacteria of two different groups (*Xenorhabdus* and *Photorhabdus*) of the family Enterobacteriaceae (Rahn, 1937) Ewing *et al.*, 1980. Symbiotic bacteria of *Steinernema* belong to genus *Xenorhabdus* Thomas and Poinar (1979) and *Heterorhabditis* belong to *Photorhabdus* Akhurst *et al.* (1996), which were differentiated on the basis of phenotypic, molecular and biochemical test (Shahina *et al.*, 2004; Kazmi *et al.*, 2013). Bacteria associated with IJs of indigenous species of EPNs were isolated from the haemolymph of dead *G. mellonella* larvae by the hanging drop technique (Poinar and Thomas, 1966). The bacteria were cultured on different media (Nutrient, MacConkey and NBTA agar medium) at 25–28°C. For purification, single colonies of bacteria were subcultured on new agar medium plates.

Phenotypic observation was that *G. mellonella* larvae dead by steinernematid nematodes became greyish-yellow, whereas those dead by *Heterorhabditids* became brick red. Symbiotic bacteria of *Heterorhabditis* species were luminescent, while symbiotic bacteria of *Steinernema* species were non-luminescent when observed in the darkroom. Moreover, bacterial isolates have particular morphological characteristics regarding the coloration of bacterial colonies on different media. Analysis of biochemical tests revealed that species of both symbiotic bacteria had a mutuality association with species of both EPN genera.

Shahina *et al.* (2004) reported three species of symbiotic bacteria, namely *P. luminescens* (Boemare *et al.*, 1993), *X. nematophila* and *X. bovenii* (Akhurst, 1980), as the first records of Pakistan. These bacterial symbionts were associated with four EPNs, namely *S. pakistanense* (*X. nematophila*), *S. asiaticum* (*X. nematophila*), *S. feltiae* (*X. bovienii*) and *H. indica* (*P. luminescens*). Later, the bacterial symbionts of three isolates of EPNs, namely *S. abbasi* (*X. indica*), *S. siamkayai* (*Xenorhabdus stockiae*) and *H. bacteriophora* (*P. luminescens*), were identified through phenotypic, SEM and biochemical tests (Kazmi *et al.*, 2013). Molecular methods based on 16S rRNA gene sequences have been used to distinguish several groups within the genus *Photorhabdus* and *Xenorhabdus* (Brunel *et al.*, 1997).

19.4.23 Virulence of heat-tolerant indigenous entomopathogenic nematodes against insect pests

Heat tolerance of IJs of indigenous species of EPNs at different temperatures (20–38°C) was analysed, as it was a major factor influencing nematode mortality and the survival of the IJs. *S. pakistanense* was the most heat-/temperature-tolerant species, with maximum survival rate (95%) at 38°C (Shahina *et al.*, 2005; Salma and Shahina, 2012, 2013). *S. siamkayai* was also a warmth-adapted nematode species and survived (95%) at high temperature (38°C) (Mehreen and Shahina, 2013).

19.4.24 Mass production of entomopathogenic nematodes

EPNs are exceedingly virulent, causing debilitation and sterility (death) of a number of insects, and can be cultured easily *in vivo* and *in vitro* (Ehlers, 2001). They can be mass produced through *in vivo* and *in vitro* methods on artificial diet, and have a great potential to reproduce (Tabassum and Shahina, 2004b; Salma and Shahina, 2012, 2013; Salma, 2013).

In vitro *technology*

The solid-state culture technique is common practice for the production of large quantities of nematodes.

CHICKEN OFFAL MEDIA. Mass rearing of IJs (*Steinernema* and *Heterorhabditis*) was carried out using chicken offal medium on a porous foam substrate and symbiotic bacteria in 500 ml Erlenmeyer flasks in the laboratory according to the method given by Bedding (1984). One to two million to 5–7 million IJs of *Steinernema* and *Heterorhabditis* per 500 ml flask (containing 80 g chicken offal medium) were produced, respectively. The IJs were stored in distilled water with a drop of Triton X-100 (a wetting agent) in a tissue culture flask in a growth chamber at 15–20°C. The infectivity of the IJs reared on monoxenic solid artificial medium remained for up to 2 months (Tabassum and Shahina, 2004b). A long-term storage procedure of EPN species of *Steinernema* and *Heterorhabditis* has been developed by Shahina for use as biopesticides. By this method, 50% EPN infectivity was maintained for 6 months at 10–15°C. One US international patent (US 2011/0197289) was achieved in this regard during 2011.

DIFFERENT MEDIA. *In vitro*, four different media, namely soya flour, wheat flour, lipid media and maize flour, on assemblage culture on *G. mellonella* larvae and lipid modified media were used for the mass culture of eight nematode species. Soya flour medium gave the highest population of nematodes as compared to the other media. The minimum multiplication of nematodes was found in maize flour medium. *S. pakistanense* produced the maximum number of IJs as compared to other nematode species. The production of IJs increased significantly in assemblage media as compared to being cultured separately (Salma and Shahina, 2012).

In vivo *technology*

Standard methods of *in vivo* culturing of nematodes on *G. mellonella* were followed for the maintenance of nematode strains. EPN species belonging to *Steinernema* and *Heterorhabditis* were cultured *in vivo* on three insect species, namely greater wax moth larvae, red flour beetle and pulse beetle, for mass production, exposed to six different dose concentrations of IJs. Maximum numbers of IJs were found on *G. mellonella* larvae. *S. pakistanense* produced an enormous yield of IJs as compared to the other nematode species in all three insects (Salma and Shahina, 2012). Soya flour medium (*in vitro*) and *G. mellonella* larvae (*in vivo*) were exposed to different temperatures. Maximum production of all species was found at high temperatures (32 and 38°C), except for *S. feltiae*, which produced at low temperature (20°C), due to its cold-tolerant habitat. Moreover, production of IJs was significantly higher in *G. mellonella* larvae as compared to soya flour medium. A national patent (IPO Patent No 141447) was obtained for mass production research of EPNs during 2013.

19.5 Application of Entomopathogenic Nematodes in Vegetable and Arable Crop Protection

Protection methodology has been developed for the prevention of crop losses due to pests in the field, preharvest losses, and during storage, postharvest losses. Indigenous isolates of *Steinernema* and *Heterorhabditis* and their symbiont bacteria have been evaluated for their virulence against several insect pests of important crops, stored grain pests, root-knot and other plant parasitic nematodes (Shahina *et al.*, 1999, 2009, 2010b, 2011, 2013b; Shahina and Salma, 2009, 2010, 2011a,b; Shahina and Tabassum, 2010b) under laboratory and field conditions (Table 19.3).

Table 19.3. Major target pests for entomopathogenic nematode/symbiotic bacteria from Pakistan.

Pest group	Scientific name	Common name	Life stages	Nematode/bacteria species[a]
Lepidoptera	*Helicoverpa armigera*	American bollworm	Larva	Sp, Sa, Sf, Hi
	Earias insulana	Spotted bollworm	Larva, pupa, adult	Sp, Sa, Sf, Hi
	E. vittella	Spotted bollworm	Larva, pupa, adult	Sp, Sa, Sf, Hi
	Papillio demoleus	Lemon butterfly	Larva, pupa, adult	Sp
	Leucinodes orbonalis	Brinjal fruit borer	Larva, pupa, adult	Sp
	Etiella zinckenella	Pod borer	Larva, pupa, adult	Sp
	Chilo infuscatellus	Sugarcane shoot borer	Larva	Sp
	Scirpophaga nivella	Sugarcane top borer	Larva	Sp
	Tryporyza incertulus	Rice stem borer	Larva	Sp
	Cnaphalocrocis medinalis	Rice leaf roller	Larva	Sp
	Spodoptera exigua	Lesser armyworm	Larva	Sp
	Pectinophora gossypiella	Pink bollworm	Larva	Sp, Sa, Sf, Hi
	Galleria mellonella	Wax moth larvae	Larva	Hb, Ss, Sp, Sa, Hi, Sab, Sf, Xn, Pl
Coleoptera	*Holotrichia consanguinea*	White grub	Larva	Sp
	Rhynchophorus ferrugineus	Red palm weevil	Egg, larva, pupa, adult	Hb, Ss, Sp, Sa, Hi, Sab, Sf
	Lasioderma serricorne	Cigarette beetle	Larva, pupa, adult	Hb, Ss, Sp, Sa, Hi, Sab, Sf
	Callosobruchus chinensis	Pulse beetle	Larva, adult	Hb, Ss, Sp, Sa, Sab, Hi, Sf
	Sitophilus oryzae	Rice weevil	Pupa, adult	Hb, Ss, Sp, Sa, Sab, Hi, Sf
	Tribolium castaneum	Red flour beetle	Larva, adult	Hb, Ss, Sp, Sa, Hi, Sf
Diptera	*Bactrocera zonata*	Mango fruit fly	Maggot, pupa, adult	Sp, Sa, Sf, Hi
Hemiptera	*Drosicha mangiferae*	Mango mealybug	Adult	Sp
	Planococcus ficus	Vine mealybug	Adult	Xn
Blattodea	*Periplaneta americana*	Cockroach	Adult	Sp
Isoptera	*Macrotermes* spp.	Subterranean termite	Adult	Sp, Pl
	Microtermes spp.	Subterranean termite	Adult	Sa, Ss, Hb
Tylenchida	*Meloidogyne* spp.	Root-knot nematodes	Egg, larva, adult	Sp, Hi
	Helicotylenchus dihystera	Spiral nematode	Adult	Sp
	Hemicriconemoides mangiferae	Sheath nematode	Adult	Sp
	Rotylenchulus reniformis	Reniform nematode	Adult	Sp
	Tylenchorhynchus brevilineatus	Stunt nematode	Adult	Sp
Aphelenchida	*Aphelenchoides sacchari*	Sugarcane nematode	Adult	Sp

Notes: [a]Hb = *Heterorhabditis bacteriophora*; Hi = *Heterorhabditis indica*; Pl = *Photorhabdus luminescens*; Sa = *Steinernema asiaticum*; Sab = *Steinernema abbasi*; Sf = *Steinernema feltiae*; Sp = *Steinernema pakistanense*; Ss = *Steinernema siamkayai*; Xn = *Xenorhabdus nematophila*.

19.5.1 Field and greenhouse environment

Tomato (*Solanum lycopersicum* L.)

To evaluate the effect of EPNs against root-knot nematodes on tomato plant growth parameters, field experiments were carried out. EPNs *S. pakistanense* and *H. indica*, at 5000 IJs/plant in water suspension, and root-knot nematode *Meloidogyne incognita*, at 500 eggs/plant, were applied to tomato cv Moneymaker seedlings.

Greenhouse trials were also carried out to determine the potential of *S. pakistanense* and *H. indica* against root-knot nematode *M. incognita* on tomato plant seedlings. EPNs at 2000 IJs/plant in 10 ml water, together with 500 *M. incognita* eggs, plant growth and other parameters were recorded after 30 days.

In field and greenhouse trials, *S. pakistanense* was more persuasive than *H. indica* against *M. incognita*. Both trials gave more or less similar results. Divergence was significant between treated and untreated plants on different parameters of plants. The plant growth parameters per tomato plants were higher in EPN-treated plants than in treatment inoculated with root-knot nematodes (RKN) only. A significant difference in the female population was found for control plants applied with RKN treatment as compared to plants treated with *H. indica* and *S. pakistanense* with RKN and 1 week prior to RKN inoculation. The total population of nematodes was reduced in plants treated with RKN, *H. indica* and *S. pakistanense*, respectively. The root-knot nematode population/plant was lowered significantly in treated plants than in the control (Shahina and Salma, 2012; Salma, 2013).

The results of field and greenhouse experiments revealed that the population of root-knot nematodes was suppressed by the application of EPNs. It proves that EPNs release some quantities of nematode antagonistic metabolites that protect roots from further penetration of root-knot nematodes.

Other greenhouse experiments on EPNs (*S. asiaticum* (Pakistani isolate), *S. glaseri*, *S. asiaticum*, *H. indica* and *H. bacteriophora* (UK isolates) were conducted to study the effect on the life cycle of *M. incognita* and its development in a tomato (cv Moneymaker) host plant. The results showed all the EPN species at 5000 IJs/plant suppressed the penetration of root-knot nematode (*M. incognita*) in tomato plant roots after 1 month. *S. asiaticum* at 5000 concentration gave best results with less penetration comparable to others (Khan, 2009; Javed *et al.*, 2012). Dead as well as live IJs of the same EPN species reduced the population of *M. incognita* in tomato (Khan *et al.*, 2010).

Okra (*Abelmoschus esculentus* Moench)

Field and greenhouse experiments were carried out to evaluate the effect of EPN nematodes (*S. pakistanense* and *H. indica*) on different insect pests (mealybug, spotted bollworm, American bollworm) and root-knot nematodes on okra. *S. pakistanense* gave significantly better results by suppressing the population of these insect pests. The highest mortality of pests was observed at the highest dose (300 IJs/ml) after 24 h in spotted bollworms (*Earias vittella* and *Earias insulana*), root-knot nematode (*M. incognita*), American bollworm (*Helicoverpa armigera*) and mealybug (*Planococcus ficus*) in descending order (Shahina and Mehreen, 2012).

Brinjal (*Solanum melongena* L.)

An experimental trial was carried out to assess the potential of EPN species *S. glaseri*, *H. indica*, *H. bacteriophora* and *S. asiaticum* (EPNs at 2000/plant) at different time intervals (before, simultaneously and after the inoculation of RKN) against *M. incognita* (at 1000/plant) on a host plant brinjal cv Purple Queen using plant root weight and nematode development parameters (number of galls, egg masses, number of females and reproduction factors). *S. asiaticum* was found most effective at all application times with root-knot nematodes, with the least number of galls and number of egg-mass formation when applied before the inoculation of RKNs. *S. asiaticum* was also the best treatment in reducing the reproduction factors of root-knot nematodes (Aatif, 2013).

Mango (*Mangifera indica* L.)

For the biocontrol of mango fruit fly (*Bactrocera zonata*), experiments were conducted *in vitro* and *in vivo*. Outdoor application of *H. indica* at 5000 IJs/100 ml of water gave better results than other EPN species, namely *S. pakistanense*, *S. asiaticum* and *S. feltiae*, to control mango fruit fly maggots (Bhatti *et al.*, 2005; Bhatti, 2011).

Cotton (*Gossypium hirsutum* L.)

Four EPN isolates, namely, *S. pakistanense*, *S. asiaticum*, *S. feltiae* and *H. indica*, were evaluated to test their infectivity at 1000 and 2000 juveniles/ml of water against the cotton bollworm complex (*H. armigera*, *E. insulana*, *E. vitella* and *Pectinophora gossypiella*) in the field. *S. pakistanense* caused the highest mortality at both dose levels as compared to other EPN species tested in spotted bollworm (*E. insulana* and *E. vitella*), followed by American bollworm and pink bollworm. *H. indica* was the second highest species, showing significant results in infectivity at both doses to all the four insect pests, followed by *S. asiaticum*, whereas mortality was very low to *S. feltiae* at both doses. The results revealed that *S. pakistanense* was highly virulent against all insect pest species of cotton.

EPNs belonging to the genera *Heterorhabditis* Poinar, 1976, and *Steinernema* Travassos, 1927, have a worldwide distribution, including Pakistan, and are found in all soil types and ecologically diverse habitats. They are proven to be exceptionally safe, effective and non-polluting biocontrol agents against the pests of major crops.

IJs of EPNs are capable enough to infect and kill insect pests within 24–48 h with their symbiotic association with bacteria (*Xenorhabdus* and *Photorhabdus*), which produce toxins; IJs are also able to seek out, locate and infect the host by penetration (Shahina *et al.*, 2014c).

Bermuda grass (*Cynodon dactylon* L.)

Live or dead infective stage juveniles of *S. pakistanense* showed efficacy against plant parasitic nematodes in bermuda grass. Dead IJs decreased the population of *Tylenchorhynchus* and *Helicotylenchus* significantly as compared to live IJs. The population of plant parasitic nematodes *Aphelenchoides*, *Hemicriconemoides* and *Rotylenchulus* decreased significantly after 15–30 days. There was no effect on the free-living soil nematode population (Shahina and Tabbasum, 2009).

19.6 Conclusion and Future Prospects

In a short span of time, EPNs have come to be regarded as promising biological control agents or biopesticides of a wide variety of agricultural crop pests and the main element in IPM programmes globally. Many years of field and laboratory experience should serve as a model for the control of agricultural pests through EPNs as an alternative to chemical pesticides that are otherwise hazardous to the environment and human health. Thousands of researchers worldwide are now exploring the potential use of EPNs as biological control to meet the new standards that now demand almost residue-free vegetables and fruit.

The development of proper application products in the control of efficacy, monitoring and timing of application, involvement of farmers in monitoring and decision making is necessary to prevent failures and to make farmers confident with the product; growing markets for new products are usually those that are considered reliable by farmers for managing more species of pests.

IPM is a dynamic system that is adaptable to diverse management approaches. In these approaches, pest management decisions are taken by the individual but are influenced by the diversity of public and private values. In the international market, the withdrawal of approval for the use of agrochemicals on many agricultural crops is increasing the market opportunities for biopesticide products. The most critical factors in reducing the efficacy of EPNs are desiccation tolerance or low humidity level and long-term preservation; therefore, optimized methods are required to meet the growing demand of EPNs in IPM programmes.

Acknowledgements

The authors are thankful to Dr Tarique Hassan Askary, one of the editors of the book, for providing the opportunity to contribute this chapter. We are thankful to the Pakistan Science Foundation, Government of Pakistan, for the financial support through PSF Projects SKU-AGR, 217, SKU-AGR, 285 and PSF-NSLP Project No 65 to carry out the research work on EPNs. The authors are highly indebted to the Ministry of Science and Technology, Government of Pakistan, for providing the support and guidance for Project PSF/ILP/SKU/Bio-047, approved under the Pak-China 17th Protocol on Science and Technology, for the molecular characterization, mass production and formulation of EPNs.

References

Aatif, H.M. (2013) Efficacy of entomopathogenic nematodes, their associated bacteria and toxins on the development of *Meloidogyne incognita* on brinjal. PhD thesis, Faculty of Agriculture, University of Agriculture, Faisalabad, Pakistan, 143 pp.

Ahmad, M., Ali, T., Khan, A.S. and Hamid, A. (2005) Perceived credibility of private sector in introducing IPM technologies with special reference to sugarcane crop in District Faisalabad, Pakistan. *Entomology* 27, 53–56.

Ahmed, I. (2009) Integrated pest management: experiences in Pakistan. *The Blessed Tree* 34–39.

Akhurst, R. (1980) Morphological and functional dimorphism in *Xenorhabdus* spp., bacteria symbiotically associated with insect pathogenic nematodes *Neoaplectana* and *Heterorhabditis*. *Journal of General Microbiology* 121, 303–309.

Akhurst, R.J., Mourant, R.G., Baud, L. and Boemare, N.E. (1996) Phenotypic and DNA relatedness study between nematode symbionts and clinical strains of the genus *Photorhabdus* (Enterobacteriaceae). *International Journal of Systematic Bacteriology* 46, 1034–1041.

Amarasinghe, L.D., Hominick, W.M., Briscoe, B.R. and Reid, A.P. (1994) Occurrence and distribution of entomopathogenic nematodes in Sri Lanka. *Journal of Helminthology* 68, 277–286.

Anis, M., Shahina, F., Reid, A.P. and Maqbool, M.A. (2000) Re-description of *Heterorhabditis indica* Poinar *et al.*, 1992 (Rhabditida: Heterorhabditidae) from Pakistan. *Pakistan Journal of Nematology* 18, 11–27.

Anis, M., Shahina, F., Reid, A.P. and Rowe, J. (2002a) *Steinernema asiaticum* sp. n. (Rhabditida: Steinernematidae) from Pakistan. *International Journal of Nematology* 12, 220–231.

Anis, M., Shahina, F. and Maqbool, M.A. (2002b) Identification of *Steinernema feltiae* Filipjev, 1934 (Rhabditida: Steinernematidae) from Pakistan. *Pakistan Journal of Nematology* 20, 65.

Askary, T.H. (2010) Nematodes as biocontrol agents. In: Lichtfouse, E. (ed.) *Sociology, Organic Farming, Climate Change and Soil Science*. Springer, The Netherlands, pp. 347–378.

Bedding, R.A. (1984) Large-scale production storage and transport of the insect parasitic nematodes *Neoaplectana* spp. and *Heterorhabditis* spp. *Annals of Applied Biology* 104, 117–120.

Bedding, R.A. and Akhurst, R.J. (1975) A simple technique for the detection of insect parasitic rhabditid nematodes in soil. *Nematologica* 21, 109–110.

Bhatti, M.I. (2011) Biological control of red palm weevil, fruit fly and armyworm through EPNs. PhD thesis. National Nematological Research Centre, University of Karachi, Karachi, Pakistan, 266 pp.

Bhatti, M.I., Firoza, K. and Shahina, F. (2005) *Bio-control of Mango Fruit Fly by Entomopathogenic Nematodes*. National Nematological Research Centre, University of Karachi, Karachi. Pakistan Agricultural Research Council, Islamabad, Agricultural Linkages Program (ALP), pp. 16 (in Urdu).

Boemore, N.E., Boyer-Giglio, M.H., Thaler, J.O. and Akhurst, R.J. (1993) The phages and bacteriocins of *Xenorhabdus* spp. symbiont of the nematode *Steinernema* spp. and *Heterorhabditis* spp. In: Bedding, R., Akhurst, R. and Kaya, H. (eds) *Nematodes and the Biological Control of Insect Pests*. CSIRO Publications, East Melbourne, Australia, pp. 13–145.

Brunel, B., Givaudan, A., Lanois, A., Akhurst, R.J. and Boemare, N. (1997) Fast and accurate identification of *Xenorhabdus* and *Photorhabdus* species by restriction analysis of PCR-amplified 16S rRNA genes. *Applied and Environmental Microbiology* 63, 574–580.

Denno, R.F., Gruner, D.S. and Kaplan, I. (2008) Potential for entomopathogenic nematodes in biological control: a meta-analytical synthesis and insights from trophic cascade theory. *Journal of Nematology* 40, 61–72.

Ehlers, R.U. (1996) Current and future use of nematodes in biocontrol: practice and commercial aspects with regard to regulatory policy issues. *Biocontrol Science and Technology* 6, 303–316.

Ehlers, R.-U. (2001) Mass production of entomopathogenic nematodes for plant protection. *Applied Microbiology and Biotechnology* 56, 623–633.

Elawad, A.S., Ahmad, W. and Reid, A.P. (1997) *Steinernema abbasi* sp. n. (Nematoda: Steinernematidae) from the Sultanate of Oman. *Fundamental and Applied Nematology* 20, 435–442.

Ewing, W.H., Farmer, J.J. III and Brenner, D.J. (1980) Proposal of Enterobacteriaceae fam. nov., nom. rev. to replace Enterobacteriaceae Rahn 1937, nom. fam. cons. (Opin. 15, Jud. Comm. 1958), which lost standing in nomenclature on 1 January 1980. *International Journal of Systematic Bacteriology* 30, 674–675.

FAO (2001) *Policy and Strategy for Rational Use of Pesticides in Pakistan*. FAO, The United Nations, Islamabad, Government of Pakistan.

Filipjev, I.N. (1934) Miscellanea Nematologica. 1. Eine neue Art der Gattung *Neoplectana* Steiner nebst Bemerkunge über die systematische stellung der letzeren. *Magazine de Parasitologie de l'Institute Zoologique de l'Academie de l'USSR* 4, 229–240.

Georgis, R., Koppenhöfer, A.M., Lacey, L.A., Bélair, G., Duncan, L.W., *et al.* (2006) Successes and failures in the use of parasitic nematodes for pest control. *Biological Control* 38, 103–123.

Government of Pakistan (2000) Report of National Commission on Agriculture, Islamabad, Pakistan. Available at: http://www.parc.gov.pk/index.php/en/about-narc/95-narc/national-integrated-pest-management-program (accessed 31 October 2016).

Hominick, W.M., Reid, A.P., Boham, A.P. and Briscoe, B.R. (1996) Entomopathogenic nematodes: biodiversity, geographical distribution and the convention on biological diversity. *Biocontrol Science and Technology* 6, 317–331.

James, C. (2007) *Global Status of Commercialized Biotech/GM Crops*: 2007. ISAAA Brief No 37. ISAAA, Ithaca, New York.

Javed, N., Khan, S.A., Haq, I.U., Atiq, M. and Kamran, M. (2012) Effect of *Steinernema glaseri* and *Heterorhabditis indica* on the plant vigour and root-knot nematodes in tomato roots at different densities and time of applications. *Pakistan Journal of Zoology* 44, 1165–1170.

Kaya, H.K. and Koppenhöfer, A.M. (1999) Biology and ecology of insecticidal nematodes. In: Polavarapu, S. (ed.) *Workshop Proceedings: Optimal Use of Insecticidal Nematodes in Pest Management*. Rutgers University, New Jersey, pp. 1–8.

Kaya, H.K., Koppenhöfer, A.M. and Johnson, M. (1999) Natural enemies of entomopathogenic nematodes. *Japanese Journal of Nematology* 28, 13–21.

Kazmi, A.R. (2012) Studies on biodiversity of entomopathogenic nematodes, symbiotic bacteria and their insecticidal properties. PhD thesis. National Nematological Research Centre, University of Karachi, Karachi, Pakistan, 206 pp.

Kazmi, A.R., Tabassum, K.A. and Shahina, F. (2013) Characterization of symbiotic bacteria associated with *Steinernema abbasi* (strain 507), *S. siamkayai* (strain 157) and *Heterorhabditis bacteriophora* (strain 1743). *Pakistan Journal of Nematology* 31, 11–19.

Khaliq, A. (2005) Pakistan. In: Peter, A.C.O. (ed.) *Non-pesticide Methods for Controlling Diseases and Insect Pests*. Asian Productivity Organization, Tokyo, Japan.

Khan, M.A., Iqbal, M., Ahmad, I. and Soomro, M.H. (2002) Economic evaluation of pesticide. Use externalities in the cotton zones of Punjab, Pakistan. *The Pakistan Development Review* 41, 683–698.

Khan, M.A., Iqbal, M. and Ahmed, I. (2007) Environment-friendly cotton production through implementing integrated pest management approach. *The Pakistan Development Review* 46, 1119–1135.

Khan, S.A. (2009) Screening of tomato cultivars against root-knot nematodes and the biological management. PhD thesis. Faculty of Agriculture, University of Agriculture, Faisalabad, Pakistan, 152 pp.

Khan, S.A., Javed, N., Kamran, M. and Atif, H.M. (2010) Suppression of *Meloidogyne incognita* by live and dead entomopathogenic nematodes in tomato. *Pakistan Journal of Nematology* 28, 135–141.

Koppenhöfer, A.M. (2000) Nematodes. In: Lacey, L.A. and Kaya, H.K. (eds) *Field Manual of Techniques in Invertebrate Pathology*. Kluwer Academic Publishers, Dordrecht, The Netherlands, pp. 283–301.

Manzoor, A. (1991) *Agricultural Pests of Pakistan*. Muqtadera National Language, Islamabad, 322 pp. (in Urdu).

Mason, J.M. and Wright, D.J. (1997) Potential for the control *Plutella xylostella* larvae with entomopathogenic nematode. *Journal of Invertebrate Pathology* 70, 234–242.

Mehreen, G. (2013) Distribution and phylogenetic relationship (molecular identification) of entomopathogenic nematodes. PhD thesis. National Nematological Research Centre, University of Karachi, Karachi, Pakistan, 301 pp.

Mehreen, G. and Shahina, F. (2013) Intraspecific variations and phylogenetic relationships among heat tolerant strains of *Steinernema siamkayai* (Rhabditida: Steinernematidae) from Pakistan. *Pakistan Journal of Nematology* 31, 105–123.

Mráček, Z. (2002) Use of entomoparasitic nematodes (EPANS) in biological control. In: Upadhyay, R.K. (ed.) *Advances in Microbial Control of Insect Pests*. Kluwer Academic/Plenum Publishers, New York, pp. 235–264.

Norton, G.W. and Mullen, J. (1994) *Economic Evaluation of IPM Programs*. Virginia Cooperative Extension Service Publication, Delhi.

Oerke, E.C. (2006) Crop losses to pests. *Journal of Agricultural Science* 144, 31–43.

Oerke, E.C., Dehne, H.W., Schonbeck, F. and Weber, A. (1994) *Crop Production and Crop Protection – Estimated Losses in Major Food and Cash Crops*. Elsevier Science, Amsterdam.

Poinar, G.O. Jr (1976) Description and biology of a new insect parasitic rhabditoid, *Heterorhadbitis bacteriophora* n. gen., n. sp. (Rhabdtoida; Heterorhabditidae n. fam.). *Nematologica* 21, 463–470.

Poinar, G.O. and Thomas, G.M. (1966) Significance of *Achromobacter nematophilus* Poinar and Thomas (Achromobacteriacae: Eubacteriales) in the development of the nematode, DD136 (*Neoaplectana* sp.: Steinernematidae). *Parasitology* 56, 385–390.

Poinar, G.O., Karunakar, G.K. and David, H. (1992) *Heterorhabditis indicus* n. sp. (Rhabditida: Nematoda) from India: separation of *Heterorhabditis* spp. by infective juveniles. *Fundamental and Applied Nematology* 15, 467–472.

Poswal, M.A. and Williamson, S. (1998) *Stepping off the Cotton Pesticide Treadmill: Preliminary Findings from a Farmer's Participatory Cotton IPM Training Project in Pakistan*. CABI Bioscience Centre, Rawalpindi, Pakistan.

Rahn, O. (1937) New principles for the classification of bacteria. *Zentralbl Bakterioi Parasitenkd Infektionskr Hyg* 96, 273–286.

Salma, J. (2013) Mass production of entomopathogenic nematodes for biological pest suppression. PhD thesis. National Nematological Research Centre, University of Karachi, Pakistan.

Salma, J. and Shahina, F. (2012) Mass production of eight Pakistani strains of entomopathogenic nematodes (Steinernematidae and Heterorhabditidae). *Pakistan Journal of Nematology* 30, 1–20.

Salma, J. and Shahina, F. (2013) Survival analysis of infective juveniles of Pakistani EPN strains (Steinernematidae and Heterorhabditidae) for mass production. *Pakistan Journal of Nematology* 31, 171–178.

Shahina, F. (2002) Entomopathogenic nematodes in Pakistan. In: Shahina, F., Firoza, K. and Shahzad, S. (eds) *Proceedings of the National Symposium on Nematology*. National Nematological Research Centre, University of Karachi, Karachi, Pakistan, pp. 7–9.

Shahina, F. and Maqbool, M.A. (1996) Isolation of entomopathogenic nematodes (Heterorhabditidae and Steinernematidae) from Pakistan. *Pakistan Journal of Nematology* 14, 135–136.

Shahina, F. and Mehreen, G. (2012) *Biological Control of Root-knot Nematode of Okra (*Abelmoschus esculentus *L.)*. National Nematological Research Centre, University of Karachi, Karachi, Pakistan, 14 pp. (in Urdu).

Shahina, F. and Mehreen, G. (2014a) A geographical survey of endemic entomopathogenic nematodes (Steinernematidae and Heterorhabditidae) as inferred from ribosomal DNA sequence of the ITS1-5.8S-ITS2 region from Pakistan. In: Rafique, M., Ibupoto, K.A., Hasan, S.A. and Mahmood, K. (eds) *Proceedings of Biodiversity of Pakistan.* Pakistan Museum of Natural History, Pakistan Science Foundation, Ministry of Science and Technology, Pakistan, pp. 479–495.

Shahina F. and Mehreen, G. (2014b) Phylogenetic relationship of Pakistani entomopathogenic nematode strains on the basis of ITS-1, 5.8S and ITS-2, D2-D3/ 28S (LSU) rDNA region and 12S rDNA mitochondrial gene. *Pakistan Journal of Nematology* 32, 173–194.

Shahina, F. and Mehreen, G. (2015) Phylogenetic analysis and biological studies of entomopathogenic nematode *Steinernema abbasi* isolates. *Pakistan Journal of Nematology* 33, 1–16.

Shahina, F. and Salma, J. (2009) Laboratory evaluation of seven Pakistani strains of entomopathogenic nematode against a stored product insect pest, pulse beetle (*Callosobruchus chinensis* L.). *Journal of Nematology* 41, 255–260.

Shahina, F. and Salma, J. (2010) Laboratory evaluation of seven Pakistani strains of entomopathogenic nematode against a stored product insect pest, rice weevil (*Sitophilus oryzae* L.). *Pakistan Journal of Nematology* 28, 295–305.

Shahina, F. and Salma, J. (2011a) Pakistani strains of entomopathogenic nematode as a biological control agent against stored grain pest, *Tribolium castaneum*. *Pakistan Journal of Nematology* 29, 25–34.

Shahina, F. and Salma, J. (2011b) Virulence of Pakistani strains of entomopathogenic nematodes against *Lasioderma serricorne*, a stored product insect pest. *International Journal of Nematology* 21, 79–85.

Shahina, F. and Salma, J. (2012) *Biological Control of Root-knot Nematode of Tomato (*Solanum lycopersicum *L.)*. National Nematological Research Centre, University of Karachi, Karachi, Pakistan, 17 pp. (in Urdu).

Shahina, F. and Tabassum, K.A. (2009) Suppression of plant parasitic nematodes in bermuda grass using live or dead entomopathogenic nematodes *Steinernema pakistanense*. *Pakistan Journal of Nematology* 27, 167–178.

Shahina, F. and Tabassum, K.A. (2010a) *Steinernema carpocapsae* (Nematoda: Steinernematidae), a new entomopathogenic nematode isolate from Pakistan. *Pakistan Journal of Nematology* 28, 59–64.

Shahina, F. and Tabassum, K.A. (2010b) Virulence of *Steinernema pakistanense* against different insect species in laboratory condition. *Pakistan Journal of Nematology* 28, 279–284.

Shahina, F., Anis, M., Zainab, S. and Maqbool, M.A. (1998) Entomopathogenic nematodes in soil samples collected from Sindh, Pakistan. *Pakistan Journal of Nematology* 16, 41–50.

Shahina, F., Anis, M., Maqbool, M.A. and Soomro, M.H. (1999) Mortality response of some insects by entomopathogenic nematode *Steinernema* BS266. *Pakistan Journal of Nematology* 17, 125–128.

Shahina, F., Anis, M., Reid, A.P., Rowe, J. and Maqbool, M.A. (2001) *Steinernema pakistanense* sp. n. (Rhabditida: Steinernematidae) from Pakistan. *International Journal of Nematology* 11, 124–133.

Shahina, F., Manzar, H. and Tabassum, K.A. (2004) Symbiotic bacteria *Xenorhabdus* and *Photorhabdus* associated with entomopathogenic nematodes in Pakistan. *Pakistan Journal of Nematology* 22, 117–128.

Shahina, F., Manzar, H. and Bhatti, M.I. (2005) Effect of salt and temperature stress on survival and infectivity of infective juveniles of entomopathogenic nematodes. Proceedings of International Symposium on Biodiversity of Nematodes in Pakistan. *Pakistan Journal of Nematology* 23, 103–109.

Shahina, F., Nguyen, K.B., Kazmi, A.R., Tabassum, K.A. and Firoza, K. (2007) Morphometrics and comments on *Steinernema abbasi* Elawad, Ahmad & Reid, 1997 and *Heterorhabditis bacteriophora* Poinar, 1976 isolates found in Pakistan. *International Journal of Nematology* 17, 213–216.

Shahina, F., Salma, J., Mahreen, G., Bhatti, M.I. and Tabassum, K.A. (2009) Rearing of *Rhynchophorus ferrugineus* in laboratory and field conditions for carrying out various efficacy studies using EPN. *Pakistan Journal of Nematology* 27, 219–228.

Shahina, F., Yan, X., Han, R. and Mahreen, G. (2010a) Identification and phylogenetic analysis of entomopathogenic nematode (Steinernematidae and Heterorhabditidae) from Pakistan as inferred from ribosomal DNA sequences of the ITS1-5. 8s-ITS2 region. *Pakistan Journal of Nematology* 28, 75–93.

Shahina, F., Kazmi, A.R., Tabassum, K.A., Mahreen, G. and Salma, J. (2010b) Evaluation of *Xenorhabdus nematophila* from *Steinernema asiaticum* against wax moth larvae and vine mealy bug. *International Journal of Nematology* 20, 203–210.

Shahina, F., Tabassum, K.A., Salma, J. and Mehreen, G. (2011) Biopesticidal effect of *Photorhabdus luminescens* against *Galleria mellonella* larvae and subterranean termite (Termitidae: Macrotermes). *Pakistan Journal of Nematology* 29, 35–43.

Shahina, F., Tabassum, K.A., Salma, J. and Mehreen, G. (2012) Re-description of *Steinernema litorale* Yoshida, 2004 (Nematoda: Steinernematidae) from Pakistan on the basis of morphometric, ITS1-5.8S-ITS2; 28S rDNA (D2/D3) and mitochondrial rDNA region. *International Journal of Nematology* 22, 81–93.

Shahina, F., Tabassum, K.A., Mehreen, G. and Salma, J. (2013a) *Steinernema maqbooli* n. sp. (Rhabditida: Steinernematidae) a species of the 'bicornutum' group from Pakistan. *International Journal of Nematology* 23, 59–72.

Shahina, F., Tabassum, K.A. and Soomro, M.H. (2013b) *Entomopathogenic Nematodes and their Bacterial Symbionts from Pakistan: Taxonomy, Application, Nematodes and Bacteria*. LAP Lambert Academic Publishing, Germany.

Shahina, F., Xun, Y., Lihong, Q., Richou, H., Mehreen, G., Tabassum, A.K. and Salma, J. (2014a) A new entomopathogenic nematode, *Steinernema bifurcatum* n. sp. (Rhabditida: Steinernematidae) from Punjab, Pakistan. *Nematology* 16, 821–836.

Shahina, F., Firoza, K. and Tabassum, K.A. (2014b) Nematode fauna of Pakistan: an update. In: Rafique, M., Ibupoto, K.A., Hasan, S.A. and Mahmood, K. (eds) *Biodiversity of Pakistan*. Pakistan Museum of Natural History, Pakistan Science Foundation, Ministry of Science & Technology, Pakistan, pp. 117–128.

Shahina, F., Tabassum, K.A. and Habib, M.A. (2014c) Potential of EPN in management of cotton bollworms in Pakistan. *Pakistan Journal of Nematology* 32, 85–90.

Shahina, F., Tabassum, K.A., Ali, S., Solangi, G.S., Mehreen, G. and Salma, J. (2015) *Steinernema balochiense* n. sp. (Rhabditida: Steinernematidae) a new entomopathogenic nematode Pakistan. *Zootaxa* 3904, 387–402.

Shahina, F., Tabassum, K.A., Salma, J., Mehreen, G. and Knoetze, R. (2016) *Heterorhabditis pakistanense* n. sp. (Nematoda: Heterorhabditidae) a new entomopathogenic nematode Pakistan. *Journal of Helminthology* 1–14. DOI: 10.1017/S0022149x16000158.

Shapiro-Ilan, D.I., Gouge, D.H. and Koppenhöfer, A.M. (2002) Factors affecting commercial success: case studies in cotton, turf and citrus. In: Gaugler, R. (ed.) *Entomopathogenic Nematology*. CAB International, Wallingford, UK, pp. 333–356.

Shapiro-Ilan, D.I., Gouge, D.H., Piggott, S.J. and Fife, J.P. (2006) Application technology and environmental considerations for use of entomopathogenic nematodes in biological control. *Biological Control* 38, 124–133.

Soomro, M.H., Shahina, F., Ahmad, I. and Maqbool, M.A. (2002) An alternative to pesticides in Pakistan. In: Shahina, F., Firoza, K. and Shahzad, S. (eds) *Proceedings of the National Symposium on Nematology*. National Nematological Research Centre, University of Karachi, Karachi, Pakistan, pp. 1–5.

Stock, S.P., Somsook, V. and Reid, A.P. (1998) *Steinernema siamkayai* n. sp. (Rhabditida: Steinernematidae) an entomopathogenic nematode from Thailand. *Systematic Parasitology* 41, 105–113.

Stock, S.P., Pryor, B.M. and Kaya, H.K. (1999) Distribution of entomopathogenic nematodes (Steinernematidae & Heterorhabditidae) in natural habitats in California. *Biodiversity and Conservation* 8, 535–549.

Tabassum, K.A. (2010) Taxonomic studies on plant parasitic nematodes of sugar cane fields with special reference to entomopathogenic nematodes. PhD thesis. National Nematological Research Centre, University of Karachi, Karachi, Pakistan, 304 pp.

Tabassum, K.A. and Shahina, F. (2004a) Re-description of *Steinernema feltiae* Filipjev, 1934 (Nematoda: Steinernematidae) from Pakistan. *Pakistan Journal of Nematology* 22, 1–8.

Tabassum, K.A. and Shahina, F. (2004b) *In vitro* mass rearing of different species of entomopathogenic nematodes in monoxenic solid culture. *Pakistan Journal of Nematology* 22, 167–175.

Tabassum, K.A., Shahina, F. and Abid, B. (2005) Occurrence of entomopathogenic nematodes (EPN) in Pakistan. Proceedings of International Symposium on Biodiversity of Nematodes in Pakistan. *Pakistan Journal of Nematology* 23, 99–102.

Tabassum, K.A., Kazmi, A.R. and Shahina, F. (2007) A re-description of *Steinernema siamkayai* Stock *et al.*, 1998 (Rhabditida: Steinernematidae) from Pakistan. *International Journal of Nematology* 17, 217–224.

Tariq, M.I., Afzal, S., Hussain, I. and Sultana, N. (2007) Pesticides exposure in Pakistan: a review. *Environment International* 33, 1107–1122.

Thomas, G.M. and Poinar, G.O. Jr (1979) *Xenorhabdus* gen. nov., a genus of entomopathogenic and nematophilic bacteria of the family Enterobacteriaceae. *International Journal of Systematic Bacteriology* 29, 352–360.

Travassos, L. (1927) Sobre o genera *Oxysomatium. Boletín Biologico* 5, 20–21.

Vashisth, S., Chandel, Y.S. and Sharma, P.K. (2013) Entomopathogenic nematodes – a review. *Agricultural Reviews* 34, 163–175.

Weiser, J. (1955) *Neoaplectana carpocapsae* n. sp. (Anguillulata: Steinernematinae), novy cizopasnik housenek obalece jablecneho, *Carpocapsa pomonella* L. *Vestnik Ceskoslovaenske Zoologicke Spolecynosti* 19, 44–52.

White, G.F. (1927) A method for obtaining infective nematode larvae from cultures. *Science* 66, 302–303.

WHO (1990) *Public Health Impact of Pesticides Used in Agriculture*. WHO, Geneva, Switzerland, 128 pp.

Wouts, W.M., Mracek, Z., Gerdin, S. and Bedding, R.A. (1982) *Neoplectana* Steiner, 1929 a junior synonym of *Steinernema* Travassos, 1927 (Nematoda: Rhabditida). *Parasitology* 4, 147–154.

Yoshida, M. (2004) *Steinernema litorale* n. sp. (Rhabditida: Steinernematidae), a new entomopathogenic nematode from Japan. *Nematology* 6, 819–838.

20 Status of Entomopathogenic Nematodes in Integrated Pest Management Strategies in South Africa

Justin L. Hatting,[1*] and Antoinette P. Malan[2]

[1]*South African Agricultural Research Council-Small Grain Institute (ARC-SGI), Bethlehem, South Africa;* [2]*Department of Conservation Ecology and Entomology, Faculty of AgriSciences, Stellenbosch University, South Africa*

20.1 Introduction

The implementation of biological control in an integrated pest management (IPM) system is not only aimed at suppressing specific target organisms, but also supports the sustainability and resilience of crop and soil management practices. Ironically, the active life of a chemical product may actually be prolonged through reduced selection pressure brought about by the biological component. In a purely biological approach, the ultimate aim would be to arrive at, and capitalize on, the suppressive service/s provided by the environment as experienced in natural ecosystems.

South Africa has a diverse agricultural landscape, with cultivated land covering an area of 2.76 million hectares (Mha), of which 82% is used for commercial farming. Coherently, the geographical locale of these industries is governed by climate, which is an equally diverse phenomenon (Schulze, 1997). Total annual rainfall ranges from less than 150 mm in the west to more than 500 mm over much of the eastern parts, exceeding 1000 mm over parts of the escarpment and along the south-eastern coastal belt. Maximum temperatures during summer can occasionally exceed 40°C (in the north-western and north-eastern low-lying interior), while winter night-time temperatures can drop below freezing over much of the plateau.

Amid such varying climatic conditions, insect pests constitute a major biotic stress factor across many agricultural industries, with several hundred species being acknowledged as major pests (Prinsloo and Uys, 2015). Chemical insect control in South Africa dates back to the late 1800s (Giliomee, 2013), and it is currently an industry with >100 active ingredients registered under the Fertilizers, Farm Feeds, Agricultural Remedies and Stock Remedies Act 36 of 1947 (CropLife South Africa, s.d.). However, given the many challenges that are associated with chemical pesticides (London and Myers, 1995; Quinn *et al.*, 2011), the South African government has, through legislation, either banned or limited the use/sale of several insecticides since the late 1970s. Recent interventions include the withdrawal of monocrotophos (2005), chlorpyrifos (products for home use, 2010), endosulfan (2012), aldicarb (2012) and methyl bromide (2015), fuelling the need for alternative remedies.

Biological control *per se* is not a new concept in South Africa, with this approach (widely employed against invasive plant species) dating back more than a century (Moran *et al.*, 2013). Attempts at insect biocontrol, through the use of parasitoids and/or predators, has an equally rich history, with classic examples including the potato tuber moth, *Phthorimaea operculella* (Zeller) (Lepidoptera:

*Corresponding author. E-mail: hattingJ@arc.agric.za

Gelechiidae), diamond back moth, *Plutella xylostella* (Linnaeus) (Lepidoptera: Plutellidae) and African bollworm, *Helicoverpa armigera* (Hübner) (Lepidoptera: Noctuidae), as well as gramineous lepidopteran stalk borers and forestry-related pests (Neuenschwander *et al.*, 2003, and references within). Awareness of and interest in entomopathogenic nematodes (EPNs) is more recent (early 1950s onwards), with an overview provided here and by Malan and Hatting (2015).

20.2 Integrated Pest Management in South Africa

IPM strategies are currently implemented against several insect pests in South Africa, of which noteworthy targets include: false codling moth (FCM), *Thaumatotibia leucotreta* Meyrick (Lepidoptera: Tortricidae), on citrus; potato leaf miner, *Liriomyza huidobrensis* (Blanchard) (Diptera: Agromyzidae), in potato; grapevine mealybug, *Planococcus ficus* (Signoret) (Hemiptera: Pseudococcidae), on grapes; sugarcane stalk borer, *Eldana saccharina* Walker (Lepidoptera: Pyralidae), in sugarcane; fruit flies (Diptera: Tephritidae) in citrus; and codling moth, *Cydia pomonella* (L.) (Lepidoptera: Tortricidae), in apples and pears. Although none of these strategies currently include EPNs as a standard component, notable progress is being made against some pests.

The present chapter provides an up-to-date account of the status of the research that has been conducted so far on the potential use of EPNs in IPM systems being implemented across diverse agricultural industries of South Africa.

20.3 Application of Biological Control Agents in Integrated Pest Management in South Africa

20.3.1 What biological methods can offer to integrated pest management

The use of an IPM strategy, in which a biological remedy is applied towards the end of the crop growth period (i.e. close to harvest), could serve to address issues pertaining to maximum residue limits (MRLs) and natural enemy mortality. A case in point relates to the African bollworm, *H. armigera*, on wheat. This pest is generally observed on (mature) wheat in the soft dough stage, when populations of many parasitoids and predatory insects, which are commonly associated with aphids, reach their peak. The application of a non-specific pyrethroid at such time could have a severe impact on natural enemy abundance.

The use of EPNs as a biological control agent in IPM also offers potential in the form of recycling, with a long-term residual effect acting against the soil stages of the target pest. Under laboratory conditions, a local nematode, *Steinernema yirgalemense* Nguyen, Tesmafariam, Gozel, Gaugler and Adams (Rhabditida: Steinernematidae), was found to cause 100% mortality after a period of 4 months compared to the zero persistence observed with a second nematode, *Heterorhabditis zealandica* Poinar. When comparing these results to those that were found with *S. yirgalemense* in subsequent field trials, it was evident that the environmental conditions that prevailed in the one vineyard were less conducive to *S. yirgalemense* persistence than were those in a second vineyard, only a few kilometres away (Le Vieux and Malan, 2015).

Manrakhan *et al.* (2014) studied the effect of a naturally occurring EPN population on FCM, finding fruit infestation to be significantly lower in a citrus block where EPNs were conserved, compared to a section of the orchard that was subjected to nematicide applications. Conservation biological control (Barbosa, 1998) and the management of naturally occurring EPN populations in orchards can also boost the benefits that EPNs offer in terms of an IPM approach.

20.3.2 Products containing entomopathogenic nematodes in South Africa

Currently, no locally produced EPN product is available on the South African market. Interest in the mass production and formulation of indigenous EPN species/strains is, however, evident from recent work conducted by Fasemore (2012), Ferreira (2012), Van Zyl (2012), Ferreira *et al.* (2014a, 2016b), Van Zyl and Malan (2014a,b, 2015), Ramakuwela (2015), Ramakuwela *et al.* (2015) and Ramakuwela *et al.*(2016). In South Africa, EPN-based products constitute an 'agricultural remedy', and, as such, are governed by the Fertilizers, Farm Feeds, Agricultural Remedies and Stock Remedies Act 36 of 1947. According to this Act, an 'agricultural remedy' means any chemical substance, or biological remedy, or any mixture, or combination, of any substance, or remedy, that is intended, or offered, to be used for the destruction, control, repelling, attraction, or prevention of any undesired microbe, alga, nematode, fungus, insect, plant, vertebrate, or invertebrate, or any product thereof. The use of EPNs, however, is still in its infancy in this country, as it has largely been impeded by the lack of locally produced/formulated (indigenous) products.

20.4 An Outline of Entomopathogenic Nematode Studies at South African Research Centres

Locally, interest in EPNs dates back to the early 1950s, when EPNs were first recorded from the black maize beetle, *Heteronychus arator* Fabricius (Coleoptera: Scarabaeoidea), collected from a maize (*Zea mays*) field near Grahamstown in the Eastern Cape province (Harington, 1953). No attempts at applying these organisms as a means of pest control were, however, made until the late 1980s, when above-ground applications against larval stages of the sugarcane borer, *E. saccharina*, were initiated by the South African Sugar Association Experimental Station (SASEX) (Spaull, 1988). An investigation by the Agricultural Research Council (ARC-Infruitec) Stellenbosch, into the biological control of the banded fruit weevil, *Phlyctinus callosus* (Schönerr) (Coleoptera: Curculionidae), facilitated soil collection and trapping from 1993 to 1994 from deciduous fruit orchards in the Western Cape province, yielding a heterorhabditid (Basson, 1993) that was later confirmed to be *Heterorhabditis bacteriophora* Poinar (Grenier *et al.*, 1996a,b). A revived interest in applied research into EPNs ensued during the early 2000s (Hatting and Kaya, 2001), with research starting in earnest in 2003 at the ARC-Small Grain Institute (ARC-SGI) near Bethlehem, Free State province, and continuing a year later at Stellenbosch University, in the Western Cape province. Since the description of the first new EPN species for South Africa in 2006 as *Steinernema khoisanae* Nguyen, Malan and Gozel (Rhabditida: Steinernematidae) (Nguyen *et al.*, 2006), several other descriptions and records of occurrence have followed, as are listed in Table 20.1.

The above-mentioned species descriptions emanate from extensive surveys that were conducted across several of the nine South African provinces. Area-wide surveys conducted in this field include those of Malan *et al.* (2006), Molotsane *et al.* (2007) and Hatting *et al.* (2009), while more crop-specific surveys relate to citrus (Malan *et al.*, 2011) and sugarcane (Spaull, 1988, 1991; Pillay *et al.*,

Table 20.1. Entomopathogenic nematodes (*Heterorhabditis* and *Steinernema*) reported from South Africa (species in alphabetical order).

Species	Authority	Distribution (cf. Fig. 20.1)	Reference
Heterorhabditis bacteriophora	Poinar	1	Malan *et al.*, 2006, 2011; Hatting *et al.*, 2009
H. noenieputensis[a]	Ferreira, Van Reenen, Pagès, Tailiez, Malan and Dicks	2	Malan *et al.*, 2014
H. safricana[a]	Malan, Nguyen, De Waal and Tiedt	3	Malan *et al.*, 2008
H. zealandica	Poinar	4	Malan *et al.*, 2006
Steinernema beitlechemi[a]	Çimen, Půža, Nermut', Hattıng, Ramakuwela and Hazir	5	Çimen *et al.*, 2016a
S. biddulphi[a]	Çimen, Půža, Nermut', Hattıng, Ramakuwela and Hazir	6	Çimen *et al.*, 2016b
S. citrae[a]	Stokwe, Malan, Nguyen, Knoetze and Tiedt	7	Stokwe *et al.*, 2011
S. fabii[a]	Abate, Malan, Tiedt, Wingfield, Slippers and Hurley	8	Abate *et al.*, 2016
S. innovationi[a]	Çimen, Lee, Hatting, Hazir and Stock	9	Çimen *et al.*, 2015
S. jeffreyense[a]	Malan, Knoetze and Tiedt	10	Malan *et al.*, 2016a
S. khoisanae[a]	Nguyen, Malan and Gozel	11	Nguyen *et al.*, 2006
S. nguyeni[a]	Malan, Knoetze and Tiedt	12	Malan *et al.*, 2016b
S. sacchari[a]	Nthenga, Knoetze, Berry, Tiedt and Malan	13	Nthenga *et al.*, 2014
S. tophus[a]	Çimen, Lee, Hatting, Hazir and Stock	14	Çimen *et al.*, 2014
S. yirgalemense	Nguyen, Tesmafariam, Gozel, Gaugler and Adams	15	Malan *et al.*, 2011

Note: [a]Type specimen.

2009). Notably, the most prevalent species recorded across all surveys was *H. bacteriophora*. The principal objective of these surveys was to obtain indigenous nematodes for development as bioinsecticide(s) against key South African insect pests. The current occurrence of EPN species in the different provinces of South Africa is indicated below (Fig. 20.1).

The bacterial symbionts associated with South African EPNs comprise: *Xenorhabdus khoisanae* Ferreira, Van Reenen, Gozel, Malan and Dicks (Enterobacteriales: Enterobacteriaceae) with *S. khoisanae* (Ferreira, 2012; Ferreira *et al.*, 2013a); *Photorhabdus heterorhabditis* Ferreira, Van Reenen, Endo, Tailiez, Pagès, Spröer, Malan and Dicks (Enterobacteriales: Enterobacteriaceae) with *H. zealandica* (Ferreira *et al.*, 2014b); and *Photorhabdus luminescens* subsp. *noenieputensis* Ferreira, Van Reenen, Pagès, Tailiez, Malan and Dicks (Enterobacteriales: Enterobacteriaceae) with *Heterorhabditis noenieputensis* (Ferreira *et al.*, 2013b). The bacteria associated with *S. yirgalemense* was confirmed as being *Xenorhabdus indica* (Ferreira *et al.*, 2016a), which was previously described from *Steinernema abbasi* (syn. *S. termophylum*) (Elawad *et al.*, 1997). Notably, the species of bacteria associated with *H. zealandica* from South Africa was identified as *P. heterorhabditis*, while that which was isolated from New Zealand (and Florida) was identified as *Photorhabdus temperata* Fischer-Le Saux, Viallard, Brunel, Normand and Boemare (Enterobacteriales: Enterobacteriaceae) (Ferreira *et al.*, 2014a).

20.5 Application of Entomopathogenic Nematodes in Vegetable Crop Protection

20.5.1 Field environment

To date, no field-related research has been conducted in South Africa on the use of EPNs for the control of vegetable crop pests.

20.5.2 Greenhouse environment

The African bollworm, *H. armigera*, is a truly polyphagous pest in southern Africa (Annecke and Moran, 1982; Prinsloo and Uys, 2015), with some 38 commodities having chemical insecticide registrations listed (CropLife South Africa, s.d.). Research

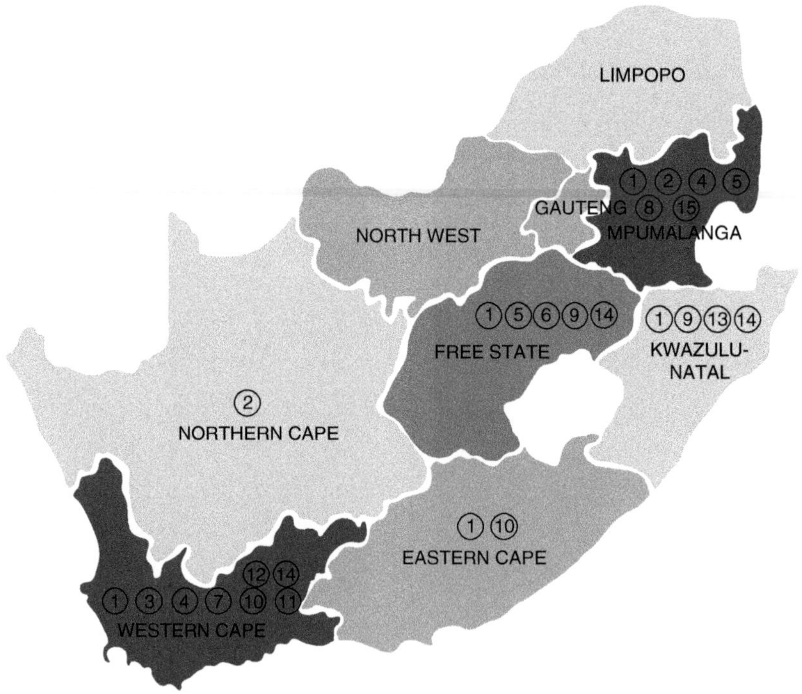

Fig. 20.1. Distribution of entomopathogenic nematodes in South Africa (see Table 20.1 for numerical key descriptions).

conducted by ARC-SGI has considered the use of entomopathogenic fungi against (above-ground) larvae of *H. armigera* (Hatting, 2012), but, as with many other lepidopteran pests, *H. armigera* has a soilborne pupal stage that presents an opportunity for control with EPNs. The focus on biological control has gained global attention, given the extensive reporting that has occurred on the development of resistance to commonly used insecticides. According to the Arthropod Pesticide Resistance Database (Michigan State University, s.d.), this insect has shown resistance to at least 48 insecticide-active ingredients, including dichlorodiphenyltrichloroethane (DDT). The South African scenario raises particular concern, as a large proportion of registered insecticides belong to the synthetic pyrethroid group.

ARC-SGI research has considered the application of EPNs inside their nematode-killed (carrier) hosts (Jansson *et al.*, 1993), which has, in the past, demonstrated improved nematode dispersal (Shapiro-Ilan and Glazer, 1996), infectivity (Shapiro-Ilan and Lewis, 1999), survival (Perez *et al.*, 2003) and efficacy (Shapiro-Ilan *et al.*, 2003). Two South African populations of *H. bacteriophora* (populations SGI 22 and SGI 173) (Hatting *et al.*, 2009), as well as population SGI 148, *Steinernema tophus* Çimen, Lee, Hatting, Hazir and Stock (Çimen *et al.*, 2014), were applied against the pre-pupal and pupal stages of *H. armigera*, using final-instar *Tenebrio molitor* L. (Coleoptera: Tenebrionidae) as carrier host. These isolates originate from the Free State province of South Africa, which is an area in a climatic zone that is defined as 'humid subtropical with summer rainfall and cool (warmest month <22°C)' (Hatting *et al.*, 2009), and where the soil is typically expected to harbour *H. armigera*, given the major crops, such as soybeans, wheat, maize, sunflower, apples and selected vegetables, that are typically cultivated in the region. Briefly, the methodology entailed the following procedure: the establishment, in the greenhouse, of bean seedlings of *Phaseolus vulgaris* L. (Fabales: Fabaceae) in pots, the once-off inoculation of soil by means of an EPN-infected *T. molitor* larva, and the release of two *H. armigera* final-instar larvae per pot after 4, 8, 12 and 16 weeks (i.e. indicating the approximate IJ ages). Eight days after each release, all insects were removed, the mortality noted, and both IJ yield, and day of IJ emergence, recorded (ARC-SGI, unpublished).

The *in vivo* production and application of EPNs via *T. molitor* proved successful against *H. armigera*, with mean pooled mortalities of 80% ± 10% and 78% ± 11% being recorded with 2- and 4-week-old IJs, respectively. A noticeable decline was, however, evident from the 8th week onwards, with only 8% ± 2% (pooled) mortality being recorded with 16-week-old IJs. IJ production in *H. armigera* was generally higher for SGI 173, with a pooled average of 70,564 IJs being produced over the entire duration of the trial, compared with the 23,669 and 7202 IJs produced by SGI 22 and SGI 148, respectively. The study also revealed that the ability of EPNs to produce offspring, as a fitness trait, was correlated negatively with IJ age. In all three isolates, this ability deteriorated, with only SGI 173 producing some offspring (1664 IJs/cadaver) following infection with 16-week-old IJs.

Although surveys to quantify the (natural) level of pre-pupal and pupal stage parasitism have not yet been conducted locally, the phenomenon was explored over a 5-year period in the Lower Rio Grande Valley, Texas, where infection of 9 and 12% in fall armyworm and corn armyworm, respectively, was noted (Raulston *et al.*, 1992). Other studies reported on augmentative attempts that were made against such noctuid species as *Heliothis virescens* (Fabricius) (Bell and Hardee, 1994; Bell, 1995) and *Helicoverpa zea* (Boddie) (Cabanillas and Raulston, 1995, 1996; Feaster and Steinkraus, 1996). Recently, a trial under seemingly challenging environmental conditions, against the soil stages of *H. armigera* on chickpea, found up to 70% moth suppression with *Steinernema* sp. (Rhabditida: Steinernematidae) at a rate of 6×10^9 IJs/ha (Hussain *et al.*, 2014). The data from the ARC-SGI study suggest time-mediated fitness among the IJs, as noted, for mortality and IJ production, as well as for the duration of IJ emergence. Although a similar 'worst-case scenario' (i.e. with no alternative insect host being present in the soil) is unlikely to occur under natural field conditions, the potential deterioration in IJ fitness over time should be taken into account when considering EPN applications against noctuid soil stages.

20.6 Application of Entomopathogenic Nematodes in Arable Crop Protection

20.6.1 Sugarcane

Sugarcane is cultivated mainly in the east of South Africa from Pondoland, Eastern Cape, throughout the coastal areas of KwaZulu-Natal, through neighbouring Swaziland into southern Mpumalanga; covering an area of *c*.340,000 ha locally (Conlong and Way, 2015). This area supports almost 23,900 registered sugarcane growers who annually produce, on average, 18.8 million tons (Mt) of sugarcane (South African Sugar Association, s.d.). Some 35 insect species are listed as being pests of sugarcane in southern Africa (Conlong and Way, 2015), with an overview of their biology and control being provided by Carnegie and Conlong (1994).

The first attempt to control an insect pest in South Africa using EPNs was undertaken by Spaull (1988) against the sugarcane stalk borer, *E. saccharina* Walker (Lepidoptera: Pyralidae). The larval stage of *E. saccharina* is the most injurious insect pest of sugarcane, *Saccharum* spp., in southern Africa (Leslie, 2004; Goebel and Sallam, 2011), causing physical stalk damage, reduced sucrose levels and/or poor plant vigour (Goebel and Way, 2003). To date, control of *E. saccharina* in South Africa has been based on an IPM approach encompassing cultural, genetic, chemical and biological strategies (Carnegie, 1981; Keeping, 2006; Conlong and Rutherford, 2009; Leslie, 2009; Rutherford and Conlong, 2010; Conlong and Way, 2015; Goble *et al.*, 2017).

During the initial study, Spaull (1988) applied (via mistblower) 100,000–1,000,000 IJs, suspended in 200 ml of water, per infested stalk, with up to 56% larval mortality recorded. Although this level of control can be considered as being relatively good, the author expressed concern over the excessive IJ rate and water volume (equivalent to 26,000 l/ha) required for conducting the procedure. In subsequent trials, a surfactant was added, with the number of IJs being reduced to 87,000 in only 57 ml of water (and with the application being performed during the late afternoon), with up to 45% mortality being recorded (Spaull, 1990). Although being a reduction from the 26,000 l/ha previously used (Spaull, 1988), the current 7400 l/ha was still considered to be impractical. At an even lower volume of *c*.3000 l/ha, mortality dropped to <20%, leaving little hope of achieving adequate control with a low water volume. Attention was then shifted to evaluating the performance of more 'cold-tolerant' EPN strains, in the hope of fewer IJs having to be used per hectare. However, the two new strains that were tested (*Heterorhabditis* (strain Widenham) and *Steinernema* (strain Emoyeni)) did not outperform the original Hsp1 strain, with mortalities ranging from 6 to 11% for all three strains when 100,000 IJs were applied in 37 ml of water (plus 0.1% Triton X100) per stalk (Spaull, 1991). Notably, a typical dose-response was observed, rather than a temperature-linked response, with mortalities increasing from 11% to 36% when the concentration of Hsp1 was increased from 100,000 to 500,000 IJs/stalk.

Given the challenges associated with the above-ground application of EPNs to the cane foliage against *E. saccharina*, the delivery of IJs directly to the stubble and/or root stool shortly after harvesting was also considered as a potential strategy, especially in the older fields, where infestation was known to increase progressively. ARC-SGI, in collaboration with the South African Sugar Research Institute (SASRI), initiated laboratory assays to: (i) verify the pathogenicity of several newly collected indigenous EPNs (Hatting *et al.*, 2009; Çimen *et al.*, 2014, 2015) against *E. saccharina*; (ii) measure the potential impact of sugarcane sap (sucrose), as osmolyte, on the survival of IJs; and (iii) investigate the movement and ability of IJs to infect *E. saccharina* inside infested sugarcane stalks. Briefly, the methodologies entailed the use of a piece of filter paper in a Petri dish assay, exposing final-instar *E. saccharina* to five *H. bacteriophora* strains (SGI 32, SGI 43, SGI 180, INF 61, SASRI 75), three *S. tophus* (R 343, SASRI 356, SASRI 426), two *Steinernema innovationi* Çimen, Lee, Hatting, Hazir and Stock (SGI 35, SASRI 198) and one *S. khoisanae* (R 293); the exposure of five EPN strains (SGI 32, SGI 35, SGI 43, SASRI 75, SASRI 426) to sugarcane (cultivar N12) sap concentrations of 50% (diluted with sterile distilled water) and 100% (undiluted sap), with survival checks being undertaken after 24, 48 and 72 h; and the artificial infestation of sugarcane stalks with mid-instar *E. saccharina* larvae by way of vertically drilled holes, followed by topical inoculation with 1 ml EPN (SGI 35) suspension/stalk (Malan and Hatting, 2015).

The six strains, SGI 35, SGI 43, R 293, R 343, SASRI 75 and SASRI 356, caused 100% mortality, with positive recycling in the host. Strains SGI 32, SGI 180 and INF 61 caused only 33% mortality, with no recycling being recorded for SGI 32 and INF 61. An LC_{90} of 44 IJs/larva (fiducial limits: 26–617) was calculated for SGI 35. Seemingly, *E. saccharina* can be considered a susceptible host, as other studies have also reported 100% mortality, especially in the case of steinernematids (Pillay *et al.*, 2009).

EPNs differed in their ability to survive in pure (undiluted) sap, with only SGI 35 showing >80% survival after 72 h. Other studies have also reported different levels of EPN sensitivity to various osmolytes, as was demonstrated with *Steinernema carpocapsae* (Weiser) Wouts, Mráček, Gerdin and Bedding (Glazer and Salame, 2000; Shamseldean *et al.*, 2004; Yan *et al.*, 2010). Seemingly, the selection of EPN species/strains, based on their ability to tolerate the sucrose-rich environment within a sugarcane plant, should be considered when targeting *E. saccharina* and other borer species, such as *Diatraea saccharalis* (Fabricius) (Lepidoptera: Crambidae) (Bellini and Dolinski, 2012).

On day 7 post-application, 65% (n = 37 stalks) of the treated stalks were found to harbour dead *E. saccharina*, of which 14 (58%) larvae showed positive EPN recycling. Control mortality was 5%. These data support the notion of using EPNs to target *E. saccharina* larvae inside the stubble directly after harvest, while the cut wounds are still 'fresh' and relatively uncontaminated. Moreover, the low water volume of only 1 ml/stalk supports the practicality of adopting such a strategy under field conditions. Additional research aimed at optimizing the dose (IJs/ml), the formulation and the application method appears to be warranted.

20.6.2 Grapevine

In South Africa, grapes, *Vitis vinifera* L. (Vitales: Vitaceae), are mostly cultivated in the Western Cape province (mainly in the Hex River and Olifants River valleys, as well as in the Berg River area) and in the Northern Cape province (along the lower Orange River), with 'secondary' production occurring in the provinces of Mpumalanga and Limpopo (Allsopp *et al.*, 2015). In 2013, South Africa exported 260,000 Mt of fresh table grapes (see IndexMundi, s.d.), while 525.6 million l of wine (natural, fortified and sparkling) were exported in the same year (South African Wine Industry Statistics, s.d.). Grapevines are host to 35 insect pests, of which the families Margarodidae, Phylloxeridae, Pseudococcidae, Cicadellidae, Noctuidae and Curculionidae are of particular importance. The grapevine mealybug, *P. ficus* (Signoret) (Hemiptera: Pseudococcidae), is considered a key pest, given its ability to act as a vector for the grapevine leaf roll virus and as a contributor towards honeydew production, thus facilitating sooty mould growth.

The grapevine mealybug, which closely resembles the citrus mealybug, *Planococcus citri* (Risso), was first recorded as a pest in South Africa in 1930 (Allsopp *et al.*, 2015). This mealybug can hibernate on grapevine roots, thereby presenting a window for control by EPNs. Bioassays were conducted to investigate the dose-response of *P. ficus* to *H. zealandica*, *S. yirgalemense* and a commercially produced *H. bacteriophora*, with LC_{50}, LC_{90} values of 19, 82; 13, 80; and 36, 555 IJs/insect, respectively. The movement of both *H. zealandica* and *S. yirgalemense* through a soil profile was confirmed, using a soil column assay, with mealybug mortalities of 82% and 95%, respectively (Le Vieux and Malan, 2013b). These authors also contained *P. ficus* in pierced Eppendorf tubes, buried at a depth of 150 mm in the soil, to which *S. yirgalemense* was applied, with up to 50% mortality being recorded after an exposure period of 48 h in the vineyard. Also, up to 70% IJ persistence was measured after 12 weeks in one vineyard. Notably, of the cadavers that were present 6 days after nematode infection, 49% of the *H. zealandica* and 60% of the *S. yirgalemense* infected cadavers were still intact, suggesting the presence of scavenger deterrent factors (Le Vieux and Malan, 2015). Olfactometry tests showed the higher attraction of *S. yirgalemense* to damaged *V. vinifera* roots than to *P. ficus*, further underscoring the potential use of EPNs against this pest (Le Vieux and Malan, 2015).

20.7 Other Applications of Entomopathogenic Nematodes in South Africa

20.7.1 Parks and urban tree plantings, municipal forests

To date, no research has been conducted in this regard.

20.7.2 Orchards

Pome fruits

Apples and pears are produced mainly in the Western and Eastern Cape provinces, with some apple orchard expansion also being evident in the Free State and Mpumalanga provinces. During the 2009/10 season, apples contributed approximately 34% (R2.9 billion) of the total gross value for deciduous fruits (R8.8 billion) from South Africa (Republic of South Africa. Department: Agriculture, Forestry and Fisheries, 2011). Notably, the European export market accounted for 42% (128,000 t) of total South African apple exports in 2010. This market, as with many others around the globe, is 'sensitive' to pesticide residues, underscoring the need for alternative, environmentally sound pest control practices.

CODLING MOTH, *CYDIA POMONELLA* (L.) (LEPIDOPTERA: TORTRICIDAE). The principal insect pest affecting apples and pears in South Africa is the codling moth, *C. pomonella* (L.) (Lepidoptera: Tortricidae) (Barnes, 1991; Giliomee *et al.*, 2015; Pringle *et al.*, 2015). Interest in targeting this insect with EPNs resides in the fact that *C. pomonella* overwinters as a diapausing larval population from June to August (i.e. wintertime in the southern hemisphere). With the onset of spring, the larvae again turn into pupae, from which the moths exit in late spring/early summer. These moths lay their eggs on the young fruit, wood and/or leaves, from which the larvae hatch and infest the fruit. The frass-filled tunnels render the fruit unmarketable (Welter, 2008). In autumn, directly prior to harvest, the final larval instar moves from the fruit to such cryptic habitats as pruning wounds, spurs and the trunk, as well as to debris around the tree, where they enter diapause inside silken cocoons (Riedl *et al.*, 1998; Cossentine *et al.*, 2004; Pringle *et al.*, 2015). IPM strategies include the use of chemical insecticides, pheromone traps (monitoring, mating disruption, 'attract and kill'), the sterile insect technique and such horticultural practices as fruit thinning and clean picking at harvest (Pringle *et al.*, 2015). Odendaal *et al.* (2015) provide a current review on the control of codling moth in South Africa, with special emphasis on the use of EPNs.

One strategy of targeting *C. pomonella* larvae with EPNs entails the treatment of bulk bins prior to the fruit being placed in cold storage, and before the empty bins (as a source of reinfestation) are moved into the orchards for harvesting. To test the potential of EPNs as a bin treatment, artificially infested 'mini' bins were treated with *H. zealandica* at a rate of 100 IJs/ml (De Waal *et al.*, 2010). Maximum mortality was achieved when the bins were pre-wet for at least 1 min, and then maintained at maximum humidity (tarping; >95% relative humidity (RH)) post-treatment for at least 3 days. The use of such adjuvants as Reversal 10™ further enhanced the efficacy of the procedure, thus demonstrating the potential of EPNs for bin disinfection using EPNs under optimal conditions. Another strategy that was investigated by De Waal *et al.* (2011b) entailed using *H. zealandica* in combination with mulches (pine chips, wheat straw, pinewood shavings, blackwood and apple wood chips) to control diapausing larvae. Larvae were confined in mesh cages filled with different mulches, yielding up to 88% mortality (i.e. with the pinewood shavings), although again requiring high humidity (>95%) for at least 3 days. At low temperatures (<15°C), the levels of control remained <48%, as opposed to the higher mortality that was recorded at temperatures between 20 and 25°C.

In an attempt to screen a relatively large number of EPN isolates against *C. pomonella*, laboratory assays were performed with *H. zealandica*, *Steinernema citrae* Stokwe, Malan, Nguyen, Knoetze and Tiedt, *S. khoisanae* (J96, SF87), *S. yirgalemense* and *Steinernema jeffreyense* Malan, Tiedt and Knoetze. At optimum laboratory conditions, the codling moth was found to be highly susceptible to all nematode isolates, with 78–100% mortality being recorded at a (low) concentration of 50 IJs/insect. At suboptimal temperature regimens (i.e. 10 h at 17°C and 14 h at 12°C), however, EPN performance was severely compromised, causing <3% host mortality (De Waal *et al.*, 2011a). Overall, the main constraints remain the need to maintain adequate moisture levels (i.e. measured as water activity (aw); De Waal *et al.*, 2011b) and suitable temperatures. In order to address the aw-factor (i.e. amount of free water), De Waal *et al.* (2013) investigated the addition of a superabsorbent polymer, Zeba®, on the performance of *H. zealandica*, thus enabling the EPNs to infect codling moth larvae at an aw of ≥0.92 (aw_{50} = 0.94 and aw_{90} = 0.96). The addition of Zeba® to nematode suspensions improved the level of control obtained at 60% and 80% RH in the laboratory, as well as enhancing the survival, and the infection ability, of EPNs in the field.

Field trials were conducted by Odendaal *et al.* (2016a, b) using cages filled with apple bark as a containment method, with the application of exotic *Steinernema feltiae* (Filipjev) Wouts, Mráček, Gerdin and Bedding, as well as *H. bacteriophora* and an endemic *S. yirgalemense*, being done by means of dipping the cages rather than spraying, so as to prevent biological pollution. After a 24 h exposure to field conditions, as experienced in the Western Cape province, it was concluded that both *S. feltiae* and *S. yirgalemense* were capable of giving >70% control, with *S. feltiae* being the preferred candidate for codling moth control and *H. bacteriophora* the least. In a field spray trial,

S. jeffreyense proved to be more effective than was either *S. yirgalemense* or *H. bacteriophora* (Odendaal *et al.*, 2015). Comparing results from the field with simulated results in the laboratory, it was concluded that relative humidity was the most important environmental factor influencing field results.

BANDED FRUIT WEEVIL, *PHLYCTINUS CALLOSUS* (SCHÖNHERR) (COLEOPTERA: CURCULIONIDAE). The banded fruit weevil, *P. callosus* (Schönherr) (Coleoptera: Curculionidae), which is indigenous to South Africa, is an important pest of apples, although also being noted on peaches, nectarines, grapevine and berries, as well as on various broad-leaved weeds and ornamentals (Bredenhand *et al.*, 2010; Prinsloo and Uys, 2015). This weevil is well adapted to the Mediterranean climate of the Western Cape province, where apples can suffer up to 40% damage (Witt *et al.*, 1995). The eggs are laid either on plant material or in organic matter on/just below the soil surface, with the subterranean larval stages feeding on the host plant roots (Barnes and Pringle, 1989; Pringle *et al.*, 2015). The majority of the larval stages tend to stay in the top 10 cm of soil during the winter months (Barnes, 1989). Pupation lasts approximately 14 days, with the adults emerging during late spring and early summer to migrate, as flightless adults, up the tree trunks to reach the available fruit (Barnes, 1989; Barnes and Giliomee, 1992). Chemical control of *P. callosus* is challenged by the development of tolerance to pyrethroids, with an indication of the occurrence of cross-tolerance to acephate (Barnes *et al.*, 1994, 1996). Trunk barriers are only used for monitoring purposes, as such use is very labour-intensive.

A protracted subterranean larval stage (lasting from 2 to 5 months, depending on the season) presents an opportunity for control using EPNs. Assays with two *H. bacteriophora* isolates and one *H. zealandica*, at a concentration of 400 IJs/insect and a 4-day exposure time against both adults and larvae, were performed by Ferreira and Malan (2014). The authors concerned recorded mortalities between 41% and 73% against larvae (Fig. 20.2), and between 13% and 35% against adults, with *H. zealandica* causing the highest mortality.

A strategy whereby EPNs are applied to the soil during winter and/or early spring would seem to be the most suitable, although low soil temperatures at such time could impede EPN activity/performance. A low-temperature-tolerant EPN should, therefore, be sought and tested under field conditions in South Africa.

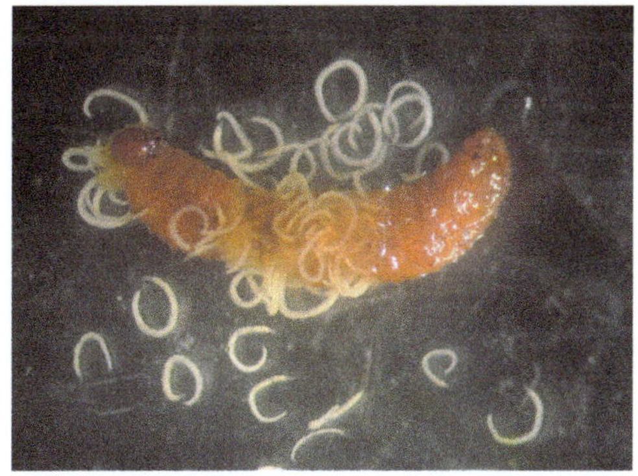

Fig. 20.2. Larval stage of *Phlyctinus callosus* infected with *Heterorhabditis noenieputensis*.

OBSCURE MEALYBUG, *PSEUDOCOCCUS VIBURNI* (SIGNORET) (HEMIPTERA: PSEUDOCOCCIDAE). The obscure mealybug, *Pseudococcus viburni* (Signoret) (Hemiptera: Pseudococcidae), is one of the most common, and serious, pests of apples and pears in South Africa, with two other species, *Pseudococcus calceolariae* (Maskell) (citrophilus mealybug) and *Pseudococcus longispinus* (Targioni Tozzetti) (long-tailed mealybug), also being recognized as acting in this capacity (Wakgari and Giliomee, 2004; Pringle *et al.*, 2015). In an attempt to identify the isolates of EPNs with superior virulence against *P. viburni* and two other mealybug species, several EPN isolates were screened in laboratory assays, with both *H. zealandica* and *S. yirgalemense* being found to be capable of reproducing successfully in *P. viburni* (Stokwe, 2009; Van Niekerk and Malan, 2012; Le Vieux and Malan, 2013a; Stokwe and Malan, 2016). In addition, adult and intermediate *P. viburni* were found to be more susceptible to EPN infection than were the crawlers, most likely as a result of the latter's small size (Stokwe, 2009; Bastidas *et al.*, 2014). Field-collected, mealybug-infested apples sprayed with EPNs have given an indication of their ability to locate and infect *P. viburni* on the fruit surface, as well as in the ovary and calyx of the fruit concerned (Stokwe, 2009). This ability supports the notion of using EPNs against arboreal populations of *P. viburni* on apples (Fig. 20.3), which is a strategy that is likely to benefit from the use of surfactants to alleviate humidity constraints (see the section on mealybugs below).

Fig. 20.3. Adult mealybug infected with *Heterorhabditis bacteriophora*.

Citrus

Citrus is a major export commodity (with an annual European Union (EU) market value of *c*.3 billion Rand) that is produced mainly in the provinces of Limpopo, the Eastern Cape, Mpumalanga, the Western Cape and KwaZulu-Natal. Approximately 65% of total fruit production is exported, despite the varying MRLs imposed by international markets. IPM remains the key strategy, incorporating both chemical and biorational approaches (Carpenter *et al.*, 2007; Moore and Hattingh, 2012). Although almost 60 insect species are listed as being pests of citrus, the quarantine pests, namely the FCM, *T. leucotreta*, and various fruit flies (Diptera: Tephritidae), remain the most important, due to their possible exportation as 'internal' fruit invaders (Grout and Moore, 2015).

FALSE CODLING MOTH, *THAUMATOTIBIA LEUCOTRETA* MEYRICK (LEPIDOPTERA: TORTRICIDAE). Soil stages in the life cycle of both the FCM and fruit flies support the potential employment of EPNs in targeting such pests. Female FCMs lay their eggs on the surface of the fruit, from where the larvae burrow into the fruit and develop through five instars. The fifth instar then exits the fruit and drops to the ground, where it penetrates the top layer of soil and spins a cocoon in which it pupates (Daiber, 1980, 1989). Pupation lasts *c*.8–10 days, depending on the prevailing temperature, after which they emerge from the soil as adult moths. The FCM is multivoltine, producing up to six generations per year (Newton, 1998), with no winter diapause. The stages presenting a possible window for control with EPNs include final-instar larvae, pre-pupae, pupae (cocooned) and emerging moths. Six indigenous EPN species were shown to be highly virulent against final-instar FCM larvae, with *S. yirgalemense* causing 100% mortality at a concentration as low as 50 IJs/insect (Malan *et al.*, 2011). Notably, emerging moths were also prone to infection, potentially aiding in the long-distance dissemination of the EPNs. Semi-field trials by these authors, using caged FCM larvae, yielded promising results, with *H. zealandica* causing >80% control and persistence over 49 days (Malan & Moore, 2016). Large-scale field trials with (imported) commercially formulated *H. bacteriophora* are currently under way, with promising results for future implementation.

FRUIT FLIES (DIPTERA: TEPHRITIDAE). Three species of fruit fly (Diptera: Tephritidae) are noted as being important citrus pests in South Africa, namely Asian fruit fly, *Bactrocera invadens* Drew, Tsurutu and White, Mediterranean fruit fly (Medfly), *Ceratitis capitata* (Wiedemann), and Natal fruit fly, *Ceratitis rosa* Karsch. These species, however, are not restricted to citrus, also affecting subtropical fruit, deciduous and/or stone fruit, together with the Marula fruit fly, *Ceratitis cosyra* (Walker), and five-spotted fruit fly, *Ceratitis quinaria* (Bezzi) (De Meyer *et al.*, 2008; Prinsloo and Uys, 2015). As in the case of the FCM, the final-instar fruit fly larva exits the fruit and drops to the ground where it burrows a few millimetres into the topsoil to pupate.

Laboratory assays were performed with local isolates of *H. bacteriophora*, *H. zealandica* and *S. khoisanae* to investigate their ability to infect pupariating larvae, pupae and adults of *C. capitata* and *C. rosa* (Malan and Manrakhan, 2009). Both pupariating larvae and adult flies (Fig. 20.4) were susceptible to nematode infection, whereas no infection was recorded against pupae *per se*. Similarly, pupal resistance to EPN infection was also recently reported for the Queensland fruit fly, *Bactrocera tryoni* (Froggatt) (Langford *et al.*, 2014). Interestingly, Malan and Manrakhan (2009) found that some pupariating larvae, infected with nematodes, still managed to pupate, but gave rise to malformed puparia (trapping the nematodes inside). Pupariating larvae of *C. capitata* were generally more susceptible to infection than were those of *C. rosa*. Highest infectivity against larvae of *C. capitata*

Fig. 20.4. Adult *Ceratitis capitata* infected with *Steinernema khoisanae*.

and *C. rosa* was recorded with *H. bacteriophora* and *H. zealandica*, respectively. In contrast, the adults of both species were highly susceptible to infection by the steinernematid, *S. khoisanae*.

Seemingly, the phenology of the particular fly species governs the optimal time of application, as was pointed out by Herz *et al.* (2006) in targeting the cherry fruit fly, *Rhagoletis cerasi* L. (Diptera, Tephritidae). Ideally, EPNs need to be present in the soil shortly before, or at the time of, larval drop, thus enabling them to infect their target prior to the completion of pupation.

MEALYBUGS (HEMIPTERA: PSEUDOCOCCIDAE). Four mealybug species of the seven that are noted as pests on citrus in southern Africa are of particular importance, namely citrus mealybug, *P. citri*, oleander mealybug (*Paracoccus burnerae* (Brain)), Karoo thorn mealybug (*Nipaecoccus viridis* (Newstead)) and the long-tailed mealybug, *P. longispinus* (Grout and Moore, 2015). Local isolates of EPNs were evaluated as potential control agents against adult female *P. citri*, with *S. yirgalemense* and *H. zealandica* causing >90% mortality (Van Niekerk and Malan, 2012). Although being highly susceptible to EPN infection, arboreal populations of *P. citri* might be able to escape such infection, due to the limited amount of surface moisture that tends to be available under field conditions. Following application, EPNs have a limited window of opportunity in which to locate and infect their host, as the level of available free water (aw) gradually decreases. An aw-bioassay indicated *S. yirgalemense* to be relatively tolerant to low levels of free water, as well as being able to locate, and to infect, *P. citri* more quickly (within 30 min) than were *H. zealandica*. The results also showed the first 2–4 h post-application to be the most decisive time for establishing successful infection of mealybugs (Van Niekerk and Malan, 2012). Subsequent research by Van Niekerk and Malan (2013) was aimed at investigating the addition of adjuvants to EPN suspensions so as to improve EPN performance. An aqueous suspension of *H. zealandica*, amended with 0.3% Zeba®, significantly increased *P. citri* mortality at 80% RH, with a temperature cycle starting at 22°C for 14 h and continuing at 11°C for 11 h. The same polymer formulation increased the mortality of *P. citri* by *S. yirgalemense* by 21% at 60% RH and by 27% at 80% RH (Van Niekerk and Malan, 2014b). Clearly, the mixing of EPNs with surfactants, gels, polymers and/or other adjuvants remains an area that is actively explored (see review by Shapiro-Ilan *et al.*, 2012), with it potentially proving crucial in the targeting of such pests as mealybugs on citrus and apples in South Africa, and elsewhere.

20.7.3 Production of garden mushroom and oyster mushroom

Although the use of EPNs against such mushroom pests as fungus gnats (Diptera: Sciaridae) is well documented abroad (Richardson and Grewal, 1991; Gouge and Hague, 1995; Harris *et al.*, 1995;

Grewal, 2007), the approach stills remains unexplored in South Africa. Potential also exists for the use of EPNs against sciarids affecting local pine tree nurseries (Hurley *et al.*, 2006).

20.7.4 Poultry production

The use of entomopathogens against house fly adults and larvae in poultry houses has been considered in South Africa, although focusing only on the fungus *Beauveria bassiana* (Balsamo) Vuillemin (Hypocreales: Cordycipitaceae) and on the bacterium *Bacillus thuringiensis* Berliner (Bacilliales: Bacilliaceae) var. *israelensis* (Bti) (Mwamburi *et al.*, 2009, 2011). Worldwide, however, many other insect pests, including fleas, darkling beetles, dermestid beetles, blowflies, moths and cockroaches (Axtell, 1999), might still be targeted with EPNs (Tóth *et al.*, 2005; Maketon *et al.*, 2010; Ali *et al.*, 2011; Alves *et al.*, 2012).

20.7.5 Application of entomopathogenic nematodes against stable flies

Although the use of EPNs against stable flies, *Stomoxys calcitrans* (L.) (Diptera: Muscidae), is not a current practice in South Africa, the adoption of such an approach warrants consideration. In a first step towards the establishment of an integrated fly management programme, the phenology of stable flies was investigated at a large feedlot near Heidelberg, Gauteng province, from October 2012 to September 2013 (Evert, 2014). In the study, the flies showed a typical late-summer activity pattern, peaking in numbers in January through February, and declining in number from early March. Limited to no fly activity was found to occur during the winter months from May to June 2013. The numbers collected in traps were correlated with fly counts on cattle in order to calculate a future threshold for control. Preliminary observations of the influence of temperature, wind speed and rainfall were also noted. The susceptibility of muscid fly larvae to EPNs, which was recently demonstrated by Ramakuwela (2015), serves as a departure point for the inclusion of EPNs in an integrated stable fly management programme in South Africa.

20.8 Interactions of Entomopathogenic Nematodes with Pesticides

Testing EPN compatibility with pesticides is important, when considering their use in an IPM system. Enabling the tank mixing of EPNs with other control products in a specific IPM programme could lead to increased control of the target pest, with greater cost-effectiveness, and with a reduction in the amount of application time required (Koppenhöfer and Grewal, 2005).

Both chemical and biorational remedies are commonly used in South Africa as part of an IPM strategy in citrus. A study conducted by Van Niekerk and Malan (2014a) measured the tolerance of two EPN species (*H. zealandica* and *S. yirgalemense*) to aqueous solutions of Cyperfos 500 EC® (chlorpyrifos and cypermethrin), Cryptogran™ (*Cryptophlebia leucotreta* granulovirus), Helicovir™ (nucleopolyhedrovirus), Nu-Film-P® (poly-1-P-menthene) and Zeba® (starch-g-polypotassium salt), in terms of infectivity and survival. *H. zealandica* proved to be highly compatible with all products tested, with no significant increase occurring in nematode mortality. The products concerned also did not affect the ability of *H. zealandica* to infect mealworm larvae after exposure to products over a 24 h period. Significant increases in the mortality of *S. yirgalemense* were recorded after 12 h of exposure to Nu-Film-P® and Cyperfos 500 EC®, and after 6 h of exposure to Nu-Film-P®. Le Vieux and Malan (2015) investigated the survival and infectivity of the IJs of exotic *H. bacteriophora* and *S. feltiae* after exposure of the IJs to time intervals of 0, 6, 12 and 24 h to the recommended dosage of imidacloprid. No negative results were obtained for either the survival or the infectivity of the IJs of both nematode species.

Contract research conducted by ARC-SGI (J.L. Hatting, 2012, unpublished results) for the South African wool industry (Cape Wools SA) investigated the compatibility of two strains of *Heterorhabditis* (strains NWGA 1 and NWGA 30) with the chemicals Zipdip® (triazophos) and Dazzle® (diazinon), which have been registered for use against ectoparasites, including blowfly (*Lucilia cuprina* Wiedemann) (Diptera: Calliphoridae), on sheep in South Africa. The above-mentioned substances were mixed with water at a chemical:water ratio of 1 l:1000 l and 1 l:150 l, respectively. The IJs were exposed to the chemical pesticides by means of 'washing' the IJs three times (in a series of successive transfers) in each solution mentioned for a period of 30 s. After the third and final transfer, 2 ml of 1000 IJs/ml (in chemical solution) was pipetted into 35 mm Petri dishes. The Petri dishes were then sealed and incubated at room temperature for 72 h. Survival was determined every 24 h by means of observing, at random, 100 IJs and recording how many were either dead or alive at the time of observation. Best compatibility was noted with the chemical insecticide Dazzle®. Compatibility with Zipdip® was observed for a period of only 24 h, after which IJ viability dropped to *c.*50% (at 72 h). Likewise, the infectivity of the nematodes (to mealworm larvae) exposed to Zipdip® was found to be severely inhibited. The combined use of EPNs with a sublethal dose of Dazzle® (resulting in an even smaller effect on the EPNs) should be considered. At such a dose, the stress factor imposed on the target pest (*L. cuprina*) by the chemical should inhibit feeding, while rendering the host more susceptible to the IJs. A mixture of IJs (in 0.01% Break-Thru® surfactant) with Dazzle® at <10% of the field-recommended rate warrants further investigation.

20.9 Technical Restrictions for Nematode Applications

As South Africa is regarded as a water-scarce country, one of the main aims of agricultural production is, understandably, the conservation and preservation of water. Most of the orchards countrywide are irrigated by means of drip irrigation, as there has been a move away from the use of overhead and sprinkler irrigation. The application of EPNs through such systems warrants closer consideration. Moreover, South Africa boasts a relatively high UV radiation index, with Cape Town experiencing values between 7 and 10 during the summer months of October through to February (World Health Organization, s.d.). Rapid inundation of the soil profile after (topical) application is thus crucial, given the general sensitivity of EPNs to UV radiation as well as desiccation. In future, the use of EPN formulations containing UV protectants when targeting arboreal species will most likely be called for under the prevailing harsh summer conditions.

20.10 Entomopathogenic Nematodes – Potential Threat to Ecosystems

Currently, the soil application of exotic EPNs in large commercial field trials in South Africa is being undertaken, at least in part, to assess the potential of its, as yet, unknown impact on the environment. This holds true especially in the case of *S. feltiae*, on which no local impact study has yet been done in South Africa. In general, a minimal amount of work has been undertaken on the ecology and the ecosystem services of EPNs after inundative outdoor application.

The predatory beetle, *Cryptolaemus montrouzieri* Mulsant (Coleoptera: Coccinellidae), is a commercially produced biocontrol agent that is used periodically against mealybugs in South Africa. The compatibility of two endemic EPNs with *C. montrouzieri* (and selected chemical/biorational insecticides) was investigated by Van Niekerk and Malan (2014a). These authors reported high levels of larval susceptibility to *H. zealandica* and *S. yirgalemense*, with the adult beetles being found to be twice as susceptible to *S. yirgalemense* as compared to *H. zealandica*. Accordingly, in an IPM system where other biological agents are used, the application of EPNs should be fine-tuned so as not to interfere with the performance of other natural enemies.

Le Vieux and Malan (2015) touched on different ecological aspects pertaining to the use of EPNs. The parameters concerned included: (i) the persistence levels involved (which could be either short or long term at the individual vineyard level within the same environment); (ii) whether, and to what extent, the associated bacteria could act as a deterrent for ants; and (iii) the result of damaged grapevine roots being more attractive to EPNs than they are to the target pest itself.

The only research with regard to the physiology of EPNs that has been conducted so far is that which has been undertaken by Hill *et al.* (2015). Their results showed inherent genetic divergence in terms of thermal specialization between *S. yirgalemense* and *H. zealandica*. *S. yirgalemense* showed higher temperature tolerance in comparison to such tolerance shown by *H. zealandica* (LT_{50} = 40.8°C and 36.7°C, respectively), with weaker low-temperature tolerance in comparison to that of *H. zealandica* (LT_{50} = –2.4°C and –9.7°C, respectively). The results obtained in this respect suggest that the two species occupy thermal niches divergent to each other. The gaining of more knowledge on the physiology of EPN species should lead to a better understanding of the occurrence and functioning of agricultural ecosystems in different geographical regions of South Africa.

20.11 Conclusions

Investigations into the occurrence and distribution of EPNs in South Africa only started in earnest during this century, as opposed to other regions of the world such as Europe and the USA. Despite the limited number of large-scale surveys that have been conducted in this country so far, ten new species have been described since 2006, with at least three additional species descriptions in preparation. This scenario underscores the outstanding potential for this hitherto unexplored biological resource, of which the benefits have yet to be fully discovered. Commercial farmers are already geared up for the outdoor use of EPNs and, because of the current unavailability of locally produced commercial products, exotic EPNs are being imported and applied on a large scale. The main concern of scientists with regard to the import and application of millions of nematodes in an agricultural setting is that of biological pollution and the risks that such actions could possibly hold in terms of 'masking' the occurrence (and possible replacement) of endemic species (Abate *et al.*, 2017)..

It is acknowledged that in most cases, EPNs cannot be expected to be used as a stand-alone control measure, but that they should be applied in combination with other control measures employed in an IPM system. Even 50% field control of the targeted insect/s using EPNs would be regarded as sufficient to suppress the next generation of insects that are otherwise likely to escape chemical control, with possible crop damage, and/or develop pesticide resistance.

Acknowledgements

We would like to acknowledge the following South African institutions (in alphabetical order) for financial support: Agricultural Research Council, Citrus Research International, HORTGRO Science, Winetech and the Winter Cereal Trust. Photograph credits, Figs 20.2–20.4: University of Stellenbosch.

References

Abate, B.A., Malan, A.P., Tiedt, L.R., Wingfield, M.J., Slippers, B. and Hurley, B.P. (2016) *Steinernema fabii* n. sp. (Rhabditida: Steinernematidae), a new entomopathogenic nematode from South Africa. *Nematology* 18, 235–255.

Abate, B.A., Wingfield, M.J., Slippers, B. and Hurley, B.P. (2017) Commercialization of entomopathogenic nematodes: Should import regulations be revised? *Biocontrol Science and Technology*, DOI: 10.1080/09583157.2016.1278200

Ali, S.S., Sirvastava, M. and Shankar, P. (2011) First report on susceptibility of khapra beetle (*Trogoderma granarium*) against *Steinernema masoodi* and its *in vivo* production. *Trends in Biosciences* 4, 140–141.

Allsopp, E., Barnes, B.N., Blomefield, T.L. and Pringle, K.L. (2015) Grapevine. In: Prinsloo, G.L. and Uys, V.M. (eds) *Insects of Cultivated Plants and Natural Pastures in Southern Africa*. Entomological Society of Southern Africa, Pretoria, South Africa, pp. 420–437.

Alves, V.S., Nerves, P.M.J.D., Alves, L.F., Moino, A. and Holz, N. (2012) Entomopathogenic nematodes (Rhabditida: Heterorhabditidae and Steinernematidae) screening for lesser mealworm *Alphitobius diaperinus* (Coleoptera: Tenebrionidae) control. *Revista Colombiana de Entomología* 38, 76–80.

Annecke, D.P. and Moran, V.C. (1982) *Insects and Mites of Cultivated Plants in South Africa*. Butterworths, Durban, South Africa.

Axtell, R.C. (1999) Poultry integrated pest management: status and future. *Integrated Pest Management Reviews* 4, 53–73.

Barbosa, P. (1998) *Conservation Biological Control*. Academic Press, London.

Barnes, B.N. (1989) Embryonic and immature stages of *Phlyctinus callosus* Boh. (Coleoptera, Curculionidae) – aspects of biology and behavior with respect to control in deciduous fruit orchards. *Journal of the Entomological Society of Southern Africa* 52, 165–178.

Barnes, B.N. and Giliomee, J.H. (1992) Fruit-feeding behavior of banded fruit weevil, *Phlyctinus callosus* (Schönherr) (Col, Curculionidae), in apple orchards. *Journal of Applied Entomology – Zeitschrift für Angewandte Entomologie* 113, 407–415.

Barnes, B.N. and Pringle, K.L. (1989) Oviposition by the banded fruit weevil, *Phlyctinus callosus* (Schoenherr) (Coleoptera, Curculionidae), in deciduous fruit orchards in South Africa. *Bulletin of Entomological Research* 79, 31–40.

Barnes, B.N., Knipe, M.C. and Calitz, F.J. (1994) Trunk barriers provide effective control of banded fruit weevil on apples and nectarines. *Deciduous Fruit Grower* 44, 327–340.

Barnes, B.N., Knipe, M.C. and Calitz, F.J. (1996) Latest results with trunk exclusion barriers for weevil control on apples (Jongste resultate met stamsperbande vir kalanderbeheer op appels). *Deciduous Fruit Grower* 46, 284–287.

Barnes, M.M. (1991) Codling moth occurrence, host race formation, and damage. In: Van der Geest, L.P.S. and Evenhuis, H.H. (eds) *Tortricid Pests: Their Biology, Natural Enemies and Control*. Elsevier, Amsterdam, pp. 313–327.

Basson, S. (1993) Biological control of *Phlyctinus callosus* by entomogenous nematodes. *Proceedings of the 11th Symposium of the Nematological Society of Southern Africa*, held at Rustenburg, South Africa. Phytophylactica 25, 303.

Bastidas, B., Edgar, P. and San-Blas, E. (2014) Size does matter: the life cycle of *Steinernema* spp. in micro-insect hosts. *Journal of Invertebrate Pathology* 121, 46–55.

Bell, M.R. (1995) Effects of an entomopathogenic nematode and nuclear polyhedrosis virus on emergence of *Heliothis virescens* (Lep: Noctuidae). *Journal of Entomological Science* 30, 243–250.

Bell, M.R. and Hardee, D.D. (1994) Tobacco budworm: possible use of various entomopathogens in large area pest management. *Proceedings of the Beltwide Cotton Production Conference, Jan. 5–8, 1994*. Beltwide Cotton Production, San Diego, California, pp. 1168–1171.

Bellini, L.L. and Dolinski, C. (2012) Foliar application of entomopathogenic nematodes (Rhabditida: Steinernematidae and Heterorhabditidae) for the control of *Diatraea saccharalis* in greenhouse. *Ciências Agrárias, Londrina* 33, 997–1004.

Bredenhand, E., Van Hoorn, A., May, F., Ferreira, T. and Johnson, S. (2010) Evaluation of techniques for monitoring banded fruit weevil, *Phlyctinus callosus* (Schoenherr) (Coleoptera: Curculionidae), infestation in blueberry orchards. *African Entomology* 18, 205–209.

Cabanillas, H.E. and Raulston, J.R. (1995) Impact of *Steinernema riobravis* (Rhabditida: Steinernematidae) on the control of *Helicoverpa zea* (Lepidoptera: Noctuidae) in corn. *Journal of Economic Entomology* 88, 58–64.

Cabanillas, H.E. and Raulston, J.R. (1996) Effects of furrow irrigation on the distribution and infectivity of *Steinernema riobravis* against corn earworm in corn. *Fundamental and Applied Nematology* 19, 273–281.

Carnegie, A.J.M. (1981) Combatting *Eldana saccharina* Walker: a progress report. *Proceedings of the International Society of Sugar Cane Technologists* 55, 116–119.

Carnegie, A.J.M. and Conlong, D.E. (1994) Biology, pest status and control measure relationships of sugarcane insect pests. *Proceedings of the Second Sugar Cane Entomology Workshop of the International Society of Sugar Cane Technologists*. SASA Experiment Station, Mount Edgecombe, South Africa.

Carpenter, J., Bloem, S. and Hofmeyer, H. (2007) Area-wide control tactics for the false codling moth *Thaumatotibia leucotreta* in South Africa: a potential invasive species. In: Vreysen, M.J.B., Robertson, A.S. and Hendrichs, J. (eds) *Area-wide Control of Pest Insects*. Springer Science & Business Media, New York, pp. 351–359.

Çimen, H.C., Lee, M.-M., Hatting, J., Hazir, S. and Stock, P.S. (2014) *Steinernema tophus* sp. n. (Nematoda: Steinernematidae), a new entomopathogenic nematode from South Africa. *Zootaxa* 3821, 337–353.

Çimen, H.C., Lee, M.-M., Hatting, J., Hazir, S. and Stock, S.P. (2015) *Steinernema innovationi* n. sp. (Panagrolaimomorpha: Steinernematidae), a new entomopathogenic nematode species from South Africa. *Journal of Helminthology* 89, 415–427.

Çimen, H., Půža, V., Nermuť, J., Hatting, J., Hazir, S. and Ramakuwela, T. (2016) *Steinernema beitlechemi* n. sp., a new entomopathogenic nematode (Nematoda: Steinernematidae) from South Africa. *Nematology* 18, 439–453.

Çimen, H., Půža, V., Nermut, J., Hatting, J., Ramakuwela, T. (2016b) *Steinernema biddulphi* n. sp., a new entomopathogenic nematode (Nematoda: Steinernematidae) from South Africa. *Journal of Nematology* 48, 148–158.

Conlong, D.E. and Rutherford, R.S. (2009) Conventional and new biological and habitat interventions for integrated pest management systems: review and case studies using *Eldana saccharina* Walker (Lepidoptera: Pyralidae). In: Peshin, R. and Dhawan A.K. (eds) *Integrated Pest Management: Innovation–Development Process*. vol. 1. Springer, Dordrecht, The Netherlands, pp. 241–260.

Conlong, D.E. and Way, M.J. (2015) Sugarcane. In: Prinsloo, G.L. and Uys, V.M. (eds) *Insects of Cultivated Plants and Natural Pastures in Southern Africa*. Entomological Society of Southern Africa, Pretoria, South Africa, pp. 156–176.

Cossentine, J.E., Sholberg, P.L., Jensen, L.B.J., Bedford, K.E. and Sheperd, T.C. (2004) Fumigation of empty fruit bins with carbon dioxide to control diapausing codling moth larvae and *Penicillium expansum* Link. ex Tom spores. *HortScience* 39, 429–432.

CropLife South Africa [s.d.] Agricultural remedies database. Available at: www.croplife.co.za/ (accessed 6 October 2016).

Daiber, C.C. (1980) A study of the biology of the false codling moth *Cryptophlebia leucotreta* (Meyr.): the adult and generations during the year. *Phytophylactica* 12, 187–193.

Daiber, C.C. (1989) The false codling moth, *Cryptophlebia leucotreta* (Meyr.) (Lepidoptera, Tortricidae), in Southern Africa. *Journal of Plant Diseases and Protection* 96, 71–80.

De Meyer, M., Robertson, M.P., Peterson, A.T. and Mansell, M.W. (2008) Ecological niches and potential geographical distributions of Mediterranean fruit fly (*Ceratitis capitata*) and Natal fruit fly (*Ceratitis rosa*). *Journal of Biogeography* 35, 270–281.

De Waal, J.Y., Malan, A.P., Levings, J. and Addison, M.F. (2010) Key elements in the successful control of diapausing codling moth, *Cydia pomonella* (L.) (Lepidoptera: Tortricidae) in fruit bins with a South African isolate of *Heterorhabditis zealandica* (Rhabditida: Heterorhabditidae). *Biocontrol Science and Technology* 20, 489–502.

De Waal, J.Y., Malan, A.P. and Addison, M.F. (2011a) Efficacy of entomopathogenic nematodes (Rhabditida: Heterorhabditidae and Steinernematidae) against codling moth, *Cydia pomonella* (Lepidoptera: Tortricidae) in temperate regions. *Biocontrol Science and Technology* 20, 489–502.

De Waal, J.Y., Malan, A.P. and Addison, M.F. (2011b) Evaluating mulches together with *Heterorhabditis zealandica* (Rhabditida: Heterorhabditidae) for the control of diapausing codling moth larvae, *Cydia pomonella* (L.) (Lepidoptera: Tortricidae). *Biocontrol Science and Technology* 21, 225–270.

De Waal, J.Y., Addison, M.F. and Malan, A.P. (2013) A superabsorbent polymer formulation for improved efficacy of *Heterorhabditis zealandica* (Rhabditida: Heterorhabditidae) control of codling moth larvae, *Cydia pomonella* (L.) (Lepidoptera: Tortricidae). *Biocontrol Science and Technology* 23, 62–78.

Elawad, S., Ahmad, W. and Reid, A.P. (1997) *Steinernema abbasi* sp. n. (Nematoda: Steinernematidae) from the Sultanate of Oman. *Fundamental and Applied Nematology* 20, 435–442.

Evert, M.M. (2014) The temporal distribution and relative abundance of stable flies (*Stomoxys calcitrans*) (Diptera: Muscidae) in a feedlot near Heidelberg, Gauteng, South Africa. MSc thesis. North-West University, Potchefstroom, South Africa.

Fasemore, O.A. (2012) An investigation of the possiblities of scaling-up and mass-producing entomopathogenic nematodes (EPNs). PhD dissertation. University of the Witwatersrand, South Africa.

Feaster, M.A. and Steinkraus, D.C. (1996) Inundative biological control of *Helicoverpa zea* (Lepidoptera: Noctuidae) with the entomopathogenic nematode *Steinernema riobravis* (Rhabditida: Steinernematidae). *Biological Control* 7, 38–43.

Ferreira, T. (2012) Characterisation of nematode symbiotic bacteria and the *in vitro* liquid culture of *Heterorhabditis zealandica* and *Steinernema yirgalemense*. PhD dissertation. Stellenbosch University, South Africa.

Ferreira, T. and Malan, A.P. (2014) Potential of entomopathogenic nematodes for the control of the banded fruit weevil, *Phlyctinus callosus* (Schönherr) (Coleoptera: Curculionidae). *Journal of Helminthology* 88, 293–301.

Ferreira, T., Van Reenen, C.A., Endo, A., Spröer, C., Malan, A.P. and Dicks, L.M.T. (2013a) Description of *Xenorhabdus khoisanae* sp. nov., a symbiont of the entomopathogenic nematode *Steinernema khoisanae*. *International Journal of Systematic and Evolutionary Microbiology* 63, 3220–3224.

Ferreira, T., Van Reenen, C.A., Pagès, S., Tailliez, P., Malan, A.P. and Dicks, L.M.T. (2013b) Description of *Photorhabdus luminescens* subsp. *noenieputensis* subsp. nov., a symbiotic bacterium associated with a new *Heterorhabditis* species related to *Heterorhabditis indica*. *International Journal of Systematic and Evolutionary Microbiology* 63, 1853–1858.

Ferreira, T., Addison, M.F. and Malan, A.P. (2014a) *In vitro* liquid culture of a South African isolate of *Heterorhabditis zealandica* for the control of insect pests. *African Entomology* 22, 80–92.

Ferreira, T., Van Reenen, C.A., Endo, A., Spröer, C., Malan, A.P. and Dicks, L.M.T. (2014b) Description of *Photorhabdus heterorhabditis* sp. nov., a symbiont of the entomopathogenic nematode *Heterorhabditis zealandica*. *International Journal of Systematic and Evolutionary Microbiology* 64, 1540–1545.

Ferreira, T., Van Reenen, C.A., Tailliez, P., Pagès, S., Malan, A.P. and Dicks, L.M.T. (2016a) First report of the symbiotic bacteria, *Xenorhabdus indica*, associated with the entomopathogenic nematode *Steinernema yirgalemense*. *Journal of Helminthology* 90, 108–112.

Ferreira, T., Addison, M.F. and Malan, A.P. (2016b) Development and population dynamics of *Steinernema yirgalemense* and growth characteristics of its associated *Xenorhabdus* symbiont in liquid culture. *Journal of Helminthology* 90, 364–371.

Giliomee, J.H. (2013) Entomology in South Africa: Where do we come from, where are we now and where are we going? *South African Journal of Science* 109, 1–3. Available at: www.sajs.co.za/issue/2013/109/1-2/january-february-2013 (accessed 6 October 2016).

Giliomee, J.H., Blomefield, T.L., Barnes, B.N. and Pringle, K.L. (2015) Pear. In: Prinsloo, G.L. and Uys, V.M. (eds) *Insects of Cultivated Plants and Natural Pastures in Southern Africa*. Entomological Society of Southern Africa, Pretoria, South Africa, pp. 366–375.

Glazer, I. and Salame, L. (2000) Osmotic survival of the entomopathogenic nematode *Steinernema carpocapsae*. *Biological Control* 18, 251–257.

Goble, T., Almeida, J. and Conlong, D. (2017) Microbial control of sugarcane insect pests. In: Lacey, L.A. (ed.) *Microbial Control of Insect and Mite Pests: From Theory to Practice*. Academic Press, Amsterdam, pp. 299–312.

Goebel, F.R. and Sallam, N. (2011) New pest threats for sugarcane in the new bioeconomy and how to manage them. *Current Opinion in Environmental Sustainability* 3, 81–89.

Goebel, F.R. and Way, M.J. (2003) Investigation of the impact of *Eldana saccharina* (Lepidoptera: Pyralidae) on sugarcane yield in field trials in Zululand. *Proceedings of the South African Sugar Technologists' Association* 77, 256–265.

Gouge, D.H. and Hague, N.G.M. (1995) Glasshouse control of fungus gnats, *Bradysia pauper*, on fuchsias by *Steinernema feltiae*. *Fundamentals in Applied Nematology* 18, 77–80.

Grenier, E., Bonifassi, E., Abad, P. and Laumond, D. (1996a) Use of species specific satellite DNAs as diagnostic probes in the identification of Steinernematidae and Heterorhabditidae entomopathogenic nematodes. *Parasitology* 113, 483–489.

Grenier, E., Laumond, C. and Abad, P. (1996b) Molecular characterization of two species-specific tandemly repeated DNAs from entomopathogenic nematodes *Steinernema* and *Heterorhabditis* (Nematoda: Rhabditida). *Molecular and Biochemical Parasitology* 83, 47–56.

Grewal, P.S. (2007) Mushroom pests. In: Lacey, L. and Kaya, H. (eds) *Field Manual of Techniques in Invertebrate*

Pathology. Springer, Dordrecht, The Netherlands, pp. 457–461.

Grout, T.G. and Moore, S.D. (2015) Citrus. In: Prinsloo, G.L. and Uys, V.M. (eds) *Insects of Cultivated Plants and Natural Pastures in Southern Africa*. Entomological Society of Southern Africa, Pretoria, South Africa, pp. 448–499.

Harington, J.S. (1953) Observation on the biology, the parasites and the taxonomic position of the maize beetle – *Heteronychus san-helenae* Blanch. *South African Journal of Science* 50, 11–14.

Harris, M.A., Oetting, R.D. and Gardner, W.A. (1995) Use of entomopathogenic nematodes and a new monitoring technique for control of fungus gnats, *Bradysia coprophila* (Diptera: Sciaridae), in floriculture. *Biological Control* 5, 412–418.

Hatting, J. (2012) Comparison of three entomopathogenic fungi against the bollworm, *Helicoverpa armigera* (Hübner) (Lepidoptera: Noctuidae), employing topical vs *per os* inoculation techniques. *African Entomology* 20, 91–100.

Hatting, J.L. and Kaya, H.K. (2001) Entomopathogenic nematodes: prospects for biological control in South Africa. In: *Proceedings of the 13th Entomological Congress, Entomological Society of Southern Africa, 2–5 July*. Entomological Society of Southern Africa, Pietermaritzburg, South Africa, pp. 28–29.

Hatting, J., Stock, P.S. and Hazir, S. (2009) Diversity and distribution of entomopathogenic nematodes (Steinernematidae, Heterorhabditidae) in South Africa. *Journal of Invertebrate Pathology* 102, 120–128.

Herz, A., Köppler, K., Vogt, H., Elias, E., Peters, A. and Katz, P. (2006) Biological control of the cherry fruit fly, *Rhagoletis cerasi* L. (Diptera, Tephritidae) by use of entomopathogenic nematodes: first experiences towards practical implementation. In: Markus, B. (ed.) *Ecofruit – 12th International Conference on Cultivation Technique and Phytopathological Problems in Organic Fruit-Growing: Proceedings to the Conference from 31st January to 2nd February 2006 at Weinsberg, Germany*. Available at: http://orgprints.org/13795/1/ecofruit_12th_12.pdf (accessed 10 October 2016).

Hill, M.P., Malan, A.P. and Terblanche, J.S. (2015) Divergent thermal specialisation of two South African entomopathogenic nematodes. PeerJ PrePrints. Available at: https://dx.doi.org/10.7287/peerj.preprints.1152v1 (accessed 6 October 2016).

Hurley, B.P., Govender, P., Coutinho, T.A., Wingfield, B.D. and Wingfield, M.J. (2006) Fungus gnats and other Diptera in South African forestry nurseries and their possible association with the pitch canker fungus. *South African Journal of Science* 103, 43–46.

Hussain, M.A., Ahmad, R. and Ahmad, W. (2014) Evaluation of *Steinernema masoodi* (Rhabditida: Steinernematidae) against soil-dwelling life stage of *Helicoverpa armigera* (Lepidoptera: Noctuidae) in laboratory and microplot study. *Canadian Journal of Plant Protection* 2, 4–8.

IndexMundi [s.d.] South Africa fresh table grapes exports by year. Available at: www.Indexmundi.com/agriculture/?country=za&commodity=grapes&graph=exports (accessed 6 October 2016).

Jansson, R.K., Lecrone, S.H. and Gaugler, R. (1993) Field efficacy and persistence of entomopathogenic nematodes (Rhabditida: Steinernematidae, Heterorhabditidae) for control of sweet potato weevil (Coleoptera: Apionidae) in southern Florida. *Journal of Economic Entomology* 86, 1055–1063.

Keeping, M.G. (2006) Screening of South African sugarcane cultivars for resistance to stalk borer, *Eldana saccharina* Walker (Lepidoptera: Pyralidae). *African Entomology* 14, 277–288.

Koppenhöfer, A.M. and Grewal, P.S. (2005) Compatibility and interactions with agrochemicals and other biocontrol agents. In: Grewal, P.S., Ehlers, R.-U. and Shapiro-Ilan, D.I. (eds) *Nematodes as Biocontrol Agents*. CAB International, Wallingford, UK, pp. 363–378.

Langford, E.A., Nielsen, U.N., Johnson, S.N. and Riegler, M. (2014) Susceptibility of Queensland fruit fly, *Bactrocera tryoni* (Froggatt) (Diptera: Tephritidae), to entomopathogenic nematodes. *Biological Control* 69, 34–39.

Leslie, G. (2004) Pests of sugarcane. In: James, G. (ed.) *Sugarcane*. Blackwell Science, Oxford, UK, pp. 78–100.

Leslie, G.W. (2009) Estimating the economic injury level and the economic threshold for the use of alpha-cypermethrin against the sugarcane borer, *Eldana saccharina* Walker (Lepidoptera: Pyralidae). *International Journal of Pest Management* 55, 37–44.

Le Vieux, P.D. and Malan, A.P. (2013a) An overview of the vine mealybug (*Planococcus ficus*) in South African vineyards and the use of entomopathogenic nematodes as potential biocontrol agents. *South African Journal of Enology and Viticulture* 34, 34–45.

Le Vieux, P.D. and Malan, A.P. (2013b) The potential use of entomopathogenic nematodes to control *Planococcus ficus* (Signoret) (Hemiptera: Pseudococcidae). *South African Journal of Enology and Viticulture* 34, 296–306.

Le Vieux, P.D. and Malan, A.P. (2015) Prospects for using entomopathogenic nematodes to control the vine mealybug, *Planococcus ficus*, in South African vineyards. *South African Journal of Enology and Viticulture* 36, 59–70.

London, L. and Myers, J. (1995) Critical issues for agrichemical safety in South Africa. *American Journal of Industrial Medicine* 27, 1–14.

Maketon, M., Hominchan, A. and Hotaka, D. (2010) Control of American cockroach (*Periplaneta americana*) and German cockroach (*Blattella germanica*) by entomopathogenic nematodes. *Revista Colombiana de Entomología* 36, 249–253.

Malan, A.P. and Hatting, J. (2015) Entomopathogenic nematode exploitation: case studies in laboratory and field applications from South Africa. In: Campos-Herrera R. (ed.) *Nematode Pathogenesis of Insects and Other Pests. Sustainability in Plant and Crop Protection: Ecology and Applied Technologies for Sustainable Plant and Crop Protection.* Springer International Publishing, Switzerland, pp. 475–506.

Malan, A.P. and Manrakhan, A. (2009) Susceptibility of the Mediterranean fruit fly (*Ceratitis capitata*) and the Natal fruit fly (*Ceratitis rosa*) to entomopathogenic nematodes. *Journal of Invertebrate Pathology* 100, 47–49.

Malan, A.P. and Moore, S.D. (2016) Evaluation of local entomopathogenic nematodes for the control of false codling moth, *Thaumatotibia leucotreta* (Meyrick, 1913), in a citrus orchard in South Africa. *African Entomology* 24, 489–501.

Malan, A.P., Knoetze, R. and Moore, S.D. (2011) Isolation and identification of entomopathogenic nematodes from citrus orchards and their biocontrol potential against false codling moth. *Journal of Invertebrate Patholology* 108, 115–125.

Malan, A.P., Knoetze, R. and Tiedt, L.R. (2014) *Heterorhabditis noenieputensis* n. sp. (Rhabditida: Heterorhabditidae), a new entomopathogenic nematode from South Africa. *Journal of Helminthology* 88, 138–151.

Malan, A.P., Knoetze, R. and Tiedt, T.R. (2016a) *Steinernema jeffreyense* n. sp. (Rhabditida: Steinernematidae), a new entomopathogenic nematode from South Africa. *Journal of Helminthology* 90, 262–278.

Malan, A.P., Knoetze, R. and Tiedt, T.R. (2016b) *Steinernema nguyeni* n. sp. (Rhabditida: Steinernematidae), a new entomopathogenic nematode from South Africa. *Nematology* 18, 571–590.

Malan, A.P., Nguyen, K.B. and Addison, M.F. (2006) Entomopathogenic nematodes (Steinernematidae and Heterorhabditidae) from the southwestern parts of South Africa. *African Plant Protection* 12, 65–69.

Malan, A.P., Nguyen, K.B., De Waal, J.Y. and Tiedt, L. (2008) *Heterorhabditis safricana* n. sp. (Rhabditida: Heterorhabditidae), a new entomopathogenic nematode from South Africa. *Nematology* 10, 381–396.

Manrakhan, A., Daneel, J.-H. and Moore, S.D. (2014) The impact of naturally occurring entomopathogenic nematodes on false codling moth, *Thaumatotibia leucotreta* (Lepidoptera: Tortricidae), in citrus orchards. *Biocontrol Science and Technology* 24, 241–245.

Michigan State University [s.d.] Insecticide Resistance Action Committee. Arthropod Pesticide Resistance Database. Available at: www.pesticideresistance.com/ (accessed 6 October 2016).

Molotsane, R., Ngoma, L. and Gray, V. (2007) A biogeographic survey of entomopathogenic nematodes of Gauteng and the North West Province of South Africa. *South African Journal of Plant and Soil* 24, 244.

Moore, S.D. and Hattingh, V. (2012) A review of current pre-harvest control options for false codling moth in citrus in southern Africa. *South African Fruit Journal* 11, 82–85.

Moran, V.C., Hoffmann, J.H. and Zimmermann, H.G. (2013) 100 years of biological control of invasive alien plants in South Africa: history, practice and achievements. *South African Journal of Science* 109 (9/10), 1–6.

Mwamburi, L.A., Laing, M.D. and Miller, R. (2009) Interaction between *Beauveria bassiana* and *Bacillus thuringiensis* var. *israelensis* for the control of house fly larvae and adults in poultry houses. *Poultry Science* 88, 2307–2314.

Mwamburi, L.A., Laing, M.D. and Miller, R. (2011) Laboratory and field evaluation of formulated *Bacillus thuringiensis* var. *israelensis* as a feed additive and using topical applications for control of *Musca domestica* (Diptera: Muscidae) larvae in caged-poultry manure. *Environmental Entomology* 40, 52–58.

Neuenschwander, P., Borgemeister, C. and Langewald, J. (2003) *Biological Control in IPM Systems in Africa*. CAB International, Wallingford, UK.

Newton, P.J. (1998) False codling moth *Cryptophlebia leuctreta* (Meyrick). In: Bedford, E.C.G., Van den Berg, M.A. and De Villiers, E.A. (eds) *Citrus Pests in the Republic of South Africa*. Dynamic Ad, Nelspruit, South Africa, pp. 194–200.

Nguyen, K.B., Malan, A.P. and Gozel, U. (2006) *Steinernema khoisanae* n. sp. (Rhabditida: Steinernematidae), a new entomopathogenic nematode from South Africa. *Nematology* 8, 157–175.

Nthenga, I., Knoetze, R., Berry, S., Tiedt, L.R. and Malan, A.P. (2014) *Steinernema sacchari* n. sp. (Rhabditida: Steinernematidae), a new entomopathogenic nematode from South Africa. *Nematology* 16, 475–494.

Odendaal, D., Addison, M.F. and Malan, A.P. (2015) Control of codling moth (*Cydia pomonella*) with special emphasis on using entomopathogenic nematodes. *African Entomology* 23, 259–274.

Odendaal, D., Addison, M.F. and Malan, A.P. (2016a) Entomopathogenic nematodes for the control of codling moth (*Cydia pomonella* L.) in field and laboratory trials. *Journal of Helminthology* 90, 615–623.

Odendaal, D., Addison, M.F. and Malan, A.P. (2016b) Evaluation of above ground application of entomopathogenic nematodes for the control of diapausing codling moth (*Cydia pomonella* L.) under natural conditions. *African Entomology* 24, 61–74.

Perez, E.E., Lewis, E.E. and Shapiro-Ilan, D.I. (2003) Impact of host cadaver on survival and infectivity of entomopathogenic nematodes (Rhabditida: Steinernematidae and Heterorhabditidae) under desiccating conditions. *Journal of Invertebrate Pathology* 82, 111–118.

Pillay, U., Martin, L.A., Rutherford, R.S. and Berry, S.D. (2009) Entomopathogenic nematodes in sugarcane in South Africa. *Proceedings of the South African Sugar Technologists' Association* 82, 538–541.

Pringle, K.L., Barnes, B.N. and Blomefield, T.L. (2015) Apple. In: Prinsloo, G.L. and Uys, V.M. (eds) *Insects of Cultivated Plants and Natural Pastures in Southern Africa*. Entomological Society of Southern Africa, Pretoria, South Africa, pp. 350–365.

Prinsloo, G.L. and Uys, V.M. (2015) *Insects of Cultivated Plants and Natural Pastures in Southern Africa*. Entomological Society of Southern Africa, Pretoria, South Africa.

Quinn, L.P., De Vos, B.J., Fernandes-Whaley, M., Roos, C., Bouwman, H., *et al.* (2011) Pesticide use in South Africa: one of the largest importers of pesticides in Africa. In: Stoytcheva, M. (ed.) *Pesticides in the Modern World – Pesticides Use and Management*. Available at: https://www.environment.gov.za/sites/default/files/docs/pesticide_usein_sa.pdf (accessed 6 October 2016).

Ramakuwela, T. (2015) Development of a new entomopathogenic nematode species, *Steinernema innovationi*: biological characterization and mass production. PhD dissertation. University of KwaZulu-Natal, Pietermaritzburg, South Africa.

Ramakuwela, T., Hatting, J.L., Laing, M.D., Hazir, S. and Thiebaut, N. (2015) Effect of storage temperature and duration on survival and infectivity of *Steinernema innovationi* (Rhabditida: Steinernematidae). *Journal of Nematology* 47, 332–336.

Ramakuwela, T., Hatting, J.L., Laing, M.D., Hazir, S. and Thiebaut, N. (2016) *In vitro* solid-state production of *Steinernema innovationi* with cost analysis. *Biocontrol Science and Technology* 26, 792–808.

Raulston, J.R., Pair, S.D., Loera, J. and Cabanillas, H.E. (1992) Prepupal and pupal parasitism of *Helicoverpa zea* and *Spodoptera frugiperda* (Lepidoptera: Noctuidae) by *Steinernema* sp. in cornfields in the lower Rio Grande Valley. *Journal of Economic Entomology* 85, 1666–1670.

Republic of South Africa. Department: Agriculture, Forestry and Fisheries (2011) A profile of the South African apple market value chain 2011. Available at: www.nda.agric.za/docs/AMCP/Applemvcp2011-12.pdf (accessed 6 October 2016).

Richardson, P.N. and Grewal, P.S. (1991) Comparative assessment of biological (Nematoda: *Steinernema feltiae*) and chemical methods of control for the mushroom fly, *Lycoriella auripila* (Diptera: Sciaridae). *Biocontrol Science and Technology* 1, 217–228.

Riedl, H., Blomefield, T.L. and Giliomee, J.H. (1998) A century of codling moth control in South Africa: II. Current and future status of codling moth management. *Journal of South African Society of Horticultural Science* 8, 32–54.

Rutherford, R.S. and Conlong, D.E. (2010) Combating sugarcane pests in South Africa: from researching biotic interactions to bio-intensive integrated pest management in the field. *Proceedings of the International Society of Sugar Cane Technologists* 27, 1–17.

Schulze, R.E. (1997) *South African Atlas of Agrohydrology and Climatology*. Water Research Commission, Pretoria, Report TT82/96. 276 pp.

Shamseldean, M.M., El-Sadawy, H. and Allam, S.F.M. (2004) Comparative safety of entomopathogenic nematodes on honeybee workers. *Egyptian Journal of Biological Pest Control* 14, 147–153.

Shapiro-Ilan, D.I. and Glazer, I. (1996) Comparison of entomopathogenic nematode dispersal from infected hosts versus aqueous suspension. *Environmental Entomology* 25, 1455–1461.

Shapiro-Ilan, D.I. and Lewis, E.E. (1999) Comparison of entomopathogenic nematode infectivity from infected hosts versus aqueous suspension. *Environmental Entomology* 28, 907–911.

Shapiro-Ilan, D.I., Lewis, E.E., Tedders, W. and Son, Y. (2003) Superior efficacy observed in entomopathogenic nematodes applied in infected-host cadavers compared with application in aqueous suspension. *Journal of Invertebrate Pathology* 83, 270–272.

Shapiro-Ilan, D.I., Han, R. and Dolinksi, C. (2012) Entomopathogenic nematode production and application technology. *Journal of Nematology* 44, 206–217.

South African Sugar Association [s.d.] Cane growing in South Africa. Available at: www.sasa.org.za/sugar_industry/CaneGrowinginSA.aspx (accessed 6 October 2016).

South African Wine Industry Statistics [s.d.] SAWIS. Available at: www.sawis.co.za/info/download/Engels_2014.pdf (accessed 6 October 2016).

Spaull, V.W. (1988) A preliminary evaluation of the entomogenous nematodes to control the African sugarcane stalk borer *Eldana Saccharina* (Lepidoptera: Pyralidae). *Proceedings of the South African Sugar Technologists' Association* 62, 120–123.

Spaull, V.W. (1990) Field tests to control the pylarid, *Eldana saccharina*, with an entomogenous nematode, *Heterorhabditis* sp. *Proceedings of the South African Sugar Technologists' Association* 64, 103–106.

Spaull, V.W. (1991) *Heterorhabditis* and *Steinernema* species (Nematoda: Rhabditida) for the control of a sugarcane stalk borer in South Africa. *Phytophylactica* 23, 213–215.

Stokwe, N.F. (2009) Entomopathogenic nematodes: characterization of a new species, long-term storage and control of obscure mealybug, *Pseudococcus viburni* (Hemiptera: Pseudococcidae) under laboratory conditions. MSc thesis. Stellenbosch University, Stellenbosch, South Africa.

Stokwe, N.F. and Malan, A.P. (2016) Susceptibility of the obscure mealybug, *Pseudococcus viburni* (Signoret) (Pseudococcidae), to South African isolates of entomopathogenic nematodes. *International Journal of Pest Management*, 62, 119–128.

Stokwe, N.F., Malan, A.P., Nguyen, K.B., Knoetze, R. and Tiedt, L. (2011) *Steinernema citrae* n. sp. (Rhabditida:

Steinernematidae), a new entomopathogenic nematode from South Africa. *Nematology* 13, 567–587.

Tóth, E.M., Márialigeti, K., Fodor, A., Lucskai, A. and Farkas, R. (2005) Evaluation of efficacy of entomopathogenic nematodes against larvae of *Lucilia sericata* (Meigen, 1826) (Diptera: Calliphoridae). *Acta Veterinaria Hungarica* 53, 65–71.

Van Niekerk, S. and Malan, A.P. (2012) Potential of South African entomopathogenic nematodes (Heterorhabditidae and Steinernematidae) for control of the citrus mealybug, *Planococcus citri* (Pseudococcidae). *Journal of Invertebrate Pathology* 111, 166–176.

Van Niekerk, S. and Malan, A.P. (2013) Adjuvants to improve control of *Planococcus citri* (Hemiptera: Pseudococcidae) using entomopathogenic nematodes (Heterorhabditidae and Steinernematidae). *Journal of Helminthology* 89, 189–195.

Van Niekerk, S. and Malan, A.P. (2014a) Compatibility of *Heterorhabditis zealandica* and *Steinernema yirgalemense* with agrochemicals and biological control agents. *African Entomology* 22, 49–56.

Van Niekerk, S. and Malan, A.P. (2014b) Evaluating the efficacy of a polymer-surfactant formulation to improve control of *Planococcus citri* (Risso) (Hemiptera: Pseudococcidae) under simulated natural conditions. *African Plant Protection* 17, 1–8.

Van Zyl, C. (2012) The *in vivo* production of *Heterorhabditis zealandica* and *Heterorhabditis bacteriophora*. MSc thesis. Stellenbosch University, Stellenbosch, South Africa.

Van Zyl, C. and Malan, A.P. (2014a) The role of entomopathogenic nematodes as biological control agents of insect pests, with emphasis on the history of their mass culturing and *in vivo* production. *African Entomology* 22, 235–249.

Van Zyl, C. and Malan, A. (2014b) Optimisation of inoculation techniques for *in vivo* mass culture of entomopathogenic nematodes through nematode and insect host manipulation. *African Entomology* 24, 405–416.

Van Zyl, C. and Malan, A.P. (2015) Cost-effective culturing of *Galleria mellonella* and *Tenebrio molitor* and nematode production in various hosts. *African Entomology* 23, 361–375.

Wakgari, W.M. and Giliomee, J.H. (2004) Insect pests of astringent persimmon in the Western and Eastern Cape provinces of South Africa. *African Plant Protection* 10, 1–5.

Welter, S.C. (2008) Codling moth. In: Resh, H.C.R.T. (ed.) *Encyclopedia of Insects*. Elsevier, Amsterdam, The Netherlands, pp. 189–199.

Witt, A.B.R., Little, R.M. and Crowe, T.M. (1995) The effectiveness of helmeted guineafowl *Numida meleagris* (Linnaeus 1766) in controlling the banded fruit weevil *Phlyctinus callosus* (Schönherr, 1826), and their impact on other invertebrates in apple orchards in the Western Cape province, South Africa. *Agriculture Ecosystems & Environment* 55, 169–179.

World Health Organization (s.d.) (Ultraviolet radiation and the INTERSUN Programme). Available at: http://www.who.int/uv/intersunprogramme/activities/uv_index/en/index3.html (accessed 6 October 2016).

Yan, X., Liu, X., Han, R., Chen, S., De Clerc, P. and Moens, M. (2010) Osmotic induction of anhydrobiosis in entomopathogenic nematodes of the genera *Heterorhabditis* and *Steinernema*. *Biological Control* 53, 325–330.

21 Status of Entomopathogenic Nematodes in Integrated Pest Management Strategies in Italy

Eustachio Tarasco,[1]* Adriano Ragni[1] and Giovanna Curto[2]

[1]Dipartimento del Suolo, della Pianta e degli Alimenti (DISSPA), University of Bari 'Aldo Moro' Bari, Italy; [2]Servizio Fitosanitario Regione Emilia Romagna, Nematology Laboratory, Bologna, Italy

21.1 Introduction

Pest management of harmful insects by applying entomopathogenic nematodes (EPNs) has been developed in Italy since the early 1980s by Deseö (1982), Deseö *et al.* (1984) and Triggiani (1983), continuing in the 1990s (Boselli *et al.*, 1997) and developing strongly since the 2000s in vegetable crops in the open field and in the greenhouse, in orchards and ornamentals.

Laboratory tests and surveys of native EPN species and strains have promoted this biocontrol that has been improved in Italy not only in organic farming but also in integrated pest management (IPM) in agriculture, in forestry and in urban parks frequented by people.

The first data on EPNs in Italy go back to the first samplings carried out in the 1980s (Deseö *et al.*, 1988). Subsequent surveys conducted in recent decades (Tarasco and Triggiani, 1997, 2005, 2007; Triggiani *et al.*, 2004; Clausi and Vinciguerra, 2005, 2007, 2008; Rappazzo *et al.*, 2005, 2011; Susurluk *et al.*, 2007; Tarasco *et al.*, 2008, 2009, 2015; Clausi *et al.*, 2011) have contributed to the discovery of many other EPN strains and new species. Discovery and knowledge of indigenous EPN populations in Italy has continued with the increase of the availability of commercial products and their use in agricultural and forest pest management. Commercial products containing *Heterorhabditis bacteriophora*, *Heterorhabditis megidis*, *Steinernema feltiae* and *Steinernema carpocapsae* are applied for the effective control of several agricultural insect pests belonging to the Heteroptera, Lepidoptera, Coleoptera, Diptera and Hymenoptera orders (Table 21.1).

In Italy, EPN formulations have been tested effectively and applied to control *Tomicus piniperda* L. (Coleoptera: Scolytidae) (Triggiani, 1983) and *Thaumetopoea pityocampa* Denis and Schiffermüller (Lepidoptera: Thaumetopoeidae) on pine (Triggiani and Tarasco, 2002); *Curculio elephas* Gyll. (Coleoptera: Curculionidae), *Curculio glandium* Mars. (Coleoptera: Curculionidae) and *Pammene castanicola* Trematerra (Lepidoptera: Tortricidae), *Pammene fasciana* L. (Lepidoptera: Tortricidae), *Cydia splendana* Hüb. (Lepidoptera: Tortricidae) and *Cydia fagiglandana* (Zeller) (Lepidoptera: Tortricidae) on chestnut (Clausi and Vinciguerra, 2007; Curto *et al.*, 2009b; Clausi *et al.*, 2014); *Curculio nucum* L. (Coleoptera: Curculionidae) on hazelnut (Blum *et al.*, 2009); *Cossus cossus* L. (Lepidoptera: Cossidae) and *Zeuzera pyrina* L. (Lepidoptera: Cossidae) on orchards (Deseö, 1982; Deseö *et al.*, 1984), *Synanthedon tipuliformis* (Clerck) and *Synanthedon spuleri* Fuchs (Lepidoptera: Sesiidae) on persimmon (Caruso *et al.*, 2014); *Hoplocampa brevis* Klug. (Hymenoptera: Tenthredinidae) (Curto *et al.*, 2007) and *Cydia pomonella* L. (Lepidoptera: Tortricidae) (Reggiani *et al.*, 2008; Curto *et al.*, 2009a) on pear; *Capnodis tenebrionis* L. (Coleoptera: Buprestidae) on apricot (Marannino *et al.*, 2003);

*Corresponding author. E-mail: eustachio.tarasco@uniba.it

Table 21.1. List of entomopathogenic nematodes (EPNs) available in the market and recommended against target insect species.

Class insect Order and Family	Species	Stage susceptible to EPNs	Application site	Cultivation	EPN species recommended
Coleoptera					
Buprestidae	*Capnodis tenebrionis*	Larva	Soil	Stone fruit, mainly apricot	*Heterorhabditis bacteriophora*, *Steinernema carpocapsae*, *S. feltiae*
Cerambycidae	*Saperda carcharias*	Larva	Wood tunnels	Poplar	*S. carpocapsae*, *S. feltiae*
Chrysomelidae	*Chaetocnema* spp., *Altica* spp.	Larva	Soil	Potato, beet, vegetables	*S. carpocapsae*
	Diabrotica virgifera virgifera	Larva	Soil	Maize	*H. bacteriophora*, *S. carpocapsae*
	Xanthogaleruca luteola	Larva	Leaves	Elm	*S. carpocapsae*
Curculionidae	*Otiorhynchus* spp.	Larva, pupa	Soil	Ornamental, strawberry, small fruit	*H. bacteriophora*, *S. carpocapsae*, *S. feltiae*
	Rhynchophorus ferrugineus	Larva, pupa	Plant	Palm	*S. carpocapsae*, *S. feltiae*
	Temnorhinus mendicus	Larva, pupa	Soil	Sugarbeet	*H. bacteriophora*, *S. carpocapsae*
	Curculio elephas, *C. glandium*, *C. nucum*	Larva, pupa	Soil	Chestnut, hazelnut	*H. bacteriophora*, *S. carpocapsae* *S. feltiae*
	Tomicus piniperda	Larva, pupa	Bark, stumps	Pine	*H. bacteriophora*, *S. carpocapsae* *S. feltiae*
Diptera					
Agromyzidae	*Liriomyza* spp.	Larva	Leaves	Ornamental, vegetables	*S. feltiae*
Sciaridae	*Lycoria* spp., *Lycoriella* spp., *Bradysia* spp. and *Platosciara* spp.	Larva	Soil	Ornamental, mushrooms	*S. feltiae*
Hemiptera					
Tingidae	*Corythuca ciliata*	Adult overwintering	Bark	Plane tree	*H. bacteriophora*, *S. carpocapsae*
Hymenoptera					
Tenthredinidae	*Hoplocampa brevis*	Larva	Soil, leaves	Pear	*H. bacteriophora*, *S. feltiae*
	Caliroa varipes	Larva	Soil, leaves	Oak	*H. bacteriophora*, *S. feltiae*
Lepidoptera					
Castniidae	*Paysandisia archon*	Larva	Crown, leaves insertion	Palm	*S. carpocapsae*, *S. feltiae*
Cossidae	*Cossus cossus*, *Zeuzera pyrina*	Larva	Wood tunnels	Orchards	*S. carpocapsae*, *S. feltiae*
Gelechiidae	*Tuta absoluta*	Larva	Leaves	Tomato, potato, pepper, aubergine	*H. bacteriophora*, *S. carpocapsae*, *S. feltiae*
Noctuidae	*Agrotis* spp., *Spodoptera* spp.	Larva	Soil, leaves	Ornamental, vegetables	*S. carpocapsae*, *S. feltiae*

Pyralidae	*Euzophera bigella*	Larva	Trunk, branches	Orchards	*S. carpocapsae*, *S. feltiae*
Sesiidae	*Synanthedon* spp.	Larva	Trunk, bark	Orchards, small fruit	*S. carpocapsae*, *S. feltiae*
Thaumetopoeidae	*Thaumetopoea pityocampa*	Larva	Insect nest	Pine	*S. feltiae*
Tortricidae	*Pammene fasciana*, *Cydia splendana*, *C. fagiglandana*	Larva	Soil	Chestnut	*H. bacteriophora*, *S. carpocapsae*, *S. feltiae*
	C. pomonella, *C. molesta*, *C. funebrana*	Larva	Trunk, branches	Apple, pear, peach, apricot, plum	*S. carpocapsae*, *S. feltiae*
Thysanoptera					
Thripidae	*Frankliniella occidentalis*	Juveniles	Leaves	Ornamental, vegetables	*S. feltiae*
Class Gasteropoda					
Order and Family					
Pulmonata					
Milacidae		Juveniles	Soil	Various	*Phasmarhabditis hermaphrodita*
Limacidae		Juveniles	Soil	Various	*P. hermaphrodita*
Arionidae		Juveniles	Soil	Various	*P. hermaphrodita*

Saperda carcharias L. (Coleoptera: Cerambycidae) on poplar (Curto *et al.*, 2003); *Xanthogaleruca luteola* Müller (Coleoptera: Chrysomelidae) on elm (Tarasco and Triggiani, 2007); *Corythucha ciliata* Say (Hemiptera: Tingidae) on plane tree (Tarasco and Triggiani, 2006); *Caliroa varipes* Klug. (Hymenoptera Tenthredinidae) on oak (Curto *et al.*, 2008); *Paysandisia archon* Burm. (Lepidoptera: Castniidae) (Nardi *et al.*, 2009) and *Rhynchophorus ferrugineus* Oliver (Coleoptera: Curculionidae) (Clausi and Vinciguerra, 2008; Nardi *et al.*, 2011; Triggiani and Tarasco, 2011) on palm; *Otiorhynchus sulcatus* F. (Coleoptera: Curculionidae) on ornamental plants and strawberry (Curto *et al.*, 2001); *Temnorhinus mendicus* Gyll. (Coleoptera: Curculionidae) on sugarbeet (Boselli *et al.*, 1997); *Rhytidoderes plicatus* Oliv. (Coleoptera: Curculionidae) on savoy cabbage (Tarasco and Triggiani, 2002); *Diabrotica virgifera virgifera* LeConte (Coleoptera: Chrysomelidae) on maize (Toepfer *et al.*, 2012); Lepidoptera noctuids on artichokes (Ippolito and Triggiani, 1995); *Liriomyza trifolii* (Burgess) (Diptera; Agromyzidae) on flowering plants (Colombo and Locatelli, 1985); Diptera Sciaridae genera *Lycoria* Meigen, *Lycoriella* Frey, *Bradysia* Winnertz and *Platosciara* Berg, in nurseries of plant material and in mushroom cultivation (Curto, 2006); and slugs and snails in horticulture with the specific nematode *Phasmarhabditis hermaphrodita* Sch. (Rhabditida: Rhabditidae) (Gengotti *et al.*, 2006).

The present chapter is focused on the description of the current strategies of EPN application on the key pests in Italy, with important information on the commercial products and their timing and distribution methods (Curto, 2014).

21.2 Integrated Pest Management in Italy

IPM began in Italy during the 1980s, as a reaction to plant protection management as calendar-based pest control with the use of synthetic chemicals. The change occurred not only because of the ever greater environmental awareness of consumers but also and especially because of the legislative action with both European and national acts that progressively have restricted the use of chemical pesticides (Franceschini *et al.*, 2009). Finally, the European Union (EU) framework Directive 128/2009 for the sustainable use of pesticides has decreed IPM as the sole method of disease management beside biological control management. Since 1 January 2014, the Directive allows the use of chemicals only as a part of IPM, to limit their application to only when it is unavoidable. In Italy, the Directive was implemented by Legislative Decree No 150 on 14 August 2012 (Italian Republic, 2012). The national law has introduced the drafting of a national plan of action for the sustainable use of plant protection products (Plan). The Plan was adopted on 26 November 2012 and defined objectives, evaluation methods, mode and timetable for the reduction of the risks and impacts of the use of pesticides on human health, environment and biodiversity. In addition, it promotes the development and introduction of either integrated control and production methods or techniques of alternative control, in order to reduce the dependence on chemical plant protection products, also in relation to the need to ensure economically sustainable production, satisfying the requirements of quality established under the current rules. The objectives of the Plan shall cover the following areas: (i) protection of the users of plant protection products and of the population affected; (ii) consumer protection; (iii) preservation of aquatic environment and of drinking water; and (iv) conservation of biodiversity and ecosystems. According to the law, the Italian regions are responsible for the implementation of integrated control (compulsory regime from 2014). Each region provides technical support for weather and disease monitoring and data processing, technical assistance and formation for integrated control and organic farming drafting, and weekly technical bulletins of information for specialized operators through adequate media. Economic aid with European and national tools is directed solely to farmers who respect those laws. The application of IPM is increasing greatly in Italy as a result of the effects of these new rules: in 2012, 50% of orchards were already being managed by IPM (Caponero, 2012).

21.3 Application of Biological Control Agents in Integrated Pest Management in Italy

Biocontrol agents are very successful in Italy and have been accepted by farmers, obviously in organic agriculture, but in IPM too, when populations of harmful insects increase out of control or show resistance to chemicals; furthermore, EPNs are applied successfully against juvenile stages of borer insects tunnelling roots, woody parts of trees or living in the soil.

In Italy, products containing EPNs are sold by various companies that not always produce them directly: in some cases, these companies sell EPNs reared by foreign factories. The products, the selling company, the EPN species and the key pests, indicated on the product brochure, currently available in Italy are listed in Table 21.2.

21.4 An Outline of Studies on Entomopathogenic Nematodes in Italy

Surveys to assess the dissemination of EPNs have been carried out throughout much of Italy, applying a biogeographic methodology to define Italian EPN biodiversity (Tarasco *et al.*, 2015). Currently, almost 140 indigenous strains of EPNs have been extracted from different natural and anthropic surroundings. They belong to the following species: *H. bacteriophora*, *Heterorhabditis downesi*, *H. megidis*, *Steinernema feltiae*, *Steinernema affine*, *Steinernema kraussei*, *Steinernema apuliae*, *Steinernema ichnusae*, *S. carpocapsae*, *Steinernema vulcanicum*, *Steinernema* 'isolate S.sp.MY7' of '*S. intermedium* group', *Steinernema arenarium* and *Oscheius onirici*. Steinernematids are more widespread and biodiverse than heterorhabditids, and *S. feltiae* and *H. bacteriophora* are the most commonly encountered species. EPN strains and species are found mainly in natural sites rather than in cultivated soil. *S. kraussei*, *H. downesi* and *H. megidis* were collected only in Sicily, whereas two of the species recently described – *S. apuliae* and *S. vulcanicum* – are known to be isolated only in Italy, and appear to be endemic; *S. ichnusae*, firstly isolated from Sardinia only, seems to be more widespread in the Mediterranean area. *O. onirici*, a new species recently described, was the latest EPN species found and was isolated in a cave in Tuscany (Torrini *et al.*, 2015). This study on the ecology of EPNs demonstrates that they are particularly present in two natural habitats, pinewoods and broadleaf woodland, where there are the highest values of biodiversity indexes.

21.5 Application of Entomopathogenic Nematodes in Vegetable Crop Protection

21.5.1 Vine weevil (*Otiorhynchus sulcatus*) control in nurseries, ornamental plants and strawberry

The vine weevil completes one generation per year and overwinters as young larva in the soil; the last larval stage is reached next spring and the adult flies in June. The adult population consists almost exclusively of parthenogenetic females that lay eggs at different times, at the base of plants. In summer, the eggs hatch and the newborn larvae can destroy the root system.

Damage is caused mainly by the larvae, which penetrate the roots, digging niches and tunnels; in addition, the larvae feed on the roots and can break them. Infested plants show a stunted growth, with final wilting and withering. Adults cause mild and slightly harmful erosions on the edge of the leaves.

EPNs from the genus *Heterorhabditis* have shown the best infectivity on vine weevils, although acceptable outcomes are obtained with formulated products based on *S. kraussei*, *S. feltiae* and, to a lesser extent, *S. carpocapsae* (Curto *et al.*, 2001). Formulations based on *H. bacteriophora* are marketed in Italy in boxes containing 50 million infective juveniles (IJs). Application doses correspond to 500,000 IJs/m^2 (one pack for 250 m^2) or 25,000–40,000 IJs/plant (one package for 1250–2000 m^2).

The best result in managing *O. sulcatus* infestation is obtained treating the insect pest at the first larval stage, which is the most susceptible, when it also starts its feeding activity, and the application is done in late summer and when the soil temperature is between 18 and 22°C. Spring treatments on overwintering mature larvae and newly formed pupae could be successful only with soil temperatures above 15°C and with an application rate of 500,000 nematodes/m^2. EPN treatment can be distributed to soil by either drip irrigation or a watering can, that is, for small plantations, on ornamental nurseries and strawberry crops.

21.6 Application of Entomopathogenic Nematodes in Arable Crop Protection

Western corn rootworm is one of the most severe pests of maize, endemic in the maize area in the USA. Over 25 years ago, this pest has moved to Europe, causing destructive attacks to maize fields. Joint efforts were set by many countries, including Italy, to develop an EPN-based biocontrol for the western corn rootworm (Toepfer *et al.*, 2012).

The most suitable EPNs for *Diabrotica* control belong to the genus *Heterorhabditis*. Both species *H. bacteriophora* and *H. megidis* are effective, with more than 80% of pest mortality in field applications. Formulated products based on *H. bacteriophora* are marketed in Italy in packs containing 250 million nematodes. The application rate is 1.5–3.0

Table 21.2. Entomopathogenic nematode-based products in Italy.

EPN-based product	Selling company	EPN species and number per pack	Key pests
Optinem Cydia	Agrifutur	*Steinernema feltiae* (1.5 billion)	Codling moth (*Cydia pomonella*)
Optinem OS	Agrifutur	*Heterorhabditis bacteriophora* (50 million)	Vine weevil (*Otiorhynchus sulcatus*) and other Coleoptera in the soil
Optinem Cydia SC	Agrifutur	*S. carpocapsae* (1.5 billion)	Codling moth (*C. pomonella*)
Optinem Sciar	Agrifutur	*S. feltiae* (50 million)	Juveniles of Sciarid flies
Heterorhabditis megidis	Andermatt Biocontrol AG	*H. megidis* (5 million, 50 million, 250 million)	Vine weevil (*O. sulcatus*), weevils
H. bacteriophora	Andermatt Biocontrol AG	*H. bacteriophora* (5 million, 50 million, 250 million)	Weevils, vine weevil (*O. sulcatus*), hazelnut borer (*Curculio nucum*), chestnut moth (*Cydia splendana*), beech seed moth (*Cydia fagiglandana*), western corn rootworm (*Diabrotica virgifera virgifera*)
S. feltiae	Andermatt Biocontrol AG	*S. feltiae* (5 million, 50 million, 250 million)	Juveniles of Sciarid flies, codling moth (*C. pomonella*), thrips juveniles and pupae, leaf miners (leaf applications)
S. carpocapsae	Andermatt Biocontrol AG	*S. carpocapsae* (5 million, 50 million, 250 million); different pack for palm applications	Codling moth (*C. pomonella*), noctuid moths (*Agrotis* spp.), palm borer (*Paysandisia archon*), red palm weevil (*Rhynchophorus ferrugineus*), crane fly (*Tipula* spp.)
Nemopak HB	Bioplanet	*H. bacteriophora* (50 million, 500 million)	Vine weevil (*O. sulcatus*) and other Coleoptera in the soil
Nemopak SC palms	Bioplanet	*S. carpocapsae* (50 million + 500 cc of liquid distributors)	Red palm weevil (*R. ferrugineus*), palm borer (*P. archon*)
Nemopak SF	Bioplanet	*S. feltiae* (50 million)	Flies juveniles (Diptera Sciaridae, Foridae, Agromizidae, Muscidae), moth juveniles (Lepidoptera Noctuidae, Agrotidae, Cossidae, Sesidae)
Nemapom	CBC Biogard Italia	*S. feltiae* (1.5 billion)	Codling moth (*C. pomonella*), other Lepidoptera of orchards (*Cydia* spp., *Synanthedon* spp., etc.), flatheaded woodborer (*Capnodis tenebrionis*)
Nemastar	CBC Biogard Italia	*S. carpocapsae* (50 million, 500 million)	Codling moth (*C. pomonella*), other Lepidoptera of orchards (*Cydia* spp., *Synanthedon* spp., etc.), palm borer (*P. archon*), red palm weevil (*R. ferrugineus*), noctuid and crambid moths (*Spodoptera* spp., *Agrotis* spp., *Duponchelia fovealis*, etc.), summer chafer (*Amphimallon solstitialis*), *Rhizotrogus aestivus*, flatheaded woodborer (*C. tenebrionis*), adults of mole cricket (*Gryllotalpa gryllotalpa*), crane fly (*Tipula paludosa*, *Tipula oleracea*)
Nemaplus	CBC Biogard Italia	*S. feltiae* (50 million, 500 million)	Codling moth (*C. pomonella*), other Lepidoptera of orchards (*Cydia* spp., *Synanthedon* spp., etc.), *Frankliniella occidentalis*, juveniles of Sciarid flies (*Bradysia* spp.), Diptera Ephrytidae shore fly (*Scatella stagnalis*, *Scatella tenuicosta*), leek moth (*Acrolepiopsis assectella*), asparagus borer (*Parahypopta caestrum*), leaf miners (*Liriomyza* spp.), tomato borer (*Tuta absoluta*)

Nematop	CBC Biogard Italia	*H. bacteriophora* (50 million, 500 million)	Vine weevil (*O. sulcatus*)
Capsanem	Koppert Italia	*S. carpocapsae* (2 × 25 million, 2 × 250 million)	Noctuid moths, crane fly (*Tipula* spp.), large pine weevil (*Hylobius abietis*), flatheaded woodborer (*C. tenebrionis*)
Entonem	Koppert Italia	*S. feltiae* (50 million, 2 × 250 million)	Vine weevil (*O. sulcatus*) and juveniles of Sciarid flies (soil applications), thrips juveniles and pupae (leaf applications)
Larvanem	Koppert Italia	*H. bacteriophora* (50 million, 2 × 250 million)	Weevils, vine weevil (*O. sulcatus*)
Larvanem - M	Koppert Italia	*H. megidis* (50 million)	Weevils, vine weevil (*O. sulcatus*)
Scia-Rid	Koppert Italia	*S. feltiae* (2 × 250 million)	Flies juveniles (Diptera Sciaridae, Foridae)
Nemax F	Serbios	*S. feltiae* (250 million), box of 6 packs	Codling moth (*C. pomonella*), Oriental fruit moth (*Cydia molesta*), plum fruit moth (*Cydia funebrana*), chestnut moth (*C. splendana*), beech seed moth (*C. fagiglandana*), peach twig borer (*Anarsia lineatella*), quince moth (*Euzophera bigella*)
Nemax H	Serbios	*H. bacteriophora* (250 million), box of 6 packs	Vine weevil (*O. sulcatus*), western corn rootworm (*D. virgifera virgifera*)
Xedanema	Xeda Italia	*S. feltiae* (250 million)	Codling moth (*C. pomonella*), Oriental fruit moth (*C. molesta*), plum fruit moth (*C. funebrana*), chestnut moth (*C. splendana*), beech seed moth (*C. fagiglandana*), peach twig borer (*A. lineatella*), quince moth (*E. bigella*), red palm weevil (*R. ferrugineus*)

billion IJs/ha. Nematodes are applied in the soil, along rows, at maize sowing, a few weeks before the hatch of *D. virgifera virgifera* eggs. EPN distribution is carried out by water. Soil humidity is crucial for the success of application and depends on soil texture and weather conditions. It is recommended to apply at least 20 hl of water/ha when combined with rain or irrigation, otherwise applying at least 30–40 hl/ha.

21.7 Applications of Entomopathogenic Nematodes in Parks, Urban Tree Plantings and Municipal Forest

21.7.1 Experience in control of red palm weevil (*Rhynchophorus ferrugineus*) in an Italian central-eastern region

R. ferrugineus (Olivier) (Coleoptera: Curculionidae), the red palm weevil (RPW), is an alien invasive pest devastating palms in the Mediterranean area, native to southern India and detected for the first time in Italy in 2004 (Sacchetti *et al.*, 2006). This pest is now widespread in the Italian peninsula, where it affects ornamental palms and has killed more than 30,000 palm exemplars since its introduction.

The Marche region (located in central-eastern Italy) has implemented an IPM protocol for the management of RPW, aimed at containing the diffusion of this insect pest in a demarcated area according to the Commission Decision 2010/467/EU (Nardi *et al.*, 2011). Integrated management of RPW is performed on infested plants and consists of the following steps: (i) research by visual inspection of the presence of the pest on susceptible hosts; (ii) mass trapping of weevil adults in the buffer zone by the use of commercial traps containing the aggregation pheromone of the target insect attractant; (iii) pruning the top of the tree canopy in a spherical shape; (iv) washing of the crown (head of palm) with water at high pressure; and finally (v) application of either chemical insecticides or EPNs. All tested IPM methods have shown to be promising tools to control *R. ferrugineus*. The application of *S. carpocapsae* is done at the rate of 50 million IJs/palm tree, applied in 15–20 l water of suspension and sprayed at low pressure over the palm top at 30, 60 and 180 days after spherical pruning, to prevent new weevil infestations. *S. carpocapsae* can be applied alone or in combination with chitosan or azadirachtin. The overall efficacy of the integrated approach (spherical pruning plus treatment application) in the control of *R. ferrugineus* exceeds 70%. The effectiveness of IPM with just *S. carpocapsae* application ranges between 71 and 79%, compared to 86–93% of chemical (chlorpyrifos + pyrethroids or acetamiprid + bifentrin) applications. The addition of chitosan to EPN treatments does not seem to increase their efficacy (59–81%). It is observed that *S. carpocapsae* is able to parasitize all the weevil stages (larva, pupa and adult), as already reported by Triggiani and Tarasco (2011) in laboratory tests, even if the mortality of adults is lower than larvae. The IPM strategy has been tested to be applicable and effective in controlling the spreading of *R. ferrugineus* when applied with quarantine measures.

21.7.2 Control of the sawfly, *Caliroa varipes*, on oak in urban parks

C. varipes is a Hymenoptera Tenthredinidae, which at Italian latitudes completes two generations per year and overwinters as mature larva cocooned in the soil. The adults, originating from overwintering larvae, fly from April to May. The females lay more than ten eggs under the lower epidermis of the leaf; the larvae emerge in mid-May and develop during this month, living gregariously and eating the leaf parenchyma. When mature, they drop and bury themselves in the soil, where they cocoon and pupate, giving a second generation by the end of June.

Severe infestation of *C. varipes* can be treated with a suspension of *H. bacteriophora* by the last 10 days of May, when the sawfly larvae of the first generation are ready to drop. The soil under the most damaged trees has to be irrigated by 5 mm of water, then a suspension of *H. bacteriophora* at a dose of 5 billion IJs/ha is applied on the soil surface, by means of a watering can, while most of the mature larvae are dropping. Finally, the soil is covered with a plastic net (1 mm × 1 mm mesh). EPN application to the soil, at the most suitable time, is successful in controlling the first generation of *C. varipes*. The irrigation supplied before the EPN treatment and the cool and shadowed habitat under the oak trees promote the penetration of *H. bacteriophora* into sawfly mature larvae before pupation (Fig. 21.1).

As a result, more than 70% decrease in *C. varipes* population has been recorded. A further treatment at the end of the second generation (overwintering) can prevent *C. varipes* infestation the following year (Curto *et al.*, 2008).

21.7.3 Control of pine processionary moth (*Thaumetopoea pityocampa*) in pinewoods

The pine processionary moth (PPM), *T. pityocampa* (Denis and Schiffemüller) (Lepidoptera: Notodontidae), is a highly destructive pine tree pest, infecting pinewoods, especially *Pinus nigra* and *Pinus halepensis* (Masutti and Battisti, 1990; Hodar *et al.*, 2002; Stastny *et al.*, 2006). The pest is univoltine and widespread throughout the Mediterranean region. Their larvae are ready to pupate in the spring, and leave the tree in a single file, head-to-tail procession, searching for soil sites suitable for pupation. Adults emerge and are airborne throughout the summer. Very good results have been achieved by controlling PPM larvae with gel formulations based on native and/or commercial strains of *S. feltiae*, *S. carpocapsae* and *H. bacteriophora* (Triggiani and Tarasco, 2002). EPNs were injected in the PPM overwintering nests at doses corresponding to 100,000–300,000 IJs/30 cc of gel suspension. Nest temperatures around 12–15°C represent the best conditions for the EPN application against overwintering larval stages of *T. pityocampa*. All EPN species are able to kill the larvae, with the best performances obtained by *S. feltiae* native or commercial strains. The EPN treatment can be applied to the nest on small surfaces, on ornamental nurseries and small trees.

Fig. 21.1. *Caliroa varipes* larvae parasitized by *Heterorhabditis bacteriophora*. Photograph courtesy of G. Curto.

21.8 Applications of Entomopathogenic Nematodes in Orchards

21.8.1 Codling moth (*Cydia pomonella*) control on pear and apple orchards

Codling moth is one of the most harmful insects of Pomaceae species; larvae directly damage the fruit, digging tunnels towards the seeds and causing severe yield losses, both in quality and in quantity.

In Italy, *C. pomonella* completes three generations per year and overwinters as cocooned mature larva, under the bark of the trunk of pear and apple trees and secondarily in the soil at the base of plants. In recent years, a specific application strategy of EPNs against this insect has been developed. *S. feltiae* and *S. carpocapsae*, the most suitable species, are applied in orchards against overwintering larvae in autumn (Curto *et al.*, 2010). EPNs are distributed on the trunks and lower branches of apple and pear trees. *C. pomonella* larvae overwinter in the bark crevices, protected by a slight cocoon. Nematodes can easily penetrate this weak defence and colonize the larva (Fig. 21.2). The most suitable application rate is 1.5 billion IJs in 1500 l of water/ha, applied by an atomizer machine, with the higher nozzles closed and without filters on. Treatment timing should be decided according to the weather

Fig. 21.2. *Cydia pomonella* larvae parasitized by *Steinernema feltiae*. Photographs courtesy of G. Curto.

forecast: in fact, EPNs are completely effective with temperatures ≥12°C (although *S. feltiae* is active even at temperatures below 12°C) and a prolonged wetting state of the bark. The application of nematodes is more effective when carried out at the beginning of a rainfall, because the trunks can dry out quickly just after rain and the nematodes can die. EPNs dragged to the ground by the rain remain active for 2 weeks after their application. In the case of treatments done in the absence of rainfall, it is recommended that high volumes of water are sprayed on the trees before and after the EPN application, during cloudy, not windy, foggy weather and with high relative humidity. The new protocol for the use of EPNs against codling moth has allowed more than 1500 ha to be treated throughout northern Italy.

On the conizozythe use of EPNs in spring against overwintering larvae of codling moth does not give good results because the moth larvae enter in pupation stage that nematodes are unable to penetrate.

The treatments with EPNs can be included advantageously in an IPM strategy of the orchards, in order to reduce the populations of *C. pomonella* and have a smaller population to be controlled in the following spring/summer.

21.8.2 Control of the sawfly, *Hoplocampa brevis*, in pear orchard

H. brevis (Hymenoptera, Tenthredinidae) causes severe damage to pear fruit in Italy, particularly in organic orchards, resulting in heavy economical losses and proving to be a difficult problem to solve.

H. brevis completes one generation per year and overwinters as a mature larva cocooned in the soil at 5–10 cm deep (Pollini, 1998), pupating in the springtime. The flight of the adults lasts for 10–12 days, and for up to 25 days with low temperatures. The females lay eggs singly in the flower buds using their ovipositor; eggs hatch after a week and the newborn larvae enter a small pear and dig a tunnel toward the central part of the fruit (primary attack), then leave it burrowing another one (secondary attack) and finally drop in the soil. Sawfly populations live aggregated, so the infestation is generally only placed in part of the orchard (Antropoli *et al.*, 1994).

One foliar treatment of *S. feltiae* at a dose of 2.5 billion IJs/ha, applied during the sawfly larvae migration from the first to the second fruit, succeeded in colonizing and controlling them in the fruit, with the same effectiveness of rotenone. However, EPNs have been demonstrated as being unable to enter the hole of the primary attack, while the secondary one is an excellent pathway for EPN penetration into the fruit, and into the sawfly larva in the tunnel. Therefore, the importance of foliar application is not so consistent in decreasing the number of worm-eaten fruit at pear harvest, but could be effective in decreasing the overall sawfly population in the orchard. In this regard, after rainfall or an irrigation of 5 mm of water, one soil application of 5 billion IJs/ha of either *H. bacteriophora* or *S. carpocapsae* at the base of the trees, just before the drop of mature *H. brevis* larvae, has been effective in reducing sawfly adults the following spring (Curto *et al.*, 2007).

21.8.3 Control of *Capnodis tenebrionis* on stone-fruit trees

A further interesting use of EPNs is represented by treatments against neonate larvae of *C. tenebrionis* (Coleoptera: Buprestidae); an important and common insect pest of Rosaceae trees in several countries of the Mediterranean Basin. Infestations on stone-fruit trees are often so severe as to destroy the orchards totally as a consequence of larval feeding on the roots. Egg laying occurs in summer on the trunk base or nearby in the soil. A survey on the eclosion mode pointed out that neonate larvae, wherever eggs are laid, always come into contact with the soil environment, so justifying EPN application. The susceptibility of newly hatched larvae to EPN species was assessed in two laboratory bioassays by using multiwell tissue culture plates and cherry plum potted plants, respectively. Indigenous Italian strains of *S. carpocapsae* and *H. bacteriophora* proved to be very effective in controlling *C. tenebrionis* larvae (Marannino *et al.*, 2003). The application dose of 50 million IJs/plant against *C. tenebrionis* can be added to IPM strategies with the aim of reducing the populations of insect pests in orchards and in plant nurseries.

21.8.4 Control of the clearwing moths (*Synanthedon tipuliformis* and *Synanthedon spuleri*), on persimmon in northern Italy

Clearwing *S. spuleri* is a moth widespread in northern Italy, causing heavy damage to persimmon orchards. Larvae develop under the bark, at the fork of large branches; their feeding breaks off the sap vessels, causing decay of the plant, yellowing of leaves and darkening of the leaf veins. Clearwings (*S. spuleri* and *S. tipuliformis*) complete one generation per year, and the adults fly from May to September (Pollini, 2007); currant clearwing (*S. tipuliformis*) feeds on blackcurrant, redcurrant, gooseberry, hazelnut and persimmon, where its life cycle is similar to the *S. spuleri* cycle (Caruso *et al.*, 2014).

One application of *S. feltiae* carried out in September–October just before or soon after rainfall, with temperatures ≥10–12°C, is effective at 1.5 billion IJs in 12 hl water/ha. Treatment is sprayed by an atomizer, at a less than 15 bar pressure and removing the nozzles' filters. EPN efficacy in the control of *Synanthedon* spp. reaches a larval mortality of about 60%; this result can be considered satisfactory because of the length of the adult flight and the lack of insecticides authorized on persimmon (Caruso *et al.*, 2014).

21.8.5 Control of carpophagous Lepidoptera (*Pammene fasciana*, *Cydia splendana* and *Cydia fagiglandana*) in chestnut

P. fasciana (chestnut leaf roller), *C. splendana* (chestnut moth) and *C. fagiglandana* (beech seed moth) are the main Lepidoptera species of chestnut in Italy. They directly damage the growing and ripening fruit and cause sizeable quantitative and economic losses.

All these species complete one generation per year and overwinter as mature larvae in the soil, but the adults of *P. fasciana* fly from June to September, with a peak in mid-July: those of *C. fagiglandana* fly from the end of July to the beginning of October, with a maximum in August, and those of *C. splendana* from the end of August to September. Therefore, from late summer to the end of autumn, after fruit damage, mature larvae of the three species bury themselves in the soil for overwintering: the first to leave the fruit is *P. fasciana*, the second is *C. fagiglandana* and the last one is *C. splendana*. In northern Italy, the most representative species of carpophagous insects in chestnut groves is *C. splendana*, and to a lesser extent *C. fagiglandana*; both could be assessed as just one harmful pest (Martini *et al.*, 1998).

The application of EPN suspension on the ground showed encouraging results in decreasing fruit damage at harvest. The most suitable EPN species is *H. bacteriophora*; the EPN treatment is carried out by the end of May, before insect pupation, at the first useful rainfall or distributing high water volumes to the soil, with a temperature ≥12°C. The treatment jet is directed towards the areas of burr accumulation in the chestnut grove, generally at the bottom of a slope; the recommended dose is 1.5 billion IJs/ha mixed with different volumes of water, depending on the soil moisture; the most suitable equipment for EPN distribution can be a launch connected with either a tractor power pump with a pressure <15 bars, or to a water tank placed aloft.

Autumnal EPN application can be effective only until the temperature stays above 10°C. Furthermore, farmers in that season are very busy with chestnut harvesting and processing, and usually prefer to apply the EPN suspension in spring. Anyway, the EPN strategy allows a considerable decrease in moth damage to the fruit (57%), regarded as being satisfactory by chestnut growers (Curto *et al.*, 2009b).

21.9 Applications of Entomopathogenic Nematodes in Production of Garden Mushroom and Oyster Mushroom

21.9.1 Sciarid flies (Diptera: Sciaridae) control in nursery of plant material and in mushroom cultivation

Sciarid flies (Diptera: Sciaridae) can cause damage in young plant material such as seedlings, rootstock and cuttings, and in mushroom cultivation; they belong mainly to the genera *Lycoria*, *Lycoriella*, *Bradysia* and *Platosciara*, and are known as 'flies of peat and mushroom'.

Diptera Sciaridae go through seven stages; namely, egg, four larval stages, pupa and adult. Adults tend to gather in shady and moist areas and are not harmful. Eggs are laid in the soil. Larvae are whitish, up to 5 mm long, living just below the surface of substrate and feeding on organic matter, young roots, mycelium, stems and leaves in contact with the soil: this reduces the uptake of water and nutrients, causing plant death. Indirect damage can be caused by larvae that transmit mites, nematodes, viruses and fungal spores. Pupation takes place in the soil as well. With temperatures above 24°C, reproduction is continuous, the cycle lasts 3–4 weeks and the number of generations per year can be high; with lower temperatures, the life cycle is completed after 8 weeks.

Effective results in the control of Sciarid larvae are obtained by EPNs belonging exclusively to *S. feltiae* species (Curto, 2006). Formulations based on this nematode are marketed in Italy in packs containing 50 million IJs. The doses correspond to

300,000–600,000 IJs/m^2 (one package for 80–150 m^2), 10,000–20,000 IJs/l of soil, or increasing doses, depending on the diameter of the vessel (Table 21.3).

EPN treatments start at the beginning of the crop cycle, before the increase of fly populations. A single application is generally sufficient to contain the infestation, but a second one could be necessary after about 6 weeks. If the application is delayed, the population of Sciarids can reach a high level and the effectiveness of treatments decreases. Optimum temperatures for maximum activity of *S. feltiae* correspond to 15–20°C; temperatures below 10°C and above 30°C nullify the action of this nematode.

21.10 Interactions of Entomopathogenic Nematodes with Pesticides

Generally, EPNs can be combined with chemical insecticides and fungicides, showing in most cases a good interaction; in some cases, insecticides can have a synergistic effect, when sublethal doses weaken insects' defences (Koppenhofer *et al.*, 2003) or paralyse the insects (Nishimatsu and Jackson, 1998), so that the EPNs can attack and penetrate them easily.

In Italy, studies on the compatibility of EPNs with pesticides were carried out by Italian scientists in the 1980s and 1990s and published in Barbarossa *et al.* (1996), Rovesti *et al.* (1988) and Rovesti and Deseö (1990, 1991). Currently, this research topic has not been faced systematically in Italy. Nevertheless, for the first time, the additive interaction between EPNs and *Bacillus thuringiensis aizawai* to control noctuid moths (*Spodoptera exigua* and *Autographa gamma*) has been demonstrated in open field trials (Lanzoni *et al.*, 2014).

21.11 Technical Restrictions for Nematode Applications

The application of EPNs in different Italian agricultural habitats needs particular recommendations in order to succeed in overcoming the limitations that are related to the characteristics of nematodes. EPNs are effective at soil temperatures higher than 10–12°C, in the presence of moisture and when protected from UV light. For this reason, the most effective treatments are made, in Italy, in spring or in autumn, in the late afternoon, irrigating the soil to maintain optimal humidity for 3–5 days (Curto, 2014). Foliar application and treatment on insects shielded in bark have poor efficacy and are frequently ineffective because of the great sensitivity of nematodes to UV light and dehydration. Nevertheless, experiments conducted in Italy have allowed the use of EPNs to be extended also to the field, which is an unusual habitat for them. The control of overwintering larvae of *C. pomonella*, placed in the nooks of bark, was achieved successfully with EPN applications carried out in autumn, moistening the surface of the trunks with 1500 l of water/ha 1 h before the treatment. The same volume of water is required to apply EPNs when treating the trunks directly with sprayers, with a working pressure of lower than 20 bar and with nozzles with a diameter >0.5 mm (Reggiani *et al.*, 2008). The control of *X. luteola* was achieved successfully in the Apulia region (south Italy) by foliar application of EPNs in gel suspensions with IdrosorbSR 200 and Xanthan gum (Triggiani and Tarasco, 2008).

It is well known that EPNs are able to move in galleries excavated by xylophagous insects and to kill them. Deseö (1982; Deseö *et al.*, 1984) tested, in the early 1980s, an application method to place nematodes in the proximity of holes made in the tree trunk by insects. Cotton wool sticks soaked in EPN suspensions were inserted in holes dug by the larvae of *C. cossus* and *Z. pyrina*. Poplars attacked by *S. carcharias* were treated by spraying the trunks with EPN suspensions (Curto and Ricci, 2003).

Table 21.3. Doses of *Steinernema feltiae* for the control of Diptera: Sciaridae in pots.

Pot diameter (mm)	IJs/pot	Number of pots treated with a pack of 50 million IJs
14	25,000	2000
18	40,000	1250
22	60,000	830
26	80,000	625
30	100,000	500

21.12 Entomopathogenic Nematodes – Potential Threat to Ecosystems

The environmental impact of EPN applications in soil ecosystems was studied in Italy by Clausi and Vinciguerra (2007). Nematodes were applied at the rate of 500,000 larvae/m^2 in two uncultivated chestnut groves on Mount Etna (Sicily). In each site there were three plots; one plot as the control and two others treated, respectively, with *H. megidis* and *S. feltiae* commercial products. In one site, just *H. megidis* was applied since *S. feltiae* was naturally present. The survey of soil macrofauna was performed by the use of pitfall traps. Data comparison does not reveal any decreasing or change in the composition of the macroflora in treated and untreated plots. In one site, there was a great decrease of Coleoptera in the plots treated with EPNs, but a similar trend was observed also in untreated plots of the other site. Therefore, the variation observed was considered as normal fluctuations of populations without effect due to treatments.

21.13 Conclusions

In Italy, the use of EPNs has been implemented both in guidelines for organic farming and in those for integrated management agriculture, according to the evidence of their effectiveness obtained in field trials performed for 2 or more years. EPNs have to be applied when the temperature and wetting of substrate can facilitate the search for the host. Therefore, their field applications should be carried out in rainy periods in autumn and in spring. Another key factor for the effectiveness of EPN treatments is to target the susceptible host stages (larvae and, in case of Coleoptera, also pupae). According to these rules, it has been possible to include the use of EPNs in the IPM guidelines for the control of Lepidoptera as codling moth in orchards of apple, pear and walnut, currant clearwing in persimmon, acorn moth and beech moth in chestnut, and for the control of Coleoptera as vine weevil on strawberry fields.

Italian farmers are aware of how to store and handle EPN-based products. The acceptance of farmers is generally very positive for these products. Those positive results were obtained by a broad network of technical assistance for the implementation of integrated agriculture, present in many Italian regions. The technical assistance provides farmers with useful information on effective treatment by EPNs, implementing mathematical models to calculate the days of the highest pest presence in a susceptible stage, installing pheromone traps to assess the density of the pest population, and taking advantage of an efficient network for meteorological forecasting.

Future research plans aim to extend the knowledge on the presence and biodiversity of EPNs in areas where those studies have not yet been performed. It is also planned to undertake molecular studies to select EPNs according their response to abiotic stresses (temperature, water, etc.). All these data will provide useful tools to design additional new IPM strategies that include EPNs as biocontrol agents, together with other natural control means, for a lower use of chemicals in agriculture.

Acknowledgements

The authors express their gratitude to the Italian Ministry of Education, University and Research and to the Emilia Romagna region for financial support.

References

Antropoli, A., Faccioli, G. and Pasqualini, E. (1994) Relationship between adult captures of *Hoplocampa brevis* Klug. (Hymenoptera, Tenthredinidae) and larval infestation. *Bulletin OILB-SROP* 17, 1–4.

Barbarossa, B., Frulla, R., Grazzi, G. and Rovesti, L. (1996) Compatibility of *Steinernema carpocapsae* with pesticides. *Bulletin OILB-SROP* 19, 136–139.

Blum, B., Kron Morelli, R., Vinotti, V. and Ragni, A. (2009) Control of *Curculio nucum*, the hazelnut borer, by entomopathogenic nematodes. *Acta Horticulturae* 845, 567–570.

Boselli, M., Curto, G.M. and Tacconi, R. (1997) Field efficacy of entomopathogenic nematodes against the sugarbeet weevil *Temnorhinus* (= *Conorrhynchus*) *mendicus* Gyll. (Coleoptera: Curculionidae). *Biocontrol Science and Technology* 7, 231–238.

Caponero, A. (2012) Lotta integrata obbligatoria L'Italia è (quasi) pronta. *Terra e Vita* 9, 38–43.

Caruso, S., Vergnani, S., Reggiani, A. and Curto, G. (2014) Valutazione di nematodi entomopatogeni per il controllo della sesia del kaki in Emilia-Romagna [Evaluation of entomopathogenic nematodes for the control of currant clearwing (*Synanthedon tipuliformis*) on persimmon]. *Proceedings of 'Giornate Fitopatologiche'*, Vol 1, 18–21 March, 2014, Chianciano Terme, Italy. CLUEB – Cooperativa Libraria Universitaria Editrice Bologna, Bologna, Italy, pp. 133–136.

Clausi, M. and Vinciguerra, M.T. (2005) I nematodi entomopatogeni in un progetto per lo sviluppo sostenibile dei castagneti. *Nematologia Mediterranea* 33, 91–94.

Clausi, M. and Vinciguerra, M.T. (2007) Efficacy and environmental impact of entomopathogenic nematodes used against nut insect pests in some chestnut woods on Etna (Italy). *IOBC/WPRS Bulletin* 30, 33–37.

Clausi, M. and Vinciguerra, M.T. (2008) Prove di lotta biologica contro il Punteruolo rosso delle palme per mezzo di Nematodi entomopatogeni. In: *La ricerca scientifica sul punteruolo rosso e gli altri fitofagi delle palme in Sicilia*, Vol 1. Regione Siciliana Assessorato Agricoltura E Foreste Dipartimento Interventi Infrastrutturali Servizio XI - Servizi allo Sviluppo, Italy, pp. 165–168.

Clausi, M., Leone, D., Vinciguerra, M.T., Rappazzo, G. and Tarasco, E. (2014) Laboratory tests on the biocontrol of

chestnut insect pests on Etna (Sicily, Italy) by means of entomopathogenic nematodes. *Redia* 97, 133–136.
Clausi, M., Longo, A., Rappazzo, G., Tarasco, E. and Vinciguerra, M.T. (2011) *Steinernema vulcanicum* n. sp. (Rhabditida: Steinernematidae), a new entomopathogenic nematode species from Sicily (Italy). *Nematology* 13, 409–423.
Colombo, M. and Locatelli, D.P. (1985) Laboratory evaluation of the activity of *Steinernema feltiae* Filip. and *Heterorhabditis* spp. on *Liriomyza trifolii* (Burgess) and *Opogona sacchari* (Bojer) infesting cultivated flowering plants. *La Difesa delle Piante* 8, 263–269.
Curto, G. (2006) I principali fitofagi degli apparati radicali di piante ornamentali e forestali in vivaio: efficaci metodi di difesa biologica con nematodi entomopatogeni. A Review. *Petria* 16(3), 295–319.
Curto, G. (2014) Nematodi entomopatogeni (*Steinernema*, *Heterorhabdits*). In: Butturini, A. and Galassi, T. (eds) *Difesa fitosanitaria in produzione integrata*. Edagricole, Bologna, Italy, pp. 283–291.
Curto, G. and Ricci, M. (2003) Saperda del Pioppo. In: *Abstracts of Convegno Nazionale 'Nematodi quali agenti biologici di controllo su insetti di rilevanza sia per l'agricoltura che per la salute dell'uomo'*, 26 September 2003, Perugia, Italy.
Curto, G., Boselli, M. and Ricci, M. (2001) Winter field trial against blackvine weevils *Otiorhynchus sulcatus* in potted *Euonymus japonica* using the cold active nematode *Steinernema kraussei* N0093 strain. In: Griffin C.T., Burnell, A.M., Downes, M.J. and Mulder, R. (eds) *Developments in Entomopathogenic Nematode/Bacterial Research*. COST Action 819 – European Commission EUR 1969, Office for Official Publications of the European Communities L-2985, Luxembourg, pp. 285–290.
Curto, G., Boselli, M., Vergnani, S. and Reggiani, A. (2007) Effectiveness of entomopathogenic nematodes in the control of sawflies (*Hoplocampa brevis*) in pear orchards. *IOBC/WPRS Bulletin* 30, 13–17.
Curto, G., Caruso, S., Reggiani, A. and Vergnani, S. (2009a) Efficacia dei nematodi entomopatogeni nel contenimento delle larve di *Cydia pomonella* L. nell'Italia settentrionale. *Frutticoltura* 3, 45–49.
Curto, G., Caruso, S., Reggiani, A. and Vergnani, S. (2010) Difesa del melo dalla carpocapsa con nematodi. *L'Informatore Agrario* 39, 48–50.
Curto, G., Reggiani, A., Dallavalle, E. and Bariselli, M. (2009b) Control of carpophagous Lepidoptera in chestnut by means of entomopathogenic nematodes. *IOBC/ WPRS Bulletin* 45, 345–348.
Curto, G., Santi, R., Vai, N., Barani, A., Vezzadini, S. and Ricci, M. (2003) Controllo biologico di larve di *Saperda carcharias* (L.) in un pioppeto industriale. *Proceedings National Meeting 'Nematodi quali agenti biologici di controllo su insetti di rilevanza sia per l'agricoltura che per la salute'*, 23 September 2003, Perugia, Italy (as CD file).
Curto, G., Vai, N. and Dallavalle, E. (2008) Control of *Caliroa varipes* (Klug, 1814) (Hymenoptera, Tenthredinidae) with entomopathogenic nematodes, in an urban park of Bologna suburbs. *Redia* 91, 173–176.
Deseö, K.V. (1982) Prove di lotta col nematode entomopatogeno *Neoaplectana carpocapsae* Weiser contro i rodilegno *Cossus cossus* L. e *Zeuzera pyrina* L. (Lepidoptera; Cossidae). [Trials with *Neoaplectana carpocapsae* Weiser (Nematoda) in the control of two treeborers *Cossus cossus* L. and *Zeuzera pyrina* L. (Lepidoptera; Cossidae)]. *Proceedings of 'Giornate Fitopatologiche'*, Vol 3, 16–19 March 1982, Sanremo, Italy. CLUEB – Cooperativa Libraria Universitaria Editrice Bologna, Bologna, Italy, pp. 3–10.
Deseö, K.V., Fantoni, P. and Lazzari, G.L. (1988) Presenza di nematodi entomopatogeni (*Steinernema* spp., *Heterorhabditis* spp.) nei terreni agricoli d'Italia. [Natural occurrence of Enthomopathogenic Nematodes (*Steinernema* spp. and *Heterorhabditis* spp.) in farmland soil in Italy]. *Proceedings of 'Giornate Fitopatologiche'*, Vol 2, 16–20 May, 1988, Lecce, Italy, pp. 269–280.
Deseö, K.V., Grassi, S., Foschi, F. and Rovesti, L. (1984) Un sistema di lotta biologica contro il Rodilegno giallo (*Zeuzera pyrina* L., Lepidoptera Cossidae). [Biological methods for the control of the Leopard Moth (*Zeuzera pyrina* L., Lepidoptera Cossidae)]. *Proceedings of 'Giornate Fitopatologiche'*, Vol 2, 26–29 March 1984, Sorrento, Italy, pp. 403–414.
Franceschini, S., Ladurner, E. and Benuzzi, M. (2009) General perspective of biocontrol industry. *Insect Pathogens and Insect Nematodes IOBC/WPRS Bullettin* 45, 17–25.
Gengotti, S., Censi, D. and Curto, G. (2006) Alternative alla metaldeide per il controllo dei gasteropodi. *L'Informatore Agrario* 12, 56–59.
Hodar, J.A., Zamora, R. and Castro, J. (2002) Host utilisation by moth and larval survival of pine processionary caterpillar *Thaumetopoea pityocampa* in relation to food quality in three *Pinus* species. *Ecological Entomology* 27, 292–301.
Ippolito, R. and Triggiani, O. (1995) Efficacia di *Heterorhabditis* e *Steinernema feltiae* Fili. (=*bibionis*) (Nematoda) nei riguardi di larve di Lepidotteri Nottuidi su carciofo. *Informatore Fitopatologico* 9, 55–60.
Italian Republic (2012) DECRETO LEGISLATIVO 14 agosto 2012, n. 150 Attuazione della direttiva 2009/128/CE che istituisce un quadro per l'azione comunitaria ai fini dell'utilizzo sostenibile dei pesticidi. (12G0171) (GU n.202 del 30-8-2012 – Suppl. Ordinario n. 177).
Koppenhofer, A.M., Cowles, R.S., Cowles, E.A., Fuzy, E.M. and Kaya, H.K. (2003) Effect of neonicotinoid insecticide synergists on entomopathogenic nematode fitness. *Entomologia Experimentalis et Applicata* 106, 7–18.
Lanzoni, A., Ade, G., Martelli, R., Radeghieri, P. and Pezzi, F. (2014) Technological aspects of *Steinernema carpocapsae* spray application alone or mixed with

Bacillus thuringiensis aizawai in spinach crop. *Bulletin of Insectology* 67, 115–123.

Marannino, P., Tarasco, E. and De Lillo, E. (2003) Biological notes on larval hatching in *Capnodis tenebrionis* (L.) (Coleoptera Buprestidae) and evaluation of entomopathogenic nematodes in controlling neonate larvae. *Redia* 86, 101–105.

Martini, A., Baronio, P., Baldassari, N. and Rocchetta, G. (1998) I lepidotteri tortricidi del castagno (*Pammene fasciana* (L.), *Cydia fagiglandana* (Zel.) e *Cydia splendana* (Hb.)) valutati come un'unica entità di danno. *Bollettino Istituto Entomologia 'G. Grandi' Università di Bologna* 52, 105–114.

Masutti, L. and Battisti, A. (1990) *Thaumetopoea pityocampa* (Den. and Schiff.) in Italy. Bionomics and perspectives of integrated control. *Journal of Applied Entomology* 110, 229–234.

Nardi, S., Ricci, E., Lozzi, R., Marozzi, F., Ladurner, E., *et al.* (2009) Use of entomopathogenic nematodes for the control of *Paysandisia archon* Burmeister. *IOBC/WPRS Bulletin* 45, 375–378.

Nardi, S., Ricci, E., Lozzi, R., Marozzi, F., Ladurner, E., *et al.* (2011) Control of *Rhynchophorus ferrugineus* (Olivier, 1790) according to EU Decision 2007/365/CE in the Marche region (Central-Eastern Italy). *OEPP/EPPO Bulletin* 41, 103–115.

Nishimatsu, T. and Jackson, I.J. (1998) Interaction of insecticides, entomopathogenic nematodes and larvae of the western corn rootworm (Coleoptera: Chrysomelidae). *Journal of Economic Entomology* 91, 410–418.

Pollini, A. (1998) *Manuale di entomologia applicata*. Edagricole, Bologna, Italy, pp. 1255–1257.

Pollini, A. (2007) *La difesa delle piante da frutto. Sintomi, diagnosi e terapia*. IV edizione. Edagricole, Bologna, Italy, 233 pp.

Rappazzo, G., Avanzato, C., Petronio, G., Clausi, M., Tarasco, E., *et al.* (2011) Bacterial symbionts of entomopathogenic nematodes of the genus *Steinernema* from South Italy: phenotypic features and evolutionary implications. *IOBC/WPRS Bulletin* 66, 321–325.

Rappazzo, G., Clausi, M. and Vinciguerra, M.T. (2005) Molecular identification of entomopathogenic nematodes found in pristine chestnut growing areas on the foothills of Etna (Sicily). *Nematologia Mediterranea* 33, 101–104.

Reggiani, A., Curto, G., Vergnani, S., Caruso, S. and Boselli, M. (2008) Effectiveness of entomopathogenic nematodes in the control of *Cydia pomonella* overwintering larvae in Northern Italy. *IOBC/WPRS Bulletin* 31, 287–293.

Rovesti, L. and Deseö, K.K. (1990) Compatibility of chemical pesticides with entomopathogenic nematodes, *Steinernema carpocapsae* Weiser and *S. feltiae* Filipjev (Nematoda: Steinernematidae). *Nematologica* 36, 237–245.

Rovesti, L. and Deseö, K.K. (1991) Compatibility of pesticides with entomopathogenic nematodes, *Heterorhabditis heliothidis*. *Nematologica*, 37, 113–122.

Rovesti, L., Heinzpeter, E.W., Tagliente, F. and Dese, K.V. (1988) Compatibility of pesticides with the entomopathogenic nematode *Heterorhabditis bacteriophora* Poinar (Nematoda: Heterorhabditidae). *Nematology* 34, 462–476.

Sacchetti, P., Camera, A., Granchietti, A., Rosi, M.C. and Marzialetti, P. (2006) Identificazione, biologia e diffusione del curculionide *Rhynchophorus ferrugineus* (Olivier). *Informatore Fitopatologico* 6, 35–40.

Stastny, M., Battisti A., Petrucco-Toffolo E., Schlyter, F. and Larsson, S. (2006) Host-plant use in the range expansion of the pine processionary moth, *Thaumetopoea pityocampa*. *Ecological Entomology* 31, 481–490.

Susurluk, I.A., Tarasco, E., Triggiani, O. and Ehlers, R.U. (2007) Molecular characterisation of entomopathogenic nematodes isolated in Italy by PCR–RFLP analysis of the ITS region of the ribosomal DNA repeat unit. *Nematologia Mediterranea* 35, 23–28.

Tarasco, E. and Triggiani, O. (1997) Survey of *Steinernema* and *Heterorhabditis* (Rhabditida: Nematoda) in Southern Italian soils. *Entomologica* 31, 117–123.

Tarasco, E. and Triggiani, O. (2002) Biocontrol of *Rhytidoderes plicatus* Oliv. (Coleoptera, Curculionidae) in potted savoy cabbage with entomopathogenic nematodes. *Entomologica* 36, 157–164.

Tarasco, E. and Triggiani, O. (2005) Entomopathogenic nematodes of forest habitats in Southern Italy and their pathogenicity for overwintering larvae of *Thaumetopoea pityocampa* (Den. et Schiff.) (Lepidoptera, Thaumetopoeidae). In: Lieutier, F. and Ghaioule, D. (eds) *Entomological Research in Mediterranean Forest Ecosystems*. INRA, France, pp. 257–263.

Tarasco, E. and Triggiani, O. (2006) Evaluation and comparison of entomopathogenic nematodes and fungi to control *Corythucha ciliata* Say (Rhynchota Tingidae). *Redia* 89, 51–54.

Tarasco, E. and Triggiani, O. (2007) Entomopathogenic nematodes and forest insects in Italy. *IOBC/WPRS Bulletin* 30, 63–65.

Tarasco, E., Clausi, M., Rappazzo, G., Panzavolta, T., Curto, G., *et al.* (2015) Biodiversity of entomopathogenic nematodes in Italy. *Journal of Helminthology* 89, 359–366.

Tarasco, E., Clausi, M., Rappazzo, G., Vinciguerra, M.T., Longo, A. and Triggiani, O. (2009) Could Italy be considered a favourite place in Europe for EPN biodiversity? *IOBC/WPRS Bulletin* 45, 387–389.

Tarasco, E., Mráček, Z., Nguyen, K.B. and Triggiani, O. (2008) *Steinernema ichnusae* sp. n. (Nematoda: Steinernematidae), a new entomopathogenic nematode from Sardinia Island (Italy). *Journal of Invertebrate Pathology* 99, 173–185.

Toepfer, S., Knuth, P., Glas, M. and Kuhlmann, U. (2012) Successful application of entomopathogenic nematodes for the biological control of western corn rootworm larvae in Europe – a mini review. *Proceedings International Conference on the German Diabrotica Research Program*, 14–16 November 2012, Berlin, Germany. Julius-Kühn-Archiv 2014 No 444. Julius Kühn Institut, Quedlinburg, Germany, 204 pp.

Torrini, G., Mazza, G., Carletti, B., Benvenuti, C., Roversi, P.F., *et al*. (2015) *Oscheius onirici* sp. n. (Nematoda: *Rhabditidae*): a new entomopathogenic nematode from an italian cave. *Zootaxa* 3937(3), 533–548.

Triggiani, O. (1983) Sensibilità del *Tomicus (Blastophagus) piniperda* L. (Coleoptera: Scolytidae) a Nematodi della famiglia Steinernematidae e Heterorhabditidae. *Entomologica* 18, 215–223.

Triggiani, O. and Tarasco, E. (2002) Efficacy and persistence of entomopathogenic nematodes in controlling larval populations of *Thaumetopoea pityocampa* (Lepidoptera: Thaumetopoeidae). *Biocontrol Science and Technology* 12, 747–752.

Triggiani, O. and Tarasco, E. (2008) Foliar application of EPNs to control *Xantogaleruca luteola* Mueller (Coleoptera, Chrysomelidae): insect pathogens and insect parasitic nematodes. *IOBC/WPRS Bulletin* 31, 294–296.

Triggiani, O. and Tarasco, E. (2011) Evaluation of the effects of autochthonous and commercial isolates of Steinernematidae and Heterorhabditidae on *Rhynchophorus ferrugineus*. *Bulletin of Insectology* 64, 175–180.

Triggiani, O., Mráček, Z. and Reid, A. (2004) *Steinernema apuliae* sp. n. (Rhabditida: Steinernematidae): a new entomopathogenic nematode from southern Italy. *Zootaxa* 460, 1–12

22 Status of Entomopathogenic Nematodes in Integrated Pest Management Strategies in Poland

JOLANTA KOWALSKA*

Institute of Plant Protection – National Research Institute, Poznań, Poland

22.1 Introduction

In Poland, biological protection has a long tradition. Polish researchers have developed technology to produce many biological products, some of which have been produced and used on a larger scale (Lipa and Pruszyński, 2010). Biological agents (BAs) have been applied in the control of the most severe greenhouse pests (such as aphids, mites, greenhouse whiteflies and thrips). However, their efficacy depends on numerous abiotic factors and, therefore, results are not always as satisfactory as would be expected (Sosnowska and Fiedler, 2009).

Poland is a very important producer of vegetables in Europe. On an area of 5000 ha, greenhouse production of vegetables has been estimated at: 2500 ha of tomato, about 1500 ha of cucumber and about 1000 ha of other vegetables (Smolenski, 2007). In 2012, the total area of vegetable production under cover was 5305 ha, of which horticulture accounted for 1689 ha and tunnels 3573 ha (Fig. 22.1).

Entomopathogenic nematodes (EPNs) have been used successfully for many years to control pests, mainly in vegetable crops. They are safe from the point of view of human health and the environment, and their application is established, although new targets are still identified and further studies are needed (e.g. Kucharska *et al.*, 2009; Askary, 2010; Wojciechowska *et al.*, 2013). EPNs use insects as a means of transport in order to reach new habitats and hosts. These relationships have a unique character and have been described in many papers (e.g. Kaya, 1990; Gaugler, 2002; Kowalska, 2006). This phenomenon is important for the protection of plants. EPNs infect many different types of soil insects, including the larval forms of butterflies, moths, beetles and flies, as well as adult crickets and grasshoppers (Friedmann, 1990). They have chemoreceptors, are motile and active in host seeking and are highly virulent, thus killing their host within a few days. EPNs can be cultured easily *in vitro*; have a high reproductive potential; have a broad host range; are safe for plants, animal health and the environment; can be applied easily using standard spray equipment; have the potential to be recycled in the environment; are compatible with many chemical pesticides; are easy to mass produce; and have long-term efficacy (Dubey *et al.*, 2011). This chapter focuses on the above-mentioned attributes in relation to Polish research.

22.2 Integrated Pest Management in Poland

From 1 January 2014, the Republic of Poland has put into effect the Regulation of the Minister of Agriculture and Rural Development principles of integrated pest management (IPM). General IPM principles are specified in Annex III to Directive 2009/128/EC. IPM involves the careful consideration of all available plant protection methods and the application of appropriate measures to limit the growth of pest populations. These measures include applications of plant protection products (PPPs) and other forms of intervention at economically and ecologically reasonable levels. It is a complex

*E-mail: J.Kowalska@iorpib.poznan.pl

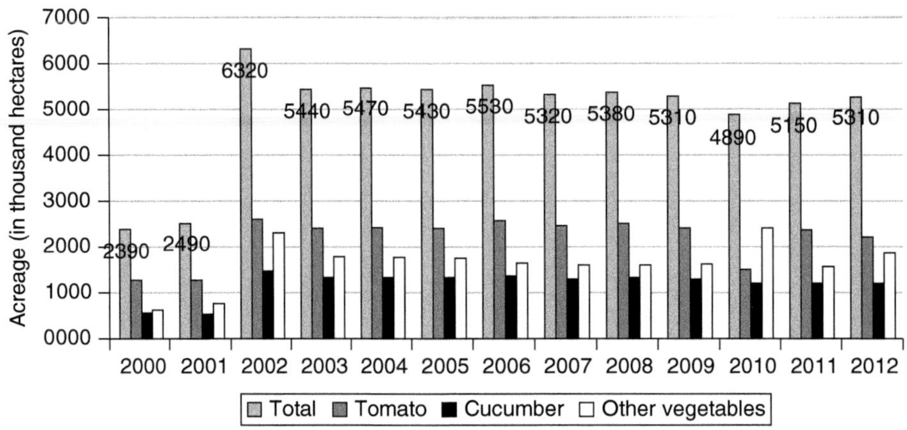

Fig. 22.1. Acreage of vegetables cultivated under cover (glasshouse and plastic tunnels) in Poland in 2000–2012. (From Filipiak, 2014.)

method of protecting plants against pests, involving all available plant protection methods, in particular non-chemical methods, so as to minimize the risk to the health of humans, animals and the environment. IPM emphasizes a healthy crop with the least possible disruption to agroecosystems, and encourages natural pest control. The principles of IPM are consistent with those of organic farming, for which additional restrictions apply, especially when it comes to nitrogen fertilization and synthetic pesticides.

Pest control should be also achieved by many agrotechnical methods (rotation, use of resistant varieties, balanced fertilization, creating favourable conditions for beneficial organisms) and the use of phytosanitary hygiene measures and PPPs in a way that minimizes the risk of resistance in pests. The lists of pesticides approved for integrated crop production are published in the Plant Protection Recommendations issued by the Institute of Plant Protection – National Research Institute (IPP-NRI) in Poznań. Recommendations for IPM in horticulture have been developed by the Institute of Horticulture in Skierniewice.

Biological pest control methods are a valuable tactic in pest management programmes, have been applied worldwide for many years and are an integral part of IPM. The manipulation of beneficial organisms remains a very important tool in IPM programmes for insect pests. EPNs from the families Steinernematidae and Heterorhabditidae are generally considered as beneficial nematodes and may be used in the IPM of agroecosystems. EPNs are vectors of insecticidal bacteria. Many abiotic and biotic factors influence their survival and virulence. Knowledge on EPN–insect–bacteria interactions ensures the increasing efficacy of treatments towards pest control (Kowalska, 2006).

22.3 What Can Biological Methods Offer to Integrated Pest Management?

Species and specific strains of antagonistic (i.e. pathogenic, parasitic and predaceous) organisms can limit populations of some important pests and pathogens effectively and safely, and can be integrated with other control measures. The most important factor in decisions on the widespread use of specific biological agents is the technology of their mass production and formulation. Knowledge about their application under specific conditions of some crop cultures is growing continuously. Equally important is the knowledge on mechanisms of the regulation of antagonistic organisms by environmental factors. The organisms used in plant protection may be classified into two groups, i.e. microorganisms and macroorganisms. EPNs, as well as parasitic insects and predatory mites, belong to the latter group. In Poland, the biological agents applied in crop protection are used mainly in greenhouse vegetable, mushroom and oyster mushroom production.

The scope of applications for biological methods in Poland has been influenced adversely by the stipulations of Directive 91/414 as, because of the dramatic increase in registration requirements and failure to adopt specific regulations for biological agents, many biological and other non-chemical preparations have been withdrawn (Table 22.1).

Biological and biotechnical agents were registered in Poland based on the Act on plant protection of 12 July 1995. Thus 44 biological agents were registered (as in 2000). The next Polish act of 2003, which entered into force prior to Poland's accession to the EU, adapted principles of introduction for pest control agents in Poland to EU regulations. The introduction of macroorganisms to turnover was not regulated. Thus, Poland does not have macroorganism registration regulations, which means that presently any preparation containing such living organisms may be used.

In 2000, out of 42 biopreparations, only 4 contained EPNs; 3 of *Steinernema feltiae* and 1 of *Heterorhabditis megidis* were put on the market. In 2009, 41 preparations containing living organisms were commercially available, with only 4 containing invasive nematode juveniles. All the leading European companies have representatives in Poland; for example, Becker Underwood (UK), Koppert (The Netherlands), Biobest (Belgium) and Polish products of Owinema.

At present, the number of products using EPNs has increased (Table 22.2; Fig. 22.2). However, a problem arises with defining the species, products and current applications in Poland.

22.4 An Outline of Studies on Entomopathogenic Nematodes in Polish Research Centres

The present chapter is focused on a wide range of Polish studies on EPNs. In the 1960s, the Institute of Ecology, Polish Academy of Sciences, was founded and studies on nematodes from the family *Steinernematidae* were initiated. Research was conducted on insect defence reactions to contact with invasive nematode stages. Most studies concerned the mortality of the Colorado potato beetle (*Leptinotarsa decemlineata*), the quantitative and qualitative changes in different haemocyte groups and the numbers of bacterial phagocytes (*Xenorhabdus nematophilus*). It was found that after direct injection of *X. nematophilus* to the insect haemocoel, the number of haemocytes increased within 18 h. In the invasion of nematode larvae to the insect haemocoel, the number of haemocytes had already increased considerably within 6 h. Tests using *L. decemlineata* showed that changes in haemocyte numbers depended on the physiological status of the insects. The greatest haemocyte increase was recorded in adult beetles after winter diapause, beetles feeding in summer responded the weakest, while no defence response was found in those wintering in soil (Jaworska, 2012).

Research also concerned the pathogenicity of nematodes towards various hosts, leading to practical applications of EPNs in biological and integrated pest control in forests and crop fields. Next, the studies focused on the health status of the environment, including industrial areas and soils contaminated with heavy metals and their effect on the activity, reproduction and infection potential of EPN larvae (Jaworska *et al.*, 1997).

Another Polish research centre investigating nematodes and their role in plant protection was the Agricultural University of Poznań. Research analysed naturally parasitized populations of bark beetles, *Tomicus piniperda* and *Tomicus minor*, with mainly *Parasitaphelenchus* spp. identified in insects and their galleries (Michalski and Tomalak, 1984; Tomalak *et al.*, 1984). Soon, studies also covered parasitic nematodes and their relationships with bark beetles, with four new parasitic nematode species described (*Sulphuretylenchus pseudoundulatus*,

Table 22.1. Number of different products admissible in plant production in 2006–2015.

Class of products	Plant protection recommendations developed by IPP-NRI for:				
	2006/2007	2008/2009	2010/2011	2012/2013	2014/2015
Plant growth regulators	62	53	48	51	57
Plant resistance stimulators	8	9	7	–	1
Attractants and repellents	24	24	12	7	8
Pheromones	17	16	11	–	–
Biological and biotechnical agents	64	57	39	25	21
Total	175	159	117	83	87

Table 22.2. Biopesticides based on entomopathogenic and slug parasitic nematodes in Poland.

Pests	Name of biopesticides	Nematode species
Otiorhynchus sulcatus Larvae of Cusrculionidae	LARVANEM, B-GREEN	*Heterorhabditis bacteriophora*
Otiorhynchus spp.	Heterorhabditis–System, Nemasys HM Exhibitline hm	*H. megidis*
Sciarid fly (e.g. *Sciara analis*), thrips (e.g. *Frankliniella occidentalis*), *Liriomyza* spp.	Entonem, Steinernema–System, SCIA-RID, Nemasys F, Exhibiline sf	*Steinernema feltiae*
Caterpillars, *Gryllotalpa gryllotalpa*, larvae of Tipulidae	Capsanem, Exhibitline sc,	*S. carpocapsae*
Otiorhynchus spp.	Nemasys L, Exhibitline srb	*S. krausei*
Slug	Nemaslug, Phasmarhabditis–System	*Phasmarhabditis hermaphrodita*

S. nopimingi, *S. posterruteri*, *Parasitylenchus caudapapilli*) (Tomalak *et al*., 1989). The Institute of Plant Protection began research into nematodes from the genera *Steinernema* and *Heterorhabditis* in 1988, initially to select species and strains, as well as the conditions required, for the effective control of *Sciaridae* in greenhouses and mushrooms (Tomalak, 2012). Continued studies showed that *S. feltiae* might infest *Sciaridae* larvae effectively in organic and mineral substrates (Tomalak *et al*., 2005; Kowalska and Komosa, 2006). The use of nematodes in mushroom growing is limited due to the environmental conditions of the mushroom substrate, affecting *S. feltiae* activity negatively (Tomalak and Lipa, 1991). Extensive studies on the genetic improvement of *S. feltiae* aimed at increased nematode efficacy in controlling selected pests, resulting in the selected ScP strain of *S. feltiae* with an enhanced ability to find and infest *Lycoriella ingenua* larvae in substrate with high CO_2 and ammonia contents (Tomalak, 1994a), thus reducing the dose of nematodes from $5.6–18.5 \times 10^6$ to 1×10^6 larvae/m^2 culture. This strain was subjected to cyclic selection passages and used to produce biopreparation by Owiplant (Owińska, near Poznań). By extensive selection, a series of point mutations were induced and characterized, responsible for morphological or behavioural changes in infective *S. feltiae* larvae. Described mutations have not been used in practice. The next genetic improvement covered other species from the genera *Steinernema* and *Heterorhabditis*, used to control *Frankliniella occidentalis* in greenhouse crops (Tomalak, 1994b).

Fig. 22.2. Healthy (white colour) and *Heterorhabditis megidis* infected larvae of *Otiorhynchus sulcatus*. (From Tomalak, 2008a.)

Subsequent research at the renamed IPP-NRI, Poznań, concerned the use of EPNs to control Polish beetle species (*Melolontha* spp. and *Amphimallon solstitiale*) resistant to infections caused by EPNs (see Section 22.4.3, this chapter, on other applications of EPNs in Poland). Studies at IPP-NRI focused on the potential applications of EPNs in areas with minimum or no chemical pest control, e.g. recreation or urban green areas (see Section 22.4.3, this chapter, on other applications of EPNs in Poland).

The Catholic University of Lublin was another research centre investigating EPNs and the factors determining insect infestation by isolated nematodes and their use against grubs.

According to the studies of Tumialis *et al.* (2014a), the first record of *Steinernema kraussei* from Poland was demonstrated. The nematode was isolated from coniferous woodlands in four localities in central Poland. Preliminary identification of the species was done based on morphometric measurements. To confirm a nematode species of the genus *Steinernema*, the result was supported by the description of the internal transcribed spacer (ITS) region. Other research of Tumialis *et al.* (2014b) concerned a study to determine the effect of different initial doses of the infective juveniles (IJs) (50 IJs, 200IJs and 1000 IJs) of *H. megidis* (Rhabditida: Heterorhabditidae), strain IsM15/09, on the recovery, final yields and per cent of final yields in *Galleria mellonella* larvae. The per cent of recovery was not related directly to the initial dose. Final yields also did not change with the initial dose. However, the per cent of yields were highly negatively correlated with the initial dose of nematodes, and was the highest with a dose of 50 IJs. An additional point of the study was to investigate whether the nematodes were able to produce progeny from one hermaphroditic individual. The results showed that the invasive larvae resumed growth and transformed into hermaphroditic individuals that reproduced without cross-fertilization.

22.4.1 Application of entomopathogenic nematodes for the protection of vegetables

The present chapter's aims are to indicate the usefulness of EPNs in vegetable crops cultivated both in the field and under cover.

Field conditions

The increasing area of organic farms in Poland, the growing public awareness of environmental issues and requirements concerning the culture of fresh vegetables and fruit and their quality lead to increased biological control in agriculture. The efficacy of EPNs was confirmed in vegetable fields in the control of *Psila rosae* larvae (Jaworska, 1995) and *Anthomyidae* sp. larvae in cabbage crops (Jaworska, 1992b), as well as in the control of beetles (e.g. Colorado potato beetle) in field crops (Sandner, 1990). Various EPN species have been described in organic potato growing (Erlichowski, 2007), also stressing the strong effect of abiotic and biotic factors affecting the efficacy of the treatment (Kowalska, 2002a). The efficacy of the use of nematodes was confirmed for *Ceutorhynchus quadridens* (Jaworska, 1992a) and *Lilioceris merdigera* (Jaworska and Dudek, 1992). High infestation of caterpillars was observed in some butterflies, e.g. *Pieris brassicae* and *Mamestra brassicae* (Stanuszek, 1974).

Greenhouse conditions

EPNs from the families Steinernematidae and Heterorhabditidae are used extensively against thrips in greenhouse vegetable growing (mainly cucumber and tomato) (Fiedler and Sosnowska, 2009). Imago of *F. occidentalis* and its two nymph stages feed on the green organs, flowers and fruit of infested plants. Non-feeding pre-pupae and pupae develop mainly in the soil and substrate (approximately 40% complete life cycle). Studies concerned the selection of *S. feltiae* to produce a strain with increased efficacy to control the soil stages of *F. occidentalis* in crop growing. The commercial strain *S. feltiae* SN reached a marginal efficacy of 20.7%: the hybridization of geographically distant strains enhanced the efficacy of *S. feltiae* and reduced the emergence of pests pupating in the soil by almost 75% (Tomalak, 1998). An increased dose from 100 to 300 invasive larvae (IJs)/cm^2 provided a similar efficacy, showing no need to exceed the commonly recommended dose of 100 larvae/cm^2.

In tunnel crops, Sciaridae and Scatella develop in large numbers due to the high moisture content and decomposing organic matter in the substrate and mycelium. The most common Sciaridae include the genera *Bradysia* and *Lycoriella*. Biological control usually involves EPNs *S. feltiae* actively searching for Sciaridae larvae in the substrate, used at a dose of 50 million IJs/100–150 m^2 substrate in watering, spraying or irrigation. The application should be repeated two or three time during plant vegetation (Piątkowski, 2009).

22.4.2 Application of entomopathogenic nematodes for arable crop protection

The present chapter also focuses on EPNs and their practical usefulness in arable crops in which the main pests are snails and slugs. Biological control against snails and slugs is applied using *Phasmarhabditis hermaphrodita* juveniles. The greatest damage in field cultures is caused by *Deroceras reticulatum*, *Arion lusitanicus*, *Arion rufus* and *Arion distinctus*, for example, and the greatest losses are observed in winter rape, winter wheat, potato, cabbage, sugarbeet, carrot, pea and clover. The damage to crops occurs at different development phases, depending on the species. In a study by Jaworska (1996), *Deroceras* sp. and *Arion* sp. were exposed to infective larvae of *Steinernema carpocapsae* and *Heterorhabditis bacteriophora*. Although the progress of infection was slow, the next nematode generations developed in dead slugs. These studies were continued in successive years on *Helix pomatia* L., *Deroceras agreste* and *D. reticulatum* in field trials applying nematodes *S. feltiae* and *H. megidis*. Another experiment on the non-chemical protection of broad bean concerned another pest, *Sitona lineatus*. None of the nematodes was fully effective against slugs, while *H. megidis* was effective against *Sitona* when applied to soil, reducing the root nodule damage by pest larvae (Jaworska, 2004). The sensitivity of *Sitona* beetles to other nematode species was confirmed earlier (Jaworska and Wiech, 1988; Jaworska and Ropek, 1994).

Losses in potato, beet and cereal fields are also caused by caterpillars of *Agrotis* spp. Laboratory experiments assessed the natural sensitivity of older caterpillar stages (L_5 and L_6) of *Agrotis exclamationis* to infections by larvae of *Steinernema bicornutum* Sb360 and *H. megidis* 01PL – nematodes isolated from natural habitats in Poland (Kowalska and Jakubowska, 2006). Both nematode species were applied at 25 and 50 larvae/cm^2, with high caterpillar mortality recorded after the application of *H. megidis* 01PL (92.5 and 95% for 25 and 50/cm^2 IJs, respectively). For *S. bicornutum* Sb360, caterpillar mortality was 42.5 and 72.5%, depending on the dose.

22.4.3 Other applications of entomopathogenic nematodes in Poland

Targets other than agricultural targets have also been indicated for EPNs. This chapter focuses on municipal green zones, orchards and others, where EPNs can be applied with success.

Parks, urban tree plantings and municipal forests

Studies showed that EPNs introduced, for example, as soil sprays under trees may effectively reduce populations of butterflies, Hymenoptera and beetles – pests of urban trees and shrubs – during the pest species' pupation in soil (Tomalak, 2012; Fig. 22.3). A single application of *S. feltiae* and *H. megidis* reduced pest numbers in successive seasons. In the city of Poznań, the most important insect pests include, for example, *Operophtera* spp., as well as *Altica quercetorum*, *Agelastica alni*, *Phytodecta* spp., *Pyrrhalta viburni*, *Caliroa annulipes*, *Arge* spp., *Nematus* spp., *Croesus* spp. and *Hemichroa* spp. (Tomalak, 2002, 2003, 2006, 2008a). The pest control treatment should be performed under trees immediately before the pests descend to the soil. Recorded efficacy was 90–100%.

Research at IPP-NRI also concerned the use of EPNs to control soil pests, e.g. grubs. The susceptibility of different cockchafer species was determined depending on their development cycle, the presence of alternative hosts in the soil (Kowalska, 2000) and the potential application of nematodes with carbosulfane in chemical grub control (Kowalska, 2002b).

The larvae of *A. solstitiale* causing damage in municipal greens, for example, frequently coexist with the larvae of other insects, e.g. *Strophosoma faber*. Laboratory studies analysed the effect of the presence of an alternative host insect for nematodes used to control grubs. Both insect species were infested by EPNs, with *S. faber* being the more sensitive host for nematodes. At the same time, infested weevil larvae were reserves of infective larvae of the next nematode generation (Kowalska, 2000).

Several studies provided no positive or repeatable results for the control of beetles from the genera

Fig. 22.3. The larvae of sawflies *Hemichroa crocea* (pest occurring on urban trees), which went into the soil to pupate, infected by *Heterorhabditis megidis*. (From Tomalak, 2008a.)

Amphimallon and *Melolontha* (Kowalska, 2003, 2004), while an experiment in a forest nursery confirmed the efficacy of *H. megidis* to control the cockchafer grub population by approximately 50% at a dose of 100 infective larvae/cm^2 (Kowalska, 2001). Lower doses of 50 and 25 larvae/cm^2 were not effective. Younger development stages (L_1–L_2) of these cockchafers showed greater susceptibility to nematode infection. Grubs actively defend themselves against infective nematode larvae by leaving the locality where the larvae are applied, or actively destroy it (Kowalska, 2004). Among the genera tested, *H. megidis* was the most effective.

Studies on biological crop protection against insect pests, for example *Melolontha melolontha*, *Melolontha hippocastani* and *Cameraria ohridella*, were conducted in the Lublin research centre (Kreft *et al.*, 2004, 2005, 2011), showing that *C. ohridella* pupae after wintering on *Aesculus hippocastanum* leaves were sensitive to infection by EPNs (Skrzypek *et al.*, 2005). Nematodes from the genus *Steinernema* were most effective in infesting pupae in cocoons at three doses of 5, 25 and 50 larvae/insect at 15°C and 20°C (Kreft *et al.*, 2005). Promising laboratory results led to field trials, in which nematodes were applied in autumn and spring at 5 larvae/cm^2, reducing the number of pest butterflies in spring. The results recorded at IPP-NRI, confirmed for larvae of *Melolontha* spp., were unsatisfactory due to their active defence (destroying penetrating infective larvae) and exceptionally strong immune system, resulting in the encapsulation of larvae penetrating the haemocoel (Kowalska, 2001).

Grub larval stage L_1 is the most susceptible to infestation by EPNs, while L_3 is the least sensitive (Kreft *et al.*, 2011). Among the nematode species isolated throughout Poland, the larvae of *Steinernema arenarium* and *H. megidis* were the most effective, in contrast to *S. feltiae* and *H. bacteriophora* (Kreft *et al.*, 2011).

Experiments were conducted in a forest nursery using preparations of Polish strains *S. arenarium* and *H. megidis* at 50 and 150 million larvae/acre. The efficacy of the nematodes in limiting grub populations was confirmed within the next 2 months after treatment. Due to the physico-chemical properties of the nursery soils, the survival rates of infective larvae were low and the treatment had to be repeated. In a study by Kowalska (2001) in a 5-year forest culture, *H. megidis* at 100 larvae/cm^2 reduced 1-year (L_1) grubs of *M. melolontha* – it may be an alternative to chemical insecticides. Extensive research at IPP-NRI also confirmed the efficacy of *H. megidis* against *Phyllopertha horticola* (Kowalska, 2003).

Studies also concerned populations of EPNs isolated from grubs of *Melolontha* spp., collected from forests differing in light, moisture and temperature conditions. Grubs were infested by *Diplogasteroides magnus*, *Koerneria* sp. and *Rhabditis* sp., with most infested grubs collected from shaded locations with the highest moisture content (Kreft *et al.*, 2012).

The use of EPNs (*Steinernema* and *Heterorhabditis* spp.) against *Hylobius abietis* is currently being studied in many countries. In 2011, similar studies began in Poland to develop a new biological control method of *H. abietis* using native nematode isolates. The first laboratory experiments were on the biological activity of three native isolates: two of *S. feltiae* and one of *S. kraussei*, applied at 100 invasive larvae per 1 larva of *H. abietis*. No statistical differences were found between larvae mortality and extent of infection by the tested isolates. In all variants, mortality and infection were extensive, amounting to 92–100% and 80–86%, respectively. The results indicated the different activity of nematode isolates within one species (Tumialis *et al.*, 2013). The use of EPNs such as *Steinernema* and *Heterorhabditis* spp. against the large pine weevil, *H. abietis*, has been mainly studied in Poland, aimed at evaluating the effectiveness of commercially produced biopreparations that contained EPNs and were applied against the large pine weevil. The treatments consisted of spraying soil around *Pinus sylvestris* stumps with an aqueous suspension of *S. carpocapsae*, *S. feltiae*, *H. bacteriophora*, *Heterorhabditis downesi* and *H. megidis* at a dose of 3.5 million IJs/1000 ml of water per stump. The roots of experimental stumps were analysed 4 weeks subsequent to the treatments. After roots debarking, *H. abietis* larvae were isolated, and the percentage of insects parasitized by EPNs was assessed. All tested nematodes showed the ability to parasitize *H. abietis* larvae overwintering in *P. silvestris* stumps. Such conclusions demonstrated the insecticidal effectiveness of biopreparations containing the mentioned nematodes against *H. abietis* larvae. Sprays with the use of tested biopreparations at a dose of 3.5 million invasive nematode larvae/stump applied to soil and litter around the stumps resulted in the increased mortality of large pine weevil larvae. The highest extensity of parasitism was shown by *H. downesi* and *S. carpocapsae* nematode species (Skrzecz *et al.*, 2014).

Orchards

EPNs are also used to control selected pests in orchards. A method was developed to introduce water suspension of the invasive nematode *H. megidis* to soil. The preparation was introduced in autumn to soil under the tree where the larvae of *Hoplocampa minuta* pupate. Similar attempts to use EPNs were also made for apple worm, by spraying tree trunks and soil under the canopy (Majewski, 2011). The applicability of EPNs, particularly *H. megidis*, to protect orchards against sawflies was confirmed in extensive studies by Tomalak (2004, 2006), e.g. assessing the application of EPNs (*Steinernema* and *Heterorhabditis*) against larvae and pupae, transforming and wintering in soil, from the most important orchard pests (Hymenoptera: Symphyta). Nematodes were introduced to soil at 1 or 5 × 10^5 invasive larvae/m^2. The results showed effective control of *Hoplocampa* spp. and *Nematus ribesii*.

Production of mushrooms and oyster mushrooms

The successful use of infective larvae of EPNs against Sciaridae larvae was described by Tomalak (2000). Larvae feeding on mushroom mycelium may damage young fruiting bodies. Treatments with *S. feltiae* at 0.5–1.0 million invasive larvae/m^2 are repeated three to six times during the production cycle (due to high carbon dioxide contents), with the first treatment before the substrate cover. The infectivity of this nematode against *Lycoriella solani* may be modified. Insect parasitizing rates by *S. feltiae* changed with changes in water content in the substrate cover, ranging from 1.0:0.5 to 1.0:2.5 by weight of substrate dry matter to contained water, and decreasing until complete substrate saturation (Tomalak, 1998). Tomalak conducted extensive research on oyster mushroom and the effect of its mycelium on the viability and activity of EPNs *S. feltiae*, *H. megidis* and *H. bacteriophora*, while also assessing the infectivity of these nematodes against Diptera developing in oyster mushroom mycelium (Tomalak, 2008b). Laboratory tests showed no significant effect of mycelium on the viability of these larvae. However, at a large number of *S. feltiae* larvae in the substrate, most *Heterorhabditis* spp. migrated, which indicated a repellent effect of oyster mushroom mycelium on these nematodes. The infectivity of the nematode species tested was comparable to that observed in garden mushroom substrate.

Poultry production

EPNs may also be used against the larvae and adult forms of *Alphitobius diaperinus*, a pest often found in poultry houses. These noxious insects feed on animal feed and are carriers of many diseases, e.g. salmonellosis, Marek's disease and leukaemia. They may also cause serious damage to wooden elements of poultry houses. The most effective nematodes, *S. feltiae*, were applied on pine sawdust used in poultry houses. However, the invasiveness of EPNs decreases with an increase in the ammonia content from chicken droppings (Pezowicz *et al.*, 2013).

Application of entomopathogenic nematodes against stable flies

EPNs were also used in stables to control the population of the housefly, *Musca domestica*, and improve animal welfare. Commercial products with *S. feltiae*, *S. carpocapsae*, *H. megidis* and *H. bacteriophora* varied in their virulence, as housefly larvae were killed more effectively by nematodes of the family Steinernematidae than Heterorhabditidae, with the highest efficacy observed for *S. feltiae* and *S. carpocapsae* (Wojciechowska *et al.*, 2013).

22.5 Interactions of Entomopathogenic Nematodes with Pesticides

The occurrence and activity of EPNs depend on many abiotic and biotic factors determining its further development in soil. Numerous studies show that nematodes tolerate chemical soil contamination (Kamionek, 1992; Jaworska *et al.*, 1997). However, certain insecticides, particularly carbamates and organophosphorus compounds, are toxic to the larvae of invasive nematodes. Dzięgielewska, in her study (2012), compared the frequency and species composition of Steinernematidae and Heterorhabditidae in orchards protected and not protected chemically. The greatest numbers of soil samples were collected from apple and apple–plum orchards, with soil reaction being mostly neutral. Soil samples were collected three times in a season (spring, summer, autumn) and nematodes were determined using the standard method with *G. mellonella*. Unprotected orchards had a greater species diversity and greater frequency in samples, the latter being dependent on the sampling site and presence of potential hosts. The greatest species variety was recorded in orchards colonized by *Cydia pomonella* and *Hoplocampa* spp.

The interactions of pesticides with infective larvae of EPNs were assessed in studies on the application of carbosulfate with *H. megidis* 01PL larvae, which reduced cockchafer grub populations effectively (Kowalska, 2001). Carbosulfate had no effect on the viability of nematode larvae. Additionally, increased larvae penetration of insects was observed (Kowalska, 2002b), similar to that reported in *G. melonella* caterpillars by Bednarek *et al.* (2000) for *S. feltiae* larvae after their incubation with carbosulfate. Other studies assessed the effect of azadirachtin on the viability of *S. feltiae* larvae. Its dose of 100 µl/ml water had no effect on increased mortality or infectiveness towards *G. melonella*.

22.6 Technical Restrictions for Nematode Applications

Technical requirements for the use of biological agents are extremely important because of their impact on the final effectiveness of applicable factors. The present chapter focuses on this aspect. EPNs are applied when mixed with water, sprayed on plants or soil/substrate at an average concentration of 500–1000 nematodes/ml. To facilitate flow, nozzles need to be over 0.5 mm in diameter. Mechanical damage to the nematodes by high pressure and temperature, and filters, needs to be prevented, as losses of *S. feltiae* caused by liquid temperature increasing to 38.4°C were reported (Chojnacki, 2007). At temperatures ranging from 4 to 30°C, the mobility of *S. feltiae* larvae is constant (Chojnacki, 2010). Nematodes are sometimes applied through irrigation systems, although this is effective only at high water velocity, as some nematodes may be deposited in the system at slow flow of the suspension.

In Poland, trials were conducted on the technical parameters limiting the efficacy of EPN applications; for example, *S. feltiae* larvae in liquid sprayed by two types of sprayer (TeeJet XR 8002 and XR 11002). Both the type of sprayer and the outlet height influenced non-uniform nematode larvae concentrations in sprayed liquid (Chojnacki, 2008).

Later studies investigated the effect of a sprayer mesh filter on losses of EPNs. Many companies recommend filter replacement, which requires filtered water to be mixed with nematodes, or equipment cleaning. An experiment on *S. feltiae* larvae using a 100-mesh filter and flow rate of 20 dm^3/min showed no nematode losses for 900 larvae/ml (Chojnacki, 2012).

22.7 Entomopathogenic Nematodes in Agriculture – Potential Threat to Protected Beetle Species

The application of nematode-based biopreparations in biological pest control has many positive and negative effects, with particular concerns over selectivity in colonizing and killing potential hosts by nematodes. Ecosystems with EPNs are inhabited by pests, beneficial insects and legally protected species. Selected aspects from this topic are presented in this chapter.

Beetles are the most studied group of hosts for nematodes; thus, nearly 50% of papers on nematodes are devoted to Coleoptera. One of them is the hermit beetle (*Osmoderma eremita*), endangered and strictly protected under the Bern Convention. In contrast, the marbled rose chafer, *Protaetia lugubris*, despite its rarity, it is not a protected species. Its habitat preferences are similar to those of the hermit beetle, which makes it potentially endangered similarly as *O. eremita* (Kucharska *et al.*, 2009).

22.8 Conclusions and Future Prospects

Research on EPNs in Poland began in the 1980s, and in this short period EPNs have become widely accepted as biological agents for the control of soil-dwelling and foliage pests of many crops in glasshouses, in fields, in shelterbelt urban areas and in orchards. In Poland, about 30% of vegetables are produced under IPM. Growers have now clearly noted the specific advantages of EPNs in greenhouses. EPNs have proved their efficacy when used alone or in integration with other management practices. Using EPNs to control the fungus gnat in glasshouses has already become a standard procedure, as has control of the black vine weevil in the open field. However, EPNs in the control of Scarabaeidae are still not proven efficacious, as noted by the Polish research centres. Additionally, new target pests have been identified, as in mushroom or poultry production. Thus, in a bid to reduce pesticide use, the future role of biological and integrated control is expected to increase (Sosnowska and Fiedler, 2009).

Acknowledgements

I would like to thank the Institute of Plant Protection – National Research Institute in Poznań for the opportunity to develop my passion for research. Thanks to my staff from the Department of Ecology and Environmental Protection for work on common projects.

References

Askary, T.H. (2010) Nematodes as biocontrol agents. In: Lichtfouse, E. (ed.) *Sociology, Organic Farming, Climate Change and Soil Science*. Springer, The Netherlands, pp. 347–378.

Bednarek, A., Popowska-Nowak, E., Pezowicz, E. and Kamionek, M. (2000) Influence of insecticides, carbosulfan and carbofuran on mortality and pathogenity of entomopathogenic nematodes and fungus. *International Organisation for Biological Control Bulletin* 23, 115–116.

Chojnacki, J. (2007) Zjawiska termiczne w opryskiwaczu w aspekcie aplikacji biologicznych środków ochrony roślin. *Inżynieria Rolnicza* 8, 37–42.

Chojnacki, J. (2008) Wpływ wysokości opryskiwania na równomierność osadzania biologicznych środków ochrony roślin – owadobójczych nicieni. *Journal of Research and Applications in Agricultural Engineering* 53, 32–34.

Chojnacki, J. (2010) Wpływ temperatury otoczenia na intensywność poruszania się owadobójczych nicieni. *Journal of Research and Applications in Agricultural Engineering* 55, 37–39.

Chojnacki, J. (2012) Wpływ filtra siatkowego zamontowanego w instalacji opryskiwacza na straty owadobójczych nicieni. *Journal of Research and Applications in Agricultural Engineering* 57, 37–40.

Dubey, J., Tiwary, B.N. and Ganguly, S. (2011) Biopesticidal nematodes and molecular techniques involved for their characterization. In: Dhingra, H.K., Jha, P.N. and Bajpai, P. (eds) *Current Topics in Biotechnology and Microbiology*. Lap Lambert Academic Publishing Ag and Co Kg, Dudweller Landstr, Germany, pp. 7–36.

Dzięgielewska, M. (2012) Występowanie nicieni owadobójczych z rodziny Steinernematidae i Heterorhabditidae w sadach chronionych i niechronionych chemicznie. *Progress in Plant Protection/Postępy w Ochronie Roślin* 52, 415–420.

Erlichowski, T. (2007) *Skład gatunkowy, szkodliwość i zwalczanie drutowców (Coleoptera: Elateridae) w uprawie ziemniaka*. PhD thesis. Plant Breeding and Acclimatization Institute (IHAR) – National Research Institute, Bonin, Poland, 94 pp.

Fiedler, Ż. and Sosnowska, D. (2009) Aktualny stan ochrony roślin warzywnych w uprawach szklarniowych przed szkodnikami z wykorzystaniem czynników biologicznych. *Progress in Plant Protection/Postępy w Ochronie Roślin* 49, 1474–1479.

Filipiak, T. (2014) *Zmiany na rynku warzyw i w gospodarstwach warzywniczych w Polsce po integracji z Unią Europejską*. SGGW, Warsaw, 184 pp.

Friedmann, M.J. (1990) Commercial production and development. In: Gaugler, R. and Kaya, H.K. (eds) *Entomopathogenic Nematodes in Biological Control*. CRC Press, Boca Raton, Florida, pp. 153–172.

Gaugler, R. (2002) *Entomopathogenic Nematology*. CAB International, Wallingford, UK.

Jaworska, M. (1992a) Influence of the *Steinernema feltiae* (Filipjev) and *Heterorhabditis bacteriophora* (Poinar) on the larvae of cabbage stem weevil – *Ceutorhynchus quadridens* (Penz). *Materiały Konferencji Naukowej* PAN Skierniewice 31, 63–67.

Jaworska, M. (1992b) *Badania nad możliwością wykorzystania owadobójczych nicieni (Nematoda, Rhabditida) do zwalczania śmietki kapuścianki i cebulanki w uprawie kalafiorów i cebuli.* Habilitation thesis. Agricultural University in Cracow, Poland.

Jaworska, M. (1995) Entomopathogenic nematodes for control of carrot fly (*Psila rosae* F.) *Entonematologia* II 4, 1–9.

Jaworska, M. (1996) Możliwości biologicznego zwalczania pomrowów. *Progress in Plant Protection/Postępy w Ochronie Roślin* 36, 23–25.

Jaworska, M. (2004) Możliwości niechemicznej ochrony bobu (*Vicia faba* L.) przed szkodnikami. *Progress in Plant Protection/Postępy w Ochronie Roślin* 44, 755–757.

Jaworska, M. (2012) Entomopatogeniczne nicienie w ochronie roślin. In: Skrzecz, I. and Sierpińska, A. (eds) *Kierunki rozwoju patologii owadów w Polsce*. Instytut Badawczy Leśnictwa, Poland, pp. 92–101.

Jaworska, M. and Dudek, B. (1992) Susceptibility of *Lilioceris merdigera* to entomopathogenous nematodes. *Materiały Konferencji Naukowej* PAN Skierniewice 31, 68–73.

Jaworska M. and Ropek D. (1994) Influence of host-plant on the susceptibility of *Sitona lineatus* L. to *Steinernema carpocapsae* (Weiser). *Journal of Invertebrate Pathology* 64, 96–99.

Jaworska, M. and Wiech, K. (1988) Susceptibility of the clover root weevil, *Sitona hispidulus* (Col. Curculionidae) to *Steinernema feltiae*, *S. bibionis* and *Heterorhabditis bacteriophora*. *Journal of Applied Entomology* 106, 372–376.

Jaworska, M., Jasiewicz, Cz. and Gorczyca, A. (1997) Wpływ zanieczyszczenia metalami ciężkimi gleb ogrodów działkowych śląska na aktywność mikroorganizmów glebowych. *Progress in Plant Protection/Postępy w Ochronie Roślin* 37, 276–278.

Kamionek, M. (1992) *Wpływ pestycydów na nicienie entomofilne*. PhD thesis. University of Life Sciences – SGGW, Warsaw.

Kaya, H.K. (1990) Soil ecology. In: Gaugler, R. and Kaya, H.K. (eds) *Entomopathogenic Nematodes in Biological Control*. CRC Press, Boca Raton, Florida, pp. 93–115.

Kowalska, J. (2000) Effect of alternative host *Strophosoma faber* (Herbst) on efficacy of the entomopathogenic nematode *Steinernema glaseri* in control of

Amphimallon solstitiale grubs in the soil. *Journal of Plant Protection Research* 40, 244–248.

Kowalska, J. (2001) Próba zastosowania nicieni owadobójczych oraz metody integrowanej w zwalczaniu pędraków chrabąszcza majowego *Melolontha melolontha* L. w uprawie leśnej. *Sylwan* 2, 89–95.

Kowalska, J. (2002a) Ekologiczne uwarunkowania liczebności wybranych szkodników glebowych. *Sylwan* 12, 97–103.

Kowalska, J. (2002b) Wpływ karbosulfanu na żywotność larw inwazyjnych nicienia owadobójczego *Heterorhabditis megidis* 01Pl. *Progress in Plant Protection/ Postępy w Ochronie Roślin* 42, 472–475.

Kowalska, J. (2003) Śmiertelność pędraków *Phyllopertha horticola* przy łącznym stosowaniu *Heterorhabtitis megidis* i diazynonu. *Progress in Plant Protection/ Postępy w Ochronie Roślin* 43, 735–740.

Kowalska, J. (2004) Wpływ chrabąszcza majowego *Melolontha melolontha* na liczebność aktywnych larw infekcyjnych w populacjach nowych izolatów *Steinernema* spp. *Progress in Plant Protection/ Postępy w Ochronie Roślin* 44, 857–860.

Kowalska, J. (2006) Wzajemne powiązania pomiędzy nicieniami owadobójczymi, owadami i bakteriami oraz ich wykorzystanie w praktyce. *Wiadomości Parazytologiczne* 52, 93–98.

Kowalska, J. and Jakubowska, M. (2006) Ocena wrażliwości *Agrotis exclamationis* L. na wybrane gatunki nicieni owadobójczych. *Progress in Plant Protection/Postępy w Ochronie Roślin* 46, 349–351.

Kowalska, J. and Komosa, A. (2006) Ocena skuteczności *Steinernema feltiae* w zwalczaniu ziemiórek (Diptera: Sciaridae) w podłożu z włókna drzewnego w uprawie pomidora szklarniowego. *Progress in Plant Protection/ Postępy w Ochronie Roślin* 46, 476–479.

Kreft, A., Skrzypek, H. and Kazimierczak, W. (2004) Effectiveness of infection of *Cameraria ohridella* (Lepidoptera: Gracillariidae) larvae by entomopathogenic nematodes *Steinernema* spp. and *Heterorhabditis* sp. (Nematoda: Rhbditida). *Annaless Maria Curie-Skłodowska University*, sectio C, 59, 23–30.

Kreft, A., Skrzypek, H. and Kazimierczak, W. (2005) Susceptibility of *Cameraria ohridella* (Lepidoptera: Gracillariidae) pupae to entomopathogenic nematodes *Steinernema* spp. and *Heterorhabditis* sp. (Nematoda: Rhbditida). *Annaless Maria Curie-Skłodowska University*, sectio C, 60, 47–54.

Kreft, A., Skrzypek, H., Kazimierczak, W., Lech-Pituch, E. and Skrabucha, A. (2011) Efektywnos´c´ poraz˙ania larw chraba˛szcza majowego (*Melolontha melolontha* L.) przez entomopatogeniczne nicienie. *Progress in Plant Protection/Postępy w Ochronie Ros´lin* 51, 771–775.

Kreft, A., Kazimierczak, W., Sierpińska, A., Skrzypek, H., Skrabucha A., *et al.* (2012) Charakterystyka nicieni wyizolowanych z pędraków chrabąszcza majowego *Melolontha* sp. (Coleoptera: Scarabaeidae). *Progress in Plant Protection/Postepy w Ochronie Roślin* 52, 1093–1097.

Kucharska, K., Kucharski, D. and Pezowicz, E. (2009) Nicienie entomopatogeniczne w rolnictwie – potencjalne zagrożenie dla chronionych gatunków chrząszczy. *Annals of Warsaw University of Life Sciences – SGGW. Animal Science* 46, 205–209.

Lipa J.J. and Pruszyński S. (2010) Stan wykorzystania metod biologicznych w ochronie roślin w Polsce i na świecie. *Progress in Plant Protection/Postępy w Ochronie Roślin*, 50, 1033–1043.

Majewski, A. (2011) Organizmy pożyteczne w sadach. Pasożyty szkodników. *Sad Nowoczesny* 3, 60–61.

Michalski, J. and Tomalak, M. (1984) Zniekształcenia niektórych narządów wewnętrznych *Tomicus piniperda* L. (Coleoptera: Scolytidae) wywołane przez nicienie (Nematoda). *Roczniki Akademii Rolniczej w Poznaniu* 42, 43–49.

Pezowicz, E., Mazurkiewicz, A. and Tumialis, D. (2013) Wrażliwość imago i larw pleśniakowca lśniącego *Alphitobius diaperinus* (Panzer 1797) w ściółce z trocin na wybrane gatunki i szczepy nicieni Steinernematidae i Heterorhabditidae w warunkach laboratoryjnych. *Annals of Warsaw University of Life Sciences – SGGW, Animal Science* 52, 161–166.

Piątkowski, J. (2009) Ziemiórki i brzegówki w uprawach pod osłonami. *Hasło Ogrodnicze* 2, 90–93.

Sandner, H. (1990) *Pasożyty w służbie człowieka*. Państwowe Wydawnictwo Naukowe Warszawa, pp. 67–127.

Skrzecz, I., Tumialis, D., Pezowicz, E. and Sowińska, A. (2014) Evaluation of biological activity of biopreparations containing nematodes from the genera *Steinernema* and *Heterorhabditis* used for reducing large pine weevil *Hylobius abietis* L. population in pine *Pinus sylvestris* L. stumps. *Folia Forestalia Polonica*, Series A 54, 196–201.

Skrzypek, H., Kreft, A., Kazimierczak, W. and Wilden, T. (2005) Wstępne badania nad nową metodą ograniczania liczebności zimujących w liściach poczwarek *Cameraria ohridella* (Lepidoptera: Gracillariidae) przez różne szczepy entomopatogenicznych nicieni. *Uprawa i ochrona drzew. Czasopismo Międzynarodowego Towarzystwa Uprawy i Ochrony Drzew* 14, 5–9.

Smolenski, T. (2007) Produkcja warzyw pod osłonami – co było, jest i będzie. *Owoce Warzywa Kwiaty* 18, 4–5.

Sosnowska, D. and Fiedler, Ż. (2009) Biological control of pest on vegetable plants in greenhouses in Poland. In: Stojnić B. (ed.) *Proceedings of VI Congress of Plant Protection* (Book II). Zlatibor, 23–27 November 2009. International Organization for Biological Control of Noxious Animals and Plants - East Palearctic Regional Section (EPRS IOBC), pp. 96–103.

Stanuszek, S. (1974) *Neoplectana feltiae pieredarum* n. ecotype (Nematoda: Rhabditoidea) – a parasite of *Pieris brassicae* L. and *Mamestra brassicae* L. in Poland. Morphology and biology. *Zeszyty Problemowe Postępów Nauk Rolniczych* 154, 361–391.

Tomalak, M. (1994a) Selective breeding of *Steinernema feltiae* (Filipjev) (Nematoda: Steinernematidae) for improved efficacy in control of a mushroom fly, *Lycoriella solani* Winnertz (Diptera: Sciaridae). *Biocontrol Science and Technology* 4, 187–198.
Tomalak, M. (1994b) Genetic improvement of *Steinernema feltiae* for integrated control of the Western Flower, *Frankliniella occidentalis*. *International Organisation for Biological Control Bulletin* 17(3), 17–20.
Tomalak, M. (1998) *Selekcja i mutageneza w genetycznym doskonaleniu nicieni owadobójczych dla celów biologicznego zwalczania szkodników*. Habilitation thesis. Institute of Plant Protection, Poznań, Poland.
Tomalak, M. (2000) Wykorzystanie nicienie owadobójczych w ochronie roślin. *Ochrona Roślin* 9, 2–3.
Tomalak, M. (2002) Potential new targets for entomopathogenic nematodes in biological control of urban pests. *Nematology* 4, 199.
Tomalak, M. (2003) Biocontrol potential of entomopathogenic nematodes against winter moths (*Operophtera brumata* and *O. fagata*) (Lepidoptera:Geometridae) infesting urban trees. *Biocontrol Science and Technology* 13, 517–527.
Tomalak, M. (2004) Susceptibility of *Amphimallon solstitialae* to entomopathogenic nematodes. COST 850 Biocontrol Symbioses WG3 meeting (2–3 May 2004, Merelbeke, Belgium), Abstracts, 8.
Tomalak, M. (2006) Potencjał nicieni owadobójczych w biologicznym zwalczaniu błonkówek atakujących drzewa owocowe i parkowe. *Progress in Plant Protection/ Postępy w Ochronie Roślin* 46, 249–255.
Tomalak, M. (2008a) Organizmy pożyteczne w środowisku rolniczym. In: Tomalak, M. and Sosnowska, D. (eds) Organizmy pożyteczne w środowisku rolniczym. Instytut Ochrony Roślin – PIB w Poznaniu, Poland, 95 pp.
Tomalak, M. (2008b) Problemy ochrony upraw boczniaka (*Pleurotus* spp.) przed szkodnikami. *Progress in Plant Protection/Postępy w Ochronie Roślin* 48, 978–987.
Tomalak, M. (2012) Nicienie zwia̧zane z owadami i ich znaczenie w ochronie ros´lin – historia badan´ prowadzonych w os´rodkach naukowych Poznania. In: Skrzecz, I. and Sierpin´ska, A. (eds) *Kierunki rozwoju patologii owadów w Polsce*. Instytut Badawczy Les´nictwa, Poland, pp. 102–110.
Tomalak, M. and Lipa, J.J. (1991) Factors affecting entomophilic activity of *Neoaplectana feltiae* in mushroom compost. *Entomologia et Applicata* 59, 105–110.
Tomalak, M., Michalski, J. and Grocholski, J. (1984) The influence of nematodes on the structure of genitalia of *Tomicus piniperda* L. (Coleoptera: Scolytidae). *Journal of Invertebrate Pathology* 43, 358–362.
Tomalak, M., Welch, H.E. and Galloway, T.D. (1989) Nematode parasites of bark beetles (Scolytidae) in southern Manitoba, with description of three species of *Sulphuretylenchus* Rühm (Nematoda: Allantonematidae). *Canadian Journal of Zoology* 67, 2497–2505.
Tomalak, M., Piggott, S. and Jagdale, G. (2005) Glasshouse application. In: Grewal, P.S., Ehlers, R-U. and Shapiro-Ilan, D.I. (eds) *Nematodes as Biological Control Agents*. CAB International, Wallingford, UK, pp. 147–166.
Tumialis, D., Skrzecz, I., Mazurkiewicz, A., Pezowicz, E. and Góral, K. (2013) Wrażliwość larw *Hylobius abietis* (L.) na rodzime gatunki i szczepy nicieni entomopatogenicznych. *Sylwan* 157, 769–774.
Tumialis, D., Gromadka, R., Skrzecz, I., Pezowicz, E., Mazurkiewicz, A. and Popowska-Nowak, E. (2014a) *Steinernema kraussei* (Steiner, 1923) (Rhabditida: Steinernematidae) – the first record from Poland. *Heliminthologica* 51, 162–166.
Tumialis, D., Pezowicz, E., Mazurkiewicz, A., Skrzecz, I., Popowska-Nowak, E. and Petrykowska, A. (2014b) The effect of initial dose on the recovery and final yields of *Heterorhabditis megidis* (Rhabditida: Heterorhabditidae) in larvae of the great wax moth, *Galleria mellonella*. *Acta Parasitologica* 59, 213–218.
Wojciechowska, I., Kamionek, M. and Morytz, B. (2013) Biologiczne metody zwalczania muchy domowej w stajniach. *Annals of Warsaw University of Life Sciences – SGGW. Animal Science* 52, 227–236.

23 Status of Entomopathogenic Nematodes in Integrated Pest Management Strategies in China

Congli Wang* and Chunjie Li

Key Laboratory of Mollisols Agroecology, Northeast Institute of Geography and Agroecology, Chinese Academy of Sciences, Harbin, China

23.1 Introduction

Entomopathogenic nematodes (EPNs) have been the subject of more and more research and commercial interest for the past four decades. They are considered as important biological control agents of soil pests over the world due to the high demand for ecologically sustainable agriculture. The advantages for using EPNs include their wide range of insect hosts, the fact that they are non-toxic and non-polluting to the environment and humans, are mass produced easily and have the ability to kill insects rapidly. The two families, Steinernematidae and Heterorhabditidae, are major effective biocontrol insecticides that are associated symbiotically with bacteria of the genera *Xenorhabdus* and *Photorhabdus*, respectively (Kaya and Gaugler, 1993; Askary, 2010). Infective juveniles (IJs) first find their host and penetrate into the insect through natural openings; the symbiotic bacteria are then released from the EPNs, multiply and cause death of the insect within 1 to 2 days (Wouts, 1991); the nematodes then feed on the bacteria and produce more infective juveniles.

Research on EPNs in China began in the early 1980s. Great achievements with EPNs as biocontrol agents in Chinese agriculture have been accomplished, due to more governmental funding input and an increasing number of trained EPN scientists since the 1980s. These were demonstrated with the numerous publications associated with EPN studies, including journals, theses, patents, meetings, reviews and others, and have been increasing in China since 1980 (Fig. 23.1). One of the important milestones was an international collaborative programme established among the Guangdong Entomological Institute in Guangzhou, the Institute of Biological Control, Chinese Academy of Agricultural Sciences in Beijing, and the Division of Entomology, CSIRO, Australia. The programme was supported by the Australian Centre for International Agricultural Research during the mid-1980s and early 1990s (Bedding, 1990; Kaya *et al.*, 2006). This international collaboration accelerated EPN field application and basic research in China.

A series of review papers on EPN research advances were published. These research advances included various aspects such as the application of *Steinernema* spp. and *Heterorhabditis* spp. (Han, 1994; Hao *et al.*, 2001; Chen *et al.*, 2002; Chu *et al.*, 2002; Liu *et al.*, 2002; Wang *et al.*, 2002; Jin *et al.*, 2003; Wu *et al.*, 2014c; Yan *et al.*, 2014a), EPN taxonomy, symbiotic bacteria and the association between EPNs and their bacteria (Tu, 1996; Li and Guo, 2005; You *et al.*, 2006; Xia *et al.*, 2006; Qiu and Han, 2007; Xue *et al.*, 2009; Qiu *et al.*, 2010; Wang *et al.*, 2013), cold tolerance and antifreeze protein (Jing *et al.*, 2002; Wang *et al.*, 2011), anhydrobiosis (Yan and Han, 2010), nematode development and behaviour studies (Li and Han, 2007), EPN association with chemical nematicides (Li and Han, 2001), EPNs against pests on fruit trees and forest (Sun *et al.*, 2003), EPNs against root gnat, *Bradysia odoriphaga*, on Chinese chive (Zhang *et al.*, 2013a,b; Bai *et al.*, 2015), and so on.

*Corresponding author. E-mail: clwangjb@yahoo.com

 Biocontrol Agents: Entomopathogenic and Slug Parasitic Nematodes (eds. M.M.M. Abd-Elgawad, T.H. Askary and J. Coupland)

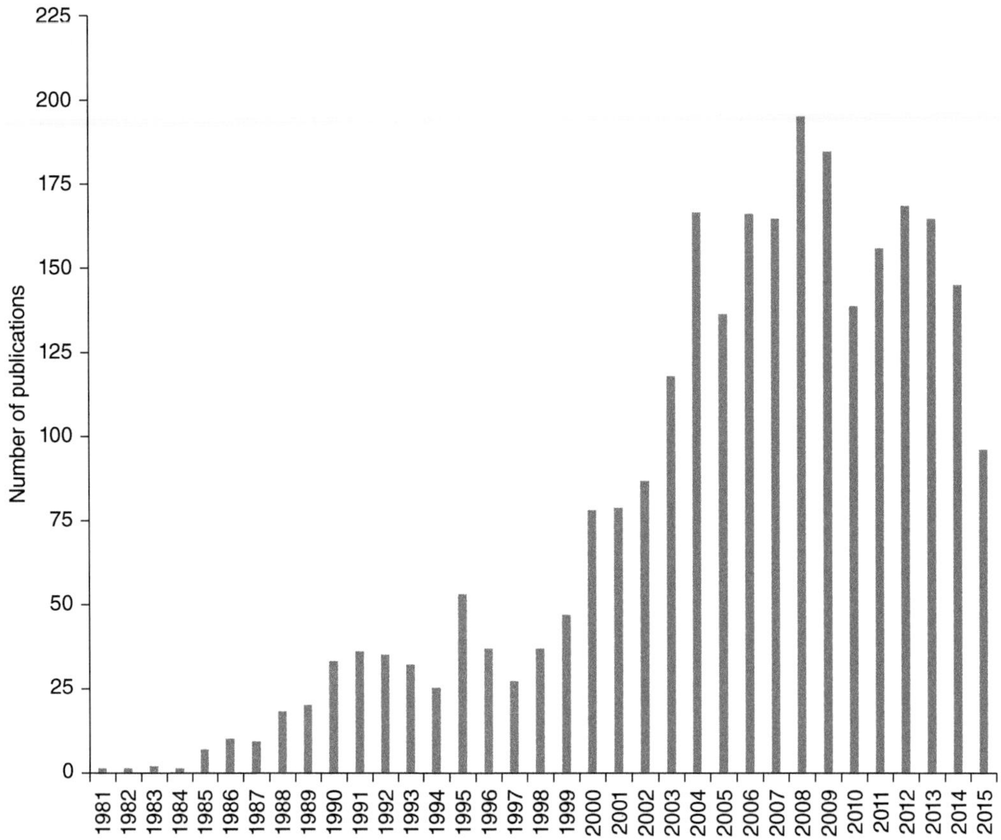

Fig. 23.1. The number of publications associated with entomopathogenic nematode studies in China from 1981 to 2015. The data, including journals, theses, meetings, reviews, patents and others in Chinese, were collected from the website http://oversea.cnki.net

In this chapter, we summarize the current status of EPNs as biocontrol agents against important Chinese agricultural pests in integrated pest management (IPM) and some important research advances of EPNs in controlling pests in China. The present drawbacks in research on selected EPN–insect systems and potential threats to ecosystems are discussed.

23.2 Integrated Pest Management in China

The concept of 'integrated control' emerged in China in the mid-1950s (Ma, 1976). However, until 1975, the Ministry of Agriculture and Forest organized one plant protection meeting to determine the definite strategy of 'Prevention first and then Integrated Pest Control' (Qiu and Yang, 2007; Liu *et al.*, 2014). The aim was to change the situation of the single chemical control strategy of killing all pests, which resulted in the chemical resistance of pests, environmental pollution and other associated ecological damage problems (Qiu and Yang, 2007). Since then, IPM has been the foundation of sustainable agricultural development in China. The Chinese government invested a large amount of money and set up different important research projects (for example: 973 projects, 863 projects and the State Key Projects) to explore new IPM strategies to control pests. Through these projects, a series of achievements on pest control were obtained, including a national and provincial network and IPM systems in important crops (rice, wheat, maize, cotton, vegetables, etc.) (Wu *et al.*, 2009). These IPM systems include the monitoring and forecasting of important pests, ecological regulation, chemical control with low toxicity, biological control, crop resistance, transgenic techniques, biological engineering, information technology (global positioning systems (GPS), geographic information systems (GIS) and computer science) and others (Wu *et al.*, 2009). Recently, with the increasing realization of environmental pollution, including soil, water and air, due to chemical pesticides, accompanied with serious health problems in China, green ecological agriculture came to the public's attention, resulting in farmers turning to the biological control of pests.

23.3 Application of Biological Control Agents in Integrated Pest Management in China

23.3.1 What can biological methods offer to integrated pest management?

Biological control can be traced back in history to 304 AD, with the use of yellow citrus ants (*Oecophylla smaragdina*) against citrus fruit tree damage (Liu *et al.*, 2014). After the foundation of the People's Republic of China, research on biological control began, and increased steadily (Liu *et al.*, 2014). A special issue on the progress of the biological control of pests in important economic and commercial crops in China was reviewed (Liu *et al.*, 2014). The identified natural enemies of pests that include predators, parasitoids and entomopathogenic pathogens (bacteria, viruses, fungi) were described in detail in each crop. EPNs used against pests were not reported in these reviews since studies on EPNs did not raise awareness in China until the late 1970s, as less success against pests of field crops was obtained with EPNs than with other natural enemies of pests.

23.3.2 A list of available products containing entomopathogenic nematodes in China

Two available commercial products containing EPNs can be found on the market. One is 'Lvcaobao', developed by the Guangdong Entomological Institute and produced by Guangzhou Lvenyuan Biotechnology Co, Ltd (http://baike.haosou.com/doc/6643607.html); the other is 'Naima', developed by the Northeast Institute of Geography and Agroecology, Chinese Academy of Sciences, and produced by Henan Jiyuan Baiyun Industry Co, Ltd (http://www.keyunnpv.cn/Products/tiandishengwunongyao/330273.html) to control chive root gnat (*B. odoriphaga*) and other hidden pests below ground. The formula of 'Nema' is powder and the application methods include spraying to the root surface or mixing nematodes with irrigation water.

23.4 An Outline of Studies on Entomopathogenic Nematodes in Chinese Research Centres

The major applications of EPNs in China are in the control of pests on fruit trees, forest, ornamental and vegetable crops and house pests. The efficient EPN strains against pests isolated from China include *Heterorhabditis bacteriophora* H06 (Taishan I), *Steinernema* sp. X-7, *Steinernema glaseri* C85011, *Steinernema feltiae* Beijing, *Steinernema* sp. CB-2y etc. (Feng and Han, 2005). However, nematode efficacy against soil pests has been variable and not stable, and practical application is limited to a small scale, due to the high cost of large-scale application to field crops (wheat, rice, soybean, potato and groundnut, etc.). Furthermore, although research on EPNs in China covers various areas from basic taxonomy to application technology, more information is required for practical use. Accordingly, research on understanding the interactions among EPNs, symbiotic bacteria, insect, plant and/or ecological environment in the soil and other related aspects are/is being increased. Basic knowledge gained from these endeavours will benefit the practical application of EPNs in the field.

23.5 Application of Entomopathogenic Nematodes in Vegetable Crop Protection

23.5.1 Application of entomopathogenic nematodes in Chinese chive

Root gnat (*B. odoriphaga*) is one of the important soil pests on Chinese chive (*Allium tuberosum*), a favoured vegetable in Asia. The chive gnat also attacks shallot, garlic, cabbage, radish, melon, celery and other vegetables, causing significant reduction in yield as well as product quality (Ma *et al.*, 2013). The larvae feed on the bulbs and stems, which causes plants to be stunted or even killed. Multiple overlapping generations of the pest in one year make it hard to control. Frequent application of insecticides against *B. odoriphaga* has caused insect resistance to insecticides. Previously used effective but highly toxic insecticides, phorate (3911), methamidophos and parathion have threatened health and caused environmental problems in China (Sun *et al.*, 2004). Therefore, seeking EPNs as an alternative biocontrol agent against the chive gnat is now a matter of urgency in the development of sustainable agriculture.

Yang and Zhang (1990a) found that the strain *Heterorhabditis* sp. D1 could parasitize the larvae and pupae of root gnat. Zhang *et al.* (1994a) identified that two strains, *Steinernema bibionis* Otio and *Heterorahbiditis* sp. CB15, among six screened strains had greater parasitic ability, and an 87% parasitism rate of chive gnat was obtained for the Otio strain and 93% for the CB15 strain at day 5 after treatments. Field trials showed approximately 69% control efficacy with the two strains (Zhang *et al.*, 1994a). The root gnat was determined to be sensitive to the strain *Heterorhabditis indica* LN2 out of 27 screened EPNs (Sun *et al.*, 2004). Further studies indicated that 85% and 88% larval mortality occurred when 200 and 400 IJs/insect were used (Sun *et al.*, 2004). LN2 caused high larval mortality of root gnat when the soil temperature was at 25–30°C and soil moisture at 10–15% (Sun *et al.*, 2004). Yang *et al.* (2004) screened a highly virulent strain, *S. feltiae* PS4, originally from Australia, which caused 82.3% larval mortality of chive gnat at day 8 after the application of 60 IJs/ml at 25°C, and 85.7% mortality rate 12 days after the application of 120 IJs/ml at 15°C. Field application of *S. feltiae* PS4 resulted in 55% pest control, slightly less than the application of the insecticide phoxim, and plant damage decreased 71% (Yang *et al.*, 2004). Ma *et al.* (2013) identified nine promising virulent species/strains (*Steinernema ceratophorum* HQA-87; *Steinernema hebeiense* JY-82; *S. feltiae* JY-90, JY-17; *Steinernema litorale* HXY-68; *H. indica* ZZ-68; *H. bacteriophora* NY-63, HQ-94; and *Heterorhabditis megidis* LFS-10) against *B. odoriphaga* out of 145 EPN isolates from 13 species collected from northern China. Pot experiments indicated that the promising isolates caused a 78–94% reduction of the chive gnat population at an application rate of 75 IJs/cm^2, while field experiments showed that *S. feltiae* JY-90 had the strongest suppression of *B. odoriphaga* at 14 days after EPN application. *S feltiae* JY-17 and *S. hebeiense* JY-82 reduced the population of *B. odoriphaga* significantly, which was similar to treatment with the insecticide phoxim (Ma *et al.*, 2013).

In the authors' group, local isolates of *H. bacteriophora* from Harbin, in the north-east of China, had been shown to tolerate relatively lower soil temperatures (–5 to –13°C) and be efficient against the chive gnat (*B. odoriphaga*) in laboratory tests (Li *et al.*, 2011a, 2013). Field application of *H. bacteriophora* (Hb) in Ningxia Hui Autonomous Region increased yield by approximately 25–64%, and plant quality was also improved (Pan *et al.*, 2013; Zhang *et al.*, 2014). Inconsistent EPN control in the field was observed, suggesting environmental factors affected control efficiency. To dissect the infection process, a powerful transparent, semisolid pluronic gel that mimicked three-dimensional soil to allow nematodes to move freely (Wang *et al.*, 2009a) was used to study EPN host-habitat location, which is a very important step in searching for hosts (Li *et al.*, 2015). We found that one isolate, Hb-HBN, showed the strongest attraction to chive roots among the species/isolates of EPN tested. Surprisingly, attraction to a chive root following storage for up to 30 days was much less than for Hb-HBN that were stored for longer, with an optimal attraction observed for *H. bacteriophora* that were stored for 75–90 days. This might be associated with phased infectivity. Our studies indicate host-habitat root exudates play an important role in directing EPN localization, which provides a better understanding of EPN foraging behaviour (Li *et al.*, 2015). Tritrophic interactions among EPNs, chive and root gnat are being investigated with this valuable pluronic gel system to gain more insight in improving stable control in the field. Figures 23.2 and 23.3 show laboratory and field work in our research team.

23.5.2 Application of entomopathogenic nematode in other vegetables

Steinernema spp. and *Heterorhabiditis* spp. were also applied to control other pests in associated vegetables (Table 23.1) (Chu *et al.*, 2002). One of the well-known examples was that the striped flea beetle (*Phyllotreta striolata*), an economically important pest of cabbage and other brassicas, could be controlled by EPNs (Yan *et al.*, 2013, 2014a). They also infect aubergines, melons and legume vegetables. Li and Wang (1990) found that *S. feltiae* A24 and *Heterorhabditis* sp. 86H-1 caused 100% larval parasitism rate of *P. striolata*. *S. feltiae* and *Steinernema carpocapsae* have been proved to be promising biocontrol agents against *P. striolata*, and more than 70% control was obtained in field trials (Li and Wang, 1990; Hou *et al.*, 2001; Wang and Han, 2008). The parasitism

Fig. 23.2. The laboratory work associated with entomopathogenic nematodes (EPNs) in the Northeast Institute of Geography and Agroecology, Chinese Academy of Sciences (NIGA-CAS).

(a) EPN application for chive root gnat (*Bradysia odoriphaga*) control on chive

(b) EPN application for onion maggot (*Delia antiqua*) control on garlic

Fig. 23.3. The greenhouse and field application of entomopathogenic nematodes (EPNs) in Shandong province (a) and in Jiangsu province (b).

Table 23.1. Entomopathogenic nematodes against important pests in vegetables.

Nematode	Insect	Reference
Steinernema feltiae A24, *Heterorhabditis* sp. 86H-1	Striped flea beetle (*Phyllotreta striolata*)	Li and Wang, 1990; Hou *et al.*, 2001; Wang and Han, 2008
S. carpocapsae All, *Heterorhabditis indica* LN2	Striped flea beetle (*P. striolata*)	Yan *et al.*, 2013
S. carpocapsae A24	Daikou leaf beetle (*Phaedon brassicae*)	Pan *et al.*, 1999
S. carpocapsae A11	Diamondback moth (*Plutella xylostella*)	Jin *et al.*, 2013
S. carpocapsae Mex, *H. indica* LN2	Black cutworm (*Agrotis ipsilon*)	Yan *et al.*, 2014b
S. carpocapsae NC116, *S. feltiae* SF-SN, *H. bacteriophora* H06	Black cutworm (*A. ipsilon*)	Wu *et al.*, 2015

rate of *S. carpocapsae* Agriotos to the third- or fourth-instar larvae of cabbage beetle (*Colaphellus bowringi*) reached 90% in pot tests (Wei *et al.*, 2000). Yan *et al.* (2013) identified that both *S. carpocapsae* All and *H. indica* LN2 decreased *P. striolata* populations and increased cabbage yield when applied at a rate of at least 0.75×10^9 IJs/ha as compared to the control and insecticide treatments alone.

23.6 Application of Entomopathogenic Nematodes in Arable Crop Protection

EPNs have not been widely used against pests on field crops, such as rice, wheat and maize. The rice water weevil (*Lissorhoptrus oryzophilus*) is a quarantine pest worldwide. It occurs in more than nine provinces in China. The larvae cause much more damage, as they feed on roots and prune them, than the adult weevils feeding on leaves. The yield loss could reach 30%, and up to 70–80% with strong infestation. *S. feltiae* Otio was identified as highly virulent to the rice water weevil, and the mortality rate was affected by time and dose (Sun *et al.*, 2006; Li *et al.*, 2007a).

The Asian corn borer (*Ostrinia furnacalis*) is a serious pest of maize, sorghum and cotton in China. The larval mortality and lethal speed of *O. furnacalis* were affected by exposure time to the strain of *S. feltiae* Beijing (Yang *et al.*, 1991). He *et al.* (1991) found 56.7% mortality at 10 days after the application of *S. feltiae* Agriotos into the whorl leaves of maize in laboratory tests. In field trials, 90.5% and 80.4% larval mortality were obtained at 6 days after releasing nematodes at a rate of 2000 IJs/plant into summer maize in 1989 and 1990, respectively. The larval mortality of *O. furnacalis* reached 92.5%, 100% and 100%, respectively, after using 50, 100 and 200 IJs of *S. carpocapsae* A24 strain in laboratory bioassays (Sun and Li, 2004). When 300–4000 IJs were released into whorl leaves of maize in the field, 70–81.8% control was obtained (Sun and Li, 2004).

The grub damage on groundnut (*Holotrichia parallela* and *Holotrichia oblita*) and soybean (*Holotrichia diomphalia*) was controlled effectively by *Heterorhabditis* spp. (Li *et al.*, 1993, 2007b; Xu *et al.*, 2008) and on beet with *S. glaseri* sp. NC32 (Li *et al.*, 1995). Guo *et al.* (2013, 2015) determined that both *Steinernema longicaudum* X-7 and *H. bacteriophora* H06 effectively decreased the population of the white grub larvae in the groundnut field, and the yield increased more than 50% compared to the control plants (water treatment) and more than 15% compared to chlorpyrifos-treated plants. The laboratory tests demonstrated that the larval mortality of *H. diomphalia* reached 76% with the application of 580 IJs/cm^2 of the strain *S. carpocapsae* A24 sprayed on the soil surface and 100% mortality rate with 120 IJs/cm^2 of the strain *H. bacteriophora* H06 (Zhang *et al.*, 1996).

23.7 Other Applications of Entomopathogenic Nematodes in China

23.7.1 Parks and urban tree plantings, municipal forests

A very successful example of using EPNs to control pests in trees was that of *S. carpocapsae* effectively killing carpenterworm (*Holcocerus insularis*) that attacked ash (*Fraxinus pennsylvannica*), the Chinese scholar tree (*Sophora japonica*) and willows (*Salix* spp.), mostly in northern China (Yang and Zhang, 1990b; Yang *et al.*, 1993; Kaya *et al.*, 2006). The carpenterworm bores into the tree trunk and makes tunnels to the exterior, through which it expels boring dust. Approximately 30–80% of shade trees were infested and 5% killed annually. Instead of toxic chemicals, a suspension of *S. feltiae* Agriotos was injected into tunnels made by the carpenterworm and 97% larval mortality and 94% decrease in survival bores were obtained in two big cities, Tianjin and Shijiazhuang (Yang *et al.*, 1992). Several other cities in China used *S. carpocapsae* to control carpenterworm in infested street trees (Yang *et al.*, 1993; Kaya *et al.*, 2006). In addition, *S. feltiae* was used effectively to suppress the moth *Zeuzera multistrigata* in Fujian province and the moth *Cossus mongolicus* in Gansu province that attacked forest trees (Yang *et al.*, 1990; Zhang *et al.*, 1991).

Long-horned beetles (*Cerambycidae*) are another group of serious pests with wide host ranges. Within this family, the citrus long-horned beetle (*Anoplophora chinensis*) and Asian long-horned beetle (*Anoplophora glabripennis*) are native to China and Asia and are destructive pests attacking street and park trees, forest trees, orchards, ornamentals, landscape and other species of trees. Huang *et al.* (1997) found that *S. feltiae* Beijing and *S. carpocapsae* MK could parasitize older larvae, pupae and adults of *A. chinensis*, and 94–100% larval mortality was obtained after 4–6 days of EPN application. Field application of the two strains resulted in 90% control of the pest after application of 1×10^4 IJs/beetle (Huang *et al.*, 1997). Injection of 7500 IJs/ml of *S. carpocapsae* Beijing to tree holes infected by *A. glabripennis* reduced the infection rate by 86.4% (Liu *et al.*, 1998a). Zhang *et al.* (2007) identified two strains (*H. bacteriophora* NJ and *S. carpocapsae* All) that caused above 80% control of the larvae in street trees in the city of Harbin, Heilongjiang province. Li *et al.* (2011c) reported that the pathogenicity of *Heterorhabditis* sp. ZH to *A. glabripennis* was high and could be a potential biocontrol agent against Asian long-horned beetles. Laboratory tests showed 94.6% adult mortality after 10 days of application of *H. bacteriophora* ZH and 100% larval mortality after only 4 days of EPN application (Li *et al.*, 2011c).

The bamboo weevil (*Cyrtotrachelus longimanus*) is a damaging pest of bamboo shoots, an economically important resource in southern China. Liu *et al.* (1989) screened and identified three strains (KG, NC52 and 85011) of *S. glaseri* that resulted in 80% mortality of weevil larvae and more than 70% control of the insect when injected at a rate of 2000 IJs/shoot. When the combination of *S. glaseri* 85011 with a low amount of chemical dimethoate was injected into infested bamboo shoots, control efficacy was increased to above 80%, and 50% of the infested bamboo grew normally (Liu *et al.*, 1993).

The Japanese pine sawyer, *Monochamus alternatus*, is an important forest pest, as well as the principal vector of the pinewood nematode, *Bursaphelenchus xylophilus*, in China. Zhang *et al.* (2012) explored a time–dose–mortality (TMD) model of the control function of *S. carpocapsae* All (Sc-All) and *H. bacteriophora* WG (Hb-WG) against *M. alternatus* in laboratory tests. When 6×10^4 IJs/ml dose was used, 82% larval mortality was obtained for Sc-All and 74% for Hb-WG, and the lethal time (LT_{50}) of Sc-All was 3.98 days and 4.06 days for Hb-WG (Zhang *et al.*, 2012). These studies provide a basis for field control of *M. alternatus* by using EPNs.

23.7.2 Orchards

The peach fruit moth (*Carposina niponensis*, also called *Carposina sasakii*) has been one of the most damaging pests in various fruit trees, including apple, peach, pear, apricot, plum and date, in China since the 1950s (Huang *et al.*, 1976; Jiang *et al.*, 1990; Yang, 1991; Li *et al.*, 2012). The pest has been spreading quickly with the increasing planting areas of fruit trees through the whole country in recent years. Yield loss could be up to 65% in the apple orchard, and US$1.7 billion economic losses occur per year (Wang, 1993). *C. niponensis* has one generation a year in most apple growing areas in China. A second generation may occur in warmer areas. The use of toxic or persistent insecticides such as BHC, DDT, phoxim and folithion had caused serious environmental pollution and ecological damage, and insecticide resistance occurred until the 1980s (Wang, 1993; Chu *et al.*, 2002). Beginning in 1981, EPNs have been applied to control *C. niponensis* in the apple orchards in many provinces, such as in Henan, Shandong, Hebei and Liaoning (Zhou *et al.*, 1989; Yang *et al.*, 2000; Chu *et al.*, 2002).

Li *et al.* (1984) reported that the larval mortality of *C. niponensis* reached 92–95% after the application of *S. carpocapsae* in laboratory assays, and mortality of the emerged insect larvae in the soil was up to 98% when applied at the rate of 46–91 × 10^4 IJs/m^2 in the apple orchard. This was better control than treatments with the chemical insecticides hexachlorocyclohexane and phoxim. Field application in apple orchards with *H. bacteriophora* Taishan I showed that larval mortality was 92% and fruit infestation was controlled to 0.3% (Li *et al.*, 1990). *Steinernema* sp. CB-2Y killed 88% of overwintering larvae (Zhou and Yang, 1991). The application of 1 × 10^9 IJs of *Steinernema* sp. CZ-88/667 m^2 in a Zhengzhou apple orchard decreased fruit damage compared with treatments of chemical insecticides when soil temperature was 25–30°C and soil moisture was 10–20%. The control could last up to 1 month (Liu *et al.*, 1998c). When 1.5 × 10^9 IJs/ha of *S. carpocapsae* Agriotes were applied to 299 ha of apple orchard in Shandong province at a soil temperature of 26.2–27.1°C, air temperature of 25.2–29.5°C and relative humidity (RH) of 80–89%, fruit damage by peach fruit moth was controlled to 0.4–0.5%. Fruit damage was increased to 2.4% at higher temperatures (soil temperature 26.5–29°C, air temperature 29.5–34°C) and decreased RH (76.8%) (Yang *et al.*, 2000). Li *et al.* (2012) found that one application of 1.5 × 10^8 IJs/667m^2 of *H. bacteriophora* H06 in a date orchard at the stage of emergence of overwintering adults resulted in 55% control, and 63% for one application at the stage of emergence of older larvae out of fruit. Control efficacy increased to 90% when combining two applications together at the two stages (Li *et al.*, 2012).

Applications of EPNs against insects in fruit trees were reviewed by Sun *et al.* (2003). The strain CB-2Y of *Steinernema* sp. caused 93–100% larval mortality of plum fruit sawflies, *Hoplocampa fulvicornis*, at 4 days after EPN application in laboratory assays, 70% mortality at 7 days after field application and 75% mortality at 18 days (Liu, 1993). Pot experiments showed 87–92% larvae mortality of pear weevil, *Rhynchites foveipennis*, at 4 weeks after application of 6–12 × 10^5 IJs/m^2 of *S. felitae*, while 80–83% larval mortality was obtained at 4 weeks after field application of 6–8 × 10^5 IJs/m^2 (Liu *et al.*, 1991). The lichi stem borer (LSB) (*Arbela dea*) is a dominant boring insect pest on lichi and other fruit and forest trees in south China. When *S. carpocapsae* Agriotis with a dose of 1000–3000 IJs/ml was injected into holes of lichi trees infected by LSB with a mass application in 260 hm^2 of lichi plantations, 85–100% mortality of LSB was obtained (Xu and Xie, 1997). The red-necked long-horn beetle (RLB, *Aromia bungii*) causes 30–100% damage in fruit trees each year in China (Liu *et al.*, 1997). Peach and almond infected by RLB was controlled effectively by *Steinernema* spp. (Liu *et al.*, 1997, 1998b). Other studies with EPNs controlling lichi longicorn beetle (*Aristobia testudo*) (Xu *et al.*, 1995), *Chelidonium arentatum* and *Lepidarbela dea* in lichi and other fruit tree orchards were conducted (Luo *et al.*, 1997). The studies on EPNs against pests in trees are listed in Table 23.2.

23.7.3 Production of edible mushroom

Mushroom is a much loved table food in China. Sciaridae, Mucetophilidae and Cecidomyiidae are very important pests on cultivated edible mushroom (Yang *et al.*, 2010a). Natural enemies, such as insects, bacterial and plant toxins, were reported to control these pests (Yang *et al.*, 2010a). However, little information is available on EPN controlling pests on mushroom crops. Yang *et al.* (2010b) evaluated the efficacy of *S. feltiae* with 100 IJs/m^2 against sciarid flies in *Agaricus bisporus* mushroom beds in two locations (Beijing and Fujian province) with two methods of trapping pests (yellow sticky trap and light trap). The yellow sticky trap method indicated 69% reduction of pest trapping rate after 19 days of application of *S. feltiae* in Beijing and 82% reduction after 33 days following treatment in Fujian province (Yang *et al.*, 2010b). Light trap assays in the Beijing location showed 25.6% reduction of trapped pest rate after 12 days of application of *S. feltiae* compared with the control (Yang *et al.*, 2010b).

23.7.4 Application of entomopathogenic nematodes against stable flies

There are few studies on the control of house pests by EPNs in China. Xu *et al.* (1994) demonstrated that *S. feltiae* Otio was a promising strain to infect the housefly (*Musca domestica*), a very serious pest on animal farms in China. Spraying approximately 1 × 10^7 IJs/m^2 on pig manure reduced the number of housefly larvae by 50% (Xu *et al.*, 1994). In laboratory assays, the virulence of *S. feltiae* Otio to housefly larvae and adults was affected by temperature (optimum range: 15–25°C), moisture (optimum RH 22%) and storage duration of nematodes (Wang *et al.*, 1998). The larval lethal rate of *M. domestica* reached 95% when 1000 IJs/fly were applied in the laboratory assays (Zhang *et al.*, 1994b).

Table 23.2. Entomopathogenic nematodes against important pests on trees.

Nematode species	Insect	Reference
Steinernema feltiae	Citrus long-horned beetle (*Anoplophora chinensis*); Asian long-horned beetle (*Anoplophora glabripennis*); Lichi stem borer (LSB, *Arbela dea*); Walnut moth (*Atrijuglans hetaohei*); Pear weevil (*Rhynchites foveipennis*)	Xu and Xie, 1997; Sun *et al*., 1990; Huang *et al*., 1997; Liu *et al*., 1991; Wang *et al*., 1997
S. carpocapsae	Citrus long-horned beetle (*A. chinensis*); Litchipara-stem borer (*Lepidarbela dea*); Lichi longicorn beetle (*Aristobia testudo*); Asian long-horned beetle (*A. glabripennis*); Carpenterworm (*Holcocerus insularis*)	Yang and Zhang, 1990b; Yang *et al*., 1993; Huang *et al*., 1997; Luo *et al*., 1997; Liu *et al*., 1998a; Kaya *et al*., 2006; Zhang *et al*., 2007
Steinernema sp. CB-2Y, CZ-88, Agriotis	Peach fruit moth (*Carposina niponensis*); Plum fruit sawflies (*Hoplocampa fulvicornis*)	Zhou and Yang, 1991; Liu *et al*., 1998c; Yang *et al*., 2000; Liu, 1993
Steinernema spp.	Red-necked long-horn beetle (RLB, *Aromia bungii*); Japanese pine sawyer (*Monochamus alternatus*)	Liu *et al*., 1997, 1998b; Zhang *et al*., 2012
S. caudatum	Litchipara-stem borer (*Lepidarbela dea*); Lichi longicorn beetle (*A. testudo*)	Luo *et al*., 1997
S. glaseri KG, NC52 and 85011	Bamboo weevil (*Cyrtotrachelus longimanus*)	Liu *et al*., 1989
Heterorhabditis bacteriophora ZH	Asian long-horned beetle (*A. glabripennis*)	Li *et al*., 2011c
H. bacteriophora H06	Peach fruit moth (*Carposina niponensis*)	Li *et al*., 2012

23.8 Interactions of Entomopathogenic Nematodes with Pesticides and Others

In order to improve control efficiency, the combination of EPNs with insecticides or impact of chemical pestcides on EPNs was studied. For example, a comparison of larval mortality of the chive root gnat (*B. odoriphaga*) caused by EPNs and chemical insecticides, or a combination of EPNs with insecticides, was studied by a few laboratories (Sun *et al.*, 2004; Sun and Li, 2007; An *et al.*, 2012; Ma *et al.*, 2012; Zhao *et al.*, 2013; Wu *et al.*, 2014a). These studies suggested that the effect of insecticides on EPN survival and infectivity was dependent on the chemicals and EPN species/strains. For example, phoxim, chlorpyrifos and emamectin benzoate affected the survival of *H. indica* LN2 but had no effect on the survival of *H. bacteriophora* H06 (Zhao *et al.*, 2013). The optimum release time without affecting survival and infectivity on both LN2 and H06 was 10, 10 and 20 days after the application of phoxim, chlorpyrifos and emamectin benzoate, respectively (Zhao *et al.*, 2013). Using a combination of *S. carpocapsae* strain SF-SN with a low concentration of chlorpyrifos, imidacloprid and beta-cypermethrin increased the lethality to *B. odoriphaga* by 34, 21 and 55%, respectively, compared to treatment with any one insecticide alone (Wu *et al.*, 2014a).

When *S. carpocapsae* All with 4000 IJs/ml were mixed with pesticides of chlorpyrifos and imidacloprid, the control of *Rhabdoscelus lineaticollis*, a new invasive pest attacking palms and sugarcanes, reached 89%, compared to 72% for chlorpyrifos only, 2.5% for imidacloprid only and 28–53% for EPNs used alone (Wang *et al.*, 2007a). Zhang *et al.* (2006) screened the interactions between *S. longicaudum* X-7 and insecticides and found that *S. longicaudum* X-7 had a synergistic interaction with the diflubenzuron insecticide, and had additive mortality with flufenoxuron, chlorbenzuron, matrine, chlorfluazuron or tebufenozide insecticides against the J3s of the target grub *Holotrichia ovata*.

Wu *et al.* (2014b) found that the combination of *H. bacteriophora* H06 (Taishan I) with groundnut root nodule bacteria (*Bradyrhizobium* sp.) not only increased the control effect of groundnut white grubs significantly but also increased groundnut leaf thickness, weight and area, and pod and kernel weight when compared with the treatments with nematodes or rhizobia alone.

23.9 Technical Restrictions for Nematode Applications

Although great progress has been achieved, there are still numerous factors that affect control efficacy and limit commercial application. These factors include not only application techniques, pest search strategies, desiccation tolerance and high temperature tolerance but also cost, shelf life, handling, mixing, coverage, and so on. Currently, investigation of the natural occurrence of nematodes, screening nematodes for adaption to local climate and the identification of symbiotic bacteria from EPNs with traditional and molecular techniques has broadened the knowledge of EPNs and improved control efficiency.

The natural occurrence of EPNs was investigated (Chen *et al.*, 2006; Xing *et al.*, 2009; Ma *et al.*, 2010; Wang *et al.*, 2014). The infectivity of the EPNs was affected by temperature and moisture (Yu *et al.*, 2003). Screening desiccation-resistant EPN strains is one of the most important strategies for long-term storage and large-scale application (Yan and Han, 2010). Chen *et al.* (1999) found that *S. carpocapsae* CB-16 was resistant to desiccation at 25°C, RH 50.5% for 9 h. Xing *et al.* (2009) demonstrated 80% survival rate for *Steinernema kraussei* and 73.7% for *Steinernema bicornutum* in those same conditions. The survival rate was low for *Heterorhabditis* spp. when isolated EPN species in Gansu province were treated in the same conditions as above. Both species, *S. kraussei* and *S. bicornutum*, showed 17% penetration rate and 100% lethal rate to wax moth after recovery from 9 h dehydration (Xing *et al.*, 2009). Li *et al.* (2011b) further confirmed that *Steinernema* spp. had a higher survival rate than *Heterorhabditis* spp. under desiccation. Lu *et al.* (2014) identified one isolate of *S. litorale* that caused 86% larval mortality of *Galleria mellonella* after being treated for 96 h at 10°C, but zero mortality was caused by other tested nematodes. Different combinations of EPNs and symbiotic bacteria greatly influenced the infectivity, dry weight and biochemical compositions of EPNs, and the cultural methods and the composition of cultural medium also influenced the biochemical compositions of nematodes (Qiu *et al.*, 2003).

Screening of highly toxic strains of symbiotic bacteria and testing their insecticidal activity is another important strategy to improve control efficiency. Wang *et al.* (2000) identified *Xenorhabdus bovienii* A54 as highly toxic to insects from six tested strains isolated from EPN symbionts. Larval mortality of *G. mellonella* reached 93% when cell-free culture from toxic symbiotic bacteria was injected into insect larvae (Wang *et al.*, 2000). When fed artificial diet with the toxin, the larval mortality of *O. furnacalis* and *Helicoverpa armigera* were 100% and 90% within 120 h, respectively (Wang *et al.*, 2000). Seven insecticidal toxins isolated from a highly virulent strain, *Xenorhabdus poinarii* 5-5B, of symbiotic bacteria were highly toxic to *O. furnacalis*, *Tenebrio molitor*, *Plutella xylostella*, *Mythimna separata* and *Latphygma exigua* (Wang *et al.*, 2008a). The insecticidal activity of metabolites of *Xenorhabdus nematophila* YL001 and *X. bovienii* YL002 associated with *Steinernema* spp. was evaluated by Wang *et al.* (2009b). The results showed that both symbiotic bacteria had highly injectable toxicity for *G. mellonella*, with 100% mortality at 48 h after injection of fermented liquid through cell-free aqueous filtrates at different cultural times. Stomach toxicity, growth inhibition activities and anti-feeding activities were lower

(Wang *et al.*, 2009b). Jin *et al.* (2010) determined that ten strains of symbiotic bacteria caused over 75% anti-feeding activity in the second-instar larvae of diamondback moth (*P. xylostella*), as measured by using the leaf-disc test.

Molecular technology was applied to the identification of EPNs (Li *et al.*, 2002; Zhou *et al.*, 2002; Qiu *et al.*, 2003), symbiotic bacteria of EPNs (Wang *et al.*, 2007b, 2008b; Fang *et al.*, 2008; Yang *et al.*, 2008; Lu *et al.*, 2014), differential expression analysis of development-related genes of EPNs (Xia *et al.*, 2008) and construction of cDNA library and analysis of expressed sequence tags for IJs of *S. feltiae* (Ma *et al.*, 2008). Further, toxic proteins from symbiotic bacteria were identified and isolated (Cui *et al.*, 2003, 2005; Yang *et al.*, 2007). Studies of oral toxicities from extracellular and intracellular proteins extracted from symbiotic bacteria *X. nematophila* showed various inhibitory activities against neonatal larvae of *H. armigera* (Yang *et al.*, 2007). The inhibition percentages reached to 97–98% for intracellular proteins and 87–96% for extracellular proteins (Yang *et al.*, 2007). Cui *et al.* (2003, 2005) identified one gene, *xptA2*, with insecticidal activity against *H. armigera* and the blasted sequence associated with oral toxicity to insects. These toxic proteins have a promising future for controlling insects.

23.10 Entomopathogenic Nematodes – Potential Threat to Ecosystems

Currently, there is no knowledge available about the potential threat of EPN presence to ecosystems or protected beetles in China.

23.11 Conclusion and Future Prospects

In summary, the major achievement of the application of EPNs in China is obvious in the control of pests on fruit trees, forest, ornamental and vegetable crops and house pests. The field application of EPNs and active research associated with using EPNs against pests indicate EPNs as one of the effective biocontrol agents in China. Increasing the intensive research on topics such as nematode formulation, spraying methods, screening virulent nematodes, etc., and the increasing demands for the realization of the development of sustainable agriculture will stimulate the application of EPNs not only on the larger scale but also against more pests in China.

In order to achieve zero growth for pesticide application in 2020, the Chinese Agricultural Ministry has decided to continue to carry out professional control systems and green prevention and control field bases. In 2015, the Chinese government established 218 demonstration experimental bases throughout the country, and promoted green control strategies for rice, wheat, maize, potato, cotton, groundnuts, vegetables, apples, citrus, tea and other crops. Also, another series of application funding from the Chinese Academy of Sciences and other plant protection agents or companies are available for biological control. EPNs as potential control organisms have started to receive more attention and funding sources for scientists to conduct more application studies in China. We have obtained more funding from the Chinese Academy of Sciences to apply EPNs to field control of below-ground pests on vegetables including chive, garlic, onion, tomato and cabbage. The combination of EPNs with low toxic chemicals, biofertilizers and other integrated control methods will be applied to the field. Finally, a few demonstration fields with thousands of acres under EPN control will be established and on-site observation tours will be held for farmers, national and local extension scientists, universities and research institute government sectors in order to make more people realize the importance of biological control with EPNs to environmental and human safety. With increasing government investment, the realization of environment protection and the development of sustainable agriculture, more achievements for EPN application in China could be obtained in the near future.

Acknowledgements

The authors thank Teresa Mullen's helpful comments on this chapter and grant support from 'Science and Technology Service Network Initiative (STS) project funded by Chinese Academy of Sciences (KFJ-SW-STS-143-9).

References

An, L., Jia, L., Yuan, W., Wang, X., Chen, S., *et al.* (2012) Effect of entomopathogenic nematodes on controlling *Bradysia odoriphaga* Yang et Zhang and soil nematode community. *Journal of Agro-Environment Science* 31, 898–903.

Askary, T.H. (2010) Nematodes as biocontrol agents. In: Lichtfouse, E. (ed.) *Sociology, Organic Farming, Climate Change and Soil Science*. Springer, The Netherlands, pp. 347–378.

Bai, G., Ma, H., Wang, X., Wu, L., Shen, G., *et al.* (2015) Research progress on controlling *Bradysia odoriphaga* by entomopathogenic nematodes. *China Plant Protection* 35, 25–33.

Bedding, R.A. (1990) Logistics and strategies for introducing entomopathogenic nematode technology into developing countries. In: Gaugler, R. and Kaya, H.K. (eds) *Entomopathogenic Nematodes in Biological Control*. CRC Press, Boca Raton, Florida, pp. 233–246.

Chen, S., Yang, H., Luan, X., Yang, X. and Jian, H. (1999) The anti-desiccation ability of entomopathogenic nematodes. *Chinese Journal of Biological Control* 15, 100–103.

Chen, S., Huang, W. and Chen Q. (2002) Research advances on entomopathogenic nematodes. *Journal of Southwest Forestry College* 22, 63–66.

Chen, S., Li, X. and Shi, J. (2006) Occurrence of entomopathogenic nematodes in Hebei Province. *Chinese Journal of Biological Control* 22, 40–44.

Chu, D., Cong, B., Liu, H. and Meng, Y. (2002) Research progress of EPN application in China. *Journal of Laiyang Agricultural College* 19, 132–135.

Cui, L., Qiu, L., Xin, Z., Fang, Y. and Pang, Y. (2003) Importance BP of five genes presented in *Xenorhabdus nematophilus* BP toxin gene cluster to its insecticidal activity. *Acta Microbiolgica Sinica* 43, 747–752.

Cui, L., Wang, Q., Qiu, L. and Pang, Y. (2005) Insecticidal activity of toxin gene *xptA2* in *Xenorhabdus nematophila* BP. *Acta Phytophylacica Sinica* 32, 407–410.

Fang, X., Guo, Q., Yi, X., Wang, R. and Zhang, X. (2008) Molecular identification and cultivation characteristics of entomopathogenic bacteria. *Journal of Northwest A & F University* 36, 99–204.

Feng, S. and Han, R. (2005) Advances in storage and formulation of entomopathogenic nematodes *Steinernema* and *Heterorhabditis*. *Plant Protection* 31, 20–25.

Guo, W., Yan, X., Zhao, G. and Han, R. (2013) Efficacy of entomopathogenic *Steinernema* and *Heterorhabditis* nematodes against white grubs (Coleoptera: Scarabaeidae) in peanut fields. *Journal of Economic Entomology* 106, 1112–1117.

Guo, W., Yan, X., Zhao, G., Chen, J. and Han, R. (2015) Efficacy of entomopathogenic *Steinernema* and *Heterorhabiditis* nematodes against *Holotrichia oblita*. *Journal of Pest Science* 88, 359–368.

Han, R. (1994) Advances in the research of entomopathogenic nematodes *Steinernema* and *Heterorhabditis* in China. *Entomologia Sinica* 1, 346–364.

Hao, D., Gao, Z., Zhuang, Q. and Li, H. (2001) Advance of entomopathogenic nematodes. *Journal of Forestry Research* 12, 257–262.

He, K., Zhou, D. and Yang, H. (1991) Biological control of Asian corn borer with entomopathogenic nematode, *Steinernema feltiae* Agriotos. *Chinese Journal of Biological Control* 7, 1–6.

Hou, Y., Pang, X. and Liang, G. (2001) On the partial application of *Steinernema carpocapsae* strain A24 against striped flea beetle *Phyllotreta striolata*. *Acta Phytophylacica Sinica* 2, 151–156.

Huang, J., Ding, M., Lin, Z., Gao, M., Huang, H. and Jian, H. (1997) Study on control capacity of 11 strains of nematodes on *Anoplophora chinensis*. *Forest Pest and Disease* 2, 3–5.

Huang, K.S., Wang, Y.Z., Yie, Z.C., Zhang, L.Y. and Su, Z.C. (1976) Effect of light cycle and temperature on the diapause of *Carposina niponensis*. *Acta Entomologica Sinica* 19, 149–156.

Jiang, Y., Piao, C., Zhang, S., Wu, S. and Zhao, F. (1990) On the damage and 'on-tree' action threshold of peach fruit borer to apple. *Acta Phytophylacica Sinica* 17, 359–364.

Jin, Y., Han, R. and Cong, B. (2003) General situation of the application research on entomopathogenic nematodes. *Natural Enemies of Insects* 25, 175–184.

Jin, Y., Han, R. and Cong, B. (2010) Characteristics of insect anti-feedants from entomopathogenic bacteria *Xenorhabdus nematophilus* strain All against *Plutella xylostella*. *Chinese Journal of Biological Control* 26, 132–137.

Jin, Y., Han, R. and Cong, B. (2013) Screening of adjuvants for entomopathogenic nematodes of *Steinernema carpocapsae* A11 strain in foliar application. *Plant Protection* 39, 194–200.

Jing, X., Hao, S. and Kang, L. (2002) Cold adaptation in insects: progress in antifreeze protein research. *Acta Entomological Sinica* 45, 679–683.

Kaya, H.K. and Gaugler, R. (1993) Entomopathogenic nematodes. *Annual Review Entomology* 38, 181–206.

Kaya, H.K., Aguillera, M.M., Alumai, A., Choo, H.Y., de la Torre, M., *et al.* (2006) Status of entomopathogenic nematodes and their symbiotic bacteria from selected countries or regions of the world. *Biological Control* 38, 134–155.

Li, A., Zhang, Y., Zhang, H., Sun, F., Zhang, X. and Sun, R. (2012) Studies on dynamic occurrence and biological control of peach fruit moth in date orchard. *China Fruits* 2, 53–56.

Li, C., Tan, G., Wang, Y. and Xu, Y. (2011a) Occurrence of entomopathogenic nematodes resources and overwintering conditions in Heiloingjiang Province. *Plant Protection* 37, 120–123.

Li, C., Wang, Y., Xu, Y., Zhang, Z., Meng, J., *et al.* (2013) Efficacy of a mixed formulation of *Heterorhabditis bacteriophora* Zt (Heilongjiang isolate) and *Steinernema carpocapsae* All against *Bradysia odoriphaga*. *Chinese Journal of Applied Entomology* 50, 235–241.

Li, C., Wang, Y., Hu, Y., Hua, C. and Wang, C. (2015) Three dimensional study of wounded plant roots recruiting entomopathogenic nematodes with Pluronic gel as a medium. *Biological Control* 89, 68–74.

Li, H. and Han, R. (2007) Advances in research on the infective behaviors of entomopathogenic nematodes. *Chinese Bulletin of Entomology* 44, 637–642.

Li, J., Han, R., Cao, L., Liu, X., Peng, T. and Li, L. (2002) Molecular identification of entomopathogenic nematode species (*Steinernema* and *Heterorhabditis*) by RFLP analysis of PCR amplified rDNA internal transcribed spacer (ITS) region. *Natural Enemies of Insects* 24, 97–104.

Li, J., Qian, H., Dong, H. and Cong, B. (2007a) Laboratory test on the infectivity of entomopathogenic nematodes to *Lissorhoptrus oryzophilus*. *Chinese Journal of Biological Control* 23, 188–190.

Li, J., Sun, C., Kang, Y. and Ma, J. (2007b) Control effect of entomopathogenic nematodes to peanut grubs. *Agrochemicals* 46, 62–63.

Li, J., Wang, Z., Yan, A., Wang, Q., Huang, D., *et al.* (2011b) Anti-desiccation ability of several entomopathogenic nematode strains and their pathogenicity to *Grapholitha molesta*. *Journal of Agricultural University of Hebei* 34, 90–92.

Li, N., Wang, Z., Yan, A., Huang, D. and Bi, R. (2011c) Virulence of eleven strains of entomopathogenic nematodes to *Anoplophora glabripennis*. *Journal of Northeast Forestry University* 39, 68–69, 100.

Li, Q. and Han, R. (2001) Research advance on interactions of entomopathogenic nematodes and other insecticidal: a review. *Natural Enemies of Insects* 23, 170–180.

Li, S., Lian, J., Qiu, L., Li, Y. and Cai, F. (1984) Study of control of peach fruit moth using entomopathogenic nematodes. *Chinese Fruits* 4, 31–36.

Li, S., Liu, J. and He, D. (1990) Studies on the utilization of entomopathogenic nematode Taishan I for peach fruit borer control. *Acta Phytophylacica Sinica* 17, 237–240.

Li, S., Ding, W. and Han, F. (1993) The application of entomopathogenic nematode Taishan No. 1 against peanut grubs. *Acta Phytophylacica Sinica* 20, 55–59.

Li, X. and Guo, F. (2005) Research advance on entomopathogenic nematodes and associated symbiotic bacteria. *Shanxi Journal of Agricultural Sciences* 5, 82–84.

Li, X. and Wang, G. (1990) A preliminary study on the application of entomopathogenic nematodes against striped flea beetle. *Acta Phytophylacica Sinica* 17, 229–231.

Li, X., Mei, S., Wu, H., Huo, Y., Song, J., *et al.* (1995) On the application of an entomopathogenic nematode against larger black cockchafer in beet fields. *Plant Protection* 4, 14–16.

Liu, H., Chang, Q. and Ma, F. (1998a) Field test on control of *Anopiophora glaripennis* (Motsch) and *Melanophila decastigma* Fabr. with *Steinernema carpocapsae*. *Ningxia Agricultural and Forestry Technology* 5, 15–17.

Liu, N., Zhang, Z., Zheng, L. and Chen, J. (1989) Study on the entomopathogenic nematodes for biological control of bamboo weevils, *Cyrtotrachelus longimanus*. *Natural Enemies of Insects* 11, 44–50.

Liu, N., Liang, M., Zheng, L. and Liang, Y. (1993) The bionomics of bamboo weevil and its control by entomopathogenic nematode. *Acta Phytophylacica Sinica* 20, 343–347.

Liu, Q. (1993) Infectivity of entomopathogenic nematodes *Steinernema* sp. CB-2Y strain to sawfly *Hoplocampa* spp. *Natural Enemies of Insects* 2, 84–87.

Liu, Q., Wang, Y. and Zhou, H. (1997) A study on the application of entomopathogenic nematodes for controlling larvae of red longicorn beetle (RLB). *Acta Agriculturae Boreali- Sinica* 12, 97–101.

Liu, Q., Wang, Y., Tong, F., Zhang, W. and Xu, L. (1998b) Study on application techniques of *Steinernema* nematodes against RNL. *Journal of China Agricultural University* 3, 17–21.

Liu, Q., Zhao, Y., Yan, Y., Wang, Y. and Lazer, I.G. (2002) Progress of applied research on entomopathogenic nematodes in biological control in China. *Journal of China Agricultural University* 7, 65–69.

Liu, S., Rao, A. and Vinson, B.S. (2014) Biological control in China: past, present and future – an introduction to this special issue. *Biological Control* 68, 1–5.

Liu, X., Wu, X., Zhao, C., Wei, G. and Deng, T. (1991) Study on control of *Rhynchites foveipennis* with *Steinernema feltiae*. *Chinese Journal of Biological Control* 7, 166–168.

Liu, Z., Li, Y., Zong, J., Zhang, G. and Chai, F. (1998c) Studies on the application of entomopathogenic nematodes CZ-88 for control of peach fruit moth. *Journal of Fruit Science* 15, 53–57.

Lu, Y., Han, L., Zhao, K., Song, J. and Zhu, M. (2014) Isolation and identification of entomopathogenic nematodes and their symbiotic bacteria from Harbin (Northeast China) and infectivity of entomopathogenic nematodes in low temperature. *Chinese Journal of Biological Control* 30, 558–563.

Luo, Q., Tan, C., Chen, Z., Li, Z. and Li, X. (1997) A study on biological control of litchi para-stem borer (*Metarbelidae*) and longicorn beetle (*Cerambycidae*) by nematodes. *Journal of South China Agricultural University* 18, 25–30.

Ma, J., Zhang, J., Li, X., Li, M. and Chen, S. (2008) Construction of the cDNA library from IJs of *Steinernema feltiae* G26 and analysis of express sequence tags. *Acta Agriculturae Boreali-Sinica* 23, 72–76.

Ma, J., Chen, S., Zou, Y., Li, X., Han, R., *et al.* (2010) Natural occurrence of entomopathogenic nematodes in North China. *Russian Journal of Nematology* 18, 117–126.

Ma, J., Li, X., Geng, Y. and Chen, S. (2012) Impact of chemical pesticide on the infection of entomopathogenic nematodes to *Bradysia odoriphaga*. *Nematology Research in China* 4, 186–187.

Ma, J., Chen, S., Moens, M., Han, R.C. and De Clercq, P. (2013) Efficacy of entomopathogenic nematodes (Rhabditida: *Steinernematidae* and *Heterorhabiditidae*) against the chive gnat, *Bradysia odoriphaga*. *Journal of Pest Science* 86, 551–561.

Ma, S. (1976) On the integrated control of agricultural insect pests. *Acta Entomologica Sinica* 19, 129–139.

Pan, F., Zhang, S., Chen, Y., Meng, J., Wu, D., *et al.* (2013) Management of entomopathogenic nematode on *Bradysia odoriphaga* in Ningxia. *Anhui Agricultural Science Bulletin* 19, 83–86.

Pan, K., Lu, X., Liu, N., Liu, X. and Li, Q. (1999) Studies on sensitivity of *Steinernema carpocapsae* to *Phaedom brassicae* Baly. *Natural Enemies of Insects* 21, 93–95.

Qiu, S. and Yang, H. (2007) Biological control, an important technology in integrated pest management. *Plant Protection* 33, 1–6.

Qiu, X. and Han, R. (2007) Resources of entomopathogenic nematodes and advances in taxonomic techniques. *Acta Entomologica Sinica* 50, 286–296.

Qiu, X., Han, R. and Xu, Z. (2003) rDNA ITS polymorphism analysis of entomopathogenic nematodes. *Natural Enemies of Insects* 25, 130–141.

Qiu, X., Cao, L. and Han, R. (2010) Advances in molecular biology of mutualism in symbiotic bacteria associated with entomopathogenic nematode. *Chinese Bulletin of Entomology* 47, 824–833.

Sun, R. and Li, A. (2004) Study on biological control of Asian corn borer (*Ostrinia furnacalis*) with entomopathogenic nematodes. *Shandong Agricultural Sciences* 2, 60–61.

Sun, R. and Li, A. (2007) Studies on combination effect of entomopathogenic nematode H06 and insecticide against *Bradysia odoriphaga*. *Chinese Journal of Pesticide Science* 9, 66–70.

Sun, R., Li, A. and Yang, X. (2003) Studies on application of entomopathogenic nematodes to control fruit pests. *Deciduous Fruits* 2, 13–14.

Sun, R., Li., A., Han, R., Cao, L. and Liu, X. (2004) Factors affecting the control of *Bradysia odoriphaga* with entomopathogenic nematode *Heterorhabditis indica* LN2. *Natural Enemies of Insects* 26, 150–155.

Sun, W., Cong, B., Dong, H. and Qian, H. (2006) Study on the efficacy of *S. feltiae* Otio on rice water weevil in laboratory. *Journal of Anhui Agricultural Science* 34, 1620–1621.

Sun, Y., Wang, G. and Tian, M. (1990) A study of the outbreak regularity of *Atrijuglans hetaohei* Yang and its control techniques. *Acta Agriculturae Boreali-Sinica* 18 (S), 102–104.

Tu, L. (1996) Research advance of taxonomy of entomopathogenic nematodes in China. *Biological Bulletin* 31, 10–11.

Wang, C., Lower, S. and Williamson, V.M., (2009a) Application of pluronic gel to the study of root-knot nematode behavior. *Nematology* 11, 453–464.

Wang, G. and Han, R. (2008) Biological control of striped flea beetle, *Phyllotreta striolata* Fabricius. *Chinese Journal of Biological Control* 24, 91–93.

Wang, G., Han, R., Chen, J. and Cao, L. (2007a) Combined efficacy of entomopathogenic nematode *Steinernema carpocapsae* All and pesticide against *Rhabdoscelus lineaticollis* (Heller). *Chinese Journal of Biological Control* 23, 218–222.

Wang, H., Cao, Y., Li, T. and Zong, X. (1998) The infection assays of entomopathogenic nematodes against house fly maggots. *Shandong Journal of Animal Science and Veterinary Medicine* 3, 2–4.

Wang, H., Cong, B., Dong, H. and Qian, H. (2008a) Isolation and purification of insecticidal proteins from *Xenorhabdus poinarii* 5-5B. *Agrochemicals* 47, 29–43.

Wang, H., Wang, Q., Li, G. and Lu, X. (2002) Advance on entomopathogenic nematodes. *Journal of Agricultural University of Hebei* 25 (S), 219–223.

Wang, H., Luan, J.B., Dong, H., Qian, H.T. and Cong, B. (2014) Natural occurrence of entomopathogenic nematodes in Liaoning (Northeast China). *Journal of Asia-Pacific Entomology* 17, 399–406.

Wang, J. (1993) Control of the peach fruit moth, *Carposina niponensis*, using entomopathogenic nematodes. In: Bedding, R., Akhurst, R. and Kaya, H. (eds) *Nematodes and the Biological Control of Insect Pests*. CSIRO Publications, East Melbourne, Australia, pp. 59–65.

Wang, L., Pan, Y., Yang, H. and Huang, D. (2000) Screening of high toxic strains from *Xenorhabdus* spp. and *Photorhabdus* spp. and assays of insecticidal activities. *Acta Entomologica Sinica* 43, 98–103.

Wang, L., Xu, Y. and Li, C. (2011) The research of entomopathogenic nematode cold tolerance. *System Sciences and Comprehensive Studies in Agriculture* 27, 490–494.

Wang, L., Qiu, X., Wu, C., Chen, J. and Han, R. (2013) Advances in the species-specific colonization of entomopathogenic nematodes by their symbiotic bacteria. *Journal of Environmental Entomology* 35, 232–241.

Wang, R., Sun, Y. and Yin, K. (1997) Studies on *Steinernema feltiae* Beijing against *Atrijuglans hetaohei* Yang. *Shanxi Agricultural Science* 1, 5–7.

Wang, W., Peng, X., Qiu, L. and Pang, Y. (2008b) Molecular identification of symbiotic bacteria of four recently described species of entomopathogenic nematode. *Acta Scientiarum Naturalium Universitatis Sunyatseni* 47, 52–55.

Wang, X., Dong, H. and Cong, B. (2007b) Determined the taxonomic status of *Xenorhabdus* and *Photorhabdus* bacteria by 16S rDNA. *Chinese Bulletin of Entomology* 44, 259–262.

Wang, Y., Zhang, Q. and Zhang, X. (2009b) Studies on insecticidal activity of metabolites of symbiotic bacteria associated with entomopathogenic nematodes. *Journal of Northwest A & F University* 37, 165–170.

Wei, H., Li, F., Wan, L., Lin, K., Wu, D. and Zhao, F. (2000) Sensitivity of entomopathogenic nematode, *Steinernema feltiae* to vegetable beetle, *Colaphellus bowringi* Baly. *Acta Agriculturae Universitatis Jianxiensis* 22, 243–244.

Wouts, W.M. (1991) *Steinernema* (*Neoaplectana*) and *Heterorhabditis* species. In: Nickle, W.C. (ed.) *Manual of Agricultural Nematology*. Dekker, New York, pp. 855–897.

Wu, H., Xin, L., Gong, Q., Zhang, K., Cao, G. and Sun, R. (2014a) Evaluation of the effects of infection by different entomopathogenic nematodes and chemical pesticides on *Bradysia odoriphaga*. *Chinese Journal of Applied Entomology* 51, 1060–1068.

Wu, H., Zhang, K., Gong, Q. and Sun, R. (2014b) Synergism of entomopathogenic nematodes and rhizobia on peanut cultivation. *Journal of Nuclear Agricultural Sciences* 28, 1885–1890.

Wu, H., Fan, K., Xin, L., Cao, G. and Sun, R. (2015) Virulence and control effect of entomopathogenic nematodes against *Agrotis ipsilon* (Rottenberg). *Journal of Plant Protection* 42, 244–250.

Wu, K., Lu, Y. and Wang, Z. (2009) Advance in integrated pest management of crops in China. *Chinese Bulletin of Entomology* 46, 831–836.

Wu, W., Yin, J., Cao, Y., Xiao, C. and Li, K. (2014c) Entomopathogenic nematodes and their application in China. *Chinese Journal of Biological Control* 30, 817–822.

Xia, Y., Han, R. and Li, R. (2006) The symbiosis between entomopathogenic nematodes and their bacteria. *Chinese Bulletin of Entomology* 43, 283–287.

Xia, Y., Qiu, X. and Han, R. (2008) The differential expression of development-related genes of entomopathogenic nematode. *Chinese Journal of Biological Control* 24, 22–29.

Xing, Y., Qian, X., Liu, C. and Li, H. (2009) Anti-desiccation ability of several entomopathogenic nematodes in Gansu Province. *Guizhou Agricultural Sciences* 37, 90–91.

Xu, J. and Xie, R. (1997) Mass application of entomopathogenic nematode against lichi stem borer. *Acta Phytophylacica Sinica* 24, 150–154.

Xu, J., Han, R., Liu, X. and Cao, L. (1994) Some factors influencing the control effectiveness of entomopathogenic nematodes to house fly, *Musca domestica* Vicina maggots in pig farm. *Natural Enemies of Insects* 16, 21–27.

Xu, J., Han, R., Liu, X., Cao, L. and Yang, P. (1995) The application of codling moth nematode against the larvae of lichi longicorn beetle. *Acta Phytophylacica Sinica* 22, 12–16.

Xu, Y., Qian, X., Li, C., Wang, Y. and Liu, C. (2008) Biocontrol efficacy of entomopathogenic nematode *Heterorhabditis bacteriophora*–NJ against larvae of *Holotrichia diomphalia* Bates. *System Sciences and Comprehensive Studies in Agriculture* 24, 106–109.

Xue, R., Wang, X., Li, M. and Zeng, F. (2009) Research progress of toxic protein from entomopathogenic nematode symbiotic bacteria. *Biotechnology Bulletin* 4, 21–24.

Yan, X. and Han, R. (2010) Anhydrobiosis of entomopathogenic nematodes. *Chinese Journal of Biological Control* 26, 90–95.

Yan, X., Han, R., Moens, M., Chen, S. and De Clercq, P. (2013) Field evaluation of entomopathogenic nematodes for biological control of striped flea beetle, *Phyllotreta striolata* (Coleoptera: Chrysomelidae). *BioControl* 58, 247–256.

Yan, X., Guo, W., Zhao, G. and Han, R. (2014a) Research advances in subterranean pest control by entomopathogenic nematodes. *Journal of Environmental Entomology* 36, 1018–1024.

Yan, X., Wang, X., Han, R. and Qiu, X. (2014b) Utilisation of entomopathogenic nematodes, *Heterorhabditis* spp. and *Steinernema* spp., for the control of *Agrotis ipsilon* (Lepidoptera Noctuidae) in China. *Nematology* 16, 31–40.

Yang, B., Yang, H., Liu, Z., Yang, X. and Yuan, J. (2007) Comparison of toxicity between extracellular and intracellular proteins of entomopathogenic nematode symbiotic bacteria against *Tenebrio armigera*. *Chinese Journal of Biological Control* 23, 33–38.

Yang, H. (1991) Control of peach fruit moth using entomopathogenic nematodes. *Chinese Journal of Biological Control* 3, 114.

Yang, H. and Zhang, G. (1990a) Infectivity of the entomopathogenic nematode, *Heterorhabditis* sp. D1 to *Bradysia odoriphaga*. *Chinese Journal of Biological Control* 6, 110–112.

Yang, H. and Zhang, Y. (1990b) Biological control of *Holcocerus insularis* with Steinernema feltiae. *Journal of Chinese Landscape Architecture* 1, 55.

Yang, H., Zhang, S., Zhang, G., Huang, J. and Zheng, H. (1990) Control of leopard moth, *Zeuzera multistrigata* Leuconota on casuarina trees by plugging sponge carrying *Steinernema feltiae*. *Chinese Journal of Biological Control* 6, 167–169.

Yang, H., Zhang, G. and Zhang, S. (1991) Relationship between host exposure time of *Steinernema feltiae* and the mortality and killing speed of *Ostrinia furnacalis*. *Chinese Journal of Biological Control* 7, 154–156.

Yang, H., Zhang, G., Zhang, S., Lin, Z., Li, G., *et al.* (1992) Study on the application of entomopathogenic nematode *Steinernema feltiae* Filitjev against *Holococerus insularis* Staudinger. *Acta Phytophylacica Sinica* 19, 37–40.

Yang, H., Zhang, G., Zhang, S. and Jian, H. (1993) Biological control of tree borers (Lepidoptera: Cossidae) in China with the nematode *Steinernema carpocapsae*. In: Bedding, R., Akhurst, R. and Kaya, H. (eds) *Nematodes and the Biological Control of Insect Pests*. CSIRO Publications, East Melbourne, Australia, pp. 33–40.

Yang, H., Yang, X. and Zhang, J. (2010a) Research and application of biological control of insect pests on cultivated edible mushroom. *Chinese Journal of Biological Control* 26, 1–6.

Yang, P., Liu, J., Liu, N. and Han, R. (2000) Large-scale application of the nematode *Steinernema carpocapsae* Agriotes against peach fruit borers. *Natural Enemies of Insects* 22, 79–82.

Yang, X., Jian, H., Yang, H., Liu, Z. and Yuan J. (2004) Using entomopathogenic nematodes for control of chive gnat, *Bradysia odoriphaga* Yang et Zhang. *Acta Phytophylacica Sinica* 31, 33–37.

Yang, X., Liu, Z., Zhang, R., Yang, H., Yuan, J. and Jian, H. (2008) Isolation and identification of seven symbiotic bacteria from entomopathogenic nematodes. *Microbiology* 35, 225–229.

Yang, X., Yang, H., Shi, Y., Wei, B. and Zhang, J. (2010b) Biological control of sciarid flies with the entomopathogenic nematode *Steinernema feltiae*. *Acta Edulis Fungi* 17, 93–96.

You, J., Liang, S. and Han, R. (2006) Research advances on symbiosis and pathogenicity of symbiotic bacteria associated with entomopathogenic nematodes. *Chinese Journal of Biological Control* 22, 168–173.

Yu, X., Wang, D., Liu, J., Xie, P., Lu, Z. and Liu, X. (2003) Infectivity of *Steinernema carpocapsae* Ohio to some insect pests and influence of temperature and humidity under laboratory condition. *Jiangsu Journal of Agricultural Science* 19, 3–17.

Zhang, B., Wang, X., Chen, X. and Li, L. (1994a) Study on control of chive root gnat using entomopathogenic nematodes. *Science and Technology of Tianjin Agriculture and Forestry* 2, 4–6.

Zhang, G., Yang, H. and Zhang, S. (1994b) Laboratory tests on the infectivity of the nematode, *Steinernema carpocapsae* on house fly, *Musca domestica*. *Chinese Journal of Biological Control* 2, 89.

Zhang, G., Cheng, G., Zhang, S. and Yan, L. (2007) Preliminary study of the control and prevention of *Anoplophora glabripennis* with insect pathogenic nematode in Harbin. *Forest By-Product and Specialty in China* 89, 34–36.

Zhang, H., Yin, S., Zhang, M. and Chen, Z. (2013a) Research progress on control techniques of *Bradysia odoriphaga*. *Journal of Henan Agricultural Sciences* 42, 6–9.

Zhang, L., Zhao, K., Li, G., Song, J. and Li, Q. (1996) Effect of nematodes against grub *Holotrichia diomphalia* Bates (Coleoptera: Melolonthidae). *Journal of Northeast Agricultural University* 27, 240–242.

Zhang, S., Xu, Y. and Pan, F. (2013b) The research progress of *Bradysia odoriphaga*. *Anhui Agricultural Science Bulletin* 19, 82–84.

Zhang, X., Jiang, Z. and Xie, X. (1991) Control of carpenterworm with entomopathogenic nematode in Lanzhou area. *Chinese Journal of Biological Control* 1, 44.

Zhang, Y., Yang, Z., Wang, X., Wu, C., Ma, S. and Lu, Z. (2012) Laboratory test on control of Japanese pine sawyer by two entomopathogenic nematodes species. *Scientia Silvae Sinicae* 48, 64–69.

Zhang, Y., Xu, Y., Wu, D., Pan, F., Xiao, L. and Cao, Y. (2014) Management efficiency of *Bradysia odoriphaga* with entomopathogenic nematode. *Anhui Agricultural Science Bulletin* 20, 70–73.

Zhang, Z., Cao, L., Liu, X., Wang, G., Xu, Z. and Han, R. (2006) Screening of insecticides as synergists for entomopathogenic nematodes *Steinernema longicaudum* X-7. *Chinese Bulletin of Entomology* 43, 68–73.

Zhao, G., Guo, W., Yan, X., Liu, Y., Chen, J. and Han, R. (2013) Effects of chemical insecticides used in chive fields on the survival and infectivity of entomopathogenic nematodes. *Journal of Environmental Entomology* 35, 458–465.

Zhou, S., Zai, Y., Wang, R., Sun, Y. and Lu, S. (1989) Application of entomopathogenic nematodes to control peach fruit moth. *Liaoning Agricultural Science* 5, 30–33.

Zhou, Y. and Yang, H. (1991) A comparative study on the virulence against *Carposina nipponensis* between a newly isolated insect nematode *Steinernema* sp. (CB-2Y) and *Steinernema feltiae* Agriotos. *Chinese Journal of Biological Control* 7, 58–60.

Zhou, Y., Jian, H. and Wang, Z. (2002) RAPD analysis of 11 strains of *Steinernema carpocapsae*. *Acta Parasitology et Medica Entomologica Sinica* 9, 85–91.

24 Status of Entomopathogenic Nematodes in Integrated Pest Management Strategies in Egypt

Mahfouz M.M. Abd-Elgawad*

Phytopathology Department, National Research Centre, Giza, Egypt

24.1 Introduction

Since ancient times, Egypt has been known as a gift of the river Nile. Agriculture was, and still is, the main activity of the Egyptian population and therefore others will truly claim that it is a gift of the Egyptians. In ancient Egypt, natural control of agricultural pests had included factors that destroyed or limited the spread of these pests; even without human intervention. Natural conditions had exclusively accomplished partial reduction in many pest populations through different non-anthropogenic factors. These included nutritional factors such as non-availability of pest food due to drought or lack of their hosts; weather conditions such as high or low temperatures, humidity, wind activity and rainfall; biological factors included natural enemies, for example predators, parasites, parasitoids and soilborne pathogens (e.g. fungi, bacteria and viruses) that can kill or cause diseases of these pests; and finally, topographic factors such as the presence of deserts, mountains and lakes that do not support favourable habitat for many pest species. All these factors could naturally limit the spread of agricultural pests, but they did not negate the fact that Egyptians, like other peoples, had over time applied different methods to combat agricultural pests. Currently, they have been trying, whenever possible, to maintain, develop and/or invent cultural methods and practices in order to maximize their usefulness and/or reduce their negative impacts, if any. In fact, many pest species seriously damage agricultural outputs and significantly reduce crop yields, and therefore some benchmark publications have been reporting their damage, economic losses and/or control measures in Egypt (e.g. Abd-Elsalam, 1993; Anon., 2012, 2014; Abd-Elgawad, 2014a) and elsewhere (e.g. Peshin and Pimentel, 2014; Pimentel and Peshin, 2014).

Pest management has gradually developed, but to different degrees in various regions of the world. Nevertheless, tremendous strides have been made worldwide since the mid-20th century due to the continuous developments in synthetic pesticides. In many regions, the management of insect pests has generally evolved from a heavy dependence on prophylactic pesticide use to a more rational utilization of these compounds within integrated pest management (IPM) schemes. IPM typically decreases pesticide use by monitoring both pest and beneficial organisms to evaluate crop loss as the basis for management decisions (Timmer and Duncan, 1999). Nevertheless, the voluminous amounts of synthetic pesticides used in many countries remain important and cause serious threats to the environment, especially natural biological control processes, wildlife, groundwater contamination, resource depletion and human health and safety (Edwards, 1993; Pimentel and Peshin, 2014).

Growing dissatisfaction with these pesticides has enhanced interest in organic methods for crop and pest management. Moreover, when considering the costs and benefits of various management paradigms, Campos-Herrera *et al.* (2010) speculated that adding more chemical pesticides did not necessarily entail increased output per unit area. In the Mediterranean Basin, Spanish and Italian citrus growers use fourfold and 15-fold more amounts of

*E-mail: mahfouzian2000@yahoo.com

pesticides than do Greek growers, but obtain similar citrus production per hectare (Gutiérrez *et al.*, 2005). Nevertheless, pest management is of utmost importance in almost all agriculture systems, and the development of sustainable pest management schemes demands a substantial comprehension of how populations of pests and their natural enemies behave in specific crop systems/habitats (Abd-Elsalam, 1993; Rechcigl and Rechcigl, 2000; Horowitz and Ishaaya, 2004; Lacey and Shapiro-Ilan, 2008).

In Egypt, a significant redirection in the types and choice of applicable pesticides is under way with a clear aim to avoid the health hazards and environmental issues resulting from traditional chemical pesticides (Abd-Elgawad, 2008a). Regrettably, some pesticides banned in developed countries are still used in less developed ones such as Egypt (Abd-Elgawad and Askary, 2015). On the contrary, in developed countries, issues of pesticide resistance, safety, health and environmental concerns, which have led to the banning of some effective fumigant and granular pesticides and the development of pest biotypes that break host plant resistance, have materialized hopes to integrate the use of biocontrol agents such as entomopathogenic nematodes (EPNs) in pest management plans (Dolinski *et al.*, 2012). Technical advances in EPN formulation, shelf life, *in vitro* and *in vivo* production and application have supplied further scope to integrate the use of EPNs in pest management schemes (Cruz-Martínez *et al.*, 2017). Unfortunately, useful EPN applications and techniques are more frequently unexploited and poorly understood in many developing than developed countries.

Therefore, I think it is important that EPN technology and application reaches further than the EU, USA, Canada, Australasia and the likes. Because EPNs can be adopted and produced in relatively unsophisticated facilities, I believe that publications that take the techniques to the developing world should be promoted and current opportunities discussed. My standpoint is that the use of EPNs and/or the bacterial associates must be a way forward in crop protection/pest management in Egypt. Therefore, in this chapter, the status of EPNs in IPM programmes in Egypt has been reviewed, and the current hindrances and possible solutions of managing pests using EPNs alone or within the IPM concept have also been reviewed. For example, a prototype, established in my institute, The National Research Centre, for economic and adequate EPN production has been presented. Its expansion into other Egyptian governorates is especially timely since Egypt is now experiencing continuous increase of areas under organic agricultural management to meet the demands of the European Union (EU), USA, Australian and Japanese markets for fruit, vegetables and medicinal plants; the EU is the major importer of Egyptian produce (Abd-Elgawad and Spiridonov, 2014).

24.2 Integrated Pest Management in Egypt

As is the case in many other countries, pest control represents one of the most important factors affecting the agricultural production process in Egypt because it helps to protect the productivity of crops and thus achieve rewarding return for the farmer, cooperative and agricultural company. The pest control process has evolved into the use of a range of modern technologies through accurate combinations of available control measures depending on the maximum utilization of physical and biological means. This process, called IPM, is always achieved through an environmental, economic and social perspective. The process is based on the management of pest population levels in order to at least maintain the pest species below the economic threshold level. IPM programmes avoid the use of chemical pesticides, except in the case of necessity. IPM schemes as administered by the Egyptian Ministry of Agriculture and Land Reclamation (MALR) consider the most appropriate pesticide(s) against the target pest, as well as the optimal pesticide concentration and time of application to maximize the benefits and reduce the damages and costs (Anon., 2012, 2014). These programmes aim to preserve the activity of natural enemies, without harm to human health and the environment, while increasing the competitiveness of the export for safe agricultural products. This latter objective is achieved through a commitment to maximum permissible limits for residues of chemical pesticides in Egyptian agricultural products.

With the advent of the 21st century, biopesticide use has been increasing steadily as successions of pests were shown to be manageable through biocontrol techniques mostly by native and, to a much lesser degree, introduced natural enemies in Egypt. Many farmers and growers learned that their produce could be improved without the use of harmful chemical insecticides but we hope to increase significantly this number of knowledgeable growers.

In fact, numerous plant species especially related to Egyptian fruit and vegetables grown for the fresh market often need no more than a few well-timed pesticide treatments to manage arthropod pests and soilborne diseases. Those educated growers have also realized that serious outbreaks of some pest species are almost always associated with the disruption of biocontrol agents through the unnecessary or intensive use of chemical pesticides. Eventually, they know that IPM is necessary for sustainable agriculture. On the contrary, for Egyptian growers who have not recognized the benefits of IPM, agricultural extensions have been trying to enhance their knowledge and skills in the field of IPM. Moreover, the Egyptian International Centre for Agriculture (EICA) periodically offers an international training course on integrated pest management and cotton technology for trainees from the agricultural section and its relevant fields, nominated from countries of Africa, Asia, Latin America and Eastern Europe, with a minimum of 3 years of experience in the field (http://eicaeg.org/ipm1.html). The major goal of the course is to enhance the participants' knowledge and skills in the discipline of IPM and cotton technology. The course conveys Egyptian experience in the field of IPM and cotton technology at both the research and application levels, and teaches the principles and gains of the adapted technologies applied in the diagnosis of the major pests and soilborne diseases affecting agriculture based on Egyptian practices. It provides adequate knowledge to design IPM and cotton technology programmes relevant to the ecological conditions of the participating countries, solve issues that stem from pest resistance to pesticides, and presents the constraints facing IPM and cotton technology. Programmes established in the countries of the trainees are also discussed during the country report sessions, to reach more sound IPM schemes. On the other hand, workshops are usually held, especially in remote governorates, from time to time to address IPM as a means of conscious intervention in an integrated manner not detrimental to the environment to ensure the control of pests below the economic limit.

24.3 Application of Biological Control Agents in Integrated Pest Management in Egypt

Egyptian growers are used to implement different agricultural practices such as ploughing, crop rotation, fallowing, use of healthy planting stocks, resistant cultivars and trap or antagonistic crops, physical control, proper time of planting and harvesting, sanitation, chemical applications and general care of host crops, as well as regulatory control. These practices could affect different pest species adversely, even implicitly. Moreover, quite a number of biopesticides registered in Egypt are recommended by the Egyptian ad hoc committee for agricultural pesticides to be used against specific pests. They may be compatible with one or more of these agricultural practices (Anon., 2014); i.e. these multiple tactics also consider environmental, economic, human health and ecological impacts to comply with IPM programmes. In this context, prophylactic measures including chemical insecticides may also be used as preventative or curative tactics. Nevertheless, the implementation of these practices varies from one grower/company to another, ranging from tactful and adequate use of them to the partial or full neglect of applying them effectively and efficiently.

Recently, management methods of pests have matured into two environmentally acceptable groups in Egypt (Abd-Elgawad and Spiridonov, 2014). One group is currently employing a more rational application of chemical compounds within IPM programmes. Such programmes typically reduce synthetic and/or non-synthetic chemical pesticide use by monitoring both pest and beneficial organisms, and predicting crop loss as the basis for management decisions. The other group is known as the organic farming system. The general principles in the International Federation of Organic Agriculture Movements (IFOAM) basic standards state that the management of pests, diseases and weeds should occur through the application of biological and cultural means in organic farming. Cropping practices (e.g. rotations, intercropping and diversity) and soil fertility practices are central to such a management (Altieri, 1999). So, the organic standards permit the use of inputs such as biologicals when non-synthetic chemicals for preventative measures are not sufficient. On the other hand, there is a third group that still relies heavily on methods implementing synthetic chemicals to combat pests in Egypt. Hence, these Egyptian farmers'/growers' awareness of adequate pest control measures need to be developed, not only to avoid yield losses but also to ensure high quality of the crops grown through the use of environmentally friendly alternatives to chemical pesticides. This is especially important since strategies based solely on cultural and tillage practices are frequently not sufficient in controlling agricultural pests due to pest abundance and their broad host spectrum in Egypt. Because of the importance of mass media in the enlightenment of and providing of culture to the public, continuous efforts should be adopted by both governmental and private sectors to increase the number of people interested in the activities and goals of biological control tactics within IPM programmes, especially against major pest species, hoping to reap clean and safe agricultural products for humans and the environment in Egypt. Therefore, this chapter addresses EPNs as attractive, but unfortunately not common, biocontrol agents in IPM in Egypt.

Due to growing satisfaction with biocontrol agents, the share of biologicals is generally increasing in different segments of the Egyptian pesticide market. For example, commercial products such as bio-fly and bio-weevil are recommended against the white fly and red palm weevil, respectively. Also, fungal diseases responsible for failure of the graft process in grapevine were treated with commercial biofungicides Bio-Zeid (*Trichoderma album*), Rhizo-in (*Bacillus subtilis*) and Bio-Arc (*Bacillus megaterium*), which could reduce the percentage of disease incidence by 40.06%, 26.92% and 25.18 %, respectively (Abo Rehab *et al.*, 2013). Likewise, commercial bionematicides for plant parasitic nematodes included products such as Micronema and Nemaless (Abd-Elgawad *et al.*, 2008; Abd-Elgawad and Vagelas, 2015) and bioacaricides such as Biosect (*Beauveria bassiana*) against mites (Anon, 2014). Yet, because of the continuous isolation and introduction of new biocontrol agents and formulations in different market segments and differences in the susceptibility of pest species to biopesticide efficacies/formulations, it is difficult to provide up-to-date information (Cruz-Martínez *et al.*, 2017). Therefore, conclusive recommendations, of the ad hoc committee for agricultural pesticides, adopted by MALR to combat agricultural pests are published biannually (e.g. Anon., 2012, 2014). Each publication records economically important agricultural pests, their host crops and recommended synthetic and non-synthetic chemicals, as well as biocontrol agents, and other biorational compounds to control each pest species; within an IPM programme, if applicable. It is imperative that experimentations for each of these recommended biocontrol agents have proved its efficacy against the targeted pest species, and consequently that the committee has approved it either singly or within an IPM programme, according to the method of application and guidelines provided (Anon., 2012, 2014).

24.4 An Outline of Infrastructure and Studies on Entomopathogenic Nematodes in Egypt

There are three sectors within which entomonematological research, teaching and extension fall. These sectors are: Egyptian universities, the Agricultural Research Centre (ARC) of MALR and the National Research Centre (NRC) of the Ministry of Scientific Research. Egyptian universities include mainly faculties of agriculture in which nematology is taught to undergraduate and postgraduate students. Most students enrolled for postgraduate studies in nematology and EPNs come from the Entomology, Plant Protection and Plant Pathology Institutes and Departments of the ARC and NRC, as well as faculties of agriculture and sciences. Currently, other Egyptian institutes have begun to initiate nematology research sections; for example, The City for Scientific Researches and Technology Applications (CSAT) in Alexandria.

There are over 100 (with PhD degrees) trained nematologists in Egypt, mainly distributed among the universities, ARC and NRC. All of the PhD nematologists, and more junior nematologists (MSc) and graduate students, have developed The Egyptian Society of Agro-Nematology (Abd-Elgawad, 2014b). Currently, the numbers of nematologists are enough to advance the level of nematode education and training in Egypt, but a few drawbacks interrupt the course of progress, especially in EPN application. These include lack of coordination and cooperation between specialists, insufficient awareness of EPN techniques, especially by small-scale farmers, shortage in scientific instruments and limited financial resources and equipment. Although the nematology laboratories of these sectors have some highly sophisticated scientific equipment, which is made available through US Agency for International Development (USAID) projects and other local and foreign funds, some of the equipment is overcommitted. Most laboratories are still in need of more facilities; for example, electronic microscopes, new incubators, centrifuges and misting chambers for nematode extraction. While nematology laboratories of the faculties of agriculture and their staff teach and provide training courses to undergraduate and graduate students, the laboratories of

NRC and ARC provide training for agricultural engineers working in the fields of plant protection and phytopathology, and facilities to carry out research work for their staff members. Also, NRC and ARC have strong links to the extension-related services.

24.4.1 Background of entomopathogenic nematode research and diversity in Egypt

The real start on EPN research in Egypt began in the 1970s when the late Dr Tayseer El-Kifl worked on the biological control potential of *Neoaplectana* (= *Steinernema*) *carpocapsae* against the cotton leafworm, *Spodoptera littoralis*, one of the most economically important insect pests in Egypt (Abd-Elgawad and Spiridonov, 2014). This work was part of her Master and PhD research programmes (El-Kifl, 1980, 1984) and was followed by that of Abdel-Kawy (1985). Their work depended mainly on foreign/imported, not indigenous, EPNs. Although some foreign EPNs have been introduced to Egypt, they have been used experimentally; not as a classical biological control. On the other hand, the first fundamental Egyptian survey to isolate indigenous EPNs was published by Shamseldean and Abd-Elgawad (1994a). Since then, researchers in Egypt have been isolating and laboratory rearing EPNs, mostly on greater wax moth, *Galleria mellonella*, larvae (Table 24.1). They have been testing their biocontrol potential against main insect pests in Egypt, such as *S. littoralis*, the black cutworm, *Agrotis ipsilon*, and the hairy rose beetle, *Tropinota squalida* (e.g. El-Kifl and Sammour, 1989; Hussein, 2004; Shamseldean and Atwa, 2004a,b; Abdel-Razek and Abd-Elgawad, 2007; Shamseldean *et al.*, 2010). Nevertheless, more native species with a diversity of target traits should be sought to overcome behavioural and environmental barriers and optimize field performance.

Abd-Elgawad (2001) suggested minute and tentative definitions of terms for EPN classification based on Egyptian key insect pests as differential hosts. His study demonstrated inter- and intraspecific variation in the virulence and reproductive potential of newly isolated Egyptian EPNs. In fact, Egyptian researchers have been trying to control many economically important insect species (Table 24.2), and therefore nematode–host matching should continuously be in parallel with new

Table 24.1. Occurrence of entomopathogenic nematodes and their symbiotic bacteria in Egypt.

Nematode	Bacterial symbionts	Soil texture	Habitat	Reference
Heterorhabditis spp. *Steinernema* spp.	*Photorhabdus* spp. *Xenorhabdus* spp.	Mainly sandy	Cultivated land	Shamseldean and Abd-Elgawad, 1995; Salama and Abd-Elgawad, 2001; Hussein, 2004; Hussein and Abou El-Souud, 2006; Abd-Elgawad, 2008b
Heterorhabditis bacteriophora	*Photorhabdus luminescens*	Sandy, loamy and clay	Mango and guava trees	Abd-Elgawad and Nguyen, 2007
H. egyptii	*Photorhabdus* sp.	Silty	Mango trees	Abd-Elgawad and Ameen, 2005
H. indica	*P. luminescens akhurstii*	Sandy	Citrus and tomato-planted soil	Abu-Shady *et al.*, 2011; Abd-Elgawad, 2012; Abd El-Zaher and Abd-Elgawad, 2012; Abouelhag and El-Sadawy, 2012
H. taysearae	*Photorhabdus* sp.	Sandy	Hibiscus shrub soil	Shamseldean *et al.*, 1996; Abd-Elgawad, 2012
Steinernema abbasi	*Xenorhabdus indica*	Sandy and loamy	Cultivated land	Abu-Shady *et al.*, 2011; Abouelhag and El-Sadawy, 2012
S. carpocapsae	*X. nematophila*	Sandy	Cultivated land	Abdel-Kawy *et al.*, 1996; Shamseldean *et al.*, 1996; Abu-Shady *et al.*, 2011; Abouelhag and El-Sadawy, 2012
S. kushidai	*X. japonica*	Sandy	Hibiscus shrub	Shamseldean and Atwa, 2004a

Table 24.2. Potential target pests for entomopathogenic nematodes in Egypt.

Order Family	Scientific name/ common name	Commodity	Entomopathogenic nematode species	References
Coleoptera Curculionidae	*Rhynchophorus ferrugineus*/ the red palm weevil	Palm trees	H sp., Hb, Sg, Sc,	Shamseldean and Abd-Elgawad, 1994b; El-Bishry *et al.*, 2000; Abd-Elgawad, 2001; Shamseldean, 2004; Shamseldean and Atwa, 2004b
Coleoptera Scarabaeidae	*Temnorhynchus baal* Reiche/strawberry scarab beetle	Strawberry	Sc, Sk, S sp., Sg, Hb	Shamseldean and Atwa, 2004a
	Tropinota squalida Scop./hairy rose beetle	Vegetables Ornamentals	Hb, Hi	Abdel-Razek and Abd-Elgawad, 2013
Lepidoptera Cossidae	*Zeuzera pyrina* L./ the leopard moth	Fruit trees, grape, apple and olive	Sc, Hb, H sp.	Abdel-Kawy *et al.*, 1992, 1996; Shamseldean, 2000
Lepidoptera Noctuidae	*Sesamia cretica*/the sugarcane stem borers	Sugarcane, maize	Sc, H sp., Sar, Sr, Hb, Ht	Saleh *et al.*, 1999
	Spodoptera littoralis Biosduval/cotton leafworm	Greenhouse Vegetables Field crops	Hb, Sc, Sr, Sa, Hi, H sp.	Abbas and Saleh, 1998; Shamseldean *et al.*, 1999; Abd-Elgawad, 2001; Abd-Elgawad and Abdel-Razek, 2005; Abdel-Razek and Abd-Elgawad, 2007; Saleh *et al.*, 2015
	Agrotis ipsilon Hufnagel/black cutworm	Greenhouse Vegetables	Hb	El-Kifl and Sammour, 1989; Hussein, 2004
Lepidoptera Plutellidae	*Plutella xylostella* L./ the diamondback moth	Cabbage	Ht, Hb, Sc	Saleh, 1995; Abd-Elgawad and Abdel-Razek, 2008
Pieridae	*Artogia (pieris) rapae* L./cabbage worm	Cabbage	Sc, Sr, H sp.	Azazy and Mousa, 2005
Lepidoptera Pyralidae	*Hellula undalis* Fabricius/cabbage webworm	Cabbage	Sc, Sr, H sp.	Azazy and Mousa, 2005
	Galleria mellonella L./greater wax moth	Bee wax	Hb, Hi, Sc, Sr, H sp.	Shamseldean and Abd-Elgawad, 1995; Shamseldean *et al.*, 1996; Abd-Elgawad, 2001; Abd-Elgawad, 2008b
Sesiidae	*Synanthedon myopaeformis* (Borkhausen)/red-belted clearwing	Apple	Hb, Sc	Abdel-Kawy *et al.*, 1996
Tylenchida Meloidogynidae	*Meloidogyne* spp./ root-knot nematodes	Greenhouse Vegetables Field crops	Hb, Sc, S sp., Sr, H sp., Sf	Abd-Elgawad and Aboul-Eid, 2002; Keila *et al.*, 2008; Abd-Elgawad, 2008b; Ornat *et al.*, 2011

Notes: Sa = *Steinernema abbasi*; Sar = *Steinernema arenarium*; Sc = *Steinernema carpocapsae*; Sf = *Steinernema feltiae*; Sg = *Steinernema glaseri*; Sk = *Steinernema karii*; Sr = *Steinernema riobrave*; S sp. = *Steinernema* species; Hb = *Heterorhabditis bacteriophora*; Hi = *Heterorhabditis indica*; Ht = *Heterorhabditis taysearae*; H sp. = *Heterorhabditis* species.

EPN isolates (Abd-Elgawad, 2001). For some of these insects, especially those that do not live in soil (e.g. the red palm weevil), it is unclear whether control with EPNs would ever be practical or economically feasible in Egypt (e.g. Abd-Elgawad, 1996; El-Bishry *et al.*, 2000). Egyptian researchers are trying to solve such problems and to be acquainted with the new advances and techniques in entomopathogenic nematology through joint scientific research projects with developed countries, especially the USA. Therefore, several mutual Egyptian–USA projects have been established aiming at cooperation in various aspects of technology transfer, knowledge sharing and eventually advancement of entomopathogenic nematology. Entomopathogenic nematologists from American institutions such as Rutgers University, University of Arizona and University of Florida have contributed to such programmes. This cooperation has also resulted in mutual publications with their Egyptian counterparts (e.g. Abd-Elgawad and Nguyen, 2007; Abu-Shady *et al.*, 2011; Campos-Herrera *et al.*, 2013).

Heterorhabditis spp. were found apparently to be more common than *Steinernema* spp. in Egyptian fauna (Shamseldean and Abd-Elgawad, 1994a; Abd-Elgawad *et al.*, 2013; Abd-Elgawad, 2014c). The identification of indigenous EPNs and the soil texture and habitats from where they were isolated are summarized in Table 24.1. Many other heterorhabditids, and some steinernematids, have been isolated from Egyptian soils but have yet to be identified (Abd-Elgawad, 2012; Abd-Elgawad and Spiridonov, 2014). While *Heterorhabditis taysearae* has originated from Egypt as a new species having several geographical isolates, other species such as *H. egyptii*, isolated from Egypt (Abd-Elgawad and Ameen, 2005), may be species inquirendae because its molecular identification is incomplete (Abd-Elgawad *et al.*, 2013).

Many other aspects of bioassays were conducted; for example, to compare inoculum levels of EPNs, efficacy of EPN species/isolates or their efficacy related to other insecticides. Diapausing larvae of the European corn borer, *Ostrinia nubilalis* (Lepidoptera: Pyralidae), were exposed under laboratory conditions to inoculum levels of EPNs (Saleh and Elkifl, 1994). *S. carpocapsae* proved to be the most virulent to the pest larvae, followed by *Heterorhabditis bacteriophora*, while *Steinernema glaseri* was the least virulent. Yet, *H. bacteriophora* reproduced better and produced more nematode-infective juveniles (IJs) in the larvae than did *S. carpocapsae*. Inside the maize stalk, *H. bacteriophora* succeeded in invading the pest larvae. Also, two apple tree borers, *Zeuzera pyrina* (Lepidoptera: Cossidae) and *Synanthedon myopaeformis* (Lepidoptera: Sesiidae) were controlled by EPNs in two apple orchards (Abdel-Kawy *et al.*, 1996). In addition to the untreated check, 24 treatments (4 nematodes × 3 application techniques × 2 concentrations) were conducted in each of two orchards. The four nematodes included imported and native *S. carpocapsae* and *H. bacteriophora* with concentrations of 800 or 1600 IJs/ml water each. The techniques included spraying active galleries with IJ suspension, injecting the galleries with nematode suspension, or blocking the gallery opening with a small cotton ball previously soaked in IJ suspension. The active selected galleries were those producing fresh frass, where the insects in the galleries were considered dead when galleries were not producing frass, or by dissecting small branches to see the dead insects. Up to 88.89 and 63.33% mortalities were observed in *Z. pyrina* and *S. myopaeformis*, but the spraying technique had the least efficacy. The native nematodes (*S. carpocapsae* S2 and *H. bacteriophora* S1) were comparable to, and in some cases better than, the imported nematodes (*S. carpocapsae* All and *H. bacteriophora* HP88). EPNs could control *A. ipsilon* (Lepidoptera: Noctuidae) on maize plants in the field (Ragab and Saleh, 2000). Nematode suspension containing four Egyptian isolates of *H. bacteriophora* was applied through drip irrigation in a tomato field and increased the yield 60.2% compared to the untreated control (Abd-Elgawad and Aboul-Eid, 2001). Such a substantial effect of EPNs on tomato growth and yield is possibly due to killing the insect pests. Among 13 dead larvae of the Egyptian cotton leafworm (*S. littoralis*) and the green worm (*Spodoptera exigua*) collected samples from the tomato field, infective stages of *H. bacteriophora* could be extracted from an *S. littoralis* larva using the White nematode trapping technique (Woodring and Kaya, 1988). Meanwhile, the cotton leafworm-infected tomato fruits were 7, 3, 8.5 and 9% in oxamyl, EPNs, *Serratia marcescens*, and untreated check, respectively. It is likely that the reduction in the percentage of infested fruit within *H. bacteriophora*-treated plots was due to the insect nematodes, since *S. littoralis* was a highly susceptible host (e.g. Shamseldean *et al.*, 1994). Moreover, small larvae of both insect species were detected from the field at the time of nematode application. It is possible that the nematodes attacked the insects near the soil surface, since *H. bacteriophora* IJs are cruisers seeking their hosts, and/or during trying to pupate under the soil surface. A month after nematode treatment, contrary to the untreated control, the *Meloidogyne incognita* population level in the same field soil was undetectable (Abd-Elgawad and Aboul-Eid, 2001). Shamseldean and Atwa (2004a) obtained promising results from testing five EPN species against the scarab beetle, *Temnorhynchus baal* (Reiche), a novel insect pest of strawberry in Egypt, under laboratory and field conditions. *Plutella xylostella* population densities and their percentage reduction at weekly intervals on two cabbage cultivars after the application of *H. bacteriophora* or *S. carpocapsae* at three different concentrations under field conditions were recorded (Abd-Elgawad and Abdel-Razek, 2008). The percentage of insect mortality was affected by both the tested cultivars and the EPN species. When *S. carpocapsae* S2 was chosen for single application against *Cassidavittata* in a sugarbeet field, it could kill 65% of the larvae, 92% of the pupae and 57.3% of the adults within a week (Saleh *et al.*, 2009). Also, Saleh *et al.* (2015) tested the effect of several additives or adjuvants on the performance of two Egyptian isolates of EPNs namely, *H. bacteriophora* BA1 and *S. carpocapsae* BA2 to control larvae of *S. littoralis* and *A. ipsilon* infesting maize plants. They found some adjuvants to have no side effects against EPN infectivity or survival, while others significantly enhanced EPN efficacy in both semi-field and field tests. The combined effect of more than one adjuvant was more efficient than single adjuvants. Moreover, including formulation adjuvants in the EPN suspension on maize plants gave better results in the case of *S. littoralis* infesting whorl maize leaves than in the case of *A. ipsilon* targeted at the stem base of maize plants (Saleh *et al.*, 2015). Eventually, Abd-Elgawad (2001) stressed that EPNs were capable of combating many Egyptian insect pests in the laboratory but end-users needed to know more about the characteristics of nematode species and the proper use of each one to reduce the efficacy gap between laboratory and field results, and consequently reach a sounder pest management.

Abd-Elgawad and Spiridonov (2014) outlined the current status of EPN research and application as a biopesticide in Egypt and Russia. For optimized EPN application, one should select a proper daytime and a proper period to target the susceptible stage of the pest, select the application rate, method and the proper nematode species to match the most susceptible pest stage, and take into account that the storage of formulated nematode species varies (Dreves and Lee, 2015). Interestingly, Abd-Elgawad and Spiridonov (2014) suggested that growers should also be advised to consider ways to target specific zones within their fields that have the insect pests for EPN application, to minimize wasted money and unprofitable expense. So, instead of random EPN application, nematodes should be sprayed deliberately on insect aggregates (Abd-Elgawad, 2014c).

The influence of four Egyptian isolates of EPNs, each applied at 0.55×10^9 IJs/acre, on natural soil populations of plant parasitic nematodes infesting watermelon in a newly reclaimed area in Egypt was studied (Abd-Elgawad and Aboul-Eid, 2002). The three isolates, *Heterorhabditis* sp. EA1, *H. bacteriophora* EB5 and *H. bacteriophora* EGG, reduced *M. incognita* populations significantly ($P \leq 0.05$) at 7 and 21 days after treatments. Contrary to the untreated check, a reduction in the *Tylenchorhynchus* spp. numbers was a progressive trend 7 and 21 days after treatment by *Heterorhabditis* sp. EA1 and *Heterorhabditis indica* EI3. *Criconemella* spp. were undetected 7 and 21 days after treatment by *H. bacteriophora* EB5, but their population levels were undetectable 7 and 49 days after treatment by *H. indica* EI3 and *H. bacteriophora* EGG, respectively. Significant ($P \leq 0.05$), differences in the spiral nematode population levels occurred only in plots treated by *H. indica* EI3 and *H. bacteriophora* EGG. Therefore, Abd-Elgawad and Aboul-Eid (2002) discussed differences in EPN efficacy and their possible control of phytonematodes.

Moreover, EPN surveys are still in progress, hopefully to isolate promising new species/strains, since various regions in Egypt have different environmental and edaphic factors. Based on such variability in climate, soil types and crops, an EPN survey divided Egypt into four natural geographic regions: (i) northern (Mediterranean Sea coast and Nile Delta); (ii) middle (Nile valley); (iii) southern; and (iv) Sinai Peninsula regions (Abu-Shady *et al.*, 2011). Overall, a total of 1000 soil samples were collected. Of these, more than 180 samples (18%) were EPN positive. About 100 isolates have been characterized molecularly to species level. Internal transcribed spacer (ITS) rDNA sequence analysis was considered as an initial screening/diagnostic approach. Molecular analysis revealed that 58% of these EPN positive samples contained nematodes in the family Heterorhabditidae, and the remaining 42% of the samples yielded nematodes in the Steinernematidae. All recovered *Heterorhabditis* isolates were identified as *H. indica*. Among steinernematids, isolates were placed mainly into two species: *Steinernema abbasi* and *S. carpocapsae*. One sample in this survey from the Aswan governorate (southern region) yielded a *Steinernema* species that, in addition to several EPN isolates from other Egyptian laboratories, needed further taxonomic study (Hunt and Nguyen, 2016). This nematode isolate is a member of the *glaseri* group and is closely related to *Steinernema arenarium*. Species in the *glaseri* group have been reported from Mediterranean countries, including Italy, Morocco and Israel. Abu-Shady *et al.* (2011) presumed that the isolate was a novel species; however, further studies are warranted to confirm its complete identification. On the other hand, Abu-Shady *et al.* (2011) concluded that EPN fauna in Egypt did not seem to be very diverse with respect to the number of species. Although factors such as climate and soil types limit species diversity, EPN species such as *S. abbasi* and *H. indica*, which are adapted to hot and dry climate conditions, prevail and are widely distributed in different Egyptian habitats (Abu-Shady *et al.*, 2011).

Numerous parasites and pathogens infect domestic and farm animals in Egypt, causing considerable economic losses in animal and poultry production. For example, tick-related diseases and injuries represent serious problems in Egypt. As obligatory blood-sucker arthropods, several tick species feed and damage domestic and farm animal's skins, resulting in losses in the production of eggs, meat and/or milk. In addition, these tick species are vectors of serious bacteria, viruses and other pathogens. The main measures used to control ticks and tick-borne pathogens are applications of chemical pesticides; for example, acaricides, which have side effects on the environment and on human health. Hence, Egyptian researchers have been trying to replace such pesticides with safe alternatives. In this respect, although EPNs proved effective against some ticks, such as *Hyalomma dromedrii* (El-Sadawy and Habeeb, 1998) and *Boophilus annulatus* (El-Sadawy and Abdel-Shafy, 2007), research has hardly exceeded the experimentation stage in the laboratory and the field. On the other hand, no information is available on EPN application in the production process of garden mushroom and oyster mushroom in Egypt, probably because mushroom cultivation still occupies very small areas in Egypt. Due to the insufficient contribution of EPNs in IPM programmes in Egypt, the attractive attributes that qualify EPNs as biological control agents in Egyptian agricultural systems are presented. I should point out that Egyptian EPNs do not differ in these traits from worldwide ones in general.

24.4.2 Desirable traits of entomopathogenic nematodes as attributes for inclusion in integrated pest management programmes

Rationale

It is well known that the success of any biocontrol agent(s) against pests depends on its relevant key traits, such as environmental tolerance, host-finding ability, rapid speed of killing the host, high reproduction efficiency, ability for *in vitro* mass culture, synchronization with the target pest species and ease of application. Taking such attributes into consideration, insecticidal nematodes (Steinernematidae and Heterorhabditidae) are the most successful example of a biological pesticide used against soil-dwelling and wood-boring insect pests (e.g. Kaya and Gaugler, 1993; Grewal *et al.*, 2005a; Lacey and Georgis, 2012). They possess virtually all the attributes of an ideal biological control agent. They are extraordinarily lethal, with most insects being killed within 36 h. Dozens of important insect pests are susceptible, yet EPNs are environmentally benign (Gaugler and Kaya, 1990; Grewal, 2012a,b). Despite their high pathogenicity to insects, their lack of threat to the environment led the US Environmental Protection Agency to exempt nematodes from registration requirements; nearly all countries provide almost similar exemption. This high degree of safety means that, unlike chemicals, or even *Bacillus thuringiensis*, nematode applications do not require masks or other safety equipment; and re-entry time, residues, pollution and pollinator toxicity are not issues. Nematodes do not require specialized application equipment, as they are compatible with existing equipment, including pressurized, mist, electrostatic, fan and aerial sprayers. Application via irrigation has further improved grower acceptance.

Merits of their biology

The biology of EPNs offers them as an attractive bioinsecticidal alternative to chemicals. They are obligate parasites of insects. EPNs function cooperatively with mutualistic bacteria, which are essential to complement their life cycle as biocontrol agents of insect pests. The specific associations between EPNs and their symbiotic bacteria are well established and limited to *Xenorhabdus* for *Steinernema* and *Photorhabdus* for *Heterorhabditis* (Boemare *et al.*, 1993). The nematode IJs secure a safe harbour for the bacteria, as intestinal symbionts in soil, and then enable these bacteria to enter the insect haemolymph. To reciprocate, the symbionts defend, through antibiotic preservation, against insect cadaver secondary invaders and supply EPNs with essential nutrients inside the host cadaver that are sufficient for nematode reproduction (Bai *et al.*, 2013). These IJs enter the host insect either via its natural apertures (spiracles, oral cavity and anus), or they can in some cases penetrate the insect cuticle (Dowds and Peters, 2002). The nematode IJs retain the cuticle of the second stage as a sheath. The sheath is especially important because IJs are considered the only free-living stage of EPNs that can live outside the insect host. They complete their development and live for several generations inside their host. The nematodes reproduce efficiently; therefore, each of the host cadavers eventually turns into an insect bag filled with tens or hundreds of thousands of IJs. When food is depleted, IJs exit from the bag, i.e. the host cadaver (Grewal and Georgis, 1999). Since IJs typically leave the host cadaver to search for a new host to infect, it is presumed that such a sheath provides protection against unfavourable environmental conditions. Moreover, Womersley (1990) suggested that this sheath might help to minimize the rate of water loss from a nematode body subjected to desiccation. Infective juveniles do not feed, and can survive in the soil for several months (Gaugler, 2002; Grewal *et al.*, 2005a). The nematode species *Steinernema anomali*, *Steinernema feltiae* and *H. bacteriophora* were found to tolerate freezing, with a lower lethal temperature of −14, −22 and −19°C, respectively (Brown and Gaugler, 1995). On the other hand, *Heterorhabditis zealandica* can avoid freezing (Wharton and Surry, 1994). Its sheath blocks innoculative freezing and permits extensive supercooling to −32°C, but exsheathed IJs freeze above −6°C and die. *H. zealandica* IJs exert such tolerance through withstanding extracellular ice formation or by evading freezing at sub-zero temperatures by supercooling (Wharton *et al.*, 1984).

Their foraging strategy

The behavioural ecology of EPNs has been studied and should be exploited to achieve their optimum utilization (e.g. Lewis *et al.*, 1992, 2006; Kaya and Gaugler, 1993; Abd-Elgawad and Spiridonov, 2014). Investigations on the foraging strategies of EPNs, in terms of ambush- and cruise-foraging strategies, highlighted the importance of nematode–host matching for their skillful application. Gaugler *et al.* (1997) implemented this strategy to make recommendations on which EPN species should be utilized in biocontrol approaches under field conditions. They reported cruise-foraging species as the best suited nematodes to find immobile hosts, including those deep in soil or in other cryptic habitats. On the contrary, they excluded ambush foragers as unsuitable for the task of locating immobile insects, but rather moving hosts as good targets for ambushers. Nevertheless, more field studies indicated that this distinction between the two, ambush- and cruise-foraging strategies, might be oversimplified (Griffin, 2012). For example, while there are obviously main dissimilarities in the habitual foraging mode of species, Griffin (2012) stresses that classification as an ambusher does not hinder success against distant cryptic insects. The ability of *S. carpocapsae* to control the large pine weevil supports her point of view. This weevil, *Hylobius abietis*, lives under the bark of tree roots, while *S. carpocapsae* poured at the soil surface can successfully infect the larvae and pupae of *H. abietis* within the roots at depths of up to 40 cm in the soil (Dillon *et al.*, 2006). Under field conditions, Dillon *et al.* (2006) found a decrease in parasitization rates with the depth of the host location in soil. Nevertheless, such a decrease was comparable for all EPN species tested, whether they were ambush, cruise or intermediate foragers. Such results imply that EPNs use a wide range of host-seeking strategies. Yet, if the foraging strategy is shown exclusively in the form of only one of the two extremes for a nematode species, then cruiser nematodes will actively seek out hosts; they are more effective at infecting stationary hosts. On the other hand, ambushers remain stationary and infect passing hosts. So, the latter foraging strategy is appropriate for attacking fast-moving hosts. However, cruisers and ambushers are actually end points along a continuum, and therefore I would assume that numerous nematodes are intermediate. The latter nematodes are capable of foraging within the range between cruising and ambushing approaches for finding and locating their hosts (Lewis, 2002; Dillman *et al.*, 2012). Moreover, other EPN species of *Steinernema* exhibit jumping, an uncommon action among tiny, tender-bodied and limbless organisms (Campbell and Kaya, 1999; Dillman *et al.*, 2012). For foraging strategies of EPNs, jumping is intended to follow a short-range, host-seeking approach, which expedites attachment to the host when the host is in close proximity (Campbell and Gauger, 1993). Bal *et al.* (2014) compared the lateral dispersal of the ambusher, *S. carpocapsae*, and the cruiser, *H. bacteriophora*, nematodes from infected host cadavers and concluded that ambush foragers must employ a long-distance dispersal strategy to maximize reproductive success in the absence of hosts. They counted numbers of both baited *G. mellonella* larvae killed and the IJs penetrated in dead baits to compute the percentage of IJs dispersed from the source cadavers, based on the emergence potential and penetration efficiency of the two nematode species. They recorded that vegetation enhanced dispersal of both species, but more so for *H. bacteriophora.* About 4% of the *S. carpocapsae* population dispersed faster than the fastest *H. bacteriophora*, reaching 30–61 cm, compared to only 2% of the *H. bacteriophora* population dispersing this far. This use of 'sprinters' for long-distance dispersal may represent an adaptive dispersal strategy by the otherwise ambush forager *S. carpocapsae* in the absence of hosts (Bal *et al.*, 2014).

Olfaction frames entomopathogenic nematode–host interactions

Other investigations have proved that EPNs are attracted to CO_2 as well as to a number of other odorants (e.g. Hallem *et al.*, 2011; Dillman *et al.*, 2012). Hallem *et al.* (2011) documented that the broad host range of these parasitic nematodes resulted from their ability to respond to the universally produced signal, carbon dioxide (CO_2), as well as a wide array of odours, including host-specific odours that were identified using thermal desorption-gas chromatography-mass spectroscopy. Dillman *et al.* (2012) recorded that the nematodes, *Steinernema riobrave*, *S. carpocapsae*, *H. bacteriophora*, *Steinernema scapterisci*, *S. glaseri* and *Oscheius carolinensis*, chosen on the basis of both phylogenetic and behavioural diversity, varied in their response to the odour blends released by live potential hosts and to individual host-derived odorants. EPNs utilized the very common host cue CO_2 and insect-specific odorants for locating the host insect. The relative significance of CO_2 versus host-specific odorants differs for various parasite–host collections and for various host-seeking manners. Numerous host-derived odours activate host-seeking actions in a species-specific manner (Dillman *et al.*, 2012). Yet, although EPNs have developed specialized olfactory arrays that likely serve to adequate host selection, a comparison of host virulence with host-evoked chemotaxis and jumping approaches demonstrated that some EPNs were attracted to invertebrate species that were not favourable hosts. Hence, appeal to non-host insects in the non-attendance of insect hosts might secure a survival meritor, even longevity, to EPNs by serving dispersal to more favourable environmental conditions (Dillman *et al.*, 2012). Occasionally, however, olfactory preferences can possibly push EPNs to pursue non-hosts or dead-end hosts. Eventually, host selection is a multiple process that can be divided into several steps: host location, host attachment, host recognition and host penetration (Kaya and Gaugler, 1993). Host attraction represents only a component of sequential steps that constitute the selection process. Other components can mediate successful host recognition and penetration, as well as avoid the fatal decision to infect an inappropriate host (e.g. Kaya and Gaugler, 1993; Abd-Elgawad, 2001; Dillman *et al.*, 2012).

Relative ease of their mass production and application

In vivo and *in vitro* (solid or liquid fermentation) approaches for EPN production are available. For laboratory experiments and small-scale field experimentations, *in vivo* production of EPNs offers the proper approach (Shapiro-Ilan *et al.*, 2012). Typically, *in vivo* production fits best where a lack of capital, scientific expertise or infrastructure exists. Therefore, *in vivo* production is utilized currently and adequately in niche markets and with smallholding farmers. On the contrary, the *in vitro* approach is operated when large-scale production is requested at acceptable quality and reasonable cost. On the other hand, enhanced efficacy in EPN applications can be eased via improved delivery mechanisms (e.g. cadaver application) or optimization of spray equipment (Abd-Elgawad *et al.*, 2003; Shapiro-Ilan *et al.*, 2012). Authentic advances have been achieved in promoting EPN formulations

(Grewal, 2002; Abd-Elgawad *et al.*, 2003; Lacey and Georgis, 2012; Shapiro-Ilan *et al.*, 2012; Cruz-Martínez *et al.*, 2017). Such advances were especially necessary for above-ground applications of EPNs that have lagged behind those of soil and subterranean implementation. Foliar and above-ground applications face obstacles, especially in terms of EPN sensitivity to desiccation, ultraviolet radiation and temperature extremes. Solving these problems is under way through mixing EPNs with surfactants or polymers, or with sprayable gels. Also, bait formulations and insect host cadavers can enhance EPN persistence and reduce the quantity of nematodes required per unit area (Shapiro-Ilan *et al.*, 2012). Anti-desiccants have been shown to increase the effectiveness of EPNs against certain pest species (Lacey and Georgis, 2012). Eventually, more fundamental and applied research on different aspects of EPN safety, biology, compatibility and/or interactions with other biotic and abiotic factors, mass production, formulation, storage, quality and application methods and technology to enhance their immediate and carryover effect under different conditions will allow us to adapt the best way to use them more efficiently. Genetic improvements may also be used to enhance EPN efficacy under various ecosystems.

24.5 Entomopathogenic Nematode Compatibility and Interactions with Other Chemicals and Control Agents

Entomopathogenic species of nematodes are frequently applied to sites and ecosystems that are traditionally prepared for other inputs. These latter may interact with EPNs, including chemical pesticides, surfactants (e.g. wetting agents), growth regulators, fertilizers, biorational compounds, biocontrol agents (e.g. bacteria, fungi and viruses) and soil amendments (Divya and Sankar, 2009). In order to save time, effort and costs, it is economically attractive to tank mix or add EPNs together with one or more of such inputs. Admittedly, the sensitivity of EPN IJs varies greatly to different inputs, according to their active ingredients/composition, concentrations and degree of proximity or contact with the IJs. Moreover, we can link EPN sensitivity, tolerance or resistance to insecticides with their capability to oppose environmental conditions such as extreme temperatures and osmotic stresses. For example, the mortality of IJs mixed with insecticides was higher at elevated temperatures (Laznik and Trdan, 2013).

On the other hand, researchers have been testing EPNs with other pest control agents for various purposes: (i) for example, to control different pest species or different stages of just one species; and (ii) to investigate the output of their combinations, for example their interactions may have additive, synergistic or antagonistic effects on pest mortality. Such studies may apply all control agents, including EPNs, simultaneously or within a short time interval of each other. EPNs may also be tank mixed with other control agents. The patterns of interactions between EPNs and other agents may range from effects on control agent viability and/or virulence, control agent recycling in the target pest, to effects on the pests' susceptibility to the control agents (Koppenhöfer and Grewal, 2005). Also, Stuart and Gaugler (1996) presumed that the possible genetic adaptation of EPNs to different biotic and abiotic factors influences nematode compatibility and interactions with other agrochemicals and pesticides.

24.5.1 Compatibility with chemicals

It is favourable that some EPN species can readily reproduce and persist in the environment; they may have direct and/or indirect adverse effects on plant pathogens such as phytonematode populations; they can indirectly serve in enhancing soil quality; and they are compatible with a wide range of non-synthetic and synthetic chemicals and biological pesticides used in IPM programmes (Lacey and Georgis, 2012). Generally, IJs can tolerate short-term exposure, i.e. 2–8 h (Divya and Sankar, 2009) and up to 24 h (Laznik *et al.*, 2012; Laznik and Trdan, 2013), to a wide variety of chemicals and biologicals. Koppenhöfer and Grewal (2005) listed herbicides, insecticides, fungicides, growth regulators and fertilizers that may or may not be tank mixed and used together against field pests. This offers a cost-effective alternative to manage more than one field pest/problem simultaneously. For example, Kulkarni *et al.* (2013) reported that IJs of *S. carpocapsae* were compatible with some modern chemical insecticides (Imidacloprid and Thiomethoxam) and biopesticides (Neemgold, Spinosad and Agropest Bt.), even in concentrations higher than the recommended doses in terms of survival and further infectivity, and thus could be used in combination for pest management. Nematodes are also compatible with most inorganic fertilizers when they are applied inundatively

(Bednarek and Gaugler, 1997; Divya and Sankar, 2009). For organic manure, EPNs should be tested carefully since their biota and other components may vary from one composted manure to another. Some of these components may be incompatible, especially of fresh manure, which can be injurious to EPNs. Generally, mixing EPNs with other compatible agricultural inputs enables us to utilize EPNs in IPM systems. Again, caution should be exercised since up to a definite concentration limit for each chemical may be tolerated by the nematodes. In other words, EPNs exposed to higher concentrations above such a limit may be damaged. Therefore, factors such as the application volume and system used should be considered carefully (Alumai and Grewal, 2004). On the other hand, some agrochemicals are not compatible with EPNs and therefore an appropriate time interval between the applications should be pursued. The length of such an interval relies on the persistence of the agrochemicals in the target substrate (Grewal *et al.*, 1998). Koppenhöfer and Grewal (2005) reported that 1 and 2 weeks post-application of chemical insecticides and chemical nematicides, respectively, was usually recommended before nematode application. Furthermore, even when a specific chemical pesticide is not deemed compatible, the use of both agents (chemical and nematode) can be implemented by time spacing between applications (e.g. 1–2 weeks).

24.5.2 Interactions with other biocontrol agents

The grouping of EPNs with other control agents was sometimes found to be synergistic, achieving higher mortality of the host insect species than either agent alone. Koppenhöfer and Kaya (1997) proved additive and synergistic interaction when EPNs were combined with *B. thuringiensis* for the control of scarab grubs. EPNs showed synergism with the neonicotinoid insecticide imidacloprid (Koppenhöfer and Kaya, 1998; Koppenhöfer *et al.*, 2000; Polavarapu *et al.*, 2007; Koppenhöfer and Fuzy; 2008). This type of synergistic interaction of some pesticides with EPNs, and consequently enhancing EPN efficacy in inundative release, was also proved by Nishimatsu and Jackson (1998). On the contrary, an antagonism occurred between imidacloprid and *Steinernema scarabaei* when applied together against the scarab European chafer, *Rhizotrogus majalis* (Cappaert and Koppenhöfer, 2003). Therefore, the strategy of synergistic effect of EPNs with other insecticide(s) should be used on a workable basis against scarab larvae. On the other hand, Ansari *et al.* (2008) found that EPNs (either *H. bacteriophora*, *S. feltiae* or *Steinernema kraussei*) applied 1 or 2 weeks after adding the fungus *Metarhizium anisopliae* could achieve 100% mortality of the black vine weevil, *Otiorhynchus sulcatus*, larvae, even when both the fungus and nematodes were applied at less than the recommended rates. They concluded that EPNs and *M. anisopliae* could function together synergistically in potted Emerald Gold plant, *Euonymus fortune*, under greenhouse conditions and might fulfil a perfect and economically feasible method for control of *O. sulcatus* larvae.

Some authors tested both chemicals and bioinsecticides. Laznik and Trdan (2013) found that azadirachtin (a secondary metabolite related to the limonoid group and present in neem seeds) and pirimicarb (a selective carbamate insecticide) did not affect the viability of *H. bacteriophora* and *S. feltiae*, but both compounds did affect the viability of *S. carpocapsae* and *S. kraussei*. The active ingredients azadirachtin, toxin of *B. thuringiensis* var. *kurstaki* and imidacloprid did not affect the viability of the domestic strain *S. feltiae* B30. By contrast, these insecticides significantly reduced the number of living IJs of a commercial product of the same EPN species. Lufenuron and abamectin were the only insecticides tested by Laznik and Trdan (2013) that influenced the mortality of *H. bacteriophora*, suggesting that *H. bacteriophora* was the species most tolerant to the tested insecticides. Moreover, the ingredients abamectin and lufenuron have a well-known nematicide effect, which may explain the relatively high mortality values, whereas *S. carpocapsae* and *S. kraussei* proved to be the most sensitive species (Laznik and Trdan, 2013).

Also, applying the bacterium *B. thuringiensis* and the nematode *S. carpocapsae*, both at half rate, against the diamondback moth larvae, *P. xylostella* (Lepidoptera: Plutellidae), on watercress led to 58% control. This percentage was significantly better than that of each applied at full rate (Baur *et al.*, 1998). The authors reported that the application of EPNs without the bacterium twice would probably be ineffective in obtaining such a percentage. Yet they recommended that EPNs could be exploited as constituents in IPM of *P. xylostella* in order to increase efficacy while alleviating resistance to *B. thuringiensis*. Also on watercress, Baur *et al.* (1997) showed that additives could generally enhance EPN persistence and efficacy, but the enhancement was likely not sufficient to raise the feasibility of above-ground applications of EPNs to control the diamondback moth larvae. Schroer and Ehlers (2005) and Schroer *et al.* (2005) demonstrated parallel results for *S. carpocapsae* and a combination of the nematode and *B. thuringiensis* with a polymer. Similarly, less than 50% of intended *P. xylostella* populations were suppressed with high concentrations of EPNs (Somvanshi *et al.*, 2006; Nyasani *et al.*, 2008). Interestingly, Lello *et al.* (1996) found that grand hydraulic nozzles delivered so great a number of IJs on to plant shoots that EPNs attained up to 98% mortality.

24.6 Egyptian Scenario for Entomopathogenic Nematodes in Integrated Pest Management

As any country, Egypt has its own fauna and flora that should not be damaged or managed improperly. Therefore, at the state level, recent governmental regulations have been established for critical examinations of any foreign/imported live microorganism including EPNs and/or their symbiotic bacteria. These regulations are issued to avoid also the introduction of exotic species that might displace native species. On the other hand, research has admittedly been carried out on different facets of indigenous EPNs in Egypt in order to use them primarily as biological control agents to replace, at least partially, hazardous chemical insecticides. These programmes encompass a variety of approaches, such as studies on EPN applications, enhancement of their effectiveness, their interactions with other soil microorganisms/insecticides and integration of biocontrol with other management techniques (e.g. Saleh and Sammour, 1995; Shamseldean *et al.*, 1999; Salama and Abd-Elgawad, 2002; Abdel-Rahman *et al.*, 2008, 2010, 2012; Abd-Elgawad *et al.*, 2010; Shamseldean, 2010; Hussein and Abdel-Aty, 2012; Atwa *et al.*, 2013; Abd-Elgawad and Spiridonov, 2014). However, the global scenario for EPN compatibility and interactions with other chemicals and control agents should be cautiously used as a tentative guide when no data are available as essential milestones for applying Egyptian EPN strains under field conditions. This is especially important since inter- and intraspecific variation in virulence and reproductive potential of newly isolated Egyptian nematodes have also been reported (Abd-Elgawad, 2001). Moreover, the laboratory percentage mortality of key insect species exposed to indigenous EPN isolates in Egypt was perfect, with significant numbers of emerging nematode IJs (Table 24.3). Despite the high efficacy shown, nematodes have very limited usage for such pests on a large scale in Egypt. That is because the definitive test of host suitability is efficacy under field conditions where a minimum of three solid field trials to establish host suitability is required (Shapiro-Ilan *et al.*, 2002). An analysis of nematode host suitability (based on field efficacy) for some of the most extensively studied insect pests in Egypt is illustrated in Table 24.4 (Abd-Elgawad and Spiridonov, 2014). The table includes only host–nematode combinations with at least three refereed publications on field efficacy to confirm reliability. Reliable outcomes should come from multi-site and multi-season tests. The best matches tend to be for nematodes that have high virulence toward hosts in a protected environment or cryptic habitat, e.g. leaf miner, black cutworm, borers, etc. Nematode applications to environments exposed to temperature extremes, ultraviolet radiation or desiccation, e.g. some *Spodoptera* spp., are prone to failure. On the other hand, Abd-Elgawad and Nguyen (2007) found variations ($P \leq 0.05$) in heat and desiccation tolerance among four Egyptian heterorhabditid isolates extracted from distant regions, which could stimulate additional surveys for EPNs in areas characterized by extreme environmental conditions to overcome such limitations.

Eventually, EPNs could be experimented with on a large-scale in Egypt but applied by end-users on smaller areas. This is because of both general and specific reasons. The common reasons centre on factors such as the limited nematode market within the confines of high product cost and restricted availability, limited spatial modelling to predict the nematodes' persistence and fates, and suboptimum ease-of-use and efficacy (see Coupland *et al.*, Chapter 2, this volume). In addition, in Egypt, many growers are unaware of EPNs as biocontrol agents and their application techniques. Yet experimentation has proven the usefulness of EPNs in IPM in Egypt through two main approaches; that is, EPNs as the only pesticide in addition to common agricultural practices for specific insect species, or EPNs with other pesticide(s).

Table 24.3. Mortality of insects exposed to Egyptian heterorhabditid nematode isolate EGB1 and numbers of emerging nematode infective juveniles (IJs). (Laboratory tests adapted from Shamseldean *et al.*, 1994).

Insect host			
Common name	Scientific name	Mortality (%)	Mean number of IJs/insect cadaver
Greater wax moth	*Galleria mellonella*	100	>250,000
Egyptian cotton leafworm	*Spodoptera littoralis*	100	>100,000
Common mole cricket	*Gryllotalpa gryllotalpa*	100	>50,000
Greasy or black cutworm	*Agrotis ipsilon*	100	>100,000
Spin bollworm	*Earias insulana*	100	>50,000
Pink bollworm	*Pectionophora gossypiella*	100	>250,000
Pink borer of sugarcane	*Sesamia cretica*	100	>75,000
Grape fruit worm	*Polychrosis botrana*	100	>10,000
Honeydew moth	*Cryptoblabes gnidiella*	100	>10,000
Apple stem borer	*Zeuzera pyrina*	100	>250,000
Fruit fly (larvae)	*Ceratitis capitata*	100	No emergence
Fruit fly (pupae)	*C. capitata*	100	No emergence
Looper caterpillars	*Chlorissa* spp.	100	>10,000

Table 24.4. An analysis of host suitability for entomopathogenic nematodes against various insect pests. (Adapted from Shapiro-Ilan *et al.*, 2002.)[a]

Pest	Nematode species	Host suitability[b] (% suppression)	Number of references in analysis
White grub (*Tropinota squalida*)	*Heterorhabditis bacteriophora*	Good (68)	3
Leaf miner (*Liromyza trifolii*)	*Steinernema carpocapsae*	Good (68)[c]	3
Black cutworm (*Agrotis ipsilon*)	*S. carpocapsae*	Excellent (86)	7
Diamondback moth (*Plutella xylostella*)	*S. carpocapsae*	Fair (56)	5
Corn earworm (*Helicoverpa zea*)	*S. riobrave*	Excellent (90)[d]	4
Borers (*Synanthedon* spp.)	*S. feltiae*	Excellent (86)	4
Spodoptera spp.	*S. carpocapsae*	Poor (27)[e]	3
Tomato leaf miner (*Tuta absoluta*)	*S. feltiae*	Fair (45)[f]	3

Notes: [a]Only pests (by genus or species) with at least three refereed publications on the field efficacy of each nematode species were included, and only if the rate of nematode application was not excessive (i.e not >125 nematodes/cm^2). [b]Host suitability ratings, excellent, good, fair and poor, are based on suppression levels of 80–100, 60–79, 40–59, and <40%, respectively. The suppression levels were calculated by averaging results from field trials in the associated references. [c]Indoor applications in high humidity. [d]Soil applications. [e]Above-ground applications. [f]Foliar applications (Jacobson and Martin, 2011).

24.6.1 Scenario for entomopathogenic nematode production and end-users

Although some foreign EPN species/strains have been imported and applied (e.g. Saleh and Ragab, 1999; Abdel-Rahman *et al.*, 2008; Ibrahem, 2015), indigenous EPN strains have proved relatively more tolerable for environmental conditions and were effective in field and laboratory tests (e.g. Abd-Elgawad, 2001; Abd-Elgawad and Nguyen, 2007; Abdel-Rahman *et al.*, 2010; Saleh *et al.*, 2015; Table 24.3). Such references have been citing both indigenous and foreign EPNs for application, but indigenous nematodes are more appreciated because they are more adaptive and less expensive, without any risk to fauna and flora than imported species/strains. The tolerance and high efficacy demonstrated by native EPNs have nominated them for further experimentations and development. A few of these EPNs are being commercialized locally, e.g. NOBUG© in Egypt, but on a very limited scale (Abd-Elgawad and Spiridonov, 2014).

The active ingredient, *S. carpocapsae* or *H. bacteriophora*, of this product is placed on moist polyether-polyurethane sponge to control mainly the caterpillar and pupal stages of lepidopterous insect pests by *S. carpocapsae*, or to control insect borers of trees and shrubs by *H. bacteriophora*. Usually, a 5-l package is diluted by 300 l of water per 1 feddan (feddan = 4200 m^2) for arthropod pest control in such fields as tomato, pepper, lettuce and cabbage for *S. carpocapsae*, or trees of pear, grape, peach and cherry for *H. bacteriophora*. EPN species associated with an Egyptian company may vary over time according to the market need.

EPNs are produced through two approaches in Egypt, mainly by *in vivo* methods but to a lesser degree through *in vitro* solid culture. For example, Abd-Elgawad (2012) has worked out a simple course to *in vitro* culture nematode alternately with *in vivo*. The reason behind this alternation is to enhance EPN production quantity via *in vitro* method and to restore or revive its virulence and infectivity to the maximal possible standard through *in vivo* reproduction using economic techniques adequate for Egyptian conditions. This system is being optimized to reach the maximal possible standard of EPN virulence and production volume; that is, the *in vivo* method is complementary to the *in vitro* solid medium technique in order to avoid defects of either method if used alone. Professor R. Gaugler collaborated with Egyptian scholars through a programme to scale up *in vivo* methods (Gaugler *et al.*, 2002). Thereafter, some improvements were added to LOTEK (Gaugler *et al.*, 2002), and NEMA NRC was established as a simple, inexpensive, largely automated insecticidal nematode production apparatus in Egypt developed from the basic LOTEK system in the USA. Apparent improvements, for example, included installing a key, to close the apparatus during non-use status, and four feet, each inside a vessel with water in it, to carry it and prevent ants as well as undesirable organisms from reaching the trays. Nevertheless, each of the two, *in vivo* and *in vitro*, methods may be used singly when needed; that is, if we have a limited area to be treated, it is better to get the most virulent EPNs through *in vivo* culture where NEMA NRC can produce small or medium amounts of insecticidal nematodes. On the contrary, for vast field areas, the need is to employ *in vitro* culture to get sufficient biopesticide. Both *in vivo* and *in vitro* culturing are operated, and further production improvements are being thought of through the following sequential steps: nematode and host selection, introducing further refinements to the apparatus design, scale up and evaluation. Briefly, existing *in vivo* nematode harvest systems rely on infected host cadavers being supported over a reservoir of water (Gaugler *et al.*, 2002; Fig. 24.1). This system does not depend on nematode migration or a reservoir. It uses shallow perforated holding trays and a timed misting system (Fig. 24.2). Exposure to free water induces the IJs to emerge from the cadaver; the EPNs are washed off into a collecting tray, then cleaned and concentrated by a cross-filtration system. Because the system is largely automated, it is projected to produce significant labour savings. A small harvester and separator have been built at the National Research Centre (NRC) in Egypt. Progress has already been made in inoculation and harvest techniques and postharvest EPN concentration and cleaning. The parameters of such an *in vivo* system that may be further optimized include tube length and diameter, flow rate and initial suspension concentration. These and other strain-specific parameters are being adjusted to meet the local and cheap facilities and needs and optimized for the best production capability. It is presumed that the organic sector can benefit significantly from and in return help in the development and spread of EPN utilization in Egypt through this system. This manifold system can provide real opportunities for the production of biocontrol agents even as a cottage industry.

The challenge now for significant and commercial EPN production is to encourage the private and public sectors to participate financially in upgrading the approaches of pest control while drawing growers' attention to their proper applications and relevant techniques. This may be through increasing governmental money allocated to the scientific research sector, enhancing academic–industry partnerships and raising the awareness of growers to use effective EPNs. Admittedly, other biocontrol agents have been recognized earlier and consequently possess better market segments than EPNs in Egypt (Anon., 2012, 2014). Expanding EPN utilization is hindered substantially by the limited knowledge of these EPNs of the extension personnel who advise Egyptian farmers, as well as by the lack of awareness of EPN technology by resource-poor farmers. The end-users of EPNs in Egypt need to know more about the characteristics of nematode species and the proper use of each one to reach a sounder pest management. Therefore,

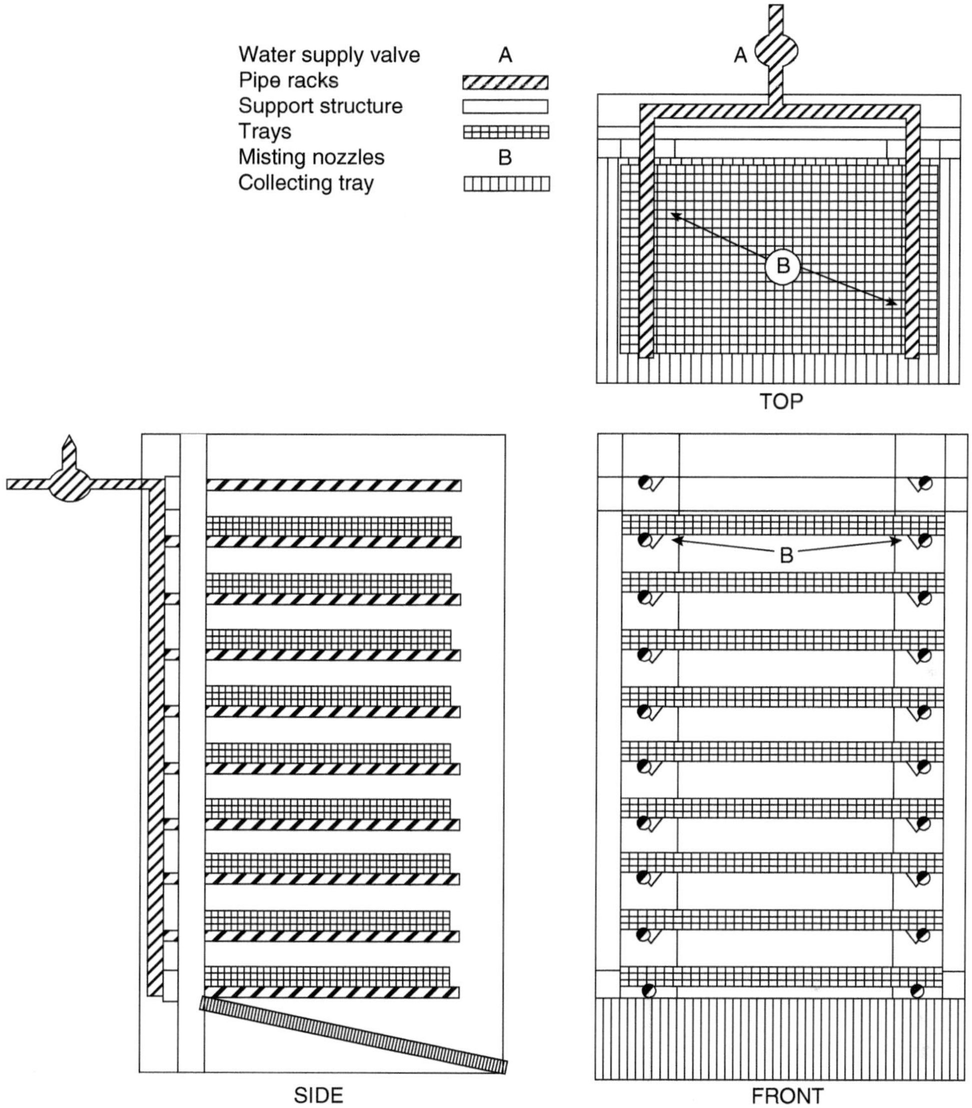

Fig. 24.1. The basic prototype for *in vivo* mass production of entomopathogenic nematodes and their symbiotic bacteria. (Courtesy of Randy Gaugler.)

programmes should be adopted to raise the awareness of farmers and extensions of EPNs as bioinsecticides. They should aim to inform (and train, if applicable) farmers and extension officers of the importance, mode of action, application and impact of EPNs on insect pests, crops and soil health. Moreover, end-users should be aware that the strength of such a production system in Egypt lies in its potential to produce locally adapted nematode strains for the control of the above-mentioned key pests, instead of a 'one size fits all' product in foreign fermenters.

On the other hand, if economic factors are not favourable, even a strategy involving the most suitable match of nematode to target pest is doomed to failure. Economic factors include the grower's perceived need to control the pest, the relative cost of nematodes compared with other management options, the value of the commodity (e.g. per hectare) and the overall importance of the commodity

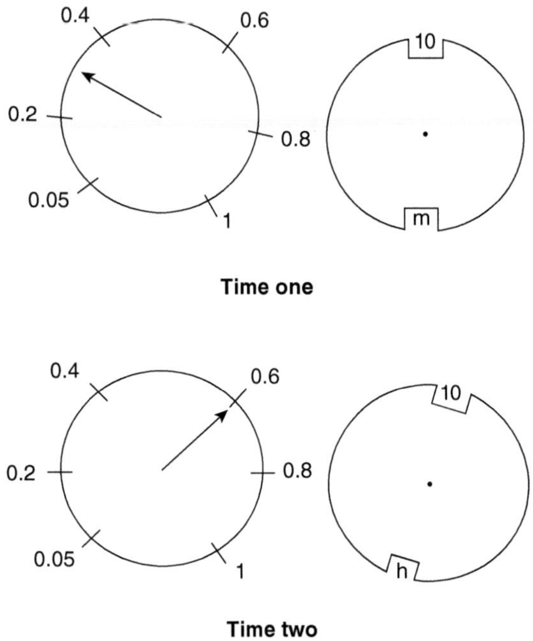

Fig. 24.2. Four scales of the timer are set to provide a continuous 3-min misting cycle at 6-h intervals (Abd-Elgawad, unpublished). Time one = 0.3 × 10 = 3 min (a periodic misting cycle). Time two = 0.6 × 10 h = 6 h (a 6-h interval without misting).

in the agricultural market (Shapiro-Ilan *et al.*, 2002). The exemption of EPNs from Environmental Protection Agency (EPA) pesticide registration has in parallel benefitted commercialization efforts over almost the entire world. Niche markets tend to be amenable to EPN usage, not only because the crop value is high but also because the commodity occupies a very small segment of the agricultural market. Therefore, attempts to unravel the complexity in terms of the above-mentioned factors that hinder the significant share of EPNs in the pesticide market in Egypt need investigation into why merchandisers are reluctant to trade in EPNs. This is especially important since the relevant sectors are expected to have a larger role to play. Such a new dogma of biologicals to replace chemical insecticides by environmentally friendly biopesticides should be expanded to non-high-value crops. For instance, major row crops in Egypt such as maize, cotton, potatoes, soybeans and wheat are often unreachable for EPN marketing because the crop value is low and the market segment is huge. In addition, although organic farming in Egypt should be attractive for EPN application, some practical problems have been found (Abd-Elgawad and Spiridonov, 2014). For example, Egyptian farmers stopped growing maize, as it could not be sold organically, while others reported that tomato was too difficult to produce organically. This has narrowed the possibility where EPNs could be used to combat pests of maize and tomato. Main EPN-susceptible maize and tomato pests proven to be good EPN targets include the cotton leafworm, *S. littoralis* (Abbas and Saleh, 1998; Abdel-Razek and Abd-Elgawad, 2007), the black cutworm, *A. ipsilon* (Shamseldean *et al.*, 1994; Hussein, 2004), the green worm, *S. exigua* (Abd-Elgawad and Aboul-Eid, 2001) and the sugarcane stem borers, *Sesamia cretica* (Saleh *et al.*, 1999).

Farms only become certified organic when a certain target crop is present – this is the reason for the general lack of drastic change towards EPN applications on organic farms, since pests of these crops might not be good targets for EPNs. Also, some organic farmers manage pests based on a 'wait-and-see' strategy and do not consider insect pests as a major problem unless they abundantly exceed the economic threshold level. Pest management strategy in Egyptian organic farms is based on the use of a variety of control types, including natural sprays (such as garlic extracts), copper sulfate and various types of pheromone traps, as well as biocontrol agents such as *B. thuringiensis*, *Trichogramma* spp., *B. bassiana*, predators and EPNs. Unfortunately, almost all references to organic agriculture in Egypt do not mention EPNs, due to their limited application, lack of information, or both (Abd-Elgawad and Spiridonov, 2014).

A twofold goal, that is controlling plant parasitic nematodes and insect pests simultaneously by EPNs in Egypt (Abd-Elgawad and Aboul-Eid, 2001, 2002; Keila *et al.*, 2008), may encourage EPN usage, and consequently contribute to such an expansion to more end-users in the future. This is especially important since biochemical and molecular evidence has recently proved that EPNs induce systemic resistance in plants (Jagdale *et al.*, 2009). Yet, to achieve the first goal necessitates optimal and relevant application tactics of EPNs; for example, delivery of the dauer stage juveniles near the plant roots for possible effective phytonematode control (Abd-Elgawad *et al.*, 2013). Abd-Elgawad (2001) discussed matching nematode and insect species to achieve optimal field performance in the light of the nematode–host selection process and the characteristics of commercially available nematode species. For best matching, the nematode and host together must produce the highest IJ yield per cost unit per gram of host (Kaya and Gaugler, 1993).

Another important point is to set perfect quality control and standardization for EPN applications by technical know-how experts and users. Also, application rates should be sound. Furthermore, feedback supplied by end-users is of utmost importance since self-regulation and product refinements by EPN producers should be set up accordingly. Enhanced reliability of the EPN industry will probably be attained through an industry-created convention on standards for EPN quality (Grewal, 2002). On the other hand, to meet all such issues in kind, some Egyptian producers have adopted a cost-effective scenario to avoid the high cost of EPN goods compared to chemical pesticides; i.e. the producer is at the same time also the distributor. Furthermore, the producer, the National Research Centre (NRC), has nominated some of its employees with experience in EPN application. Those latter are well prepared to go everywhere in Egypt to apply EPNs, if applicable. Therefore, those experts can recognize EPN issues such as the viability of IJs, contamination and nematode fate/persistence. On the other hand, being responsible for producing, distributing, applying and following up EPNs all by the same company will increase profit margins. For such a multiple service offered by NRC, the utilization of EPNs seems more attractive and less costly than for other countries. For example, EPN production was recorded by Abd-Elgawad (2012) to be more expensive by only approximately 20% when its price was compared to the absolute price of chemical insecticide to control the black cutworm in Egypt. A solid media mass production for the production of *S. carpocapsae* at the standard application rate of 1.05×10^9 nematodes/feddan costs L.E. 120. On the contrary, grower cost for Hostathion 40 EC (emulsifiable concentrate), a chemical insecticide, is L.E. 100. So, Hostathion is about 20% less expensive than EPNs. The increase in value of the dollar relative to Egyptian pound currency and the increase in commodity prices since 2012 may inflate this estimate further. Yet, if we take into account the insecticide damage/pollution to the environment generated from soil, plants, animals and water, it is apparent that such adverse effects and their impact on health and social services exceed many times the original price of a biopesticide. Moreover, since manpower is to be found in Egypt on a large scale with cheap wages, EPN treatment of 1 feddan costs L.E. 300 (= US$40 or US$95.24/ha). Such costs cover product, distributor and application expenses (M. Saleh, Egypt, 2015, personal communication). In another example, a bioinsecticide only costs the grower US$52.33/ha compared to US$10–20/ha for chemical insecticides in the USA (Georgis, 2002). In spite of this, enlightening end-users of the direct and indirect benefits gained by using EPNs for the effective control of insect pests, especially in protected environments or soil habitats, as well as the safety of humans and the environment, seems imperative to widen the EPN market in Egypt.

24.6.2 Entomopathogenic nematodes with non-pesticide agricultural practices

Since pest management is an integral part of crop protection to ensure high quantitative and qualitative crop yields, Egyptian farmers and the government always try to adjust agricultural practices to achieve this objective. These practices include mainly irrigation, fertilization and tillage, but other agricultural traditions are also exercised for specific crops (Abd-Elsalam, 1993; Anon., 2012, 2014). These traditions may include host-plant resistance, biological methods and interference approaches. All such practices/traditions contribute in one way or another to crop protection/pest management. Briefly, tillage exposes the soil stages of the insect pests to be killed by direct sun rays, irrigation may partially suffocate these stages and/or maintain plant health/vitality, and fertilization strengthens the plant to tolerate the damage of the pests. On the formal level, an outstanding example of adjusting special practices for crop protection is that Egyptian law prohibits the irrigation of berseem (Egyptian clover) after 10 May annually in order to prevent the spread/movement of the cotton leaf worm (*S. littoralis*), especially from berseem to cotton and other cultivated crops. This prevention has been reducing the efforts and pesticides needed to combat the pest. Judicious and timely application of EPNs, or even chemicals when necessary, as a complementary approach to one or more of such practices/traditions may alleviate the pest problem(s) and/or keep the insect population below the threshold level. In other words, control measures of insect pests can be divided conveniently into two major categories: i.e. cultural and chemical control measures. Abd-Elgawad (2013) stressed that none of these measures should be relied upon exclusively for pest management in Egypt. Rather, when practicalities and economics permit, each management procedure should be considered for use in conjunction with other relevant available tactics for arthropod pest control and used in an integrated programme for pest management.

Recent technological advances have backed IPM development in Egypt through enhancing conventional techniques. Thus, more options for pest management and control measures are available. In this respect, Abd-Elgawad (2014d) has obtained good potato yield by exploiting organic wastes, to minimize pollution and reduce the addition of chemical fertilizers, insecticides and fungicides in preparing and mulching potato through novel soil amendments/composts and pest management technology. Also, predictive computer models have been utilized as a decision support system, especially for protected crops (greenhouses, tunnels, etc.). Moreover, current biotechnology is providing transgenic plants for pest resistance in economically important crops. For example, the transgenic tomato cv. Moneymaker overexpressing Bt (Cry 2Ab) gene was produced in Egypt (Saker *et al.*, 2011). The plants induced mortality of *Helicoverpa armigera* and *Phthorimaea operculella* when fed on Bt tomato. The authors suggested that the transgenic tomato plants could be used for management of the target lepidopteran insect pests endemic in Egypt. Furthermore, semiochemicals and selective pesticides have been adopted by Egyptian MARL (Anon., 2012, 2014) to reduce the impact on non-target species.

Agricultural practices may have direct and indirect effects on soil properties, and consequently on soil microbiota such as EPNs. In order to analyse such effects, we should link them to other elements, such as the food web of organisms living in the soil and the climatic features and edaphic factors. In Egypt, such nested biotic and abiotic factors impacting soil microbiota need to be understood more in order to realize a better eco-friendly strategy of pest management, and consequently widen the framework of non-chemical options, as well as the judicious use of pesticides. In this context, an increase in the efficacy of EPNs has been possible experimentally when such biocontrol agents are integrated with other components/management methods in Egypt (e.g. Dimetry *et al.*, 1996; Salama and Abd-Elgawad, 2002; Abdel-Rahman *et al.*, 2008, 2010, 2012).

Like environmental conditions and edaphic factors, anthropogenic activities in terms of agricultural practices affect EPN occurrence, diversity, reproduction and persistence. In Egypt, Abd-Elgawad (2014c) speculated that the undisturbed habitat of the perennial mango had likely contributed to the occurrence of EPNs, while crop, pest and natural enemy removal at the end of the tomato growing season, as well as soil ploughing and preparation for the new crop, and because of the usual chemical applications on tomato, no EPNs were detected at three surveyed tomato fields. Hence, Abd-Elgawad (2014c) concluded that such anthropogenic interference might hinder the reliability of EPNs as conservation biological control agents, especially in seasonal crops like many vegetables, maize, broad bean and cotton. This does not negate the possible use of EPNs in a conservation approach, especially in undisturbed areas like fruit trees, turf and landscapes (Lewis *et al.*, 1998). Yet those latter, turf and landscapes, are found in limited areas in Egypt. EPN treatment for some of these areas was actually done at least once, but with no follow-up due to inadequate awareness as well as lack of available fund and/or specialists. So, practically, augmentation biocontrol, in terms of short-term inundative release, is the most common and implemented tactic in many agricultural areas in Egypt. However, since the above-mentioned, as well as ongoing, studies are proving the common occurrence of important EPNs in a variety of soil ecosystems in Egypt, conservation biocontrol as longer-term strategies may be more appreciated (Lewis *et al.*, 1998; Stuart *et al.*, 2006), as recently demonstrated elsewhere (Stuart *et al.*, 2008; Duncan *et al.*, 2013). Stuart *et al.* (2008) speculated that, under adequate conditions, conservation biological control strategies centred on EPNs could supply long-term arthropod pest control and be more workable and less expensive than augmentation. Although EPN conservation strategies remain undeveloped and have largely been a matter of speculation (Lewis *et al.*, 1998; Stuart *et al.*, 2006, 2008), Duncan *et al.* (2013) found that modifying orchard planting sites could conserve EPNs, reduce weevil herbivory and increase citrus tree growth, survival and fruit yield. Therefore, in order to encourage the use of EPN conservation strategies in Egypt, we should create and/or maintain optimal conditions for EPN persistence. These conditions might include undisturbed habitats (Abd-Elgawad, 2014c), some forms of organic manure mulches (Bednarek and Gaugler, 1997; Duncan *et al.*, 2007), certain soil types (Shamseldean and Abd-Elgawad, 1994a,b; Duncan *et al.*, 2003, 2013) and non-antagonistic factors (Kaya and Koppenhöfer, 1996; Denno *et al.*, 2008).

Since EPNs should be applied as soon as a pest population starts to inflict injuries, reapplying nematodes, usually at 7-day intervals, depends on the success of the first EPN release. Therefore, the fate/persistence of released EPNs is influenced by soil moisture and type, environmental condition and percentage of active EPNs released. Hence, definite agricultural practices are necessary to ensure optimum EPN efficacy. Usually, irrigation of the intended area before and, if needed, after EPN application is imperative to offer the moist conditions necessary for EPN movement to find hosts, as well as to avoid exposing the nematode to desiccation. Also, maximum efficacy is obtained when EPNs are applied at high relative humidity, moderate ambient and soil temperature (10–30°C), and just before sunset (Divya and Sankar, 2009).

24.6.3 Scenario for entomopathogenic nematodes with other pesticides

Essentially, for a specific pesticide, caution should be exercised so that we decide compatibility on a case-by-case basis, since its compatibility with an EPN may vary between pesticides, even within one chemical group (Rovesti and Deseo, 1990). Likewise, compatibility with each candidate product of EPNs for an IPM programme should be tested separately. Attention should be paid to the fact that the compatibility of an EPN with pesticides is not only species specific but also strain specific (Laznik *et al.*, 2012; Laznik and Trdan, 2013). Furthermore, pesticides can influence the occurrence or distribution of endemic EPNs directly; e.g. by applying nematode-incompatible pesticides or indirectly through, for example, affecting soil variables. Ultimately, it is of prime importance to define the grade to which EPNs may be affected by the chemicals involved in IPM programmes.

Egyptian researchers have been investigating the different scenarios for EPNs with other pesticides in IPM programmes. The separate and combined actions of an Egyptian (S2) and an imported (All) strain of *S. carpocapsae*, as well as three insecticides on *S. littoralis* larvae, were tested and compared for their efficacy (Saleh and Sammour, 1995). They reported that selecron was incompatible with the EPNs tested. It was more harmful for the nematodes at higher concentrations and/or longer exposure period. On the contrary, muthrin and nudrin were convenient to EPNs, especially in a short exposure period. Also, locally available plant extracts against the greasy cutworm, *A. ipsilon*, and their side effects on the EPN *S. carpocapsae*, were investigated by Dimetry *et al.* (1996). Fetoh *et al.* (2009) found that the two biopesticides, spinosad and proclaim, used separately against *A. ipsilon* in infested strawberry fields, were more effective than either *S. carpocapsae* or *H. bacteriophora*. Spinosad was more effective than proclaim. As a new approach of IPM, the mixing of either nematode with spinosad or proclaim significantly increased *A. ipsilon* mortality under laboratory and field conditions. Atwa (2013) determined the efficacy of two Egyptian strains, *Steinernema* sp. (EBN1e) and *H. bacteriophora* (EBN10k), with two concentrations (500 IJs and 1000 IJs) mixed with recommended doses of seven chemical insecticides and four chemical fungicides against fifth-instar larvae of *S. littoralis*. EPNs–pesticides complexes were tested for two exposure times (48 and 96 h), after which the nematodes were washed several times in distillated water and the susceptibility of the fifth-instar larvae of *S. littoralis* to the treated nematodes was then examined. *H. bacteriophora* (EBN10k) was found to be highly sensitive to Penconazole, Imidacloprid, Chlorfluazuron, Thiocyclam, Methiocarb, Methomyl, Trimiltox Forte and Benomyl, but only sensitive to Diafenthiuron, Mancozeb and Captan, with significant variations in EPN survival for the two exposure times. Methiocarb, Methomyl, Penconazole, Thiocyclam and Mancozeb were the most sensitive among the tested insecticides after 96 h exposure time. While *Steinernema* sp. (EBN1e) was highly sensitive to Thiocyclam, Chlorfluazuron and Methiocarb and sensitive or moderately sensitive to Diafenthiuron, Penconazole, Imidacloprid, Trimiltox Forte, Methomyl, Benomyl, Captan and Mancozeb. Generally, the nematode response could be categorized as highly sensitive, sensitive, moderately sensitive and tolerant. Also, *Steinernema* sp. (EBN1e) was more tolerant to the insecticides/fungicides used than *H. bacteriophora* (EBN10k). On the other hand, the insecticides/fungicides used have different ($P \leq 0.05$) effects on each nematode strain (Atwa, 2013). Furthermore, the steinernematid showed better reproduction rate than the heterorhabditid strain in all treatments (Atwa *et al.*, 2013). Differences ($P \leq 0.05$) were also found in EPN reproduction between the two nematode concentrations or between the two exposure times relative to the chemicals tested.

24.7 Conclusions and Future Prospects

IPM as an eco-friendly-based pest control strategy is still in progress in Egypt. The strategy follows the general concept of utilizing such natural factors as natural enemies and weather to induce pest mortality. Its involved control tactics try to maintain these factors for sustainable agriculture. IPM in Egypt also uses chemical pesticides where natural control factors indicate a need based on systematic monitoring of the targeted pest(s). Typically, an ideal IPM programme takes into account all available pest control actions and examines the potential compatibility or interaction among various control tactics, weather, cultural practices, other pests and the intended crop. Admittedly, many commercial products (with biocontrol agents as the active ingredients) are available on the Egyptian pesticide market (Anon., 2014), such as Protecto 9.4% WP, Dipel 2 × 6.4% WP, Delfin 85% WG, or Dipel DF 6.4% WG (*B. thuringiensis*), Bio Arc 6% Powder (*B. megaterium*), Bio Zeid 2.5 Powder (*T. album*), Biosect (*B. bassiana*) and Rhizo N Powder (*B. subtilis*). Yet only a few EPN species are being commercialized locally; e.g. NOBUG© in Egypt. They are distributed on a narrow commercial scale due to costs and reliability (Abd-Elgawad and Spiridonov, 2014). In order to cut EPN expenses and operate efficiently, the EPN producer works also as the distributor and applicator. On the other hand, several reports have accurately analysed the factors responsible for limiting the expansion of EPN application, recommended approaches for widening EPN usage and identified research priorities (Grewal *et al.*, 2005b; Grewal, 2012a; Abd-Elgawad and Spiridonov, 2014; see Coupland *et al.*, Chapter 2, this volume). More financial, informative, regulatory and technical support is needed in order to exploit the potential for increased pest management through EPN applications in Egypt and elsewhere. Interestingly, EPNs can be adopted and produced in relatively unsophisticated facilities, and consequently their techniques should be promoted and current opportunities exploited so that wider utilization of EPNs and/or their bacterial associates must be the way forward in crop protection/pest management programmes in Egypt. In this respect, the current distinguished alternative *in vivo* and *in vitro* production system of EPNs established at NRC (Figs 24.1 and 24.2) should be expanded to other Egyptian areas, while drawing growers' attention to their proper handling and application, as well as their attractive merits. Optimistically, such a system may help to expand the limited use of EPNs on high-value speciality crops to major ones.

Widening EPN application in Egypt is especially important since many economically important insect pests have also proved to be susceptible to indigenous nematodes, with possibly unexplored vast genetic variability among their isolates (Tables 24.1–24.4). Egyptian EPN isolates may be more adaptive and less expensive, without any risk to Egyptian fauna and flora than imported strains. Furthermore, they are capable of seeking and killing their insect hosts, especially in soil and cryptic habitats where many chemical pesticides fail to reach. As indicated above, insecticidal nematodes (Steinernematidae and Heterorhabditidae) have additional merits in their biology, foraging strategy and ease of formulation and application. The IJs are compatible with most agrochemicals under field conditions (Shapiro-Ilan and Gaugler, 2015) and may be tank mixed with most chemical herbicides and fungicides, as well as many insecticides such as bacterial or fungal products (Koppenhöfer and Grewal, 2005). In some cases, adding chemical or biocontrol agents to EPNs had no side effects (Dimetry *et al.*, 1996) or results in synergistic levels of insect mortality (Nishimatsu and Jackson, 1998; Fetoh and Azzazy, 2004). Therefore, optimal chances are there to integrate EPNs with other pesticides that can increase EPN pathogenicity/effectiveness in proper IPM programmes (Grewal, 2012a,b). On the contrary, since there is another, but harmful, group of nematodes (i.e. phytonematodes), adding nematicides such as aldicarb and carbofuran with EPNs in an IPM programme should be avoided. EPNs can be used with some other chemicals by allowing proper time spacing between both applications. Similarly, some inert ingredients, additives or adjuvants used in EPN formulations can be toxic to nematodes (Krishnayya and Grewal, 2002), while others enhanced EPN efficacy significantly (Saleh *et al.*, 2015).

Improving EPN efficacy for prophylactic usage is another main challenge I see ahead of us in Egypt. This may be achieved successfully through slow-released formulations of EPNs, such as EPN-infected cadavers (Abd-Elgawad *et al.*, 2003). EPNs in traps formulated to lure and kill key pests have also been applied successfully (Grewal, 2012a). Moreover, EPN experimentation against serious pests of other commodities such as poultry and farm animal produce, for example *H. dromedrii* (El-Sadawy and Habeeb, 1998) and *B. annulatus* (El-Sadawy and Abdel-Shafy, 2007), showed promising results in Egypt, but with no commercial applications.

Eventually, recent advances in EPN research, especially in areas of expanding EPN conservation strategies (Duncan *et al.*, 2013), EPN identification and genomics (Hunt and Nguyen, 2016; DeLuca and Abd-Elgawad, Chapter 6, this volume), population ecology (see Hussaini, Chapter 5, this volume) and the molecular mechanism of EPN–bacteria symbiosis and pathogenicity (see Baiocchi *et al.*, Chapter 25, this volume) are expected to provide the needed support for more successful output and commercial application of EPNs in Egypt and elsewhere.

Acknowledgements

I would like to thank Drs Randy Gaugler, Larry Duncan, David Shapiro-Ilan, Raquel Campos-Herrera, Tarique Hassan Askary, Mohamed F.M. Eissa, Mostafa Hammam, Mahmoud Saleh and Nadia Dimetry for the information provided. This work was supported in part by US-Egypt Science and Technology Joint Fund (cycle 17; project no. 172) and Egyptian In-House, National Research Centre Project No 11030133 entitled 'Pesticide alternatives against soilborne pathogens attacking legume cultivation in Egypt'.

References

Abbas, M.S.T. and Saleh, M.M.E. (1998) Comparative pathogenicity of *Steinernema abbasi* and *S. riobravae* to *Spodoptera littoralis* (lepidoptera: noctuidae). *International Journal of Nematology* 8, 43–45.

Abd-Elgawad, M.M.M. (1996) The Indian red palm weevil: modernization of the pest control methods (in Arabic). *Agriculture and Development in the Arab Homeland* 15, 36–45.

Abd-Elgawad, M.M.M. (2001) Entomopathogenic nematode-host matching. *Egyptian Journal of Agronematology* 5, 91–104.

Abd-Elgawad, M.M.M. (2008a) The current status of phytonematode management in Egypt with special reference to applicable nematicides. *Egyptian Journal of Agronematology* 6, 33–46.

Abd-Elgawad, M.M.M. (2008b) Economic production of biopesticides using techniques adequate for local conditions. Basic Scientific Research Council, Academy of Scientific Research and Technology, Giza, Egypt. Final report (Febraury), 88 pp.

Abd-Elgawad, M.M.M. (2012) Developing bioinsecticide production technology through Egyptian–Chinese complementary approach. Convention projects for scientific and technological cooperation, Egyptian Academy of Scientific Research and Technology, Final report, 81 pp.

Abd-Elgawad, M.M.M. (2013) Phytonematode damage, economic threshold and management with special reference to Egypt. *Egyptian Journal of Agronematology* 12, 159–176.

Abd-Elgawad, M.M.M. (2014a) Yield losses by phytonematodes: challenges and opportunities with special reference to Egypt. *Egyptian Journal of Agronematology* 13, 75–94.

Abd-Elgawad, M.M.M. (2014b) Nematology education and training in Egypt and neighbouring countries. *Journal of Nematology* 46, 130–131.

Abd-Elgawad, M.M.M. (2014c) Spatial patterns of *Tuta absoluta* and heterorhabditid nematodes. *Russian Journal of Nematology* 22, 89–100.

Abd-Elgawad, M.M.M. (2014d) Integrated management of nematode pests and pre- and post-harvest diseases of economically important crops. A report of in-house National Research Centre Project No 10120604, 62 pp.

Abd-Elgawad, M.M.M. and Abdel-Razek, A.S. (2005) Susceptibility of the cotton leafworm, *Spodoptera littoralis* (Lepidoptera: Noctuidae) to different heterorhabditid nematodes and their bacterial symbionts. *Egyptian Journal of Agricultural Research* 2, 457–472.

Abd-Elgawad, M.M.M. and Abdel-Razek, A.S. (2008) Effect of entomopathogenic nematodes on populations of the diamondback moth, *Plutella xylostella* Linn. (Lepidoptera: Plutellidae) infesting cabbage. *Egyptian Journal of Agronematology* 6, 207–215.

Abd-Elgawad, M.M.M. and Aboul-Eid, H.Z. (2001) Effects of oxamyl, insect nematodes and *Serratia marcescens* on a polyspecific nematode community and yield of tomato. *Egyptian Journal of Agronematology* 5, 79–89.

Abd-Elgawad, M.M.M. and Aboul-Eid, H.Z. (2002) Effects of entomopathogenic nematodes on a polyspecific nematode community infecting watermelon plants in Egypt. *International Journal of Nematology* 12, 41–45.

Abd-Elgawad, M.M.M. and Ameen, H.H. (2005) *Heterorhabditis egyptii* n. sp. (Rhabditida: Heterorhabditidae) from Egypt. *Egyptian Journal of Agricultural Research* 2, 855–870.

Abd-Elgawad, M.M.M. and Askary, T.H. (2015) Impact of phytonematodes on agriculture economy. In: Askary, T.H. and Martinelli, P.R.P. (eds) *Biocontrol Agents of Phytonematodes*. CAB International, Wallingford, UK, pp. 3–49.

Abd-Elgawad, M.M.M. and Nguyen, K.B. (2007) Isolation, identification and environmental tolerance of new heterorhabditid populations from Egypt. *International Journal of Nematology* 17, 116–123.

Abd-Elgawad, M.M.M. and Spiridonov, S.E. (2014) Entomopathogenic nematode application in Egypt and Russia: challenges and opportunities. *Egyptian Journal of Agronematology* 13, 1–22.

Abd-Elgawad, M.M.M. and Vagelas, I.K. (2015) Nematophagous bacteria: field application and commercialization. In: Askary, T.H. and Martinelli, P.R.P. (eds) *Biocontrol Agents of Phytonematodes*. CAB International, Wallingford, UK, pp. 296–306.

Abd-Elgawad, M.M.M., Aboul-Eid, H.Z. and Mohamed, M.A. (2003) Entomopathogenic nematode virulence and persistence in relation to their enzymatic activity and formulations. *Egyptian Journal of Agricultural Research – The National Research Centre* 1, 663–674.

Abd-Elgawad, M.M.M., Mohamed, M.M.M. and El-Gamal, N.G.S. (2008) Development of safe chemical and biological formulations for control of nematodes in cucumber. *Egyptian Pharmaceutical Journal* 7, 41–50.

Abd-Elgawad, M., Mohamed, M., El-Mougy, N.S., El-Gamal, N. and Abdel-Kader, M.M. (2010) Protective treatments against soilborne pathogens in citrus orchards. *Journal of Plant Protection Research* 50, 512–519.

Abd-Elgawad, M.M.M., Jian H., Qiao, Y. and Hammam, M.M.A. (2013) Entomopathogenic nematodes and their potential against insects and phytonematodes in Egypt. In: Pridannikov, M.V., Udalova, Zh.V. and Butorina, N.N. (eds) *Proceedings of the 10th International Nematological Symposium*. Russian Society of Nematologists, Golitsyno – Bolshie Vyazemy, Moscow, Russia, pp. 98–101.

Abd-Elsalam, A.L. (1993) *Insect Pests in Egypt and Arab Countries and Their Controlling Methods* (in Arabic). Parts 1 and 2. Modem Egyptian Press, Cairo, 612 and 781 pp.

Abd El-Zaher, F.H. and Abd-Elgawad, M.M.M. (2012) Assessment of the antagonistic potential for the bacterial symbionts of entomopathogenic nematodes. *Journal of Applied Sciences Research* 8, 4590–4599.

Abdel-Kawy, A.M. (1985) Biological control of some lepidopterous pests by the nematode *Neoaplectana carpocapsae* and the physiological changes associated with it. PhD thesis. Faculty of Agriculture, Cairo University, Cairo, 117 pp.

Abdel-Kawy, A.M., El-Bishry, M.H. and El-Kifl, T.A.H. (1992) Controlling the leopard moth *Zeuzera pyrina* by three entomopathogenic nematode species in the field. *Faculty of Agriculture Bulletin, Cairo University* 43, 769–780.

Abdel-Kawy, A.M., Saleh, M.M.E. and El-Azzazi, A.M. (1996) Using entomopathogenic nematodes in the control of two apple borers. *Egyptian Journal of Biological Pest Control* 6, 147–151.

Abdel-Rahman, R.M., Abd El-Razzik, M.I., Osman, E.A. and El-Badawey, S.S. (2008) Laboratory evaluation of three Egyptian Heterorhabditid (Heterorhabditidae) and one imported Steinernematid nematodes (Steinernematidae) against mealybugs *Planococcus citri* (Risso) and *Icerya aegyptiaca* (Douglas) (Homoptera: Pseudococcidae and Margarodidae). *Egyptian Journal of Agricultural Research* 86, 659–668.

Abdel-Rahman, R.M., Abd El-Razzik, M.I., Osman, E.A. and El-Badawey, S.S. (2010) Enhancement effect of oleyl-polypeptide as an additive to entomopathogenic nematodes for controlling *Ferrisia virgata* (Cockerell). *Journal Egyptian–German Society of Zoology* 60 E, 1–13.

Abdel-Rahman, R.M., Abd El-Razzik, M.I., Osman, E.A. and Mangoud, A.A.H. (2012) Efficacy of the entomopathogenic nematodes and their based-product on some species of mealybugs (Hemiptera: Pseudococcidae) in Egypt. *Egyptian Academic Journal of Biological Sciences* 5, 193–196.

Abdel-Razek, A.S. and Abd-Elgawad, M.M.M. (2007) Investigation on the efficacy of entomopathogenic nematodes against *Spodoptera littoralis* (Biosd.) and *Galleria mellonella* (L.). *Archives of Phytopathology and Plant Protection* 40, 414–422.

Abdel-Razek, A.S. and Abd-Elgawad, M.M.M. (2013) Effects of entomopathogenic nematodes on suppressing hairy rose beetle, *Tropinota squalida* Scop. (Coleoptera: Scarabaeidae) population in cauliflower field in Egypt. *International Journal of Biological, Life Science and Engineering* 7, 28–32.

Abo Rehab, M.E.A., Korra, A.K.M., Kamhawy, M.A.M. and Youssef, K.Y.A (2013) Fungal species associated with graft union on grapevine, its impact on graft failure process and attempted solutions in Egypt. *International Journal of Agriculture and Forestry* 3, 52–59.

Abouelhag, H.A. and El-Sadawy, H.A. (2012) Natural associations between symbionts *Photorhabdus* spp. and *Xenorhabdus* spp. and bacteria related to *Ochrobactrum anthropi*, *Bacillus pumilus* and *Enterobacter cloacae*. *Life Science Journal* 9, 4783–4790.

Abu-Shady, N.M., Shamseldean, M.M., Abd-Elbary, N.A. and Stock, S.P. (2011) Diversity and distribution of entomopathogenic nematodes (Heterorhabditidae and Steinernematidae) in Egypt. *Journal of Nematology* 43, 224.

Altieri, M.A. (1999) The ecological role of biodiversity in agroecosystems. *Agriculture, Ecosystems & Environment* 74, 19–31.

Alumai, A. and Grewal, P.S. (2004) Tankmix-compatibility of the entomopathogenic nematodes, *Heterorhabditis bacteriophora* and *Steinernema carpocapsae*, with selected chemical pesticides used in turfgrass. *Biocontrol Science and Technology* 14, 725–730.

Anon. (2012) *Recommendations Adopted to Combat Agricultural Pests*. Agricultural Pesticides Committee, Egyptian Ministry of Agriculture and Land Reclamation,

Media Support Center Press, Dekernes, Dakahlia, Egypt (in Arabic), 319 pp.

Anon. (2014) *Recommendations Adopted to Combat Agricultural Pests*. Agricultural Pesticides Committee, Egyptian Ministry of Agriculture and Land Reclamation, Media Support Center Press, Dekernes, Dakahlia, Egypt (in Arabic), 296 pp.

Ansari, M.A., Shah, F.A. and Butt, T.M. (2008) Combined use of entomopathogenic nematodes and *Metarhizium anisopliae* as a new approach for black vine weevil, *Otiorhynchus sulcatus*, control. *Entomologia Experimentalis et Applicata* 129, 340–347.

Atwa, A.A. (2013) Susceptibility of *Spodoptera littoralis* (Boisd.) to treated entomopathogenic rhabdities, *Heterorhabditis bacteriophora* and *Steinernema* sp. by different pesticides. *Journal of Biopesticides* 6, 149–159.

Atwa, A.A., Shamseldean, M.M. and Yonis, F.A. (2013) The effect of different pesticides on reproduction of entomopathogenic nematodes. *Turkish Journal of Entomology* 37, 493–502.

Azazy, A.M. and Mousa, G.M. (2005) Biological control of cabbage webworm, *Hellula undalis* and cabbage worm, *Artogia rapae* using entomopathogenic nematodes on cabbage seedlings. *Third International Conference for Pest Control*. Plant Protection Research Institute, Agricultural Research Centre, Giza, Egypt, 11 pp.

Bai, X., Adams, B.J., Ciche, T.A., Clifton, S., Gaugler, R., *et al.* (2013) A lover and a fighter: the genome sequence of an entomopathogenic nematode *Heterorhabditis bacteriophora*. *PLoS ONE* 8(7), e69618, DOI: 10.1371/journal.pone.0069618.

Bal, H.K., Taylor, R.A.J. and Grewal, P.S. (2014) Ambush foraging entomopathogenic nematodes employ 'sprinters' for long-distance dispersal in the absence of hosts. *Journal of Parasitology* 100, 422–432.

Baur, M.E., Kaya, H.K., Gaugler, R. and Tabashnik, B. (1997) Effects of adjuvants on entomopathogenic nematode persistence and efficacy against *Plutella xylostella*. *Biocontrol, Science and Technology* 7, 513–525.

Baur, M.E., Kaya, H.K., Tabashnik, B.E. and Chilcutt, C.F. (1998) Suppression of diamondback moth (Lepidoptera: Plutellidae) with an entomopathogenic nematode (Rhabditida: Steinernematidae) and *Bacillus thuringiensis* Berliner. *Journal of Economic Entomology* 91, 1089–1095.

Bednarek, A. and Gaugler, R. (1997) Compatibility of soil amendments with entomopathogenic nematodes. *Journal of Nematology* 29, 220–227.

Boemare, N.E., Akhurst, R.J. and Mourant, R.G. (1993) DNA relatedness between *Xenorhabdus* spp. (Enterobacteriaceae), symbiotic bacteria of entomopathogenic nematodes, a proposal to transfer *Xenorhabdus luminescens* to a new genus, *Photorhabdus* gen. nov. *International Journal of Systematic Bacteriology* 43, 249–255.

Brown, I.M. and Gaugler, R. (1995) Cold tolerance of steinernematid and heterorhabditid nematodes. *Journal of Thermal Biology* 23, 75–80.

Campbell, J.F. and Gauger, R. (1993) Nictation behaviour and its ecological implications in the host search strategies of entomopathogenic nematodes (Heterorhabditidae and Steinernematidae). *Behaviour* 126, 155–169.

Campbell, J.F. and Kaya, H.K. (1999) How and why a parasitic nematode jumps. *Nature* 397, 485–486.

Campos-Herrera, R., Stuart, R.J., EL-Borai, F.E., Gutiérrez, C. and Duncan L.W. (2010) Entomopathogenic nematode ecology and biological control in Florida citrus. In: Ciancio, A. and Mukerji, K.G. (eds) *Integrated Management of Arthropod Pests and Insect Borne Diseases*, Vol 5. Springer Science+Business Media BV, New York, pp. 101–130.

Campos-Herrera, R., Pathak, E., El-Borai, F.E., Schumann, A., Abd-Elgawad, M.M.M., and Duncan, L.W. (2013) New citriculture system suppresses native and augmented entomopathogenic nematodes. *Biological Control* 66, 183–194.

Cappaert, D.L. and Koppenhöfer, A.M. (2003) *Steinernema scarabaei*, an entomopathogenic nematode for control of the European chafer. *Biological Control* 28, 379–386.

Cruz-Martínez, H., Ruiz-Vega, J., Matadamas-Ortíz, P.T., Cortés-Martínez, C.I. and Rosas-Diaz, J. (2017) Formulation of entomopathogenic nematodes for crop pest control: a review. *Plant Protection Science* 53, 15–24.

Denno, R.F., Gruner, D.S. and Kaplan, I. (2008) Potential for entomopathogenic nematodes in biological control: a meta-analytical synthesis and insights from trophic cascade theory. *Journal of Nematology* 40, 61–72.

Dillman, A.R., Guillermin, M.L., Lee, J.H., Kim, B., Sternberg, P.W. and Hallem, E.A. (2012) Olfaction shapes host–parasite interactions in insect-parasitic nematodes. *Proceedings of the National Academy of Sciences of the United States of America* 109, E2324–E2333.

Dillon, A.B., Ward, D., Downes, M.J. and Griffin, C.T. (2006) Suppression of the large pine weevil *Hylobius abietis* (L.) (Coleoptera: Curculionidae) in pine stumps by entomopathogenic nematodes with different foraging strategies. *Biological Control* 38, 217–226.

Dimetry, N.Z., Ismail, I.A., El-Gengaihi, S. and Saleh, M.M.E. (1996) The use of extracts of locally available plants against the greasy cutworm, *Agrotis ipsilon* (Hufn.) and their side effects on the entomopathogenic nematode, *Steinernema carpocapsae*. *Egyptian Journal of Biological Pest Control* 6, 115–120.

Divya, K. and Sankar, M. (2009) Entomopathogenic nematodes in pest management. *Indian Journal of Science and Technology* 2, 53–60.

Dolinski, C., Choo, H.Y. and Duncan, L.W. (2012) Grower acceptance of entomopathogenic nematodes: case studies on three continents. *Journal of Nematology* 44, 226–235.

Dowds, B.C.A. and Peters, A. (2002) Virulence mechanisms. In: Gaugler, R. (ed.) *Entomopathogenic Nematology*. CAB International, Wallingford, UK, pp. 79–98.

Dreves, A.J. and Lee, J. (2015) Pacific Northwest insect management handbook: entomopathogenic-nematodes. Available at: http://pnwhandbooks.org/insect/ipm/entomopathogenic-nematodes (accessed 12 October 2016).

Duncan, L.W., Graham, J.H., Dunn, D.C., Zellers, J., McCoy, C.W. and Nguyen, K. (2003) Incidence of endemic entomopathogenic nematodes following application of *Steinernema riobrave* for control of *Diaprepes abbreviatus*. *Journal of Nematology* 35, 178–186.

Duncan, L.W., Graham, J.H., Zellers, J., Bright, D., Dunn, D.C., *et al.* (2007) Food web responses to augmenting the entomopathogenic nematodes in bare and animal manure-mulched soil. *Journal of Nematology* 39, 176–189.

Duncan, L.W., Stuart, R.J., El-Borai, F.E., Campos-Herrera, R., Pathak, E., *et al.* (2013) Modifying orchard planting sites conserves entomopathogenic nematodes, reduces weevil herbivory and increases citrus tree growth, survival and fruit yield. *Biological Control* 64, 26–36.

Edwards, C.A. (1993) The impact of pesticides on the environment. In: Pimentel, D. and Lehman, H. (eds) *The Pesticide Question, Environment, Economics and Ethics*. Chapman and Hall, New York, pp. 13–46.

El-Bishry, M.H., El-Sebay, Y. and Al-Elimi, M.H. (2000) Impact of the environment in date palm infested with *Rhynchophorus ferrugineus* on five entomopathogenic nematodes (Rhabditida). *International Journal of Nematology* 10, 75–80.

El-Kifl, T.A.H. (1980) Utilization of the nematode *Neoaplectana carpocapsae* in the biological control of cotton leafworm, *Spodoptera littoralis*. MSc thesis. Faculty of Agriculture, Cairo University, Egypt.

El-Kifl, T.A.H. (1984) Factors affecting potentialities of entomogenous nematodes in the biological control of insect pests. PhD thesis. Faculty of Agriculture, Cairo University, Egypt.

El-Kifl, T.A.H. and Sammour, E.A. (1989) Possible use of the endoparasitic nematode *Neoaplectana caprocpsae* (Weiser) combined with the insecticide hostathion for controlling the cutworm, *Agrotis ipsilon* (Hufn). *Bulletin of the Entomological Society of Egypt, Economic Series* 17, 215–223.

El-Sadawy, H.A. and Abdel-Shafy, S. (2007) Laboratory and field studies on entomopathogenic nematodes as a biocontrol agent for the cattle tick *Boophilus annulatus* (Acari: Ixodidae). *Acarologia* 47, 25–31.

El-Sadawy, H.A.E. and Habeeb, S.M. (1998) Testing some entomopathogenic nematodes for the biocontrol of *Hyalomma dromedrii* Koch (Acarina: Ixodidae). *Journal of the Union of Arab Biologists – Cairo* 10A, 1–11.

Fetoh, B.A. and Azzazy, A.M. (2004) Integration of three biocontrol agents for the control of the cabbage worm, *Pieris rapae*. *Egyptian Journal of Biological Pest Control* 14, 225–230.

Fetoh, B.A., Khaled, A.S. and El-Nagar, T.F.K. (2009) Combined effect of entomopathogenic nematodes and biopesticides to control the greasy cut worm, *Agrotis ipsilion* (Hufn.) in the strawberry fields. *Egyptian Academic Journal of Biological Sciences* 2, 227–236.

Gaugler, R. (ed.) (2002) *Entomopathogenic Nematology*. CAB International, Wallingford, UK.

Gaugler, R. and Kaya, H.K. (eds) (1990) *Entomopathogenic Nematodes in Biological Control*. CRC Press, Boca Raton, Florida.

Gaugler, R., Lewis, E. and Stuart, R.J. (1997) Ecology in the service of biological control: the case of entomopathogenic nematodes. *Oecologia* 109, 483–489.

Gaugler, R., Brown, I., Shapiro-Ilan, D.I. and Atwa, A. (2002) Automated technology for *in vivo* mass production of entomopathogenic nematodes. *Biological Control* 24, 199–206.

Georgis, R. (2002) The Biosys Experiment: an insider's perspective. In: Gaugler, R. (ed.) *Entomopathogenic Nematology*. CAB International, Wallingford, UK, pp. 357–372.

Grewal, P. (2002) Formulation and application technology. In: Gaugler, R. (ed.) *Entomopathogenic Nematology*. CAB International, Wallingford, UK, pp. 265–288.

Grewal, P.S. (2012a) Entomopathogenic nematodes as tools in integrated pest management. In: Abrol, D.P. and Shankar, U. (eds) *Integrated Pest Management: Principles and Practice*. CAB International, Wallingford, UK, pp. 162–236.

Grewal, P.S. (2012b) From IPM to ecosystem management: the case of urban lawn. In: Abrol, D.P. and Shankar, U. (eds) *Integrated Pest Management: Principles and Practice*. CAB International, Wallingford, pp. 450–488.

Grewal, P.S. and Georgis, R. (1999) Entomopathogenic nematodes. In: Hall, F.R. and Menn, J.J. (eds) *Methods in Biotechnology*, Vol 5: *Biopesticides: Use and Delivery*. Humana Press, Totowa, New Jersey, pp. 271–299.

Grewal, P.S., Weber, T.A. and Betterley, D.A. (1998) Compatibility of *Steinernema feltiae* with chemicals used in mushroom production. *Mushroom News* 46, 6–10.

Grewal, P.S., Ehlers, R.-U. and Shapiro-Ilan, D.I. (eds) (2005a) *Nematodes as Biocontrol Agents*. CAB International, Wallingford, UK.

Grewal, P.S., Ehlers, R.-U. and Shapiro-Ilan, D.I. (2005b) Critical issues and research needs for expanding the use of nematodes in biocontrol. In: Grewal, P.S., Ehlers, R.-U. and Shapiro-Ilan, D.I. (eds) *Nematodes as Biocontrol Agents*. CAB International, Wallingford, UK, pp. 479–489.

Griffin, C.T. (2012) Perspectives on the behavior of entomopathogenic nematodes from dispersal to reproduction: traits contributing to nematode fitness and biocontrol efficacy. *Journal of Nematology* 44, 177–184.

Gutiérrez, C., Campos-Herrera, R., Jiménez, J., Escuer, M., Labrador, S. and Rivera, Y. (2005) Control de plagas en agricultura ecológica: eficacia de los nematodos entomopatógenos. (Pest control in ecological agriculture: efficacy of entomopathogenic nematodes). *Alimentaría* (April) 4, 62–67.

Hallem, E.A., Dillman, A.R., Hong, A.V., Zhang, Y., Yano, J.M., *et al.* (2011) A sensory code for host seeking in parasitic nematodes. *Current Biology* 21, 377–383.

Horowitz, A.R. and Ishaaya, I. (2004) *Insect Pest Management: Field and Protected Crops*. Springer, Berlin.

Hunt, D.J. and Nguyen, K.B. (2016) *Advances in Entomopathogenic Nematode Taxonomy and Phylogeny*, Vol. 12, Nematology Monographs and Perspectives. Brill, Leiden, The Netherlands, 437 pp.

Hussein, M.A. (2004) Utilization of entomopathogenic nematodes for biological control of some lepidopterous pests. PhD thesis. Faculty of Science, Ain Shams University, Cairo.

Hussein, M.A. and Abdel-Aty, M.A. (2012) Formulation of two native entomopathogenic nematodes at room temperature. *Journal of Biopesticides* 5(S), 23–26.

Hussein, M.A. and Abou El-Souud, A.B. (2006) Isolation and characterization of two heterorhabditid and one steinernematid nematodes from Egypt. *International Journal of Nematology* 16, 7–12.

Ibrahem, H.I.H. (2015) Entomopathogenic nematodes as biocontrol agents of *Tuta absoluta* on tomato plants. MSc thesis. Faculty of Agriculture, Cairo University, Egypt.

Jacobson, R.J. and Martin, G.A. (2011) A potential role for entomopathogenic nematodes within IPM of *Tuta absoluta* (Meyrick) on organic tomato crops. *IOBC/WPRS Bulletin* 68, 71–74.

Jagdale, G.B., Kamoun, S. and Grewal, P.S. (2009) Entomopathogenic nematodes induce components of systemic resistance in plants: biochemical and molecular evidence. *Biological Control* 51, 102–109.

Kaya, H.K. and Gaugler, R. (1993) Entomopathogenic nematodes. *Annual Review of Entomology* 38, 181–206.

Kaya, H.K. and Koppenhöfer, A.M. (1996) Effects of microbial and other antagonistic organism and competition on entomopathogenic nematodes. *Biocontrol Science and Technology* 6, 357–372.

Keila, A.M., Shamseldean, M.M. and Bekhiet, M.A. (2008) Evaluation of entomopathogenic nematodes as biocontrol agents of the root-knot nematode, *Meloidogyne javanica*. *Egyptian Journal of Agronematology* 6, 283–301.

Koppenhöfer, A.M. and Fuzy, E.M. (2008) Early timing and new combinations to increase the efficacy of neonicotinoid entomopathogenic nematode (Rhabditida: Heterorhabditidae) combinations against white grubs (Coleoptera: Scarabaeidae). *Pest Management Science* 64, 725–735.

Koppenhöfer, A.M. and Grewal, P.S. (2005) Compatibility and interactions with agrochemicals and other biocontrol agents. In: Grewal, P.S., Ehlers, R.U. and Shapiro-Ilan, D.I. (eds) *Nematodes as Biocontrol Agents*. CAB International, Wallingford, UK, pp. 363–381.

Koppenhöfer, A.M. and Kaya, H.K. (1997) Additive and synergistic interaction between entomopathogenic nematodes and *Bacillus thuringiensis* for scarab grub control. *Biological Control* 8, 131–137.

Koppenhöfer, A.M. and Kaya, H.K. (1998) Synergism of imidacloprid and an entomopathogenic nematode: a novel approach to white grub (Coleoptera: Scarabaeidae) control in turfgrass. *Journal of Economic Entomology* 91, 618–623.

Koppenhöfer, A.M., Brown, I.M., Gaugler, R., Grewal, P.S., Kaya, H.K. and Klein, M.G. (2000) Synergism of entomopathogenic nematodes and imidacloprid against white grubs: greenhouse and field evaluation. *Biological Control* 19, 245–251.

Krishnayya, P.V. and Grewal, P.S. (2002) Effect of neem and selected fungicides on viability and virulence of the entomopathogenic nematode *Steinernema feltiae*. *Biocontrol Science and Technology* 12, 259–266.

Kulkarni, N., Paunikar, S. and Mishra, V.K. (2013) Tolerance of entomopathogenic nematode, *Steinernema carpocapsae* to some modern insecticides and biopesticides. *Annals of Entomology* 31, 129–134.

Lacey, L.A. and Georgis, R. (2012) Entomopathogenic nematodes for control of insect pests above and below ground with comments on commercial production. *Journal of Nematology* 44, 218–225.

Lacey, L.A. and Shapiro-Ilan, D.I. (2008) Microbial control of insect pests in temperate orchard systems: potential for incorporation into IPM. *Annual Review of Entomology* 53, 121–144.

Laznik, Ẑ. and Trdan, S. (2013) The influence of insecticides on the viability of entomopathogenic nematodes (Rhabditida: Steinernematidae and Heterorhabditidae) under laboratory conditions. *Pest Management Science* 70, 784–789.

Laznik, Ẑ., Vidrih, M. and Trdan, S. (2012) Effect of different fungicides on viability of entomopathogenic nematodes *Steinernema feltiae* (Filipjev), *S. carpocapsae* Weiser and *Heterorhabditis downesi* Stock, Griffin and Burnell (Nematoda: Rhabditida) under laboratory conditions. *Chilean Journal of Agricultural Research* 72, 62–67.

Lello, E.R., Patel, M.N., Matthews, G.A. and Wright, D.J. (1996) Application technology for entomopathogenic nematodes against foliar pests. *Crop Protection* 15, 567–574.

Lewis, E.E. (2002) Behavioral ecology. In: Gaugler, R. (ed.) *Entomopathogenic Nematology*. CAB International, Wallingford, UK, pp. 205–223.

Lewis, E.E., Gaugler, R. and Harrison, R. (1992) Entomopathogenic nematode host finding: response to host contact cues by cruise and ambush foragers. *Parasitology* 105, 309–319.

Lewis, E.E., Campbell, J.F. and Gaugler, R. (1998) A conservation approach to using entomopathogenic nematodes in turf and landscapes. In: Barbosa, P. (ed.) *Conservation Biological Control*. Academic Press, New York, pp. 235–254.

Lewis, E.E., Campbell, J., Griffin, C., Kaya, H. and Peters, A. (2006) Behavioral ecology of entomopathogenic nematodes. *Biological Control* 38, 66–79.

Nishimatsu, T. and Jackson, J.J. (1998) Interaction of insecticides, entomopathogenic nematodes, and larvae of the western corn rootworm (Coleoptera: Chrysomelidae). *Journal of Economic Entomology* 91, 410–418.

Nyasani, J.O., Kimenju, J.W., Olubayo, F.M. and Wilson, M.J. (2008) Laboratory and field investigations using indigenous entomopathogenic nematodes for biological control of *Plutella xylostella* in Kenya. *International Journal of Pest Management* 54, 355–361.

Ornat, C., Abd-Elgawad, M.M.M. and Sorribas, F.J. (2011) Native isolates of entomopathogenic nematodes in Spain and Egypt and development of *Meloidogyne javanica* on tomato. *Nematropica* 41, 411.

Peshin, R. and Pimentel, D. (2014) *Integrated Pest Management; Experiences with Implementation, Global Overview*, Vol 4. Springer, Dordrecht, The Netherlands.

Pimentel, D. and Peshin, R. (2014) *Integrated Pest Management: Pesticide Problems*, Vol 3. Springer Science+Business Media, New York.

Polavarapu, S., Koppenhöfer, A.M., Barry, J.D., Holdcraft, R.J. and Fuzy, E.M. (2007) Entomopathogenic nematodes and neonicotinoids for remedial control of oriental beetle, *Anomala orientalis* (Coleoptera: Scarabaeidae), in highbush blueberry. *Crop Protection* 26, 1266–1271.

Ragab, Z.A. and Saleh, M.M.E. (2000) Control of *Agrotis ipsilon* (Lepidoptera: Noctuidae) on corn plants with entomopathogenic nematodes (Nematoda: Steinernematidae, Heterorhabditidae). *Egyptian Journal of Applied Science* 15, 232–238.

Rechcigl, J.E. and Rechcigl, N.A. (2000) *Insect Pest Management: Techniques for Environmental Protection*. Lewis Publishers, Boca Raton, Florida.

Rovesti, L. and Deseo, K.V. (1990) Compatibility of chemical pesticides with entomopathogenic nematodes *Steinernema carpocapsae* Weiser and *S. feltiae* Filipjev (Nematoda: Steinernematidae). *Nematologica* 36, 237–245.

Saker, M.M., Salama, H.S., Salama, M., El-Banna, A. and Abdel Ghany, N.M. (2011) Production of transgenic tomato plants expressing Cry 2Ab gene for the control of some lepidopterous insects endemic in Egypt. *Journal of Genetic Engineering and Biotechnology* 9, 149–155.

Salama, H.S. and Abd-Elgawad, M.M.M. (2001) Isolation of heterorhabditid nematodes from palm tree planted areas and their implication in the red palm weevil control. *Anzeiger für Schädlingskunde Pflanzenschutz Umweltschutz* 74, 43–45.

Salama, H.S. and Abd-Elgawad, M.M.M. (2002) Activity of heterorhabditid nematodes at high temperature and in combination with cytoplasmic polyhedrosis virus. *Anzeiger für Schädlingskunde Pflanzenschutz Umweltschutz* 75, 78–80.

Saleh, M.M.E (1995) Efficiency of the Egyptian entomopathogenic nematode, *Heterorhabditis taysearae* in controlling the cabbage-worm, *Pieris rapae* (L.) (Lepidoptera: Pieridae). *Egyptian Journal of Biological Pest Control* 5, 99–112.

Saleh, M.M.E. and Elkifl, T.A.H. (1994) Virulence of certain entomopathogenic nematodes to the European corn borer, *Ostrinia nubilalis*. *Egyptian Journal of Biological Pest Control* 4, 125–131.

Saleh, M.M.E. and Ragab, Z.A. (1999) Susceptibility of *Spodoptera littoralis* (Boisd) and *Agrotis ipsilon* (Hufn.) larvae to Egyptian and imported entomopathognic nematodes. *Egyptian Journal of Applied Sciences* 14, 213–223.

Saleh, M.M.E. and Sammour, E.A. (1995) Interactions of three insecticides and two entomopathogenic nematodes against *Spodoptera littoralis* (Boised) larvae. *Egyptian Journal of Biological Pest Control* 6, 71–74.

Saleh, M.M.E., Matter, M.M. and Hussein, M.A. (1999) Efficiency of entomopathogenic nematodes in controlling *Sesamia cretica* (Lepidoptera: Noctuidae) in Egypt. *Bulletin of the National Research Centre, Egypt* 25, 181–188.

Saleh, M.M.E, Draz, K.A.A., Mansour, M.A., Hussein, M.A. and Zawrah, M.F.M. (2009) Controlling the sugar beet beetle *Cassida vittata* with entomopathogenic nematodes. *Journal of Pest Science* 82, 289–294.

Saleh, M.M.E., Hussein, M.A., Hafez, G.A., Hussein, M.A., Salem, H.A. and Metwally, H.M.S. (2015) Foliar application of entomopathogenic nematodes for controlling *Spodoptera littoralis* and *Agrotis ipsilon* (Lepidoptera: Noctuidae) on corn plants. *Advances in Applied Agricultural Science* 3, 51–61.

Schroer, S. and Ehlers, R.-U. (2005) Foliar application of the entomopathogenic nematode *Steinernema carpocapsae* for biological control of diamondback moth larvae (*Plutella xylostella*). *Biological Control* 33, 81–86.

Schroer, S., Sulistydanto, D. and Ehlers, R.-U. (2005) Control of *Plutella xylostella* using polymer-formulated *Steinernema carpocapsae* and *Bacillus thuringiensis* in cabbage fields. *Journal of Applied Entomology* 129, 128–204.

Shamseldean, M.M. (2000) Laboratory bioassays and field applications of entomopathogenic nematode isolates and/or species belong to the genera *Heterorhabditis* and *Steinernema* against the leopard moth *Zeuzera pyrina* L. (Lepidoptera: Cossidae). *Journal of Agricultural Science, Mansoura University* 25, 8143–8150.

Shamseldean, M.M. (2004) Laboratory trials and field applications of Egyptian and foreign entomopathogenic nematodes used against the red palm weevil, *Rhynchophorus ferrugineus* (Oliv.). *International Journal of Nematology* 14, 51–63.

Shamseldean, M.M. (2010) Nematodes and their use in biological control of insect pests in the Arab countries (in Arabic). In: Abu-Gharbieh, W.A., Al-Hazmi, A.S.,

Stephan, Z.A. and Dawabah, A.A. (eds) *Plant Nematodes in Arab Countries*. Darwael for Publishing, Arab Society of Plant Protection, Amman, pp. 927–969.

Shamseldean, M.M. and Abd-Elgawad, M.M. (1994a) Natural occurrence of insect pathogenic nematodes (Rhabditida: Heterorhabditidae) in Egyptian soils. *Afro-Asian Journal of Nematology* 4, 151–154.

Shamseldean, M.M. and Abd-Elgawad, M.M. (1994b) Laboratory evaluation of six Egyptian isolates of heterorhabditid nematodes for control of the red palm weevil. *Egyptian Journal of Applied Science* 9, 670–679.

Shamseldean, M.M. and Abd-Elgawad, M.M. (1995) Enhanced biological control potential of entomopathogenic nematodes through genetic engineering for thermotolerance. NARP collaborative project # H-035, Egyptian Ministry of Agriculture, Giza. Final report, 72 pp.

Shamseldean, M.M. and Atwa, A.A. (2004a) Laboratory and field tests of entomopathogenic nematodes against the scarab beetle *Temnorhynchus baal* (Reiche) a novel insect pest of strawberry in Egypt. *Egyptian Journal of Biological Pest Control* 14, 127–133.

Shamseldean, M.M. and Atwa, A.A. (2004b) Virulence of Egyptian Steinernematid nematodes used against the red palm weevil, *Rhynchophorusferrugineus* (Oliv.) *Egyptian Journal of Biological Pest Control* 14, 135–140.

Shamseldean, M.M., Abd-Elgawad, M.M. and El-Bishry, M.H. (1994) Comparative infectivity of native heterorhabditid nematodes and the potential use of one isolate against some economically important pests in Egypt. *Egyptian Journal of Biological Pest Control* 4, 133–139.

Shamseldean, M.M., Abou El-Sooud, A.B. and Saleh, M.M.E. (1996) Isolation of *Steinernema carpocapsae* (Waiser, 1955) Poinar, 1990 (Steinernematidae), *Heterorhabditis bacteriophora* Poinar 1976 and *Heterorhabditis indicus* Poinar *et al.*, 1992 (Heterorhabditidae) as first record from Egypt. *Egyptian Journal of Biological Pest Control* 6, 187–201.

Shamseldean, M.M., Abd-Elgawad, M.M. and Atwa, A.A. (1999) Factors affecting pathogenicity of an Egyptian strain of *Heterorhabditis indicus* infecting cotton leafworm, *Spodoptera littoralis*. *International Journal of Nematology* 9, 90–94.

Shamseldean, M.M., Keila, A.M., Bekhiet, M.A. and Anany, A.E. (2010) Evaluation of the entomopathogenic nematodes, *Steinernema* spp. and *Heterorhabditis bacteriophora* in the control of mole cricket, *Gryllotalpa africana*. *Minufiya Journal of Agricultural Research* 35, 1849–1857.

Shapiro-Ilan, D.I. and Gaugler, R. (2015) Nematodes (Rhabditida: Steinernematidae & Heterorhabditidae). In: Shelton, A. (ed.) *Biological Control: A Guide to Natural Enemies in North America.* Available at: https://biocontrol.entomology.cornell.edu/pathogens/nematodes.php (accessed 15 March 2015).

Shapiro-Ilan, D.I., Gouge, D.H. and Koppenhofer, A.M. (2002) Factors affecting commercial success: case studies in cotton, turf and citrus. In: Gaugler, R. (ed.) *Entomopathogenic Nematology*. CAB International, New York, pp. 333–356.

Shapiro-Ilan, D.I., Han, R. and Dolinksi, C. (2012) Entomopathogenic nematode production and application technology. *Journal of Nematology* 44, 206–217.

Somvanshi, V.S., Ganguly, S. and Paul, A.V.N. (2006) Field efficacy of the entomopathogenic nematode *Steinernema thermophilum* Ganguly and Singh (Rhabditida: Steinernematidae) against diamondback moth (*Plutella xylostella* L.) infesting cabbage. *Biological Control* 37, 9–15.

Stuart, R.J. and Gaugler, R. (1996) Genetic adaptation and founder effect in laboratory populations of the entomopathogenic nematode *Steinernema glaseri*. *Canadian Journal of Zoology* 74, 164–170.

Stuart, R.J., Barbercheck, M.E., Grewal, P.S., Taylor, R.A.J. and Hoy, C.W. (2006) Population biology of entomopathogenic nematodes: concepts, issues and models. *Biological Control* 38, 80–102.

Stuart, R.J., El-Borai, F.E. and Duncan, L.W. (2008) From augmentation to conservation of entomopathogenic nematodes: trophic cascades, habitat manipulation and enhanced biological control of *Diaprepes abbreviatus* root weevils in Florida citrus groves. *Journal of Nematology* 40, 7384.

Timmer, L.W. and Duncan, L.W. (1999) *Citrus Health Management*. APS Press, St Paul, Minnesota.

Wharton, D.A. and Surry, M.R. (1994) Cold tolerance mechanisms of the infective larvae of the insect parasitic nematode *Heterorhabditis zealandica* Poinar. *Cryo Letters* 15, 353–360.

Wharton, D.A., Young, S.R. and Barrett, J. (1984) Cold tolerance in nematodes. *Journal of Comparative Physiology* 154, 73–77.

Womersley, C.Z. (1990) Dehydration survival and anhydrobiotic potential. In: Gaugler, R. and Kaya, H.K. (eds) *Entomopathogenic Nematodes in Biological Control*. CRC Press, Boca Raton, Florida, pp. 117–137.

Woodring, J.L. and Kaya, H.K. (1988) Steinernematid and heterorhabditid nematodes: a handbook of techniques. *Southern Cooperative Series Bulletin* No 331, Arkansas Agricultural Experiment Station, Fayetteville, Arkansas, 30 pp.

Part V: Genetics for Enhancing Efficacy in Entomopathogenic Nematodes

25 Genetic Improvement of Entomopathogenic Nematodes for Enhanced Biological Control

Tiffany Baiocchi,[1] Mahfouz M.M. Abd-Elgawad[2] and Adler R. Dillman[3]*

[1]*Department of Biochemistry, University of California, Riverside, California, USA;* [2]*Phytopathology Department, National Research Centre, Giza, Egypt;* [3]*Department of Nematology, University of California, Riverside, California, USA*

25.1 Introduction

The world loses 32.1%, on average, of all crops due to herbivory by pests (mostly insects), viruses, weeds and crop diseases, with insect pests being a major driver of this loss (Oerke, 2006). Insecticidal nematodes of the genera *Heterorhabditis* and *Steinernema* are highly pathogenic and are used as biocontrol agents of numerous insect pests. Due to the many desirable traits of these entomopathogenic nematodes (EPNs), they have been commercialized on several continents. However, lack of consistent efficacy in the field prevents these nematodes from being more widely used; thus, researchers have been working to enhance their efficacy against arthropod pests under field conditions for decades. Two main strategies have been employed to improve EPN field efficacy and increase their usefulness in agriculture. The first is to isolate and identify new EPN species and/or populations (*e.g.*, Hunt and Nguyen, 2016). One line of reasoning is that endemic EPN species or populations will be adapted to local environmental conditions and to native pests, and may provide superior control compared to newly introduced species or populations of EPNs (Gaugler, 1988; Gaugler *et al.*, 1997a; Hiltpold, 2015). Considerable efforts have been spent surveying to extract new EPN isolates, which may lead to increased genetic variation and to the development of new nematode strains. Consequently, the first strategy to improve EPN efficacy relies on isolation and/or breeding of EPNs for enhanced insect pest suppression (see Subramanian and Muthulakshmi, Chapter 26, this volume). Traits typically targeted by this approach include increased tolerance to temperature extremes, ultraviolet light and desiccation, as well as higher host-seeking ability, virulence and resistance to nematicides. Improving these traits in EPNs has been done primarily by classical genetic techniques such as breeding and selection. However, the improved trait is not always stable, and trait gains can sometimes be lost once the selective pressure is removed (Glazer, 2014). Moreover, selection of some traits can sometimes lead to inadvertent reduction of others, or even loss of fitness (Gaugler *et al.*, 1989, 1990; Burnell, 2002). There is also the concern of inbreeding depression or other means of fitness loss during EPN mass production or as a result of continuous laboratory culture (Bilgrami *et al.*, 2006; Chaston *et al.*, 2011). Hence, the second major EPN improvement strategy, which is to use modern genetic and molecular tools to improve certain traits through genetic engineering. These tools have not yet been fully employed to improve EPN field efficacy in biological control, but genetic engineering remains a major strategy none the less (Burnell, 2002; Glazer, 2014, 2015). Progress has been made toward tool development and technology transfer from the *Caenorhabditis elegans* community,

*Corresponding author. E-mail: adlerd@ucr.edu

but the application of modern techniques to improve EPN efficacy is still in its infancy.

EPNs have been studied for and used in biocontrol programmes for more than 70 years. *Steinernema glaseri* was described in 1929 (Steiner, 1929) and was heavily studied for its potential to control the Japanese beetle (*Popillia japonica*). This nematode was quickly adopted by state-wide programmes in the 1940s and 1950s (Girth *et al.*, 1940; Glaser *et al.*, 1940; Cory and Langford, 1944, 1955; Fleming, 1968). *Steinernema carpocapsae* was identified as one of the most promising biocontrol agents because of its high virulence, broad host range, amenability to mass production on a commercial scale, and because it is exempt from government registration (Gaugler *et al.*, 1989). Many of the factors that make *S. carpocapsae* appealing as a biocontrol agent are also true of other steinernematid and heterorhabditid species. But clearly understanding the biology of both the pests to be controlled and the potential biocontrol agents is important. For example, biological control of white grubs, the root-feeding larvae of scarab beetles, will be different from a successful biocontrol programme targeting mole cricket pests (and indeed they are) (Shapiro-Ilan *et al.*, 2002; Frank and Walker, 2006). These pests have different life cycles and immune responses, and have co-evolved with different pathogens. White grubs have developed defence mechanisms that include infrequent carbon dioxide output, a dense peritrophic membrane, sieve plates over their spiracles, frequent defecation and certain behavioural and immune defences (Shapiro-Ilan *et al.*, 2002). To control white grub populations effectively using EPNs will require either using nematode species that have adapted to parasitizing white grubs or modifying an established EPN population through selection or genetic means to include characteristics that allow them to find and infect white grub larvae effectively in the field. It has been shown that cruise foragers like certain *Heterorhabditis* species and *S. glaseri* are better adapted to finding and infecting sedentary soil-dwelling insects such as white grubs than ambush foragers like *S. carpocapsae* (Gaugler *et al.*, 1997a; Lewis, 2002). Controlling mole crickets on the other hand requires a different approach, since they seem to be refractory to infection from many species of EPNs (Dillman *et al.*, 2012a), but instead are susceptible to *Steinernema scapterisci*, which is reported to be a cricket specialist (Nguyen and Smart, 1990, 1991). So while our understanding of pest insects and EPNs continues to increase, we are constantly discovering and describing new species of EPNs that may be better suited to controlling some pests than the currently used biocontrol agents. In this chapter, we discuss the strategies and future directions of genetic improvement of EPNs for biocontrol.

25.2 Traits for Improvement

There are several factors that limit the efficacy of control afforded by EPNs in the field. Broadly, these can be separated into two categories; infectivity and persistence. Infectivity includes the cascade of events beginning with host seeking and ending with the emergence of new infective juveniles (IJs) from a nutrient-depleted insect cadaver. Persistence includes all of the traits that help the EPNs survive and deal with the physical, chemical and biological components of the environment into which they are placed; everything from heat and desiccation to predator avoidance. There are also additional concerns regarding commercial production and shelf life (for more details see Askary and Ahmad, Chapter 13, Laznik and Trdan, Chapter 30, this volume), but in this chapter we will focus on traits that directly influence their performance in the field.

25.2.1 Infectivity

EPN infectivity can be influenced at every step of the infection process. First, the nematodes perform host-seeking behaviours to locate a suitable host, then they must obtain entry into the haemolymph, release their mutualistic bacteria, overcome the host immune response, kill the host, liberate host tissues and nutrients into a useable form, develop, reproduce and, eventually, when cadaver resources become limited, the rising generation of nematodes must acquire or associate with their symbiont and emerge as IJs to repeat the cycle again (Kaya and Gaugler, 1993; Lewis *et al.*, 1995, 2006). Any part of this process, and other parts not mentioned (*e.g.*, Hunt and Nguyen, 2016), could potentially be targeted for genetic improvement and could lead to a higher level of infectivity, thus increasing the efficacy of EPN biocontrol. There have already been a significant number of studies done to investigate selective improvement of host-seeking behaviours and infectivity (e.g. Gaugler *et al.*, 1989; Gaugler and Campbell, 1991; Grewal *et al.*, 1993; Tomalak, 1994; Peters and Ehlers, 1998; Hiltpold *et al.*, 2010; Bal *et al.*, 2014; Vadnal *et al.*, 2017). However, modern genetic techniques such as genetic transformation have not yet been used to enhance any part of the cascade leading to infection.

25.2.2 Persistence and survival

EPN persistence and survival is greatly influenced by the ability of a nematode to withstand desiccation and a range of temperatures. Industrial production of EPNs relies on desiccation tolerance as a means of prolonging the life of nematodes for storage and shipment. Improvements for desiccation and heat tolerance so far have relied on selective breeding as a means for enhancing survival.

Because EPNs are limnoterrestrial, like all terrestrial nematodes, they are extremely sensitive to desiccation and so desiccation tolerance is essential in prolonging the life of EPNs, not only to increase shelf life for the purpose of distribution but also to ensure their persistence in the field. Low water content induces quiescence in nematodes, and also ensures that bacterial and fungal growth is prevented. However, desiccation is a harsh treatment that results in a high mortality rate. Selection for nematodes that can withstand the desiccation process has successfully produced strains of EPNs that can survive the process with a high rate of survival (Nimkingrat *et al.*, 2013a). Along with selective breeding, hybridization has also proven to be a powerful tool in enhancing desired traits such as desiccation and heat tolerance. Some species of nematodes are more adept at surviving dry conditions and a wider range of temperatures than others. Using these particular strains of EPN for hybridization, followed by selective breeding, has proven to be successful in producing new, more highly tolerant EPN lines (Nimkingrat *et al.*, 2013b). However, there are indications that the desired genes involved with desiccation tolerance are carried by heterozygous individuals. For species of nematodes that can reproduce asexually, these traits are easily retained by maintaining the nematodes in liquid culture. For gonochoristic species (which reproduce sexually) – such as many species of *Steinernema* – maintaining a population of heterozygous individuals is not possible (Ehlers, 2001).

In addition to selective breeding, the process of slow desiccation allows for adaptation. It has been shown that allowing for adaptation can be equally as influential as selective breeding.

There are indications that a slower desiccation process provides enough time for adaptation that results in an increase in survivorship (Nimkingrat *et al.*, 2013a). Additionally, heat tolerance has shown to be improved in individuals that were allowed time to adapt to dry environments (Salame *et al.*, 2010).

Heat tolerance is also another trait that has strongly flexible genetic components. This means that through selective breading these traits can be enhanced so that the EPN lines can withstand a broader range or higher range of temperatures. The heritability of heat tolerance is relatively high for some species of nematodes (such as *Heterorhabditis bacteriophora*), but lower in others (Bal *et al.*, 2014). Through selective breeding, heat tolerance can be improved; however, laboratory selection for these attributes has been shown to have deleterious effects on other desirable characteristics (Chaston *et al.*, 2011).

Inbreeding can have depressive effects on many desired traits. Some characteristics that are most highly affected include heat tolerance and fecundity. However, trait deterioration can be overcome with outbreeding. Outbreeding utilizes an outside line of EPNs that is mixed with the existing lines for breeding, and this method has been shown to be successful at restoring trait deterioration due to inbreeding (Chaston *et al.*, 2011).

Dispersal is another trait that can be selected for in order to increase survivorship and persistence. Selection for dispersal can increase the distances EPNs will travel to find a new host, as well as the speed at which they can travel. Much like heat and desiccation tolerance, there are trade-offs that can arise when selecting for enhanced dispersal, such as decreased reproductive potential and a reduced rate of nictation (Bal *et al.*, 2014). However, these particular trade-offs, unlike the trade-offs for heat and desiccation tolerance, appear to be behaviourally based. Out of the population of individuals that have an inclination to move faster and disperse farther, the majority are males (Bal *et al.*, 2014). However, due to the asymmetry of the sex ratio in this population, the reproduction potential is reduced. Additionally, these *sprinter* individuals have reduced nictation frequencies. Although the ability to nictate is not lost, these individuals have adopted a more cruiser-type method of finding a new host. Unlike the trade-offs that occur in selection for heat and desiccation tolerance, these reductions in reproduction potential and nictation frequency are not something that can be fixed with outbreeding, as they are behavioural aspects that are linked closely with breeding for the characteristic of enhanced dispersal (Bal *et al.*, 2014).

Overall selective breeding and hybridization have proven to be powerful tools in assisting with enhancements of EPN characteristics such as heat and desiccation tolerance and dispersal. Inbreeding can have deleterious effects, leading to reduced fitness, but can be overcome with outbreeding to restore the desired traits. However, the issue of maintaining desired characteristics will prove to be the biggest challenge for species of EPNs that reproduce sexually. Yet this particular issue may prove to be overcome through the utilization of molecular methods for genetic improvement.

25.3 Molecular Methods for Genetic Improvement

The idea of using recombinant DNA research to improve biological control has been around in the literature since at least the mid-1980s (Beckendorf and Hoy, 1985). The simplified idea is that once a useful gene is identified, one that could enhance the persistence or field efficacy of biological control agents, this gene could then be added to other species to confer similar advantages (Hoy, 1992). This has already been done successfully in several systems, especially in agricultural systems where genetic engineering is widespread and increasing. For example, the bacterium *Bacillus thuringiensis* has been studied extensively for its insecticidal activity and has been shown to produce many toxins that are highly effective at killing insect pests. Some of the insecticidal toxins from *B. thuringiensis* have been inserted into the genomes of many important agricultural crop plants to reduce crop loss and damage by insect herbivory (Vaeck *et al.*, 1987; Glare and O'Callaghan, 2000; Romeis *et al.*, 2006). Entomopathogenic fungi are another group where genetic engineering has led to enhanced biocontrol of insect pests. Cuticle-degrading enzymes and toxins are important for fungal infection of insects, many of which are encoded by single genes that are amenable to genetic manipulation (St Leger and Wang, 2010). Multiple copies of a cuticle-degrading protease gene were inserted into the genome of the entomopathogenic fungus *Metarhizium anisopliae* and were expressed constitutively, leading to reduced survival in infected insect hosts (St Leger *et al.*, 1996).

Genetic enhancement has benefitted from descriptive exploration of natural proteins and gene products. Myriad arachnid, scorpion and bacterial proteins and toxins have been considered for the genetic enhancement of biocontrol using genetic engineering (Edwards and Gatehouse, 2007; Whetstone and Hammock, 2007). Scorpion toxins have even been inserted into entomopathogenic fungi and have improved their ability to kill insect hosts (Wang and St Leger, 2007; Pava-Ripoll *et al.*, 2008). Certainly, more candidate biocontrol genes and toxins will come to light as we continue to discover and sequence more species. However, natural proteins and genes are not the only resource available for genetic enhancement as synthetic genes and proteins have also been used. For example, some of the chitinase genes in the entomopathogenic fungus *Beauvaria bassiana* do not have recognizable chitin-binding domains. So researchers have constructed novel hybrid chitinase genes where they fused endogenous functional chitinase genes from *B. bassiana* with chitin-binding domains from other species, which resulted in increased killing of insect hosts by the transgenic fungi (Fan *et al.*, 2007). These examples from other systems open exciting possibilities for the genetic improvement of EPNs, even if researchers only repeat what has been done in entomopathogenic fungi or plants, and with some creativity, perhaps even greater improvements can be realized.

The development of molecular and genetic tools for EPNs and technology transfer from the powerful genetic model *C. elegans* has been slow. Many of the powerful tools that allow for genetic engineering (genetic transformation) and the functional characterization of genes (RNAi) were developed in *C. elegans* 20 or more years ago. (Kimble *et al.*, 1982; Fire *et al.*, 1991; Mello *et al.*, 1991; Broverman *et al.*, 1993; Mello and Fire, 1995). There are encouraging examples of successful genetic transformation of EPNs in the literature (Hashmi *et al.*, 1995a, 1997, 1998; Vellai *et al.*, 1999), but these results were published 15 or more years ago and researchers in the field have not developed these methods further, nor have transgenic EPNs become commercially available. Hashmi *et al.* (1995a, 1998) reported transforming *H. bacteriophora* with a copy of the *C. elegans hsp70A* gene and that this resulted in dramatically increased heat tolerance in the transgenic nematodes under laboratory conditions. The transgenic nematodes had normal growth and development and did not appear to differ in virulence from the parental strain (Hashmi *et al.*, 1998). Although the transgenic nematodes and parental strain did not differ in persistence in the field (Gaugler *et al.*, 1997b; Wilson *et al.*, 1999), these studies have gone a long way in showing that transgenic techniques can be applied to EPNs successfully.

Steinernema feltiae has also been transformed successfully, given the stress resistance gene trehalose-phosphate synthase 1 from yeast (Vellai *et al.*, 1999). These transgenic nematodes showed an increase in stress tolerance, further demonstrating the potential of molecular methods in improving EPN traits. Surprisingly, there have not been additional published reports of transgenic success in EPNs. Certainly, there is interest in the field to use molecular genetics to explore gene function in EPNs and to enhance their efficacy in biocontrol, but perhaps the perceived difficulty of these techniques has prevented further developments. There are now multiple ways to develop transgenic strains of nematodes, including microinjection of genetic constructs into the gonad (Evans, 2006) and microparticle bombardment (Praitis *et al.*, 2001). Although microinjection has been used to make transgenic EPNs (Hashmi *et al.*, 1995a), this technique has not been adopted by many EPN researchers. Perhaps microparticle bombardment, which is done on many nematodes and with multiple genetic constructs at the same time, will have more success in the EPN community.

Being able to introduce exogenous or modified genes into EPNs in a consistent and reliable manner is paramount to our ability to employ molecular methods in genetic improvement of EPNs in biocontrol. Other techniques, such as RNAi, will be extremely useful in determining the molecular function of the genes (Fire *et al.*, 1991, 1998). As with genetic transformation, although RNAi has been applied successfully in *H. bacteriophora* (Ciche and Sternberg, 2007), it has not been adopted by the field at large nor has it been developed further. Before we can make significant progress using molecular methods for genetic improvement of EPNs, these techniques need to be developed and adopted by the EPN community of researchers. Progress has been made toward developing these tools with the sequencing of the *H. bacteriophora* genome and multiple *Steinernema* congeners (Bai *et al.*, 2013; Dillman *et al.*, 2015). The huge amount of data released by these sequencing projects both regarding gene presence and expression should further enhance and encourage transgenic approaches by providing a large list of candidate parasitism and survival genes that could be used to enhance EPNs for biological control.

25.3.1 Genomic analyses of *Heterorhabditis bacteriophora*

The 77 Mb *H. bacteriophora* genome sequence has revealed numerous putative parasitism and survival genes that could be manipulated by transgenic approaches (Table 25.1) (Bai *et al.*, 2013). But it has also raised many questions since the *H. bacteriophora* genome contains more than 10,000 proteins of unknown function. *H. bacteriophora*'s obligate association with the bacterium *Photorhabdus luminescens* has shaped the content of its genome as the nematode relies on the bacteria for nutrients and protection. The bacteria produce an arsenal of enzymes and proteases to overcome insect host immunity, degrade host tissues and make them available for the developing nematodes, and prevent opportunistic fungi and bacteria from making use of the nutrient-rich insect cadaver. Unlike *Steinernema*, axenic (bacteria-free) *H. bacteriophora* nematodes are unable to kill insect hosts, suggesting that the bacteria do most of the heavy lifting on host entry (Han and Ehlers, 2000; Eleftherianos *et al.*, 2010). *H. bacteriophora* may rely mostly on *Photorhabdus* for dealing with the insect host post-infection, but it is still the nematode that must locate hosts to infect, gain entry into the haemolymph and persist in the soil until a new host is found, leaving

Table 25.1. Comparison of the sequenced entomopathogenic nematode genomes as previously reported. The *Steinernema monticolum* genome was not included. (From Bai *et al.*, 2013; Dillman *et al.*, 2015).

	Heterorhabditis bacteriophora	*Steinernema carpocapsae*	*Steinernema scapterisci*	*Steinernema feltiae*	*Steinernema glaseri*
Genome size (Mb)	77.0	85.6	79.4	82.4	92.9
N50 (bp)	312,328	299,566	90,783	47,472	37,444
No of scaffolds	1,263	1,578	2,877	5,839	7,515
No of predicted genes	21,250	28,313	31,378	33,459	34,143
GC content (%)	32.2	45.53	47.98	46.99	47.63

plenty of room for genetic improvements to enhance field efficacy. Most of the insulin/IGF-1 signalling pathway genes were found in the *H. bacteriophora* genome (Bai *et al.*, 2013). These genes play a critical role in dauer formation and are important regulators of stress resistance and innate immunity in *C. elegans*, and are thought to perform similar functions in *H. bacteriophora*. The 19 genes found to be conserved in the insulin/IGF-1 signalling pathway are all potential candidates for genetic manipulation or enhancement, as increasing IJ longevity and/or stress resistance may lead to better field efficacy and pest control.

Another gene family explored in the *H. bacteriophora* genome is the G protein-coupled receptor (GPCR) family. There are at least 82 predicted GPCRs in the *H. bacteriophora* genome, and these are important because they could be sensory receptors functioning in olfaction and host-seeking behaviour (Robertson and Thomas, 2006; Thomas and Robertson, 2008; Srinivasan *et al.*, 2013). The abundance and diversity of GPCRs can reflect the niche inhabited by a nematode, and in the case of EPNs, putative olfactory receptors are of high value as they could be used to enhance host seeking or adjust the specificity of hosts being sought. Increasing our knowledge of how GPCRs are used in host seeking could be critical to improving or altering the host seeking of EPNs.

Any genes that play a role in the symbiotic association between EPN and the insect-pathogenic bacteria they carry could potentially be used to enhance the biological control potential of EPNs. *H. bacteriophora* seems to have a reduced or modified immune response compared to *C. elegans*. It has far fewer C-type lectin domain-containing proteins (9 compared to 133), which function in the immune response of *C. elegans* to bacterial infection (Schulenburg *et al.*, 2008; Bai *et al.*, 2013). This reduction in C-type lectin domain-containing proteins may be related to the association between *H. bacteriophora* and *P. luminescens* (Ciche *et al.*, 2008). But it is also not known to what extent bacterial infection might impede on the efficacy of EPNs as biocontrol agents. More research needs to be done, but the immune response of EPNs is yet another area that could be harnessed to improve field efficacy against insect pests.

Secreted products are another area that is rife with potential for genetic enhancement of EPNs. The field efficacy of EPNs will benefit from our increased understanding and characterization of the existing secreted products and their role in invasion and infection. One can imagine improving host killing by EPNs simply by adding additional copies of genes already present, similar to the transgenic inclusion of multiple endogenous cuticle-degrading proteases in entomopathogenic fungi (St Leger *et al.*, 1996). Another possible route of improvement is by the addition of new secreted products, like the transgenic addition of a scorpion venom gene into the genome of the entomopathogenic fungus *M. anisopliae* (Pava-Ripoll *et al.*, 2008). *H. bacteriophora* has fewer than 30 predicted protease and protease inhibitors in its secretome (Bai *et al.*, 2013). The presence of such a small number of putatively secreted proteases and protease inhibitors may reflect the heavy reliance of the nematode on *P. luminescens* for immune suppression and tissue degradation of the insect host. However, there are non-canonical secretion signals that are poorly understood and it is possible that *H. bacteriophora* has a larger secretome than is predicted by the abundance of signal peptides (Bennuru *et al.*, 2009, 2011). EPN-secreted proteases are known to influence penetration of the nematode into the host haemolymph (Abuhatab *et al.*, 1995), tissue degradation of insect hosts (McKerrow *et al.*, 2006), as well as immune suppression (Balasubramanian *et al.*, 2009).

The genomic sequence of *H. bacteriophora* provides a long list of candidate genes that could be used to improve infectivity and/or survival. The successful transformations and use of RNAi in *H. bacteriophora*, coupled with this archive of new genes to be studied, is expected to lead to significant advances in the application of molecular genetics to improve the field efficacy of EPNs in biological control.

25.3.2 Genomic analyses of *Steinernema*

The draft genomes of five species of *Steinernema* (*S. carpocapsae*, *S. feltiae*, *S. glaseri*, *Steinernema monticolum* and *S. scapterisci*) have recently been sequenced and annotated, and have revealed numerous genes that could be involved in the entomopathogenic lifestyle and could be used in programmes to improve traits important for biological control (Table 25.1) (Dillman *et al.*, 2015). Similar to what has been found in the *H. bacteriophora* genome, more than 10,000 predicted proteins (~37% of the predicted proteome) seem to have no orthologues with other animals, or even other nematodes (Dillman *et al.*, 2015). Determining the function of these orphan proteins in EPNs could yield insight into genes important for infection or survival and persistence, and thus be useful to future transgenic endeavours in EPNs. The genomes of these EPNs are similar in size but differ considerably in nucleotide prevalence (G+C content), which may influence the application and use of recombinant DNA techniques for genetic enhancement. Codon usage preferences can influence gene expression and regulation (Rao *et al.*, 2013), which could affect technology transfer from *C. elegans* to the EPNs, with technology developed in the *C. elegans* model being applied more easily to *H. bacteriophora* due to their closer ancestry and similar nucleotide prevalence (Table 25.1) (van Megen *et al.*, 2009; Bai *et al.*, 2013; Dillman *et al.*, 2015), although this needs to be further explored experimentally.

In contrast to the *H. bacteriophora* genome, the sequenced species of *Steinernema* have an abundance of predicted protease and protease inhibitors with signal peptides. The genome sequence revealed *Steinernema*-specific expansions of serine and metalloproteases (Dillman *et al.*, 2012b, 2015). Proteases are known to be important in invasion and host-killing for *Steinernema*, making the repertoire of proteases and protease inhibitors an important group of proteins for investigation in future selection and recombinant studies (Abuhatab *et al.*, 1995; Balasubramanian *et al.*, 2009). Proteases in steinernematids have been shown to play an important role in suppressing insect host immunity, as well as tissue degradation (e.g. Balasubramanian *et al.*, 2009, 2010; Toubarro *et al.*, 2009, 2010). Protease inhibitors are also expanded in the sequenced steinernematids, and there are functional studies showing that protease inhibitors play an important role in nematode evasion of host immunity (Milstone *et al.*, 2000; Zang and Maizels, 2001), although the mechanism remains poorly understood. One provocative possibility is that host range and specificity of EPNs may be influenced by their armoury of secreted products, and that using transformation the host range and/or specificity could be altered by the addition or removal of certain secreted products from the secretome. Not enough is known about the evolution of insect immunity, but as more genomes are being studied, it seems that insect immunity could differ dramatically between orders and that niche partitioning among EPNs could be based on individual species' abilities to overcome or avoid the immune response of certain hosts (Elsik, 2010; Gerardo *et al.*, 2010).

Fatty acid- and retinol-binding (FAR) proteins are another interesting gene family that has been highlighted in the genomes of *Steinernema*, as they seem to be expanded in these species (Dillman *et al.*, 2015). FAR proteins are thought to play a key role in parasitism by functioning in the sequestration of host retinoids, as well as by contributing to immune evasion or suppression, although their exact functional role is not well understood (Garofalo *et al.*, 2002; Kennedy *et al.*, 2013). They are particularly interesting because they seem to be involved in nematode parasitism of animals, insects and plants (Hao *et al.*, 2010; Iberkleid *et al.*, 2013; Kennedy *et al.*, 2013), which makes understanding their mechanistic function important for both biocontrol as well as disease treatment and prevention.

The availability of the genomic sequence data and putative proteomes provides a large number of genes that could be useful in increasing the infectivity of EPNs. We have mentioned proteases, protease inhibitors, FAR proteins and GPCRs, although there are certainly many more to be mined from these data. There is also an abundance of genes that could lead to increased IJ persistence in the soil, as well as increased survival. In addition to genes involved in the insulin/IGF-1 signalling pathway, there are known stress-tolerance genes like heat-shock proteins, trehalose-related molecules and pathways, as well as all of the orthologues and paralogues that have been expanded within the EPNs. What remains is for these genes to be functionally tested, perhaps using biochemical or RNAi techniques, and applied in recombinant DNA experiments. The field has been advanced by the sequencing of these genomes, but whether this leads to actual improvements in the field efficacy of EPN biocontrol remains to be seen.

25.4 Troubleshooting

As in crop breeding (e.g. Abd-Elgawad, 1991), success in the genetic improvement of EPNs for enhanced biological control generally depends on three main phases. These are hereditable genetic variation of the desired trait(s), adequate screening and improvement techniques, and the capacity of the improved strain in terms of its efficacy, fitness and stability in laboratory, greenhouse and, more importantly, under field conditions. Eventually, to materialize the net gain of such an improvement process, its cost should be justified by the advantages or profits obtained (e.g. Hoy, 1985).

Although selective breeding and mutagenesis have been classical approaches for genetic improvement of EPNs, they often have substantial problems in some or all of the above-mentioned categories. One difficulty is that genetic refinement of EPNs to enhance their biological control potential through selective breeding is likely only if the desired alleles are found in the gene pools of the examined nematode populations. For example, insufficient genetic variation in resistance to ultraviolet light (UV) resulted in rejecting selective breeding as an option to improve UV tolerance in *S. carpocapsae* (Gaugler *et al.*, 1989). It is well established that sexual reproduction, mutations and the recombination of linked genes (crossing over) are major tools of the phenotypic variability of EPNs in nature. While sexual reproduction offers recombinations between genes located in different chromosomes, crossing over can perform the recombination of alleles present on the same chromosome. Therefore, broad surveys of diverse habitats to isolate novel EPNs to widen the genetic variation and consequently develop new nematode strains are considered excellent troubleshooting tools. Selective breeding is the preferred approach to improve a polygenic trait where each gene encodes only a small effect on the same phenotype. Optimally, the selected improvements will have high heritability (mathematically calculated as h^2). If the targeted trait has low heritability (h^2), then, a programme centred on the introduction and selection of new mutations affecting this trait would be preferable (Glazer *et al.*, 1991).

Nevertheless, many researchers (e.g. Gaugler *et al.*, 1990; Hastings, 1994; Burnell, 2002; Grewal *et al.*, 2011; Glazer, 2014, 2015) reported that traits improved in this way were not stable and that the phenotypes selected for tended to revert gradually to the unselected state following relaxation of the selection pressure. A wise strategy to troubleshoot these problems is to let nature perform the selection process, when possible. For instance, EPN strains endemic to areas with climate extremes will be naturally acclimated through selection to survive and reproduce in those conditions. Interestingly, EPN strains with these phenotypes have been extracted in programmes aimed at collecting cold-tolerant and warm-tolerant strains (Glazer *et al.*, 1996; Mracek *et al.*, 1998; Burnell, 2002; Abd-Elgawad and Nguyen, 2007). If natural selection is not applicable, the default strategy, to hinder the reverse of the targeted trait, is to cryopreserve the improved strains and to reapply selection pressure at constant intervals (Burnell, 2002).

Another issue that can result from applying artificial selection for enhancement of nematode strains is the occurrence of unintentional alteration of traits that may affect the fitness of the improved nematode strain adversely. Sometimes, the genes encoding for the targeted trait do not segregate independently. It is possible that some targeted genes have unknown linkage, which may lead to such problems. Therefore, more knowledge and tools for genetic improvement of EPNs are required. Certainly, modern genetic techniques such as RNA sequencing (RNA-seq) and transcriptomics can contribute to our understanding of these processes and provide needed genetic and molecular information (Glazer, 2014). For example, exposure of *H. bacteriophora* to desiccation stress could lead to increased desiccation tolerance. On relaxation of selection pressure, the tolerance gained was lost again during *in vivo* production, although this desired tolerance was positively retained in liquid-cultured EPNs (Anbesse *et al.*, 2013). Since the transcriptome is continually changing, RNA-seq can be utilized to detect a snapshot of RNA existence and quantity from a genome at a fixed moment in time (Chu and Corey, 2012), which may enable us to follow and understand the molecular basis of a beneficial EPN trait. Nevertheless, such reported results (e.g. Anbesse *et al.*, 2013) cannot be interpreted reliably and conclusively through a single technique, but a combination of genetic approaches to improve EPNs as bioinsecticides holds promise for solving this and other related issues. In this respect, other modern genetic techniques such as those related to genome editing, e.g. transcription activator-like effector nucleases (TALENs) and clustered regularly interspaced short palindromic repeats that rely on a protein called Cas9 (CRISPR-Cas9), may also be used to modify traits (Glazer, 2014). For instance, TALENs are synthetic restriction enzymes that can cut DNA strands at a specific sequence. Transcription activator-like effectors (TALEs) can be engineered quickly to bind practically any desired DNA sequence (Boch, 2011). By combining such an engineered TALE with a DNA cleavage domain (which cuts DNA strands), one can engineer restriction enzymes that are specific for any desired DNA sequence. When these artificial enzymes are inserted in to the intended cells, they can be utilized for genome editing in situ (https://en.wikipedia.org/wiki/Transcription_activator-like_effector_nuclease).

The continuous progress in developing molecular tools should be directed to fundamental and applied EPN genetics, creating novel useful approaches while fine-tuning current methodology.

For example, fine-tuning of transgenic techniques can be applied successfully to EPNs through trial and error, as may occur in fixing the appropriate settings in order to pull needles for microinjection of EPNs (Hashmi *et al.*, 1999). On the other hand, Hashmi *et al.* (1995b) could validate a new and efficient genetic transformation system for EPNs, using arrays of micromechanical piercing structures, which is quick and easy to use. Such modern and other novel genetic techniques are desperately needed to address the challenges concerning improvement of EPNs for enhanced biocontrol of arthropod pests.

25.5 Conclusions and Future Prospects

The increase in both EPN species descriptions and in general EPN research is exciting and bodes well for the future of biocontrol using EPNs, and is encouraging for the prospects of genetic improvement of EPNs. Historically, the significant strain improvements made to EPN biocontrol have come through the discovery of new species or through classical genetics. However, there are promising reports of successful transgenic methods being applied in EPNs, but they remain few and have not been followed up by more recent studies (Hashmi *et al.*, 1998; Vellai *et al.*, 1999). Genetic improvement of EPNs via selection will continue to be important; however, transgenic organisms have become more common and they allow for improvements that simply are not possible with conventional methods. The advances that have been made and the continued development of molecular tools will be used to address difficult applied and fundamental questions.

Better field efficacy is needed if EPNs are to provide pest control at a level that is competitive with other means. Ultimately, this problem is likely to be solved as EPN biology continues to be a vibrant and growing field of research, attracting more and new researchers to this field. This continued growth and enthusiasm for EPN biology will be achieved by emphasizing not only the applied aspects of EPN research but also their incredible usefulness as models for basic biology (Akhurst and Dunphy, 1993; Emelianoff *et al.*, 2008; Dillman and Sternberg, 2012). EPNs remain a powerful system for the study of evolution, symbiosis, parasitism and development. They are also useful models of animal parasitism, since they are easier to work with than most vertebrate-parasitic species. There remain many questions that are unknown or underexplored, leaving EPN biology a field that is pregnant with possibility and that has plenty of room for more investigators to apply their skill and creativity to studying their interesting biology.

Acknowledgements

We thank the community of nematode researchers for the archive of information that has guided and influenced our own research. Tiffany Baiocchi and Adler Dillman were supported by funds from the University of California, Riverside. Mahfouz Abd-Elgawad was supported in part by US-Egypt Science and Technology Joint Fund (cycle 17; project no. 172) and Egyptian In-House, National Research Centre Project No 11030133 entitled 'Pesticide alternatives against soilborne pathogens attacking legume cultivation in Egypt'.

References

Abd-Elgawad, M.M.M. (1991) A new rating scale for screening plant genotypes against root-knot and reniform nematodes. *Anzeiger für Schädlingskunde Pflanzenschutz Umweltschutz* 64, 37–39.

Abd-Elgawad, M.M.M. and Nguyen, K.B. (2007) Isolation, identification and environmental tolerance of new heterorhabditid populations from Egypt. *International Journal of Nematology* 17, 116–123.

Abuhatab, M., Selvan, S. and Gaugler, R. (1995) Role of proteases in penetration of insect gut by the entomopathogenic nematode *Steinernema glaseri* (Nematoda, Steinernematidae). *Journal of Invertebrate Pathology* 66, 125–130.

Akhurst, R.J. and Dunphy, G.B. (1993) Tripartite interactions between symbiotically associated entomopathogenic bacteria, nematodes, and their insect hosts. In: Beckage, N.E., Thompson, S.N. and Federici, B. (eds) *Parasites and Pathogens of Insects.* Academic Press, New York, pp. 1–23.

Anbesse, S., Sumaya, N.H., Dörfler, A.V., Strauch, O. and Ehlers R.-U. (2013) Selective breeding for desiccation tolerance in liquid culture provides genetically stable inbred lines of the entomopathogenic nematode *Heterorhabditis bacteriophora*. *Applied Microbiology and Biotechnology* 97, 731–739.

Bai, X.D., Adams, B.J., Ciche, T.A., Clifton, S., Gaugler, R., *et al.* (2013) A lover and a fighter: the genome sequence of an entomopathogenic nematode *Heterorhabditis bacteriophora*. *Plos One* 8(7), e69618.

Bal, H.K., Michel, A.P. and Grewal, P.S. (2014) Genetic selection of the ambush foraging entomopathogenic nematode, *Steinernema carpocapsae* for enhanced dispersal and its associated trade-offs. *Evolutionary Ecology* 28, 923–939.

Balasubramanian, N., Hao, Y.J., Toubarro, D., Nascimento, G. and Simoes, N. (2009) Purification, biochemical and molecular analysis of a chymotrypsin protease with prophenoloxidase suppression activity from the entomopathogenic nematode *Steinernema carpocapsae*. *International Journal for Parasitology* 39, 975–984.

Balasubramanian, N., Toubarro, D. and Simoes, N. (2010) Biochemical study and *in vitro* insect immune suppression by a trypsin-like secreted protease from the nematode *Steinernema carpocapsae*. *Parasite Immunology* 32, 165–175.

Beckendorf, S.K. and Hoy, M.A. (1985) Genetic improvement of arthropod natural enemies through selection, hybridization or genetic engineering techniques. In: Hoy, M.A. and Herzog, D.C. (eds) *Biological Control in Agricultural IPM Systems.* Academic Press, Orlando, Florida, pp. 167–187.

Bennuru, S., Semnani, R., Meng, Z., Ribeiro, J.M.C., Veenstra, T.D. and Nutman, T.B. (2009) *Brugia malayi* excreted/secreted proteins at the host/parasite interface: stage- and gender-specific proteomic profiling. *Plos Neglected Tropical Diseases* 3, e525.

Bennuru, S., Meng, Z., Ribeiro, J.M., Semnani, R.T., Ghedin, E., *et al.* (2011) Stage-specific proteomic expression patterns of the human filarial parasite *Brugia malayi* and its endosymbiont *Wolbachia*. *Proceedings of the National Academy of Sciences of the United States of America* 108, 9649–9654.

Bilgrami, A.L., Gauger, R., Shapiro-Ilan, D.I. and Adams, B.J. (2006) Source of trait deterioration in entomopathogenic nematodes *Heterorhabditis bacteriophora* and *Steinernema carpocapsae* during *in vivo* culture. *Nematology* 8, 397–409.

Boch, J. (2011) TALEs of genome targeting. *Nature Biotechnology* 29, 135–136.

Broverman, S., Macmorris, M. and Blumenthal, T. (1993) Alteration of *Caenorhabditis elegans* gene-expression by targeted transformation. *Proceedings of the National Academy of Sciences of the United States of America* 90, 4359–4363.

Burnell, A. (2002) Genetics and genetic improvement. In: Gaugler, R. (ed.) *Entomopathogenic Nematology*. CAB International, Wallingford, UK, pp. 241–264.

Chaston, J.M., Dillman, A.R., Shapiro-Ilan, D.I., Bilgrami, A.L., Gaugler, R., *et al.* (2011) Outcrossing and crossbreeding recovers deteriorated traits in laboratory cultured *Steinernema carpocapsae* nematodes. *International Journal of Parasitology* 41, 801–809.

Chu, Y. and Corey, D.R. (2012) RNA sequencing: platform selection, experimental design, and data interpretation. *Nucleic Acid Therapeutics* 22, 271–274.

Ciche, T.A. and Sternberg, P.W. (2007) Postembryonic RNAi in *Heterorhabditis bacteriophora*: a nematode insect parasite and host for insect pathogenic symbionts. *BMC Developmental Biology* 7, 101, DOI: 10.1186/1471-213X-7-101.

Ciche, T.A., Kim, K.S., Kaufmann-Daszczuk, B., Nguyen, K.C. and Hall, D.H. (2008) Cell invasion and matricide during *Photorhabdus luminescens* transmission by *Heterorhabditis bacteriophora* nematodes. *Applied and Environmental Microbiology* 74, 2275–2287.

Cory, E.N. and Langford, G.S. (1944) The Japanese beetle in Maryland. *Maryland University Agricultural Extension Bulletin* 88, 24.

Cory, E.N. and Langford, G.S. (1955) The Japanese Beetle Retardation Program in Maryland. *Maryland University Agricultural Extension Bulletin* 156, 20.

Dillman, A.R. and Sternberg, P.W. (2012) Entomopathogenic nematodes. *Current Biology* 22, R430–431.

Dillman, A.R., Guillermin, M.L., Lee, J.H., Kim, B., Sternberg, P.W. and Hallem, E.A. (2012a) Olfaction shapes host–parasite interactions in parasitic nematodes. *Proceedings of the National Academy of Sciences of the United States of America* 109, E2324–2333.

Dillman, A.R., Mortazavi, A. and Sternberg, P.W. (2012b) Incorporating genomics into the toolkit of nematology. *Journal of Nematology* 44, 191–205.

Dillman, A.R., Macchietto, M., Porter, C.F., Rogers, A., Williams, B., *et al.* (2015) Comparative genomics of *Steinernema* reveals deeply conserved gene regulatory networks. *Genome Biology* 16(1), 200, DOI: 10.1186/s13059-015-0746-6.

Edwards, M.G. and Gatehouse, A.M.R. (2007) Biotechnology in crop protection: towards sustainable insect control. In: Vurro, M. and Gressel, J. (eds) *Novel Biotechnologies for Biocontrol Agent Enhancement and Management.* Springer, Dordrecht, The Netherlands, pp. 1–24.

Ehlers, R.-U. (2001) Mass production of entomopathogenic nematodes for plant protection. *Applied Microbiology and Biotechnology* 56, 623–633.

Eleftherianos, I., Ffrench-Constant, R.H., Clarke, D.J., Dowling, A.J. and Reynolds, S.E. (2010) Dissecting the immune response to the entomopathogen *Photorhabdus*. *Trends in Microbiology* 18, 552–560.

Elsik, C.G. (2010) The pea aphid genome sequence brings theories of insect defense into question. *Genome Biology* 11, 106, DOI:10.1186/gb-*2010-11-*2-106.

Emelianoff, V., Chapuis, E., Le Brun, N., Chiral, M., Moulia, C. and Ferdy, J.B. (2008) A survival-reproduction trade-off in entomopathogenic nematodes mediated by their bacterial symbionts. *Evolution* 62, 932–942.

Evans, T.C. (2006) Transformation and microinjection. In: Community, T.C.E.R. (ed.) *WormBook.* DOI: 10.1895/wormbook.1.108.1.

Fan, Y.H., Fang, W.G., Guo, S.J., Pei, X.Q., Zhang, Y.J., *et al.* (2007) Increased insect virulence in *Beauveria bassiana* strains overexpressing an engineered chitinase. *Applied and Environmental Microbiology* 73, 295–302.

Fire, A., Albertson, D., Harrison, S.W. and Moerman, D.G. (1991) Production of antisense RNA leads to effective and specific-inhibition of gene-expression in *C. elegans* muscle. *Development* 113, 503–514.

Fire, A., Xu, S., Montgomery, M.K., Kostas, S.A., Driver, S.E. and Mello, C.C. (1998) Potent and specific genetic interference by double-stranded RNA in *Caenorhabditis elegans*. *Nature* 391, 806–811.

Fleming, W.E. (1968) *Biological Control of the Japanese Beetle*. Technical Bulletin 1383. Agricultural Research Service, US Department of Agriculture, Washington, DC, 78 pp.

Frank, J.H. and Walker, T.J. (2006) Permanent control of pest mole crickets (Orthoptera: Gryllotalpidae: *Scapteriscus*) in Florida. *American Entomologist* 52, 138–144.

Garofalo, A., Klager, S.L., Rowlinson, M.C., Nirmalan, N., Klion, A., *et al.* (2002) The FAR proteins of filarial nematodes: secretion, glycosylation and lipid binding characteristics. *Molecular and Biochemical Parasitology* 122, 161–170.

Gaugler, R. (1988) Ecological considerations in the biological control of soil-inhabiting insects with entomopathogenic nematodes. *Agriculture, Ecosystems and Environment* 24, 351–360.

Gaugler, R. and Campbell, J.F. (1991) Selection for enhanced host-finding of scarab larvae (Coleoptera, Scarabaeidae) in an entomopathogenic nematode. *Environmental Entomology* 20, 700–706.

Gaugler, R., Campbell, J.F. and McGuire, T.R. (1989) Selection for host-finding in *Steinernema feltiae*. *Journal of Invertebrate Pathology* 54, 363–372.

Gaugler, R., Campbell, J.F. and McGuire, T.R. (1990) Fitness of a genetically improved entomopathogenic nematode. *Journal of Invertebrate Pathology* 56, 106–116.

Gaugler, R., Lewis, E. and Stuart, R.J. (1997a) Ecology in the service of biological control: the case of entomopathogenic nematodes. *Oecologia* 109, 483–489.

Gaugler, R., Wilson, M. and Shearer, P. (1997b) Field release and environmental fate of a transgenic entomopathogenic nematode. *Biological Control* 9, 75–80.

Gerardo, N.M., Altincicek, B., Anselme, C., Atamian, H., Barribeau, S.M., *et al.* (2010) Immunity and other defenses in pea aphids, *Acyrthosiphon pisum*. *Genome Biology* 11(2), R21, DOI: 10.1186/gb-2010-11-2-r21.

Girth, H.B., McCoy, E.E. and Glaser, R.W. (1940) Field experiments with a nematode parasite of the Japanese beetle. *New Jersey Department of Agriculture Circulation,* 317, 21.

Glare, T.R. and O'Callaghan, M. (2000) *Bacillus thuringiensis: Biology, Ecology, and Safety*. John Wiley & Sons Ltd, Chichester, UK.

Glaser, R.W., McCoy, E.E. and Girth, H.B. (1940) The biology and economic importance of a nematode parasite of insects. *Journal of Parasitology* 26, 479–495.

Glazer, I. (2014) Genetic improvement and breeding of EPN: the race for the "super nematode". *Journal of Nematology* 46, 168.

Glazer, I. (2015) Improvement of entomopathogenic nematodes: a genetic approach. In: Campos-Herrera, R. (ed.) *Nematodes Pathogenesis of Insects and Other Pests*. Springer International Publishing, Neuchâtel, Switzerland, pp. 29–56.

Glazer, I., Gaugler, R. and Segal, D. (1991) Genetics of the nematode *Heterorhabditis bacteriophora* strain HP88: the diversity of beneficial traits. *Journal of Nematology* 23, 324–333.

Glazer, I., Kozodoi, E., Hashmi, G. and Gaugler, R. (1996) Biological characteristics of the entomopathogenic nematode *Heterorhabditis* sp. IS-5: a heat tolerant isolate from Israel. *Nematologica* 42, 481–492.

Grewal, P.S., Tomalak, M., Keil, C.B.O. and Gaugler, R. (1993) Evaluation of a genetically selected strain of *Steinernema feltiae* against the mushroom sciarid *Lycoriella mali*. *Annals of Applied Biology* 123, 695–702.

Grewel, P.S., Bai, X. and Jagdale, G.B. (2011) Longevity and stress tolerance of entomopathogenic nematodes. In: Perry, R.N. and Wharton, D.A. (eds) *Molecular and Physiological Basis of Nematode Survival*. CAB International, Wallingford, UK, pp. 157–181.

Han, R. and Ehlers, R.U. (2000) Pathogenicity, development, and reproduction of *Heterorhabditis bacteriophora* and *Steinernema carpocapsae* under axenic *in vivo* conditions. *Journal of Invertebrate Pathology* 75, 55–58.

Hao, Y.J., Montiel, R., Abubucker, S., Mitreva, M. and Simoes, N. (2010) Transcripts analysis of the entomopathogenic nematode *Steinernema carpocapsae* induced *in vitro* with insect haemolymph. *Molecular and Biochemical Parasitology* 169, 79–86.

Hashmi, S., Hashmi, G. and Gaugler, R. (1995a) Genetic transformation of an entomopathogenic nematode by microinjection. *Journal of Invertebrate Pathology* 66, 293–296.

Hashmi, S., Ling, P., Hashmi, G., Reed, M., Gaugler, R. and Trimmer, W. (1995b) Genetic transformation of nematodes using arrays of micromechanical piercing structures. *Biotechniques* 19, 766–770.

Hashmi, S., Abuhatab, M.A. and Gaugler, R.R. (1997) GFP: green fluorescent protein a versatile gene marker for entomopathogenic nematodes. *Fundamental and Applied Nematology* 20, 323–327.

Hashmi, S., Hashmi, G., Glazer, I. and Gaugler, R. (1998) Thermal response of *Heterorhabditis bacteriophora* transformed with the *Caenorhabditis elegans* hsp70 encoding gene. *Journal of Experimental Zoology* 281, 164–170.

Hashmi, S., Hashmi, G. and Gaugler, R. (1999) Transformation of nematodes by microinjection. In: Lacal, J.C., Feramisco, J. and Perona, R. (eds) *Methods and Tools in Biosciences and Medicine*

Microinjection. Birkhäuser Verlag, Basel, Switzerland, pp. 233–240.

Hastings, I.M. (1994) Introduction to quantitative genetics: inbreeding, heritability estimates, artificial selection. In: Burnell, A.M., Ehlers, R.-U. and Masson, J.P. (eds) *Genetics of Entomopathogenic Nematode–Bacterium Complexes*. European Commission Publications, Luxembourg, pp.120–129.

Hiltpold, I. (2015) Prospects in the application technology and formulation of entomopathogenic nematodes for biological control of insect pests. In: Campos-Herrera, R. (ed.) *Nematode Pathogenesis of Insects and Other Pests: Ecology and Applied Technologies for Sustainable Plant and Crop Protection.* Springer International Publishing, Neuchâtel, Switzerland, pp. 187–206.

Hiltpold, I., Baroni, M., Toepfer, S., Kuhlmann, U. and Turlings, T.C.J. (2010) Selection of entomopathogenic nematodes for enhanced responsiveness to a volatile root signal helps to control a major root pest. *Journal of Experimental Biology* 213, 2417–2423.

Hoy, M.A. (1985) Improving establishment of arthropod natural enemies. In: Hoy, M.A. and Herzog, D.C. (eds) *Biological Control in Agricultural IPM Systems*. Academic Press, Orlando, Florida, pp. 151–166.

Hoy, M.A. (1992) Biological control of arthropods: genetic engineering and environmental risks. *Biological Control* 2, 166–170.

Hunt, D.J. and Nguyen, K.B. (2016) *Advances in Entomopathogenic Nematode Taxonomy and Phylogeny*, Vol. 12, Nematology Monographs and Perspectives. Brill, Leiden, The Netherlands, 437 pp.

Iberkleid, I., Vieira, P., Engler, J.D., Firester, K., Spiegel, Y. and Horowitz, S.B. (2013) Fatty acid and retinol-binding protein, Mj-FAR-1 induces tomato host susceptibility to root-knot nematodes. *Plos One* 8(5), e64586, DOI: 10.1371/journal.pone.0064586.

Kaya, H.K. and Gaugler, R. (1993) Entomopathogenic nematodes. *Annual Review of Entomology* 38, 181–206.

Kennedy, M.W., Corsico, B., Cooper, A. and Smith, B.O. (2013) The unusual lipid-binding proteins of nematodes: NPAs, nemFABPs and FARs. In: Kennedy, M.W. and Harnett, W. (eds) *Parasitic Nematodes: Molecular Biology, Biochemistry, and Immunology.* CAB International, Wallingford, UK, pp. 397–412.

Kimble, J., Hodgkin, J., Smith, T. and Smith, J. (1982) Suppression of an amber mutation by micro-injection of suppressor transfer-RNA in *C. elegans*. *Nature* 299, 456–458.

Lewis, E.E. (2002) Behavioral ecology. In: Gauger, R. (ed.) *Entomopathogenic Nematology.* CAB International, Wallingford, UK, pp. 205–223.

Lewis, E.E., Grewal, P.S. and Gauger, R. (1995) Hierarchical order of host cues in parasite foraging strategies. *Parasitology* 110, 207–213.

Lewis, E.E., Campbell, J., Griffin, C., Kaya, H. and Peters, A. (2006) Behavioral ecology of entomopathogenic nematodes. *Biological Control* 38, 66–79.

McKerrow, J.H., Caffrey, C., Kelly, B., Loke, P. and Sajid, M. (2006) Proteases in parasitic diseases. *Annual Review of Pathology – Mechanisms of Disease* 1, 497–536.

Mello, C. and Fire, A. (1995) DNA transformation. In: Shakes, D.C. and Epstein, H.F. (eds) *Methods in Cell Biology*, Vol 48. Academic Press, San Diego, pp. 451–482.

Mello, C.C., Kramer, J.M., Stinchcomb, D. and Ambros, V. (1991) Efficient gene-transfer in *C. elegans* – extrachromosomal maintenance and integration of transforming sequences. *Embo Journal* 10, 3959–3970.

Milstone, A.M., Harrison, L.M., Bungiro, R.D., Kuzmic, P. and Cappello, M. (2000) A broad spectrum Kunitz type serine protease inhibitor secreted by the hookworm *Ancylostoma ceylanicum*. *Journal of Biological Chemistry* 275, 29391–29399.

Mracek, Z., Becvar, S., Kindlmann, P. and Webster, J.M. (1998) Infectivity and specificity of Canadian and Czech isolates of *Steinernema kraussei* (Steiner, 1923) to some insect pests at low temperatures in the laboratory. *Nematologica* 44, 437–448.

Nguyen, K.B. and Smart, G.C. (1990) *Steinernema scapterisci* n. sp. (Rhabditida, Steinernematidae). *Journal of Nematology* 22, 187–199.

Nguyen, K.B. and Smart, G.C. (1991) Pathogenicity of *Steinernema scapterisci* to selected invertebrates. *Journal of Nematology* 23, 7–11.

Nimkingrat, P., Uhlmann, F., Strauch, O. and Ehlers, R.-U. (2013a) Desiccation tolerance of dauers of entomopathogenic nematodes of the genus *Steinernema*. *Nematology* 15, 451–458.

Nimkingrat, P., Strauch, O. and Ehlers, R.U. (2013b) Hybridisation and genetic selection for improving desiccation tolerance of the entomopathogenic nematode Steinernema feltiae. *Biocontrol Science and Technology* 23, 348–361.

Oerke, E.C. (2006) Crop losses to pests. *Journal of Agricultural Science* 144, 31–43.

Pava-Ripoll, M., Posada, F.J., Momen, B., Wang, C. and Leger, R.S. (2008) Increased pathogenicity against coffee berry borer, *Hypothenemus hampei* (Coleoptera: Curculionidae) by *Metarhizium anisopliae* expressing the scorpion toxin (AaIT) gene. *Journal of Invertebrate Pathology* 99, 220–226.

Peters, A. and Ehlers, R.-U. (1998) Evaluation and selection for enhanced nematode pathogenicity against *Tipula* spp. In: Simoes, N., Boemare, N. and Ehlers, R.-U. (eds) *Pathogenicity of Entomopathogenic Nematodes Versus Insect Defence Mechanisms: Impact on Selection of Virulent Strains.* European Commission Publications COST 819, Brussels, pp. 225–241.

Praitis, V., Casey, E., Collar, D. and Austin, J. (2001) Creation of low-copy integrated transgenic lines in *Caenorhabditis elegans*. *Genetics* 157, 1217–1226.

Rao, Y.S., Chai, X.W., Wang, Z.F., Nie, Q.H. and Zhang, X.Q. (2013) Impact of GC content on gene expression pattern in chicken. *Genetics Selection Evolution* 45, 9, DOI: 10.1186/1297-9686-45-9.

Robertson, H.M. and Thomas, J.H. (2006) The putative chemoreceptor families of *C. elegans*. *WormBook* 6, 1–12.

Romeis, J., Meissle, M. and Bigler, F. (2006) Transgenic crops expressing *Bacillus thuringiensis* toxins and biological control. *Nature Biotechnology* 24, 63–71.

Salame, L., Glazer, I., Chubinishvilli, M.T. and Chkhubianishvili, T. (2010) Genetic improvement of the desiccation tolerance and host-seeking ability of the entomopathogenic nematode *Steinernema feltiae*. *Phytoparasitica* 38, 359–368.

Schulenburg, H., Hoeppner, M.P., Weiner, J. and Bornberg-Bauer, E. (2008) Specificity of the innate immune system and diversity of C-type lectin domain (CTLD) proteins in the nematode *Caenorhabditis elegans*. *Immunobiology* 213, 237–250.

Shapiro-Ilan, D.I., Gouge, D.H. and Koppenhofer, A.M. (2002) Factors affecting commercial success: case studies in cotton, turf and citrus. In: Gaugler, R. (ed.) *Entomopathogenic Nematology*. CAB International, Wallingford, UK, pp. 333–356.

Srinivasan, J., Dillman, A.R., Macchietto, M.G., Heikkinen, L., Lakso, M., *et al.* (2013) The draft genome and transcriptome of *Panagrellus redivivus* are shaped by the harsh demands of a free-living lifestyle. *Genetics* 193, 1279–1295.

St Leger, R.J. and Wang, C. (2010) Genetic engineering of fungal biocontrol agents to achieve greater efficacy against insect pests. *Applied Microbial Biotechnology* 85, 901–907.

St Leger, R.J., Joshi, L., Bidochka, M.J. and Roberts, D.W. (1996) Construction of an improved mycoinsecticide over-expressing a toxic protease. *Proceedings of the National Academy of Sciences of the United States of America* 93, 6349–6354.

Steiner, G. (1929) *Neoaplectana glaseri* n. g., n. sp. (Oxyuridae), a new nemic parasite of the Japanese beetle (*Popillia japonica* Newm.). *Journal of the Washington Academy of Sciences* 29, 436–440.

Thomas, J.H. and Robertson, H.M. (2008) The *Caenorhabditis* chemoreceptor gene families. *BMC Biology* 6, 42, DOI: 10.1186/1741-7007-6-42.

Tomalak, M. (1994) Selective breeding of *Steinernema feltiae* (Filipjev) (Nematoda, Steinernematidae) for improved efficacy in control of a mushroom fly, *Lycoriella solani* Winnertz (Diptera, Sciaridae). *Biocontrol Science and Technology* 4, 187–198.

Toubarro, D., Lucena-Robles, M., Nascimento, G., Costa, G., Montiel, R., *et al.* (2009) An apoptosis-inducing serine protease secreted by the entomopathogenic nematode *Steinernema carpocapsae*. *International Journal for Parasitology* 39, 1319–1330.

Toubarro, D., Lucena-Robles, M., Nascimento, G., Santos, R., Montiel, R., *et al.* (2010) Serine protease-mediated host invasion by the parasitic nematode *Steinernema carpocapsae*. *Journal of Biological Chemistry* 285, 30666–30675.

Vadnal, J., Ratnappan, R., Keaney, M., Kenney, E., Eleftherianos, I., O'Halloran, D. and Hawdon, J.M. (2017) Identification of candidate infection genes from the model entomopathogenic nematode *Heterorhabditis bacteriophora*. *BioMed Central Genomics* 18, 8. doi:10.1186/s12864-016-3468-6.

Vaeck, M., Reynaerts, A., Hofte, H., Jansens, S., Debeuckeleer, M., *et al.* (1987) Transgenic plants protected from insect attack. *Nature* 328, 33–37.

van Megen, H., van Den Elsen, S., Holterman, M., Karssen, G., Mooyman, P., *et al.* (2009) A phylogenetic tree of nematodes based on about 1200 full-length small subunit ribosomal DNA sequences. *Nematology* 11, 927–950.

Vellai, T., Molnar, A., Laktos, L., Banfalvi, Z., Fodor, A. and Saringer, G. (1999) Transgenic nematodes carrying a cloned stress resistance gene from yeast. In: Glazer, I., Richardson, P., Boemare, N. and Coudert, F. (eds) *Survival of Entomopathogenic Nematodes*. Office for Official Publications of the European Communities, Luxembourg, pp. 105–119.

Wang, C. and St Leger, R.J. (2007) A scorpion neurotoxin increases the potency of a fungal insecticide. *Nature Biotechnology* 25, 1455–1456.

Whetstone, P.A. and Hammock, B.D. (2007) Delivery methods for peptide and protein toxins in insect control. *Toxicon* 49, 576–596.

Wilson, M., Xin, W.M., Hashmi, S. and Gaugler, R. (1999) Risk assessment and fitness of a transgenic entomopathogenic nematode. *Biological Control* 15, 81–87.

Zang, X. and Maizels, R.M. (2001) Serine proteinase inhibitors from nematodes and the arms race between host and pathogen. *Trends in Biochemical Sciences* 26, 191–197.

26 Breeding Entomopathogenic Nematodes for Enhanced Insect Pest Suppression

S. Subramanian* and M. Muthulakshmi

Department of Nematology, Tamil Nadu Agricultural University, Coimbatore, India

26.1 Introduction

Entomopathogenic nematodes (EPNs) have successful commercial application against many pests, such as scarab larvae in turf and lawns, insect pests in protected horticulture, mole crickets in lawn and turf, black vine weevil in nursery plants, fungus gnats in mushroom production and diaprepes root weevil in citrus (Lacey and Georgis, 2012). Yet biopesticides based on EPNs and their symbiotic bacteria are aimed exclusively for relatively small niche markets. Therefore, it is clear that the EPN market share should be increased. A promising approach to attain such an increase should be *via* developing new strains of nematodes and bacteria suitable for biological control of insect pests in unprotected horticulture and other agriculture systems (Glazer, 2015). EPN strains with improvements in their pathogenicity, host range, environmental tolerance and shelf life are essentially necessary to meet the current agricultural/horticultural pest problems. Actually, many of such EPN attributes have been identified (Gaugler, 2002; Grewal *et al.*, 2005) and promising relevant approaches are in progress (Campos-Herrera, 2015).

The increased knowledge of the basic biology of EPNs and their associated symbiotic bacteria has led to awareness among researchers of the potential for the genetic improvement of these nematodes and bacteria. The genetic studies carried out on *Heterorhabditis* and *Steinernema* have used both conventional and new techniques. The former approach, that is, selective breeding, inbreeding and mutagenesis, is still utilized effectively and will be reviewed herein. Also, recent genetic improvements of EPNs for enhanced biological control have been summarized (Glazer, 2015; see Baiocchi *et al.*, Chapter 25, this volume). Apart from phylogenetics and diagnostics, molecular approaches were used more extensively in genetic studies of the symbiotic bacteria *Xenorhabdus* and *Photorhabdus* than in the studies of their hosts, *Steinernema* and *Heterorhabditis*, respectively (Dowds, 1994; Burnell, 2002; Campos-Herrera, 2015).

26.2 Methodology for Genetic Improvement

26.2.1 Classical genetics

The beneficial traits of microorganisms are often inherited polygenically. Three main approaches have been in use in classical genetics: selective breeding, cross-breeding and mutagenesis. Many scientists use 'cross-breeding' to express the reproduction between two distinct but conspecific microorganisms (usually two strains/isolates related to the same species), while 'hybridization' is used to express this reproduction but between two species. With EPNs, the two terms are mostly used as synonyms referring to the reproduction between two distinct but conspecific nematode strains (Shapiro-Ilan *et al.*, 1997).

Selective breeding

For the polygenically inherited traits of EPNs, artificial selection of beneficial attributes is the most adequate method for genetic improvement (Burnell, 2002). Therefore, the efficacy of selective breeding

*Corresponding author. E-mail: subbunema@yahoo.co.in

in the genetic improvement of beneficial traits has been well established for *Heterorhabditis bacteriophora*; the heritability values for the traits of host finding, ultraviolet tolerance and heat tolerance were high (Glazer *et al.*, 1991), and therefore, were more responsive to process the techniques of selective breeding. These values are usually deducted from inbred lines or the degree of resemblance between relatives. It is also very important that the gene pool of the intended trait(s) is wide enough for the selection process (Abd-Elgawad and Nguyen, 2007). Early selective breeding programmes have been applied successfully to *Steinernema* for the traits of host finding and improved pest control efficiency (Tomalak, 1994). Griffin and Downes (1994) reported the successful selection of *Heterorhabditis* for improved infectivity at low temperatures. A recent record of previous breeding programmes, in terms of selection breeding, hybridization and genetic transformation, intended for genetic improvement of EPNs has been published (Glazer, 2015).

Crucially, characterization of beneficial traits and their distribution among the utilized EPN species and strains can be the cornerstone for success in nematode application in a definite microclimate and/or special geographic area. Thus, developing predictive power on such desirable attributes can assist in rational decisions *via* EPN application. Moreover, characterization of relative desirable trait qualities among nematode strains/species can assist in developing strain improvement programmes, e.g. *via* hybridization and selective breeding (Shapiro-Ilan *et al.*, 1997; Mukuka *et al.*, 2010; Shapiro-Ilan *et al.*, 2014a). For example, Shapiro-Ilan *et al.* (2014a) found no correlation between freeze and desiccation tolerance among ten EPN species; most of them were of recent commercial application. Those authors, therefore, reported that other newly isolated or uncharacterized species could not be examined to predict these two traits of stress. Yet the screening of the possibly existing variation among the ten EPN species, which included numerous strains, could help in selecting the proper nematode for addressing desiccation or freezing tolerance, as well as for further comparisons (Strauch *et al.*, 2004; Shapiro-Ilan *et al.*, 2014a).

Examples of selective breeding programmes for enhancing EPN efficacy are presented in Table 26.1. The reader should consider that in some instances the selected trait in the laboratory did not translate into enhanced EPN efficacy under field conditions. Nevertheless, selective breeding is likely to remain an important component of EPN genetic improvement programmes, due to its potential efficacy in improving general nematode performance (Campos-Herrera, 2015).

Basic steps necessary for a selective breeding programme in the field of EPNs were suggested (Gaugler *et al.*, 1989; Burnell, 2002; as in Fig. 26.1). Gaugler *et al.* (1989) proposed that a clarification of potential heritability of a desirable trait might be approached using a number of ecologically distinct strains *via* assessing their phenotypic variability for that trait (Gaugler *et al.*, 1989). It should also be possible to analyse heritability by obtaining gravid first generation of EPN females individually to pick up their progeny and utilize a variety of full sib families. This technique will have merit because each of these sib families is derived from an individual infective juvenile (IJ).

Genotypic changes can be introduced into organisms by mutagenesis, but mutagenesis suffers from the drawback that only one or a small number of genes can be altered at any one time. An additional problem is that mutagenesis usually has the effect of inactivating genes, so it is not a very useful approach for the introduction of new genetic information. Constitutive expression of certain stress-response genes may be a means of protecting EPNs or their symbionts from environmental stresses.

Hybridization

Hybridization is another non-molecular avenue to improve EPN populations/strains. It aims at moving beneficial traits *via* cross-breeding described above from one nematode population/strain to another. Shapiro-Ilan *et al.* (1997) proved the ability to transfer a desirable trait, heat tolerance, between two different populations of the same EPN species; that is, reproduction between the heat-tolerant population *H. bacteriophora* IS-5 (Glazer *et al.*, 1996) and the laboratory *H. bacteriophora* HP88 population. The hybrid nature of the offspring was verified with a mutant resulting from *H. bacteriophora* HP88 (Hp-dpy-2) as a marker (Koltai *et al.*, 1994), as well as *via* backcrossing. Screening the offspring for heat tolerance was carried out *via* assessing the IJs survival after 2 h exposure at 40°C. Shapiro-Ilan *et al.* (1997) found that survival of the hybrid progeny showed a higher rate than that of *H. bacteriophora* HP88, and similar to that of

Table 26.1. Examples of promising isolate characterization and selection breeding courses for species of *Heterorhabditis* and *Steinernema*.[a]

Entomopathogenic nematode species	Intended trait	Heritability value (h^2)	Approach	Outputs	Fitness change	Stability	Notes	References
Heterorhabditis bacteriophora	Heat tolerance	0.68	Four selection rounds	Increase from 38.5 to 39.2°C	NT	NT	High heritability	Ehlers *et al*., 2005
H. bacteriophora	Heat and desiccation tolerance	ND	Exposure at 37°C or low relative humidity (25 ± 0.5°C, 40–57% relative humidity)	Strain EG_4 showed the best heat tolerance, but strains EG_2, EG_3 and EG_4 had higher tolerance to desiccation than others	NT	NT	Promising preliminary characterization of isolates for selection breeding	Abd-Elgawad and Nguyen, 2007
H. bacteriophora	Desiccation tolerance in osmotic solution	ND	Six selection rounds	Reduced value of water activity	No change	NT		Mukuka *et al*., 2010
Steinernema carpocapsae	Increased dispersal	0.60	Capturing the fastest and furthest reaching IJs exiting from an infected *Galleria mellonella*	21-37 fold increase in the per cent of IJs dispersing from the source cadaver	Reduced reproduction capacity and nictation ability	NT	The selected lines had more males (72 %) than the foundation population (44 %)	Bal *et al*., 2014
S. diaprepesi	Relatively dry soil	ND	Three cycles of migration *via* soil to infect insect larvae 10 cm distant	Survival for 7 days in the absence of food was greatest at 6% soil moisture	NT	NT	Promising preliminary characterization of isolates for selection breeding	El-Borai *et al*., 2016
S. feltiae	Host-finding ability	0.50	Thirteen rounds in host-finding bioassay	20-27 fold increase in host finding	Better host penetration and reproduction; loss in storage stability	Gradual decrease	Used high-variation foundation strain	Gaugler and Campbell, 1989; Gaugler *et al*., 1990

Note: [a]The table is not meant to be a comprehensive record (see also Grewal *et al*., 2005; Glazer, 2015). ND = not determined; NT = not tested.

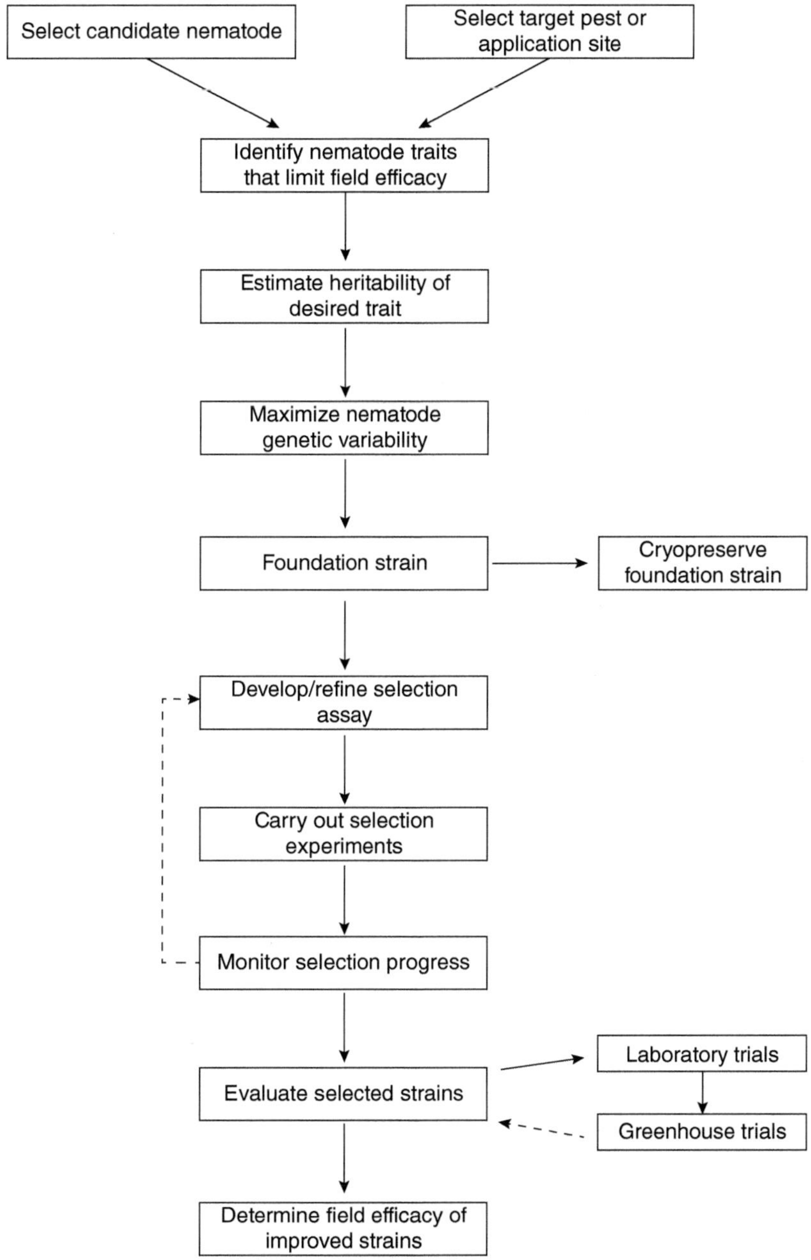

Fig. 26.1. A schematic illustration to design a selection breeding course for entomopathogenic nematodes. (From Gaugler *et al.*, 1989; Burnell, 2002.)

H. bacteriophora IS-5, after six passages through *Galleria mellonella* last-instar larvae. Clearly, *H. bacteriophora* IS-5 and the hybrid populations killed these insect larvae more rapidly than *H. bacteriophora* HP88 at 32°C. Yet at 10°C, the two heat-tolerant populations demonstrated sensitivity to cold storage.

The life cycle of heterorhabditid nematodes is not identical to that of the steinernematids, since the first heterorhabditid generation is totally hermaphrodite but subsequent generations include females, males and hermaphrodites (Strauch *et al.*, 1994). On the other hand, only amphimictic females and males characterize steinernematid species, except *Steinernema hermaphroditum*, which has hermaphrodites also in the first adult generation (Stock *et al.*, 2004). Therefore, contrary to most steinernematids, special care must be exercised to ensure that progeny nematodes in controlled heterorhabditid crosses arise from intended mating rather than self-replication. In order to overcome this obstruction, marker mutations could be utilized (Shapiro-Ilan *et al.*, 1997; Shapiro-Ilan *et al.*, 2014b). Also, Shapiro-Ilan *et al.* (2005) utilized hybridization to obtain superior virulence and environmental tolerance in *Steinernema carpocapsae* populations. Both hybridization and selection, as non-molecular methods, may be used in combination to produce elite EPN populations that can withstand environmental extremes (Mukuka *et al.*, 2010).

Mutagenesis

The biology and short length of the EPN life cycle render them as tractable organisms for strain improvement through mutagenesis as well. Routine screening of nematode IJs for a desirable trait can be quite time-consuming, while lethal screens can offer a comfortable approach of mutant selection. This latter is quite suitable for screening strains with enhanced stress tolerances. If a dose mortality curve is acquired for the parental strain for an environmental stress such as temperature extreme, UV tolerance, desiccation and nematicides, then such ecological factors that lead to 100% mortality of the parental strain can be utilized to screen for resistant mutants. Likewise, behavioural screens, as in strains with differences in their host range, can also be made to ease mass screening for mutant selection (Burnell, 2002). The behaviour of *S. carpocapsae* demonstrated a sensitive thermotactic regime (Burman and Pye, 1980). A thermal gradient is commonly used whereupon the nematodes will accumulate at their favourable temperature. Cryophilic mutant strains were obtained from *Heterorhabditis* sp. K122 *via* using the mutagen ethylmethane sulfonate (EMS) (Nugent, 1996). All the acquired cryophilic mutants demonstrated an enhanced infection potential for *Tenebrio molitor* larvae at 9°C but not at 20°C. Mutants of *Heterorhabditis megidis* that tolerate desiccation have been obtained after EMS mutagenesis (O'Leary and Burnell, 1997). Burnell (2002) concluded that the EMS mutagenesis should generally have more application in EPN strain improvement plans, especially for strain improvements against environmental stresses.

Other protocols, for example transposon mutagenesis, have been developed to obtain a working system for EPNs. Molecular parasites capable of replicating their DNA sequences into new sites within their host's genome are known as transposons. Yet their major problems are the availability of manpower and resources, in addition to research focus (Burnell, 2002). Insertion of a transposon into the coding region of a nematode gene usually causes a gene mutation. The resulting mutated gene may further be identified on a polyacrylamide gel.

26.2.2 Genetic engineering

The impact of genetic engineering on strain improvement is seen most clearly in genetically engineered microorganisms (GEMs). GEMs were initially used as research tools, but are now widely used in the biotechnology industry for the production of a variety of enzymes and pharmaceutical products, such as hormones, growth factors and antibiotics (Puhler, 1993). Novel techniques in the field of engineering EPNs have been suggested (Glazer, 2014).

EPNs and their symbiont bacteria both have taxonomically related genera that are used as model organisms in molecular genetics (*Caenorhabditis elegans* and *Escherichia coli*, respectively). Thus, it has frequently been suggested that the techniques that have been developed for these model organisms could be readily adapted for use with EPNs and their symbiont bacteria, and that genetic improvement of EPNs and/or their symbionts by transgenic means will be achieved in the near future. We believe, however, that the rate of progress is likely to be relatively slow as there are many inherent difficulties to be resolved, the major one being the paucity of basic research knowledge on the biochemistry and genetics of the traits that might be changed in EPNs and their bacteria. In many cases, we do not know what genes might be usefully inserted into these organisms for the purposes of strain improvement. Many desirable phenotypes are likely to result from the concerted action of an array of genes and will, therefore, be less amenable to genetic engineering. Thus, it will be necessary to identify single genes whose products directly confer an advantageous phenotype to the transgenic nematode or bacterium, or to target genes coding for key regulatory enzymes in critical metabolic pathways (Hope, 1994).

26.3 Targets for Breeding Programmes of Entomopathogenic Nematodes

The main targets for breeding programmes in *Heterorhabditis* and *Steinernema* are almost the same as those for genetic improvements. These targets are increased efficacy, resistance to environmental extremes, the development of anhydrobiotic strains and increased suitability of *Heterorhabditis* for culture in liquid fermenters. For genetic improvement, these targets may also be addressed as infectivity, persistence and survival (see Baiocchi *et al.*, Chapter 25, this volume).

26.3.1 Efficacy

Efficacy can be broken down into many components, such as dispersal activity in the soil, host-finding ability, host specificity, host penetration, avoidance, or suppression of the immune response of the host, successful packaging and transmission of the symbiont bacterium. Such complex physiological and behavioural attributes are likely to result from the action of many interacting genes. Thus, selective breeding would be an appropriate means by which genetic improvements could be made to phenotypes considered to be important in EPN efficacy. Selection experiments are most likely to succeed if a single trait is selected or if two traits, at most, are co-selected. Several laboratory studies have shown that, under apparently favourable conditions for infection, only a proportion (in some cases as low as 13%) of individual IJs in a population are infective at any one time (Curran and Heng, 1992). Although culture conditions can influence infectivity significantly, the proportion of IJs in a population that is infective at any one time under defined conditions appears to be species and strain related (Curran, 1993). This suggests that the control of the proportion of IJs in the population that is actively infective has a genetic basis and may be amenable to modification by genetic means. If activation into an infective state depends on the perception of environmental cues by the IJ, then mutagenesis would seem to be the most appropriate means of modifying such a sensory activation system.

26.3.2 Resistance to environmental extremes

Improvements to EPN physiology and biochemistry that would increase the survival time or the infective capacity of IJs in unfavourable environments (e.g. at low or high temperatures, in conditions of low soil moisture or on foliar surfaces) would provide new opportunities for the use of EPNs in biological control. On the other hand, physiological and behavioural variations between EPN species that conformed to documented differences in habitat preference, such as the water potential of inhabited soil, were reported by El-Borai *et al.* (2016), who speculated that such a behavioural character was probably favourable in leading the engineering or selection breeding of EPN species for utilization under different environmental systems.

Many of the complex physiological traits that are responsible for the physiological tolerances of animals are polygenic in nature; thus, the most effective means of genetically improving such traits is by selective breeding. However, in situations where a key regulatory gene may be involved in controlling the expression of several genes in a coordinated pathway, mutagenesis may be appropriate. Griffin and Downes (1994) have reported the successful selection of *Heterorhabditis* for improved infectivity at low temperatures. and O'Leary and Burnell (1995) have isolated desiccation-resistant strains of *Heterorhabditis* by mutagenesis. In cases where biochemical or molecular genetic studies clearly indicate that the product of a known gene can affect the environmental physiology of EPNs, then the construction of transgenic nematodes may yield improved strains. For example, the chilling tolerance of tobacco plants has been modified successfully by altering the degree of fatty acid unsaturation of phosphatidylglycerol in plants transformed with the gene coding for glycerol-3-acyltransferase (Murata *et al.*, 1992). EPNs are sensitive to inactivation by UV radiation (Gaugler *et al.*, 1992) and this, in combination with their poor desiccation tolerance, limits their usefulness for the biological control of insect pests on foliar surfaces. The mechanism of UV tolerance and damage repair in prokaryotes and eukaryotes is well understood at the molecular level (Strid *et al.*, 1994), so this trait may be amenable to genetic manipulation by transgenic means in EPNs. Selective breeding or mutagenesis would seem to be the most promising methods for the genetic improvement of anhydrobiotic potential in EPNs.

Nematode variability of genetic bases – concerning tolerance to desiccation, heat and hypoxia, as well as fitness in terms of reproduction potential and virulence – among different *S. carpocapsae* populations could be estimated *via* phenotypic differences. Pronounced variations were noted among these natural populations for such tolerance traits. North Carolina populations demonstrated more pronounced stress tolerance than other populations extracted from Ohio. Interestingly, important differences could be found among populations extracted from the same locality. The percentage of IJ survival was 37–82% after exposure to 40°C for 2 h. Difference in IJ survival after exposure to hypoxic regime or osmotic desiccation was as high as threefold. Moreover, some of the tested populations performed better than the commonly used 'All' strain in these tolerance traits. Strikingly, a population designated KMD33 demonstrated superiority to the 'All' strain across the tested traits. Reproductive potential and virulence varied greatly among populations. Yet these two traits were less variable than the traits of tolerance to desiccation, heat and hypoxia. Eventually, the pronounced variability in tolerance attributes among these populations proposed the utilization of a selection technique to improve the above-mentioned traits genetically (Somasekhar *et al.*, 2002). Topics related to enhancing EPNs' performance as effective bioinsecticides against environmental extremes and addressing their relevant physiology and biochemistry, survival biology, behavioural ecology and formulation, application and production technology have been reported (e.g. Glazer *et al.*, 1997; Gaugler, 2002; Grewal *et al.*, 2006; Campos-Herrera, 2015).

26.4 Entomopathogenic Bacteria

The main targets for genetic improvement of *Xenorhabdus* and *Photorhabdus* are pathogenicity, host specificity, symbiont specificity and resistance to environmental extremes such as cold or desiccation and control of phase variation.

26.4.1 Pathogenicity

Most pathogenicity work performed with EPNs has used the wax moth, *G. mellonella*, as the target insect. These insects are extremely sensitive to EPNs; for example, when *G. mellonella* larvae are injected with *Xenorhabdus* or *Photorhabdus* spp., the LD_{50} is less than 3 (Jackson *et al.*, 1995). However, this does not reflect the situation with other insect species. For example, the LD_{50} for *Xenorhabdus bovienii* is 40,000 for *Tipula oleracea* (Wulff *et al.*, 1993) and *Xenorhabdus nematophilus* was relatively non-pathogenic to *Popillia japonica* (Yeh and Alm, 1992).

There are two main classes of insect-pathogenic bacteria: spore formers and non-spore formers. *Xenorhabdus* and *Photorhabdus* fit into the class of non-spore formers. This group is usually pathogenic as a result of the production of extracellular enzymes and/or lipopolysaccharides (LPSs) that destroy the haemocytes and internal organs of the insect after the bacteria have reached the haemocoel (Lysenko, 1985). Little work has been done on characterizing the virulence determinants of *Xenorhabdus* and *Photorhabdus*. They both secrete lipases, lecithinases and proteases. Clarke and Dowds (1995) have shown that the secreted lipase contributes to the virulence of *Photorhabdus* sp. towards *G. mellonella*. Virulence might be improved by increased toxin activity or specificity, or improved avoidance by cells of various components of the insect immune response. However, much basic research is needed to elucidate the mechanisms of virulence.

26.4.2 Host specificity

It is clear from the different LD_{50} values mentioned above that *Xenorhabdus* and *Photorhabdus* have different degrees of virulence for different insect species. Further work needs to be carried out to assess their virulence towards a wide range of both insect pests and beneficial insects. In addition, the biochemical basis of this difference in the sensitivity of different insect species has not been investigated. In the absence of this information, there is no possibility of genetically engineered bacteria for specificity for certain insect pests while ensuring that beneficial insects are unharmed. One possible type of structure of importance or bacterial adhesion to host tissue is the fimbriae on the bacterial surface (Brehelin *et al.*, 1993). In other bacterial species, fimbriae are known to play a major role in host specificity because of their role in adhesion to surfaces. A strain of *S. carpocapsae* (G-13) selectively bred for improved host finding showed an increased fitness concerning host penetration and reproduction potential and a lack of fitness for storage stability (Gaugler *et al.*, 1990).

The G13 strain of *S. carpocapsae* (Weiser), previously selected for enhanced host finding of wax moth larvae, *G. mellonella* L., was subjected to selection for the same trait against scarab larvae. Twenty rounds of selection produced a 3.7-fold increase, from 18.4 to 68.7%, in nematode (S20 strain) location of non-diapausing larvae of the Japanese beetle, *P. japonica* Newman. However, the overall improvement over wild types (Foundation and All strains) after 33 rounds (13 against *G. mellonella* and 20 against scarabs) was 153- and 72-fold, respectively. IJs had difficulty locating diapausing *P. japonica* larvae, and selection did not result in any enhancement in finding these larvae. Selection for host finding of scarabs resulted in a 24.6% gain in host finding of *G. mellonella*, suggesting selection was for a general feature present at different levels in different hosts. The use of potassium hydroxide (KOH) to absorb CO_2 resulted in the complete inhibition of the S20 strain response to insect hosts, verifying that nematodes have been selected for enhanced sensitivity to CO_2. S20 nematodes were significantly less storage stable than the Foundation strain (Gaugler and Campbell, 2000).

26.4.3 Symbiont specificity

Studies have shown that it is possible, in some cases, to grow *Steinernema* and *Heterorhabditis* on bacteria other than their own specific symbionts (Gerritsen and Smits, 1993). When symbionts were exchanged between *Heterorhabditis* species and strains, only some nematode bacterial combinations were compatible (Gerritsen and Smits, 1993), and in some cases, bacterial strains that could support nematode growth were not packaged by the nematodes, resulting in non-pathogenic IJs. Han *et al.* (1991) have shown that by exchanging bacteria between compatible *Heterorhabditis* strains, the pathogenicity of these strains was increased or decreased, depending on the bacteria on which they were grown. Thus, although there is a degree of specificity for the bacterial strain that will support the growth of, and be carried by, a given nematode strain, some scope exists for the improvement of EPN pathogenicity by making different combinations of bacteria and nematodes.

26.4.4 Resistance to extremes in the environment

EPNs are ineffective at the low temperatures prevailing in temperate climates in spring and autumn. Work on the adaptation of *Photorhabdus* and *X. nematophilus* to low temperatures (Leisman *et al.*, 1995), and a review of the differences between psychrophiles and mesophiles and of low temperature shock in other species of bacteria (Clarke and Dowds, 1994), suggests that many genes are involved in low temperature adaptation. These might include genes coding for cold shock and outer membrane proteins, and are expected to include enzymes involved in desaturation of membrane fatty acids. By analogy to other stress responses, it is possible that a small number of regulatory genes are involved in controlling expression of the adaptive genes, suggesting that mutations in the regulatory genes might yield cold-active mutant strains. However, it is not known whether such strains would display low temperature virulence as well as low temperature growth.

Another stress adaptation that might be improved upon usefully is the response to desiccation. It is known that the nematode is sensitive to drying, but the sensitivity of the bacteria has not yet been determined. A range of adaptations of bacteria to desiccation have been demonstrated or hypothesized, including trehalose production, membrane structure and composition and capsule formation (Potts, 1994). Nothing is known about these characteristics in *Xenorhabdus* and *Photorhabdus*, except that these bacteria have fluid membranes typical of psychrotrophs (Clarke and Dowds, 1994) and that the primary forms synthesize capsules (Brehelin *et al.*, 1993).

Temperature extremes in terms of cold and heat sensitivity may severely restrict the biocontrol potential of EPNs in definite environments and/or seasons. *Steinernema feltiae* strain SN and its symbiotic bacterium *X. bovenii* successfully underwent a course of selection for improved cold tolerance *via* continuous passage through *G. mellonella* larvae at 15°C (Grewal *et al.*, 1996). Such a selection led to increased IJ virulence at low temperatures. At 8°C, the nematode attained the best increase of ninefold. The results indicated that such functional traits were correlated genetically along a continuously distributed ecological variable, but the fitness function was not always constant. Trade-offs in nematode-induced mortality and rapid kill of insect host were observed at the higher temperature *via* six passages of artificial selection (Grewal *et al.*, 1996). Emelianoff *et al.* (2008) stressed that these symbiotic bacteria played an important role in the survival–reproduction trade-offs of EPNs. Bal *et al.* (2014) examined *S. carpocapsae* for enhanced dispersal and reported its associated trade-offs.

26.4.5 Phase variation

Xenorhabdus and *Photorhabdus* display a form of phenotypic instability known as phase variation. Thus, a culture grown from a single colony of primary phase (phase I) generates secondary forms (phase II) at a rate that is highly dependent on the strain. This has proved to be a major nuisance in the industrial production of the nematodes because the yield of nematodes from growth on secondary-form bacteria is only a fraction of that from primary forms (Akhurst, 1980). In addition, the primary form is preferentially retained by the infective-stage nematode. The reason for these differences may be that the secondary form fails to produce a range of enzymes and secondary metabolites made by the primary form, including extracellular protease, lipase, antibiotics, fimbriae, capsule and many other products. Possibly because of the conservation of energy by the secondary form, it can recommence growth following starvation much more rapidly than the primary form, indicating a more efficient nutrient uptake system. The secondary form also has higher levels of respiratory enzymes (Smigielski *et al.*, 1994). These authors suggest that the secondary phase may be better adapted to conditions in soil as free-living bacteria. It seems apparent that mutating *Xenorhabdus* and *Photorhabdus* into a stable primary form would be advantageous to the industrial production of these bacteria. However, the role of the secondary form in the life cycle or in adapting to changes in the environment needs to be elucidated before we can be certain that such mutants would be advantageous when applied in the field. Many pathogens, and possibly most bacterial populations, undergo various forms of phase variation, all of which assist the bacteria in adapting to the variety of conditions that they encounter (Brunham *et al.*, 1993). It is unlikely that such a complex change in metabolism is without major adaptive implications for *Xenorhabdus* and *Photorhabdus* as well. There may also be problems with the technology of generating a mutant that is completely locked into the primary phase. Work is accumulating that suggests that there are multiple regulatory controls affecting the phase shift. There may be a single regulatory gene controlling a number of other regulators, but so far there is no evidence for this. The inactivation of phase-variant characteristics seems to be regulated at a post-translational level for lipase and protease (Wang and Dowds, 1993), but might be regulated by antisense RNA in the case of bioluminescence, as suggested by Hosseini (1994). Hosseini (1994) examined the phase-variant characteristics of *Photorhabdus luminescens* and found factors such as low-salt media and rifampicin to affect some, but not all, of these characteristics (different sets in each case). Thus, any postulated primary level of control of the phase-variant phenotypes must have a variable effect on the different characteristics, if indeed such an overall control exists. The reader can find more details about *Xenorhabdus* spp. in Javed *et al.*, Chapter 11, this volume, and *Photorhabdus* spp. in Abd-Elgawad, Chapter 12, this volume.

26.5 Risks of Release of Genetically Modified Entomopathogenic Nematodes and/or their Symbiont Bacteria

The field release of genetically modified EPNs and symbiont bacteria may be beneficial in terms of improved pest control, harmful to new target pests and a reduced use of less selective and more toxic pesticides. The regulatory authorities will assess the risks posed by the release of genetically engineered organisms (GEOs).

The possible ecological risks posed by the introduction into the environment of GEOs are:

- the transfer of foreign genes to other organisms in the ecosystem, especially to deleterious organisms such as pathogens and pests
- the possibility that the GEOs will survive better than the parent strain in some ecological systems, with unknown, but possibly negative, outcomes (Burnell, 2002).

It seems likely that these symbiotic bacteria may not be released from the nematodes or insects into the soil in significant numbers and/or that the bacteria may not survive for significant lengths of time in the free state in soil. *Photorhabdus* can receive genes from *E. coli* by conjugation, and gene transformation among other bacterial species may occur (Dowds, 1994). A specific risk from such events might be the generation of insect-pathogenic species of bacteria from non-pathogens, with a possibly broader host range. The most obviously damaging target would be beneficial insect species, such as pollinators, but the infection of other species is not impossible. For example, a strain of *Photorhabdus* has been isolated from human wounds (Colepicolo *et al.*, 1989), where the bacteria clearly survive (see Abd-Elgawad, Chapter 12, this volume).

Ehlers and Peters (1995) have evaluated the possible environmental impacts and risks of the inundative release of non-engineered EPNs for the control of insect pests. Their salient observations are:

- that *Steinernema* and *Heterorhabditis* are limited to the soil environment and their host spectrum is limited to arthropods, particularly insects
- all available evidence indicates that the application of EPNs has an insignificant impact on non-target hosts, including beneficial insects (the majority of beneficial insects do not have soil-dwelling larvae)
- following inundative release, EPN numbers decline rapidly
- EPN population dynamics are influenced by soil-dwelling nematophagous antagonists such as fungi and invertebrate predators.

26.6 Conclusions

Breeding of EPNs for enhanced insect pest suppression is one of the most important aspects that can improve EPNs performance as bioinsecticides. The procedures followed in classical genetics in breeding EPNs and their symbiotic bacteria for enhanced insect pest suppression can be implemented for EPN genetic improvement in parallel to genetic engineering. Numerous desirable traits are probably controlled polygenically. So, selective breeding to address such traits would be more adequate. Better understanding of basic and applied genetics of such traits in both EPNs and their symbionts should facilitate our endeavours to produce 'super' nematodes as bioinsecticides.

Due to the technical difficulties involved in the engineering of EPNs and their symbionts, it is likely that, at least in the initial stages, transgenic EPNs and bacteria will differ from their parent strains by only a single gene or by small numbers of genes. Thus, it is possible that the release of such genetically modified organisms (GMOs) would pose less threat than the release of non-native EPN species, since the latter would result in the release of a greater amount of novel genes into the ecosystem. As Tiedje *et al.* (1989) point out, however, the important consideration is the character of the phenotypic changes expressed after the manipulation of the organism's genotype and not simply how many genes have been added or deleted.

Numerous desirable traits have been targeted for breeding and results reported (e.g. Campos-Herrera *et al.*, 2015; Table 26.1). Actually, utilizing superior nematode strains may contribute to the attractive attributes mentioned above, or even novel traits that could be conceived. Successful EPN breeding for a desired trait usually has at least a twofold goal. For example, a nematode strain with higher efficacy (e.g. host-finding ability, environmental tolerance, virulence) against insect pest(s) will need fewer EPNs per unit area, which is a cost-effective approach too. Likewise, the more yield *via* EPN mass production, gained by a nematode strain, or typical IJs produced, the better cost-efficiency it takes into consideration (Shapiro-Ilan *et al.*, 2014b). Initially, surveys can be carried out to discover new strains and widen the nematodes' genetic pool. Then, the above-mentioned breeding techniques should be implemented correctly. Such an implementation should consider the adequate breeding technique for each case, utilize strain(s) that may possess superior levels of desired traits, which can then be screened to compare the resulting strain(s) to the existing ones, and assess the efficacy gap between laboratory and field conditions. Therefore, the screening process is usually attained by initial selection of the best nematode candidates in the laboratory. The laboratory-based selections should target various desired traits, such as nematode efficacy to control a definite pest(s) under simulated actual conditions. In the case of selecting strain(s) with high mass production, the strain's reproductive capacity should be examined under full-scale production systems. Finally, assuring the desired insect pest suppression under field conditions is crucial and should be confirmed using the nematode strain(s) in its commercial/ultimate formulation.

Acknowledgements

We are very pleased to place on record the help and services provided by the staff and students of the Department of Nematology, Tamil Nadu Agricultural University, Coimbatore, Tamil Nadu, India, during the preparation of this chapter.

References

Abd-Elgawad, M.M.M. and Nguyen, K.B. (2007) Isolation, identification and environmental tolerance of new heterorhabditid populations from Egypt. *International Journal of Nematology* 17(2), 116–123.

Akhurst, R.J. (1980) Morphological and functional dimorphism in *Xenorhabdus* spp., bacteria symbiotically associated with the insect pathogenic nematodes *Neoaplectana* and *Heterorhabditis*. *Journal of General Microbiology* 121, 303–309.

Bal, K.H., Michel, P.A. and Grewal, S.P. (2014) Genetic selection of the ambush foraging entomopathogenic nematode, *Steinernema carpocapsae* for enhanced dispersal and its associated trade-offs. *Evolutionary Ecology* 28, 923–939.

Brehelin, M., Cherqui, A., Drif, L., Luciani, J., Akhurst, R. and Boemare, N. (1993) Ultrastructural study of surface components of *Xenorhabdus* sp. in different cell phases and culture conditions. *Journal of Invertebrate Pathology* 61, 188–191.

Brunham, R.C., Plummer, F.A. and Stephens, R.S. (1993) Bacterial antigenic variation, host immune response, and pathogen-host coevolution. *Infection and Immunity* 61, 2273–2276.

Burman, M. and Pye, A.E. (1980) *Neoaplectana carpocapsae*: movement of nematode populations on a thermal gradient. *Experimental Parasitology* 49, 258–265.

Burnell A. (2002) Genetics and genetic improvement. In: Gaugler, R. (ed.) *Entomopathogenic Nematology*. CAB International, Wallingford, UK, pp. 333–356.

Campos-Herrera, R. (ed.) (2015) *Nematode Pathogenesis of Insects and Other Pests*. Springer International Publishing, Switzerland, 521 pp.

Campos-Herrera, R., El-Borai, F.E. and Duncan, L.W. (2015) Modifying soil to enhance biological control of belowground dwelling insects in citrus groves under organic agriculture in Florida. *Biological Control* 84, 53–63.

Clarke, D.J. and Dowds, B.C.A. (1994) Cold adaptation in *Photorhabdus* spp. In: Burnella, M., Ehlers, R.-U. and Massoni, P. (eds) *Genetics of Entomopathogenic Nematode–Bacterium Complexes*. European Commission Publication EUR 15681 EN, Luxembourg, pp. 170–177.

Clarke, D.J. and Dowds, B.C.A. (1995) Virulence mechanisms of *Photorhabdus* sp. strain K122 towards wax moth larvae. *Journal of Invertebrate Pathology* 66, 149–155.

Colepicolo, P., Cho, K.W., Poinar, G.C. Jr and Hastings, J. (1989) Growth and luminescence of the bacterium *Xenorhabdus luminescens* from a human wound. *Applied and Environmental Microbiology* 55, 2601–2606.

Curran, J. (1993) Post application biology of entomopathogenic nematodes in soil. In: Bedding, R.A., Akhurst, R.J. and Kaya, H.K. (eds) *Nematodes and the Biological Control of Insect Pests*. CSIRO Publications, Melbourne, Australia, pp. 67–77.

Curran, J. and Heng, I. (1992) Comparison of three methods for estimating the number of entomopathogenic nematodes present in soil samples. *Journal of Nematology* 24, 172–176.

Dowds, B.C.A. (1994) Molecular genetics of *Xenorhabdus* and *Photorhabdus. Proceedings of the VIth International Colloquium on Invertebrate Pathology and Microbial Control. Society for Invertebrate Pathology*, Montpellier, France, pp. 95–100.

Ehlers, R.-U. and Peters, A. (1995) Entomopathogenic nematodes in biological control: feasibility, perspectives and possible risks. In: Hokkanenh, M.T. and Lynchi, M. (eds) *Biological Control: Benefits and Risks*. Cambridge University Press, Cambridge, UK, pp. 119–136.

Ehlers, R.-U., Oestergaard, J., Hollmer, S., Wingen, M. and Strauch, O. (2005) Genetic selection for heat tolerance and low temperature activity of the entomopathogenic nematode–bacterium complex *Heterorhabditis bacteriophora–Photorhabdus luminescens. BioControl* 50, 699–716.

El-Borai, F., Killiny, N. and Duncan, L.W. (2016) Concilience in entomopathogenic nematode responses to water potential and their geospatial patterns in Florida. *Frontiers in Microbiology* 7, 356, DOI: 10.3389/fmicb.2016.00356.

Emelianoff, V., Chapuis, E., Le Brun, N., Chiral, M., Moulia, C. and Ferdy, J.B. (2008) A survival–reproduction trade-off in entomopathogenic nematodes mediated by their bacterial symbionts. *Evolution* 62, 932–942.

Gaugler, R. (ed.) (2002) *Entomopathogenic Nematology*. CAB International, Wallingford, UK, 388 pp.

Gaugler, R. and Campbell, J.F. (1989) Selection for host-finding in *Steinernema feltiae. Journal of Invertebrate Pathology* 54, 363–372.

Gaugler, R. and Campbell, J.F. (2000) Selection for enhanced host-finding of scarab larvae (Coleoptera: Scarabaeidae) in an entomopathogenic nematode. *Environmental Entomology* 20, 700–706.

Gaugler, R., McGuire, T. and Campbell, J. (1989) Genetic variability among strains of the entomopathogenic nematode *Steinernema feltiae. Journal of Nematology* 21, 247–253.

Gaugler, R., Campbell, J.F. and McGuire, T.R. (1990) Fitness of a genetically improved entomopathogenic nematode. *Journal of Invertebrate Pathology* 56, 106–116.

Gaugler, R., Bednarek, A. and Campbell, J.F. (1992) Ultraviolet inactivation of heterorhabditid and steinernematid nematodes. *Journal of Invertebrate Pathology* 59, 155–160.

Gerritsen, L.J.M. and Smits, P.H. (1993) Variation in pathogenicity of recombinations of *Heterorhabditis* and

Xenorhabdus luminescens strains. *Fundamental and Applied Nematology* 16, 367–373.

Glazer, I. (2014) Genetic improvement and breeding of EPN: the race for the "super nematode". *Journal of Nematology* 46, 168.

Glazer, I. (2015) Improvement of entomopathogenic nematodes: a genetic approach. In: Campos-Herrera, R. (ed.) *Nematodes Pathogenesis of Insects and Other Pests*. Springer International Publishing, Switzerland, pp. 29–56.

Glazer, I., Gaugler, R. and Segal, D. (1991) Genetics of the nematode *Heterorhabditis bacteriophora* strain HP88: the diversity of beneficial traits. *Journal of Nematology* 23, 324–333.

Glazer, I., Kozodoi, E., Hashmi, G. and Gaugler, R. (1996) The entomopathogenic nematode *Heterorhabditis bacteriophora* IS-5; a heat tolerant isolate from Israel. *Nematologica* 42, 481–492.

Glazer, I., Salame, I. and Segal, D. (1997) Genetic enhancement of nematicide resistance in entomopathogenic nematodes. *Biocontrol Science and Technology* 7, 499–512.

Grewal, P.S., Gaugler, R. and Wang, Y. (1996) Enhanced cold tolerance of the entomopathogenic nematode *Steinernema feltiae* through genetic selection. *Annals of Applied Biology* 129, 335–341.

Grewal, P.S., Ehlers, R.-U. and Shapiro-Ilan, D.I. (eds) (2005) *Nematodes as Biocontrol Agents*. CAB International, Wallingford, UK, 505 pp.

Grewal, P.S., Bornstein-Forst, S., Burnell, A.M., Glazer, I. and Jagdale, G.B. (2006) Physiological, genetic, and molecular mechanisms of chemoreception, thermobiosis, and anhydrobiosis in entomopathogenic nematodes. *Biological Control* 38, 54–65.

Griffin, C.T. and Downes, M.J. (1994) Selection of *Heterorhabditis* sp. for improved infectivity at low temperatures. In: Burnella, M., Ehlers, R.-U. and Masson, J.P. (eds) *Genetics of Entomopathogenic Nematode–Bacterium Complexes.* European Commission Publications, Luxembourg, pp. 143–149.

Han, R., Wouts, W.M. and Li, L. (1991) Development of *Heterorhabditis* spp. strains associated with different *Xenorhabdus luminescens* isolates. *Journal of Invertebrate Pathology* 58, 27–32.

Hope, I.A. (1994) Genetic transformation of *Caenorhabditis elegans.* In: Burnella, M., Ehlers, R.-U. and Masson, J.P. (eds) *Genetics of Entomopathogenic Nematode–Bacterium Complexes.* European Commission Publications, Luxembourg, pp. 14–22.

Hosseini, P.K. (1994) Regulation of bioluminescence in *Photorhabdus* luminescens HM, primary and secondary forms. MSc thesis, University of Wisconsin, Milwaukee, Wisconsin.

Jackson, T.J., Wang, H., Nugent, M.J., Griffin, C.T., Burnell, A.M. and Dowds, B.C.A. (1995) Isolation of insect pathogenic bacteria, *Providencia rettgeri*, from *Heterorhabditis* spp. *Journal of Applied Bacteriology* 78, 237–244.

Koltai, H., Glazer, I. and Segal, D. (1994) Phenotypic and genetic analysis of two mutants of *Heterorhabditis bacteriophora* strain HP88. *Journal of Nematology* 26, 32–39.

Lacey, L.A. and Georgis, R. (2012) Entomopathogenic nematodes for control of insect pests above and below ground with comments on commercial production. *Journal of Nematology* 44(2), 218–225.

Leisman, G.B., Wakau, J. and Forst, S.A. (1995) Characterization and environmental regulation of outer membrane proteins in *Xenorhabdus nematophilus*. *Applied and Environmental Microbiology* 61, 200–204.

Lysenko, O. (1985) Non-spore forming bacteria pathogenic to insects: incidence and mechanisms. *Annual Review of Microbiology* 39, 673–695.

Mukuka, J., Strauch, O., Hoppe, C. and Ehlers, R.-U. (2010) Improvement of heat and desiccation tolerance in *Heterorhabditis bacteriophora* through crossbreeding of tolerant strains and successive genetic selection. *BioControl* 55, 511–521.

Murata, N., Ishizaki-Nishizawa, O., Higashi, S., Hayashi, H., Tasaka, Y. and Nishida, I. (1992) Genetically engineered alteration in the chilling sensitivity of plants. *Nature* 356, 710–713.

Nugent, M.J. (1996) An investigation of cold tolerance and the genetic improvement of the entomopathogenic nematode *Heterorhabditis*. PhD thesis. National University of Ireland, Maynooth, Ireland.

O'Leary, S.A. and Burnell, A.M. (1995) The genetic improvement of desiccation tolerance in the insect parasitic nematode *Heterorhabditis*. Proceedings of the 22nd International Symposium of the European Society of Nematologists held at Ghent, Belgium, 7–14 August, 1994. *Nematologica* 5, 451–458.

O'Leary, S.A. and Burnell, A.M. (1997) The isolation of mutants of *Heterorhabditis megidis* (Strain UK211) with increased desiccation tolerance. *Fundamental and Applied Nematology* 20, 197–205.

Potts, M. (1994) Desiccation tolerance of prokaryotes. *Microbiological Reviews* 58, 755–805.

Puhler, A. (1993) *Genetic Engineering of Microorganisms*. VCH, Weinheim, Germany.

Shapiro-Ilan, D.I., Glazer, I. and Segal, D. (1997) Genetic improvement of heat tolerance in *Heterorhabditis bacteriophora* through hybridization. *Biological Control* 8, 153–159.

Shapiro-Ilan, D.I., Stuart, J.R. and McCoy, W.C. (2005) Targeted improvement of *Steinernema carpocapsae* for control of the pecan weevil, *Curculio caryae* (Horn) (Coleoptera: Curculionidae) through hybridization and bacterial transfer. *Biological Control* 34, 215–221.

Shapiro-Ilan, D.I., Brown, I. and Lewis, E. (2014a) Freezing and desiccation tolerance in entomopathogenic nematodes: diversity and correlation of traits. *Journal of Nematology* 46, 27–34.

Shapiro-Ilan, D.I., Han, R. and Qiu, X. (2014b) Production of entomopathogenic nematodes. In: Morales-Ramos, J., Rojas, G. and Shapiro-Ilan, D.I. (eds) *Mass Production of Beneficial Organisms: Invertebrates and Entomopathogens*. Academic Press, San Diego, California, pp. 321–356.

Smigielski, A.J., Akhurst, R.J. and Boemare, N.E. (1994) Phase variation in *Xenorhabdus nematophilus* and *Photorhabdus luminescens*: differences in respiratory activity and membrane energization. *Applied and Environmental Microbiology* 60, 120–125.

Somasekhar, N., Grewal, P.S. and Klein, M.G. (2002) Genetic variability in stress tolerance and fitness among natural populations of *Steinernema carpocapsae*. *Biological Control* 23, 303–310.

Stock, S.P., Griffin, C.T. and Chaerani, R. (2004) Morphological and molecular characterization of *Steinernema hermaphroditum* n. sp. (Nematoda: Steinernematidae), an entomopathogenic nematode from Indonesia, and its phylogenetic relationships with other members of the genus. *Nematology* 6, 401–412.

Strauch, O., Stoessel, S. and Ehlers, R.U. (1994) Culture conditions define automictic or amphimictic reproduction in entomopathogenic rhabditid nematodes of the genus *Heterorhabditis*. *Fundamental and Applied Nematology* 17, 575–582.

Strauch, O., Oestergaard, J., Hollmer, S. and Ehlers, R.U. (2004) Genetic improvement of the desiccation tolerance of the entomopathogenic nematode *Heterorhabditis bacteriophora* through selective breeding. *Biological Control* 31(2), 218–226.

Strid, A., Chow, W.S. and Anderson, J.M. (1994) UV-B damage and protection at the molecular level in plants. *Photosynthesis Research* 39, 475–489.

Tiedje, J.M., Colwell, R.K., Grossman, Y.N., Hodson, R.E. Lenski, R.E., *et al*. (1989) The planned introduction of genetically engineered organisms: ecological considerations and recommendations. *Ecology* 70, 298–315.

Tomalak, M. (1994) Selective breeding of *Steinernema feltiae* (Filipjev) (Nematoda, Steinernematidae) for improved efficacy in control of a mushroom fly *Lycoriella solani* Winnertz (Diptera, Sciaridae). *Biocontrol Science and Technology* 4, 187–198.

Wang, H. and Dowds, B.C.A. (1993) Phase variation in *Xenorhabdus luminescens*: cloning and sequencing of the lipase gene and analysis of its expression in primary and secondary phases of the bacterium. *Journal of Bacteriology* 175, 1665–1673.

Wulff, A., Peters, A. and Ehlers, R.-U. (1993) Pathogenicity of the *Steinernema feltiae–Xenorhabdus bovienii* complex to *Tipula oleracea*. *IOBCIWPRS Bulletin* 17, 99–102.

Yeh, T. and Alm, S.R. (1992) Effects of entomopathogenic nematode species, rate, soil moisture, and bacteria on control of Japanese beetle (Coleoptera: Scarabaeidae) larvae in the laboratory. *Journal of Economic Entomology* 85, 2144–2148.

Part VI: Slug Parasitic Nematodes

27 Slug Parasitic Nematodes: Biology, Parasitism, Production and Application

Jiří Nermuť* and Vladimír Půža

Biology Centre CAS, Institute of Entomology, České Budějovice, Czech Republic

27.1 Introduction

Slugs are widely distributed pests that cause damage in a large range of crops, especially in moist areas all around the world. One of the most important slug pests in the Americas, Europe and Australia is the grey garden slug, *Deroceras reticulatum* (family Agriolimacidae). *Arion vulgaris* (family Arionidae), an invasive species, causes damage to many crops in different parts of Europe. The slug pests are also reported from other families such as Limacidae and Milacidae. Recently, some changes in agricultural practices like cultivation reduction and the use of cover crops have contributed to the increase of slug populations. Therefore, farmers and researchers around the entire world direct their attention to the methods of slug regulation, by chemical, biological and other means. Biological control is gaining attention these days, as this method is non-hazardous, cost-effective and generally safe to non-target organisms. Among the biocontrol agents, parasitic nematodes, *Phasmarhabditis hermaphrodita* have attracted attention in the past two decades among scientists and farm growers (Askary, 2010; Askary *et al.*, 2012). Unfortunately, there is no or very low efficacy of this parasite on the big species such as *A. vulgaris* (Senoldová and Rotrekl, 2005) or *Achatina fulica* (Williams and Rae, 2015), and many others. Because there are no other bioagents applicable against harmful terrestrial molluscs, these dismal conditions generate the need to find another, more effective and environmentally safe way for slug and snail control. Nematodes, as the most common parasites of molluscs (Grewal *et al.*, 2003), are one of the best choices. Nematodes belonging to several families, namely Agfidae, Alaninematidae, Alloionematidae, Angiostomatidae, Cosmocercidae, Diplogasteridae, Mermithidae and Rhabditidae, are known as common parasites of many slugs and snails belonging to class Gastropoda (Fig. 27.1; Table 27.1).

Despite this fact, the knowledge of their biology and ecology is very poor, and so far only one species, *P. hermaphrodita* (family Rhabditidae), has been introduced in the market. This nematode can be mass produced in modified solid media (Bedding, 1984), but liquid media in air-lift fermenters up to 20,000 l can also be used (Wilson and Grewal, 2005). Harvested nematodes are then formulated in vermiculite, polymer or clay and distributed for field applications where a variety of equipment ranging from simple watering to tractor sprayers are used for application in soil. Even though other rhabditid, cephalobid (Charwat and Davies, 1999) or alloionematid nematodes have received attention as potential biocontrol agents for slugs (Laznik *et al.*, 2010; Nermut', unpublished), *P. hermaphrodita* is the only commercially produced slug parasitic nematode used by growers.

Parasitic nematodes are to be found in all organs of the general body cavity, pallial cavity and also foot muscle (Morand *et al.*, 2004). They are able to infect slugs through the body wall, orally, or during mating of two individuals. The life cycles of slug parasitic nematodes are usually very varied. Generally, there

*Corresponding author. E-mail: Jirka.Nermut@seznam.cz

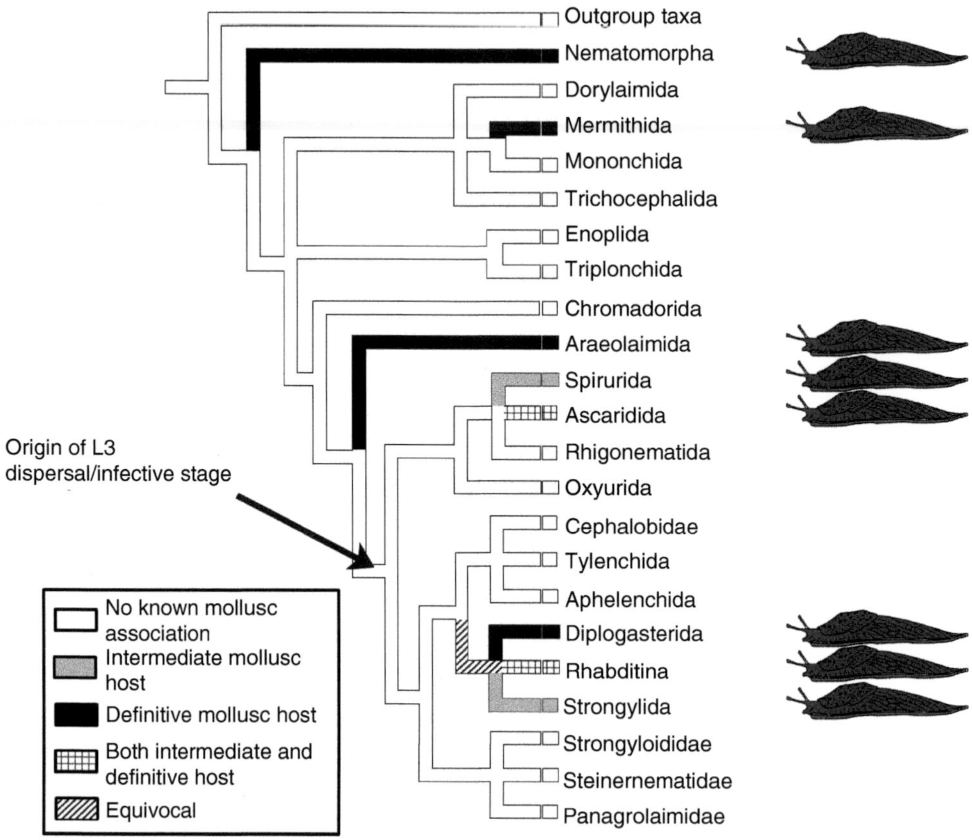

Fig. 27.1. Phylogenetic distribution of mollusc host utilization among the Nematoda. (Phylogenetic tree is modified according to Grewal *et al*., 2003.)

are described saprobic, necromenic and parasitic cycles (Wilson and Grewal, 2005). However, many slug parasitic nematodes are able to switch among these cycles; for example, *P. hermaphrodita* or *Alloionema appendiculatum* are able to use all three cycles according to the environmental conditions. Unfortunately, the biology and ecology of most species is unknown. We only have more information on the parasitism and influence on mortality, feeding activity or reproduction of host for three nematodes *P. hermaphrodita*, *A. appendiculatum* and *Nemhelix bakeri*, while the studies of other species are at very early stages.

This chapter provides thorough information on the biology and ecology of slug parasitic nematodes, besides assessing the potential for their commercial use in the biological control of slug pests, with special emphasis on production, storage and techniques of application.

27.2 Biology and Parasitism of Slug Parasitic Nematodes

27.2.1 Biology and ecology

Nematodes are usually a thread-like, cylindrical, generally fusiform, rarely sac-like and almost ubiquitous organism. Their life strategies range from free-living bacteriophages to invertebrate or vertebrate, but also plant, parasites. The size of nematodes is from microscopic to giant worms like *Placentonema gigantissima*, which can be more than 8 m long. But mollusc parasitic nematodes usually never exceed several millimetres, except for Mermithids. The life cycle of most nematodes consists of several stages: egg, four larval stages and adults. In parasitic species, the infective stages are usually third-stage larvae and adults. But the general biology of nematodes is very well known and we will therefore focus mainly on some interesting aspects of slug parasitic

Table 27.1. Selected parasitic nematodes of terrestrial molluscs.

Family	Species	Mollusc host	References
Agfidae	*Agfa flexilis* Dujardin	*Limax cinereoniger* Wolf (Limacidae)	Morand *et al.*, 2004
	Agfa morandi Ribas and Casanova	*Limax* sp. Linnaeus (Limacidae)	Ribas and Casanova, 2002
	Agfa tauricus Korol and Spiridonov	*Krynickillus melanocephalus* Kaleniczenko (Agriolimacidae)	Korol and Spiridonov, 1991
		Limacus flavus Linnaeus (Limacidae)	Korol and Spiridonov, 1991
		Oxychilus deilus Bourguignat (Zonitidae)	Korol and Spiridonov, 1991
Alaninematidae	*Alaninema venmansi* Théodorides	*Amphidromus contrarius* Müller (Camaenidae)	Théodorides, 1957
	Alaninema njoroensis Puylaert	Leaf-veined slugs (Athoracophoridae)	Ivanova *et al.*, 2013b
	Alaninema ngata Ivanova *et al.*	Leaf-veined slugs (Athoracophoridae)	Ivanova *et al.*, 2013b
Alloionematidae	*Alloionema appendiculatum* Schneider	Agriolimacidae, Arionidae, Helicidae, Hygromiidae and Succineidae	Nermuť *et al.*, 2015
	Neoalloionema tricaudatum Ivanova *et al.*	*Cyclophorus* sp. Montfort (Cyclophoridae)	Ivanova *et al.*, 2016
Angiostomatidae	*Angiostoma aspersae* Morand	*Cantareus aspersus* Müller (Helicidae)	Morand, 1986
	Angiostoma limacis Dujardin	Arionidae	Mengert, 1953
	Angiostoma sp.	Arionidae, Agriolimacidae, Limacidae, Milacidae, Zonitidae and Rhytididae	Morand *et al.*, 2004
Cosmocercidae	*Cosmocercoides dukae* Holl	North American slugs and snails	Morand *et al.*, 2004
	Nemhelix bakeri Morand and Petter	European helicid snails	Morand and Petter, 1986
Diplogasteridae	*Hugotdiplogaster neozelandia* Morand and Barker	*Athoracophorus bitentaculatus* Quoy and Gaimard (Athoracophoridae)	Morand and Barker, 1995
	Diplogasterids	*Limacus flavus* Linnaeus (Limacidae)	Charwat and Davies, 1999
		Cernuella virgata da Costa (Hygromiidae)	Charwat and Davies, 1999
		Theba pisana Müller (Helicidae)	Charwat and Davies, 1999
Mermithidae	*Mermis nigrescens* Dujardin	*Deroceras reticulatum* Müller (Agriolimacidae)	Chitwood and Chitwood, 1937
	Hexamermis albicans Siebold	*Deroceras berytensis* Bourguigant	Chitwood and Chitwood, 1937
	Mermithids	*Succinea pfeifferi* Risso	Mienis, 1986
		Veronicella ameghini Gambetta (Vaginulidae)	Dundee, 1977
		Limacus flavus Linnaeus (Limacidae)	Thiengo, 1995
		Sarsinula marginata Semper (Vaginulidae)	Thiengo, 1995

Continued

Table 27.1. Continued.

Family	Species	Mollusc host	References
Rhabditidae	*Phasmarhabditis hermaphrodita* Schneider	Limacidae, Agriolimacidea, Arionidae, Vaginulidae	Rae *et al.*, 2007
	Phasmarhabditis neopapillosa Mengert	Limacidae, Agriolimacidea, Arionidae, Vaginulidae	Mengert, 1953
	Caenorhabditis briggsae Dougherty and Nigon	*Deroceras* sp. Rafinesque (Agriolimacidae)	Nermut' (unpublished)
	Caenorhabditis elegans Maupas	*Limax* sp. Linnaeus (Limacidae)	Ross *et al.*, 2012
	Rhabditis gracilicauda de Man	*Arion owenii* Davies (Arionidae)	Ross *et al.*, 2010a
	Rhabditis axei Cobbold	*Achatina achatina* Linnaeus (Achatinidae)	Odaibo *et al.*, 2000

nematodes biology and behaviour. Most of the mollusc parasitic nematodes spend at least a part of their life cycle in the soil, and the soil environment is the place where most molluscs are usually infected with nematodes. Naturally, there are some exceptions, see *Mermis nigrescens* or *N. bakeri* below. Nematodes that infect molluscs directly in the soil need some mechanism for how to find and approach the mollusc. Slug and snail movement is very slow, but still too fast for the nematodes. So, how can nematodes catch and infect slugs when their natural density is very low? Quite easily: during their long evolution, mollusc parasitic nematodes adopted some tricky behaviour. As we know, at least *P. hermaphrodita* (Rae *et al.*, 2006; Nermut' *et al.*, 2012) and *A. appendiculatum* (Nermut', unpublished), but very probably other nematodes also, readily react to some host-associated cues. The cue can be CO_2 or other volatile compounds produced by the living slug, its mucus, faeces, etc. Nematodes react very strongly to all of them, not only to alive slug. Of course it is not surprising, we know that *P. hermaphrodita* is able to complete the whole life cycle also on slug faeces and other organic material (Nermut' *et al.*, 2014a), and the same is true for *A. appendiculatum* (Mengert, 1953; Nermut', unpublished). But this behaviour also gives the nematode another advantage. Molluscs, especially slugs, have a homing behaviour. It means that the animal uses the same shelter every day, and this place is usually covered by a large amount of mucus and faeces. Therefore, if the nematodes readily react to these cues, they highly increase their chance to meet a suitable host. It is interesting to note that *P. hermaphrodita* can react readily not only to water-soluble cues, as do other nematodes, but also to volatile cues (Nermut' *et al.*, 2012) that can be related to its habitat, soil surface and organic matter where its slug hosts also live.

As mentioned above, at least *P. hermaphrodita* is able to complete its life cycle on different organic matter in the soil. But is the development and quality of this slug parasitic nematode affected by the substrate on which it grows? It certainly is. But of interest is that the quality of the growing substrate is expressed mostly in the yield of dauer juveniles (DJs), not in the quality of progeny (Nermut' *et al.*, 2014a). Very similar results were also found in experiments with *A. appendiculatum* (Nermut', unpublished), while in entomopathogenic nematodes (EPNs), for example, the quality of substrate influenced not only the yield but also the quality of dauers (infective juveniles). The reason for this difference could be that while EPNs are true parasites, mollusc parasitic nematodes still retain both parasitic and free-living life cycles, and the ability to produce full quality juveniles in a wide range of conditions is a big advantage that helps them to survive in various environments.

As was mentioned above, knowledge on the biology of mollusc parasitic nematodes is very poor in general, and prevalence and natural occurrence is not the exception. This state is a consequence of the regrettably low interest of the scientific community in this topic. Despite only anecdotal data being available, it seems that the occurrence of many mollusc parasitic nematodes is patchy, and in the case of *A. appendiculatum* and *P. hermaphrodita*, prevalence varies from zero to 100%, depending on the habitat and host species (Charwat and

Davies, 1999; Morand *et al.*, 2004; Ross *et al.*, 2010a; Nermut', unpublished). Detailed biology and basic information on the morphology of eight mollusc parasitic families are introduced below under different sections, and are also summarized in tabular form in Table 27.1.

27.3 Morphology of Different Families of Slug Parasitic Nematodes

27.3.1 Agfidae

A brief description

Knowledge on the family Agfidae is still very poor, and the position of this family in the system of nematodes is uncertain. Actually, there are three described nematodes in the family Agfidae (Morand *et al.*, 2004): *Agfa flexilis*, a parasite of the slug *Limax cinereoniger* (Limacidae), *Agfa morandi*, parasitic in *Limax* sp. (Limacidae), and *Agfa tauricus*, a parasite of *Krynickillus melanocephalus* (Agriolimacidae), *Limacus flavus* (Limacidae) and *Oxychilus deilus* (Zonitidae). All of these species are exclusive parasites of slugs (Ribas and Casanova, 2002) and are described from the salivary glands or genital tract of the mollusc host, even though Ivanova *et al.* (2013a) stated that the occurrence of *Agfa* in genital tracts is probably erroneous information.

All species are known, for example, from Europe (Morand *et al.*, 2004; Ivanova *et al.*, 2013a) and South Africa (Ross *et al.*, 2012) or North America (Ross *et al.*, 2010a). *A. flexilis* is very common in France, Belgium and the UK (Ivanova *et al.*, 2013a), while *A. tauricus* has so far only been reported from Crimea and Bulgaria and is seemingly restricted to southern regions of Eastern Europe.

MORPHOLOGY. Elongated pharynx with a valvated basal bulb; oral opening with three lips; stoma small and cylindrical; males with bursa and six pairs of pedunculate papillae; adult nematodes 2–4 mm long (Morand *et al.*, 2004).

27.3.2 Alaninematidae

A brief description

The Alaninematidae belonging to group Panogrolaimorpha is a very poorly known family that includes only three species; *Alaninema venmansi* from Niki-Niki (South Timor), isolated from *Amphidromus contrarius* (Gastropoda: Camaenidae), *Alaninema njoroensis* from Njoro in Kenya and *Alaninema ngata* from New Zealand, isolated from leaf-veined slugs (Gastropoda: Athoracophoridae). These nematodes live in the digestive tract or pallial cavity of their hosts, but detailed biology is unknown. According to present knowledge, it seems that this family is restricted to the southern hemisphere.

MORPHOLOGY. Small pharynx with a basal valveless bulb; oral opening circular; nerve ring posterior to the pharynx; males without bursa; nematodes 3–8 mm long (Morand *et al.*, 2004).

27.3.3 Alloionematidae

A brief description

The family Alloionematidae consists of two genera: *Alloionema*, with only one species, *A. appendiculatum*; and *Rhabditophanes*, with three or four valid species, *Rhabditophanes schneideri*, *Rhabditophanes cobbi*, *Rhabditophanes aphodii* and *Rhabditophanes insolitus*. Nematodes belonging to the genus *Rhabditophanes* are free-living, saprophytic nematodes occurring in moulds, rotting wood, excrements and the frass of bark beetles. In cow dung, they are associated with *Aphodius* beetles (Andrassy, 1983), but *R. schneideri* has also been reported as phoretic on cave pseudoscorpions (Čurčič *et al.*, 2008).

A. appendiculatum is a common larval parasite of many terrestrial molluscs that was first described by Schneider (1859) from the body of the slug *Arion ater*. This nematode has both a parasitic and a free-living life cycle (Mengert, 1953). Its third-stage larvae (DJs) invade the foot muscle of snails and slugs, and after moulting into fourth-stage larvae, they remain encysted inside the host's tissue. These fourth-stage larvae are able to leave slugs to mature and reproduce in the soil (parasitic generation). The progeny of the parasitic generation make a free-living, saprophytic generation that can persist in suitable organic material for a long time (currently, the culture has been held in laboratory conditions for 4 years; Nermut', unpublished data). The development of these saprophytic forms is very fast: the whole cycle is complete within 3 or 4 days. When the source of food is depleted, new DJs are produced and spread in the soil to infect hosts (Fig. 27.2). All stages of both saprophytic and parasitic generations are bacteriophagous and freely associated with many bacteria; for example,

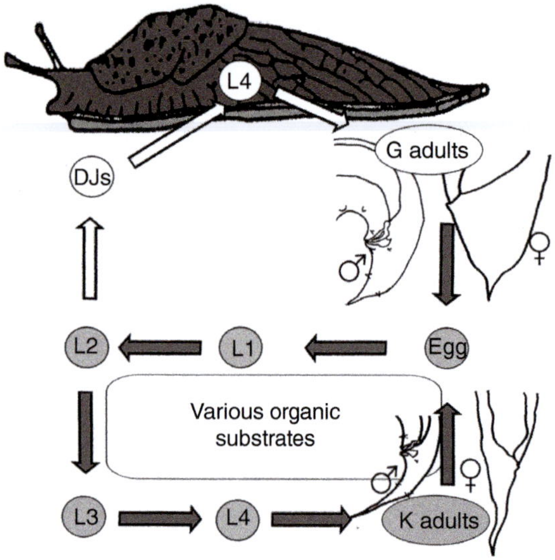

Fig. 27.2. The life cycle of *Alloionema appendiculatum*. *Notes*: DJs = dauer juveniles; L1 = first larval stage; L2 = second larval stage; L3 = third larval stage (called DJs in some special cases); L4 = fourth larval stage; G adults = parasitic adults; K adults = saprophytic adults.

Acinetobacters sp., *Pseudomonas* sp., *Neisseiria* sp. (Nermut', unpublished).

A. appendiculatum is a parasite of molluscs belonging to the families Agriolimacidae, Arionidae, Helicidae, Hygromiidae and Succineidae. The prevalence of *A. appendiculatum* ranges from less than 0.01% in *Cantareus aspersus* to up to 100% in some arionid slugs (Morand *et al.*, 2004; Nermut', unpublished). Despite the low prevalence in natural populations of *C. aspersus*, this nematode is very frequent in heliculture conditions, where *A. appendiculatum* causes a high mortality of farmed snails (Cabaret *et al.*, 1988). A high mortality of *A. vulgaris* (syn. *Arion lusitanicus*) was observed by Laznik *et al.* (2010) and studied thoroughly by Nermut' *et al.* (2014b), who did not find significant increase of mortality of *A. vulgaris* after infection with *A. appendiculatum*, except in stressful conditions (high water content in a rearing substrate).

Alloionema has been reported from Europe, Asia, North America (Andrassy, 1983; Ross *et al.*, 2010a) and Australia (Charwat and Davies, 1999).

MORPHOLOGY. Saprophytic (S) adults are smaller than parasitic (P) ones, body length: male 930 μm (S) or 1500 μm (P), female 1300 μm (S) or 2000 μm (P), DJs 450 μm. Pharynx has a basal valvated bulb and median bulb without valves; oral opening is triangular; males bursa absent (Fig. 27.3). Tail digitate (P) or filiform (S) (Morand *et al.*, 2004).

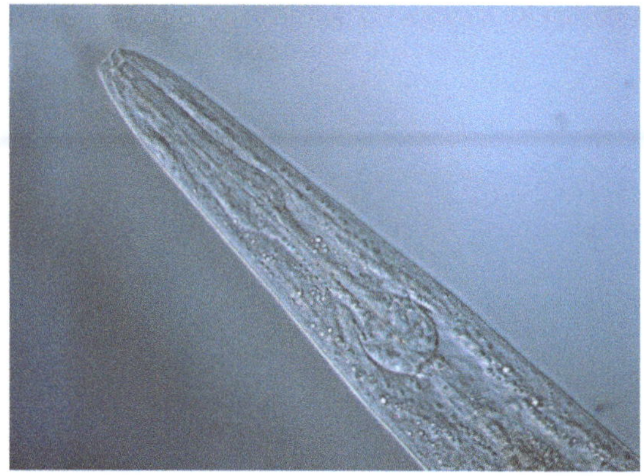

Fig. 27.3. *Alloionema appendiculatum*: female head part with pharynx.

27.3.4 Angiostomatidae

A brief description

The Angiostomatidae family has worldwide distribution. They are known from Europe, Africa, New Zealand and South-east Asia. Actually, there are 17 described species in two genera, *Angiostoma* and *Aulacnema*, 13 from terrestrial molluscs and 4 from reptiles and amphibians (Ross *et al.*, 2011), but it seems that the amphibian host of *Angiostoma plethodontis* is infected only by ingesting parasitized molluscs. Angiostomatids usually parasitize the digestive tract of their host, but *Angiostoma aspersae* has been found in the pallial cavity of *C. asperses* (Helicidae) and *Angiostoma gladicola* has been recovered from the hepatopancreas of *Megaustenia* sp. (Helicarionidae). Furthermore, *Angiostoma carettae* lives in the lungs of sea turtles. Except for the two mollusc families mentioned above, *Angiostoma* also parasitize the families Arionidae, Agriolimacidae, Limacidae, Milacidae, Zonitidae and Rhytididae.

Among the various relationships between nematodes and molluscs, *Angiostoma* is truly parasitic (Grewal *et al.*, 2003). Nematodes parasitize their host as adults or fourth-stage larvae. Eggs are released from the host's body with its faeces or mucus, and rapidly develop outside until they reach the infective stage (first–second stage in ovoviviparous species, or third stage in oviparous species). Infective larvae are swallowed by the host with food, or they actively penetrate the mantle cavity (Morand, 1986, 1988).

MORPHOLOGY. Pharynx with a valvated basal bulb; median bulb without valves; oral opening circular or triangular; stoma cylindrical; male bursa present but can be reduced sometimes; bursa with ten or less pairs of pedunculate papillae; adults 2–4 mm long (Morand *et al.*, 2004).

27.3.5 Cosmocercidae

A brief description

The nematodes in the family cosmocercidae are usually parasites of reptiles and amphibians, but two genera are known as mollusc parasitic, namely *Nemhelix* and *Cosmocercoides*. *Cosmocercoides dukae* parasitize the pallial cavity of many North American slugs and snails. *N. bakeri* and some other species of the genus parasitize the reproductive organs of European helicid snails.

C. dukae inhabits the pulmonary cavity of some terrestrial molluscs. The females lay third-stage larvae that leave the pulmonary cavity to the outside environment. Later, they start looking for new hosts and infect them again via the pulmonary cavity, but third-stage larvae are also able to penetrate the genital tract and infect the snail's eggs. Thanks to this strategy, the mollusc's hatchlings are already infected (Morand and Faliex, 1994).

Under natural conditions, *N. bakeri* is frequently associated with *Helix aspersa*. This nematode lives and reproduces in the genital tract of its host. Infection of the new host (e.g. *H. aspersa*) by *N. bakeri* occurs only during mating, when the parasite is exchanged along with the spermatophores (Morand, 1988). This means that the mollusc's juveniles are always free of infection. *N. bakeri* and other mollusc parasitic nematodes reduce the fecundity of their hosts (Morand, 1989), but their potential for mollusc biocontrol is questionable.

MORPHOLOGY. Pharynx with basal valvated bulb and median valveless bulb; subtriangular oral opening; males without bursa; 14–15 pairs of papillae; parasitic adults 2–6 mm long (Morand *et al.*, 2004).

27.3.6 Diplogasteridae

A brief description

Diplogasterids are usually free-living soil nematodes, but some species, especially juveniles, can be associated with insects. The only species that is known to be associated with molluscs is *Hugotdiplogaster neozelandia*. This nematode has parasitic adults that are located in the genital tract of its host, *Athoracophorus bitentaculatus* (Athoracophoridae) in New Zealand (Morand and Barker, 1995). Unfortunately, there is no other information about the life cycle and biology of this nematode. However, Charwat and Davies (1999) isolated some diplogasterids in *Lehmania flava*, *Cernuella virgata* and *Theba pisana*, but the impact of these nematodes on molluscs was very low or negligible.

MORPHOLOGY. Pharynx with basal valvated bulb and median valveless bulb; male bursa small or absent in the case of *H. neozelandia*; nematodes less than 2 mm long (Morand *et al.*, 2004).

27.3.7 Mermithidae

A brief description

Mermithids are frequent parasites of insects in aquatic and terrestrial habitats; for example, *Romanomermis culicivorax* parasitizing mosquito larvae, or *M. nigrescens*, which are quite frequent parasitic nematodes of grasshoppers in the Czech Republic (Mrác˘ek and Weiser, 1988). Findings of mermithids in molluscs are also quite common, but it seems that these nematodes use molluscs only as facultative paratenic hosts (Morand *et al.*, 2004). Several species are known to infect molluscs; for example, *M. nigrescens* in the grey garden slug, *D. reticulatum* (Agriolimacidae), *Hexamermis albicans* in *Deroceras berytensis* and *Succinea pfeifferi*. Some other not determined mermithids were found in *Veronicella ameghini* (Vaginulidae) in North America, *L. flavus* (Limacidae), *Sarsinula marginata* (Vaginulidae) in Brazil and *Deroceras panormitanum* (Agriolimacidae) in the UK.

M. nigrescens has parasitic larvae and free-living adults that lay eggs on plants, especially on leaves, usually early in the morning or during the night, when there is high ambient humidity. Eggs that are very resistant to dry conditions and UV radiation are able to remain on plants for the whole season, or for even more than 1 year. When the eggs are eaten by the suitable host, invasive larvae hatch and penetrate into the haemocoel via the gut wall and develop for several weeks. When the larva is 'grown up', it leaves the host by penetrating the body wall. The host is usually infected with some pathogens via the opening and dies shortly afterward. Emerged larvae develop into post-parasites in the soil, and adults mate later. The whole development in the soil can take several months, and therefore the whole life cycle can take more than 1 year (Fig. 27.4).

The life cycle of other species can be very different; for example, in some *Hexamermis* sp. nematodes, the host is infected with pre-parasitic larvae and not with eggs, as in the previous case (Mráček and Weiser, 1988).

MORPHOLOGY. Instead of a pharynx, there is a slender pharyngeal tube, head with two, four or six papillae, anus and bursa absent, several rows and caudal papillae, very long nematodes, 1–10 cm or more (Morand *et al.*, 2004).

27.4 Rhabditidae

A brief description

The rhabditidae is a large family consisting of many bacteriophagous free-living, phoretic and parasitic nematodes that are often associated with insects. But several species belonging to the genera *Rhabditis*, *Caenorhabditis* and *Phasmarhabditis* are also parasites of terrestrial molluscs.

The slug parasitic nematode, *P. hermaphrodita* (Schneider) Andrassy, is an almost cosmopolitan species capable of killing many slug and snail species, such as *D. reticulatum* (Fig. 27.5). The infective juveniles (IJs), also called dauer juveniles (DJs), infect slugs in the area beneath the mantle surrounding the shell. They usually cause a disease with characteristic symptoms, particularly a swelling of the mantle. It has been reported that *P. hermaphrodita* acts as a vector to transport the bacteria *Moraxella osloensis* into the shell cavity of *D. reticulatum* (Tan and Grewal, 2001a). The IJs move through the soil and locate and infect the slug by penetrating through a natural opening at the back side of the mantle. After entering the host, the bacterial cells are released by IJs. The bacteria start multiplying there and the nematodes feed on those bacteria, developing into self-fertilizing hermaphrodites (Askary, 2010; Askary *et al.*, 2012). The infection often leads to the death of the slug within 7–21 days (Fig. 27.6). New DJs, which are released from the slug cadaver, spread into the soil and look for new hosts (Wilson

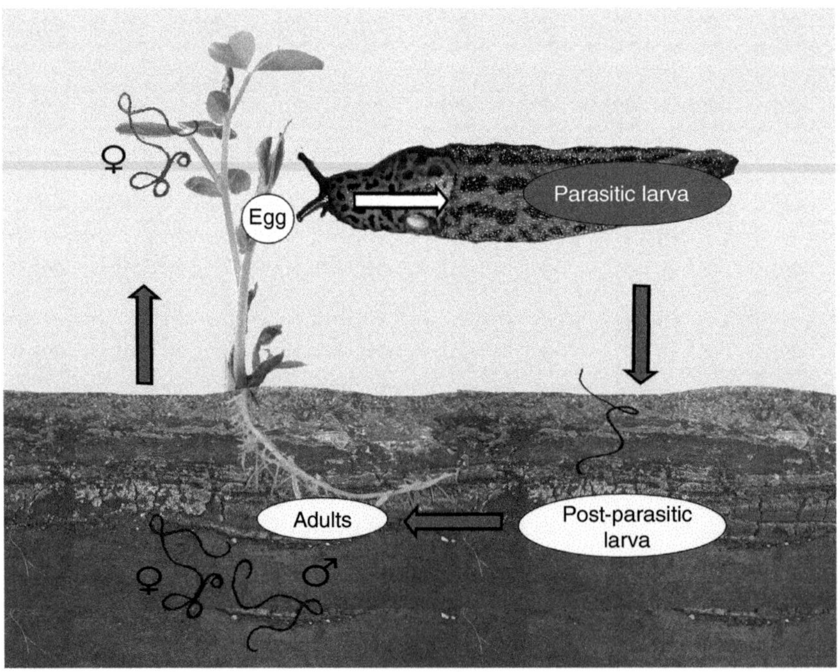

Fig. 27.4. The life cycle of *Mermis nigrescens*.

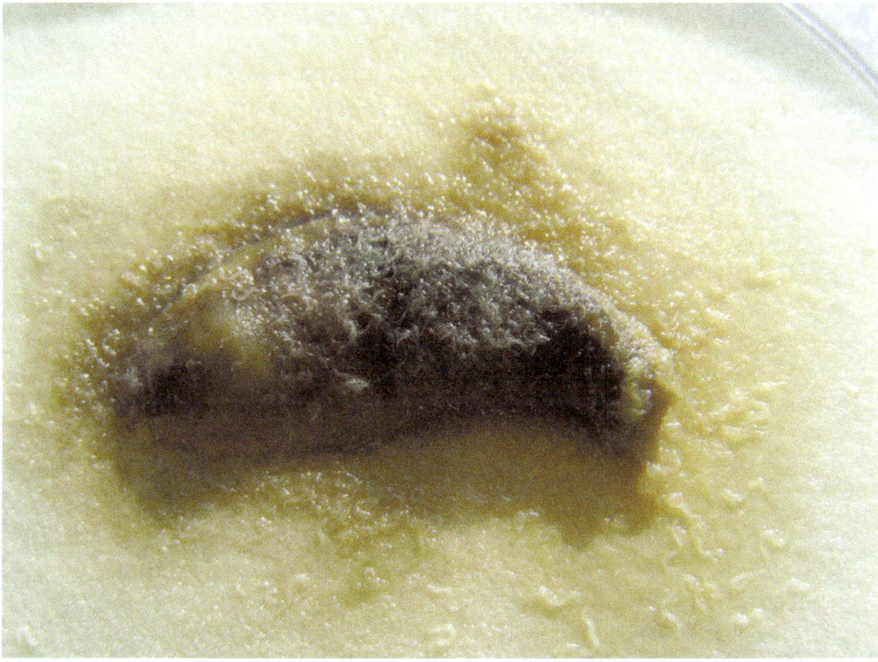

Fig. 27.5. Grey garden slug, *Deroceras reticulatum*, covered with adults of *Phasmarhabditis hermaphrodita* several days after infection in laboratory conditions.

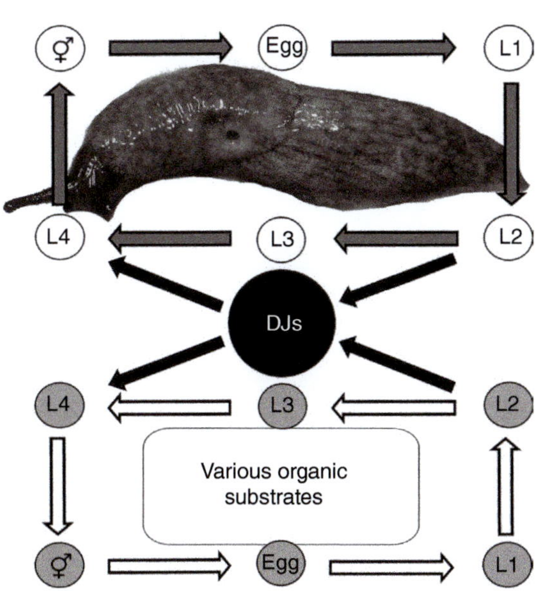

Fig. 27.6. The life cycle of *Phasmarhabditis hermaphrodita*.

Notes: DJs = dauer juveniles; L1 = first larval stage; L2 = second larval stage; L3 = third larval stage (called DJs in some special cases); L4 = fourth larval stage.

et al., 1993). Apart from the parasitic cycle, *P. hermaphrodita* also has a necromenic life cycle (Mengert, 1953), and has been shown to reproduce on dead earthworms (Rae *et al.*, 2009), leaf litter (MacMillan *et al.*, 2009) and slugs or slug faeces homogenates (Tan and Grewal, 2001b), which are very strong attractants for this nematode (Rae *et al.*, 2006; Nermut' *et al.*, 2012).

P. hermaphrodita is a bacteriophagous nematode that does not live in a strict association with only one species of bacteria, as do EPNs, but is associated with, and able to feed on, many bacterial species (Wilson *et al.*, 1995a; Rae *et al.*, 2010) that are common in its habitat. Bacterial species also significantly influence *P. hermaphrodita* progeny production (Wilson *et al.*, 1995b; Nermut' *et al.*, 2014a) and are responsible for the pathogenicity of the nematode–bacteria complex towards their hosts (Wilson *et al.*, 1995b).

Except for *P. hermaphrodita*, there are many other rhabditid parasites of slugs. Quite frequent species are *Phasmarhabditis papillosa* Andrassy and *Phasmarhabditis neopapillosa*, which, unike *P. hermaphrodita*, have a frequent occurrence of males (Hooper *et al.*, 1999). Sometimes, *Caenorhabditis briggsae* is found in *Deroceras* sp. (Nermut', unpublished), *Caenorhabditis elegans* in *Limax* sp. (Ross *et al.*, 2012) or *Rhabditis gracilicauda* in *Arion owenii* (Ross *et al.*, 2010a), *Rhabditis axei* in *Achatina achatina* (Odaibo *et al.*, 2000) and some other rhabditids in many other molluscs (Charwat and Davies, 1999).

MORPHOLOGY. Nematodes usually small, less than 2 mm long. Typical rhabditoid pharynx with valvated basal bulb and median bulb without valves. Oral opening triangular. Males usually with bursa and nine pairs of pedunculate papillae (Morand *et al.*, 2004).

27.5 Parasitism

The utilization of molluscan hosts is probably very lucrative for nematodes. It seems that molluscs have been acquired as hosts by nematodes on a number of occasions (up to four times), even though there are only two major groups of nematodes that frequently use molluscs as hosts: Metastrongyloids and Rhabditids. Metastrongyloids use molluscs usually as intermediate hosts, while Rhabditids use molluscs as definitive hosts (Grewal *et al.*, 2003).

Metazoan phylogenetic analyses suggest that ancestral nematode lineage originated in the mid or late Proterozoic (Giribet, 2002; Zrzavý *et al.*, 1998), a long time prior to the evolution of arthropods and molluscs. The parasitism of mollusc by Mermithids, one of the oldest group of Nematoda, suggests that this parasitic relationship among nematodes and molluscs occurred very early, probably shortly after the Cambrian species radiation. According to Poinar (1993), mollusc parasitism arose first, followed by the parasitism of arthropods, while the free-living rhabditid, *Alloionema*- and *Pellioditis*-like nematodes were ancestral. This theory is supported by, for example, the apparent similarity of life cycles of *P. hermaphrodita* and EPNs (Heterorhabditidae and Steinernematidae), or by the fact that most of the nematodes that are using molluscs as definitive hosts are bacterial feeders that are able to produce DJs (usually third-stage larvae) and still maintain both free-living and parasitic life cycles (Grewal *et al.*, 2003; Ross *et al.*, 2010b).

In general, there are three types of association of nematodes and molluscs: molluscs as (i) paratenic, (ii) intermediate and (iii) definitive hosts. Paratenic hosts serve only for transferring parasites from one host to another while no development takes place (Anderson, 2000). This strategy is typical for vertebrate parasitic nematodes; for example, *Galba corvus* (Lymnaeidae) is a paratenic host for third-stage larva of *Anguillicola crassus*, whose definitive host is the eel, *Anguilla anguilla* (Moravec, 1996).

Intermediate hosts are those that serve the nematodes for completing part of their life cycle. The invasive stage (usually first-stage larvae) infects the mollusc and develops inside its body into the infective stage. Then, the mollusc is usually swallowed by the final vertebrate host, and the cycle is completed: again, a strategy typical for vertebrate parasites, *Muellerius capillaris*, for example, the lung parasite of sheep that use different slugs and snails as intermediate hosts. This type of association is also known in the family Angiostomatidae: *Angiostoma limacis* parasitizing salamanders are known from the digestive tract of *A. ater* (Morand and Spiridonov, 1989). From the biocontrol point of view, the most interesting type of association is 'molluscs as definitive hosts' (see Table 27.1), because the mollusc often dies at the end of the nematode's life cycle. Three different levels can be seen in this association. The first level is represented by the nematodes that have parasitic juveniles, while the adults are free living. A good example of this association is the Mermithids, for example *M. nigrescens* or *Mermis albicans* (see above), that are able to kill the host when leaving its body, but also *A. appendiculatum* (see above), whose ability to kill the host is still a little questionable.

Another level of this association is typical for the nematodes that complete the whole cycle in the body of the mollusc host, but never kill it. This includes a vast amount of nematodes, such as *A. flexilis*, *R. axei* or *H. neozelandia*, etc.

The last level is occupied by the true killers: nematodes that cause significant mortality of infected molluscs. *Daubaylia potomaca*, *P. hermaphrodita*, *P. papillosa* and *P. neopapillosa* are typical members of this group.

The effect of nematode parasitism on the mollusc host has only been studied in three nematodes, *P. hermaphrodita*, *A. appendiculatum* and *N. bakeri*. While the influence of *P. hermaphrodita* on the mortality of slugs is well known, the information on other nematodes is not so plentiful, and is sometimes contradictory. *A. appendiculatum* is able to cause the death of *C. aspersus* and, according to Laznik *et al.* (2010), also *A. vulgaris*, but a later, thorough study on *A. vulgaris* (Nermut' *et al.*, 2014b) does not support this statement. According to the study of Morand (1988), *N. bakeri* is responsible for the decreasing number of eggs produced in *C. aspersus*. Mixed infections with two or more nematodes are very frequent; however, the effect on the mortality, fecundity or fitness of molluscs after mixed infection is unclear. But infection with *A. appendiculatum* can be lower if the *C. aspersus* had previously been infected with *M. capillaris* (Cabaret and Morand, 1990).

27.6 Mass Production

Many nematodes can be more or less large-scale produced *in vivo* or *in vitro* in solid or liquid medium. Some of them can be produced on nematode growth medium (NGM) with appropriate bacteria, Wouts' agar, dog food agar, chicken offal medium, etc. (Stock and Goodrich-Blair, 2012). *In vivo* methods are usually more time-consuming and the results (yields) are not as consistent as in the case of *in vitro* methods.

In slug parasitic nematodes, there are two species especially that can be produced easily, *P. hermaphrodita* and *A. appendiculatum*. The first of them is produced commercially for use in biocontrol, and the other for scientific purposes only. *A. appendiculatum* can be produced very easily on homogenized pig kidneys placed on 2% agar plates at 15°C. The culture under this condition can survive for many years in saprophytic form. Only the source of food (kidney) must be changed every second week (Nermut' *et al.*, 2014b). But this nematode can, in fact, also be mass produced in slightly modified medium, according to Bedding (1984).

Various methods for the *in vitro* production of *P. hermaphrodita* were developed in the 1990s by Wilson (e.g. Wilson *et al.*, 1993). Wilson showed that *P. hermaphrodita* could grow in a xenic culture in solid foam chip culture according to Bedding (1984), and also in liquid cultures. Actually, this species is the only commercially mass produced slug parasitic nematode. The technology that is used for producing *P. hermaphrodita* is a slightly modified method used for the mass production of EPNs. The nematodes are cultured monoxenically, to ensure consistency of quality and predictability (Lunau *et al.*, 1993). An appropriate bacterium is extracted from the nematodes and, subsequently, sterile nematode eggs are prepared and applied to the medium pre-inoculated with the bacterium. The nematodes are produced routinely in air-lift fermenters, up to 20,000 l in suitable medium that allows approximately 100,000 DJs/ml of the culture to be yielded. When the maximum yield is obtained, the nematodes are concentrated by centrifuge and later formulated in vermiculite, polymer of clay (Wilson and Grewal, 2005) that slightly dehydrates and immobilizes the nematodes, which saves energy reserves more effectively in this state. The formulated nematodes are packed into 8 μm high-density polyethylene bags that allow exchange of air but which retain water. This final product can be stored in a refrigerator for up to 6 months (Morand *et al.*, 2004).

27.7 Application

The only commercial product based on mollusc parasitic nematodes is Nemaslug© (Becker Underwood) with bioagent *P. hermaphrodita*. Therefore, this section will target mostly the use of this nematode against noxious slugs, mainly the grey garden slug, *D. reticulatum*, and other members of Agriolimacidae and, to a lesser extent, also other species from the families Arionidae, Limacidae, Milacidae and others in different crops. Experiments with other nematode species, *A. appdendiculatum* for example (Laznik *et al.*, 2010; Nermut' *et al.*, 2014b) and some other rhabditids (Charwat and Davies, 1999), had already been done, but the results and the potential of these nematodes for use in biocontrol are a little questionable and too far from practical impact.

The general recommendation is to apply *P. hermaphrodita* on wet soil at a dose of 300,000/m^2 and water the soil surface shortly after application. The best time for application is early evening, when the soil temperature is about 15°C. Nematode efficacy can be increased with cultivation of the soil immediately after application (Hass *et al.*, 1999a). The nematodes are then better protected against UV radiation and drying. Nozzles and filters should have holes at least 1 mm wide and the pressure should not be exceed 5 bar. It is necessary to avoid the application of *P. hermaphrodita* in areas that have been treated with toxic chemicals, e.g. pellets with active ingredient methiocarb used against noxious slugs. Combination with metaldehyde is possible, because this compound does not affect nematodes at the concentration recommended for slug control.

There are two different strategies for application. The common strategy is to apply the nematodes over the whole soil surface, even though some other strategies for local applications have been developed. *D. reticulatum* slugs tend to avoid places treated with *P. hermaphrodita* (Wilson *et al.*, 1999). Therefore, there were some ideas to apply the nematodes only around individual plants or in narrow bands centred on crop rows. Unfortunately, the assumption of protecting plants in this way, with a lower amount of nematodes, was not confirmed. There is no benefit associated with band application as opposed to uniform application (Hass *et al.*, 1999b). The number of *P. hermaphrodita* DJs decreases in time and the repellent effect to slugs subsides.

A very interesting idea on how to reduce the dose of nematodes and retain efficacy against slugs has been published by Grewal *et al.* (2001). The principle is to apply nematodes at a dose of $0.6 \times 10^6/m^2$ only under artificial shelters that are used by slugs during the day. This method provides almost the same effect as uniform application of $0.3 \times 10^6/m^2$.

Early application of *P. hermaphrodita* before germination and drilling does not provide any benefits because the slugs infected with nematodes reduce their feeding activity almost immediately after infection (Glen *et al.*, 2000). Repeated application of lower than the recommended dose can be very effective. In Brussels sprouts, the application of 50,000 DJs/m^2 is repeated three times in 1-month intervals. This represents a 50% reduction of the previously recommended single application, while the efficacy is almost the same as in the case of using metaldehyde pellets (Ester *et al.*, 2003).

P. hermaphrodita can also be used successfully in integrated pest management systems in combination with some slug pellets. The assumption is that the nematode is better at killing slugs under the soil surface, while slug pellets are more effective on the surface (Rae *et al.*, 2007). Ester and Geelen (1996) stated that application of the standard rate of nematodes with methiocarb at a dose of 80 g of active ingredient per square metre provides the best protection to sugarbeet. What more laboratory bioassays indicate is that metaldehyde has no effect on the mortality of *P. hermaphrodita*. From this point of view, methiocarb is more dangerous, because this carbamate has a very strong negative effect on the survival of beneficial nematodes (Nermut' and Mráček, 2010), *P. hermaphrodita* included (Wilson *et al.*, 2000).

P. hermaphrodita is applied to many plants in greenhouses, vegetables, ornamentals and in arable crops; for example, *Cymbidium* sp., lettuce, cabbage, Brussels sprouts, *Asparagus* sp., oilseed rape, wheat or sugarbeet, and many others. The most common pests are *Deroceras* sp. and *Arion* sp., but also *Lehmania valentiana*. Repeated uniform application on the soil surface is usual, but *P. hermaphrodita* can also be applied successfully in the plastic tunnels or pots that are frequently used in greenhouses. The nematodes have a future in arable crops, especially on organic farms, where no cheap chemicals like methiocarb or metaldehyde can be applied, and nematodes are therefore the only solution to the slug problem.

27.8 Conclusion and Future Prospects

Biological pest control is a very strong trend of the past decades. The main advantages consist of the reduction in using hazardous chemical compounds, with their negative influence on environment, human or farm animal health and agricultural product quality. Unlike insect pests, terrestrial slugs and snails are a slightly neglected group of harmful organisms. Until now, only one biological agent has been produced, the nematode *P. hermaphrodita*. This slug parasitic nematode is actually widely used against noxious molluscs in many crops and many states around the world. And the amount of nematodes sold will probably increase in the next few years. The problem that does not allow mass use in agriculture is the very high price of this nematode on the market, and also the limited effect on some big (and often invasive) mollusc species. In the Czech Republic, the price of 30 million DJs is €40. This means that the cost of 1 ha application can be about €4,000, which is an absolutely unacceptable price for most growers. Naturally, government programmes financially support using bioagents, including nematodes, and this support can be 25–70% of the price of the bioagents; but despite this fact, the price remains very high and therefore *P. hermaphrodita* can be used only in the most profitable crops or in horticulture. In the Czech Republic, this nematode is actually used only in the hobby segment of private gardens, because this market segment is not dependent on economic profit, and therefore the higher price is not a problem. According to the official statistics of the Ministry of Agriculture of the Czech Republic, only 30 million DJs (1 package) were used in wineries and 570 million DJs (19 packages) were used in vegetables: altogether, 20 packages in 1 year (2013), really not much. Fortunately, sales in the hobby market are much higher. The solution to high price can be seen in integrated pest management systems. The use of nematodes in combination with bait pellets is possible and thus allows the dose not only of this very costly bioagent but also of the chemical pesticides to be reduced significantly (Ester and

Geelen, 1996; Wilson *et al.*, 2000). It is also possible to enhance the efficacy of nematodes with soil cultivation (Hass *et al.*, 1999a) or by using repeated applications of low doses (Ester *et al.*, 2003) or other methods (Rae *et al.*, 2007).

As stated above, the use of other mollusc parasitic nematodes in biocontrol is probably possible, but the knowledge of and research on these animals are at a very early stage and we cannot therefore expect to see new mollusc parasitic nematodes on the market in the near future. Particularly in the case of species such as *A. vulgaris* and *A. fulica*, or others, which are invasive in many countries, a bioagent with the ability to control populations of these molluscs is very desirable. We have to hope that intensive targeted research will soon bring about a good solution.

Acknowledgements

Preparation of this chapter was supported by the Czech Ministry of Education, Youth and Sport and Amvis in the Program KONTAKT II, Project LH 12 105.

References

Anderson, R.C. (2000) *Nematode Parasites of Vertebrates: Their Development and Transmission*. CAB International, Wallingford, UK.

Andrassy, I. (1983) *A Taxonomic Review of the Sub-order Rhabditina (Nematoda: Secernentea).* Office de la Recherche Scientifique et Technique Outre-Mer, Paris.

Askary, T.H. (2010) Nematodes as biocontrol agents. In: Lichtfouse, E. (ed.) *Sociology, Organic Farming, Climate Change and Soil Science*. Springer, The Netherlands, pp. 347–378.

Askary, T.H., Khan, A.A., Waliullah, M.I.S., Banday, S.A., Iqbal, U. and Mir, M.M. (2012) Slug pest management through nematodes in agricultural and horticultural crops. In: Boeri, F. and Chung, J.A. (eds) *Nematodes: Morphology, Functions and Management Strategies*. Nova Publishers, New York, pp. 197–211.

Bedding, R.A. (1984) Large scale production, storage and transport of the insect parasitic nematodes *Neoaplectana* spp. and *Heterorhabditis* spp. *Annals of Applied Biology* 104, 117–120.

Cabaret, J. and Morand, S. (1990) Single and dual infections of the land snail *Helix aspersa* with *Muellerius capillaris* and *Alloionema appendiculatum* (Nematoda). *Journal of Parasitology* 76, 579–580.

Cabaret, J., Morand, S., Aubert, C. and Yvore, P. (1988) Snail farming – a survey of breeding management, hygine and parasitism of the garden snail, *Helix aspersa* Müller. *Journal of Molluscan Studies* 54, 209–214.

Charwat, S.M. and Davies, K.A. (1999) Laboratory screening of nematodes isolated from South Australia for potential as biocontrol agents of helicid snails. *Journal of Invertebrate Pathology* 74, 55–61.

Chitwood, B.G. and Chitwood, M.B. (1937) Snails as hosts and carriers of nematodes and Nematomorpha. *The Nautilus* 50, 130–135.

Čurčič, B.P.M., Sudhaus, W., Dimitijevič, R.N., Makarov, S.E. and Tomič, V.T. (2008) *Rhabditophanes schneideri* (Rhabditida) phoretic on a cave pseudoscorpion. *Journal of Invertebrate Pathology* 99, 254–256.

Dundee, D.S. (1977) Mermithids in introduced molluscs. *Journal of Parasitology* 63, 590.

Ester, A. and Geelen, P.M.T.M. (1996) Integrated control of slugs in a sugar beet crop growing in a rye cover crop. In: Henderson, I.F. (ed.) *Slug and Snail Pests in Agriculture.* Monograph No 41, British Crop Protection Council, Thornton Health, UK, pp. 445–450.

Ester, A., van Rozen, K. and Molendijk, L.P.G. (2003) Field experiments using the rhabditid nematode *Phasmarhabditis hermaphrodita* or salt as control measures against slugs in green asparagus. *Crop Protection* 22, 689–695.

Giribet, G. (2002) Current advances in the phylogenetic reconstruction of metazoan evolution. A new paradigm for the Cambrian explosion? *Molecular Phylogenetics and Evolution* 24, 345–357.

Glen, D.M., Wilson, M.J., Brain, P. and Stroud, G. (2000) Feeding activity and survival of slugs, *Deroceras reticulatum*, exposed to the rhabditid nematode, *Phasmarhabditis hermaphrodita*: a model of dose response. *Biological Control* 17, 73–81.

Grewal, P.S., Grewal, S.K., Taylor, R.A.J. and Hammond, R.B. (2001) Application of molluscicidal nematodes to slug shelters: a novel approach to economic biological control of slugs. *Biological Control* 22, 72–80.

Grewal, P.S., Grewal, S.K., Tan, L. and Adams, B.J. (2003) Parasitism of molluscs by nematodes: types of associations and evolutionary trends. *Journal of Nematology* 35, 146–156.

Hass, B., Glen, D.M., Brain, P. and Hughes, L.A. (1999a) Targeting biocontrol with the slug-parasitic nematode *Phasmarhabditis hermaphrodita* in slug feeding areas: a model study. *Biocontrol Science and Technology* 9, 587–598.

Hass, B., Hughes, L.A. and Glen, D.M. (1999b) Overall versus band application of the nematode *Phasmarhabditis hermaphrodita* with and without incorporation into soil, for biological control of slugs in winter wheat. *Biocontrol Science and Technology* 9, 579–586.

Hooper, D.J., Wilson, M.J., Rowe, J.A. and Glen, D.M. (1999) Some observations on the morphology and protein profiles of the slug-parasitic nematodes *Phasmarhabditis hermaphrodita* and *P. neopapillosa* (Nematoda: Rhabditidae). *Nematology* 1, 173–182.

Ivanova, E.S., Panayotova-Pencheva, M.S. and Spiridonov, S.E. (2013a) Observations on the nematode fauna of terrestrial molluscs of the Sofia area (Bulgaria) and the Crimea peninsula (Ukraine). *Russian Journal of Nematology* 21, 41–49.

Ivanova, E.S., Spiridonov, S.E., Clark, W.C., Tourna, M., Wilson, M.J. and Barker, G.M. (2013b) Description and systematic affinity of *Alaninema ngata* n. sp. (Alaninematidae: Panagrolaimorpha) parasitising leaf-veined slugs (Athoracophoridae: Pulmonata) in New Zealand. *Nematology* 15, 859–870.

Ivanova, E., Van Luc, P. and Spiridonov, S. (2016) *Neoalloionema tricaudatum* gen. n., sp. n. (Nematoda: Alloionematidae) associated with a cyclophorid snail in Cuc Phuong Natural Park, Vietnam. *Nematology* 18, 109–120.

Korol, E.N. and Spiridonov, S.E. (1991) *Angiostoma kimmeriensis* sp. n. and *Agfa tauricus* sp. n. – parasitic Rhabditida (Nematoda) from Crimean terrestrial molluscs. *Helminthologia* 28, 179–182.

Laznik, Z., Ross, J.L. and Trdan, S. (2010) Massive occurrence and identification of the nematode *Alloionema appendiculatum* Schneider (Rhabditida: Alloionematidae) found in Arionidae slugs in Slovenia. *Acta Agriculturae Slovenica* 95, 43–49.

Lunau, S., Stoessel, S., Schmidt-Peisker, A.J. and Ehlers, R.-U. (1993) Establishment of monoxenic inocula for scaling up *in vitro* cultures of the entomopathogenic nematodes *Steinernema* spp. and *Heterorhabditis* spp. *Nematologica* 39, 385–399.

MacMillan, K., Haukeland, S., Rae, R., Young, I., Crawford, J., *et al.* (2009) Dispersal patterns and behaviour of the nematode *Phasmarhabditis hermaphrodita* in mineral soils and organic media. *Soil Biology and Biochemistry* 41, 1483–1490.

Mengert, H. (1953) Nematoden und Schnecken. *Zeitschrift für Morphologie und Ökologie der Tiere* 48, 311–349.

Mienis, H.K. (1986) First record of the parasitic nematode, *Hexamermis albicans*, from a slug in Israël. *Zoology in the Middle East* 1, 156.

Morand, S. (1986) *Angiostoma aspersae* n. sp. (Nematoda, Angiostomatidae) parasite de *Helix aspersa* Müller (Gastropoda, Helicidae). *Bulletin Museum National Histoire Naturelle, Paris* 11, 111–115.

Morand, S. (1988) Cycle évolutif de *Nemhelix bakeri* Morand et Petter (Nematoda, Cosmocercidae) parasite de l'appareil génital de *Helix aspersa* (Gastropoda, Helicidae). *Canadian Journal of Zoology* 66, 1796–1802.

Morand, S. (1989) Deux nouveaux nématodes Cosmocercidae parasites des escargots terrestres: *Cepaea nemoralis* L. et *Cepaea hortensis* Müller. *Bulletin Museum National Histoire Naturelle, Paris* 11, 563–570.

Morand, S. and Barker, G.M. (1995) *Hugotdiplogaster neozelandia* n. gen., n. sp. (Nematoda: Diplogasteridae), a parasite of the New Zealand endemic slug, *Athoracophorus bitentaculatus* (Quoy and Gaimard, 1832) (Gastropoda: Athoracophoridae). *New Zealand Journal of Zoology* 22, 109–113.

Morand, S. and Faliex, E. (1994) Study on the life-cycle of a sexually transmitted nematode parasite of a terrestrial snail. *Journal of Parasitology* 80, 1049–1052.

Morand, S. and Petter, A.J. (1986) *Nemhelix bakeri* n. gen., n. sp. (Nematoda: Cosmocercidae) parasite de l'appareil génital de *Helix aspersa* (Gastropoda: Helicidae) en France. *Canadian Journal of Zoology* 64, 2008–2011.

Morand, S. and Spiridinov, S. (1989) Redescription de trois espèces d'Angiostomatidae (Nematoda: Rhabditidae), parasites de Gastropodes Pulmones Stlommatophores, et description du cycle evolutif de deux d'entre elles. *Bulletin du Museum National d'Histoire Naturelle, Paris*, 11, 367–385.

Morand, S., Wilson, M.J. and Glen, D.M. (2004) Nematodes (Nematoda) parasitic in terrestrial gastropods. In: Barker, G.M. (ed.) *Natural Enemies of Terrestrial Molluscs*. CAB International, Wallingford, UK, pp. 525–557.

Moravec, F. (1996) Aquatic invertebrates (snails) as new paratenic hosts of *Anguillicola crassus* (Nematoda: Dracunculoidea) and the role of paratenic hosts in the life cycle of this parasite. *Diseases of Aquatic Organisms* 27, 237–239.

Mráček, Z. and Weiser, J. (1988) *Parazitické hlístice hymzu*. Academia, Praha, Czech Republic.

Nermuť, J. and Mráček Z. (2010) The influence of pesticides on the viability and infectivity of entomopathogenic nematodes (Nematoda: Steinernematidae). *Russian Journal of Nematology* 18, 141–148.

Nermuť, J., Půža, V. and Mráček, Z. (2012) The response of *Phasmarhabditis hermaphrodita* (Nematoda: Rhabditidae) and *Steinernema feltiae* (Nematoda: Steinernematidae) to different host-associated cues. *Biological Control* 61, 201–206.

Nermuť, J., Půža, V. and Mráček, Z. (2014a) The effect of different growing substrates on the development and quality of *Phasmarhabditis hermaphrodita* (Nematoda: Rhabditidae). *Biocontrol Science and Technology* 24, 1026–1038.

Nermuť, J., Půža, V. and Mráček, Z. (2014b) Some observations on morphology and ecology of mollusc-parasitic nematode *Alloionema appendiculatum*. *47th Annual Meeting of SIP, Program and Abstracts*, 3–7 August 2014, Mainz, Germany, pp. 125–126.

Nermuť, J., Půža V. and Mráček, Z. (2015) Re-description of the slug-parasitic nematode *Alloionema appendiculatum* Schneider, 1859 (Rhabditida: Alloionematidae). *Nematology* 17, 897–910.

Odaibo, A.B., Dehinbo, A.J., Olofintoye, A.O. and Falode, O.A. (2000) Occurrence and distribution of *Rhabditis axei* (Rhabditida; Rhabditidae) in African giant snails in southwestern Nigeria. *Helminthologia* 37, 233–235.

Poinar, G.O. Jr (1993) Origins and phylogenetic relationships of the entomophilic rhabditids, *Heterorhabditis* and *Steinernema*. *Fundamental and Applied Nematology* 16, 333–338.

Rae, R.G., Robertson, J.F. and Wilson, M.J. (2006) The chemotactic response of *Phasmarhabditis hermaphrodita* (Nematoda: Rhabditida) to cues of *Deroceras reticulatum* (Mollusca: Gastropoda). *Nematology* 8, 197–200.

Rae, R., Verdun, C., Grewal, P., Robertson, J.F. and Wilson, M.J. (2007) Biological control of terrestrial molluscs using *Phasmarhabditis hermaphrodita* – progress and prospects. *Pest Management Science* 63, 1153–1164.

Rae, R.G., Robertson, J.F. and Wilson, M.J. (2009) Chemoattraction and host preference of the gastropod parasitic nematode *Phasmarhabditis hermaphrodita*. *Journal of Parasitology* 95, 517–526.

Rae, R.G., Tourna, M. and Wilson, M.J. (2010) The slug parasitic nematode *Phasmarhabditis hermaphrodita* associates with complex and variable bacterial assemblages that do not affect its virulence. *Journal of Invertebrate Pathology* 104, 222–226.

Ribas, A. and Casanova, J. (2002) *Agfa morandi* sp. n. (Rhabditida, Agfidae), a parasite of *Limax* sp. (Gastropoda, Limacidae). *Parasitology Research* 88, 745–747.

Ross, J.L., Ivanova, E.S., Severns, P.M. and Wilson, M.J. (2010a) The role of parasite release in invasion of the USA by European slugs. *Biological Invasions* 12, 603–610.

Ross, J.L., Ivanova, E.S., Spiridonov, S.E., Waeyenberge, L., Moens, M., Graeme, W.N. and Wilson, M.J. (2010b) Molecular phylogeny of slug-parasitic nematodes inferred from 18S rRNa gene sequences. *Molecular Phylogenetics and Evolution* 55, 738–743.

Ross, J.L., Malan, A.P. and Ivanova, E.S. (2011) *Angiostoma margaretae* n. sp (Nematoda: Angiostomatidae), a parasite of the milacid slug *Milax gagates* Draparnaud collected near Caledon, South Africa. *Systematic Parasitology* 79, 71–76.

Ross, J.L., Ivanova, E.S., Sirgel, W.F., Malan, A.P. and Wilson, M.J. (2012) Diversity and distribution of nematodes associated with terrestrial slugs in the Western Cape Province of South Africa. *Journal of Helminthology* 86, 215–221.

Schneider, A. (1859) *Monographie der Nematoden*. Druck und Verlag von Georg Reimer, Berlin, Germany.

Senoldová, P. and Rotrekl, J. (2005) Rozdílná účinnost parazitických hlístic *Phasmarhabditis hermaphrodita* na slimáčky rodu *Deroceras* a plzáky rodu *Arion*.; Differentiated effects of parasitic nematodes *Phasmarhabditis hermaphrodita* on slugs of the genus *Deroceras* and snails of the genus *Arion*. *Acta Universitatis Agriculturae et Silviculturae Mendelianae Brunensis* 53, 171–175.

Stock, S.P. and Goodrich-Blair, H. (2012) Nematode parasites, pathogens and associates of insects and invertebrates of economic importance. In: Lacey, L.A. (ed.) *Manual of Techniques in Invertebrate Pathology.* Academic Press, London, pp. 373–426.

Tan, L. and Grewal, P.S. (2001a) Pathogenecity of *Moraxella osloensis*, a bacterium associated with the nematode *Phasmarhabditis hermaphrodita*, to the slug *Deroceras reticulatum*. *Applied and Environmental Microbiology* 67, 5010–5016.

Tan, L. and Grewal, P.S. (2001b) Infection behaviour of the rhabditid nematode *Phasmarhabditis hermaphrodita* to the grey garden slug *Deroceras reticulatum*. *Journal of Parasitology* 87, 1349–1354.

Théodoridès, J. (1957) Sur un nématode nouveau parasite d'un mollusque gastéropode d'Indonésie *Alaninema venmansi* n.g. n. sp. (Nematoda, Drilonematidae). *Annales de Parasitologie* 32, 267–270.

Thiengo, S.C. (1995) Presence of mermithid larvae (Nematoda, Mermithidae) in *Limax flavus* and *Sarasinula marginata* (Mollusca, Gastropoda) in Brazil. *Memorias Instituto Oswaldo Cruz* 90, 347.

Williams, A.J. and Rae, R. (2015) Susceptibility of the Giant African snail (*Achatina fulica*) exposed to the gastropod parasitic nematode *Phasmarhabditis hermaphrodita*. *Journal of Invertebrate Pathology* 127, 122–126.

Wilson, M.J. and Grewal, P.S. (2005) Biology, production and formulation of slug-parasitic nematodes, In: Grewal, P.S., Ehlers, R.-U. and Shapiro-Ilan, D.I. (eds) *Nematodes as Biocontrol Agents*. CAB International, Wallingford, UK, pp. 421–429.

Wilson, M.J., Glen, D.M. and George, S.K. (1993) The rhabditid nematode *Phasmarhabditis hermaphrodita* as a potential biological-control agent for slugs. *Biocontrol Science and Technology* 3, 503–511.

Wilson, M.J., Glen, D.M., George, S.K. and Pearce, J.D. (1995a) Selection of a bacterium for the mass-production of *Phasmarhabditis hermaphrodita* (Nematoda, Rhabditidae) as a biocontrol agent for slugs. *Fundamental and Applied Nematology* 18, 419–425.

Wilson, M.J., Glen, D.M., Pearce, J.D. and Rodgers, P.B. (1995b) Monoxenic culture of the slug parasite *Phasmarhabditis hermaphrodita* (Nematoda, Rhabditidae) with different bacteria in liquid and solid-phase. *Fundamental and Applied Nematology* 18, 159–166.

Wilson, M.J., Hughes, L.A., Jefferies, D. and Glen, D.M. (1999) Slugs (*Deroceras reticulatum* and *Arion ater* agg.) avoid soil treated with the rhabditid nematode *Phasmarhabditis hermaphrodita*. *Biological Control* 16, 170–176.

Wilson, M.J., Major, B., Hughes, L.A. and Glen, D.M. (2000) Integrating biological and chemical control of slugs. In: *Proceedings of Brighton Crop Protection Conference – Pests and Diseases*. British Crop Protection Council, Alton, Hants, UK, pp. 1007–1012.

Zrzavý, J., Mihulka, S., Kepka, P., Bezděk, A. and Tietz, D. (1998) Phylogeny of the Metazoa based on morphological and 18S ribosomal DNA evidence. *Cladistics* 14, 249–285.

28 The Discovery and Commercialization of a Slug Parasitic Nematode

DAVID M. GLEN[1]* AND JAMES COUPLAND[2]

[1]Styloma Research Consulting, Cheddar, UK; [2]FarmForest Research Inc, Almonte, Ontario, Canada

28.1 Introduction

Terrestrial slugs and snails are gastropod molluscs belonging to the Order Styllomatophora. They are important pests of agriculture and horticulture throughout the world (Godan, 1983; South, 1992; Barker, 2002). Terrestrial slugs have evolved from snails by losing the hard calcareous protective shell (Barker, 2001). Without a bulky shell, slugs are able to move readily through small cracks and channels in the soil, enabling them to search for food and to shelter from unfavourable conditions within the soil. Slugs are especially important as pests in regions with moist mild climates, such as Western Europe. The main pest species in Europe belong to the families Agriolimacidae (mainly *Deroceras reticulatum*), Arionidae and Milacidae. They are important pests of a wide range of crops in agriculture and horticulture, as well as in home gardens. They can survive in soil that is cultivated but populations are reduced by cultivation (Glen and Symondson, 2003), and the severity of slug problems tends to increase wherever conservation tillage methods (low-till or, especially, no-till methods) of growing crops are introduced (Glen and Symondson, 2003) to protect soils from severe erosion (Crovetto, 1996; El Titi, 2003). Agronomic changes such as the increased growing of autumn-sown cereals (Glen and Moens, 2002) and the increased areas of oilseed rape (Moens and Glen, 2002) have also increased the severity of slug problems. Moreover, slugs and snails have become pests in new regions of the world as a result of spread by humans (Barker, 2002).

The development of chemical control has not matched the increasing status of slugs and snails as pests (Henderson and Triebskorn, 2002). These authors stress that chemical molluscicides have limited efficacy for economic and technical reasons. Economically, the world market for molluscicides is too fragmented to justify commercial R&D expenditure on primary molluscicide screens. Technically, contact-action molluscicides need to be water soluble in order to penetrate the mucous layer surrounding the body of slugs and snails, but water-soluble compounds are leached away rapidly in the wet conditions under which slugs are most active and troublesome as pests. For this reason, most chemical molluscicides are applied in bait pellets (Bailey, 2002; Henderson and Triebskorn, 2002) to the soil surface. However, molluscicidal bait pellets have limited efficacy, especially in situations where slugs cause crop damage beneath the soil surface; for example, asparagus spears early in the growing season (Port and Ester, 2002), potatoes close to harvest (Beer, 1989) and winter wheat at establishment (Glen and Moens, 2002). Chemicals also have unwanted side effects and are subject to restrictions. The European Union has recently banned the use of methiocarb bait pellets for slug control because they are perceived to be hazardous for grain-eating farm birds such as finches and sparrows (Clarke, 2014). Although metaldehyde (the most widely used molluscicide) is harmless to beneficial insects (Samsøe-Petersen *et al.*, 1992) and earthworms (Edwards *et al.*, 2009), its use in bait pellets can result in accidental poisoning of mammals such as dogs (e.g. Buhl *et al.*, 2013). Metaldehyde has also recently been found to leach into drinking water supplies in higher than permissible concentrations

*Corresponding author. E-mail: davidmglen@btopenworld.com

  Biocontrol Agents: Entomopathogenic and Slug Parasitic Nematodes (eds. M.M.M. Abd-Elgawad, T.H. Askary and J. Coupland)

(Kay and Grayson, 2014), which has necessitated the industry to promote stewardship methods (Anon., 2015) and to develop additional water purification processes to remove residues (e.g. Nabeerasool *et al.*, 2015).

Cultural measures are valuable for slug control but their usefulness tends to be restricted. For example, cultivations can kill large numbers of slugs but many still survive, especially in wet conditions (Glen and Symondson, 2003). The preparation of fine, firm seedbeds greatly reduces the risk of slug damage to seedling crops, but it may not be possible for farmers to prepare such seedbeds when soils are very wet or dry, especially on heavy soils (Glen and Moens, 2002). Furthermore, as noted above, cultivation operations may be undesirable because they can result in soil erosion, especially in very wet or dry soil (Glen and Symondson, 2003).

For all the above reasons, there is a clear need for more effective and less environmentally damaging methods of controlling damage by slugs and snails, especially methods that are reliably effective in wet weather and in situations where slugs cause severe damage below the soil surface.

28.2 Preliminary Investigations

The discovery and commercialization of slug parasitic nematodes began with a proposal by one of the authors (D.M. Glen) in early 1987 for funding to survey for potential biological control agents for use against slug pests in the UK. The proposal for funding was based on the need for more effective and selective slug control, and was also partly inspired by the observations (Glen *et al.*, 1988; Christian *et al.*, 1999) in a long-term study of slug numbers on arable land in Oxfordshire, England, that the abundance of three species of slugs (*D. reticulatum*, *Arion intermedius* and *Arion silvaticus*) had declined by more than 90% in one year (1985), despite the weather being generally wet and suitable for slug activity, survival and breeding, and without any changes in farming practice that could explain such a dramatic reduction in numbers. Numbers then remained low in subsequent years. This suggested that a biological agent could have been responsible. However, few such potential biological control agents were recorded in the literature (Port and Port, 1986), and it was unlikely that any of them would meet the needs for a biological control agent that could be used in a commercially appropriate manner. These commercial criteria included effectiveness against a wide range of pest species of slugs, ease of production, handling and storage, suitable persistence after application and lack of harmful effects on non-target organisms.

The proposal for research funding was submitted from Long Ashton Research Station (LARS), University of Bristol (one of the research stations of what was then the UK Agricultural and Food Research Council (AFRC)), to the Agricultural Genetics Company (AGC). AGC had recently been set up to exploit the discoveries of AFRC science. Funding was secured and Michael J. Wilson, a new graduate in microbiology from the University of Bristol, was appointed to begin a search for any organisms that could have potential as biological control agents. At all stages, the research at LARS was done in close collaboration with AGC and its (then) subsidiary companies, MicroBio and Axis Genetics.

Michael Wilson started by collecting slugs (mainly *D. reticulatum*) in traps baited with bran in rough grassland at Long Ashton Research Station and keeping them in crowded conditions in plastic containers in the laboratory at 10°C. This temperature was chosen because it was representative of the relatively cool conditions of spring and autumn in the UK under which slugs were most troublesome as pests. Within a short time, Wilson found that *D. reticulatum* in some containers were infested by nematodes and dying. When nematodes from these slugs were transferred to containers of apparently healthy *D. reticulatum*, these slugs also showed signs of nematode infestation, most noticeably a swollen mantle (Fig. 28.1), followed by death, with the nematodes spreading and breeding over the entire cadaver (Fig. 28.2). The nematodes were then isolated and grown in culture with nutrient medium, following Hooper (1986), and then modified as described in the following section.

The nematodes were identified (Andrassy, 1983) as belonging to the Sub-Order Rhabditina and, with the help of David Hooper from Rothamsted Research Station, as belonging to the genus *Phasmarhabditis*. Andrassy (1983) lists two morphologically identical species of *Phasmarhabditis* that differ only in the number of males present in the population. Males and females are equally abundant in *Phasmarhabditis neopapillosa*, whereas males are extremely rare in *Phasmarhabditis hermaphrodita*. The initial isolate was *P. hermaphrodita*, but *P. neopapillosa* was also isolated later from slugs collected at LARS. Nematodes of both species were examined using light microscopy and electron microscopy. Protein profiles were also studied using isoelectric focusing (Hooper *et al.*, 1999). No differences were found between the two species using all of these methods.

AGC was interested in making further investigations of these slug parasitic nematodes, partly because their subsidiary company, MicroBio, was already producing insect parasitic nematodes on a large scale for commercial use, based on research done at the Glasshouse Crops Research Institute, Littlehampton, UK. Insect parasitic nematodes were being produced using liquid culture or by solid culture in bags or trays of foam chips impregnated with growth medium.

Fig. 28.1. The grey field slug, *Deroceras reticulatum*, in the foreground shows a swollen mantle characteristic of parasitism by *Phasmarhabditis hermaphrodita*. An unparasitized *D. reticulatum* is shown in the background for comparison.

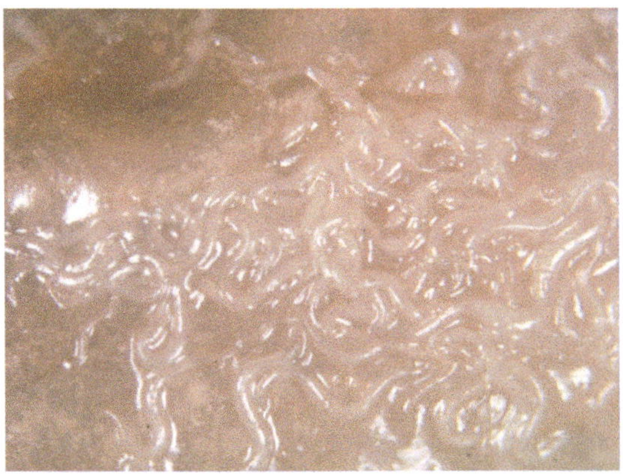

Fig. 28.2. *Phasmarhabditis hermaphrodita* feeding on a cadaver of the grey field slug, *Deroceras reticulatum*.

Another important reason for AGC's interest was that although *P. hermaphrodita* and *P. neopapillosa* had previously been recorded as being associated with slugs, they had not been recorded as parasites capable of causing disease and killing slugs. Thus, there was a potential opportunity for a patentable biocontrol product based on these nematodes. Both *P. hermaphrodita* and *P. neopapillosa* had been recorded from a number of locations in France and Germany as non-feeding infective juveniles or dauer larvae (third-stage juveniles with mouth and anus closed, inside a retained second-stage cuticle) (Fig. 28.3) living in the body cavity of slugs and snails, but later stages were not found in live slugs and snails (Maupas, 1900; Morand, 1988). Although these species were not considered to be parasites of slugs, they were thought to show some association with slugs, in that the infective juveniles entered the bodies of living slugs, where they remained until after the slug died, then grew and reproduced on the decaying slug cadaver, eventually producing infective juveniles once again when the food supply became depleted (Mengert, 1953). Infective juveniles were believed to survive in the soil ready to repeat the cycle by entering the body of slugs when the opportunity arose.

Wilson *et al.* (1993a) show how *P. hermaphrodita* acts as a slug parasite. It enters into the shell cavity (containing the vestigial shell) just below the dorsal surface of the mantle and induces a bacterial

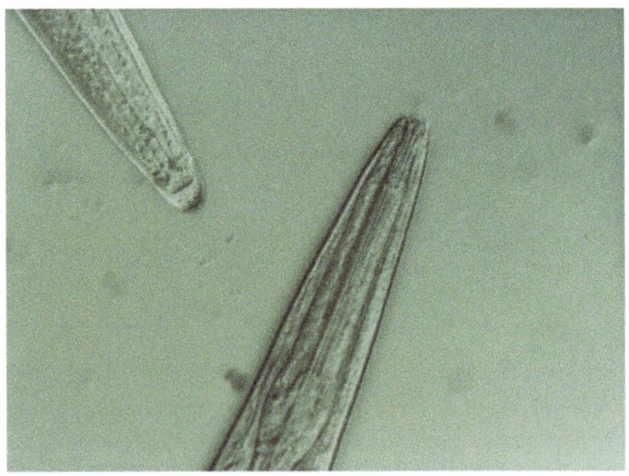

Fig. 28.3. Third-stage juvenile with open mouth (right) and infective third-stage (dauer) juvenile (left) with retained cuticle of the second juvenile stage and closed mouth.

infection in the fluid inside the shell cavity. This causes the mantle to swell in a characteristic way. Shortly after this, the slug dies and nematodes spread and reproduce all over the cadaver before once again producing infective juveniles.

28.3 Culturing, Harvesting and Storing *Phasmarhabditis hermaphrodita*

These nematodes are obligate bacterial feeders. Under favourable conditions, they develop through four juvenile stages into autogamous hermaphrodite adults that are mainly oviparous, with eggs laid in an advanced stage of development (Maupas, 1900). However, when food supply is depleted and toxic by-products build up, they give rise to third-stage infective juveniles (dauer) that are able to survive outside the host and are able to infect slugs (Wilson *et al.*, 1993a). For this reason, effective methods of rearing and storing large numbers of infective juveniles are necessary prerequisites for the use of *P. hermaphrodita* as a biocontrol agent. Techniques for rearing and storing infective juveniles were initially tested and developed as described by Wilson *et al.* (1993b). First, *P. hermaphrodita* taken from slugs were cultured xenically (i.e. with an unknown mixture of bacteria) in small numbers on kidney-based medium in Petri dishes. When larger numbers were required for experimental purposes, they were reared in foam-chip cultures similar to those used for insect parasitic nematodes (Bedding, 1984), which provide a plentiful air supply for the nematodes. *P. hermaphrodita* were then grown in small flasks of aerated liquid cultures, with baffled flasks giving better results than unbaffled flasks, presumably because the baffles improved oxygen absorption by the medium (Wilson *et al.*, 1993b).

28.3.1 Selection of pathogenic bacterial isolates

Many bacterial isolates were found to be associated with *Phasmarhabditis* nematodes. Xenic cultures are undesirable in commercial practice because they may contain bacteria pathogenic to the nematodes, or ones that produce toxic metabolites or ones that inhibit the growth of other, more beneficial bacteria. It was desirable to grow *P. hermaphrodita* in monoxenic culture (with a single strain of bacterium) as recommended for insect parasitic nematodes by Poinar and Hansen (1986).

The bacteria tested for monoxenic cultures (Wilson *et al.*, 1995a,b) were obtained from a range of sources, including the intestines of surface-sterilized nematodes from thriving xenic foam-chip cultures, from slugs (*D. reticulatum*) infected with *P. hermaphrodita*, together with isolates from the free-living nematode, *Caenorhabditis elegans*, and the insect parasitic nematode, *Steinernema feltiae*. Nine bacterial isolates were injected into the body cavity of *D. reticulatum*. Of these, only two isolates were found to be pathogenic.

First- and second-stage juveniles of *P. hermaphrodita* were axenized, then these sterile juveniles were introduced to monoxenic cultures in Petri dishes. Next, those bacterial isolates that supported good growth were used to culture nematodes in liquid medium in 250 ml conical flasks (Wilson *et al.*, 1995a,b). One of the isolates that was pathogenic to slugs when injected (*Aeromonas hydrophila*) did not support nematode growth, whereas the other (*Pseudomonas fluorescens*, isolate 140) supported good nematode growth but produced nematodes that were not pathogenic to *D. reticulatum*. Of the bacteria tested, *Providencia rettgeri* was found to give the best yields of infective juvenile *P. hermaphrodita* in liquid culture. However, when tested in bioassays for their ability to infect and kill slugs (*D. reticulatum*), infective juveniles produced in cultures with *P. rettgeri* gave inconsistent results. Nematodes produced on *Moraxella osloensis* or *P. fluorescens* (isolate 141) were pathogenic. Because of this and the finding that *M. osloensis* gave only slightly lower yields of infective juveniles than *P. rettgeri*, *M. osloensis* was selected as the preferred bacterial species for further investigations leading towards commercial production of *P. hermaphrodita* nematodes as biocontrol agents for slugs.

These studies clearly demonstrated that interactions between the nematode, the bacterium and the host slug were important in determining the value of *P. hermaphrodita* as a biocontrol agent.

28.3.2 Commercial production

In commercial production, *P. hermaphrodita* are reared in monoxenic liquid cultures with *M. osloensis* in fermenters. Yields of infective larvae (Glen *et al.*, 1996) are influenced by the composition of the growth medium and the environment within the fermenter, where yields in excess of 100,000 infective juveniles/ml were achieved in early commercial production. Glen *et al.* (1996) stated that infective juveniles were harvested from the fermentation medium by centrifugation, followed by repeated washing in water, concentrated to give between 0.1 × 10^6 and 2.0 × 10^6 nematodes/g and mixed with calcium montmorillonite clay to produce a water-dispersible friable formulation containing 0.05 × 10^6 to 1.8 × 10^6 larvae/g (wet weight), which is stored in bags of 8 μm high-density polyethylene. *P. hermaphrodita* can be maintained in this formulation, under refrigeration, for 6 months. When required, the formulation is mixed with water and applied to soil using conventional spray equipment or a watering can fitted with a rose.

28.4 Host Range Studies

Laboratory bioassays (Wilson *et al.*, 1993a) to test the exposure of slugs to different concentrations of nematodes were initially done in plastic tubes. However, the survival of untreated slugs was sometimes poor in these tubes, so a soil-based assay was used with boxes containing soil aggregates (each aggregate 12–24 mm across), which allowed slugs to move readily through the soil, which was either treated with infective juveniles of *P. hermaphrodita* or left untreated. Boxes of soil containing slugs and nematodes were kept at 10°C (a temperature typical of the autumn, when slugs are particularly troublesome as crop pests). These studies showed a clear dose response, with the percentage mortality of slugs (*D. reticulatum*) increasing on a probit scale in relation to nematode dose. Other species of slugs were also shown to be susceptible to nematode infection, including *Deroceras panormitanum*, *A. intermedius*, *Arion distinctus*, *Arion ater*, *A. silvaticus*, *Tandonia sowerbyi* and *Tandonia budapestensis*. All the slugs tested were considered to be adults, except for *A. ater*, which were juveniles about the same size as adult *D. reticulatum*. The snails *Helix aspersa* and *Monacha cantiana* were also shown to be killed by *P. hermaphrodita*. Other helicid snails (*Theba pisana*, *Trochoidea elegans*, *Cochlicella acuta* and *Cernuella virgata*) in southern France were also shown by Coupland (1995) to be susceptible. However, subsequent studies have shown that the larger species of slugs and snails are only likely to be killed by *P. hermaphrodita* when they are in the smaller juvenile stages (Speiser *et al.*, 2001).

Tests showed that the host range appeared to be confined to molluscs (Wilson *et al.*, 2000) because other invertebrates, including earthworms and a range of insect species, were not affected by exposure to high doses of the nematodes in laboratory tests. Moreover, as *P. hermaphrodita* is rapidly killed at 35°C (Wilson *et al.*, 1993b), it is not able to survive in homoeothermic animals.

28.5 Initial Field Experiments in Biocontrol

28.5.1 Mini-plot trials

Following the encouraging results of laboratory bioassays, it was important to establish as quickly as possible whether *P. hermaphrodita* would be effective in controlling slug damage to plants under field conditions. At this stage (early 1989), only small quantities of these nematodes were being produced in xenic cultures at LARS. Therefore, the initial outdoor replicated trials were done in mini-plots (each 0.7 × 0.7 m) of soil at LARS (Wilson *et al.*, 1994b). The mini-plots were separated from each other by solid barriers below soil level and by slug-proof copper-mesh barriers between the plots above soil level. Two experiments were done in 1989. Between the experiments, the soil was removed and the plots and underlying gravel were treated to kill any remaining nematodes before the plots were filled with fresh soil. Slugs (*D. reticulatum*) were introduced to the soil before the start of each trial. In the first trial, infective juveniles of *P. hermaphrodita* were applied as a drench at a rate equivalent to 2.1 × 10^6/m^2 and compared with untreated plots and with plots treated with methiocarb pellets at the recommended rate of 0.55 g product/m^2 (Draza, 4% a.i., Bayer; a standard chemical treatment at that time for controlling slug damage). Immediately after treatment, Chinese cabbage seedlings were planted in each plot. Chinese cabbage was chosen as a test crop because it was attractive to slugs, easily grown and it was representative of a wide range of crops and ornamental plants that were susceptible to slug damage as seedlings. In the second trial, six doses of infective juveniles were tested, ranging from 1 × 10^4 to 2 × 10^6/m^2. Chinese cabbage seedlings were also planted in this trial, along with wheat seeds.

In the first mini-plot experiment, the plots treated with *P. hermaphrodita* showed significantly less damage than the untreated plots. From 3 weeks after treatment until the end (6 weeks after treatment), there was also significantly less damage on plots treated with *P. hermaphrodita* than on methiocarb-treated plots. In the second experiment, slug damage decreased as nematode dose increased from 1×10^4 to $8 \times 10^4/m^2$ but then showed little or no further reduction in damage at higher doses. Doses of $8 \times 10^4/m^2$ and above gave similar protection to methiocarb pellets. As the nematodes used in both field experiments were reared in mixed bacterial cultures, it is possible that variation in the bacterial population make-up in the nematode growth medium may have affected their pathogenicity to slugs. This did not explain the lack of dose response above $8 \times 10^4/m^2$ in the second experiment, which was puzzling. However, similar findings of weak or no dose response have been reported for insect parasitic nematodes by Georgis and Gaugler (1991) in their review of field trials.

Following completion of these two mini-plot trials in 1989, it was clearly important to eliminate any possible variability in the bacterial population used to cultivate *P. hermaphrodita*. Thus, no further trials were initiated until a suitable bacterium had been selected, as described in the preceding section.

28.5.2 First field trial in winter wheat

Slugs are one of the most important pests of winter wheat in Western Europe, especially so in the UK because the mild damp climate is highly favourable for slugs (Port and Port, 1986; Glen and Moens, 2002). The most severe damage is done at early crop establishment, below ground, when slugs destroy seed embryos or the growing points of young seedlings. Although damage can be reduced by appropriate cultural control measures and by the use of chemical molluscicides applied as slug pellets, these measures tend to be less effective in wet weather conditions, i.e. at the time when slugs are most damaging as pests. In addition, slug pellets have unwanted side effects, including toxicity to non-target organisms and leaching of active ingredients into water supplies (as described in the introduction to this chapter). Thus, there is a definite need for more effective and selective methods of controlling slug damage to winter wheat.

The first winter wheat trial, at LARS in autumn 1991 (Wilson *et al.*, 1994a), was an important milestone because of the above, but also for two other important reasons: (i) it was the first field experiment that used infective juveniles of *P. hermaphrodita* reared in monoxenic liquid cultures by MicroBio; (ii) it was the first test of these nematodes against a natural field population of slugs.

The wheat crop was shallow sown following a crop of oilseed rape, into a cloddy seedbed. These are all factors that increase the risk of slug damage (Glen and Moens, 2002). There were five replicates of seven treatments. Five doses of nematodes ($1 \times 10^4/m^2$, $3 \times 10^4/m^2$, $1 \times 10^5/m^2$, $3 \times 10^5/m^2$, $1 \times 10^6/m^2$) were compared with untreated plots and plots treated with methiocarb pellets (Draza, Bayer) at 0.55 g/m^2. The nematodes were suspended in 4 l of tap water per plot (6 m × 6 m) using a knapsack sprayer fitted with a coarse nozzle. All treatments were applied just after drilling, at around sunset to minimize exposure of the nematodes to ultraviolet rays.

The soil was relatively moist at the time of drilling and nematode application, but subsequent high winds resulted in rapid drying of the soil surface. Thus, weather conditions were not ideal for nematode activity and survival in the first week after treatment, but subsequent rainfall resulted in improved conditions. The number of wheat plants that survived slug attack increased linearly and strongly with the log nematode dose, while the percentage of leaf area destroyed declined linearly with the log dose.

The impact of treatment with *P. hermaphrodita* on slug numbers, activity and on the prevalence of nematode infection was assessed by two methods: refuge trapping and soil sampling. *D. reticulatum* was the predominant species found together with nine other species (Wilson *et al.*, 1994a). Virtually no dead or moribund slugs were found on the soil surface after treatment with *P. hermaphrodita*, in contrast to plots treated with methiocarb, where large numbers of slugs showing symptoms of poisoning were found on the soil surface soon after treatment. One and two weeks after treatment, there were significantly more slugs in traps on plots treated with the two highest doses of nematodes than on untreated plots. During this period, more of the trapped slugs showed signs of nematode infection than was expected from the prevalence of infection in slugs in the soil samples, suggesting that the presence of *P. hermaphrodita* caused the slugs to alter their behaviour, perhaps trying to avoid contact with these nematodes, as shown by Wilson *et al.* (1999). After 6 weeks, there were significantly fewer slugs in traps on plots treated with the highest nematode dose. Slug numbers in soil samples were not affected significantly by nematode dose until 5 weeks after treatment, when there was a significant inverse relationship between slug numbers and dose. However, by then, slug numbers on untreated plots had declined to low levels similar to those found on the plots treated with the highest dose of *P. hermaphrodita*. The rather weak effect of treatment with *P. hermaphrodita* on slug numbers, combined with the strong effect of such treatment in reducing slug damage, suggested that its efficacy as a biocontrol agent was mainly through the inhibition of feeding by slugs rather than its effects in killing slugs. Laboratory studies have confirmed the importance of feeding inhibition for the success of this nematode as a biocontrol agent (Glen *et al.*, 2000a).

28.5.3 First trials in lettuce

In 1992 and 1993, infective juveniles were produced in sufficient numbers in monoxenic liquid cultures with *M. osloensis* by MicroBio Ltd for two trials to be carried out at LARS to investigate the biocontrol of slugs in lettuce grown in polythene tunnels (Wilson *et al.*, 1995c). Such tunnels were known to provide excellent conditions for slug population growth in spring and autumn, with suitable temperatures and high humidity. Thus, they provided an excellent test of the biocontrol potential of *P. hermaphrodita*. Both lettuce trials compared treatments with infective juveniles applied as a drench to soil at four dose rates (3×10^4, 1×10^5, 3×10^5 and $1 \times 10^6/m^2$) before planting lettuce seedlings. These treatments were compared with untreated plots and with plots treated with methiocarb pellets (0.55 g/m^2) evenly spread on the soil surface immediately after planting. In the first trial, in spring 1992, nematode treatments were applied 1 day before planting. Plots treated with methiocarb pellets showed significantly reduced slug damage to lettuce seedlings (measured over the first 3 weeks after planting), but the nematode treatments did not. In the second experiment in spring 1993, in a different polythene tunnel, nematode treatments were applied 4 days before planting. In this second experiment, the methiocarb treatment and the treatment with the second highest dose of nematodes reduced slug damage significantly. In both experiments, increasing the nematode dose reduced the numbers of slugs in lettuce plants significantly at harvest. Such contamination is economically important because it can result in rejection of a batch of lettuce. Thus, reduction in the number of slugs in harvested lettuce is as economically important as reduction in damage to the seedlings.

28.6 Commercialization

28.6.1 Product launch

Research on the biocontrol of slugs using *P. hermaphrodita* remained confidential until publication of an international patent (Wilson *et al.*, 1993c). Following this, a commercial product (Nemaslug®) was launched by MicroBio in 1994 for the control of slugs in the home garden market in the UK. This stimulated worldwide interest in the news media. The home garden market was chosen for initial launch because it could be supplied with the relatively small quantities available from commercial production at that stage. Each pack of Nemaslug contained 6×10^6 infective juveniles of *P. hermaphrodita*. The recommendation was for this to be applied to 20 m^2 of moist soil ($3 \times 10^5/m^2$) using a watering can fitted with a rose.

28.6.2 Early commercial trials

From 1993 onwards, MicroBio conducted many European grower trials with the commercial formulation of *P. hermaphrodita* (Glen *et al.*, 1996). Initial trials were set up in order to demonstrate efficacy against common European slug species (predominantly *D. reticulatum*) in comparison to chemical molluscicides sold on the home garden market (based on metaldehyde). A single treatment with 3×10^5 nematodes/m^2 provided control equivalent to metaldehyde-based formulations in a number of different trial locations across Europe, with a range of climatic conditions and soil types. However, reduced product efficacy was observed when applied to heavy, waterlogged soils. Here, it is likely that the nematodes were unable to penetrate the upper soil horizons and quickly died when the soil surface dried out.

Subsequent commercial trials focused on both fine-tuning the label recommendation for the home garden market and also to demonstrate efficacy in high-value horticultural crops. Consistent slug control was demonstrated in all but the most extreme conditions. However, the application of *P. hermaphrodita* to mulched borders, or thatched lawns, resulted in a significant reduction in slug control. This indicated that, as is also the case for entomopathogenic nematodes (EPNs), it is essential that the nematodes are applied directly to moist, bare soil in order to maximize response. As previously mentioned, larger slugs and snails were found to be less sensitive to the application of *P. hermaphrodita* than either their juvenile counterparts or smaller species.

In horticulture, T-tape irrigation has become an increasingly popular method of irrigating mulched crops such as strawberries. A T-tape grower trial in strawberries, commissioned through the UK high street supermarket chain, Sainsbury's, indicated that the efficacy of *P. hermaphrodita* (applied at 3 $\times 10^4/m^2$ through double-line T-tape) was comparable to chemical pelleted baits in terms of the percentage reduction in fruit damage. The numbers of carabid beetles, which are slug predators, remained unchanged following the use of *P. hermaphrodita* in this trial (Glen *et al.*, 1996).

Grower trials demonstrated that the control of plant damage could be sustained for 5–8 weeks following an application of *P. hermaphrodita*. A good example was a series of UK field trials on iceberg lettuce. Slug control is critical in this high-value, often irrigated, crop.

Growers routinely use two to three applications of pelleted molluscicide throughout the crop's 9- to 10-week period of field growth. It is essential that pelleted baits do not become lodged within leaf bases, and that slugs are controlled before they move from the soil into the developing plant both to feed and to seek refuge. These field trials concluded that a single application of *P. hermaphrodita* approximately 4 weeks after planting resulted in the effective control of plant damage until harvest. A combination of the early inhibition of slug feeding, followed by slugs burrowing into the soil before death, makes *P. hermaphrodita* an ideal slug control agent for such a market.

28.7 Further Studies to Extend the Commercial Use of *P. hermaphrodita*

Following the product launch, the efficacy of *P. hermaphrodita* was quickly demonstrated in small-scale field trials against slugs in lettuce in Switzerland (Speiser and Andermatt, 1994, 1996). Further research at LARS in collaboration with MicroBio aimed to extend the use of *P. hermaphrodita* from home gardens to commercial agriculture and horticulture. One project, funded jointly by MicroBio and the UK Ministry of Agriculture, Fisheries and Food, focused on its use in arable crops (principally winter wheat and winter oilseed rape). These studies showed that the incorporation of nematodes into the surface layers of soil could help to prevent the loss of nematode efficacy in drier soil conditions (Wilson *et al.*, 1996). It was possible to reduce the overall dose of *P. hermaphrodita* applied without reduction in efficacy by targeting applications to specific slug feeding areas (Hass *et al.*, 1999a,b). However, despite these findings, the overall cost of applying these nematodes was simply too high for their use in slug control in arable crops and to be commercially viable.

Another major EU-funded project entitled 'Novel technologies for the integrated control of slug damage in key horticultural crops' investigated the use of *P. hermaphrodita* in combination with a range of other control measures in a variety of horticultural crops including salad vegetables, asparagus and Brussels sprouts. This involved seven collaborating organizations in France, The Netherlands, Spain and Switzerland, in addition to the UK (Glen *et al.*, 2000b; Speiser *et al.*, 2000, 2001; Iglesias *et al.*, 2001, 2003; Ester *et al.*, 2003, 2005; Vernavá *et al.*, 2004). During the course of this project, more than 30 new strains of *P. hermaphrodita* capable of killing *D. reticulatum* were isolated from The Netherlands, Spain, Switzerland and the UK, demonstrating the widespread distribution of this species in Europe. The product containing *P. hermaphrodita* was tested in asparagus, Brussels sprouts and a wide range of salad crops, with mostly positive results. In general, two or more applications of *P. hermaphrodita* were found to give better control of slug damage compared with a single application. Some aspects of these investigations are described below.

28.7.1 Influence of timing of application on nematode efficacy

In initial mini-plot field experiments, protection from slug damage by *P. hermaphrodita* was found to increase over the first 2 weeks after application (Wilson *et al.*, 1994b). Since severe damage by slugs to many crops is done soon after sowing or planting, this suggests that it may be beneficial to apply nematodes to the soil before susceptible crops are sown. However, field experiments showed little or no benefit from nematode application 1 or 2 weeks prior to sowing a crop of cereals or rape, compared with application at the time of drilling or planting (e.g. Wilson *et al.*, 1995d). In these experiments, application to moist soil appeared to be more important for achieving the effective control of slug damage than the precise timing of application.

Because crops such as oilseed rape provide slugs with an excellent source of food and shelter, slug numbers usually increase considerably in these crops, with a consequent high risk of damage to following crops, such as winter wheat. This suggested that it might be possible to use the nematode to prevent the build-up of slug populations in oilseed rape. However, in two trials where the nematode was applied to oilseed rape at the time of drilling, soil sampling at intervals during the growing season revealed no effect on slug population increase in this crop (Wilson *et al.*, 1995d).

28.7.2 Value of incorporating nematodes into soil after application

P. hermaphrodita, like its slug hosts, is very dependent on moisture for survival. Thus, nematodes are normally likely to be applied to moist soil where slugs are likely to cause damage. However, the surface of arable soils tends to dry rapidly following cultivation, and although there may be sufficient moisture below the surface for slug activity, nematodes applied to the soil surface under these conditions will become desiccated and die shortly afterwards. Poor performance of the nematode in some field trials has been associated with dry soil conditions (Wilson *et al.*, 1995d, 1996). Trials were set up to investigate the possibility of incorporating nematodes into the top 2–10 cm of surface soil in order to improve their reliability (Wilson *et al.*, 1996). When nematodes were applied to moist soil, incorporation had no effects on nematode efficacy. However, when nematodes were applied to a dry seedbed immediately after drilling winter wheat, nematodes left on the soil surface were ineffective, whereas nematodes incorporated into the soil using a tined implement working to a depth of 5 or 10 cm, reduced slug damage significantly. Thus, shallow incorporation of nematodes was beneficial in dry soil conditions and had a neutral effect in moist soil. In many crops, nematodes could be incorporated readily into soil during seedbed preparation.

28.7.3 Exploiting the repellent effects of nematodes

In the first winter wheat field trial (Glen *et al.*, 1994; Wilson *et al.*, 1994a), it was noted that slug behaviour appeared to have changed on nematode-treated plots, such that more slugs were recorded beneath refuge traps placed on the soil surface of these plots, compared to untreated plots. Moreover, a higher percentage of slugs recorded beneath the traps showed symptoms of nematode infection than slugs recorded in soil samples. Laboratory studies have since demonstrated that slugs avoid feeding and resting on soil treated with nematodes (Wilson *et al.*, 1999). This finding suggested that it might be possible to apply nematodes only to soil immediately surrounding susceptible plants or along crop rows, thus reducing the dose of nematodes required for effective protection. Repellence probably explains why *P. hermaphrodita* is effective in reducing slug damage to Brussels sprouts, which develop well above soil level: Ester *et al.* (2005) found a substantial reduction in damage when *P. hermaphrodita* was applied as an overhead spray to this crop.

28.7.4 Effects on snails of conservation interest

Although several snail species have been shown to be killed by *P. hermaphrodita* in the laboratory, in two field experiments where nematodes were applied to plots adjacent to field margins, no adverse effects were detected on snails living in field margins (Wilson *et al.*, 2000). Experiments on the horizontal movement of the nematodes in soil showed that the vast majority of nematodes moved less than 1 cm from the site of placement (Wilson *et al.*, 2000). These results indicate that the application of nematodes to arable soil is unlikely to result in nematodes spreading to semi-natural habitats in sufficient numbers to affect the slugs and snails living there.

28.8 Conclusions

This project indicates the amount of effort involved both in finding a nematode biological control agent and in bringing this agent to commercialization. While EPNs have been discovered and commercialized widely for use against insect pests, the discovery against more poorly understood but equally important agricultural pests (such as molluscs) has progressed more slowly. It may be that as chemical pesticides become more unfavourable in the public eye, more research into biological solutions may be more supported. Although slug parasitic nematodes are currently too expensive for widespread use in broad-acre agriculture, this could change as a result of advances in the technologies of production, storage and application. Commercial treatments have already been made in potatoes (Brown *et al.*, 2011), a relatively high-value broad-acre crop. Moreover, the recent discovery of *P. hermaphrodita* in the cadavers of invasive slugs in the USA (De Ley *et al.*, 2014) opens up the possibility for their commercial use in this important market. Modelling studies strongly indicate that the success of *P. hermaphrodita* as a biocontrol agent against slugs not only is dependent on the application dose of these nematodes but also the density of the slug population is highly important (Wilson *et al.*, 2004). This is an area where further research is warranted in order to improve the predictability and success of biocontrol using *P. hermaphrodita*.

Acknowledgements

We thank all collaborators and funders of the research described in this chapter for their invaluable contributions.

References

Andrassy, I. (1983) *A Taxonomic Review of the Sub-order Rhabditina (Nematoda: Secernentina).* ORSTOM, Paris, 241 pp.

Anon. (2015) http://www.getpelletwise.co.uk/news/metaldehyde_stewardship (accessed 17 October 2016).

Bailey, S.E.R. (2002) Molluscicidal baits for control of terrestrial gastropods. In: Barker, G.M. (ed.) *Molluscs as Crop Pests.* CAB International, Wallingford, UK, pp. 33–54.

Barker, G.M. (2001) Gastropods on land: phylogeny, diversity and adaptive morphology. In: Barker (ed.) *The Biology of Terrestrial Molluscs.* CAB International, Wallingford, UK, pp. 1–146.

Barker, G.M. (2002) *Molluscs as Crop Pests.* CAB International, Wallingford, UK.

Bedding, R.A. (1984) Large scale production, storage and transport of the insect parasitic nematodes *Neoaplectana* spp. and *Heterorhabditis* spp. *Annals of Applied Biology* 104, 117–120.

Beer, G.J. (1989) *Levels and Economics of Slug Damage in Potato Crops 1987 and 1988.* Monograph - British Crop Protection Council 1989 No 41, pp. 101–105.

Brown, A.P., Barker, A., Hopkins, A. and Nelson, D. (2011) Application of *Phasmarhabditis hermaphrodita* (Nemaslug®) to commercial broad acre crops. IOBC/WPRS Bulletin 64, 99–104.

Buhl, K.J., Berman, F.W. and Stone, D.L. (2013) Reports of metaldehyde and iron phosphate exposures in animals and characterization of suspected iron toxicosis in dogs. *Journal of the American Veterinary Medical Association* 242, 1244–1248.

Christian, D.G., Bacon, E.T.G., Brockie, D., Glen, D.M., Gutteridge, R.J. and Jenkyn, J.F. (1999) A long-term study of straw disposal methods and direct-drilling or cultivations on winter cereals grown on a clay soil. *Journal of Agricultural Engineering Research* 73, 297–309.

Clarke, A. (2014) EU votes to ban methiocarb slug pellets. *Farmers Weekly*, 22 January 2014.

Coupland, J.B. (1995) Susceptibility of helicid snails to isolates of the nematode *Phasmarhabditis hermaphrodita* from southern France. *Journal of Invertebrate Pathology* 66, 207–208.

Crovetto, C. (1996) *Stubble Over the Soil.* American Society of Agronomy, Madison, Wisconsin.

De Ley, I.T.T., McDonnell, R.D., Lopez, S., Paine, T.D. and De Ley, P. (2014) *Phasmarhabditis hermaphrodita* (Nematoda: Rhabditidae), a potential biocontrol agent isolated for the first time from invasive slugs in North America. *Nematology* 16, 1129–1138.

Edwards, C.A., Arancon, N.Q., Vasko-Bennett, M., Little, B. and Askar, A. (2009) The relative toxicity of metaldehyde and iron phosphate-based molluscicides to earthworms. *Crop Protection* 28, 289–294.

El Titi, A. (2003) *Soil Tillage in Agroecosystems.* CRC Press, Boca Raton, Florida.

Ester, A., Rozen, K. van and Molendijk, L.P.G. (2003) Field experiments using the rhabditid nematode *Phasmarhabditis hermaphrodita* or salt as control measures against slugs in green asparagus. *Crop Protection* 22, 689–695.

Ester, A., Huiting, H. and Rozen, K. van (2005) Slug parasitic nematodes in vegetable crops. Bulletin OILB/SROP 28, 39–42.

Georgis, R. and Gaugler, R.E. (1991) Predictability in biological control using entomopathogenic nematodes. *Journal of Economic Entomology* 84, 713–720.

Glen, D.M. and Moens, R. (2002) Agriolimacidae, Arionidae and Milacidae as pests in West European cereals. In: Barker, G.M. (ed.) *Molluscs as Crop Pests.* CAB International, Wallingford, UK, pp. 271–300.

Glen, D.M. and Symondson, W.O.C. (2003) Influence of soil tillage on slugs and their natural enemies. In: El Titi, A. (ed.) *The Role of Soil Tillage in Agroecosystems.* CRC Press, Boca Raton, Florida, pp. 207–227.

Glen, D.M., Wiltshire, C.W. and Milsom, N.F. (1988) Effects of straw disposal on slug problems in cereals. In: *Environmental Aspects of Applied Biology*, Vol 2. Aspects of Applied Biology 17, Association of Applied Biologists, Wellesbourne, UK, pp. 173–179.

Glen, D.M., Wilson, M.J., Pearce, J.D. and Rodgers, P.B. (1994) Discovery and investigation of a novel nematode parasite for biological control of slugs. In: *Proceedings of the Brighton Crop Protection Conference, Pests and Diseases.* British Crop Protection Council, Alton, Hants, UK, pp. 617–624.

Glen, D.M., Wilson, M.J. and Hughes, L. (1996) Exploring and exploiting the potential of the rhabditid nematode

Phasmarhabditis hermaphrodita as a biocontrol agent for slugs. In: Henderson, I.F. (ed.) *Slug and Snail Pests in Agriculture*. Monograph No 66, British Crop Protection Council, Thornton Heath, UK, pp. 271–280.

Glen, D.M., Wilson, M.J., Brain, P. and Stroud, G. (2000a) Feeding activity and survival of slugs, *Deroceras reticulatum*, exposed to the rhabditid nematode, *Phasmarhabditis hermaphrodita*: a model of dose–response. *Biological Control* 17, 73–81.

Glen, D.M., Wiltshire, C.W., Hughes, L., Ester, A., Rozen, van K., *et al.* (2000b) The use of slug-parasitic nematodes and other techniques for control of slug and snail damage in horticultural crops. In: *Proceedings of the Brighton Crop Protection Conference, Pests and Diseases*. British Crop Protection Council, Alton, Hants, UK, pp. 345–350.

Godan, D. (1983) *Pest Slugs and Snails*. Springer-Verlag, Berlin.

Hass, B., Hughes, L.A. and Glen, D.M. (1999a) Overall versus band application of the nematode, *Phasmarhabditis hermaphrodita*, with and without incorporation into soil, for biological control of slugs in winter wheat. *Biocontrol Science and Technology* 9, 579–586.

Hass, B., Glen, D.M., Brain, P. and Hughes L.A. (1999b) Targeting biocontrol with the slug-parasitic nematode *Phasmarhabditis hermaphrodita* in slug feeding areas: a model study. *Biocontrol Science and Technology* 9, 587–598.

Henderson, I. and Triebskorn, R. (2002) Chemical control of terrestrial gastropods. In: Barker, G.M. (ed.) *Molluscs as Crop Pests*. CAB International, Wallingford, UK, pp. 1–31.

Hooper, D.J. (1986) Culturing nematodes and related experimental techniques. In: Southey, J.F. (ed.) *Laboratory Methods for Work with Plant and Soil Nematodes*. MAFF Reference Book No 402. HMSO, London, pp. 133–137.

Hooper, D.J., Wilson, M.J., Rowe, J.A. and Glen, D.M. (1999) Some observations on the morphology and protein profiles of the slug-parasitic nematodes *Phasmarhabditis hermaphrodita* and *P. neopapillosa* (Nematoda: Rhabditidae). *Nematology* 1, 173–182.

Iglesias J., Castillejo, J. and Castro, R. (2001) Field test using the nematode *Phasmarhabditis hermaphrodita* for biocontrol of slugs in Spain. *Biocontrol Science and Technology* 11, 93–98.

Iglesias, J., Castillejo, J. and Castro, R. (2003) The effects of repeated applications of the molluscicide metaldehyde and the biocontrol nematode *Phasmarhabditis hermaphrodita* on molluscs, earthworms, nematodes, acarids and collembolans: a two-year study in north-west Spain. *Pest Management Science* 59, 1217–1224.

Kay, P. and Grayson, R. (2014) Using water industry data to assess the metaldehyde pollution problem. *Water and Environment Journal* 28, 410–417.

Maupas, E. (1900) Modes et formes de reproduction des nematodes. *Archives de Zoologie* 8, 464–642.

Mengert, H. (1953) Nematoden und Schneken. *Zeitschrift für Morphologie und Ökologie Tiere* 41, 311–349.

Moens, R. and Glen, D.M. (2002) Agriolimacidae, Arionidae and Milacidae as pests in West European oilseed rape. In: Barker, G.M. (ed.) *Molluscs as Crop Pests*. CAB International, Wallingford, UK, pp. 425–439.

Morand, S. (1988) Contribution á l'étude d'un systeme hôtes-parasites: nematodes associés á quelques mollusques terrestres. Doctorate thesis. University of Rennes, France, 335 pp.

Nabeerasool, M.A., Campen, A.K., Polya, D.A., Brown, N.W. and Dongen, B.E. van (2015) Removal of metaldehyde from water using a novel coupled adsorption and electrochemical destruction technique. *Water* 7, 3057–3071.

Poinar, G.O. Jr and Hansen, E.L. (1986) Associations between nematodes and bacteria. *Helminthological Abstracts Series B* 33, 61–81.

Port, C.M. and Port, G.R. (1986) The biology and behavior of slugs in relation to crop damage and control. *Agricultural Zoology Reviews* 1, 255–299.

Port, G. and Ester, A. (2002) Gastropods as pests in vegetable and ornamental crops in Western Europe. In: Barker, G.M. (ed.) *Molluscs as Crop Pests*. CAB International, Wallingford, UK, pp. 337–352.

Samsøe-Petersen, L., Bieri, M. and Büchs, W. (1992) Interpretation of laboratory measured effects of slug pellets on soil dwelling invertebrates. *Aspects of Applied Biology* 31, 87–96.

South, A. (1992) *Terrestrial Slugs, Biology, Ecology and Control*. Chapman and Hall, London, 428 pp.

Speiser, B. and Andermatt, M. (1994) Biokontrolle von Schneken mit Nematoden. *Agrarforschung* 1, 115–118.

Speiser, B. and Andermatt, M. (1996) Field trials with *Phasmarhabditis hermaphrodita* in Switzerland. In: Henderson, I.F. (ed.) *Slug and Snail Pests in Agriculture*. Monograph No 66. British Crop Protection Council, Thornton Heath, UK, pp. 419–424.

Speiser, B., Glen, D., Bohan, A., Aalten, M., Hughes, L., *et al.* (2000) 'Novel Slug Control' – combined efforts to devise strategies of slug control for organic and integrated farming. In: Alföldi, T., Lockeretz, W. and Niggli, U. (eds) *The World Grows Organic. Proceedings of 13th International IFOAM Scientific Conference*. Basel, Switzerland, pp. 28–31.

Speiser, B., Zaller, J.G. and Neudecker, A. (2001) Size-specific susceptibility of the pest slugs *Deroceras reticulatum* and *Arion lusitanicus* to the nematode biocontrol agent *Phasmarhabditis hermaphrodita*. *BioControl* 46, 311–320.

Vernavá, M.N., Phillips-Aalten, P.M., Hughes, L.A., Rowcliffe, H., Wiltshire, C.W. and Glen, D.M. (2004) Influences of preceding cover crops on slug damage and biological control using *Phasmarhabditis hermaphrodita*. *Annals of Applied Biology* 145, 279–284.

Wilson, M.J., Glen, D.M. and George, S.K. (1993a) The rhabditid nematode *Phasmarhabditis hermaphrodita* as a potential biological control agent for slugs. *Biocontrol Science and Technology* 3, 503–511.

Wilson, M.J., Glen, D.M., George, S.K. and Butler, R.C. (1993b) Mass cultivation and storage of the rhabditid nematode *Phasmarhabditis hermaphrodita*, a biocontrol agent for slugs. *Biocontrol Science and Technology* 3, 513–521.

Wilson, M.J., Glen, D.M. and Pearce, J.D. (1993c) Biological Control of Molluscs. International Patent Number WO 93/00816.

Wilson, M.J., Glen, D.M., George, S.K., Pearce, J.D. and Wiltshire, C.W. (1994a) Biological control of slugs in winter wheat using the rhabditid nematode *Phasmarhabditis hermaphrodita*. *Annals of Applied Biology* 125, 377–390.

Wilson, M.J., Glen, D.M., Wiltshire, C.W. and George, S.K. (1994b) Mini-plot field experiments using the rhabditid nematode *Phasmarhabditis hermaphrodita* for biocontrol of slugs. *Biocontrol Science and Technology* 4, 103–113.

Wilson, M.J., Glen, D.M., Pearce, J.D. and Rodgers, P.B. (1995a) Monoxenic culture of the slug parasite *Phasmarhabditis hermaphrodita* (Nematoda: Rhabditidae) with different bacteria in liquid and solid phase. *Fundamental and Applied Nematology* 18, 159–166.

Wilson, M.J., Glen, D.M., George, S.K. and Pearce, J.D. (1995b) Selection of a bacterium for the mass production of *Phasmarhabditis hermaphrodita* (Nematoda: Rhabditidae) as a biocontrol agent for slugs. *Fundamental and Applied Nematology* 18, 419–425.

Wilson, M.J., Glen, D.M., George, S.K. and Hughes, L.A. (1995c) Biocontrol of slugs in protected lettuce using the rhabditid nematode *Phasmarhabditis hermaphrodita*. *Biocontrol Science and Technology* 5, 233–242.

Wilson, M.J., Hughes, L.A. and Glen, D.M. (1995d) Developing strategies for the nematode, *Phasmarhabditis hermaphrodita*, as a biological control agent for slugs in integrated crop management systems. In: McKinlay, R.G. and Atkinson, D. (eds) *Integrated Crop Protection: Towards Sustainability?* British Crop Protection Council Monograph No 63, Farnham, Surrey, UK, pp. 33–40.

Wilson, M.J., Hughes, L.A., Hamacher, G.M., Barahona, L.D. and Glen, D.M. (1996) Effects of soil incorporation on the efficacy of the rhabditid nematode, *Phasmarhabditis hermaphrodita*, as a biological control agent for slugs. *Annals of Applied Biology* 128, 117–126.

Wilson, M.J., Hughes, L.A., Jefferies, D. and Glen, D.M. (1999) Slugs (*Deroceras reticulatum* and *Arion ater* agg.) avoid soil treated with the rhabditid nematode *Phasmarhabditis hermaphrodita*. *Biological Control* 16, 170–176.

Wilson, M.J., Hughes, L.A., Hamacher, G.M. and Glen, D.M. (2000) Effects of *Phasmarhabditis hermaphrodita* on non-target molluscs. *Pest Management Science* 56, 711–716.

Wilson, M.J., Glen, D.M., Hamacher, G.M. and Smith, J.U. (2004) A model to optimise biological control of slugs using nematode parasites. *Applied Soil Ecology* 26, 179–191.

29 *Phasmarhabditis*: The Slug and Snail Parasitic Nematodes in North America

Irma Tandingan De Ley,[1,3*] Rory McDonnell,[2] Timothy D. Paine[3] and Paul De Ley[1]

[1]*Department of Nematology, University of California, Riverside, California, USA;* [2]*Department of Crop and Soil Science, Oregon State University, Corvallis, Oregon, USA;* [3]*Department of Entomology, University of California, Riverside, California*

29.1 Introduction

Invasive snails and slugs are among the most serious pest threats to agriculture and horticulture in the USA. They feed directly on many crops, vector plants (e.g. orchardgrass choke) and human pathogens (e.g. *Escherichia coli*), and lower crop quality through faecal and mucus contamination. Management relies heavily on molluscicides, with metaldehyde generally regarded as the active ingredient standard practice. Although it can be effective, it is very toxic to mammals, and under high relative humidity and low temperatures, activity is lower and gastropods may recover. Iron phosphate and sodium ferric ethylenediaminetetraacetic acid (EDTA) are often used in high humidity or where there are concerns about pet safety. Both materials act slower than metaldehyde, and iron phosphate may be toxic to non-targets such as earthworms. Methiocarb is a restricted-use material and is toxic to mammals including humans, birds, bees and aquatic taxa. Other control options (e.g. copper barriers) are impractical for large areas. There is hence an urgent need for more effective and safe strategies such as biological control.

The nematode, *Phasmarhabditis hermaphrodita* (Schneider) Andrássy, was recently discovered for the first time in the USA, in addition to *Phasmarhabditis papillosa* (Schneider) Andrássy and new species *Phasmarhabditis californica* (Tandingan De Ley *et al.*, 2016). The importance of this discovery in terms of gastropod management is evidenced by the number of pest gastropod species that attack crops throughout the USA that are reportedly killed by *P. hermaphrodita*, and by the success of this agent (Nemaslug®) in Europe in reducing damage and inhibiting feeding of slugs on high-value and agricultural crops and ornamentals. Furthermore, this discovery opens possibilities for a potentially more sustainable and safe strategy (biological control), either alone or as an integrated pest management (IPM) component for pest gastropods. However, the US Department of Agriculture-Animal and Plant Health Inspection Service (USDA-APHIS) has regulatory measures in place for the interstate movement or release of a local or foreign strain of an invertebrate biological control agent (IBCA).

This chapter is focused on the importance of invasive gastropods in North America; the common tactics used for their management and the search for and discovery of *P. hermaphrodita* in North America, which could be an effective, efficient and safe alternative; the discovery of multiple *Phasmarhabditis* species in California and species diagnostics; and the implications, future priorities and prospects for establishing the biological control potential of other *Phasmarhabditis* species.

29.2 Invasive Gastropods in North America

The invasive malacofauna in North America is dominated by species that have been introduced

*Corresponding author. E-mail: Irma.deley@ucr.edu

  Biocontrol Agents: Entomopathogenic and Slug Parasitic Nematodes (eds. M.M.M. Abd-Elgawad, T.H. Askary and J. Coupland)

both intentionally and accidentally from Europe (Robinson, 1999; Cowie and Robinson, 2003), with some representatives from subtropical to tropical regions (McDonnell *et al.*, 2008). In terms of invasive slugs and semi-slugs, at least 29 exotic species representing nine families (Agriolimacidae, Arionidae, Ariophantidae, Boettgerillidae, Limacidae, Milacidae, Philomycidae, Testacellidae and Veronicellidae) are known from North America. Furthermore, the three snails (*Lissachatina fulica* Férussac, *Euglandina rosea* Férussac and *Pomacea canaliculata* Lamarck) that are listed among the world's most damaging invasive species (Lowe *et al.*, 2004) are known from and are problematic in North America (Lach and Cowie, 1999; Cowie, 2001). Although large sums of money are spent by government agencies (e.g. US Department of Agriculture (USDA)) to prevent additional species from entering the USA, new invasive gastropod species continue to be discovered (*Tandonia kusceri* Wagner (Gerber, 2013); and *Boettgerilla pallens* Simroth (McDonnell *et al.*, 2014)).

When exotic snails and slugs become established in North America, they can cause a wide range of economic and environmental damage. Many species, including *Deroceras reticulatum* Müller, are direct pests of agricultural and horticultural production and can cause substantial yield losses (Byers *et al.*, 1985; Hammond and Stinner 1987; Hammond *et al.*, 1999; Byers, 2002; Hammond and Byers, 2002; Sakovich, 2002). In addition to reducing the yield at harvest, gastropods also decrease the quality rating of produce by surface scarring, soiling with mucus and contaminating with faecal matter (Fig. 29.1). For highly valuable and vulnerable commodities (e.g. strawberries), such contamination greatly reduces market value. Other species, such as *Limacus flavus* L., are serious pests of stored produce, including apples and pears (Godan, 1983) (Fig. 29.2). There is also significant cost associated with gastropod eradication programmes; for example, eradication of a single, relatively small population of *L. fulica* in southern Florida in the 1960s and 1970s cost approximately US$1 million (USDA, 2011).

Invasive gastropods have also resulted in severe biodiversity loss. The introduction of the predatory snail, particularly *E. rosea*, to control *L. fulica* is widely regarded as being an ecological disaster (Civeyrel and Simberloff, 1996). There is no convincing evidence that the control programmes have been a success, but there is compelling evidence that *E. rosea* has had a major negative impact on endemic Pacific island snails, including the endemic Hawaiian achatinelline tree snails (Hadfield and Kay, 1981; Hadfield and Mountain, 1981; Hadfield, 1986; Hadfield and Miller, 1989; Hadfield *et al.*, 1993; Stone, 1999), and likely on other species, including the speciose endemic Hawaiian family Amastridae (Griffiths *et al.*, 1993). Of the ~750 species of endemic land snails described from Hawaii (Cowie, 1996), Solem (1990) estimated that only 25–35% remained. Furthermore, McDonnell *et al.* (2016a) recently reported that the commercially sold biocontrol agent *Rumina decollata* L. was not an effective snail predator and raised doubts about its efficacy as a biological control agent of *Cornu aspersum* Müller in Californian citrus.

Snails and slugs are also important vectors of human disease. In Hawaii, *Parmarion martensi* Simroth is a major vector of *Angiostrongylus cantonensis* (Hollingsworth *et al.*, 2007). This nematode causes eosinophilic meningoencephalitis, a serious debilitating disease that is potentially fatal in humans

Fig. 29.1. Damage of *Deroceras reticulatum* on Angel trumpet (*Brugmansia* sp.).

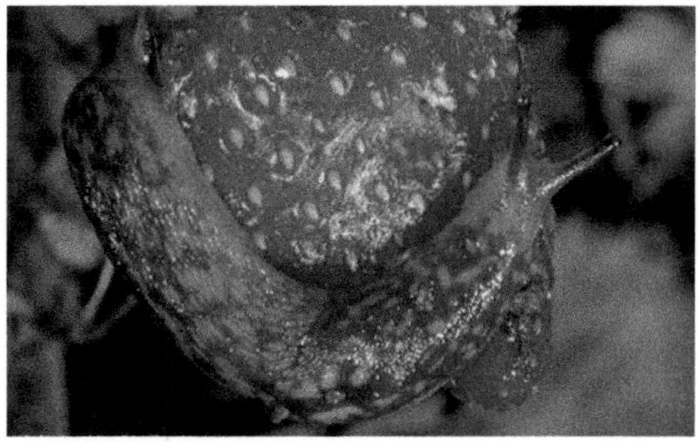

Fig. 29.2. Damage of *Limacus flavus* on strawberry fruit (courtesy of Jack Kelly Clark).

(Aguiar *et al.*, 1981; Lindo *et al.*, 2004). Furthermore, Sproston *et al.* (2006), based on the persistence of *E. coli* O157 on the external surface and in the faeces of *D. reticulatum*, concluded that slugs had the potential to act as novel vectors of *E. coli* O157. The following year, Raloff (2007) suggested that slugs were partially responsible for the highly publicized recalls of faecally contaminated spinach and other salad crops grown in California during 2006 (*Los Angeles Times*, 16 September 2006). The contamination sickened hundreds of consumers, thirty of whom developed acute kidney failure, and five people died. The presence of gastropods in horticultural produce destined for interstate transport in the USA can also be very costly for growers. In the USA, a grower/shipper exporting to another state generally must have a holding area that is certified as mollusc free before being allowed to ship (Parrella *et al.*, 1985). For example, beginning in 1984, the US states of Alabama, Arkansas, Louisiana, Mississippi, Oregon, Tennessee and Virginia imposed quarantines on any material from California infested with *C. aspersum*, while Florida went even further and established a quarantine prohibiting all plant-eating molluscs. Regulated material from California can be accepted in Florida only if authorized officials have inspected the shipment, and each shipment must be accompanied by a special certificate stating that the articles have been inspected and meet Florida's requirements. The discovery of a plant-eating mollusc such as *C. aspersum* in a Californian shipment of regulated material in Florida results in the supplier having to pay the costs of returning the regulated materials or the destruction of the shipment, and their shipping permit may be revoked, all at great cost to the supplier.

In North America, as in other parts of the world, a suite of strategies are used to manage invasive gastropod pests. However, there is a heavy reliance on chemical molluscicides such as metaldehyde (generally regarded as the current active ingredient standard practice), iron phosphate, chelated iron and methiocarb (emergency use treatment), but many of these compounds have damaging non-target effects (Bailey, 2002; Langan *et al.*, 2004; Langan and Shaw, 2006). For example, metaldehyde is toxic to mammals, gastropods can recover from metaldehyde poisoning under wet conditions (Bailey, 2002) and there is evidence that prolonged exposure can result in increased tolerance (Salmijah *et al.*, 2000). Furthermore, recent research has shown that earthworms can decrease the efficacy of molluscicides by pulling the bait underground, thereby making it inaccessible to surface-foraging snails and slugs (Gavin *et al.*, 2012). Cultural control options also play an important role in managing these pests in North America (Barker, 2002). Copper can provide an effective barrier to gastropod movement in citrus orchards (Sakovich, 2002), but it is expensive. Traps have also been used traditionally (Clements and Murray, 1991; Hagnell *et al.*, 2006). In terrestrial cropping systems, light to moderate tillage can help reduce gastropod abundance, but this is contrary to the philosophy of conservation tillage (Hammond and Byers, 2002). In recent years, there have been attempts to develop new control options for snail and slug pests in North America. For example, McDonnell *et al.* (2016b) recently identified clove bud oil drenches as a potentially useful approach to managing *C. aspersum* in potted plants, but cost might be a limiting factor to widespread use. There is hence an urgent need for the development of more effective, safe and sustainable control strategies.

29.3 *Phasmarhabditis hermaphrodita* and Development of Nemaslug in the UK

In a search for slug biological control agents, *P. hermaphrodita* was rediscovered from *D. reticulatum* in the UK (Wilson *et al.*, 1993c) and patented for use in gastropod biological control (Wilson *et al.*, 1993b). Its biological control potential was studied extensively by determining the optimum production, viability and infectivity of *P. hermaphrodita* when grown in monoxenic cultures with bacteria (Wilson *et al.*, 1995b,c). From over 100 bacterial isolates, 9 belonging to 8 species were tested. *Moraxella osloensis* proved the most promising and was used in the commercialization of biological control product Nemaslug® by MicroBio Ltd, UK, a company that was later acquired by Becker Underwood and finally by BASF Agricultural Specialties. The product has been approved for application on vegetables, high-value crops and field crops in the UK, Ireland, France, The Netherlands, Belgium, Germany, Denmark, Norway, Finland, Poland, Spain, the Czech Republic, Italy and Switzerland (Rae *et al.*, 2007).

29.4 Early Gastropod Nematode Surveys in North America

Nematode surveys have previously been conducted on terrestrial slugs, snails, or both, in northern California (Kaya and Mitani, 2000) and central Maine (Gleich *et al.*, 1977), and multi-state surveys covering nine states in the USA and two in Canada

(Grewal *et al.*, 2000), and 11 states (including Hawaii) all in the USA (Ross *et al.*, 2010). These surveys were aimed at finding nematodes for the biological control of gastropods, including *Phasmarhabditis* in particular, or examining gastropods that are intermediate hosts of a meningeal nematode, *Parelaphostrongylus tenuis* Dougherty (Gleich *et al.*, 1977). Native and invasive gastropod taxa were collected from different urban and rural agricultural habitats.

Nematodes were recovered from gastropods by direct dissection of washed specimens (Ross *et al.*, 2010); examination of moribund or dead slugs (Grewal *et al.*, 2000; Kaya and Mitani, 2000); or examining extracts from White traps (Kaya and Stock, 1997; Kaya and Mitani, 2000) or cut-up and digested specimens from a modified Baermann set-up (Gleich *et al.*, 1977). Most nematodes recovered with these techniques belonged to free-living Rhabditidae (Rhabditomorpha), with six true parasites from Protostrongylidae, Agfidae, Angiostomatidae and Cosmocercidae (Gleich *et al.*, 1977; Ross *et al.*, 2010). For two decades, Nemaslug® has not been approved for sale and application in the USA, despite its reported success in Europe, because, based on these surveys, *P. hermaphrodita* was not known to occur in North America.

29.5 Multiple *Phasmarhabditis* Species in California

During recent surveys in search for potential biological control agents of gastropods affecting speciality crops in California, gastropods were collected throughout the state, reared on organic carrots, and over 700 that died were transferred to agar in nematode emergence assays. Of these, more than 40% had associated nematodes, with more nematode occurrence in southern California (57.7%) than in northern California (27.2%). Nematodes were identified using reverse taxonomy (De Ley *et al.*, 2005; Kanzaki *et al.*, 2012) by sequencing the D2-D3 domains of the large subunit (LSU) and/or near complete small subunit (SSU) rRNA genes. The nematodes belong to at least 12 major groups that include Rhabditidae (*Oscheius dolichura* Schneider and *Oscheius tipulae* Lam and Webster; *Caenorhabditis elegans* Maupas), Diplogasteridae (*Protodiplogasteroides* sp. or *Pseudodiplogasteroides* sp.), Cosmocercoidea (*Cosmocercoides* sp.) and Alloionematidae (*Alloionema appendiculatum* Schneider and a new Alloionema species, *A. similis* (Holovachov *et al.*, 2016)). *P. hermaphrodita* was isolated for the first time in North America from cadavers of the invasive slugs, *D. reticulatum*, *Deroceras laeve* and *Lehmannia valentiana* (Tandingan De Ley, *et al.*, 2014) from three counties (Table 29.1). This is the first confirmed record of this nematode in North America, and a new host association with *L. valentiana*. This discovery opens up possibilities for commercialization of the local strain in the USA, or market expansion for the UK strain of Nemaslug® by BASF Agricultural Specialties. In either case, the nematode has to be widely prevalent in the state where the biocontrol agent will be introduced.

The USDA-APHIS regulates the introduction and use of an IBCA in the USA under the authority of the Plant Protection Act (PPA) of 2000. APHIS requires a Plant Protection and Quarantine (PPQ) 526 Permit for the importation, interstate movement and environmental release of biological control organisms and biopesticides:

1. Use of local strain: 'A PPQ 526 permit is typically not required for the interstate movement or release into the environment of domestically isolated microorganisms that are not plant pests, and are widely prevalent (= established throughout their ecological range) in the continental United States.' For the US strain of *P. hermaphrodita* to be used and marketed outside California, its widespread presence in other states has to be established, thus requiring additional intensive surveys in the target markets.
2. Use of a foreign strain: 'A PPQ 526 permit is required for the importation, and interstate movement of microbial biological control organisms of foreign origin.' Therefore, importation of the foreign strain of *P. hermaphrodita* with bacteria *M. osloensis*, i.e. the commercialized product Nemaslug®, into or for commercialization in the USA also requires a permit.

During the same California statewide surveys, *P. papillosa* was also recovered from a cadaver of *D. reticulatum*, and this constituted the first record of this nematode species in the Americas (Tandingan De Ley *et al.*, 2016). In addition, *P. californica*, isolated from four invasive slug taxa *D. reticulatum*, *D. laeve*, *L valentiana*, and *Arion hortensis* agg.,[1] was described as new to science (Table 29.1) and was the most widely distributed of the three species in three counties.

The discovery of multiple *Phasmarhabditis* species in California suggests that these nematodes may have become more widespread in recent years, possibly due to the intercontinental, intercountry and interstate movement of plants and planting materials with commercial establishments such as nurseries and garden centres likely acting as foci for the spread of invasive gastropods and their

Table 29.1. *Phasmarhabditis* species, hosts and sampling locations in California.

Phasmarhabditis spp	Host	Location
P. hermaphrodita	*Deroceras reticulatum*	Eureka, Humboldt Co
	D. laeve	Sonoma, Sonoma Co
	Lehmannia valentiana	Eureka, Humboldt Co
	Unidentified, possibly *D. reticulatum*	Oakland, Alameda Co
P. californica	*D. reticulatum*	Eureka, Humboldt Co;
	D. reticulatum	McKinleyville, Humboldt Co
	Arion hortensis	Eureka, Humboldt Co
	D. laeve	Eureka, Humboldt Co
	D. laeve	Salinas, Monterey Co
	L. valentiana	Eureka, Humboldt Co
	Unidentified, possibly *D. reticulatum*	Oakland, Alameda Co
P. papillosa	*D. reticulatum*	San Diego, San Diego Co

natural enemies. Given the success of *P. hermaphrodita* in Europe, determination of the biological control potential of the three species in California should be a research priority.

29.6 Species Diagnostics

It is widely recognized that *Phasmarhabditis* is a problematic genus/taxon with a number of poorly described species. There are currently seven species isolated from terrestrial gastropods (Table 29.2); one from marine habitat [*P. nidrosienses* (Allgén 1949) Andrássy, 1983; Germany and UK], one from littoral detritus [*P. valida* (Sudhaus, 1974) Andrássy, 1983; New Zealand] and another from leaf litter on the forest floor (*P. huizhouensis*; Huang *et al.*, 2015). However, in a recent revision, *P. nidrosienses* and *P. valida* were moved to a new genus *Buetschlinema* (Sudhaus, 2011). Among the gastropod species, *P. papillosa* is poorly described and will require neotypes to be collected and designated as such. The decision for designating *Phasmarhabditis* new species 2 (apud Tandingan De Ley *et al.*, 2014) as *P. papillosa* will need to be reconsidered should there be an available neotype, e.g. from Germany, properly characterized using current acceptable taxonomic standards (Tandingan De Ley *et al.*, 2016) for comparison with the California isolate.

Generally, morphometric values are highly variable, and measurements are affected by the source of specimens like synthetic media for growth, food source, host or substrate. Nematodes that are isolated or grown on slugs tend to have bigger measurements (Hooper *et al.*, 1999) e.g. 10% . Variations may also be attributed to human error and method of choice as well as substandard quality of specimens (e.g. robust specimens like mature female *Phasmarhabditis* could be flattened between microscope slide and cover glass). These are major considerations when using only morphometrics for species diagnostics. Like most of Rhabditidae, juvenile stages are impossible to differentiate between genera or species and therefore additional information and in particular diagnostics using established and tested molecular barcoding genes cannot be overemphasized. Following is a polytomous key to show a combination of biological, morphological and morphometric characters (Table 29.2), which can be used in combination with genetic data.

Among the seven gastropod-associated species, two have hermaphroditic reproductive strategy: *P. californica* and *P. hermaphrodita* so the nematode tends toward gonochorism (see binomial key). When specimens come from agar culture with bacterial lawn, the former can be distinguished from the latter by its shorter body (1501 ± 117 μm) and tail length (83 ± 8.9 μm), with tail constricted at third of its length while the latter has a longer body (1542 ± 161 μm) and tail (99 ± 10.4 μm) with constriction at about half of its length. In addition, the former has three autapomorphies in its concatenated SSU and D2-D3 sequences versus the latter with 20 (Table 29.3).

Among the recently characterized gonochoristic species, three share the same male papillar arrangement of three pairs precloacal, two pairs adcloacal, one pair postcloacal and three pairs near the posterior end (3/2/+1+3). *P. apuliae* can be distinguished from *P. bonaquaense* and *P. huizhouensis* by having a long, conical tail (152 ± 16.7 μm), and longest male, 2.1 mm (1.9-2.4 mm). Both *P. bonaquaense* and *P. huizhouensis* females have short cupula-shaped tails with spicate terminus but they differ in male body length, with the former being shorter at 1.8 mm (1.4-2.1 mm) and the latter the shortest at

Table 29.2. Polytomous key for gastropod *Phasmarhabditis* species.[a] All measurements are in µm (except where indicated otherwise) and in the form: mean ± s.d. (range).

Species	Reproductive Strategy	Body length	Body length	Tail shape	Tail length	Spicule length	Papillar arrangement[b]	Hosts/substrate	References
		♀	♂	♀	♀	♂	♂		
Phasmarhabditis apuliae	amphimictic, Gonochoristic both sexes frequent	2623 ± 163 (2262-2848)	2096 ± 128 (1898-2363)	Long, tapering	152 ± 16.7 (125-187)	81 ± 4.3 (73-87)	3/2/+1+3	dead slugs (*Milax sowerbyi* Férussac and *M. gagates* Draparnaud)	Nermut' *et al.*, 2016a
P. bonaquaense	amphimictic, Gonochoristic both sexes frequent	2349 ± 186 (1878-2626)	1829 ± 224 (1414-2121)	short, cupula-shaped, with long filiform terminus	85.3 ±12.9 (66.5-109.5)	77 ± 4.8 (68.4-85.9)	3/2/+1+3	dead slugs (*Malacolimax tenellus* Müller	Nermut' *et al.*, 2016b
P. californica	Hermaphrodite	1501 ± 117 (1298–1757)	–	Short, conoid, constricted at one-third its length	83 ± 8.9 (66–94)	–	–	1% agar culture (from *Deroceras reticulatum* Müller ITD272)	Tandingan De Ley *et al.*, 2016
P. hermaphrodita, USA	Hermaphrodite	1542 ± 161 (1284–1721)	–	Conoid, wider part forming 47 (50–74)% of tail length, hyaline tail tip short	99 ± 10.4 (85–117)	–	–	1% agar culture (from *D. reticulatum* Müller ITD 046)	Tandingan De Ley *et al.*, 2014
P. hermaphrodita,[c] UK	Hermaphrodite ♂ rare	1354 ± 115 (1186–1525)	–	Conical, 3 ABW, tapers to filiform terminus	104 ± 8.6 (82–113)	53.0[d]	4/1/+3+1[d]	Bacterial culture (from *D. reticulatum* Müller)	Hooper *et al.*, 1999
P. huizhouensis	Gonochoristic ♂ common	1875 ± 341 (1333–2341)	1282 ± 248 (908–1669)	Short cupula-shaped, spicate tail constricted ¾ along its length[e]	91 ± 6.5 (81–106)	69.96 ± 5.62 (61.07–81.88)	3/2/+1+3	NGM culture (from rotting leaves)	Huang *et al.*, 2015

Continued

Table 29.2. Continued.

Species	Reproductive Strategy	Body length ♀	Body length ♂	Tail shape ♀	Tail length ♀	Spicule length ♂	Papillar arrangement[b] ♂	Hosts/substrate	References
P. neopapillosa,[c] UK	Gonochoristic	2227 ± 190	1585 ± 90	Similar to *P. hermaphrodita*	157 ± 15.3	68 ± 2.8	4/1/+3+1	*D. reticulatum* Müller	Hooper *et al.*, 1999
		(1817–2449)	(1432–1771)		(141–174)	(60–71)			
P. papillosa, Germany	Gonochoristic	3 mm	–	Cupula-shaped, spicate		–	9 pairs (arrangement unclear)	–	Schneider, 1866
P. papillosa,[f] Germany	Gonochoristic	(1.6–3.4 mm)	1.2–2.4 mm	Cupula-shaped, spicate		–	3/1+1/+1+3[g]	–	Andrássy, 1983
P. papillosa, USA	Gonochoristic ♂ common	1590 ± 195	1233 ± 184	Long, cupula-shaped spicate tail constricted halfway along its length	106 ± 16.5	(46.7–133.6)	3/1/+2+3	Agar culture (from *D. reticulatum* Müller ITD510)	Tandingan De Ley *et al.*, 2016
		(1202–1932)	(1011–1565)		(73–130)	(46.7–133.6)			
P. tawfiki, Egypt	Gonochoristic	1716 + 347	1337 ± 159	Elongate conoid, filiform terminus	144	62.9 ± 8.2	2/1/6	Helix (= *Eobania vermiculata*) Müller; *Limax flavus* L.	Azzam, 2003
		(1150–2370)	(980–1535)			(54–75)			

Notes: [a]*Phasmarhabditis huizhouensis* was not isolated from a gastropod but was included here because of its highly similar morphological characteristics with *Phasmarhabditis papillosa*. [b]Papillar arrangement or Ray formula = number of pre-cloacal/adcloacal/post-cloacal genital papillae. [c]Characters and morphometrics based on Hooper *et al*. (1999). [d]Based on Sudhaus's (1976) description of a single male. [e]Based on Fig. 1b of Huang *et al*. (2015). [f]Characters and morphometrics based on Andrássy (1983). [g]Based on Fig. 1b of Mengert (1953). ABW = anal body width; NGM = nematode growth medium (isolated from rotting leaves).

Table 29.3. List of autapomorphies for *Phasmarhabditis* based on the secondary structure models of concatenated nearly complete small sub-unit (SSU) and D2-D3 domains of the large sub-unit (LSU) ribosomal DNA sequences.[a]

Species	Gene, nucleotide position	Transition	Transversion
Phasmarhabditis californica n. sp. (SSU = 1; LSU = 2)	SSU, 429	T -> C	
	LSU, 123		A -> T
	LSU, 211	A -> G	
Phasmarhabditis hermaphrodita (SSU = 6; LSU = 14)	SSU, 184	T -> C	
	SSU, 679		A -> T
	SSU, 725		T -> A
	SSU, 786	T -> C	
	SSU, 964		T -> A
	SSU, 1256		C -> A
	SSU, 1570	T -> C	
	LSU, 122	C -> T	
	LSU, 130	G -> A	
	LSU, 138	C -> T	
	LSU, 153	T -> C	
	LSU, 162	G -> A	
	LSU, 169	T -> C	
	LSU, 183	G -> A	
	LSU, 187	G -> A	
	LSU, 188	G -> A	
	LSU, 214	A -> G	
	LSU, 334		A -> T
	LSU, 370	G -> A	
	LSU, 404	C -> T	
Phasmarhabditis huizhouensis (SSU = 24; LSU = 7)	SSU, 169		A -> C
	SSU, 272		C -> A
	SSU, 357	T -> C	
	SSU, 373	G -> A	
	SSU, 375		A -> T
	SSU, 505		A -> T
	SSU, 506		T -> A
	SSU, 555		T -> A
	SSU, 592		T -> G
	SSU, 601		A -> T
	SSU, 606	A -> G	
	SSU, 611		G -> T
	SSU, 613	T -> C	
	SSU, 614	A -> G	
	SSU, 623	C -> T	
	SSU, 627	T -> C	
	SSU, 640		G -> C
	SSU, 642		T -> A
	SSU, 683	T -> C	
	SSU, 947		A -> T
	SSU, 955	T -> C	
	SSU, 963		T -> G
	SSU, 1410		A -> C
	SSU, 1607	G -> A	
	D2D3, 59		T -> A
	D2D3, 191		T -> G
	D2D3, 192	T -> C	
	D2D3, 194	A -> G	

Continued

Table 29.3. Continued.

Species	Gene, nucleotide position	Transition	Transversion
	D2D3, 326	G –> A	
	D2D3, 344		A –> C
	D2D3, 363		C –> A
Phasmarhabditis neopapillosa	SSU, 157		C –> A
(SSU = 4; LSU = NA)	SSU, 161	G –> A	
	SSU, 1380	C –> T	
	SSU, 1576	C –> T	
Phasmarhabditis papillosa	SSU, 035	G –> A	
(SSU = 8; LSU = 1)[b]	SSU, 036	A –> G	
	SSU, 126	G –> A	
	SSU, 189	G –> A	
	SSU, 609	A –> G	
	SSU, 625	T –> C	
	SSU, 717	T –> C	
	SSU, 718		A –> T
	LSU, 178	G –> A	

Notes: [a]Total of 2334 positions containing 1725 (SSU) and 609 (LSU) nucleotide positions; [b]based on CLUSTAL (Jeanmougin *et al.*, 1998) and MAFFT-aligned (Katoh *et al.*, 2009) D2-D3 data set of LSU. NA = not available.

1.3 mm (0.91-1.7 mm). Two species *P. neopapillosa and P. hermaphrodita*, both with papillar arrangement of 4/1/+3+1, also have very similar long and conical female tails. However, *P. hermaphrodita* female body is longer (1.8 mm; 1.5-2.4 mm) than *P. neopapillosa* (1.6 mm (1.4-1.8 mm). Additionally, having very rare male in the case of *P. hermaphrodita* is an important character to separate the two. Two other populations/species have unique papillar arrangement: A population designated as *P. papillosa* has 3/1/+2+3, while *P. tawfiki* has 2/1/6.

The use of papillar arrangement in the key is an example of how morphological data, when available, can better inform our attempt at species diagnostics. In such case, the light micrographs and/or scanning electron micrographs SEM proved to be more informative than either written descriptions or drawings that are difficult to verify without access to original specimens. These illustrations, although providing "some" useful information, may not be enough to draw an accurate conclusion as to the species' identity when the drawings don't support the written descriptions (cf. papillar arrangements on Fig 29.3 c and d, ventral and lateral view of the male posterior end are not similar in Azzam, 2003). However, the use of reverse taxonomy, by sequencing the D2-D3 domain of the rDNA would be the faster and accurate approach, followed by more morphological studies if need be, and if said sequences are unique.

Key to *Phasmarhabditis* species

1. Species **no known male or rare**, reproductive strategy hermaphroditic.
- Body shorter (1501 ± 117 µm); tail short (83 ± 8.9 µm), conoid, constricted at 1/3 its length* .... *P. californica*
- Body longer (1542 ± 161 µm); tail longer (99 ± 10.4 µm) with constriction at about 1/2 of its length* .... *P. hermaphrodita*
- Species with **males quite common**, reproductive strategy gonochoristic........ 2............2

2. Males with **papillar arrangement** of 2 pairs precloacal, 2 pairs adcloacal, 1 pair postcloacal, 3 pairs near the posterior end of the tail **(3/2/+1+3)**
- **Female tail long** (152 ± 16.7 µm), **conical, male body longest**, 2.1 mm (1.9-2.4 mm)** .... *P. apuliae*
- **Female tail short, cupula-shaped** with long **spicate** terminus
  - **Male body shorter**, 1.8 mm (1.4-2.1 mm)** .... *P. bonaquaense*
  - **Male body shortest**, 1.3 mm (0.91-1.7 mm)* .... *P. huizhouensis*
- Males with papillar arrangement of 4 pairs precloacal, 1 pair adcloacal, 3 pairs postcloacal, 1 pair near the posterior end of the tail **(4/1/+3+1)**
  - (Specimens from slug) Female body longer, 1.8 mm (1.5-2.4 mm) (*male extremely rare*

Fig. 29.3. Infection of *Phasmarhabditis hermaphrodita* on *Lehmannia valentiana*. (a) An infected *L. valentiana* with exposed vestigial shell and cavity; (b) highly constricted body behind the mantle at 5 days after exposure (DAE); and (c) decomposing cadaver covered with the infective juveniles (IJs) of *P. hermaphrodita* 7 DAE.

but included for comparison)**......... *P. hermaphrodita*, UK

- (Specimens from slug) Female body shorter, 1.6 mm (1.4-1.8 mm)**......... *P.neopapillosa*, UK

- Males with papillar arrangement of 3 pairs precloacal, 1 pair adcloacal, 2 pairs postcloacal, 3 pairs near the posterior end of the tail (3/1/+2+3)*.....................cf *P. papillosa*, US
- Males with papillar arrangement of 2 pairs precloacal, 1 pair adcloacal, 6 pairs distributed postcloacal towards the posterior end of the tail (2/1/6)**....................................*P. tawfiki*

* specimens for study obtained from agar culture

** specimens for study obtained from slug substrate

29.7 Implications and Future Prospects

29.7.1 Geographic distribution

P. hermaphrodita and *P. californica* were both recovered from the cooler counties within the ecoregion of the Northern California Coast, while *P. papillosa* was isolated only in San Diego, Southern California Coastal region (USDA ecoregions). Nematode species or strains that have established in their new geographic range may be more virulent and effective against slug species prevalent in the same area than in another (Coupland, 1995). There is a need to undertake tests to determine the optimum temperature range for *Phasmarhabditis* species because this will be important for determining the optimum control efficiency and effectivity in states or countries that have highly diverse ecohabitats. For example, *P. hermaphrodita* infects at temperature as low as 5°C, but dies rapidly at 26–35°C (Wilson *et al.*, 1993c). Similar information will be needed for the US strain of this species, as well as *P. californica* and *P. papillosa*, in order to determine their viability in areas outside their geographic range. *P. hermaphrodita* and other species have been reported from different countries and continents with diverse climate, including Chile (France and Gerding, 2000), Iran (Karimi *et al.*, 2003), Egypt (Azzam, 2003; Genena *et al.*, 2011), South Africa (Ross *et al.*, 2012), New Zealand (Wilson *et al.*, 2012) and Norway (Ross *et al.*, 2015), South China (Huang *et al.*, 2015), Italy (Nermut' *et al.*, 2016a) and Czech Republic (Nermut' *et al.*, 2016b), suggesting that the genus is widely distributed. More intensive surveys in the USA and elsewhere will be essential to map the biogeography and distribution of known species and to recover new taxa.

29.7.2 Bacterial choice

To date, 40 bacterial species representing over 100 single colonies have been recovered from the authors' nematode collection, each identified by sequencing the 16SrDNA gene (Jackson *et al.*, 2015; unpublished data). Preliminary bioassays have been completed to determine the pathogenicity of six isolates that have the most frequent association with the three US *Phasmarhabditis* spp. and across the other available rhabditid cultures. These bacteria will be screened further in liquid or solid phase to determine which isolates best support growth, viability and infectivity of dauer juveniles of the three *Phasmarhabditis* species and identify the best nematode–bacteria combination (Wilson *et al.*, 1995a) for subsequent macrocosm experiments.

29.7.3 Biological control potential

Phasmarhabditis hermaphrodita are able to parasitize and kill several slug species, the wide spread pest of many agricultural and horticultural crops (Askary, 2010; Askary *et al.*, 2012). Our initial experiments show that xenic *P. hermaphrodita* causes mortality to *L. valentiana* starting 3 days after exposure (DAE). The slugs exhibit swelling of the mantle, followed by constriction of the area behind the mantle, exposure of the shell cavity, extrusion of the vestigial shell and rupture of the integument (Fig. 29.3). At 5 DAE, the slugs are dead or in varying degrees of symptom development and decomposition, with hundreds to thousands of nematodes swarming either in the 'pool' of the shell cavity or over the cadaver. Further trials are needed to establish the biological control potential of US *P. hermaphrodita* in monoxenic culture with selected candidate bacteria against other invasive slug and snail species. Symptom development on *D. reticulatum* is very similar to *L. valentiana*, with characteristic swelling of the mantle, girdling behind the mantle, exposure of the cavity with numerous developing nematodes, and nematodes eventually decomposing the cadaver leaving only the vestigial shell (Fig 29.4).

29.7.4 Host range

Host range will be determined to document the effects of candidate nematode species on gastropod targets and non-targets (Grewal *et al.*, 2003). *P. hermaphrodita* is known to cause mortality and feeding inhibition

Fig. 29.4. Infection of *Phasmarhabditisherm aphrodita* on *Derocerasreticulatum*. (a) An infected *Deroceras reticulatum* with typical swollen mantle and constriction behind it; (b) exposed cavities and a ruptured integument; and (c) developing nematodes on the cadaver. Photographs of different specimens taken 3-4 days after exposure to *Phasmarhabditis hermaphrodita aphrodita* (ITD272).

Table 29.4. Susceptibility of terrestrial gastropods to *Phasmarhabditis* species recovered in California. (Revised from Rae *et al.*, 2007.)

Gastropods	Family	Species	*Phasmarhabditis* spp.	Reaction	References
Slugs	Agriolimacidae	*Deroceras laeve* (Müller)	*Phasmarhabditis californica*	Susceptible	Tandingan De Ley *et al.*, 2016
			P. hermaphrodita		Grewal *et al.*, 2003; Tandingan De Ley *et al.*, 2014
		D. panormitanum (Lessona and Pollonera)	*P. hermaphrodita*	Susceptible	Wilson *et al.*, 1993a; Ross *et al.*, 2015
		D. reticulatum (Müller)	*P. californica*	Susceptible	Tandingan De Ley *et al.*, 2016
			P. hermaphrodita		Wilson *et al.*, 1993a; Tandingan De Ley *et al.*, 2014; Ross *et al.*, 2015
			P. papillosa		Tandingan De Ley *et al.*, 2016
	Arionidae	*Arion ater* (L.)	*P. hermaphrodita*	Susceptible, juvenile	Wilson *et al.*, 1993a; Rae *et al.*, 2007; Ross *et al.*, 2015
		A. distinctus (Mabille)	*P. hermaphrodita*	Susceptible	Wilson *et al.*, 1993a; Iglesias and Speiser, 2001
		A. empiricorum Férussac	*P. hermaphrodita*	Susceptible	Mengert, 1953
		A. fuscus (Müller)	*P. hermaphrodita*	Susceptible	Ross *et al.*, 2015
		A. hortensis (Férussac)	*P. hermaphrodita*	Not susceptible (UK isolate); Susceptible (US isolate)	Iglesias and Speiser, 2001; Grewal *et al.*, 2003
		A. hortensis agg.[b]			Tandingan De Ley *et al.*, 2014
		A. hortensis agg.[b]	*P. californica*	Susceptible	Tandingan De Ley *et al.*, 2016
		A. intermedius (Normand)	*P. hermaphrodita*	Susceptible	Wilson *et al.*, 1993a
		A. lusitanicus (Mabille)	*P. hermaphrodita*	Susceptible, juvenile	Glen *et al.*, 2000; Speiser *et al.*, 2001; Grimm, 2002
		A. rufus/*A. ater* hybrid	*P. hermaphrodita*	Susceptible	Ross *et al.*, 2015
		A. silvaticus (Lohmander)	*P. hermaphrodita*	Susceptible	Wilson *et al.*, 1993a
		A. subfuscus Draparnaud	*P. hermaphrodita*	Not susceptible	Grewal *et al.*, 2003
		A. vulgaris Moquin-Tandon	*P. hermaphrodita*	Susceptible	Ross *et al.*, 2015
	Limacidae	*Lehmannia valenciana* (Férussac)[a]	*P. californica*	Susceptible	Tandingan De Ley *et al.*, 2016
			P. hermaphrodita		Tandingan De Ley *et al.*, 2014
		Limax maximus (L.)	*P. hermaphrodita*	Not susceptible	Grewal *et al.*, 2003
	Milacidae	*Tandonia budapestensis* (Hazay)	*P. hermaphrodita*	Susceptible	Wilson *et al.*, 1993a
		T. sowerbyi (Férussac)	*P. hermaphrodita*	Susceptible	Wilson *et al.*, 1993a
	Veronicellidae	*Leidyula floridana* (Leidy)	*P. hermaphrodita*	Susceptible	Grewal *et al.*, 2003

Continued

Table 29.4. Continued

Gastropods	Family	Species	*Phasmarhabditis* spp.	Reaction	References
Snails	Helicidae	*Cepaea hortensis* (Müller)	*P. hermaphrodita*	Susceptible	Wilson *et al.*, 2000
		Cepaea nemoralis (L.)	*P. hermaphrodita*	Not susceptible	Wilson *et al.*, 2000
		Cernuella virgata (da Costa)	*P. hermaphrodita*	Susceptible	Coupland, 1995
		Cochlicella acuta (Müller)	*P. hermaphrodita*	Susceptible	Coupland, 1995
		Helix aspersa (Müller)	*P. hermaphrodita*	Susceptible, juvenile	Glen *et al.*, 1996
		Monacha cantiana (Montagu)	*P. hermaphrodita*	Susceptible	Wilson *et al.*, 2000
		Theba pisana (Müller)	*P. hermaphrodita*	Susceptible	Coupland, 1995
	Lymnaeidae	*Lymnaea stagnalis* (L.)	*P. hermaphrodita*	Susceptible	Wilson *et al.*, 1993b; Morley and Morrit, 2006
	Physidae	*Physa fontinalis* (L.)	*P. hermaphrodita*	Not susceptible	Wilson *et al.*, 2000
	Pomatiasidae	*Pomatias elegans* (Müller)	*P. hermaphrodita*	Not susceptible	Wilson *et al.*, 2000
	Zonitidae	*Oxychilus helveticus* (Blum)	*P. hermaphrodita*	Not susceptible	Wilson *et al.*, 2000
	Clausiliidae	*Clausilia bidentata* (Strüm)	*P. hermaphrodita*	Not susceptible	Wilson *et al.*, 2000
	Endodontidae	*Discus rotundatus* (Müller)	*P. hermaphrodita*	Not susceptible	Wilson *et al.*, 2000

Notes: [a]New record for *Phasmarhabditis hermaphrodita*; [b]new record for *Phasmarhabditis*.

in at least fifteen slug species belonging to five families (Agriolaimacidae, Arionidae, Milacidae and Veronicellidae) with new host associations, *L. valentiana* (Limacidae) and *A. hortensis* (Arionidae) (Tandingan De Ley *et al.*, 2014); and seven snail species belonging to two families (Helicidae and Lymnaeidae) (Table 29.4). Interestingly, the serious pest *A. hortensis* agg. was susceptible to *P. hermaphrodita* US isolate but not to UK isolate, highlighting the importance of establishing the host range of local *Phasmarhabditis* species and isolates.

Among the three *Phasmarhabditis* species, the new species *P. californica* has the highest per cent recovery in California (Tandingan De Ley *et al.*, 2016). All four slug hosts (*D. laeve*, *D. reticulatum*, *L. valentiana* and *A. hortensis* agg.; see Table 29.1) were the most abundant taxa in California nurseries and garden centres. Although studies showed Nemaslug® as safe to non-targets such as the earthworm *Lumbricus terrestris* L. (Grewal and Grewal, 2003), this result contradicts the recovery of a yet undescribed but genetically distinct *Phasmarhabditis* EM434 from an unidentified earthworm in the Bronx, New York (http://www.nyu.edu/projects/fitch/WSRN/strains/em434.html) and another nematode provisionally identified as *Phasmarhabditis* from *L. terrestris* at the University of Illinois Urbana-Champaign campus, Illinois (Zaborski *et al.*, 2001).

29.7.5 Nursery and field tests

Following microcosm studies in the laboratory, the biological control potential of the chosen candidate(s) will be verified in release trials. Studies will include different doses; novel strategies and rates of application; trials in different agricultural, horticultural and urban systems; and possibilities for use in integrated pest management. Initial markets to use a potential *Phasmarhabditis* species as a commercial product will likely be home gardens and nurseries (Wilson *et al.*, 2012).

29.7.6 Novel formulations and delivery

P. hermaphrodita was packaged as infective juveniles (IJs) that were partially dehydrated on to an inert carrier such as clay or vermiculite, or in some cases in liquid formulation, but both had poor survival because of lack of oxygen and difficulty in storage and handling (Wilson, 2004). Nemaslug® is currently formulated as a water-dispersible polymer gel (Pieterse *et al.*, 2016) in two-pack sizes and is effective against small to medium-sized slugs for preventative control, knock-down application, or in an overall population management scheme to break the slug life cycle in highly infested areas (https://agriculture.basf.com/en/Crop-Protection/Nemaslug-Global.html; accessed 21 October 2016). Different nematode strains or species grown in different bacteria may prove more lethal to invasive or native pest slugs and snails. Therefore, finding an optimum *Phasmarhabditis*–bacteria partnership for the local market will contribute towards efficiency and likely lower slug management cost.

The use of 'nemabags', root dips (Grewal *et al.*, 2005), and 'tea bags' (Menzler-Hokkanen and Hokkanen, 2003) all aim to effect a slow release, low rate delivery system. The adaptability of local strains and species to these known approaches will be important, especially in small-scale applications like home gardens.

Biological control successes with Nemaslug® in glasshouse or field applications are well documented (see extensive review by Rae *et al.*, 2007). Damage reduction has been demonstrated in agricultural crops, for example oilseed rape and winter wheat (Speiser and Andermatt, 1996; Wilson *et al.*, 1994; Wilson *et al.*, 1995a), as well as horticultural, ornamental and other high-value crops, for example asparagus (Ester *et al.*, 2003b), Brussels sprouts (Ester *et al.*, 2003c), cabbages (Grubisis *et al.*, 2003), orchids (Ester *et al.*, 2003a) and *Impatiens* and *Hosta* (Grewal *et al.*, 2001), using different methods and rates of application (see extensive review by Rae *et al.*, 2007). The recommended dose for the commercially available strain is 30 IJs/cm^2. However, split applications of lower doses (Glen *et al.*, 2000; Ester *et al.*, 2003b,c), spot application of the product around plants (courgettes, Glen *et al.*, 2000) and under artificial slug shelters (Grewal *et al.*, 2001), drip application (Glen *et al.*, 1996) or repeated low doses of band applications (Ester *et al.*, 2003b,c) reduced plant damage, with biological control efficacy comparable to, and sometimes better than, the control molluscicides (metaldehyde and/or methiocarb).

Biological control applications of local *P. hermaphrodita* isolates and of other *Phasmarhabditis* species therefore merit closer investigation, particularly in organic production systems of high-value crops such as ornamentals, strawberries and salad greens.

29.8 Conclusions

Invasive gastropods are among the most damaging pests of agriculture and horticulture in North America, with many species feeding directly on and reducing the yield and quality ratings of a wide range of crops (Byers, 2002; Hammond and Byers,

2002; Sakovich, 2002). Some species are known to vector human pathogens such as the potentially fatal *A. cantonensis* (Hollingsworth *et al.*, 2007), and others have resulted in severe biodiversity loss on Pacific islands (Cowie, 2001). Although a wide range of management strategies are available, growers continue to struggle to control these pests. Molluscicides such as metaldehyde are currently the most common management tool, and although they can be effective, there is a risk of serious non-target effects (Bailey, 2002) and increasing gastropod tolerance to metaldehyde (Salmijah *et al.*, 2000). Many cultural control options (e.g. copper barriers) are impractical for large areas, and previous attempts at biological control have often ended in ecological disaster (Cowie, 2001) or questionable efficacy (McDonnell *et al.*, 2016).

P. hermaphrodita, marketed as Nemaslug® for over 20 years in 14 European countries (Wilson *et al.*, 1993a; Rae *et al.*, 2007), has proven successful by reducing damage and inhibiting slug feeding on a number of field crops (Wilson *et al.*, 1994) and horticultural and ornamental high-value crops (Ester *et al.*, 2003a,b,c), with efficacy oftentimes comparable to, and sometimes better than, molluscicides. Since *P. hermaphrodita* can also interact synergistically with some chemical molluscicide, therefore a combination of multiple tactics can be prepared to achieve a satisfactory result (Askary *et al.*, 2012). However, it has not been possible to market this product in the USA because the nematode had not been recovered in the USA until recently. The discovery of multiple species of *Phasmarhabditis* for the first time in the USA (Tandingan De Ley *et al.*, 2014, 2016) opens possibilities for a more sustainable and safe strategy (biological control) for managing pest gastropods, either alone or as an integrated pest management (IPM) component (Ester *et al.*, 2003b,c).

However, the USDA-APHIS requires permits for interstate movement or release of a local or foreign strain of an IBCA unless proven that the organism is widely prevalent or established in the target area of mainland USA. As such, state and nationwide surveys are required for either market expansion of an IBCA of local or foreign origin. Additional surveys will provide data on biogeography and ecology, and as more species are recovered, additional molecular barcodes and rRNA sequences will be made available for fast, reliable and efficient species diagnostics and phylogenetic studies. Furthermore, the biological control potential of each of the three *Phasmarhabditis* species and the best combination of nematode and bacteria will need to be determined. Host range tests will verify the IBCA's effects on targets and non-targets. Greenhouse and field tests of the best candidate will be required to establish optimum effective doses on different crops, considering the benefits of low but repeated application (Glen *et al.*, 2000; Ester *et al.*, 2003b,c) and using the most efficient delivery system (Menzler-Hokkanen and Hokkanen, 2003; Grewal *et al.*, 2005).

In closing, the advent of high-throughput genome sequencing technologies has revolutionized our approach to biological research from fundamental to applied sciences and has improved our understanding of parasite biology dramatically (Forrester and Hall, 2014; Blaxter, 2011). As the cost for sequencing becomes less prohibitive, more genome projects are completed. These result in massive data sets including, for example, draft genomes and transcriptomes that reveal stage-specific gene families that are likely involved in the parasitism of, for example, entomopathogenic nematodes *Steinernema carpocapsae* and four congeners (Dillman *et al.*, 2015). Whole-genome sequencing and RNA sequencing of specific stages of *P. hermaphrodita* and other candidate species will inform our understanding of *Phasmarhabditis* parasitism and improve future research direction on how best to manage pest gastropods using these nematodes.

Acknowledgements

The project on gastropod biological control was made possible by a California Department of Food and Agriculture (CDFA) Specialty Crop Block Grant (SCB12059) to the authors.

Endnote

[1]Species complex comprising *Arion hortensis* Férussac, *Arion distinctus* Mabille and *Arion owenii* Davies. Diagnostics rely mainly on combined morphological and molecular data (e.g. mitochondrial cytochrome c oxidase I and 16S rRNA sequences).

References

Aguiar, P.H., Morera, P. and Pascual, J. (1981) First record of *Angiostrongylus cantonensis* in Cuba. *American Journal of Tropical Medicine and Hygiene* 30, 963–965.

Allgén, C. (1949) Über einige südschwedische Brackwasser-und Erdnematoden. *Kungliga Fysiografiska Sällskapets i Lund Förhandlingar* 19, 3–19.

Andrássy, I. (1983) *A Taxonomic Teview of the Suborder Rhabditina (Nematoda: Secernentia)*. ORSTOM, Paris.

Askary, T.H. (2010) Nematodes as biocontrol agents. In: Lichtfouse, E. (ed.) *Sociology, Organic Farming, Climate Change and Soil Science.* Springer, The Netherlands, pp. 347–378.

Askary, T.H., Khan, A.A., Waliullah, M.I.S., Banday, S.A., Iqbal, U. and Mir, M.M. (2012) Slug pest management through nematodes in agricultural and horticultural crops. In: Boeri, F. and Chung, J.A. (eds) *Nematodes: Morphology, Functions and Management Strategies*. Nova Publishers, New York, USA, pp. 197–211.

Azzam, K.M. (2003) Description of the nematode *Phasmarhabditis tawfiki* n. sp. isolated from Egyptian terrestrial snails and slugs. *Journal of the Egyptian German Society of Zoology* 42, 79–87.

Bailey, S.E.R. (2002) Molluscicidal baits for control of terrestrial gastropods. In: Barker, G.M. (ed.) *Molluscs as Crop Pests*. CAB International, Wallingford, UK, pp. 33–54.

Barker, G.M. (2002) *Molluscs as Crop Pests.* CAB International, Wallingford, UK.

Blaxter, M. (2011) Nematodes: the worm and its relatives. *PLoS Biology* 9(4): e1001050, DOI: 10.1371/journal.pbio.1001050.

Byers, R.A. (2002) Agriolimacidae and Arionidae as pests in lucerne and other legumes in forage systems of northeastern North America. In: Barker, G.M. (ed.) *Molluscs as Crop Pests*. CAB International, Wallingford, UK, pp. 325–336.

Byers, R.A., Templeton, W.C., Mangan, R.L., Bierlein, D.L., Campbell, W.F. and Donley, H.J. (1985) Establishment of legumes in grass swards: effects of pesticides on slugs, insects, legume seedling numbers and forage yield and quality. *Grass Forage Science* 40, 41–48.

Civeyrel, L. and Simberloff, D. (1996) A tale of two snails: is the cure worse than the disease? *Biodiversity and Conservation* 5, 1231–1252.

Clements, R.O. and Murray, P.J. (1991) Comparison between defined-area slug traps and other methods of trapping slugs in cereal fields. *Crop Protection* 10, 152–154.

Coupland, J. (1995) Susceptibility of helicid snails to isolates of the nematode *Phasmarhabditis hermaphrodita* from southern France. *Journal of Invertebrate Pathology* 66, 207–208.

Cowie, R.H. (1996) Variation in species diversity and shell shape in Hawaiian land snails: in situ speciation and ecological relationships. *Evolution* 49, 1191–1202.

Cowie, R.H. (2001) Can snails ever be effective and safe biocontrol agents? *International Journal of Pest Management* 47, 23–40.

Cowie, R.H. and Robinson, G.D. (2003) Pathways of introduction of non-indigenous land and freshwater snails and slugs. In: Ruiz, G.M. and Carlton, J.T. (eds) *Invasive Species: Vectors and Management Strategies*. Island Press, Washington, DC, pp. 93–122.

De Ley, P., Tandingan De Ley, I., Morris, K., Abebe, E., Mundo-Ocampo, M., *et al.* (2005) An integrated approach to fast and informative morphological vouchering of nematodes for applications in molecular barcoding. *Philosophical Transactions of the Royal Society of London. B Biological Science* 360, 1945–1958.

Dillman, A.R., Macchietto, M., Porter, C.F., Rogers, A., Williams, B., *et al.* (2015) Comparative genomics of *Steinernema* reveals deeply conserved gene regulatory networks. *Genome Biology* 16, 200, DOI: 10.1186/s13059-015-0746-6.

Ester, A., van Rozen, K. and Hazendonk, A. (2003a) Efficacy of pesticides to control *Lehmannia valentiana* (Férussac) in orchids (Cymbidium) in greenhouse experiments, In: Henderson, I.F. (ed.) *Slugs and Snails: Agricultural, Veterinary and Environmental Perspectives*. BCPC Symposium Proceedings No 80. British Crop Protection Council, Alton, Hants, UK, pp. 89–94.

Ester, A., van Rozen, K. and Molendijk, L.P.G. (2003b) Field experiments using the rhabditid nematode *Phasmarhabditis hermaphrodita* or salt as control measures against slugs in green asparagus. *Crop Protection* 22, 689–695.

Ester, A., Huiting, H.F., Molendijk, L.P.G. and Vlaswinkel, M.E.T. (2003c) The rhabditid nematode *Phasmarhabditis hermaphrodita* Schneider as a potential biological agent to control field slugs *Deroceras reticulatum* (Müller) in Brussels sprouts. In: Henderson, I.F. (ed.) *Slugs and Snails: Agricultural, Veterinary and Environmental Perspectives*. BCPC Symposium Proceedings No 80. British Crop Protection Council, Alton, Hants, UK, pp. 313–318.

Forrester, S.J. and Hall, N. (2014) The revolution of whole genome sequencing to study parasites. *Molecular and Biochemical Parasitology* 195, 77–81.

France, A. and Gerding, M. (2000) Discovery of *Phasmarhabditis hermaphrodita* in Chile and its pathological differences with the U.K. isolate in slug control. In: Proceedings of the Society of Nematologists 39th Annual Meeting, Quebec City, Quebec, Canada, June 24–28 2000. *Journal of Nematology* 32, 430.

Gavin, W.E., Mueller-Warrant, G.W., Griffith, S.M. and Banowetz, G.M. (2012) Removal of molluscicidal bait pellets by earthworms and its impact on control of the gray field slug in western Oregon grass seed fields. *Crop Protection* 42, 94–101.

Genena, M.A.M., Mostafa, F.A.M., Fouly, A.H. and Yousef, A.A. (2011) First record for the slug parasitic nematode, *Phasmarhabditis hermaphrodita* (Schneider) in Egypt. *Archives of Phytopathology and Plant Protection* 44, 340–345.

Gerber, J. (2013) First American record of the exotic slug *Tandonia kusceri* (Gastropoda: Milacidae). *Nautilus* 128, 59–63.

Gleich, J.G., Gilbert, F.F. and Kutscha, N.P. (1977) Nematodes in terrestrial gastropods from Central Maine. *Journal of Wildlife Diseases* 13, 43–46.

Glen, D.M., Wilson, M.J., Hughes, L., Cargeeg, P. and Hajjar, A. (1996) Exploring and exploiting the potential of the rhabditid nematode *Phasmarhabditis hermaphrodita* as a biocontrol agent for slugs, In: Henderson, I.F. (ed.) *Slugs and Snails: Agricultural, Veterinary and Environmental Perspectives*. BCPC Symposium Proceedings No 66. British Crop Protection Council, Alton, Hants, UK, pp. 271–280.

Glen, D.M., Wiltshire, C.W., Hughes, L., Ester, A., Van Rozen, K., *et al.* (2000) The use of slug-parasitic nematodes and other techniques for control of slug and snail damage in horticultural crops. In: *Proceedings Brighton Crop Protection Conference – Pests and Diseases*. British Crop Protection Council, Alton, Hants, UK, pp. 345–350.

Godan, D.G. (1983) *Pest Slugs and Snails*. Biology and Control. Springer-Verlag, Berlin.

Grewal, S.K. and Grewal, P.S. (2003) Survival of earthworms exposed to the slug-parasitic nematode *Phasmarhabditis hermaphrodita*. *Journal of Invertebrate Pathology* 82, 72–74.

Grewal, S.K., Grewal, P.S., Brown, I., Tan, L., Hammond, R.B. and Gaugler, R. (2000) First North American survey for the recovery of nematodes associated with mollusks. In: Proceedings of the Society of Nematologists 39th Annual Meeting, Quebec City, Quebec, Canada, June 24–28 2000. *Journal of Nematology* 32, 432.

Grewal P.S., Grewal, S.K., Taylor, R.A.J. and Hammond, R.B. (2001) Application of molluscicidal nematodes to slug shelters: a novel approach to economic biological control of slugs. *Biological Control* 22, 72–80.

Grewal, S.K., Grewal, P.S. and Hammond, R.B. (2003) Susceptibility of North American native and non-native slugs (Mollusca: Gastropoda) to *Phasmarhabditis hermaphrodita* (Nematoda: Rhabditidae). *Biocontrol Science and Technology* 13, 119–125.

Grewal, P.S., Ehlers, R.U. and Shapiro-Ilan, D.I. (2005) Critical issues and research needs for expanding the use of nematodes in biocontrol. In: Grewal, P.S., Ehlers, R.U. and Shapiro-Ilan, D.I. (eds) *Nematodes As Biocontrol Agents*. CAB International, Wallingford, UK, pp. 479–489.

Griffiths, O., Cook, A. and Wells, S.M. (1993) The diet of the carnivorous snail *Euglandina rosea* in Mauritius and its implications for threatened island gastropod faunas. *Journal of Zoology* 229, 79–89.

Grimm, B. (2002) Effect of the nematode *Phasmarhabditis hermaphrodita* on young stages of the pest slug *Arion lusitanicus*. *Journal of Molluscan Studies* 68, 25–28.

Grubisis, D., Ostrec, L.J. and Dusak, I. (2003) Biological control of slugs in vegetable crops in Croatia, In: Henderson, I.F. (ed.) *Slugs and Snails: Agricultural, Veterinary and Environmental Perspectives*. BCPC Symposium Proceedings No 80. British Crop Protection Council, Alton, Hants, UK, pp. 115–120.

Hadfield, M.G. (1986) Extinction in Hawaiian achatinelline snails. *Malacologia* 27, 67–81.

Hadfield, M.G. and Kay, E.A. (1981) The multiple villainies of *Euglandina rosea* (or its human proponents). *Hawaiian Shell News* 29, 5–6.

Hadfield, M.G. and Miller, S.E. (1989) Demographic studies on Hawaii's endangered tree snails: *Partulina proxima*. *Pacific Science* 43, 1–16.

Hadfield, M.G. and Mountain, B.S. (1981) A field study of a vanishing species, *Achatinella mustelina* (Gastropoda, Pulmonata), in the Waianae Mountains of Oahu. *Pacific Science* 34, 345–358.

Hadfield, M.G., Miller, S.E. and Carwile, A.H. (1993) The decimation of endemic Hawai'ian tree snails by alien predators. *American Zoologist* 33, 610–622.

Hagnell, J., Schander, C., Nilsson, M., Ragnarsson, J., Valstar, H., *et al*. (2006) How to trap a slug: commercial versus homemade slug traps. *Crop Protection* 25, 212–215.

Hammond, R.B. and Byers, R.A. (2002) Agriolimacidae and Arionidae as pests in conservation tillage soybean and maize cropping in North America. In: Barker, G.M. (ed.) *Molluscs as Crop Pests*. CAB International, Wallingford, UK, pp. 301–314.

Hammond, R.B. and Stinner, B.R. (1987) Seedcorn maggots (Diptera: Anthomyiidae) and slugs in conservation tillage systems in Ohio, USA. *Journal of Economic Entomology* 80, 680–684.

Hammond, R.B., Beck, T., Smith, J.A., Amos, R., Barker, J., *et al*. (1999) Slugs in conservation tillage corn and soybeans in the eastern corn belt. *Journal of Entomological Science* 34, 467–478.

Hollingsworth, R.G., Kaneta, R., Sullivan, S.J., Bishop, H.S., Qvarnstrom, Y., *et al*. (2007) Distribution of *Parmarion* cf. *martensi* (Pulmonata: Helicarionidae), a new semi-slug pest on Hawaii Island, and its potential as a vector for human angiostrongyliasis. *Pacific Science* 61, 457–467.

Holovachov, O., Boström, S., Tandingan De Ley, I., McDonnell, R.J., Alvarado, S., *et al*. (2016) *Alloionema similis* n. sp., a genetically divergent sibling species of *A. appendiculatum* Schneider, 1859 (Rhabditida: Alloionematidae) from invasive slugs in California, USA. *Systematic Parasitology* 93, 877–898.

Hooper, D.J., Wilson, M.J., Rowe, J.A. and Glen, D.M. (1999) Some observations on the morphology and protein profiles of the slug- parasitic nematodes, *Phasmarhabditis hermaphrodita* and *P. neopapillosa* (Nematoda: Rhabditidae). *Nematology* 1, 173–182.

Huang, R.E., Ye, W., Ren, X. and Zhao, Z. (2015) Morphological and molecular characterization of *Phasmarhabditis huizhouensis* sp. nov. (Nematoda: Rhabditidae), a new rhabditid nematode from South, China. *PLoS ONE* 10(12): e0144386, DOI: 10.1371/journal.pone.0144386.

Iglesias, J. and Speiser, B. (2001) Consumption rate and susceptibility to parasitic nematodes and chemical molluscicides of the pest slugs *Arion hortensis* s. s. and *A. distinctus*. *Journal of Pest Science* 74, 159–166.

Jeanmougin, F., Thompson, J.D., Gouy, M., Higgins, D.G. and Gibson, T.J. (1998). Multiple sequence alignment with Clustal X. *Trends in Biochemical Science* 23, 403–405. Available at: http://www.clustal.org/clustal2/ (accessed 1 November 2016).

Kanzaki, N., Giblin-Davis, R.M., Scheffrahn, R.H., Taki, H., Esquivel, A., *et al*. (2012) Reverse taxonomy for elucidating diversity of insect-associated nematodes: a case study with termites. *PLoS ONE* 7(8), e43865, DOI: 10.1371/journal.pone.0043865.

Karimi, J., Kharazi-Pakadel, A. and Robert, S.J. (2003) Report of pathogenic nematodes of slugs, *Phasmarhabditis hermaphrodita* (Nematoda: Rhabditida) in Iran. *Journal of Entomological Society of Iran* 22, 77–78.

Katoh, K., Asimenos, G. and Toh, H. (2009). Multiple Alignment of DNA Sequences with MAFFT. *Methods in Molecular Biology* 537, 39–64. In: Posada, D. (ed.) *Bioinformatics for DNA Sequence Analysis*. Springer Science, New York. Available at: http://mafft.cbrc.jp/alignment/software/ (accessed 1 November 2016).

Kaya, H. and Mitani, D. (2000) Molluscicidal nematodes for biological control of pest slugs. *Slosson Report* 1999–2000, pp. 1–4. Available at: http://slosson.ucdavis.edu/newsletters/Kaya_200029038.pdf (accessed 1 November 2016).

Kaya, H.K. and Stock, S.P. (1997) Techniques in insect nematology, In: Lacey, L.A. (ed.) *Manual of Techniques in Insect Pathology*. Academic Press, San Diego, California, pp. 281–324.

Lach, L. and Cowie, R.H. (1999) The spread of the introduced freshwater apple snail *Pomacea canaliculata* (Lamarck) (Gastropoda: Ampullariidae) on O'ahu, Hawaii. *Bishop Museum Occasional Papers* 58, 66–71.

Langan, A.M. and Shaw, E.M. (2006) Responses of the earthworm *Lumbricus terrestris* (L.) to iron phosphate and metaldehyde slug pellet formulations. *Applied Soil Ecology* 34, 184–189.

Langan, A.M., Taylor, A. and Wheater, C.P. (2004) Effects of metaldehyde and methiocarb on feeding preferences and survival of a slug predator (*Pterostichus melanarius* (F.): Carabidae, Pterostichini). *Journal of Applied Entomology* 128, 51–55.

Lindo, J.F., Escoffery, C.T., Reid, B., Codrington, C., Cunningham-Myrie, C. and Eberhard, M.L. (2004) Fatal autochthonous eosinophilic meningitis in a Jamaican child caused by *Angiostrongylus cantonensis*. *American Tropical Medicine and Hygiene* 70, 425–428.

Lowe, S., Browne, M., Boudjelas, S. and De Poorter, M. (2004) 100 of the World's Worst Invasive Alien Species. A selection from the Global Invasive Species Database. Published by The Invasive Species Specialist Group (ISSG), a specialist group of the Species Survival Commission (SSC) of the World Conservation Union (IUCN). Available at: http://www.issg.org/pdf/publications/worst_100/english_100_worst.pdf (accessed 20 October 2016).

McDonnell, R.J., Hansen, A., Paine, T.D. and Gormally, M.J. (2008) A record of the invasive slug *Veronicella cubensis* (Pfeiffer, 1840) in California. *The Veliger* 50, 81–82.

McDonnell, R.J., Robinson, D.G., Barr, N., Tandingan De Ley, I., De Ley, P. and Paine, T.D. (2014) First report of the invasive slug *Boettgerilla pallens* Simroth, 1912 (Boettgerillidae) in the United States. *American Malacological Bulletin* 32, 209–210.

McDonnell, R.J., Santangelo, R., Paine, T. and Hoddle, M. (2016a) The feeding behaviour of *Rumina decollata* (Subulinidae: Gastropoda) raises questions about its efficacy as a biological control agent for *Cornu aspersum* (Helicidae: Gastropoda). *Biocontrol Science and Technology* 26, 331–336.

McDonnell, R.J., Yoo, J., Patel, K., Rios, L., Hollingsworth, R., *et al.* (2016b) Can essential oils be used as novel drench treatments for the eggs and juveniles of the pest snail *Cornu aspersum* in potted plants? *Journal of Pest Science* 89, 549–555.

Mengert, H. (1953) Nematoden und schnecken. *Zeitschrift für Morphologie und Ökologie der Tiere* 41, 311–349.

Menzler-Hokkanen, I. and Hokkanen, H.M.T. (2003) Entomopathogenic nematode delivery systems for biological control of pests on major outdoor crops: the case of oilseed rape. In: *Abstracts for the 36th Annual Meeting of the Society of Invertebrate Pathology Program*, 60 pp.

Morley, N.J. and Morrit, D. (2006) The effects of the slug biological control agent, *Phasmarhabditis hermaphrodita* (Nematoda) on non-target aquatic molluscs. *Journal of Invertebrate Pathology* 92, 112–114.

Nermuť J., Půža V. and Mráček Z. (2016a) *Phasmarhabditis apuliae* n. sp. (Nematoda: Rhabditidae), a new rhabditid nematode from milacid slugs. *Nematology* 18, 1095–1112.

Nermuť, J., Půža, V., Mekete, T. and Mráček, Z. (2016b) *Phasmarhabditis bonaquaense* n. sp. (Nematoda: Rhabditidae), a new slug-parasitic nematode from the Czech Republic. *Zootaxa* 4179, 530–546.

Parrella, M.P., Robb, K.L. and Morishita, P. (1985) Snails and slugs in ornamentals. *California Agriculturist* 39, 6–8.

Pieterse, A., Malan, A.P. and Ross, J.L. (2016) Nematodes that associate with terrestrial molluscs as definitive hosts, including *Phasmarhabditis hermaphrodita* (Rhabditida: Rhabditidae) and its development as a biological molluscicide. *Journal of Helminthology*, DOI: 10.1017/S0022149X16000572.

Rae, R., Verdun, C., Grewal, P.S., Robertson, J.F. and Wilson, M.J. (2007) Biological control of terrestrial molluscs using *Phasmarhabditis hermaphrodita*-progress and prospects. *Pest Management Science* 63, 1153–1164.

Raloff, J. (2007) Lettuce liability. Programs to keep salad germfree raise wildlife and conservation concerns. *Science News* 172, 362–364.

Robinson, D.G. (1999) Alien invasions: the effects of the global economy on non-marine gastropod introductions into the United States. *Malacologia* 41, 413–438.

Ross, J.L., Ivanova, E.S., Severns, P.M. and Wilson, M.J. (2010) The role of parasite release in invasion of the USA by European slugs. *Biological Invasions* 12, 603–610.

Ross, J.L., Ivanova, E.S., Sirgel, W.F., Malan, A.P. and Wilson, M.J. (2012) Diversity and distribution of nematodes associated with terrestrial slugs in the

Western Cape Province of South Africa. *Journal of Helminthology* 86, 215–221.

Ross, J.L., Ivanova, E.S., Hatteland, B.A., Brurberg, M.B. and Haukeland, S. (2015) Survey of nematodes associated with terrestrial slugs in Norway. *Journal of Helminthology*, DOI: 10.1017/S0022149X15000784.

Sakovich, N.J. (2002) Integrated management of *Cantareus aspersus* (Müller) (Helicidae) as a pest of citrus in California. In: Barker, G. (ed.) *Molluscs as Crop Pests*. CAB International, Wallingford, UK, pp. 353–360.

Salmijah, S., Chan, M.K., Hong, B.H., Maimon, A. and Ismail, B.S. (2000) Development of resistance in *Achatina fulica* Fer. and *Bradybaena similaris* Fer. towards metaldehyde. *Plant Protection Quarterly* 15, 2–5.

Schneider, A. (1866) *Monographie der nematoden*. Verlag von Georg Reimer, Berlin.

Solem, A. (1990) How many Hawaiian land snail species are left? And what we can do for them. *Bishop Museum Occasional Papers* 30, 27–40.

Speiser, B. and Andermatt, M. (1996) Field trials with *Phasmarhabditis hermaphrodita* in Switzerland. In: Henderson, I.F. (ed.) *Slugs and Snails: Agricultural, Veterinary and Environmental Perspectives*. BCPC Symposium Proceedings No 66. British Crop Protection Council, Alton, Hants, UK, pp. 419–424.

Speiser, B., Zaller, J.G. and Newdecker, A. (2001) Size-specific susceptibility of the pest slugs *Deroceras reticulatum* and *Arion lusitanicus* to the nematode biocontrol agent *Phasmarhabditis hermaphrodita*. *Biocontrol* 46, 311–320.

Sproston, E.L., Macrae, M., Ogden, I.D., Wilson, M.J. and Strachan N.J.C. (2006) Slugs: potential novel vectors of *Escherichia coli* O157. *Applied Environmental Microbiology* 72, 144–149.

Stone, R. (1999) Keeping paradise safe for the natives. *Science* 285, 1837.

Sudhaus W. (1974) Zur Systematik, Verbreitung, Ökologie und Biologie neuer und wenig bekannter Rhabditiden (Nematoda). 2. Teil. *Zoologische Jahrbücher (Systematik)* 101, 417–465.

Sudhaus, W. (1976) Vergleichende Untersuchungen zur Phylogenie, Systematik, Ökologie, Biologie und Ethologie der Rhabditidae (Nematoda). *Zoologia* 43, 1–229.

Sudhaus, W. (2011) Phylogenetic systematization and catalogue of paraphyletic 'Rhabditidae' (Secernentea, Nematoda). *Journal of Nematode Morphology and Systematics* 14, 113–178.

Tandingan De Ley, I., McDonnell, R., Lopez, S., Paine, T.D. and De Ley, P. (2014) *Phasmarhabditis hermaphrodita* (Nematoda: Rhabditidae), a potential biocontrol agent isolated for the first time from invasive slugs in North America. *Nematology* 16, 1129–1138.

Tandingan De Ley, I., Holovachov, O., McDonnell, R., Bert, W., Paine, T.D. and De Ley, P. (2016) Description of *Phasmarhabditis californica* n. sp. and first report of *P. papillosa* (Nematoda: Rhabditidae) from invasive slugs in the USA. *Nematology* 18, 175–193.

USDA (United States Department of Agriculture) (2011) *Pest Alert. Giant African Snails:* A Foreign Threat to U.S Agriculture. United States Department of Agriculture Animal and Plant Health Inspection Service, 2 pp.

Wilson, M.J. (2004) Nematode containing formulation. World Intellectual Property Organisation, International Patent Publication Number: WO/2004/066733.

Wilson, M.J., Glen, D.M. and George, S.K. (1993a) The rhabditid nematode *Phasmarhabditis hermaphrodita* as a potential biological control agent for slugs. *Biocontrol Science and Technology* 3, 503–511.

Wilson, M.J., Glen, D.M. and Pearce, J.D. (1993b) Biological control of molluscs. World Intellectual Property Organisation, International Patent Publication Number: WO/93/00816.

Wilson, M.J., Glen, D.M., George, S.K. and Butler, R.C. (1993b) Mass cultivation and storage of the rhabditid nematode *Phasmarhabditis hermaphrodita*, a biocontrol agent of slugs. *Biocontrol Science and Technology* 3, 513–521.

Wilson, M.J., Glen, D.M., George, S.K., Pearce, J.D. and Wiltshire, C.W. (1994) Biological control of slugs in winter wheat using the rhabditid nematode *Phasmarhabditis hermaphrodita*. *Annals of Applied Biology* 125, 377–390.

Wilson, M.J., Hughes, L.A. and Glen, D.M. (1995a) Developing strategies for the nematode, *Phasmarhabditis hermaphrodita*, as a biological control agent for slugs in integrated crop management systems. In: McKinlay, R.G. and Atkinson, D. (eds) *Integrated Crop Protection: Towards Sustainability*. BCPC Symposium Proceedings No 63. British Crop Protection Council, Alton, Hants, UK, pp. 33–40.

Wilson, M.J., Glen, D.M., Pearce, J.D. and Rodgers, P.B. (1995b) Monoxenic culture of the slug parasite *Phasmarhabditis hermaphrodita* (Nematoda: Rhabditidae) with different bacteria in liquid and solid phase. *Fundamental and Applied Nematology* 18, 159–166.

Wilson, M.J., Glen, D.M., Pearce, J.D. and Rodgers, P.B. (1995c) Selection of a bacterium for the mass production of *Phasmarhabditis hermaphrodita* (Nematoda: Rhabditidae) as a biocontrol agent for slugs. *Fundamental and Applied Nematology* 18, 419–425.

Wilson, M.J., Hughes, L.A., Hamacher, G.M. and Glen, D.M. (2000) Effects of *Phasmarhabditis hermaphrodita* on non-target molluscs. *Pest Management Science* 56, 711–716.

Wilson, M.J., Burch, G., Tourna, M., Aalders, L.T. and Barker, G.M. (2012) The potential of a New Zealand strain of *Phasmarhabditis hermaphrodita* for biological control of slugs. *New Zealand Plant Protection* 65, 161–165.

Zaborski, E.R., Gittenger, L.A.S. and Roberts, S.J. (2001) A possible *Phasmarhabditis* spp. (Nematoda: Rhabditidae) isolated from *Lumbricus terrestris* (Oligochaeta: Lumbricidae). *Journal of Invertebrate Pathology* 77, 282–287.

Part VII: Commercialization and Future Prospects

30 Compatibility between Entomopathogenic Nematodes and Phytopharmaceuticals

Žiga Laznik* and Stanislav Trdan

University of Ljubljana, Biotechnical Faculty, Department of Agronomy, Ljubljana, Slovenia

30.1 Introduction

Entomopathogenic nematodes (EPNs) are soil organisms that typically live with bacteria in mutualistic symbiosis (Kaya and Gaugler, 1993). The nematodes enter insects and release a symbiotic bacterium into their bodies. By releasing toxins, the bacterium causes the death of its host (Gaugler, 2002). Nematodes are classified as beneficial organisms because they can be used to suppress economically harmful insects (Gaugler, 2002; Askary, 2010; Laznik and Trdan, 2011).

The importance of EPNs was discovered in the 1930s in the USA. In 1923, Glaser and Fox discovered a nematode that parasitized the Japanese beetle, *Popillia japonica* Newman (Gaugler and Kaya, 1990). The discovery was quite exceptional, yet the aforementioned manner of plant protection was forgotten until the 1960s, primarily due to the intensive use of chemical insecticides (Koppenhöfer and Kaya, 2002). The concept of biotic plant protection was revived when the public became aware of the toxicity of chlorinated hydrocarbons, which had been used extensively until then. The biotic suppression of harmful insects using EPNs was thus revived. Three decades ago, this was limited to laboratory work, whereas now it represents an applied science used in agriculture (Laznik and Trdan, 2011).

30.2 Factors that Influence the Survival of Entomopathogenic Nematodes

The survival of EPNs is influenced by internal factors, such as genetics, physiology and behaviour, and external factors that are divided into abiotic and biotic factors (Laznik and Trdan, 2011). Abiotic factors include temperature, moisture, soil texture, ultraviolet light and phytopharmaceuticals (Glazer, 2002). The biotic factors that are favourable for nematodes create a suitable environment for the survival of EPNs. Among these is the presence of EPN hosts and plants. Biotic factors can also have antagonistic effects on nematodes (Glazer, 2002).

Nematodes can survive adverse environmental conditions when in a dormant state (Barrett, 1991). Organisms in a state of dormancy can be in diapause (inactive). The state of inactivity refers to a delay in the development of an organism. Development does not continue until certain requirements are met, even when favourable environmental conditions have been re-established. Inactivity is an optional response to adverse environmental conditions. In a state of dormancy, metabolism slows but can return to its original condition after favourable environmental conditions are re-established. The environmental conditions that affect EPNs adversely are lack of water, extreme temperatures, lack of oxygen and osmotic stress (Barrett, 1991). Each of these factors influences the survival of nematodes. The majority of EPNs can only enter a state of dormancy outside the host in the infective juvenile (IJ) stage (Glazer, 2002).

IJs have two external epidermis layers (Campbell and Gaugler, 1991). There are structural differences in the epidermis between different EPN species. The epidermis plays an important role in retaining water in the body, thereby preventing dehydration (Timper and Kaya, 1989). During dormancy, IJs

*Corresponding author. E-mail: Ziga.Laznik@bf.uni-lj.si

have closed oral and anal apertures, thus preventing harmful microbial organisms and chemicals from entering. They do not feed in this period, and their survival depends on their internal sources of energy (reserves). The survival of IJs therefore depends on their metabolism and the initial level of energy reserves. Energy reserves are necessary for supporting the physiological and behavioural processes that accompany the adjustment to environmental stress (Glazer, 2002).

30.3 Phytopharmaceuticals

The nematodes from the families Steinernematidae and Heterorhabditidae can survive even if exposed to certain phytopharmaceuticals (Rovesti and Deseö, 1990; Head *et al.*, 2000; Glazer, 2002; Koppenhöfer *et al.*, 2002; Krishnayya and Grewal, 2002; De Nardo and Grewal, 2003; Schroer *et al.*, 2005; Laznik *et al.*, 2012; Laznik and Trdan, 2014; Ulu *et al.*, 2016). IJs, however, are susceptible to several nematicides that are used in agriculture (Rovesti and Deseö, 1990). It is advisable that certain phytopharmaceuticals only be used 1 or 2 weeks after nematode application (Koppenhöfer and Grewal, 2005).

EPNs can be mixed with different phytopharmaceuticals (Fig. 30.1). Nematodes and other preparations for plant protection can be applied synchronously or at short intervals. Because of their ease of use, nematodes and various phytopharmaceuticals can also be mixed in mixing chambers, which increases the possibility of interaction due to a higher concentration of nematodes and phytopharmaceuticals. The mixing of nematodes is performed primarily to increase the efficiency of suppressing insect pests on plants and different groups of harmful organisms synchronously (Laznik *et al.*, 2012; Laznik and Trdan, 2014; Ulu *et al.*, 2016). The types of interactions that emerge when nematodes and phytopharmaceuticals are mixed can differ considerably (Koppenhöfer and Grewal, 2005).

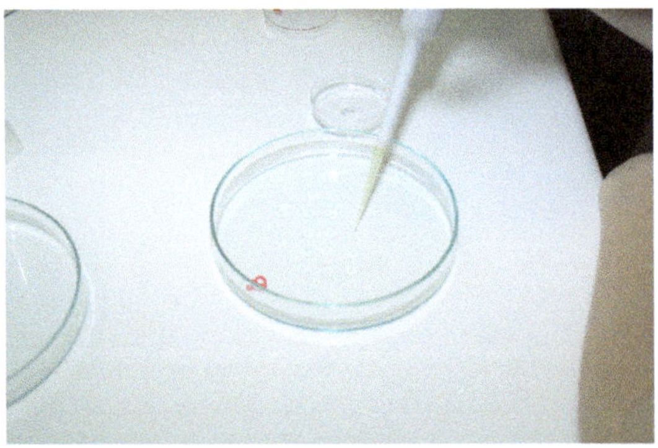

Fig. 30.1. Testing the compatibility of EPNs to phytopharmaceuticals in laboratory investigation.

IJs of EPNs can survive short exposures (from 2 to 24 h) to numerous chemical and bioinsecticides, fungicides, herbicides, fertilizers and growth regulators. Therefore, they can be mixed together and can also be applied to plants synchronously (Rovesti and Deseö, 1990; De Nardo and Grewal, 2003; Laznik and Trdan, 2014). This provides a low-cost alternative for plant protection and enables the use of nematodes in integrated pest management (IPM). The actual concentration of chemicals to which nematodes are exposed is different, due to the quantity of applications (Alumai and Grewal, 2004). Incompatibility between phytopharmaceuticals and EPNs can be controlled using an optimal time interval between applications, the length of which depends on the persistence of the active ingredient in the target substrate. Though specific information on the optimal time interval of application is limited, it is usually recommended that nematodes be applied 1 or 2 weeks after the application of chemical insecticides and nematicides (Koppenhöfer and Grewal, 2005).

30.4 Factors that Influence the Compatibility between Entomopathogenic Nematodes and Phytopharmaceuticals

The majority of previous studies examining the compatibility of EPNs and chemicals were carried out *in vitro*, with direct exposure of EPNs to phytopharmaceuticals (Rovesti and Deseo, 1990; Gordon *et al.*, 1996; De Nardo and Grewal, 2003; García del Pino and Morton, 2010; Laznik and Trdan, 2014). There were considerable differences in the compatibility between EPNs and phytopharmaceuticals from the same chemical group that contained EPNs as the active ingredient, which rendered these data unreliable. Therefore, each of these products should be tested individually for use in an IPM system. Sweeping generalizations about compatibility based on data from different races of nematodes are undesirable (Laznik and Trdan, 2014).

30.4.1 Entomopathogenic nematode species

The findings of the studies examining the compatibility and non-compatibility of different EPN species/strains and the active ingredients of agrochemicals are presented in Tables 30.1–30.10.

Numerous studies have shown that the compatibility between EPNs and the studied phytopharmaceuticals depends largely on both the species and the strain of EPN (Rovesti *et al.*, 1988; Krishnayya and Grewal, 2002; De Nardo and Grewal, 2003; Laznik *et al.*, 2012; Laznik and Trdan, 2014). Laznik and Trdan (2014) reported that the active ingredients azadirachtin and pirimicarb did not influence the mortality of the EPN species *Steinernema feltiae* and *Heterorhabditis bacteriophora*. However, the aforementioned active ingredients did influence the mortality of the EPN species *Steinernema carpocapsae* and *Steinernema kraussei*. The compatibility between the active ingredient azadirachtin and the nematode species *S. feltiae* was also reported in the study conducted by Grewal *et al.* (1998).

30.4.2 Entomopathogenic nematode strains

Laznik *et al.* (2012) recently reported that the compatibility between EPNs and phytopharmaceuticals (fungicides) was not only a species-specific characteristic but also a strain-specific characteristic. Laznik and Trdan (2014) reached similar conclusions. The active ingredients of azadirachtin, i.e. the toxin of the bacterium *Bacillus thuringiensis* var. *kurstaki* and imidacloprid, had no effect on the strain *S. feltiae* B30. However, the aforementioned active ingredients considerably reduced the number of live infective larvae in the commercial preparation whose active ingredient was *S. feltiae*. Laznik and Trdan (2014) did not confirm any strain-specific characteristic of the different strains (strain C101, commercial strain) of *S. carpocapsae* with respect to the compatibility of the insecticides studied. In the study of Laznik and Trdan (2014), azadirachtin, a toxin of *B. thuringiensis* var. *kurstaki*, and imidacloprid reduced the number of IJs of the species *S. carpocapsae*, which was incongruent with similar studies in which the stated active ingredients did not have any effect on the survival or infectiveness of this species (Barbarossa *et al.*, 1996; Grewal *et al.*, 1998; Koppenhöfer *et al.*, 2003; Alumai and Grewal, 2004). On the basis of the data, we hypothesize that the strain C101 and the commercial preparation with the active ingredient *S. carpocapsae* are genetically more similar than other aforementioned strains of *S. carpocapsae*. This supports the theory proposed by Stuart and Gaugler (1996), who discovered the possibility that EPNs might adapt genetically to different biotic and abiotic factors, which can influence the results.

Table 30.1. Compatibility of entomopathogenic nematode *Steinernema arenarium* with the active ingredients of agrochemicals.

	+ (compatible interaction)	– (non-compatible interaction)
F		Fipronil (García del Pino and Jové, 2005)

30.4.3 Temperature

Temperature influences the mortality of EPNs significantly when used in combination with phytopharmaceuticals (Laznik *et al.*, 2012; Laznik and Trdan, 2014). The influence of temperature on EPNs depends on the nematode species. In general, IJs become slow at temperatures below 10–15°C, and inactive at temperatures above 30–40°C. The majority of IJs parasitize insects most effectively at temperatures between 20 and 30°C. The exceptions include the species *S. feltiae*, which is most effective at a temperature interval between 12 and 25°C, and *Steinernema riobrave*, which behaves in the same manner in the temperature interval between 25 and 35°C (Koppenhöfer, 2000). Higher temperatures increase metabolic activity and the consumption of energy reserves, consequently shortening the lifespan of nematodes (Koppenhöfer and Kaya, 2002). Temperatures above 32°C negatively influence the reproduction, growth and survival of different organisms, including nematodes.

In the research conducted by Laznik *et al.* (2012), temperature had an important influence on the mortality of IJs when the nematodes were mixed with fungicides. At higher temperatures, the mortality of the IJs was higher (average number of living IJs at 15°C (2398.9 ± 128.3); 20°C (1947.4 ± 93.8); and 25°C (1720.2 ± 87.9)). Among the EPNs studied (*S. feltiae*, *S. carpocapsae* and *Heterorhabditis downesi*), the opposite pattern was only found for *H. downesi* strain 3173 (average number of living IJs at 15°C (1119.0 ± 171.9); 20°C (2491.4 ± 111.0); and 25°C (2199.9 ± 187.2)). This strain was isolated in Hungary (Tóth, 2006), where it is known to be active at lower temperatures (12°C) (Lola-Luz *et al.*, 2005). The sensitivity, resistance or tolerance of EPNs to fungicides has been connected to their ability to withstand osmotic stresses;

Table 30.2. Compatibility of entomopathogenic nematode *Steinernema carpocapsae* with the active ingredients of agrochemicals.

	+ (compatible interaction)	– (non-compatible interaction)
A	Atrazine (Rovesti and Deseö, 1990)	Abamectin (Laznik and Trdan, 2014; Sabino *et al.*, 2014)
	Azadirachtin (Poinar *et al.*, 1990)	Alachlor (Rovesti and Deseö, 1990)
		Aldicarb (Rovesti and Deseö, 1990)
		Aluminum (Alumai and Grewal, 2004)
		Amitraz (Rovesti and Deseö, 1990)
		Azadirachtin (Laznik and Trdan, 2014)
B	Benomyl (Zimmerman and Cranshaw, 1990)	Bendiocarb (Zimmerman and Cranshaw, 1990)
	Bt (Poinar *et al.*, 1990)	Beta cyfluthrin (Negrisoli *et al.*, 2010)
	Bt *kurstaki* (Barbarossa *et al.*, 1996)	Binapacryl (Prakasa Rao *et al.*, 1975)
	Bt *san diego* (Poinar *et al.*, 1990)	Bt *kurstaki* (Laznik and Trdan, 2014)
	Bt *tenebrionis* (Sterk *et al.*, 1999)	
	Bupirimate (Rovesti and Deseö, 1990)	
C	Carbaryl (Alumai and Grewal, 2004)	Carbendazin (Rovesti *et al.*, 1988)
	Carbendazin (Rovesti *et al.*, 1988)	Chlorfenvinphos (Prakasa Rao *et al.*, 1975)
	Carbofuran (Rovesti and Deseö, 1990)	Chlorpyrifos (Zimmerman and Cranshaw, 1990; Sabino *et al.*, 2014)
	Carbosulfan (Rovesti and Deseö, 1990)	Cypermethrin (Negrisoli *et al.*, 2010)
	Carboxin + thiram (Rovesti and Deseö, 1990)	
	Chlorantranilprole (Sabino *et al.*, 2014)	
	Chloridazone (Rovesti *et al.*, 1988)	
	Chlorothalonil (Rovesti *et al.*, 1988)	
	Chlorpyrifos (Alumai and Grewal, 2004; Negrisoli *et al.*, 2010)	
	Chlorpyriphos-ethyl (Rovesti and Deseö, 1990)	
	Chlorsulfuron (Rovesti and Deseö, 1990)	
	Chlorthal-dimethyl (Rovesti and Deseö, 1990)	
	Chlortoluron (Rovesti and Deseö, 1990)	
	Clopyralid (Rovesti and Deseö, 1990)	
	Cycloxydim (Barbarossa *et al.*, 1996)	
	Cyfluthrin (Rovesti and Deseö, 1990)	
	Cymoxanil + folpet (Rovesti and Deseö, 1990)	
	Cyproconazole (Barbarossa *et al.*, 1996)	
D	Dalapon (Rovesti and Deseö, 1990)	Deltamethrin (Negrisoli *et al.*, 2010)
	Deltamethrin (Rovesti and Deseö, 1990; Sabino *et al.*, 2014)	Diazinon (Rovesti and Deseö, 1990; Zimmerman and Cranshaw, 1990)
	Deltamethrin + triazofos (Negrisoli *et al.*, 2010)	Difenoconazole (Barbarossa *et al.*, 1996)
	Diafenthiuron (Barbarossa *et al.*, 1996)	Diflubenzuron (Negrisoli *et al.*, 2010)
	Diazinon (Zimmerman and Cranshaw, 1990)	Dodine (Rovesti and Deseö, 1990)
	Dicamba (Zimmerman and Cranshaw, 1990)	
	Dicofol (Rovesti and Deseö, 1990)	
	Diflubenzuron (Rovesti and Deseö, 1990)	
	Dimethylamino-propyl (Sabino *et al.*, 2014)	
	Dinocap (Rovesti and Deseö, 1990)	
	Diphenadim (Rovesti and Deseö, 1990)	
	Dithianon (Rovesti and Deseö, 1990)	
	DNOC (Rovesti and Deseö, 1990)	
E	Endosulfan (Das and Divakar, 1987)	
	Espinosad (Negrisoli *et al.*, 2010)	
	Exitiazox (Rovesti and Deseö, 1990)	

Continued

Table 30.2. Continued.

	+ (compatible interaction)	– (non-compatible interaction)
F	Fenarimol (Rovesti and Deseö, 1990)	Fenamiphos (Rovesti and Deseö, 1990)
	Fenithion (Prakasa Rao *et al.*, 1975)	Fenitrothion (Prakasa Rao *et al.*, 1975)
	Fenpropimorph (Rovesti and Deseö, 1990)	Flubenzimine (Rovesti and Deseö, 1990)
	Fentin-acetate (Rovesti and Deseö, 1990)	Formothion (Prakasa Rao *et al.*, 1975)
	Fipronil (García del Pino and Jové, 2005)	
	Flazifop butyl (Rovesti *et al.*, 1988)	
	Fluroxypyr (Barbarossa *et al.*, 1996)	
	Fluvalinate (Barbarossa *et al.*, 1996)	
	Folpet (Rovesti and Deseö, 1990)	
	Fomesafen (Rovesti and Deseö, 1990)	
	Fonofos (Rovesti and Deseö, 1990)	
	Fosethyl (Barbarossa *et al.*, 1996)	
G	Gamacyhalothrin (Negrisoli *et al.*, 2010)	
	Glyphosate (Rovesti and Deseö, 1990)	
H	Halofenozide (Alumai and Grewal, 2004)	Heptenophos (Barbarossa *et al.*, 1996)
	Haloxyfop (Barbarossa *et al.*, 1996)	
I	Imidacloprid (Koppenhöfer *et al.*, 2003; Laznik and Trdan, 2014)	
	Ioxynil (Rovesti and Deseö, 1990)	
L	Lambda cyhalothrin + chlorantranilprole (Sabino *et al.*, 2014)	Lambda cyhalothrin (Negrisoli *et al.*, 2010; Laznik and Trdan, 2014)
	L-flamprop isopropyl (Rovesti and Deseö, 1990)	Lecithin (Barbarossa *et al.*, 1996)
	Lenacil (Rovesti and Deseö, 1990)	Lufenuron (Laznik and Trdan, 2014)
	Lindane (Rovesti and Deseö, 1990)	
	Lufenuron (Negrisoli *et al.*, 2010)	
M	Mannose (Rovesti and Deseö, 1990)	Malathion (Das and Divakumar, 1987)
	MCPP (Rovesti and Deseö, 1990)	Mercurious chloride (Zimmerman and Cranshaw, 1990)
	Mefenoxam (Alumai and Grewal, 2004)	Methamsodium (Rovesti and Deseö, 1990)
	Mercoprop-p (Barbarossa *et al.*, 1996)	Methil parathion (Negrisoli *et al.*, 2010)
	Metalaxyl + folpet (Rovesti and Deseö, 1990)	Methomyl (Rovesti and Deseö, 1990)
	Metamitron (Sterk *et al.*, 1999)	Methoxyfenozide (Negrisoli *et al.*, 2010)
	Metazachlor (Barbarossa *et al.*, 1996)	Monocrotophos (Prakasa Rao *et al.*, 1975)
	Metrobuzin (Rovesti and Deseö, 1990)	
N	Neem (Rovesti *et al.*, 1988)	
O		Oxyfluorfen (Rovesti and Deseö, 1990)
P	Penconazole (Sterk *et al.*, 1999)	Paraquat (Rovesti and Deseö, 1990)
	Pendimethalin (Rovesti and Deseö, 1990)	Parathion (Rovesti and Deseö, 1990)
	Pentachloronitrobenzene (Zimmerman and Cranshaw, 1990)	Permethrin (Negrisoli *et al.*, 2010)
	Phenthioate (Prakasa Rao *et al.*, 1975)	Phorate (Rovesti and Deseö, 1990)
	Phosalone (Prakasa Rao *et al.*, 1975; Barbarossa *et al.*, 1996)	Phosalone (Prakasa Rao *et al.*, 1975; Barbarossa *et al.*, 1996)
	Phosmet (Sterk *et al.*, 1999)	Pirimicarb (Laznik and Trdan, 2014)
	Phosphamidon (Rovesti and Deseö, 1990)	Propargite (Rovesti and Deseö, 1990)
	Prochloraz (Rovesti *et al.*, 1988)	Pymetrozine (Laznik and Trdan, 2014)
	Procyidone (Rovesti and Deseö, 1990)	Pyridimine phosphate (Prakasa Rao *et al.*, 1975)
	Propamocarb (Rovesti and Deseö, 1990)	
	Propiconazole (Zimmerman and Cranshaw, 1990)	
	Propyzadime (Rovesti and Deseö, 1990)	
	Pyrazophos (Rovesti and Deseö, 1990)	
R	Rapeseed oil (Barbarossa *et al.*, 1996)	
S	Spinosad (Alumai and Grewal, 2004)	2-4-D sodium salt (Gupta and Siddiqi, 1999)
	Sulfur (Rovesti and Deseö, 1990)	

Continued

Table 30.2. Continued.

	+ (compatible interaction)	– (non-compatible interaction)
T	Tebuconazole (Sterk *et al.*, 1999)	Triazofos (Negrisoli *et al.*, 2010)
	Teflubenzuron (Sterk *et al.*, 1999)	Trichlorfon (Alumai and Grewal, 2004)
	Terbufos (Rovesti and Deseö, 1990)	Triflumuron (Negrisoli *et al.*, 2010)
	Thiamethoxam (Koppenhöfer *et al.*, 2003; Alumai and Grewal, 2004)	Trifluralin (Rovesti and Deseö, 1990)
	Thiamethoxam + lambda cyhalothrin (Sabino *et al.*, 2014)	
	Thiram (Rovesti and Deseö, 1990)	
	Tolclofos-methyl (Rovesti *et al.*, 1988)	
	Tolylfuanide (Barbarossa *et al.*, 1996)	
	Triflumuron (Sabino *et al.*, 2014)	
	Triforine (Rovesti and Deseö, 1990)	
V	Vamidothion (Prakasa Rao *et al.*, 1975)	*Verticillium lecanii* (Barbarossa *et al.*, 1996)
W	White oil (Rovesti and Deseö, 1990)	

Table 30.3. Compatibility of entomopathogenic nematode *Steinernema feltiae* with the active ingredients of agrochemicals.

	+ (compatible interaction)	– (non-compatible interaction)
A	Atrazine (Rovesti and Deseö, 1990)	Abamectin (Head *et al.*, 2000; Laznik and Trdan, 2014)
	Azadirachtin (Grewal *et al.*, 1998; Laznik and Trdan, 2014)	Acephate (De Nardo and Grewal, 2003)
	Azoxystrobin (Krishnayya and Grewal, 2002; Laznik *et al.*, 2012)	Alachlor (Rovesti and Deseö, 1990)
		Aldicarb (Rovesti and Deseö, 1990)
		Amitraz (Rovesti and Deseö, 1990)
		Anilazine (Hassan *et al.*, 1994)
		Azadirachtin (Krishnayya and Grewal, 2002)
B	Benomyl (Zimmerman and Cranshaw, 1990)	Bendiocarb (Zimmerman and Cranshaw, 1990)
	Bentazone (Hassan *et al.*, 1994)	Bt (De Nardo and Grewal, 2003)
	Bt (De Nardo and Grewal, 2003)	
	Bt *kurstaki* (Laznik and Trdan, 2014)	
	Bupirimate (Rovesti and Deseö, 1990)	
	Buprofezin (Hassan *et al.*, 1994)	
C	Carbaryl (Zimmerman and Cranshaw, 1990)	Calcium hypochlorite (Grewal *et al.*, 1998)
	Carbendazin (Rovesti and Deseö, 1990)	Chlorpyriphos (Zimmerman and Cranshaw, 1990)
	Carbofuran (Rovesti and Deseö, 1990)	Chlorpyriphos-ethyl (Rovesti and Deseö, 1990)
	Carbosulfan (Rovesti and Deseö, 1990)	Cinameldehyde (Krishnayya and Grewal, 2002)
	Carboxin + thiram (Rovesti and Deseö, 1990)	Cyfluthrin (Sterk *et al.*, 1999)
	Chloridazon (Rovesti and Deseö, 1990)	Cypermethrin (Cuthbertson *et al.*, 2007)
	Chlorothalonil (Rovesti *et al.*, 1989; Grewal *et al.*, 1998)	
	Chlorsulfuron (Rovesti and Deseö, 1990)	
	Chlorthal-dimethyl (Rovesti and Deseö, 1990)	
	Chlortoluron (Rovesti and Deseö, 1990)	
	Ciopyralid (Hassan *et al.*, 1994)	
	Clethodim (De Nardo and Grewal, 2003)	
	Clopyralid (Rovesti and Deseö, 1990)	
	Cyhexatin (Rovesti and Deseö, 1990)	
	Cymoxanil + folpet (Rovesti and Deseö, 1990)	
	Cyproconazole (Sterk *et al.*, 1999)	
	Cyromazine (Sterk *et al.*, 1999)	

Continued

Table 30.3. Continued.

	+ (compatible interaction)	– (non-compatible interaction)
D	Dalapon (Rovesti and Deseö, 1990)	Deltamethrin (Rovesti and Deseö, 1990; Head *et al.*, 2000)
	Deltamethrin (Rovesti and Deseö, 1990; Head *et al.*, 2000)	Diazinon (Rovesti and Deseö 1990; Zimmerman and Cranshaw, 1990)
	Diazinon (Zimmerman and Cranshaw, 1990)	Dodine (Rovesti and Deseö, 1990)
	Dicamba (Zimmerman and Cranshaw, 1990)	
	Dicofol (Rovesti and Deseö, 1990)	
	Diflubenzuron (Grewal and Richardson, 1993)	
	Dimethoate (Head *et al.*, 2000; Cuthbertson *et al.*, 2007)	
	Dinocab (Rovesti and Deseö, 1990)	
	Diphenamid (Rovesti and Deseö, 1990)	
	Dithianon (Rovesti and Deseö, 1990)	
	DNOC (Rovesti and Deseö, 1990)	
E	Endosulfan (Rovesti and Deseö, 1990)	Ethofumesate (Sterk *et al.*, 1999)
	Etridiazole (De Nardo and Grewal, 2003)	
	Exitiazox (Rovesti and Deseö, 1990)	
F	Fenarimol (Rovesti and Deseö, 1990)	Fatty acid potassium salts (Hassan *et al.*, 1994)
	Fenpropimorph (Rovesti and Deseö, 1990)	Fenamiphos (Rovesti and Deseö, 1990)
	Fluazifop butyl (Rovesti and Deseö, 1990)	Fenoxycarb (De Nardo and Grewal, 2003)
	Flufenoxuron (Sterk *et al.*, 1999)	Fenpropathrin (Hassan *et al.*, 1994)
	Flutdioxonil (De Nardo and Grewal, 2003)	Fentin-acetate (Rovesti and Deseö, 1990)
	Flutolanil (De Nardo and Grewal, 2003)	Flubenzimine (Rovesti and Deseö, 1990)
	Fluvalinate (Sterk *et al.*, 1999)	Fluoxypyr (Sterk *et al.*, 1999)
	Folpet (Rovesti and Deseö, 1990)	Formaldehyde (Grewal *et al.*, 1998)
	Fomesafen (Rovesti and Deseö, 1990)	
	Fonofos (Rovesti and Deseö, 1990)	
G	Glyphosate (Rovesti and Deseö, 1990)	
H	Hexaconazole (Hassan *et al.*, 1994)	Haloxyfop (Sterk *et al.*, 1999)
		Heptenophos (Head *et al.*, 2000)
		Hydrogen peroxide and peroxyacetic acid mixture (Krishnayya and Grewal, 2002)
I	Imidacloprid (Koppenhöfer *et al.*, 2003; Cuthbertson *et al.*, 2007)	Isoproturon (Sterk *et al.*, 1999)
	Ioxynil (Rovesti and Deseö, 1990)	
	Iprodione (De Nardo and Grewal, 2003)	
L	Lenacil (Rovesti and Deseö, 1990)	Lambda-cyhalotrin (Sterk *et al.*, 1999; Laznik and Trdan, 2014)
	Lindane (Rovesti and Deseö, 1990)	Lecithin (Sterk *et al.*, 1999)
	L-flamprop-isopropyl (Rovesti and Deseö, 1990)	Lufenuron (Laznik and Trdan, 2014)
M	Mannose (Rovesti and Deseö, 1990)	Mancozeb (Laznik *et al.*, 2012)
	MCPP (Rovesti and Deseö, 1990)	Mercurious chloride (Zimmerman and Cranshaw, 1990)
	Mefenoxam (De Nardo and Grewal, 2003)	Metamidophos (Hassan *et al.*, 1994)
	Metalaxyl + folpet (Rovesti and Deseö, 1990; Laznik *et al.*, 2012)	Metamitron (Sterk *et al.*, 1999)
	Methoprene (Grewal *et al.*, 1998)	Methansodium (Rovesti and Deseö, 1990)
	Metrobuzin (Rovesti and Deseö, 1990)	Methomyl (Rovesti and Deseö, 1990)
N	Neem (Rovesti *et al.*, 1988)	Neem oil (Krishnayya and Grewal, 2002)
	Neem oil (Krishnayya and Grewal, 2002)	
O		Oxyfluorfen (Rovesti and Deseö, 1990)

Continued

Table 30.3. Continued.

	+ (compatible interaction)	– (non-compatible interaction)
P	Penconazole (Sterk *et al*., 1999)	Paraquat (Rovesti and Deseö, 1990)
	Pendimethalin (Rovesti and Deseö, 1990)	Parathion (Rovesti and Deseö, 1990)
	Pentachloronitrobenzene (Zimmerman and Cranshaw, 1990)	Phosphamidon (Rovesti and Deseö, 1990)
	Phorate (Rovesti and Deseö, 1990)	Procymidone (Hassan *et al*., 1994)
	Phosmet (Sterk *et al*., 1999)	Propyzamide (Rovesti and Deseö, 1990)
	Pirimicarb (Laznik and Trdan, 2014)	Pymetrozine (Laznik and Trdan, 2014)
	Prochloraz (Rovesti and Deseö, 1990)	
	Prochloraz-manganese (Grewal *et al*., 1998)	
	Propamocarb (Rovesti and Deseö, 1990)	
	Propargite (Rovesti and Deseö, 1990)	
	Propiconazole (Zimmerman and Cranshaw, 1990)	
	Pyrazophos (Rovesti and Deseö, 1990)	
	Pyridine-carboxamide (Laznik *et al*., 2012)	
Q		Quizalofop-ethyl (Hassan *et al*., 1994)
S	Spinosad (De Nardo and Grewal, 2003)	Sethoxydim (Hassan *et al*., 1994)
	Streptomyces (De Nardo and Grewal, 2003)	Sulfur (Hassan *et al*., 1994)
T	Terbufos (Rovesti and Deseö, 1990)	Tebuconazole (Sterk *et al*., 1999; Laznik *et al*., 2012)
	Thiamethoxam (Koppenhöfer *et al*., 2003)	Teflubenzuron (Sterk *et al*., 1999)
	Thiophanate-methyl (De Nardo and Grewal, 2003)	Thiobendazole (Grewal *et al*., 1998)
	Thiram (Rovesti and Deseö, 1990)	Tralkoxydim (Hassan *et al*., 1994)
	Tolcofos-methyl (Rovesti and Deseö, 1990)	Tridemorph (Hassan *et al*., 1994)
	Tradimenol (Hassan *et al*., 1994)	Triforine (Hassan *et al*., 1994)
	Trichlorfon (Head *et al*., 2000)	
	Trichoderma harzianum (De Nardo and Grewal, 2003)	
	Triflumizole (De Nardo and Grewal, 2003)	
	Triforine (Rovesti and Deseö, 1990)	
W	White oil (Rovesti and Deseö, 1990)	

Table 30.4. Compatibility of entomopathogenic nematode *Steinernema glaseri* with the active ingredients of agrochemicals.

	+ (compatible interaction)	– (non-compatible interaction)
B		Beta cyfluthrin (Negrisoli *et al*., 2010)
C	Chlorpyrifos (Negrisoli *et al*., 2010)	
	Cypermethrin (Negrisoli *et al*., 2010)	
D	Deltamethrin (Negrisoli *et al*., 2010)	
	Deltamethrin + triazofos (Negrisoli *et al*., 2010)	
	Diflubenzuron (Negrisoli *et al*., 2010)	
E	Espinosad (Negrisoli *et al*., 2010)	
G	Gamacyhalothrin (Negrisoli *et al*., 2010)	
I	Imidaclorid (Koppenhöfer *et al*., 2003)	
L	Lufenuron (Negrisoli *et al*., 2010)	Lambda cyhalothrin (Negrisoli *et al*., 2010)
M	Mepronil (Lee *et al*., 1999)	Methil parathion (Negrisoli *et al*., 2010)
	Methoxyfenozide (Negrisoli *et al*., 2010)	
N	Neem (Rovesti *et al*., 1988)	
P	Permethrin (Negrisoli *et al*., 2010)	Permethrin (Negrisoli *et al*., 2010)
T	Thiamethoxam (Koppenhöfer *et al*., 2003)	
	Tolclofos-methyl (Lee *et al*., 1999)	
	Triazofos (Negrisoli *et al*., 2010)	
	Triflumuron (Negrisoli *et al*., 2010)	

Table 30.5. Compatibility of entomopathogenic nematode *Steinernema kraussei* with the active ingredients of agrochemicals.

	+ (compatible interaction)	– (non-compatible interaction)
A		Abamectin (Laznik and Trdan, 2014)
		Azadirachtin (Laznik and Trdan, 2014)
B		Bt *kurstaki* (Laznik and Trdan, 2014)
I		Imidacloprid (Laznik and Trdan, 2014)
L		Lambda-cyhalotrin (Laznik and Trdan, 2014)
		Lufenuron (Laznik and Trdan, 2014)
P		Pirimicarb (Laznik and Trdan, 2014)
		Pymetrozine (Laznik and Trdan, 2014)

Table 30.6. Compatibility of entomopathogenic nematode *Steinernema yirgalemense* with the active ingredients of agrochemicals.

	+ (compatible interaction)	– (non-compatible interaction)
C	*Cryptophlebia leucotreta* GV (van Niekerk and Malan, 2014)	Chlorpirifos (van Niekerk and Malan, 2014)
H	*Helicoverpa armigera* nucleopolyhedrovirus (HearNPV) (van Niekerk and Malan, 2014)	

Table 30.7. Compatibility of entomopathogenic nematode *Heterorhabditis amazonensis* with the active ingredients of agrochemicals.

	+ (compatible interaction)	– (non-compatible interaction)
A		Abamectin (Sabino *et al.*, 2014)
C	Chlorantranilprole (Sabino *et al.*, 2014)	Chlorpyrifos (Sabino *et al.*, 2014)
D	Deltamethrin (Sabino *et al.*, 2014)	
	Dimethylamino-propyl (Sabino *et al.*, 2014)	
L	Lambda cyhalothrin + chlorantranilprole (Sabino *et al.*, 2014)	
T	Thiamethoxam + lambda cyhalothrin (Sabino *et al.*, 2014)	
	Triflumuron (Sabino *et al.*, 2014)	

Table 30.8. Compatibility of entomopathogenic nematode *Heterorhabditis bacteriophora* with the active ingredients of agrochemicals.

	+ (compatible interaction)	– (non-compatible interaction)
A	Abamectin (Laznik and Trdan, 2014)	Acephate (Rovesti *et al.*, 1988)
	Amitraz (Rovesti *et al.*, 1988)	Alachlor (Rovesti *et al.*, 1988)
	Atrazine (Rovesti and Deseö, 1990)	Aldicarb (Rovesti *et al.*, 1988)
	Azadirachtin (Laznik and Trdan, 2014)	Aluminum tris (Alumai and Grewal, 2004)
B	Benomyl (Zimmerman and Cranshaw, 1990)	Bendicarb (Zimmerman and Cranshaw, 1990)
	Bt (Poinar *et al.*, 1990)	Bt *kurstaki* (Laznik and Trdan, 2014)
	Bt *kurstaki* (Poinar *et al.*, 1990)	
	Bt *san diego* (Poinar *et al.*, 1990)	
	Bupirmate (Rovesti *et al.*, 1988)	

Continued

Table 30.8. Continued.

	+ (compatible interaction)	– (non-compatible interaction)
C	Carboxin + thiram (Rovesti *et al.*, 1988)	Carbaryl (Zimmerman and Cranshaw, 1990)
	Chloropyriphos-ethyl (Rovesti *et al.*, 1988)	Carbendazin (Rovesti *et al.*, 1988)
	Chlorothalonil (Rovesti *et al.*, 1988)	Carbofuran (Rovesti *et al.*, 1988; Rovesti and Deseö, 1990)
	Chlorpyrifon (Alumai and Grewal, 2004)	Carbosulfan (Rovesti *et al.*, 1988; Rovesti and Deseö, 1990)
	Chlorsulfuron (Rovesti *et al.*, 1988)	Chloridazon (Rovesti *et al.*, 1988)
	Chlorthal-dimethyl (Rovesti *et al.*, 1988)	Chlorpyrifos (Zimmerman and Cranshaw, 1990)
	Chlortoluron (Rovesti *et al.*, 1988)	Cymoxanil + folpet (Rovesti *et al.*, 1988)
	Clopyralid (Rovesti *et al.*, 1988)	
	Copper sulfate (Rovesti *et al.*, 1988)	
	Cyhexatin (Rovesti *et al.*, 1988)	
D	Dalapon (Rovesti *et al.*, 1988)	Diazinon (Rovesti and Deseö, 1990; Zimmerman and Cranshaw, 1990)
	Diazinon (Rovesti *et al.*,1988; Rovesti and Deseö, 1990)	Dodine (Rovesti *et al.*, 1988)
	Dicamba (Zimmerman and Cranshaw, 1990)	
	Dicofol (Rovesti *et al.*, 1988)	
	Difubezuron (Rovesti *et al.*, 1988)	
	Dinocap (Rovesti *et al.*, 1988)	
	Diphenamid (Rovesti *et al.*, 1988)	
	Dithianon (Rovesti *et al.*, 1988)	
	DNOC (Rovesti *et al.*, 1988)	
E	Endosulfan (Das and Divakumar, 1987)	
	Exitazon (Rovesti *et al.*, 1988)	
F	Fenarimol (Rovesti *et al.*, 1988)	Fenamiphos (Rovesti *et al.*, 1988)
	Fenpropimorph (Rovesti *et al.*, 1988)	Fentin-acetate (Rovesti *et al.*, 1988)
	Fipronil (García del Pino and Jové, 2005)	Flubenzimine (Rovesti *et al.*, 1988; Rovesti and Deseö, 1990)
	Fluazifop butyl (Rovesti *et al.*, 1988)	
	Folpet (Rovesti *et al.*, 1988)	
	Fomesafen (Rovesti *et al.*, 1988)	
	Fonofos (Rovesti *et al.*, 1988)	
G	Glyphosate (Rovesti *et al.*, 1988; Rovesti and Deseö, 1990)	
H		Halofenozide (Alumai and Grewal, 2004)
I	Imidacloprid (Koppenhöfer and Kaya, 1998; Alumai and Grewal, 2004)	Isofenphos + phoxim (Poinar *et al.*, 1990)
	Ioxynil (Rovesti *et al.*, 1988)	
	Isofenphos + phoxim (Poinar *et al.*, 1990)	
L	Lambda-cyhalotrin (Laznik and Trdan, 2014)	Linuron (Rovesti *et al.*, 1988; Rovesti and Deseö, 1990)
	Lenacil (Rovesti *et al.*, 1988)	Lufenuron (Laznik and Trdan, 2014)
	L-flampop-isopropyl (Rovesti and Deseö, 1990)	
	Lindane (gamma HCH) (Rovesti *et al.*, 1988)	
	Linuron (Rovesti *et al.*, 1988; Rovesti and Deseö, 1990)	
M	Mannose (Rovesti *et al.*, 1988)	Mercurious chloride (Zimmerman and Cranshaw, 1990)
	MCPP (mecoprop) (Rovesti *et al.*, 1988)	Metalaxyl + folpet (Rovesti *et al.*, 1988)
	Mefenoxam (Alumai and Grewal, 2004)	
	Mepronil (Lee *et al.*, 1999)	
	Methamsodium (Rovesti *et al.*, 1988)	
	Methomyl (Rovesti *et al.*, 1988; Rovesti and Deseö, 1990)	
	Metrobuzin (Rovesti *et al.*, 1988)	

Continued

Table 30.8. Continued.

	+ (compatible interaction)	– (non-compatible interaction)
N		Neem (Prakasa Rao *et al.*, 1975)
O	Oxyfluorfen (Rovesti *et al.*, 1988)	
P	Pentachloronitrobenzene (Zimmerman and Cranshaw, 1990)	Paraquat (Rovesti *et al.*, 1988; Rovesti and Deseö, 1990)
	Pirimicarb (Laznik and Trdan, 2014)	Parathion (Rovesti *et al.*, 1988)
	Phenmedipham (Rovesti *et al.*, 1988)	Pendimethalin (Rovesti *et al.*, 1988)
	Phosphamidon (Rovesti *et al.*, 1988; Rovesti and Deseö, 1990)	Phorate (Rovesti *et al.*, 1988)
	Propargite (Rovesti *et al.*, 1988; Rovesti and Deseö, 1990)	Phosphamidon (Rovesti *et al.*, 1988; Rovesti and Deseö, 1990)
	Propiconazole (Rovesti *et al.*, 1988)	Prochloraz (Rovesti *et al.*, 1988)
	Pymetrozine (Laznik and Trdan, 2014)	Procymidone (Rovesti *et al.*, 1988)
	Pyrazophos (Rovesti *et al.*, 1988)	Propamocarb (Rovesti *et al.*, 1988)
		Propargite (Rovesti *et al.*, 1988; Rovesti and Deseö, 1990)
		Propyzamide (Rovesti *et al.*, 1988)
S	Spinosad (Alumai and Grewal, 2004)	
	Sulfur (Rovesti *et al.*, 1988)	
T	Terbufos (Rovesti *et al.*, 1988; Rovesti and Deseö, 1990)	Terbufos (Rovesti *et al.*, 1988; Rovesti and Deseö, 1990)
	Thiamethoxam (Koppenhöfer *et al.*, 2003; Alumai and Grewal, 2004)	Thiamethoxam (Koppenhöfer *et al.*, 2003; Alumai and Grewal, 2004)
	Thiram (Rovesti *et al.*, 1988)	Trichlorfon (Alumai and Grewal, 2004)
	Tolclofos-methyl (Rovesti *et al.*, 1988; Lee *et al.*, 1999)	Trifluralin (Rovesti *et al.*, 1988; Rovesti and Deseö, 1990)
	Triforine (Rovesti *et al.*, 1988)	
W	White oil (Rovesti *et al.*, 1988)	

Table 30.9. Compatibility of entomopathogenic nematode *Heterorhabditis indica* with the active ingredients of agrochemicals.

	+ (compatible interaction)	– (non-compatible interaction)
B		Beta cyfluthrin (Negrisoli *et al.*, 2010)
C	Chlorpyrifos (Negrisoli *et al.*, 2010)	Chlorpyrifos (Negrisoli *et al.*, 2010)
	Cypermethrin (Negrisoli *et al.*, 2010)	Cypermethrin (Negrisoli *et al.*, 2010)
D	Deltamethrin (Negrisoli *et al.*, 2010)	
	Deltamethrin + triazofos (Negrisoli *et al.*, 2010)	
	Diflubenzuron (Negrisoli *et al.*, 2010)	
E		Espinosad (Negrisoli *et al.*, 2010)
G	Gamacyhalothrin (Negrisoli *et al.*, 2010)	
L	Lufenuron (Negrisoli *et al.*, 2010)	Lambda cyhalothrin (Negrisoli *et al.*, 2010)
M		Methil parathion (Negrisoli *et al.*, 2010)
		Methoxyfenozide (Negrisoli *et al.*, 2010)
P		Permethrin (Negrisoli *et al.*, 2010)
T		Triazofos (Negrisoli *et al.*, 2010)
		Triflumuron (Negrisoli *et al.*, 2010)

Table 30.10. Compatibility of entomopathogenic nematode *Heterorhabditis zealandica* with the active ingredients of agrochemicals.

	+ (compatible interaction)	– (non-compatible interaction)
C	Chlorpirifos (van Niekerk and Malan, 2014) *Cryptophlebia leucotreta* GV (van Niekerk and Malan, 2014)	
H	*Helicoverpa armigera* nucleopolyhedrovirus (HearNPV)(van Niekerk and Malan, 2014)	

however, there is insufficient information on this interesting biological phenomenon (Thurston *et al.*, 1994; Finnegan *et al.*, 1999). Similar conclusions were obtained in the research conducted by Laznik and Trdan (2014), who tested several EPN species (*S. feltiae*, *S. carpocapsae*, *S. kraussei* and *H. bacteriophora*) with different insecticides (Fig. 30.2).

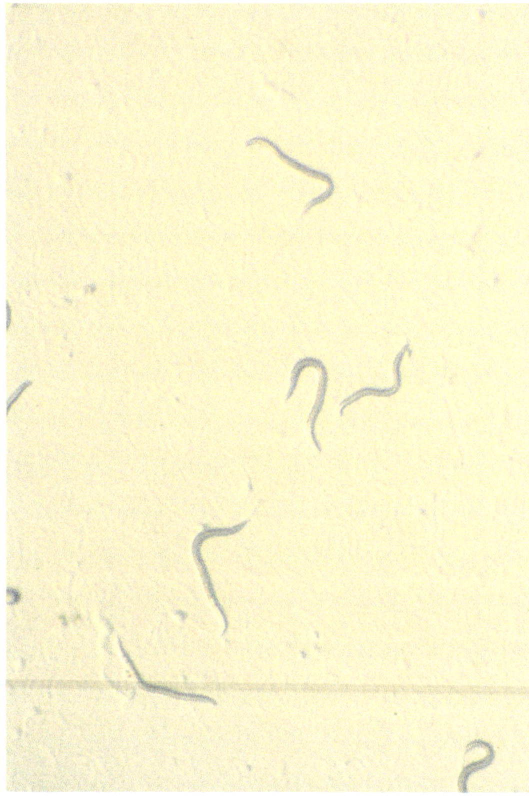

Fig. 30.2. *Steinernema feltiae* survived exposure to the phytopharmaceuticals tested.

30.5 The Use of Entomopathogenic Nematodes and Selected Phytopharmaceuticals in Integrated Pest Management

On the basis of the results of previous studies, we believe that the use EPNs in combination with phytopharmaceuticals can be an important component of an IPM system. EPNs can be applied using the same equipment that is used for phytopharmaceuticals, and therefore offers a cost-effective alternative for suppressing mites, insects and other disease-causing agents. In the suppression of harmful organisms, synchronous application saves time and money (Georgis, 1990).

The active ingredients azoxystrobin (Anand *et al.*, 2008), propamocarb (Urban and Lebeda, 2007) and sulfur (Bassino *et al.*, 1976) are known for their efficiency against powdery mildew (*Erysiphe cichoracearum* DC) and mildew on cucumbers (*Pseudoperonospora cubensis* ((Berk. and MA Curtis) Rostovzev)). Previous studies have shown that EPNs, when applied appropriately, effectively suppress certain pests on cucumbers, for example, western flower thrips (*Frankliniella occidentalis* Pergande) (Ebssa *et al.*, 2004; Trdan *et al.*, 2007) and greenhouse whiteflies (*Trialeurodes vaporariorum* (Westwood)) (Laznik *et al.*, 2011). The synchronous application of fungicides and EPNs in cases where mixed suspensions are applied to suppress specific harmful pests is thus justified. EPNs have proved to be very efficient biotic agents for suppressing some harmful pests, such as Colorado potato beetles (*Leptinotarsa decemlineata* (Say)) (Laznik *et al.*, 2010) and cereal leaf beetles (*Oulema melanopus* (L.)) (Laznik *et al.*, 2011). The synchronous application of fungicide and EPNs in such examples is justifiable for controlling the aforementioned pests. Similar conclusions can be drawn when dealing with the active ingredients metiram, copper hydroxide and mancozeb. These compounds specifically control potato blight (*Phytophthora infestans* (Mont.) de Bary), lettuce downy mildew (*Bremia lactucae* Regel), wheat leaf blotch (*Septoria tritici* Thüm.) and early blight of tomatoes

(*Alternaria solani* Sorauer) (Milus, 1994; Stepanović *et al.*, 2009; Stevenson, 2009), as well as some insect pests such as the Colorado potato beetle, cabbage armyworm (*Mamestra brassicae* (L.)), barley wireworm (*Agriotes fuscicollis* Miwa), cereal leaf beetle and the tomato leaf miner (*Tuta absoluta* (Povolny)). Previous research has shown that EPNs are very effective biological agents for the control of these insect pests (Zhao *et al.*, 1996; Batalla-Carrera *et al.*, 2010; Laznik *et al.*, 2010). However, the future implementation of the synchronous application of EPNs and fungicides must be supported by field experiments because the results of laboratory experiments cannot be applied uncritically to conditions that hold for the environment.

30.6 Conclusions

EPNs are used to control a wide range of foliar insect pests (Laznik and Trdan, 2011). When applied under favourable conditions, these nematodes can be as effective as chemical insecticides (Laznik *et al.*, 2011). EPNs can be combined with other agricultural chemicals and control agents for various purposes. First, nematodes and other control agents may be applied synchronously or within a short time interval of each other to control different pest species or stages of a pest. Nematodes may be combined with other control agents to achieve better control of a single pest through additive or, preferably, synergistic effects on pest mortality (De Nardo and Grewal 2003; Laznik *et al.*, 2012). For convenience, nematodes may also be tank mixed with other control agents, thereby increasing the chances of interactions due to the higher concentration of both agents. Nematode–insecticide combinations in tank mixes could offer a cost-effective alternative to foliar IPM systems. The use of EPNs in combination with insecticide could offer a solution for overcoming the problem of insecticide resistance. However, prior to an ecologically based IPM approach involving nematode–agrochemical combinations in tank mixes, the compatibility of these nematodes with new and routinely used agrochemicals needs to be established.

Acknowledgements

This chapter was completed under the programme Horticulture (P4-0013) funded by the Slovenian Research Agency. Part of the researches presented in this chapter was funded within Professional Tasks from the Field of Plant Protection, a programme funded by the Ministry of Agriculture, Forestry, and Food of Phytosanitary Administration of the Republic of Slovenia, and within bilateral cooperation in science and technology between Slovenia and Hungary in the period 2008–2011. We thank Sanja Ljubi, Polona Pust, Nika Bajc, Urška Držaj, Jaka Rupnik and Žan Tegelj for technical assistance.

References

Alumai, A. and Grewal, P.S. (2004) Tank mix-compatibility of the entomopathogenic nematodes, *Heterorhabditis bacteriophora* and *Steinernema carpocapsae*, with selected chemical pesticides used in turfgrass. *Biocontrol Science and Technology* 14, 725–730.

Anand, T., Chandrasekaran, A., Pillai-Kuttalam, S., Senthilraja, G., Raguchander, T. and Samiyappan, R. (2008) Effectiveness of azoxystrobin in the control of *Erysiphe cichoracearum* and *Psedoperonospora cubensis* on cucumber. *Journal of Plant Protection Research* 48, 147–158.

Askary, T.H. (2010) Nematodes as biocontrol agents. In: Lichtfouse, E. (ed.) *Sociology, Organic Farming,* Climate Change and Soil Science. Springer, The Netherlands, pp. 347–378.

Barbarossa, B., Frulla, R., Grazzi, G. and Rovesti, L. (1996) Compatibility of *Steinernema carpocapsae* with pesticides. *International Organization for Biological and Integrated Control Bulletin* 19, 136–139.

Barrett, J. (1991) Anhydrobiotic nematodes. *Agricultural Zoology Reviews* 4, 161–176.

Bassino, J.P., Bailly, R., Blanc, M. and Pascal, C. (1976) Powdery mildew of melon. The efficiency of various products. *Defense des Vegetaux* 31, 151–157.

Batalla-Carrera, L., Morton, A. and García del Pino, F. (2010) Efficacy of entomopathogenic nematodes against the tomato leafminer *Tuta absoluta* in laboratory and greenhouse conditions. *BioControl* 55, 523–530.

Campbell, L.R. and Gaugler, R. (1991) Role of the sheath in desiccation tolerance of two entomopathogenic nematodes. *Nematologica* 37, 324–332.

Cuthbertson, A.G.S., Mathers, J.J., Northing, P., Luo, W. and Walters, K.F.A. (2007) Establishing effect of commonly used insecticides for aphid control on the infectivity of the entomopathogenic nematode *Steinernema feltiae* using a streamlined screening method. *Journal of Environmental Research and Development* 2, 1–5.

Das, J.N. and Divakar, B.J. (1987) Compatibility of certain pesticides with DD-136 nematode. *Plant Protection Bulletin* 39, 20–22.

De Nardo, E.A.B. and Grewal, P.S. (2003) Compatibility of *Steinernema feltiae* (Nematoda: Steinernematidae) with pesticides and plant growth regulators used in glasshouse plant production. *Biocontrol Science and Technology* 13, 441–448.

Ebssa, L., Borgemeister, C. and Poehling, H.M. (2004) Effectiveness of different species/strains of entomopathogenic nematodes for control of western flower thrips (*Frankliniella occidentalis*) at various concentrations, host densities, and temperatures. *Biological Control* 29, 145–154.

Finnegan, M.M., Dawnes, J.D., O'Regan, M. and Griffin, C.T. (1999) Effect of salt and temperature stresses on the survival and infectivity of *Heterorhabditis* spp. IJs. *Nematology* 1, 69–78.

García del Pino, F. and Jové, M. (2005) Compatibility of entomopathogenic nematodes with fipronil. *Journal of Helminthology* 79, 333–337.

García del Pino, F. and Morton, A. (2010) Synergistic effect of the herbicides glyphosate and MCPA on survival of entomopathogenic nematodes. *Biocontrol Science and Technology* 20, 483–488.

Gaugler R. (2002) *Entomopathogenic Nematology*. CAB International, Wallingford, UK, 373 pp.

Gaugler, R. and Kaya, H.K. (1990) *Entomopathogenic Nematodes in Biological Control*. CRC Press, Boca Raton, Florida, 365 pp.

Georgis, R. (1990) Formulation and application technology. In: Gaugler, R. and Kaya, H.K. (eds.) *Entomopathogenic Nematodes in Biological Control*. CRC Press, Boca Raton, Florida, pp. 173–194.

Glazer, I. (2002) Survival biology. In: Gaugler, R. (ed.) *Entomopathogenic Nematology*. CAB International, Wallingford, UK, pp. 169–187.

Gordon, R., Chippett, J. and Tilley, J. (1996) Effects of 2 carbamates on infective juveniles of *Steinernema carpocapsae* All strain and *Steinernema feltiae* Umea strain. *Journal of Nematology* 28, 310–317.

Grewal, P.S. and Richardson, P.N. (1993) Effect of application rates of *Steinernema feltiae* (Nematoda Steinernematidae) on biological control of the mushroom fly *Lycoriella auripila* (Diptera: Sciaridae). *Biocontrol Science and Technology* 3, 29–40.

Grewal, P.S., Weber, T.A. and Betterley, D.A. (1998) Compatibility of *Steinernema feltiae* with chemicals used in mushroom production. *Mushroom News* 46, 6–10.

Gupta, P. and Siddiqi, M.R. (1999) Compatibility studies on *Steinernema carpocapsae* with some pesticidal chemicals. *Indian Journal of Entomology* 61, 220–225.

Hassan, S.A., Bigler, F., Bogenschütz, H., Boller, E., Brun, J., *et al.* (1994) Results of the sixth joint pesticide testing programme of the IOBC/WPRS-Working Group 'Pesticides and Beneficial Arthropods'. *Entomophaga* 39, 107–119.

Head, J., Walters, K.F.A. and Langton, S. (2000) The compatibility of the entomopathogenic nematode, *Steinernema feltiae*, and chemical insecticides for the control of the South American leafminer, *Liriomyza huidobrensis*. *Biocontrol* 45, 345–353.

Kaya, H.K. and Gaugler R. (1993) Entomopathogenic nematodes. *Annual Review of Entomology* 38, 181–206.

Koppenhöfer, A.M. (2000) Nematodes. In: Kaya H.K. (ed.) *Field Manual of Techniques in Invertebrate Pathology*. Kluwer Academic Publishers, The Netherlands, pp. 283–301.

Koppenhöfer, A.M. and Grewal, P.S. (2005) Compatibility and interactions with agrochemicals and other biocontrol agents. In: Grewal, P.S., Ehlers, R.-U. and Shapiro-Ilan, D.I. (eds) *Nematodes as Biocontrol Agents*. CAB International, Wallingford, UK, pp. 363–381.

Koppenhöfer, A.M. and Kaya, H.K. (1998) Synergism of imidacloprid and an entomopathogenic nematode: a novel approach to white grub (Coleoptera: Scarabaeidae) control in turfgrass. *Journal of Economic Entomology* 91, 618–623.

Koppenhöfer, A.M. and Kaya, H.K. (2002) Entomopathogenic nematodes and insect pest management. *Microbial Biopesticides* 15, 277–305.

Koppenhöfer, A.M., Cowles, R.S., Cowles, E.A., Fuzy, E.M. and Baumgartner, L. (2002) Comparison of neonicotinoid insecticides as synergists for entomopathogenic nematodes. *Biological Control* 24, 90–97.

Koppenhöfer, A.M., Cowles, R.S., Cowles, E.A., Fuzy, E.M. and Kaya, H.K. (2003) Effect of neonicotinoid insecticide synergists on entomopathogenic nematode fitness. *Entomologia Experimentalis et Applicata* 106, 7–18.

Krishnayya, P.V. and Grewal, P.S. (2002) Effect of neem and selected fungicides on viability and virulence of the entomopathogenic nematode *Steinernema feltiae*. *Biocontrol Science and Technology* 12, 259–266.

Laznik, Ž. and Trdan, S. (2011) Entomopathogenic nematodes (Nematoda: Rhabditida) in Slovenia: from tabula rasa to implementation into crop production systems. In: Perveen, F. (ed.) *Insecticides – Pest Engineering*. Intech, Rijeka, Croatia, pp. 627–656.

Laznik, Ž. and Trdan, S. (2014) The influence of insecticides on the viability of entomopathogenic nematodes (Rhabditida: Steinernematidae and Heterorhabditidae) under laboratory conditions. *Pest Management Science* 70, 784–789.

Laznik, Ž., Tóth, T., Lakatos, T., Vidrih, M. and Trdan, S. (2010) Control of the Colorado potato beetle (*Leptinotarsa decemlineata* Say) on potato under field conditions: a comparison of the efficacy of foliar application of two strains of *Steinernema feltiae* (Filipjev) and spraying with thiametoxam. *Journal of Plant Diseases and Protection* 117, 129–135.

Laznik, Ž., Žnidarčič, D. and Trdan, S. (2011) Control of *Trialeurodes vaporariorum* (Westwood) adults on glasshouse-grown cucumbers in four different growth substrates: an efficacy comparison of foliar application of *Steinernema feltiae* (Filipjev) and spraying with thiamethoxam. *Turkish Journal of Agriculture and Forestry* 35, 631–640.

Laznik, Ž., Vidrih, M. and Trdan, S. (2012) Effect of different fungicides on viability of entomopathogenic nematodes *Steinernema feltiae* (Filipjev), *S. carpocapsae* Weiser and *Heterorhabditis downesi* Stock, Griffin & Burnell (Nematoda: Rhabditida) under laboratory conditions. *Chilean Journal of Agricultural Research* 72, 62–67.

Lee, D.W., Choo, H.Y. and Choi, E.J. (1999) Effect of aqueous solution of pesticides on survival of entomopathogenic nematodes. *Korean Journal of Turfgrass Science* 13, 213–222.

Lola-Luz, T., Downes, M. and Dunne, R. (2005) Control of black vine weevil larvae *Otiorhyncus sulcatus* (Fabricius) (Coleoptera: Curculionidae) in grow bags outdoors with nematodes. *Agriculture and Forest Entomology* 7, 121–126.

Milus, E.A. (1994) Effect of foliar fungicides on disease control, yield and test weight of soft red winter wheat. *Crop Protection* 13, 291–295.

Negrisoli, A.S. Jr, Garcia, M.S. and Barbosa-Negrisoli, C.R.C. (2010) Compatibility of entomopathogenic nematodes (Nematoda: Rhabditida) with registered insecticides for *Spodoptera frugiperda* (Smith, 1797) Lepidoptera: Noctuidae) under laboratory conditions. *Crop Protection* 29, 545–549.

Poinar, G.O. Jr, Thomas, G.M. and Lighthart, B. (1990) Bioassay to determine the effect of commercial preparations of *Bacillus thuringiensis* on entomogenous rhabditoid nematodes. *Agriculture, Ecosystems and Environment* 30, 195–202.

Prakasa Rao, P.S., Das, P.K. and Padhi, G. (1975) Note of compatibility of DD-136 (*Neoaplectana dutkyi*), an insect parasitic nematode, with some insecticides and fertilizers. *Indian Journal of Agricultural Sciences* 45, 275–277.

Rovesti, L. and Deseö, K.V. (1990) Compatibility of chemical pesticides with entomopathogenic nematodes *Steinernema carpocapsae* Weiser and *S. feltiae* Filipjev (Nematoda: Steinernematidae). *Nematologica* 36, 237–245.

Rovesti, L., Heinzpeter, E.W., Tagliente, F. and Deseö, K.V. (1988) Compatibility of pesticides with the entomopathogenic nematode *Heterorhabditis bacteriophora* Poinar (Nematoda: Heterorhabditidae). *Nematologica* 34, 462–476.

Sabino, P.H.S., Sales, S.F., Guevara, E.J., Moino, A. Jr and Filgueiras, C.C. (2014) Compatibility of entomopathogenic nematodes (Nematoda: Rhabditida) with insecticides used in the tomato crop. *Nematoda* 1, e03014.

Schroer, S., Ziermann, D. and Ehlers R.-U. (2005) Mode of action of a surfactant-polymer formulation to support performance of the entomopathogenic nematode *Steinernema carpocapsae* for control of diamond-back moth larvae (*Plutella xylostella*). *Biocontrol Science and Technology* 15, 601–613.

Stepanović, M., Vuksa, P., Potočnik, I., Milijašević, S., Rekanović, E. and Todorović, B. (2009) Integrated management of viral sensitivity of *Alternaria solani* isolates to some fungicides. *Acta Horticulturae* 808, 377–380.

Sterk, G., Hassan, S.A., Baillod, M., Bakker, F., Bigler, F., *et al.* (1999) Results of the seventh joint pesticide testing programme carried out by the IOBC/WPRS-Working Group 'Pesticides and Beneficial Organisms'. *Biocontrol* 44, 99–117.

Stevenson, W.R. (2009) Late blight control strategies in the United States. *Acta Horticulturae* 834, 83–86.

Stuart, R.J. and Gaugler, R. (1996) Genetic adaptation and founder effect in laboratory populations of the entomopathogenic nematode *Steinernema glaseri*. *Canadian Journal of Zoology* 74, 164–170.

Thurston, G.S., Ni, Y. and Kaya, H.K. (1994) Influence of salinity on survival and infectivity of entomopathogenic nematodes. *Journal of Nematology* 26, 345–351.

Timper, P. and Kaya, H.K. (1989) Role of the 2nd stage cuticle of entomogenous nematode in preventing infection by nematophagous fungi. *Journal of Invertebrate Pathology* 54, 314–321.

Tóth, T. (2006) Collection of entomopathogenic nematodes for the biological control of insect pests. *Journal of Fruit and Ornamental Plant Research* 14, 225–230.

Trdan, S., Žnidarčič, D. and Vidrih, M. (2007) Control of *Frankliniella occidentalis* on glasshouse-grown cucumbers: an efficacy comparison of foliar application of *Steinernema feltiae* and spraying with abamectin. *Russian Journal of Nematology* 15, 25–34.

Ulu, T.C., Sadic, B. and Susurluk, I.A. (2016) Effects of different pesticides on virulence and mortality of some entomopathogenic nematodes. *Research Report* 13, 111–115.

Urban, J. and Lebeda, A. (2007) Variation of fungicide resistance in Czech populations of *Psedoperonospora cubensis*. *Journal of Phytopathology* 155, 143–151.

van Niekerk, S. and Malan, A.P. (2014) Compatibility of biological control agents and agrochemicals to entomopathogenic nematodes *Steinernema yirgalemense* and *Heterorhabditis zealandica*. *African Entomology* 22, 49–56.

Zhao, K.J., Zhang, L.K., Song, J., Zhang, L.H., Li, G.X., *et al.* (1996) On the application of steinernematid nematodes against eight lepidopterous and coleopterous insect pests. *Acta Phytophylacica Sinica* 23, 20–24.

Zimmerman, R.J. and Cranshaw, W.S. (1990) Compatibility of three entomogenous nematodes (Rhabditida) in aqueous solutions of pesticides used in turfgrass maintenance. *Journal of Economic Entomology* 83, 97–100.

31 Strategies for Making Entomopathogenic Nematodes Cost-effective Biocontrol Agents

M. Nagesh,[1*] Tarique Hassan Askary,[2] Balachander Manohar,[3] Shailesha Nanjundaiah Arakalagud[3] and Rajan[4]

[1]*Division of Plant Protection, ICAR-CPRI, Shimla, Himachal Pradesh, India;* [2]*Division of Entomology, Sher-e-Kashmir University of Agricultural Sciences and Technology, Srinagar, India;* [3]*ICAR-National Bureau of Agriculturally Important Insects (NBAII), Bengaluru, India;* [4]*Indian Council of Agricultural Research, New Delhi, India*

31.1 Introduction

Entomopathogenic nematodes (EPNs) belonging to Steinernematidae and Heterorhabditidae are now used to control several agriculturally important insect pests of the different orders. The first nematode (*Steinernema carpocapsae*) used successfully in the control of an insect pest was from Australia, and the Commonwealth Scientific and Industrial Research Organization (CSIRO) was the first in the world to use EPNs commercially against black vine weevil in ornamentals and against the currant borer moth in blackcurrants (Georgis, 2002). There are several species of EPNs used around the world against a variety of pests in niche markets (e.g. fungus gnats in nurseries, hydroponics and mushroom; weevils on ornamentals, strawberries, cranberries, citrus and bananas; scarabs of turf, ornamentals and blueberries; cutworms, webworms, billbugs and mole crickets; termites in wooden articles and trees; peach borer moth in apples and carpenter worm in shade trees in China and fig trees in the USA) (Friedman, 1990; Poinar, 1990; Kaya and Gaugler, 1993; Akhurst, 1996; Gaugler, 2002; Grewal *et al.*, 2005). Experts from the Organisation for Economic Cooperation and Development (OECD) and European countries participating in the European Commission's Cooperation in the Field of Science and Technical Research (COST) Action 819, along with 12 other participants, reviewed the potential safety issues associated with the introduction and commercial use of non-indigenous nematodes for insect biological control. The consensus view of the participants was that EPNs possessed specific biological and ecological features that made their use in biological control exceptionally safe to the environment, production and application personnel, end-users and public, animal and aqua health. Therefore, it was recommended that EPNs should not be subject to any kind of registration. This biosafety recommendation has not only intensified research on EPNs but also encouraged the industry to join hands for farm-oriented output.

Steinernematid and heterorhabditid nematode-based products are commercially available for insect control. Innovations in production through liquid fermentation have enabled these nematodes to become competitive with chemical insecticides in

*Corresponding author. E mail: nageshm.nbaii@gmail.com

niche markets. Substantial improvement in the stability of formulations and better shelf life have been developed by immobilizing or partially desiccating large numbers of nematodes on specific carriers such as clay, polyacrylamide and alginate gels (see Askary and Ahmad, Chapter 13, this volume). Also, the large-scale application of nematodes is now feasible due to the ease of mixing and applying these formulations. Paradoxically, the uptake of these environmentally safe and pest-effective bioagents is obviously constrained by uneconomical costs and temporal and spatial availability, compared to synthetic pesticides. Currently, the costs involved in the application of recommended doses of EPN products that are available in the market are substantially high across countries. There is an urgent need to address the issue of making EPN products more cost-effective and available.

In this chapter, we discuss the basic events involved in transforming EPNs into value-added products in order to evolve strategies for making EPNs cost-effective biocontrol agents and a part of the regular commercial supply chain. Further, we have compared the status of the EPN market scenario in the West with that of developing countries, taking the Indian scenario as an example, based on certain parameters that we assumed to have a role in rationalizing costs.

31.2 Diversity in Entomopathogenic Nematodes and their Potential

More than 30 nematode families are known to host taxa that parasitize or are associated with insects (Nickle, 1972; Poinar, 1990; Kaya and Stock, 1997). However, because of their biocontrol potential, research has concentrated on seven families: Mermithidae, Allantonematidae, Neotylenchidae, Sphaerularidae, Rhabditidae, Steinernematidae and Heterorhabditidae, the latter two currently receiving the most attention as control agents of soil insect pests. Besides insect hosts, *Phasmarhabditis hermaphrodita* (Schneider), a member of the family Rhabditidae, is reported to suppress several slug species, and has been developed as a biological molluscicide (Wilson *et al.*, 1993; Glen and Wilson, 1997; Wilson and Gaugler, 2000).

EPNs belong to the families Steinernematidae and Heterorhabditidae, and have been studied extensively as biological control agents in the management of the soil-dwelling stages of insect pests (Kaya and Gaugler, 1993; Askary, 2010). These nematodes occur naturally in soil and their infective juveniles (IJs) actively infect a broad range of insects and possess attributes like high virulence, quick kill in 24–72 h, high fecundity, host-searching/-foraging ability, safety to non-targets, dispersal capability, ease of mass production and exemption from US Environmental Protection Agency (EPA) registration.

31.3 Desirable Attributes of Entomopathogenic Nematodes to Develop Them as Commercial Products

1. Environmentally safe: EPNs, like all biological control organisms, are considered pesticides. However, EPNs are exempted from federal pesticide registration, Central Insecticides Board and the Registration Committee (CIB and RC). EPNs and their associated bacterial symbionts have been proved to be safe to warm-blooded vertebrates, including humans (Poinar *et al.*, 1982; Boemare *et al.*, 1996). We at National Bureau of Agricultural Insect Resources (ICAR-NBAIR), formerly National Bureau of Agriculturally Important Insects ICAR-(NBAII), have documented the safety and/or the minimal adverse effects of EPNs to many aboveground non-target invertebrates and their effects on the soil microfauna (earthworms, collembolans, etc.) and flora in the past 15 years. Somasekhar *et al.* (unpublished) found no decrease in the population of free-living nematodes up to 60 days after one application of native and exotic strains of *Heterorhabditis* species in turfgrass, but a significant reduction in the population of plant parasitic nematodes was observed. According to the recommendations of the OECD and the European Commission's COST workshop, nematodes are multicellular organisms and should not be regulated as microorganisms (Ehlers and Hokkanen, 1996). However, introduction of non-indigenous nematode species should be regulated, and recommendations were included in the Food and Agriculture Organization (FAO) Code of Conduct for the Import and Release of Biological Control (FAO, 1996).

2. Wide host range: In laboratory tests, *S. carpocapsae* alone infected more than 250 species of insects from over 75 families in 11 orders (Poinar, 1975). The nematodes attack a far wider spectrum of insects in the laboratory, where host contact is assured, environmental conditions are optimal and no ecological or behavioural barriers to infection exist (Kaya and Gaugler, 1993; Gaugler *et al.*, 1997). EPNs are especially efficacious against insects in soil and cryptic habitats (Lacey *et al.*, 2001).
3. Amenability to commercial production systems: EPNs are amenable to scale-up production, formulations and field delivery systems (Shapiro-Ilan *et al.*, 2012).
4. Compatibility with other agrochemicals: EPNs are often applied to sites and ecosystems that routinely receive other inputs that may interact with nematodes, including chemical pesticides, surfactants (e.g. wetting agents), fertilizers and soil amendments. Infective juveniles of EPNs are tolerant of short exposures (10–12 h) to most agrochemicals, including herbicides, fungicides, acaricides and insecticides (Hussaini *et al.*, 2001; Nagesh *et al.*, 2015), and therefore can often be tank mixed. However, some pesticides can reduce nematode infectivity and survival. Some pesticides act synergistically with EPNs and therefore improve their efficacy per dose applied and acceptability in integrated pest management (IPM) packages.

Despite environmental incentives and strong research efforts, the commercialization of EPNs has been slow to evolve, unlike other microbial biocontrol agents. Fermentation and post-production processes are more complex and expensive to operate than synthetic chemical processes. Significantly, even with the ever-growing regulatory, biodiversity and intellectual property (IP) requirements, the demand for biological control products is growing.

Commercialization essentially addresses the transformation of (temporal and spatial) demand to qualitative and quantitative supply. The principles and approaches to transform a nematode to a commercial product are necessarily similar to that of the development of a commercial product for a fungal or bacterial bioagent. Unlike microbials, being multicellular organisms, EPNs inherently have commercially undesirable features like short shelf life, absence of life stages resistant to adverse environment, typical asexual reproduction of short life cycles, etc., consequently demanding special protocols during production, shelf-life, packaging, storage, transport and delivery systems and adding costs. As a result, despite enormous potential for EPN products, their prohibitive costs are limiting their commercial use and benefit to farmers.

31.4 Status of the World Market for Entomopathogenic Nematodes in Comparison with Biopesticides

The first report on the global status of EPN production in comparison to the production of biopesticides was by Shapiro-Ilan and Grewal (2008). Significantly, EPN production expressed in terms of US dollars was fluctuating between 2003 and 2009 and ranging between US$14 million and US$19 million, while the production of total biopesticides recorded an increasing trend from US$195 million to US$257 million. The biopesticides market is projected to reach US$6.6 billion by 2020, and is expected to grow at a compound annual growth rate (CAGR) of 18.8% from 2015 to 2020 (http://www.marketsandmarkets.com/Market-Reports/biopesticides-267.html and http://www.marketsandmarkets.com/PressReleases/biopesticide.asp). The market for EPNs is expected to increase by 25% by 2020. To reach these markets, the product costs of EPNs will have to decrease further in the future. One possibility to reduce application costs related with the use of EPNs is the inoculative application to cause long-term effects on pest populations (Ehlers, 2003).

A sample survey made for EPN products that are currently available in different continents and their costs is presented in Table 31.1. Notably, the number of products/nematodes, their formulations, dosages, target pests and crops are variable across the continents and among countries. Further, commercialization and supply chains of EPNs seem to be systematic and stable in Western countries compared to the developing countries, including India, where the demand and market for EPNs is in infancy (Table 31.2).

Table 31.1. Status of entomopathogenic nematode (EPN) products available in various countries across different continents.

Continent: AFRICA
EGYPT

Species of entomopathogenic nematode	Target crop	Target pest	Formulation	*Dosage/acre	Name of product	Weight/IJs per pack	Name of company and country	Cost
Steinernema carpocapsae or Heterorhabditis bacteriophora	Tomato, pepper, lettuce and cabbage	Plant borers	EPN product is placed on moist polyether-polyurethane sponge	5-l package is diluted in 300 l of water per 1 feddan (feddan = 4200 m^2)	NOBUG	400 g	The National Research Centre (NRC), Egypt	US$40

SOUTH AFRICA

Species of entomopathogenic nematode	Target crop	Target pest	Formulation	*Dosage/acre	Name of product	Weight/IJs per pack	Name of company and country	Cost
H. bacteriophora	Citrus	Citrus, all varieties false codling moth (*Thaumatotibia leucotreta*)	Wettable powder	5×10^8 IJs/ha applied as 10 IJs/cm^2 1 million IJs/tree or 10 IJs/cm^2	Cryptonem	50 million or 250 million or 500 million	River Bioscience, South Africa	–
	Apple	Apples/codling moth Various banded fruit weevil		30 IJs/cm^2 5000 IJs/ml drench by injection with lance into soil				
	Strawberries	Strawberries: black vine weevil Berry crops, ornamentals: black vine weevil		50 IJs/cm^2 or 0.5 million IJs/m^2 10 IJs/cm^2				
	Mushroom	Mushrooms/sciarids						

Continent: ASIA
INDIA

Species of entomopathogenic nematode	Target crop	Target pest	Formulation	*Dosage/acre	Name of product	Weight/IJs per pack	Name of company and country	Cost
H. indica	Arecanut, cardamom, groundnut, sugarcane, potato, maize	White grubs, termites	WP	4 kg	Soldier	1 kg	Multiplex	US$15–20
	Aubergine	Ash weevil	WP	4 kg				
	Sweet potato	Sweet potato weevil	WP	4 kg				

Continued

Table 31.1. Continued.

Species of entomopathogenic nematode	Target crop	Target pest	Formulation	*Dosage/acre	Name of product	Weight/IJs per pack	Name of company and country	Cost
H. indica	Arecanut, cardamom, groundnut, sugarcane, potato, maize	White grubs, termites	WP	8 kg	Armour	1 kg	Ponalab	US$15–20
	Aubergine	Ash weevil	WP	8 kg				
	Sweet potato	Sweet potato weevil	WP	8 kg				
H. indica	Arecanut, cardamom, groundnut, sugarcane, potato, maize	White grubs, termites	WP	8 kg		1 kg	Camson Biotech	US$15–20
	Aubergine	Ash weevil	WP	8 kg				
	Sweet potato	Sweet potato weevil	WP	8 kg				
H. bacteriophora	Arecanut, cardamom, groundnut, sugarcane, potato, maize	White grubs, termites	WP	8 kg	Grubcide	1 kg	Sri Biotech	–

PAKISTAN

Species of entomopathogenic nematode	Target crop	Target pest	Formulation	Dosage/acre	Name of product	Weight/IJs per pack	Name of company and country	Cost
S. pakistanense	Okra and cotton	American, spotted, pink bollworm and root-knot nematodes	Sponge	10 million IJs	PakNema-1	200 g	NNRC, Pakistan	US$20
S. maqbooli	Tomato	Root-knot nematodes	Sponge	10 million IJs	PakNema-2	200 g	NNRC, Pakistan	US$20
S. bifurcatum	Vegetables	Pupal stages in soil and termite	Sponge	10 million IJs	PakNema-3	200 g	NNRC, Pakistan	US$20
S. balochiense	Mango	Fruit fly pupa	Sponge	10 million IJs	PakNema-4	200 g	NNRC, Pakistan	US$20

Continent: EUROPE
CZECH REPUBLIC

Species of entomopathogenic nematode	Target crop	Target pest	Formulation	*Dosage/acre	Name of product	Weight/IJs per pack	Name of company and country	Cost
S. feltiae	Planting substrate, mushrooms, flowers, greenhouses, ornamentals, vegetables, fruit	Sciaridae – larvae	Inert carrier, infection larvae mixed with inert carrier or on clay	0.5 million/m^2	Entonem	5/50/500 million IJs	Biocont Laboratory spol. s r. o.	US$5.7/40
H. bacteriophora	Planting substrates, ornamentals, greenhouses, strawberries, fruit	Curculionidae, genus Otiorhynchus	Inert carrier, infection larvae mixed with inert carrier or on clay	0.5 million/m^2	Larvanem	5/50/500 million IJs	Biocont Laboratory spol. s r. o.	US$5.65/40
S. feltiae	Planting substrate, mushrooms, flowers, greenhouses, ornamentals, vegetables, fruit	Sciaridae – larvae	Inert carrier, infection larvae mixed with inert carrier or on clay	0.5 million/m^2	Nemaplus	18 million IJs	e-nema GmbH	No data
Phasmarhabiditis hermaphrodita	Ornamentals, vegetables, fruit	Slugs (Limacidae, Agriolimacidae, Arionidae)	Inert carrier, infection larvae mixed with inert carrier or on clay	0.3 million/m^2	Nemaslug	12/30 million DJs	BASF Agricultural Specialities Ltd	US$24/52
H. bacteriophora	Planting substrates, strawberries	Curculionidae, genus Otiorhynchus	Inert carrier, infection larvae mixed with inert carrier or on clay	0.5 million/m^2	Nematop	10 million IJs	e-nema GmbH	–
P. hermaphrodita	Ornamentals, vegetables, fruit, planting substrates	Slugs (Limacidae, Agriolimacidae, Arionidae)	Inert carrier, infection larvae mixed with inert carrier or on clay	0.3 million/m^2	Phasmar habditis-System	12/30 million DJs	Biobest NV	–
S. feltiae	Planting substrates, mushrooms, flowers, greenhouses, ornamentals, vegetables, fruit	Sciaridae – larvae	Inert carrier, infection larvae mixed with inert carrier or on clay	0.5–1 million/m^2	Steinernema-System	5/50 million IJs	Biobest NV	–

Continued

Table 31.1. Continued.

GERMANY

Species of entomopathogenic nematode	Target crop	Target pest	Formulation	*Dosage/acre	Name of product	Weight/IJs per pack	Name of company and country	Cost
S. feltiae	Apple	Overwintering larvae of codling moth (*Cydia pomonella*) and red-belted clearwing (Synanthedon myopaeformis) Autumn treatment with nemapom® nematodes reduces overwintering larvae and pest occurrence in the following spring. Larvae die within 48 h after application	Clay powder	Young plantations: 750 million IJs/1000 l of water/m trunk. It is sufficient to treat only the lower part (1 m) Bigger and older trees: 1.5 million IJs/1500 l water. Treat full crown and take care that the older branches and arms are well treated at their base.	Nemapom	5 million IJs	e-nema GmbH, Germany	US$5
S. feltiae	Mushroom	Glasshouse sciarids	Clay powder	50 million IJs/100 m^2 in 20–50 l of water or for individual treatment, 10,000 IJs/l of potting compost. In case of a higher content of composted green waste or perlite into the substrate mixture, the dosage of nematodes should be increased by 50%	Nemaplus	5 million IJs	e-nema GmbH, Germany	US$5
H. bacteriophora	Turf grass, Vegetables	Chafer grubs (billbugs, Japanese beetle, northern and southern mask chafer, Oriental beetle, Asiatic garden beetle)	Clay powder	0.5 million nematodes/m^2	Nemagreen	5 million IJs	e-nema GmbH, Germany	US$5
H. bacteriophora	Strawberries, fruit tree nurseries	White grubs, black vine weevil (*Otiorhynchus sulcatus*), effective against larva and pupa	Clay powder	Soil: 0.5 million IJs/m^2 Container culture: 10,000 IJs/l of soil Strawberries: 25,000 IJs/plant	Nematop	5 million	e-nema GmbH, Germany	–

S. carpocapsae		*Capnodis tenebrionis*, leatherjackets, mole crickets (very efficient against adult mole crickets but less efficient against young larvae) and black cutworms	Clay powder	500 million IJs/1000 m^2 in 1200 l of water	Nemastar	5 million IJs	e-nema GmbH, Germany	US$5

ITALY

Species of entomopathogenic nematode	Target crop	Target pest	Formulation	Dosage/acre	Name of product	Weight/IJs per pack	Name of company and country	Cost
S. feltiae	Orchards	*Cydia pomonella*	Wettable powder	1.5 million/ha	OPTIMEN F	1.5 billion	Agrifutur Srl (Italy)	–
H. bacteriophora	Strawberry plants	*Otiorhynchus sulcatus* and other soil beetles	Wettable powder	0.2–0.4 million/m^2	OPTINEM H	50 million	Agrifutur Srl (Italy)	–
S. carpocapsae	Orchards	*C. pomonella* *C. molesta* *Euzophera bigella*	Wettable powder	1.5 billion/ha	OPTIMEN C	1.5 billion	Agrifutur Srl (Italy)	–
H. bacteriophora	Strawberry plants, hazel, chestnut tree, maize	*O. sulcatus, Balaninus nucum, C. splendana, C. fagiglandana, Diabrotica virgifera virgifera*	Wettable powder	0.5 million/m^2	*H. bacteriophora*	5/50/250 million	Andermatt Biocontrol AG (Swiss Confederation)	–
S. feltiae	Pot plants, young plants, sowings, scions, mushroom beds	*Sciaridae, C. pomonella,* thrips	Wettable powder	0.5–1 million/m^2	*S. feltiae*	5/50/250 million	Andermatt Biocontrol AG (Swiss Confederation)	–
S. carpocapsae	Orchards, horticultural crops, palm trees	*C. pomonella, Agrotis spp., Rhynchophorus ferrugineus, Tipula sp., Gryllotalpa gryllotalpa, Duponchelia fovealis, Paysandisia archon*	Wettable powder	0.5–1 million/m^2 Rhynchophorus: 10–30 million/palm tree	*S. carpocapsae*	5/50/250 million	Andermatt Biocontrol AG (Swiss Confederation)	–

Continued

Table 31.1. Continued.

Species of entomopathogenic nematode	Target crop	Target pest	Formulation	Dosage/acre	Name of product	Weight/IJs per pack	Name of company and country	Cost
H. bacteriophora	Strawberry plants and horticultural crops	*O. sulcatus and other soil beetles*	Wettable powder	0.2–0.4 million/m^2	*NEMAPACK HB*	50/250 million	Bioplanet s.c.a. (Italy)	–
S. carpocapsae	Orchards	*C. pomonella*	Wettable powder	1.5 billion/ha	*NEMAPACK SC*	50/500 million	Bioplanet s.c.a. (Italy)	–
S. feltiae	Pot plants, young plants, sowings, scions, mushroom beds	*Sciarids, phorids, Agromyza spp., muscids, nottuids, Agrotis spp., Cossus spp., sesids*	Wettable powder	0.3–0.6 million/m^2	*NEMAPACK SF*	50/500 million	Bioplanet s.c.a. (Italy)	–
S. carpocapsae	Palm trees	*R. ferrugineus, P. archon*	Wettable powder	17/25 million/tree	*NEMAPACK SC Palme*	50 million nematodes with 500 cc of liquid applicator	Bioplanet s.c.a. (Italy)	–
S. carpocapsae	Palm trees, orchards, horticultural crops, lawns	*R. ferrugineus, P. archon, Cydia spp., Synanthedon spp., Spodoptera spp., Agrotis spp., D. fovealis, Amphimallon solstitialis, Rhizotrogus aestivus, C. tenebrionis, G. grillotalpa, Tipula paludosa, T. oleracea*	Wettable powder	Cydia: 1.5 billion/ha; Rhynchophorus: up to 50 million/tree; Lepidoptera and Gryllotalpa: 0.5 million/m^2; Amphimallon: 1–1.5 million/m^2; Capnodis: 1–3 million/tree; Tipula: 0.5–1 million/m^2	*Nemastar*	50/500 million	Biogard Divisone CBC (Europe) Srl (Italy)	–
H. bacteriophora	Strawberry plants and horticultural crops	*O. sulcatus*	Wettable powder	0.5 million/m^2	*Nematop*	50/500 million	Biogard Divisone CBC (Europe) Srl (Italy)	

S. feltiae	Orchards, ornamentals, horticultural crops	*Cydia spp., Synanthedon spp., Frankliniella occidentalis, Bradysia spp., Acrolepiopsis assectella, Ephrytidae, Scatella stagnalis, Scatellatenui costa, Parahypopta caestrum, Lyriomiza spp., Tuta absoluta*	Wettable powder	Cydia: 1.5 billion/ha; Frankliniella and Bradysia: 0.125 million/m^2; Acrolepiopsis: 1 million/m^2; Lyriomiza and Tuta: 0.25 million/m^2	*Nemaplus*	50/500 million	Biogard Divisone CBC (Europe) Srl (Italy)	–
S. feltiae	Orchards	*Cydia spp., Synanthedon spp. C. tenebrionis*	Wettable powder	Cydia, Synanthedon: 1.5 billion/ha; Capnodis: 1–3 million/tree	*Nemapom*	1.5 billion	Biogard Divisone CBC (Europe) Srl (Italy)	–
S. carpocapsae	Orchards, ornamentals, horticultural crops	*Noctuids, D. fovealis, Tipula spp., Hylobius abietis, A. solstitialis, R. aestivus, C. tenebrionis, G. grillotalpa*	Wettable powder	0.5 million/m^2	*Capsanem*	50/500 million	Koppert Italia Srl (Italy)	–
S. feltiae	Ornamentals, pot plants, young plants, sowings, scions, mushroom beds	*Sciarids and thrips*	Wettable powder	0.5 million/m^2	*Entonem*	50/500 million	Koppert Italia Srl (Italy)	–
H. bacteriophora	Strawberry plants and horticultural crops	*O. sulcatus*	Wettable powder	0.5/1 million/m^2	*Larvanem*	50/500 million	Koppert Italia Srl (Italy)	–
H. bacteriophora	Horticultural crops, lawns	*Phyllopertha horticola, Serica brunnea, Hoplia spp., Aphodius spp.*	Wettable powder	0.5 million/m^2	*Terranem*	50/500 million	Koppert Italia Srl (Italy)	–

Continued

Table 31.1. Continued.

Species of entomopathogenic nematode	Target crop	Target pest	Formulation	Dosage/acre	Name of product	Weight/IJs per pack	Name of company and country	Cost
S. feltiae	Chestnut trees, orchards, palm trees	*C. pomonella, Synanthedon spuleri, C. molesta, E. bigella, Cydia funebrana, Anarsia lineatella, C. splendana, C. fagiglandana, R. ferrugineus*	Wettable powder	1/1.5 billion/ha 15/50 million/palm tree	*Nemax F*	250 million	Serbios Srl (Italy)	–
H. bacteriophora	Strawberry, maize	*O. sulcatus and D. virgifera virgifera*	Wettable powder	1/1.5 billion/ha	*Nemax H*	250 million	Serbios Srl (Italy)	–
S. feltiae	Orchards, palm trees	*C. pomonella, C. molesta, C. funebrana, C. splendana, C. fagiglandana, A. lineatella, E. bigella, R. ferrugineus*	Wettable powder	1/1.5 billion/ha 15/50 million/palm tree	*Xedanema*	250 million	Xeda Italia (Italy)	–

POLAND

Species of entomopathogenic nematode	Target crop	Target pest	Formulation	*Dosage/acre	Name of product	Weight/IJs per pack	Name of company and country	Cost
H. bacteriophora	Strawberry, home gardens	Larvae of *O. sulcatus, grubs of P. horticola, Melolontha spp., A. solstitialis*	WP Clay	Low incidence of insect pests: 50 IJs/cm^2 High incidence: 100 IJs/cm^2	LARVANEM	50 million IJs or 2 × 250 million IJs	Koppert	US\$28 US\$178
H. bacteriophora	Turfgrass, golf courses	Grubs of *P. horticola*	WP Clay	5 million IJs/10 m^2	B-GREEN	5 million IJs or 50 million IJs or 500 million IJs	Biobest	US\$43 – –

H. megidis	Ornamental plants, strawberries	Larvae of *O. sulcatus*	WP Clay	500,000 IJs/m^2– 1 million IJs/m^2	Heterorhabditis-System	50 million IJs or 500 million IJs	Biobest	US$27 US$165
H. megidis	Orchards	*O. sulcatus*	WP Clay	Interventional: 250 million IJs/250 m^2 Preventive: 250 million IJs/500 m^2	Nemasys HM	50 million IJs 250 million IJs	Syngenta Bioline Becker Underwood	– US$97
H. megidis	Ornamental plants in pots; raspberry plants; strawberry plants	*O. sulcatus*	WP Clay	Interventional: 1 million IJs/m^2 Preventive: 500,000 IJs/m^2	Exhibilinehm	50 million IJs or 250 million IJs or 500 million IJs	Syngenta Bioline	No data
S. feltiae	Greenhouse crops	*Thrips tabaci, P. horticola*	WP Clay	Soil application: 0.25–1.0 million IJs/m^2 Foliar application: 0.25 million IJs/m^2	Entonem	50 million IJs or 500 million IJs	Koppert	US$22 US$121
S. feltiae	Trees in nurseries, tomato plants, sweet pepper, aubergine, cucumber, ornamental plants	*F. occidentalis, Sciaridae, Diptera*	WP Clay	50 million IJs/m^2	Steinernema-System	5 million IJs or 50 million IJs or 250 million IJs or 1250 million IJs	Biobest	No data US$25 No data No data
S. feltiae	Plants grown in greenhouses	*F. occidentalis, Sciaridae, Diptera*	WP Clay	2 million IJs/m^2	SCIARID	2 × 250 million IJs	Koppert	US$105
S. feltiae	Plants grown in greenhouses	*F. occidentalis, Sciaridae, Diptera*	WP Clay	Sciaridae: 0.5–1 million IJs/m^2 F. occidentalis: 0.125–0.25 million IJs/m^2	Nemasys F	50 million IJs or 250 million IJs	Syngenta Bioline	US$22 US$66
S. feltiae	Ornamental plants	*F. occidentalis, T. tabaci*	WP Clay		Exhibitline Sf	50 million IJs or 250 million IJs or 1250 million IJs	Syngenta Bioline	US$41 US$160 US$263
S. carpocapsae	Orchards	Noctuidae – larvae Pyralidae – larvae Tipulidae – larvae Coleoptera – larvae *G. gryllotalpa*	WP Clay	Interventional: 50 million IJs/100m^2 Preventive: 500 million IJs/1000 m^2 or 0.5 million/m^2	Capsanem	50 million IJs or 2 × 250 million IJs	Koppert	US$28 US$149

Continued

Table 31.1. Continued.

Species of entomopathogenic nematode	Target crop	Target pest	Formulation	*Dosage/acre	Name of product	Weight/IJs per pack	Name of company and country	Cost
S. carpocapsae	Turfgrass	*Agrostis* spp., *Tipula* spp.	WP Clay		Exhibitline Sc	50 million IJs or 250 million IJs	Syngenta Bioline	No data No data
S. kraussei	Ornamental plants in pots	*O. sulcatus*	WP Clay	Interventional: 250 million/250 m^2 or 250 million/ 400–500 m^2 plants grown in containers Preventive: 250 million/500 m^2 or 250 million/800–1000 m^2 plants grown in containers	Nemasys L	50 million IJs or 250 million IJs	Syngenta Bioline	US$32 US$94
S. kraussei	Ornamental plants, strawberry plants and others	*O. sulcatus*	WP Clay	1 million/m^2	Exhibitline Srb		Syngenta Bioline	No data
P. hermaphrodita	Agricultural and horticultural crops	Snails	WP Clay	50,000/m^2 150,000/m^2 300,000/m^2	Nemaslug	30 million IJs or 250million IJs	Becker Underwood	US$27 US$113
P. hermaphrodita	Agricultural and horticultural crops	Snails	WP Clay	12 million/40 m^2 30 million/100 m^2	Phasmarhabditis-System	12 million IJs or 30 million IJs	Biobest	US$11 US$36

RUSSIA

Species of entomopathogenic nematode	Target crop	Target pest	Formulation	*Dosage/ acre	Name of product	Weight/IJs per pack	Name of company and country	Cost
S. feltiae	Ornamentals (especially roses) in greenhouses and in pots); tomatoes and cucumbers in greenhouses and open plots; decorative trees of high commercial value on open terrain	Western flower thrips, wireworm, insects in tunnels beneath the bark, Melolontha chafers and other soil-inhabiting insects	Water suspension <8 atm no sieves	4 kg/ha	Entonem F	8 kg	OOO Biometodika, Russia	Approx. US$100/1 kg of suspension

S. carpocapsae	Ornamentals (especially roses) in greenhouses and in pots); tomatoes and cucumbers in greenhouses and open plots; decorative trees of high commercial value on open terrain	Western flower thrips, wireworm, insects in tunnels beneath the bark; soil-inhabiting insects	Water suspension <8 atm; no sieves	4 kg/ha	Nemabact	5 kg	OOO Biometodika, Russia	Approx. US$100/1 kg of suspension

Continent: SOUTH AMERICA
BRAZIL

Species of entomopathogenic nematode	Target crop	Target pest	Formulation	*Dosage/acre	Name of product	Weight/IJs per pack	Name of company and country	Cost
Deladenus siricidicola This nematode is mostly used in Brazil (>20 lakh ha)	Pine	*Sirex noctilio* (pine wasp)	Gel	20 ml (1 million nematodes) Each dose is enough to treat 10 infected trees	Deladenus Siricidicola (the product name is same as nematode species)	–	Produced by Embrapa Floresta (Federal Research Corporation in Brazil) using also a cottage system that they copy from Australia. This nematode has not been registered so far (there is no commercial name) but despite this, Embrapa has produced and sold it. This is the most successful programme in Brazil so far.	US$2 (approx.)
H. indica LPP30 or *H. baujardi* LPP7	Guava	*Conotrachelus psidii*	Infected juveniles	2.5 × 108 IJs/ha	Infected larvae of *Galleria mellonella* (there is no commercial name, growers know it as infected larvae)	–	UENF (Universidade Estadual Norte Fluminense) (state university in Brazil)	US$30 (approx.)

Continued

Table 31.1. Continued.

Species of entomopathogenic nematode	Target crop	Target pest	Formulation	*Dosage/acre	Name of product	Weight/IJs per pack	Name of company and country	Cost
S. braziliense	Sugarcane	*Sphenophorus levis*	Sponge	1–2 × 10^8 IJs/ha	Bio Steinernema (the only product registered in Brazil)	500 mg	Bio Controle (a private company in Brazil)	US$30 (approx.)

Notes: *denotes the recommended dose as per the label claim of the company and not by reference of published data. WP = wettable powder. Koppert BV from The Netherlands has a branch in Poland known as Koppert Polska; Biobest from Belgium offers their products by local distributors of agricultural products e.g. Hortico, BioPartner; Syngenta Bioline Ltd is registered in England. In Poland, the local distributor of agricultural products, e.g. APS Beneficials, offers Syngenta products. Becker Underwood has been acquired by BASF Crop Protection from England and Wales. Products with EPNs are sold by different agricultural and horticultural centres in Poland, e.g. AZeco.pl.

IJs = Infective juveniles. DJs = Daeur juveniles**.** Infective juveniles are active and ready to infect the insect host while Daeur juveniles are inactive 3rd instar juveniles which cannot directly infect the insect host but need to tranform to infective juveniles.

Table 31.2. Comparative status of entomopathogenic nematodes (EPNs) for integrated pest management in Western countries (USA, Europe) and in India.

Parameters	USA, Europe	India (efforts of ICAR-NBAII)
Number of pests covered	24	8
Area covered	3 continents[a]	20,000 ha
Number of crops	>20	4–6
Number of *Heterorhabditis* spp. developed as commercial products	5–7	2
Number of *Steinernema* spp. developed as commercial products	10–12	4
Technologies evolved for EPN production, formulations, storage, etc.	8	1
Number of companies	13–15	4–6 (total of 12 acquired technologies)
Production systems	*In vitro* and *in vivo*	*In vivo*
Downstream processing	Mechanized	Semi-mechanized
Formulations that are available on the market	Cadaver, sponge, Water Dispersible Granule (WDG) and Wettable Powder (WP)	WP
Shelf life of formulations evolved	46 months	8–10 months
Supply chain established	Network, Internet, on board	Dealers
Application methods	Mechanized	Manual
Shopping	On Web/Internet	Personal

Notes: Nagesh *et al*., 2013. [a]3 continents = Asia, Europe and North America.

31.5 Strategies for Making Entomopathogenic Nematodes Cost-effective Biocontrol Agents

31.5.1 Factors contributing to the costs

The transformation of a living nematode into an end product in the form of final commercial-scale formulation involves costs. Essentially and ideally, the costs should cover the expenditure on inputs, services, fees, taxes, intellectual property costs, royalties, etc., involved from the first to the last step in the flow chart mentioned below (Fig. 31.1). Therefore, the critical stages and the purposes that involve the recurring and non-recurring costs can be identified and analysed, before arriving at the strategies to make EPNs cost effective biocontrol agents. Fixed costs involving infrastructure (buildings, machinery, vehicle purchase, etc.) are dependent primarily on the scale of operation and the location of the establishment, besides several other related costs.

Among the different stages, transformation, production, post-production processing, formulations, packaging, transportation, storage (shelf life) and delivery systems are critical events that involve cost factors. These are the critical stages that need to be examined for rationalization of the cost factors to make biological control agents cost-effective. Expenditure on IP, royalty, registration and other certification charges are generally fixed and charged on actual basis. These costs also add to the project outlay and are therefore to be considered as an investment.

31.5.2 Cost-effective production to scale

EPNs can be mass produced by *in vivo* or *in vitro* methods for laboratory tests or for commercial production (Shapiro-Ilan *et al.*, 2012). Large-scale commercial yields of EPNs require economies of scale (i.e. decreasing costs in yields with an increasing scale of operation) (Friedman, 1990). Without such economies of scale, the commercial production of nematodes will be confined to cottage industries. Mass production of EPNs has evolved from the first large-scale *in vitro* solid media production by Glaser *et al.* (1940), to the *in vivo* production by Dutky *et al.* (1964), to the three-dimensional solid media *in vitro* process (Bedding, 1981, 1984), to the *in vitro* liquid fermentation production methods (Friedman, 1990).

The basic method for small-scale *in vivo* production was described by Kaya and Stock (1997), whereas a large-scale method to produce nematodes

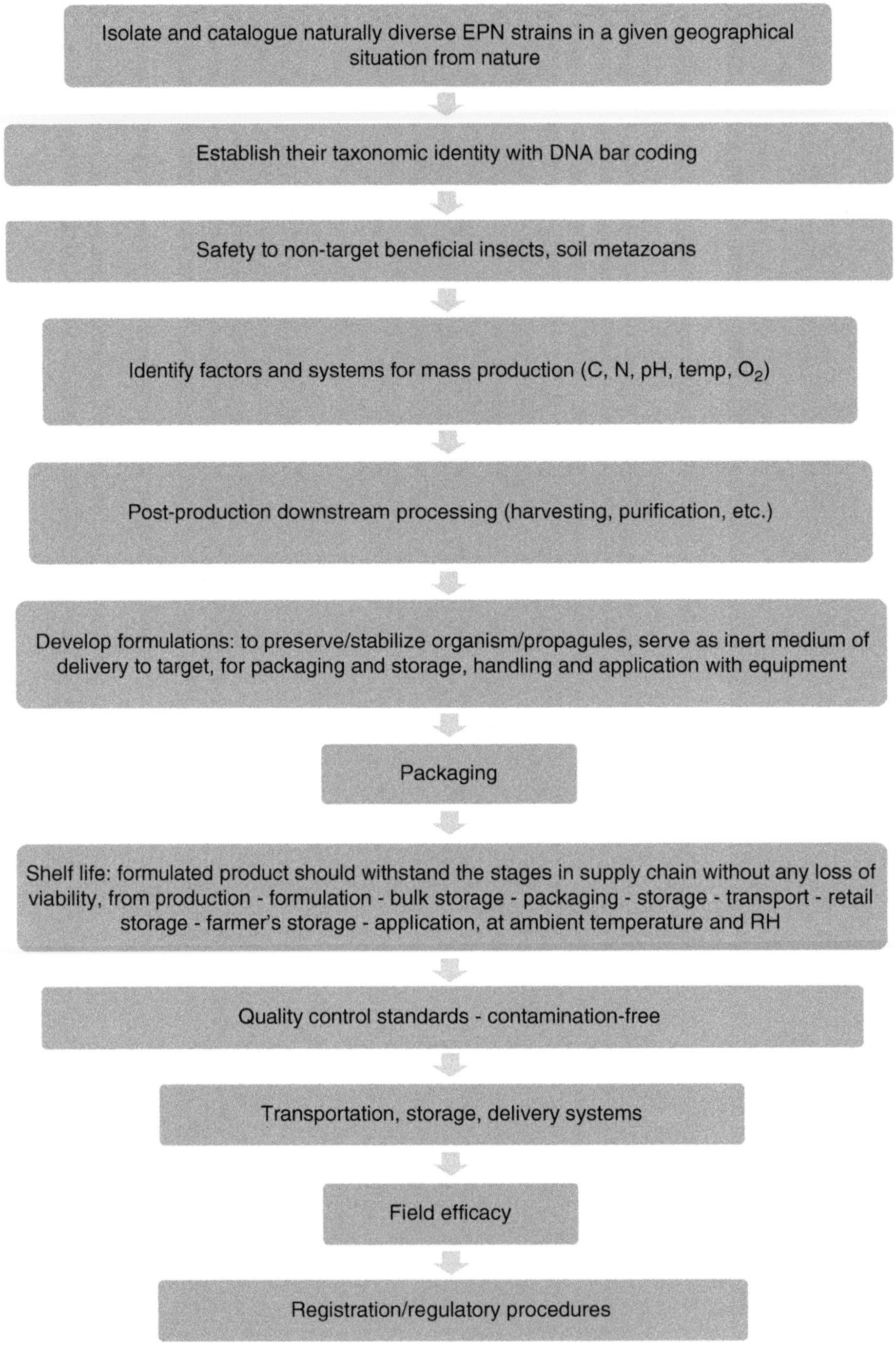

Fig. 31.1. Critical steps involved in transforming EPNs into value-added end products for commercial use.

using insect hosts was described by Gaugler *et al.* (2000). In *in vivo* method of production, the larvae of wax moth, *Galleria mellonella*, are most commonly used the world over, because of their commercial availability. Depending on the nematode species to be multiplied, yields to the tune of 1.0×10^5 to 6×10^5 IJs/larva of *G. mellonella* can be achieved (Nagesh *et al.*, 2013). In some Western countries, larvae of the mealworm beetle, *Tenebrio molitor*, are also used, which yield much higher numbers of juveniles per larva. However, its use is restricted as it is also a stored grain pest and is quarantined in many countries. In some other countries, larvae of another stored grain pest, *Corcyra cephalonica*, is also commonly used for laboratory rearing of nematodes.

In vivo production is labour-intensive, lacks economies of scale and is costly, but it is also simple and reliable and results in high-quality nematodes (Gaugler *et al.*, 2000). This technique of nematode production can be promoted as a cottage industry in developing countries.With these leads, we have standardized and validated *in vivo* processes for large-scale production of *Heterorhabditis indica*, *Heterorhabditis bacteriophora*, *Steinernema abbasi* and *S. carpocapsae* and developed a distinct commercial EPN industry in India in the past decade. Using the *in vivo* process, we could achieve a scale-up of 6–12 t of wettable powder (WP) formulation/month, with a quality-assured shelf life of 10–12 months, depending on the nematode species. The costs of *in vivo* production primarily encompass the recurring expenditure on insect diet ingredients, plastic ware, water, electricity and human resources on an actual basis. Diet ingredients include semi-processed cereals, milk powder, honey, glycerol, vitamins, etc., as feed for larvae for 3 weeks. Ideally, one larva of *G. mellonella* is observed to consume about 80–100 g of diet to grow into a pre-pupating larva. There is therefore a necessity to find relatively low cost dietary alternatives to these components to reduce the overall costs. Among other recurring expenses in *in vitro* production, the costs of electricity for temperature and RH maintenance, and fresh water for harvesting the juveniles in aqueous suspension are important. The use of alternative energy sources like solar and wind energy for generating electric power plays a vital role in reducing these costs in the industry and increasing profitability. It has been universally observed that *in vivo* production is simple, demanding low investment on infrastructure and producing good-quality nematodes with longer shelf life, but is certainly time- and space-consuming and labour-intensive, which has a direct bearing on commercial-scale production. Therefore, ideally, *in vitro* production systems should be replicated spatially, with the involvement of rural and peri-urban artisans and a semi-skilled work force.

Early workers on EPN production and commercialization have realized that *in vivo* production is constrained by the costs involved, the timely availability of host insects and the time required per production cycle, and have thus explored the possibilities of monoxenic *in vitro* production. *In vitro* production was standardized initially on solid media. The first successful commercial-scale monoxenic culture was developed by bedding in 'solid' culture (Bedding, 1981, 1984). Nematodes are cultured on a crumbed polyether polyurethane sponge impregnated with emulsified beef fat and pigs' kidneys, along with symbiotic bacteria. Using this method, approximately 6×10^5 to 10×10^5 IJs/g of medium were achieved. Since then, this method has been used commercially in Australia, China and the USA. In a scale-up mode, Friedman reported that the solid culture method was economically feasible up to a production level of approximately 10×10^{12} nematodes/month (Friedman, 1990). Labour costs increase significantly for nematode production beyond this level. The advantages of the solid media method are that capital costs are low, limited expertise is required (but need more skills than the *in vivo* method) and the logistics of production are flexible. Because of limited economies of scale and levels of contamination, this production method is best suited for countries that have low labour costs or for serving high-value markets.

Friedman reported a liquid fermentation technique for the large-scale production of nematodes (Friedman, 1990). The liquid fermentation method has economies of scale because the proportion of labour and capital costs decreases in scale as operating costs increase. Among the different methods, liquid fermentation technology has the lowest production cost. Therefore, liquid production is the method of choice for larger companies with multiple products in developed countries (Nickle *et al.*, 1988). A number of nematode species have been produced successfully in 75,000–80,000 l bioreactors, including *S. carpocapsae*, *Steinernema riobrave*, *Steinernema kushidai*, *Steinernema feltiae*, *Steinernema glaseri*,

Steinernema scapterisci, *H. bacteriophora* and *Heterorhabditis megidis*, with yield capacity as high as 250,000 IJs/ml, depending on the nematode species. Costs on infrastructure for production, post-production processing and annual maintenance are proportionally high, depending on the scale of the fermentor. Relatively, recurring or production working costs on ingredients/raw material, power supply and working staff are low and economical. However, the quality and vigour of nematode juveniles produced through *in vitro* production have been reported to be inferior to those obtained through *in vivo* production.

31.5.3 Postharvest downstream processing

Infective juveniles of EPNs, once mass produced *in vivo* or *in vitro*, primarily pass through the postharvest processes including harvesting, repeated filtration to remove substrate material, purification and concentration in aqueous or semi-dry form for developing into appropriate formulations. These events require one-time investment on equipment, including appropriate centrifuges, filtration facilities, harvesters, containers, water circulation systems, etc., and running or recurring costs are on power, water and labour costs.

31.5.4 Cost-effective formulations

The development of appropriate formulations of EPNs is a critical event during their commercial use and endurance in the production–supply chain. Formulations involve costs, and may influence the costs on packaging too. Therefore, while developing formulations, their universality of use, ease of application, storage and transport also need to be considered, besides their ability to sustain EPN viability and shelf life. Several formulations have been developed for storage and application of EPNs, including aqueous suspensions, sponge impregnation, alginate gels, granules, wettable dispersible granules, wettable powders, etc. In the simplest type of formulation, the nematodes are impregnated on to moist carrier substrates, providing substantial interstitial spaces leading to increased gas exchange. Such carriers include polyether polyurethane sponge, cedar shavings, peat vermiculite, etc. Nematodes held on the sponge need to be hand-squeezed into water before application, whereas from the other carriers they may be applied directly to the soil as mulch. Due to labour-intensive application, limited economy of scale and constant refrigeration requirements, these formulations are only applicable in small home lawn and garden markets. Further, activated charcoal has been used to restrict nematode movement (Yukawa and Pitt, 1985). Kaya and Nelsen (1985) were the first to report the encapsulation of EPNs with calcium alginate. This discovery subsequently led to the development of a commercial nematode product that used thin sheets of calcium alginate spread over a plastic screen to trap nematodes (Georgis, 1990). For application, nematodes are released from the alginate gel matrix by dissolving it in water with the aid of sodium citrate. The alginate-based *S. carpocapsae* products were the first to possess shelf life of about 3–4 months at room temperature, and led to an increased acceptability of nematodes in the high- and medium-value niche markets. However, the time-consuming extraction steps, transportation and the problematic disposal of containers rendered this formulation unsuitable for large-scale use. Although gel-based, wheat flour and attapulgite clay chip based formulations were evolved, none of these formulations offered any advantage over the alginate gels. A significant advancement was made with the advent of a water-dispersible granule (WDG) formulation in which IJs were encased in 10–20 mm diameter granules consisting of mixtures of various types of silica, clays, cellulose, lignin and starches.

Steinernematids, especially *S. carpocapsae*, can be maintained for up to 5 months at room temperature or up to 12 months under refrigeration (Georgis, 1990; Georgis and Hague, 1991). Formulated or unformulated, most heterorhabditids do not have a long shelf life and suffer high mortalities in storage.

Hussaini and Nagesh (2000–2008) at NBAIR attempted to develop wettable powder formulations of Heterorhabditids and Steinernematids with improved shelf life, scalability and ease of application. Based on these studies, Nagesh *et al.* (2010) developed novel WP formulations for EPNs, with a shelf life of 10–12 months, and validated their efficacy through All India Co-ordinated Research Project (Biological Control of Crop Pests, Diseases and Weeds) and the All India Network Project (Whitegrubs and Other Soil Arthropods) trials and front line demonstrations (FLDs). These efforts effectively established a niche market for EPNs and a viable alternative to chemical pesticides for managing several soilborne arthropods.

31.5.5 Packaging, storage and shelf life of entomopathogenic nematodes

The loss of viability during storage (storage stability) is one of the most challenging barriers to overcome in the case of higher organisms, including EPNs. In general, EPNs do not have a long shelf life. The IJs of EPNs are not a resting stage, and juveniles are metabolically active and use energy (lipid and glycogen) reserves while in formulation (Shapiro-Ilan and Lewis, 1999). These nematodes invariably require an appropriate amount of moisture and aeration for their viability during storage. In some formulations, nematodes are forced into an anhydrobiotic stage and reduce their metabolic activity before being packaged into formulations. Some of the formulations are required to be stored and transported at regulated temperatures in refrigerated containers, obviously increasing their costs.

31.5.6 Quality assurance is another important factor that plays a vital role in market price

Improved reliability, quality assurance and customer satisfaction are important components for the health and success of the nematode industry. Many corporate companies invest a great amount of funds into these components and standards for quality. Gaugler *et al.* (2000) suggested that the EPN cottage industry lacked rigorous quality control and consumers were rarely able to provide feedback on the quality of EPNs.

31.5.7 Regulations – essential component of commercialization

The nematode–bacterium complex has been exempted from registration in the USA by the EPA (Nickle *et al.*, 1988) and is regulated by the Animal and Plant Health Inspection Service. These regulations vary among countries. Indigenous nematodes are exempted from registration in many European countries, Australia and the USA, while in other countries they are subject to similar registration procedures as for a chemical pesticide. The importation and use of non-indigenous and transgenic nematode species are subject to strict regulations in most countries. Some countries consider foreign strains of endemic species to be exotic, and this can be a major obstacle to the commercialization of EPNs. Guidelines for the introduction of exotic beneficial nematodes, proposed by Nickle *et al.* (1988), offer a basis from which national guidelines can be developed.

31.6 Production and Commercialization of Entomopathogenic Nematodes in Developing Countries

Comparatively, the EPN market and its commercialization in the West is more robust and evolved than in a developing country like India, based on certain factors, including the number of crops and pests covered, the number of EPN species that have been introduced into the supply chain and market, the extent of mechanization in production and post-production processes, the choice of formulations that are available in the market and the type of supply chain established. Significantly, EPN production systems in the West have almost all transformed to *in vitro* from *in vivo*, while in countries like India, the *in vivo* system continues to be the mainstay. Also, the number of patent-compliant EPN technologies are higher (8–10) in the West, while India currently has only one technology related to EPNs that is awaiting patent grant and has been transferred to industry for commercialization of the production systems. Awareness of EPN products and their market networking are much advanced in the West, while the systems are lacking or evolving in India and many other countries. Crop diversity and the number and diversity of insect pests across a country like India have potentiated the applicability and necessity of wider application of EPNs. For instance, the acreage under selected field, fruit and vegetable crops (National Horticulture Board, 2011; http://www.fnbnews.com/; www.indiastat.com) and rough estimates on damage due to soil pests (primarily, white grubs) at 10% of the area correspond to: 0.765 million ha (Mha)/0.0765 Mha for banana; 0.119 Mha/0.0119 Mha for grape; 0.0778 Mha/0.00778 Mha for cardamom; 0.410 Mha/0.041 Mha for groundnut; 6.89 Mha/0.689 Mha for arecanut; and 5.25 Mha/0.525 Mha for sugarcane. Nagesh *et al.* (2015) demonstrated the potential of different species of EPNs in the management of soilborne insect pests in these crops (Table 31.3). Further, in order to meet the field requirements, production and formulation technologies were transferred to industry during 2010 and 2015. Ten private firms are actively involved in the production and supply of EPNs to the tune of 586 metric tons, which in our estimation can

Table 31.3. Potential of entomopathogenic nematodes in controlling soilborne arthropods in India, as established by the National Bureau of Agriculturally Important Insects. (From Nagesh et al., 2015.)

Family	Scientific name	Commodity	States	Nematode
Curculionidae	*Basilepta fulvicornae*	Cardamom	Kerala, Karnataka, TN	Sa, Sc, Hi
Curculionidae	*Cylas formicarius*	Sweet potato	Karnataka, TN, Odisha	Sc, Sa, Hi, Sg
Scarabaeidae	*Holotrichia* spp., *Anomala* spp.	Ginger, turmeric	Karnataka, TN, Kerala, Odisha	Sc, Sa, Hi, Sg
Scarabaeidae	*Holotrichia* spp.	Potato	Karnataka, TN	Sc, Sa, Hi
Scarabaeidae	*Holotrichia* spp., *Anomala* spp.	Fodder grasses	Karnataka, TN, Odisha	Sc, Sa, Hi, Hb, Sg
Scarabaeidae	*Holotrichia* spp., *Anomala* spp.	Maize	Karnataka, TN, Maharashtra	Sc, Sa, Hi, Hb, Sg
Scarabaeidae	*Holotrichia* spp., *Anomala* spp.	Redgram (pigeon pea)	Karnataka, TN	Sc, Hi, Sa
Scarabaeidae	*Holotrichia* spp., *Anomala* spp.	Groundnut	Karnataka, TN	Sc, Hi, Sa
Curculionidae	*Otiorhynchus sulcatus*	Nursery	Karnataka, TN	Sc, Hi
Curculionidae	*Myllocerus subfasciatus*	Aubergine	Karnataka, TN, AP, Maharashtra	Sb, Hi
Curculionidae	*Odoiporus longicollis*	Banana	TN	Sb, Hi, Sc
Curculionidae	*Cosmopolites sordidus*	Banana	Karnataka, TN	
Scarabaeidae	*Exomala* spp.	Turfgrass	Karnataka	Sc, Sg, Hb
Scarabaeidae	*Leucopholis leucophora, Leucopholis burmeisteri, Leucopholis coneophora*	Arecanut	Karnataka, Kerala, Maharashtra, TN	Sg, Sc, Hi, Hb
Scarabaeidae	*Anomala* sp.	Vegetables	Karnataka, Maharashtra	Sg, Sc, Sa, Hi, Hb
Lepidoptera	*Herpetogramma phaeopteralis*	Turfgrass	Karnataka, Maharashtra	Sg, Sc, Sa, Hi, Hb
Thysanopotera	*Frankliniella occidentalis*	Polyhouse vegetable and ornamental crops	Karnataka, Maharashtra	Sa, Sf, Hi

Notes: TN = Tamil Nadu; AP = Andhra Pradesh; Sa = *Steinernema abbasi*; Sc = *Steinernema carpocapsae*; Sf = *Steinernema feltiae*; Sg = *Steinernema glaseri*; Hb = *Heterorhabditis bacteriophora*; Hi = *Heterorhabditis indica*.

hardly meet the demand of the projected area of white grub damage in a single crop (Table 31.4). Therefore, this inequitable demand and supply affects the costs and quality of EPN products on the market immensely, to the disadvantage of the farmer in a developing country like India.

31.7 Conclusion

Developing countries like India across the continents have the capability to produce significant levels of EPN formulations at the current costs of production to meet projected EPN requirements. However, the actual production levels are low due to dependence on low investment *in vivo* technologies that are low in production and productivity, lacking in infrastructure of established supply chain, storage facility, etc., thus making EPN production systems and products not cost-effective. The ability to expand the market to multiple pest control applications on wider areas and crops is expected automatically to enhance the commercial production potential to scale, and also ensure cost benefits from production to scale in the case of EPNs. Commercial preference and demand for the EPN product depend on various factors such as its relative efficacy over the synthetic chemicals available, label claims on number of crops and pests, quality, shelf-life and safety as per the label claim, timely availability for immediate field use, and dose and method of application which have a bearing on the price of the product. The wider the crop and pest number in the label claim, the product should be more cost-effective. Similarly, low doses of formulation and simple methods of application (not requiring special equipment, care etc.) contribute to reduction in costs. Strategically, there is an urgent need to establish a demand-driven EPN supply chain with private–public partnership at both the local and the global level, taking guidelines from other microbial biocontrol agents, such as *Bacillus thurengiensis*, *Trichoderma* species, *Metarhizium anisopliae*, *Beauveria bassiana*, pseudomonads, etc., to achieve the cost-effective availability of EPNs on temporal and spatial bases. This can be achieved through the exchange and replication of information across countries, and a paradigm shift in the technologies related to *in vitro* production to scale, formulation technologies and the mechanization of different production, post-production processes and delivery systems involved in the development of the products. Further, *in vitro* production systems in the West are validated for a few species of EPNs, which need to be expanded to other potential species of heterorhabditids and steinernematids through intensive research. The incorporation of improvised isolates with attributes such as desiccation and high-temperature tolerance can enhance uptake in rain-fed and arid crop conditions. EPNs can become cost-effective bioagents by their effective incorporation as a component in IPM packages with wider label claims. Last, but not least, a farmer–public–private partnership to establish a supply chain would make EPNs cost-effective and market-sustainable bioagents in times to come.

Table 31.4. List of Department of Scientific and Industrial Research (DSIR) and recognized companies to whom National Bureau of Agricultural Insect Resources (NBAIR) have commercialized and licensed entomopathogenic nematode technologies. (From personal communications Nagesh, M., Hussaini, S. S. and Shylesha, A. N., ICAR-NBAIR, Bengaluru, India with the licensed companies, 2012–2014).

Technology	Licensed to	Production per annum (tons) January 2016
Novel insecticidal wettable powder formulations of *Heterorhabditis indica* strain NBAII HI1 and *H. bacteriophora* strain NBAII HB5 for the biological control of white grubs and other insect pests, and the methods thereof for their preparation and use	Multiplex, Bengaluru	400
	Camson Biotech, Bengaluru	114
	Farmer, Ghaziabad, Uttar Pradesh	Not known
	Allwin Industries, Indore, Madhya Pradesh	Start up
	Sri Biotech Laboratories India Ltd, Hyderabad	Start up
	Sri Venkateshwara Chemicals (KN Biosciences), Hyderabad	Not Known
	Ponalab, Bengaluru	30
	Rauf Research Foundation, Sirsi, Karnataka	30
	Eco Agri Biotech, Kottayam, Kerala	12
	UAS, Shimoga, Karnataka	Start up

Acknowledgements

We would like to thank Drs Jolanta Kowalska, J. Nermut, Mahfouz M.M. Abd-Elgawad, Sergei Spiridinov, Shahina Fayaz, Luis Grarrigós Leite, Vanessa Andaló, Justin Hatting, A.P. Malan and Eustachio Tarasco, for the information provided and/or help in preparing the manuscript.

References

Akhurst, R.J. (1996) From then to now – a brief review of entomopathogenic nematodes and their symbiotic bacteria. *Proceedings of IInd International Symposium on Entomopathogenic Nematodes and their Symbiotic Bacteria*, pp. 3–8.

Askary, T.H. (2010) Nematodes as biocontrol agents. In: Lichtfouse, E. (ed.) *Sociology, Organic Farming,* Climate Change and Soil Science. Springer, The Netherlands, pp. 347–378.

Bedding, R.A. (1981) Low cost *in vitro* mass production of *Neoaplectana and Heterorhabditis* species (Nematoda) for field control of insect pests. *Nematologica* 27, 109–114.

Bedding, R.A. (1984) Large scale production, storage and transport of the insect-parasitic nematodes *Neoaplectana* spp. *and Heterorhabditis* spp. *Annals of Applied Biology* 104, 1104–1120.

Biopesticides Market (2015) Biopesticides Market by Type (Bioinsecticides, Biofungicides, Bioherbicides, and Bionematicides), Origin (Beneficial insects, Microbials, and Biochemical), Mode of application, Formulation, Crop type & Region - Global forecast to 2020 http://www.marketsandmarkets.com/Market-Reports/biopesticides-267.html (accessed 1 December 2016).

Boemare, N.E., Laumond, C. and Mauleon, H. (1996) The nematode-bacterium complexes: biology, life cycle, and vertebrate safety. *Biocontrol Science and Technology* 6, 333–345.

Dutky, S.R., Thompson, J.V. and Cantwill, G.E. (1964) A technique for the mass propagation of the DD 136 nematode. *Journal of Insect Pathology* 6, 417–422.

Ehlers, R.-U. (2003) Entomopathogenic nematodes in the European biocontrol market. *Communications in Agricultural and Applied Biological Sciences* 68, 3–16.

Ehlers, R.-U. and Hokkanen, H.M.T. (1996) Insect biocontrol with non-endemic entomopathogenic nematodes (*Steinernema* and *Heterorhabditis* spp.): conclusions and recommendations of a combined OECD and COST workshop on scientific and regulatory policy issues. *Biocontrol Science and Technology* 6, 295–302.

FAO (1996) International standards for phytosanitary measures. Code of conduct for the import and release of exotic biological control agents. FAO Publication No. 3. pp.1–19.

Friedman, M.J. (1990) Commercial production and development. In: Gaugler, R. and Kaya, H.K. (eds) *Entomopathogenic Nematodes in Biological Control*. CRC Press, Boca Raton, Florida, pp.153–172.

Gaugler, R. (2002) *Entomopathogenic Nematology.* CAB International, Wallingford, UK.

Gaugler, R, Lewis, E.E. and Stuart, R.J. (1997) Ecology in the service of biological control: The case of entomopathogenic nematodes. *Oecology* 109, 483–489.

Gaugler, R., Grewal, P.S. and Kaya, H.K. (2000) Quality assessment of commercially produced entomopathogenic nematodes. *Biological Control* 17, 100–109.

Georgis, R. (1990) Formulation and application technology. In: Gaugler, R. and Kaya, H.K. (eds) *Entomopathogenic Nematodes in Biological Control*. CRC Press, Boca Raton, Florida, pp. 173–191.

Georgis, R. (2002) The Biosys experience: an insider's perspective. In: Gaugler, R. (ed.) *Entomopathogenic Nematology*. CAB International, Wallingford, UK, pp. 357–372.

Georgis, R. and Hague, N.G.M. (1991) Nematodes as biological pesticides. *Pesticides Outlook* 2, 29–32.

Glaser, R.W., McCoy, E.E. and Girth, H.B. (1940) The biology and economic importance of a nematode parasitic in insects. *Journal of Parasitology* 26, 479–495.

Glen, D.M. and Wilson, M.J. (1997) Slug-parasitic nematodes as biocontrol agents for slugs. *Agro Food Industry Hi-Tech* 8, 23–27.

Grewal, P.S., Ehlers, R.-U. and Shapiro-Ilan, D.I. (2005) *Nematodes as Biocontrol Agents*. CAB International, Wallingford, UK.

Hussaini, S.S., Singh, S.P. and Shakeela, V. (2001) Compatibility of entomopathogenic nematodes (Steinernematidae, Heterorhabditidae: Rhabditida) with selected pesticides and their influence on some biological traits. *Entomon* 26, 37–44.

Kaya, H.K. and Gaugler, R. (1993) Entomopathogenic nematodes. *Annual Review of Entomology* 38, 181–206.

Kaya, H.K. and Nelson, C.E. (1985) Encapsulation of steinernematid and heterorhabditid nematodes with calcium alginate; a new approach for insect control and other applications. *Environmental Entomology* 14, 572–574.

Kaya, H.K. and Stock, S.P. (1997) Techniques in insect nematology. In: Lacey, L.A. (ed.) *Manual of Techniques in Insect Pathology*. Academic Press San Diego, California, pp. 281–324.

Lacey, L.A., Frutos, R., Kaya, H.K. and Vails, P.P. (2001) Insect pathogens as biological control agents. Do they have a future? *Biological Control* 21, 230–248.

Nagesh, M., Javeed, S. and Rabindra, R.J. (2010) Development of novel insecticidal wettable powder formulations of *Heterorhabditis indica* strain NBAII HI1 and *Heterorhabditis bacteriophora* strain NBAII HB5 for the biological control of white grubs and other insect pests, and the methods thereof for their preparation and use. Indian Patent Application Number: 3490/CHE/2010.

Nagesh, M., Shailesha, A.N., Patil, J. and Nikhita, R.P. (2013) Improvised *in vivo* production and formulations of entomopathogenic nematodes, and current status of commercialization in India. Emerging Opportunities for the Mass Production and Quality Assurance of Invertebrates. 13th IOBC-MRQA Workshop, Bengaluru, India.

Nagesh, M., Shailesha, A.N., Patil, J. and Jalali, S.K. (2015) Genetic diversity, biology and utilization of entomopathogenic nematodes (EPN) against cryptic pests. Consolidated Research Project Report 2012–2015. ICAR-NBAIR, Bengaluru, India.

Nickle, W.R. (1972) A contribution to our knowledge of the Mermithidae (Nematoda). *Journal of Nematology* 4, 113–146.

Nickle, W.R., Drea, J.J. and Coulson, J.R. (1988) Guidelines for introducing beneficial insect-parasitic nematodes into the United States. *Annals of Applied Nematology* 2, 50–56.

Poinar, G.O. Jr (1975) *Entomogenous Nematodes: A Manual and Host List of Insect–Nematode Associations*. Brill, Leiden, Germany.

Poinar, G.O. Jr (1990) Taxonomy and biology of Steinernematidae and Heterorhabditidae. In: Gaugler, R. and Kaya, H.K. (eds) *Entomopathogenic Nematodes in Biological Control*. CRC Press, Boca Raton, Florida, pp. 23–61.

Poinar, G.O. Jr., Thomas, G.M., Presser, S.B. and Hardy, J.L. (1982) Inoculation of entomogenous nematodes, *Neoaplectana* and *Heterorhabditis* and their associated bacteria, *Xenorhabdus* spp., into chicks and mice. *Environmental Entomology* 11, 132–138.

Shapiro-Ilan, D.I. and Grewal, P. (2008) Entomopathogenic nematodes and insect management. In: Capinera, J.L. (ed) *Encyclopedia of Entomology*. 2nd edition. Springer, Dordrecht, The Netherlands, pp. 1336–1340.

Shapiro-Ilan, D.I. and Lewis, E.E. (1999) Comparison of entomopathogenic nematode infectivity from infected hosts versus aqueous suspension. *Environment and Ecology* 28, 907–911.

Shapiro-Ilan, D.I., Han, R. and Dolinksi, C. (2012) Entomopathogenic nematode production and application technology. *Journal of Nematology* 44, 206–217.

Wilson, M. and Gaugler, R. (2000) *Phasmarhabditis hermaphrodita* (Rhabditida: Rhabditidae). In: *Biological Control: A Guide to Natural Enemies in North America*. Cornell University.

Wilson, M.J., Glen, D.M. and George, S.K. (1993) The rhabditid nematode *Phasmarhabditis hermaphrodita* as a potential biological control agent for slugs. *Biocontrol Science and Technology* 3, 503–511.

Yukawa, T. and Pitt, J.M. (1985) Nematode storage and transport. *World Patent No.* WO 85/03412.

32 Future Thrusts in Expanding the Use of Entomopathogenic and Slug Parasitic Nematodes in Agriculture

Tarique Hassan Askary,[1]* Jiří Nermuť,[2] M. Jamal Ahmad[1] and Mushtaq Ahmad Ganai[1]

[1]*Division of Entomology, Sher-e-Kashmir University of Agricultural Sciences and Technology, Srinagar, India;* [2]*Institute of Entomology, Biology Centre CAS, České Budějovice, Czech Republic*

32.1 Introduction

Entomopathogenic (*Steinernema* and *Heterorhabditis* sp.) and slug parasitic (*Phasmarhabditis* sp.) nematodes are important group of beneficial organisms inhabiting the soil. The symbiotic association of these nematodes with bacteria, i.e. the nematode–bacterium complex, makes their status as potential biopesticides, as well as a viable pest management alternative (Askary, 2010). Furthermore, these nematodes have the potential to affect the diversity of native fauna in soil ecosystems, even though they do not have any direct parasite–host or predator–prey relationship. These attributes have drawn the attention of research practitioners working in different laboratories of the world to exploit these beneficial nematodes and their bacterial symbionts against the insects and slug pests of various crops of agricultural importance. In the past three decades, such nematodes have been formulated and their products commercialized in the global market under different names (Table 32.1). However, a few of them have been withdrawn from the market due to their poor efficacy against target pests under field conditions, probably because of their poor adaptability to environmental conditions, or production or formulation problems.

Nowadays, these beneficial nematodes, as well as their respective bacterial symbionts, are being viewed as exciting subjects for basic research in the areas of ecology, biodiversity, evolution and molecular genetics. However, the share of nematode-based biopesticides in the global market is too low, for various reasons, and therefore the vast majority of extensive research is still focused on their suitable formulation, greater shelf life, enhanced efficacy, enhanced carryover effect and, finally, low product cost.

In this chapter, we highlight the issues related to the expansion and commercialization of two beneficial nematodes, i.e. entomopathogenic and slug parasitic. Their limitations and valuable suggestions to overcome them have been discussed in detail.

32.2 Commercially Available Nematodes

In recent years, entomopathogenic nematode (EPN) products containing *Steinernema* and *Heterorhabditis* have been prepared successfully by different companies of the world and commercialized in the international market. The most promising species that have been formulated successfully and exploited as commercial products are *Steinernema carpocapsae*, *Steinernema feltiae*, *Steinernema riobrave*, *Steinernema scapterisci*, *Steinernema kushidai*, *Heterorhabditis megidis* and *Heterorhabditis bacteriophora* (Table 32.1). The products, under different names, are being sold by both developed as well as developing countries of the world, such as the USA, the UK, Germany, Switzerland, The Netherlands, India, Japan, Poland, Austria and Canada. As far as slug-parasitic nematodes are concerned, the only species

*Corresponding author. E-mail: tariq_askary@rediffmail.com

 Biocontrol Agents: Entomopathogenic and Slug Parasitic Nematodes (eds. M.M.M. Abd-Elgawad, T.H. Askary and J. Coupland)

Table 32.1. Commercial products of entomopathogenic and slug parasitic nematodes prepared by different countries.

Nematode	Product name	Country
Steinernema carpocapsae	Boden-Niitzlinge	Germany
	Mioplant	Austria
	Nemabakt	Russia
	Biosafe	Columbia
	Biovector	Columbia
	Green Commandos	India
	Bouncer	India
	CAPSANEM	Netherlands
	Carpocapsae-System	Canada
	Sanoplant	Switzerland
	Helix	Canada
	X-GNAT	USA
S. feltiae	Nemasys	UK
	Entonem	Netherlands
	Owinema SC	Poland
	Exhibit	Switzerland
	Stealth	UK
	Magnet	USA
S. scapterisci	Proactant Ss	USA
S. riobrave	Biovector	Columbia
S. kushidai	SDS biotech	Japan
Heterorhabditis bacteriophora	Otinem	Switzerland
	Soil Commandos	India
	E-Nema Gmbh	Germany
H. megidis	Nemasys	UK
	Larvanem	Netherlands
	NovoNem	Germany
H. indica	Soldier	India
Phasmarhabditis hermaphrodita	Nemaslug	UK

is *Phasmarhabditis hermaphrodita*, commercially available (Wilson *et al.*, 1993) and considered most successful in European countries (Rae *et al.*, 2007). *P. hermaphrodita* are being mass produced by the company BASF (previously Becker Underwood UK Ltd) and sold under the trade name of Nemaslug® (Fig. 32.1).

This product is now commercialized in different European countries, namely Belgium, the Czech Republic, Denmark, Finland, France, Germany, the UK, Ireland, Italy, The Netherlands, Norway, Poland, Spain, Sweden and Switzerland, and total sales are around €1 million per annum. The shelf life of Nemaslug® is comparatively short as compared to other EPNs (Ester and Wilson, 2005). However, sales in the USA, Australia and some other countries are not allowed because *P. hermaphrodita* has never been isolated (until a short time ago) in these states and the introduction of exotic species is strictly forbidden. Now, the recent discovery of *P. hermaphrodita* in the cadavers of invasive slugs in the USA (Tandingan De Ley *et al.*, 2014) can been seen as ray of hope for their commercial use in this important market (see Glen and Coupland, Chapter 28, this volume).

32.3 Constraints in the Expansion of Entomopathogenic and Slug Parasitic Nematodes

There are numerous factors responsible for limiting the expansion of beneficial nematodes (EPNs and slug parasitic) that are used against target pests of agricultural importance. The constraints on buyers/users of both EPNs and slug parasitic nematodes are more or less similar. They may vary from the little experience of the users to the poor efficacy of the nematodes. High product cost and their low availability on the market are other limitations. We have explained these under the four subheadings below.

Fig. 32.1. Package of the only commercially available slug parasitic nematode *Phasmarhabditis hermaphrodita*, Nemaslug©.

32.3.1 High price of nematode products

The prime concern in the utilization of nematodes as biocontrol agents is the high price of their products, which ultimately has a negative impact on the cost–benefit ratio. For example, the product price on the market of the slug parasitic nematode *P. hermaphrodita* is actually too high for large-scale use in agriculture (at least in the Czech Republic) (see Nermut' and Půža, Chapter 27, this volume). In fact, the final high price is due to the cost incurred in formulation, storage and transport, and thus nematode-based biocontrol agents are expensive when compared with standard chemical pesticides. The high cost of equipment, labour and insects in the case of *in vivo* production, and also the high start-up costs or high capital investment at the beginning of *in vitro* production, are some important reasons why the final price of the nematode product is high (Shapiro-Ilan *et al.*, 2012, 2014; also see Askary and Ahmad, Chapter 13, this volume). The long process of nematode production (10–14 days for the completion of a single run in the production process), controlled temperature (2–10°C) for storage at all times, whether formulated or not, the requirement for proper aeration and for shipping on ice, often via overnight or priority mail, make the product expensive (Grewal *et al.*, 2005).

32.3.2 Limited product demand–supply–availability

The limiting factors with the companies involved in the production of EPNs or slug parasitic nematodes for supply to retailers, and in turn from retailers to end-users, are the lack of proper storage facilities, problems in shipping the product, the poor shelf life of the products and the seasonal nature of demand. These drawbacks push the manufacturers to go into limited production. On the other hand, continuous refrigeration and the poor shelf life of the EPN products result in their fall in demand by retailers. The buyer/user also faces the same problem, i.e. to keep the product cool until ready for use (see Askary and Ahmad, Chapter 13, this volume). The ultimate results can be seen in the form of limited demand, and so limited production and supply by the manufacturers, and thus limited availability of the nematode product on the market.

32.3.3 Insufficient knowledge of end-users

Lack of knowledge among users applying nematodes against target pests in the home lawn, kitchen garden or field often results in unsatisfactory results. The limited experience and knowledge of farm advisors and extension personnel finally leads to improper handling and mixing and application of nematode products by the end-users. They are generally not aware of the fact that storing or retaining nematode products for a long time before use will decrease the pathogenic ability of infective juveniles (IJs). Users are also unaware that while applying EPNs, there should be sufficient moisture in the soil and the soil temperature should be >55°F but <90°F (Dreves and Lee, 2015). This is essential for the proper activity of the nematodes and is why light irrigation is advised pre- and post-application of EPNs in soil. This is advantageous in two ways, i.e. washing the nematodes into the soil and decreasing the surface temperature of the soil. However, over-irrigation should be avoided, as this may cause lack of oxygen in saturated soil, which will inhibit the nematodes' activity. Foliar application of EPNs needs high humidity, and both cold and warm temperatures have a negative effect on nematode efficacy (Grewal *et al.*, 2005). Due to the above-mentioned reasons, it is advised that nematodes should be applied in the morning or evening, so that they may be protected from desiccation and ultraviolet rays.

EPN users do not have a proper knowledge of nematode application equipment. If a sprayer is being used to apply EPNs, fine-meshed filters (<50 mesh) should be removed and the nozzle of the sprayer should be checked for any clogging during application. Also, the sprayer tank should not be agitated excessively, as mechanical stress due to repeated recirculation may increase the temperature of the pump and nozzle. EPNs are sensitive to temperature, and when the temperature goes above 86°F, it becomes detrimental to the nematodes as they will be damaged and their viability will be decreased.

32.3.4 Low efficacy of nematodes

The points discussed in the above subsection are related directly or indirectly with the efficacy of EPNs and slug parasitic nematodes. The efficacy of nematodes largely depends on optimum soil temperature and moisture, and therefore its application in the early (morning) or late hours (evening) is recommended because the nematodes will be protected from scorching sunlight and the adverse effect of ultraviolet rays. However, the foremost point to which the user should pay attention is to check the expiry date of the supplied/purchased nematode product and the colour and odour of the nematodes, as a mild odour is a sign of live nematodes whereas a strong smell like ammonia indicates dead nematodes. If the product is vermiculite based or in sponge, the viability of the nematodes can be checked by soaking a small amount of product in water and then placing a droplet of this sample on a slide or in a small, clear glass bowl. When observing the sample with a 15X hand lens or microscope, the live motile and non-motile nematodes will appear as 'S' shaped and 'J' shaped (*S. carpocapsae*), respectively, whereas dead nematodes will appear to be straight and unmoving (Miles *et al.*, 2012). While applying nematodes through a sprayer or watering can, if the nematode suspension is not agitated from time to time, it will result in fast sedimentation of nematodes in the spray tank and thus become a reason for low nematode efficacy against target pests. Low efficacy is also a result of unmatched nematode species/strain and target pests. Lack of information on the compatibility of nematodes with insecticides or chemical fertilizers, particularly in those cases where they are toxic to nematodes, often results in low nematode efficacy. Not testing the compatibility of nematodes with insecticides or chemical fertilizers (in the case of integrated application) or not following the application instructions given on the EPN product label regarding the specific listing of chemicals that are toxic to nematodes will ultimately result in poor nematode efficacy.

32.4 Factors That Can Enhance and Expand the Use of Entomopathogenic and Slug Parasitic Nematodes

There are several ways to improve the use of EPNs and slug parasitic nematodes and decrease the price. We strongly believe that thrusts in the fields described briefly below will lead to the expansion of nematodes used as biocontrol agents against crop insect pests and noxious slugs.

32.4.1 Lowering the price of nematode products

As mentioned above under constraints (Section 32.3), the factors responsible for the high price of nematode products are the high costs incurred in formulation, storage and transport; therefore, researchers should pay attention to these areas in order to bring about a cheap product that would compete with chemical pesticides. Grewal *et al.* (2005) are of the opinion that increasing the efficiency of nematode production by making an improvement in the media constituents and optimizing the processing conditions with an aim to cut down the process time will finally contribute to lowering production costs. Furthermore, large-scale production, a large market with easy accessibility and a longer nematode shelf life will definitely bring down the product price. However, all this depends on a high demand for nematode products on the part of growers/users, and this in turn is related to educating users about the advantages of beneficial nematode products by extension functionaries or through advertisements or the distribution of pamphlets.

32.4.2 Increasing the demand-based supply of nematode products

The limited market supply and low availability of nematode products can be overcome by high nematode production, stable formulation, enhanced nematode shelf life and at least optimum profit margins of the manufacturers and distributors related to this business. Nematodes can be produced on a large scale by using a fermenter without any major technical hurdles, as has been demonstrated by Biosys with the production of up to 80,000 l (Grewal and Georgis, 1999), which is enough to treat 6000–8000 ha (depending on nematode species), if applied at 2.5 billion nematodes/ha (Grewal *et al.*, 2005). Nematode resistance to temperature extremes and osmotic desiccation and extending the longevity of nematodes (at least 1 year) in the final product are in great demand, because this could enable transport and long storage without very costly refrigeration. If success is gained in these areas, the availability of nematode products will increase in the market, and the final outcome will be the enhanced use of beneficial nematodes in agriculture.

32.4.3 Improving the knowledge of end-users for enhancing the use of nematode products

Users should be aware of the fact that nematode products are sensitive to temperature and moisture and still lack stable formulation, and therefore the products have a limited shelf life. Hence, it is advised to use EPN products within 1–2 days of arrival so as to obtain a satisfactory result. Nematodes have a tendency to sediment quickly in the tank mixture. Therefore, for the better application of EPNs, the tank mixture should be stirred so that the mixture may be homogenized properly. However, excessive agitation should be avoided as it increases the temperature of the spray tank and may also cause physical damage to the nematodes, which would finally result in reduced nematode efficacy. Therefore, increase in temperature above 30°C in the application equipment should be avoided (Grewal, 2002). Among the equipment used, pumps with lower capacity, such as diaphragm or roller pumps, are considered better for application (see Askary and Ahmad, Chapter 13, this volume). To enlarge the range of use of these amazing nematodes, future research should be targeted predominantly at the genetic enhancement of the nematode to increase tolerance to high temperature and desiccation.

Other ways to improve EPN or slug parasitic nematode *Phasmarhabditis* application are increasing the field efficacy and the implementation of new application strategies. The use of *P. hermaphrodita* is limited mostly to small slug species (e.g. genus *Deroceras*) and young stages of the larger species (e.g. genera *Arion*, *Milax*, *Limax*) (see Nermut' and Půža, Chapter 27, this volume). According to our experience with gardeners, application unfortunately very often fails. The reason is usually not the low quality of the product but poor or no respect to producers' protocol and application at the inappropriate time. Therefore, the first step to great efficacy must be based on a strict respect of the producers' application protocol and the right timing, which is dependent on the slug developmental stages and the environmental conditions. When we are sure of excellent application, we can begin to think about novel strategies. The most common is band application (also for EPNs) around the plants. This method provides good protection, but the number of nematodes decreases after some time and repeat application is necessary. Therefore, the economic advantage is questionable (Hass *et al.*, 1999a). Despite this, it is possible to recommend this method in short cultures or in plants that can tolerate lower levels of damage (Hass *et al.*, 1999b). Other good methods are the repeated application of a low dose (Ester *et al.*, 2003) or application in artificial slug shelters placed among plants (Grewal *et al.*, 2001).

One of the most important areas that need attention in order to enhance and expand these nematodes is the study of their bacterial associates. As we know, *P. hermaphrodita* is able to grow in association with many bacteria (Wilson *et al.*, 1995a,b; Rae *et al.*, 2010). Nowadays, this nematode is produced in monoxenic culture with the bacteria *Moraxella osloensis*, which provides very good results in both mass production and pathogenicity to slugs (Askary, 2010; Askary *et al.*, 2012), even though *Providencia rettgeri*, for example, provides much better results in mass production. Unfortunately, the results of *P. rettgeri* in pathogenicity were inconsistent, and therefore *M. osloensis* was chosen for commercial production (Wilson *et al.*, 1995b). Nevertheless, this *Phasmarhabditis* characteristic (or ability?) gives us the opportunity to find other bacterial associates that could be more effective, both in the mass production of nematodes and in the control of slug populations, especially the large species (e.g. family Arionidae, Limacidae) that these days are almost resistant to the *Moraxella–Phasmarhabditis* combination. Furthermore, it seems that the isolation of *P. hermaphrodita* is not as complicated and time-consuming as stated in the past, because the method for isolation of this nematode has been developed (Nermut', 2012). This trap is not absolutely selective, but the skilled nematologist is able to use it for 'large-scale' screening of many soil samples. This idea opens the way to obtaining many isolates of *P. hermaphrodita* and its bacterial associates. All that is needed is to make a start with screening and testing.

The study of bacterial associates is very close to enhancing the efficiency of mass producing *P. hermaphrodita*. Currently, this nematode is produced in a monoxenic culture in huge fermenters in growth medium that strongly influences the growth and quality of bacteria, and via this also the quality and production of the nematode (Ehlers and Shapiro-Ilan, 2005). The composition of the medium and the conditions in the fermenters certainly create enough space to enhance the method, but also *P. hermaphrodita* can probably be enhanced, similarly to EPNs (e.g. Glazer *et al.*, 1997; Ehlers and Shapiro-Ilan, 2005).

Novel application strategies that could be introduced on the market in the future will probably be based on slow-release systems that will allow the rapid decrease of dose. Nematodes can be applied in 'tea bags' with super-absorbent gel (Menzler-Hokkanen and Hokkanen, 2003), but root dips and the application of infected hosts have also been studied in EPNs (Grewal *et al.*, 2005) and could probably also be used for *P. hermaphrodita* or other slug parasitic nematodse. Further, the nematode species to be applied should be selected precisely and matched with the most susceptible pest stage. Advances in mechanization and alternative production routes i.e. application of EPNs in infected host cadavers has been suggested to improve efficiency and economy of scale (Shapiro-Ilan *et al.*, 2016).

Models to optimize the use of nematodes (application and dose) can be very useful in agricultural practice. Some models have been developed, but their use in practice is a little inconvenient. One of the first models estimated the influence of various doses on the mortality and feeding activity of the grey garden slug, *Deroceras reticulatum* (Glen *et al.*, 2000). Unfortunately, this model does not work with exposure time and therefore the information received is not very accurate. Another model developed works with time (Hass *et al.*, 1999b), but only the model published by Wilson *et al.* (2004) combines both the exposure time and dose with the size of the slug population in the field. Estimation of the size of slug population is a little complicated, which means that the use of this model in practice is questionable. Nevertheless, Choi *et al.* (2004) developed interesting models for predicting the population dynamics of *D. reticulatum*, and therefore the use of descriptive models to optimize the use of *P. hermaphrodita* could be more practical and could help to increase the efficacy and decrease the price of its application. An excellent advanced model was published by Schley and Bees (2006), which showed that both short-term and high-level as well as less dramatic but sustained control of slugs was possible by varying the timing of application. Only time will show if the models take root and really provide the expected enhancement of pestiferous slug control. But growers will certainly welcome every user-friendly method for managing the slug populations in their fields.

32.5 Conclusion

The use of entomopathogenic and slug parasitic nematodes seem more meaningful in the context of organic farming, where no chemicals are sprayed on plants and the products are cost-effective and competitive. While planning to include these beneficial nematodes in integrated pest management programmes, concerted efforts are needed, both with respect to the nematode species and the insect/slug and its environment. There are a few companies that are marketing EPNs successfully; however, revenues and market share have been limited to certain markets, with little opportunity to expand to new markets. In the past decade, there has been a significant increase in the demand of biopesticide products, but the market share in global revenues for nematodes has been flat as compared to other biopesticide products. The factors that limit nematodes gaining entry into markets and achieving a significant market share are production cost, shelf life, handling, mixing, coverage, new caution signal-based pesticides, compatibility and profit margins for manufacturers and distributors (Lacey and Georgis, 2012). Researchers should pay attention to isolating indigenous nematodes in order to expand the genetic base of indigenous nematode strains, identifying those adapted to low as well as high temperature. In the present era, these beneficial nematodes and their bacterial symbionts are an exciting subject for basic research in ecology, biodiversity, evolution, biochemistry and molecular genetics. The molecular interaction between nematodes and their symbiotic bacteria needs to be explored. These endeavours would ensure EPNs and slug parasitic nematodes are more effective biopesticides, besides ensuring that nematodes and their symbionts gain prominence as unique and intrinsically interesting biological systems.

There is a research need toward the production costs, enhancement of product availability, efficacy, ease-of-use and carryover effect of nematodes (Shapiro-Ilan *et al.*, 2012). A better and efficient formulation is needed that can be fulfilled by searching out better and efficient adjuvants, surfactants and anti-caking substances, as well as anti-desiccant and UV retardant properties. In addition, researchers will have to work out the proper dose for a particular nematode strain, as well as the method, time and frequency of applications for an effective control of targeted pests.

Genetic improvement can play an important role in strengthening and expanding the market share of nematodes in the current crop–pest complex, and this may be accomplished through selection regimes and changes in bacterial payload that will result in developing stable formulations, improving nematode searching abilities, producing more virulent nematodes and increasing tolerance to environmental extremes. Moreover, other advancements are also needed, such as to strengthen the efficacy and the field host range beyond the current target insects, crops and environment (Lacey and Georgis, 2012). When achieved, the advances in these fields will reduce the great reliance on chemical pesticides, and will bring stability to crop production and productivity.

Growers aim to check crop losses in order to earn maximum profit, and researchers are also trying the same for growers, but their aim is to accomplish the task without or least disturbing the ecological balance (Askary and Martinelli, 2015). In order to incorporate nematodes as biocontrol agents with existing IPM modules for different crops, the application technology and quality of commercial nematode products need to be standardized for aerial and soil insect pests in various types of formulations. However, biological control cannot be a substitute for chemical control, which is meant for quick results, but it is environmentally safe and free from health hazards and so is widely accepted among growers (Askary, 2015). In certain cases, EPNs or slug parasitic nematodes may replace chemical pesticides when used alone or in conjugation with them, but are unlikely to dominate most pest control scenarios. In spite of this, it can be concluded that these beneficial nematodes definitely have a place of their own in crop insect and slug pest management programmes.

Acknowledgements

This work is supported in part by All India Coordinated Research Project (AICRP), ICAR on Biological Control of Crop Pests and Weeds running in the Division of Entomology, Sher-e-Kashmir University of Agricultural Sciences and Technology of Kashmir (SKUAST-K), Srinagar, Jammu and Kashmir, India. Trade names or commercial products of nematodes mentioned in this chapter are only aimed to provide specific information to readers and do not imply recommendation or endorsement.

References

Askary, T.H. (2010) Nematodes as biocontrol agents. In: Lichtfouse, E. (ed.) *Sociology, Organic Farming, Climate Change and Soil Science*. Springer, The Netherlands, pp. 347–378.

Askary, T.H. (2015) Limitations, research needs and future prospects in the biological control of phytonematodes. In: Askary, T.H. and Martinelli, P.R.P. (eds) *Biocontrol Agents of Phytonematodes*. CAB International, Wallingford, UK, pp. 446–454.

Askary, T.H. and Martinelli, P.R.P. (2015) *Biocontrol Agents of Phytonematodes*. CAB International, Wallingford, UK.

Askary, T.H., Khan, A.A., Waliullah, M.I.S., Banday, S.A., Iqbal, U. and Mir, M.M. (2012) Slug pest management through nematodes in agricultural and horticultural crops. In: Boeri, F. and Chung, J.A. (eds) *Nematodes: Morphology, Functions and Management Strategies*. Nova Publishers, New York, pp. 197–211.

Choi, Y.H., Bohan, D.A., Powers, S.J., Wiltshire, C.W., Glen, D.M. and Semenov, M.A. (2004) Modelling *Deroceras reticulatum* (Gastropoda) population dynamics based on daily temperature and rainfall. *Agriculture, Ecosystems and Environment* 103, 519–525.

Dreves, A.J. and Lee, J. (2015) Entomopathogenic nematodes. Available at: http://insect.pnwhandbooks.org/ipm/entomopathogenic-nematodes (accessed 9 March 2016).

Ehlers, R.-U. and Shapiro-Ilan, D.I. (2005) Mass production. In: Grewal, P.S., Ehlers, R.-U. and Shapiro-Ilan, D.I. (eds) *Nematodes as Biocontrol Agents*. CAB International, Wallingford, UK, pp. 65–78.

Ester, A. and Wilson, M.J. (2005) Application of slug-parasitic nematodes. In: Grewal, P.S., Ehlers, R.-U. and Shapiro-Ilan, D.I. (eds) *Nematodes as Biocontrol Agents*. CAB International, Wallingford, UK, pp. 431–444.

Ester, A., van Rozen, K. and Molendijk, L.P.G. (2003) Field experiments using the rhabditid nematode *Phasmarhabditis hermaphrodita* or salt as control measures against slugs in green asparagus. *Crop Protection* 22, 689–695.

Glazer, I., Salame, L. and Segal, D. (1997) Genetic enhancement of nematicide resistance in entomopathogenic nematodes. *Biocontrol Science and Technology* 7, 499–512.

Glen, D.M., Wilson, M.J., Brain, P. and Stroud, G. (2000) Feeding activity and survival of slugs, *Deroceras reticulatum*, exposed to the rhabditid nematode, *Phasmarhabditis hermaphrodita*: a model of dose response. *Biological Control* 17, 73–81.

Grewal, P.S. (2002) Formulation and application technology. In: Gaugler, R. (ed.). *Entomopathogenic Nematology*. CAB International, Wallingford, UK, pp. 265–287.

Grewal, P.S. and Georgis, R. (1999) Entomopathogenic nematodes. In: Hall, F.R. and Menn, J. (eds) *Methods in Biotechnology: Biopesticides: Use and Delivery*. Humana Press, Totowa, New Jersey, pp. 271–299.

Grewal, P.S., Grewal, S.K., Taylor, R.A.J. and Hammond, R.B. (2001) Application of molluscicidal nematodes to slug shelters: a novel approach to economic biological control of slugs. *Biological Control* 22, 72–80.

Grewal, P.S., Ehlers, R.-U. and Shapiro-Ilan, D.I. (2005) Critical issues and research needs for expanding the use of nematodes in biocontrol. In: Grewal, P.S., Ehlers, R.-U. and Shapiro-Ilan, D.I. (eds) *Nematodes as Biocontrol Agents*. CAB International, Wallingford, UK, pp. 479–489.

Hass, B., Hughes, L.A. and Glen, D.M. (1999a) Overall versus band application of the nematode *Phasmarhabditis hermaphrodita* with and without incorporation into soil, for biological control of slugs in winter wheat. *Biocontrol Science and Technology* 9, 579–586.

Hass, B., Glen, D.M., Brain, P. and Hughes, L.A. (1999b) Targeting biocontrol with the slug-parasitic nematode *Phasmarhabditis hermaphrodita* in slug feeding areas: a model study. *Biocontrol Science and Technology* 9, 587–598.

Lacey, L.A. and Georgis, R. (2012) Entomopathogenic nematodes for control of insect pests above and below ground with comments on commercial production. *Journal of Nematology* 44, 218–225.

Menzler-Hokkanen, I. and Hokkanen, H.M.T. (2003) Entomopathogenic nematode delivery systems for biological control of pests on major outdoor crops: the case of oilseed rape. *The Society of Invertebrate Pathology Program and Abstracts for the 36th Annual Meeting, Burlington, Vermont*, USA, 60 pp. (Abstract).

Miles, C., Blethen, C., Gaugler, R., Shapiro-Ilan, D.I. and Murray, T. (2012) Using entomopathogenic nematodes for crop insect pest control. *Pacific North West Extension Publications*, USA, pp. 1–9.

Nermut', J. (2012) The persistence of *Phasmarhabditis hermaphrodita* (Rhabditida: Rhabditidae) in different substrates. *Russian Journal of Nematology* 20, 61–64.

Rae, R., Verdun, C., Grewal, P., Robertson, J.F. and Wilson, M.J. (2007) Biological control of terrestrial molluscs using *Phasmarhabditis hermaphrodita* – progress and prospects. *Pest Management Science* 63, 1153–1164.

Rae, R.G., Tourna, M. and Wilson, M.J. (2010) The slug parasitic nematode *Phasmarhabditis hermaphrodita* associates with complex and variable bacterial assemblages that do not affect its virulence. *Journal of Invertebrate Pathology* 104, 222–226.

Schley, D. and Bees, M.A. (2006) The role of time delays in a non-autonomous host–parasitoid model of slug biocontrol with nematodes. *Ecological Modelling* 193, 543–559.

Shapiro-Ilan, D.I., Han, R. and Dolinksi, C. (2012) Entomopathogenic nematode production and application technology. *Journal of Nematology* 44, 206–217.

Shapiro-Ilan, D.I., Han, R. and Qiu, X. (2014) Production of entomopathogenic nematodes. In: Morales-Ramos, J.A., Rojas, M.G. and Shapiro-Ilan, D.I. (eds) *Mass Production of Beneficial Organisms: Invertebrates and Entomopathogens.* Academic Press, Amsterdam, pp. 321–355.

Shapiro-Ilan, D.I., Morales-Ramos, J.A. and Rojas, M.G. (2016) *In vivo* production of entomopathogenic nematodes. *Methods in Molecular Biology* 1477, 137–158.

Tandingan De Ley, I., McDonnell, R.J., Lopez, S., Paine, T.D. and De Ley, P. (2014) *Phasmarhabditis hermaphrodita* (Nematoda: Rhabditidae), a potential biocontrol agent isolated for the first time from invasive slugs in North America. *Nematology* 16, 1129–1138.

Wilson, M.J., Glen, D.M. and George, S.K. (1993) The rhabditid nematode *Phasmarhabditis hermaphrodita* as a potential biological-control agent for slugs. *Biocontrol Science and Technology* 3, 503–511.

Wilson, M.J., Glen, D.M., George, S.K. and Pearce, J.D. (1995a) Selection of a bacterium for the mass-production of *Phasmarhabditis hermaphrodita* (Nematoda, Rhabditidae) as a biocontrol agent for slugs. *Fundamental and Applied Nematology* 18, 419–425.

Wilson, M.J., Glen, D.M., Pearce, J.D. and Rodgers, P.B. (1995b) Monoxenic culture of the slug parasite *Phasmarhabditis Hermaphrodita* (Nematoda, Rhabditidae) with different bacteria in liquid and solid-phase. *Fundamental and Applied Nematology* 18, 159–166.

Wilson, M.J., Glen, D.M., Hamacher, G.M. and Smith, J.U. (2004) A model to optimise biological control of slugs using nematode parasites. *Applied Soil Ecology* 26, 179–191.

Index

Note: Page numbers in **bold** type refer to **figures**
Page numbers in *italic* type refer to *tables*